Parryella filifolia (upper), *Astragalus mollissimus* var. *mogollonicus* (lower) [Photos by MF Wojciechowski]

Legumes of Arizona

An Illustrated Flora and Reference

m m a pope

Legumes of Arizona

An Illustrated Flora and Reference

Editors
Michelle M. McMahon, University of Arizona, Tucson, Arizona
Matthew B. Johnson, University of Arizona, Tucson and Boyce Thompson Arboretum, Superior, Arizona
Kirsten N. Lake, University of Arizona, Tucson, Arizona
Martin F. Wojciechowski, Arizona State University, Tempe, Arizona
Mark W. Bierner, University of Texas, Austin, Texas and University of Arizona, Tucson, Arizona

BRIT PRESS

Fort Worth Botanic Garden | Botanical Research Institute of Texas
Center for Botanical Exploration & Discovery
Fort Worth, Texas

Legumes of Arizona: An illustrated flora and reference

Printed in Korea
ISBN: 13-978-1889878-77-5

Published by BRIT Press, Fort Worth Botanic Garden | Botanical Research Institute of Texas, Fort Worth, Texas, U.S.A.

Cover design
Rebecca Horn
Book layout and production
Rebecca Horn, rhorngraphics

Front cover
Desert ironwood, *Olneya tesota*, grows in the driest conditions yet flowers profusely and beautifully, as shown here. The illustrator, Margaret Pope, is a scientific illustrator and founder of the Sonoran Desert Florilegium Program (art-botanical.org).

Please cite this book as follows:
Legumes of Arizona Editorial Committee. 2024. Legumes of Arizona: An illustrated flora and reference. Sida, Bot. Misc. 64. Botanical Research Institute of Texas, Fort Worth, Texas, U.S.A.

Distribution of copies by:
BRIT Press, Fort Worth Botanic Garden | Botanical Research Institute of Texas
1700 University Drive
Fort Worth, Texas 76107-3400, U.S.A.
Telephone: 1.817.332.4441
Fax: 1.817.332.4112
Email: press@brit.org
Websites: britpress.org, shopbritpress.org

BRIT PRESS

Botanical Miscellany, No. 64
ISSN: 0883-1475
28 June 2024

THE DESERT LEGUME PROGRAM

The Desert Legume Program (DELEP) is a research and service program focused on arid land legumes. Seeds of more than 1400 taxa, from 67 countries across six continents have been gathered, processed, and are preserved as living germplasm. These resources, i.e., sets of living seeds, have been provided to researchers around the world to study systematics, genomes, microbial interactions, toxic secondary metabolites, to name a few. In Arizona, DELEP has analyzed many of the collected taxa for their horticultural potential in hot dry conditions, in the greenhouse, with respect to frost tolerance, and for any tendencies to escape. These results are provided here, in the discussions, for many Arizona species.

DELEP is unique in focusing on wild species of Fabaceae from dry regions. This group of plants was selected for several reasons. The Fabaceae is the third largest family of vascular plants worldwide and is well-represented in the floras of dry regions. Wild Fabaceae from dry regions are understudied in terms of their economic potential. As a group, legumes are second only to the cereal grains in terms of their importance to humanity. Legumes are utilized in many ways, including as food crops, forage crops, green manure, timber, sources of medicines and industrial compounds, for erosion control, for shade and shelter, and as landscape and amenity plants. Many species of legumes have the ability to convert atmospheric nitrogen into a plant-available form through symbiosis with bacteria on their roots.

The Desert Legume Program was established in 1988 by Dr. R. Phillip Upchurch as a joint project of the University of Arizona College of Agriculture, Life and Environmental Sciences, and Boyce Thompson Arboretum. Since July 2019, DELEP has been managed through the University of Arizona Herbarium in the School of Plant Sciences.

MISSION STATEMENT

- To acquire and preserve in perpetuity seeds of legumes native to the arid and semiarid lands of the world;
- To learn more about the nature and utility of these unique species;
- To share this germplasm with professionals and laypersons having a legitimate interest; and
- To aid in the preservation and conservation of desert legume biodiversity through both *in situ* and *ex situ* means.

DELEP plays an important role in conserving legume biodiversity. Maintaining biological diversity is a growing concern worldwide as human populations increase and more natural areas are diverted for human use. As development of natural lands continues, it is inevitable that many species may be lost without conservation efforts. Collecting and maintaining seed germplasm is one of the most basic and most important means of preserving species in the face of habitat loss.

For more information or to contribute to this mission, please visit https://cals.arizona.edu/desertlegumeprogram/.

ACKNOWLEDGMENTS

Legumes of Arizona: An Illustrated Flora and Reference was made possible by the generous support of H.B. and Jocelyn Wallace and Linda Wallace Gray through grants from the Wallace Research Foundation. Major support was also provided by Janet Spadora, Cliff Douglas at Harris Cattle Company/Arid Zone Trees, and Ron Gass at Mountain States Wholesale Nursery. The project was made possible by a longterm partnership between Boyce Thompson Arboretum and the University of Arizona College of Agriculture, Life and Environmental Sciences.

We are deeply grateful for the generous donation of original illustrations by artists associated with the Art Institute at the Arizona-Sonora Desert Museum, the illustration program at the Denver Botanical Gardens, and the Desert Botanical Garden in Phoenix. Their participation in this project has made it truly unique.

The following are gratefully acknowledged for their assistance with field work that contributed directly or indirectly to the project: Catherine Babcock, Patti Baynham, Cass Blodgett, Kenneth Coppola, Martha Johnson, Michael Chamberland, William Kendall, Elaine Moisan, Lacey Pacheco, Jeff Payne, Lorrie Polley, Patricia A. Rorabaugh, Gary Selinger, Mark Siegwarth, Kim Stone, Marceline VandeWater.

The editors acknowledge the following for insightful discussion on various aspects of the flora and of the project: C. David Bertelsen, Mima Falk, George Ferguson, James T. Verrier, Julie Crawford, and Sabra Tonn. Technical assistance was provided by Joe Black and Kai Lepley.

Extensive reviews were generously provided by C. David Bertelsen, Brad Boyle, Erin Riggs, and Glenn Rink.

Thank you to Thomas C. Brennan, Andrew T. Holycross, and the Arizona Game and Fish Department for granting permission to adapt David Brown's biotic community chapter from *A Field Guide to Amphibians and Reptiles in Arizona* for this project. Thank you to the University of Michigan Herbarium for the use of nine figures, and to The American Society of Plant Taxonomists for the use of one figure. The United States Forest Service and the Arizona State Land Department provided collecting permits.

TABLE OF CONTENTS

Clockwise from lower left: *Mimosa biuncifera, Senna covesii, Coursetia glandulosa, Dalea formosa, Mimosa dysocarpa* (Photos by M. McMahon), *Lupinus kingii* (Photo by M.F. Wojciechowski)

PREFACE

Legumes and humans have a long-standing relationship. As people around the world settled down, invented agriculture, and chose the best local plants for feeding themselves and their communities, they repeatedly domesticated two ubiquitous plant groups: legumes and grasses. In Mesopotamia, early farmers worked with lentils and wheat. Further east, they used soybeans and rice. Peas and oats in the Mediterranean, lima bean and maize in South America, and common bean and maize in North America — by pairing foods from these two groups, humans were able to obtain complete nutrition, and civilizations were born. Legumes remain mission critical: around 20% of current worldwide crop production involves a plant in the legume family (FAO, 2018 data).

Arizona sits at a sweet spot for plant diversity. Four North American deserts and the great Colorado Plateau, studded or bounded by rugged mountains extending into alpine zones, traversed by major, minor, and intermittent river systems — these elements combine to support a uniquely diverse landscape. Likewise, urban landscapes depend on distinctive palettes of arid adapted and mostly frost tolerant taxa.

Not surprisingly, natural legume diversity in Arizona is quite high — more than one in ten Arizona species belong to Fabaceae, the legume family. World-wide, there are nearly 20,000 legume species, making it third in size only to the asters (Asteraceae) and orchids (Orchidaceae). Found at nearly all elevations, the legumes particularly dominate in arid and semi-arid tropical and subtropical regions, and they are well represented throughout the temperate zones too. Legumes form a large component of Arizona's flora, yet they have not been documented in a monograph since the classic flora, *Arizona Flora* by Kearney and Peebles (1960).

Native, naturalized, and commonly cultivated legumes from Arizona are the focus of this volume. Uses and results of horticultural trials are here combined for the first time with keys, descriptions, maps, original illustrations, and photos. Our aim with this project is to provide a true reference for botanists, gardeners, naturalists, and anyone who appreciates our diverse southwestern region.

Led by the Desert Legume Program (DELEP), this project is a manifestation of the long collaboration between DELEP and Boyce Thompson Arboretum. Additional partners include the College of Agriculture, Life and Environmental Sciences at the University of Arizona, the University of Arizona Herbarium, and the Wallace Research Foundation. Most importantly, this is a product of immense volunteer contributions: 45 authors, each an expert on their groups or topics, plus 28 scientific illustrators, four photographers, one map maker, and an extraordinarily generous copy editor combined to create the contents of this volume.

Boyce Thompson Arboretum

Founded in 1924 by mining magnate Col. William Boyce Thompson, Boyce Thompson Arboretum (BTA) is the oldest and largest botanical garden in Arizona. The Arboretum comprises about 400 acres near Superior, Arizona, in a magnificent natural setting within the Sonoran Desert. The mission of Boyce Thompson Arboretum is to inspire appreciation and stewardship of desert plants, wildlife, and ecosystems through education, research, and conservation.

BTA features over 4000 plant species from dry regions around the world. A network of trails provides access to the garden areas as well as adjacent natural habitats. Ecogeographic collections recreate plant communities that can be found in parts of North and South America, Africa, Asia, Australia, and Europe. The Arboretum has extensive collections of cacti, agaves, yuccas, aloes, legumes, eucalyptus, and oaks. The last three collections are accredited by the Plant Collections Network. The Arboretum also features a demonstration garden, hummingbird/butterfly garden, herb garden, and a rose collection. In the late 2010s the Wallace Desert Garden was relocated to BTA from the former home of H. B. and Jocelyn Wallace. This involved transplanting over 5000 plant specimens, many of which were mature trees and cacti — a truly monumental task that is likely the largest garden relocation ever attempted.

BTA provides tours, presentations, workshops, and other educational opportunities for the public, year-round. Visitors can enjoy a shaded picnic area and a well-stocked gift shop as well as a nursery selling many desert-adapted plants. Due to the extensive garden areas and adjacent natural riparian habitat, the Arboretum is a popular birding destination. The Arboretum has also been involved in research since its founding and has published the semi-technical journal Desert Plants since 1979.

The Arboretum is managed by a non-profit corporation, Boyce Thompson Southwestern Arboretum, Inc. For several decades, it also had close relationships with the Arizona State Parks and the University of Arizona. Former BTA executive director and Head of the UA Department of Arboretum Affairs, Dr. Mark Bierner, initiated the Legumes of Arizona project under the combined research objectives of BTA and DELEP. Although BTA is now independent of the

UA and the State Parks system, BTA and DELEP remain strong partners in promoting arid land plant research, education, and outreach.

Editors

Michelle M. McMahon, University of Arizona, Tucson, Arizona
Matthew B. Johnson, University of Arizona, Tucson and Boyce Thompson Arboretum, Superior, Arizona
Kirsten N. Lake, University of Arizona, Tucson, Arizona
Martin F. Wojciechowski, Arizona State University, Tempe, Arizona
Mark W. Bierner, University of Texas, Austin, Texas and University of Arizona, Tucson, Arizona

Copy editing and scientific review

Mary Welch-Keesey, University of Arizona, Tucson, Arizona

Illustration Coordination

Margaret Pope, Tucson, Arizona

Contributors

Authors

Jason Alexander, University of California, Berkeley, California
Peter W. Ball, University of Toronto at Mississauga, Mississauga, Ontario, Canada
Stephen W. Bieda, III, National Weather Service, Amarillo, Texas
Benjamin Brandt, Tucson, Arizona
Steven L. Broich, Oregon State University, Corvallis, Oregon
David E. Brown, Arizona State University Biocollections, Tempe, Arizona
Richard D. Conway, Arivaca, Arizona
Alfonso Delgado-Salinas, Universidad Nacional Autónoma de México, Mexico City, Mexico
John E. Ebinger, Eastern Illinois University, Charleston, Illinois
Ashley N. Egan, Utah Valley University, Orem, Utah
Teague Embrey, Tucson, Arizona
Frank T. Farruggia, Smithsonian Institution, National Museum of Natural History, Washington, DC
Richard Stephen Felger, University of Arizona, Tucson, Arizona
Gabriel Flores-Franco, Universidad Autónoma del Estado de Morelos, Cuernavaca, Morelos, Mexico
Peter W. Fritsch, Botanical Research Institute of Texas, Fort Worth, Texas
Douglas H. Goldman, United States Department of Agriculture, Natural Resources Conservation Service, Greensboro, North Carolina
Rosaura Grether, Universidad Autónoma Metropolitana, Unidad Iztapalapa, Mexico City, Mexico
Leslie Gunatilaka, University of Arizona, Tucson, Arizona
Richard R. Halse, Oregon State University, Corvallis, Oregon
Héctor M. Hernández, Universidad Nacional Autónoma de México, Mexico City, Mexico
Colin Hughes, University of Zurich, Zurich, Switzerland
Philip Jenkins, University of Arizona, Tucson, Arizona
Matthew B. Johnson, University of Arizona, Tucson and Boyce Thompson Arboretum, Superior, Arizona
Alexander Krings, North Carolina State University, Raleigh, North Carolina
Kirsten N. Lake, University of Arizona, Tucson, Arizona
Matt Lavin, Montana State University, Bozeman, Montana
Ries Lindley, University of Arizona, Tucson, Arizona
L.J.G. van der Maesen, Netherlands Centre for Biodiversity Naturalis Wageningen University, Wageningen, the Netherlands
Brigitte Marazzi, Natural History Museum of Canton Ticino/Lugano, Ticino, Switzerland
Kathryn Mauz, Evergreen, Colorado
Michelle M. McMahon, University of Arizona, Tucson, Arizona
Guy L. Nesom, Academy of Natural Sciences of Drexel University, Philadelphia, Pennsylvania

Camille F. Nowell, California Academy of Sciences, San Francisco, California
Lourdes Rico-Arce, Royal Botanic Gardens, Kew, Richmond, United Kingdom and Comisión Nacional para el Conocimiento y Uso de la Biodiversidad, Mexico City, Mexico
Teresa Sholars, Mendocino College Coast Campus (College of the Redwoods), Fort Bragg, California
David S. Seigler, University of Illinois at Urbana-Champaign, Urbana
Beryl B. Simpson, University of Texas at Austin, Billie L. Turner Plant Resources Center, Austin, Texas
Ernest Small, Ottawa Research and Development Centre, Agriculture and Agri-Food Canada, Ottawa, Canada
Lawrence R. Stritch, U.S. Forest Service, Washington, D.C.
Leticia Torres-Colín, Universidad Nacional Autónoma de México, Mexico City, Mexico
Michael A. Vincent, Miami University, Oxford, Ohio
Mary Welch-Keesey, University of Arizona, Tucson, Arizona
Martin F. Wojciechowski, Arizona State University, Tempe, Arizona
Michael Woods, Troy University, Troy, Alabama
Richard P. Wunderlin, University of South Florida, Tampa, Florida

Illustrators

Susan E. Ashton, Phoenix, Arizona
Marsha Bennett, Phoenix, Arizona
Chris Bondante, Tucson, Arizona
Kathleen Bailey Cray, Tucson, Arizona
Karen Gengle, Phoenix, Arizona
John H. Gerdes, Tucson, Arizona
Linda Gerstle, Denver, Colorado
Rebecca Hall, Denver, Colorado
Gabriele Henn, Phoenix, Arizona
Christine Hubbell, Denver, Colorado
Elaine Hultgren, Phoenix, Arizona
Matthew B. Johnson, Tucson, Arizona
Marjorie Leggitt, Denver, Colorado
Jo Ann Loza, Phoenix, Arizona
Ardys Lurtsema, Tucson, Arizona
Deborah C. Nance, Denver, Colorado
Joyce Riney Peters, Phoenix, Arizona
Margaret Pope, Tucson, Arizona
Randall C. Raak, Denver, Colorado
Kim Ratelle-Collins, Palominas, Arizona
Lynn Reves, Phoenix, Arizona
Gillian Rice, Phoenix, Arizona
Katherine Salwei, Palominas, Arizona
Daniela Siroky, Phoenix, Arizona
Sandy Turico, Gilbert, Arizona
Marceline VandeWater, Phoenix, Arizona
Barbara Vucich, Denver, Colorado
Gigi Wilson, Phoenix, Arizona

Photographers

Jillian H. Cowles, Vail, Arizona
Chris Kline, Superior, Arizona
Martin F. Wojciechowski, Tempe, Arizona
Chris Bondante, Tucson, Arizona

Maps

Rob Hastings, Tucson, Arizona

ADDENDUM

As this book was going to press, Becklund and Ayers (Systematic Botany 47:1012–1024) published a revision of *Errazurizia rotundata* (Wooton) Barneby as the new genus and species, *Pictarena rotundata* (Wooton) Becklund.

Becklund, L.E. & T.J. Ayers. 2022. The Phylogeny of *Errazurizia* (Fabaceae: Amorpheae) and description of the new monotypic genus *Pictarena*. Syst. Bot. 47:1012–1024.

Palo verdes with brittlebush, saguaros, and organ pipes in Organ Pipe Cactus National Monument. Photo: M.J. Sanderson

Arizona's diversity of legumes is unparalleled—clovers in alpine meadows;
locoweeds in pine forests; locusts in oak woodlands;
palo verdes, ironwoods, and mesquites in cactus-studded deserts—the state's biomes
and biodiversity invite exploration.

INTRODUCTION

Regional biodiversity depends on countless factors, perhaps most notably geology and climate. Also, of course, individual plant species never exist in a vacuum and are instead components of biotic communities. With that in mind, we start this volume with discussions of geology, biotic communities, and climate of Arizona. Additional sections follow, providing information on conservation efforts, the history of botanical collecting in Arizona, and the research value of bioprospecting arid land legumes, as well as brief summaries of legume classification and characteristics. Our aim is to develop a context for the legume diversity described in the rest of the volume, in which we present treatments and at least one illustration for each genus, maps for each native or naturalized species, and photographs representing most genera. Detailed comments regarding creation of the treatments are at the end of the introductory material.

ARIZONA GEOLOGY

Richard D. Conway

Introduction

Arizona! Just the word evokes images and ideas of geology. Rocks and landscapes with colors and forms that delight and challenge the imagination. From the Grand Canyon and the Petrified Forest in the north to the Sky Island ranges in the southeast, from the eastern high plateaus of the White Mountains covered with pine forests and meadows to the ragged desolation of the scorched deserts in the southwest, the diversity and extravagance of Arizona's geology is unique on the planet.

Arizona has had a long and complex geologic history, revealing many of the times and events that created western North America. Part of this record is remarkably well preserved in places like the Grand Canyon, but all the places possess their stories and mysteries and keep geologists coming back to learn more, hoping to discover their secrets.

The map in Figure 1 shows Arizona's landforms as an elevation model. In the north is a very large area of high undulating terrain with few features except for the great incision of the Colorado River. Trending northwest across the state is a belt of complex mountainous landscapes. In the southwest is a vast tract of low elevation desert scattered with small craggy mountain ranges. The southeast corner of the state is an area where large mountain ranges rise from deserts floors to over 9,000 ft (2,740 m) in elevation.

The map is labeled to show physiographic provinces, which are regions geographers use to organize landscapes by their geomorphology. For Arizona these are the Colorado Plateau, the Basin and Range and the Arizona Transition Zone. The Colorado Plateau is a region including the Four Corners area of Utah, Colorado, New Mexico and Arizona that has been an uplifted and a stable part of the Earth's crust for millions of years, its layered rocks little deformed since their origins. It is surrounded by geologic mayhem where the rocks have been twisted, broken, compressed, buckled, stretched and melted.

The Basin and Range Province is one those active areas where the crust was stretched, thinned, and broken into faulted blocks. This physiographic region extends from southern Idaho and Oregon through Nevada and eastern California deep into Mexico. It is best developed in central Nevada where it resembles, in the words of an early geologist, an army of caterpillars resolutely marching north out of Mexico. While the Basin and Range evolved the stable Colorado Plateau crust remained high and intact.

The Transition Zone Province is a rugged mountainous broken mass of crustal blocks containing pieces of both the Colorado Plateau and the Basin and Range. In central Arizona, the Mogollon Rim surmounts the Transition Zone and defines the southern edge of the Colorado Plateau. This landscape is an ecological and cultural barrier between northern and southern Arizona and is frequently called the Central Highlands.

Of course, all of this is the landscape expression of the rocks. Figure 2a shows the rocks in a simplified geologic map of the state and is accompanied by an explanation of the colors and symbols (Fig. 2b). Notice that the explanation is arranged in geologic time order. Spend a few minutes studying the patterns of colors, checking what kind of rock they symbolize and their geologic age. Figure 3 is the geologic time scale accompanied by a summary of important Arizona geologic events.

How can we approach, in a few pages, the long and complex history of such a place? Geologic summaries often take a historical approach, proceeding from earliest times, reading the rock record as an account of events interpreted from the rocks. We will take the physiographic approach instead, and cruise drone-like over the landscapes, descending to take a closer look at places and features that characterize and highlight Arizona's geology. This is a landscape approach, but it is more than that because we also want to see ancient landscapes. When we fly across Black Mesa we want to see

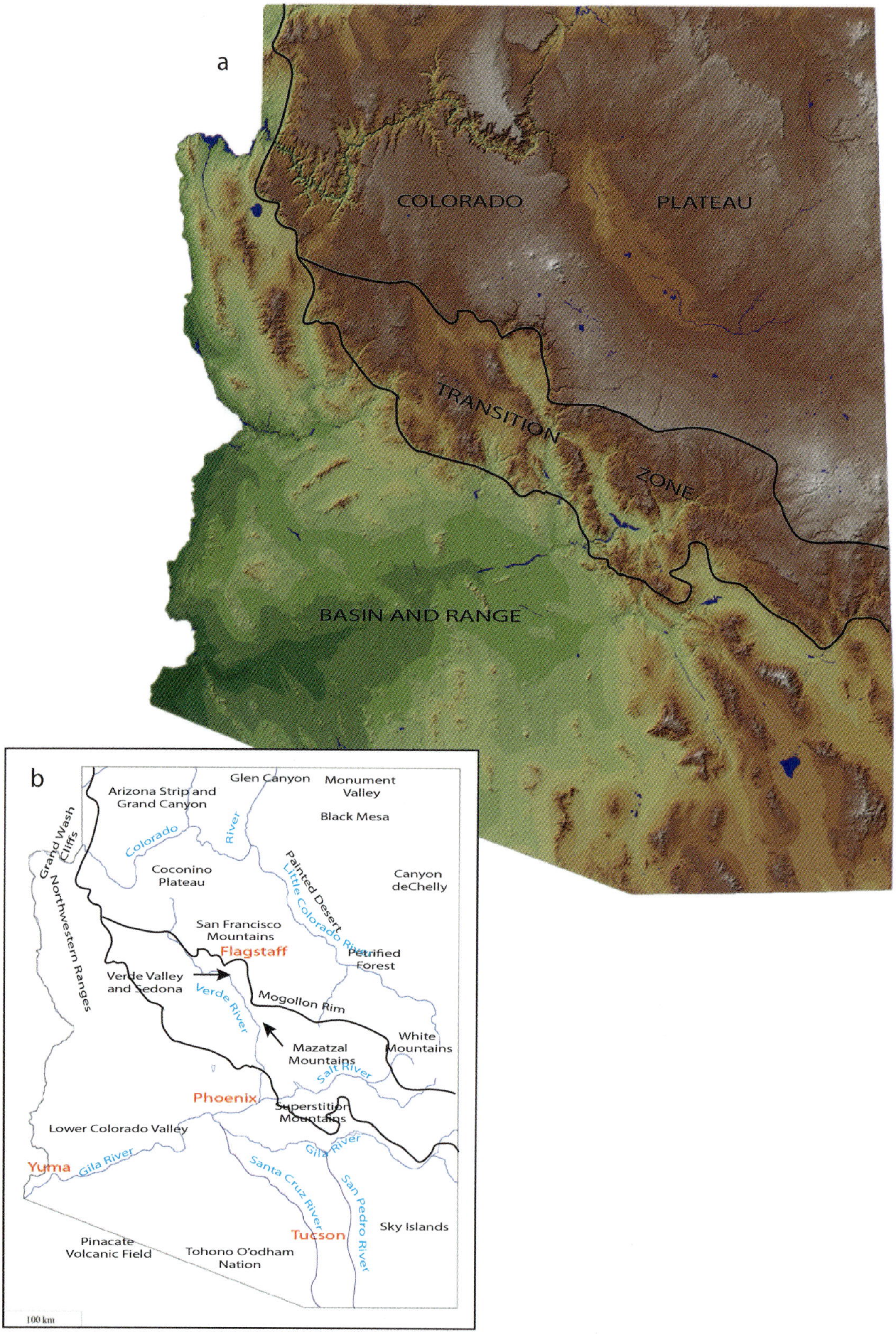

Figure 1. Physiographic regions and major landmarks of Arizona. **a.** colors indicate elevation; **b.** major cities, streams, and areas.

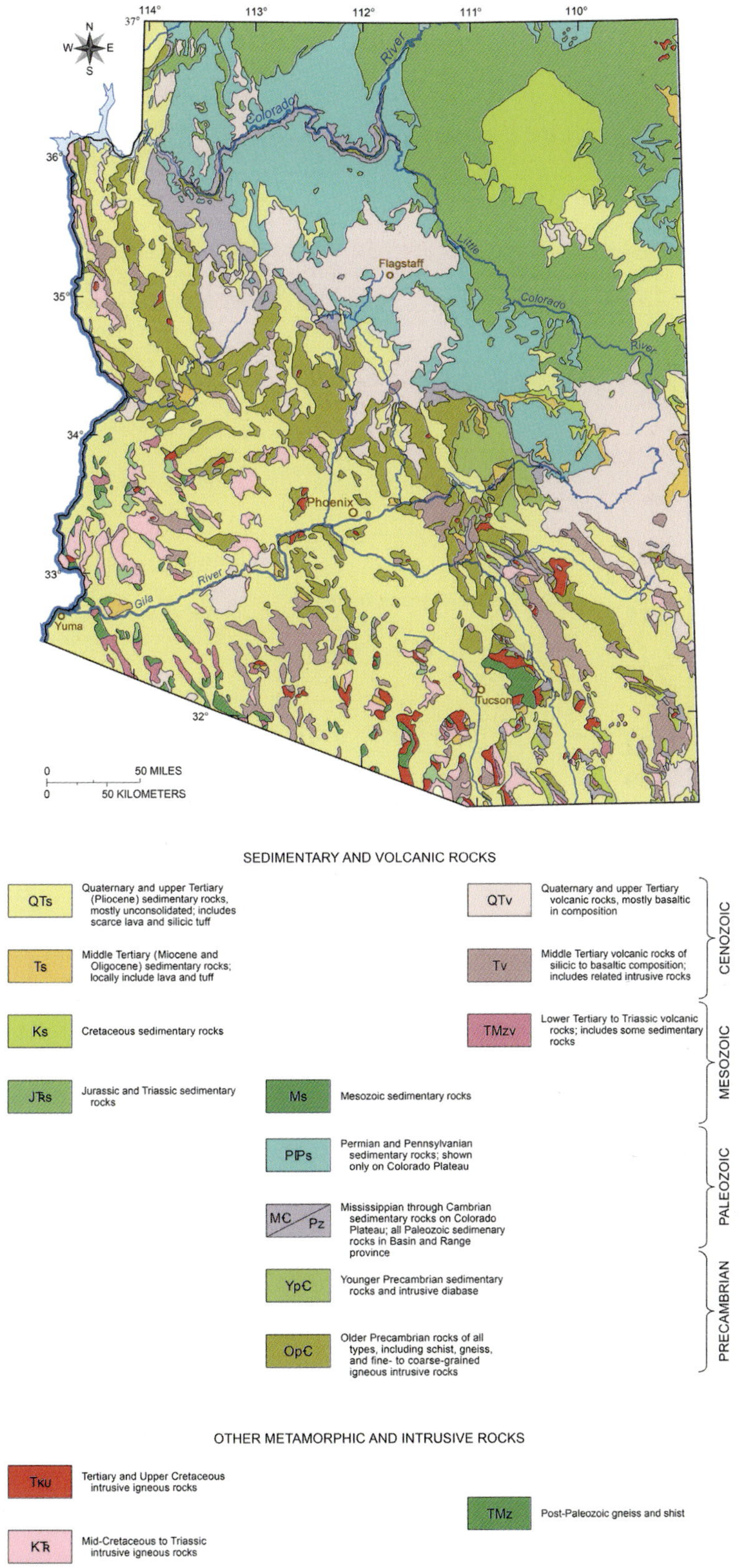

Figure 2. a. Geologic map of Arizona; **b.** legend. Adapted from Cooper et al. (1969).

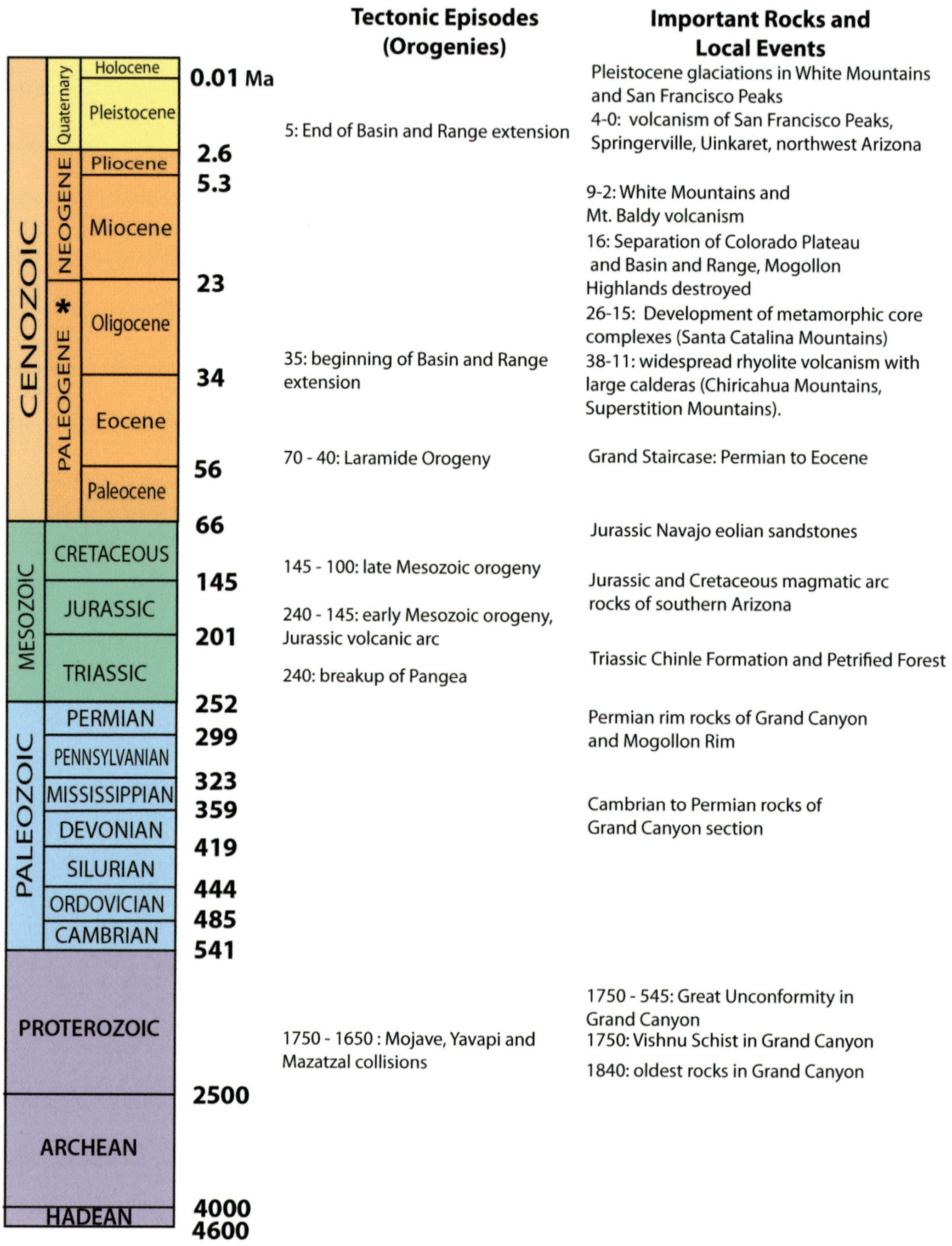

Figure 3. Geological time scale and important events for Arizona. Numbers represent millions of years ago (Ma). *The Tertiary Period, which has been replaced by the Paleogene and Neogene Periods, is still widely used and appears in the text.

in our mind a vast warm equatorial plain of meandering rivers and dense tropical forests. When we look at the red rocks towering above Sedona, we try to visualize a fluctuating coastline that transforms from beach to sand dunes to arid coastal salt-flats and back again in reverse order. We want to look at a modern landscape and visualize ancestral landscapes within it and read their stories.

Arizona's biological diversity ultimately comes down to two things—rocks and climate. An interface is created between them that modulates the interaction of the two and creates the surface of our world. This is the soil. In some places it is just weathered mineral grains, in other places a complex soil profile, and in several places it is absent. We will generally and very briefly describe the soils of the areas we visit. A convenient way of classifying soils that relates closely to geography uses soil temperature and precipitation. Figure 4 shows the seven soil groups that occur in Arizona and can be mapped. Each area includes many mapped soil associations, geographic soil bodies that are segments of the soil mantle. As you might imagine, the pattern of soil associations is extremely intricate, so we will restrict our survey to the seven soil groups.

We will begin our survey in southern Arizona where the rocks and landscapes reveal a long history beginning with the assembly of the continent through episodes of mountain building and the growth of North America to the creation of the present-day Basin and Range landforms. Next, we will observe the Colorado Plateau, a land of crustal tranquility. There is hardly a better place to interpret and visualize ancient landscapes. Finally, we will examine the Transition Zone, which is an amazing mash-up of the stories of the other two regions.

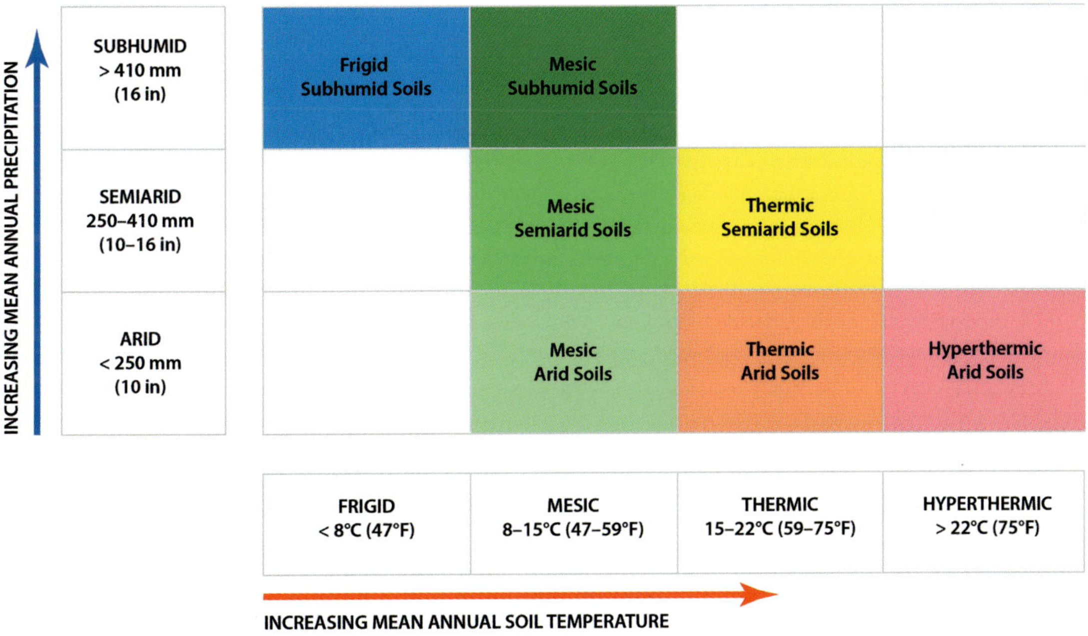

Figure 4. Soils of Arizona. Classification scheme is based on soil temperature and precipitation. Based on information from Hendricks (1985).

BASIN AND RANGE

Southeast Arizona

The Sky Islands region is a world where intense deserts shimmer just a few miles from peaks that reach into conifer forests. It is an area in southeast Arizona, southwest New Mexico, and northern Sonora, Mexico, that is sometimes called the Madrean Sky Islands, named after the Madrean pine-oak woodlands that occur in the Sierra Madre Occidental of Mexico. The ranges are islands because they are isolated from one another and have evolved different ecosystems.

Each range is an uplifted and eroded fault block that rises above the surrounding basins and often presents a spectacular display of its piece of geologic history. Some like the Santa Catalina and Santa Rita Mountains near Tucson are over 9,000 ft (2,740 m) and contain deep-seated intrusive igneous and metamorphic rocks which were created during times of mountain building and major crustal deformation called orogenies. Others, like the Whetstone Mountains exhibit thousands of feet of limestone deposited in warm shallow Paleozoic continental seas. The Chiricahua Mountains are the largest of the Sky Islands and expose the material of a very violent volcanic eruption during Cenozoic time. The Pinaleño Mountains along the New Mexico border are the highest Sky Island range at 10,724 ft (3,268 m) and contain rocks that are 1,800 million years old. Within the westernmost range, the 7,734 ft (2,357 m) Baboquivari Mountains, there are intrusive igneous rocks formed near the core of a Mesozoic volcanic arc.

It is because of uplift and erosion that the ranges of the Basin and Range Province expose rocks that reveal episodes in the geologic history of the West. Each range tells its own story. Sometimes the stories in different ranges are separated by many millions of years and some overlap in time. By piecing together the records from different ranges geologists have been able to assemble a history for the evolution of southern Arizona from about 1,650 million years ago to now. Because of the depth of their history, the Sky Islands are a logical place in our account to summarize this history and place it in the chronicle of western North America.

About 240 million years ago the supercontinent Pangea began breaking up and North America commenced its westward drift. The western margin of the continent began colliding with Pacific Ocean tectonic plates, driving the denser oceanic plates beneath the overriding western margin of the continent. This process, called subduction, applies compressional forces to the continental plate as the subducting oceanic crust sinks beneath it into the mantle, melts, and sends magma up into the continental crust. The ascending magma forms masses of intrusive igneous rocks within the crust and great arcs of volcanoes near the edge of the continent. The compressional forces cause faulting where slabs of rock are thrust over one another, thickening the continental crust. Heat and pressure create metamorphic rocks.

The American West is part of a very long belt of mountains, called the American Cordillera, which extends from Alaska to the southern tip of South America and Antarctica. The Andes have much in common with the North American Cordillera and help us visualize the evolution of western North America. The development of the cordillera was very protracted.Vast tracts of ocean floor disappeared beneath the continent, huge arcs of volcanoes grew along its margin, and exotic fragments of distant oceanic islands migrated in and became part of the growing continent. There was a time when the Pacific coastline was where eastern California is today. There are rocks in California, Oregon, and Washington that began their journeys in Alaska and others that came from Baja California. There are rocks just west of Phoenix that were once ocean floor. Through the Mesozoic, mountain building migrated eastward forming mountain ranges that have long since vanished. The compressional orogenic history culminated with the Laramide Orogeny, 80 to 50 million years ago, and the development of the Rocky Mountains.

The modern landscapes began to develop about 15 million years ago. With the formation of the San Andreas Fault plate boundary about 30 million years ago, western North American changed from a compressional orogen to an extensional one, and western North America experienced an episode of crustal extension that stretched and nearly rifted the West from North America. As the crust stretched, rock masses migrated westward and some are now found many miles west of their original locations. Deeper levels of the crust stretched and thinned like taffy, in some cases by 100%, while the shallower brittle levels of the crust broke into blocks. Some blocks subsided, slid, and tilted along high-angle faults to form fault basins, called grabens, often leaving the adjacent blocks thousands of feet higher. Some blocks dropped down between two border faults. Many blocks slid and tilted along a single fault that flattened with depth and were moved horizontally along a nearly flat fault plane. These faults are all called normal because their motion is driven by gravity. The upper block slides down along the fault plane during a time when compressional forces are relaxed or have become tensional. The high-angle border faults commonly create straight-line topographic boundaries with the mountains of the upthrown block rising abruptly from the basin. The structural result of these processes is called block-faulting. Normal faulting is the opposite of the thrust faulting that occurs during compressional mountain building. Strike-slip faulting is another type that occurs when one crustal mass slides horizontally alongside of another. The modern San Andreas fault is the iconic example and is just the latest expression of the tectonics that have dominated the western margin of the North American continent for many millions of years.

Look closely at the landforms map (Fig. 1) and you can see a swarm of sub-parallel ranges south of the edge of the Transition Zone. As you study this pattern, you can almost imagine the ranges breaking away from the Colorado Plateau mass and from one another as they drift westward! The consequences of this extension were often more than the formation of fault blocks. In some areas like the Santa Catalina Mountains near Tucson, magmatic upwelling was activated, and the overlying rocks were intensely deformed by heating and shearing to form metamorphic core complexes.

The fault basins rapidly filled with sediment washed from the ranges and created the distinctive Basin and Range landforms. The Sky Islands are a special example of this kind of landscape because of the high elevation of the ranges and vertical relief. They are also islands in the sense that they rise above a sea of lowlands between the Sierra Madre Occidental and the Rocky Mountains.

Beneath the valley floors are thousands of feet of sediments that were washed from the ranges. They document the vigorous erosion of the mountains and filling of the fault basins as they subsided during 10 million years of extension, faulting, and subsidence. Erosion continued as faulting waned about 5 million years ago and the modern landscape began to emerge. Pouring from the deepening canyons with their sediment load, the streams flooded across the valley floors and built alluvial fans out into the valleys. Overlapping fans formed aprons of sediments, called bajadas that gently slope away from the mountain front to the valley floor. The abrupt mountain front at the head of the bajadas is a unique feature that we will consider when we get to Southwest Arizona.

At first the basins were closed and water drained to their centers, evaporated, and formed playa lakes. By the early Pleistocene the separate basins became integrated into the Santa Cruz and San Pedro rivers, becoming part of the Colorado River system. During Pleistocene time climates fluctuated from semiarid to cool and moist. Streams alternately cut valleys and then filled them, created alluvial fans and then dissected them. The alluvial fans and stream terraces above the valley floors and the channel deposits of the valleys preserve a record of the rivers' activity from the Pliocene, through the Pleistocene to modern times. During glacial advances when cool, moist pluvial conditions prevailed south of the ice front, lakes were abundant. Willcox Playa is a remnant of pluvial Lake Cochise, Arizona's only pluvial lake.

The sands and silts of the valley floors as well as the fans and terraces are mantled with Thermic Semiarid Soils in which grow Sonoran Desertscrub and grasslands with abundant mesquite woodlands. At higher elevations the rocky surfaces are weathered and eroded from the underlying rocks, commonly granitic and volcanic. Here the soils are Mesic Subhumid and support oak, juniper, and pinyon of the Madrean Evergreen Woodlands. The angular rocks of the canyons and talus slopes within the mountains grade to the rounded material of the alluvial fans. Picking through the cobbles and boulders on a fan looking for different rock types just might save you a hike into the mountains!

Some of the Sky Islands rise to elevations where the freeze-thaw process helped create a rugged precipitous topography steep with rocky slopes, cliffs, and abundant young talus. Frigid Subhumid Soils support Rocky Mountain conifer forests. Monsoon-season flash floods are essential agents in creating these landscapes. Virtually all the erosion takes place during flash flood events.

Lower Colorado River Valley

West and north of the Baboquivari Mountains, the westernmost Sky Island range, is a seemingly endless desert. This expanse is the heart of the Sonoran Desert region of the Basin and Range Province. Arid land geologic processes like wind and intense thunderstorm-driven running water have created a place of extremes. The Gila and Colorado Rivers pass through the region on their way to the Gulf of California leaving many of the basins with internal drainage. The soils are all Hyperthermic Arid and have developed on the Basin and Range block-fault geomorphology. Our geologic region includes the Arizona Uplands with saguaros and paloverdes on the rocky slopes and bajadas, and the Lower Colorado Valley with shrubs like creosote bush and many cacti.

In contrast to the Sky Islands, the mountainous areas of southwestern Arizona are smaller, often with large distances between them. Aprons of overlapping alluvial fans forms extensive bajadas gracefully sloping away from the mountains and seamlessly connecting the contrasting landforms, often concealing the border faults that separate the blocks. Talus cones on the mountain slopes are widespread. Many are old, formed during the Pleistocene and are now being eroded away. Because of the sparse vegetation, flash flood events are even more effective in this region.

Landscape development in arid and semiarid block-faulted regions creates a unique landform called a pediment. The high rugged topography of the mountains does not gradually change to the smooth gently sloping floor of the valley, as in a more humid area. An abrupt change in slope occurs where streams flow out of canyons at the mountain front. As the streams emerge from the front, they "whip" laterally back and forth over time scouring the surface and creating a gently sloping planar surface called a pediment. Bedrock occurs at the surface or beneath a thin veneer of alluvium. Basin-ward the bedrock is covered with thicker and thicker alluvial deposits forming bajadas. Through time the mountain front retreats, maintaining the abrupt front and the mountainous area becomes smaller and smaller eventually disappearing beneath sand and gravel. It is common to see an isolated crag surmounting a long, elegant bajada, the last remains of a mountain range.

Border Ranges West to Yuma

Some ranges westward from the Sky Islands to Yuma display glimpses of the granitic and metamorphic rocks of the early Proterozoic continental basement. Others expose Mesozoic to early Tertiary sedimentary, metamorphic, and igneous rocks that formed during the development of the Cordilleran mountain belt. Their rocks demonstrate the intensity of deformation and transformation that occurs near the axis of an orogenic zone.

Widespread middle- to late-Tertiary volcanic rocks form ragged and hostile terrains. An exceptional area of basaltic volcanism is the Pinacate Volcanic Field. It is a 580-square mile, or 150,000-hectare (ha), oval just south of the Mexican border that exemplifies Quaternary volcanism in southwestern Arizona. Lava flows, cinder cones and other explosive features look like they happened yesterday. This volcanism began a few million years ago and the most recent eruption occurred about 12,000 years ago. Another large basaltic field west of Gila Bend contains shield volcanoes similar to the Hawaiian volcanoes.

Phoenix Area

Driving I-10 from Tucson to Phoenix takes you from the Sky Island ranges out into the realm of wind. The drive takes you across Quaternary stream deposits of the Santa Cruz River that merge with those of the San Pedro and then, in the middle of Phoenix, with the Gila and Salt rivers. The urban area and agricultural lands have been developed on a huge accumulation of sediments from the weathering and erosion of all of southeast Arizona and parts of southwest New Mexico. These sediments are over 11,000 ft (3,400 m) thick beneath Gilbert. Rising from this plain are rocks that range from early Proterozoic granitic and metamorphic rocks to middle Tertiary basalt. South Mountain is an urban metamorphic core complex.

Superstition Mountains

Just east of Phoenix are the spectacular volcanic landscapes of the Superstition Mountains. The pinnacles, towers, and impenetrable canyons are carved from welded tuff. Around 25 million years ago very violent eruptions of viscous rhyolite magma spread ash and pyroclastic flow deposits over 2,500 square miles (650,000 ha). The material of many of the pyroclastic flows was still molten when it came to rest and fused into the welded tuffs from which the pinnacles and canyons have been eroded. Often, big rhyolite eruptions climax with the catastrophic collapse of the summit area into the evacuated magma chamber forming a caldera. Although they are no longer visible, several calderas were formed during the Superstition Mountain eruptions. Splendid views of these rocks can be seen on the drive between Apache Junction and Lake Roosevelt.

Northwestern Ranges

The western ranges north of the Gila River seem scattered randomly across great tracts of alluvial desert etched with the graceful patterns of flowing water. When viewed from altitude in a commercial jet, there is no sense of the north-south order expected from Basin and Range protocols. Part of this is because of complications in the compressional deformation of this part of the Cordillera during Mesozoic mountain building. Later Basin and Range extensional patterns were influenced by the structures created and several unusual northeast-trending ranges formed. They contain metamorphic core complexes and major exposures of early Proterozoic rocks.

Basin and Range order returns west of the Colorado Plateau and could hardly be more dramatic. Several ranges with the stature of the Sky Islands stand just west of the plateau's edge. The great escarpment of the Grand Wash Cliffs defines the western break of the Colorado Plateau. Here the rocks of the Basin and Range dropped 15,000 ft (4,500 m), sliding westward and breaking away from the plateau along the Grand Wash Fault between 16 and 9 million years ago. Today the faulted blocks stand high as north-trending ranges separated by deep basins. The Hualapai Basin is an internally drained playa basin and was filled with a mile (2.5 km) of salt because subsidence progressed during extension. It was in this corridor of basins that evidence critical to understanding the history of the Grand Canyon was discovered and debated.

COLORADO PLATEAU

The Colorado Plateau is 130,000 square miles (33,000,000 ha) of plateaus, canyons, valleys, badlands, deserts, peaks, escarpments, buttes, towers, hoodoos, and more. This breathtaking array of landforms includes parts of Utah, Colorado, New Mexico, and Arizona. From the inner gorges of the Grand Canyon to the active dunes of the Moenkopi Plateau, a record spanning 1.84 billion years of history is present.

The many layers of rock exposed in the Grand Canyon give evidence that the plateau did not exist before 80 million years ago. At this time a continental sea and its fringing flood plains retreated in response to plate tectonic action far to the west. This Mesozoic orogeny is recorded west and south of the Plateau as we have seen in the mountains of southern Arizona. By 80 million years ago things began to change and the effects of the orogeny began to migrate east. This is called the Laramide Orogeny and ultimately the deformation and uplifts of the Rocky Mountains and their intervening basins occurred. The Colorado Plateau was uplifted with deformation limited to only a few monoclinal folds and early movement along major faults. Remarkably, the Colorado Plateau and its rocks remained intact and relatively undeformed. Why the Plateau endured and why it was uplifted as a stable platform, as if there was a planetary scale mechanical scissor-lift beneath it, is a major question for geologists. However it happened, the Plateau area has been uplifted about 7,000 ft (2,000 m) since Laramide time.

The boundaries that define the Colorado Plateau are sometimes spectacular and sometimes only a line on a map. To explore the Colorado Plateau, we will begin along its southern boundary, move westward to explore some major areas and features, and look more closely at the plateau's history.

White Mountains

The southeast corner of the plateau is buried beneath a large area of late Cenozoic volcanism. Volcanic activity from 9 to 2 million years ago built the White Mountain Volcanic Field that was an expanse of lava flows, cinder cones, and ash deposits. The massive Mount Baldy stratovolcano and its long, spreading flows of basalt and andesite was then constructed on the older volcanics. It may have reached an elevation of 13,000 ft (4,000 m). The volcano has since been deeply eroded and today Mount Baldy has an elevation of 11,409 ft (3,477 m). During the last Ice Age, it hosted several glaciers that formed classic glacial landforms like cirques, u-shaped valleys, and moraines.

As cinder cones were eroded away and the lava flows smoothed, the modern landscape evolved and uplift caused streams to cut deep canyons. Today Mount Baldy surmounts the sprawling White Mountains landscape, covered with mixed conifer and ponderosa pine forests. Deeply eroded drainages descend into woodlands of pinyon, juniper, and oak. The last episode of volcanism ended about 300,000 years ago and produced the Springerville Volcanic Field where there are many well preserved volcanic features.

The soils of the higher White Mountains are Frigid Subhumid and transition south and west into Mesic Subhumid and Mesic Semiarid Soils of the Transition Zone. The soils become Mesic Arid northward across the Springerville Volcanic Field and onto the Colorado Plateau.

Mogollon Rim to the Grand Wash Cliffs

The southern edge of the Plateau continues northwestward across the western White Mountains to the Mogollon Rim northeast of Payson. This great erosional escarpment rises with grandeur above the Transition Zone and Basin and Range, sharply defining the southern boundary of the plateau. Capping the rim is the Permian Kaibab Limestone, Toroweap Formation, and the Coconino Sandstone, the same rocks that occur at the rim of the Grand Canyon. From their 8,000-ft (2,400-m) elevation these formations are inclined gently northward and disappear beneath the younger rocks of the Painted Desert. The modern rim marks a moment in the northward retreat of the southern edge of the Colorado Plateau. To its south lies the disarray of faulted and subsided blocks of the Transition Zone.

The ponderosa and mixed conifer forest of the rim grades northward into pinyon-juniper woodlands as the Frigid Subhumid Soils transition to Mesic Semiarid Soils then to the Mesic Arid Soils landscapes of the Moenkopi Plateau and the Painted Desert.

Following the Mogollon Rim northwest, Sedona soon appears in a breathtaking setting of red-rock cliffs, towers, and pinnacles. It is a distillation of the drama of the Colorado Plateau, but is technically part of the Transition Zone and will be described later. The boundary continues northwest and then becomes obscured by volcanic rocks until it turns northward and becomes the Grand Wash Cliffs. Here the Plateau is in direct contact with the Basin and Range across a fault that dropped the rocks 15,000 ft (4,600 m) as the crust extended, subsided and moved westward. The Colorado River emerges in a most stately fashion from the Grand Canyon through the Grand Wash Cliffs, turns south and continues its journey.

Northwest Plateaus

The northwestern part of the Colorado Plateau in Arizona is a landscape of plateaus, seven of them, separated from one another by faults and monoclinal folds. North of Flagstaff near Cameron, U.S. Route 89 crosses the Painted Desert. These nearly horizontal, multihued sedimentary rocks are part of a broad swath of Triassic and Jurassic rocks that sweeps across the northeast corner of Arizona. We will talk about them shortly. North of Cameron the dip of the beds steepen and they gather together as a narrow and ragged escarpment that curves in from the east and parallels the highway for many miles. This is the Echo Cliffs monocline and exposes steeply upturned and eroded Triassic rocks and the overlying Navajo Sandstone. West of the highway the rocks have been uplifted to form the Marble Platform that is floored by the Permian Kaibab Limestone. Thousands of feet of Mesozoic rock have been stripped off the surface including the rocks of the Echo Cliffs that plunge eastward off the monocline and disappear beneath the Painted Desert.

The Marble Platform is the easternmost of the seven plateaus. To its west the land rises abruptly onto the surface of the Kaibab Upwarp. The eastern edge of this structure is defined by the East Kaibab Monocline. U.S. Route 89A makes a steep twisting ascent of the monocline exposing fossiliferous Permian limestones in the roadcuts. The upwarp is easily seen from space as a dark elongate teardrop of Rocky Mountain conifer forests and boreal and subalpine conifer forests at the highest elevations.

It is across this structural and topographic barrier that the Colorado has cut the Grand Canyon. North of the canyon the upwarp is called the Kaibab Plateau and rises to an elevation of about 8,700 ft (2,700 m). The North Rim of the Grand is at 8,200 ft (2,500 m), 1,200 ft (370 m) higher than the South Rim. The Coconino Plateau is the broad region south of the canyon. The connection between the east side of the Coconino Plateau and the Kaibab Plateau is evident on maps and space imagery.

The Coconino Plateau is 5,000 square miles (1,000,000 ha) of sprawling uplands and drainages that range from ponderosa pine forests to Great Basin sage and grasslands. Most of it is above 5,000 ft (1,520 m) and gently descends westward into the dissected drainage of Havasu/Cataract Creek. It then ascends again to the uplands of the Havasupai and Hualapai Indian reservations and its western edge at the Grand Wash Cliffs. The southern third is covered by the volcanic fields of the San Francisco Peaks near Flagstaff. Frigid Subhumid Soils of the San Francisco Mountains transition to Mesic Semiarid Soils and to Thermic Semiarid Soils that occur in the lower reaches of Havasu Creek. The Permian limestones that form the rim of the Grand Canyon are just beneath the surface over much of the Coconino Plateau.

The high landscapes north of the canyon rim continue westward from the Kaibab Plateau as a series of uplifted plateaus separated by faults to the edge of the Colorado Plateau and the Grand Wash Cliffs. North of the Grand Canyon the Kanab Plateau reaches 5,700 ft (1,740 m) and west of it is the Uinkaret Plateau which reaches 8,028 ft (2,446 m). This is the site of the Quaternary Uinkaret Volcanic Field. Uinkaret volcanoes erupted basalt most recently from 600,000 to about 1,000 years ago. Some flows descended into the Grand Canyon forming dams and temporary lakes. The westernmost Shivwits Plateau rises to about 7,000 ft (2,130 m) and is also capped by an area of Quaternary volcanism.

The high plateaus are mantled with Mesic Semiarid and Arid Soils that host Great Basin Conifer Woodlands with Rocky Mountain Woodlands at their highest elevations. The intervening drainages descend into Great Basin Desertscrub and Grasslands. Near the canyon of the Colorado River, Mesic Arid Soils with Mohave Desertscrub are encountered.

Grand Canyon

In contrast to our aerial survey, floating through the Grand Canyon is a very intimate experience. Raft trips start at Lee's Ferry and are immediately in Marble Canyon with Permian rocks at river level. In only 23 miles the cliffs of the Mississippian Red Wall Limestone are reached. At river mile 63 the Great Unconformity is reached. This is a profound and worldwide gap in the rock record that occurs at the beginning of the Paleozoic. Here the Cambrian Tapeats Sandstone, 525 million years old, rest on 1,150-million-year-old sandstone of the late Proterozoic Grand Canyon Supergroup, a series of sediments and volcanics deposited from 1,200 to 740 million years ago. The unconformity can easily be seen from the South Rim at a level where the horizontal rocks of the Paleozoic overlie the tilted rocks of the Proterozoic Era. At mile 77 the river descends through another major unconformity and enters the Inner Gorge and the oldest rocks exposed in the canyon, the granites, schists and gneisses of the "Vishnu Basement Rocks." These 1,700-million-year-old early Proterozoic rocks are the oldest rocks in Arizona. The oldest dated rock is the Elves Chasm Gneiss, 1,840 million years old. This reach of the river, a mile below the canyon rim, flows through narrow dark gorges of glistening black schist. Brilliant dikes of pink granite seem to leap out of the water and ascend the canyon walls. In this very hot, arid place a remarkable array of cactus, agaves, and many other desert plants ornament the walls. Rising high above are the ramps and cliffs of the Paleozoic rocks. At river mile 120 is Blacktail Canyon where travelers make a pilgrimage into a slot canyon shrine. Here the 525-million-year-old Tapeats sandstone rests on 1,750-million-year-old Vishnu Schist. For some it can be a spiritual experience to span with your hand the 1,200 million years of missing historical record.

The western section of the canyon exposes mostly Paleozoic rocks. A remarkable exception is seen at Lava Rapid. On the rim is Vulcan's Throne, a cinder cone on the brink of the Uinkaret Plateau that sent flows of basalt into the canyon and dammed the river about 70,000 years ago. Downstream from Vulcan's Throne the river passes through Vishnu Basement Rocks and Cambrian rocks with 3,500 ft (1,070 m) of Paleozoic rocks rising to the rim. At river mile 277 the river flows from the Colorado Plateau into the Basin and Range, turns south and makes its way to the Gulf of California. The history of the Colorado River has been a focus of controversy ever since scientists began writing about it in the middle of the nineteenth century. Only in recent years has its very complex story been revealed. Simple questions about the age of the Canyon are always met with complicated answers. It began cutting the present canyon no earlier than 5 or 6 million years ago, influenced by the development of the Basin and Range, the uplift of the Colorado Plateau, and ancient drainage systems including an early one that drained to the northeast across the plateau region 70 million years ago.

San Francisco Peaks Volcanic Field

Volcanism has been central to the geologic history of Arizona. The state is splashed with volcanic rocks that erupted during Cenozoic time. Even the Colorado Plateau has experienced volcanism, although not nearly as much as surrounding areas. The peaks of the San Francisco Field are the volcanic set piece of the Colorado Plateau. Humphreys Peak is a deeply eroded stratovolcano that rises to 12,633 ft (3,850 meters) above its retinue of volcanic features. East of the high peaks is a field of cinder cones that includes Sunset Crater which erupted in A.D. 1064. More than 600 volcanoes are spread over 1,800 square miles (466,000 ha) of the southern Coconino Plateau. Volcanism began about 6 million years ago. The magma types that have erupted range from very viscous high-silica magmas that create the rock rhyolite and form volcanic domes to intermediate-silica types that form andesite and are characteristic of stratovolcanoes like those of the Pacific Rim of Fire, to low-silica lava that forms cinder cones and flows of fluid basalt. The high peaks are part of a stratovolcano that began erupting about 400,000 years ago, but the basalt cinder cones, east of the peaks are the youngest and most common type of volcano.

Mt. Humphreys is one of two areas in Arizona that supported glaciers during the Pleistocene. The other, you will recall, is Mt. Baldy. Evidence for the glaciation of Mt. Humphreys includes several small cirques that occupy the summit area. The Inner Basin, which drains the southern slopes, has the U-shape of a glaciated valley and is bordered by moraines. The glacier was only about 1.5 miles long, but its glacial deposits form an important source of water for Flagstaff.

The San Francisco Peaks terrain is part of a large area of Frigid Subhumid Soils and Rocky Mountain conifer forests that includes the Mogollon Rim. It is nearly surrounded by upland tracts of Mesic Semiarid Soils that include the Coconino Plateau and the slopes of the Mogollon Rim. Mt. Humphreys is host to Alpine Tundra.

Arizona Strip

From Lee's Ferry the river traveler floats through thousands of feet of stratified rocks to the ancient crystalline core of the continent. A few miles downstream from Lee's Ferry the Navajo Bridge is anchored to the Permian Kaibab Limestone and carries U.S. Route 89A across Marble Canyon. A west-bound traveler crosses the bridge and drives across the Marble Platform, which quickly conceals the canyon. To the north an immense brilliant red escarpment called the Vermillion Cliffs rises nearly 3,000 ft (910 m) to the Paria Plateau and extends westward from the East Kaibab Monocline and Kaibab Plateau. This is the Arizona home of the California condor. The cliffs are a stunning display of Jurassic sandstones and shales that tell of vast sand seas that covered parts of Arizona, Utah, New Mexico, Colorado, and Wyoming. The Vermillion Cliffs are one part of an extraordinary landscape that extends from the Kaibab Plateau north into Utah. It is called the Grand Staircase. Whatever image the name Grand Staircase evokes, it will likely fall short of the surpassing vista that unfolds from the north end of the Kaibab Plateau. The landscape displays, step by step, the rocks and history through the Mesozoic and into the Cenozoic. The Vermillion Cliffs are the second step. Ascending above them are the Jurassic White Cliffs, the Cretaceous Grey Cliffs, and the Cenozoic Pink Cliffs of Bryce Canyon. The ascent culminates at Cedar Breaks National Monument in Utah along the lofty 10,000 ft (3,050 m) high edge of the Markagunt Plateau after passing through about 16,000 ft (4,880 m) of stratified rocks from the depths of the Grand Canyon. Cedar Breaks is not in Arizona, but it would be difficult to find a place that offers a better opportunity to attempt to grasp the planetary grandeur of the Basin and Range and its companion the Colorado Plateau.

Glen Canyon

A few miles above Marble Canyon the Colorado River is released from Lake Powell through the turbines of Glen Canyon Dam. This is one of those places that concentrates the spectacular red rock scenery of a region into one dazzling display. The waters of Lake Powell drowned Glen Canyon and this mystical place now exists only in memories, words, and photographic images. The main Colorado canyon and its maze of tributary slot canyons are carved from the Jurassic cross-bedded eolian sandstones of the Navajo Formation. Today the Navajo sandstone forms the vertical cliffs and appealing inlets seen in Arizona. In Utah younger eolian sandstones and shallow-sea sandstones and siltstones sport alternating red and white stripes.

Black Mesa, The Little Colorado River, Painted Desert, and Petrified Forest

The highlands of the Mogollon Rim and the Kaibab Plateau, supported by late Paleozoic rocks, form a great arc embracing the northeast corner of Arizona. This arc is the southern rim of the Colorado Plateau and from their heights these rocks disappear northward beneath successively younger rocks. At the center of this concentric pattern is Black Mesa. This area is defined by the rocks and inseparable from them. Encircling the Cretaceous rocks of Black Mesa is a ring of older Triassic and Jurassic rocks that create the classic landscapes in the land of the Little Colorado River. This modest looking river drains a large area of the Colorado Plateau and releases its load of red sand and mud into the Colorado River above Grand Canyon.

Eastward from the San Francisco Peaks basalt fields and the Echo Cliffs to the Little Colorado River is the windswept country of the Painted Desert. Its landscapes are eroded into desert plains and low plateaus with little vegetation to hold the soil. Modern sand dunes move across the sandstones and shales of ancient, arid times. What little soil there is is Mesic Arid and supports Great Basin Desertscrub and Semidesert Grasslands.

Interstate 40 east of Flagstaff crosses the complex and multihued landscape of the Painted Desert and enters the Petrified Forest west of Holbrook. Of all the wildly colored layers of rock, perhaps the most intense and vibrant are those of the Triassic Chinle Formation. The colors and landforms are so compelling that Georgia O'Keeffe chose to spend her life among them where they occur at Ghost Ranch in New Mexico. During late Triassic time a river system, perhaps as large as today's Amazon River, drained mountains to the south and east. North America was near the equator and rich and diverse ecosystems thrived and evolved. The forests and stream-side habitats were filled with a remarkable array of plants and animals that included large trees and the early ancestors of the dinosaurs. In Petrified Forest National Park huge tree trunks litter the desert. Hidden from the casual sightseer are bone quarries that have revised our understanding of the origin of dinosaurs.

North of the Petrified Forest is a 1,000-square-mile (259,000-ha) field of 300 explosive volcanic features called Hopi Buttes that formed from 4 to 8 million years ago. The buttes are black basaltic volcanic necks eroded from volcanic pipes and ring-shaped ash deposits resembling coliseums.

Northeastern Arizona

The culmination of the westward stratigraphic and topographic ascent from the valley of the Little Colorado River is Black Mesa with pinyon-juniper woodlands growing on Mesic Semiarid Soils. Its eastern margin is a bold escarpment of Cretaceous sandstone and its western margin slopes into Mesic Arid Soils with their Great Basin Desertscrub and Semidesert Grasslands of the Little Colorado River Valley. Beneath the pinyon and juniper woodlands are Cretaceous sandstones and shales containing many coal seams that formed along meandering river systems that drained westward across warm lush coastal lowlands into a shallow continental sea. Thick layers of organic material were deposited along with fine sands and muds. The carbon-rich beds were compressed and transformed into coal layers that were mined by Hopi residents for centuries and by Peabody Western Coal Company from the 1960s through 2019.

The escarpment of Black Mesa looks eastward across the Four Corners region of Arizona, New Mexico, Colorado, and Utah. Along the New Mexico border the Carrizo and Chuska Mountains ascend from the Chinle Valley to nearly 10,000 ft (3,050 m) and host pinyon-juniper and mixed conifer and ponderosa forests on Frigid Subhumid Soils near the highest elevations. The large monoclinal uplifts of the Monument Upwarp and Defiance Uplift are scattered with the unusual volcanic features of the 25- to 30-million-year-old Navajo Volcanic Field. Ship Rock in New Mexico and Agathla north of Black Mesa are the classic examples. Commonly called volcanic necks, they are volcanic conduits filled with the shattered debris of very vigorous eruptions and are accompanied by dikes and other intrusive features. Erosion of the surrounding rocks has created dark spires, towers, and walls.

North of Black Mesa is Monument Valley, the topographic expression of the Monument Upwarp. As with the Kaibab Upwarp the up-folding brings older rocks to the surface and exposes them to erosion. Here they are Permian eolian sandstones and shales deposited when much of the western interior of the continent was an arid plain. The vertical towers, spires, and buttes are eroded from resistant cross-bedded sandstone that is nearly pure quartz dune sand. They rest on broad sloping bases of easily eroded shales. The quartz sandstones are very brittle and when they were bent during the upward flexing of the rocks they broke into an orderly system of cracks called joints. The joints focused weathering and erosion and began the disassembly of the plateau because it separated into mesas which were then reduced to buttes, spires, and pinnacles.

Several streams flowing west off the border highlands have cut deeply into the Triassic and Permian rocks creating one of the Colorado Plateau's most iconic places—Canyon de Chelly. The canyon is cut in highly cross-bedded eolian sandstones and shales, the same rocks we saw in Monument Valley. Vertical cliffs rise directly from the valley floor with its meandering stream. Over-hanging rock shelters with abundant evidence of extensive human use are eroded in the walls of sandstone. The photographer Edward Curtis spent much time here and because of his work, Canyon de Chelly has come to represent the Colorado Plateau for much of the world.

ARIZONA TRANSITION ZONE

Plant your feet on an airy ledge of Coconino Sandstone and look across a boundless pine-covered wilderness that fades away far to the south. You are on the Mogollon Rim near the town of Payson and your view is of a long mountainous belt that trends northwest across the central part of the state. As an ecological region it is often called the Central Highlands and it contains some of the most rugged and remote country in Arizona.

The region is called the Transition Zone because it separates the Colorado Plateau from the Basin and Range and contains features of both. Mountains such as the plateau-like Sierra Ancha and the rugged peaks of the Mazatzal Mountains are bordered by elongate valleys like the Verde Valley and Tonto Basin. Familiar rocks from the Colorado Plateau are mixed with rocks found in the ranges of southeast Arizona. The sense of soaring heights of the landscape is reflected in the rocks. Geologists would say the region is structurally high because deeply buried Precambrian rocks, like those in the gorges of the Grand Canyon have been uplifted and eroded to create some mountain peaks of the Transition Zone.

From the ponderosa pines that grow in the Frigid Subhumid Soils of the Rocky Mountain Conifer Forests along the Mogollon Rim, the terrain descends to Mesic Subhumid and Mesic Semiarid Soils. Slopes are covered with juniper and pinyon pine of the Great Basin Conifer Woodlands. Some areas of Interior Chaparral occur below the rim and cover higher ridges and mountainous divides. Higher elevation valleys like the Verde Valley are Semidesert Grasslands with Arizona Uplands Desertscrub. Lower elevation valleys like the Tonto Basin are Arizona Uplands Desertscrub growing on Thermic Semiarid Soils. Small enclaves of Madrean Evergreen Woodlands are scattered throughout the zone.

During the late Cretaceous about 80 million years ago when the Laramide Orogeny was transforming the interior West, a mountain range called the Mogollon Highlands was uplifted to heights greater than the rim it was named after. Sediments were shed northward off this range across the much lower area that would later become the Colorado Plateau. When the Basin and Range extension occurred the Mogollon Highlands broke up, foundered, and created the southern edge of the high Colorado Plateau. As the ancient rim of the Colorado Plateau began to retreat to the north, its eroded sediments were carried southward. When you look south across this region from the Mogollon Rim you are seeing the immense rubble pile of the destruction, each fragment containing its part of the remarkable story.

The uplands of the eastern Transition Zone are concealed beneath Tertiary volcanics and the White River Volcanic Field. The deeply dissected landscape crests in forests of ponderosa pine and mixed conifers. The canyons descend into pinyon, juniper, and oak woodlands. The Blue River, the Black River, and the San Francisco River flow through deep valleys and canyons to join the Gila River. This wild and very remote country was selected for the successful reintroduction of the Mexican grey wolf.

The Salt River drains the northern part of these highlands and is deeply entrenched where it flows toward the Gila River. Travelers on Arizona Highway 77 are amazed to discover the Salt River Canyon and a vista that could easily be of the Grand Canyon. Here the same rocks that appear beneath the Great Unconformity as the Grand Canyon Supergroup are exposed in the Salt River Canyon walls.

In the central part of the Transition Zone, its landforms seem to merge with those of the Basin and Range. This is one of those places that reminds you that landscapes are gradational and lines on a map are not. The Mazatzal Mountains with their flanking Tonto Basin are a range that seems to be one of the Sky Island swarm. Among its peaks are 1,650-million-year-old rocks that record the collision and accretion of island arcs to the southern margin of the growing North American continent. This event is called the Mazatzal Orogeny and was a major moment in the early growth of the continent.

Sedona dwells at the foot of the Colorado Plateau and its rocks tell Plateau stories. The red and white layered rocks are Triassic through Jurassic sandstones, siltstones, and shales. The red colors indicate coastal plain river deposits and the white color indicates coastal dunes and beaches. Scanning up through the layers is like fast forwarding through hundreds of thousands and millions of years of a fluctuating coastline when ocean beach becomes coastal salt flats or dunes and sediment is washed off the continent from as far away as Appalachia. Crowning the top of the escarpment are Permian limestones and the Coconino Sandstone.

Sedona is part of the Verde Valley, which was created by northward retreat of the Mogollon Rim, block faulting, and basaltic volcanism. Jerome sits on an upthrown fault block that overlooks Sedona and is the site of fabulously rich deposits of copper, gold, and silver that were created 1,800 million years ago on the seafloor in an island arc setting near the southern margin of the ancestral continent.

West of Sedona the Transition Zone becomes less of a distinct region. The western plateaus of the Grand Canyon region prevail as the Transition Zone abuts the Grand Wash trough and the basins along the Basin and Range detachment corridor.

CONCLUSION

Our metaphorical drone revealed to us a dazzling and thought-provoking montage of landscapes and gave us glimpses of ancient scenes in the creation of those landscapes. Now, after pausing in its flight, our drone rises vertically, its view becoming the painter's pallet of the Colorado Plateau. It flies higher and the plateau becomes an immense dish, its western rim washed by the waves of ranges of the Basin and Range, and its eastern rim buttressed by the Rocky Mountain massifs. It flies higher and the North American Cordillera is revealed as a strikingly distinct part of the continent, the precision of its coastal edge matched by the Rocky Mountain front rising above the Great Plains. When all of North America comes into view we see the American West is in the middle of the immense range of mountains that extends from Alaska southward across nearly half a planet.

The planetary order becomes most apparent when the landscapes and seascapes of the earth's crust are seen from space. We realize the Sky Islands, the San Francisco Peaks, the Navajo Sandstone, and the Grand Canyon did not just happen. Rather each are manifestations of the functioning of a planetary system that has been operating and evolving for 4.6 billion years. The challenge of making sense of a tiny yet extraordinarily complex portion of the earth's crust—as defined by the boundaries of the state of Arizona—is overwhelming to both the career geologist and the curious, but it leads both to profound wonder and a simple exclamation. Arizona!

BIOTIC COMMUNITIES

Introduction

Arizona lies in a nexus for plant diversity, with extreme climate and geology and a combination of temperate and tropical elements. All four North American deserts are represented—Great Basin, Mojave, Chihuahuan, and Sonoran. Within a day's drive from any corner of the state, one can relax in alpine tundra. The diversity of abiotic factors naturally results in tremendous diversity of biotic communities.

In this book, we use the classification system for biotic communities, or biomes, originally published as "Biotic Communities of the American Southwest: United States and Mexico" by David E. Brown (ed.) in Desert Plants (1982). This classification system considers climate, physiognomy, distribution, dominant and common plant species, and characteristic vertebrates. Below, these communities are briefly defined in terms of climate, physiognomy, and typical plants. Figure 5 provides a visual reference for the approximate locations of these communities in Arizona. Distribution statements for our native and naturalized plants include a list of biotic communities in which they are typically found.

COMMUNITIES OF ARIZONA

Adapted from *A Field Guide to Amphibians and Reptiles in Arizona*, 2006, by permission from Thomas C. Brennan, Andrew T. Holycross, and the Arizona Game and Fish Department.

Riparian Communities

Arizona's major biotic communities are intersected by numerous creeks and rivers that stretch from high in the mountains down through the woodlands and across the low deserts. Scores of specialized plants and animals make their homes in and along these streams and rivers. Referred to as "riparian," these biological communities contribute substantially to southwestern biodiversity and are critical components of complex southwestern ecosystems. Characterized by organisms with high water requirements such as tall broad-leaved trees, fish, and frogs, riparian corridors often starkly contrast with the surrounding biotic community. River corridors in low desertscrub communities are often lined with mesquite trees, desert willow, seepwillow, reeds, saltcedar, and cottonwoods. At middle elevations cottonwoods and willows become more prevalent and are joined by alder, ash, walnut, and sycamore trees. Arizona's highest elevation streams are lined by a variety of trees including maples, elders, alders, and willows. Dams, groundwater pumping, and water diversions have dried long stretches of many southwestern rivers and introduced species have altered many remaining riparian communities by displacing native plants and animals. It has been estimated that less than 10% of Arizona's riparian communities remain in their natural form. Only through aggressive restoration and recovery efforts can we help conserve these elements of Arizona's natural heritage.

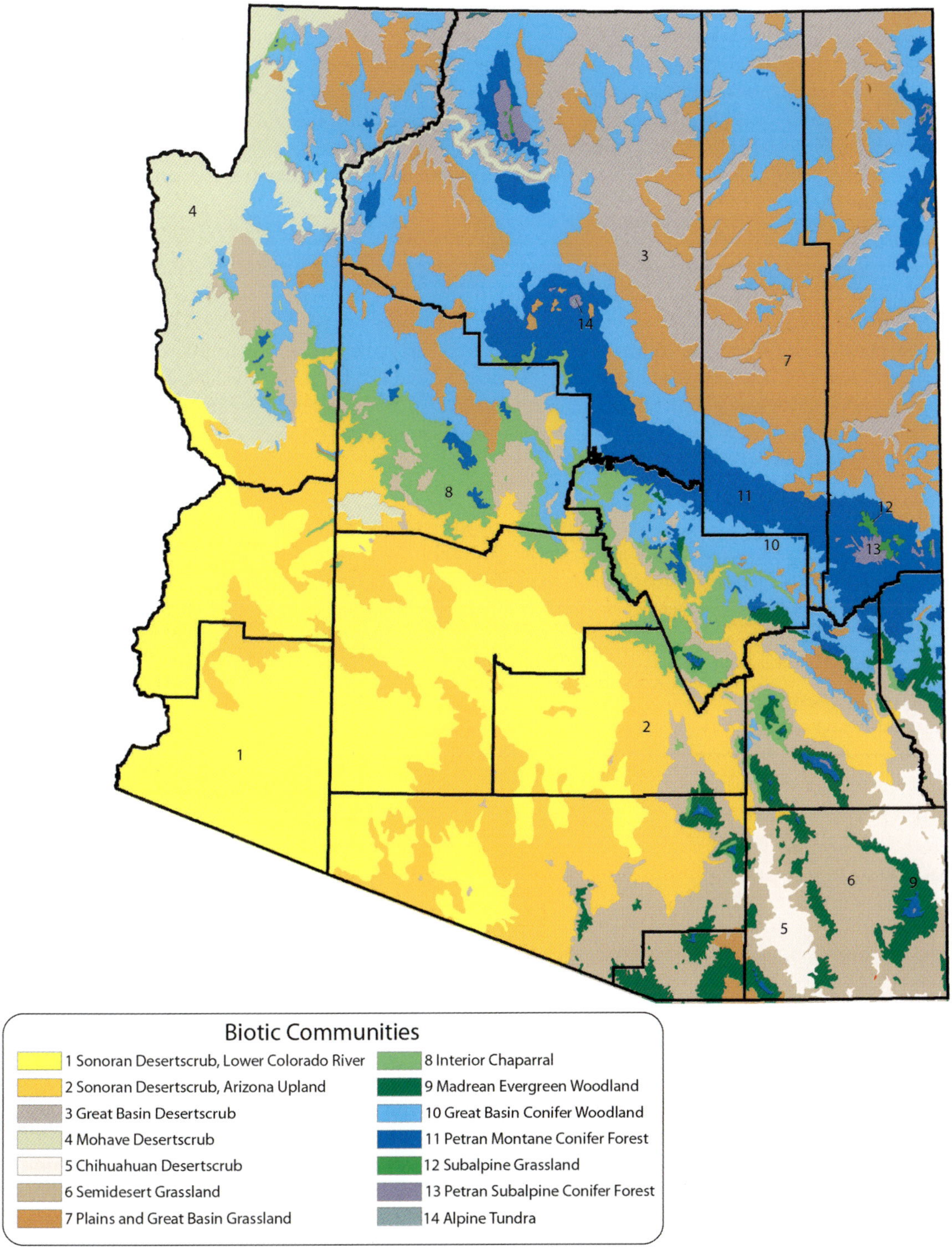

Figure 5. Biotic communities of Arizona. Adapted from original map created by David E. Brown, Thomas C. Brennan, and Andrew T. Holycross.

Sonoran Desertscrub

Lower Colorado River Desertscrub subdivision

Topography in this community generally consists of brushy flatlands transected by dry washes (24–400 m or 80–1,300 ft). This flat desert is found below the Arizona Upland Desertscrub subdivision of the Sonoran Desert. A combination of low annual rainfall (typically less than 200 mm or 8 in) and high temperatures make this Arizona's driest biotic community. Some regions regularly go a year or more without significant rainfall. Plants adapted to this xeric environment include creosote bush, white bursage, and desert saltbush. Washes are lined by legume trees such as ironwood, blue palo verde, and smoke tree.

Arizona Upland Desertscrub subdivision

Typically found on low mountains, hills, and bajadas (300–1,050 m or 980–3,500 ft), this community is above the Lower Colorado River subdivision of the Sonoran Desert and below Interior Chaparral or Semidesert Grassland. Average annual rainfall ranges from 200–425 mm (8–17 in), about 30–60% of which falls in summer. Saguaro, various cholla, various prickly-pear, fishhook pincushion, and fishhook barrel characterize this cactus-rich desert. The dominant trees are foothill paloverde, blue paloverde, ironwood, velvet mesquite, and cat-claw acacia. Common shrubs include brittlebush, triangle-leaf bursage, and jojoba. Invasive non-native grasses now carpet much of this landscape and carry fire through a community that evolved in its absence. Consequently, large sections of this picturesque community are being converted to non-native grass and scrublands.

Chihuahuan Desertscrub

In Arizona this community is generally found in flatlands and on hillsides and low sloping bajadas (900–1,200 m or 3,000–4,000 ft). It is often below Madrean Evergreen Woodland and adjacent to or mixed with Semidesert Grassland. Mean annual precipitation ranges from 200 mm to more than 300 mm (12 in), more than half of which falls between May and September. It is a warm temperate desert characterized by hot summers and cold winters. Temperatures can reach freezing (or below) for more than 100 nights per year in some areas. Conspicuous plants include creosote bush, tarbush, whitethorn acacia, ocotillo, viscid acacia, allthorn, mariola, and western honey mesquite.

Mohave Desertscrub

Topography in this community generally includes flatlands, plains, low hills, and bajadas (300–1,300 m or 980–4,270 ft). Mohave Desertscrub is often found below Great Basin Desertscrub, Great Basin Conifer Woodland, or Interior Chaparral and is sometimes adjacent to or mixed with Sonoran Desertscrub. Annual rainfall is low, averaging less than 200 mm (7 in) at most localities. Conspicuous shrubs in this xeric environment include creosote bush, desert holly, white burrobrush, Mohave yucca, blackbrush, and shadscale. The Joshua tree is found only in Mohave Desertscrub.

Great Basin Desertscrub

Topography in this community generally consists of plateaus, low hills, and mesas (1,200–2,200 m or 3,940–7,200 ft). This desert community is often adjacent to or below Great Basin Conifer Woodland and Plains and Great Basin Grassland and sometimes above Mohave Desertscrub (e.g., in much of Grand Canyon). Low annual precipitation (averaging less than 250 mm, or 10 in) is uniformly scattered throughout the year and much of it arrives as snow. Great Basin Desertscrub extends further north than any other North American desert. The presence of cold-temperate vegetation sets this desert apart from our other desert communities. Conspicuous plants include big sagebrush, Bigelow sagebrush, shadscale, winterfat, and blackbrush.

Semidesert Grassland

In Arizona this community is most often found in low valleys and on rolling hills (1,100–1,700 m or 3,600–5,600 ft). Typically, these grasslands are situated below Interior Chaparral or woodlands and are adjacent to or surrounded by desertscrub communities. Average annual rainfall ranges from 250–450 mm (10–18 in), most of which falls during summer. Drought and overgrazing have contributed to substantial shrub invasion and desertification in much of Arizona's Semidesert Grassland. Consequently, many grassland communities in Arizona now resemble desertscrub. Winters are relatively mild with freezing temperatures occurring on less than 100–150 nights in an average year. Conspicuous plants include tobosa, hairy grama, palmilla, sotol, soaptree yucca, and a variety of agave, cholla, and prickly-pear.

Plains and Great Basin Grassland

Topography in this community generally consists of flat and open plains, but also includes low mesas and rolling hills (1,500–2,300 m or 5,000–7,550 ft). Usually below Great Basin Conifer Woodland and above Great Basin Desertscrub, in southeastern Arizona it adjoins Madrean Evergreen Woodland. Average annual rainfall ranges from 300–500 mm (12–20 in), much of which comes from summer thunderstorms. This community has evolved with regular cycles of lightning-caused summer fires that were carried across large areas by wind. Fire suppression and cattle grazing have reduced the spread of natural fires in this community. These disruptions, along with extended droughts, have resulted in the invasion of weedy annuals, shrubs and Great Basin Conifer Woodland. Conspicuous plants now include four-wing saltbush, snakeweed, Mormon tea, and soapweed yucca. The grasses that used to dominate this community include blue grama, galleta, Indian ricegrass, and alkalai sacaton.

Interior Chaparral

In Arizona, this community is found on mountain slopes below the Mogollon Rim (1,050–2,150 m or 3,450–7,000 ft). It is usually situated above Sonoran Desertscrub or Semidesert Grassland, and below Great Basin Conifer Woodland or Petran Montane Conifer Forest. Average annual rainfall ranges from 350–600 mm (14–24 in), virtually none of which falls from April–June. Interior Chaparral intermittently receives snow at its higher elevations. Vegetation consists of dense, difficult to penetrate, stands of large evergreen shrubs such as shrub live oak, manzanita, birchleaf mountain mahogany, skunkbush sumac, desert ceanothus, silktassels, hollyleaf buckthorn, cliffrose, Arizona rosewood, and desert olive.

Madrean Evergreen Woodland

This biotic community most often develops on mountain slopes and hills, but sometimes extends into adjacent grassland valleys (1,300–2,200 m or 4,250–7,200 ft). This woodland usually occurs below Petran Montane Conifer Forest and above Semidesert Grassland or rarely desertscrub communities. In Arizona, average annual rainfall ranges from 400–600 mm (16–24 in). Mild winters and wet summers facilitate the growth of many oaks, including Emory oak, Arizona white oak, gray oak, and Mexican blue oak. Other conspicuous trees found in this woodland include alligator-bark juniper, border pinyon, and madrone. Rainbow cactus, barrel cactus, Schott's yucca, banana yucca, sotol, Palmer's agave, and Parry's agave are the most noticeable plants in the understory.

Great Basin Conifer Woodland

Sometimes referred to simply as Pinyon-Juniper Woodland, this community generally occupies foothills and lower mountain slopes, but above the Mogollon Rim it is found on the plains and hills of the Colorado Plateau (1,200–2,300 m or 4,000–7,500 ft). Below the Mogollon Rim this community typically occurs adjacent to or above Interior Chaparral and below Petran Montane Conifer Forest. Above the Mogollon Rim it is usually adjacent to Plains and Great Basin Grassland or Great Basin Desertscrub. Average annual precipitation ranges from 250–500 mm (10–20 in). This woodland is well-adapted to a cold climate and usually experiences freezing temperatures more than 150 nights per year. Consistently low temperatures exclude evergreen oaks and other warm-temperate vegetation. This relatively low diversity community is dominated by pinyon and junipers, including Rocky Mountain juniper, Utah juniper, single-leaf pinyon, and Rocky Mountain pinyon. The understory is comprised of grasses and shrubs such as big sage.

Petran Montane Conifer Forest

This community is found on the slopes and peaks of our higher mountains, as well as across some of our high plateaus (2,000–3,050 m or 6,500–10,000 ft). It generally occurs above Interior Chaparral or Great Basin Conifer Woodland and below Petran Subalpine Conifer Forest. These forests are among the driest in North America with average annual rainfall ranging from 460–760 mm (18–30 in). This community is dominated by ponderosa pine, but also includes Arizona white pine, Douglas fir, Gambel oak, and New Mexican Locust. Frequent low intensity fires once maintained open stands of mature forest characterized by large ponderosa pine with an open, grassy understory. Active fire suppression and removal of fine fuels by overgrazing has decreased the frequency and spread of these fires. These changes in natural fire cycles have substantially increased the density of woody vegetation (understory shrubs and young ponderosa pines). Consequently, fire no longer maintains these forests, but results in catastrophic conflagrations and conversion of forest to shrubby grasslands or Interior Chaparral.

Petran Subalpine Conifer Forest

Restricted to the slopes of Arizona's tallest mountains (2,450–3,800 m or 8,000–12,500 ft), usually above Petran Montane Conifer Forest and below Alpine Tundra. This community is very cold and relatively wet, receiving 635–1,000 mm (25–39 in) of annual precipitation via summer monsoon storms and winter snows. The growing season is usually fewer than 75 days per year, during which nighttime frosts are not uncommon. Engelmann spruce and corkbark fir dominate and are joined by aspen and Douglas fir at the lower elevations. Dry sunny slopes are populated by white pine and bristlecone pine.

Alpine Tundra

In Arizona, this community only occurs on the high peaks of San Francisco Mountain (3,500–3,862 m or 11,500–12,633 ft) near Flagstaff. It occurs above Petran Subalpine Conifer Forest. The growing season here is very brief. Severely cold temperatures and the scarcity of liquid water prevent trees from growing. Dominant growth consists of low-growing shrubs, herbaceous plants, lichens, and mosses. Some examples are golden avens, Parry's lousewort, moonwort, wild candytuft, orange sneezeweed, strawberry, and alpine speedwell. The small parcel of Alpine Tundra on San Francisco Mountain is home to the San Francisco groundsel, a small, yellow-flowered plant found nowhere else in the world. The lower elevations of Alpine Tundra are penetrated by elements of Petran Subalpine Conifer Forest, including Engelmann spruce, corkbark fir, and bristlecone pine.

CLIMATE AND WEATHER OF ARIZONA

Stephen W. Bieda, III

Arizona is a semi-arid/arid state covering an area of approximately 297,850 km^2 (115,000 mi^2). The state contains diverse climate types due to the widely varying elevations (21 to 3852 m, 69 to 12,640 ft) in it. The southwestern corner is very arid because the elevation is quite low and usually on the downward (lee) side of mountain regions. Locations further north and east gradually (and sometimes dramatically) increase in elevation, which allows these regions to capture more moisture from passing storm systems due to ascent of air masses that create instability and convection. As a result of these dramatic elevational shifts (Fig. 1) and the resulting increase in precipitation, the state has far lusher vegetation than similarly located subtropical latitude deserts (e.g., the Sahara, Jacobs et al. 2005).

The most noticeable aspect of Arizona's climate is the mild to hot mean daily average temperatures (Fig. 6) and the rather high amounts of precipitation that fall over the region when compared to other desert climates (Sheppard et al. 2002, Jacobs et al. 2005). Average annual precipitation ranges from 75 mm (3.0 in) in the lowest desert region of southwest Arizona to 480 mm (18.9 in) in the Mogollon Rim country of east central Arizona and over 900 mm (35.4 in) on the highest mountain peaks (Fig. 7). Another notable aspect of Arizona's climate is the temperature extremes that have been recorded at -40° and 53°C (-40° and 127°F). Some locations in southwestern Arizona may remain frost-free in mild winters while freezing temperatures can occur even in summer at the highest elevations in the state. The frequency, duration, and intensity of winter freezes help to limit the northward and altitudinal distributions of plants with subtropical origins. Available soil moisture is a major factor determining the lower elevational limits of montane species.

The region is characterized by two periods of precipitation coincident with the northward track of the subtropical ridge (which brings the North American Monsoon Season) and the polar jet stream. This bimodal rainfall pattern contributes greatly to the diversity of plant species found in the state because it favors both cool-season and warm-season plant species. What follows is a discussion of seasonal interactions of these important weather features and their influence on the amount of precipitation that may result.

Winter Precipitation and ENSO Cycle

During the winter months (December to February), approximately a third to a half of the annual average precipitation occurs. The polar jet stream occurs further south in these winter months and brings moisture to the west coast of the United States and to Arizona. Frontal boundaries (zones where air masses meet) crossing Arizona are often cold enough to bring snow to northern and higher elevation regions of Arizona, and rarely bring snow to the southern deserts. However, during these months the amount of precipitation that falls and the year-to-year seasonal temperature swings are highly variable and strongly influenced by coincident phases of the El Niño Southern Oscillation (ENSO).

The phases of the ENSO cycle have an important role in determining whether winter precipitation is above or below average during any given year (Fig. 8). When ENSO is in its El Niño phase, the storm track tends to be persistently west to east and frequently directed at the state of Arizona. This situation allows for wetter than normal conditions to occur during the winter months and aids in water table recharge via snowpack in northern Arizona. However, when ENSO is in its La Niña phase, the jet stream becomes more variable and is displaced further north. The overland route the jet stream then takes normally presents Arizona with much drier than normal conditions, often increasing the threat of drought and wildfires in the region.

Summer Precipitation and the North American Monsoon System

The North American Monsoon System (NAMS) is an important contributor to rainfall in the state of Arizona. The monsoon season, which runs from June 15 through September 30, is characterized by a significant increase in the dewpoint

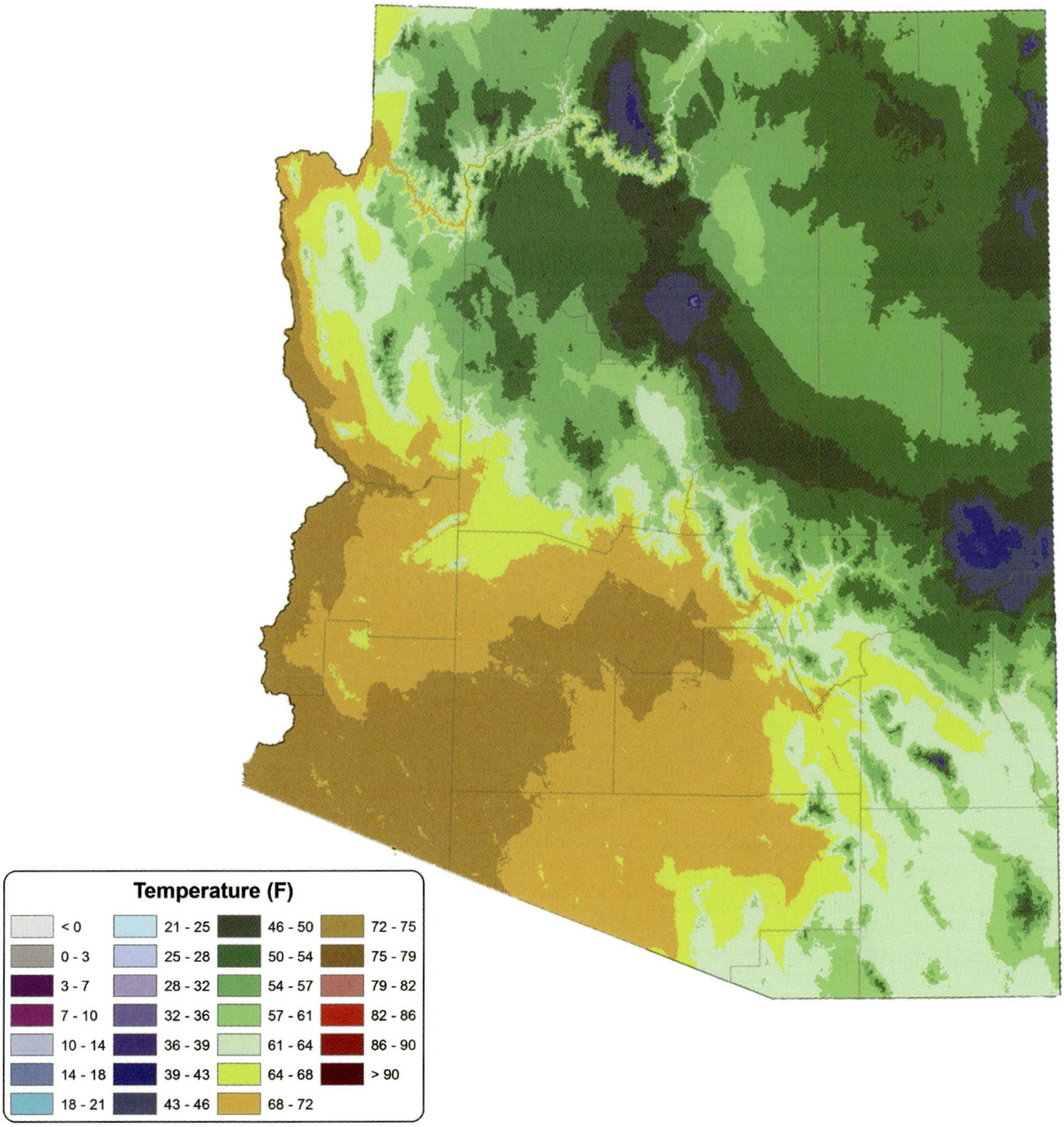

Figure 6. Temperature variation across Arizona. Thirty-year average annual temperatures 1981–2010. Excerpted from country map, copyright © 2015, PRISM Climate Group, Oregon State University. Downloaded 12/2020.

and relative humidity. Thunderstorms may form, bringing localized heavy rainfall that can result in flash flooding. Some storms can become severe, with damaging winds and hail. Outflow winds from thunderstorms can generate extensive dust storms. Tornadoes are rare in Arizona but do occur.

Variability of monsoon precipitation is related to the positioning and seasonal evolution of upper-level atmospheric features on interannual and longer time scales, namely an upper-level monsoon/subtropical ridge (Carleton 1986, Carleton et al. 1990, Douglas et al. 1993, Watson et al. 1994, Adams and Comrie 1997, Comrie and Glenn 1998, Grantz et al. 2007, Bieda et al. 2009). If this ridge migrates anomalously farther north or northeast of its climatological location, for example northwestern Texas during late July and early August, the core NAMS region experiences enhanced upper-level easterlies and an increase in rainfall amounts and coverage in Arizona. If the upper-level monsoon ridge is instead anomalously farther south, Arizona experiences a weaker (or non-existent) tropical influence and convection is

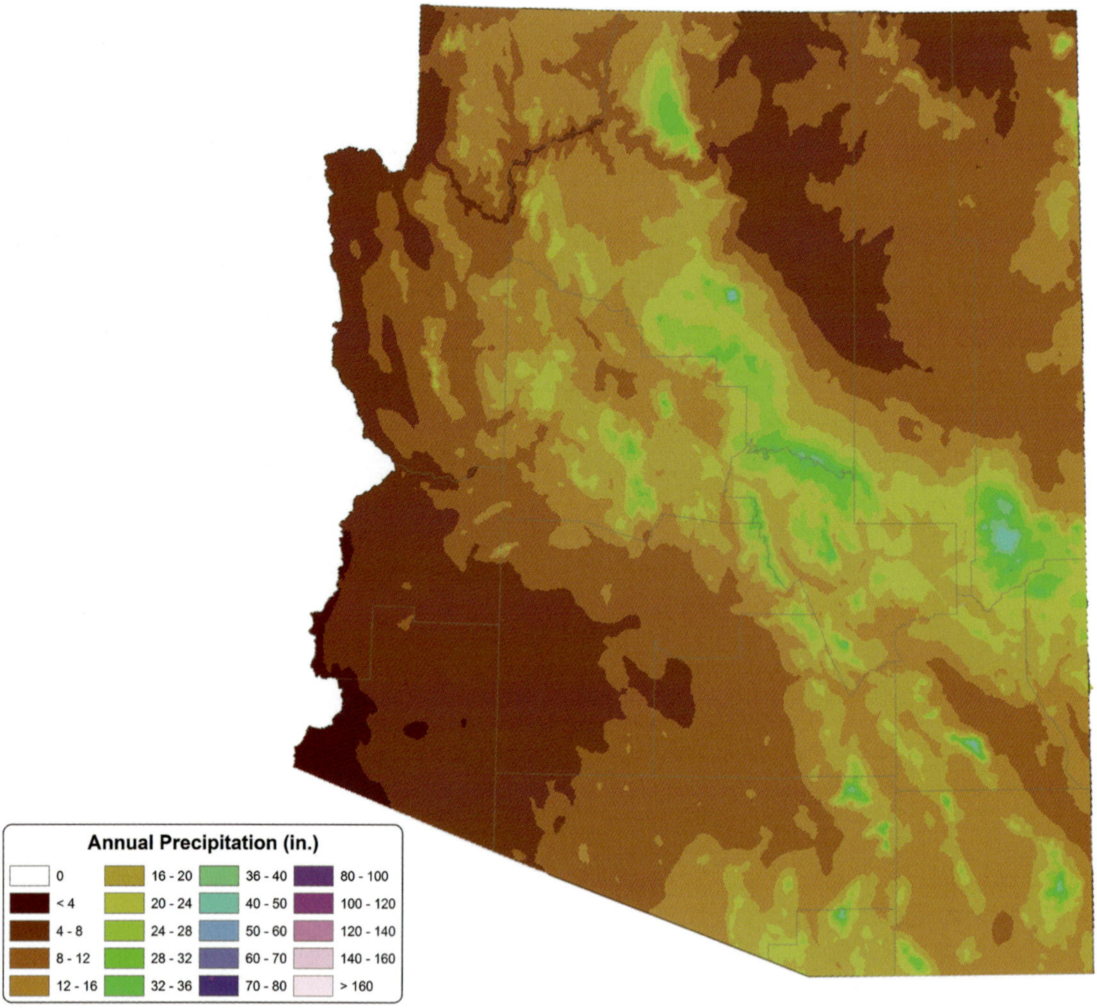

Figure 7. Precipitation variation across Arizona. Thirty-year average annual precipitation 1981–2010. Excerpted from country map, copyright © 2013, PRISM Climate Group, Oregon State University. Downloaded 12/2020.

normally suppressed. This relationship is out of phase with precipitation anomalies over the central United States (Carleton 1986, Carleton et al. 1990, Douglas et al. 1993, Watson et al. 1994, Adams and Comrie 1997, Comrie and Glenn 1998, Grantz et al. 2007).

The strength of the monsoon anticyclone during the monsoon onset period in Arizona (late June/early July) is significantly related, in a statistical sense, to the state of the Pacific Sea Surface Temperatures (SSTs). If an El Niño prevails in the eastern and central tropical Pacific and the eastern North Pacific has warmer-than-normal SST anomalies, a southward displacement of the monsoon ridge is likely to occur, resulting in less precipitation in the state of Arizona compared to average (Castro et al. 2001, Castro et al. 2007). On the other hand, when the opposite condition occurs and a La Niña prevails in the eastern and northern tropical Pacific in conjunction with a North Pacific horseshoe-shaped negative SST anomaly, the monsoon ridge is displaced northward. In this case, more precipitation occurs than average.

Another, less reliable source of moisture is provided by tropical cyclonic storms from the eastern Pacific and Gulf of California. The remnants of hurricanes and tropical storms move into the state in the summer and autumn once in approximately 2.5 years and can bring a modest 15%-25% of precipitation to Arizona (Ritchie et al. 2011). Extensive flooding has resulted from these storms. The moisture delivered by tropical storms in the autumn can contribute to abundant wildflower growth during the following spring.

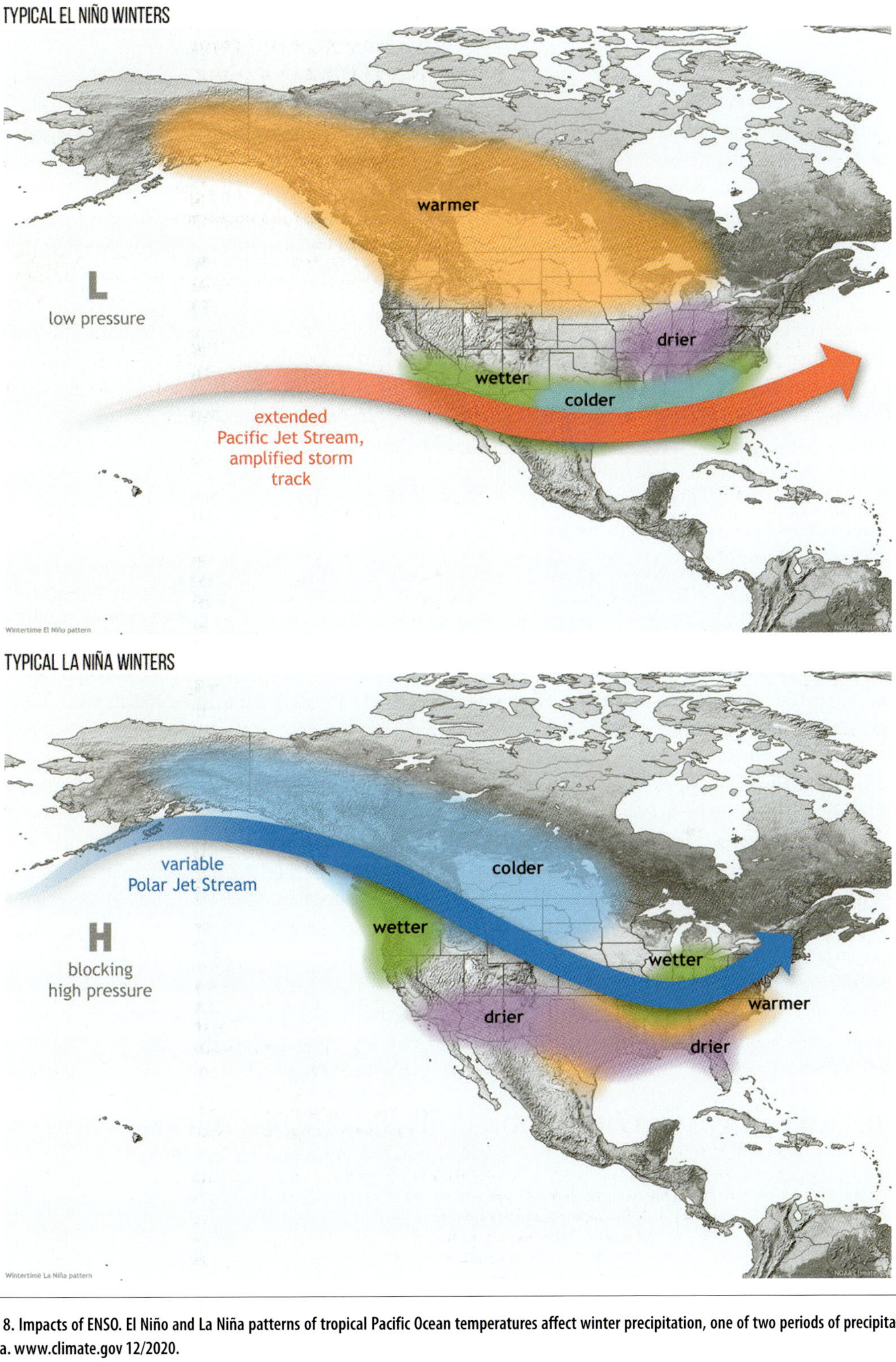

Figure 8. Impacts of ENSO. El Niño and La Niña patterns of tropical Pacific Ocean temperatures affect winter precipitation, one of two periods of precipitation in Arizona. www.climate.gov 12/2020.

CONSERVATION IN ARIZONA

Matthew B. Johnson

While native plant communities remain in much of Arizona, there has been substantial alteration of the vegetation since Anglo-European settlement. Ranching, farming, mining, and urbanization have all impacted the native vegetation of the state to varying degrees. Vigorous fire suppression efforts during much of the 20th century have altered plant community structure and led to a buildup of fuel loads leading to destructive wildfires. This, in combination with invasive species such as buffelgrass (*Cenchrus ciliaris*), red brome (*Bromus rubens*), and cheatgrass (*Bromus tectorum*) have changed fire regimes over large areas of Arizona. In addition to altering fire frequency and intensity, these and other invasive species such as tamarisk (*Tamarix* spp.) and Sahara mustard (*Brassica tournefortii*) also displace indigenous species.

A large portion of the land in Arizona is administered by various governmental agencies including the Bureau of Land Management, Bureau of Indian Affairs, Tribal governments, Forest Service, National Park Service, Department of Defense, Fish and Wildlife Service, Arizona State Land Department, and Arizona State Parks. These agencies have conservation policies as a part of their respective missions and provide varying degrees of protection for the plants on lands they manage. Additional lands are protected in municipal and county parks. Non-governmental organizations such as The Nature Conservancy and the Audubon Society protect unique habitats and rare plants through purchase of land. Many private landowners manage their property for conservation purposes, and some have established conservation easements on their land to protect it into the future.

Various municipalities have enacted ordinances requiring open space to be set aside, preserving native plants in place, as well as the use of native plants in landscaping. Wash protection ordinances and wildlife corridors also preserve habitat. In Pima County, the Sonoran Desert Conservation Plan is an ambitious and far-sighted effort to purchase land for open space and to preserve important plant and wildlife habitats.

Organizations such as the Arizona Native Plant Society, Coalition for Sonoran Desert Protection, Sky Island Alliance, Sierra Club, and others promote the conservation of natural habitats and the plants and animals that depend on them. Botanical gardens in the state display native plants and work to educate the public about their importance. Volunteer groups such as the Sonoran Desert Weedwhackers are working to eliminate invasive species such as buffelgrass and fountain grass (*Cenchrus setaceus*) through regular work days and educational programs to raise public awareness of the threats posed by these plants.

Arizona is fortunate to have extensive areas of public land where plants are protected and people can visit natural habitats. The diverse natural environments of the state are a major economic asset. Visitors and residents alike with an interest in outdoor activities can enjoy the beauty and diversity of the state's natural areas year round. Despite the large areas of protected vegetation in Arizona, threats from invasive species, urbanization, and climate change pose significant challenges for conservation of the state's native plants.

Information on rarity and protected status is included in the discussion sections for taxa of conservation concern. Additionally, the Arizona Game and Fish Department maintains the Arizona Heritage Data Management System and the Navajo Nation maintains the Navajo Natural Heritage Program. These contain extensive data on rare plants. Other land management agencies also maintain information on rare plants occurring on lands under their jurisdiction.

A HISTORY OF PLANT COLLECTION (AND LEGUME DISCOVERY) IN ARIZONA

Kathryn Mauz

When botanical collectors first came to the region that is now Arizona, it was still called New Mexico, itself a province of Old Mexico occupied by peoples both indigenous and descended from Spanish occupation. Exploring parties of the nineteenth century endured all the region's extremes of climate with only the shelter and clothing that could be strapped to a mule train. They carried some food and hunted when they were able, and their livestock routinely endured hunger where forage was scarce. They could carry some water, but were continuously concerned with finding it on the landscape even as they collected observations, animals, plants, and data that produced original images, maps, and scientific records of the lands traversed. Even the leaders of these expeditions were occasionally uncertain of their prospects, sometimes frustrated, and at times inspired by the wonders around them. Just 34 years after the region was first surveyed in 1851, eminent botanist Asa Gray traveled through northern Arizona's canyonlands with relative ease—by rail and wagon—concluding, "Altogether, there is nothing like this cañon. The far-famed Yosemite is more beautiful and more varied, but not more magnificent, nor half so strange and weird" (Gray 1885, p. 517). Gray had the benefit of knowing not only where he was and where he was going, but how to get back again. He also enjoyed early outposts of hospitality that assured food, water, and a comfortable place to sleep. Gray's impressions of the scenery and flora of the country echoed those of many earlier explorers in their more sanguine moments and reflect why it is that Arizona's plant life continues to fascinate people and to draw them to its varied terrain.

Plant collection in North America was two centuries old by the time a botanically-minded explorer reached the boundaries of modern Arizona. The East, the South, the Pacific Northwest and West Coast, the high latitudes, even many parts of modern-day Mexico had already been visited by naturalists if not also broadly explored (Brendel 1879, 1880). Arriving at the eastern extent of the Colorado Desert in May 1832, Thomas Coulter described the location of the ford of the Río Colorado "not a mile" below the junction of the Gila River. There, he told of "a broad plain now pretty well covered with poplar and brushwood" (Coulter 1835, p. 63), producing at least a botanical description if not specimens from this boundary river. He had journeyed overland from Monterey, California, to the ford, where he remained "stationary" for 10 days (Coulter 1835, pp. 61–62) before retracing his route to the coast. Just how far Coulter crossed into what is today Arizona may be debated, but that stay surely provided opportunity to explore the surroundings.

Several of Coulter's new species (e.g., *Prosopis pubescens* Benth., *Dalea mollis* Benth.) occupy ranges that include southwestern Arizona, but some type localities were evidently reported incorrectly in the protologues and remain ambiguous (Coville 1895). Certain other collections are plants of the Sonoran Desert region. Because Coulter had earlier collected in Sonora proper and because today's Arizona was then still part of Mexico, specimens that are botanically Sonoran but reported from "Sonora Alta" or just "Mexico" could possibly have originated in Arizona (McVaugh 1943). It has been suggested that one of the new species, *Brickellia coulteri* A. Gray, although reported from "California," was more likely collected near Yuma and, therefore, may have been the earliest Arizona plant collection (Coville 1895).

Nearly a decade and a half elapsed between Coulter's foray—arguably the first by a botanist to reach the state—and the next expedition that included a naturalist among its members. In the fall of 1846, early in the war between the United States and Mexico, topographical engineer William H. Emory traveled across the plains and desert Southwest to San Diego in the company of a military expedition commanded by Stephen W. Kearny under the banner of "the Army of the West." All along this journey, Emory noted plants and vegetation in his journal and regarded the strange landscape as both newcomer and foreigner.

The expedition entered the headwaters of the Gila River and crossed into today's Arizona about the third week of October. The party for the most part followed (and mapped for the first time) the Gila's course, arriving at its mouth at the Colorado River one month later. Along the Gila, Emory indicated additional "vegetable curiosities" among the cacti and arboreal legumes, observing, "the mezquite, acacia, prosopis odorata, and prosopis glandulosa, now form the principal growth. Under the name mezquite, the voyageur comprises all of the acacia and prosopis family" (Emory 1848, p. 65). In his journal, Emory noted "a new green-barked acacia, covering the plains above the river bed" (Emory 1848, p. 67)—a reference to *Cercidium floridum* Benth. ex A. Gray [*Parkinsonia florida* (Benth. ex A. Gray) S. Watson], for which Emory's collections became part of the type series. The "mezquite" rarely failed to figure in his descriptions as the journey continued.

Emory first mentioned the huge "pitahaya" cactus, later described botanically as *Cereus giganteus* Engelm. [*Carnegiea gigantea* (Engelm.) Britton & Rose], along the middle Gila (Emory 1848, p. 72). Later, deep within saguaro country, Emory contemplated his surroundings: "my thoughts went back to the States ... I was struck most forcibly with the fact that not one object in the whole view, animal, vegetable, or mineral, had any thing in common with the products of any State in the Union, with the single exception of the cotton wood, which is found in the western States, and seems to grow wherever water flows ..." (Emory 1848, p. 77). As the sometimes arduous overland campaign went on, Emory added more legumes to his collections, including *Sesbania macrocarpa* Muhl. ex Raf. [*Sesbania herbacea* (Mill.) McVaugh], an "unknown" *Dalea*, and *Prosopis emoryi* Torr., published as a new species by John Torrey (1848, p. 139). (The latter name is superseded by *Prosopis pubescens* Benth., described just two years earlier from Thomas Coulter's "California" specimen.) Still other collections from the Gila within Arizona were among the type material for species both new to science and characteristic of this region: *Kallstroemia grandiflora* Torr. ex A. Gray, *Obione polycarpa* Torr. [*Atriplex polycarpa* (Torr.) S. Watson], *Obione lentiformis* Torr. [*Atriplex lentiformis* (Torr.) S. Watson], *Tessaria borealis* Torr. & A. Gray [*Tessaria sericea* (Nutt.) Shinners], *Tetradymia ramosissima* Torr., *Hymenoclea monogyra* Torr. & A. Gray [*Ambrosia monogyra* (Torr. & A. Gray) Strother & B. G. Baldwin], *Baccharis emoryi* A. Gray, and *Sphaeralcea emoryi* Torr. ex A. Gray.

Several "peacetime" expeditions through the territory ensued in the 1850s—political, exploratory, or economic in their objectives—that included naturalists among their members. Prominent among these was the United States and Mexican Boundary Survey, which included several parties operating through this vast desert region from 1848 to 1855. Although the Leguminosae was one of the largest families presented in the Botany of the Boundary (Torrey 1859), few of the collections reported were made within modern-day Arizona. (More collections were made than appeared in Torrey's account, however, and those of Wright and Thurber, in particular, were addressed elsewhere by Asa Gray.) Unlike the surgeon-naturalists who accompanied other topographical surveys, the Boundary Survey parties included persons designated as botanists who also functioned as assistants to the surveyors. Charles Wright collected on many legs of the Survey but worked in the grasslands of southeastern Arizona on just one excursion to Santa Cruz, Sonora, in 1851. Within modern Arizona, he collected plants in the San Pedro, Babocomari, and Sonoita Valleys, including the type of *Cassia wrightii* A. Gray [*Chamaecrista serpens* (L.) Greene var. *wrightii* (A. Gray) H.S. Irwin & Barneby]. Charles Parry and George Thurber each made collections on the lower Gila River and elsewhere in "Sonora." Parry collected the type specimen for the spring-blooming *Penstemon puniceus* A. Gray var. *parryi* A. Gray [*Penstemon parryi* (A. Gray) A. Gray], while Thurber collected the types of *Petalonyx thurberi* A. Gray along with the charismatic desert trees *Olneya tesota* A. Gray and *Holacantha emoryi* A. Gray [*Castela emoryi* (A. Gray) Moran & Felger] on the lower Gila River. Arthur Schott, who had worked on the survey of the Río Grande section, also collected in the Gila and Santa Cruz River Valleys in 1854–1855, and some of his collections reported from "Sierra del Pajarito" may well have come from the Arizona side of the international boundary line that was then being marked. He collected the type of *Dalea schottii* Torr. [*Psorothamnus schottii* (Torr.) Barneby] from "banks of the Colorado" near Yuma (Torrey 1859, p. 53).

In northern Arizona, the expedition led by topographical engineer Lorenzo Sitgreaves explored down the Zuñi, Little Colorado, and Colorado Rivers from late September to the end of November 1851. The route included an overland passage through the San Francisco Peaks region to the head of the Bill Williams River (today, one of the volcanic summits there bears the name of the party's leader). Climbing into the mountains after leaving the Little Colorado, Sitgreaves observed the succession of forest cover: "... the cedar gave place to the nut-bearing pine; and this, when near the summit, to a pine of larger growth, with long leaves. ... The ground was covered with fresh grass, and well timbered with tall pines, mingled, after attaining a certain altitude, with aspens of a brilliant yellow" (Sitgreaves 1853, pp. 10–11). This language contrasts considerably with his unoptimistic descriptions of the desert landscapes before and after this portion of the journey. Along the Colorado, Sitgreaves further noted aspects of the surrounding flora, in which the arboreal legumes were clearly conspicuous: "Some large cotton-wood trees grow directly upon the river banks, but the growth of the rest of the valley is small consisting chiefly of mezquit, tornilla, willow, and a singular tree with a smooth, pale green bark, and leaves so diminutive as to require a close proximity to discern them" (Sitgreaves 1853, p. 21).

Naturalist Samuel Woodhouse (1853) prepared a general description of the region, naming plants in their habitats along with animals seen, heard, or collected. Despite the late season, the botanical collections reported from the Arizona section of the expedition (including a handful of legumes) represented both taxonomic breadth and variety of form (see Torrey 1853). Several new plants were described from these collections, all characteristic members of the pine forest or plateau floras on and above the Mogollon Rim: four species of *Eriogonum*, a new genus in *Acanthochiton wrightii* Torr. [*Amaranthus acanthochiton* J. D. Sauer], plus *Teloxys cornuta* Torr. [*Chenopodium graveolens* Willd.], *Heuchera sitgreavesii* Rydb. [*Heuchera versicolor* Greene], and *Lupinus sitgreavesii* S. Watson [see *Lupinus argenteus* Pursh var. *argenteus*].

Two of the Pacific railroad surveys crossed Arizona territory. These were led by Boundary Survey veteran Amiel W. Whipple on the 35th parallel (northern Arizona), 1853–1854, and by John G. Parke on the 32nd parallel (southern Arizona), 1854 and 1855. Lately employed on the Boundary Survey in New Mexico and Texas, John M. Bigelow served as the naturalist to Whipple's expedition. The party crossed northern Arizona late in the year, following the Zuñi and Little Colorado Rivers to camp at the San Francisco Peaks by Christmas day 1853, then spent about three weeks in the valley of the Río Verde before descending the Bill Williams River to the "Colorado Grande" in February 1854. Bigelow's report remarked upon the quality and abundance of grass (especially the *grama)* on the plateaus, the many novel forms of cacti (*Echinocactus whipplei* Engelm. & J. M. Bigelow [*Sclerocactus whipplei* (Engelm. & J. M. Bigelow) Britton & Rose] was described from specimens collected on the Little Colorado), the vast tracts of dark green pinyon pine, the varied chapparal flora, and the forest resources of the mountain ranges. Though evidently impressed by the geography, Bigelow lamented, "this was the most unpropitious season of the whole year for the collection of herbaceous plants, and must account for the meagreness of my collections in this part of our journey" (Bigelow 1856, p. 6).

By the time the expedition approached the valley of the Colorado River, before crossing to California, Bigelow had collected the type specimen for *Sophora arizonica* S. Watson [*Dermatophyllum arizonicum* (S. Watson) Vincent] and reported collecting a series of springtime annuals, among which were "fine specimens" of *Hosackia puberula* Benth. [*Acmispon oroboides* (Kunth) Brouillet] (Bigelow 1856, p. 13) and *Lupinus sparsiflorus* Benth. (Torrey 1857). Indeed, Torrey remarked, "The most interesting region of this part of the route is the valley of William's river ... some of the most remarkable plants of the collection were found here" (Torrey 1857, p. 59). It was on the Bill Williams fork that the arboreal legumes featured prominently in Bigelow's descriptions: "The timber of this valley is composed almost exclusively of cotton-wood, or alamo, mezquite, 'green-barked acacia,' of Major Emory's report, curly mezquite (*Strombocarpus pubescens*), two other leguminous trees, and some willows. One of the leguminous trees is the *Olneya Tesota* of Dr. Gray ... The two mezquites are rather small in this valley, seldom attaining more than fifteen or twenty feet in height. The *Tesota*, 'green-barked acacia,' and the other leguminous tree, grow fully as large as the mezquites" (Bigelow 1856, p. 11). To these could be added *Dalea spinosa* A. Gray [*Psorothamnus spinosus* (A. Gray) Barneby] and *Parkinsonia microphylla* Torr. (with *Phoradendron californicum* Nutt.) collected on the Bill Williams fork (Torrey 1857). As Bigelow had implied, however, few herbaceous legumes were collected between late summer on the high plains and springtime on the California coast. Many new species were named from the collections of the expedition, including some now familiar plants from Arizona: *Canotia holacantha* Torr., *Burrielia lanosa* A. Gray [*Eriophyllum lanosum* (A. Gray) A. Gray], *Acanthogonum rigidum* Torr. [*Chorizanthe rigida* (Torr.) Torr. & A. Gray], *Penstemon microphyllus* A. Gray, *Penstemon spectabilis* Thurb. ex A. Gray, *Dasylirion bigelovii* Torr. [*Nolina bigelovii* (Torr.) S. Watson], and *Crossosoma bigelovii* S. Watson.

The railroad surveys led by John Parke crossed southern Arizona from February to March 1854 and June to August 1855, following the Gila River and the dry course of the Santa Cruz, then the "emigrant road" east to the San Pedro Valley, across the dry "Playa de los Pimas" (Willcox Playa), and north of the Dos Cabezas on to New Mexico. Parke's (1855) first journal is spare on botanical detail, and that is frequently confined to greasewood, mesquite, "sage," and "bunch-grass" (often *grama*). With the party was physician and naturalist Adolphus Heermann who, although better known as an ornithologist (Fischer 2001), collected a number of plants within the future borders of Arizona (Antisell 1856). Of those collections reported by Torrey (1856), all were from the second survey during a more favorable time of year. Predictably, among those collections were a number of legumes: *Hosackia puberula* Benth. [*Acmispon oroboides* (Kunth) Brouillet] in the Chiricahua Mountains; *Amorpha fruticosa* L., *Acacia constricta* Benth. [*Vachellia constricta* (Benth.) Seigler & Ebinger], *Acacia cuspidata* Schltdl. [*Acaciella angustissima* (Mill.) Britton & Rose], and *Cassia lindheimeriana* Scheele [*Senna lindheimeriana* (Scheele) H.S. Irwin & Barneby] on the San Pedro River; and the mesquites *Prosopis pubescens* Benth. and *Prosopis glandulosa* Torr. on the lower Gila River.

The Colorado River itself, a milestone that earlier parties had been eager simply to cross, became the subject of exploration a few years later when Whipple Survey veteran Joseph C. Ives led an expedition by steamboat from the river's mouth to the head of navigation at Black Canyon, then overland east across the Colorado Plateau. Travel with the full party began at Fort Yuma on 11 January 1858. Little mention is made of vegetation in Ives' (1861) narrative, only occasionally remarking upon common riparian trees or riverside stands of marsh or cane. Uncharacteristically, he wrote effusively about the Mohave Valley and, later on, the now-drowned Cottonwood Valley: "... the belt of bottom land is narrow, and dotted with graceful clusters of stately cottonwood in full and brilliant leaf. The river flows sometimes through green meadows, bordered with purple and gold rushes, and then between high banks, where rich masses of foliage overhang the stream" (Ives 1861, p. 78). The overland portion of Ives' explorations began on March 23, traversing the plateaus and minor canyons of northern Arizona. A sometimes difficult month elapsed before the party arrived in the parklands around the San Francisco Peaks, where the verdant spring flora figured in each entry of Ives' report. Their ensuing excursions into the mesas around the Painted Desert in early May 1858 are memorialized in a handful of geographic features named for explorers in the party.

John S. Newberry served as geologist and naturalist to the expedition and, according to Ives, collected extensively along the route. That the overland trip was made at a favorable time of the year—unlike the passages of the Sitgreaves and Whipple expeditions—presented opportunity for a truly representative collection from the plateau region that included many springtime annuals (see Gray et al. 1861). From the banks of the Colorado River, collections of by-then-familiar woody legumes were made, along with collections of *Abutilon newberryi* S. Watson [*Horsfordia newberryi* (S. Watson) A. Gray], *Eriogonum deflexum* Torr., and *Antirrhinum filipes* A. Gray, each of which would be named as new. J. C. Ives (1861, p. 131) indicated that "collections and field-notes" from the river passage were sent downriver with the steamboat when his party left for the plateaus, and those packages would then have gone around to California and back to New York by sea; among them, presumably, were these. New among Newberry's collections from the pine-forested volcanic region were *Euphorbia lurida* Engelm. and *Trifolium longipes* Nutt. var. *pygmaeum* A. Gray. From the mesalands of northern Arizona came new species in *Phacelia ivesiana* Torr., *Physaria newberryi* A. Gray, A. Gray, *Sophora stenophylla* A. Gray, and *Trifolium subcaulescens* A. Gray [*Trifolium gymnocarpon* Nutt.], along with previously known beans and locoweeds from new localities.

Numerous collections were made in Arizona in the 1860s by physician-naturalists stationed at frontier military posts in the recently declared Arizona Territory. Among these men was Elliott Coues, who arrived at Ft. Whipple from Zuñi in the summer of 1864 and remained there for just over a year. Although Coues would become best known as an ornithologist, his collections included many plants destined for George Engelmann's herbarium. In the summer of 1865, Coues collected in the vicinity of the post with newly arrived Edward Palmer and made a round trip on the Colorado River to Ft. Yuma before being transferred to the East Coast (Cutright and Broadhead 1981). *Cassia covesii* A. Gray [*Senna covesii* (A. Gray) H.S. Irwin & Barneby], from near Ft. Whipple, was a new species among the Coues and Palmer collections. Palmer served briefly as physician before being transferred to Camp Lincoln (later Camp Verde), whereupon surgeon Charles Smart took up the medical post at Ft. Whipple. Smart made many collections around Verde Mesa, the Verde River, and the Mazatzal Mountains, before transferring to Camp Lowell in Tucson in July 1867. He served there for two years and made trips to other military posts in southern Arizona (Mauz 2011). Among Smart's plant collections, possibly numbering 300 from Arizona (McVaugh 1956), were the type specimens of *Isocoma tenuisecta* Greene, *Filago arizonica* A. Gray [*Logfia arizonica* (A. Gray) Holub], and *Stylocline arizonica* Coville.

Edward Palmer's military service in Arizona lasted for just two years, 1865–1867, during which he sent hundreds of plant collections, some co-attributed to Coues, to George Engelmann for determination. He collected in the Verde Valley and then around Camp Grant, on the San Pedro River, before seeking a medical discharge and leaving the state for California. Over his long career as a professional collector, Palmer returned to collect in Arizona under the sponsorship of different government agencies (including the Department of Agriculture and the Smithsonian Institution) from 1869–1870, 1876–1877, and in 1890. His travels to all corners of the territory produced hundreds more specimens delivered to the National Herbarium, Gray Herbarium, and a few others (McVaugh 1956). Among those new species described from Palmer's Arizona collections were *Agave palmeri* Engelm., *Lupinus palmeri* S. Watson [*Lupinus argenteus* Pursh var. *palmeri* (S. Watson) Barneby], and several *Astragalus* including *Astragalus allochrous* A. Gray.

In 1869, John Wesley Powell led the first scientific expedition down the Colorado River through the Grand Canyon. No botanist accompanied the party and, although he was broadly trained in the natural sciences, Powell's recorded observations on natural history were principally geological. When he observed the vegetation, his remarks were generally about the *lack of it* on uplands of "naked rock," the strips of riparian vegetation (willow, cottonwood, boxelder, hackberry, "oak glens" and "aspens") alongside valleys or canyon floors, and special occurrences of "mosses" and "fern decked alcoves" about desert springs on sandstone walls (e.g., Powell 1875, pp. 72, 79, 178). In August 1869, floating through Marble Canyon in northern Arizona, Powell named a series of cliff-side springs "Vasey's Paradise, in honor of the botanist [George Vasey] who traveled with us last year [1868, in Colorado Territory]" (Powell 1875, p. 76). Just upstream of the confluence with the "Colorado Chiquito" (Little Colorado River), he saw "*mesquite* shrubs, or small trees, with finely divided leaves and pods, somewhat like the locust" (Powell 1875, p. 77, italics original). At least a few specimens from the expedition in this part of Arizona made their way to botanists back East. Among these were seeds of two species of amaranth (*Amaranthus leucocarpus* S. Watson and *A. powellii* S. Watson) and of a salt-bush (*Atriplex powellii* S. Watson) that were grown out and the plants described as new.

Near the end of the era of continent-spanning exploring and surveying expeditions were the "Geographical Surveys West of the One Hundredth Meridian" (1871–1879) directed by George M. Wheeler. Branches of the Survey first entered Arizona in 1871, 1872, and 1873. In 1871, parties crossed the northern plateau and descended as far south as the Gila River, Camp Grant, and Camp Lowell in Tucson (Wheeler 1872). In his summary of the botany from the southwestern Surveys, Joseph T. Rothrock (1878) indicated that surgeon-naturalist Walter J. Hoffman and geologist Oscar Loew had each contributed plant collections from Arizona. The territories of New Mexico and Arizona were visited again under Rothrock in 1874. That survey began in Santa Fe, crossed the White Mountains (Camp Apache), and made the circuit of military camps in southern Arizona Territory: Camp Goodwin on the Gila River, Camp Grant in the Sulphur Springs Valley (and an ascent of neighboring Mt. Graham), Camp Bowie in the Chiricahua Mountains, Camp Crittenden on Sonoita Creek, and Camp Lowell in the Rillito Valley near Tucson, before returning to New Mexico. The expedition produced hundreds of plant collections from Arizona (see Rothrock 1878, including the Leguminosae by Sereno Watson), of which many were new species. Among those well-known from the region today are *Nolina microcarpa* S. Watson and *Bouteloua rothrockii* Vasey.

By the 1880s, the major river valleys and biomes of Arizona had been crossed and sampled by botanists, resulting in perhaps thousands of collections representing Arizona's flora deposited in herbaria. The second International Boundary Survey was yet to come in the 1890s, to be led by Edgar Mearns, who collected in Arizona while he was the surgeon at Fort Verde, 1884–1888, and added *Hosackia mearnsii* Britton [*Acmispon mearnsii* (Britton) Brouillet] to the flora. Otherwise, botanical collection fell primarily to individuals working independently or under the sponsorship of government agencies or scientific institutions. Eastern taxonomic botanists continued to desire new species, but also sought collections from new areas as they worked to compile data on the ranges and habitats of species already known. The collectors who supplied them were many. One observer wrote, "this territory is fast becoming the favorite haunt of our more intrepid collectors, for its physical features, while most forbidding so far as comfort is concerned, favor the growth of very peculiar plants" (Anonymous 1882, p. 8). In Arizona, several were not only prolific (aided greatly by the long-awaited arrival of two transcontinental railways within the decade) but became notorious for their output and discoveries: Edward L. Greene, Marcus E. Jones, John G. and Sarah Lemmon (*Lupinus lemmonii* C. P. Sm., see *Lupinus argenteus* Pursh var. *argenteus*), Cyrus G. Pringle (*Dalea pringlei* A. Gray), Henry H. Rusby (*Astragalus rusbyi* Greene), William F. Parish, and George R. Vasey. Still other collectors, professional and casual alike, sent specimens back East that today are both important records from new localities and witnesses to landscapes that have since been radically altered (Mauz 2011). Because the topographical surveys had often preferentially followed river valleys, this and ensuing decades saw new emphasis on the scattered Madrean and desert mountain ranges in the southern half of the Territory. Even in the early twentieth century, Philadelphia botanist Edwin Bartram remarked, "the surrounding mountains, or at least their foothills, are of easy access; and, as the representation of Arizona plants in most of our larger eastern herbaria is quite meagre, the ambitious collector finds an almost unlimited field in which to work" (Bartram 1922, p. 237).

Plant collections from the early expeditions had been exported to eastern institutions where established botanists identified the specimens and described new species. In the mid-19th century, these were the herbaria of John Torrey (now at NY), Asa Gray (GH), George Engelmann (now at MO), the Academy of Natural Sciences of Philadelphia (PH), and the fledgling herbarium at the Smithsonian Institution (US). With the opening of the University of Arizona in 1891, the Territory's first herbarium soon followed under its founder, professor of botany James W. Toumey. Its second curator, John J. Thornber, added substantially to collections from all parts of Arizona over his 40-year career. The herbarium at Tucson (ARIZ) has since grown to be one of the largest in the Southwest and the single-largest repository of Arizona plant specimens, and other institutional herbaria in Arizona (e.g., ASC, ASU) are important holdings of the state's biodiversity (Holmgren et al. 1990). Through the 20th century, faculty and students from the state's universities, staff of the Arizona Agricultural Experiment Station, scientists at the Carnegie Institution's Desert Botanical Laboratory, botanists with government agencies, and visiting investigators from elsewhere have increased the scope of Arizona plant collections (see the bibliographies in Moore and Cole 2004, for floristic treatments; Ewan 1936, for assorted taxonomic and historical works). The regions' scientists drew upon these specimens and reports as they began to synthesize the plant resources of Arizona taxonomically (e.g., Tidestrom and Kittell 1941, Kearney and Peebles 1942) and geographically (e.g., Shreve 1951, Brown 1994), and continue to do so in support of modern challenges in phylogenetic analyses, plant conservation, resource management, and environmental change.

BIOPROSPECTING LEGUMES OF ARIZONA

Leslie Gunatilaka

Plants of the family Fabaceae (Leguminosae), commonly known as legumes, are only second to those of the family Poaceae (Gramineae; cereals) in their economic importance. In addition to utility of their high-quality protein rich seeds (e.g., beans and peas), legumes are extensively used as sources of timber, fiber, gums, resins, tannins and other colorings, and agrochemicals including insecticides and molluscicides. Many legumes also find applications as medicinal plants. As a result of these practical uses, some legumes have been extensively investigated for their biochemical compounds (secondary metabolites) especially those responsible for their claimed medicinal properties.

Bioprospecting or biodiversity prospecting refers to the process of searching for potentially useful genetic resources and biochemical compounds in nature (Eisner 1994, Reid et al. 1993). As legume seeds (pulses) are rich sources of proteins, it is important to understand legume biodiversity for the purpose of bioprospecting such genetic resources for additional supplies of plant proteins. Bioprospecting efforts are underway in many parts of the world to identify and evaluate underexploited legume food sources as alternate protein crops for the future (Vijayakumari et al. 1994, Jaiwal and Singh 2003). Medicinal uses of some legumes have prompted bioprospecting these and other legumes for their biologically active secondary metabolites (natural products) with potential applications as pharmaceuticals.

However, despite the availability of a variety of native legume species of Arizona and a diverse collection of non-native legumes in the Desert Legume Program (DELEP), only a handful of bioprospecting studies have been conducted on legumes of Arizona.

Natural Products of the Leguminosae

Chemical investigations of plants of the family Fabaceae have resulted in numerous natural products, many of which are unique to legumes. *Phytochemical Dictionary of the Leguminosae* contains reports on the occurrence of natural products in 2,678 species classified in 385 genera of this plant family (Harborne 1994). Legumes are particularly rich in phenolic compounds including flavonoids, isoflavonoids, quinones, furanocoumarins, and xanthones. Other common natural products such as amino acids and peptides, alkaloids, and terpenoids are also known to occur in legumes. However, bioprospecting studies on legumes of Arizona have been directed towards screening, isolation, characterization, and evaluation of antimicrobial flavonoids and anticancer triterpenoid saponins, and these classes of natural products will be considered here with special reference to their pharmaceutical potential.

Bioprospecting Legumes for Phenolic Compounds

Phenolic plant pigments, including flavonoids and their analogues, are responsible for the colors of flowers, fruits, and sometimes leaves. Approximately 4,000 flavonoids and 870 isoflavonoids are known from the plant kingdom (Bruneton 1999) and many of these natural products also occur in legumes. A variety of plant-animal interactions are influenced by flavonoids and isoflavonoids. The colors of flowers and fruits help to attract pollinators and seed dispersers. Some flavonoids, e.g., kaempferol (Fig. 9), which occur in many legumes of Arizona, including species of the genera *Astragalus, Caesalpinia, Desmodium, Lathyrus, Medicago, Phaseolus, Prosopis, Senna, Trifolium, Vicia*, and segregates of the genus *Acacia*, protect these plants against harmful ultraviolet B (UV-B) radiation that is predominant in the United States Southwest. The flavonoids, apigenin and luteolin (Fig. 9) act as signaling molecules in legume-rhizobium bacteria interactions, assisting legumes in fixation of atmospheric nitrogen. Some isoflavonoids, e.g., (–)-medicarpin (Fig. 9), are known to induce defense against fungal attack in alfalfa (*Medicago sativa*), a major legume crop of Arizona.

As a part of an ongoing screening of arid land plants for potential biomedical uses, scientists at the UA Southwest Center for Natural Products Research and Commercialization (Natural Products Center) have investigated a number of native and non-native legumes available from DELEP. Bioactivity-guided fractionation of an extract of aerial parts of *Eysenhardtia texana*, a legume obtained from DELEP, afforded two new natural products, flavanone **1** and flavanone **2** (Fig. 10) with antibacterial activity against *Staphylococcus aureus* and antifungal activity against *Candida albicans* (Wächter et al. 1999). Based on this discovery an international patent application has been filed by the Research Development Foundation of Texas (Hoffmann et al. 2000).

There is a growing body of evidence to suggest that flavonoids also have the ability to inhibit multidrug-resistance (MDR) pumps in bacteria. In a bioprospecting project to discover phenolic natural products that enhance antibiotic activity against model pathogenic bacteria, G. Belofsky and colleagues (2004) of the University of Tulsa, Oklahoma, have investigated *Dalea versicolor*, a legume native to the Sonoran Desert of Arizona and Mexico. Their work resulted in the isolation and characterization of three phenolic compounds exhibiting direct or synergistic activity toward the human pathogen *Staphylococcus aureus* and the opportunistic pathogen *Bacillus cereus*. These compounds were identified as flavanone **3**, isoflavanone **4**, and the phenolic natural product chalcone **5** (Fig. 11). Of these, **3** and **4** are new natural products. In addition to its antimicrobial activity, chalcone **5** was found to potentiate the activity of known antibiotics—erythromycin and tetracycline.

Bioprospecting Legumes for Triterpenoid Glycosides

Legumes are known to contain a wide range of terpenoid natural products (Harborne 1994). Over two dozen gibberellin-type diterpenoid growth regulators have been encountered in legumes, including in *Phaseolus* and *Pisum* species. Triterpenoid glycosides, also known as triterpenoid saponins, occur widely in fodder legumes such as *Medicago* species. Recognizing that desert plant species were relatively poorly characterized for bioactive natural products, a bioprospecting effort was undertaken by the UA Natural Products Center (NPC) in collaboration with the Boyce Thompson Institute for Plant Research, New York, and University of Texas M.D. Anderson Cancer Center, Houston, to evaluate plant extracts from legume species native to arid and semiarid regions of the world. Screening of extracts derived from source plant material obtained from DELEP for anticancer activity identified the Australian species *Acacia victoriae* as a promising legume for bioprospecting (Mujoo et al. 2001). Anticancer activity-guided fractionation of an extract derived from seed and pods of this legume afforded two new triterpenoid saponins, avicin **D** and avicin **G** (Fig. 12), as natural products responsible for the anticancer activity (Jayatilake et al. 2003).

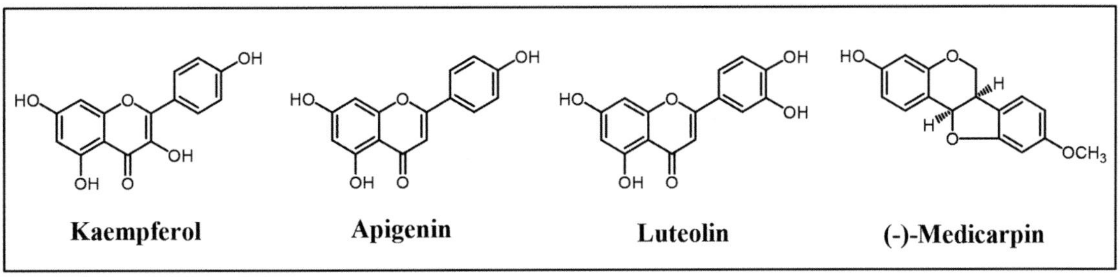

Figure 9. Structures of kaempferol, apigenin, luteolin and (−)-medicarpin.

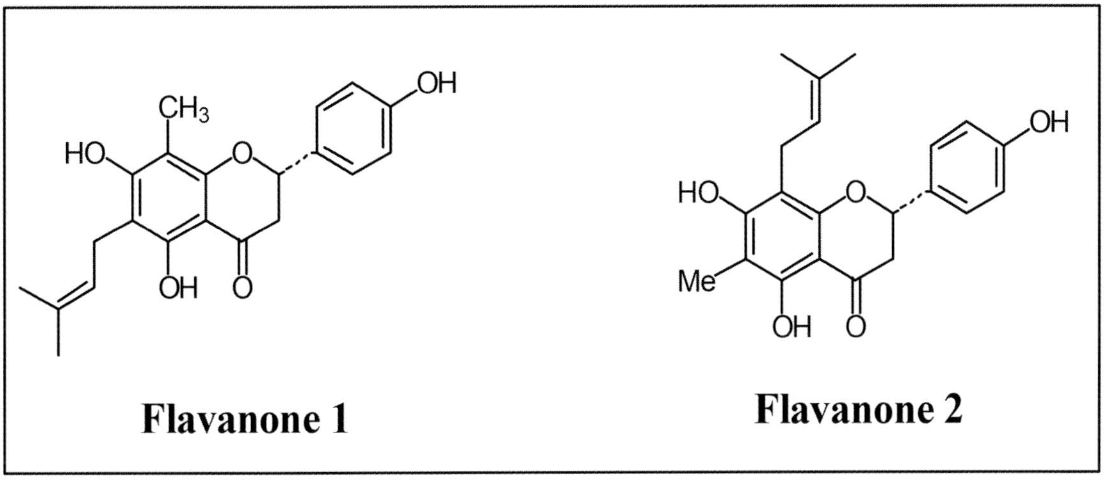

Figure 10. Structures of flavanone **1** and flavanone **2**.

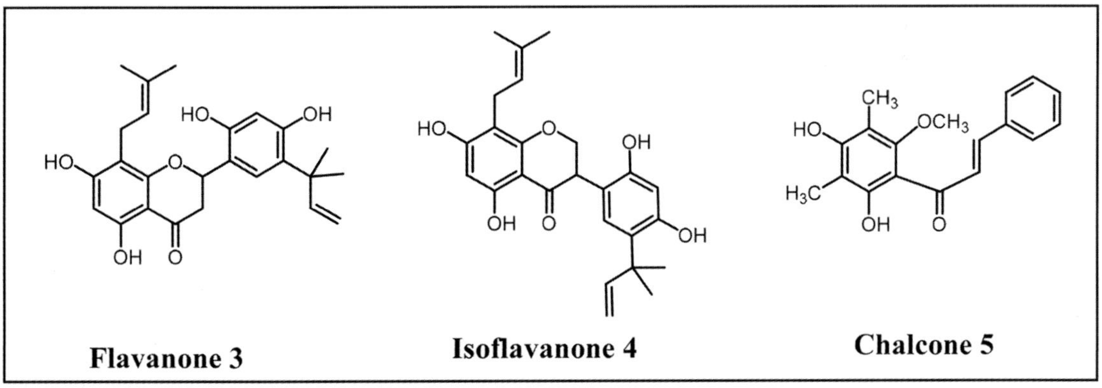

Figure 11. Structures of flavanone **3**, isoflavanone **4**, and chalcone **5**.

In addition to the originally observed cytotoxic activity, the fraction of the extract of *Acacia victoriae* containing a mixture of avicins has been shown to inhibit chemically induced mouse skin carcinogenesis suggesting these natural products could prove to be useful in suppressing the development of human skin cancer and other epithelial malignancies (Hanausek et al. 2001). Subsequent studies have demonstrated avicins **D** and **G** to exhibit cytoprotective, anti-mutagenic, anti-inflammatory, and anti-oxidant activities (Wang et al. 2010). It has recently been shown that avicins can bind to glucocorticoid receptor (GR) and induce its nuclear translocation. This and other studies have led to the classification of avicin **D** as a non-steroidal selective glucocorticoid receptor modulator (Haridas et al. 2011). The promise of

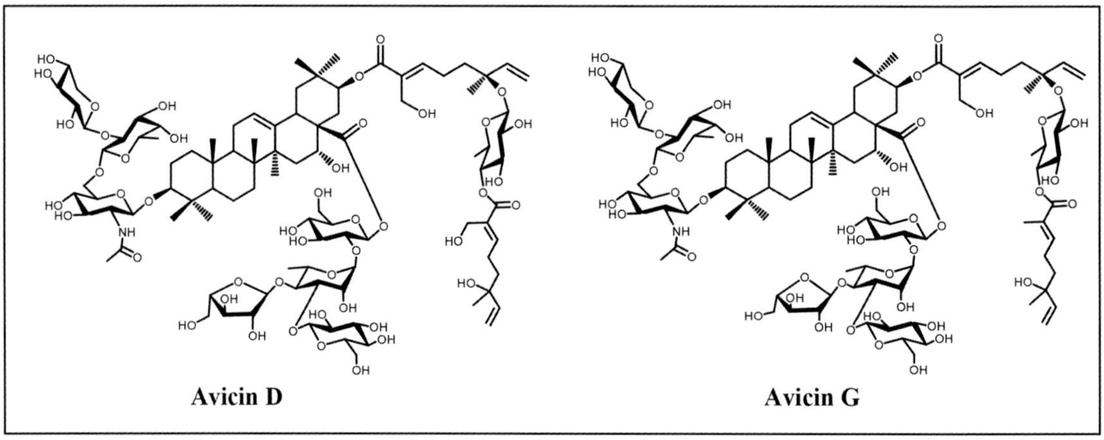

Figure 12. Structures of avicin **D** and avicin **G**.

avicin **D** as a potential anticancer drug prompted a multicenter phase I clinical study and is the subject of a recent patent (Gutterman and Haridas, 2016).

THE LEGUME FAMILY

Classification

The legume family, Fabaceae (or Leguminosae), is easily distinguished from other flowering plant families by its suite of characteristics, as described under morphology, below. Classification within the family, however, has undergone a few substantial changes. Until relatively recently, three subfamilies were recognized: Papilionoideae, Caesalpinioideae, and Mimosoideae. These groupings were largely based on the forms of the flowers, described below.

We now know that these subfamilies do not accurately capture the evolutionary history of the family. Thanks to the ease of generating more and more data, mostly from DNA sequences, as well as advances in our ability to analyze massive amounts of data carefully and thoroughly, a new picture has emerged. The data support a single common ancestor for all the species in Papilionoideae, and the same for the mimosoids. When we look at the caesalpinioids, though, we have major problems! A very few share a common ancestor with the mimosoids, but the rest are more and more distant cousins.

Why do these ancestors matter? The last several decades have seen a resurgence of an ideal promoted by Darwin: "Our classifications will come to be, as far as they can be so made, genealogies ..." (Darwin 1859). Classification systems that are based on genealogies, or phylogeny, and not on some other arbitrary characteristic, are far more stable and powerful for making predictions about traits or understanding evolutionary processes. The old Caesalpinioideae do not share a single common ancestor to the exclusion of the mimosoids, so we needed a new classification scheme.

To establish new subfamilies that accurately reflect ancestry, the Legume Phylogeny Working Group identified six common ancestors on the most recent phylogeny of the family (LPWG 2017) and used these to name new subfamilies: Cercidoideae, Detarioideae, Duparqetioideae, Dialioideae, Caesalpinioideae, Papilionoideae (Fig. 13). Mimosoideae, as a subfamily, is no longer recognized. Mimosoid species share a common ancestor with each other, so the group is a clade (light green semi-oval, Fig. 13), but the clade is nested within the newly circumscribed Caesalpinioideae (green semi-oval, including light and dark green, Fig. 13). We still use "mimosoid" to refer to members of this relatively young clade (Fig. 13) and its flower form, a form which is pretty unusual for this part of the angiosperm tree of life.

Morphology and Characteristics

The legume family is well distinguished from other families of flowering plants by several characteristics—most notably, the fruit or legume itself. Familiar to anyone who eats green beans or opens peanut shells, the legume is nonetheless a peculiar fruit because it is composed of a single carpel that splits on two sides (or, if it doesn't split open, it still has sutures on both sides) (Fig. 14). Other non-legume fruits made of only one carpel split on one side only, like milkweed follicles. An easy way to see that the legume is a single carpel is to carefully open a green bean and note that the seeds are all attached on one side (Fig. 14). Vegetatively, with a few exceptions, members of the legume family can also be readily

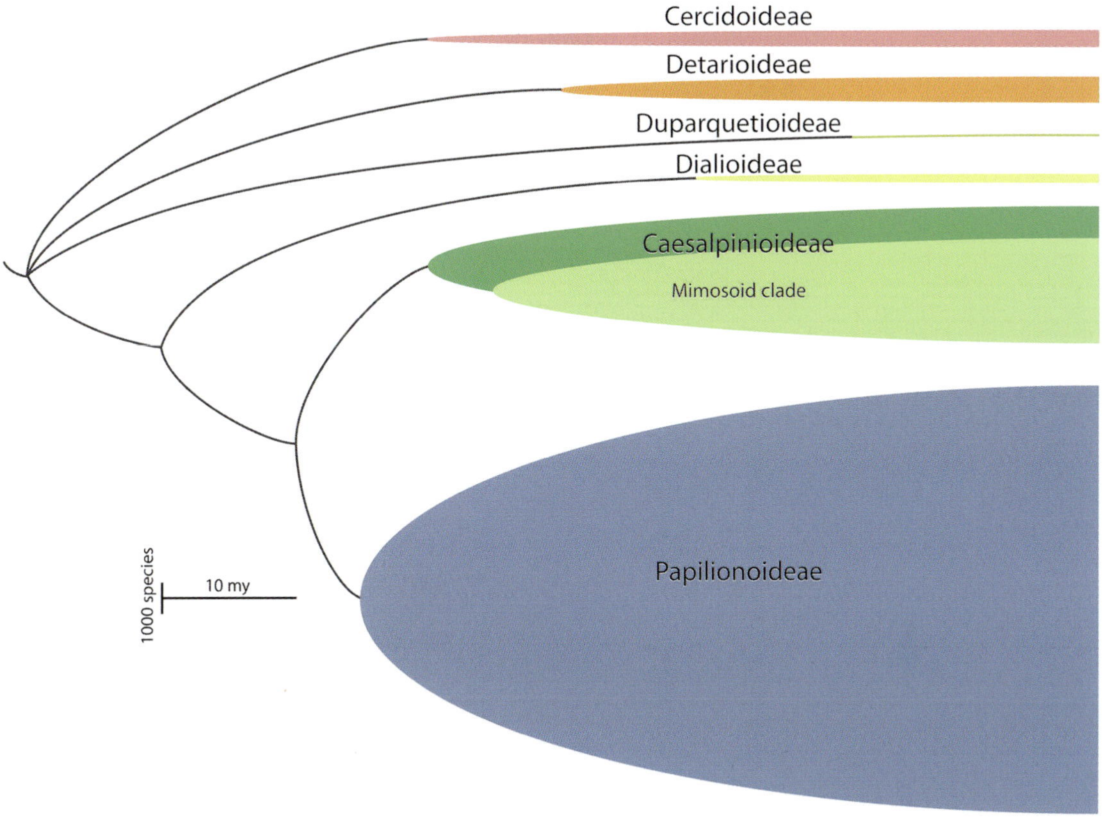

Figure 13. Phylogeny of Fabaceae. Each of the colored groups with names ending in "-oideae" are the new subfamilies, based on LPWG 2017. Note that the Mimosoid group, a group of genera that includes *Acacia*, *Mimosa*, and relatives, and that was formerly recognized as a subfamily, is now a clade nested within the larger Caesalpinioideae. Relative sizes, in number of species, is indicated by the height of the colored semi-ovals; approximate ages of the groups and of their common ancestors (where the lines join) are indicated on the horizontal axis. Ages and sizes are approximated from the Angiosperm Phylogeny Website (Stevens 2017).

distinguished by alternate compound leaves with stipules and pulvini (Fig. 15). The stipules in many (particularly in our arid and semi-arid habitats) form spines or spiny hooks and are therefore likely to be familiar in a less positive way than the fruits. Pulvini are also famous in their own right; these areas of swollen tissue at the base of leaves, and sometimes of leaflets as well, can allow quick movements in response to stimuli, e.g., the rapid folding of leaflets in the "sensitive plant," a popular horticultural mimosoid.

Three distinct flower forms are found in the family: papilionoid, caesalpinoid, and mimosoid (Fig. 16). In all, five sepals are fused, at least at the base, but the petal arrangement is distinct for each type. In the mimosoid flower form, found throughout the mimosoid group, the small flowers are radially symmetric, and the five petals are usually at least basally fused. Ten or more stamens are either free or also fused at the base and are often well exserted. Long showy stamens combined with congested inflorescences, such as round balls or elongated spikes containing many small flowers, give this group a distinctive look as seen in the "fairy dusters" (*Calliandra*) or "acacias" (*Acacia*, *Vachellia*, *Mariosousa*, etc.).

Papilionoid group members almost all have papilionoid flowers, which are bilaterally symmetric, larger, and with more conspicuous petals than in the mimosoids. In a papilionoid flower the five free petals are not all the same, so they have special names. The upper petal is called the banner and often has an eyespot or nectar guides, and/or it changes color following pollination. The two lateral petals are called wings and are often directed a bit outward. The two lower petals, the keel petals, commonly adhere lightly or are fused to each other along their outer margins, enclosing the reproductive parts. The keel petals can have folds near the base that catch on folds on the corresponding wing petal. In those cases, the four lower petals therefore operate a bit like a trick platform. Bees land to access nectar around the base of the ovary, thereby depressing the entire set of petals and revealing the stamens and pistil, perhaps getting their bellies dusted with pollen in the process. Most papilionoid flowers have 10 stamens; the upper stamen (closest to the banner) is free and the other nine are partly fused, forming a sheath around the ovary.

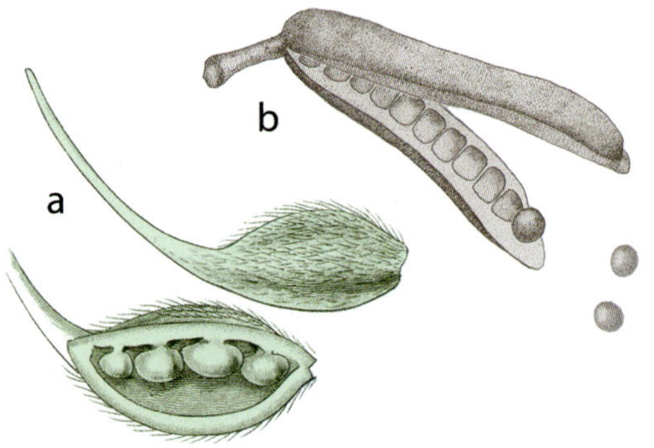

Figure 14. Fruits of Fabaceae. **a.** young fruit showing seed attachments on upper suture only (*Geoffroea decorticans*, excerpted from illustration in Hooker, 1833); **b.** mature fruit showing the typical dehiscence along two sutures, by which the pod splits into two valves, despite comprising a single carpel (*Ebenopsis ebano,* from illustration by Daniela Siroky).

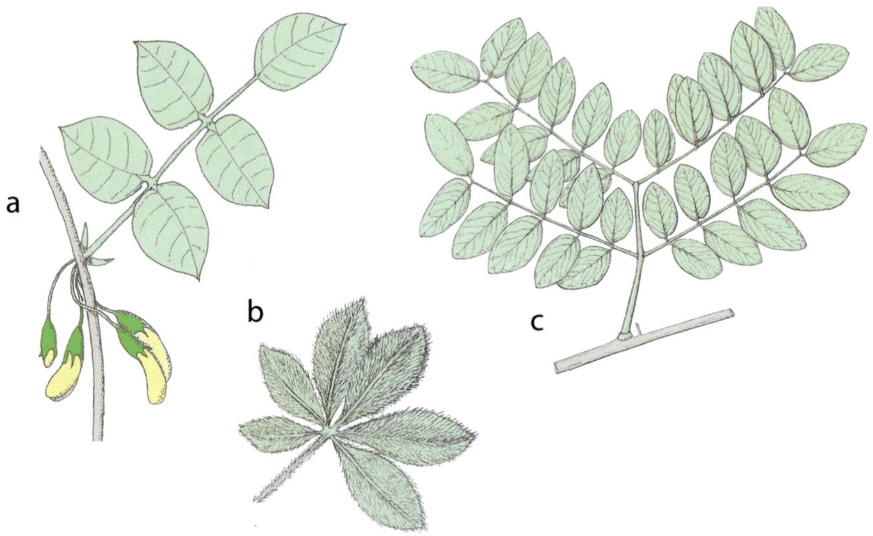

Figure 15. Leaves of Fabaceae. **a.** pinnately compound leaf with odd number of leaflets (*Nissolia schottii*, from illustration by Barbara Vucich); **b.** palmately compound leaf (*Lupinus huachucanus*, from illustration by Joyce Riney Peters); **c.** bipinnately compound leaf with even number of primary leaflets and even number of secondary leaflets and pulvinus at base of the leaf (*Coulteria platyloba*, from illustration by Marsha Bennett).

Flowers that have the caesalpinoid flower form are far more variable, which makes sense—it is now clear that they do not form a natural group (see Classification section, above). Nevertheless, they share pentamery (five-partedness) with all members of the family and exhibit a range of symmetries from radial to bilateral, as commonly seen with the papilionoids. However, an easy way to distinguish the papilionoid flowers from caesalpinoid flowers is through the position of the upper petal: slightly inside the others in the caesalpinoid form, but outside the others in the papilionoid form (Fig. 16).

Other differences between the subfamilies include characteristics of the leaves. Papilionoids generally have once-compound leaves with an odd number of leaflets; twice-compound leaves with an even number of leaflets are common throughout the rest of the family (Fig. 15).

INTRODUCTION TO THE FLORA AND REFERENCE

Overview of Fabaceae in Arizona

Native, naturalized, and commonly cultivated legume diversity is rich here in Arizona, with strong representation of three of the six new subfamilies. Most Arizona legumes are from the largest subfamily, the Papilionoideae, with 503

Figure 16. Flowers of Fabaceae. The generalized mimosoid flower form has fused sepals (green), fused petals (yellow), and often many showy stamens (style not shown). The generalized caesalpinoid flower has bilateral symmetry, five free petals, with the upper petal (white) slightly inside of the neighboring lateral petals (yellow). The generalized papilionoid flower also has bilateral symmetry and five petals, all free at least at their bases. The upper petal (pink) is called the banner and is outside of the neighboring two lateral petals (blue), which are called wings. The lowest pair (pale pink), often lightly touching at their lower margins, surround the stamens and the style and are called keel petals. Illustrations by Matthew B. Johnson.

genera worldwide and 57 in Arizona. The other two subfamilies represented in Arizona are the new Caesalpinioideae, (30 genera in Arizona, including those formerly in the Mimosoideae; see Classification section, above), and Cercidoideae (2 genera in AZ). The other three subfamilies are exclusively tropical (LPWG 2017).

Most of the genera and species in Arizona are native, but several of the commonly cultivated genera are here only as guests (Table 1).

Table 1. Arizona legume diversity.

	Native	Naturalized	Cultivated (only)	Total
Genera	59	17	21	89
Species	289	31	61	381
All taxa (including var. & subsp.)	354	36	64	454

Diversity of Arizona legumes is also quite high when compared to other, nearby, floras (Table 2). For example, counting all native and naturalized taxa (including varieties and subspecies), we have 390 taxa spread across our 295,000 km^2. This makes our per-area legume diversity higher than California, New Mexico, and Texas (although this is sensitive to floristic updates). Utah exceeds the diversity in Arizona, and in both Arizona and Utah, the cause is simple: *Astragalus*. With over 3,000 species worldwide, it is the largest genus in the world. The New World center of diversity for this immense genus is the intermountain West, and it is well represented in Arizona. With 79 native species, more than 25% of Arizona legumes species are in *Astragalus*.

Table 2. Comparison of native and naturalized Fabaceae in Arizona to other state floras.

	AZ	CA[1]	NM[2]	TX[3]	UT[4]
Genera	68	50	56	74	42
Species	320	405	262	367	246
Total taxa	390	541	322	398	324
Area (x 1000 km^2)	295	425	316	679	220
Diversity[5]	1.32	1.27	1.02	0.59	1.47

[1]Baldwin 2012, [2]Allred 2012, [3]Johnston 1990, [4]Welsh et al. 2003, [5]Number of total taxa per unit area (1000 km^2).

TAXONOMIC TREATMENTS

The main text of this volume provides taxonomic treatments for all the native, naturalized, and cultivated legume taxa in the state. Each treatment was provided by a taxonomic expert to ensure up-to-date nomenclature and taxonomic concepts. Of course, names will continue to change; experts at local herbaria should always be consulted for changes made after publication. Following are a few notes that may assist in the use of these treatments.

Descriptions include some botanical terminology (see glossary), but technical terms have been reduced whenever possible. Information given in descriptions for genera is generally not repeated in descriptions for individual taxa. Descriptions, especially for genera, represent the taxon's full range of variation, not just the range of variation found in Arizona. However, uncommon features or measurement not found in Arizona are highlighted through the use of brackets. Members of the editorial committee supplemented treatments with information on ecology and on horticultural, agricultural, and other uses, where appropriate.

Keys

The key to the genera accommodates legume species occurring in Arizona, including the genera and species that are commonly cultivated in Arizona. Keys within genera likewise focus on Arizona species, and keys within species are included only when more than one variety or subspecies occurs here. Every attempt has been made to ensure that the key couplets include more than one set of characteristics, but this is particularly challenging in some cases, e.g., when distinguishing the remarkably large number of locoweeds (*Astragalus*) in Arizona.

Geographic Distributions

Geographic distributions were estimated using location data derived from herbarium vouchers in the Southwest Environmental Information Network (SEINet; swbiodiversity.org). These data were curated by the editors to exclude material of dubious identity or surprising range extensions that editors were not able to verify. The resulting dot maps that accompany each taxon may be considered conservative. They are intended to provide a quick visual idea of where the taxon is found and how often it has been collected, as an estimate of range and density. The county lists as presented in the treatments are likely to be accurate, although new populations certainly await discovery among the topographically complex regions of Arizona.

Each treatment indicates the origin of the taxa. Native plants are those that occur naturally in Arizona, not as a consequence of human activity. Naturalized plants include those taxa from outside Arizona that nevertheless have become established here, reproducing without human fostering. Naturalized taxa may or may not be considered weeds; this designation is usually reserved for a rampantly reproducing species, often posing some threat to agriculture or to natural habitats. Of all the Arizona legumes, only one member is formally designated as a noxious weed by the USDA: *Prosopis velutina*—and it is one of our cherished Sonoran Desert natives! This volume also includes information on taxa commonly cultivated in the state, including many natives that are also cultivated. For all, the full range of the species is provided at the county level within Arizona and with less granularity outside the state.

Each treatment also indicates the biotic communities in which to expect the taxa; these communities are described in the chapter on Biotic Communities.

Nomenclature

The scientific name of each species (or variety or subspecies) is followed by the "authority" i.e., the name of the author(s) who originally published that taxon name. If the species name is transferred to a new genus or other rank, the name of the original author(s) is retained and placed in parentheses, followed by the name of the author(s) who made the taxonomic transfer. For example, *Dalea spinosa* A. Gray (the "desert smoke tree") was first described and named in 1854 by Asa Gray. In 1977 Rupert C. Barneby taxonomically revised the genus *Dalea* (Barneby 1977), and transferred the species to the genus *Psorothamnus*, with the new, correct name becoming *Psorothamnus spinosus* (A. Gray) Barneby.

Nomenclatural change is inevitable as we learn more about plant groups, especially their evolutionary, or phylogenetic, relationships. Indeed, perhaps particularly in the plant world, synonyms are common—these are sets of two or more valid names applied to the same biological entity, or taxon. In the above example, "*Dalea spinosa* A. Gray" became a synonym of the new name, "*Psorothamnus spinosus* (A. Gray) Barneby." These are important because different resources may refer to the same thing, but by different names! Field guides, the last comprehensive flora for Arizona (Kearney and Peebles 1960), online resources such as USDA Plants, the International Plant Name Index, Tropicos—all of these can disagree. As of this publication, the names included here can be considered the "right" or preferred names for these Arizona taxa, but new discoveries after publication will result in still more changes. The best resource is always your local herbarium (e.g., the University of Arizona Herbarium, the Arizona State University Herbarium, the Deaver Herbarium at Northern Arizona University), where experts are tasked with keeping up to date on this all-important question of "what's the right name?"

Limited synonymy is provided in this publication. In general, many of the names used for legumes in the *Arizona Flora* (Kearney and Peebles 1960) are no longer recognized but are listed here as synonyms. One genus from the 60-year-old flora, *Krameria* Loefl., is no longer considered part of the legume family (Simpson 1989). Other synonyms are given solely to help clarify nomenclature or for general interest. However, no attempt is made to provide complete synonymy. Selected common names (e.g., "blue palo verde" for *Parkinsonia florida* S. Watson) are included in the treatments and in the index.

Horticulture and Cultivation

The legume family offers a varied palette of native and introduced plants that are suitable for landscape use in Arizona. Plants in this family come in a wide variety of shapes, sizes, and forms, and they can fill many functions in the landscape. Shade trees can provide relief from the intense desert sun. Various shrubs and trees can be planted as screens, barriers, windbreaks, and hedges and provide food and shelter to wildlife. Legumes are ornamental with attractive growth habits, bark, and foliage. Many taxa produce masses of colorful and sometimes fragrant flowers. Numerous taxa tolerate high temperatures and low humidity and are drought resistant once established. There are few pest and disease problems with legumes in arid landscape environments. Many legumes are able to fix atmospheric nitrogen, minimizing or eliminating the need for supplemental fertilizers. The attractiveness of these plants and their ability to survive the rigors of harsh environments and thrive with a minimum of care makes them well adapted for landscaping and urban forestry in Arizona and other regions of the world with similar climates.

A considerable number of legume species are commercially grown in Arizona for landscape use. Many of these are native to the state. Some have been introduced from Texas and Mexico, while others are native to the Mediterranean Region, Africa, Asia, Australia, and South America. Additional species are likely to become available in the future as they are evaluated for their landscape potential. Many of these plants are readily available from nurseries that carry low water-use plants. Some recent introductions and plants with limited demand can be found at specialty nurseries or botanical garden plant sales.

In general, plants in cultivation under favorable growing conditions will respond by growing more rapidly than their wild counterparts and may achieve a larger size than is typical for the species unless growth is restricted by winter freezing or some other factor. Eventual mature size is an important consideration when selecting a planting site. Selecting the proper plant for the site can avoid considerable effort and expense to correct mistakes. Some plants that are normally deciduous in the wild may remain in leaf year-round in cultivation under favorable conditions of moisture and temperature. Specific information on horticultural characteristics, along with landscape applications and limitations, are included in the discussion sections for both native taxa that are grown as landscape plants and taxa that are found in Arizona only as cultivated plants.

Sensitivity to freezing is a significant limiting factor for survival of many species of legumes in Southwest landscapes. Other factors that can influence cold hardiness in plants include duration and intensity of the freeze, the individual genotype of the plant, maturity of the plant, whether the plant is growing or is dormant and hardened off, and protection provided by adjacent plants and buildings. Some plants with freeze-sensitive stems and foliage exhibit rapid recovery the following spring while other species recover slowly. There is a need for further selection of germplasm that exhibits greater cold tolerance among many taxa.

Propagation of most legumes is by seeds. With many of these plants, scarification of the seeds provides for rapid and uniform germination; however, some legume seeds will germinate well without scarification. Some nurseries have made selections of plants with particular desirable characteristics, such as growth form, flowering traits, or lack of thorns/spines. These are typically propagated by vegetative cuttings or through tissue culture to maintain uniformity. Care of legumes in a nursery setting is fairly standard, subject only to specific limitations such as temperature tolerances. Where appropriate, these limitations are noted in the individual discussions.

KEY TO THE GENERA

Below is a dichotomous key to the legumes of Arizona, focusing on the traits as found in our taxa. For example, not all *Errazurizia* species have 0 petals, but it keys out there because the only Arizona representative, *Errazurizia rotundata,* generally lacks petals. Every effort has been made to incorporate variation between individuals, but in some cases, it may be necessary to follow multiple paths, particularly if the specimen in hand is poorly developed. Usually, flowers and leaves are required for success; for some couplets, fruits are needed as well. All characteristics require no more than a typical 10x hand lens, and technical terms are defined in the glossary.

Major Groups

1. Flowers radially symmetric; petals inconspicuous, basally fused; stamens 10-many, exserted well beyond petals; leaves bipinnate or simple (phyllodes) ____ **Mimosoid Clade**
1′ Flowers weakly to strongly bilaterally symmetric; petals usually conspicuous, not basally fused (free); stamens 1–10, often not exserted much beyond petals except in *Caesalpinia* and *Erythrostemon*; leaves pinnate or bipinnate, 1-foliolate (appearing simple), 2-foliolate, or palmate.
  2. Corolla papilionoid ("pea-like") with lower pair of petals usually fused into a keel, uppermost petal (banner) outermost in bud, or corolla lacking in *Errazurizia rotundata* and *Parryella*; stamens enclosed by keel except in *Amorpha*, some *Dalea*, *Errazurizia*, *Eysenhardtia*, and *Parryella*, filaments all or 9 usually fused; leaves 1-foliolate (appearing simple), palmate or pinnate (including 3-foliolate) ____ **Papilionoideae**
  2′ Corolla not papilionoid, with petals free, uppermost petal (banner) innermost in bud; stamens not usually enclosed by lower petals (keel) (except *Cercis*), filaments free; leaves 1-foliolate (appearing simple), 2-foliolate, pinnate or bipinnate.
    3. Leaves pinnate or bipinnate; corolla yellow to orange to red, or absent ____ **Caesalpinioideae (excluding Mimosoid Clade)**
    3′ Leaves 1-foliolate or 2-foliolate; corolla white, pink, pink-purple, red-purple, blue-purple ____ **Cercidoideae**

Caesalpinioideae (excluding Mimosoid Clade)

1. Inflorescences arising from trunks and older branches; corolla absent ____ **Ceratonia**
1′ Inflorescences terminal or axillary at distal ends of stems; corolla present, often showy.
  2. Leaves all pinnate; corolla generally yellow.
    3. Stems armed; fruits dehiscent along suture in mid-region of valves ____ **Haematoxylum**
    3′ Stems unarmed; fruits dehiscent along suture at edge of valves or from distal end.
      4. Plants herbaceous perennials or shrubs; bracteoles absent; valves of fruit not elastically dehiscent and not coiling upon dehiscence ____ **Senna**
      4′ Plants annuals or delicate herbaceous perennials; bracteoles 2; valves of fruits elastically dehiscent and coiling upon dehiscence ____ **Chamaecrista**
  2′ Leaves generally bipinnate (both pinnate and bipinnate in some; if leaf interpreted as pinnate, see *Parkinsonia*); corolla yellow to orange, less often red or with orange to red marks or pale greenish.
    5. Flowers 3–5 mm; sepals and petals alike in color and texture; trees, generally armed ____ **Gleditsia**
    5′ Flowers more than 5 mm; sepals and petals not alike in color and texture; herbaceous perennials, shrubs or trees, armed or unarmed.
      6. Plants armed.
        7. Plants with stipular spines at nodes or thorns in leaf axils or twigs thorn-tipped; calyx with lobes subequal ____ **Parkinsonia** (in part)
        7′ Plants with prickles; calyx with lobes strongly unequal.
          8. Plants with slender prickles; corolla with filaments red, 31–56 mm; fruits stiffly papery and elastically dehiscent ____ **Caesalpinia pulcherrima** (in part)
          8′ Plants with stout prickles; corolla with filaments pale yellow or greenish yellow, 11–13 mm; fruits leathery and indehiscent ____ **Tara** (in part)
      6′ Plants unarmed.
        9. Trees with 2 pinnae per leaf ____ **Parkinsonia** (in part)
        9′ Herbaceous perennials, shrubs, or trees with 3 or more pinnae per leaf.
          10. Herbaceous perennials or if shrubs then leaves with 3 pinnae per leaf.
            11. Leaflets with orange (black on drying) glandular punctate trichomes on lower (abaxial) surface ____ **Pomaria**
            11′ Leaflets lacking glandular-punctate trichomes on lower (abaxial) surface ____ **Hoffmannseggia**
          10′ Shrubs or trees with 4 or more pinnae per leaf.
            12. Leaves odd bipinnate.
              13. Bark of stems and trunks exfoliating in thin, irregular plates; fruits woody, indehiscent ____ **Libidibia**
              13′ Bark of stems and trunks not exfoliating; fruits stiffly papery, elastically dehiscent ____ **Erythrostemon**
            12′ Leaves even bipinnate.
              14. Leaves with 2 pairs of pinnae; leaflets 2–8 mm ____ **Parkinsonia** (in part)
              14′ Leaves with 2–9 pairs of pinnae; leaflets 6–100 mm.
                15. Fruits 1–1.3 cm wide, constricted between seeds ____ **Tara** (in part)
                15′ Fruits 1.5–3.6 cm wide; not constricted between seeds.
                  16. Leaves with 4–9 pairs of pinnae; leaflets 6–12(–25) mm; lower calyx lobe with an entire margin; filaments 31–56 mm; fruits elastically dehiscent ____ **Caesalpinia pulcherrima** (in part)
                  16′ Leaves with 2–3(–4) pairs of pinnae; leaflets 30–100 mm; lower calyx lobe with a glandular-fringed margin; filaments 7–9 mm; fruits indehiscent ____ **Coulteria**

Cercidoideae

1. Leaves 1-foliolate, orbicular to ovate to cordate to subreniform ____ **Cercis**
1′ Leaves 1-foliolate and 2-lobed, or 2-foliolate, suborbicular to ovate to lanceolate ____ **Bauhinia**

Mimosoid Clade

1. Stamens 10 (or fewer).
  2. Plants unarmed; inflorescences heads.
    3. Perennial herbs, subshrubs or shrubs to less than 3 m height; stipules bristle-like or awl-shaped; sterile flowers with staminodia present at base of inflorescences; fruits 2.5–4 mm wide ______ **Desmanthus**
    3′ Shrubs or small trees 2 to 18 m height; stipules lanceolate or ovate sometimes; sterile flowers absent; fruits 11–21 mm wide ______ **Leucaena**
  2′ Plants armed (or if unarmed as in some *Prosopis*, then inflorescences racemes); inflorescences heads, spikes, or racemes.
    4. Plants armed with prickles; fruits valvately dehiscent detaching from the rim or separating into segments that detach from rim, lacking pulpy mesocarp ______ **Mimosa**
    4′ Plants armed with nodal spines or stipular spines (some stems with armament minute); fruits indehiscent, with pulpy mesocarp ______ **Prosopis**
1′ Stamens more than 10.
  5. Filaments fused proximally, free distally.
    6. Plants armed.
      7. Leaflets 2–6 pairs per pinna; inflorescences cylindrical spikes; fruits woody ______ **Ebenopsis**
      7′ Leaflets 5–21 pairs per pinna; inflorescences globose heads; fruits stiffly papery ______ **Havardia**
    6′ Plants unarmed.
      8. Trees (or shrubs 2+ m); leaves 10–32 cm with (13–)17–76 pairs of leaflets per pinna; petiole with a gland; inflorescences axillary or terminal corymbs or subglobose racemes; fruits not or tardily dehiscent.
        9. Flowers with filaments white proximally and pink distally; fruits with rim not separating from valves ______ **Albizia**
        9′ Flowers with filaments white throughout; fruits with rim separating from valves ______ **Lysiloma**
      8′ Herbs, subshrubs, or shrubs to 2 m; leaves to 10(–13) cm with (3–)4–20(–27) pairs of leaflets per pinna (if more than 15 pairs of leaflets per pinna then the plants herbs or low subshrubs); petiole without a gland; inflorescences axillary globose heads; fruit elastically dehiscent.
        10. Plants herbaceous or shrubby; leaflets, 4–27 pairs per pinna, overlapping, or if not overlapping, then herbaceous and not found in Sonoran Desertscrub; fruits usually erect or ascending, linear-lanceolate or oblanceolate, not translucent or translucent; widespread (*Calliandra californica* cultivated only) ______ **Calliandra**
        10′ Plants shrubby; leaflets, 3–7 pairs per pinna, not overlapping; fruits usually pendent, linear, translucent; Sonoran Desertscrub only, localized in Pima County ______ **Zapoteca**
  5′ Filaments free.
    11. Leaves simple (phyllodes) ______ **Acacia**
    11′ Leaves bipinnate.
      12. Plants armed with prickles or stipular spines (some stems with armament sparse or minute).
        13. Stems armed with prickles ______ **Senegalia**
        13′ Stems armed with stipular spines ______ **Vachellia**
      12′ Plants unarmed (stipules of *Mariosousa millefolia* scarcely indurate and appearing spine-like, to 6.5 mm).
        14. Trees with bark becoming fissured, not exfoliating or papery ______ **Parasenegalia**
        14′ Subshrubs or shrubs, or if trees then bark uniformly smooth, exfoliating and papery.
          15. Petiole and rachis lacking glands; inflorescences globose to slightly elongated heads; flowers pedicellate; stamens 175–250 per flower; fruits tan, brown, or purplish ______ **Acaciella**
          15′ Petiole and rachis with glands (absent from some leaves); inflorescences cylindrical spikes; flowers sessile; stamens 50–100 per flower; fruits pale straw colored ______ **Mariosousa**

Papilionoideae

1. Stipules sometimes obscure, gland-like; plants herbaceous (or subshrubs in *Acmispon rigidus*); leaves pinnately compound to palmately compound, sometimes both types on the same plant (rarely some or most 1-foliolate) ______ **Acmispon**
1′ Stipules conspicuous, obscure or absent, not gland-like; plants various; leaves either pinnately compound or palmately compound (including 3-foliolate), not both types on the same plant (rarely 1-foliolate or leaves or leaflets absent).
  2. Leaves with 3, 4, or 5 broadly ovate to nearly rounded leaflets, distal-most leaflet terminal, lateral leaflets alternate; cultivated only ______ **Dalbergia**
  2′ Leaves not as above; native or cultivated.
    3. Leaves consistently with 1 leaflet or leaflets absent ______ **Group A**
    3′ Leaves consistently with more than 1 leaflet.
      4. Leaves palmately compound (rachis absent) ______ **Group B**
      4′ Leaves pinnately compound (rachis present but sometimes obscure).
        5. Leaves even pinnate ______ **Group C**
        5′ Leaves odd pinnate.
          6. Leaves with more than 3 leaflets ______ **Group D**
          6′ Leaves with 3 leaflets.
            7. Plants vines or with stems trailing ______ **Group E**
            7′ Plants not vines ______ **Group F**

Group A. Leaves with 1 leaflet or leaflets absent

1. Herbage glandular.
  2. Woody shrubs or small trees; fruit valves glandular ______ **Psorothamnus** (in part)
  2′ Herbaceous or suffrutescent; fruit valves not glandular ______ **Ladeania juncea** (in part)

1′ Herbage not glandular.
3. Fruits loments or resembling loments.
4. Fruits loments; plants not rhizomatous; unarmed; native **Desmodium** (in part)
4′ Fruits resembling loments; plants strongly rhizomatous; stems thorn-tipped; naturalized **Alhagi**
3′ Fruits neither loments nor loment-like.
5. Plants twining vines; cultivated only **Hardenbergia**
5′ Plants not vines; native.
6. Plants perennial (and annual); commonly with 3 to many leaflets, rarely unifoliolate **Astragalus** (in part)
6′ Plants annual; leaves unifoliolate.
7. Corolla yellow; fruits inflated **Crotalaria sagittalis**
7′ Corolla pinkish; fruits laterally compressed **Sphinctospermum**

Group B. Leaves palmately or pseudo-palmately compound

1. Leaflets 2 **Zornia reticulata**
1′ Leaflets 3 or more.
2. Leaflets more than 3.
3. Herbage not glandular **Lupinus**
3′ Herbage glandular **Pediomelum**
2′ Most leaves with 3 leaflets.
4. Stipules dimorphic, those at lowest, leafless nodes papery and connate into sheath, those at upper, leafy nodes often large and leaf-like **Thermopsis**
4′ Stipules not dimorphic, usually small, triangular, often fused to petiole.
5. Herbage glandular **Ladeania** (in part)
5′ Herbage not glandular **Trifolium** (in part)

Group C. Leaves even-pinnate

1. Leaves with terminal tendrils.
2. Style folded; stipules leaflet-like, often larger than leaflets **Pisum**
2′ Style not folded; stipules generally not leaflet-like, sometimes ± as large as leaflets.
3. Style pubescent on upper side **Lathyrus**
3′ Style evenly pubescent just proximal to stigma and/or tufted on lower side **Vicia**
1′ Leaves without terminal tendrils.
4. Fruits loments, strongly laterally compressed.
5. Leaves with 20–50 leaflets; fruits with surface not reticulate **Aeschynomene**
5′ Leaves with 2–4 leaflets; fruits with surface reticulate **Zornia**
4′ Fruits not loments, more or less cylindrical.
6. Plants herbaceous or suffrutescent; fruits not constricted between seeds; corolla yellow-orange with purple-maroon spots or orange-grenadine-red without purple maroon spots **Sesbania**
6′ Plants woody; fruits constricted between seeds; corolla whitish to pink or purplish.
7. Unarmed; rachis extending past distal leaflet pair into a conspicuous mucro; inflorescences with abundant stipitate glands **Coursetia glandulosa**
7′ At least some stems armed; rachis not extending past distal leaflet pair into a conspicuous mucro; inflorescences not glandular or with sparse stipitate glands **Olneya**

Group D. Leaves odd-pinnate with more than 3 leaflets

1. Plants twining vines or with stems trailing.
2. Corolla azure, purple, rose, or white; cultivated **Wisteria**
2′ Corolla yellow; native.
3. Inflorescences with bracts present; fruits samaroid, indehiscent **Nissolia**
3′ Inflorescences with bracts absent; fruits not samaroid, elastically dehiscent **Hosackia alamosana**
1′ Plants not vines.
4. Herbage glandular.
5. Fruits with stiff, hooked bristles **Glycyrrhiza**
5′ Fruits without bristles.
6. Petals 0 or 1.
7. Petal 1. **Amorpha**
7′ Petals 0.
8. Stems stiff; strongly pubescent **Errazurizia**
8′ Stems pliant; not pubescent **Parryella**
6′ Petals 5.
9. Woody shrubs or small trees.
10. Corolla white; southeastern Arizona and cultivated **Eysenhardtia**
10′ Corolla dark blue or purple; western and northern Arizona, and cultivated **Psorothamnus** (in part)
9′ Herbs or suffrutescent shrubs.
11. Leaflets without lineoles; flowers usually sessile **Dalea** (in part)
11′ Leaflets with sinuous, pallid lineoles; flowers pedicellate **Marina** (in part)
4′ Herbage not glandular (or some with glandular hairs only).
12. Fruits winged (samara-like) or fleshy (drupe-like).

13. Bark not exfoliating; leaflets 2.5–5.4 cm; fruits winged (samara-like) ____ **Tipuana**
13′ Bark exfoliating in strips or flakes; leaflets 0.3–1.6 cm; fruits fleshy (drupe-like) ____ **Geoffroea**
12′ Fruits neither winged (samara-like) nor fleshy (drupe-like).
14. Fruits with firm stiff hairs ____ **Robinia neomexicana**
14′ Fruits without firm stiff hairs.
15. Fruits loments.
16. Fruits flattened ____ **Hedysarum**
16′ Fruits angled in cross-section ____ **Securigera**
15′ Fruits not loments.
17. Large shrubs or trees.
18. At least some stems with spinescent stipules; fruits dehiscent.
19. Leaves 2–10 cm; leaflets 5–21 mm; fruits cylindrical ____ **Olneya**
19′ Leaves 20–30 cm; leaflets 25–65 mm; fruits laterally compressed ____ **Robinia pseudoacacia**
18′ Stipules not spinescent (or stipules absent); fruits indehiscent.
20. Corolla white, yellowish white, or pink; fruits black at maturity; seeds brown to black ____ **Styphnolobium**
20′ Corolla lilac to violet, violet-blue; fruits grey to light brown, brown; seeds red, orange, or buff ____ **Dermatophyllum**
17′ Herbs or low shrubs.
21. Fruits inflated.
22. Leaflet margins partly or almost entirely with teeth; stipules with teeth ____ **Cicer**
22′ Leaflets entire; stipules entire.
23. Plants spreading by roots; corollas red; style sparsely hairy below stigma; adventive, restricted to a small area of Navajo County ____ **Sphaerophysa**
23′ Plants not both spreading by roots and with red corollas; style glabrous; native, statewide ____ **Astragalus** (in part)
21′ Fruits not inflated.
24. Leaflets 5.
25. Leaflets > to >> 2 cm; corolla pink or purple ____ **Cologania angustifolia** (in part)
25′ Leaflets < 2 cm; corolla yellow ____ **Lotus**
24′ Leaflets more than 5.
26. Fruits with 1 seed.
27. Low herbs; leaflets with lineoles; banner 3.7–5 mm ____ **Marina calycosa**
27′ Shrubs to 1.2 m; leaflets without lineoles; banner 6–7 mm ____ **Indigofera**
26′ Fruits with 1–many seeds.
28. Stipules spinescent ____ **Peteria**
28′ Stipules herbaceous.
29. Fruits not constricted between the seeds.
30. Inflorescences scapose racemes; keel conspicuously porrect; pod oblong-ellipsoid or bladdery, and deeply sulcate on ventral surface but not or scarcely so on dorsal surface ____ **Oxytropis**
30′ Not both scapose racemes and with a porrect keel; pod highly heterogeneous, but not more deeply sulcate on ventral surface than on the dorsal one ____ **Astragalus** (in part)
29′ Fruits constricted between the seeds (not always apparent in young fruit).
31. Fruits constricted in all directions ____ **Sophora**
31′ Fruits constricted laterally only.
32. Fruits with transverse creases between the seeds; seeds quadrangular to rounded, not noticeably longer than broad ____ **Coursetia** (in part)
32′ Fruits without transverse creases between the seeds; seeds rectangular to oblong or oval, usually noticeably longer than broad ____ **Tephrosia**

Group E. Leaves odd-pinnate, 3-foliolate; vines or stems trailing

1. Flowers resupinate ____ **Clitoria** (in part)
1′ Flowers not resupinate.
2. Fruits loments ____ **Desmodium** (in part)
2′ Fruits not loments.
3. Keel slightly to strongly twisted or coiled.
4. Herbage with hooked hairs ____ **Phaseolus**
4′ Herbage without hooked hairs ____ **Vigna** (in part)
3′ Keel not twisted or coiled.
5. Flowers pink, bluish-purple, or red purple.
6. Flowers 1–3(–4) in leaf axils; keel shorter than the wings ____ **Cologania** (in part)
6′ Flowers (2–)5–18(–30), commonly fasciculate on inflorescence axis; keel subequal to wings or longer ____ **Galactia**
5′ Flowers yellow, orange, salmon or maroon or a combination thereof.
7. Wings longer than the other petals; style hooked and S-shaped; leaflets not resin-dotted ____ **Macroptilium**
7′ Wings not conspicuously longer than the other petals; style not hooked or S-shaped; leaflets resin-dotted ____ **Rhynchosia**

Group F. Leaves odd-pinnate, 3-foliolate; not vines

1. Flowers resupinate ____ **Clitoria** (in part)
1′ Flowers not resupinate.
2. Herbage glandular ____ **Dalea jamesii**
2′ Herbage not glandular.

3. Fruits loments.
 4. Beak in mature fruit elongated, hooked or spiraled; rare, localized in Huachuca Mts and Canelo Hills ____ **Stylosanthes**
 4′ Beak in mature fruit short, not hooked or spiraled; widespread ____ **Desmodium** (in part)
3′ Fruits not loments.
 5. Shrubs or small trees; corolla red ____ **Erythrina**
 5′ Herbaceous; corolla not red.
  6. Fruits inflated ____ **Crotalaria pumila**
  6′ Fruits not inflated.
   7. Leaflets 4–17.2 cm; fruits 5–30 cm.
    8. Inflorescence 8–25 cm; fruits 10–30 cm; seeds 18–25 ____ **Vigna** (in part)
    8′ Inflorescence 1.8–4.9 cm; fruits 5 cm; seeds 2–10 ____ **Cologania angustifolia** (in part)
   7′ Leaflets 0.5–5 cm; fruits 0.15–1.5 cm.
    9. Fruits coiled or sickle-shaped, mostly more than 2-seeded, prickly (reniform, 1 or 2 seeded and not prickly in *Medicago lupulina*) ____ **Medicago**
    9′ Fruits straight or nearly so; 1–4 seeded; never prickly.
     10. Inflorescences elongate racemes ____ **Melilotus**
     10′ Inflorescences racemes, head-like or compact spike-like ____ **Trifolium** (in part)

TREATMENTS

ACACIA Miller. Gard. Dict. Abr., ed. 4. 1754. * [from the Greek, *ake*, a point, referring to the spiny stipules of some early described species]

John E. Ebinger and David S. Seigler

Shrubs or trees, rarely vines. **Stems** lightly to highly branched; twigs rounded to angular or ridged, glabrous to pubescent; short shoots usually absent; stipular spines and prickles rarely present. **Leaves** alternate; [bipinnate or] more commonly modified to polymorphic phyllodes (enlarged, flattened petiole lacking a blade or leaflets); foliar glands normally present along the phyllode margins and/or apex; stipules normally present, commonly early deciduous, rarely spinose. **Inflorescences** loosely- to densely-flowered globose heads to cylindrical spikes, solitary to clustered in the leaf axis or in pseudo-racemes or pseudo-panicles; peduncles usually short, glabrous to pubescent, mostly not elongating in fruit; floral bracts present. **Pedicels** usually absent. **Flowers** mimosoid, radially symmetric, [4- or] 5-merous; calyx cup-shaped, lobes triangular, glabrous to pubescent; corolla cup-shaped, membranous, lobes triangular, glabrous to pubescent; stamens 20–150, 2–10 mm, usually exserted, mostly yellow to gold or creamy white; anthers attached dorsally, mostly lacking anther glands; ovary 1, sessile to short-stalked; style thread-like; stigma thread-like. **Fruits** erect to pendulous, usually dehiscing along both sutures, linear to oblong, straight to curved, mostly flat, dry, papery to leathery, glabrous to pubescent; stipe mostly short; apex sometimes beaked. **Seeds** mostly arranged in a single row, usually with a pulpy, arillate funiculus, pleurogram usually present. $x = 13$.

Species 987 (11 cultivated in the flora, 3 of these rarely adventive): 982 in Australia and Pacific, 2 in Indian Ocean area and 10 in Asia (7 of these also in Australia).

Acacia is the second largest genus in the Fabaceae and has been widely exploited for many purposes. The seeds of several Australian species were an important food resource for the Aboriginal peoples and some of these have potential for development as tree food crops for dry regions. A few species also produce an edible gum. The foliage of some species is suitable as livestock fodder. Wood has been used for construction, carpentry, turnery, and weaponry. It was used as fuel and to make charcoal. The bark of some species has been used for tanning leather. Some *Acacia* species have medicinal uses. Still others are planted for erosion control, as windbreaks, and as landscape and ornamental plants.

SELECTED REFERENCES: Nielsen 1992, Orchard and Wilson 2001a, 2001b.

1. Flowers in elongated, cylindrical spikes.
 2. Leaves (phyllodes) deltoid to triangular, one margin nearly straight the other angled to curved ____ **4. Acacia cultriformis** (in part)
 2′ Leaves (phyllodes) linear to obovate to oblong, margins similar.
  3. Leaves (phyllodes) round in cross section to linear or narrowly elliptic, 25–110 mm ____ **1. Acacia aneura**
  3′ Leaves (phyllodes) flat in cross section, usually obovate to oblong, 15–25 mm ____ **3. Acacia craspedocarpa**
1′ Flowers in globose heads.
 4. Leaves (phyllodes) deltoid to triangular, one margin nearly straight, the other rounded or angled ____ **4. Acacia cultriformis** (in part)
 4′ Leaves (phyllodes) various shapes but not deltoid or triangular, margins similar.
  5. Globose heads small, commonly 3–5 mm across.
   6. Leaves (phyllodes) pinnately veined, midvein obvious ____ **5. Acacia jennerae**
   6′ Leaves (phyllodes) with numerous parallel veins.
    7. Stems pendulous; leaves (phyllodes) with numerous close parallel veins ____ **6. Acacia pendula**
    7′ Stems erect to spreading; leaves (phyllodes) with 5–12 obvious, parallel veins ____ **7. Acacia redolens**
  5′ Globose heads larger, 6–13 mm across.
   8. Leaves (phyllodes) pinnately veined, midvein obvious.

9. Twigs ridged; fruits constricted between the seeds, stipe 8–15 mm ______ **9. Acacia saligna**
9′ Twigs not ridged; fruits not constricted between the seeds, stipe 1–3 mm ______ **8. Acacia salicina**
8′ Leaves with parallel veins.
10. Leaves less than 60 mm ______ **11. Acacia tetragonophylla**
10′ Leaves more than 100 mm.
11. Leaves flat in cross section, linear, veins 8–22, obvious ______ **10. Acacia stenophylla**
11′ Leaves round in cross section, narrowly linear with closely spaced veins ______ **2. Acacia coriacea**

1. Acacia aneura F. Mueller ex Bentham. Linnaea 26:627. 1855. * [from Greek meaning without nerves, in reference to the obscure veins of the phyllodes] MULGA Cultivated

Shrubs or trees, erect, to 18 m. **Stems** slender, spreading; bark dark gray and smooth, becoming fissured with age; twigs round, not zig-zagged, glabrous, dark purple, resinous; prickles absent. **Leaves** (phyllodes) alternate; round in cross section to flat, linear to narrowly elliptic, margins similar, straight to slightly curved, puberulent, 25–110 × 0.7–8 mm; striate with obvious parallel veins; apex narrowly acute to acuminate, apiculate, usually not hooked; gland solitary, small, just above the 1–2 mm pulvinus. **Inflorescences** densely flowered cylindrical spikes, 7–30 mm, solitary in upper leaf axis; peduncles 3–10 mm. **Pedicels** absent. **Flowers** yellow; calyx 0.5–1.1 mm, usually glabrous; corolla 0.9–1.8 mm, glabrous; filaments 3–4 mm. **Fruits** 25–100 × 6–16 mm, straight, flattened, usually not constricted between the seeds, stiffly papery, glabrous, usually resinous, light brown; stipe 2–7 mm; apex acute. **Seeds** oblong to oval, flattened, 3–6 × 2–4.5 mm; aril creamy white, small, forming a cap on the seed.

Flowering Apr–Jun. Native across c Australia.

Acacia aneura is widespread in dry areas of all Australian mainland states except Victoria and is found on a variety of soil types in various habitats from plains to mountains. It is common and often a dominant element of some plant communities including mulga woodland. *Acacia aneura* occupies over 150 million hectares in Australia (Miller 1994). Annual rainfall across its range is 150–550 mm (Thomson 1992). *Acacia aneura* is reported to nodulate (Allen and Allen 1981). Ten varieties are described in Orchard and Wilson (2001b) and additional varieties will likely be described in the future. These are distinguished mainly on differences in the phyllodes and fruits. At least two of these appear to be present in Arizona horticulture. However, because the species is wide-ranging and highly variable, the varieties are not easily separated. For this reason they are not formally treated here.

Acacia aneura is a long-lived, drought-tolerant tree. In Australia, due to its abundance and extensive distribution, it has been heavily utilized. It is a desirable fodder tree, especially during drought, though quality varies and plants have been overexploited in some areas (NRC 1979, Cunningham et al. 1992). Miller (1994) discusses the role of *Acacia aneura* as a maintenance forage during drought including the presence of tannins in the leaves that can exacerbate mineral deficiencies in livestock. The plants are useful for erosion control (Cunningham et al. 1992). Seeds were gathered and ground for food by Aboriginal peoples (Urban 1990) and have potential as a food crop in dry regions (Thomson 1992). An apple-sized insect gall called the mulga apple is also eaten (Urban 1990). Devitt (1992) indicates that the seeds, insect galls, and root grubs were eaten, and that burnt leaves and twigs were employed as chewing tobacco. The wood is hard, heavy, and durable, and is used for fuel, fence posts, and by Aboriginal peoples to make a variety of tools, weapons, and musical instruments (Simmons 1981, Cunningham et al. 1992). The sapwood is pale yellow. The heartwood is dark brown to black and is widely used to make souvenirs (NRC 1979).

Mulga is increasingly being planted in Arizona landscapes. Its size and growth habit make it ideal for planting in street medians and patios. It can also be planted as a screen and as an accent plant (Jones and Sacamano 2000). Plants with silvery and gray-green foliage are available. Canopy shape varies from conical to rounded. Established plants are highly drought tolerant but growth rate is improved with irrigation. Cultivated plants in Tucson have been observed to flower from June–February. Most plants in Arizona horticulture appear to be hardy to at least -9° C. The trees are evergreen, produce minimal litter, and require little maintenance. Mulga is readily propagated by scarified seeds.

2. Acacia coriacea de Candolle. Prodr. 2:451. 1825. * [from Latin meaning leathery] WIREWOOD, DESERT OAK, DOGWOOD Cultivated

Shrubs or trees, erect to spreading, to 10 m. **Stems** slender, erect to spreading [or sometimes pendulous]; bark gray, corky and becoming furrowed with age; twigs round to slightly ridged, not zig-zagged, glabrous to lightly sericeous, dark purple, not resinous; prickles absent. **Leaves** (phyllodes) alternate; round in cross section to flat, linear, margins similar, straight, puberulent, 110–330 × 1.2–8 mm; striate with obvious close parallel veins; apex acute, apiculate, not hooked; gland solitary, 2–6 mm above the 1–2 mm pulvinus. **Inflorescences** densely flowered globose heads 6–9 mm across, solitary or short pseudo-racemes in upper leaf axils; peduncles 5–15 mm. **Pedicels** absent. **Flowers** yellow; calyx

0.9–1.6 mm, lobes puberulent; corolla 2.3–3.1 mm, lobes puberulent; filaments 3–4 mm. **Fruits** 80–340 × 7–12 mm, straight to occasionally twisted and coiled, somewhat flattened, constricted between the seeds, leathery, appressed puberulent, not resinous, reddish brown; stipe 10–25 mm; apex acuminate. **Seeds** broadly elliptic to oblong, flattened, 4–8 × 3–5.5 mm; aril bright orange, large, forming a cap on the seed.

Flowering Apr–Jun. Native across c Australia.

Acacia coriacea is found in the Australian states of New South Wales, Northern Territory, Queensland, South Australia, and Western Australia. It is widely distributed on alluvial plains and on sand dunes in arid regions and is also found in semiarid woodlands and along drainageways. Thomson (1992) indicates that the species grows in areas that receive 100–800 mm of yearly rainfall. Orchard and Wilson (2001b) describe three varieties of *Acacia coriacea*. The variety present in Arizona horticulture is var. *sericophylla* (F. Mueller) R. S. Cowan & Maslin and is the widespread inland variety. The other varieties, var. *coriacea* and var. *pendens* R. S. Cowan & Maslin, are of restricted distribution in Western Australia.

In Australia, *Acacia coriacea* is browsed by sheep and to some extent by cattle though it is not considered to be nutritious (Cunningham et al. 1992). Aboriginal people ate the green seeds as a vegetable and ground ripe seeds to make flour (Urban 1990). The plants produce large quantities of easily harvested seeds (Thomson 1992). Thomson considers var. *pendens* to have potential as a food crop for dry tropical regions. Both green and ripe seeds have been used for food, the arils are used to make a sweet beverage and also a medicinal wash, and ashes from the burnt leaves are used with chewing tobacco (Devitt 1992).

Wirewood is suitable as an informal screen and for planting in street medians. The plants are long-lived and drought tolerant once established. This species will grow in a variety of soil types from sandy to clay. Moderate growth rates can be achieved with irrigation. Cultivated plants in Arizona are damaged below ca. -6° C. Wirewood produces some leaf litter but otherwise requires little maintenance other than removal of freeze-damaged stems. Propagation is by scarified seeds.

3. Acacia craspedocarpa F. Mueller. Australas. Chem. Druggist 2:73. 1887. * [from Greek *craspedo*, a border, and *carpus*, fruit] LEATHER-LEAF ACACIA Cultivated

Shrubs or trees, rounded, branched from base, to 4 m. **Stems** slender, spreading; bark gray and smooth, becoming rough and slightly fissured; twigs round, not zig-zagged, glabrous, dark purple, resinous; prickles absent. **Leaves** (phyllodes) alternate; flat, broadly obovate to oblong, margins similar, straight, glabrous, 15–25 × 8–12 mm; usually with 3 major parallel veins, minor veins not obvious; apex obtuse to broadly acute, apiculate, not hooked; gland not obvious; pulvinus 1–2 mm. **Inflorescence** densely flowered cylindrical spikes 12–20 mm, solitary in upper leaf axils; peduncles 12–20 mm. **Pedicels** absent. **Flowers** yellow; calyx 0.8–1.5 mm, glabrous; corolla 1.6–2.4 mm, glabrous; filaments 3–4 mm. **Fruits** 30–60 × 18–25 mm, straight, flattened, not constricted between the seeds, stiffly papery, glabrous, resinous, light brown, with prominent wings; stipe very short; apex obtuse. **Seeds** oval to circular, flattened, 7–9 × 5–7 mm; aril yellow, forming a small cap on the seed.

Flowering May–Jul. Native to wc Australia.

Acacia craspedocarpa is found on plains and rocky hills, and along watercourses in desert habitats in Western Australia. The thick, reticulate-veined fruits are distinctive among cultivated acacias in Arizona.

Leather-leaf acacia is planted as a hedge or screen and makes an interesting specimen plant. It is ideally suited for street medians. With its rounded form and gray-green foliage, the plants make ideal background shrubs and contrast well with a variety of other desert landscape plants. This species is long-lived and drought tolerant. Plants seem to tolerate a range of soils including clay. Growth rates are generally slow. The flowers are dull yellow and not particularly attractive. Flowering on cultivated plants in Tucson has been recorded from April–January. Cultivated plants have survived -9° C without damage. Litter and maintenance are minimal. Scarified seeds germinate readily.

4. Acacia cultriformis A. Cunningham ex G. Don. Gen. Hist. 2:406. 1832. * [from Latin meaning knife-shaped] KNIFE-LEAF WATTLE Cultivated

Shrubs, erect, to 4 m. **Stems** slender, spreading; bark yellow to orange, becoming gray on older stems, smooth; twigs ridged, not zig-zagged, glabrous, bluish to purplish, not resinous; prickles absent. **Leaves** (phyllodes) alternate; flat, inequilateral and often deltoid to triangular, one margin nearly straight, the other angled to curved, glabrous, 10–30 × 5–15 mm; midvein obvious, minor veins not obvious; apex acute, apiculate; gland prominent, near widest part of the leaf; pulvinus 0.3–1.0 mm. **Inflorescences** densely flowered globose to slightly elongated heads 3–7 mm, in

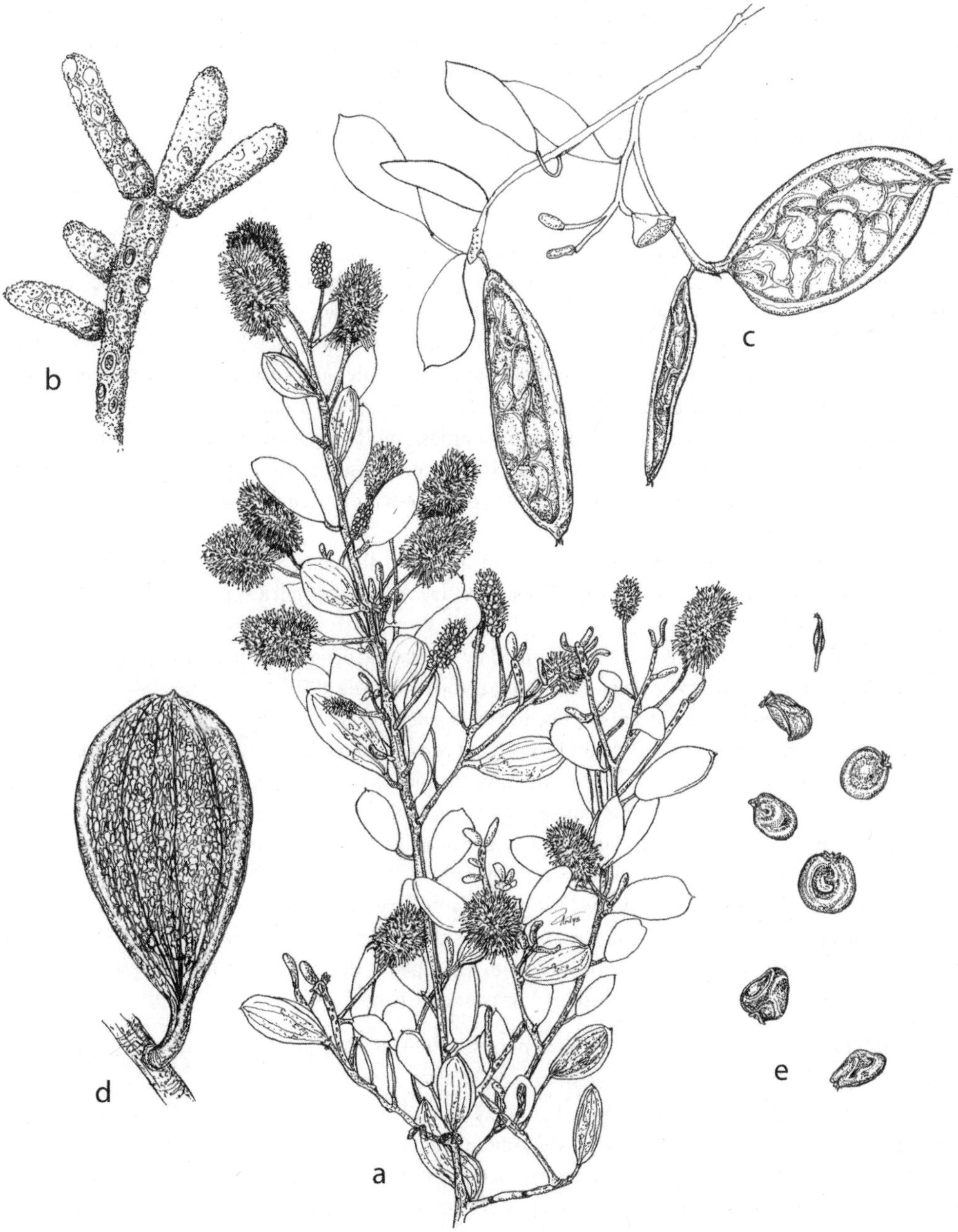

Acacia craspedocarpa: **a.** habit; **b.** developing inflorescence; **c.** fruits; **d.** phyllode detail; **e.** seeds. Illustrator: Ardys Lurtsema.

pseudo-racemes of 5–20 heads, 25–100 mm, usually solitary in upper leaf axils, glabrous; peduncles 20–50 mm. **Pedicels** absent. **Flowers** golden; calyx 0.5–0.9 mm, glabrous; corolla 1.3–1.8 mm, glabrous; filaments 2.7–3.5 mm. **Fruits** 40–90 × 5–8 mm, straight, flattened, slightly constricted between some seeds, stiffly papery, glabrous, not resinous, purplish; stipe 6–10 mm; apex long acuminate. **Seeds** oblong, flattened, 3.5–4.5 mm; aril white, club-shaped.

Flowering throughout year. Native to e Australia.

Acacia cultriformis grows as an understory plant in semiarid woodland communities in New South Wales and Queensland, Australia. It grows in sand to clay loam soils (Orchard and Wilson 2001a). This species is reported to nodulate (Allen and Allen 1981).

Knife-leaf wattle can be grown as an accent plant. With pruning, it will make a sculptural specimen (Jones and Sacamano 2000). The unusual phyllodes provide interest. The plants are typically open and somewhat rangy. Established plants should receive occasional irrigation to maintain optimum growth and appearance. Well-drained soil is preferred. Flowering is in the winter and spring, and occasionally in other seasons. Cultivated plants are hardy to -6° C. Knife-leaf wattle requires little maintenance and produces negligible litter. Propagation is by scarified seeds.

5. Acacia jennerae Maiden in A.J. Ewart and O.B. Davies, Fl. N. Terr. 333, pl. 26. 1917. * [named for A. M. Jenner, former librarian at the Royal Botanic Gardens, Sydney] COONAVITTRA WATTLE Cultivated

Shrubs or small trees, erect, to 5 m. **Stems** slender, spreading; bark orange to dull reddish, smooth, turning brown and becoming rough and fissured at base of trunk; twigs round, zig-zagged, glabrous, bluish to dark purple, resinous; prickles absent. **Leaves** (phyllodes) alternate; flat, narrowly elliptic to narrowly oblanceolate, margins similar, slightly curved, glabrous, 50–110 × 5–15(–20) mm; pinnately veined, midvein obvious, lateral veins faint; apex acute, apiculate, sometimes hooked; gland 1–3, the lowest 4–10 mm above the 1–2 mm pulvinus. **Inflorescence** densely flowered globose heads 3–5 mm across, in pseudo-racemes of 3–15 heads, 13–40 mm, solitary from the upper leaf axils; peduncles 2–5 mm. **Pedicels** absent. **Flowers** yellow; calyx 0.7–1.1 mm, glabrous; corolla 1.2–1.9 mm, glabrous; filaments 2.5–3.5 mm. **Fruits** 80–150 × 5–8 mm, straight, flattened, constricted between the seeds, leathery, glabrous, not resinous, black; stipe 3–7 mm; apex acute. **Seeds** oblong, slightly flattened, 6–8 × 3–6 mm; aril light yellow, club-shaped.

Flowering Sep–Nov. Native to w Australia.

Acacia jennerae is of sporadic distribution in arid and semiarid habitats in New South Wales, Northern Territory, Queensland, and Western Australia (Orchard and Wilson 2001a). Annual rainfall across its range varies from 200–350 mm (Thomson 1992).

The seeds of *Acacia jennerae* have been used for food by Aboriginal people in Australia and gum was also utilized (Devitt 1992). The seeds may have some potential as a food crop (Thomson 1992).

Coonavitra wattle resembles a small *Eucalyptus*, and when not in flower or fruit is not readily identified as a legume. It is being increasingly planted in Arizona. With its upright growth habit, it is well-suited for small spaces and patios. The contrasting bark and foliage are particularly attractive. In some situations, plants will profusely sucker from the roots, forming thickets if not controlled. This can be useful for screening. The suckering habit may be useful in erosion control (Jones and Sacamano 2000). The plants are tolerant of a variety of soils. This species is drought resistant, but appearance and performance are improved with occasional irrigation during drought. Flowering has been observed on cultivated plants in Tucson from July–February. Cultivated plants have survived temperatures to ca. -9° C without damage. The biggest maintenance issue with Coonavitra wattle is the tendency to form masses of root suckers. Cultivated plants in Arizona have also tended to be short-lived, with some plants declining after as little as a decade. The plants produce a modest amount of leaf letter. Scarified seeds germinate easily.

6. Acacia pendula A. Cunningham ex. G. Don. Gen. Hist. 2:404. 1823. * [from Latin *pendul*, hanging] WEEPING MYALL, BOREE Cultivated

Trees, erect, to 12 m. **Stems** slender, pendulous; bark gray, smooth, becoming gray brown, rough and fissured; twigs ridged, not zig-zagged, glabrous to lightly appressed puberulent, brown, resinous; prickles absent. **Leaves** (phyllodes) alternate; flat, narrowly elliptic, margins similar, straight to curved, glabrous, 40–140 × 3–10 mm; striate with numerous closely parallel veins with 1–3 more obvious; apex acuminate, apiculate, sometimes hooked; gland solitary, just above the 1–3 mm pulvinus. **Inflorescences** densely flowered globose heads 3–5 mm across, in pseudo-racemes of 2–7 heads, 2–9 mm, solitary in upper leaf axils; peduncles 2–8 mm. **Pedicels** absent. **Flowers** light yellow; calyx 0.4–1.1 mm, glabrous; corolla 1.2–1.9 mm, glabrous; filaments 2.5–3.5 mm. **Fruits** 80–130 × 9–12 mm, straight to curved, flattened, slightly constricted between the seeds, leathery, appressed puberulent, not resinous, dark brown; stipe 2–5 mm; apex acute. **Seeds** circular, somewhat flattened, 5–9 × 4–6 mm; aril cream, forming a small cap on the seed.

Flowering throughout year. Native to se Australia.

Acacia pendula is found on clay soils and often on floodplains in semiarid woodlands in New South Wales, Queensland, South Australia, and Victoria. Annual rainfall across its range varies from 400–650 mm (NRC 1979).

In Australia *Acacia pendula* is used for planting to form windbreaks, and is used for fuelwood and fence posts (Simmons 1981). Livestock browse the foliage, especially during drought, and it has been eliminated from portions of its range (NRC 1979, Cunningham et al. 1992). The wood is hard, heavy, and aromatic (Simmons 1981, Cunningham et al. 1992) and has been used for small wooden articles and by Aboriginal people to make boomerangs (Cunningham et al. 1992).

Weeping myall makes a beautiful shade or specimen tree for larger spaces. The rounded canopy with weeping branches and silvery gray or yellow-green foliage is particularly attractive. The plants tolerate a variety of soils. Occasional irrigation is desirable during hot dry weather. Flowering in Tucson has been observed in December and January. Jones and Sacamano (2000) report that flowering on plants in dry landscapes is rare. Cultivated plants are hardy to ca. -6° C. Leaf litter is modest. Jones and Sacamano (2000) report that weeping myall is subject to Texas root rot and damage from high winds. Propagation is by scarified seeds.

7. Acacia redolens Maslin. Nuytsia 1:327. 1974. * [from Latin *redol*, emit a scent] VANILLA-SCENTED WATTLE, ONGERUP WATTLE Cultivated and rarely adventive

Shrubs to 2 m [or small trees, erect, to 5 m]. **Stems** slender, erect to spreading; bark gray, smooth, becoming rough and shallowly fissured on old stems; twigs round, not zig-zagged, glabrous, brown to purple brown, ribs appearing resinous, fragrant when crushed (vanilla scented); prickles absent. **Leaves** (phyllodes) alternate; flat, oblanceolate, margins similar, straight, glabrous, 20–70 × 5–15 mm; striate with mostly 5–12 prominent, parallel veins; apex obtuse, apiculate, not hooked; gland solitary, at edge of 1–3 mm pulvinus. **Inflorescences** densely flowered globose heads 3–5 mm across, solitary or in short pseudo-racemes of 2–6 heads, 5–10 mm, clustered in upper leaf axils; peduncles 2–6 mm. **Pedicels** absent. **Flowers** light yellow; calyx 0.5–0.8 mm, glabrous; corolla 1.0–1.6 mm, glabrous; filaments 2–3 mm. **Fruits** 30–60 × 2–4 mm, curved, flattened, constricted between the seeds, leathery, glabrous, resinous, brown; stipe 4–8 mm; apex acuminate. **Seeds** ellipsoid to ovoid, flattened, 4.9–6.5 × 3.5–4.5 mm; aril cream-white, thickened, forming a cap on seed.

Flowering Aug–Sep. Native to coastal sw Australia. Cultivated and rarely established in the sw United States.

Acacia redolens is often found in somewhat saline or alkaline loam or clay soils in woodlands and around salt lakes (Orchard and Wilson 2001b). The shrubby form is present in Arizona horticulture.

Vanilla-scented wattle forms low dome-shaped mounds and is widely planted on freeway embankments and street medians in Arizona as a tall ground cover. It can also be used for erosion control. The plants will grow in a variety of soils including saline soils. The plants tolerate drought well but appearance is improved with occasional irrigation. Flowering in Tucson has been observed from February–June. Cultivated plants are hardy to ca. -7° C. Litter is minimal. Cultivated plants sometimes suffer extensive die out of stems, particularly during the summer. This may be reduced by planting in well-drained soil or reducing irrigation frequency. Selections with specific traits have been trademarked by commercial nurseries. Scarified seeds are easily germinated.

8. Acacia salicina Lindley in T.L. Mitchell, Three Exped. Australia 2:20. 1838. * [from Latin *salix*, willow, meaning willow-like] WILLOW-WATTLE, COOBA Cultivated and rarely adventive

Shrubs or small trees, bushy, erect, to 13 m. **Stems** slender, pendulous; bark gray and smooth, on older stems becoming gray brown, rough and fissured; twigs not ridged, slightly zig-zagged, glabrous, gray to purplish, not resinous; prickles absent. **Leaves** (phyllodes) alternate; flat, linear to narrowly elliptic, margins similar, straight to slightly curved, glabrous, 60–200 × 4–30 mm; pinnately veined, midvein obvious, lateral veins faint; apex acuminate, apiculate, sometimes hooked; glands 2–5, lowermost 0–5 mm above the 0.5–1.3 mm pulvinus. **Inflorescences** loosely flowered globose heads 8–13 mm across, in pseudo-racemes of 2–8 heads, 10–80 mm, solitary in upper leaf axils; peduncles 5–12 mm. **Pedicels** absent. **Flowers** cream to pale yellow; calyx 1.1–1.8 mm, glabrous; corolla 2.4–3.1 mm, glabrous; filaments 5–6 mm. **Fruits** 50–120 × 7–13 mm, straight, flattened, not constricted between the seeds, woody, glabrous, not resinous, gray-green; stipe 1–3 mm; apex acute. **Seeds** oblong to elliptic, flattened, 4–6.5 × 3–5 mm; aril scarlet, forming a cap on the seed.

Flowering Oct–Feb. Native to eastern half of Australia. Cultivated and rarely adventive in the sw United States.

Acacia salicina is found in parts of all Australian mainland states except for Western Australia. It prefers clay soils of low-lying areas in arid and semi-arid habitats receiving 125–500 mm annual of rainfall, where it is a component of open woodlands and riparian communities (Whibley 1980). *Acacia salicina* has become adventive and is increasing in abundance on disturbed sites around Tucson and Phoenix. It should be considered invasive in areas beyond its native distribution. The arils of many acacias, especially those that are brightly colored, serve to attract birds that disperse the seeds. *Acacia salicina* has been observed to nodulate (Allen and Allen 1981).

In Australia *Acacia salicina* can be planted to form windbreaks (Cunningham et al. 1992). The root-suckering habit can be useful in reducing soil erosion and stabilizing river banks (Whibley 1980, Cunningham et al. 1992). This species is reported to tolerate saline conditions (Simmons 1981). The foliage is reported to be browsed by livestock, but is suspected of causing poisoning under some conditions (Cunningham et al. 1992). Sheep readily eat the green pods

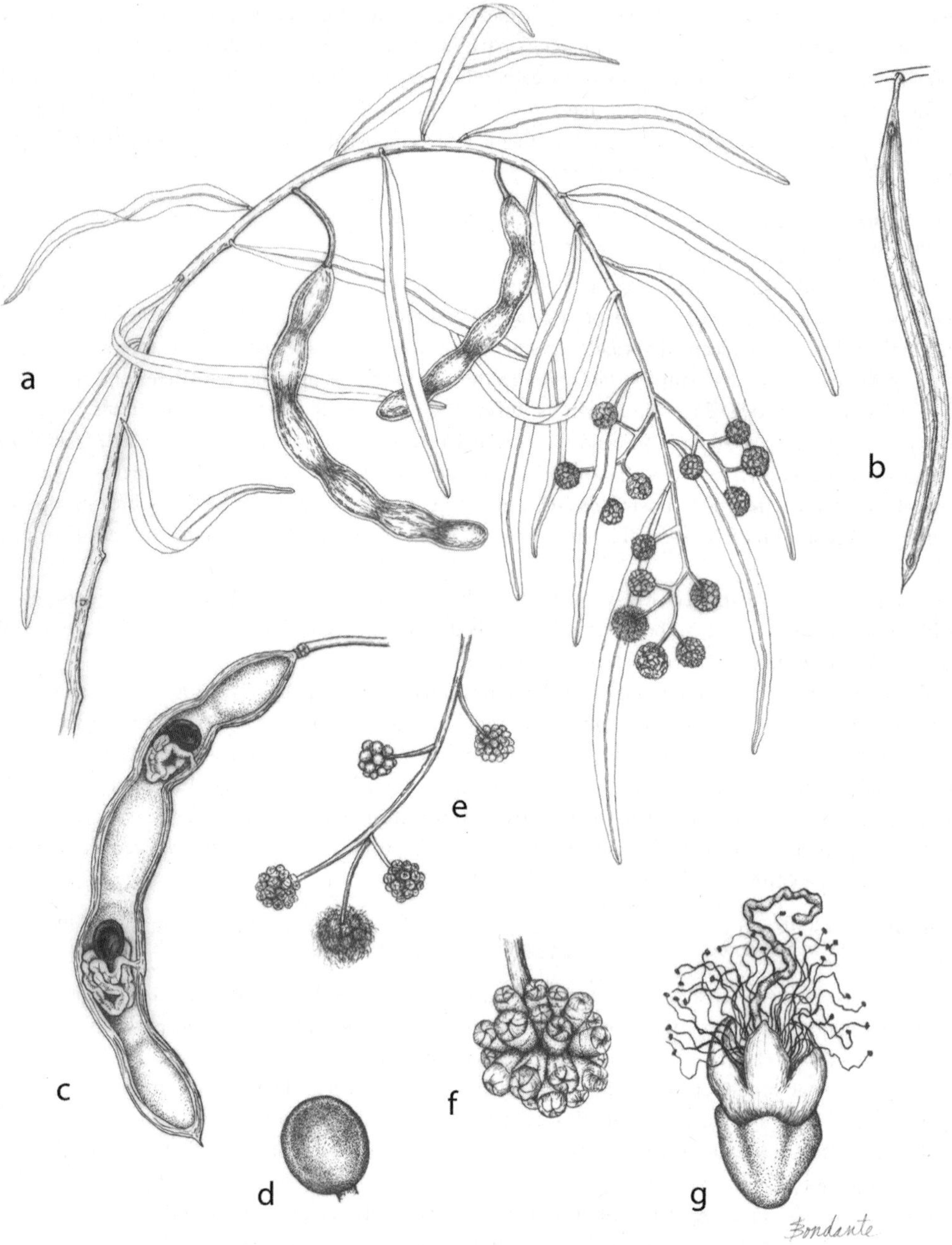

***Acacia salicina*: a. habit; b. phyllode; c. fruit valve with seed detail; d. seed; e. inflorescences; f. immature inflorescence; g. flower with developing fruit. Illustrator: Chris Bondante.**

(Cunningham et al. 1992). Wood of willow wattle is dark brown and strong (Simmons 1981, Cunningham et al. 1992) and has been used to make furniture, yokes, and other articles (Cunningham et al. 1992). The bark was used by Aboriginal people to poison fish (Simmons 1981, Cunningham et al. 1992).

Willow wattle has graceful willow-like foliage and weeping branches. It is suitable as a shade tree and for planting in patios, street medians and against buildings. Root suckers frequently develop around the parent tree. Willow wattle tolerates a wide variety of soil types. Established plants are drought tolerant. Flowering in Tucson has been recorded from September–March. The plants are hardy to -7° C. Willow wattle is subject to wind throw if improperly irrigated

(Jones and Sacamano 2000). The plants produce a moderate amount of leaf and fruit litter. Root suckers can become a maintenance issue. Scarified seeds germinate readily.

9. Acacia saligna (Labillardiere) H.L. Wendland. Comm. Acac. Aphyll. 26. 1820. * [from Latin *salignus*, of willow] GOLDEN WREATH WATTLE Cultivated

Acacia cyanophylla Lindley

Shrubs or small trees, bushy, erect, to 6 m. **Stems** slender, usually pendulous; bark gray or red brown, smooth or becoming rough and shallowly fissured on old limbs; twigs slightly ridged, slightly zig-zagged, glabrous, bluish to purplish, not resinous; prickles absent. **Leaves** (phyllodes) alternate; flat, linear to narrowly elliptic, margins similar, straight to slightly curved, glabrous, 70–250 × 6–30 mm; pinnately veined, midvein obvious, lateral veins faint; apex narrowly acuminate, apiculate, sometimes hooked; gland solitary, 0–3 mm above the 1–4 mm pulvinus. **Inflorescences** densely flowered globose heads 8–13 mm across, in pseudo-racemes of 2–10 heads, 5–40 mm, solitary in upper leaf axils; peduncles 5–15 mm. **Pedicels** absent. **Flowers** golden; calyx 1–2 mm, glabrous; corolla 2.6–3.4 mm, glabrous; filaments 5–6 mm. **Fruits** 80–140 × 5–8 mm, straight, flattened, constricted between the seeds, stiffly papery, glabrous, not resinous, dark brown; stipe 8–15 mm; apex acuminate. **Seeds** oblong, flattened, 4–6 × 2.5–4 mm; aril yellow, club-shaped, 2–3 mm.

Flowering Nov–Mar. Native to sw Australia.

Acacia saligna is native to southwestern Western Australia where it is found on various soil types in a variety of habitats, in areas receiving 300–700 mm yearly rainfall (Whibley 1980). It has been introduced elsewhere in the country and has become naturalized. It has also naturalized in other parts of the world including South Africa and California. Though *A. saligna* is known to be invasive, it has not been recorded as adventive in Arizona. *Acacia saligna* is reported to nodulate (Allen and Allen 1981).

In Australia *Acacia saligna* is useful for soil stabilization and does well in coastal areas (Whibley 1980). The plants are effective in stabilizing sand and are used for windbreaks (NRC 1980). It is fast growing and in plantations can yield 1.5–10 m^3 wood/ha per year though the wood is of low quality (NRC 1980). The plants coppice readily. The foliage is suitable for livestock fodder. NRC (1980) reports that fresh and dried foliage of *A. saligna* is used as supplementary feed for goats and sheep, and that crushed seeds have been fed to sheep without problems. An acidic gum (27% uronic acid) is exuded from damaged stems and may have potential as an ingredient in acidic foods (NRC 1979, NRC 1980).

Golden wreath wattle has a fast growth rate, but in Arizona has been short-lived and subject to considerable wind damage due to the dense canopy. It is suitable as a patio tree or screen. The plants provide shade and have dark green leaves that resemble those of some species of *Eucalyptus*. Golden wreath wattle is tolerant of a variety of soils. Periodic irrigation is desirable during hot dry weather. This species is spectacular when in flower with masses of yellow flower heads. Flowering takes place in winter and spring. Cultivated plants in Tucson have been severely damaged below -6° C. Maintenance considerations include damage from high winds, leaf litter, and the potential for root suckers. Propagation is by scarified seeds.

10. Acacia stenophylla A. Cunningham ex Bentham. London J. Bot. 1:366. 1842. * [from Greek *steno*, narrow and *phyllum*, leaf] BLACK WATTLE, RIVER COOBA, SHOESTRING ACACIA Cultivated and rarely adventive

Trees, bushy, to 20 m. **Stems** slender, mostly pendulous; bark dark gray or black, rough and furrowed; twigs round, not zig-zagged, glabrous to puberulent, dark purple, slightly resinous; prickles absent. **Leaves** (phyllodes) alternate; flat, linear, margins similar, straight to curved, usually glabrous, 150–400 × 2–10 mm; striate with 8–22 obvious parallel veins; apex narrowly acuminate, apiculate, rarely hooked; gland solitary, 0–2 mm above the 1–4 mm pulvinus. **Inflorescences** densely flowered globose head 6–10 mm across, in pseudo-racemes of 2–6 heads, 3–10 mm, solitary in upper leaf axils; peduncles 6–14 mm. **Pedicels** absent. **Flowers** creamy white to pale yellow; calyx 0.9–1.6 mm, puberulent; corolla 2–3 mm, puberulent; filaments 3.5–5.0 mm. **Fruits** 100–260 × 8–12 mm, straight, flattened, constricted between the seeds, leathery, glabrous, not resinous, dark brown; stipe 8–14 mm; apex acute. **Seeds** broadly elliptic, slightly flattened, 7–9 × 4–6 mm; aril yellow, very small.

Flowering throughout year. Native to e Australia and rarely adventive in the United States including Arizona.

Acacia stenophylla is frequently found along water courses as an understory plant beneath *Eucalyptus camaldulensis* and other species, and in seasonal swamps, typically in clay soils. It is recorded from all mainland states in Australia. This species has become adventive in disturbed sites in urban areas of Arizona and should be considered potentially invasive. Annual rainfall is 100–800 mm across its range (Thomson 1992).

The foliage of *Acacia stenophylla* is palatable to livestock (Cunningham et al. 1992). The wood is hard, heavy, and polishes well (Cunningham et al. 1992). Thomson (1992) indicates that the seeds may have some potential as a source for human food.

Despite its preference for wet areas in habitat, shoestring acacia is well-adapted to dry landscapes. It can be used as a background tree or planted against larger buildings where it is effective when planted in groups. The open canopy produces light, filtered shade. Shoestring acacia is long-lived. It tolerates a wide range of soils including saline soils. Flowering on cultivated plants in Tucson has been observed in April and from July–December. Hardiness varies from -6°– -9° C among plants grown in Tucson, though most material is hardy to -9° C. The trees produce some leaf litter and may produce root suckers. Scarified seeds germinate easily.

11. Acacia tetragonophylla F. Mueller. Fragm. 4:3. 1863. * [from Greek *tetra*, four, *gonia* an angle, and *phyllum*, leaf] DEAD FINISH, PRICKLY WATTLE Cultivated

Straggly shrubs or small trees, erect to spreading, to 5 m. **Stems** slender, spreading; bark gray, smooth or becoming rough and slightly fissured at base of large trunks; twigs round, not zig-zagged, glabrous, dark purple, not resinous; prickles absent. **Leaves** (phyllodes) clustered on short spur branches, alternate on new growth; round and 5–7 ridged in cross section, linear-subulate, margins similar, straight, glabrous, 10–60 × 0.8–1.2 mm; parallel veins; apex acuminate, sharp-pointed; gland usually absent. **Inflorescences** densely flowered globose heads 6–9 mm across, 1–5 in leaf axil; peduncles 8–25 mm. **Pedicels** absent. **Flowers** yellow; calyx of separate sepals, 0.6–1.3 mm, glabrous; corolla 1.5–2.2 mm, lobes glabrous; filaments 3–4 mm. **Fruits** 50–110 × 4–5 mm, twisted and coiled, flattened, constricted between the seeds, leathery, glabrous, not resinous, dark reddish brown; stipe 2–6 mm; apex acuminate. **Seeds** elliptic to oblong, flattened, 4–6 × 2–4 mm; aril bright yellow, encircling the seed.

Flowering Apr–Jun. Native to southern half of Australia.

Acacia tetragonophylla is widespread in arid habitats on a variety of soil types in Australia where it has been recorded from all mainland states except Victoria. It grows in regions that receive 125–300 mm of yearly rainfall (Whibley 1980). The fascicled phyllodes of older stems are unique among species of *Acacia* cultivated in Arizona. Nodulation is reported for *A. tetragonophylla* (Allen and Allen 1981).

Acacia tetragonophylla can be planted as a windbreak and for soil stabilization (Whibley 1980, Cunningham et al. 1992). In Australia the foliage is considered an important livestock browse during drought in some regions (Cunningham et al. 1992, Orchard and Wilson 2001a). The seeds have been ground for food (Urban 1990). Devitt (1992) indicates that green and ripe seeds were used for food and that root grubs were eaten. Orchard and Wilson (2001a) report that the phyllodes and root bark are used by Aboriginal people in Northern Territory to treat skin lesions and warts and for bandaging fractured bones. The bark is used to treat coughs in Western Australia. The wood is red brown, heavy, and durable, and is aromatic when freshly cut (Simmons 1981), though the small diameter limits its use.

With its stiff, sharp-tipped phyllodes and shrubby habit, dead finish makes an effective barrier. It can also be planted as an informal screen and makes a distinctive specimen plant. The dark green foliage contrasts nicely with the yellow flowers. Established plants are drought tolerant but grow slowly without supplemental irrigation. The plants require well-drained soil. Cultivated plants in Tucson have been observed to flower from January–May and in November. Dead finish is hardy to ca. -6° C. Freeze-damaged plants recover slowly. This species is moderately salt-tolerant (Simmons 1981). The sharp-tipped phyllodes are a limitation when selecting a planting site. Litter is minimal. Dead finish is easily grown from scarified seeds.

Taxa with potential for cultivation

Additional taxa of *Acacia* have been introduced into Arizona horticulture but are generally not available from commercial sources. New introductions are likely to become available over time. All of the following species have evergreen phyllodes and are native to Australia. They are rarely available in Arizona, though some are likely to become more widely planted in the future.

Acacia brachybotrya Bentham. London J. Bot. 1:347. 1842

Gray mulga forms a rounded shrub to 3 m tall, is unarmed, and has gray-green oblanceolate to obovate phyllodes. The flowers are yellow, in rounded heads. Gray mulga is hardy to -7° C.

Acacia cyclops A. Cunningham ex G. Don. Gen. Hist. 2:404. 1832

Western coastal wattle is an unarmed, rounded shrub or tree up to 6 m tall. The phyllodes are narrowly oblong to narrowly obovate. Yellow flowers are produced in globular heads. Hardy to ca. -6° C.

Acacia notabilis F. Mueller. Fragm. 1:6. 1858
Notable wattle forms a rounded to spreading, unarmed shrub to 3 m tall. The gray-green phyllodes are oblong-lanceolate, and yellow flowers are produced in rounded heads. Hardy to ca. -7° C.

Acacia papyrocarpa Bentham. Fl. Austral. 2:338. 1864
Western myall typically forms an unarmed, multiple-stemmed, umbrella-shaped shrub or tree to 7 m tall. The phyllodes are linear and silvery green. Yellow flowers are produced in round heads. Cultivated plants in Arizona appear to be hardy to at least -9° C.

Acacia pruinocarpa Tindale. Contr. New South Wales Natl. Herb. 4:73. 1968
Gidgee forms a tidy, unarmed, low-branched tree 5–10 m tall, or sometimes a multiple-stemmed shrub with relatively thick twigs. The gray bark and linear to oblong, blue-green phyllodes are attractive. The rounded flower heads are bright yellow. Gidgee is hardy to ca. -6° C.

Acacia trachycarpa E. Pritzel. Bot. Jahrb. Syst. 35:308. 1904
Acacia trachycarpa is an unarmed shrub or tree to 6 m tall with linear, sharp-tipped phyllodes. Yellow flowers are produced in cylindrical spikes. Hardy to ca. -5° C.

Acacia trineura F. Mueller. Fragm. 4:5. 1863
Three-nerved wattle forms a rounded, unarmed shrub or small tree to 5 m tall. The phyllodes are narrowly obovate or oblanceolate and dark gray-green. Rounded flower heads are yellow. Hardy to -9° C.

Acacia victoriae Bentham in T.L. Mitchell, J. Exped. Trop. Australia 333. 1848
Victoria wattle, bramble wattle, is a shrub or small tree to 5 m tall, with blue-green to yellow-green linear to lanceolate or elliptical phyllodes and short, paired stipular spines at the nodes. The flowers are cream colored and are produced in masses of rounded heads. Victoria wattle is hardy to -7° C.

ACACIELLA Britton & Rose. N. Amer. Fl. 23:96. 1928. * [diminutive form of *Acacia*]
John E. Ebinger and David S. Seigler

Subshrubs, shrubs, or small trees, less frequently perennial herbs, mostly erect. **Stems** lightly branched; twigs rounded or angular, glabrous to pubescent; short shoots absent; prickles and stipular spines absent. **Leaves** alternate; bipinnate, pinnae 2–35 pairs per leaf, mostly opposite; stipules mostly small, not obvious, early deciduous to persistent; petioles round, usually channeled above, petiolar glands absent; rachis usually channeled above, lacking glands. **Leaflets** 2–75 pairs per pinna, opposite, subsessile, membranous, glabrous to pubescent. **Inflorescences** loosely- to densely-flowered globose to slightly elongated heads, solitary to clustered in the leaf axis, commonly in terminal or axillary racemose or paniculate clusters; peduncles short, glabrous to pubescent, not elongating in fruit; flower bracts present. **Flowers** mimosoid, radially symmetric, 5-merous; calyx cup-shaped, lobes triangular, glabrous to pubescent; corolla cup-shaped, membranous, lobes triangular, glabrous to pubescent, valvate in bud; nectariferous disk present; stamens 175–250, 3–9 mm, exserted, creamy-white to pink; anthers attached dorsally, lacking a gland; ovary 1, sessile to short stipitate; style thread-like; stigma thread-like. **Fruits** tan, brown, or purplish, erect to pendulous, dehiscing along both sutures, linear to oblong, straight, flattened, dry, membranous to stiffly papery, glabrous to pubescent; stipe short; apex sometimes beaked. **Seeds** arranged in a single row, not surrounded by pulp, oval to circular, strongly flattened, smooth, pleurogram present. $x = 13$.

Species 15 (2 native in the flora): sw United States, Mexico, Central America, South America.

Various species are used by wildlife for food and are sometimes browsed by livestock. Some species may be useful for erosion control.

SELECTED REFERENCE: Rico Arce and Bachman 2006.

1. Leaflets 2–4 × 0.5–1.3 mm, only the midvein prominent beneath ____ **1. Acaciella angustissima**
1′ Leaflets 4–13 × 1.5–3.1 mm, lateral veins mostly obvious beneath ____ **2. Acaciella lemmonii**

1. Acaciella angustissima (Miller) Britton & Rose. N. Amer. Fl. 23:100. 1928. * [Latin *angustissima*, narrow]

Mimosa angustissima Miller, Gard. Dict. (ed. 8), no. 19. 1768; *Acacia angustissima* (Miller) Kuntze; *Senegalia angustissima* (Miller) Pedley

Shrubs or small trees, erect, to 12 m, sometimes somewhat shrubby with a thick woody caudex. **Stems** slender; twigs strongly zigzagged, round, glabrous to puberulent. **Leaves:** pinnae 2–32 pairs per leaf, 6–105 mm; stipules linear, to 6.5 mm, glabrous to lightly ciliate; petioles channeled above, 12–35 mm, glabrous to appressed pubescent to pilose; rachis

5–200 mm. **Leaflets** 6–75 pairs per pinna, linear to oblong, oblique and obtuse at base, broadly acute to obtuse at apex, 2–4 × 0.5–1.3 mm, usually glabrous, only the midvein prominent beneath. **Inflorescences** loosely-flowered globose heads 9–18 mm across, solitary (rarely 2–3) in the leaf axil to racemose or paniculate clusters; peduncles 1–21 mm. **Flowers** greenish-white, drying pale pink; calyx 0.6–1.2 mm glabrous; corolla 1.9–4.5 mm, glabrous; filaments 3–7.6 mm. **Fruits** 25–90 × 7–17 mm, rarely constricted between the seeds, stiffly papery, glabrous; stipe 8–12 mm; apex obtuse to acuminate, short beaked. **Seeds** 2.6–4.6 mm across, mottled gray-green and dark gray to dark brown.

Varieties 3 (2 native in the flora): widespread from the south central U.S. to Arizona with disjunct populations in Florida; also much of Mexico to Central America.

1. Plants pubescent to glabrate; pinnae 9–17 pairs per leaf ________ **1a. Acaciella angustissima** var. **angustissima**
1′ Plants glabrous or nearly so; pinnae 2–8(–9) pairs per leaf ________ **1b. Acaciella angustissima** var. **texensis**

1a. Acaciella angustissima (Miller) Britton & Rose var. **angustissima** * FERN ACACIA, PRAIRIE ACACIA, WHITE-BALL ACACIA Native, Cultivated

Acacia angulosa Bertoloni; *Acacia boliviana* Rusby; *Acacia elegans* Schlechtendal; *Acacia insignis* Hoffmannsegg; *Acacia suffrutescens* Rose; *Acaciella breviracemosa* Britton & Rose; *Acaciella ciliata* Britton & Rose; *Acaciella costaricensis* Britton & Rose; *Acaciella delicata* Britton & Rose; *Acaciella ferrisiae* Britton & Rose; *Acaciella rensonii* Britton & Rose; *Acaciella smithii* Britton & Rose; *Acaciella santanderensis* Britton & Killip; *Acaciella talpana* Britton & Rose

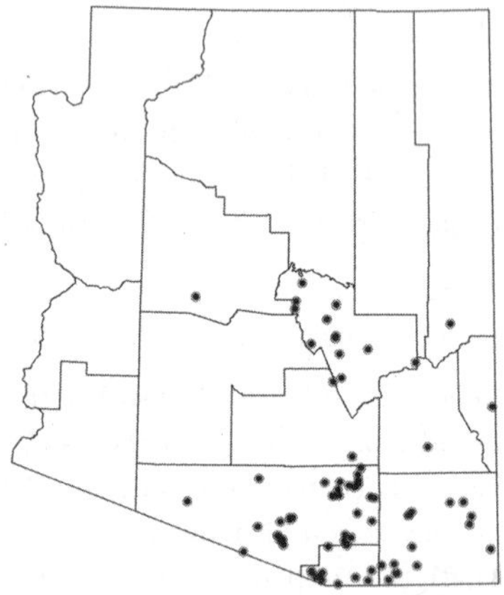

Shrubs or small trees, erect, to 2 [12] m. **Stems:** twigs glabrate to puberulent. **Leaves:** pinnae 9–17 pairs per leaf, 25–50 mm; stipules to 2.5 mm; petioles 12–35 mm, glabrate to pubescent; rachis 65–200 mm. **Leaflets** 20–40 pairs per pinna, linear, broadly acute at apex, 2.4–3.5 × 0.5–1.3 mm. **Inflorescences** 10–16 mm across, mostly in leafy racemose or paniculate clusters; peduncles 1–1.5 mm. **Flowers**: calyx 0.6–1 mm; corolla 2.5–4.5 mm; filaments 4.5–7.6 mm. **Fruits** 40–90 × 10–17 mm, not constricted between the seeds. **Seeds** dark brown.

Flowering May–Oct[all year]. Dry to relatively moist sites of slopes, arroyos, and canyons, often among rocks, disturbed sites; [sea level–]700–2100[–3000] m; (occasional in Sonoran Desertscrub), Semidesert Grassland, Interior Chaparral, open Madrean Evergreen Woodland; Cochise, Gila, Graham, Greenlee, Maricopa, Navajo, Pima, Pinal, Santa Cruz, Yavapai cos.; sw United States; Mexico; Central America; South America (n Argentina).

In Arizona, var. *angustissima* often forms low, many-stemmed plants to 1 m or more across. This is the result of the stems being killed by winter freezing. Southward in Sonora, the plants develop into small trees. In Mexico it also occurs in secondary tropical deciduous forest and dry thornscrub. The leaves are winter-deciduous. Nodulation is reported for *Acaciella angustissima* in Mexico (Rincón-Rosales et al. 2009).

Fern acacia is sometimes cultivated in Arizona gardens for its showy white flower heads that attract butterflies. Stems are damaged below ca. -5° C. The plants are readily propagated from seeds. Vines (1960) indicates that *Acaciella angustissima* can be propagated by root division.

1b. Acaciella angustissima (Miller) Britton & Rose var. **texensis** (Nuttall ex Torrey & A. Gray) L. Rico. Fl. Guerrero 25:44. 2005. * [of Texas] TEXAS WHITE-BALL ACACIA Native

Acacia texensis Nuttall ex Torrey & A. Gray, Fl. N. Amer. 1:404. 1840; *Acacia cuspidata* Schlechtendal; *Acacia hirta* Nuttall ex Torrey & A. Gray; *Acaciella texensis* (Torrey & A. Gray) Britton & Rose

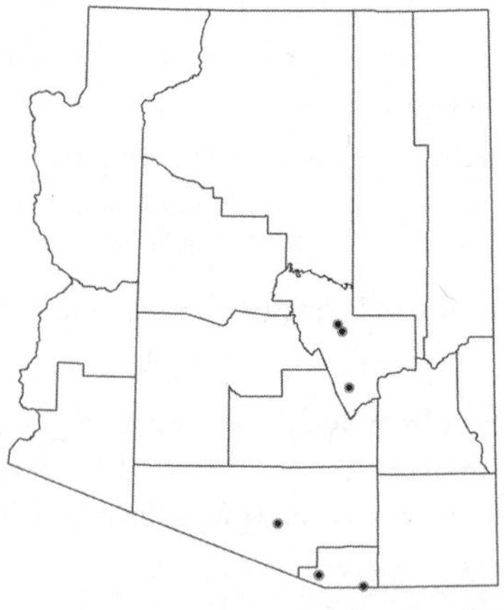

Shrubs to 1 [3] m, occasionally short shrubby with a thick woody caudex and less than 20 cm. **Stems**: twigs glabrous. **Leaves:** pinnae 2–8(–9) pairs per leaf, 6–30 mm; stipules to 4.5 mm; petioles 12–25 mm, mostly glabrous; rachis 5–75 mm. **Leaflets** 6–25 pairs per pinna, linear, acute to obtuse at apex, 2–4 × 0.5–1.3 mm. **Inflorescences** 9–14 mm across, solitary (rarely 2–3) in the leaf axil to racemose clusters; peduncles 5–18 mm. **Flowers:** calyx 0.6–1.2 mm; corolla 1.9–4.5 mm; filaments 3–6 mm. **Fruits** 25–90 × 7–17 mm, sometimes constricted between the seeds. **Seeds** dark brown.

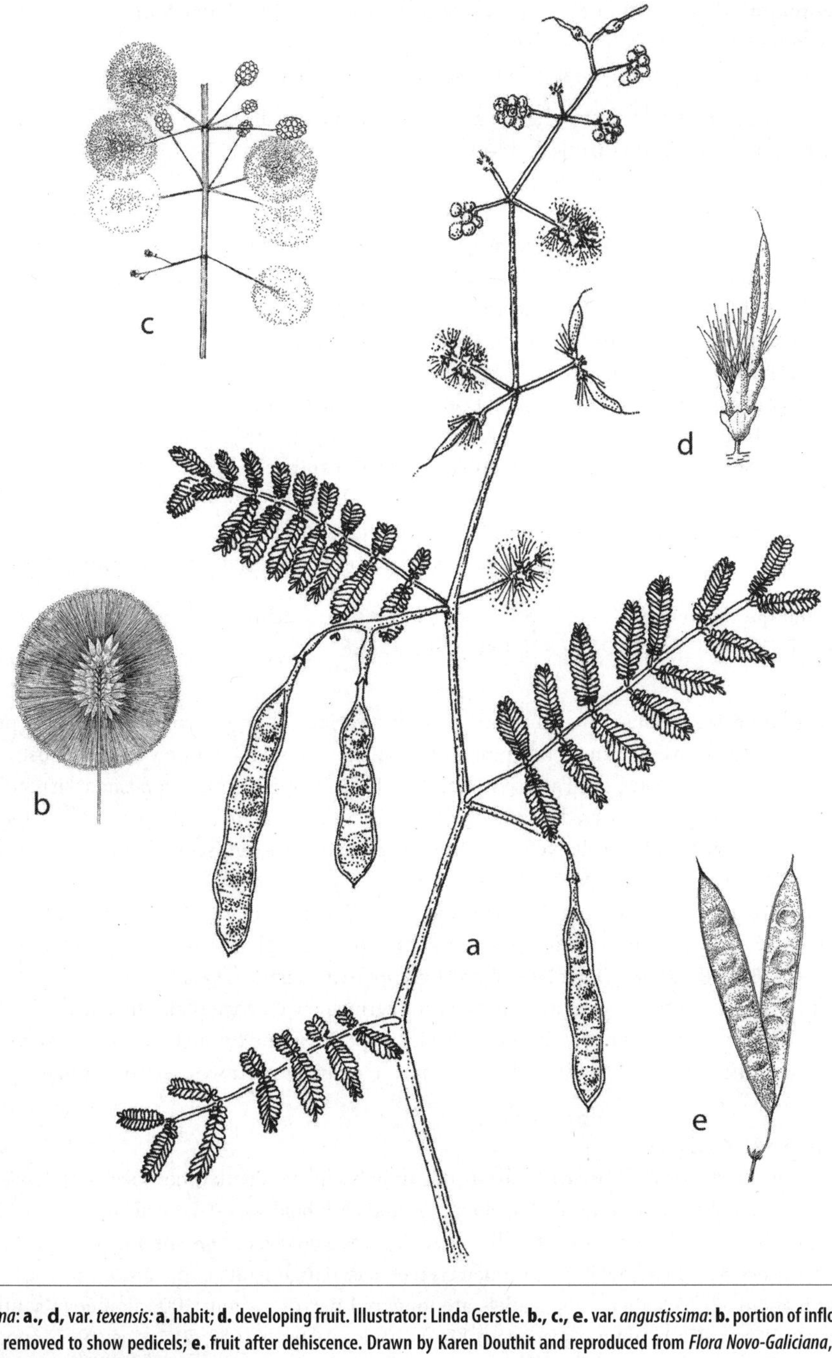

***Acaciella angustissima*: a., d, var. *texensis:* a. habit; d. developing fruit. Illustrator: Linda Gerstle. b., c., e. var. *angustissima*: b. portion of inflorescence; c. raceme with facing flowers removed to show pedicels; e. fruit after dehiscence. Drawn by Karen Douthit and reproduced from *Flora Novo-Galiciana*, vol. 5, fig. 18, with permission of the University of Michigan Herbarium.**

Flowering Jun–Oct[all year]. Dry sites on limestone, disturbed sites; [sea level–]900–1500[–3000] m; Semidesert Grassland, [thornscrub, prairie]; Gila, Pima cos.; sw United States; c Mexico.

Rico Arce and Bachman (2006) reported var. *texensis* from Gila Co. We have seen material from Gila Co.: Parker Creek, Sierra Ancha Mountains (*F. W. Gould 3682*, 23 Aug 1946, Field Museum); and Pima Co.: Santa Cruz Valley near Tucson (*C. G. Pringle s.n.* May–Jun 1891, Field Museum).

2. Acaciella lemmonii (Rose) Britton & Rose. N. Amer. Fl. 23:103. 1928. * [for John Gill Lemmon, 1832–1908, American botanist and plant explorer] Native

Acacia lemmonii Rose, Contr. U.S. Nat. Herb. 12:409. 1909; *Acaciella shrevei* Britton & Rose

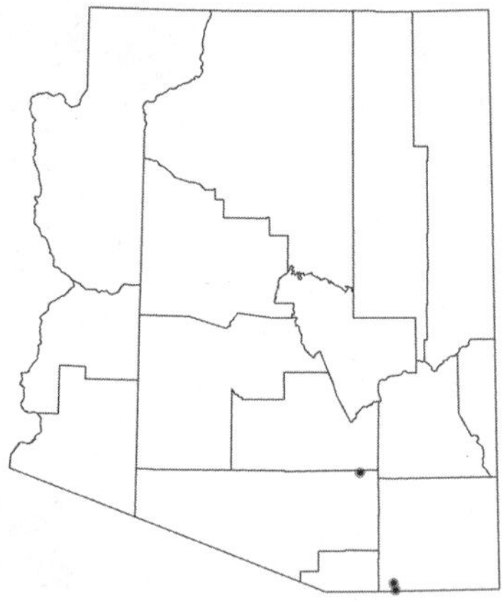

Shrubs, erect, to 3 m. **Stems** slender, twigs slightly zigzagged, round, usually pubescent. **Leaves:** pinnae 4–12 pairs per leaf, 30–70 mm; stipules linear, to 7.5 mm, glabrous to long ciliate; petioles channeled above, 15–41 mm, glabrous to appressed pubescent; rachis 25–140 mm. **Leaflets** 10–30 pairs per pinna, oblong, oblique and truncate at base, acute at apex, 4–13 × 1.5–3.1 mm, glabrous, lateral veins mostly obvious beneath. **Inflorescences** loosely-flowered globose heads 12–17 mm across, solitary (rarely 2–3) in the leaf axil or in terminal racemose clusters; peduncles 6–21 mm. **Flowers** greenish-white, drying pale pink; calyx 0.4–1.1 mm, glabrous; corolla 1.9–3.8 mm, glabrous; filaments 5.5–7.4 mm. **Fruits** 30–70 × 6–13 mm, not constricted between the seeds, stiffly papery, usually glabrous; stipe 8–12 mm; apex acuminate to obtuse, short beaked. **Seeds** 3–4.5 mm across, mottled gray-green and dark gray to dark brown.

Flowering May–Sep. Dry sites on limestone; 900–1600 m; Semidesert Grasslands, Madrean Evergreen Woodland, [Sinaloan Thornscrub]; Cochise, Pima cos.; sw United States; n Mexico.

ACMISPON Rafinesque. Atlantic J. 1(4):144–145. 1832. * [according to Rafinesque, point-hooked, in reference to the beak of the fruits] DEER VETCH, BIRD'S-FOOT TREFOIL, LOTUS
Ries Lindley

Annual or perennial herbs, or subshrubs, unarmed; vestiture various, from glabrous to noticeably pubescent, with hairs spreading to strigose. **Leaves** alternate, odd pinnate to palmately compound (rarely some or most 1-foliolate); stipules gland-like, often inconspicuous; stipels absent. **Leaflets** 1–13, margins entire, opposite or irregularly arranged. **Inflorescences** axial, umbels, 1–6 flowered; bracts generally absent or 1(–3); peduncle present, not apparent in some. **Pedicellate. Flowers** papilionoid; corolla yellow, white, or pink. **Fruits** dehiscent or indehiscent, oblong. **Seeds** 1–several.

Species ± 30 (12 native in the flora, 1 of those cultivated): sw Canada, United States, Mexico, Chile.

Acmispon comprises the majority of the New World taxa formerly placed in *Lotus* as well as some previously included in *Hosackia* (Brouillet 2008). Other New World taxa are treated under *Hosackia*.

Acmispon species may hybridize, creating plants with intermediate characteristics that are difficult to identify at the species level. This is especially true of *Acmispon oroboides*, a species found throughout most of Arizona. In southern Arizona there are morphologies that are intermediate with *A. greenei*. In northwest Arizona it may hybridize with *A. utahensis*.

The *Acmispon oroboides* Complex

Acmispon oroboides is a polymorphic species, probably due to hybridization with other species of *Acmispon*. The holotype for this species was collected prior to 1823 by Bonpland and Humboldt near Guanajuato, Mexico, 1500 kilometers south of the Arizona border with Mexico. Originally called *Tephrosia oroboides*, this specimen's name became the basionym for *Lotus oroboides*. The Bonpland type specimen is erect, has a single fruit per peduncle, and the leaves are clearly odd-pinnate with about 11–13 leaflets. The leaflets are long and narrow. About 400 kilometers south-southeast of Arizona, near Cuauhtémoc, Chihuahua, a specimen of *A. oroboides* (*Steussey 1050*, 20 August 1967) appears to be similar to the Bonpland specimen, but with fewer leaflets per leaf, and leaflet spacing more crowded. *Acmispon oroboides* in northern Sonora and Chihuahua and southern Arizona may also exhibit leaflets so long and narrow they may be called linear, and these linear leaflets may be more widely spaced, so that the leaves appear more plainly pinnate, as in the holotype.

The morphological forms discussed above are all within an expected range of variation for *Acmispon oroboides*, but in Arizona increased variation is seen in certain geographic areas due to hybridization and ecological variation (Ottley 1944). Ottley notes that north-northwestern and southern Arizona are areas where this type of variation is more intense. In southern Arizona, plants intermediate between *A. oroboides* and *A. greenei* are relatively common. A second group of plant forms intermediate to *A. oroboides* and *A. utahensis* has also been collected in Arizona, mostly north of Pima County.

Comparative Descriptions of *Acmispon greenei* and *A. utahensis* and Intermediates with *A. oroboides*

To understand the intermediate variations, a brief comparative description of typical *Acmispon greenei* and *A. utahensis* is necessary. In Arizona, *A. greenei* is most common in the southeastern part of the state in Cochise and Santa Cruz counties as well as parts of Pima, Pinal, and Graham counties. Some plants have also been reported from Gila, Mohave, Greenlee, Coconino, and Yavapai counties. The typical form is procumbent to low ascending with spreading hairs on the stems and leaves. At the base of the plant is a cluster of more closely spaced leaves that are smaller than the distal leaves. All the leaflets, basal and distal, are broader and rounder in general than those of *A. oroboides*. Intermediates to the two species also have a cluster of closely spaced leaves at the base that are broader and rounder than the distal leaves, a trait common in *A. greenei*. The distal leaves of the intermediates may have long and narrow leaflets or leaflets that are more mid-way between the two extremes of long and narrow vs. short and broad. The intermediates often have strigose hairs, the expected condition in *A. oroboides*, or may have some spreading or disorganized hairs at the distal ends. The trait of spreading hairs seems to be weakly heritable in these forms.

Acmispon utahensis in Arizona generally occupies the hilly and mountainous country of the Arizona Central Highlands as far southeast as southern Gila County as well as the upper Grand Canyon and parts of northwestern Arizona. This is an erect or ascending plant with palmate leaves. Generally, this species has 2–5 flowers per inflorescence and 2–3 bracts just beneath the flowers or fruits. *Acmispon oroboides* by contrast generally has 1–2 flowers and 0–1 bracts.

The differences in gross habit and character are more subtle in the intermediates between *Acmispon oroboides* and *A. utahensis*. The number of flowers per inflorescence in these two species overlap, with both species sometimes having two flowers per inflorescence. The strict palmate character of leaves in *A. utahensis* sometimes skews toward the very-closely spaced leaflets of the irregularly pinnate leaves in some *A. oroboides*. *Acmispon utahensis* may also hybridize with *A. rigidus*, forming intermediates with the long internodes typical of that species.

Ottley's views on the range extent of *Acmispon oroboides* in Arizona are as follows: "Many of the plants of northwestern Arizona and contiguous areas may appear to key to *Acmispon oroboides* but field and garden observations have convinced us that these plants do not belong to that species, even though they resemble it in many respects. *Acmispon oroboides* reaches, in Arizona, its farthest extension to the northwest in the Santa Catalina Mountains, Pima County."

Ottley did not say what she thought those northwest Arizona plants might be. A search of herbaria databases for plants outside Ottley's distributional boundaries for *Acmispon oroboides* revealed eight specimens with images. Four of these appear to be *A. utahensis*, each with long internodes and petioles somewhat longer than typical. Two more specimens appear to be better suited to *A. greenei*, and the remaining two appear to be *A. mearnsii*. An examination of physical specimens in the University of Arizona Herbarium collection yielded similar results, i.e., none of the specimens collected in northwest Arizona was clearly *A. oroboides*.

Ottley suggested that transplant studies and chromosome work were needed for *Lotus* (including taxa now treated as *Acmispon*). The same can still be said today. Clearly there are improvements that need to be made in species treatments, but simple, gross morphology is unlikely to improve the current systematics nearly as much as will cytology, phylogeny, molecular studies, or horticultural work.

SELECTED REFERENCES: Barneby 1989, Isely 1998, McDougall 1973, Ottley 1944.

1. Plants annual.
    2. Stigma pubescent (sometimes not noticeably so in *Acmispon strigosus*); fruits curved, coiled, or bent along distal portion.
        3. Leaf axis generally not flattened; peduncle absent; wings shorter than keel; fruits indehiscent, cylindrical with a portion distal to seeds that is attenuate and coiled or curved; seeds 2 ____ **7. Acmispon micranthus**
        3′ Leaf axis flattened; peduncle generally less than 10(–14) mm; wings longer than keel; fruits dehiscent, somewhat laterally compressed, distally curved; seeds 5–10 ____ **10. Acmispon strigosus**
    2′ Stigma glabrous; fruits ± straight over their entire length.
        4. Leaves pinnately 3-foliolate or 1-foliolate; fruits 15–35 mm ____ **1. Acmispon americanus**
        4′ Leaves irregularly pinnate, subpinnate or palmate with 2–7 leaflets; fruits 6–19 mm.
            5. Inflorescence with 2–4(–5) flowers, peduncles 2–20(–40) mm; calyx 2–3 mm, lobes shorter than tube; wings shorter than keel ____ **5. Acmispon maritimus** var. **brevivexillus**
            5′ Inflorescence with 1(–2) flowers, peduncles absent or less than 2 mm; calyx 3–6.5 mm, lobes longer than or equal to tube; wings about equal to keel.
                6. Plants prostrate, decumbent, mat-forming, noticeably pubescent; fruits 3–4.5 mm wide; abundant and widespread, 300–800 m elevation ____ **2. Acmispon brachycarpus**
                6′ Plants decumbent to erect, ± glabrous; fruits 2–3 mm wide; rare and localized in northern Coconino and Mohave cos., 700–2000 m elevation ____ **3. Acmispon denticulatus**

1′ Plants perennial.
    7. Leaves strictly palmate or appearing so.

8. Peduncles 0–6 mm (mostly absent) __________ **12. Acmispon wrightii**
8′ Peduncles 3–70 mm.
9. Peduncles 3–40 mm, inflorescence with 1–2(–3) flowers, bracts 0–1(–3); widespread in southern, central and northern Arizona __________ **8. Acmispon oroboides** (in part)
9′ Peduncles 19–70 mm, inflorescence with (1–)2–5 flowers, bracts (1–)2–3; localized in central and northern Arizona __________ **11. Acmispon utahensis**
7′ Leaves pinnate, irregularly pinnate, or subpalmate.
10. Stems with internodes mostly 4 cm or greater (some forms from arid, low-elevation areas of western Pima and Yuma cos. with internodes less than 4 cm); leaves scattered; fruits ± cylindrical __________ **9. Acmispon rigidus**
10′ Stems with internodes less than 4 cm; leaves appear crowded; fruits somewhat laterally compressed.
11. Pubescence mostly obscuring the underlying surface, strigose; fruits 3–4 mm wide __________ **6. Acmispon mearnsii**
11′ Pubescence usually not obscuring underlying surface, strigose or with spreading hairs; fruits 1.5–2.9 mm wide.
12. Plants usually with a cluster of basal leaves; stem pubescence primarily of spreading hairs or in part strigose __________ **4. Acmispon greenei**
12′ Plants usually without a cluster of basal leaves; stem pubescence primarily strigose or some plants with a few spreading hairs __________ **8. Acmispon oroboides** (in part)

1. Acmispon americanus (Nuttall) Rydberg. Bull. Torrey Bot. Club 40(2):45. 1913. * [of America] AMERICAN BIRD'S-FOOT TREFOIL, PURSH'S BIRD'S-FOOT TREFOIL, SPANISH CLOVER

Trigonella americana Nuttall, Gen. N. Amer. Pl. 2:120. 1818; *Lotus americanus* (Nuttall) Bischoff; *Lotus purshianus* Clements & E. G. Clements; *Lotus unifoliolatus* (Hooker) Bentham

Annual herbs, erect to ascending, rarely prostrate, ± glabrous to pubescent. **Stems** to 70 cm. **Leaves** pinnate (sometimes 1 foliolate); axis not flattened. **Leaflets** 1 or 3, 10–19 × 3–7 mm. **Inflorescences** 1-flowered; bracts 1; peduncle 4–20 mm. **Flowers** 4–7 mm; calyx 3–6 mm, lobes longer than tube; corolla white or yellow to pink, wings about equal to keel; stigma glabrous. **Fruit** dehiscent, somewhat laterally compressed, ± straight, 15–35 × 2–3 mm. **Seeds** 3–8.

Varieties 2 (1 native in the flora): sw Canada, w United States, nw Mexico. Variety *helleri* is restricted to the southeastern U.S.

1a. Acmispon americanus (Nuttall) Rydberg var. **americanus** * Native

Lotus purshianus Clements & E. G. Clements var. *purshianus*

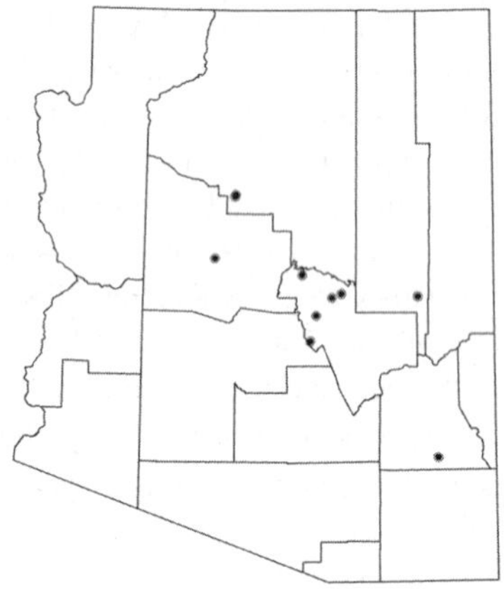

Annual herbs, with conspicuous spreading hairs especially in the distal portion.

Flowering May–Oct. Often found on moist soils of stream banks and drainages, canyons, marshy areas, springs, meadows, and less commonly on slopes; [near sea level–]1070–2260[–2400] m; Interior Chaparral, Madrean Evergreen Woodland, Petran Montane Conifer Forest; Apache, Cochise, Coconino, Gila, Graham, Navajo, Yavapai cos.; California, Idaho, Montana, Nevada, New Mexico, Oregon, Washington and disjunct eastward in the central U.S. where it is widely distributed from Texas northward to North Dakota and Wisconsin; Canada (British Columbia, Manitoba, Saskatchewan); Mexico (Baja California).

Acmispon americanus var. *americanus* is found as scattered populations at elevations of 1000–2300 m across Arizona from Cochise County to the central and north central portions of the state. It is typically found in moist sites but can occur on slopes. Seeds of *A. americanus* are eaten by a variety of birds and small rodents (Graham 1941), and the herbage is considered to be important browse for mule deer (Dixon 1934). The plants may be of some use in erosion control (Graham 1941, Stubbendieck and Conard 1989).

2. Acmispon brachycarpus (Bentham) D.D. Sokoloff. Ann. Bot. Fenn. 37(2):130. 2000. * [from Greek *brachy*, short and *karpos*, fruit] SHORT-POD LOTUS, FOOTHILL DEER VETCH; *COLCHITA* Native

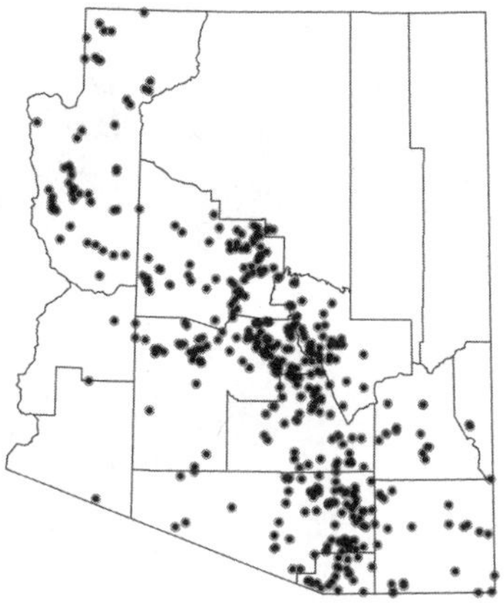

Hosackia brachycarpa Bentham, Pl. Hartw. 306. 1848[1849]; *Lotus humistratus* Greene

Annual herbs, prostrate, decumbent, mat-forming, and sometimes low ascending, noticeably pubescent, pubescence dense in new growth. Stems 10–20(–40) cm. Leaves subpinnate or palmate; axis flattened. Leaflets 3–5, irregularly arranged, 5–11 × 2–6 mm. Inflorescences 1-flowered; bracts 0; peduncle 0. Flowers 3.5–6.5 mm; calyx 3.5–6.5 mm, lobes 1–2 times the length of the tube; corolla yellow, reddening in age, wings about equal to keel; stigma glabrous. Fruit dehiscent, somewhat laterally compressed, ± straight, 9–14 × [2.5–]3–4.5 mm. Seeds 2–5.

Flowering Mar–May(–Jun). Growing in a wide range of soils from sand and clay to gravel and rock on slopes, canyons, valleys, arroyos, floodplains, and roadsides; 300–800 m; Sonoran Desertscrub, Chihuahuan Desertscrub, Mohave Desertscrub, Semidesert Grassland, Interior Chaparral, (Madrean Evergreen Woodland, Great Basin Conifer Woodland); Cochise, Coconino, Gila, Graham, Greenlee, La Paz, Maricopa, Mohave, Pima, Pinal, Santa Cruz, Yavapai, Yuma cos.; California, Idaho, Nevada, New Mexico, Oregon, Texas, Utah; Mexico (Baja California, Sonora).

Acmispon brachycarpus is distributed across Arizona in a broad band from southeast to northwest. It is absent from low-elevation deserts, higher mountains, and the Colorado Plateau. A winter-spring ephemeral, it can be common in wet years. The seeds of *A. brachycarpus* are an important food resource for a variety of birds and rodents (Graham 1941).

3. Acmispon denticulatus (Drew) D.D. Sokoloff. Ann. Bot. Fenn. 37(2):130. 2000. * [from Latin *denti*, tooth and *cula*, little] TOOTHED LOTUS, RIVER-BAR LOTUS Native

Hosackia denticulata Drew, Bull. Torrey Bot. Club 16(6):151. 1889; *Lotus denticulatus* (Drew) Greene

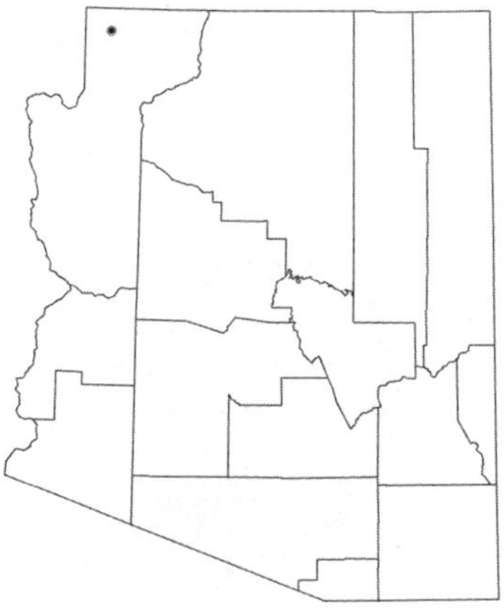

Annual herbs, decumbent to erect, ± glabrous with some scattered hairs and some dense pubescence on new growth. **Stems** 10–40 cm. **Leaves** subpinnate or palmate; axis flattened. **Leaflets** 2–4, 3–6 × 8–14 mm. **Inflorescences** with 1(–2) flowers; bracts 0; peduncles absent or less than 2 mm. **Flowers** 6–7.5 mm; calyx 3–5 mm, lobes longer than or equal to tube; corolla white, pale yellow, or pink, often darkening in age, wings about equal to keel; stigma glabrous. **Fruit** dehiscent, somewhat laterally compressed, ± straight, 6–19 × 2–3 mm. **Seeds** 2–3.

Flowering (Apr–)May–Jun(–Jul). Sandy soil of drainages and floodplains, [clay and rocky soils of valleys and slopes]; [sea level–]700–2000[–2300] m; Great Basin Conifer Woodland; Coconino, Mohave cos.; California, Oregon, Utah, Washington; Canada (British Columbia).

Primarily a species of the West Coast, *Acmispon denticulatus* is known in Arizona from 3 collections in northern Coconino and Mohave cos.

4. Acmispon greenei (Wooton & Standley) Brouillet. J. Bot. Res. Inst. Texas 2(1):390. 2008. * [for Edward L. Greene, 1843–1915, American botanist] GREENE'S BIRD'S-FOOT TREFOIL Native

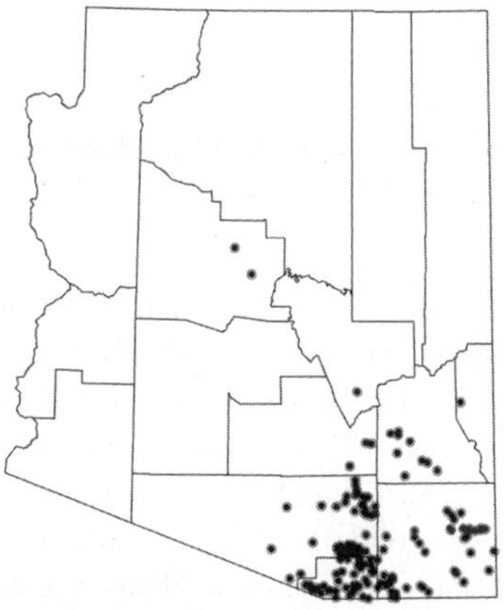

Anisolotus greenei Wooton & Standley, Contr. U.S. Natl. Herb. 16(4):135. 1913; *Hosackia greenei* (Wooton & Standley) Wiggins; *Lotus greenei* (Wooton & Standley) Ottley; *Lotus mollis* Greene; *Lotus neomexicanus* Greene (in part); *Ottleya greenei* (Wooton & Standley) D.D. Sokoloff

Perennial herbs, procumbent to low-ascending; from tap root; often with spreading hairs or in part strigose, pubescence not dense, only occasionally approaching a density that obscures the underlying surface; usually with a dense basal cluster of smaller leaves with leaflets noticeably shorter and broader than more distal leaves, these lower leaf clusters sometimes falling with age. **Stems** to 40 cm; internodes less than 4 cm. **Leaves** appearing

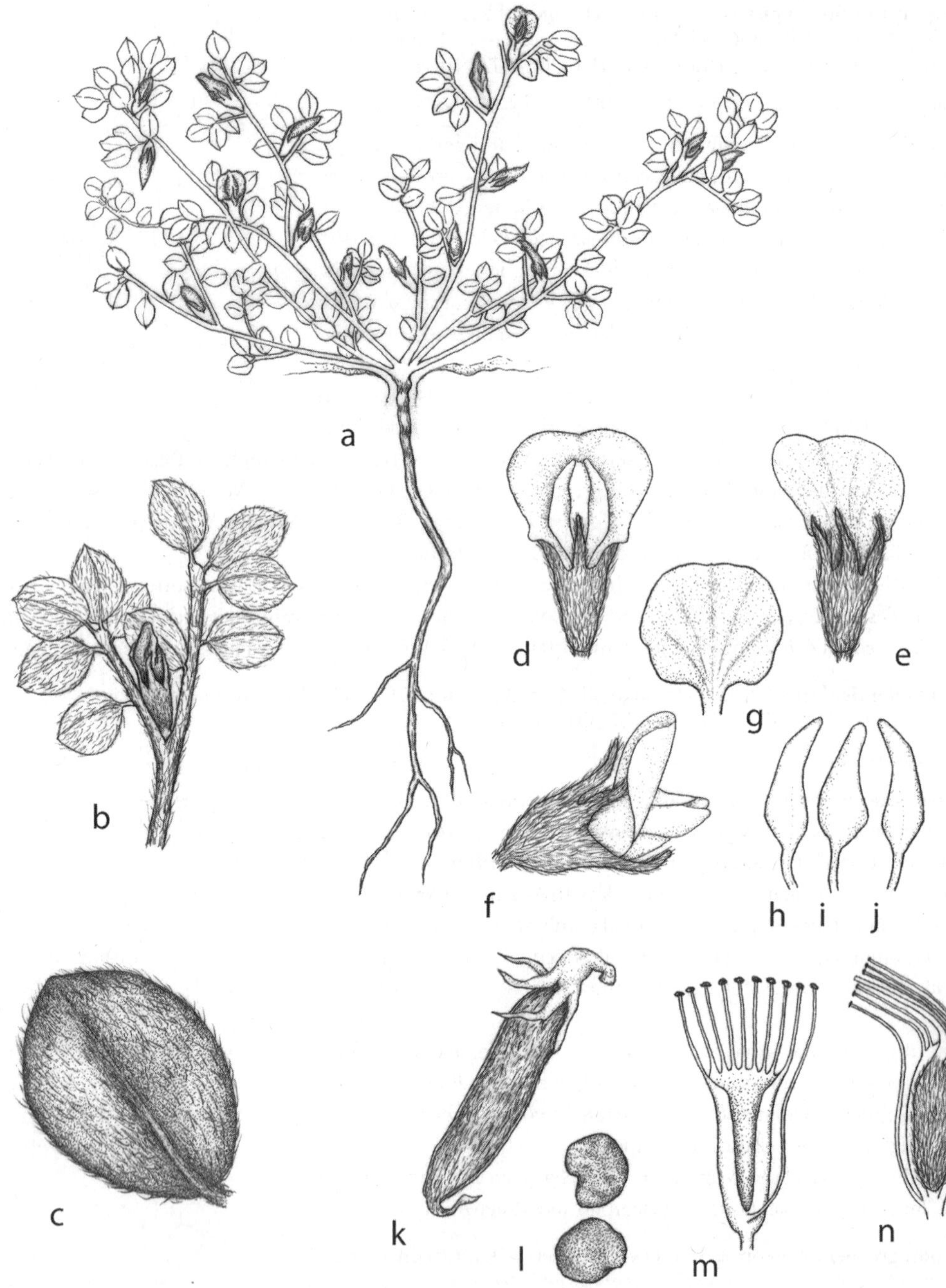

***Acmispon brachycarpus*: a. habit; b. node with leaves and flower; c. leaflet; d., e., f., flower; g., h., i., j. petals; k. fruit; l. seeds, m. stamens; n. pistil and stamens. Illustrator: Chris Bondante.**

crowded; pinnate to irregularly pinnate; axis not flattened. **Leaflets** 3–6, 3–11 × 1–4 mm. **Inflorescences** with 1–3(–5) flowers; bracts 0 or 1; peduncle 2.5–40(–50) mm. **Flowers** 10–15 mm; calyx 6–9 mm, lobes almost equal to tube; corolla yellow to yellowish-white, wings longer than keel; stigma pubescent. **Fruit** dehiscent, pubescent, laterally compressed, ± straight, 22–32 × 1.5–2.8 mm, appearing slender. **Seeds** few.

Flowering Mar–Jun(–Aug). Sandy to rocky soils of slopes, ledges, canyons, arroyos, stream channels, and roadsides; (800–)1200–1900[–2400] m; (Sonoran Desertscrub), Chihuahuan Desertscrub, (Great Basin Desertscrub), Semidesert Grassland, Plains Grassland, Madrean Evergreen Woodland, (Great Basin Conifer Woodland); Cochise,

Coconino, Graham, Greenlee, Gila, Mohave, Pima, Pinal, Santa Cruz, Yavapai cos.; New Mexico; Mexico (Chihuahua, Sonora).

Acmispon greenei is widespread and locally common at mid-elevations in southeastern Arizona. It frequently occurs in rocky sites. Disjunct populations are reported for Coconino, Mohave, and Yavapai counties. *Acmispon greenei* may hybridize with *Acmispon oroboides*. Please see the discussion following the genus description.

Seeds of *Acmispon greenei* are eaten by scaled quail, and the herbage was highly palatable to mule deer and white-tailed deer in feeding experiments (Nichol 1938).

5. Acmispon maritimus (Nuttall) D.D. Sokoloff. Ann. Bot. Fenn. 37(2):129. 2000. * [from Latin *maritim*, of the sea] COASTAL BIRD'S-FOOT TREFOIL, COASTAL LOTUS

Hosackia maritima Nuttall, Fl. N. Amer. 1(2):326. 1838; *Lotus salsuginosus* Greene

Annual herbs, prostrate to ascending, glabrous or strigose. **Stems** to 35 cm. **Leaves** irregularly pinnate; axis moderately flattened. **Leaflets** 3–7(–8), 5–15 × 2–7.5 mm. **Inflorescences** with 2–4(–5) flowers, early inflorescences with 1(–2) flowers; bracts 1; peduncles 2–20(–40) mm. **Flowers** 3–9 mm; calyx 2–3 mm, lobes shorter than tube; corolla yellow, wings about equal to or shorter than keel; stigma glabrous. **Fruit** dehiscent, somewhat laterally compressed, ± straight, 10–30 × 2 mm. **Seeds** (1–)4–8.

Varieties 2 (1 native in the flora). Variety *maritimus* is found in California and Baja California.

5a. Acmispon maritimus (Nuttall) D.D. Sokoloff var. **brevivexillus** (Ottley) Brouillet. J. Bot. Res. Inst. Texas 2(1):391. 2008. * [from Greek *brevi*, short and *vexillum*, banner] Native

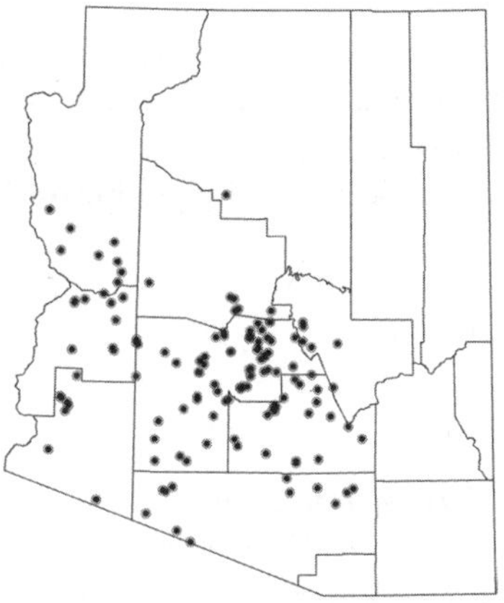

Lotus salsuginosus Greene var. *brevivexillus* Ottley, Univ. Calif. Publ. Bot. 10(3):217. 1923

Flowers 3–5 mm; wings and banner shorter than keel. **Fruit** 10–15[–18] mm, some narrowing between seeds.

Flowering [Jan–]Feb–Apr[–May]. Various soils of slopes, valleys, arroyos, and roadsides; [30–]320–1220 m; Sonoran Desertscrub; Gila, La Paz, Maricopa, Mohave, Pima, Pinal, Yavapai, Yuma cos.; California, Nevada; Mexico (Sonora).

Acmispon maritimus var. *brevivexillus* is a winter/spring annual that is widespread and often abundant across the Sonoran Desert in wet years. The species occurs in a variety of soils ranging from sandy to gravelly and rocky substrates. It is occasionally found in riparian habitats.

Both Isely and Brouillet note that the fruit of *Acmispon maritimus* var. *brevivexillus* exhibit a constriction between the seeds (Brouillet 2012, Isely 1998). This character is either subtle or inconsistent. No easily discernable constriction was noted in the specimens from the University of Arizona Herbarium.

6. Acmispon mearnsii (Britton) Brouillet. J. Bot. Res. Inst. Texas 2(1):391. 2008. * [for Edgar A. Mearns, 1856–1916, American military physician, ornithologist, and naturalist] MEARNS' BIRD'S-FOOT TREFOIL

Hosackia mearnsii Britton, Trans. New York Acad. Sci. 8(3–4):65. 1889; *Lotus mearnsii* (Britton) Britton ex Greene; *Ottleya mearnsii* (Britton) D.D. Sokoloff

Perennial herbs, from rootstocks, procumbent or decumbent often with some upper stems ascending, pubescence strigose, dense and silvery, mostly obscuring the underlying surface. **Stems** to 40 cm; internodes less than 4 cm. **Leaves** appearing crowded; pinnate or irregularly pinnate; axis not flattened. **Leaflets** 3–5, 6–16 × 3–12 mm. **Inflorescences** with (1–)2–5(–6) flowers; bracts 0–1; peduncles 20–150 mm. **Flowers** 10–18 mm; calyx 6–8 mm, lobes generally shorter than tube; corolla yellow or red-tinged, wings longer than keel; stigma pubescent. **Fruit** dehiscent, somewhat laterally compressed, ± straight, 15–37 × 3–4 mm, giving a broad appearance. **Seeds** few.

2 varieties (2 native in the flora): endemic to Arizona.

1. Plants of various soils and locations within the Colorado Plateau and the transition zone (Central Highlands) above 950 m elevation; internodes mostly longer than leaves, leaflets more than twice as long as wide; fruit 17–22 mm __________ **6b. Acmispon mearnsii** var. **mearnsii**

1′ Plants of the lacustrine deposits of the Verde River at about 850 m elevation; internodes mostly shorter than leaves, leaflets twice as long as wide or less; fruit 25–38 mm __________ **6a. Acmispon mearnsii** var. **equisolensis**

6a. Acmispon mearnsii (Britton) Brouillet var. **equisolensis** (J.L. Anderson) Brouillet * [from Latin *equi*, equal, and Greek *soleno*, channel] Native

Lotus mearnsii (Britton) Britton ex Greene var. *equisolensis* J.L. Anderson, Madroño 43(2):261–263. 1996

Stems: internodes mostly shorter than leaves. **Leaflets** twice as long as wide or less. Fruit 25–38 mm.

Flowering Feb–Apr. Lacustrine soils of the Lower Verde River valley south of Horseshoe Reservoir; ca. 850 m; Sonoran Desertscrub; Maricopa, Yavapai cos.; Arizona.

The overall plant habit is more compact than that of variety *mearnsii*.

6b. Acmispon mearnsii (Britton) Brouillet var. **mearnsii** * Native

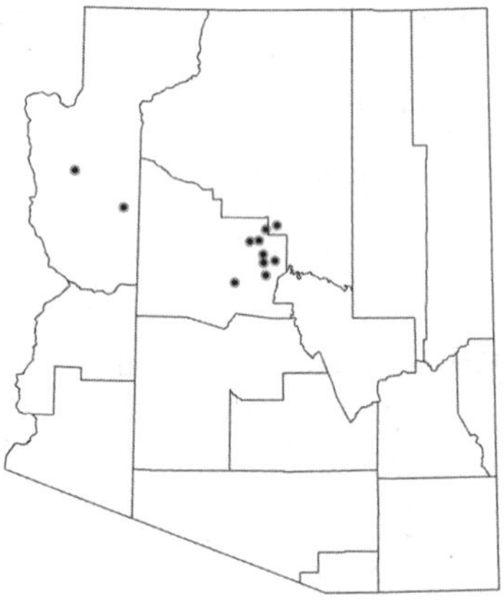

Stems: internodes mostly greater than leaves. **Leaflets** 2 or more times longer than wide. **Fruit** 17–22 mm.

Flowering Mar–Jul. Various soils and habitats; 950–2100 m; Sonoran Desertscrub, Semidesert Grassland, Plains and Great Basin Grassland, Great Basin Conifer Woodland, Petran Montane Conifer Forest; Coconino, Yavapai cos.; Arizona.

Variety *mearnsii* plants can reach about a meter in diameter.

7. Acmispon micranthus (Torrey & A. Gray) Brouillet. J. Bot. Res. Inst. Texas 2(1):391. 2008. * [from Greek *micro*, small and *anthus*, flower] SAN DIEGO BIRD'S-FOOT TREFOIL, LITTLE-FLOWER DEER VETCH Native

Hosackia micrantha Torrey & A. Gray, Fl. N. Amer. 1(2):324. 1838; *Lotus hamatus* Greene

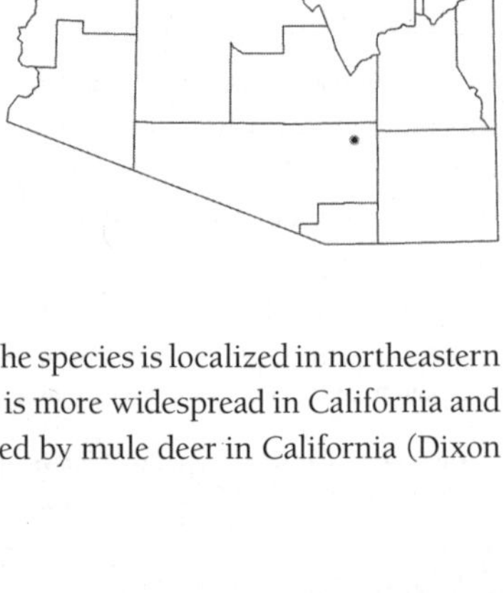

Annual herbs, prostrate to ascending, glabrous or puberulent. **Stems** 20–60 cm. **Leaves** irregularly pinnate or palmate; axis usually not flattened. **Leaflets** 4–7, 3–8 × 1–4 mm. **Inflorescences** 1–6 flowered; bracts absent; peduncle absent. **Flowers** 3–8 mm, calyx 1.5–3 mm, lobes about 1/3 length of tube; corolla 2.5–8 mm, yellowish, wings shorter than keel, stigma pubescent. **Fruit** indehiscent, cylindrical with a portion distal to seeds that is attenuate and coiled or curved, as long as or longer than seed-bearing portion, 10–15 × to 1 mm. **Seeds** 2.

Flowering Feb–Jul. Various soils of slopes, canyons, valleys, and roadsides; [near sea level–]1000–1600 m; Sonoran Desertscrub, Semidesert Grassland, Madrean Evergreen Woodland; Pima, Yavapai cos.; California; Mexico (Baja California, Sonora).

In Arizona, *Acmispon micranthus* is apparently highly localized and rare. It is known from only a few collections from the Santa Catalina and Baboquivari Mountains, and the Altar Valley in Pima County, and one collection (*R. Van Buren 261*, 01 March 1993) from along Interstate Highway 17 in Yavapai County. The species is localized in northeastern Sonora where it has been found in thornscrub as well as Sonoran Desertscrub. It is more widespread in California and Baja California in a variety of plant communities. *Acmispon micranthus* is browsed by mule deer in California (Dixon 1934).

8. Acmispon oroboides (Kunth) Brouillet. J. Bot. Res. Inst. Texas 2(1):391. 2008. * NEW MEXICO BIRD'S-FOOT TREFOIL, DEER VETCH Native

Tephrosia oroboides Kunth, Nov. Gen. Sp. (quarto ed.) 6: plates 578–579. 1823[1824]; *Lotus longebracteatus* Rydberg; *Lotus neomexicanus* Greene (in part); *Lotus oroboides* (Kunth) Ottley; *Lotus oroboides* (Kunth) Ottley var. *nanus* Isely; *Lotus oroboides* (Kunth) Ottley var. *nummularius* (M.E. Jones.) Isely; *Lotus oroboides* (Kunth) Ottley var. *plebeius* (Brandegee) Ottley; *Lotus plebeius* (Brandegee) Barneby; *Ottleya oroboides* (Kunth) D.D. Sokoloff

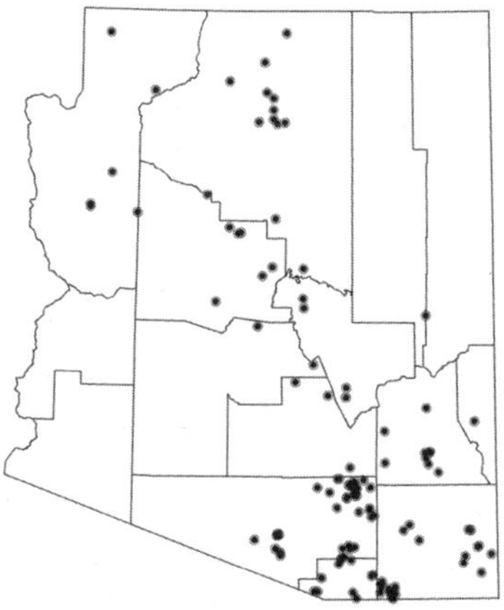

Perennial herbs, procumbent, erect or ascending, from tap root, pubescence usually strigose or some plants with a few spreading hairs, not dense, only occasionally approaching a density that obscures the underlying surface; mostly without a dense basal cluster of smaller leaves with leaflets noticeably shorter and broader than those of more distal leaves, or the cluster much reduced. **Stems** to 40 cm; internodes less than 4 cm. **Leaves** irregularly pinnate, subpalmate or strictly palmate; axis not flattened; appearing crowded. **Leaflets** 3–5(–7)[–13], 4.2–20 × 0.5–4.6 mm. **Inflorescences** with 1–2(–3) flowers; bracts 0–1(–3); peduncles 3–40 mm. **Flowers** 11–17 mm; calyx 6–9 mm, lobes shorter than or equal to tube; corolla yellow becoming orange-red with age, wings longer than keel; stigma pubescent. **Fruits** dehiscent, somewhat laterally compressed, ± straight, 17–43 × 2.2–2.9 mm, appearing slender. **Seeds** few to numerous.

Flowering [Feb–]Mar–Oct[–Nov], occasionally year-round (Bertelsen, personal communication). Found in a variety of soil types on ridges, slopes, cliffs, canyons, along streams, springs, and roadsides; [100–]900–2400[–2800] m; (Sonoran Desertscrub), Semidesert Grassland, Madrean Evergreen Woodland, (Great Basin Conifer Woodland), Petran Montane Conifer Forest; Apache, Cochise, Coconino, Gila, Graham, Greenlee, Maricopa, Mohave, Pima, Pinal, Santa Cruz, Yavapai cos.; Nevada, New Mexico, Texas, Utah; Mexico (Chihuahua, Coahuila, Durango, Hidalgo, México, Puebla, Sinaloa, Sonora).

Acmispon oroboides is widespread and often abundant in Arizona at mid- and higher elevations. At lower elevations, the plants are usually restricted to riparian habitats or other sites where moisture is available. As outlined in the discussion at the beginning of this treatment, *A. oroboides* exhibits significant morphological variation, particularly in its leaf characters. Isely (1981) separated the material found in the U.S. into three varieties: *oroboides*, *nanus*, and *nummularius*. These varieties are not recognized here as there seems to be considerable overlap of distinguishing characters. Barneby (1989) placed all material referable to *A. oroboides* from the U.S. into *Lotus plebeius*.

9. Acmispon rigidus (Bentham) Brouillet. J. Bot. Res. Inst. Texas 2(1):392. 2008. * [from Latin *rigid*, stiff, referring to the stems] DESERT ROCK PEA, BROOM LOTUS, SHRUBBY LOTUS Native, Cultivated

Hosackia rigida Bentham, Pl. Hartw. 305. 1848[1849]; *Lotus rigidus* (Bentham) Greene; *Ottleya rigida* (Bentham) D.D. Sokoloff

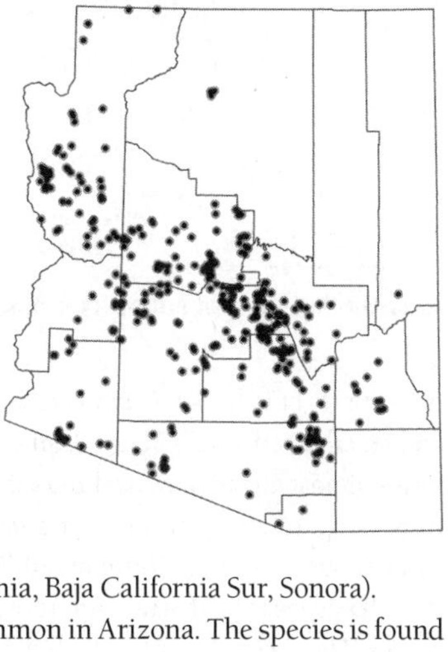

Perennial herbs, subshrubs developing a woody crown, erect to ascending, often glabrate in age, internodes generally greater than 4 cm (smaller forms in western Pima County and Yuma County have internodes less than 4 cm), few-leaved. **Stems** (20–)50–100(–150) cm. **Leaves** scattered; subpalmate at apex, irregularly pinnate below; axis not flattened. **Leaflets** 3–5, 5–22 × 1–4.5 mm. **Inflorescences** 1–3-flowered; bracts 0–3; peduncle 40–90(–130) mm. **Flowers** 12–20 mm; calyx 7–9 mm, lobes shorter than tube; corolla yellow to cream, wings longer than keel; stigma pubescent. **Fruits** dehiscent, glabrous, ± cylindrical, ± straight, 25–40 × 3.5–5 mm. **Seeds** 18–30.

Flowering Jan–Jun(–Dec). Often found on rocky soils of ridges, cliffs, slopes, canyons, arroyos, stream banks, floodplains, and roadsides; 10–1700[–2300] m; Sonoran Desertscrub, Mohave Desertscrub, Semidesert Grassland, Interior Chaparral, Madrean Evergreen Woodland, Great Basin Conifer Woodland; Coconino, Gila, Graham, La Paz, Maricopa, Mohave, Pima, Pinal, Yavapai, Yuma cos.; California, Nevada, New Mexico, Utah; Mexico (Baja California, Baja California Sur, Sonora).

Acmispon rigidus is a xerophytic subshrub that is widespread and locally common in Arizona. The species is found in many habitats. Outside Arizona, *A. rigidus* is found in coastal chaparral and Jeffrey pine forest. In the hottest and

***Acmispon oroboides*: a. habit; b. flowers; c. fruit; d. stem detail; e. leaf detail. Illustrator: Marsha Bennett.**

driest parts of its range *A. rigidus* may be reduced in all its dimensions, with internodes less than 4 cm, smaller leaves, and smaller stem size. These diminutive plants also have strigose fruit and stems, whereas the typical, more robust forms almost always have glabrous fruit and stems older than one year that are glabrate. Leaves in the typical form tend to be subpalmate at the apex and irregularly pinnate below, and in the diminutive form the leaves are subpalmate throughout. Felger and Rutman (2015) have referred all these diminutive dryland plants to *Acmispon rigidus*.

Acmispon rigidus is used for revegetation and is occasionally grown as a flowering perennial for low water-use landscapes. The green stems and showy yellow and red flowers are attractive. This species is well-suited for rock gardens or

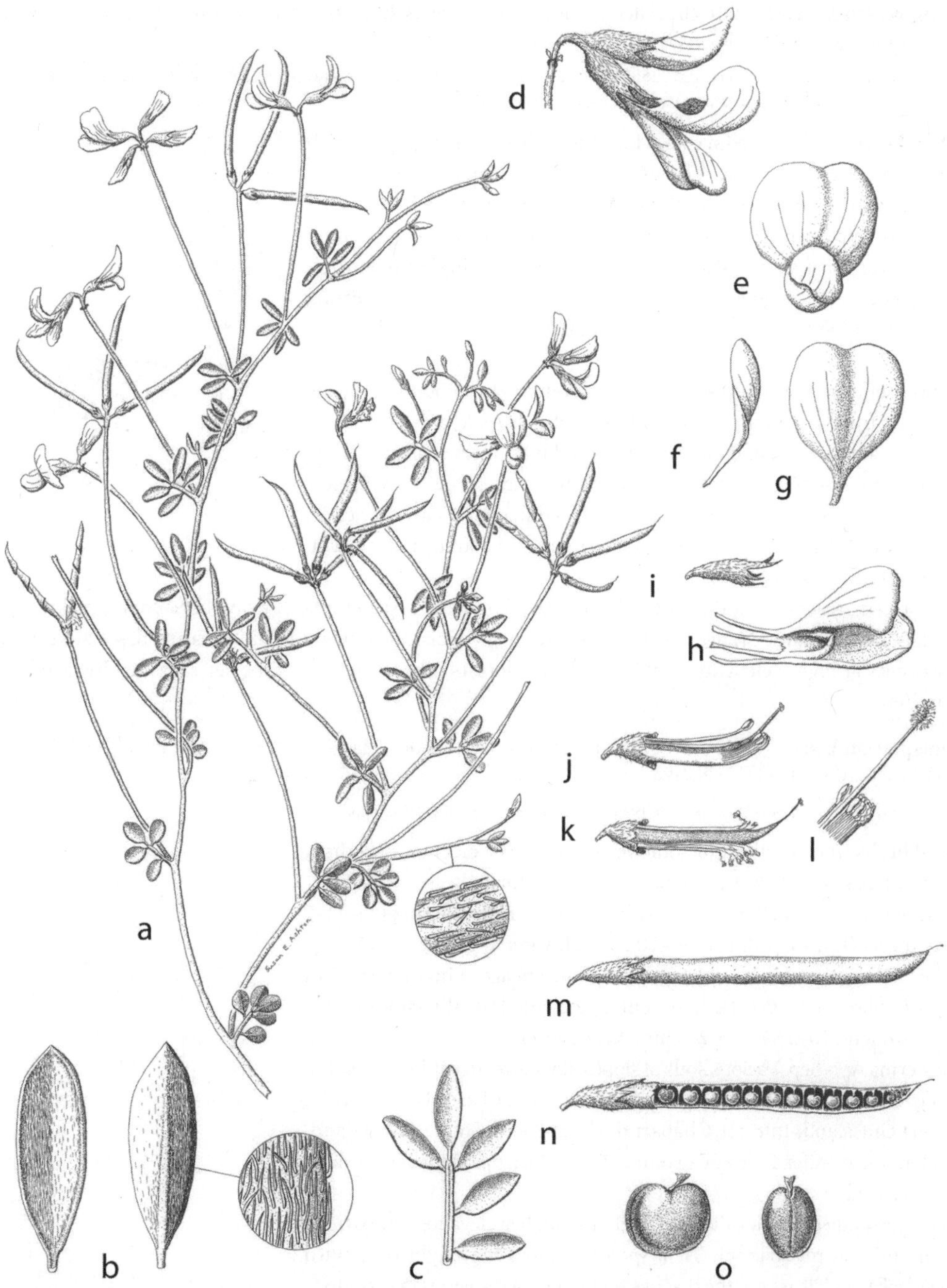

Acmispon rigidus: **a.** habit; **b.** detail of upper and lower leaflet surface; **c.** leaf; **d.** flowers (side view); **e.** flower (front view); **f., g., h.** petals; **i.** calyx; **j., k., l.** detail of stamens and pistil; **m.** fruit; **n.** fruit detail; **o.** seeds. Illustrator: Susan E. Ashton.

planting among rocks with other flowering perennials and succulents. For best appearance, desert rock pea should be planted in a sunny location. It will tolerate reflected heat. Once established, the plants require infrequent irrigation to maintain appearance. Desert rock pea tolerates temperatures to at least -12° C. The plants produce negligible litter. Old stems may be removed if desired for appearance. Propagation is by scarified seeds.

10. Acmispon strigosus (Nuttall) Brouillet. J. Bot. Res. Inst. Texas 2(1) 392. 2008. * [from Latin *strigos*, thin, meager] HAIRY LOTUS, BRISTLE DEER VETCH Native

Hosackia strigosa Nuttall, Fl. N. Amer. 1(2):326. 1838; *Lotus strigosus* (Nuttall) Greene; *Lotus strigosus* (Nuttall) Greene var. *tomentellus* (Greene) Isely; *Lotus tomentellus* Greene; *Ottleya strigosa* (Nuttall) D.D. Sokoloff

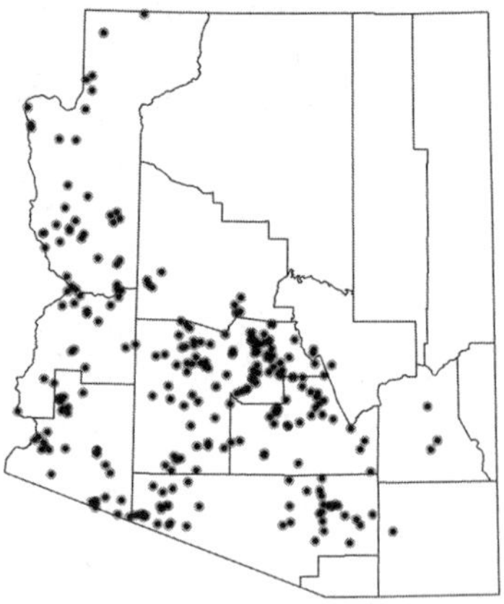

Annual herbs, prostrate sometimes mat-forming, pubescence strigose or with spreading hairs. **Stems** 3–50 cm. **Leaves** irregularly pinnate; axis flattened, ± blade-like. **Leaflets** 4–9, 3–10 × 1–4 mm. **Inflorescences** with 1–2 flowers; bracts 0–1; peduncle generally less than 10(–14) mm. **Flowers** 4–8(–10) mm; calyx 2.5–4 mm, lobes shorter than tube; corolla yellow, orange or red in age, wings longer than keel; stigma pubescent, sometimes not noticeably so. **Fruits** dehiscent, somewhat laterally compressed, distally curved, 10–20 × 1–3 mm. **Seeds** 5–10.

Flowering (Jan–)Feb–Apr(–May), peak bloom in Mar. Found in a wide range of soils from sand and clay loam to gravelly or rocky soils of ridges, slopes, valley, arroyos, floodplains, and on sand dunes; [near sea level–]100–1200[–1400] m; Sonoran Desertscrub, Mohave Desertscrub, (Semidesert Grassland, Interior Chaparral); Coconino, Gila, Graham, La Paz, Maricopa, Mohave, Pima, Pinal, Santa Cruz, Yavapai, Yuma cos.; California, Nevada; Mexico (Baja California, Sonora).

Acmispon strigosus is a widespread and often abundant winter-spring ephemeral of the Sonoran and lower Mohave Deserts. It is found on a wide range of substrates and can grow on sand dunes. In California and Baja California, it occasionally occurs at higher elevations into Jeffrey pine forests. The seeds of this species are eaten by Gambel's quail (Graham 1941).

11. Acmispon utahensis (Ottley) Brouillet. J. Bot. Res. Inst. Texas 2(1):392. 2008. * [of Utah] UTAH BIRD'S-FOOT TREFOIL, UTAH LOTUS Native

Lotus utahensis Ottley, Brittonia 5(2):108–109. 1944; *Ottleya utahensis* (Ottley) D.D. Sokoloff

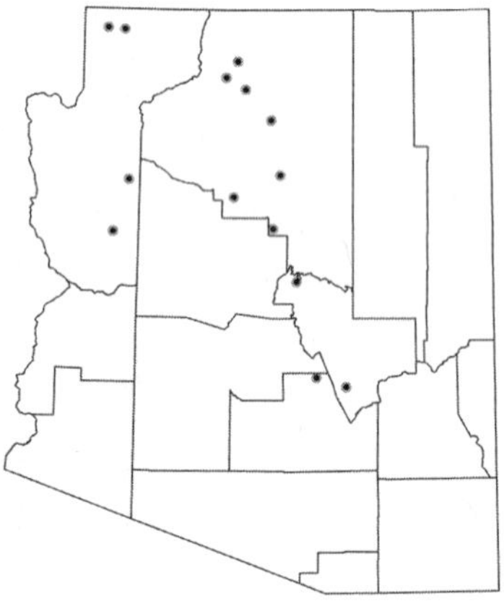

Perennial herbs from a caudex or sometimes rootstocks, erect or ascending, pubescent. **Stems** 20–40(–50) cm. **Leaves** strictly palmate, sessile or subsessile; axis not flattened. **Leaflets** 3–12, 7–22 × 1–4 mm. **Inflorescences** (1–)2–5 flowered; bracts (1–)2–3; peduncle 19–70 mm. **Flowers** 5–15 mm; calyx 4–11 mm, lobes shorter than tube; corolla yellow, orange in age, wings longer than keel; stigma pubescent. **Fruits** dehiscent, somewhat laterally compressed, strigose, ± straight, 20–30(–35) × 2–3 mm. **Seeds** several.

Flowering Apr–Sep. Various soils of slopes, flats, stream banks, roadsides, and rarely on sand dunes; 910–2710[–2850] m; (Great Basin Desertscrub), Semidesert Grassland, Interior Chaparral, Great Basin Conifer Woodland, Petran Montane Conifer Forest; Coconino, Gila, Maricopa, Mohave, Yavapai cos.; Nevada, Utah.

Acmispon utahensis is localized at mid- and higher elevations in parts of northern and central Arizona. *Acmispon utahensis* may hybridize with *Acmispon oroboides*. Please see the discussion following the genus description.

12. Acmispon wrightii (A. Gray) Brouillet. J. Bot. Res. Inst. Texas 2(1):392. 2008. * [for Charles Wright, 1811–1885, American botanist] WRIGHT'S DEER VETCH, SCRUB LOTUS Native

Hosackia wrightii A. Gray, Smithsonian Contr. Knowl. 5(6):42. 1853; *Lotus wrightii* (A. Gray) Greene; *Ottleya wrightii* (A. Gray) D.D. Sokoloff

Perennial herbs, erect or ascending, pubescent, branched throughout, internodes seldom greater than 4 cm. **Stems** 20–40 cm. **Leaves** strictly palmate. **Leaflets** 3–6, 5–20 × 0.7–2 mm. **Inflorescences** with 1–2 flowers; bracts 1–3 (incomplete leaves may subtend the flowers and appear bract-like); peduncles 0–6[–19] mm, mostly absent. **Flowers** 9–16(–18) mm; calyx 5–8 mm, lobes shorter than or equal to tube; corolla yellow, wings longer than keel; stigma pubescent. **Fruit** dehiscent, somewhat laterally compressed, ± straight, 15–30(–40) × 2–3 mm. **Seeds** few to several.

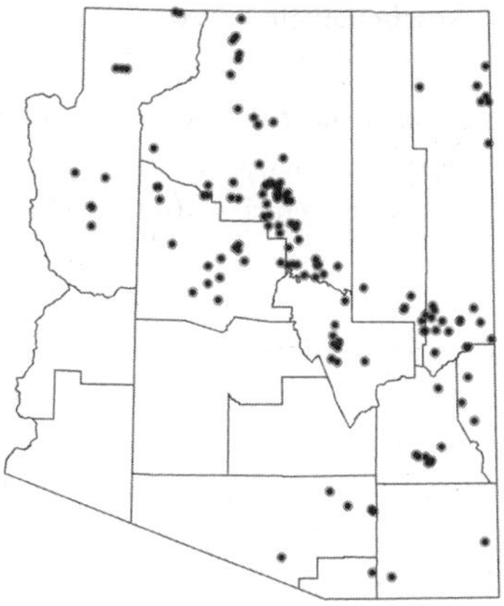

Flowering Apr–Sep[–Nov]. Various soils of slopes, canyons, valleys, drainages, stream banks, marshy areas, and roadsides; (370–)1350–2830 m; (Sonoran Desertscrub, Semidesert Grassland, Plains Grassland), Interior Chaparral, Madrean Evergreen Woodland, Great Basin Conifer Woodland, Petran Montane Conifer Forest; Apache, Cochise, Coconino, Gila, Graham, Greenlee, Maricopa, Mohave, Navajo, Pima, Pinal, Santa Cruz, Yavapai cos.; Colorado, New Mexico, Utah; Mexico (Chihuahua, Durango, Sonora).

Acmispon wrightii is widely distributed at middle to higher elevations in the state. Lower elevation records are mostly from riparian habitats. Flowering occurs during warmer months in response to moisture. Gambel's quail and scaled quail eat the seeds of *A. wrightii* (Graham 1941). Feeding experiments in Arizona have shown that the herbage is highly palatable to both mule deer and white-tailed deer (Nichol 1938). The species may have potential for erosion control and is tolerant of grazing (Graham 1941).

AESCHYNOMENE Linnaeus. Sp. Pl. 2:713. 1753. * [from Greek, *aischynomeno*, modest, sensitive, referring to the leaves which are sensitive to touch and temperature] JOINTVETCH

Matthew B. Johnson

Herbs, shrubs (climbing vines or trees); pubescence of simple hairs and bulbous-based, viscid hairs or bristles. **Stems** diverse. **Leaves** alternate or loosely clustered, even- or odd-pinnate; stipules attached basally or on underside of blade, membranous, herbaceous or leathery, persistent or deciduous; petiolate; stipels absent. **Leaflets** alternate or subopposite, subsessile, asymmetric at base, mucronate, margins entire (or serrated). **Inflorescences** racemes or panicles (or fascicled) or flowers solitary, axillary (or terminal); bracts resembling stipules, persistent or deciduous; bracteoles paired at base of calyx, appressed, deciduous (or persistent). **Pedicellate. Flowers** papilionoid, small but often conspicuous; calyx either bell-shaped with 5 subequal teeth or 2-lipped with the upper (adaxial) lip 2-lobed and the lower (abaxial) lip 3-lobed; corolla yellow, orange, or rarely red or purplish, petals short-clawed; banner often rounded and bent strongly outward; wings nearly as long as the banner, obliquely oblong or obovate with an appendage at one side of the base, free or slightly fused; keel falcate or strongly bent, fused; stamens 10, usually in 2 groups of 5 each, filaments fused for at least half their length, anthers all alike; ovary with 2–many ovules; style glabrous; stigma simple, head-like. **Fruits** loments, subsessile or stipitate, compressed, with 1–18 segments, upper (adaxial) margin straight and lower (abaxial) margin deeply notched or not, segments rectangular or "D"-shaped, separating, indehiscent or dehiscent along lower (abaxial) suture. **Seeds** 1 per fruit segment; kidney-shaped, smooth.

Species ca. 175 (1 native in the flora): tropical and warm-temperate regions of North and South America, Africa, and Asia, with one species extending into Australia.

Various species of *Aeschynomene* are used for livestock fodder, green manure, and as ornamentals (Allen and Allen 1981, Lewis et al. 2005, Rudd 1955). *Aeschynomene fluitans* is eaten as a vegetable (Allen and Allen, 1981). *Aeschynomene aspera* in Asia and *A. elaphroxylon* in Africa are sources of ambatch wood which is used for pith to make paper and pith helmets, in artificial flowers and handicrafts, and for floats and canoes (Allen and Allen 1981, Lewis et al. 2005). *Aeschynomene elaphroxylon* is also used to fill in inundated areas due to its profuse production of adventitious roots that trap and accumulate other vegetative material (Allen and Allen 1981). Wood of *A. hispida* is reported to have a density of only 40 kg/m^3, which is a third that of balsa wood (Allen and Allen 1981). Nodulation is widely reported for *Aeschynomene*. Some species have the unique trait of having both root and stem nodules (Allen and Allen 1981).

1. Aeschynomene villosa Poiret in Lamarck, Encycl. Suppl. 4(1):76. 1816. * [from Latin, *villosus*, with long, soft hairs] HAIRY JOINTVETCH

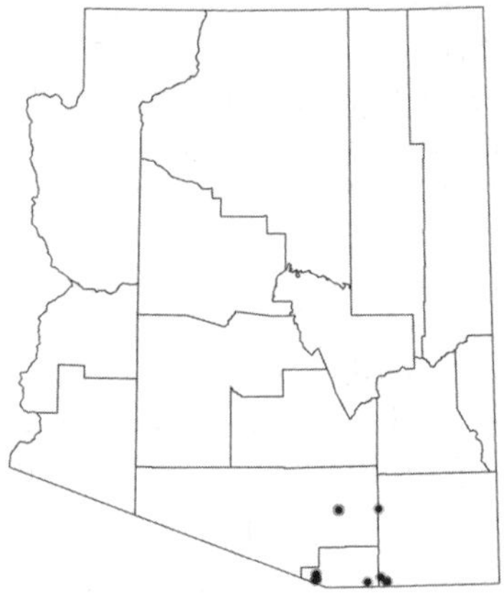

Annual or short-lived perennial herbs or becoming woody basally to 0.6 (1.2) m. **Stems** solitary or with several basal branches, delicate or robust, prostrate, decumbent or erect, hispid. **Leaves** alternate, 1.2–7 cm, even-pinnate; stipules 5–15 x 1–1.5 mm, attached on underside of blade, lanceolate with an elongated lanceolate spur, membranous, subglabrous, margins ciliate, persistent; petiole 1–4 mm. **Leaflets** 20–50, 3–15 x 1–3 mm, alternate with the distal-most leaflet terminal at the tip of the rachis, narrowly oblong, glabrous. **Inflorescences** paniculate or fascicled, 1–6 cm, 3–10(–15) flowered, hispid; bracts heart-shaped, 1.5–6 x 1–2 mm, apex acuminate, margin ciliate; bracteoles 1–4 x 0.5–1 mm, ovate-lanceolate, margin ciliate. **Pedicels** 1–9 mm, hispid. **Flowers** 3–11 mm; calyx 2–6 mm, distinctly 2-lipped, with teeth much shorter than the lips, hispid; corolla yellow or orange-yellow, sometimes with maroon streaks on the banner, banner 5–7 mm, wings 5–7 mm, keel 4–5 mm; stamens 4–7 mm. **Fruits** with a stipe 1–3 mm, with 2–6(–8) segments, segments straw-colored or pale brown, 3–4 x 2.5–4 mm, "D"-shaped with distinct notches between segments or rectangular without notches between segments, villous-hispid, enlarged bases of hairs often dark and conspicuous, margins sometimes separating from segments. **Seeds** black, 2–2.5 x 1.5–2 mm.

Varieties 3 (1 native in the flora): widespread from Arizona, Mexico, Central America, the Caribbean, and northern South America; introduced in Australia and perhaps elsewhere.

1a. Aeschynomene villosa Poiret in Lamarck var. **villosa** Native

Aeschynomene americana Linnaeus var. *villosa* (Poiret) Urban

Annual [or short-lived perennial] herbs 0.1–0.5[–1.2] m. **Stems** delicate [or robust], [prostrate or] erect. **Leaves** 1.2–4.5[–5.5] cm. **Leaflets** 4–5 x 1–1.5 mm. **Inflorescences** 1–4 cm. **Pedicels** [1–]3–5 mm. **Flowers** (3–)5(–8) mm; calyx (2.5–)3.5–4.5 mm. **Fruits** with a stipe 1–2 mm, with (2–)3–4(–5)[–7] segments, the segments 3–4 x 2.5–3.5 mm, "D"-shaped with distinct notches between segments.

Flowering Aug–Sep[–Oct]. Canyons, cienegas, and along streams in moist to dry soil; [100–]1100–1740[–2250] m; Madrean Evergreen Woodland (and Plains Grassland); Cochise, Pima, Santa Cruz cos.; Mexico (widespread from Sonora and Sinaloa to Chiapas and Vera Cruz); Central America, the Caribbean, northern South America.

Aeschynomene villosa var. *villosa* is widespread in many habitats and vegetation types across its range. In Arizona, it occurs in a few isolated populations near the International Boundary where it may be locally abundant, and it has been recently discovered in the Rincon Mountains (*D. Beckman 202*, 29 Oct 2021). The Arizona plants appear to be annuals and have slender, erect stems. Southward in more tropical habitats, *A. villosa* var. *villosa* can become more robust and can be perennial. In Sonora, this variety is found in tropical deciduous forest. *Aeschynomene villosa* has potential as a forage crop for subtropical and tropical regions with sufficient moisture. Roots of *A. villosa* var. *villosa* from Arizona are heavily nodulated.

ALBIZIA Durazzini. Mag. Tosc. 3(4):13–14. 1772. * [for Filippo degli Albizzi, 18th century Italian naturalist]
Lourdes Rico-Arce

Trees, shrubs, usually unarmed, strigillose. **Stems** usually straight, young growth with white puberulous to tomentose indument. **Leaves** alternate, bipinnate; not sensitive to the touch; stipules not spinescent, very early deciduous; petiole, rachis, and pinnae often with extrafloral nectaries. **Leaflets** opposite, usually asymmetric, the terminal pair heteromorphic. **Inflorescences** corymbs, axillary or congested, whitish, pink to yellow-green; 3 or 4 bracts, deltoid, (5–)7 mm, densely strigillose. **Flowers** mimosoid, dimorphic, in numerous species, central flowers larger than the peripheral, 5–7–8-merous, sessile; peripheral flowers small, 5-merous, calyx and corolla fused, valvate, stamens many, united into a tube (fused proximally, free distally), usually long exserted, anthers without glands. **Fruit** dehiscent or tardily dehiscent, oblong, straight, never twisted or curved, flat, margins slightly thickened, occasionally winged, valves membranous to leathery; endocarp ochre and without septa between seeds. **Seeds** 8–20 seeds per fruit; elliptic in outline, strongly biconvex, with pleurogram; aril and endosperm absent; cotyledons large, radicle curved. $x = 13$.

Species 120–140 (1 cultivated in the flora): in Africa there are ca. 36 endemic species and in the Neotropics 22. For North America, 4 are native or naturalized.

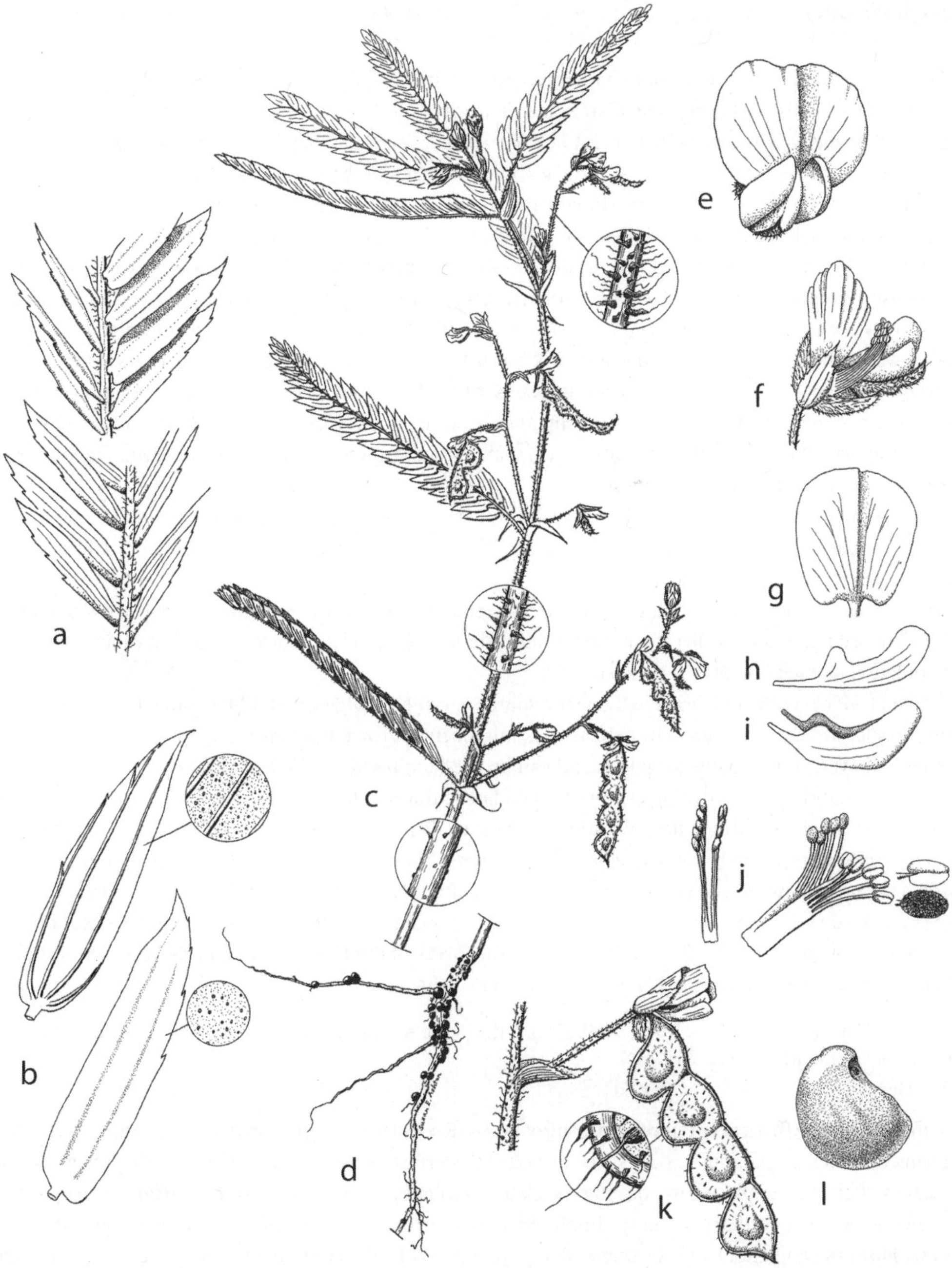

***Aeschynomene villosa*: a. leaflets; b. leaflet detail; c. habit; d. roots with nodules; e. flower; f. flower detail; g., h., i. petals; j. stamen detail; k. fruit; l. seed. Illustrator: Susan E. Ashton.**

Albizia is a pantropical genus that includes at least 470 names. Lewis and Rico Arce (2005) give a figure between 120–140 species. The type species for the genus is *Albizia julibrissin* Durazzini.

Various species of *Albizia* are used for lumber, fuelwood, livestock forage, human food, medicine, gums, tannins, paper-making, inks, dyes, soap, fish poison, insecticide, as ornamental trees, for shading crops, and for reforestation (Allen and Allen 1981, Lewis et al. 2005). *Albizia lebbeck* and *A. niopoides* are widely planted for timber production in tropical regions (NRC 1979, Rico Arce et al. 2008). Wood durability and quality varies greatly depending on the species (Allen and Allen 1981). Nodulation is widely reported for the genus (Rico Arce et al. 2008).

1. Albizia julibrissin Durazzini. Mag. Tosc. 3(4):11. 1772. * [corruption of the Persian word *gul-i abrisham*, meaning silk flower] SILK TREE, MIMOSA Cultivated

Trees to 6 [12] m. **Stems** with light, smooth bark, becoming slightly fissured with age. **Leaves** 20–32 cm; pinnae (4–)5–11 pairs, opposite, 5.5–10 cm; stipules linear lanceolate, 2 mm; petiole 3–7.5 cm with a flat, proximal or nearly so elliptic gland, glabrescent; rachis with a gland between 1 or 2 outermost (distal) pair of pinnae. **Leaflets** (13–)17–22(–36) pairs, opposite, oblong, strongly asymmetric, 0.7–1.5 cm, glabrous or nearly so, base truncate, apex acute with a small mucro, palmate venation, midvein marginal or submarginal, terminal pair of leaflets inconspicuously heteromorphic. **Inflorescences** terminal and axillary, axis densely strigillose; peduncles 3–9 cm long, densely strigillose, with a bract at the base, linear lanceolate, 2(–5) mm; flower heads 2.5–4 cm diameter, ca. 22 flowers. **Flowers** radially symmetrical, inconspicuously dimorphic; peripheral flowers with pedicels to 3 mm, strigillose, small bract at the base, quickly deciduous, 1 mm, calyx campanulate, 6 mm, 5-lobate, estrigillose towards the apex, corolla campanulate, to 12 mm, 5-lobate, strigillose towards the apex, stamens white and pink towards the apex, tube to 12 mm long, ovary 3 mm, sessile, glabrous; terminal or central flowers, sessile, without bracts, calyx 3–4.5 mm, glabrous, corolla 9–11 mm, staminal tube long exserted, to 24 mm. **Fruit** 10–20 × 1.5–2.6 cm, margin straight or slightly constricted, apex rounded with a beak 14–20 mm long, base rounded, valves membranous, glabrescent, brown or dark reddish brown, smooth or slightly rough. **Seeds** 6–8(–10) per fruit, 9 × 5 × 2 mm.

Flowering summer. Widely cultivated across the southeastern U.S. and as far north as New York and Illinois, and less frequently in the western U.S.; occasionally naturalizing in humid regions; native to Azerbaijan, China, India, Iran, Japan, Korea, Nepal, Taiwan; introduced in Greece.

Albizia julibrissin is used in Asian folk medicine. Research has demonstrated antidepressant activity in mice (Kim et al. 2007). The seeds are eaten by livestock and wildlife. In Asia, the wood is reportedly used for cabinetmaking (Vines 1960). Nodulation in *A. julibrissin* is widely reported.

Silk tree is widely cultivated, especially in the southeastern U.S. Diggs et al. (1999) state that it can become an aggressive invader in natural habitats. In Arizona it is infrequently grown for its tropical appearance and showy pink "powder-puff" flowers. It is suitable for patios and can be used to provide shade for smaller plants. Silk tree typically develops a low, dome-shaped, spreading canopy. Several horticultural selections have been developed. Silk tree can be grown in hot desert regions and at middle elevations in the state. Plants require regular irrigation to maintain health and appearance. This species exhibits considerable cold hardiness. Flowers of silk tree attract bees, butterflies, and hummingbirds. The leaves, flowers, and pods create considerable litter. The brittle wood is subject to storm damage and the trees are short-lived (Jones and Sacamano 2000). Silk tree is widely reported to be susceptible to fatal infections of Fusarium wilt. Though documented to be invasive in other areas of the U.S., *Albizia julibrissin* presents a low risk of naturalizing in Arizona. Plants are readily propagated from scarified seeds.

ALHAGI Gagnebin. Acta Helv. Phys.-Math. 2:59. 1755. * [from the Arabic *al-hajji/al-haggi*, pilgrim, said to be the Arabic name for the plant] CAMEL-THORN

Richard R. Halse

Shrub, subshrubs or suffrutescent perennial, spiny herbs. **Roots** rhizomatous, branched, creeping, deep-set. **Stems** simple to much-branched, glabrous to pubescent, armed with sterile thorn-tipped branchlets, these subtended by leaves or not. **Leaves** alternate, simple, entire, deciduous, glabrous to pubescent; stipules small. **Inflorescences** with few to several flowers; racemes, axillary, usually simple; bracteoles minute, at base of calyx; rachis rigid, thorn-tipped. **Pedicellate**. **Flowers** papilionoid; calyx campanulate, barely to evidently short lobed, lobes subequal, shorter than the calyx tube, glabrous to pubescent; corolla reddish, petals free; stamens 10, 9 united and 1 free; pistil: style thread-like, glabrous; stigma terminal. **Fruits** indehiscent, linear to curved, moniliform to torulose, resembling loments, short-stipitate, glabrous to pubescent. **Seeds** few to several, reniform to quadrangular. $x = 8$.

Species 1–5 (1 naturalized in the flora): Mediterranean Basin of northeastern Africa, southeastern Europe to Central Asia, India, China, Mongolia; naturalized in Australia, South Africa, and North America.

The number of species recognized in *Alhagi* is subject to interpretation based on robustness of growth; calyx tooth development; fruit shape; and pubescence of fruits, calyces, stems, and leaves. Shaparenko (1948/1972) and Rechinger (1984) recognized five species while Yakovlev (1979) recognized one variable species with several subspecies and varieties.

Species of *Alhagi* can serve as fodder for camels and other animals in arid areas. In the summer some species exude a sweet gummy material on the stems and branches that hardens into reddish brown lumps. These are collected by Arab

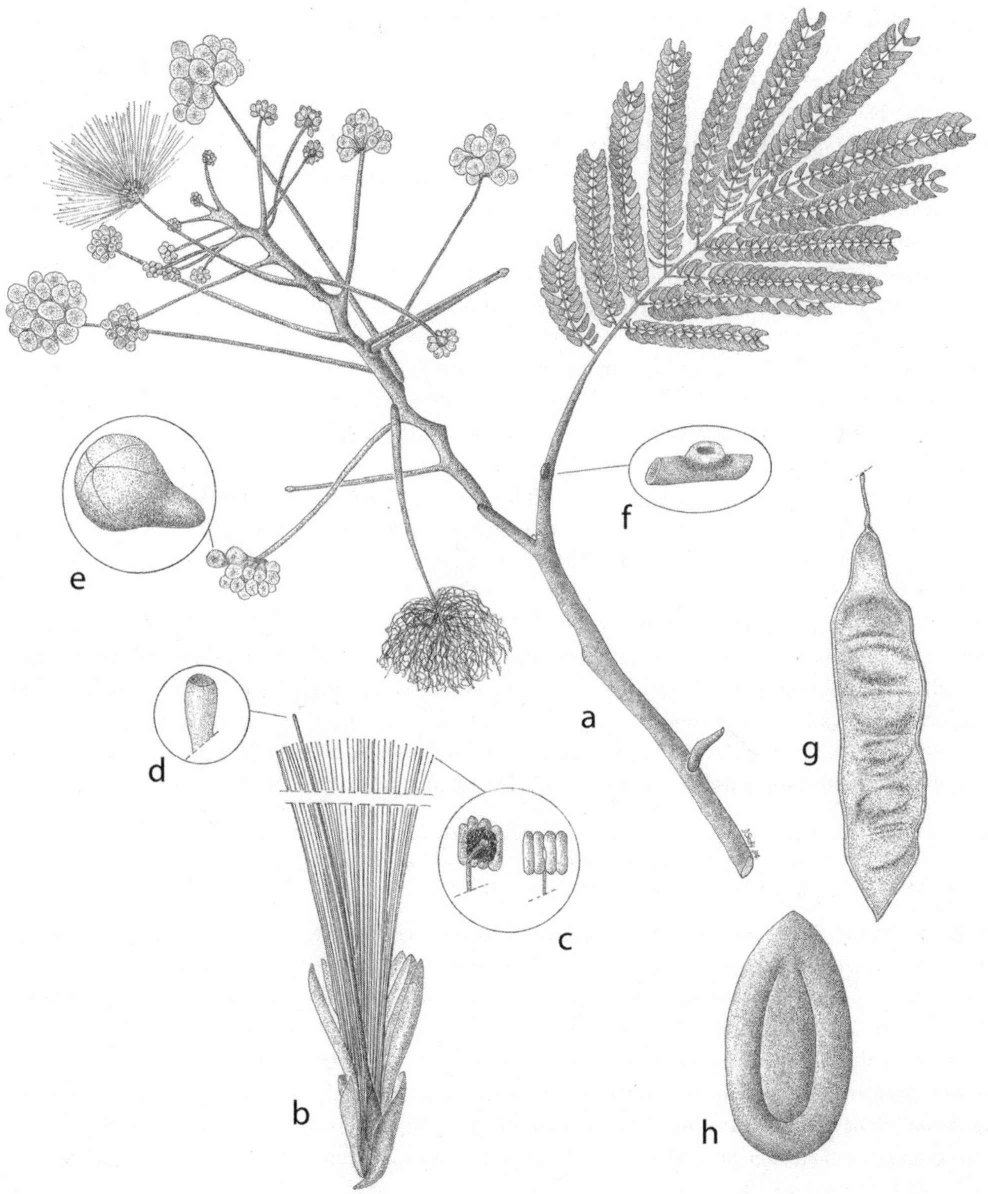

***Albizia julibrissin*: a. habit; b. flower; c. anther; d. stigma; e. flower bud; f. petiolar extrafloral nectary; g. fruit; h. seed. Illustrator: Daniela Siroky.**

nomads and used as a delicacy; they have also been used medicinally. Camel-thorn is sometimes grown in botanical gardens as a curiosity. The plants can be serious weeds in both their native and introduced habitats.

1. Alhagi maurorum Medikus. Vorles Churpfälz. Phys.-Öcon. Ges. 2:397. 1787. * CAMEL-THORN, PERSIAN MANNA, CASPIAN MANNA Naturalized

Hedysarum alhagi Linnaeus; *Alhagi pseudalhagi* (M. Bieberstein) Fischer; *Alhagi camelorum* Fischer

Subshrubs to suffrutescent herbs, 0.3–2 m. **Roots** extensively creeping, to 2.5 m. **Stems** often intricately branched, greenish, strigose to glabrate, thorn-tipped, branchlets 1–5 cm. **Leaves:** blades linear, elliptic, oblanceolate to narrowly obovate or oblong, 3–35 mm long, 1–12 mm wide, usually minutely red dotted, glabrous to pubescent; stipules 0.5–3 mm, subulate to lance-ovate, caducous; petioles short. **Inflorescences** (1–)2–7-flowered**;** racemes, axis prolonged as a thorn; bracts 0.5–1 mm, subulate. **Pedicels** 1–2 mm. **Flowers** calyx 2–3.2 mm, barely to evidently lobed, teeth low, broad, 0.2–0.5 mm, glabrous, margins sometimes puberulent, persistent; corolla 7–9.5 mm, pink to reddish purple.

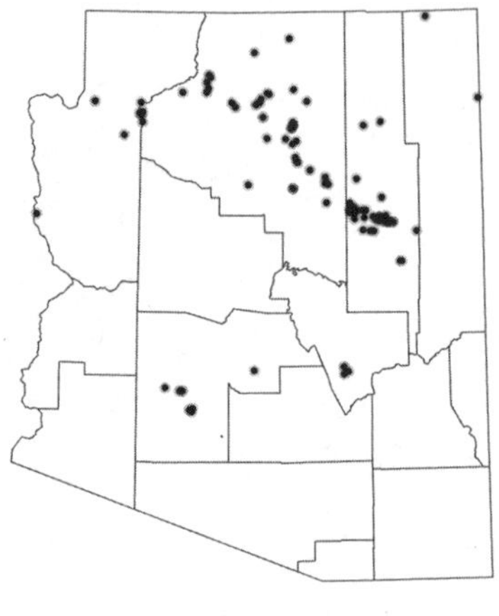

Fruits 0.5–3 cm, glabrous, stipe about as long as calyx, pods usually breaking apart between seeds. **Seeds** (1–)2–8(–10), yellowish or greenish brown to dark brown.

Flowering May–Sep(–Oct); fruiting Jul–Oct. Roadsides, ditches, disturbed agricultural areas, dunes, slopes, arroyo channels, sandy areas along rivers, alkaline meadows, playas; [190–]230–1720[–2350] m; Sonoran Desertscrub, Mohave Desertscrub, Great Basin Desertscrub and riparian communities within these biotic communities; Apache, Coconino, Gila, Maricopa, Mohave, Navajo Cos; naturalized in California, Colorado, Nevada, New Mexico, Texas, Utah, Washington; widely introduced and often naturalizing in dry regions; native to southwest Asia and the eastern Mediterranean region.

Stems and leaves of North American plants range from glabrous to densely pubescent. Young growth is almost always densely strigose; older stems and leaves may be essentially glabrous. The calyx and fruit are glabrous.

For a discussion of the choice of a specific epithet for the plants growing in North America see Isely (1998) and Barneby (1989). Awmack and Lock (2002) have divided *Alhagi maurorum* into two subspecies based in part on the pubescence of the ovary and shape of the calyx. These traits are quite variable. North American plants would belong to subsp. *maurorum*.

In Arizona, *Alhagi maurorum* is primarily found along the drainage of the Little Colorado River sporadically downstream into the Grand Canyon, but also occurs elsewhere in the state. The plants form extensive colonies and are difficult to eradicate once established. Stems have been observed emerging through asphalt pavement. Allen and Allen (1981) report nodulation for *A. maurorum*.

Camel-thorn is a state listed weed in Arizona, California, Colorado, Nevada, New Mexico, Oregon, Texas, and Washington. Eradication programs have eliminated most populations in California (DiTomaso and Healy 2007).

AMORPHA Linnaeus. Sp. Pl. 2:713. 1753. * [from Greek, *amorpho*, shapeless or deformed, in reference to the corolla which lacks wing and keel petals] FALSE INDIGO, LEAD PLANT

Michelle M. McMahon, Mary Welch-Keesey, and Matthew B. Johnson

Subshrubs or shrubs. Stems and herbage pubescent, with translucent or pale glands, pungent when bruised. **Leaves** alternate, odd-pinnate; stipules linear to awl-shaped, deciduous; petiolate; stipels linear to bristle-like, persistent. **Leaflets** (7–)11–29(–45), often irregularly arranged, entire to crenulate, with petiolules and stipels. **Inflorescences** spike-like racemes appearing terminal or forming racemose panicles; bracts present or absent. **Pedicellate. Flowers** nonpapilionoid; calyx obconic to campanulate, with 5 lobes; corolla with banner petal only, blue, white, red-purple or black-purple; stamens 10, filaments partially or mostly united into a single tube, united proximally and free distally, exserted from banner, anthers attached dorsally, subuniform, yellow or orange; ovary with 2 ovules; style sinuous, exserted beyond calyx and often beyond petal. **Fruits** obliquely obovoid to curved, subcompressed, indehiscent. **Seeds** 1 (or 2), oblong or curved. $x = 10$.

Species ca. 16 (2 native in the flora, 1 of these also cultivated): subtropical to cool temperate regions of North America from southern Canada to northern Mexico; most diverse in the southeastern U.S.

An analysis of DNA confirms the separation of our two species but suggests many other issues within the genus: several species may nest within, or experience ongoing gene flow with, *Amorpha fruticosa*; and *Amorpha* may not be monophyletic without the inclusion of *Errazurizia rotundata* and *Parryella filifolia* (Straub and Doyle 2014).

Species of *Amorpha* have a variety of uses including planting for erosion control and as windbreaks, cover crops, and ornamentals (Graham 1941, Lewis et al. 2005). Leaves of some species have been used as a substitute for tea and tobacco, in folk medicine, as an insect repellant, and in dyeing (Lewis et al. 2005). Nodulation is reported for several species in the genus (Allen and Allen 1981).

SELECTED REFERENCES: Isely 1998, Wilbur 1975.

1. Branches and leaf rachises with prickle-like glands ______ **1. Amorpha californica**
1′ Branches and leaf rachises lacking prickle-like glands ______ **2. Amorpha fruticosa**

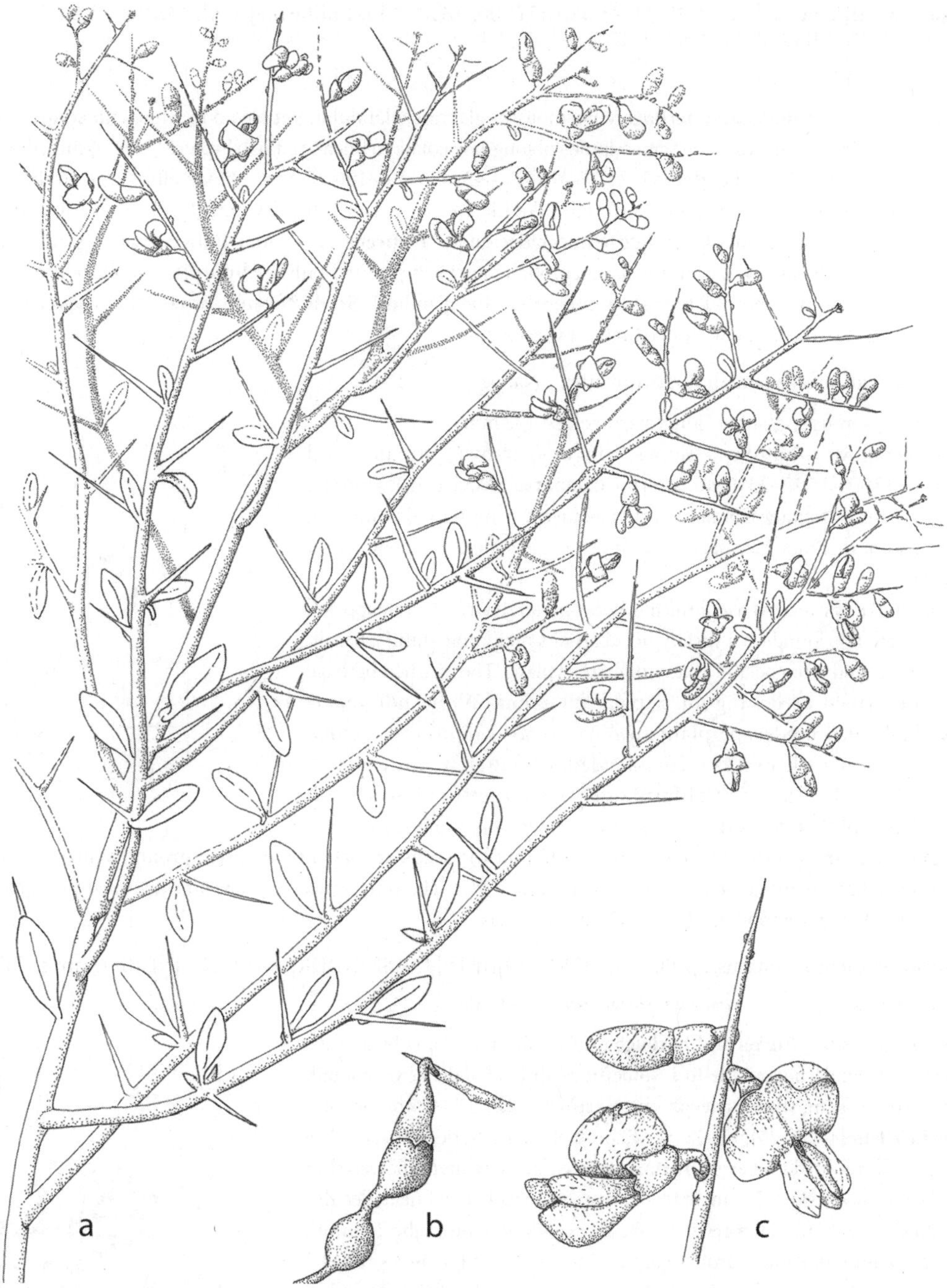

Alhagi maurorum: **a.** habit; **b.** fruit; **c.** flowers. Illustrator: Christine Hubbell.

1. Amorpha californica Nuttall. Fl. N. Amer. 1(2):306. 1838. * [of California] CALIFORNIA FALSE INDIGO, STINKING WILLOW, MOCK LOCUST

Amorpha hispidula C.K. Schneider

Shrubs to 1.5(2.5) m. **Stems and herbage** villosulous or glabrate, glandular; petiole and rachis with conical, prickle-like and sessile glands. **Leaves** 10–20 cm; stipules oblong-lanceolate to linear, reddish brown, (2–)4–6 mm, deciduous; petioles (0.5–)1–1.5(–2.5) cm. **Leaflets** 11–25, (8–)15–25(–42) × (6–)10–15(–20) mm, oblong-elliptic, green above, pallid beneath. **Inflorescences** densely many-flowered spike-like racemes, axis 5–10(–20) cm, glands prickle-like (occasionally also sessile); bracts narrowly lanceolate to linear, deciduous. **Pedicels** 0.3–1.2 mm. **Flowers** with obconic calyx 4–5 mm, tube 2–3 mm, lobes 0.5–2 mm, unequal; corolla (comprising banner only) red-purple, 5–7 × 2.5–4 mm. **Fruits** 6–8 × 3–4 mm, obliquely obovoid, the valves puberulent and glandular. **Seeds** 3–4 mm, brown or olive-green, smooth.

Varieties 2 (1 in the flora): w United States, nw Mexico.

1a. Amorpha californica Nuttall var. **californica** * Native

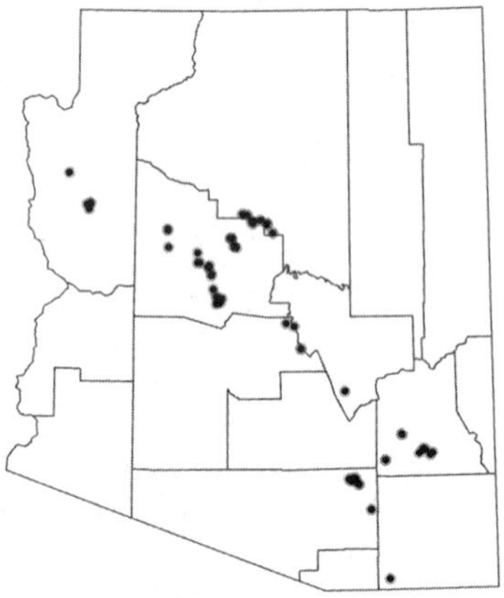

Herbage not glabrous or glabrate; longest calyx lobe 1–2 mm.

Flowering May–Jul(–Sep). Slopes, canyons, arroyos, streams, and roadsides; (1380–)1500–2350 m; Interior Chaparral, Madrean Evergreen Woodland, Petran Montane Conifer Forest; Cochise, Coconino, Gila, Graham, Maricopa, Mohave, Pima, Yavapai cos.; California, New Mexico; Mexico (Baja California).

Amorpha californica var. *californica* is widespread in Arizona but appears to be localized. It is found on a variety of substrates including granite, basalt, and sandstone, and often occurs as an understory plant. The crushed herbage is often described as having an unpleasant scent. While not generally available as a landscape plant in Arizona, *A. californica* is cultivated in California where it is recommended for butterfly gardens. It is the larval food plant for the California dogface butterfly (*Zerene eurydice*). Graham (1941) reports that the species is palatable to mule deer and white-tailed deer in experiments and also reports that the seeds are eaten by mantled ground squirrels. *Amorpha californica* has been planted for windbreaks (Graham 1941). The other variety, *Amorpha californica* var. *napensis*, is narrowly distributed west and north of San Franscisco Bay.

2. Amorpha fruticosa Linnaeus. Sp. Pl. 2:713. 1753. * [shrubby] FALSE INDIGO, BASTARD INDIGO Native, Cultivated

Amorpha fruticosa Linnaeus var. *occidentalis* (Abrams) Kearney & Peebles

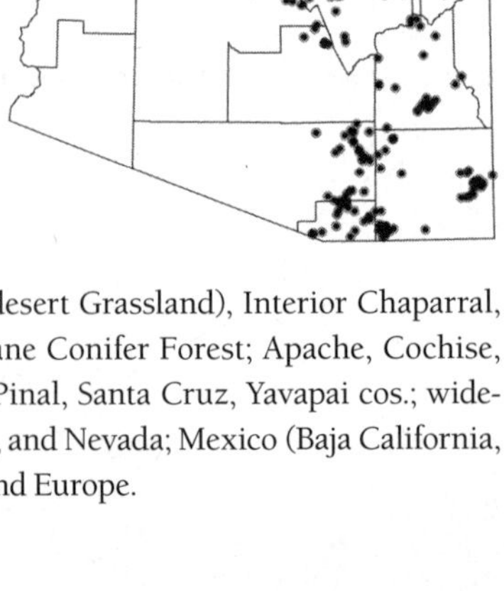

Shrubs to 3 [6] m tall. **Stems and herbage** finely puberulent or subglabrous, with scattered glands, not prickle-like, sometimes obscure, if dense very small. **Leaves** 7–20(–40) cm; stipules deciduous, narrowly linear, 2–4 mm; petioles 1–4 cm. **Leaflets** 11–21(–25), 15–42 × 7–23 mm, oblong-elliptic, ovate, or obovate, green above, pallid beneath. **Inflorescences** densely many-flowered in spike-like racemes, axis 5–20 cm; bracts narrowly lanceolate to linear, deciduous. **Pedicels** 1–2.2 mm. **Flowers** with obconic calyx 3–5 mm, tube 2–4 mm, lobes 0.2–1.2 mm, unequal; corolla (comprising banner only), 5–6 × 3.5–4.2 mm, dark red-purple to indigo. **Fruits** 5–9 × 1.5–4 mm, obliquely obovoid or incurved, glabrous or puberulent and usually glandular. **Seeds** 4–5 mm, pale green or tan, smooth.

Flowering (Apr–)May–Jul(–Sep). Generally in canyons in mountainous areas and usually along permanent or intermittent streams, drainage ditches, man-made ponds, and occasionally on moist slopes; (590–)900–2290 m; riparian communities within (upper elevations in Sonoran Desertscrub, Semidesert Grassland), Interior Chaparral, Madrean Evergreen Woodland, Great Basin Conifer Woodland, Petran Montane Conifer Forest; Apache, Cochise, Coconino, Gila, Graham, Greenlee, La Paz, Maricopa, Mohave, Navajo, Pima, Pinal, Santa Cruz, Yavapai cos.; widespread and reported for all states in the Continental U.S. except Alaska, Montana, and Nevada; Mexico (Baja California, Chihuahua, Sonora); Canada (Manitoba, Ontario, Quebec). Introduced in Asia and Europe.

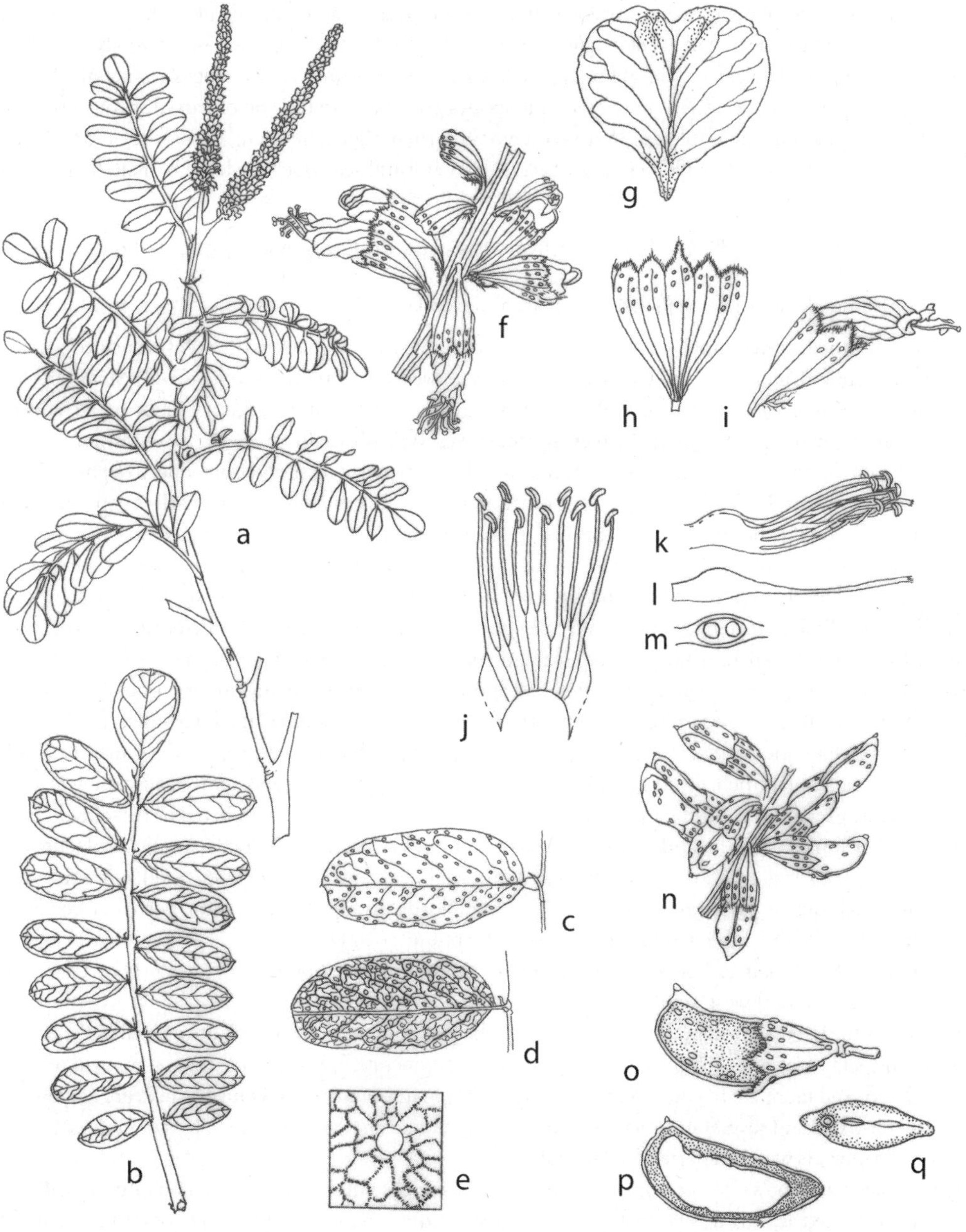

***Amorpha fruticosa*: a. habit; b. leaf; c., d. leaflets; e. leaflet surface detail; f. inflorescence section; g. banner petal; h. calyx; i. flower; j. anthers; k. anthers; l. pistil; m. ovary; n. fruits; o. fruit; p. fruit cut-away; q. seed. Illustrator: Margaret Pope.**

Amorpha fruticosa is recorded for all counties in Arizona except for Yuma County. The species is widespread and can be common or occur as isolated individuals. It is usually confined to riparian communities as an understory plant.

Burlage (1968) reports that *Amorpha fruticosa* contains alkaloids and is toxic to livestock. Compounds in *A. fruticosa* are toxic to several insect pests and repel cattle flies (Brett 1946, Roark 1947). *Amorpha fruticosa* has been extensively planted as windbreaks and is valued in Texas for erosion control plantings where roads cross water courses (Graham 1941). Graham (1941) reports that the species is an important nectar source for honey bees in the Midwestern U.S. and that bobwhite quail eat the seeds sparingly. Nodulation is widely reported for *A. fruticosa* (Allen and Allen 1981).

False indigo is occasionally grown in Arizona landscapes, especially as a butterfly plant. It is ideal for patios and other small spaces, and can be grown at colder, high elevation sites in the state than most other woody species of native legumes. The dark purple flowers with bright orange yellow anthers are attractive. It is suitable as an informal hedge or screen, though the plants are deciduous in winter. This species grows well in full sun or partial shade. It requires periodic irrigation and is not drought tolerant in desert regions. False indigo is hardy to at least -20° C and will survive extended periods of below-freezing temperatures. Several horticultural selections are known, including a white-flowered form, though these are generally not available in Arizona. Propagation is by scarified seeds.

ASTRAGALUS Linnaeus. Sp. Pl. 2:755. 1753. * [from the classical Greek word for ankle-bone] LOCOWEED, MILKVETCH

Martin F. Wojciechowski, Matthew B. Johnson, and Jason Alexander

Herbs, annual, biennial, or perennial, (sometimes partially woody); caulescent or acaulescent; unarmed (some species with spinescent petioles); hairs attached at base (basifixed) or above base, near middle (medifixed). **Stems** erect, ascending, prostrate, or decumbent; arising directly from a root-crown or from a caudex, this aerial, or subterranean, sometimes rhizomatous; usually pubescent. **Leaves** alternate, sessile, subsessile, or petiolate; commonly compound (odd-pinnate) with 3 to many petiolulate leaflets, rarely unifoliolate (appearing simple), or reduced to phyllodes; stipules present, fused to petiole or distinct, or fused to each other opposite petiole; stipels absent. **Leaflets** petiolulate or sessile; margins entire, surfaces glabrous or pubescent. **Inflorescences** with 1–95(–125) flowers; axillary or rarely terminal racemes, (sometimes spike-like, subumbellate, or solitary); bracts present, often small, membranous, subtending pedicels; bracteoles 0–2, borne on calyx base. **Pedicellate** or rarely pedicels obsolete. **Flowers** papilionoid; calyx campanulate, cylindric, subcylindric, turbinate, or obconic, lobes 5, lobes ± equal; corolla violet, blue-purple, purple, red-purple, pink-purple, pink, lilac, reddish, white, yellow, cream, ochroleucous, or greenish, petals mostly long-clawed, banner erect, wings oblong, keel suberect; stamens 10, 9 united and 1 free, anthers generally uniform or of 2 different lengths; ovary sessile or stipitate; style straight or incurved, threadlike, glabrous; stigma small, terminal; ovules 2–75(–84). **Fruits** highly diverse in orientation, size, shape, compression, texture, and dehiscence; sessile, sometimes with gynophore, subsessile, or stipitate; dehiscent throughout or apically, deciduous from calyx or from abscission layer at gynophore apex and dehiscent on ground, or persistent and dehiscent on plant; one-chambered or partially or completely two-chambered by intrusion of the dorsal (abaxial) suture, compressed laterally or dorsiventrally, to bladdery-inflated, usually straight or incurved, rarely decurved or curved inward, linear, lanceolate, oblong, ovoid, ellipsoid, circular, or almost circular, glabrous or pubescent. **Seeds** 2–75(–84); reniform, arranged in two rows, borne on slender funicles sometimes fused basally with funicular flange simulating a septum. $x = 8, 11, 12$.

Species ca. 3000 (79 native, 1 adventive in the flora): North America, South America, Europe, Asia, Africa. The genus is most diverse in southwest Asia.

Astragalus is the largest genus in the Fabaceae and the largest genus of flowering plants, as well as the largest genus in Arizona. The genus occupies a diverse array of habitats in temperate to warm temperate, semi-arid to arid shrub, woodland, desert, and montane regions. Many species are xerophytic. Numerous *Astragalus* taxa are narrow endemics to particular substrates and several Arizona taxa are of conservation concern. Nodulation by symbiotic nitrogen-fixing soil bacteria (rhizobia) is widely reported for *Astragalus*.

Some astragali are known to accumulate selenium and have potential in bioremediation of mine tailings. These species are documented as causing livestock poisoning. Species that contain selenium are reported to be malodorous, having a strong garlic-like or sulphurous odor. Various species contain toxic compounds other than selenium that can cause poisoning or "locoism" in livestock. The effects of locoism vary and the poisoning is cumulative, with some animals becoming addicted to eating the plants (Allen and Allen 1981). Other astragali are grown for pasture forage: *Astragalus chinensis*, *A. cicer*, *A. crassicarpus*, *A. falcatus*, and *A. sinicus* are considered to be among the most palatable forage species (Allen and Allen 1981).

The use of *Astragalus* species for geobotanical prospecting, using selected species in the search for minerals, is well documented. Native Arizona species that have been used for this purpose include *A. albulus*, *A. beathii*, *A. bisulcatus*, *A. crotalariae*, *A. flavus*, *A. moencoppensis*, *A. pattersonii*, *A. preussii*, and *A. toanus* (Cannon 1971). All of these are known selenium accumulators. Allen and Allen (1981) report that *A. pattersonii* and *A. preussii* have been used successfully in the search for uranium deposits. Other astragali have been used elsewhere in the world for geobotanical prospecting.

Several shrubby Old World species of *Astragalus* are the source of gum tragacanth. The gum is exuded from cuts made to the stems of the plants. Gum tragacanth has been used medicinally since ancient times and is used today for a variety of purposes including sizing textiles and paper, as a finish for rope and twine, in waterproofing and stiffening felt, and as an ingredient in cosmetics including shaving creams, lotions, nail polish, and skin and hair products (Allen and Allen 1981). Duke (1981) states that the gum is used to give body to liqueurs, as an adhesive agent for a variety of products, and as a pharmacy demulcent. The gum has been used as an emulsifier in numerous food products. In addition to the pharmacological uses just noted, gum tragacanth is used in folk medicine as a laxative as well as to treat coughing and diarrhea (Duke 1981).

The green fruits of a few species including *Astragalus crassicarpus* and *A. mexicanus* are considered suitable as vegetables for human consumption. *Astragalus cicer*, which is reported to be weakly naturalized in Arizona, and several other astragali are suitable for erosion control (Allen and Allen 1981, Duke 1981). Several species are suitable for wildflower gardens. Perhaps the most spectacular is *A. coccineus*, with its showy display of red flowers and contrasting white-hairy leaves.

Astragalus includes many taxa that exhibit considerable morphological variability, and there can be significant overlap in characters between some similar taxa. Both flowers and fruits are frequently necessary to identify a specimen, so it may be necessary to follow both halves of a couplet if characters overlap or if flowers or fruits are not present. For species with varieties that do not occur in Arizona, characters used in this key are for Arizona varieties only.

Fruit shape characters in the keys and descriptions (i.e. ellipsoid, ovate, oblong, etc.) refer to the shape of unpressed-pods in longitudinal section. In some species, the pods retain a characteristic shape even when pressed. Fruit compression characters—dorsiventrally compressed (obcompressed), laterally compressed, 3-sided compressed—refer to the shape of unpressed-pods in cross-section. Dorsiventrally compressed pods are wider than tall in cross-section and can result in a more or less rectangular cross section. The term cylindrical indicates a pod rounded in cross-section, without compression. The degree of compression of the pod in cross-section is a diagnostic fruit character and is used in the key for many taxa; unfortunately, it is not always discernable in specimens with papery-walled or immature fruits.

This treatment (key and descriptions) of the genus *Astragalus* for the *Legumes of Arizona* does not include several species that have been reported as collected in Arizona, but for which collections are either few or of uncertain validity (identity or location). These species include *Astragalus canadensis* L., *Astragalus chamissonis* (Vogel) Reiche, *Astragalus cliffordii* Welsh & Atwood, *Astragalus nidularius* Barneby, *Astragalus niveus* (Rydb.) Barneby, *Astragalus nyensis* Barneby, *Astragalus palmeri* A. Gray, *Astragalus phoenix* Barneby, *Astragalus piscator* Barneby & Welsh, *Astragalus sabulosus* M.E. Jones, and *Astragalus tenellus* Pursh. We encourage readers to contact the editors [MMM, MFW] or the herbaria at Arizona State University (ASU) or the University of Arizona (ARIZ) of any possible new collections of these species.

1. Plants annual or biennial; pubescence basifixed.
  2. Fruits 1-chambered (the septum always present at least proximally; narrow in *A. geyeri* var. *triquetrus*).
    3. Fruits slightly inflated but scarcely bladdery; banner recurved through 40°; seeds 3–7 ____ **7. Astragalus aridus**
    3′ Fruits bladdery inflated or turgid and somewhat inflated; banner recurved through 45–70°; seeds 7–21.
      4. Herbage villous-hirsutulous; fruits turgid or somewhat inflated, stiffly papery or leathery, hirsute, often villous ____ **63. Astragalus sabulonum**
      4′ Herbage strigillose; fruits bladdery-inflated, papery, strigose or strigillose (or subvillosulous or glabrous).
        5. Keel 6.2–7.5 mm long; inflorescences 10–20-flowered ____ **3. Astragalus allochrous** (in part)
        5′ Keel 3.8–6.1 mm long (up to 6.4 mm long in larger flowered variants of *A. wootonii*); inflorescences 2–10-flowered (or up to 15 in rare, robust variants of *A. wootonii*).
          6. Fruits prominently beaked, clearly differentiated from the body.
            7. Calyx lobes 0.6–1.5 mm; corolla whitish with lilac-tipped keel, rarely all pink-purple; restricted in Arizona to Mohave County ____ **34. Astragalus geyeri** (in part)
            7′ Calyx lobes 1.3–3 mm; corolla reddish-purple; widespread ____ **40. Astragalus insularis**
          6′ Fruits with beak obscure, not well-differentiated from the body.
            8. Calyx lobes 1.8–2 mm long, subulate; corolla with banner recurved through 80°; seeds 28–30; rare local endemic along the Little Colorado River in Coconino County ____ **26. Astragalus endopterus**
            8′ Calyx lobes 2–3.5 mm long, lanceolate-subulate; corolla with banner recurved through 45–50°; seeds 10–21; common and widespread ____ **78. Astragalus wootonii**
  2′ Fruits 2-chambered or partially 2-chambered.
    9. Fruits bladdery inflated (not inflated or scarcely inflated in *A. lentiginosus* var. *borreganus*), not 3-sided compressed (if ovoid or nearly globose then not laterally compressed and wider than 3 mm) ____ **43. Astragalus lentiginosus** (in part)
    9′ Fruits 3-sided compressed or plumply ovoid to nearly globose and somewhat laterally compressed, but not bladdery inflated.
      10. Banner recurved through 25°; seeds 2; fruits plumply ovoid or nearly globose and somewhat laterally compressed ____ **23. Astragalus didymocarpus**
      10′ Banner recurved through 40–45°; seeds 8–26; fruits 3-sided compressed.
        11. Fruits 6–12 mm wide; restricted in Arizona to Mohave County ____ **34. Astragalus geyeri** (in part)
        11′ Fruits (0.6–)2–3.7 mm wide; widespread.
          12. Fruits persistent, dehiscent while attached to the plant ____ **55. Astragalus nuttallianus**
          12′ Fruits deciduous, dehiscent after falling.
            13. Corollas whitish, the banner tinged purple or with purple veins; banner (4.7–)5–7 mm; wings 4.3–6.2 mm; keel beak-like; peduncles erect or spreading; fruits usually strigillose ____ **1. Astragalus acutirostris**

13′ Corollas pink-purple; banner (6–)7.3–11.2 mm; wings (5.1–)6.2–6.8(–10) mm; keel bluntly deltoid; peduncles reclining; fruits glabrous ____ **25. Astragalus emoryanus**

1′ Plants short- to long-lived perennial; pubescence basifixed or medifixed.

14. Leaflets with apex spinulose, all continuous with the rachis; racemes with 1–3 flowers ____ **41. Astragalus kentrophyta**

14′ Leaflets with apex not spinulose, jointed to the rachis or if some or all are continuous with the rachis, then racemes with 1–50 flowers (usually more than 3).

15. Pubescence medifixed.

16. Fruits stipitate, the stipe 0.7–3.3 mm.

17. Plants robust, bushy clumps, caudex woody at base, not subterranean; calyx tube (5.4–)5.7–7.4 mm long; terminal leaflets all jointed (distal leaves not reduced to a rachis); fruits ellipsoid or narrowly ellipsoid, dorsiventrally compressed or 3-sided at dehiscence ____ **2. Astragalus albulus**

17′ Plants slender and wiry, caudex subterranean with slender, rhizome-like branches, not woody; calyx tube 2.1–3.5 mm long; terminal leaflet of distal leaves continuous with the rachis (distal leaves often reduced to a subfiliform rachis); fruits subglobose and bladdery inflated ____ **14. Astragalus ceramicus** (in part)

16′ Fruits sessile (sometimes attached to a gynophore 0–0.8 mm in *A. sesquiflorus*).

18. Plants with stipules fused throughout or fused proximally and separate distally.

19. Seeds 33–70; fruits 14–40 mm, stiffly leathery to semi-woody, distinctly beaked.

20. Fruits readily deciduous, usually incurved; caudex, if present, weakly developed and not subterranean; widespread in Apache, Coconino, Mohave, Navajo, Yavapai counties ____ **4. Astragalus amphioxys** (in part)

20′ Fruits usually persistent, straight or slightly incurved; caudex well-developed and subterranean; localized in northern Apache County and rare in eastern Coconino County ____ **47. Astragalus missouriensis** (in part)

19′ Seeds 4–26; fruits 3.5–13 mm (6–20 mm in *A. humistratus*), papery, stiffly papery or leathery, beak obscure or weakly developed.

21. Stems decumbent, ascending, or erect; fruits dorsiventrally compressed, the body straight (or sometimes slightly curved in *A. sophoroides*).

22. Peduncles (3–)5–23 cm, usually surpassing the subtending leaf, with racemes well surpassing the leaves; pedicels 0.7–1.2 mm long; corolla deciduous, the banner 9–18 mm, the keel 6.5–10 mm long; seeds 6–17 ____ **31. Astragalus flavus**

22′ Peduncles (0.7–)1–3(–4) cm, shorter than the subtending leaf, with racemes not or scarcely projecting above the foliage; pedicels 0.5–0.7 mm long; corolla persistent around the fruits, the banner 8.5–10 mm, the keel 5.7–6.8 mm long; seeds 4–6 ____ **66. Astragalus sophoroides**

21′ Stems prostrate; fruits laterally compressed or 3-sided and laterally compressed, the body incurved.

23. Inflorescence axis (0.5–)1–9(–13) cm in fruit, racemes 3–30-flowered; seeds 6–26; fruits laterally compressed ____ **38. Astragalus humistratus**

23′ Inflorescence axis 0.1–0.8 cm in fruit, racemes 1–4-flowered; seeds 7–10; fruits 3-sided and laterally compressed ____ **65. Astragalus sesquiflorus**

18. Plants with stipules fused throughout or fused proximally and separate distally.

18′ Plants with stipules separate throughout.

24. Fruits leathery, stiffly leathery or slightly to semi-woody, distinctly beaked; seeds 18–70; calyx tube 5–14.3 mm.

25. Fruits dorsiventrally compressed, becoming laterally compressed distally, 9–20(–22) mm long; seeds 18–33; corolla white or ochroleucous, suffused with dull lilac or purple; wings 6–9.3 mm; keel 8.3–14.3 mm ____ **13. Astragalus castaneiformis**

25′ Fruits cylindrical to slightly dorsiventrally compressed (to laterally dilated and obtuse-angled when mature in *A. missouriensis* var. *missouriensis*), 15–40 mm; seeds 33–70; corolla pink-purple, purple, lavender, or violet-red, rarely white; wings 9–24(–25) mm; keel 8–24 mm long.

26. Fruits readily deciduous, usually incurved; caudex, if present, weakly developed and not subterranean; widespread in Apache, Coconino, Mohave, Navajo, Yavapai counties ____ **4. Astragalus amphioxys** (in part)

26′ Fruits usually persistent, straight or slightly incurved; caudex well-developed and subterranean; localized in northern Apache County and rare in eastern Coconino County ____ **47. Astragalus missouriensis** (in part)

24′ Fruits papery, cartilaginous or somewhat leathery, beak obscure or weakly developed; seeds 4–30; calyx tube 1.7–6.7 mm.

27. Fruits 3–7.5 mm, 1-chambered; seeds 4–11.

28. Inflorescences with 1–3 flowers; calyx tube not papery, 1.7–2.3 mm; fruits 3–4.8 mm; leaves 0.3–2(–3.5) cm; leaflets (3–)5–9(–11); seeds 4–6; local endemic on Kaibab and Coconino Plateaus, Coconino County ____ **21. Astragalus cremnophylax**

28′ Inflorescences with 7–26 flowers; calyx tube becoming papery and ruptured, 2.3–4 mm; fruits 4.5–7.5 mm; leaves (1.5–)2.5–9 cm; leaflets (7–)11–21(–25); seeds 7–11; widespread in Apache, Coconino, Greenlee, Navajo counties ____ **35. Astragalus gilensis**

27′ Fruits 8–37 mm, 2-chambered or nearly 2-chambered; seeds 13–28.

29. Plants caulescent, clump-forming; inflorescences with (7–)10–30 flowers ____ **8. Astragalus arizonicus**

29′ Plants acaulescent to subacaulescent, tufted or mound-forming; inflorescences with 1–9 flowers (up to 12 flowers in robust variants of *A. monumentalis*).

30. Wings deeply cleft; leaflets flat; banner recurved through 45–90°; widespread in Apache, Coconino, Gila, Graham, Mohave, Navajo, Yavapai counties ____ **12. Astragalus calycosus**

30′ Wings entire; leaflets all or mostly folded; banner recurved through 45°; localized in northern Coconino and Navajo counties near the Utah border ____ **51. Astragalus monumentalis** (in part)

15′ Pubescence basifixed (sometimes flattened and scale-like in *A. titanophilus*).

31. Terminal leaflet of at least some distal leaves continuous with the rachis (distal leaves often reduced to a subfiliform rachis).

32. Plants with stipules fused throughout or fused proximally and distinct distally.

33. Fruits bladdery-inflated __________ **14. Astragalus ceramicus** (in part)
33′ Fruits not bladdery-inflated.
34. Fruits deflexed or declined-pendulous, sessile or substipitate, the stipe 0–1.7 mm long.
35. Pedicels spreading-declined; keel 5.9–10.7(–11.4) mm; fruits declined-pendulous, the valves strigillose (var. *oblongifolius* of) __________ **46. Astragalus miser**
35′ Pedicels ascending; keel 3.5–5.4 mm; fruits deflexed, the valves glabrous __________ **77. Astragalus wingatanus** (in part)
34′ Fruits erect, ascending, or spreading, sessile.
36. Keels 5.7–7 mm long; fruits ascending to spreading, dorsiventrally compressed; Apache, Coconino, Mohave, Navajo counties __________ **48. Astragalus moencoppensis** (in part)
36′ Keels (9.5–)10.5–13.5 mm long; fruits erect, slightly laterally compressed; plants restricted to one population in Mohave Co. (var. *scidulus* of) __________ **74. Astragalus toanus**
32′ Plants with stipules distinct throughout.
37. Fruits stipitate, the stipe 3–15 mm.
38. Fruits strongly laterally compressed; corolla pink-purple; calyx purplish __________ **19. Astragalus coltonii**
38′ Fruits dorsiventrally compressed; corolla creamy white to almost white; calyx reddish-brown __________ **44. Astragalus lonchocarpus**
37′ Fruits sessile or subsessile or elevated on an obscure, stipe-like neck to 1 mm.
39. Fruits bladdery-inflated, rounded in cross-section, not compressed; plants of sand dune environments in Yuma County __________ **45. Astragalus magdalenae**
39′ Fruits laterally compressed, sharply 4-sided compressed, or cylindrical (rounded in cross section in *A. pinonis*, not bladdery inflated); plants of various substrates, not from Yuma County.
40. Fruits cylindrical or nearly so; Coconino County __________ **58. Astragalus pinonis** (in part)
40′ Fruits strongly laterally compressed or sharply 4-sided compressed; Apache, Coconino, Mohave, Navajo counties.
41. Fruits sharply 4-sided compressed, the body strongly incurved or coiled; banner (13.5–)15–19 mm; plants robust, ± leafy throughout __________ **71. Astragalus tetrapterus**
41′ Fruits laterally compressed, the body slightly curved to straight; banner 9–14.5 mm (up to 15.5 mm in robust variants of *A. episcopus*); plants wiry or rush-like, sparsely leafy to apparently leafless throughout.
42. Banner 11–14.5(–15.5) mm long; fruits sessile or tapering at base into a stout, stipe-like neck up to 0.6 mm (var. *episcopus* of) __________ **28. Astragalus episcopus** (in part)
42′ Banner 9–11.5 mm long; fruits sessile.
43. Corolla whitish, pink- to pink-purple tinged; banner recurved through 40–50°; plants from Coconino and Mohave counties (var. *lancearius* of) __________ **28. Astragalus episcopus** (in part)
43′ Corolla ochroleucous; banner recurved through 100–130°; plants from Apache, Coconino, and Navajo counties __________ **79. Astragalus xiphoides** (in part)
31′ Terminal leaflets all jointed (distal leaves not reduced to a rachis).
44. Plants with stipules fused throughout or fused proximally and distinct distally.
45. Fruits 1-chambered (the septum, if present, not over 1 mm wide in *A. tephrodes*).
46. Corolla 11–26 mm.
47. Fruits stipitate, the stipe 1–19 mm (the stipes sometimes inconspicuous and less than 1 mm in *A. hallii*).
48. Fruits ascending-erect; stipes 9–19 mm long; pedicels ascending __________ **5. Astragalus ampullarius**
48′ Fruits spreading, declined, deflexed, or pendulous; stipes 0–6 mm long; pedicels spreading to declined.
49. Fruits dorsiventrally compressed, 2–4.5 mm wide; calyx tube 2.8–5.7(–7) mm; banner 6.7–17.5 mm; seeds 5–15; rare and localized in Coconino County __________ **10. Astragalus bisulcatus**
49′ Fruits slightly to distinctly inflated, 6–12 mm wide; calyx tube 5–6.2 mm; banner 5.3–8.4 mm; seeds 24–34; Apache, Coconino, Gila, Mohave, Navajo, Yavapai counties __________ **36. Astragalus hallii**
47′ Fruits sessile.
50. Fruits green or purplish, not mottled, flattened or low-convex dorsally, the valves sparsely to densely strigillose or pilosulose, rarely glabrous __________ **70. Astragalus tephrodes** (in part)
50′ Fruits brightly purple- or red-mottled, the valves loosely strigose or villosulous __________ **80. Astragalus zionis** (in part)
46′ Corolla 3.3–11 mm.
51. Corolla white to yellowish and commonly suffused with lilac or purple.
52. Keels 3.5–5.4 mm; calyx strigose, the tube 1.5–2.6 × 1.2–1.8(–2) mm; fruits strongly laterally-compressed, glabrous __________ **77. Astragalus wingatanus** (in part)
52′ Keels (5.7–)6–7.7 mm; calyx villosulous, the tube 2.3–3.6 × 1.9–3.2 mm; fruits nearly cylindrical to dorsiventrally-compressed, sparsely villosulous.
53. Plants mat or cushion-forming with a branched caudex; inflorescence axis 0.3–1.5 cm in fruit; pods ascending, sessile, the valves strigillose __________ **15. Astragalus chuskanus**
53′ Plants diffuse, not mat- or cushion-forming, with a deeply subterranean, branched caudex; inflorescence 1–7 cm in fruit; pods spreading to declined, subsessile, the stipe-like neck 0.3–1 mm long, the valves sparsely villosulous __________ **69. Astragalus subcinereus**
51′ Corolla pink-purple to dull purplish or reddish purple (the wing tips white in some variants of *A. wingatanus*).
54. Fruits bladdery-inflated, 10–20 mm wide.
55. Keel blunt or subacute and slightly beak-like; banner recurved through 50°; pod greenish or purplish becoming straw-colored, not mottled; seeds 10–21; widespread __________ **3. Astragalus allochrous** (in part)
55′ Keel acutely triangular and beak-like; banner recurved through 90°; pod purple-mottled; seeds 21–32; Apache, Coconino, Navajo counties __________ **33. Astragalus fucatus**

54′ Fruits variously compressed, ellipsoid, oblong, ovoid, obliquely semi-ovoid, cylindrical, or ± flattened, 2.7–9 mm wide.

56. Plants acaulescent, subacaulescent, or shortly caulescent, tuft-forming ________ **22. Astragalus desperatus**

56′ Plants caulescent, not tuft-forming.

57. Fruits 5.5–8 mm; plants often selenophytic ________ **48. Astragalus moencoppensis** (in part)

57′ Fruits 8–24 mm; plants not selenophytic.

58. Keels 3.5–5.4 mm long; calyx pubescent with black hairs, the tube 1.5–2.6 × 1.2–1.8(–2) mm; pods glabrous, strongly laterally compressed, 2-sided ________ **77. Astragalus wingatanus** (in part)

58′ Keels 5–8.2 mm long; calyx pubescent with mostly white hairs, the tube 1.9–4.3 × 1.8–2.9 mm; pods strigillose or villosulous, rarely glabrous, nearly cylindrical to slightly compressed, laterally or dorsiventrally, not strongly 2-sided or laterally compressed ________ **32. Astragalus flexuosus**

45′ Fruits 2-chambered or incompletely 2-chambered.

59. Plants subacaulescent to shortly caulescent, tuft-forming; fruits bladdery-inflated, the valves hirsute-hirsutulous with spreading hairs ________ **68. Astragalus striatiflorus**

59′ Plants distinctly caulescent (sometimes tufted in *A. moencoppensis*); fruits not bladdery-inflated, the valves glabrous, strigose, strigillose, pilose, or villosulous.

60. Calyx tube 5–8.5 mm long; keel 9.6–15.4 mm long.

61. Fruits ascending to spreading and subsessile, the stipe absent or up to 0.8 mm long; calyx tube 5–6 mm long; plants from a subterranean caudex with long, rhizome-like branches; widely introduced and occasionally naturalizing, appears restricted to Coconino County ________ **16. Astragalus cicer**

61′ Fruits pendulous and stipitate, the stipe 4–9 mm long; calyx tube 6.5–8.5 mm long; plants from a subterranean caudex, the branches not long and rhizome-like; native ________ **64. Astragalus scopulorum**

60′ Calyx tube 1.8–4 mm long; keel 3.5–7 mm long.

62. Fruits ± dorsiventrally compressed.

63. Plants from a subterranean caudex, the stems decumbent to ascending, 1.5–10(–18), flexuous and slender, but not wiry; inflorescence (4–)8–22(–33) flowered, peduncle equal to or shorter than subtending leaf; pedicels spreading, becoming declined, 1–2 mm long; fruits dorsiventrally compressed ________ **17. Astragalus cobrensis**

63′ Plants from a superficial caudex, not subterranean, the stems prostrate-spreading, 5–35(–40), very slender and wiry; inflorescence 1–5(–7)-flowered, early peduncles shorter than leaves, commonly 1-flowered, later ones elongate and surpassing the leaves; pedicels ascending, 1.2–4 mm long; fruits slightly dorsiventrally compressed ________ **11. Astragalus brandegeei**

62′ Fruits 3-sided compressed.

64. Fruits glabrous, the surface lustrous ________ **24. Astragalus egglestonii**

64′ Fruits strigillose.

65. Fruits 3.2–4.1 mm wide, stipitate, the stipe 2.2–5 mm long; pedicels deflexed, becoming slightly thickened and elongating to (1.5–)2–3.2 mm long in fruit; keel 4.6–5.5 mm long ________ **62. Astragalus rusbyi**

65′ Fruits 2.3–3 mm wide, sessile or substipitate, the stipe absent or up to 0.7 mm; pedicels ascending, becoming declined in fruit, not thickened or elongating, 0.5–1.5 mm long in fruit; keel 5–6.5 mm long ________ **61. Astragalus recurvus**

44′ Plants with stipules distinct throughout.

66. Fruits 1-chambered (the septum, if present, not over 1 mm wide in *A. tephrodes*).

67. Plants distinctly caulescent, clump-forming in some taxa but not tuft- or mat-forming.

68. Fruits inflated or bladdery-inflated, round and not compressed.

69. Fruits inflated but not bladdery, becoming stiff, subsessile or stipitate, the stipe 0–8 mm; herbage malodorous (selenium-scented).

70. Pedicels deflexed; fruits fleshy, becoming leathery-woody in age ________ **59. Astragalus praelongus** (in part)

70′ Pedicels ascending; fruits thinly fleshy, becoming stiffly papery to leathery in age ________ **60. Astragalus preussii**

69′ Fruits bladdery-inflated, becoming papery, sessile; herbage not malodorous.

71. Fruits sessile on and disjointing from a linear gynophore 3.5–11 mm; banner 11–24 mm; keel 9.5–16 mm long ________ **56. Astragalus oophorus** (in part)

71′ Fruits sessile, gynophore absent; banner 5.7–9.4 mm; keel 4.7–7.5 mm long.

72. Banner 7.2–9.4 mm; keel 6.2–7.5 mm; fruits (20–)25–40(–45) mm long ________ **3. Astragalus allochrous** (in part)

72′ Banner 5.7–7.1 mm; keel 4.7–5.6 mm; fruits 6–13 mm long ________ **72. Astragalus thurberi**

68′ Fruits not inflated, strongly laterally compressed, dorsiventrally compressed, or the compression is slight (when obscure or absent, then the fruits cylindrical or rounded).

73. Fruits strongly laterally compressed.

74. Leaflets (1–)3–7, linear-filiform or linear-involute, the apex acute; fruits sessile, stipe-like neck absent, the body 4–5.5 mm wide ________ **79. Astragalus xiphoides** (in part)

74′ Leaflets 7–15 narrowly linear to linear-oblong, the apex obtuse to emarginate; fruit subsessile, the base contracted into a stout, stipe-like neck 1–1.5 mm long, the body 3.2–4.6(–5) mm wide ________ **73. Astragalus titanophilus**

73′ Fruits dorsiventrally compressed or the compression obscure to entirely absent.

75. Plants slender, weak, not clump-forming, understory plants growing up through shrubs with stems erect or reclining, not malodorous; pods stiff-papery; pedicels ascending-spreading; keels 7.5–9.7 mm ________ **58. Astragalus pinonis** (in part)

75′ Plants robust, forming moderate to very large, bushy clumps with erect to ascending stems, malodorous (selenium-scented); pods fleshy, becoming stiff-leathery to leathery-woody in age; pedicels declined or deflexed; keels 11.5–17 mm.

76. Keel apex bluntly rectangular; calyx green or yellowish, the tube 4.4–7.5 mm; banner recurved through 45°; seeds 40–75(–84); plants widespread ______ **59. Astragalus praelongus** (in part)

76′ Keel apex sharply triangular; calyx tan or whitish, the tube 6–8.8 mm; banner recurved through 50–90°; seeds 22–38; plants rare in Arizona, Mohave County and restricted to a single population around Fredonia in Coconino County, rarely collected and possibly introduced ______ **57. Astragalus pattersonii**

67′ Plants acaulescent, subacaulescent, or shortly caulescent, sometimes dwarf, or tuft- or mat-forming.

77. Corolla scarlet red, drying crimson; the banner 35–41 mm, recurved through 20–30°; keel 35–40 mm; Yuma County ______ **18. Astragalus coccineus**

77′ Corolla variously pink-purple, lilac, pale pink, dull reddish purple, or white or ochroleucous and veined or spotted with pink-purple or lilac; the banner 9.5–32 mm, recurved through 40–90°; keel 7.9–26 mm; not in Yuma County.

78. Fruits disjointing from a linear gynophore that is 5–9 mm, bladdery-inflated, ± rounded, not compressed (except in var. caulescens which is ± flattened above but rounded below ______ **56. Astragalusoophorus** (in part)

78′ Fruits sessile or subsessile, gynophore absent, dorsiventrally or laterally compressed, not bladdery-inflated.

79. Corolla 9.5–11.4 mm, dull reddish purple; seeds 8–12; fruits 5–7 mm long ______ **75. Astragalus troglodytus**

79′ Corolla 11.8–32 mm, variously pink-purple, lilac, pale pink, or white or ochroleucous and veined or spotted with pink-purple or lilac; seeds 24–46; fruits 13–65 mm long.

80. Fruits laterally compressed ______ **30. Astragalus eurylobus**

80′ Fruits dorsiventrally compressed.

81. Fruits when ripe becoming leathery to stiffly woody, but not densely hairy or mottled.

82. Inflorescences with 1–6 flowers; peduncle incurved-ascending in flower, bent downward or prostrate in fruit, the axis 0.2–2(–2.5) cm ______ **6. Astragalus argophyllus**

82′ Inflorescences with (9–)10–25(–35) flowers; peduncle incurved–ascending or erect in flower and fruit, the axis (1.5–)2–20 cm ______ **70. Astragalus tephrodes** (in part)

81′ Fruits when ripe becoming leathery and either densely hairy, villous-tomentose, or brightly purple- or red-mottled.

83. Fruits green becoming brown-straw-colored, not mottled, the valves both villous-tomentose with shorty curly hairs and densely villous-hirsute with longer, lustrous hairs that conceal the pod surface; banner recurved through 90°; leaflets 1–15 ______ **52. Astragalus newberryi**

83′ Fruits brightly purple- or red-mottled, the valves loosely strigose or villosulous, not concealing the pod surface; banner recurved through 40°; leaflets 13–25 ______ **80. Astragalus zionis** (in part)

66′ Fruits 2-chambered or incompletely 2-chambered.

84. Plants acaulescent, subacaulescent, or shortly caulescent, tuft- or clump-forming in some species.

85. Fruits dorsiventrally compressed or cylindrical.

86. Calyx pilose; herbage pilose to strigillose; inflorescences (1–)2–5(–7)-flowered, peduncles always shorter than leaves; fruit ovoid, thin and papery, 8–10 mm long; restricted to Apache County (and adjacent areas in Catron Co., New Mexico) ______ **54. Astragalus nutriosensis**

86′ Calyx villous-tomentose; herbage gray- or white-villous-tomentose; inflorescences (5–)7–45-flowered, peduncles scapiform; fruit fleshy, becoming stiff-papery, leathery, or semi-woody in age, (6–)9–24 mm; widespread ______ **50. Astragalus mollissimus**

85′ Fruits 3-sided compressed.

87. Plants low and clump-forming, but not dwarf and tuft-forming, stems from a deep-seated, subterranean caudex, the branches becoming rhizomatous ______ **42. Astragalus layneae**

87′ Plants dwarf and tuft-forming, stems from a caudex or superficial root-crown near or at surface-level, the branches not rhizomatous.

88. Banner 21.5–23.5 mm long; keels 16.5–18.5 mm; fruits 6–9 mm wide; restricted to very extreme northern Mohave County ______ **37. Astragalus holmgreniorum**

88′ Banner 11–16 mm long; keels 9–11.5 mm long; fruits 3–4(–5) mm wide; from Coconino and Navajo counties but more plants more widespread outside of Arizona ______ **51. Astragalus monumentalis** (in part)

84′ Plants distinctly caulescent, not tuft- or mat-forming.

89. Fruits 3-sided compressed or 3-sided and dorsiventrally compressed.

90. Fruits stipitate.

91. Calyx tube 2.5–3.5 mm; banner 6.3–8.5 mm; keel 5–6 mm; seeds 10–13; fruits with a stipe 1.4–2.5 mm ______ **67. Astragalus straturensis**

91′ Calyx tube 4.4–8 mm; banner 12–20 mm; keel 9.4–14.4 mm; seeds 17–32; fruits with a stipe 6–15(–17) mm ______ **29. Astragalus eremiticus**

90′ Fruits sessile.

92. Calyx tube 3.6–4.9 mm; banner 8.5–12 mm; keel 6.5–8.4 mm, with a narrowly triangular or subulate beak ______ **53. Astragalus nothoxys**

92′ Calyx tube 1.7–3 mm; banner 4.2–8 mm; keel 3.7–6.2 mm, beakless.
93. Corolla 7.5–8 mm; calyx lobes 2.8 mm; pedicels ascending; fruits ascending, 2–2.5 mm wide ____ **39. Astragalus hypoxylus**
93′ Corolla 4.2–6.2 mm; calyx lobes 0.8–1.8 mm; pedicels arched downward; fruits deflexed, 1.3–2 mm wide ____ **76. Astragalus vaccarum**

89′ Fruits inflated, turgid, dorsiventrally compressed or cylindrical, not 3-sided compressed.
94. Herbage malodorous (selenium-scented); plants growing on seleniferous substrates, forming large, bushy clumps.
95. Pedicels spreading-ascending; fruits deflexed or declined, sessile or substipitate, the stipe, if present, less than 0.8 mm; corolla bright purple with paler wing tips, rarely pure white or whitish-blue ____ **9. Astragalus beathii**
95′ Pedicels deflexed; fruits erect, spreading, or declined, sessile or stipitate, the stipe, if present up to 8 mm; corolla ochroleucous ____ **59. Astragalus praelongus** (in part)
94′ Herbage not malodorous; plants on various substrates but not restricted to seleniferous substrates, not forming large, bushy clumps.
96. Fruits 8–27 mm wide.
97. Fruits fleshy, firm, and succulent, becoming brown or blackish and wrinkled in age; the interior alveolate-spongy or pithy ____ **20. Astragalus crassicarpus**
97′ Fruits bladdery-inflated, thin-walled not fleshy or succulent, becoming straw-colored to brownish in age, not wrinkled; the interior an open cavity, not alveolate-spongy or pithy ____ **43. Astragalus lentiginosus** (in part)
96′ Fruits 4–8 mm wide.
98. Fruits cylindrical and nearly round in cross section, becoming laterally compressed when pressed; fruits sessile to short stipitate, the stipe, when present 0.5–2 mm long ____ **27. Astragalus ensiformis**
98′ Fruits dorsiventrally compressed and scarcely inflated to not inflated or, if round in cross section, then bladdery-inflated; fruits sessile or substipitate, the stipe, when present less than 0.5 mm (up to 1 mm in a rare variant of *A. mokiacensis*).
99. Fruits deciduous (all varieties excluding var. *maricopae*, which keys to couplet 100 below) ____ **43. Astragalus lentiginosus** (in part)
99′ Fruits long-persistent, the pods often found still attached to dried inflorescences for a year or more.
100. Corolla light yellow, drying ochroleucous to brownish; fruits subsessile, the base of the pod often abruptly contracted into a stipe-like base less than 0.4 mm; plants restricted to Maricopa County (flowering individuals of var. *yuccanus* are often misidentified as this taxon, which is the source of reports outside of this county) ____ **43f. Astragalus lentiginosus** var. **maricopae**
100′ Corolla pink- to red-purple (or in some populations the wings white-tipped), drying violet (flowering individuals often confused with *A. lentiginosus* var. *palans* and var. *wilsonii*); fruits subsessile, the stipe 0.5–1 mm; plants from Coconino and Mohave counties ____ **49. Astragalus mokiacensis**

1. Astragalus acutirostris S. Watson. Proc. Amer. Acad. Arts 20:360. 1885. * [from Latin *acut*, sharp and *rostrum*, a beak or snout, referring to the keel] SHARP-KEELED MILKVETCH Native

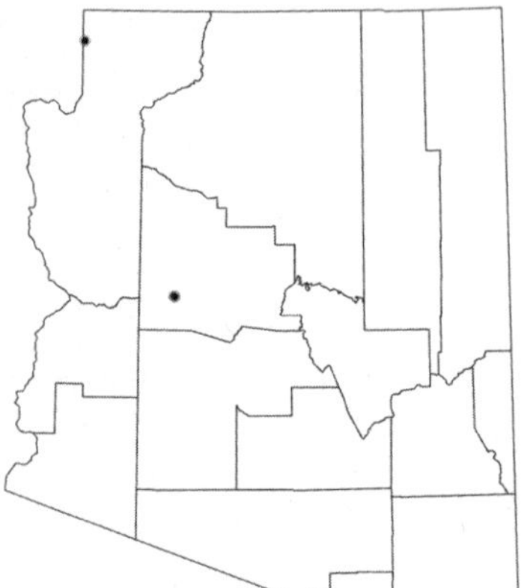

Annuals or winter annuals, caulescent, slender, 2–25(–30) cm; taproot slender; from superficial caudex; strigillose to pilosulose, hairs basifixed. **Stems** single or few to several; prostrate to decumbent or ascending. **Leaves** (1–)1.5–4.5 cm, petiolate or subsessile; stipules distinct throughout, deltoid, triangular-ovate, or lanceolate, 0.8–2.5(–3) mm, membranous. **Leaflets** (7–)9–13(–15), flat or loosely folded, oblong-oblanceolate, obovate, or cuneate, 2–10 × 2–3 mm, apex retuse, pubescent below (abaxially), glabrescent above (adaxially). **Inflorescences** with (1–)3–8 flowers; peduncle erect or spreading and incurved, (1.5–)2.5–7 cm; axis (0–)1–3.5 cm in fruit; bracts 0.7–1.3 mm; bracteoles 0. **Pedicels** spreading or declined, 0.4–0.9 mm. **Flowers**: calyx campanulate, (2.6–)2.8–3.5(–4.1) mm, strigillose; tube 1.6–2.1 × 1.4–1.9 mm; lobes lanceolate-subulate, (1–)1.2–1.5(–2.3) mm; corolla whitish, banner tinged or veins purple, (4.7–)5–7 mm; banner recurved through 45°, ovate, obovate, or nearly circular, (4.7–)5–7 mm; wings 4.3–6.2 mm; keel 4.3–5.8 mm, apex narrowly triangular, acute or subacute, often slightly beaklike; ovules 12–26. **Fruits** pendulous, spreading, or ascending and resupinate, sessile; gynophore obscure, 0.4–0.8 mm; gently incurved, lunately linear-ellipsoid, 3-sided compressed, dehiscent after falling, 12–30 × (2.2–)2.5–3.1 mm, green or purple-tinged becoming brownish straw-colored, thin, becoming papery, usually strigillose, sometimes glabrous, two-chambered. **Seeds** 12–26. $2n = 24$.

Flowering late Mar–May. Growing on slopes, often in gravelly soils; 600–1600 m; Mohave Desertscrub; Mohave, Yavapai cos.; California, Nevada; Mexico (Baja California).

Astragalus acutirostris typically occurs in creosote bush and Joshua tree communities. *Astragalus acutirostris* is technically similar to *A. pauperculus* and is often confused with western races of *A. nuttallianus*, which have persistent fruits and mostly acute leaflets (Barneby 1964a). The upwardly curved keel apex was the basis for its prior inclusion in *Oxytropis*, but the keel shape is unmatched in that genus.

2. Astragalus albulus Wooton & Standley. Contr. U.S. Natl. Herb. 16:136. 1913. * [from Latin *albulus*, whitish] CIBOLA MILKVETCH Native

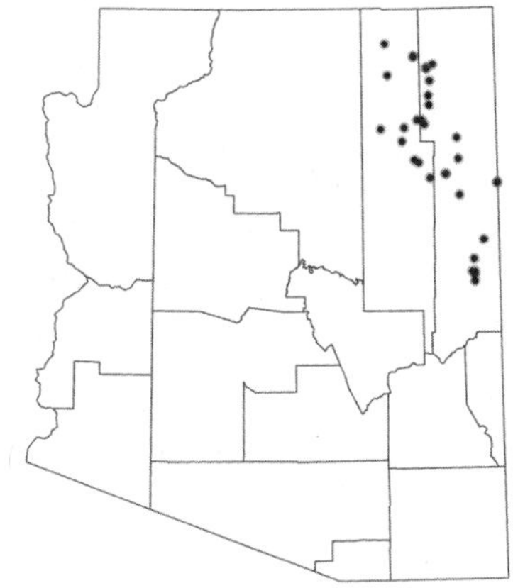

Perennials, selenophyte, forming bushy clumps, caulescent, robust, 15–85 (–120) cm; taproot shortly forked, thick, woody; from caudex, branched, becoming somewhat woody at base; strigillose to pilose, hairs medifixed. **Stems** several to numerous, erect or ascending; grey hairs. **Leaves** 3–10(–14) cm, subsessile; stipules with fused, bidentate sheathing throughout, triangular or ovate, 2.5–10 mm, scarious. **Leaflets** (7–)11–23, mostly folded, linear, linear-oblanceolate, or narrowly elliptic, (2–)4–27(–45) × 1.2–3.2 mm, apex acute to subobtuse, surfaces strigose below (abaxially), glabrous above (adaxially). **Inflorescences** 9–47-flowered; peduncle erect, 1–4(–5.5) cm; axis (2–)3–15 (–20) cm in fruit; bracts (2–)3–7 mm; bracteoles 0–2. **Pedicels** ascending, 0.3–1.2 mm. **Flowers**: calyx short-cylindric, (7.2–)7.5–11 mm, strigillose to pilose; tube (5.4–)5.7–7.4 × 3.2–4.2 mm; lobes lanceolate-subulate, (1.5–)1.8–3.6 mm; corolla whitish, keel tip faintly spotted, 13–17.2 mm; banner recurved through 45°, rhombic-oblanceolate, 13–17.2 mm; wings 12.5–15.3 mm; keel 10.4–12.8 mm, tip obtuse; ovules 5–11. **Fruits** declined or pendulous; stipe 0.7–2.5 mm; gently incurved, narrowly ellipsoid or lanceolate-ellipsoid, dorsiventrally compressed or 3-sided at dehiscence, ventral suture forming a ridge, beak laterally compressed, 9–12 × 3.3–5 mm, green becoming brown-straw-colored, papery, strigose or glabrous; one chambered. **Seeds** 5–11. $2n = 22$.

Flowering Jul–Oct. Growing in badlands, slopes, drainages, and roadsides, often in seleniferous soils; [1600–]1750–2100[–2600] m; Plains Grassland, Great Basin Conifer Woodland; Apache, Navajo Yavapai cos.; New Mexico, Texas.

Astragalus albulus, distributed mainly in New Mexico, is one of the few species of *Astragalus* that ignores the spring flowering season and flowers instead from mid-summer onward. At first sight it resembles a somewhat taller, more robust *A. flavus*, but its declined or pendulous, stipitate, dorsiventrally compressed, and subtrigonous legumes (except for the laterally compressed, cuspidate beak) easily define it as unique. It is the only milkvetch within its region that flowers and fruits from mid-summer to autumn on a regular basis. Racemes sometimes are borne from below the middle of the stem, with the portion bearing racemes much exceeding the non-flowering lower portion of the plant. *Astragalus albulus* occurs in gullied badlands; sandy clay talus under cliffs; on seleniferous substrates; with juniper, rabbitbrush, *Eriogonum,* and saltbush; and in salt-seep communities. This distinctive species would make an attractive addition to wildflower gardens.

A single record (*R. Hurd s.n.,* 14 Jun 1978) from along Interstate 17 in Yavapai Co. at 975 m elevation represents an unusual disjunction as it is far removed from the distribution of the species and at a much lower elevation than has otherwise been reported. The specimen does not match characters for *A. albulus*.

3. Astragalus allochrous A. Gray. Proc. Amer. Acad. Arts 13:366. 1878. * [from Greek *allo*, different and *chroa*, skin, of different color] HASSAYAMPA MILKVETCH Native

Biennials or perennials, short-lived, caulescent, coarse, 10–50(–60) cm; from superficial root-crown; strigillose, hairs basifixed. **Stems** several to numerous, ascending and radiating. **Leaves** 4–10 cm, shortly petiolate or subsessile; stipules usually distinct throughout, rarely fused proximally, triangular or ovate, 1.5–7 mm, papery proximally, herbaceous or submembranous distally. **Leaflets** (9–)11–21, flat or loosely folded, oblong-obovate, oblanceolate, elliptic, or narrowly oblong-elliptic, (4–)6–21 × (1.2–)2–6 mm, apex acute, obtuse, or retuse and mucronulate, surfaces strigillose below (abaxially), strigillose or glabrous above (adaxially). **Inflorescences** 10–20-flowered; peduncle incurved-ascending, 3–9(–11) cm; axis (1.5–)3.5–12 cm in fruit; bracts 1.2–2.5 mm; bracteoles 2. **Pedicels** ascending, 1–4 mm. **Flowers**: calyx campanulate or turbinate-campanulate, (3.6–)4.1–5.7 mm, strigillose; tube 2.4–3.5 × 1.9–3.6 mm; lobes

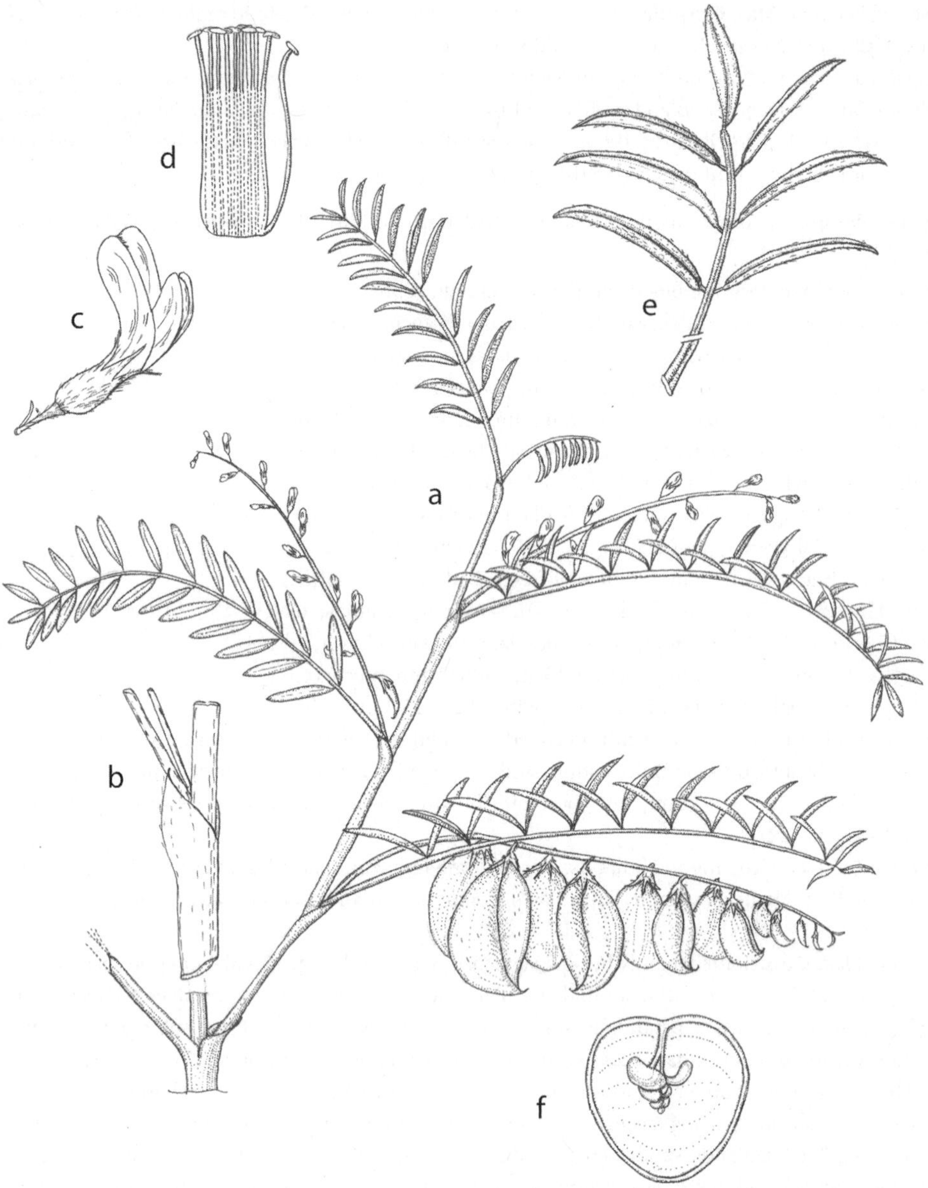

Astragalus allochrous: **a.** habit; **b.** stem node; **c.** flower; **d.** stamens; **e.** terminal leaflets; **f.** fruit cross section. Illustrator: Marsha Bennett.

lanceolate-subulate, (1.1–)1.6–2.5 mm; corolla pink-purple or reddish purple, 7.2–9.4 mm; banner recurved through 50°, broadly ovate-cuneate or obcordate, 7.2–9.4 mm; wings 6.6–8.3 mm; keel 6.2–7.5 mm, apex blunt or subacute, slightly beaklike; ovules (10–)14–21. **Fruits** spreading or declined, or humistrate and ascending; sessile; straight or slightly incurved, obliquely ellipsoid or semi-ellipsoid, bladdery-inflated, ventral suture straight, concave, or less strongly concave than the dorsal suture, (20–)25–40(–45) × 10–17 mm, beak strongly developed and clearly differentiated from body, thin, pale green or purple-cheeked becoming straw-colored, papery, translucent, surface lustrous, finely strigillose; one-chambered. **Seeds** (10–)14–21. 2*n* = 22.

Flowering (Feb–)Mar–Jul (–Sep). Plains, foothills, open valleys and along arroyos, riparian floodplains, and roadsides, on various soil types, rarely on dunes and usually absent from calcareous bedrock; 550–1900(–2300) m; Sonoran Desertscrub, (Chihuahuan Desertscrub, Mohave Desertscrub, Great Basin Desertscrub), Semidesert Grassland, (Plains Grassland), Interior Chaparral, Madrean Evergreen Woodland, Great Basin Conifer Woodland; Cochise, Coconino,

Gila, Graham, Greenlee, Maricopa, Mohave, Navajo, Pima, Pinal, Santa Cruz, Yavapai cos.; New Mexico, Texas; Mexico (Chihuahua, Sonora).

Astragalus allochrous contains swainsonine and can result in the poisoning of livestock (Allison et al. 2016). In Arizona, the species is particularly abundant in the upper Gila River Basin. It occurs in many plant communities but is most common in desert grasslands and scattered scrub-oak or juniper woodlands.

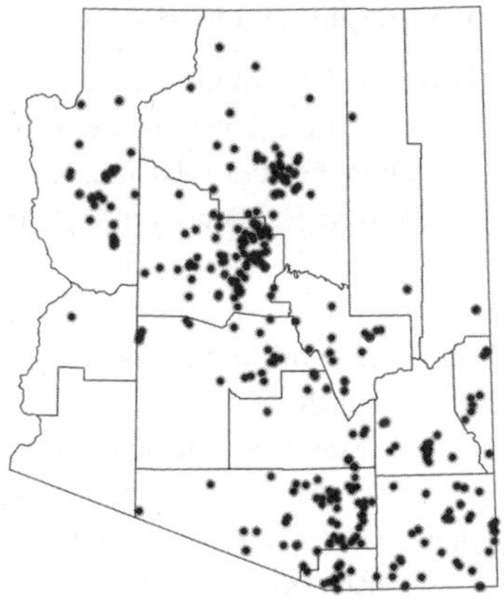

4. Astragalus amphioxys A. Gray. Proc. Amer. Acad. Arts 13:366. 1878. * [from Greek *amphi*, around or on both sides and *oxys*, sharp, referring to the fruits] CRESCENT MILKVETCH

Perennials, short-lived, sometimes flowering first year, tuft- or loosely mat-forming, subacaulescent to shortly caulescent, 2–35 cm; from weak caudex; densely strigose-strigillose, hairs medifixed. **Stems** usually several, sometimes single, usually prostrate or weakly ascending, sometimes erect, 0–20 cm, internodes often concealed by stipules. **Leaves** 2–13 cm, petiolate; stipules mostly distinct, sometimes with fused sheathing proximally, ovate or narrowly to broadly triangular, 2–13 mm, submembranous becoming papery. **Leaflets** (1–)5–21, flat or loosely folded, elliptic to obovate or oblanceolate, 3–20 × 1–9 mm, apex obtuse to acute, surfaces strigose. **Inflorescences** 2–13-flowered; peduncle ascending, (1–)2–15(–20) cm; axis (0.5–)1–6.5 cm in fruit; bracts 2.5–8 mm; bracteoles 0–2. **Pedicels** ascending, 0.6–2.5 mm. **Flowers**: calyx usually purplish, cylindric, campanulate, or cylindro-campanulate, 6.3–14.2 mm, strigose; tube (3.4–) 5.8–13.2 × 2.3–5.2 mm; lobes subulate, 1.1–3.7(–4.5) mm; corolla usually pink-purple, rarely white, 10.3–27(–28) mm; banner recurved through 40°, oblanceolate, rhombic-oblanceolate, or spatulate, 10.3–27(–28) mm; wings 9.8–24 (–25) mm; keel 8–23.6 mm, apex obtuse; ovules 42–70. **Fruits** readily deciduous, ascending; sessile; usually incurved, crescent-shaped, mostly dorsiventrally compressed, 15–40 × 5–12 mm, green, sometimes mottled, distinctly beaked, fleshy becoming stiffly leathery or slightly-woody, strigose; one-chambered with a partial, inconspicuous septum to 1 mm. **Seeds** 42–70.

Varieties 4 (4 native in the flora): w United States, n Mexico.

Astragalus amphioxys is a widely distributed species complex consisting of a series of morphological and geographical varieties, which are fairly distinctive. The complex is common in the middle Colorado Basin and adjoining regions. *Astragalus missouriensis*, widespread and common over the Great Plains and Rocky Mountain regions, is considered a closely related species; the ranges of the two are essentially allopatric, but they slightly overlap in southwestern Colorado and northwestern New Mexico. Flowering material is difficult to identify, and care must be taken in the Four Corners area of the Southwest to ensure correct identification.

1. Calyx tube (5.8–)7–13.2 mm; banner (16.2–)19–27(–28) mm; keel (13.2–)14.3–24 mm.
    2. Calyx tube (5.8–)7–10.5 mm; banner (16.2–)19–24.5 mm; keel (13.2–)14.3–18.8(–19.6) mm______ **4a. Astragalus amphioxys** var. **amphioxys**
    2′ Calyx tube 8.8–13.2 mm; banner 23–27(–28) mm; keel (17–)18–24 mm______ **4d. Astragalus amphioxys** var. **vespertinus**

1′ Calyx tube 3.4–9.3 mm; banner 10.3–16 mm; keel 8–12.7 mm.
    3. Calyx cylindro-campanulate, tube 6.1–9.3 mm; keel 11–12.7 mm; seeds 44–58; Coconino, Mohave, Yavapai counties______ **4b. Astragalus amphioxys** var. **modestus**
    3′ Calyx campanulate, tube 3.4–4.8 mm; keel 8–11.5 mm; seeds 26–36; Mohave County______ **4c. Astragalus amphioxys** var. **musimonum**

4a. Astragalus amphioxys A. Gray var. **amphioxys** * Native

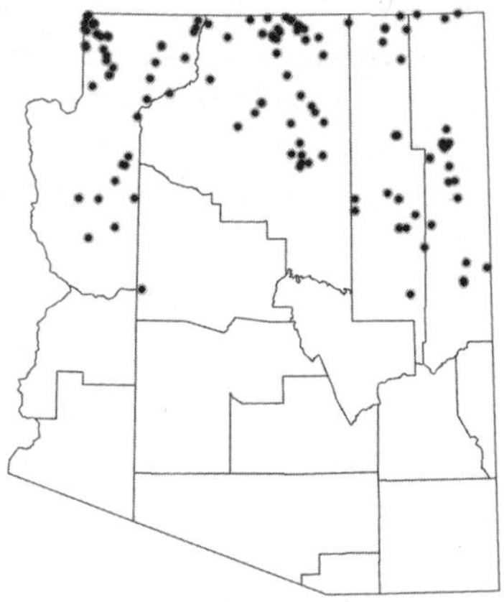

Flowers: calyx cylindric; tube (5.8–)7–10.5 × 3.2–4.2(–4.7) mm; lobes 1.5–3.7(–4.5) mm; corolla: banner (16.2–)19–24.5 mm; keel (13.2–)14.3–18.8 (–19.6) mm. **Fruits** 15–40 mm, curved. **Seeds** (42–)44–56. $2n = 22$.

Flowering late Mar–Jun(Aug–Sep). Sandy valleys, plains, gravelly hillsides, ridges, on dunes, along arroyos, rarely on gumbo clay flats; 400–2100 m; Sonoran Desertscrub, Mohave Desertscrub, Great Basin Desertscrub, Great Basin Grassland, Great Basin Conifer Woodland, Petran Montane Conifer Forest; Apache, Coconino, Mohave, Navajo, Yavapai cos.; Nevada, New Mexico, Texas, Utah; Mexico (Chihuahua).

Variety *amphioxys* exhibits high morphological variation. Across its range, var. *amphioxys* is found in creosote bush, Joshua tree, blackbrush, indigo bush (*Psorothamnus* sp.), salt desertscrub, pinyon-juniper, ponderosa pine, and mountain brush communities.

4b. Astragalus amphioxys A. Gray var. **modestus** Barneby. Leafl. W. Bot. 9:89. 1960. * [unassuming] MODEST MILKVETCH Native

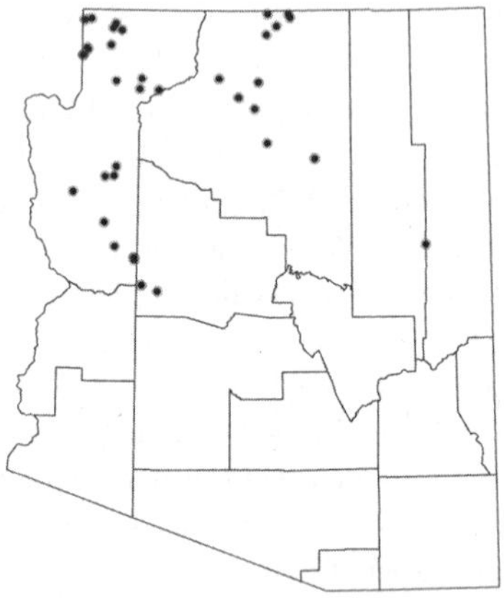

Flowers: calyx cylindro-campanulate; tube 6.1–9.3 × 2.3–3.3 mm; lobes 1.1–2.6 mm; corolla: banner 12.8–16 mm; keel 11–12.7 mm. **Fruits** 16–40 mm, crescent-shaped or incurved through 1/2 their length. **Seeds** 44–58.

Flowering Mar–May. Dry gravelly hillsides, canyons, along arroyos and roadsides, and on sandy silts, limestone, or cindery volcanic debris; (580–) 1100–1500(–2300) m; Sonoran Desertscrub, Mohave Desertscrub, Great Basin Desertscrub, Great Basin Grassland, Great Basin Conifer Woodland; Apache, Coconino, Mohave, Yavapai cos.; Nevada, Utah.

The small-flowered var. *modestus* forms the other major extreme in flower size within the species, contrasting with the large-flowered var. *vespertinus*. It is localized, forming uniform colonies within the range of var. *amphioxys* but only from valleys affluent to the Colorado River from San Juan and eastern Kane counties, Utah, and northern Coconino and Mohave counties, Arizona, with disjunct populations in Lincoln County, Nevada. Some plants from northeastern Arizona and northwestern New Mexico have similarly small flowers and might represent an eastward extension of var. *modestus*. There, however, influence of phases of *Astragalus missouriensis* cannot be ruled out. Variety *modestus* occurs in blackbrush, four-wing saltbush, Indian ricegrass, needle-and-thread grass, old man sagebrush, and mixed grass communities.

4c. Astragalus amphioxys A. Gray var. **musimonum** (Barneby) Barneby in A. Cronquist et al., Intermount. Fl. 3(B):138. 1989. * [of Sheep Mountain, Nevada] SHEEP MOUNTAIN MILKVETCH Native

Astragalus musimonum Barneby, Proc. Calif. Acad. Sci., ser. 4, 25:150, pl. 17, figs. 1–9. 1944

Flowers: calyx campanulate; tube 3.4–4.8 × 2.4–2.9 mm; lobes 1.8–2.5 mm; corolla: banner 10.3–13.4 mm; keel 8–11.5 mm. **Fruits** 15–20(–25) mm, crescent-shaped or incurved through 1/2 their length. **Seeds** 26–36.

Flowering Apr–Jun. Limestone gravels and outcrops; [1100–]1300–1400[–1800] m; Great Basin Conifer Woodland; Mohave Co.; Nevada.

Barneby (1964a) noted that *Astragalus musimonum* was known from so few examples that appraisal of species status was premature, noting simultaneously its closeness to *A. amphioxys*. In 1989 he reconsidered and combined the Sheep Mountain milkvetch as a variety within *A. amphioxys*. Isely (1998) commented on the transfer but continued to recognize it at the species level because of the sharply curved tip of the fruit, the short petal claws, and the short calyx. In Arizona, it is known from a single collection in extreme northwestern Mohave County. Across its range, var. *musimonum* is found in pinyon-juniper and blackbrush communities.

4d. Astragalus amphioxys A. Gray var. **vespertinus** (E. Sheldon) M.E. Jones. Rev. N.-Amer. Astragalus 215. 1923. * [from Latin *vespertin*, evening] EVENING MILKVETCH Native

Astragalus vespertinus E. Sheldon, Minnesota Bot. Stud. 1:150. 1894

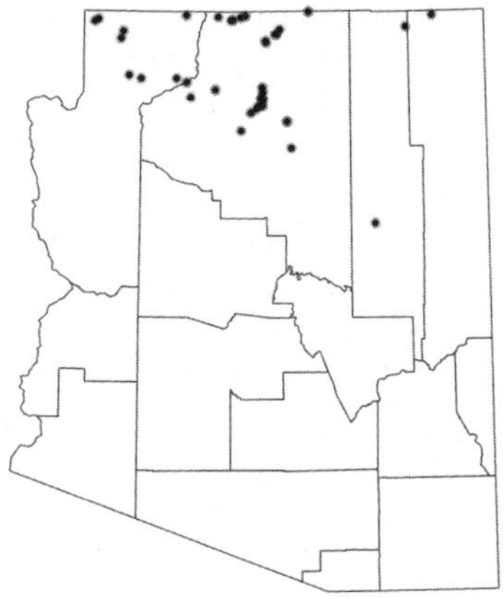

Flowers: calyx cylindric; tube 8.8–13.2 × 3.8–5.2 mm; lobes 1.3–3 mm; corolla: banner 23–27(–28) mm; keel (17–)18–24 mm. **Fruits** 30–35 mm, gently incurved. **Seeds** 50–70.

Flowering Mar–Jun. Sandy valley floors, sand dunes, ledges under cliffs, gravelly hillsides, canyons, and arroyos, usually on sandstone; 600–1800 (–2000) m; (Sonoran Desertscrub), Mohave Desertscrub, Great Basin Desertscrub, Great Basin Grassland, Great Basin Conifer Woodland; Apache, Coconino, Mohave, Navajo cos.; Colorado, New Mexico, Utah.

Variety *vespertinus* is a large-flowered, more or less northerly distributed form of the species that is, otherwise, rather poorly characterized. It occurs in creosote scrub, sagebrush, and mountain brush communities, grasslands, pinyon-juniper woodlands, riparian communities, and occasionally in upper elevation Sonoran Desert as in the Grand Canyon.

5. Astragalus ampullarius S. Watson. Amer. Naturalist 7:300. 1873. * [flask-shaped] GUMBO MILKVETCH Native

Perennials, low, slender, shortly caulescent, 2–28 cm; from deep, subterranean caudex; strigillose, hairs basifixed. **Stems** single or few, prostrate-ascending; underground for a length of 3.5–14 cm. **Leaves** 3–14 cm, petiolate; stipules with fused sheathing proximally, fused or distinct distally, ovate, deltoid, or lanceolate, 2–6 mm, membranous, apex obtuse proximally. **Leaflets** 7–15 (–19), flat, obovate, 4–15 × 2–12 mm, apex round to emarginate, surfaces strigose, sometimes glabrous on upper (adaxial) surface. **Inflorescences** 5–20-flowered; peduncle erect, 0.5–9.5 cm; axis 1.2–13 cm in fruit; bracts 1.5–3 mm; bracteoles 2. **Pedicels** ascending, 1–3 mm. **Flowers**: calyx short-cylindric, 4.8–7.5 mm, black-strigose; tube 4.2–6 × 2.5–3.5 mm; lobes broadly triangular, 0.5–1.5 mm; corolla ochroleucous, or pale pink-purple with white wing tips, 13.5–22 mm; banner recurved through 30°, elliptic or rhombic-elliptic, 13.5–20 mm; wings 10.6–18.3 mm; keel 9–12.2 mm, apex obtuse; ovules 17–25. **Fruits** ascending-erect; stipitate, stipe 9–19 mm; straight or slightly decurved, ovoid to nearly circular, round in cross section, inflated, 12–20 × 8–11 mm, thin, green, purple-dotted, becoming purplish-straw-colored, papery, glabrous or glabrate; one-chambered or nearly so with a partial septum. **Seeds** 17–25.

Flowering late Apr–Jun. Clay soils on sandstone knolls; 900–1700 m; Great Basin Conifer Woodland; Coconino, Mohave cos.; Utah.

Astragalus ampullarius is restricted to outcrops of the Triassic Chinle Formation in western Kane and eastern Washington counties, Utah, and immediately adjacent Mohave and Coconino counties, Arizona. It is often the most prevalent plant among several endemics on this formation. The dried, persistent stems and fruits of the previous year, circular-reclining, bleached and skeleton-like, are quite unlike any other species. After its discovery the species remained obscure for nearly seven decades, until rediscovered by Alice Eastwood and J. T. Howell in 1941, not that it is so rare, but because it is restricted to such a narrowly exposed geological formation, much of which is remote. The species is of conservation concern due to its limited geographic distribution.

6. Astragalus argophyllus Nuttall in J. Torrey and A. Gray, Fl. N. Amer. 1:331. 1838. * [from Greek *argo*, shining and *phyllo*, leaf] SILVER-LEAF MILKVETCH

Perennials, tufted or mat-forming, acaulescent to subacaulescent, 1.5–12 cm; from superficial caudex; densely pilose or strigillose, hairs usually basifixed. **Stems** absent to few, 0–10(–15) cm, prostrate. **Leaves** 1.5–12(–15) cm, petiole slender, often becoming persistent, internodes commonly shorter than stipules; stipules distinct throughout, lanceolate, 2–10 mm, submembranous becoming papery. **Leaflets** 7–21, flat or loosely folded, usually elliptic, rhombic-elliptic, rhombic-ovate, obovate-cuneate, elliptic-oblong, elliptic-obovate, oblanceolate, or obovate, rarely elliptic-oblanceolate, 2–15 × 1–6 mm, apex acute, obtuse, or acuminate, surfaces pilose. **Inflorescences** 1–6-flowered; peduncle incurved ascending, bent downward or prostrate in fruit, to 9 cm; axis 0.2–2(–2.5) cm in fruit; bracts 1.8–6.5 mm; bracteoles 0–2. **Pedicels** ascending, 1.2–3.8 mm. **Flowers**: calyx cylindric, 9–16.8 mm, strigillose, hairs white or black; tube 6.8–11.8 × 2.8–4.6 mm; lobes linear, 1.6–5(–5.8) mm; corolla pink-purple, often drying violet to bluish, sometimes fading, or ochroleucous-white and lilac-tinged or veined, except keel and wing tips, 15–25 mm; banner recurved through 40°, oblanceolate, broadly rhombic-oblanceolate, or spatulate, 15–24 mm; wings 12–23 mm; keel 12–20.3 mm, apex obtuse; ovules 25–43. **Fruits** ascending; sessile; straight or incurved, ovoid, lanceolate-ellipsoid, oblong-ellipsoid, or ovoid-acuminate, ± dorsiventrally compressed, 15–32(–37) × 5–12(–13) mm, fleshy, green becoming brown or straw-colored, stiffly leathery or woody, strigillose, strigillose-pilosulose, or silky-strigose; one-chambered. **Seeds** 25–43.

Varieties 4 (1 native in the flora): w United States.

Astragalus argophyllus is racially complex and varies considerably in vesture orientation, in flower size and color, and in ecology. Still, it is only with difficulty that material may be sorted into infraspecific taxa and sometimes plants cannot be satisfactorily identified.

6a. Astragalus argophyllus Nuttall var. **panguicensis** (M.E. Jones) M.E. Jones. Contr. W. Bot. 8:5. 1898. * [of the Panguitch Plateau] PANGUITCH MILKVETCH Native

Astragalus chamaeleuce A. Gray var. *panguicensis* M.E. Jones, Proc. Calif. Acad. Sci., ser. 2, 5:671. 1895

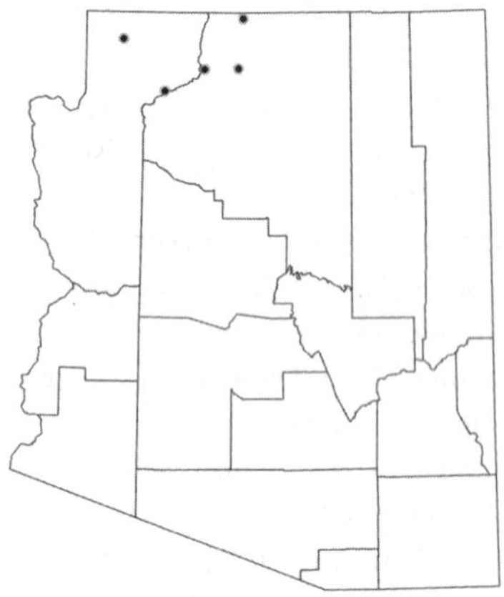

Herbage: hairs basifixed. **Stems** rarely developed, 0–5(–10) cm. **Leaves** 1.5–6(–8) cm. **Leaflets** 7–15(–17), elliptic-obovate, 2–9 mm, apex obtuse to very shortly acuminate. **Flowers**: calyx 9–12(–14) mm; tube 6.8–8.7(–10.6) × 2.8–3.4 mm; lobes 1.9–3.4(–4.2) mm; corolla pink-purple to pale tinged with lilac or dull purple: banner 15–17.5(–18.4) mm; keel 12–15.2 mm. **Fruits** lunately lanceolate-ellipsoid or narrowly and very obliquely ovoid-acuminate, 15–27 × 5–8(–9) mm, hairs ascending, silvery, to 0.5–0.8 mm.

Flowering Apr–Jul. Gravelly hillsides and benches; 1280–2690[–3200] m; Great Basin Conifer Woodland, Petran Montane Conifer Forest; Coconino, Mohave cos.; Utah.

Variety *panguicensis* is restricted to south-central Utah, with an outlying population on Mount Emma, Unikaret Plateau, in Coconino County, Arizona. Specimens keying to this variety have also been collected on the Kaibab Plateau and on Buckskin Mountain. Some collections from Mount Emma are exactly typical; others have exceptionally large flowers suggestive of the generally more northern var. *martini*. A single collection of a poorly differentiated plant with slightly larger flowers from Mt. Trumbull, and referred to as var. *alvius* S.L. Welsh, has been described (Welsh 2007). Barneby (1964a) states that the only consistent character distinguishing var. *panguicensis* from the other varieties of *Astragalus argophyllus* is the denser, persistent, silky vestiture. The species is found in ponderosa pine forest, aspen-Douglas fir-limber pine, white fir, sagebrush, and pinyon-juniper communities in Utah and Arizona.

7. Astragalus aridus A. Gray. Proc. Amer. Acad. Arts 6:223. 1864. * [from Latin *aridus*, dry] PARCHED MILKVETCH Native

Annuals, short-lived, of 2–4 months duration, caulescent, slender or coarse, 2.5–30 cm; taproot slender; from superficial root-crown; densely strigillose-pilose, hairs basifixed, white. **Stems** single or few, decumbent with ascending tips or trailing and diffuse. **Leaves** (2–)3–9(–10) cm, petiolate; stipules distinct throughout, ovate or triangular, 1.5–5(–6.5) mm, herbaceous becoming paper. **Leaflets** (7–)9–17, flat or loosely folded, oblanceolate, oblong-oblanceolate, or elliptic, 4–16 × 1.9–2.8(–4) mm, apex obtuse, truncate, or emarginate, surfaces pubescent. **Inflorescences** (3–)4–9-flowered; peduncle incurved-ascending, 2–5.5 cm; axis 1.5–5.5 cm in fruit; bracts 0.8–2.2 mm; bracteoles 0. **Pedicels** ascending, 0.2–1 mm. **Flowers**: calyx turbinate-campanulate, 3.3–4.3 mm, densely silvery pilose; tube 2.1–2.7 × 1.5–2.3 mm; lobes broadly subulate, 1–1.6 mm; corolla whitish, faintly tinged with pink-lilac or pale pink, (3.3–)3.5–6.5 mm; banner recurved through 40°, obovate-cuneate or broadly elliptic, (3.3–)3.5–6.5 mm; wings 3.6–5.4 mm; keel 3.6–5 mm, apex round or narrowly triangular and beaklike; ovules (3–)4–6(–7). **Fruits** ascending; sessile; incurved, semi-ellipsoid or lunately ellipsoid, slightly inflated but scarcely bladdery, 10–17 × 4.5–7 mm, straw-colored, thin becoming papery and opaque, canescently strigillose-pilosulose; one-chambered. **Seeds** (3–)4–6(–7).

Flowering Feb–May. Desert dunes, open sandy plains, valley floors; [-60–]100–130[–400] m; Sonoran Desertscrub; Yuma Co.; California; Mexico (Baja California, Sonora).

Astragalus aridus is a species of the Colorado Desert in southern California and adjacent Arizona. It is vaguely similar to *A. insularis*, but the flowers of *A. aridus* are smaller than those of any similar species. In Arizona, *A. aridus* is apparently restricted to extreme southern Yuma County.

8. Astragalus arizonicus A. Gray. Proc. Amer. Acad. Arts 7:398. 1868. * [of Arizona] ARIZONA MILKVETCH Native

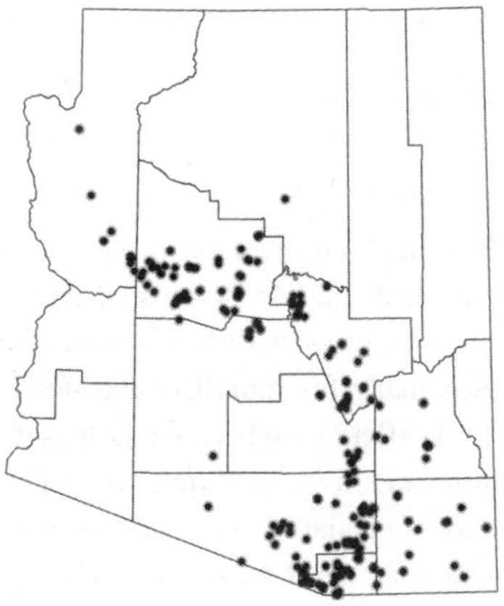

Perennials, short-lived, sometimes flowering first year, diffuse, usually slender, clump-forming, caulescent, (5–)15–40(–55) cm; from superficial caudex; silvery-canescent or silvery- to gray-strigillose, hairs medifixed. **Stems** several or numerous, decumbent or prostrate with ascending tips. **Leaves** 2–8 (–10) cm, petiolate or subsessile; stipules distinct throughout, ovate, deltoid, or triangular, 2–4 mm, submembranous becoming papery. **Leaflets** (5–)9–15(–17), flat or folded, linear, linear-oblong, linear-elliptic, narrowly lanceolate, or oval, (1.5–)3–17(–21) × 1.5–4(–5) mm, apex obtuse to mucronulate or acute, surfaces strigillose. **Inflorescences** with (7–)10–30 flowers; peduncle erect or widely diverging and incurved, (1.5–)2.5–10(–15) cm; axis (1.5–)2.5–10 cm in fruit; bracts 0.8–2(–3) mm; bracteoles 0–2. **Pedicels** ascending, 0.5–2.2 mm. **Flowers:** calyx campanulate, (4.5–)5–7 mm, strigillose; tube 3.3–4.1 × 2.1–3.1 mm; lobes broadly to slenderly subulate, (1.2–)1.4–3 mm; corolla dull pink-purple, wing tips pale or white and ± suffused with greenish yellow, keel spotted, (8.7–)9–12 mm; banner recurved through 50°, obovate-cuneate to nearly circular-flabellate, (8.7–)9–12 mm; wings (6.8–)7.1–10 mm; keel 8.2–10.4 mm, apex narrowly triangular, acute, beaklike; ovules 14–22. **Fruits** erect or ascending; sessile; straight or gently incurved, linear-oblong or linear, obtusely 3-sided compressed, 15–30 × (2.2–)2.5–4.5 mm, green, thin becoming straw-colored, papery, strigillose; two-chambered. **Seeds** 14–22. $2n = 16, 26$.

Flowering Mar–May(–Jun). Various substrates on ridges, slopes, canyons, floodplains, and roadsides; 700–1500 m; Sonoran Desertscrub, (Mohave Desertscrub), Semidesert Grassland, Interior Chaparral, (Madrean Evergreen Woodland, Great Basin Conifer Woodland); Cochise, Coconino, Gila, Graham, Maricopa, Mohave, Pima, Pinal, Santa Cruz, Yavapai cos.; New Mexico; Mexico (Sonora).

Astragalus arizonicus occurs from west-central to southeastern Arizona and into northern Sonora. The species is most common in grasslands and scrub communities on slopes. The hairs vary from definitely medifixed to those where the attachment is very near the base; careful observation and good magnification is required. *Astragalus arizonicus* is the only member of the genus, within its area of distribution, with narrowly oblong leaflets and widely-spaced flowers.

9. Astragalus beathii C.L. Porter. Madroño 6:18, pl. 3. 1941. * [for Orville A. Beath, 1884–1965, American chemist who worked on selenium in plants, the discoverer of this species] BEATH'S MILKVETCH Native

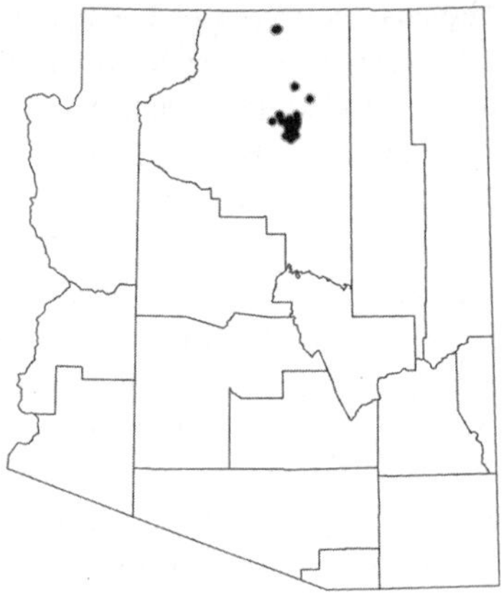

Perennials, coarse, robust, forming bushy clumps, caulescent, malodorous (selenium-scented), 25–60 cm; from superficial caudex; strigillose, sometimes glabrate proximally, hairs basifixed. **Stems** several to numerous, erect to ascending, becoming stout, hollow. **Leaves** 6–12(–15) cm, shortly petiolate; stipules distinct throughout, broadly deltoid or triangular, 3–9.5 mm, submembranous becoming pallid, papery-scarious. **Leaflets** 11–21(–23), flat, narrowly elliptic-oblanceolate to broadly obovate or oblong-obovate, 5–25 × (2.5–)3.3–6.2 mm, apex obtuse to retuse, surfaces proximally glabrous, distally pubescent below (abaxially), glabrous above (adaxially). **Inflorescences** 10–27-flowered; peduncle incurved-ascending, (2.5–)4–10 cm; axis 3.5–10 cm in fruit; bracts 2–4.2 mm; bracteoles 0–2. **Pedicels** spreading-ascending, 0.9–3 mm. **Flowers**: calyx short-cylindric, (7.7–)8.2–10.5 mm, sparsely to densely strigillose; tube purplish or deep-reddish purple, (5.3–)5.7–7.8 × (3–)3.4–4.5(–5) mm; lobes subulate, 1.8–3.1 mm; corolla bright purple (drying violet), wing tips pale or white, but rarely pure white or whitish-blue, (17–)20–25(–27) mm; banner recurved through 40°, broadly rhombic-elliptic or rhombic-ovate, (17–)20–25(–27.7) mm; wings (13.4–)15.4–19.5(–21.8) mm; keel (9.6–)11.2–16.4 mm, apex bluntly triangular; ovules 29–52. **Fruits** deflexed or declined; sessile or nearly so, stipe 0–0.6 mm; straight or slightly incurved, narrowly to plumply oblong-ellipsoid, somewhat dorsiventrally compressed, 20–39 × 7–11 mm, thick, fleshy, green or purplish-green becoming straw-colored, leathery, glabrous or minutely puberulent; nearly two-chambered, septum 1.3–2.3 mm wide. **Seeds** 29–52. $2n = 24$.

Flowering late Mar–May. Sandy flats, red clay knolls, and gullied washes in badlands, on seleniferous substrates of the Moenkopi Formation and rarely Chinle Formation; 1030–1650 m; Great Basin Desertscrub; Coconino Co.; endemic.

Astragalus beathii is a locally abundant, narrow endemic, found near Cameron, in Coconino County, where it grows in saltbush, Mormon tea, matchweed (*Gutierrezia* sp.), yucca, and galleta communities. The species is of conservation concern.

10. Astragalus bisulcatus (Hooker) A. Gray in War Department [U.S.], Pacif. Railr. Rep. 12:42. 1860. * [from Latin *bi*, two and *sulco*, furrow, with two grooves] TWO-GROOVED MILKVETCH

Phaca bisulcata Hooker, Fl. Bor.-Amer. 1:145. 1831

Perennials, coarse, sometimes slender, clump-forming, caulescent, selenophyte, 15–75 cm; from superficial caudex, branched; glabrous to mostly strigillose, hairs basifixed. **Stems** several to numerous, often erect or ascending, sometimes decumbent, diffuse, or spreading. **Leaves** 3–13.5 cm, shortly petiolate or subsessile; stipules with fused sheathing proximally, fused or distinct distally, deltoid or triangular, 2.5–10(–12) mm, submembranous becoming scarious, pallid. **Leaflets** (7–)11–35, flat or loosely folded, lanceolate-oblong to oblong, elliptic, or oblanceolate, 5–27(–32) × 1.5–11 mm, apex usually acute or obtuse, rarely emarginate, surfaces strigillose below (abaxially), glabrous or glabrate on upper (adaxial) surface. **Inflorescences** with (15–)20–80 flowers; peduncle erect, ascending, or incurved, 2.5–13 cm; axis 3–25 cm in fruit; bracts (1–)2.5–7 mm; bracteoles 0–2. **Pedicels** declined, 1–3.5 mm. **Flowers**: calyx tube red-purple to dark red, or pallid, obliquely campanulate, with a sac on one side, 3.5–9.6 mm, sparsely strigose; tube 2.8–5.7(–7) × 1.9–2.3(–2.6) mm; lobes subulate, (0.7–)1–4.5(–6) mm; corolla white, ochroleucous, dull yellowish-white, or rose-purple, keel tip often purple-spotted or suffused, (6.7–)8–17.5 mm; banner usually recurved through 45°, oblanceolate to oval-obovate, 6.7–17.5 mm; wings shorter; keel 7–13 mm, apex bluntly rounded; ovules 5–15. **Fruits** pendulous; stipe slender, 1.4–5.2(–6) mm; straight or slightly curved, linear, ellipsoid, or oblong-ellipsoid, dorsiventrally compressed, ± 3-sided, (5–)6.5–17(–20) × 2–4.5 mm, fleshy green becoming straw-colored, thinly leathery or papery, strigillose or glabrous; one-chambered. **Seeds** 5–15. $2n = 22, 24$.

Varieties 3 (1 native in the flora): w United States. Variety *haydenianus* has been reported from northeastern Arizona but more mature specimens need to be collected to verify identification.

Astragalus bisulcatus is a malodorous, primary indicator of selenium in most areas. From the western Great Plains to the western slope of the Rocky Mountains it is easily recognized by its 2-grooved fruits, where it is relatively nonvariable and presents no taxonomic problems.

10a. Astragalus bisulcatus (Hooker) A. Gray var. **bisulcatus** * Native

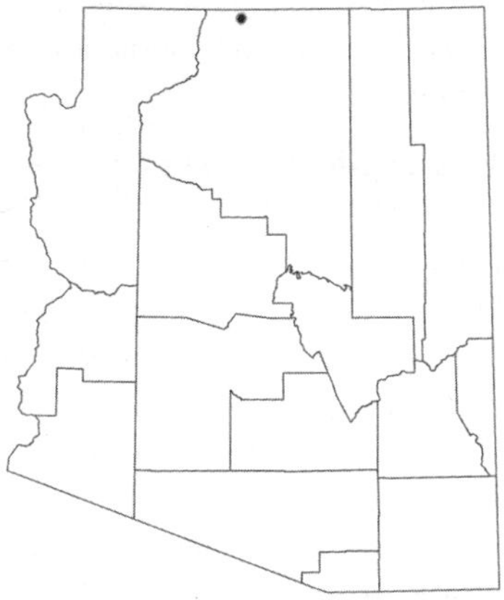

Plants usually erect, stout; herbage sparsely hairy or stems glabrous. **Leaflets** 17–29, (5–)10–25(–32) mm. **Inflorescences** 25–75-flowered; axis 5–18 cm in fruit; bracts (2–)2.5–7 mm. **Flowers**: calyx usually red-purple; tube 3.3–5.7 mm; lobes 1.5–4.5(–6) mm; corolla pink-purple, or pallid and purple- or lilac-tipped or suffused, or white, or whitish with keel tip spotted, 13–17.5 mm; banner longer than keel. **Fruits:** stipe 3–5(–6) mm; linear- or narrowly oblong-ellipsoid, (8–)10–17(–20) × 2–4.5 mm, smooth or faintly reticulate, glabrous or strigillose. **Seeds** 10–14(–15). $2n$ = 22, 24.

Flowering May–Aug. Commonly on fine-textured, saline or seleniferous substrates on slopes, benches, knolls, and canyon floors; [500–]1550–1600 [–2500] m; Great Basin Conifer Woodland; Coconino Co.; Colorado, Idaho, Kansas, Montana, Nebraska, New Mexico, North Dakota, South Dakota, Utah, Wyoming; Canada (Alberta, Manitoba, Saskatchewan).

Variety *bisulcatus*, as used here, includes the so-called "basin variants" of Barneby (1964a), which occur from southern Wyoming through western Colorado and eastern and southern Utah to northern Arizona. *Astragalus bisulcatus* has been confused with *A. racemosus*, with which it sometimes intermingles in south-central Colorado. The 2-grooved fruits of *A. racemosus* differentiate them from the ± 3-sided fruits of *A. bisulcatus*. Where sympatric, *A. bisulcatus* is also usually purple-flowered. The vars. *bisulcatus* and *haydenianus* (of Colorado and northern New Mexico) are known to contain selenium and swainsonine, both poisonous to livestock (Trelease and Beath 1949, Kingsbury 1964). Variety *bisulcatus* is known in Arizona only from the Coyote Springs Valley in northwest Coconino County; the few specimens of this plant collected from this location generally have white-ochroleucous petals with a dull purple keel tip, which may cause misidentification as var. *haydenianus*. Across its range, var. *bisulcatus* is found in short-grass prairies and plains, badlands, or in steppes, with sagebrush-grass, grass, shadscale, and less commonly with pinyon-juniper and in mountain brush communities.

11. Astragalus brandegeei Porter in T.C. Porter and J.M. Coulter, Syn. Fl. Colorado 24. 1874 (as *Astragalus Brandegei*) * [for Townshend S. Brandegee, 1843–1925, American engineer and naturalist] BRANDEGEE'S MILKVETCH Native

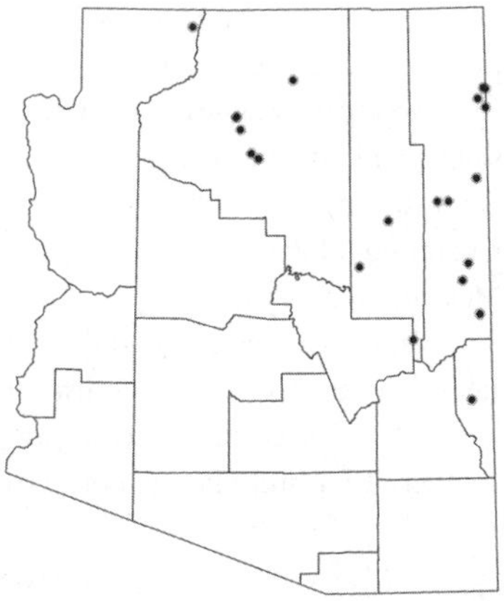

Perennials, low, delicate, often flowering first year, caulescent, 5–35(–40) cm; from superficial caudex, branched; strigillose, cinereous, hairs basifixed. **Stems** few to several, prostrate-spreading, very slender and wiry. **Leaves** (1–)2–11.5(–15) cm, petiole slender; stipules with fused sheathing proximally, fused or distinct distally, deltoid or lanceolate, 1.5–5 mm, papery proximally, herbaceous distally. **Leaflets** 5–15, flat, involute or longitudinally folded, linear-filiform to narrowly oblong, 5–27 × 0.5–2.6 mm, apex acute to obtuse, surfaces strigose below (abaxially), glabrous on upper (adaxial) upper side. **Inflorescences** with 1–5(–7) flowers, flowers widely spaced along rachis; peduncles dimorphic, early ones much shorter than leaves, commonly only 1-flowered, later ones elongate and surpassing the leaves, humistrate in fruit, 2.5–14(–17) cm; axis threadlike, 0.5–6(–8) cm in fruit; bracts 1–2 mm; bracteoles 2. **Pedicels** ascending, 1.2–4 mm. **Flowers:** calyx campanulate, 2.7–4 mm, black-strigose; tube 1.8–2.5 × 1.5–2.2 mm; lobes subulate, 0.9–2 mm; corolla ochroleucous or tinged dull lilac to violet, 4–6(–7) mm; banner abruptly recurved through 90°, ovate or nearly circular-cuneate, 4–7 mm; wings 4.7–6 mm; keel 3.5–4.5 mm, apex strongly incurved and obtuse; ovules 14–18. **Fruits** pendulous to ascending; sessile or nearly so, stipe 0–1 mm; straight or slightly decurved, obovoid to oblong-ellipsoid, slightly dorsiventrally compressed, 10–18 × 3.5–5 mm, thin, greenish or purple-dotted becoming brownish straw-colored, papery, strigose; incompletely two-chambered, septum 1.2–2 mm wide. **Seeds** 13–18. $2n$ = 22.

Flowering May–Sep. Gravelly or sandy substrates of ridges, slopes and meadows, arroyos and roadsides; 1430–2200[–2800] m; (Great Basin Desertscrub), Great Basin Grassland, Great Basin Conifer Woodland; Apache, Coconino, Navajo cos.; Colorado, New Mexico, Utah.

Astragalus brandegeei is a delicate, relatively inconspicuous plant that is seldom collected, probably because of its inconspicuous tiny flowers, threadlike peduncles, and slender stems. It is found in mixed shrublands, pinyon-juniper, or oak brush. Seeds of this species are able to germinate whenever rainfall is sufficient during the warm season (Barneby 1964a).

12. Astragalus calycosus Torrey ex S. Watson. Botany (Fortieth Parallel) 66, pl. 10, figs. 4–7. 1871. * [with a conspicuous calyx] TORREY'S MILKVETCH

Perennials, low, often diminutive, tufted or mound-forming, acaulescent, 1–12(–23) cm; from superficial caudex, branches often with well-developed thatch of persistent leaf bases; densely silvery-strigillose throughout, hairs medifixed. **Stems** usually obsolete, rarely 2, 0–2 cm, internodes concealed by stipules. **Leaves** 1–7(–12) cm, petiole slender or threadlike; stipules distinct throughout, ovate or triangular, 1.5–6 mm, submembranous becoming papery. **Leaflets** (1–)3–13, flat, obovate, oblanceolate, or elliptic, 2–19 × 1–7 mm, apex obtuse to acute, surfaces silvery strigose. **Inflorescences** with (1–)2–6(–8) flowers; peduncle scape-like, ascending or decumbent, 1–7+ cm; axis 0.2–2(–2.5) cm in fruit; bracts 0.5–2 mm; bracteoles 0 or 1, minute. **Pedicels** ascending to spreading, 0.7–3 mm. **Flowers:** calyx campanulate to short-cylindric, [5–]5.2–10.6 mm, strigose; tube 4–6.4(–6.7) × 2.2–4.5 mm; lobes subulate, 1–4.2 mm; corolla whitish, light blue, or pink-purple and wing tips whitish, 10–16.5(–20.8) mm; banner recurved through 45–90°, ovate, nearly circular, or obreniform, 10–16.5(–20.8) mm; wings 9–17(–19.7) mm, deeply cleft; keel 7.4–11.6(–12.7) mm, apex round or bluntly deltoid; ovules 13–28. **Fruits** ascending; sessile; usually straight or slightly incurved, narrowly oblong, 3-sided compressed, 8–25 × 3–4.5 mm, thin, straw-colored, becoming papery, strigose; two-chambered. **Seeds** 13–28. $2n = 22$.

Varieties 4 (2 native in the flora): w United States.

Astragalus calycosus includes two major multiracial and confluent varieties, var. *calycosus* to the north and northwest, and var. *scaposus* to the southeast; two other, minor varieties not present in Arizona.

1. Leaflets (1–)7–13, rachis usually 1+ cm; peduncle erect-ascending, 7+ cm ____ **12b. Astragalus calycosus** var. **scaposus**
1′ Leaflets 3–7, rachis to 1 cm; peduncle ascending or decumbent, 1–7 cm ____ **12a. Astragalus calycosus** var. **calycosus**

12a. Astragalus calycosus Torrey ex S. Watson var. **calycosus** * Native

Plants tuft- or mound-forming. **Leaves** 1–7 cm; rachis to 1 cm. **Leaflets** 3–7, 5–19 mm. **Inflorescences** peduncle ascending or decumbent, 1–7 cm, shorter or longer than leaves. **Flowers**: calyx lobes (1–)1.5–4.2 mm; corolla whitish to pink-purple, 10–16.5(–20.8) mm.

Flowering May–early Jul. On rock ledges, gravel, sand, and silt of ridges, slopes, and canyons; [1200–]1600–2400[–3500] m; (Great Basin Desertscrub), Great Basin Conifer Woodland, Petran Montane Conifer Forest; Apache, Coconino, Mohave, Navajo, Yavapai cos.; California, Idaho, Nevada, New Mexico, Utah, Wyoming.

Variety *calycosus* is found from east-central California to north-central Idaho, eastern Utah, and southwestern Wyoming. It occurs in mixed desert shrublands, sagebrush, pinyon-juniper, and ponderosa pine woodlands. In southern Utah and adjacent northern Arizona it passes into var. *scaposus*.

12b. Astragalus calycosus Torrey ex S. Watson var. **scaposus** (A. Gray) M.E. Jones. Zoë 4:26. 1893. * [scapose, referring to the peduncles] SCAPOSE MILKVETCH Native

Astragalus scaposus A. Gray, Proc. Amer. Acad. Arts 13:366. 1878

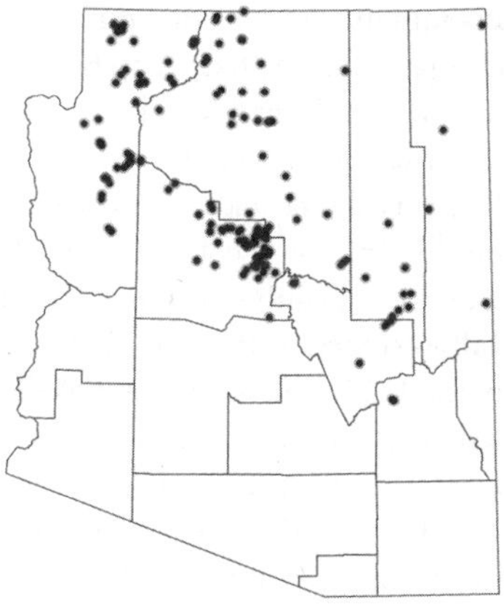

Plants growing in dense tufts. **Leaves** 1–9 cm; rachis usually 1+ cm. **Leaflets** (1–)7–13, 5–15(–19) mm. **Inflorescences** peduncle erect-ascending, 7+ cm, longer to much longer than leaves. **Flowers**: calyx lobes 1–2 mm; corolla pink-purple, wing-tips whitish, 10.5–14 mm.

Flowering Mar–Jul (Oct). Various substrates on ridges, slopes, canyons, arroyos, floodplains, and roadsides; 670–2180 m; (Sonoran Desertscrub), Great Basin Desertscrub, Great Basin Grassland, Interior Chaparral, Great Basin Conifer Woodland; Apache, Coconino, Gila, Graham, Mohave, Navajo, Yavapai cos.; Colorado, Nevada, New Mexico, Utah.

With its contrasting white/pink-purple pattern on the petals and long ascending inflorescences, var. *scaposus* is the showiest form of the species. It occurs from southeastern Nevada, across northern Arizona, to southwestern Colorado and northwestern New Mexico. It is found in juniper, pinyon-juniper, and mixed desert shrub communities.

13. Astragalus castaneiformis S. Watson. Proc. Amer. Acad. Arts 20:361. 1885 (as *Astragalus castaneaeformis*) * [from Latin *castanea*, chestnut and *forma*, shape, referring to the shape of the fruits] CHESTNUT MILKVETCH Native

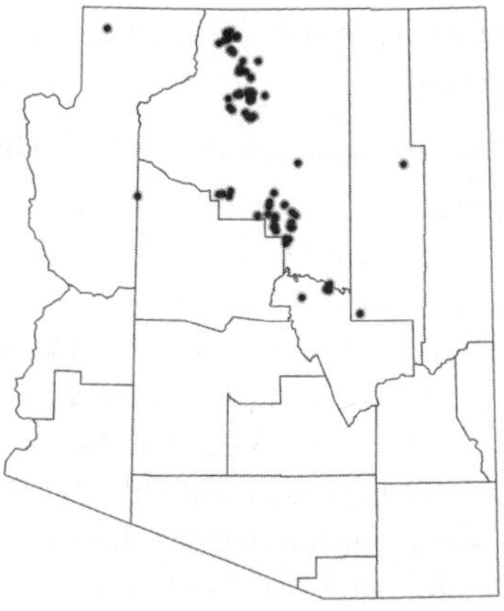

Perennials, low, dwarf, tufted, acaulescent or subacaulescent, 3–9 cm; from superficial caudex or root-crown; silky-strigose to strigillose, hairs medifixed. **Stems** absent to reduced, ascending or prostrate, radiating from root-crown, internodes mostly obscured by leaf bases, 0–3 cm. **Leaves** (1.5–)2.5–10 cm, petiolate; stipules distinct throughout, ovate, triangular, or lanceolate, 3–7 mm, submembranous. **Leaflets** 7–13, flat or loosely folded, obovate-cuneate or elliptic-oblanceolate, 3–12 × 2–5 mm, apex obtuse or subacute, surfaces strigillose. **Inflorescences** 2–10-flowered; peduncle ascending, 0.5–3(–5) cm; axis 0.1–1.5 cm in fruit; bracts 1.5–5.5 mm; bracteoles 0(–2), exceptionally conspicuous. **Pedicels** ascending, 1.7–2.4(–3) mm. **Flowers**: calyx cylindric or vase-shaped, 6–10.5 mm, strigillose, hairs black or white; tube 6.5–8 × 1.9–2.7 mm; lobes subulate, 1.3–2.5 mm; corolla white or ochroleucous, suffused with dull lilac or purple, (14.5–)15–18.5 mm; banner recurved through 40°, broadly rhombic-oblanceolate or spatulate-oblanceolate, 11–18.5 mm; wings 6–9.3 mm; keel 8.3–14.3 mm, apex obtuse; ovules 18–33. **Fruits** spreading-ascending, humistrate; sessile; scarcely to moderately incurved, obliquely ovoid, lanceolate-ovoid, or lunately ellipsoid, dorsiventrally compressed becoming laterally compressed distally, 9–20(–22) × 4–8 mm, green or purplish tinged becoming brownish or straw-colored, leathery, densely strigillose, distinctly beaked; one-chambered. **Seeds** 18–33.

Flowering Apr–early Jul. Gentle slopes and flats in dry stony soils of basaltic or, rarely, calcareous substrates; (1520–)1900–2800 m; Great Basin Conifer Woodland, Petran Montane Conifer Forest; Coconino, Gila, Mohave, Navajo, Yavapai cos.; Utah.

In Arizona, *Astragalus castaneiformis* is a species of the Kaibab and Mogollon Plateaus, Coconino County, extending southeast to the Natanes Plateau, in Gila County. To the north, it just enters Kane County, Utah, in the Buckskin Mountains. The plant grows in open ponderosa pine forests and pinyon-juniper woodlands, occasionally in pine-fir-spruce forests and rarely in oak woodlands.

14. Astragalus ceramicus E. Sheldon. Minnesota Bot. Stud. 1:19. 1894. * [an earthen pot, referring to the mottled fruits] PAINTED MILKVETCH

Phaca picta A. Gray, not *Astragalus pictus* Steudel

Perennials, slender, wiry, caulescent, 3–40(–45) cm; from deep subterranean caudex, branched, branches slender, elongate, rhizome-like; densely strigillose, ashen or silvery-canescent, hairs medifixed or basifixed. **Stems** single or few-clustered, sprawling to erect; 3–40+ cm underground. **Leaves** with slender petiole and rachis, threadlike or linear-oblong, sometimes reduced to phyllodes distally, 2–17 cm; stipules with fused sheathing proximally, mostly fused distally, deltoid or lanceolate, 1.5–9 mm, subscarious proximally, herbaceous distally. **Leaflets** (1–)3–13, usually involute, narrowly oblong to threadlike, 3–50(–80) × 0.3–4 mm, apex obtuse to retuse or acute, surfaces strigillose, hairs very slightly spreading; terminal leaflet continuous with rachis distally, at least some reduced to phyllodes. **Inflorescences** with 2–15(–25) flowers; peduncle widely diverging and incurved, 0.7–7.5 cm; axis 1–8(–15) cm in fruit; bracts 1–2.5 mm; bracteoles 0. **Pedicels** ascending to declined, 0.7–3.1 mm. **Flowers**: calyx campanulate, 3.1–6 mm, strigose; tube 2.1–3.5 × 1.6–3 mm; lobes subulate, 1–3 mm; corolla whitish or dull purplish to pink, sometimes suffused or veined with lavender, 6.3–10.8 mm; banner abruptly recurved through 85–90°, broadly obovate, nearly circular, or reniform, 6.3–10.8 mm; wings 6–10.6 mm; keel 6.4–9.1 mm, apex narrowly triangular, subacute, sometimes porrect; ovules 12–29. **Fruits** pendulous; sessile to stipitate, stipe 0–3.3 mm; straight or slightly curved, ellipsoid to ovoid, subglobose, bladdery-inflated, (10–)15–50 × 5–26 mm, bright red- or purple-mottled valves becoming purplish brown, lustrous, papery, glabrous; one-chambered. **Seeds** 12–29. $2n = 22$.

Varieties 3 (1 native in the flora): w, c United States.

Rushlike or with very few leaflets, *Astragalus ceramicus* is inconspicuous in flower but showy in fruit; it is marked by its rhizomatous habit and red-mottled, bladdery fruits. The caudex, in sandy sites, is often extensively branched and pervasive.

14a. Astragalus ceramicus E. Sheldon var. **ceramicus** * Native

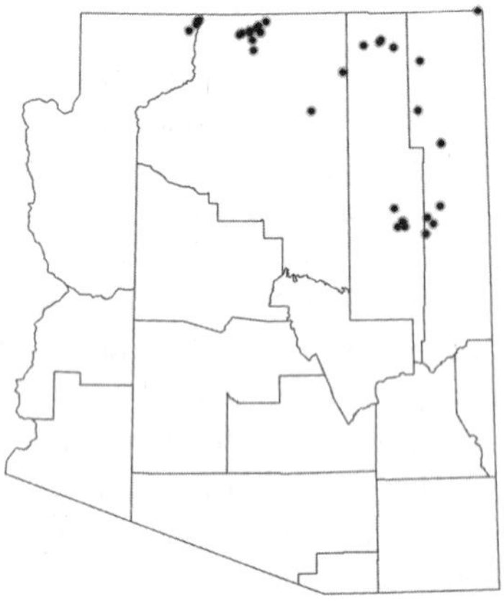

Plants with mostly medifixed hairs on stems and herbage. **Leaves** 2–12 cm. **Leaflets:** usually 1–6 pairs of lateral leaflets proximally or throughout, distally often reduced to subfiliform rachis. **Inflorescences** with 6–15(–25) flowers; axis (1–)1.5–8(–15) cm in fruit. **Flowers**: calyx 3.1–4.2 mm; tube 2.1–2.6(–3.3) mm; lobes 1–1.8 mm; corolla 6.3–8.3(–9.5) mm. **Fruits:** stipe (1–)1.5–3.3 mm; (10–)15–30 × 5–15(–22) mm. **Seeds** 12–16(–26).

Flowering late Apr–Jul. Dunes and other sandy sites of gentle slopes, valleys, canyon floors, roadsides, and other disturbed areas; 1200–2140 [–2500] m; Great Basin Desertscrub, Great Basin Grassland, Great Basin Conifer Woodland; Apache, Coconino, Mohave, Navajo, cos.; Colorado, Nevada, New Mexico, Utah.

Astragalus ceramicus var. *ceramicus* is widespread across northern Arizona but is seldom common. It is found in pinyon-juniper, sagebrush, stream bank, grassland, and mixed desert shrub communities. The roots are reportedly sweet and are eaten during the spring months by Hopi children (Kearney and Peebles 1942).

15. Astragalus chuskanus Barneby & Spellenberg. Brittonia 39:188, fig. 1. 1987. * [of the Chuska Mountains] CHUSKA MILKVETCH Native

Perennials, long-lived, mat- or cushion-forming, caulescent, 10–35(–40+) cm; from caudex, branched; strigillose to pilosulose, hairs basifixed. **Stems** numerous, prostrate, densely leafy. **Leaves** (1–)1.5–5.5 cm, petiolate; stipules with fused sheathing throughout, deltoid to ovate, 2–5.5 mm, subherbaceous becoming papery. **Leaflets** (7–)9–15, not crowded, flat, loosely folded, involute, or convex and keeled, obovate- or oblong-elliptic, 0.7–10 × 0.7–3(–4) mm, apex abruptly acuminate, surfaces pilosulose. **Inflorescences** densely 4–10-flowered; peduncle erect, 1.5–3.5(–4) cm, shorter than leaf; axis 0.3–1.2 cm in fruit; bracts 2–3.5 mm; bracteoles 0. **Pedicels** ascending-spreading, 0.9–1.5 mm. **Flowers**: calyx campanulate, 4.5–6.9 mm, pilosulose; tube

2.5–3.8 × 1.6–2 mm; lobes subulate, 1.5–3.1 mm; corolla whitish, tinged with lilac, fading pale ochroleucous, 7.3–10 mm; banner recurved through 40°, spatulate-obcordate, 7.3–10 mm; wings 7.1–8 mm; keel 5.6–6 mm, apex obtuse; ovules 4–6. **Fruits** ascending, humistrate; sessile; straight, obliquely semi-ovoid, compressed, 6 × 3 mm, pale green or straw-colored, or brownish, sometimes purple-mottled, becoming stiffly papery, strigillose; one-chambered. $2n = 24$.

Flowering late May–Jul. Sandy soil, sandstone substrates, and occasionally basalt of mesas, slopes, valleys, canyons, roadcuts, and disturbed areas; 1650–2930 m; Great Basin Conifer Woodland, Petran Montane Conifer Forest; Apache Co.; New Mexico.

Astragalus chuskanus is restricted to the Chuska Mountains and vicinity in Apache County in northeast Arizona and northwest New Mexico.

16. Astragalus cicer Linnaeus. Sp. Pl. 2:757. 1753. * [*Cicer*, the genus for chickpea] CHICKPEA MILKVETCH Adventive

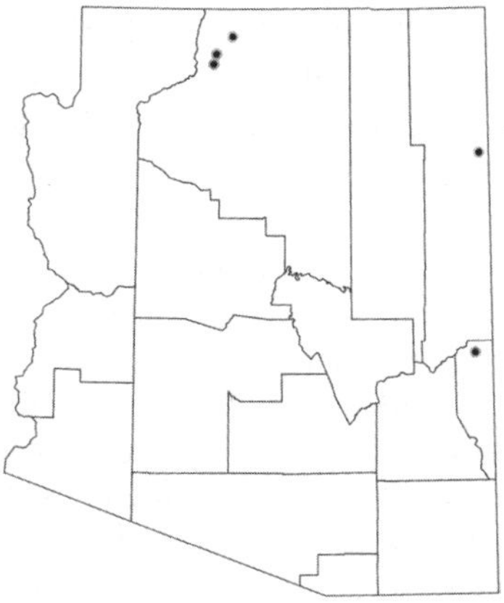

Perennials, caulescent, leafy, 30–90(–130) cm; from branched, subterranean caudex, branches long, rhizome-like; sparsely strigose-pilose or glabrate, hairs basifixed. **Stems** erect, diffuse, flexuous, becoming decumbent-clambering. **Leaves** 2–10 cm, subsessile; stipules with fused sheathing proximally, fused or distinct distally, lanceolate, 2–8 mm, becoming papery, apex sometimes acuminate or caudate. **Leaflets** 17–29(–31), flat, lanceolate-elliptic or oblong, 5–35 × 2–12(–15) mm, apex obtuse, mucronulate, or acute, surfaces strigillose, sometimes glabrescent above (adaxially). **Inflorescences** almost headlike, densely (6–)10–30-flowered, raceme shorter than leaf; peduncle incurved-ascending, 3.5–11 cm; axis (1.5–)2–5 cm in fruit; bracts 2–6.5 mm; bracteoles 0. **Pedicels** ascending, 0.3–1.5 mm. **Flowers**: calyx subcylindric, 6.5–9 mm, strigillose; tube 5–6 × 3.2–4 mm; lobes triangular-acuminate, 1.6–3 mm; corolla ochroleucous, 12.5–16.5 mm; banner not or slightly recurved, oblanceolate or spatulate, 12.5–16.2 mm; wings 11–14.5 mm; keel 9.6–10.5 mm, apex obtuse; ovules 14–26. **Fruits** ascending to spreading; subsessile, stipe to 0.8 mm; fleshy, green, broadly ovoid to nearly globose, inflated, hardly bladdery, 6–14 × 5–10(–12) mm, firm becoming leathery or stiffly papery, brown or black, pilose; two-chambered. **Seeds** 14–26. $2n = 16, 32, 48, 64$.

Flowering Jun–Sep. Various soils, often in disturbed sites; [(500–)1400–]2000–2100[–2800] m; Petran Montane Conifer Forest; Apache, Coconino, Greenlee cos.; widely introduced and occasionally naturalized in western United States and Canada (Manitoba, Saskatchewan); native to Europe and southwest Asia.

Astragalus cicer is a vigorous European species and has been spread around the American West in reclamation plantings. Its original introduction as a cover or forage crop is not recorded in taxonomic literature, but it has been in North America for at least five decades. *Astragalus cicer* is more or less equal to alfalfa in nutrient value. In Arizona, plants putatively identified as *A. cicer* (without mature fruits) have been collected from only a couple of scattered localities on the Kaibab Plateau and in Apache Co. Additional collections with fruits are needed for confirmation. Elsewhere in western North America it is found in mixed desert shrub, sedge-willow, pinyon-juniper, sagebrush, aspen, and open pine communities.

17. Astragalus cobrensis A. Gray. Smithsonian Contr. Knowl. 5(6):43. 1853. * [of Santa Rita del Cobre] COPPER MINE MILKVETCH

Perennials, low, slender, caulescent, (4–)6–20(–25) cm; from subterranean caudex; thinly strigillose or pilosulose to densely hirsutulous, hairs basifixed. **Stems** few to several, decumbent to ascending, 1.5–10(–18) cm underground, flexuous distally. **Leaves** 1.5–8(–11) cm, shortly petiolate or subsessile distally; stipules fused proximally, sometimes fused to 1/2 their length distally, triangular-lanceolate, 1.5–5 mm, papery proximally to thinly herbaceous distally, apex bidentate proximally. **Leaflets** (7–)11–19(–23), flat, broadly obovate, obovate-cuneate, broadly oblong-elliptic, or nearly circular, (1.5–)3–12(–17) × 1.5–3.8 mm, apex retuse, leaflets ± bicolored, pallid beneath (abaxially), yellowish-green and either glabrous or sparsely pubescent above (adaxially). **Inflorescences** with (4–)8–22(–33) flowers; peduncle incurved-ascending, (1–)1.5–8.5 cm, equal to or shorter than subtending leaf; axis (1–)2–6(–7.5) cm in fruit; bracts 1–2.2 mm; bracteoles 0. **Pedicels** spreading, becoming declined, 1–2 mm. **Flowers**: calyx campanulate or turbinate-campanulate, 3.2–4.5 mm, strigillose, hairs black, white, or grayish brown; tube 2.2–2.7 × 1.6–2.2 mm; lobes subulate, 1–1.8 mm; corollas whitish tinged with dull lilac, 6.5–7.8 mm; banner recurved through 85°, broadly obovate-cuneate, obovate-flabellate, or rhombic-ovate, 6.5–7.8 mm; wings 6.2–7.7 mm; keel incurved through 95–100°, semi-obovate or nearly round, 4.4–6 mm, apex blunt or sharply deltoid and obscurely porrect; ovules 10–13. **Fruits** pendulous; subsessile, stipe obscure; straight or slightly arched downward, lower (abaxial) face grooved, upper (adaxial) face convex, oblong or clavate-ellipsoid, dorsiventrally compressed, (7–)9–15 × 3–6 mm, greenish becoming straw-colored, papery, base cuneate, apex abruptly contracted into very short, triangular, cuspidate beak, surface delicately cross-reticulate, strigillose or villosulous; sometimes incompletely two-chambered, inflexed as partial or nearly complete septum 0.8–1.3 mm wide. **Seeds** 10–13. $2n = 22$.

Varieties 2 (2 native in the flora): sw United States, nw Mexico.

1. Herbage mostly sparsely strigillose, hairs appressed, fine, usually to 0.3–0.5(–0.6) mm, rarely a few longer; leaflets glabrous or glabrate above (adaxially); fruits strigillose; plants not of the Chiricahua and Peloncillo Mountains ____ **17a. Astragalus cobrensis** var. **cobrensis**
1′ Herbage pilosulose, hairs spreading and incurved-ascending, somewhat stiff, to 0.8 mm; leaflets strigillose above (adaxially); fruits white-villosulous; plants of the Chiricahua and Peloncillo Mountains, Cochise County ____ **17b. Astragalus cobrensis** var. **maguirei**

17a. Astragalus cobrensis A. Gray var. **cobrensis** * Native

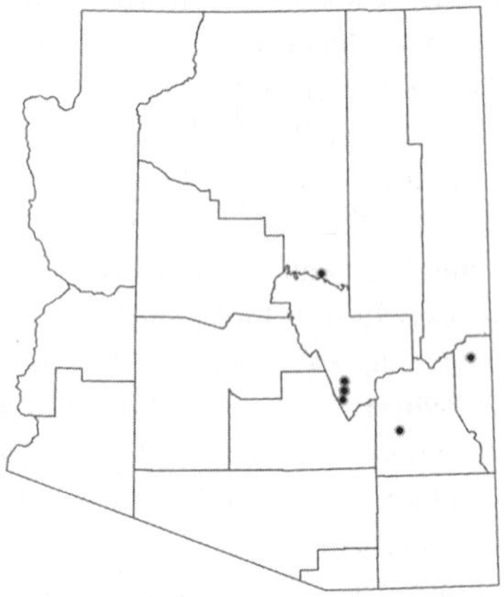

Herbage mostly sparsely strigillose, hairs appressed, fine, usually to 0.3–0.5 (–0.6) mm, rarely a few longer. **Leaflets** glabrous or glabrate above (adaxially). **Fruits** strigillose.

Flowering mid Mar–May. Dry sandy or gravelly hillsides and riparian areas, with walnut or oak in pine forests; 1600–2400 m; Interior Chaparral, Madrean Evergreen Woodland, Petran Montane Conifer Forest; Gila, Graham, Greenlee, Pima cos.; New Mexico.

Variety *cobrensis* occurs mainly within the drainage basin of the Gila River in southeastern Arizona, barely extending into the drainage of the Rio Grande in the Black Range in southwest New Mexico. A recent collection (*D. Beckman 52*, 25 April 2019, ARIZ) from the Rincon Mountains and an observation from the Santa Catalina Mountains represent the first reports of this variety in Pima County.

17b. Astragalus cobrensis A. Gray var. **maguirei** Kearney. J. Wash. Acad. Sci. 30:218. 1940. * [for Bassett Maguire, 1904–1991, American botanist and plant explorer] MAGUIRE'S MILKVETCH Native

Herbage pilosulose, hairs spreading and incurved-ascending, somewhat stiff, to 0.8 mm. **Leaflets** strigillose above (adaxially). **Fruits** white-villosulous.

Flowering Apr–May. Dry sandy or gravelly hillsides and canyons; 1670–2130 m; Madrean Evergreen Woodland; Cochise Co.; New Mexico; Mexico (Sonora).

Of the two varieties, var. *maguirei* is the more southern, occurring in the Chiricahua Mountains of southeastern Arizona, and the Peloncillo Mountains of southeastern Arizona and southwestern New Mexico. This taxon is of conservation concern.

18. Astragalus coccineus (Parry) Brandegee. Zoë 2:72. 1891. * [from Latin *coccinum*, scarlet] SCARLET MILKVETCH Native

Astragalus purshii Douglas var. *coccineus* Parry, W. Amer. Sci. 7:10. 1890

Perennials, low, tuft-forming, acaulescent, 8–15 cm; taproot woody; from root-crown; densely villous-tomentose, hairs basifixed. **Stems** absent, reduced to 1 to several crowns, with persistent thatch of stipules and petioles. **Leaves** (3–)4–10 cm, petiole stout; stipules distinct throughout, lanceolate or deltoid, 6–10 mm, papery-membranous. **Leaflets** 7–15, flat or loosely folded, obovate-cuneate to broadly oblanceolate or elliptic, (3–)5–14 × (2.3–)2.7–7.3(–9) mm, thick, apex obtuse or subacute, surfaces villous-tomentose. **Inflorescences** with (3–)5–10 flowers; peduncle ascending, (2–)4–10 cm; axis 0.5–2.5 cm in fruit; bracts 5–7 mm; bracteoles 0. **Pedicels** ascending, 3–5 mm. **Flowers**: calyx cylindric, 18–24

***Astragalus coccineus*: a. habit; b. fruit cross-section; c. fruit with seeds; d. seeds; e. leaf detail; f. flowers; g. pistil; h. pistil and stamens; i. j. k. petals. Illustrator: Joyce Riney Peters.**

mm, villous; tube (12–)13.5–16 × 4.6–6 mm; lobes subulate to lanceolate, 3.5–8 mm; corolla scarlet red, drying crimson, 35–41 mm; banner recurved through 20–30°, broadly oblanceolate, 35–40 mm; wings 34.5–37.5 mm; keel 35–40 mm, apex obtuse; ovules 34–43. **Fruits** ascending, humistrate; sessile; incurved, obliquely lanceolate-ovoid or narrowly ovoid-ellipsoid, dorsiventrally compressed, 25–40 × 9–12 mm, fleshy becoming leathery, color concealed by hairs, densely villous-tomentulose, concealing surface; one-chambered. **Seeds** 34–43.

Flowering Mar–early Jun. Open gravelly ridges and canyon benches of desert mountains and along arroyos, on decomposed granite; [50–]600–1200[–2200] m; Sonoran Desertscrub [Mohave Desertscrub, Great Basin Desertscrub, Great Basin Conifer Woodland]; Yuma Co.; California; Mexico (Baja California).

Astragalus coccineus, with its contrasting long and narrow scarlet to crimson flowers and white or ashen herbage, stands out in the genus. Only two other New World astragali have red flowers, both Mexican and apparently unrelated (*A. helleri* Fenzl and *A. sanguineus* Rydberg). Scarlet milkvetch occurs around the northern and western edges of the Mojave Desert, from Owens Valley and Death Valley to extreme northern Baja California, with an outlying station in the Kofa Mountains and Castle Dome Mountains, Arizona. It occurs in mixed warm desert shrub and pinyon-juniper or sagebrush communities. With its distinctive foliage and showy red flowers, *A. coccineus* would make an outstanding addition to wildflower gardens.

19. Astragalus coltonii M.E. Jones. Zoë 2:237. 1891 (as *Astragalus coltoni*) * [for W. F. Colton] COLTON'S MILKVETCH

Perennials, clump-forming, caulescent, 10–75 cm; from subterranean or superficial caudex, branched; strigillose, canescent, cinereous, or greenish cinereous, hairs basifixed. **Stems** several to numerous, erect or ascending. **Leaves** 2–10 cm, petiolate; stipules distinct throughout, deltoid or triangular, 1–7 mm, papery proximally, herbaceous distally, apex acuminate. **Leaflets** (1 or) 3–19, some unifoliolate distally, reduced to rachis, folded and inrolled, linear, oblong, cuneate-oblong, or ovate, (3–)4–20 × 0.3–3 mm, apex obtuse, truncate, retuse, or acute, surfaces strigose; terminal leaflet decurrent distally, not jointed to rachis. **Inflorescences** loosely (2–)5–30-flowered; peduncle erect or incurved-ascending, (4–)6.5–30 cm; axis 3–20 cm in fruit; bracts 0.5–3.2 mm; bracteoles 0. **Pedicels** spreading-declined, 0.8–2.5 mm. **Flowers**: calyx purplish, cylindric, 4.5–8 mm, strigose; tube 4–6.7 × 2.3–3.7 mm; lobes broadly subulate, 0.6–2.3 mm; corolla pink-purple, 12–19 mm; banner recurved through 35–50°, broadly rhombic-elliptic, rhombic-obovate, or broadly spatulate, 12–18.5 mm; wings 11–18.3 mm; keel 9–13 mm, apex bluntly or sharply deltoid; ovules 14–20. **Fruits** pendulous; stipe 4–11 mm; straight or gently decurved, oblong to oblanceolate, strongly flattened laterally, 19–35 × (3–)3.5–6 mm, green or reddish becoming straw-colored, stiffly papery, glabrous; one-chambered. **Seeds** 14–20.

Varieties 2 (1 native in the flora): w United States.

Colton's milkvetch, as described by Barneby (1964a), has been aligned with *Astragalus filipes* (with a similarly compressed and stipitate fruit) in section *Homalobi*. Rydberg (1929) and Jones (1923) described the stipules of *A. coltonii* as fused, but the interpretation is not strictly correct. The stipules are dimorphic, the lowest ones are papery, fused to form a bidentate sheath (by adherence to a vestigial petiole) whose structure and appearance are consistent with genuine sect. *Lonchocarpi*, while the upper ones are smaller, deltate or triangular and herbaceous.

19a. Astragalus coltonii M.E. Jones var. **moabensis** M.E. Jones. Contr. W. Bot. 8:11. 1898 (as *Astragalus coltoni*) * [of Moab] MOAB MILKVETCH Native

Astragalus canovirens (Rydb.) Barneby, Leafl. W. Bot. 4:49. 1944

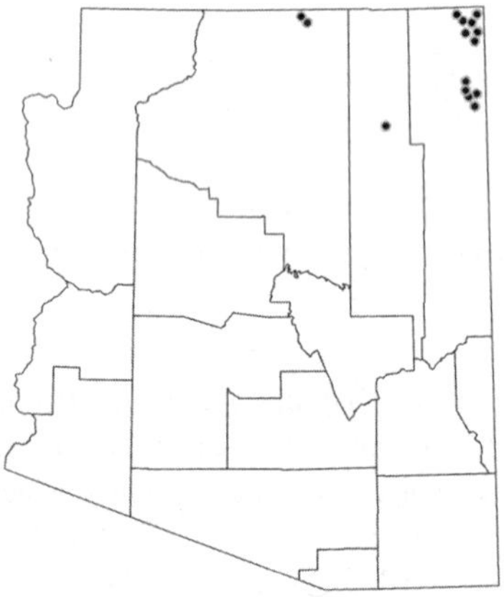

Stems from subterranean caudex; strigillose, cinereous, greenish, or canescent. **Leaves** (2–)3–9 cm. **Leaflets** (5–)9–17(–19), jointed or joint obscure in distal ones, oblong, cuneate-oblong, or ovate, (3–)5–20 mm, apex obtuse, truncate, or retuse, surfaces bright green and, usually, less densely pubescent below (abaxially). **Inflorescences** (6–)10–30-flowered; peduncle (4–)6.5–21 cm. **Fruits**: stipe 5–11 mm; 19–35 × (3–)3.5–6 mm.

Flowering Apr–Jun. Mesas, slopes, canyons and valleys, in sandy or clay soils, often with rocks or gravel, and in soil pockets in sandstone; [1400–]1700–2300[–2600] m; Great Basin Conifer Woodland, (Petran Montane Conifer Forest); Apache Co.; Colorado, New Mexico, Utah, Wyoming.

Astragalus coltonii var. *moabensis* grows in pinyon-juniper and mountain brush communities.

20. Astragalus crassicarpus Nuttall. Cat. Pl. Upper Louisiana no. 6. 1813. * [from Latin *crass*, thick and *karpos*, fruit, with thick fruits] COMMON GROUNDPLUM, BUFFALO PLUM

Perennials, caulescent, (5–)10–50(–60) cm; taproot woody; from subterranean or superficial caudex or root-crown, caudex determinate, shortly forking; strigose or strigillose, hairs basifixed. **Stems** single, few, or several, ascending or decumbent, 1–9 cm underground. **Leaves** (2–)3.5–18 cm, petiolate or subsessile; stipules distinct throughout, ovate, triangular, or lanceolate, 3–10 mm, submembranous, surfaces glabrous below (abaxially). **Leaflets** (11–)15–33, mostly flat, oblanceolate, oblong-oblanceolate, obovate, nearly circular, or broadly oval to linear-elliptic, (2–)3–24 × 2–8 mm, apex obtuse, acute, subacute, or truncate-emarginate, sometimes mucronulate, surfaces pubescent below (abaxially), glabrous or glabrescent above (adaxially). **Inflorescences** with (5–)6–25(–35) flowers; peduncle erect or ascending, (1.5–)2–9(–11) cm; axis 1–7(–9) cm in fruit; bracts 2.5–7.5 mm; bracteoles 0–2. **Pedicels** ascending or slightly arched,

2–7.5 mm. **Flowers**: calyx usually broadly or deeply campanulate or cylindro-campanulate, rarely cylindric, (6.6–)7.7–14 mm, strigillose, pilosulose, ascending-pilose, or villosulous-tomentulose; tube (5.2–)5.6–9.7 × (2.5–)3.2–5(–5.2) mm; lobes subulate or triangular-subulate, (1.3–)1.7–4.2(–5.8) mm; corolla pink-purple, red-purple, white, whitish (faintly lilac-tinged), lilac, ochroleucous, or suffused pinkish lilac, (16–)16.5–25(–27) mm; banner recurved through 45°, obovate-cuneate, broadly oblanceolate, rhombic-oblanceolate, or rhombic-elliptic, (16–)16.6–25(–27) mm; wings 12–17 mm; keel (10.7–)12–20.7 mm, apex obtuse; ovules 34–77. **Fruits** spreading or ascending, often humistrate; sessile; straight, circular, nearly circular, broadly and plumply oblong-ellipsoid, oblong-ovoid, or oblong-obovoid, ± dorsiventrally compressed, 15–40 × 12–27 mm, green, fleshy often red-tinged at first firm and succulent, becoming brown or blackish, alveolate-spongy or pithy, leathery, wrinkled on drying, glabrous; two-chambered. **Seeds** 34–77. $2n = 22$.

Varieties 5 (1 native in the flora): w North America.

The fleshy fruits of *Astragalus crassicarpus* are burnished with red or purple and resemble small plums prior to maturity. Early travelers valued the immature fruits as a summer vegetable. Prairie dogs and other rodents gather and eat the fruits (Barneby 1964a). Ultimately, the surface collapses as the mesocarp dries. The fruits are frequently broader than long (Welsh 1960).

20a. Astragalus crassicarpus Nuttall var. **cavus** Barneby. Mem. New York Bot. Gard. 13:759. 1964. * [from Latin *cavus*, hollow] HOLLOW GROUNDPLUM Native

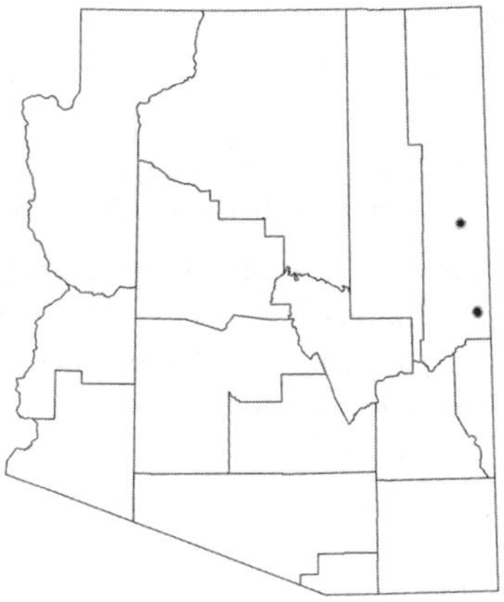

Herbage pallid green or yellowish-green. **Stems** decumbent and ascending, 5–30 cm, sometimes short, closely tuft-forming, elongating as fruit matures. **Leaves** (4–)5–15 cm. **Leaflets** (13–)17–31(–33), broadly to narrowly elliptic, oblong-oblanceolate, or obovate, (2–)4–17 mm, apex obtuse or subacute, sometimes callous-mucronulate. **Inflorescences** with 10–23(–30) flowers; peduncle 2–9(–11) cm; axis (1–)1.5–7(–9) cm in fruit. **Pedicels** 2–4 mm. **Flowers**: calyx (8.6–)9.6–12.6 mm, strigillose, pilosulose, or ascending-pilose, hairs mixed black and white or all black; tube (6.3–)7.5–9.2 × 3.8–4.5 mm; lobes (1.9–)2.3–3.6 mm; corolla whitish, faintly lilac-tinged; banner 17–25 × (8.2–)9–14 mm. **Fruits** broadly and plumply oblong-ovoid, oblong-obovoid, or nearly circular, 15–35(–40) × 12–25 mm. **Seeds** 38–51.

Flowering May–Jul. Arid sandy plains, dry stony meadows, gravelly banks, canyon benches, on granitic and various alluvial soils; (1500–)1700–2400 m; Plains Grassland, Great Basin Conifer Woodland; Apache Co.; New Mexico.

Variety *cavus* is a xeric plant, more so than its prairie and foothills counterparts to the east and north. It occurs from central New Mexico to east-central Arizona.

21. Astragalus cremnophylax Barneby. Leafl. W. Bot. 5:83. 1948. * [from Greek *cremnos*, gorge and *phylax*, watchman, referring to the habitat at the edge of the Grand Canyon] SENTRY MILKVETCH

Perennials, tuft- or cushion-forming, to 20 cm wide, acaulescent or subacaulescent, 0.5–2(–3.2) cm; from superficial caudex, branched, branches with thatch of withered, persistent stipules and leaf bases; silvery-strigillose, hairs medifixed. **Stems** obscured by stipules and leaf bases. **Leaves** 0.3–2(–3.5) cm, subsessile; stipules distinct throughout, deltoid-ovate, 1–3 mm, submembranous becoming papery. **Leaflets** (3–)5–9(–11), folded, elliptic, obovate, or nearly circular, 1.5–5 × 0.5–0.8 mm, apex obtuse or subacute, surfaces silvery-strigillose. **Inflorescences** ascending, to 0.5 cm, racemes (1–)2(–3)-flowered, shorter than leaves to scarcely exserted; axis 0.1–0.2 cm; bracts 0.6–1 mm; bracteoles 0. **Pedicels** ascending, 0.7–1 mm. **Flowers**: calyx campanulate, 2.7–3.6 mm, silky-strigillose; tube 1.7–2.3 × 1.5–1.7 mm; lobes subulate, 0.8–1.4 mm; corolla pale pink-purple, wing tips whitish, keel spotted, 3.7–8 mm; banner recurved through 45°, broadly ovate-cuneate or nearly circular-cuneate, 3.7–8 × 4–6 mm; wings 3.5–4.4 mm; keel 3.7–5.7 mm, apex bluntly deltoid, minutely beaklike; ovules 4–6. **Fruits** ascending, humistrate; sessile; straight, obliquely-ovoid, slightly flattened or basally compressed below (abaxially), 3–4.5(–4.8) × 2.5 mm, green or purple dotted, thinly fleshy, becoming straw-colored, papery, densely strigillose; one-chambered. **Seeds** 4–6.

Varieties 3 (3 native in the flora): Arizona.

Astragalus cremnophylax, one of our rarest species, consists of three (possibly more) obscure and disjunct taxa whose populations are all found within Coconino County. There are few collections of these taxa and, therefore, measurements presented herein are tentative.

1. Rachis of leaf lax, soft; plants forming soft, small mats or cushions on rock pavement; North and South Rim of the Grand Canyon _____ **21a. Astragalus cremnophylax** var. **cremnophylax**
1′ Rachis of leaf erect or ascending, stiff; plants forming tufts or prickly cushions; Buckskin Mountains, and east rim of Marble Canyon.
  2. Banner 7–8 × 5.4–6 mm, keel 5.2–5.7 mm _____ **21b. Astragalus cremnophylax** var. **hevronii**
  2′ Banner 3.7–6.3 × 4–4.6 mm, keel 3.7–4.5 mm _____ **21c. Astragalus cremnophylax** var. **myriorrhaphis**

21a. Astragalus cremnophylax Barneby var. **cremnophylax** * SENTRY MILKVETCH Native

Plants forming soft mats or cushions. **Leaves:** rachis lax, soft. **Flowers**: banner 5.3–6.1 × 4.3–4.6 mm; keel 3.7–4.3 mm.

Flowering (Apr–)May–Jun. Crevices of limestone pavement; 2150–2430 m; Great Basin Conifer Woodland; Coconino Co.

Variety *cremnophylax* is restricted to the North and South rims of the Grand Canyon where it grows in small, isolated populations in the pinyon pine belt. Travis et al. (1996) found genetic variation between plants in populations (one very low in a small South Rim population) and comparatively substantial variation between populations. Greater genetic variation is observed in the North Rim plants, which comprise a larger population.

Variety *cremnophylax* is in the Center for Plant Conservation's National Collection of Endangered Plants and is classified as endangered under the Endangered Species Act.

21b. Astragalus cremnophylax Barneby var. **hevronii** Barneby. Brittonia 44:238. 1992. * [for William Hevron, American botanist] HEVRON'S MILKVETCH, MARBLE CANYON MILKVETCH Native

Plants forming tufts or prickly cushions. **Leaves:** rachis erect or ascending, stiff. **Flowers:** banner 7–8 × 5.4–6 mm; keel 5.2–5.7 mm.

Flowering Apr–May. In Kaibab Limestone crevices; 1600–1650 m; Great Basin Desertscrub; Coconino Co.

Variety *hevronii* is known only from the eastern rim of Marble Canyon along the Colorado River. The plants are of conservation concern.

21c. Astragalus cremnophylax Barneby var. **myriorrhaphis** Barneby. Brittonia 31:463, fig. 2. 1979. * [from Greek *myrio*, numberless and *rhaphi*, needles] BUCKSKIN MOUNTAIN MILKVETCH, CLIFF MILKVETCH Native

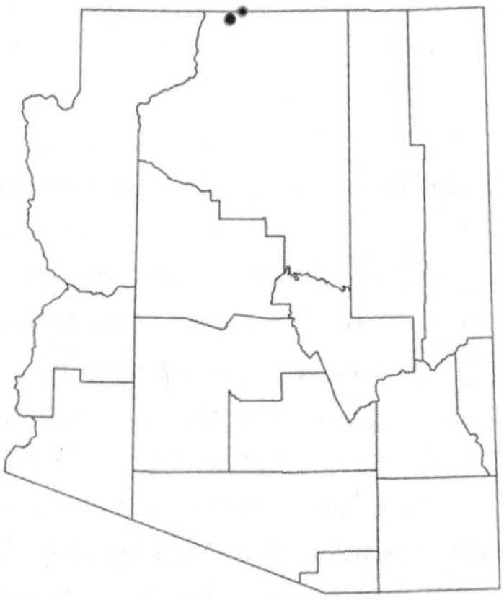

Plants forming tufts or prickly cushions. **Leaves:** rachis erect or ascending, stiff. **Flowers:** banner 3.7–6.3 × 4–4.6 mm; keel 3.7–4.5 mm.

Flowering Apr–May(–Jul). Crevices of rimrock and limestone cliffs in Kaibab Limestone; 1800–2000 m; Great Basin Conifer Woodland; Coconino Co.

Variety *myriorrhaphis* is known from several collections, all from the Buckskin Mountain, east of Fredonia. This taxon is of conservation concern.

22. Astragalus desperatus M.E. Jones. Zoë 2:243. 1891. * [from Latin *despero*, despairing] RIMROCK MILKVETCH

Perennials, tuft-forming, acaulescent, subacaulescent or shortly caulescent to 12(–15) cm; from caudex, branched; strigillose, hairs basifixed. **Stems** one to few, absent or prostrate, when developed to 8 cm, internodes often obscured by stipules. **Leaves** 1–12 cm, petiolate; stipules with fused sheathing proximally, fused or distinct distally, ovate to broadly lanceolate, 1.5–7 mm, submembranous becoming papery-scarious. **Leaflets** (3–)7–17, usually folded, elliptic to oblanceolate or obovate, 2–13 × 1–5 mm, apex acute to obtuse, surfaces strigose below (abaxially), strigose or glabrate above (adaxially). **Inflorescences** with 3–16(–18) flowers; peduncle erect or ascending, 0.5–13 cm; axis 0.4–13 cm in fruit; bracts (1–)1.5–5 mm; bracteoles 0–2. **Pedicels** ascending to declined, 0.5–1.4(–2) mm. **Flowers**: calyx campanulate, 3.5–9.4 mm, strigose-pilose; tube 2.5–7.0 × (1.3–)1.6–2.1 mm; lobes subulate, 0.8–2.6 mm; corolla pink-purple of variable intensity, 6–9 mm; banner recurved through 45°, obovate-cuneate or broadly oblanceolate, 7.4–14.5 mm; wings 6.7–13.4 mm; keel 5.9–12.5 mm, apex round to narrowly deltoid; ovules 16–28. **Fruits** declined to deflexed; sessile on a gynophore 0–1.2 mm; ± straight or lunately to very strongly incurved through 0.75(–1) spiral, obliquely ovoid to lanceolate-ellipsoid, dorsiventrally compressed, 6–19 × 3–6 mm, thin green, usually red-mottled or spotted, becoming straw-colored, papery, hirsute, hairs lustrous; one-chambered. **Seeds** 16–28.

Varieties 4 (2 native in the flora): sw United States.

Barneby (1989) recognized four varieties in the *Astragalus desperatus* complex; Isely (1998) noted those to be poorly defined. Some authors have recognized two of those varieties as separate species (*A. barnebyi* and *A. equisolensis*).

1. Flowers with calyx tube 2.5–4.2 mm, banner 7–11 mm, keel 6–9 mm, racemes mostly with 5–16(–18) flowers; Mojave and Coconino cos., Arizona ________ **22b. Astragalus desperatus** var. **desperatus**
1′ Flowers with calyx tube 4.5–7 mm, banner 12–16 mm, keel 9.5–12.5 mm; racemes mostly with 4–10 flowers; extreme northwest Coconino Co. and along the Little Colorado River in southern Navajo Co., Arizona ________ **22a. Astragalus desperatus** var. **conspectus**

22a. Astragalus desperatus M.E. Jones var. **conspectus** Barneby. Leafl. W. Bot. 5:87. 1948. * [from Latin *conspectus*, conspicuous, remarkable] Native

Astragalus barnebyi S.L. Welsh & N.D. Atwood

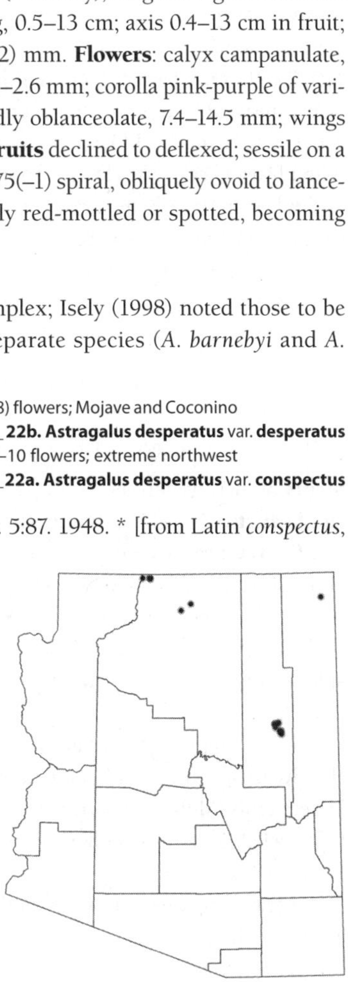

Plants dwarf and densely tufted, 3–8(–15) cm including inflorescences. **Inflorescences** mostly with 4–10 flowers, raceme not or little elongating. **Flowers** with calyx tube 4.5–7 mm, banner 12–16 mm, keel 9.5–12.5 mm. **Fruits** ± straight or lunately to very strongly incurved through 0.75(–1) spiral, obliquely ovoid to lanceolate-ellipsoid, 6–19 mm.

Flowering Apr–Jun. On sandstone rimrock in crevices and on rock ledges, discontinuously widespread and localized on Kanab Creek in nw. Coconino Co., and along Little Colorado River in s. Navajo Co.; 1500–1600 m; Great Basin Desertscrub; Apache, Coconino, Navajo cos.; Colorado, New Mexico, Utah.

22b. Astragalus desperatus M.E. Jones var. **desperatus** * Native

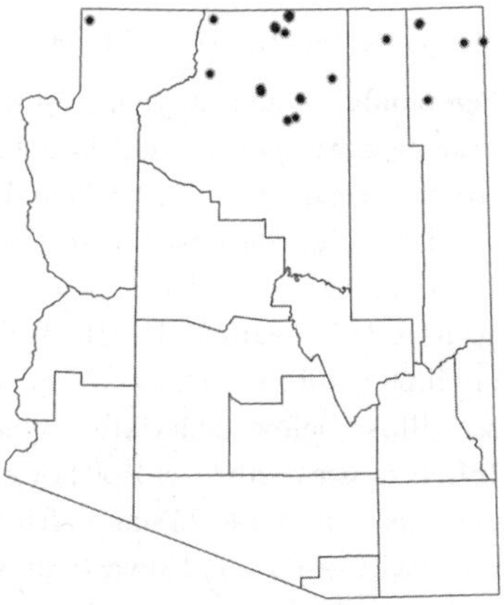

Plants shortly caulescent, 0.7–2(–2.5) cm including inflorescences. **Inflorescences** with 5–16(–18) flowers. **Flowers** with calyx-tube 2.5–4.2 mm, banner 7–11 mm, keel 6–9 mm. **Fruits** lunately curved, 11–19 mm.

Flowering Apr–Jun. On rimrock on sandy substrates, dunes, slopes, canyon floodplains, and roadsides; 1100–1900 m; Great Basin Desertscrub, Great Basin Conifer Woodland; Apache, Coconino, Mohave, Navajo cos.; Colorado, New Mexico, Utah.

23. Astragalus didymocarpus Hooker & Arnott. Bot. Beechey Voy. 334, pl. 81. 1838. * [from Greek *didymos*, twin and *karpos*, fruit, with a double fruit] TWO-SEEDED MILKVETCH

Annuals or winter annuals, caulescent, (1.5–)3–35(–45) cm; taproot slender; from superficial root-crown; sparsely to densely strigose, hairs basifixed. **Stems** single or few, erect, ascending, decumbent, or prostrate, very slender, diffuse. **Leaves** (0.8–)1–7.5 cm, petiolate or subsessile, petiole slender; stipules distinct throughout, deltoid, ovate, ovate-triangular, or lanceolate, (0.5–)1–5 mm, thinly herbaceous becoming papery-membranous. **Leaflets** (7–)9–17, flat or loosely folded, linear-oblanceolate, linear-elliptic, oblanceolate, oblong-cuneate, or cuneate, 1.5–14 × 1.3–2.2(–3.8) mm, apex deeply retuse, surfaces strigose below (abaxially), strigose, glabrate, or glabrous above (adaxially). **Inflorescences** with (5–)7–30 flowers; peduncle erect or incurved-ascending, (0.5–)1–8 cm; axis 0.2–2(–2.2) cm in fruit, not elongating; bracts 0.6–3 mm; bracteoles 0. **Pedicels** erect or ascending, straight, 0.1–0.5(–0.6) mm. **Flowers**: calyx campanulate to ovoid-campanulate, 2.5–5.7 mm, pilosulose or villosulous; tube 1.5–3.8 × (0.8–)1.2–1.8 mm; lobes subulate to subulate bristlelike, 0.8–2.5 mm; corolla whitish and tinged lavender, or pink-purple with white or pale wing tips, 2.8–10 mm; banner recurved through 25°, oblanceolate, spatulate-oblanceolate, rhombic-oblanceolate, or obovate-cuneate, 2.8–10 mm; wings 2.8–8 mm; keel 2.4–7.2 mm, apex blunt, deltoid, round, or narrowly triangular and acute or subacute, beaklike or not; ovules 2. **Fruits** erect or spreading; sessile, gynophore slender, 0–0.5 mm; straight, plumply ovoid or nearly globose, somewhat laterally compressed, (2–)2.3–4 × 1.5–3 mm, green or purple-tinged, thinly fleshy, becoming brownish, stiffly papery, surface coarsely reticulate or ridged, strigillose, strigillose-hirsutulous, pilosulose, or glabrous; ± two-chambered. **Seeds** 2.

Varieties 4 (1 native in the flora): sw United States, n Mexico.

Astragalus didymocarpus has erect flowers and fruits, which contrasts with the nodding (in late anthesis) flowers and declined fruits of *A. gambelianus*.

23a. Astragalus didymocarpus Hooker & Arnott var. **dispermus** (A. Gray) Jepson. Fl. Calif. 2:376. 1936. * [from Greek *dis*, double and *sperma*, seed, with two seeds] WICKENBURG MILKVETCH Native

Astragalus dispermus A. Gray, Proc. Amer. Acad. Arts 13:365. 1878

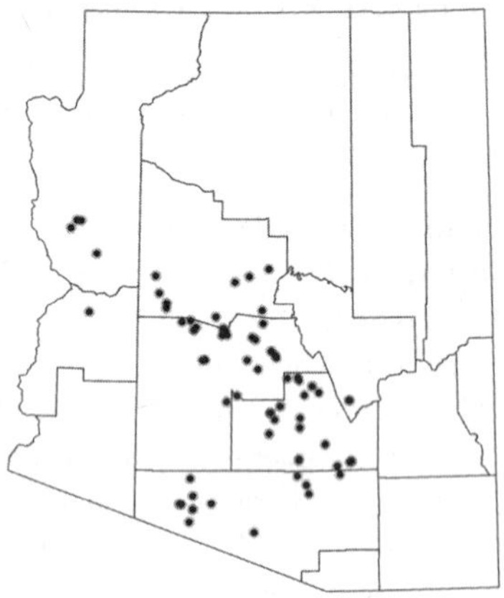

Stems diffuse, weakly ascending, or prostrate, (1.5–)4–27 cm. **Leaves** 1–3.5 cm. **Leaflets** (7–)9–15, 2–8 mm. **Inflorescences** 7–20-flowered; axis 0.3–1.2 cm. **Flowers**: calyx tube 1.7–2.3 mm; lobes subulate or subulate bristlelike, (1.3–)1.6–2.4 mm, hairs mostly white, sometimes also black; corolla (3.4–)4.3–5.4 mm; keel abruptly incurved, (3.4–)3.8–4.5 mm, apex usually bluntly deltoid, sometimes obscurely beaklike. **Fruits** 2.3–4 × 1.6–2.5 mm, deeply sulcate, strongly wrinkled transversely or diagonally, strigillose-hirsutulous. $2n = 26$.

Flowering Feb–May. Various soils of slopes, plains, canyons, arroyos, floodplains, dunes, and roadsides; [30–]350–1100(–1300) m; Sonoran Desertscrub, (Mohave Desertscrub, Semidesert Grassland, Interior Chaparral); Gila, La Paz, Maricopa, Mohave, Pima, Pinal, Yavapai cos.; California, Nevada; Mexico (Baja California).

Variety *dispermus* occurs from the western edge of the California deserts across southern Nevada to central Arizona. It is especially common in foothills of desert mountains and on arid sandy soils, with creosote bush.

24. Astragalus egglestonii (Rydberg) Kearney & Peebles. J. Wash. Acad. Sci. 29:484. 1939. * [for Willard W. Eggleston, 1863–1935, American botanist] EGGLESTON'S MILKVETCH Native

Tium egglestonii Rydberg in N.L. Britton et al., N. Amer. Fl. 24:396. 1929

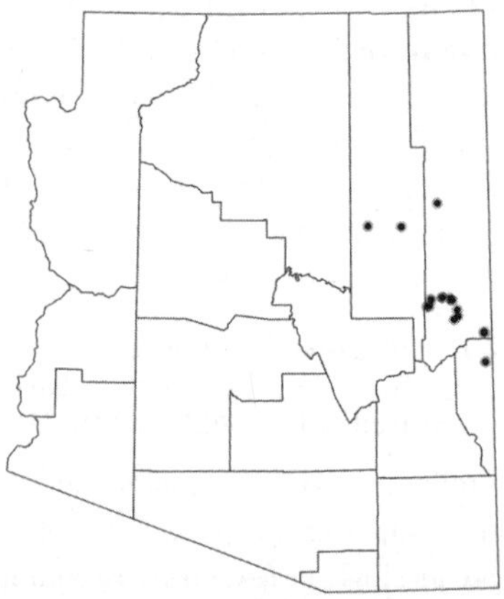

Perennials, caulescent, slender, stiff, (15–)20–55 cm; from slightly subterranean caudex; sparsely strigillose, hairs basifixed. **Stems** several, erect and strictly ascending, (15–)20–38 cm. **Leaves** (3–)4–12 cm, shortly petiolate or sessile or subsessile (distally); stipules fused proximally, distinct distally, deltoid or broadly lanceolate, 1.5–5 mm, scarious-papery proximally to herbaceous distally. **Leaflets** (17–)21–29, flat or loosely folded, oblong-oblanceolate or elliptic, 3–13 × 1.2–3.8 mm, apex obtuse to subacute, surfaces sparsely strigillose below (abaxially), glabrate or glabrous above (adaxially). **Inflorescences** with (5–)10–30 flowers; peduncle erect or ascending, (2–)4–11 cm; axis (1.5–)3–19(–25) cm in fruit; bracts 1–2 mm; bracteoles 2. **Pedicels** recurved, 0.6–1.5 mm. **Flowers:** calyx withering, persistent, becoming papery, campanulate, 3.4–4.3 mm, strigillose (hairs black and/or white); tube 2.1–2.7 × 1.6–2.2 mm; lobes subulate, 1–2 mm; corolla whitish or, sometimes, greenish-ochroleucous, 5.8–8 mm; banner abruptly recurved through 90° (or more on withering), rhombic-ovate or ovate-cuneate, 5.8–8 mm; wings 5.8–7.3 mm; keel incurved through 130°, nearly round, 4.2–5.6 mm, apex deltoid; ovules 12–16. **Fruits** pendulous, stipe 1.5–2.5 mm; straight or slightly incurved or decurved, lower(abaxial) face concave, lateral faces convex, linear-oblong or linear-oblanceolate, 3-sided compressed, 15–30 × (2.5–)3–4 mm, green, glabrous becoming straw-colored, papery, translucent, base tapering, apex cuneate, surface lustrous, finely cross-reticulate, glabrous; two-chambered, valves inflexed as complete septum 1.4–1.9 mm wide. **Seeds** 12–16.

Flowering Jul–Sep(–Oct). Slopes and roadsides, on clay or loam soils derived from basalt or granite; (1550–)1900–2700 m; Great Basin Grassland, Great Basin Conifer Woodland, Petran Montane Conifer Forest; Apache, Navajo cos.; New Mexico.

In its region *Astragalus egglestonii* is the only late-flowering species of the genus with erect stems; small, nodding flowers; and subsequently declined, 3-angled fruits (Barneby 1964a). Commonly found in meadows and open pine woods.

25. Astragalus emoryanus (Rydberg) Cory. Rhodora 38:406. 1936. * [for William H. Emory, 1811–1887, American soldier and explorer] EMORY'S MILKVETCH

Hamosa emoryana Rydberg, Bull. Torrey Bot. Club 54:327. 1927

Annuals or winter annuals (biennials), caulescent, (2.5–)4–45(–60) cm; taproot slender; from superficial caudex; sparsely to densely strigillose, hirsutulous, or subvillosulous, hairs basifixed. **Stems** few to several or numerous, diffuse, prostrate, radiating from crown. **Leaves** 1–4.5(–8) cm, shortly petiolate or subsessile; stipules mostly distinct throughout, deltoid, ovate, or triangular, 1.5–3.5(–5) mm, membranous. **Leaflets** (7–)11–19(–21), flat or folded, oval-obovate, obcordate, cuneate-oblanceolate, or elliptic-oblanceolate, 2–10(–14) × 1.6–4.1(–6.2) mm, apex obtuse or retuse, surfaces strigillose below (abaxially), strigillose or glabrous above (adaxially). **Inflorescences** (1–)2–10-flowered; peduncle reclining, (1–)2–6(–10) cm; axis (0–)0.3–2.5(–3.5) cm in fruit; bracts 0.5–1.8 mm; bracteoles 0 or 1. **Pedicels** spreading to declined, 0.5–2 mm. **Flowers**: calyx campanulate, 3.6–6 mm, strigillose; tube 1.9–3.5 × 1.7–2.5(–2.8) mm, pallid or purplish; lobes lanceolate-subulate to subulate, 1.3–2.5(–2.9) mm; corolla pink-purple, (6–)7.3–11.2 mm; banner recurved through 45°, obovate-cuneate, ovate-cuneate, or flabellate, (6–)7.3–11.2 mm; wings (5.1–)6.2–8.8(–10) mm; keel (4.5–)4.8–6.6 mm, apex bluntly deltoid; ovules (8–)10–15. **Fruits** spreading, declined, or ascending; sessile; straight or incurved through 0.25–0.5 spiral, linear-oblong to lanceolate-oblong, 3-sided compressed, deciduous, dehiscent on ground, 8–20(–22) × (2–)2.2–3.7 mm, green or purple-tinged becoming pale brown or straw-colored, papery, glabrous; ± two-chambered. **Seeds** (8–)10–15.

Varieties 2 (1 native in the flora): sw United States, ne Mexico.

Astragalus emoryanus and *A. nuttallianus* are difficult to distinguish, differing definitively only by the deciduous fruits of *A. emoryanus*. *Astragalus emoryanus* also usually has larger and showier flowers, an apically broad and commonly obtuse keel, and an elongating raceme axis. Isely (1998) provides a series of geographically restricted short keys that distinguish this from similar species.

25a. **Astragalus emoryanus** (Rydberg) Cory var. **emoryanus** * Native

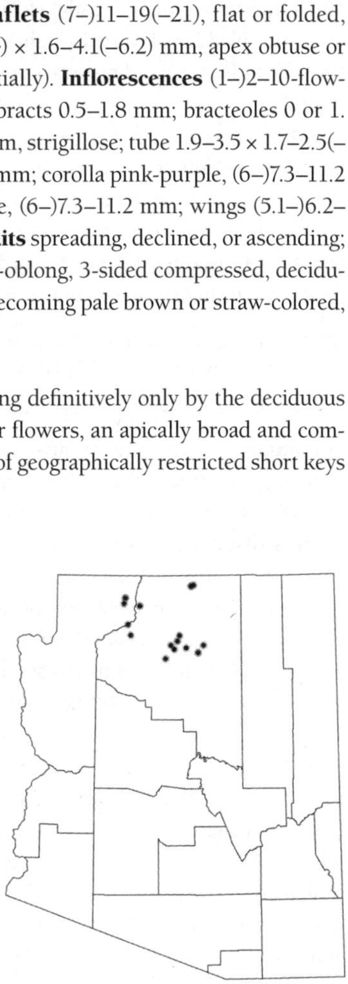

Hairs mostly appressed or subappressed. **Inflorescences** peduncle equaling or surpassing leaves. **Fruits** 12–20(–22) × (2–)2.2–3.3 mm, length 5–6.5 times width. $2n = 22, 24$.

Flowering [Feb–]Apr–Jun. Ridges, slopes, canyons, and arroyos; [600–]900–2000[–2200] m; (Mohave Desertscrub), Great Basin Desertscrub, (Great Basin Grassland), Great Basin Conifer Woodland; Coconino, Mohave cos.; New Mexico, Texas, Utah; Mexico (Chihuahua, Coahuila, Nuevo Leon, Tamaulipas).

The range of var. *emoryanus* is unusual; it occurs in Kane County, Utah, and northwestern Arizona, then skips to central New Mexico and western Texas. In the Big Bend region it is replaced by var. *terlinguensis*, but then reoccurs again in Maverick County in the Río Grande Valley and is more or less continuous to near the mouth of the Río Grande River. An isolated series of populations are found in east-central Texas and northern Mexico. In Arizona, the variety mainly occurs in Great Basin Desertscrub and pinyon-juniper woodland but has been found in grassland and in Mohave Desertscrub in the Grand Canyon. Across its range, var. *emoryanus* is found in creosote bush, mesquite, grassland, thornscrub, oak-grassland, juniper woodlands, and mixed desert shrub communities.

26. Astragalus endopterus (Barneby) Barneby. Mem. New York Bot. Gard. 13:879. 1964. * [from Greek *endon*, within and *pteron*, wing, with an inner wing] SANDBAR MILKVETCH Native

Astragalus wootonii E. Sheldon var. *endopterus* Barneby, Amer. Midl. Naturalist 41:498. 1949 (as *Astragalus wootoni*)

Annuals or biennials, slender, caulescent, 10–20 cm; with a taproot; strigillose, cinereous, hairs basifixed. **Stems** few, erect or ascending. **Leaves** 4.5–8.5 cm, petiole slender; stipules distinct throughout, deltoid or triangular, 1–2.5 mm, submembranous. **Leaflets** 7–11, flat or some folded, lanceolate, oblong, or narrowly oblong-elliptic, 8–16 mm long, apex obtuse or truncate, somewhat bi-colored with surfaces strigillose underneath (abaxially), glabrescent above (adaxially). **Inflorescences** loosely 4–7-flowered; peduncle ascending or incurved, 1.5–2 cm; axis 1.5–3 cm in fruit; bracts 1.5 mm; bracteoles 0. **Pedicels** ascending becoming declined, 1–2 mm. **Flowers**: calyx campanulate, 4 mm, strigillose; tube 2.1–2.3 × 1.7–1.8 mm; lobes subulate, 1.8–2 mm, becoming papery, ruptured; corolla whitish faintly lined pink, 6–6.3 mm; banner recurved through 80°, rhombic-oval, 6–6.3 mm; wings 5.8–6 mm; keel 4.3–4.4 mm, apex bluntly deltoid, obscurely beaklike; ovules 28–30. **Fruits** loosely spreading to declined; sessile; straight, subsymmetrically ovoid or ovoid-ellipsoid, bladdery-inflated, slightly dorsiventrally compressed, 20–27 × 11–15 mm, thin, pale green and faintly brown-mottled, becoming straw-colored, surface lustrous, finely reticulate and strigillose; one-chambered. **Seeds** 28–30.

Flowering May–Jun. Gravelly washes and sandbars of summer-dry streams, and on sand dunes; 1100–1300 m; Great Basin Desertscrub; Coconino Co.; endemic.

Astragalus endopterus is rare and local, known only from the lower Little Colorado River and Moenkopi Wash near Cameron, Coconino County. Barneby (1964a) noted the similarity to *A. wootonii,* and he and Isely (1998) reviewed the small but substantive characters that distinguish them, including the exceptionally broad funicular flange within the fruits and the higher ovule number in *A. endopterus*.

27. Astragalus ensiformis M.E. Jones. Proc. Calif. Acad. Sci., ser. 2, 5:658. 1895. * [from Latin *ensis*, sword and *forma*, form, sword-shaped] PAGUMPA MILKVETCH Native

Astragalus ensiformis M.E. Jones var. *gracilior* Barneby; *Astragalus minthorniae* (Rydberg) Jepson var. *gracilior* (Barneby) Barneby

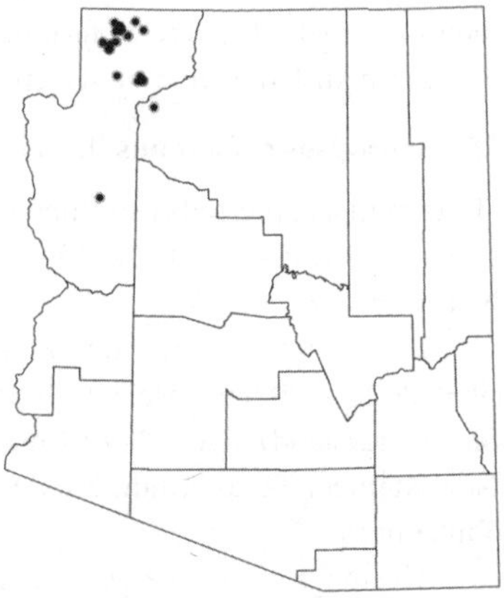

Perennials, low and rather coarse, caulescent, 8–45 cm; from shallow, subterranean caudex; strigillose, hairs basifixed. **Stems** several, decumbent or sprawling to erect; (1–)2–6(–8) cm underground. **Leaves** 4–16 cm, petiolate; stipules distinct throughout, broadly to narrowly ovate or lanceolate, 3–11 mm, papery proximally, subherbaceous distally, strongly decurrent. **Leaflets** (5–)11–23, flat, ovate to oblong, obovate, or oblanceolate, 6–24 × 1.5–13.5 mm, apex obtuse to retuse, somewhat bi-colored with surfaces strigose to sparsely so below (abaxially), yellowish-green and strigose to glabrous above (adaxially). **Inflorescences** 10–30-flowered; peduncle incurved-ascending, subhorizontal in fruit, 2.5–13(–24) cm; axis 3–14 cm in fruit; bracts 2–6 mm; bracteoles 0–2. **Pedicels** ascending to declined, 1–3.5 mm. **Flowers**: calyx short-cylindric, 5.2–7.8 mm, black-pilosulose; tube purplish, 4.5–6.5 × 2.8–3 mm; lobes subulate, 1.2–2.5 mm; corolla purplish to pink-purple, wing tips pale to white, (11–)13–17 mm; banner recurved through 45°, rhombic-ovate, 14–15 mm; wings 12.2–14 mm; keel 10.5–10.8 mm, apex obtuse; ovules 24–36. **Fruits** pendulous, but sometimes ascending to descending (or humistrate); sessile or shortly stipitate, stipe 0.5–2 mm; incurved 0.5 spiral, narrowly oblong, cylindrical and nearly round in cross section, laterally compressed when pressed, 15–32 × 4–7(–8) mm, thick and fleshy, green and red-tinged becoming brownish, becoming leathery, strigose to strigillose; two-chambered. **Seeds** 24–36.

Flowering Apr–May. Hillsides, in gravel in pinyon-juniper woodlands, canyons, and roadsides on various substrates; 1200–1700[–2400] m; Great Basin Desertscrub, Great Basin Conifer Woodland; Coconino, Mohave cos.; Nevada, Utah.

Astragalus ensiformis is found in pinyon-juniper, sagebrush, salt desert shrub, and blackbrush communities. The species is locally common on the Arizona Strip.

28. Astragalus episcopus S. Watson. Proc. Amer. Acad. Arts 10:346. 1875. * [for Francis M. Bishop, 1843–1933, American soldier and explorer] BISHOP'S MILKVETCH

Perennials, caulescent, stiffly wiry or rushlike, 20–50 cm; from a woody subterranean caudex; green or yellowish-green; strigose, hairs basifixed. **Stems** single or few, erect or ascending. **Leaves,** sparsely to apparently leafless, most reduced to rachis, 2–10 cm, petiolate; stipules distinct throughout, ovate, oblong, deltoid, or triangular, 2–13 mm, papery proximally, herbaceous distally, apex subobtuse or acuminate. **Leaflets** (0 or) 3–13 (often proximal), often involute or folded, linear to elliptic or oblong, 1–15 × 0.5–2 mm, apex acute to obtuse or emarginate, surfaces strigose; terminal leaflet decurrent distally, continuous with rachis. **Inflorescences** very loosely 4–25(–30)-flowered; peduncle erect or incurved-ascending, 6–23 cm; axis 3–30 cm in fruit; bracts 1.3–3 mm; bracteoles 0–2. **Pedicels** ascending, 1.5–3.5 mm. **Flowers**: calyx suffused with purple or very pale, 3.8–8.5(–9) mm, hairs white or black, or mixed, strigillose; tube 3–6.2 × 1.8–2.8 mm, shortly cylindric to campanulate; lobes triangular to subulate, 0.6–2.2 mm; corolla whitish, pink-tinged to pink-purple, 10–15.5 mm; banner recurved through 40–45°, rhombic-obovate or rhombic-elliptic, 9–14.5(–15.5) mm; wings 10–12.7(–14.3) mm; keel 7–12.3 mm, apex obtuse to subacute, short beaklike; ovules 8–24. **Fruits** deflexed, pendulous; sessile to tapered at base into an obscure neck, continuous with the receptacle; slightly curved or straight, oblong to lanceolate-ellipsoid or linear ellipsoid, strongly laterally compressed, 14–34 × 4–8(–9) mm, straw-colored or suffused with purple or mottled, becoming stiffly papery, glabrous or strigose; one-chambered. **Seeds** 8–24(–26). $2n$ = 22.

Varieties 2 (2 native in the flora): Arizona, Utah.

1. Calyx (3.8–)5–8.5(–9) mm; tube (3.7–)4–6.2 × 1.8–2.8 mm, at least twice as long as its diameter; banner 11–14.5(–15.5) mm, keel 7.5–12 mm; ovules 16–24 ______ **28a. Astragalus episcopus** var. **episcopus**

1′ Calyx 3.8–5.5 mm; tube 3–4.3 × 2–2.5 mm, less than twice as long as its diameter; banner 9–11.5 mm, keel 7–9 mm; ovules 8–16 ______ **28b. Astragalus episcopus** var. **lancearius**

28a. Astragalus episcopus S. Watson var. **episcopus** * Native

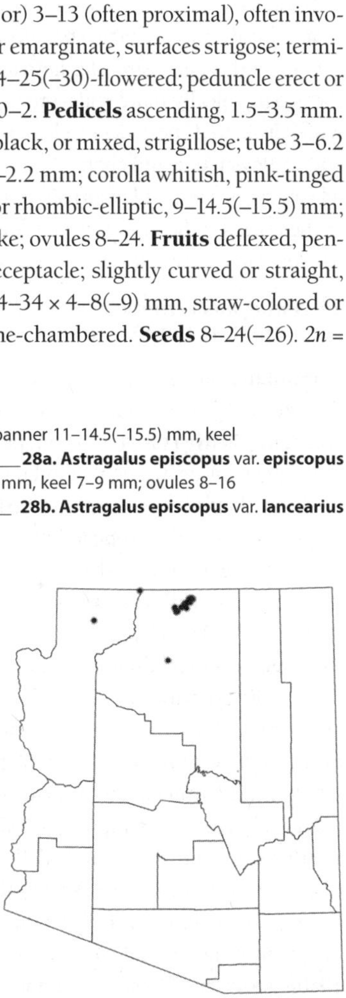

Inflorescences with (4–)6–22(–30) flowers. **Flowers**: calyx (3.8–)5–8.5(–9) mm, white strigillose; tube (3.7–)4–6.2 × 1.8–2.8 mm, at least twice as long as its diameter, campanulate to cylindric; banner 11–14.5(–15.5) mm; keel 7.5–12 mm; ovules 16–24. **Fruits** sessile or tapering at base into stout, stipe-like neck up to 0.6 mm, (15–)30–32 mm. **Seeds** 16–24(26).

Flowering May–June. Slopes, bluffs, rocky talus under cliffs, and roadsides in sandy to clay soils derived from sandstone; (900–)1200–1750(–1900) m; Great Basin Desertscrub, Great Basin Grassland, occasionally at lower elevations in Great Basin Conifer Woodland; Coconino, Mohave cos.; Utah.

Variety *episcopus* is locally plentiful in the Canyonlands area in Emery and San Juan cos., Utah, south to House Rock Valley and Kanab Creek in northern Arizona. It is found primarily in mixed desert shrub/grass and occasionally in pinyon-juniper communities.

28b. Astragalus episcopus S. Watson var. **lancearius** (A. Gray) Isely. Syst. Bot. 8:421. 1983. * [From Latin *lancearius*, lancer, for a bladelike pod] LANCER MILKVETCH Native

Astragalus lancearius A. Gray, Proc. Amer. Acad. Arts 13:370. 1878

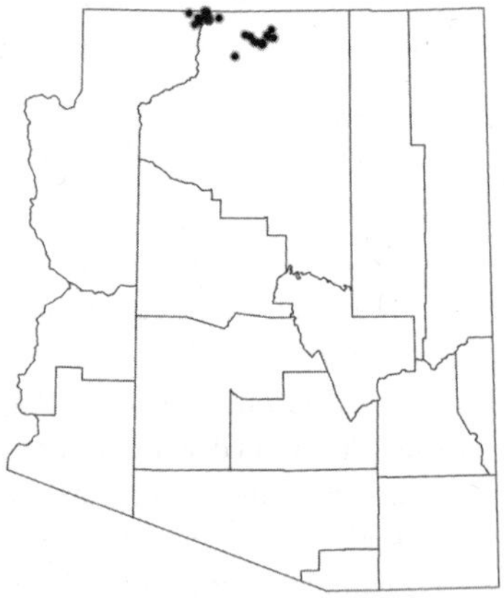

Inflorescences with (6–)10–25 flowers. **Flowers**: calyx 3.8–5.5 mm; tube 3–4.3 × 2–2.5 mm, less than twice as long as its diameter, campanulate, white- or mixed white and black strigillose; banner 9–11.5 mm; keel 7–9 mm; ovules 8–16. **Fruits** sessile, 20–34 mm. **Seeds** 8–14.

Flowering Apr–Jun. Ridges, slopes, valleys, and roadsides in various soils; 1300–1950 m; Great Basin Desertscrub, Great Basin Conifer Woodland; Coconino, Mohave cos.; Utah.

Variety *lancearius* closely resembles var. *episcopus* in size; its wiry, sparsely leafy growth habit; strigillose pubescence; and leaflets equally pubescent on both sides. *Astragalus lancearius* was first included as a variety within the geographically more widespread, contiguous *A. episcopus* by Isely (1983), who considered these taxa to be varieties differing mainly in ovule number and in flower size and shape. Variety *lancearius* occurs in sagebrush, buckwheat, serviceberry, blackbrush and pinyon-juniper communities. Localized in the valleys of the Virgin River, Kanab Creek, and Paria River in southwest Washington and southern Kane cos., Utah, var. *lancearius* is restricted to a small area near the Utah border in northern Mohave and Coconino cos., Arizona.

29. Astragalus eremiticus E. Sheldon. Minnesota Bot. Stud. 1:161. 1894. * [from the Greek *eremi*, lonely or solitary] HERMIT MILKVETCH Native

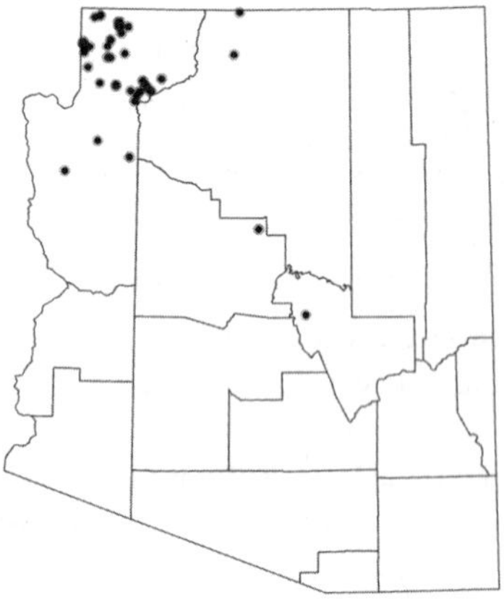

Perennials, caulescent, 20–55(–63) cm; from superficial caudex, branched; sparsely to densely strigillose, hairs basifixed. **Stems** several to many, erect or ascending. **Leaves** (3–)4.5–18 cm, petiole slender proximally; stipules distinct throughout, ovate, triangular, or lanceolate, 3–11 mm, papery-scarious proximally, subherbaceous distally. **Leaflets** 11–23, flat or folded, ovate, oblong, elliptic, or narrowly oblong to linear, 5–26 × 1.5–6 mm, apex obtuse to retuse, surfaces strigose below (abaxially), usually glabrous above (adaxially), sometimes glabrate. **Inflorescences** with (7–)10–26(–30) flowers; peduncle erect, (1.5–)2.5–17(–35) cm; axis 4–10(–15) cm in fruit; bracts 1.5–4 mm; bracteoles 0–2. **Pedicels** ascending or spreading to descending, 0.7–3.5 mm. **Flowers**: calyx short-cylindric, 5.4–9.5 mm, strigose; tube 4.4–8 × 2.2–4 mm; lobes triangular to subulate, 0.7–2.6 mm; corolla variable in color, from ochroleucous or pale yellow, to tinged pink, purplish, or pink-purple, keel spotted or not, 12–20 mm; banner recurved through 45°, rhombic-obovate or rhombic-oblanceolate, 12–20 mm; wings 10.2–18 mm; keel 9.4–14.4 mm, apex round or narrowly triangular; ovules 17–32. **Fruits** incurved to erect, rarely deflexed; stipe 6–15(–17) mm; straight, oblong to ellipsoid, 3-sided and dorsiventrally compressed, 12–27(–30) × 3.5–8(–9) mm, fleshy, green, or suffused or speckled with purple, becoming brownish straw-colored, thinly or stiffly leathery, glabrous; ± two-chambered, septum 0.5–1.2 mm wide. **Seeds** 17–32. $2n = 22, 24$.

Flowering Mar–Jul. Slopes, cliffs, canyons, and arroyos, often in gravelly limestone or igneous soils; (600–)900–1550[–2000] m; (Mohave Desertscrub), Great Basin Desertscrub, (Semidesert Grassland, Interior Chaparral), Great Basin Conifer Woodland; Apache, Coconino, Gila, Mohave, Yavapai cos.; Nevada, New Mexico, Utah.

Astragalus eremiticus is a racially complex taxon. Barneby (1964a) discussed five unnamed morphological and ecological races in place of fewer varieties that he earlier recognized (1944, 1949). Isely (1998) recognized two varieties, with var. *ampullarioides* S.L. Welsh a vigorous form of the species. That entity is considered a distinct species (*A. ampullarioides*) by Welsh (2007).

Most plants from northern Arizona, from eastern Kane County, and some from Washington County, Utah, have blue, purple, or pink petals, or petals suffused with those colors. They seem not to differ otherwise from the typically pale ochroleucous whitish flowered specimens that form the bulk of the variety away from those areas. Fruits vary from thin and almost papery to thick and leathery-woody, and from oblong to linear-oblong and not at all or not much inflated to definitely though slightly inflated. Most Arizona collections are from Mohave County, but the species has been reported in Gila County (*B. D. Parfitt 3225*, 08 Apr 1984, DES), well south of other populations. *Astragalus eremiticus* is found in creosote bush, blackbrush, Mormon tea, juniper-pinyon, live oak, and sagebrush communities.

30. Astragalus eurylobus (Barneby) Barneby. Brittonia 36:169. 1984. * [from Greek *eurybia*, wide-spreading and *lobos*, pod, with wide pods] NEEDLE MOUNTAINS MILKVETCH Native

Astragalus tephrodes A. Gray var. *eurylobus* Barneby, Mem. New York Bot. Gard. 13:643. 1964

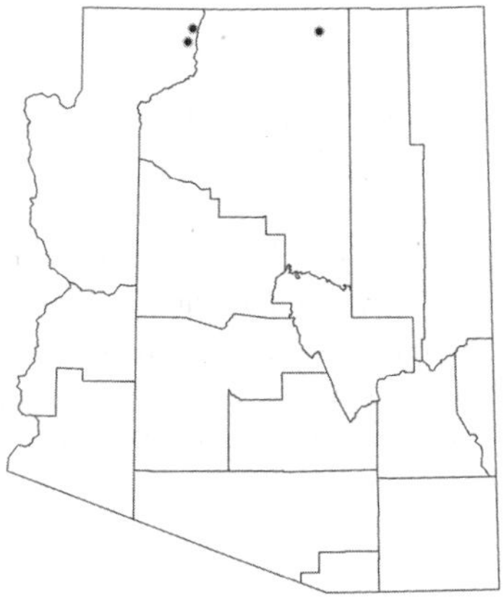

Perennials, loosely tuft-forming, subacaulescent or shortly caulescent, 10–30 cm; from caudex, branched; canescent or green, pilosulose, hairs basifixed. **Stems** absent or few, prostrate, 0.4–9 cm, internodes less than 1 cm. **Leaves** 3–24 cm, petiolate; stipules distinct throughout, deltoid or ovate, 2.5–8 mm, submembranous becoming papery. **Leaflets** 17–27, flat or loosely folded, rhombic- to elliptic-obovate or obovate-cuneate, 3–17(–27) mm, apex obtuse or emarginate, surfaces pilosulose. **Inflorescences** loosely 7–14(–26)-flowered; peduncle incurved-ascending, humistrate or procumbent in fruit, 4–18 cm; axis 2–5.5(–14) cm in fruit; bracts 1.5–8 mm; bracteoles 0(–2). **Pedicels** ascending, 0.6–3 mm. **Flowers**: calyx cylindric, 9.5–11(–12) mm, pilosulose; tube 8–9.5 × 3.9–4.4 mm; lobes subulate to triangular, 1–1.7 mm; corolla pink-purple with pallid pink-veined spot on banner, 17–20(–22) mm; banner recurved through 45°, spatulate-oblanceolate, 17–22 mm; wings 19–20.8 mm; keel 17–19 mm, apex obtuse; ovules 30–45. **Fruits** ascending, humistrate; sessile; green or purplish, straight proximally, incurved distally, obliquely oblong-ellipsoid or lanceolate-ellipsoid, laterally compressed, becoming compressed below (abaxially) on drying, 21–30(–40) × 7.5–9(–16) mm, fleshy, green or purplish, becoming semi-woody, strigillose; one-chambered. **Seeds** 30–45.

Flowering Apr–Jun. Open areas; 1300–1550[–1900] m; Great Basin Desertscrub, Great Basin Conifer Woodland; Mohave, Coconino cos.; Nevada.

Astragalus eurylobus is very similar to *A. tephrodes*, with which it was originally associated as a variety, but from which it differs in its larger, one-chambered fruits, and distribution northwest of the Grand Canyon. Overall, it is a coarser plant than *A. tephrodes*. *Astragalus eurylobus* is of conservation concern. It grows in shadscale, matchweed (*Gutierrezia* sp.), saltbush, juniper, sagebrush, and grassland communities.

31. Astragalus flavus Nuttall in J. Torrey and A. Gray, Fl. N. Amer. 1:335. 1838. * [from Latin *flavus*, yellow] YELLOW MILKVETCH

Perennials, low, loosely tufted or clump-forming, caulescent selenophyte, 5–30(–40) cm; from superficial caudex, branched, branches often impacted by clay; usually strigillose, rarely pilose-pilosulose, usually cinereous or canescent, rarely greenish, hairs medifixed. **Stems** several to numerous, decumbent to ascending or erect. **Leaves** 3–15(–18) cm, petiolate; stipules with fused sheathing, bidentate proximally throughout, deltoid or triangular, 2–10 mm, papery-scarious. **Leaflets** (5–)9–21, usually folded, linear, narrowly oblong, or oblanceolate to ovate, 3–31 × 0.5–6 mm, apex obtuse to acute, surfaces silvery strigose, rarely greenish, or glabrate or glabrous above (adaxially). **Inflorescences** 6–30-flowered; peduncle erect or ascending, (3–)5–23 cm, surpassing subtending leaves; axis 2–15 cm in fruit; bracts 1.5–5 mm; bracteoles 0. **Pedicels** ascending, 0.7–1.2 mm. **Flowers**: calyx campanulate, 4–10 mm, short-villous, pilosulose, strigillose to pilosulose; tube 3–5.2 × 2.1–3.8 mm; lobes broadly subulate to narrowly triangular, 0.9–6 mm; corolla usually whitish, ochroleucous, lemon-yellow, straw-colored, reddish lilac, reddish purple, or rarely pink-purple or suffused with pale blue, pale pink, or purple, 9–18 mm; banner recurved through 45(–90)°, broadly obovate-cuneate to rhombic-oblanceolate, 9–17.8 mm; wings (8.8–)9.3–14.4 mm; keel 6.5–10 mm, apex obtuse; ovules 6–17. **Fruits** erect; sessile; straight, oblong, dorsiventrally compressed, 7–13 × 3.5–5.5 mm, green, somewhat fleshy becoming straw-colored, leathery or stiffly papery, strigose; one-chambered. **Seeds** 6–17. $2n$ = 24, 26.

Varieties 3 (2 native in the flora): w United States.

Members of *Astragalus flavus* occupy typically fine-textured soils composed of seleniferous clays, muds, and silts. The characters used to distinguish varieties are not absolute; varieties intergrade and populations within varieties often have different features. Formal recognition of varieties may not be warranted based on quantitative differences as once thought.

1. Calyx 5.7–7.6 mm; corolla usually cream to lemon-yellow, rarely suffused with pale blue, pink, or purple, keel 8–10 mm ____ **31b. Astragalus flavus** var. **flavus**

1′ Calyx 4–6.7 mm; corolla usually whitish, sometimes tinged with pale blue or lilac, drying cream-colored, keel 6.5–8 mm ____ **31a. Astragalus flavus** var. **candicans**

31a. Astragalus flavus Nuttall var. **candicans** A. Gray. Proc. Amer. Acad. Arts 12:54. 1876. * [from Latin *candidus*, white] Native

Astragalus confertiflorus A. Gray; *Astragalus flavus* Nuttall var. *higginsii* S.L. Welsh (in part)

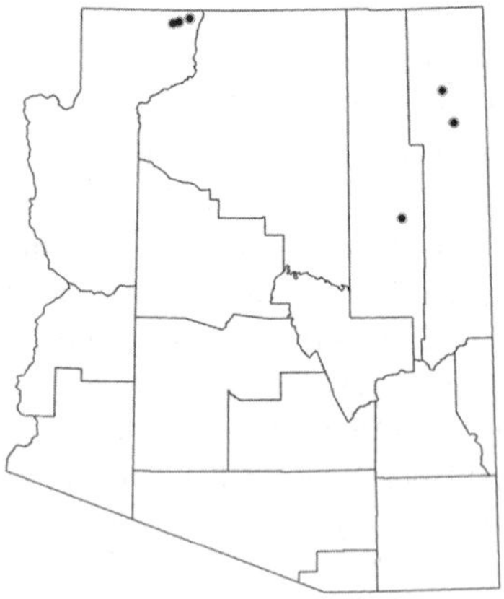

Plants Closely resembling var. *flavus*, usually of shorter duration and smaller when old. **Inflorescences** surpassing leaves; peduncle usually 3–10 cm. **Flowers**: calyx 4–6.7 mm, densely strigillose; tube 3.7–5.2 mm; lobes shorter than tube; corolla usually whitish, sometimes tinged with pale blue or lilac, drying cream-colored, 11–18 mm; keel 6.5–8 mm. **Seeds** (6–)8–14.

Flowering Apr–Jun. Habitats of var. *flavus*, but at lower elevations, plains, clay flats, knolls, and bluffs, commonly on sandstone of Triassic Chinle, Moenkopi, and other similar, fine-textured, seleniferous formations; [850–]1520–2000 m; Mohave Desertscrub, Great Basin Desertscrub; Apache, Mohave, Navajo cos.; Nevada, Utah, New Mexico.

31b. Astragalus flavus Nuttall var. **flavus** * Native

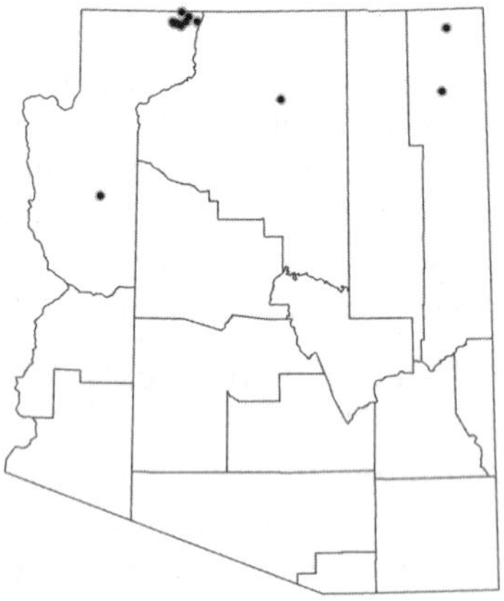

Plants forming tufts. **Inflorescences** slightly or much surpassing leaves; peduncle usually to 12 cm. **Flowers**: calyx 5.7–7.6 mm, strigose to short-villous; tube 3.2–5.2 mm; lobes 0.9–1.7 mm; corolla usually cream to lemon-yellow, rarely suffused with pale blue, pink, or purple, 11–17.8 mm, keel 8–10 mm. **Seeds** 8–17.

Flowering Apr–Jul(–Sep). Slopes, canyons, and along roads in seleniferous substrates composed of saline silts and clays derived from Triassic Chinle and Moenkopi, and from Cretaceous Mancos Shale and Tropic Shale formations, and other similarly fine-textured Tertiary formations; [800–]1500–2000[–2300] m; Great Basin Desertscrub, Great Basin Conifer Woodland; Apache, Coconino, Mohave, Navajo cos.; Colorado, New Mexico, Utah, Wyoming.

Sheep poisoning, possibly due to ingestion of selenium, attributable to var. *flavus*, is known from the lower elevations of the Uinta Basin in northeastern Utah. Variety *flavus* grows in salt desert shrub (shadscale, horsebrush, greasewood (*Sarcobatus vermiculatus*), rabbitbrush, and mat-atriplex) and pinyon-juniper communities.

32. Astragalus flexuosus (Hooker) Douglas ex G. Don. Gen. Hist. 2:256. 1832. * [from Latin *flexuosus*, much or fully bent or curved, tortuous] BENT MILKVETCH, PLIANT MILKVETCH

Phaca flexuosa Hooker, Fl. Bor.-Amer. 1:141. 1831

Perennials, caulescent, slender, 10–60 cm; from subterranean caudex, (branched); strigillose to villosulous, greenish-cinereous [or silky-canescent], hairs appressed, incurved-ascending or curly, or cinereous, hairs basifixed. **Stems** few to several, decumbent or ascending, 15–60(–70) cm, 1–20+ cm underground. **Leaves** 1.5–9 cm, shortly petiolate proximally, subsessile distally; stipules with fused sheathing proximally, fused or distinct distally, triangular or lanceolate, 1–7 mm, papery proximally, herbaceous distally. **Leaflets** (7–)9–25(–29), often folded, sometimes flat, linear, linear-oblong, oblong-oblanceolate, or obovate-cuneate, (2–)3–19 × 1–4.2 mm, apex obtuse to truncate or retuse, surfaces strigose to glabrate below (abaxially), usually glabrous above (adaxially). **Inflorescences** with (7–)10–26(–30) flowers, with raceme usually shorter than stem; peduncle incurved-ascending, widely diverging, or erect, 1.5–19 cm; axis (1–)2.5–13(–15) cm in fruit; bracts 0.6–4.5 mm; bracteoles 0–2. **Pedicels** spreading, 0.7–3.5 mm. **Flowers**: calyx campanulate, 3.3–6 mm, strigose with mostly white hairs; tube 1.9–4.3 × 1.8–2.9 mm; lobes subulate, 0.5–2 mm; corolla pink-purple to dull purplish, 7–11 mm; banner recurved through 45–90°, ovate-cuneate or rhombic-obovate, 7.2–11 mm; wings 6.9–10.5 mm; keel 5–8.2 mm, apex obtuse or acute-triangular; ovules 12–25. **Fruits** descending to spreading; sessile or obscurely stipitate, stipe 0–1.3 mm; straight or slightly incurved, linear-oblong, linear-oblanceolate, oblong-ellipsoid, lanceolate-ellipsoid, or ovoid-ellipsoid, nearly cylindrical or ± flattened, (8–)11–24 × 2.7–9 mm, green or mottled becoming straw-colored or brownish, thinly to stiffly papery, usually strigillose or villosulous, rarely glabrous; one-chambered. **Seeds** 12–25. 2*n* = 22.

Varieties 3 (2 native in the flora): North America.

1. Calyx 3.3–4.1 mm; banner abruptly recurved through 90°; keel 5–5.5 mm, nearly round, very strongly incurved through 120°; fruit 11–15(–20) × 3–4(–4.5) mm ________ **32a. Astragalus flexuosus** var. **diehlii**
1′ Calyx 3.5–6 mm; banner recurved through 45–50°; keel 7–8.2 mm, semi-obovate, incurved through 90–100°; fruit (12–)14–23 × (4–)5–9 mm ________ **32b. Astragalus flexuosus** var. **greenei**

32a. Astragalus flexuosus (Hooker) Douglas ex G. Don var. **diehlii** (M.E. Jones) Barneby. Leafl. W. Bot. 4:54. 1944. * [for I.E. Diehl, correspondent of M.E. Jones] DIEHL'S MILKVETCH Native

Astragalus diehlii M.E. Jones, Rev. N.-Amer. Astragalus 194. 1923

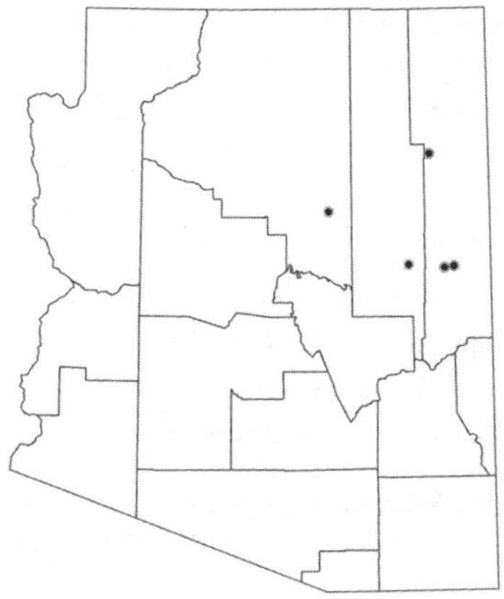

Herbage greenish cinereous. **Stems** usually obviously flexuous, 15–30(–40) cm. **Leaflets** (7–)11–17, mostly linear or narrowly oblong-oblanceolate. **Inflorescences** with 12–26 flowers; peduncle 2.5–7 cm; axis 2.5–9.5 cm in fruit. **Flowers:** calyx 3.3–4.1 mm; tube 1.9–2.3 mm; lobes 0.7–1.1(–1.4) mm; corolla: banner abruptly recurved through 90°, 7–9 mm; keel very strongly incurved through 120°, nearly round, 5–5.5 mm. **Fruits:** stipe 0 mm; oblong-ellipsoid, scarcely turgid, 11–15(–20) × 3–4(–4.5) mm, strigillose. **Seeds** 12–18.

Flowering late Apr–Jun. Slopes, valleys, canyons, arroyos, and roadsides in fine-textured substrates derived from Moenkopi (Triassic), Entrada (Jurassic), and Mancos Shale (Cretaceous) formations; 1300–2100 m; Great Basin Desertscrub, Great Basin Conifer Woodland; Apache, Coconino, Navajo cos.; Colorado, New Mexico, Utah.

32b. Astragalus flexuosus (Hooker) Douglas ex G. Don var. **greenei** (A. Gray) Barneby. Leafl. W. Bot. 9:91. 1960. * [for Edward L. Greene, 1843–1915, American botanist] GREENE'S MILKVETCH Native

Astragalus greenei A. Gray, Proc. Amer. Acad. Arts 16:105. 1881

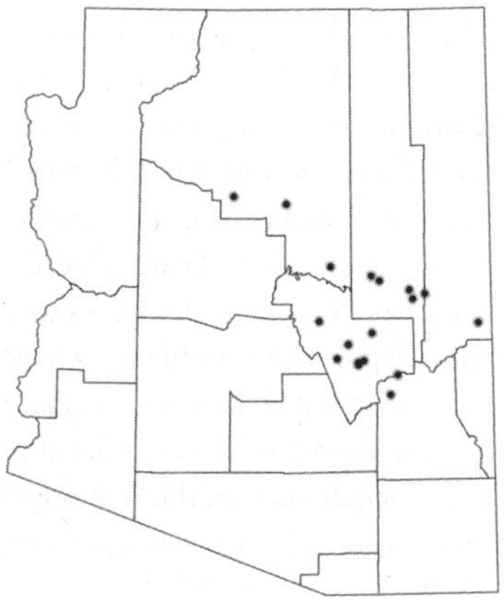

Herbage greenish cinereous. **Stems** 15–40 cm. **Leaflets** (9–)13–23, linear-oblong to obovate-cuneate. **Inflorescences** with (7–)10–30 flowers; peduncle 1.5–13.5 cm; axis (1–)2.5–10 cm in fruit. **Flowers**: calyx 3.5–6 mm; tube 2.8–4.2 mm; lobes 0.8–2 mm; corolla: banner recurved through 45–50°, 9–10.8 mm; keel incurved through 90–100°, semi-obovate, 7–8.2 mm. **Fruits**: sessile or nearly so, stipe 0–0.5 mm; lanceolate-ellipsoid to plumply ellipsoid, oblong-ellipsoid, or ovoid-ellipsoid, turgid or inflated, (12–)14–23 × (4–)5–9 mm, (base rounded when less than 5 mm wide), strigillose or villosulous. **Seeds** 16–25. 2*n* = 22.

Flowering Apr–Jun. Slopes, arroyos, badlands, and other semi-barrens; 1340–2290 m; Interior Chaparral, Great Basin Conifer Woodland, occasional in meadows in Petran Montane Conifer Forest; Apache, Coconino, Gila, Navajo cos.; New Mexico.

33. Astragalus fucatus Barneby. Leafl. W. Bot. 9:89. 1960. * [from Latin *fuco*, painted, referring to the fruit] HOPI MILKVETCH Native

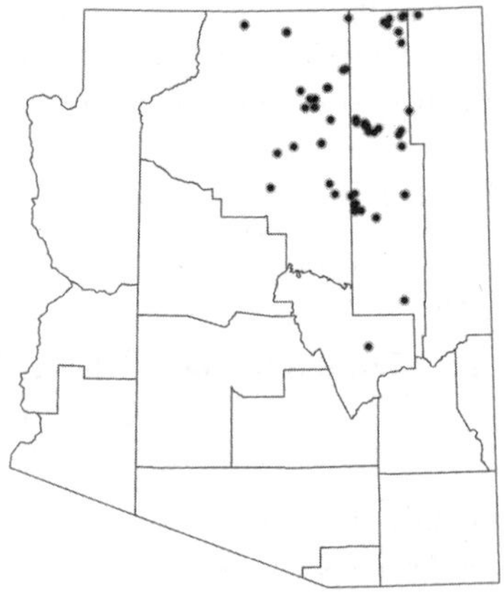

Perennials, short-lived, sometimes flowering first year, caulescent, 7–45(–70) cm; from subterranean or superficial caudex; strigillose, hairs basifixed. **Stems** single to several, ascending to erect or sprawling, flexuous, (0–)1–8 cm underground. **Leaves** 2–13 cm, petiolate; stipules with fused sheathing proximally, distinct or obscurely fused distally, broadly deltoid to lanceolate, 1–5.5 mm, papery proximally, subherbaceous or firmly papery distally. **Leaflets** 9–17, usually involute or longitudinally folded, rarely flat, oblanceolate to linear, 3–20(–25) × 0.5–4 mm, apex obtuse to retuse, surfaces strigose below (abaxially), glabrous above (adaxially). **Inflorescences** 9–27-flowered; peduncle incurved-ascending, becoming horizontal in fruit, 1–6.5 cm; axis 2–11.5 cm in fruit; bracts 0.8–2 mm; bracteoles 0. **Pedicels** ascending to declined, 0.7–3.5 mm. **Flowers:** calyx campanulate, 3.3–5.4 mm, strigose; tube 2.3–3.3 × 2.1–2.6 mm; lobes broadly subulate or triangular, 0.8–2.2 mm; corolla pink-purple, 6.4–8.7 mm; banner recurved through 90°, broadly ovate-cuneate to rhombic-flabellate, 6.4–8.2 mm; wings 6.8–7.5 mm; keel 6–7.2 mm, apex triangular and beaklike; ovules 21–32. **Fruits** spreading to declined; sessile; purple-mottled, straight, ovoid, or subglobose to ellipsoid, bladdery-inflated, 17–32 × 12–20 mm, papery, strigose; one-chambered. **Seeds** 21–32. $2n = 22$.

Flowering May–Jul(–Sep). Valleys, arroyos and dunes, typically in sand; 1300–1900(–2200) m; Apache, Coconino, Navajo cos.; Colorado, New Mexico, Utah.

Astragalus fucatus occurs in mixed sandy desertscrub habitats associated with red sandstones, in rabbitbrush, blackbrush, and juniper communities.

34. Astragalus geyeri A. Gray. Proc. Amer. Acad. Arts 6:214. 1864. * [for Charles A. Geyer, 1809–1853, German plant collector] GEYER'S MILKVETCH

Phaca annua Geyer ex Hooker, not *Astragalus annuus* de Candolle

Annuals (biennials), slender, diffuse, caulescent, (3–)6–27(–30) cm; taproot slender; from superficial root-crown; strigillose, hairs basifixed. **Stems** 1–3, prostrate to ascending or erect. **Leaves** (1.5–)2–10.5(–12) cm, petiolate; stipules distinct throughout, deltoid or lanceolate, 1.5–4 mm, thinly herbaceous or submembranous. **Leaflets** (3–)7–13, folded or flat, linear to oblong or narrowly elliptic, rarely obcordate, 3–18 × 1–5.2 mm, apex obtuse to retuse, surfaces strigose below (abaxially), strigose or glabrous above (adaxially). **Inflorescences:** loosely 2–8-flowered; peduncle incurved-ascending, 0.6–2.5 cm; axis 0.3–1.5 cm in fruit; bracts 0.7–2 mm; bracteoles 0–2. **Pedicels** ascending, 0.6–1.5 mm. **Flowers**: calyx campanulate, (2–)2.7–3.8(–4) mm, strigose; tube 1.5–2.5 × 1.5–2 mm; lobes subulate to lanceolate, 0.6–1.5 mm; corolla whitish with lilac-tipped keel, rarely all pink-purple and drying violet, 5–7.6 mm; banner recurved through 45°, ovate-cuneate or oval-cuneate, 5.2–7.6 mm; wings 4.6–6.5 mm; keel 3.8–4.8 mm, apex obtuse, deltoid; ovules 7–18. **Fruits** spreading to declined, often humistrate; sessile; straight or slightly incurved, obliquely ovoid, bladdery-inflated, 15–24 × 6–12 mm, prominently beaked and clearly differentiated from the body, thin, green or suffused purple, becoming straw-colored, papery, strigose; one-chambered or partially two-chambered. **Seeds** 7–18.

Varieties 2 (2 native in the flora): Great Basin, extending into Montana and n Arizona.

The drab flowers and inflated fruits of *Astragalus geyeri* are mostly tucked among the leaves. Of all the species of section *Inflati*, this has the largest geographic range.

1. Fruit sutures rounded dorsally (abaxially); one-chambered, not inflexed; leaflets usually linear-oblong, narrowly oblanceolate, or linear-elliptic, rarely obcordate ____ **34a. Astragalus geyeri** var. **geyeri**

1′ Fruit sutures grooved dorsally (abaxially); partially two-chambered, inflexed; leaflets broadly oblanceolate or elliptic ____ **34b. Astragalus geyeri** var. **triquetrus**

34a. Astragalus geyeri A. Gray var. **geyeri** * Native

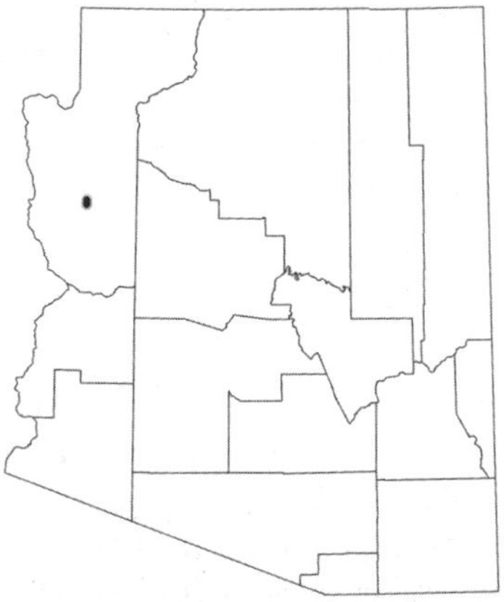

Annuals or, rarely, biennials. **Leaflets** (3–)7–13, usually linear-oblong, narrowly oblanceolate, or linear-elliptic, rarely obcordate, 3–17 mm, terminal leaflet usually longer than last pair. **Fruits**: sutures rounded dorsally (abaxially), not inflexed; one-chambered, not inflexed. **Seeds** 10–18.

Flowering Apr–Jul. Slopes, valleys, and draws; [700–]1400–1500[–1700 (–2200)] m; [Mohave Desertscrub, Great Basin Desertscrub], Great Basin Conifer Woodland; Mohave Co.; California, Idaho, Montana, Nevada, Oregon, Utah, Washington, Wyoming.

In years when growing conditions are suitable this variety may cover the ground like a planted crop. A single collection is documented for Arizona from near Hackberry, Mohave Co., in the Hualapai Mountains. Elsewhere across its wide range, var. *geyeri* is often found in desertscrub communities.

34b. Astragalus geyeri A. Gray var. **triquetrus** (A. Gray) M.E. Jones. Contr. W. Bot. 8:7. 1898. * [three-cornered] MUDDY RIVER MILKVETCH Native

Astragalus triquetrus A. Gray, Proc. Amer. Acad. Arts 13:367. 1878

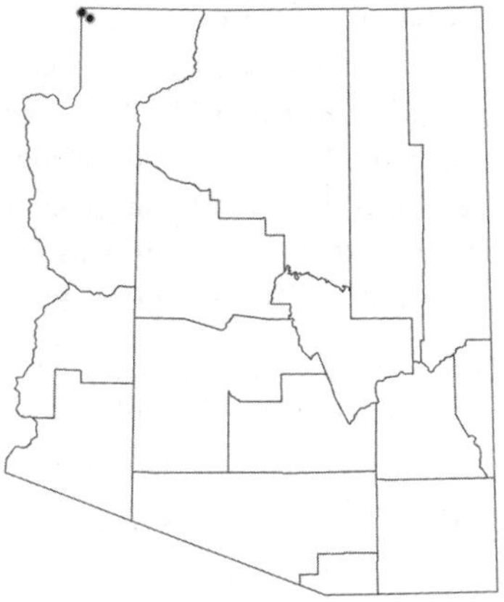

Annuals. **Leaflets** (5–)7–9, broadly oblanceolate or elliptic, 4–15 mm, terminal leaflet not conspicuously longer than last pair. **Fruits:** sutures grooved and inflexed dorsally (abaxially), 3-sided; partially two-chambered, inflexed as narrow, partial septum 0.6–1.2 mm wide. **Seeds** 7–11.

Flowering Apr–May. Sandy flats and washes; 400–730 m; Mohave Desertscrub; Mohave Co.; Nevada.

Variety *triquetrus* is a rare and local taxon from extreme northwestern Mohave County, Arizona, and southern Clark County, Nevada.

35. Astragalus gilensis Greene. Bull. Torrey Bot. Club 8:97. 1881. * [of the Gila River] GILA MILKVETCH Native

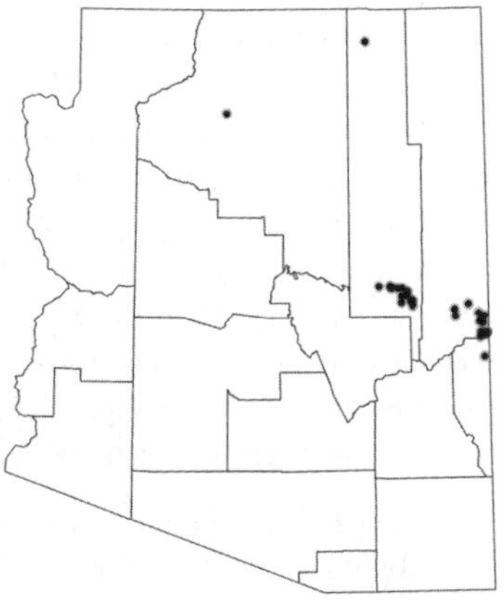

Perennials, acaulescent or nearly so, 6–25 cm; from superficial caudex, branched, tufted with thatch of persistent, withered leaf bases; ashen-strigillose, hairs medifixed. **Stems** obscured by stipules and leaf bases. **Leaves** (1.5–)2.5–9 cm, petiole slender; stipules distinct throughout, ovate or lanceolate, 4–9 mm, submembranous becoming papery. **Leaflets** (7–)11–21(–25), flat or loosely folded, elliptic, oblong-elliptic, or oblanceolate, (2–)3–12 × 2–4 mm, apex acute, obtuse, or emarginate, surfaces ashen-strigillose. **Inflorescences** densely 7–26-flowered, almost headlike cluster; peduncle subscapose, incurved-ascending, (2.5–)3.5–15(–22) cm; axis 0.5–2.5 cm in fruit, slightly elongating; bracts 1.5–4 mm; bracteoles 0–2. **Pedicels** ascending to spreading becoming declined proximally, 0.6–1.6 mm. **Flowers**: calyx campanulate, 3.7–6 mm, strigillose; tube becoming papery, ruptured, 2.3–4 × 1.6–2.4 mm; lobes subulate, 1.3–2.7 mm; corolla pink to pink-purple, wing tips white, 6.2–10.4 mm; banner recurved through 45–50°, elliptic or elliptic-oblanceolate, 6.2–10.4 mm; wings 5.8–9 mm; keel 4.3–6.8 mm, apex bluntly deltoid, sometimes beaklike; ovules 7–11. **Fruits** in oblong or circular clusters, spreading and ascending; sessile; slightly incurved, obliquely-ovoid or oblong-ellipsoid, obscurely 3-sided and somewhat laterally compressed, 4.5–7.5 × 2–4.2 mm, fleshy and firm, reddish-tinged becoming brownish, stiffly papery or somewhat leathery, strigillose; one-chambered. **Seeds** 7–11. $2n = 22$.

Flowering (Apr–)May–Sep. Dry sites on slopes and meadows beneath trees and in the open and roadsides;

1700–2700 m; Great Basin Conifer Woodland, Petran Montane Conifer Forest; Apache, Coconino, Greenlee, Navajo cos.; New Mexico.

Astragalus gilensis is reminiscent of the larger *A. troglodytus* in growth form and almost headlike inflorescences that remain compact in fruit; however, *A. troglodytus* has basifixed hairs. The species occurs along the Mogollon Rim system in east-central Arizona and west-central New Mexico, with an outlying population in that state in the Jemez Mountains. It is found in xeric ponderosa pine forests and pinyon-juniper woodlands.

36. Astragalus hallii A. Gray. Proc. Amer. Acad. Arts 6:224. 1864. * [for Elihu Hall, 1822–1882, American botanist] HALL'S MILKVETCH

Perennials, caulescent, 12–50(–55) cm; from subterranean caudex; loosely strigillose to villosulous, hairs basifixed. **Stems** several or numerous, decumbent, ascending, prostrate, or diffuse, 1–13 cm underground. **Leaves** 2–9 cm, with petiole 2–8(–12) mm or upper leaves subsessile; stipules at lower nodes small, fused into a scarious, glabrous sheath, the median and upper ones triangular or lanceolate, (1.2–)2–7 mm, subherbaceous, either strigillose or glabrescent dorsally. **Leaflets** (11–)13–27(–31), flat or involute, obovate to oblanceolate or elliptic, (1.5–)3–14(–15) × 1–7 mm, apex retuse to truncate or obtuse, surfaces strigillose below (abaxially), sparsely hairy or glabrous above (adaxially). **Inflorescences** with (7–)9–28 flowers; peduncle suberect, incurved-ascending, spreading, declined, or prostrate, 3–9.5(–11) cm; axis 1–7 cm in fruit; bracts 1.5–5 mm; bracteoles 0–2. **Pedicels** spreading-declined, 1.2–4 mm. **Flowers**: calyx campanulate, (5–)6–7(–9) mm, strigillose or villosulous; tube 5–6.2 × 1.6–2.4 mm; lobes triangular-subulate, 0.5–2.3 mm; corolla variable in color, bright or pale purple, sometimes, white-tipped or suffused with lilac or purple, 12.4–15(–18.5) mm; banner recurved through 45°, ovate-cuneate, 5.3–8.4 mm; wings (11.5–)12–17 mm; keel 9–12(–12.3) mm, apex obtuse or acute-triangular; ovules 20–34. **Fruits** spreading to declined or deflexed; subsessile or stipitate, stipe (0–)1–4.5 mm; straight or slightly incurved, oblong, oblanceolate-ellipsoid, ovoid, obovoid, or ellipsoid, slightly to distinctly inflated, (12–)13–27 × (3–)4–12 mm, green or purplish becoming straw-colored or brown, becoming papery or cartilaginous, opaque, strigillose, or villosulous; one-chambered. **Seeds** 20–34. $2n = 22$.

Varieties 2 (1 native in the flora): Rocky Mountains of Colorado and n and w New Mexico and Arizona.

36a. Astragalus hallii A. Gray var. **fallax** (M.E. Jones) Barneby. Leafl. W. Bot. 9:91. 1960. * [from Latin *fallax*, deceptive] DECEPTIVE MILKVETCH Native

Astragalus gracilentus (A. Gray) A. Gray var. *fallax* M.E. Jones, Contr. W. Bot. 8:14. 1898

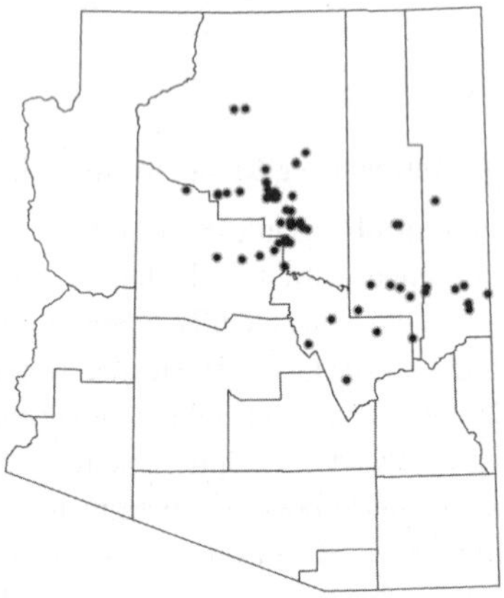

Herbage often (not consistently) cinereous, strigillose, hairs to 0.2–0.4(–0.6) mm (if 0.4–0.6 mm, then fruits distinctly inflated). **Stems** diffuse or ascending. **Leaflets**: (11–)13–25. **Inflorescences:** peduncle incurved-ascending to suberect. **Flowers**: corolla bright or pale purple, or tipped with white, or suffused with purple. **Fruits** not or seldom humistrate; stipe 0–4.5 mm; obliquely ovoid or obovoid, rarely broadly ellipsoid, usually distinctly inflated, 13–27 × 6–12 mm, usually strigillose, rarely villosulous. **Seeds** (24–)28–34.

Flowering Jun–Sep. Ridges and valleys, usually in igneous soils; 2300–3100 m; occasional in Great Basin Desertscrub and Madrean Evergreen Woodland, Great Basin Conifer Woodland, Petran Montane Conifer Forest; Apache, Coconino, Gila, Navajo, Yavapai cos.; New Mexico.

As reviewed by Barneby (1964a) var. *fallax* is multiracial and includes at least three entities of possible taxonomic significance. Variety *fallax* occurs in ponderosa pine, pinyon-juniper, oak, sagebrush, and mixed mountain brush communities. Barneby (1964a) reports this taxon as occurring in the Santa Catalina Mts., Pima Co.

37. Astragalus holmgreniorum Barneby. Brittonia 32:24, fig. 1. 1980. * [for Noel H. Holmgren, b. 1937, and Patricia K. Holmgren, b. 1940, American botanists] HOLMGREN'S MILKVETCH, PARADOX MILKVETCH Native

Perennials, dwarf, tufted, strictly acaulescent, 4–12 cm; from thickened superficial root-crown; loosely pilosulose, hairs basifixed. **Stems** absent. **Leaves** 4–21.5 cm, petiolate; stipules distinct throughout, lanceolate or lanceolate-attenuate, 2.5–8 mm, papery, surfaces pilose. **Leaflets** (5–)9–15(–17), loosely folded, broadly obovate to obcordate, 6–16 × 3.5–12 mm, apex emarginate, surfaces pilose below (abaxially), glabrous above (adaxially). **Inflorescences:** racemes 4–16-flowered; peduncle procumbent, 2–9 cm, shorter than leaf; axis 0.4–6.5 cm in fruit; bracts 4–8 mm; bracteoles 0. **Pedicels** ascending to spreading, 1–2 mm. **Flowers**: calyx cylindric, 10.5–12.5 mm, sparsely white-pilose; tube 8–9.5 ×

2.9–3.7 mm; lobes subulate, 2–3.5 mm; corolla pink-purple, 18–23.5 mm; banner recurved through 30–40°, oblanceolate, 21.5–23.5 mm; wings 17.5–20.5 mm; keel 16.5–18.5 mm, apex obtuse; ovules 30–34. **Fruits** ascending or reclining, humistrate; lunately incurved, ellipsoid, 3-sided compressed, 25–55 × 6–9 mm, green or purplish brown becoming straw-colored, leathery, glabrous; disjointing prior to dehiscence; two-chambered. **Seeds** 30–34. $2n$ = 22.

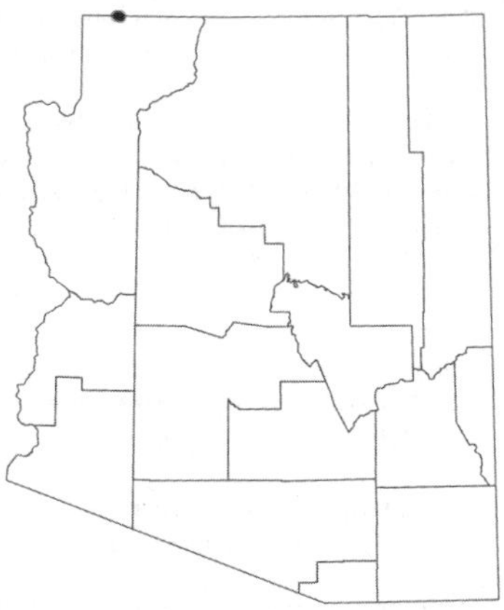

Flowering May–Jun. Rock outcrops and slopes with gravelly sandy loam soils; [800–]820–860[–900] m; Mohave Desertscrub; Mohave Co.; Utah.

Long known from collections, *Astragalus holmgreniorum* was only recently described. *Astragalus holmgreniorum* is of conservation concern. It is in the Center for Plant Conservation's National Collection of Endangered Plants and is listed as Endangered by the U.S. Fish and Wildlife Service. In Arizona, this species occurs in a very restricted area of extreme northern Mohave County.

The specific epithet commemorates Drs. Noel and Patricia Holmgren, while the common name "paradox milkvetch" is derived from their nickname Paradox for "pair o'docs" (Barneby 1989).

38. Astragalus humistratus A. Gray. Smithsonian Contr. Knowl. 5(Art. 6) 2:43. 1853. * [from Latin *humi*, ground and *stratum*, layer, referring to the low growth habit] GROUNDCOVER MILKVETCH

Perennials, mat- or clump-forming, caulescent, (4–)7–80 cm; from superficial or slightly subterranean caudex; strigose to villous, subtomentose, or pilose-pilosulose, hairs medifixed, herbage green, greenish cinereous, cinereous, or silvery-canescent. **Stems** several or numerous, prostrate, diffuse or weakly ascending **Leaves** 1–6(–7.5) cm, subsessile or proximally shortly petiolate; stipules all with fused sheathing, lower ones into a cuplike sheath, the upper ones with progressively long, free blades, 1.5–10(–11) mm, submembranous becoming scarious, surfaces pubescent below (abaxially). **Leaflets** (5–)9–17(–19), flat, loosely folded, or involute, elliptic to oblong or oblanceolate to obovate, (2–)3–17(–19) × 0.5–6 mm, apex acute or obtuse, surfaces strigose, sometimes glabrate or glabrous above (adaxially). **Inflorescences** with 3–30 flowers; peduncle widely diverging or incurved-ascending, 1–9(–9.5) cm; axis (0.5–)1–9(–13) cm in fruit; bracts 1.5–7 mm; bracteoles 0–2. **Pedicels** ascending, 0.4–2.2 mm. **Flowers**: calyx campanulate, 3.2–7.6(–8.8) mm, strigose; tube 2.1–4(–4.1) × 1.6–2 mm; lobes lanceolate-acuminate, filiform bristlelike, or subulate, (1–)1.2–4.5(–5) mm; corolla greenish to ochroleucous or purple, banner often suffused, margined, or veined purplish, pink, or lilac-purple, sometimes wing tips whitish, 5.9–11.8 mm; banner abruptly recurved through 50–85°, broadly obovate to rhombic-ovate or subquadrate, 5.9–11.8 mm; wings 5.8–10.7 mm; keel 5.1–10 mm, apex porrectly acuminate or obtuse; ovules 6–26. **Fruits** ascending to spreading; sessile; incurved, obliquely oblong-ellipsoid, lanceolate-ellipsoid, ovoid, obovoid, linear-oblong, oblanceolate, semi-ovoid, or semi-ellipsoid, laterally compressed, 6–20 × 2.5–6.5 mm, green or purple-dotted, thickly fleshy becoming straw-colored or brown, leathery or papery, strigillose or villosulous; one-chambered. **Seeds** 6–26. $2n$ = 24.

Varieties 6 (6 native in the flora): sw United States, nw Mexico.

Mostly, *Astragalus humistratus* comprises two main forms: (1) a southwestern group characterized by fruits relatively long and narrow, with many ovules, and (2) a northwestern group characterized by fruits short, plumper in silhouette, and with relatively few ovules. Varieties are spring- or autumn-flowering or, with rains, blooming both times. The prostrate mats of this species are useful in erosion control (Kearney and Peebles 1942).

1. Fruits (10–)13–20 × 3.1–6.5 mm; seeds (16–)18–26; southeast Arizona.
 2. Herbage sparsely pubescent, green or greenish cinereous; leaflets glabrous or glabrate above (adaxially) ______ **38c. Astragalus humistratus** var. **humistratus**
 2′ Herbage cinereous or silvery canescent; leaflets pubescent above (adaxially) ______ **38e. Astragalus humistratus** var. **sonorae**
1′ Fruit 6–14 × 2.5–5.7 mm; seeds 6–16; east, west Arizona.
 3. Plants with extremely fine, weak, sinuous or curly hairs; banner (3–)4–5 mm wide; fruit 2.5–3 mm wide; seeds 6–9; leaflets pubescent above (adaxially) ______ **38a. Astragalus humistratus** var. **crispulus**
 3′ Plants with straight or nearly straight hairs, sometimes shorter ones crispate or sinuous; banner 5–8.4(–9) mm wide; fruit 3–5.7 mm wide; seeds (7–)8–16; leaflets pubescent, glabrous, or glabrescent above (adaxially).
  4. Inflorescences with 3–6(–8) flowers; leaflets 2–6.5 mm; Kaibab Plateau, northern Arizona ______ **38f. Astragalus humistratus** var. **tenerrimus**
  4′ Inflorescences with (3–)7–22 flowers; leaflets (2–)3–17(–19) mm; Arizona outside of Kaibab Plateau.
   5. Herbage cinereous, canescent, or silvery; leaflets pubescent above (adaxially) ______ **38d. Astragalus humistratus** var. **humivagans**
   5′ Herbage green or greenish cinereous; leaflets glabrous above (adaxially) ______ **38b. Astragalus humistratus** var. **hosackiae**

38a. Astragalus humistratus A. Gray var. **crispulus** Barneby. Leafl. W. Bot. 4:53, figs. 24–26. 1944. * [from Latin *crispus*, with curled hairs] CURLY-HAIR MILKVETCH Native

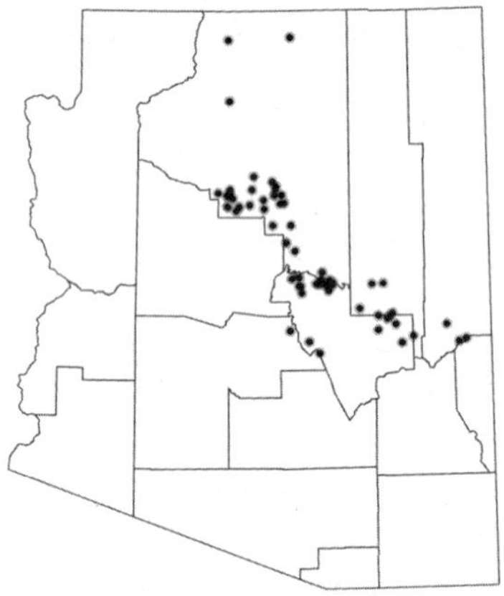

Plants gray-villous, often also subtomentose, hairs extremely fine, weak, sinuous or curly; herbage greenish cinereous. **Stems** (5–)15–55 cm. **Leaves** 1–4.5 (–5) cm; stipules 2.5–8 mm. **Leaflets** 11–15, (2–)3–14 mm, surfaces pubescent above (adaxially). **Inflorescences** with (3–)5–12 flowers; peduncle (1–)1.5–3.5(–4) cm; axis (0.5–)1–3 cm in fruit. **Flowers:** calyx 4.6–5.5 mm; tube 2.9–4 mm; lobes subulate, 1.2–2.4 mm; corolla whitish, faintly tinged with pink; banner 7–9.2 × (3–)4–5 mm. **Fruits** lunately semi-ellipsoid, incurved, 8–10 × 2.5–3 mm, villosulous. **Seeds** 6–9.

Flowering Aug–Sep. Slopes on igneous substrates; 2200–2500 m; Petran Montane Conifer Forest; Apache, Navajo cos.; New Mexico.

Variety *crispulus*, endemic to the White and San Francisco mountains of eastern Arizona and adjacent New Mexico, is sometimes sympatric with var. *humistratus*. The two are remarkably different in appearance. Variety *humivagans* extends eastward into their range generally, and perhaps exclusively, at lower elevations. Variety *crispulus* is found in xeric pine forests.

38b. Astragalus humistratus A. Gray var. **hosackiae** (Greene) M.E. Jones. Contr. W. Bot. 10:58. 1902. * [resembling species of *Hosackia*] LOTUS MILKVETCH Native

Astragalus hosackiae Greene, Bull. Calif. Acad. Sci. 1:157. 1885

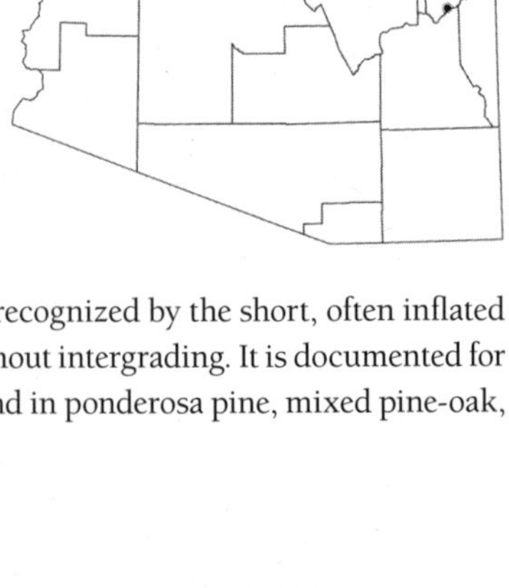

Plants sparsely strigillose or pilose-pilosulose, hairs straight or nearly so, sometimes shorter ones crispate or sinuous; herbage green or greenish cinereous. **Stems** (4–)7–36 cm. **Leaves** 1–5 cm; stipules (1.5–)2.5–8 mm. **Leaflets** 9–15, 3–9(–12) mm, surfaces glabrous above (adaxially). **Inflorescences** with 8–20 flowers; peduncle 2–6 cm; axis (0.8–)1.2–3.5 cm in fruit. **Flowers**: calyx (3.8–)4–5.5 mm; tube 2.5–3.2 mm; lobes lanceolate-acuminate or filiform bristle-like, (1–)1.4–2.5 mm; corolla greenish white, lined or suffused with dull purple, sometimes all purple with whitish wing tips; banner 6.4–8 × 5–7 mm. **Fruits** plumply semi-ovoid or obliquely ovoid, 7–10 × (3.6–)4–5.5 mm, finely strigillose. **Seeds** (7–)8–14. $2n = 24$.

Flowering May–Sep. Slopes, valleys, meadows, and roadsides, sandy to clay soils and on igneous bedrock; 1160–2750 m; Interior Chaparral, Madrean Evergreen Woodland, Great Basin Conifer Woodland, Petran Montane Conifer Forest; Apache, Coconino, Gila, Maricopa, Navajo cos.; New Mexico.

Variety *hosackiae* has the smallest flowers of these six varieties. It is easily recognized by the short, often inflated fruits. Variety *hosackiae* is locally sympatric with var. *humivagans*, ostensibly without intergrading. It is documented for Catron County, New Mexico, just east of the Arizona border. The plants are found in ponderosa pine, mixed pine-oak, oak-chaparral, and pinyon-juniper communities.

38c. Astragalus humistratus A. Gray var. **humistratus** * Native

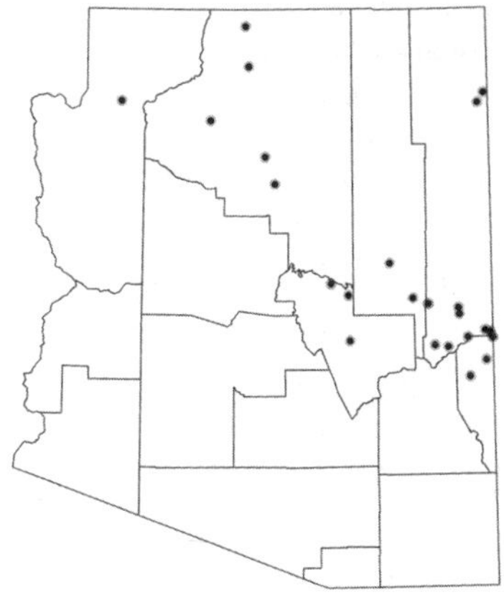

Plants sparsely pilose-pilosulose; herbage green or greenish cinereous, sparsely pubescent. **Stems** (15–)20–60 cm. **Leaves**: (1–)1.5–5.5 cm, stipules 2–9(–11) mm. **Leaflets** 11–17, 4–14 mm, surfaces glabrous or glabrate above (adaxially). **Inflorescences** with (5–)7–30 flowers; peduncle 2–8(–9.5) cm; axis 1–6 cm in fruit. **Flowers**: calyx 5.8–7.6 mm; tube 3–3.8 mm; lobes lanceolate-acuminate or filiform bristlelike, 2.5–4.5 mm; corolla greenish white to ochroleucous, often suffused or veined with purple; banner 9.5–11.8 × 6–8 mm. **Fruits** obliquely oblong-ellipsoid or lanceolate-ellipsoid, somewhat incurved, 14–18 × 4–6.5 mm, strigillose. **Seeds** (16–)18–26. $2n = 24$.

Flowering May–Sep. Rocky or gravelly slopes, canyons, floodplains, and occasionally on gravel bars in stream channels, mostly in soils derived from igneous rocks; 1300–2700 m; Great Basin Grassland, Interior Chaparral, Great Basin Conifer Woodland, Petran Montane Conifer Forest; Apache, Coconino, Gila, Greenlee, Mohave, Navajo cos.; Colorado, New Mexico, Texas; Mexico (Chihuahua).

Variety *humistratus* intergrades with var. *sonorae* in southwestern New Mexico. It grows in pine forests, pinyon-juniper communities, mesquite-juniper-saltbush and blue-grama grassland communities, and less commonly in spruce-fir forests.

38d. Astragalus humistratus A. Gray var. **humivagans** (Rydberg) Barneby. Amer. Midl. Naturalist 55:478. 1956. * [from Latin *humi*, ground and *vagans*, wandering, in reference to the growth habit] SPREADING MILKVETCH Native

Batidophaca humivagans Rydberg in N.L. Britton et al., N. Amer. Fl. 24:316. 1929

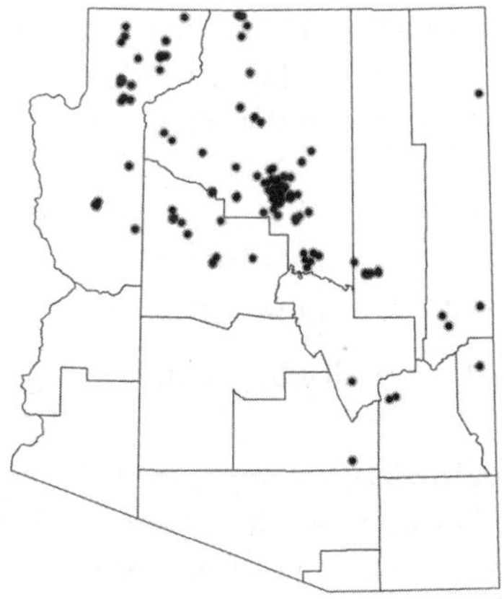

Plants densely strigose-pilose or pilosulose, hairs straight or nearly so, sometimes shorter ones crispate or sinuous; herbage cinereous, canescent, or silvery. **Stems** (6–)10–60(–80) cm, radiating from root-crown. **Leaves**: 1–6(–7.5) cm, stipules 1.5–9 mm. **Leaflets** (7–)11–17(–19), (2–)5–17(–19) mm, surfaces pubescent above (adaxially). **Inflorescences** with (3–)7–22 flowers; peduncle 2–9 cm; axis 1–9(–13) cm in fruit. **Flowers**: calyx (4.5–)5–7.4(–8.8) mm; tube (2.4–)2.7–3.7(–4.1) mm; lobes lanceolate-acuminate or filiform bristlelike, (1.4–)1.9–3.6(–5) mm; corolla greenish white or ochroleucous, lined or suffused or margined with dull purple, sometimes all purple; banner 7.2–10.2(–11.6) × 5.5–8.4(–9) mm. **Fruits** obliquely ovoid, obovoid, semi-ovoid, or oblong-ellipsoid, (6–)8–14 × 3.5–5.7 mm, densely strigillose. **Seeds** 10–16. $2n = 24$.

Flowering May–Sep. Canyon rims and walls, slopes, valleys, arroyos, and roadsides in rocky, gravelly, sandy, or clay substrates; (760–)1300–2600[–3300] m; (Great Basin Desertscrub, Semidesert Grassland, Great Basin Grassland, Interior Chaparral, Madrean Evergreen Woodland), Great Basin Conifer Woodland, Petran Montane Conifer Forest; Apache, Coconino, Gila, Greenlee, Mohave, Navajo, Pinal, Yavapai cos.; Nevada, New Mexico, Utah.

Variety *humivagans* occurs in a wide range of habitats in mountain brush, high-elevation desertscrub, galleta grasslands, sagebrush, pinyon-juniper, ponderosa pine, oak, manzanita, and aspen communities.

38e. Astragalus humistratus A. Gray var. **sonorae** (A. Gray) M.E. Jones. Contr. W. Bot. 10:58. 1902. * [of Sonora] SONORAN GROUNDCOVER MILKVETCH Native

Astragalus sonorae A. Gray, Smithsonian Contr. Knowl. 5(6):44. 1853

Plants densely strigose-pilose and strigillose; herbage cinereous or silvery canescent. **Stems** (15–)10–35 cm. **Leaves**: (1.5–)2.5–5 cm; stipules 3.5–10 mm. **Leaflets** (5–)9–17, 3–13(–15) mm, surfaces pubescent above (adaxially). **Inflorescences** with 10–26 flowers; peduncle 3–8 cm; axis 1.5–5.5 cm in fruit. **Flowers**: calyx 4.6–6.8(–8) mm; tube 2.5–3.5 mm; lobes lanceolate-acuminate or filiform bristlelike, 2–4(–4.5) mm; corolla dull purplish, purple, or banner

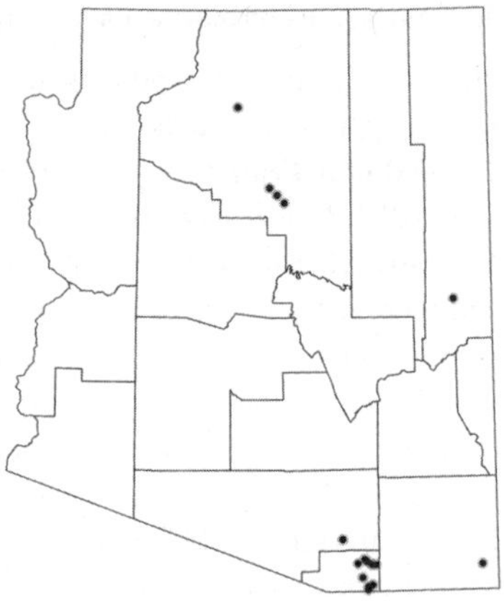

purple margined or veined; banner 8.3–10.3(–11.4) × 6–9 mm. **Fruits:** lunately linear-oblong or narrowly oblanceolate, (10–)13–20 × 3.2–4.1 mm, densely strigillose. **Seeds** (16–)18–26.

Flowering late Mar–Sep. Dry substrates, slopes, canyon floors, and arroyos, and in riparian habitats; (1100–)1400–2100(–2520); occasional in Semidesert Grassland, Plains Grassland, Madrean Evergreen Woodland, Great Basin Conifer Woodland, Petran Montane Conifer Forest; Apache, Cochise, Coconino, Pima, Santa Cruz cos.; New Mexico; Mexico (Chihuahua, Sonora).

Variety *sonorae* is similar to var. *humistratus* except for the more densely ashen or silky-pubescent foliage. In general, it is found at lower elevations and is primarily spring-flowering. It commonly grows among junipers, in ponderosa pine forests and in grasslands.

38f. Astragalus humistratus A. Gray var. **tenerrimus** M.E. Jones. Proc. Calif. Acad. Sci., ser. 2, 5:649. 1895. * [from Latin *tener*, tender or delicate] DELICATE MILKVETCH Native

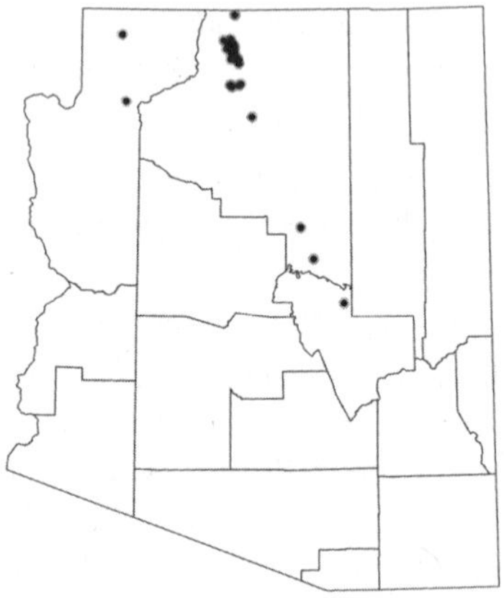

Plants forming loose mats, to 80+ cm wide, strigillose, hairs straight or nearly so, sometimes shorter ones crispate or sinuous; herbage greenish cinereous or silvery. **Stems** 6–20 cm, sometimes 1–3 cm underground. **Leaves** 1–3(–4) cm; stipules 1.5–4.5(–5.5) mm. **Leaflets** (9–)11–15, 2–6.5 mm, surfaces pubescent or glabrescent above (adaxially). **Inflorescences** with 3–6(–8) flowers; peduncle 1–3(–4) cm; axis 0.5–2 cm in fruit. **Flowers**: calyx 3.2–4.7 mm; tube 2.1–2.7 mm; lobes subulate, 1–1.2 mm; corolla whitish, veined or suffused with lilac-purple; banner 5.9–7.2 × 5.2–6.6 mm. **Fruits** lunately semi-ovoid-ellipsoid, 6–9 × 3–4.5 mm, often purple-dotted, loosely strigillose. **Seeds** 9–14.

Flowering Jul–Sep. Ridges, gentle slopes, and meadows; (1540–)2200–2800 m; Petran Montane Conifer Forest, Subalpine Conifer Forest; Coconino, Mohave cos.; endemic.

Barneby (1964a) considered variety *tenerrimus*, a diminutive Kaibab Plateau endemic, closely allied to var. *humivagans*. It grows in ponderosa pine, white fir and spruce forests, and margins of aspen stands.

39. Astragalus hypoxylus S. Watson. Proc. Amer. Acad. Arts 18:192. 1883. * [from Greek *hypo*, under or below and *xylum*, wood] HUACHUCA MILKVETCH Native

Perennials, low, caulescent, 4–11 cm; from superficial branched caudex; strigillose cinereous, hairs basifixed. **Stems** few to several, decumbent. **Leaves** 1.5–4 cm, petiole slender; stipules distinct throughout, deltoid, 1–1.5 mm, herbaceous becoming papery. **Leaflets** 11–13, flat or folded, obovate, 2–4.5 × 1–2.5(–3.5) mm, apex obtuse, surfaces strigillose below (abaxially), glabrous above (adaxially). **Inflorescences** densely 8–18-flowered; peduncle incurved-ascending, prostrate, radiating in fruit, 5–7 cm; axis 1–2.5 cm in fruit; bracts 1.3–2.5 mm; bracteoles 0. **Pedicels** ascending, 1.5 mm. **Flowers**: calyx campanulate, 6 mm, strigillose-villosulous; tube 3 × 2 mm; lobes subulate, 2.8 mm; corolla purple or bluish lavender, 7.5–8 mm; banner recurved through 45°, spatulate, 7.5–8 mm; wings 6–8 mm; keel 5.8–6.2 mm, apex bluntly deltoid; ovules 6. **Fruits** ascending; sessile; green becoming straw-colored, slightly incurved, narrowly lanceolate, 3-sided compressed, 7–9 × 2–2.5 mm, thin becoming stiffly papery, loosely strigillose; two-chambered. **Seeds** 6.

Flowering Mar–Apr. Growing in open rocky clearings and slopes on gravelly limestone substrates; 1610–1890 m; Madrean Evergreen Woodland; Cochise, Santa Cruz cos.; Mexico (Sonora).

Astragalus hypoxylus is restricted to the Huachuca and Patagonia Mountains of southeastern Arizona and adjacent northern Sonora, Mexico. It is found in oak-juniper-pinyon communities. Known only from the type collection for nearly a century, it has been collected several times since the mid-1980s. It is a species of conservation concern and is in the Center for Plant Conservation's National Collection of Endangered Plants.

40. Astragalus insularis Kellogg. Bull. Calif. Acad. Sci. 1:6. 1884. * [of islands; Cedros Island, the type locality] CEDROS MILKVETCH

Annuals or winter annuals, caulescent, often diminutive, (4–)15–55 cm; taproot slender; from superficial root-crown; strigillose, hairs basifixed. **Stems** single to few or several, erect, incurved-ascending, or prostrate. **Leaves** (2–)3–12 cm, petiole slender; stipules distinct throughout, triangular or deltoid, (1–)1.5–4(–5) mm, thinly herbaceous becoming papery. **Leaflets** (7–)11–19(–21), flat or folded, broadly to narrowly elliptic, narrowly oblong, or lanceolate, 3–19 × (1–)1.4–3(–3.9) mm, apex acute, obtuse, or emarginate, surfaces strigillose below (abaxially), strigillose, glabrescent, or glabrous above (adaxially). **Inflorescences** 3–9-flowered; peduncle incurved-ascending or widely diverging, 0.8–7 cm; axis (0.5–)1–4(–6) cm in fruit; bracts 0.6–1.3 mm; bracteoles 0. **Pedicels** horizontal becoming declined, 0.4–2.2 mm. **Flowers**: calyx campanulate or turbinate-campanulate, 3.2–4.5 mm, strigillose; tube 1.8–2.7(–2.9) × 1.8–2.3 mm; lobes subulate, 1.3–2.2(–3) mm; corolla reddish purple, 5.5–7.4 mm; banner recurved through 50–60°, ovate-cuneate, 5.5–7.4 mm; wings 5–6.5 mm; keel 4.8–6 mm, apex deltoid or triangular-acuminate, sometimes obscurely beaklike; ovules 7–12(–14). **Fruits** spreading or declined; sessile; straight, very obliquely to subsymmetrically ovoid-ellipsoid, lanceolate-ellipsoid, to subglobose, bladdery-inflated, prominently beaked, (9–)11–24 × 5–15 mm, thin, pale green and purple-suffused or -dotted, becoming straw-colored, papery, translucent, surface somewhat lustrous, sparsely strigillose; one-chambered. **Seeds** 7–12(–14). $2n = 22$.

Varieties 3 (1 native in the flora): sw United States, nw Mexico.

Astragalus insularis is a widely distributed species over the Colorado and southern Sonoran Deserts and Baja California that includes vars. *insularis* and *quentinus* M.E. Jones south of the international boundary, and var. *harwoodii* north into the Colorado Desert in southern California and southwestern Arizona; it is similar to *A. nutans*, which does not extend south of the Chuckwalla Mountains in eastern Riverside County, California.

40a. Astragalus insularis Kellogg var. **harwoodii** Munz & McBurney. Bull. S. Calif. Acad. Sci. 31:66. 1932. * [for Robert D. Harwood, 1899–1984, American botanist] HARWOOD MILKVETCH Native

Astragalus harwoodii (Munz & McBurney) Abrams

Leaflets (9–)11–19(–21). **Fruits** 14–23 × (5–)6–12(–15) mm.

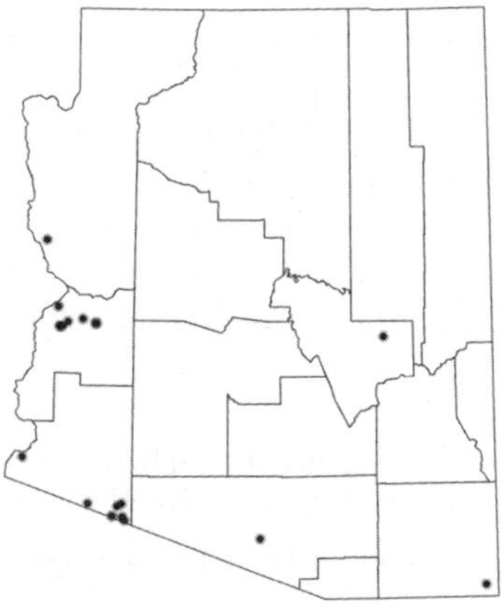

Flowering Jan–May. Open sandy flats, sandy or stony desert washes, and roadsides; [0–]30–400(–1600) m; Sonoran Desertscrub, (Madrean Evergreen Woodland); La Paz, Mohave, Pima, Yuma cos., with single reports from Cochise and Gila cos.; California; Mexico (Baja California, Sonora).

The distribution of var. *harwoodii* is centered in the southwestern portion of the Sonoran Desert where it occurs in creosote bush communities and extends in places to near the shore of the Gulf of California. Two specimens are documented from disjunct localities at much higher elevation habitats to the east: *G. E. Granfelt 67-33* (ASU) from Gila County, in Madrean Evergreen Woodland on basaltic soils at 1683 m (determined by R. C. Barneby in 1975) and *E. R. Blakeley B-1252* (DES) from Cochise County, Guadalupe Canyon at 1310 m (determined by J. H. Lehr in 1978).

41. Astragalus kentrophyta A. Gray. Proc. Acad. Nat. Sci. Philadelphia 15:60. 1864. * [from Greek *kentro*, point and *phyto*, plant, with pointed or prickly leaflets; a previous generic name] NUTTALL'S KENTROPHYTA

Kentrophyta montana Nuttall in J. Torrey and A. Gray, not *Astragalus montanus* Linnaeus

Perennials, often much branched, mat- or cushion-forming, sometimes becoming woody basally, caulescent, 5–45(–65) cm; from caudex and stolon-like, creeping stems; usually strigillose, sometimes villosulous or villous, hairs basifixed or medifixed. **Stems** few to several, prostrate, decumbent, spreading, or erect, compact to elongate **Leaves** (0.2–)0.4–2.6 cm, shortly petiolate; stipules with fused sheathing throughout or fused or distinct distally, lanceolate, 1–12 mm, scarious throughout, or herbaceous becoming papery, or stiff and thorny, apex acuminate, spinulose, or callous-mucronate. **Leaflets** (3–)5–9, grooved above (adaxially) or longitudinally folded and inrolled, decurrent, continuous with rachis, linear to narrowly elliptic or lanceolate, (1–)3–15(–17) × 0.5–1.5 mm, apex spinulose, surfaces usually strigose, rarely glabrous above (adaxially). **Inflorescences** 1–3 diminutive flowers; sessile or shortly pedunculate 0.1–0.6(–3) cm; axis to 0.5 cm in fruit; bracts 0.8–3.5 mm; bracteoles 0. **Pedicels** declined, 0.5–2 mm. **Flowers**: calyx campanulate, (2–)2.4–8.3 mm, strigose; tube 1.2–2.8(–3.3) × 1.7–2.8(–3) mm; lobes subulate, bristlelike, or spinelike, (0.5–)1.4–5 mm; corolla either white except purple-tipped keel and wings to lilac-pink, or purple with whitish wing tips, (3.9–)4–10 mm; banner recurved through 45°, broadly obovate-cuneate, flabellate, elliptic-obovate, or oblanceolate, 4–10 mm; wings 4–10 mm; keel 3–6 mm, apex obtuse; ovules 2–8. **Fruits** declined or spreading; sessile; straight or slightly curved, ellipsoid to oblong-ellipsoid, lenticular, ovoid, ovoid-lenticular, lanceolate-ellipsoid, or lanceolate-acuminate, laterally compressed, 3–10 × (0.6–)1.4–4 mm, green becoming straw-colored, papery, strigose; one-chambered. **Seeds** 2–8. $2n = 24$.

Varieties 9 (2 native in the flora): w United States and s Canada.

The varieties of *Astragalus kentrophyta* are distinguished from other astragali by the few leaflets continuous with the leafstalk and apically mucronate or spinulose at apex, and by their racemes, usually subsessile in leaf axils, bearing 1–3 tiny flowers. There are two major groups: (1) alpine or higher elevation, prostrate-matted to pulvinate-depressed plants, usually with bright purple flowers and basally-attached pubescence, with ovules 5–8, and (2) lower elevation forms that range from prostrate to bushy and ascending, with white or lavender flowers, pubescence sometimes medifixed, and ovules 2–5. With the exception of the widespread *A. kentrophyta* var. *implexus*, the varieties are largely allopatric.

1. Plants loosely mat- or clump-forming, low, to 12 cm; fruits subsymmetrically ellipsoid, lanceolate- or oblong-ellipsoid, (5–)7–10 × 2.8–4 mm; seeds 4–8 ____ **41a. Astragalus kentrophyta** var. **coloradoensis**
1′ Plants usually erect or ascending, rarely trailing, forming low, prickly bushes, 10–45(–65) cm, sometimes mat-forming; fruits narrowly ovoid-acuminate, (3.5–)4–7 × 1.5–2 mm; seeds 2–4 ____ **41b. Astragalus kentrophyta** var. **elatus**

41a. Astragalus kentrophyta A. Gray var. **coloradoensis** M.E. Jones. Contr. W. Bot. 10:63. 1902. * [of the Colorado River] CANYON KENTROPHYTA Native

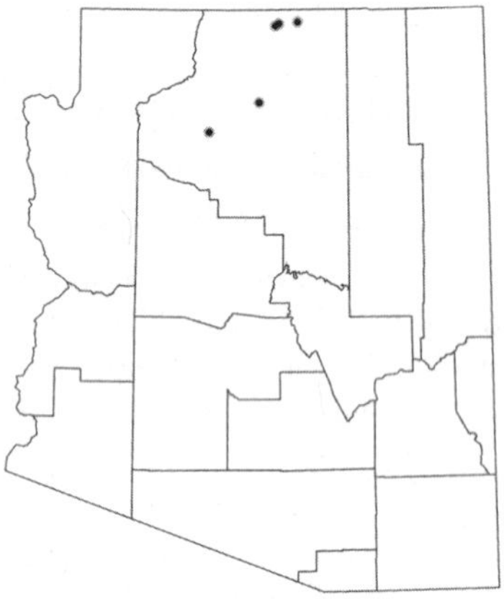

Plants loosely mat- or clump-forming, low, to 12 cm. **Stems** and herbage strigillose, hairs medifixed. **Leaves** 0.9–2.4 cm; stipules 1–12 mm. **Leaflets** usually (3–)5(–7), 4–15 mm, pubescent on both surfaces. **Inflorescence:** peduncle 0.2–0.6 cm. **Flowers**: calyx 6–8.3 mm; tube 2.4–3.3 mm; lobes subulate, spinulose, 3.4–5 mm; corolla pink-purple, 7.3–10 mm. **Fruits** subsymmetrically ellipsoid, lanceolate- or oblong-ellipsoid, (5–)7–10 × 2.8–4 mm. **Seeds** 4–8.

Flowering Apr–Jun. Sandy wash bottoms and sandy pockets on sandstone rimrock; 900–1700 m; Great Basin Desertscrub; Coconino Co.; Utah.

Variety *coloradoensis* is known from the canyons of the Colorado River and tributaries in south-central Utah and north-central Arizona where it is found in mixed desert shrub communities.

41b. Astragalus kentrophyta A. Gray var. **elatus** S. Watson. Botany (Fortieth Parallel) 77. 1871. * [from Latin *elatus*, tall, lofty] TALL KENTROPHYTA Native

Astragalus impensus (E. Sheldon) Wooton & Standley

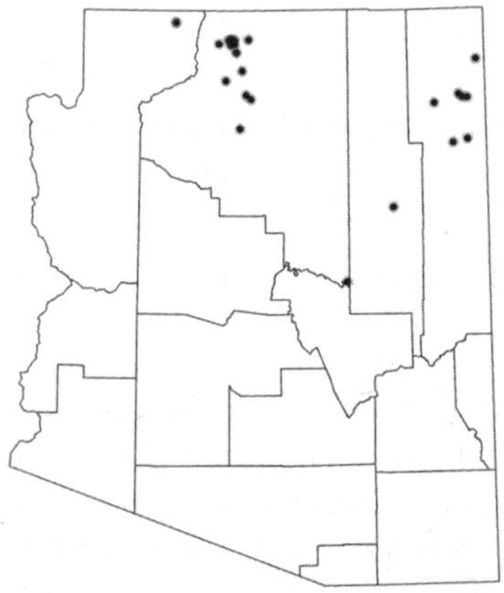

Plants usually erect or ascending, diffusely branched, rarely trailing, somewhat woody and often bushy-branched basally, forming low, prickly bushes, 10–45(–65) cm, sometimes mat-forming. **Stems** and herbage strigillose, hairs medifixed. **Leaves** (0.8–)1–2.6 cm; stipules 1–12 mm. **Leaflets** (3–)5–7, (2–)5–15(–17) mm, sometimes glabrous above (adaxially). **Inflorescence:** peduncle 0.1–0.6 cm. **Flowers**: calyx 3.4–4.4 mm; tube 1.8–2.3 mm; lobes subulate, spinulose, 1.5–2.4 mm; corolla whitish or faintly veined or tinged with purple, fading ochroleucous or, less commonly, pink-purple, 4.8–6.2 mm. **Fruits** narrowly ovoid-acuminate, (3.5–)4–7 × 1.5–2 mm. **Seeds** 2–4.

Flowering Jun–Sep. Slopes, ledges, canyons, arroyos, floodplains, and roadsides in sandy to clay or rocky soils of sandstone or limestone origin; 1500–2600[–3200] m; Great Basin Desertscrub, Great Basin Conifer

Woodland, Petran Montane Conifer Forest; Apache, Coconino, Mohave, Navajo cos.; California, Colorado, Nevada, New Mexico, Utah, Wyoming.

Both erect and prostrate forms of this variety are known, which at maturity form ascending or sprawling tangles of untidy, branched stems with prickly leaves, giving one common name "barb-wire kentrophyta." The prostrate form, typically from upper-middle elevations, simulate the var. *tegetarius*, which has, however, basifixed hairs. Across its range, var. *elatus* is found in mixed desert and salt desert shrub, juniper-pinyon, ponderosa pine, bristlecone pine, and pine-spruce communities.

42. Astragalus layneae Greene. Bull. Calif. Acad. Sci. 1:156. 1885. * [for Mary K. Layne (Mrs. Townsend S. Brandegee), 1844–1920, American botanist] LAYNE'S MILKVETCH Native

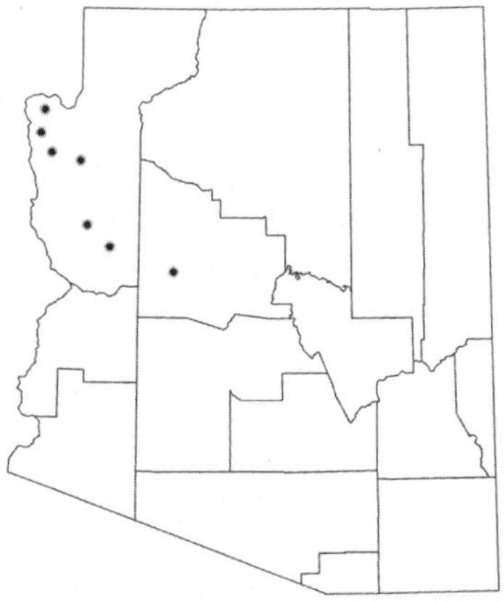

Perennials, low, coarse, short-lived, clump-forming, acaulescent or subcaulescent, 15–35 cm; from deeply subterranean caudex, branched, branches becoming rhizomatous; gray-strigillose to pilosulose, hairs basifixed. **Stems** single or few, close-set, erect or nearly so, 1–7(–10) cm underground, becoming abruptly stouter on emergence, clump-forming at irregular intervals along subterranean rhizome. **Leaves** (4–)6–16 cm, petiolate; stipules free and distinct throughout, broadly ovate-triangular to lanceolate, 3–10(–12) mm, membranous throughout or herbaceous distally. **Leaflets** (11–)13–21(–23), flat or loosely folded, ovate, obovate, rhombic-ovate, obovate-cuneate, broadly elliptic, or nearly circular, 5–18(–23) × 4.5–11 mm, apex obtuse or shallowly retuse, surfaces strigillose-pilosulose. **Inflorescences** loosely (12–)15–45-flowered; peduncle erect, 4.5–10(–14) cm; axis (3–)5–20(–30) cm in fruit; bracts 2.5–6 mm; bracteoles 2. **Pedicels** ascending becoming horizontal or slightly declined, 0.6–2.8 mm. **Flowers**: calyx deeply or, sometimes, broadly campanulate, 6–8.8 mm, densely hirsute or hirsutulous; tube 5–7.5 × (2–)2.5–3.5 mm; lobes deltoid or subulate, 1–2 mm; corolla whitish and suffused, veined, or tipped with deep purple, 12.5–17.5(–18) mm; banner recurved through 50°, broadly oblanceolate, rhombic-oblanceolate, or rhombic-spatulate, 12.5–17.5(–18) mm; wings 11.8–16.5(–17.5) mm; keel 10.4–15.4(–16.5) mm, apex obtuse; ovules 26–36. **Fruits** usually loosely ascending, incurved-ascending, or spreading, rarely declined; sessile; green or, sometimes, purplish, usually red-mottled, usually straight, sometimes crescent-incurved, very obliquely linear-ellipsoid, 3-sided compressed, (20–)30–65 × 3.5–8 mm, fleshy, green or, sometimes, purplish, usually red-mottled, becoming leathery, villosulous; partially to completely two-chambered. **Seeds** 26–36. $2n = 22, 44$.

Flowering Mar–May. Sandy flats, washes and gentle slopes or outwash fans in foothills of desert mountains and roadsides; 400–1600 m; Mohave Desertscrub; Mohave, Yavapai cos.; California, Nevada.

Astragalus layneae, named for Katharine Layne Brandegee, is the only extensively rhizomatous *Astragalus* in the Mojave Desert. The rhizomes, however, are rarely collected. This is one of the few North American *Astragalus* with a reported tetraploid chromosome number ($n = 22$), this with meiotic abnormalities in at least one population (Spellenberg 1976).

43. Astragalus lentiginosus Douglas in W.J. Hooker, Fl. Bor.-Amer. 1:151. 1831. * [from Latin *lentigo*, freckled] FRECKLED MILKVETCH

Annuals, but more often biennials or perennials, generally caulescent, (0.4–)1–100 cm; from superficial or slightly subterranean root-crown; diverse in ecology, stature, habit, and in density, distribution, and orientation of basifixed pubescence, from glabrous, to strigillose, villosulous, pilose, or tomentulose. **Stems** (solitary) few to numerous, prostrate, creeping, decumbent, ascending, or erect. **Leaves** 1–17 cm, petiolate or subsessile; stipules distinct throughout, deltoid, ovate, or lanceolate, (1–)2–7 mm, submembranous. **Leaflets** (3–)11–27(–29), flat or folded, obovate, obovate-cuneate, obcordate, oblong-elliptic, oblong-oval, elliptic, ovate, lanceolate, oblanceolate, nearly circular, or linear, (1–)2–21(–25) × 1–12 mm, apex obtuse, retuse, truncate, acute, subacute, or emarginate, surfaces glabrous, strigose, strigillose, strigillose-villosulous, pilosulose, tomentulose, or silvery-silky. **Inflorescences** with (2–)3–35(–48) flowers; peduncle erect or incurved-ascending, 0.5–11(–14) cm; axis (0.3–)0.5–18(–26) cm in fruit, not too much elongating; bracts (1.5–)2–4 mm; bracteoles usually 0. **Pedicels** ascending or spreading, 1–3 mm. **Flowers:** calyx cylindric or campanulate, (3.4–)3.8–12.5 mm, strigose, strigillose, or villosulous; tube 2.7–10 × 1.7–4(–4.4) mm; lobes subulate or triangular, 0.5–5 mm; corolla pink-purple, purple, violet, to pale lilac, white or ochroleucous, or veined, suffused, or tipped with lavender or pink, 5.5–29 mm; banner recurved 30–50°, rhombic-spatulate, 7.4–21.4 mm; wings (8–)10–27 mm; keel 6–16.4 mm, apex round; ovules (7–)10–42. **Fruits** readily deciduous when mature, except var. *maricopae*, ascending or spreading and incurved; usually sessile, rarely shortly stipitate; usually incurved, sometimes nearly straight, ovoid, obovoid, globose, lanceolate, ellipsoid, oblong, linear, linear-lanceolate, or linear- or oblong-ellipsoid, usually strongly inflated or bladdery-inflated, rarely not or scarcely so, 8–40(–48) × (3–)4–18(–19) mm, green, sometimes red- or purple-mottled or purple-speckled, usually becoming straw-colored or brownish, papery, papery-membranous, or leathery, usually glabrous, sometimes puberulent, strigose, strigillose, strigillose-villosulous, villosulous, or villous-tomentulose, interior an open cavity, not alveolae-spongy or pithy; usually two-chambered, inflexed as a complete or, rarely, incomplete septum (in beak), the funicular flange either conspicuous or minimally present. **Seeds** (7–)10–42. $2n = 22$.

Varieties 42 (12 native in the flora): w United States, Canada, Mexico.

Astragalus lentiginosus, which has many named varieties, is a widespread, polymorphic complex exhibiting a range of variation as great as occurs in many genera alone, the number and circumscription of the varieties implying active morphological and genetic differentiation (Knaus et al. 2005). The species is generally characterized by basifixed pubescence, free stipules, and typical pinnate leaves, but a sessile, early-deciduous, ovoid-acuminate, bladdery two-chambered pod with a laterally compressed, one-chambered beak is diagnostic of most forms. Historically, Rydberg (1929) placed many of the taxa, here treated as *A. lentiginosus*, in the genus *Cystium* and recognizing 33 species, but Jones (1923) treated them as a single, complex species, *A. lentiginosus*, with 18 varieties. Barneby (1964a) later recognized 36 varieties. Fruits and flowers are both needed to key most varieties successfully. Isely (1998) presents keys to the species by geographic region.

These taxa are inhabitants of the western United States, barely entering northwest Mexico and southwest Canada, regularly occurring in geologically young habitats (Knaus et al. 2005). At least 31 varieties of *Astragalus lentiginosus* have been shown to produce swainsonine and are potentially, or actually, poisonous to livestock (Cook et al. 2016).

1. Fruit not or scarcely inflated (sometimes moderately inflated in var. *wilsonii*), lengths usually more than 2 times widths.
  2. Herbage silvery and satiny; fruit silky strigillose-villosulous; La Paz and Yuma counties ______ **43c. Astragalus lentiginosus** var. **borreganus**
  2′ Herbage green or yellowish-green; fruit glabrous, strigillose, minutely pubescent, or puberulent; not in La Paz or Yuma counties.
    3. Corolla light yellow, drying ochroleucous; pods long-persistent on dried stems; Maricopa County, Arizona ______ **43f. Astragalus lentiginosus** var. **maricopae**
    3′ Corolla not light yellow to ochroleucous; pods readily deciduous; variously distributed in Arizona.
      4. Fruit papery-membranous, translucent to subtranslucent (uninflated forms of vars. *australis* and *vitreus*).
        5. Leaflets obovate, broadly elliptic, or rhombic-ovate, rarely nearly circular; south of Mogollon Rim ______ **43b. Astragalus lentiginosus** var. **australis** (in part)
        5′ Leaflets obovate-cuneate or oblong-obovate; north of the Grand Canyon ______ **43j. Astragalus lentiginosus** var. **vitreus** (in part)
      4′ Fruits stiffly papery or leathery, not translucent.
        6. Corolla pink-purple; axis of inflorescence (1–)2–12(–14.5) cm in fruit; the pedicel of the mature pod deflexed or declined; Coconino, Mohave, and Yavapai counties ______ **43h. Astragalus lentiginosus** var. **palans**
        6′ Corolla white with keel tip suffused with pink or pink-purple; axis of inflorescence 1.5–4(–13) cm in fruit; the pedicel of the pod erect or ascending; upper Verde Valley and Coconino Plateau, Coconino, Gila, and Yavapai counties ______ **43k. Astragalus lentiginosus** var. **wilsonii**

1′ Fruits inflated to strongly or bladdery-inflated, lengths usually not more than 1.5 × widths.
  7. Inflorescence in fruit: lax and open, axis (2.5–)4–18 cm.
    8. Corolla 9.1–12(–12.4) mm ______ **43e. Astragalus lentiginosus** var. **fremontii**
    8′ Corolla 11–18 mm.
      9. Fruit stiffly papery or almost leathery, usually glabrous, rarely puberulent; northwest Arizona (Mohave Co.) ______ **43a. Astragalus lentiginosus** var. **ambiguus**
      9′ Fruit thinly papery or papery-membranous, glabrous, or minutely strigillose; south- and west-central AZ.
        10. Corolla white or ochroleucous, often faintly lilac-tinged ______ **43l. Astragalus lentiginosus** var. **yuccanus**
        10′ Corolla purple, pink-purple, or lavender with white wing-tips.
          11. Plants clump-forming; corolla (13.2–)14.5–18 mm; south of Mogollon Rim ______ **43b. Astragalus lentiginosus** var. **australis** (in part)
          11′ Plants not clump-forming; corolla 11–17 mm; north, west of Colorado River.
            12. Herbage dark green, stem generally glabrous or thinly strigillose; fruit glabrous, subtranslucent ______ **43j. Astragalus lentiginosus** var. **vitreus** (in part)
            12′ Herbage cinereous or green, stems silvery-canescent or silvery or green; fruit strigillose, opaque ______ **43i. Astragalus lentiginosus** var. **stramineus**
  7′ Inflorescence in fruit: short and compact, axis 1–4(–6) cm.
    13. Plants diffuse; stems decumbent or weakly ascending; calyx lobes (2.5–)3–5 mm; Coconino Co. (Kaibab Plateau) ______ **43g. Astragalus lentiginosus** var. **oropedii**
    13′ Plants stout and coarse; stems ascending in low clumps; calyx lobes 1.5–3.2 mm; Apache, Coconino, Gila, Graham, Mohave, Navajo, Yavapai counties ______ **43d. Astragalus lentiginosus** var. **diphysus**

43a. Astragalus lentiginosus Douglas var. **ambiguus** Barneby. Mem. New York Bot. Gard. 13:945. 1964. * [from Latin *ambiguus*, doubtful] PEACH SPRINGS FRECKLED MILKVETCH Native

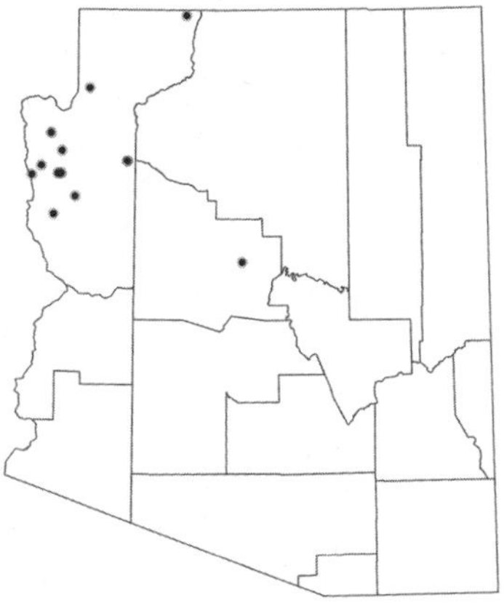

Perennials, coarse, stout, 3–5 cm, herbage thinly strigillose or nearly glabrous. **Leaves** 7–14 cm. **Leaflets** 13–21, flat, oblong-obovate or oblanceolate, (3–)7.5–15 mm, apex retuse-emarginate. **Inflorescences** loosely (7–)15–20-flowered, lax and open in fruit; peduncle 7–11 cm; axis (4–)5.5–9 cm in fruit. **Flowers**: calyx 6.7–8 mm; tube 5–5.7 mm; lobes 1.6–2.3 mm; corolla purple, (13–)15–16.7 mm; ovules 22–34. **Fruits** plumply ovoid or nearly globose, inflated but firm, 15–22 × 9–13 mm, green and non-mottled becoming straw-colored, stiffly papery or almost leathery, usually glabrous, rarely puberulent; two-chambered; beak 2–5 mm, one-chambered.

Flowering (Mar–)Apr–May(–Jun). Open limestone or granite hillsides; 1200–1500 m; Mohave Desertscrub, (Great Basin Desertscrub, Great Basin Conifer Woodland); Coconino, Mohave, Yavapai cos.; endemic.

Variety *ambiguus* is known mostly from Mohave County south or east of the Colorado River. This variety is of conservation concern.

43b. Astragalus lentiginosus Douglas var. **australis** Barneby. Leafl. W. Bot. 4:117, pl. 3, figs. 15–19. 1945. * [from Latin *australis*, southern] SOUTHERN FRECKLED MILKVETCH Native

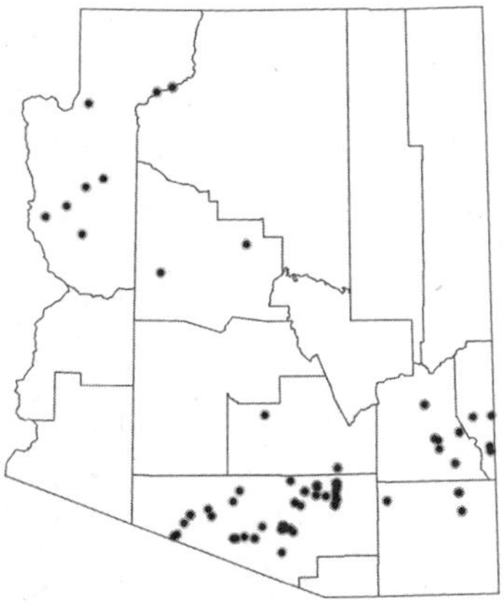

Perennials, coarse, short-lived, clump-forming, stems decumbent and ascending, 25–40(–60) cm, herbage green, strigillose-villosulous. **Leaves** (5–)6–17 cm. **Leaflets** (13–)15–21(–23), flat or loosely folded, usually obovate, broadly elliptic, or rhombic-ovate, rarely nearly circular, (4–)6–25 mm, apex emarginate or retuse. **Inflorescences** loosely (10–)15–33-flowered, lax and open in fruit; peduncle 3–7(–9) cm; axis (3.5–)4–12 cm in fruit. **Flowers:** calyx (5.3–)6.4–10.8 mm; tube (4.5–)4.8–7 mm; lobes (0.8–)1.6–3.8 mm; corolla pink-purple, sometimes pale, (13.2–)14.5–18 mm; ovules 18–22. **Fruits** narrowly to broadly ovoid-acuminate or broadly lanceolate-acuminate, slightly or greatly inflated, 12–22 × (4–)5–13(–15) mm, green or faintly mottled becoming straw-colored or purplish, thinly papery-membranous, usually glabrous, rarely sparsely strigillose, translucent; two-chambered; beak 4–8 mm, one-chambered. $2n = 22$.

Flowering Feb–May. Slopes, plains, canyons, floodplains, arroyos, and roadsides; 500–1300 m; Sonoran Desertscrub, Chihuahuan Desertscrub, Semidesert Grassland; Cochise, Coconino, Graham, Greenlee, Mohave, Pima, Pinal, Yavapai cos.; New Mexico; Mexico (Chihuahua, Sonora).

Variety *australis* is common and abundant in southeastern Arizona and southwestern New Mexico, passing to the north in Arizona into var. *yuccanus*. It is found in open desert, sandy playas and outwash fans, plains, washes in foothills of desert mountains, with creosote bush, saguaro, and in yucca-grasslands.

43c. Astragalus lentiginosus Douglas var. **borreganus** M.E. Jones, Contr. W. Bot. 8:3. 1898. * BORREGO SPRINGS MILKVETCH Native

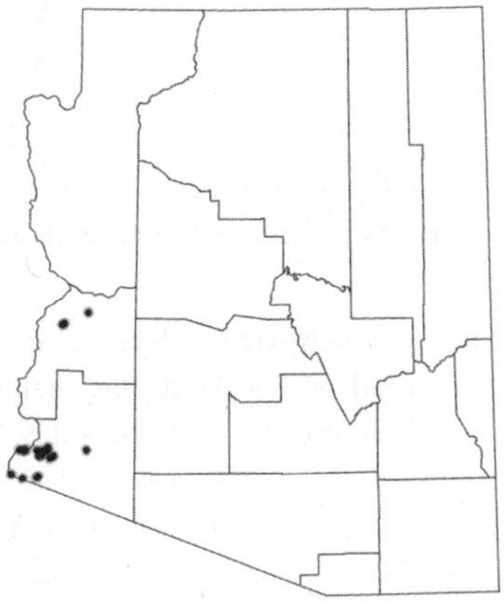

Astragalus coulteri Bentham, Pl. Hartw. 307. 1849; *Astragalus agninus* Jepson; *Astragalus arthu-schottii* A. Gray

Winter annuals, single stem either simple or branched, when persisting into a second year, forming bushy clumps of several to many erect or ascending stems, 10–30 cm, herbage densely pubescent, silvery and satiny. **Leaves** 6–10 (–16) cm. **Leaflets** (7–)15–19, flat or loosely folded, broadly obovate-cuneate to elliptic-oblanceolate, 4–14(–21) mm, apex obtuse or emarginate to retuse. **Inflorescences** loosely 13–35(–48)-flowered; peduncle 5–10 cm; axis (4.5–) 6–18(–26) cm in fruit. **Flowers:** calyx 5.2–6.6 mm; tube 4–5.1 mm; lobes 1–2.3 mm; corolla pink-purple, 12–14.8 mm; ovules (10–)13–20. **Fruits** lanceolate to ovoid-acuminate, not or scarcely inflated, slightly turgid, 15–23 × 4.5–6 mm,

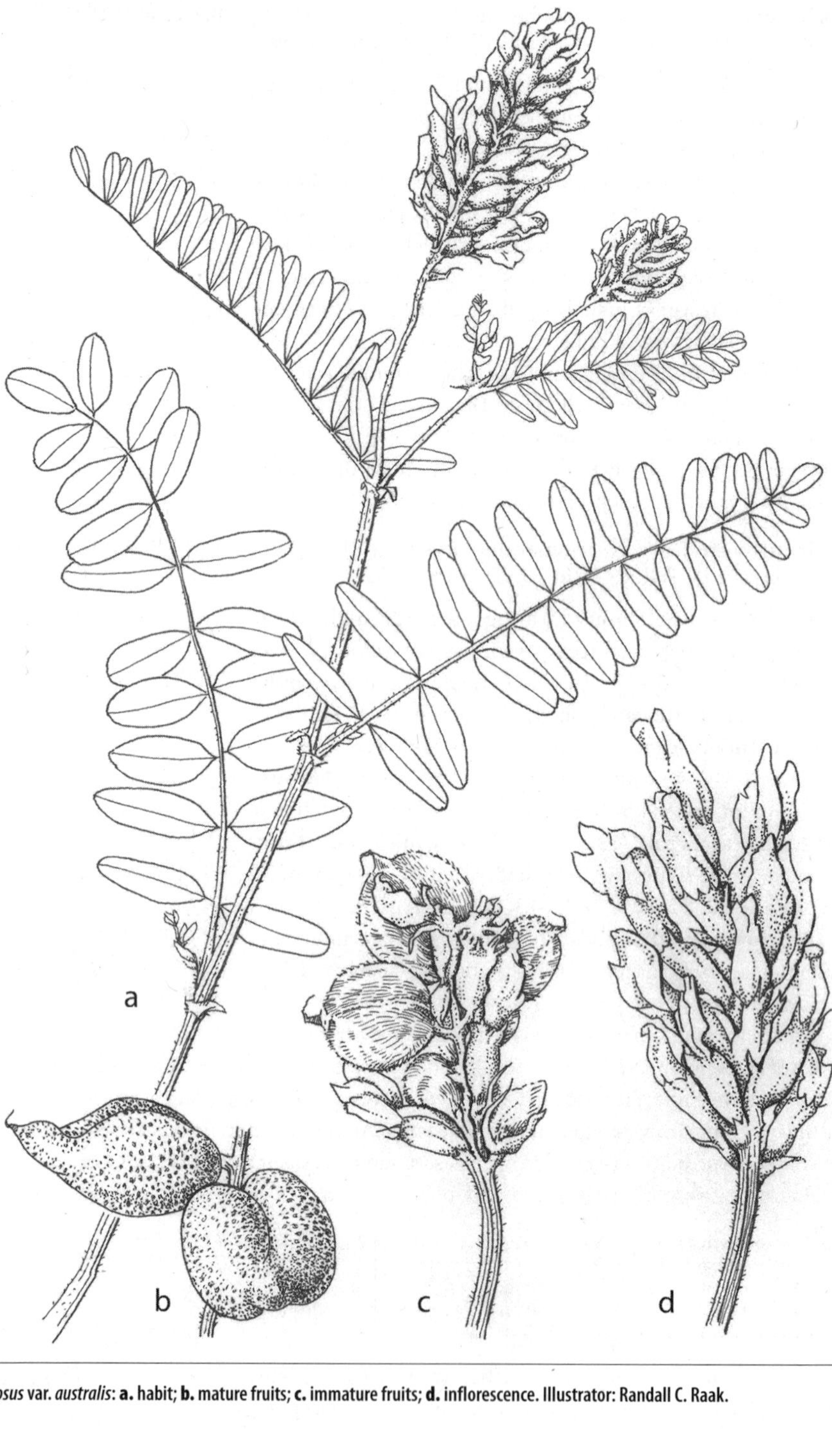

***Astragalus lentiginosus* var. *australis*: a. habit; b. mature fruits; c. immature fruits; d. inflorescence. Illustrator: Randall C. Raak.**

thin, greenish straw-colored, sometimes faintly mottled, becoming papery, silky strigillose-villosulous; ± two-chambered; beak 3–5 mm, one-chambered.

Flowering (Jan–)late Feb–May. Sandy flats and semi-stabilized dunes; 30–300 m; Sonoran Desertscrub; La Paz, Yuma cos.; California, Nevada; Mexico (Sonora).

Variety *borreganus* is of the southern Colorado Desert, extending from eastern San Diego County, California, to the Yuma Desert in extreme southwestern Yuma County, Arizona, and a few localities in La Paz County, and adjacent Sonora, Mexico.

43d. Astragalus lentiginosus Douglas var. **diphysus** (A. Gray) M.E. Jones. Proc. Calif. Acad. Sci., ser. 2, 5:673. 1895. * [from Greek *dis*, two and *phys*, bellows, with two bladders] DOUBLE BLADDER FRECKLED MILKVETCH Native

Astragalus diphysus A. Gray, Mem. Amer. Acad. Arts, n.s. 4:34. 1849; *Astragalus lentiginosus* var. *macdougalii* (E. Sheldon) M.E. Jones

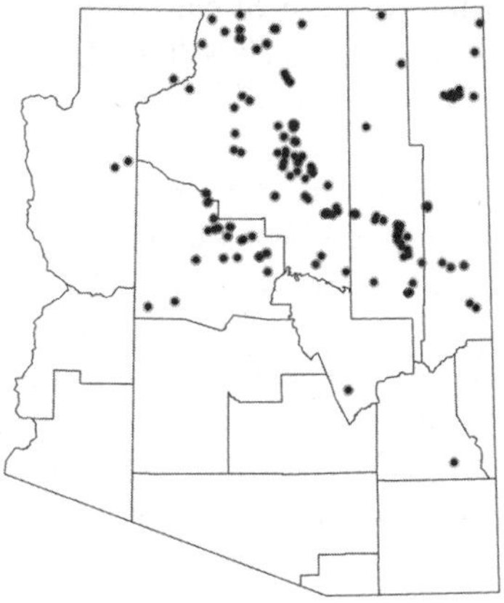

Perennials, stout and coarse, stems ascending in low clumps, (10–)15–35 (–40) cm, stems glabrous, herbage yellowish-green or dark green. **Leaves** 3–10(–14) cm. **Leaflets** (11–)15–21(–23), flat or loosely folded, oblong-oblanceolate, obovate, rhombic-obovate, elliptic, or ovate-cuneate, 4–20 mm, glabrous above (adaxially) or on both surfaces, apex truncate-emarginate. **Inflorescences** ± densely flowered, with (9–)12–24 flowers, short and compact in fruit; peduncle (1.5–)2.5–8.5 cm; axis not or little elongating, 1–4(–6) cm in fruit. **Flowers**: calyx 7–10.4 mm; tube (5.1–)5.5–8 mm; lobes 1.5–3.2 mm; corolla pink-purple, often drying violet, but fading brownish, (12.6–)14.5–19 mm; ovules 28–35. **Fruits** slightly incurved, mostly plumply ovoid or nearly globose, inflated, (10–)14–27(–30) × (6.5–)8–18 mm, green, purple- or red-tinged, or brightly mottled, becoming stiffly papery or leathery, glabrous; partially two-chambered; beak well-defined, triangular or deltoid, (3–)4–10 mm, one- or two-chambered. $2n = 22$.

Flowering Mar–Jun(–Sep). Various soils of rock outcrops, ridges, slopes, canyons, arroyos, plains, and roadsides; 970–2300 m; Great Basin Desertscrub, (Semidesert Grassland), Great Basin Grassland, (Interior Chaparral), Great Basin Conifer Woodland, Petran Montane Conifer Forest; Apache, Coconino, Gila, Graham, Mohave, Navajo, Yavapai cos.; Colorado, New Mexico, Utah.

Variety *diphysus* is widespread, ranging across the northern parts of Arizona and New Mexico, extending northward into western Colorado and Utah. Isolated collections have been made in Gila and Graham counties, Arizona, and eastern Nevada. This variety occurs in yucca-grasslands, pinyon-juniper forests, other xeric communities. Isely (1998) recognized var. *macdougalii*, here considered a form of var. *diphysus* with diminutive fruits, from the San Francisco Peaks area of Arizona.

43e. Astragalus lentiginosus Douglas var. **fremontii** (A. Gray ex Torrey) S. Watson. Botany (Fortieth Parallel) 66. 1871. * [for John C. Frémont, 1813–1890, American soldier and explorer] FRÉMONT'S FRECKLED MILKVETCH Native

Astragalus fremontii A. Gray ex Torrey in War Department [U.S.], Pacif. Railr. Rep. 4(5):80. 1857; *Astragalus fremontii* A. Gray ex Torrey subsp. *eremicus* (E. Sheldon) Abrams

Annuals, biennials, or perennials, short-lived, with a single and erect or more often diffuse, ascending stems, (4–)8–35(–40) cm, stems thinly to densely strigillose. **Leaves** (3–)4–9(–12) cm. **Leaflets** (9–)11–19, flat or loosely folded, ovate- or obovate-cuneate, broadly oblanceolate, or rhombic-elliptic, 5–19 mm, apex usually obtuse or emarginate, rarely subacute. **Inflorescences** loosely (8–)10–30-flowered, lax and open in fruit; peduncle 2.5–8.5(–10) cm; axis (2.5–)4–11(–16) cm in fruit. **Flowers**: calyx (3.4–)3.8–6(–7.9) mm; tube (2.8–)3–4.5 mm; lobes (0.6–)0.9–2(–3.5) mm; corolla usually bright purple, but rarely varying through pink-lilac to pure white, 9.1–12(–12.4) mm; ovules (17–)19–31. **Fruits** usually broadly and plumply ovoid-acuminate, rarely quite narrowly so, nearly always bladdery-inflated, 14–27(–36) × (5–)8–18 mm, pale green, often purple-freckled or purple-mottled, becoming papery-membranous, subtranslucent, glabrous or sparsely strigillose-villosulous; two-chambered or nearly so; beak (2–)3–7(–10) mm, one-chambered.

Flowering late Mar–Jul (Sep–Oct). Slopes and arroyos; 700–1500[–2800] m; Mohave Desertscrub, Great Basin Desertscrub; Mohave Co.; California, Nevada, Utah.

Variety *fremontii* is common from southeastern California across southern Nevada to southwestern Utah and northwestern Arizona at low elevations and may be very abundant when moisture is adequate in amount and in timing. To the south and west it passes into the larger-flowered var. *variabilis*; to the north in western Nevada it passes into var. *kennedyi*. Across its range, var. *fremontii* is found in braided stream gravels in riparian communities, gravelly slopes in creosote bush, Joshua tree, juniper, pinyon-juniper, and Jeffrey pine woodland communities.

43f. Astragalus lentiginosus Douglas var. **maricopae** Barneby. Leafl. W. Bot. 4:140. 1945. * [of Maricopa County, Arizona] MARICOPA MILKVETCH Native

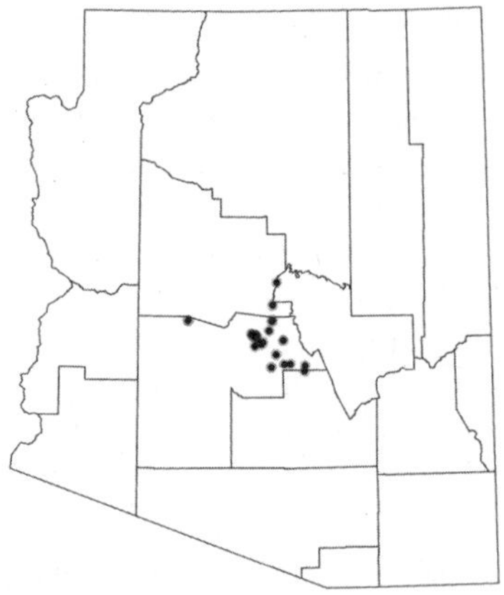

Perennials, short-lived, erect or ascending, 45–65 cm, herbage green with ochroleucous to purplish stems, subglabrate. **Leaves** (4–)6–16 cm. **Leaflets** 17–25, flat, ovate, broadly oval, oblong-elliptic, or obovate-cordate, (3–)5–22 mm, apex obtuse and mucronulate or emarginate. **Inflorescences** loosely 13–30-flowered; peduncle 5–14 cm; axis elongating, (3–)5–20 cm in fruit. **Flowers**: calyx 7.2–8.6 mm; tube 5.6–6.4 mm; lobes 1.6–2.6 mm; corolla light yellow drying ochroleucous, 15–16.5 mm; ovules 22–26. **Fruits** ascending or spreading, straight or incurved less than 90 degrees, subsessile, narrowed stipe-like base less than 0.4 mm, ascending or spreading, straight or curved, the body narrowly oblong- or linear-ellipsoid, not inflated, long persistent, 20–30 × 3.7–5 mm, somewhat fleshy, green becoming straw-colored to brownish, leathery or stiffly papery, glabrous; two-chambered; beak 3–5 mm, one-chambered; the base often abruptly contracted into a stipe-like base, less than 0.4 mm.

Flowering (Feb–)Mar–May. Stony valleys, arroyos, floodplains, and roadsides; 300–700(–1150) m; Sonoran Desertscrub, (Mohave Desertscrub); Gila, Maricopa, Yavapai cos. with one dubious report from Mohave Co.; endemic.

Variety *maricopae* is found sporadically in the area north and northeast of Phoenix, north of the Salt River in Maricopa County. It has recently been collected in adjacent Gila and Yavapai counties. In favorable years, it is common in areas where desert vegetation remains intact despite residential development. Of potential conservation concern due to highly restricted range and habitat loss.

43g. Astragalus lentiginosus Douglas var. **oropedii** Barneby. Leafl. W. Bot. 4:135. 1945. * KAIBAB PLATEAU MILKVETCH Native

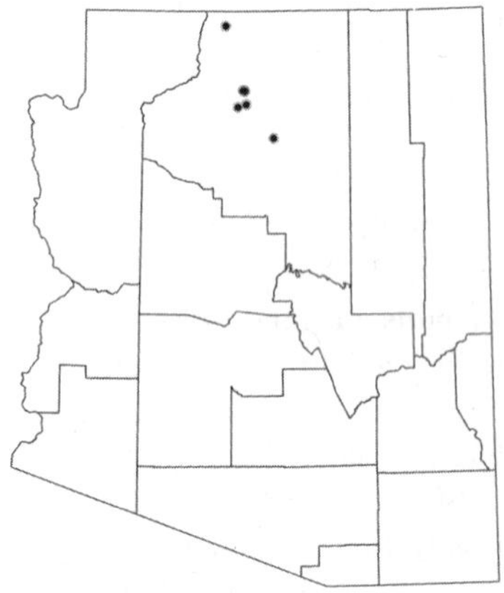

Perennials, diffuse, stems decumbent or weakly ascending, flexuous or zigzag in age, (10–)20–80 cm, stems glabrous or glabrate. **Leaves** 5–15 cm. **Leaflets** 15–21(–25), flat, broadly oblong-elliptic, ovate-oblong, or nearly circular, 5–20(–25) mm, apex rounded. **Inflorescences** shortly (8–)10–25-flowered, short and compact in fruit; peduncle (2–)3–10 cm; axis little elongating, 1.5–4(–5) cm in fruit. **Flowers**: calyx 7.5–10.8 mm; tube (4.5–)5–7.5 mm; lobes (2.5–)3–5 mm; corolla purple or pale pink-purple, (12.5–)13.2–20 mm; ovules 20–33. **Fruits** obliquely ovoid or semi-ovoid, ± strongly inflated, 13–25(–30) × 6.5–14 mm, mottled, becoming stiffly papery, glabrous; ± two-chambered; beak 5–8 mm, one-chambered.

Flowering Jun–Sep. Canyon walls and plateaus; (1600–)2100–2500 m; Great Basin Conifer Woodland, Petran Montane Conifer Forest; Coconino Co.; endemic.

Variety *oropedii* is locally common in openings in ponderosa pine forests but, apparently, confined to the Kaibab Plateau, especially the North Rim and upper levels of the northern wall of the Grand Canyon, Coconino County. Specimens with thin-textured, subdiaphanous fruits that occur within or near known localities of var. *oropedii* have been tentatively placed with var. *vitreus* (J. Alexander, Utah Valley Univ., unpublished data). However, variety *oropedii* is easily distinguished from var. *vitreus*, which is also present on the north slope of the Kaibab Plateau, by its larger flowers and more compact racemes.

43h. Astragalus lentiginosus Douglas var. **palans** (M.E. Jones) M.E. Jones. Contr. W. Bot. 8:4. 1898. * [straggling] STRAGGLING MILKVETCH Native

Astragalus palans M.E. Jones, Zoë 4:37. 1893; *Astragalus bryantii* Barneby

Perennials, sometimes short-lived, erect or incurved-ascending, (10–)15–35(–40) cm, herbage green or yellowish-green, thinly strigillose. **Leaves** (3.5–)5–11 cm. **Leaflets** 13–21(–23), flat or loosely folded, broadly obovate-cuneate, oblong-elliptic, oblong-oblanceolate, or nearly circular-obcordate, (3–)5–17(–23) mm, apex obtuse, emarginate, or subacute, upper (adaxial) surface glabrous. **Inflorescences** loosely (6–)10–28-flowered; peduncle (3.5–)5–11 cm; axis

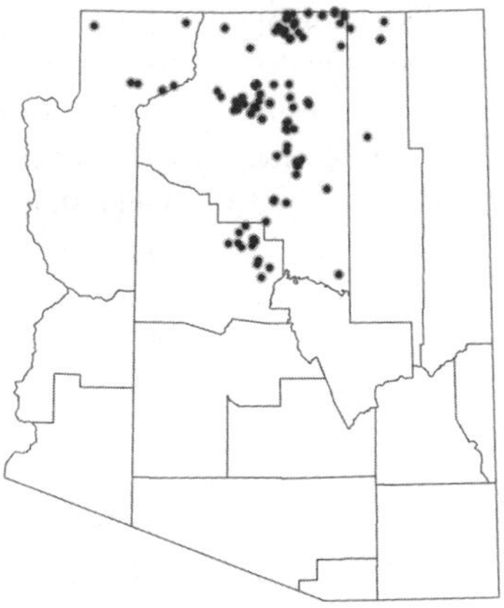

elongating or not, (1–)2–12(–14.5) cm in fruit. **Flowers**: calyx 6.3–9.4 mm; tube 4.7–6.8 mm; lobes (0.9–)1.1–3(–4) mm; corolla pink-purple, 13.5–17.5 (–18.3) mm; ovules 20–42. **Fruits** spreading, declined, or spreading-ascending, nearly straight to uniformly or hamately incurved, pedicels declined or deflexed, the body obliquely linear-lanceolate to narrowly ovoid-acuminate, not or scarcely inflated, dehiscent on ground, (12–)15–27 × 4–8.5 mm, green or brightly mottled, somewhat fleshy straw-colored then blackish, leathery or stiffly papery, not translucent, glabrous or strigillose; ± two-chambered; beak 5–8 mm, one-chambered. $2n = 22$.

Flowering (Jan–)Mar–Jun (Sep). Rocky, gravelly, and sandy substrates of ridges, slopes, canyons, arroyos, floodplains, roadsides, and on sand dunes; 700–2080 m; Great Basin Desertscrub, Great Basin Conifer Woodland; Coconino, Navajo cos.; Colorado, New Mexico, Utah. Specimens from Mohave Co. and Yavapai Co. are most likely misidentified collections of *Astragalus mokiacensis* and *A. lentiginosus* var. *wilsonii*, respectively (J. Alexander, unpublished data).

There is great variation in the pod of var. *palans*, from slender, fusiform and straight, or nearly so, and erect or ascending to incurved through a half circle. Variety *palans* occurs in salt-desert shrub, blackbrush, juniper, pinyon-juniper, and mixed desert shrub communities.

43i. Astragalus lentiginosus Douglas var. **stramineus** (Rydberg) Barneby. Leafl. W. Bot. 4:122. 1945. * [from Latin *stramineus*, straw-colored] STRAW MILKVETCH Native

Cystium stramineum Rydberg in N.L. Britton et al., N. Amer. Fl. 24:409. 1929

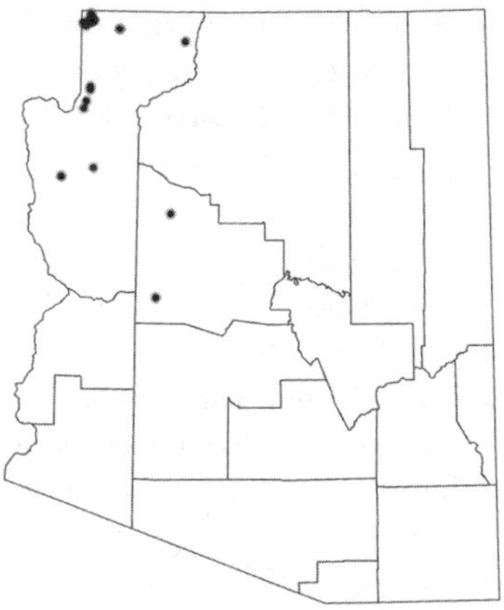

Perennials, short-lived, erect or incurved-ascending, 12–40(–65) cm, herbage strigillose, cinereous or green, stems silvery-canescent or silvery or green. **Leaves** (2.5–)4–11(–13) cm. **Leaflets** (9–)11–15, flat, broadly oblanceolate or obovate-cuneate, (3–)5–18 mm, apex openly notched. **Inflorescences** loosely or remotely 15–25(–30)-flowered, lax and open in fruit; peduncle 3.5–9 cm; axis 5–18 cm in fruit. **Flowers**: calyx 5–6.6 mm; tube 3.8–4.5 mm; lobes 1.2–2.1 mm; corolla pale purple, fading brownish or ochroleucous when dry, 11–12 mm; ovules 20–26. **Fruits** obliquely ovoid-acuminate, bladdery-inflated, (10–)17–26 × 10–16 mm, greenish, usually mottled, becoming straw-colored, somewhat stiffly papery, opaque, strigillose; two-chambered; beak (2–)3–7 mm, one-chambered.

Flowering Mar–May(–Jun). Sandy slopes, plains, arroyos, and dunes; 450–1000(–1450) m; Mohave Desertscrub, (Semidesert Grassland); Mohave, Yavapai cos.; Nevada, Utah.

Variety *stramineus* is the only member of the species with moderate-sized, pale-purple flowers known to occur in northwestern Arizona and adjacent Washington County, Utah, and southern Nevada. It is similar to the slightly smaller-flowered var. *fremontii* and to var. *variabilis* of the western Mojave Desert in California. The species is principally found in sandy and gravelly valley flats and wash habitats, extending onto disturbed sites usually with sandy soils.

43j. Astragalus lentiginosus Douglas var. **vitreus** Barneby. Leafl. W. Bot. 4:119, pl. 3, figs. 30–33. 1945. * [from Latin *vitreus*, glassy, glass-like] GLASS FRECKLED MILKVETCH Native

Perennials, stems numerous, decumbent and ascending, 15–40 cm, herbage dark green, stems generally glabrous or thinly strigillose. **Leaves** 4.5–10 cm. **Leaflets** (7–)13–19, flat, obovate-cuneate or oblong-obovate, (5–)7–17(–21) mm, apex obtuse or truncate-emarginate. **Inflorescences** loosely (10–)15–27-flowered, lax and open in fruit; peduncle 4–9.5 cm; axis 4–8.5 cm in fruit. **Flowers**: calyx 6.5–8 mm; tube 4.6–5.7 mm; lobes (1.5–)1.7–2.3 mm; corolla pink-purple or lavender with white wing tips, initially drying bluish, but occasionally corolla white with only keel or wings drying bluish in northwest Arizona populations, 13.2–17 mm; ovules 21–31. **Fruits** usually broadly ovoid, rarely lunately lanceolate-acuminate, usually strongly inflated, rarely less so, 15–25 × (7–)9–15 mm, pale green and unmottled turning pallid, papery-membranous, subtranslucent, surface lustrous, glabrous; two-chambered; beak triangular, (2–)3–5 mm, one-chambered. $2n = 22$.

Flowering Apr–Jun. Ridges, slopes, canyons, arroyos, roadsides, gullied badlands, and desert flats, on sand or clay derived from sandstone or limestone, occasionally on volcanic gravel; (620–)800–1500(–2000) m; (Mohave Desertscrub), Great Basin Desertscrub, Great Basin Grassland, Great Basin Conifer Woodland; Coconino, Mohave, Navajo cos.; Utah.

Variety *vitreus* is found in the valleys of the upper Virgin River and Kanab Creek in eastern Washington and western Kane counties, Utah, south to the northern slope of the Kaibab Plateau, and Toroweap and House Rock valleys in northern Mohave and northwestern Coconino counties, Arizona.

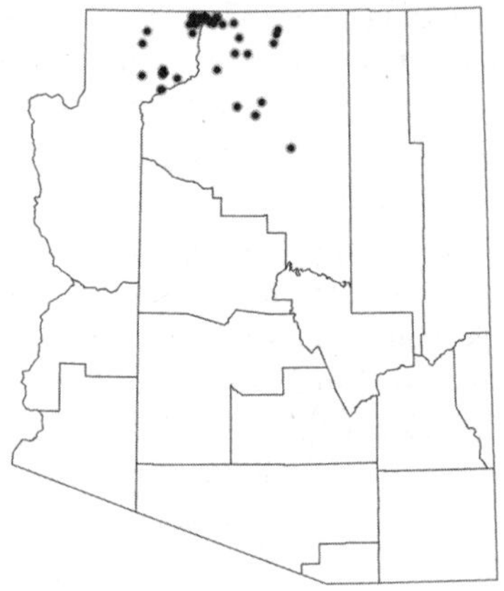

43k. Astragalus lentiginosus Douglas var. **wilsonii** (Greene) Barneby. Leafl. W. Bot. 4:139. 1945. * [for Norman C. Wilson, 19th century American explorer and plant collector who first collected this plant] WILSON'S MILKVETCH Native

Astragalus wilsonii Greene, Pittonia 3:196. 1897

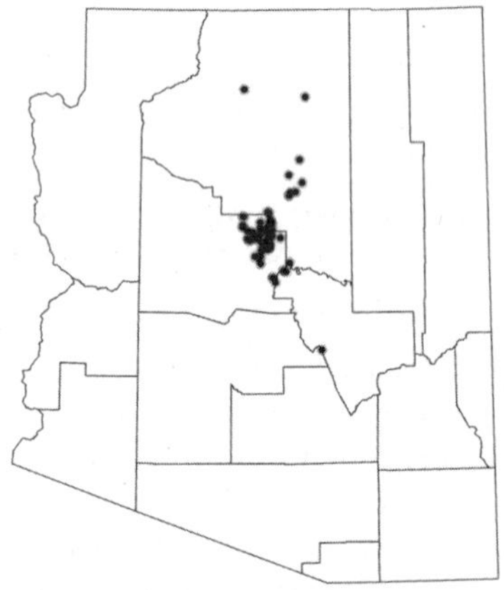

Perennials, slender to stout, prostrate or weakly-ascending stems, purplish, 20–50 cm, herbage green or subglaucescent. **Leaves** (3.5–)6–16 cm. **Leaflets** (11–)17–25, flat, ovate, broadly elliptic, or oblong-obovate, (4–)6–20(–25) mm, apex obtuse to emarginate. **Inflorescences** shortly and loosely (7–)10–17(–22)-flowered; peduncle 2.5–7(–8) cm; axis not elongating, 1.5–4 (–13) cm in fruit. **Flowers**: calyx (7–)7.7–9.4 mm; tube (5.5–)5.7–7.7 mm; lobes 1.5–3 mm; corolla white with keel tip suffused with pink or pink-purple, (14.2–)15–20 mm; ovules (29–)32–41. **Fruits** erect to ascending, rarely horizontal, straight to gently incurved, the pedicels erect or ascending, the body narrowly lanceolate-acuminate, scarcely to moderately inflated, (15–)20–35 × 3.8–7 mm, green or mottled becoming straw-colored or brownish, stiffly papery to leathery, not translucent, glabrous or puberulent; two-chambered or nearly so; beak 5–10 mm, one-chambered. $2n = 22$.

Flowering Feb–Jun (Sep). Ridges, slopes, canyons, arroyos, floodplains, and roadsides, often on volcanic substrates; 900–2200 m; (Great Basin Desertscrub), Semidesert Grassland, Interior Chaparral, Great Basin Conifer Woodland, Petran Montane Conifer Forest; Coconino, Gila, Yavapai Cos; endemic.

Variety *wilsonii* occurs on the Coconino Plateau and near the headwaters of the Verde River in southwestern Coconino, northwestern Gila, and eastern Yavapai counties. It is found in pinyon-juniper woodland, chaparral and mesquite-yucca scrub, sometimes with canotia, rarely in open ponderosa pine forests. Variety *wilsonii* resembles var. *oropedii* and var. *palans* and overlaps with var. *diphysus* (which has bright purple flowers and an inflated, almost globose pod) in its distribution area.

43l. Astragalus lentiginosus Douglas var. **yuccanus** M.E. Jones. Contr. W. Bot. 8:3. 1898. * [of Yucca, Arizona] YUCCA FRECKLED MILKVETCH Native

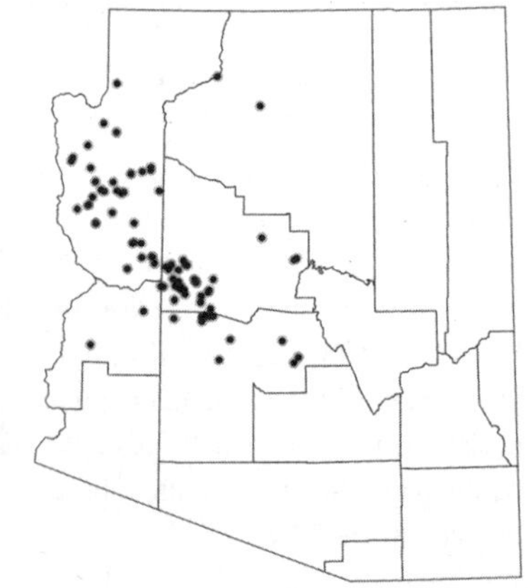

Winter annuals or short-lived perennials, robust, clump-forming, stems several to many, erect and ascending, 15–60(–100) cm, herbage strigillose or villosulous, or both. **Leaves** (5–)7–16 cm. **Leaflets** 13–21(–25), flat or loosely folded, broadly elliptic, oval, rhombic-obovate, or obovate, (4–)6–21 mm, apex retuse or emarginate. **Inflorescences** loosely (12–)15–32-flowered, lax and open in fruit; peduncle (3–)5–10 cm; axis early elongating, (4.5–)6–14 cm in fruit. **Flowers** calyx 6–7.5 mm; tube 4.4–5.7 mm; lobes (1.3–)1.5–2.4 mm; corolla white or ochroleucous, petals often faintly lilac-tinged, sometimes fading pale bluish, 11–15.6 mm, petals poorly graduated; ovules 17–26. **Fruits** plumply ovoid-acuminate or nearly globose, strongly inflated,

15–25(–32) × 10–18 mm, pale green becoming straw-colored, papery-membranous, subtranslucent, usually glabrous, rarely minutely strigillose; two-chambered; beak 3–7 mm, one-chambered. $2n = 22$.

Flowering Feb–May. Various sandy or rocky soils of slopes, plains, washes, and roadsides; 270–1440 m; Sonoran Desertscrub, Mohave Desertscrub, Semidesert Grassland, Interior Chaparral; Coconino, La Paz, Maricopa, Mohave, Yavapai cos.; Nevada.

Variety *yuccanus* occurs in west-central Arizona and barely enters Nevada in the southern tip of Clark County. This variety grows on open sandy plains, sandy or rocky washes, gulches in foothills of desert mountains, with creosote bush, and in yucca-grasslands.

44. Astragalus lonchocarpus Torrey in War Department [U.S.], Pacif. Railr. Rep. 4(5):80. 1857. * [from Greek *loncho*, lance and *karpos*, fruit, with long fruit] GREAT RUSHY MILKVETCH Native

Phaca macrocarpa A. Gray

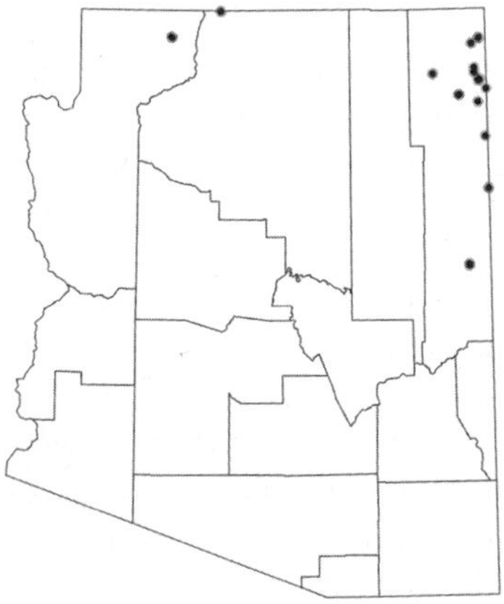

Perennials, commonly coarse and robust, sometimes slender, often densely clump-forming, caulescent, (22–)30–85(–90) cm; from shallow, subterranean caudex; strigillose or cinereous, hairs basifixed. **Stems** several to numerous, erect; 30–60 cm long, 1–8 cm underground, forming bushy to broom-like tufts. **Leaves** 2–13 cm, petiolate; stipules distinct throughout, ovate, lanceolate, or triangular, 1–9(–10) mm, stiff-papery proximally, herbaceous becoming papery distally. **Leaflets** (1 or) 3–9(–11), involute or ± flat, linear to narrowly oblanceolate, 2–36 × 0.5–4 mm, apex obtuse to acute, surfaces usually strigose, sometimes glabrous above (adaxially); terminal leaflet decurrent distally, continuous with rachis. **Inflorescences** loosely 15–50-flowered; peduncle erect or incurved-ascending, 6–24 cm; axis 3.5–45 cm in fruit. **Pedicels** spreading-declined, 1.3–4.5 mm. **Flowers**: calyx reddish-brown, cylindric, enlarged on one side, 5.8–10.3 mm, strigose; tube 5–8 × 2.7–4 mm; lobes subulate, 0.6–2.5 mm; corolla creamy-white to almost white, uniformly colored, 13–20 mm; banner recurved through 50°, broadly to narrowly rhombic-ovate, 13.5–19.5 mm; wings 14.5–19.5 mm; keel 10.5–14 mm, apex bluntly or sharply deltoid; ovules 12–26. **Fruits** pendulous; stipe 3–15 mm; often brownish, straight or slightly curved, ellipsoid to oblong, dorsiventrally compressed, convex, 22–50 × 3.3–6.2(–7.5) mm, stiffly papery, surface smooth or cross-reticulate, strigose or glabrous; one-chambered. **Seeds** 12–26. $2n = 22$.

Flowering May–Jul. Slopes, rock outcrops, canyons and arroyos, often on low-quality substrates, gullied bluffs and badlands, commonly in alkaline soils derived from sandstone; [1100–]1640–2250[–2600] m; Great Basin Desertscrub, Great Basin Grassland, Great Basin Conifer Woodland (lower elevations in Petran Montane Conifer Forest); Apache, Mohave, Navajo cos.; Colorado, Nevada, New Mexico, Utah.

The main center of distribution of *Astragalus lonchocarpus*, a relatively tall, handsome, pale-flowered species, is in the Colorado Plateau. The species is sporadic and widely dispersed in the eastern Great Basin and along the east side of the Rockies in Colorado and New Mexico. Smith et al. (1992) reported the indolizidine alkaloid swainsonine in this species, but there is little evidence that the plant is grazed by livestock. Karron (1989) reported the species to be self-compatible, and no inbreeding depression was detected in seedlings produced by self-fertilization. *Astragalus lonchocarpus* grows in alkaline and saline desertscrub, blackbrush, pinyon-juniper and occasionally in ponderosa pine-oak-juniper communities.

45. Astragalus magdalenae Green. Pittonia 1:162. 1888. * [of Bahía Magdalena, Mexico] SATINY MILKVETCH

Phaca candidissima Bentham, not *Astragalus candidissimus* Ledebour

Perennials, sometimes short-lived, flowering first season, sometimes clump-forming, caulescent, 20–70 cm; from superficial root-crown; strigillose [or strigillose-villosulous], silvery-canescent or satiny, hairs basifixed. **Stems** several or numerous, erect or ascending [diffuse], becoming woody basally. **Leaves** (1–)2–15 cm, petiolate or subsessile; stipules distinct throughout, triangular or deltoid, deciduous, (1–)2–5.5 mm, submembranous. **Leaflets** distant or crowded, (3–)9–13[–23], mostly folded or inrolled, oblanceolate, linear-oblong, oblong-obovate, obovate-cuneate, or broadly obovate, 1.5–14 × 1–2.2(–6) mm, apex usually obtuse or emarginate, rarely obtuse and apiculate, surfaces strigillose; terminal leaflet continuous with the rachis, larger than lateral. **Inflorescences** 10–37-flowered; peduncle erect or incurved-ascending, (2–)3.5–15 cm; axis 2.5–6.5 cm in fruit; bracts 1–3 mm; bracteoles 0–2. **Pedicels** ascending to spreading, 0.4–1.5 mm. **Flowers**: calyx campanulate, [4.3–]5.4–7.5(–9.2) mm, densely silky-strigillose; tube [3–]3.5–4 × 2.6–3.5 mm; lobes subulate, [1.3–]1.4–3.6(–5.2) mm; corolla pink-purple [purple], or tips whitish and margins pinkish lavender, [9.4–]10–14.2 mm; banner recurved 40°, broadly rhombic-ovate, rhombic-obovate, rhombic-elliptic, or ovate-cuneate, [9.4–]10–14.2 mm; wings 8–13.3 mm; keel [7.4–]8.5–10 mm, apex bluntly deltoid; ovules [10–]11–16 [–20]. **Fruits** spreading; sessile; pale green or purple-cheeked becoming straw-colored, straight or slightly incurved, usually broadly ovoid, ovoid-ellipsoid, or ovoid-acuminate, sometimes semi-ellipsoid or nearly circular, bladdery-inflated, [10–]20–35 × [5–]13–21 mm, thin becoming papery, translucent, strigillose; one-chambered. **Seeds** [10–]11–16[–20].

Varieties 3 (1 native in the flora): sw United States, nw Mexico.

45a. Astragalus magdalenae Greene var. **peirsonii** (Munz & McBurney) Barneby. Aliso 4:135. 1958. * [for Frank W. Peirson, 1865–1951, American botanist] PEIRSON'S MILKVETCH Native

Astragalus peirsonii Munz & McBurney, Bull. S. Calif. Acad. Sci. 31:67. 1932

Plants silvery-canescent, rarely green-cinereous. **Leaflets** distant and scattered, (3–)9–13. **Pedicels** ascending to spreading, 0.4–1.5 mm. **Flowers**: calyx 5.4–7.5(–9.2) mm; tube 3.5–4 × 2.6–3.5 mm; lobes 1.4–3.6(–5.2) mm; corolla 10–14.2 mm; keel 8.5–10 mm; ovules 11–16. **Fruits** 20–35 × 13–21 mm. **Seeds** 11–16.

Flowering Dec–Apr. Slopes and hollows in mobile dunes, usually to lee of prevailing winds; 50–300 m; Sonoran Desertscrub; Yuma Co.; California; Mexico (Sonora).

Variety *peirsonii* is locally plentiful west of Yuma, in Imperial County, and in the Borrego Valley, San Diego County. The single Arizona record (*M. Porter and K. Heil 13382*, 13 March 2002) is from 4 miles west-southwest of Yuma. The plant is endemic to very arid dune habitats and is of conservation concern. The seeds of var. *peirsonii* are among the largest of any New World *Astragalus* (Felger 2000).

46. Astragalus miser Douglas in W.J. Hooker, Fl. Bor.-Amer. 1:153. 1831. * [from Latin *miser*, wretched] WEEDY MILKVETCH Native

Perennials, low and clump-forming or taller and diffuse, caulescent or shortly caulescent, 1–35 cm; from superficial caudex, branched; strigose, strigillose-pilosulose, pilosulose, strigillose, or villous, hairs basifixed or medifixed. **Stems** usually numerous, decumbent to erect, flowering branches and shorter, sterile branches forming tuft of basal or nearly basal leaves. **Leaves** 1.5–20 cm, petiolate, sometimes shortly so distally; stipules with fused sheathing, bidentate proximally, shortly fused or distinct distally, lanceolate, 1.5–9 mm, papery-membranous throughout or, sometimes, herbaceous distally, apex acuminate. **Leaflets** 3–21, closely folded to flat, linear, oblong, elliptic, linear-elliptic, linear-oblanceolate, lanceolate, oblong-elliptic, elliptic-lanceolate, or, rarely, filiform-subulate distally, 2–30(–42) × 0.5–7 mm, apex acute, obtuse, acuminate, mucronulate, or, rarely, retuse, surfaces strigose, glabrescent, or glabrous below (abaxially), glabrous, glabrescent, or sparsely hairy above (adaxially); terminal leaflet sometimes decurrent distally, continuous with rachis. **Inflorescences** with 3–19(–25) flowers; peduncle usually incurved-ascending, 2–14 cm; axis 1–14 cm in fruit; bracts 0.6–4 mm; bracteoles usually 0. **Pedicels** spreading-declined, 0.8–3 mm. **Flowers**: calyx campanulate, (2.3–)2.4–6 mm, strigose; tube 1.7–4.2 × 1.6–3(–3.5) mm; lobes subulate, 0.5–2.6 mm; corolla lilac, pink-purple, ochroleucous, or whitish, often suffused, lined, or veined with purple, 5.3–10.6 mm; banner recurved through 40–90°, broadly ovate-cuneate to nearly circular-cuneate, (5.2–)6–13 mm; wings 5.3–10.6 mm; keel 5.9–10.7 (–11.4) mm, apex obtuse or acute-triangular, beaklike; ovules 6–19. **Fruits** declined-pendulous; sessile or subsessile, stipe 0–1 mm; straight or nearly so, linear, linear-oblong, linear-elliptic, or oblanceolate, laterally compressed, with both sutures prominent or ridge-like, 11–25 × (1.2–)2–4 mm, green, sometimes with purple speckles, becoming brown or straw-colored, papery, usually strigillose, villosulous, or glabrous, rarely with few hairs; one-chambered. **Seeds** 6–19. $2n = 22$.

Varieties 8 (1 native in the flora): w North America.

Astragalus miser, widespread in the American West along the Front Range of the Rockies, contains miserotoxin, a nitrogenous compound that is poisonous to both cattle and sheep. When identifying varieties, the presence of medifixed hairs is important. The point of attachment may be so near the base as to be almost indistinguishable from a basally attached trichome.

46a. Astragalus miser Douglas var. **oblongifolius** (Rydberg) Cronquist. Leafl. W. Bot. 7:18. 1953. * [with oblong leaflets] BAKER'S WEEDY MILKVETCH, TIMBER MILKVETCH Native

Homalobus oblongifolius Rydberg, Bull. Torrey Bot. Club 34:50. 1907

Herbage thinly to densely strigillose-pilosulose, hairs basifixed. **Stems** (1–)2–20(–24) cm. **Leaves** (2–)4–20 cm. **Leaflets** (9–)11–19(–21), linear, narrowly oblong, or elliptic to broadly oblong or oval-elliptic, (3–)5–30(–42) mm, apex mostly acute, nearly always glabrous or glabrescent above (adaxially). **Inflorescences** with 3–15(–25) flowers; axis (1–)1.5–10 cm in fruit. **Flowers**: calyx (2.8–)3.4–5.2 mm; tube 2.2–2.9 mm; lobes (0.8–)1–1.2 mm; corolla whitish or suffused or veined with dull lilac; banner (5.9–)6.5–9.5(–10.2) mm; keel 6.1–8.4 mm. **Fruits** oblanceolate, (12–)15–25 × (1.2–)2.3–4 mm; sometimes dotted or suffused with purple but not mottled, strigillose. **Seeds** 13–19.

Flowering late May–Aug. Slopes, exposed locations, and beneath the forest canopy; [1600–]2280–2580[–3500] m; Petran Montane Conifer Forest [Petran Subalpine Conifer Forest]; Coconino, Gila cos.; Colorado, Idaho, Nevada, Utah, Wyoming.

Variety *oblongifolius* forms weedy populations in the forests of Colorado and Utah where it is the only representative of the species. Across its range, this variety is found in sagebrush, oak, aspen, and spruce-fir or pine communities. The few Arizona records are from the Kaibab Plateau and Mogollon Rim.

47. Astragalus missouriensis Nuttall. Gen. N. Amer. Pl. 2:99. 1818. * [of the Missouri River] MISSOURI MILKVETCH

Perennials, low, sometimes loosely tufted or prostrate, shortly caulescent, subacaulescent, 6–20 cm, from subterranean caudex; strigose to strigillose (becoming silvery), hairs medifixed. **Stems** few to numerous, or absent in depauperate individuals, prostrate and radiating, to 10(–20) cm, internodes often concealed by stipules. **Leaves** (2–)4–14 cm, petiolate; stipules mostly distinct, sometimes with fused sheathing proximally (in var. *missouriensis*), ovate or triangular, (2–)3–9 mm, firm or submembranous becoming scarious or papery. **Leaflets** (5–)11–17(–21), flat, elliptic to obovate, 3.5–13(–17) × 3–7(–8) mm, apex acute, mucronulate, or obtuse, surfaces strigose to strigillose. **Inflorescences** with (3–)5–15 flowers; peduncle ascending, (1.5–)3.5–11 cm; axis (0.5–)1–4 cm in fruit; bracts 2.5–8 mm; bracteoles 0–2. **Pedicels** ascending to spreading, 1–3.5 mm. **Flowers**: calyx purple-tinged, cylindric or deeply campanulate, 5–13(–14.3) mm, strigillose; tube 4.1–10 × 2–4.5 mm; lobes subulate, (0.7–)1.4–4.5(–5.3) mm; corollas usually pink-purple, purple, lavender, or violet-red, rarely white, 9.5–22(–24) mm; banner recurved through 45°, rhombic-ovate, obovate-cuneate, or broadly elliptic-oblanceolate, 9.5–22(–24) mm; wings 9–17 mm; keel 8.9–17.3(–18.5) mm, apex obtuse; ovules (33–)35–55(–56). **Fruits** ascending or descending; sessile and commonly persistent on receptacle; brownish becoming black, straight or slightly incurved, subsymmetrically or obliquely oblong, oblong-ellipsoid, or ellipsoid, compressed, (11–)14–27(–30) × (4–)5–9(–10) mm, distinctly beaked, fleshy, stiffly leathery or semi- woody, strigillose, strigose, or glabrous; one-chambered or partially one-chambered. **Seeds** (33–)35–55(–56). $2n = 22$.

Varieties 4 (3 native in the flora): w North America.

From the central Canadian provinces to Trans-Pecos, Texas, and western Iowa to eastern Utah, *Astragalus missouriensis* is remarkably uniform morphologically. In the Four Corners region of the Southwest, it has undergone considerable taxonomically significant differentiation.

1. Calyx 5–5.8 mm; banner 9.5–11.8 mm; keel 8.9–10.6 mm ____ **47b. Astragalus missouriensis** var. **mimetes**
1′ Calyx 7.8–13(–14.3) mm; banner 14.5–22(–24) mm; keel (11.5–)12.8–17.3(–18.5) mm.
  2. Fruits at first nearly cylindrical or somewhat dorsiventrally compressed, straight or nearly so, subsymmetrically oblong-ellipsoid, somewhat laterally dilated and obtuse-angled, when mature apex abruptly contracted into subulate, pungent beak ____ **47c. Astragalus missouriensis** var. **missouriensis**
  2′ Fruits dorsiventrally compressed, lunately incurved, ellipsoid, contracting at apex into acute, triangular beak ____ **47a. Astragalus missouriensis** var. **amphibolus**

47a. Astragalus missouriensis Nuttall var. **amphibolus** Barneby. Amer. Midl. Naturalist 37:447. 1947. * [ambiguous, with features of both *Astragalus amphioxys* and *A. missouriensis*] MANCOS MILKVETCH Native

Plants subacaulescent to shortly caulescent. **Stems** to 10 cm. **Inflorescences** with 4–8 flowers. **Flowers**: calyx 8.5–13 mm; tube 7–10 mm; lobes 1.5–3 mm; corolla usually pink-purple, rarely white; banner (14.5–)16–22(–24) × (7.2–) 8–12.5 mm; keel (11.5–)12.8–17.3(–18.5) mm. **Fruits** sometimes deciduous, ascending to descending, dorsiventrally compressed, lunately incurved, ellipsoid, (11–)15–25 × 7–9 mm, apex dorsiventrally compressed proximal to incurved beak, strigose; one-chambered. **Seeds** 35–55.

Flowering May–Jul. On igneous or sandstone outcrops or substrates; across its range 1600–2500 m; Great Basin Desertscrub, Great Basin Conifer Woodland; Apache Co.; Colorado, New Mexico, Utah.

The fruits of var. *amphibolus* are initially dorsiventrally compressed and ultimately dehisce apically while still attached to the inflorescence (though sometimes deciduous). The fruits have a lateral ridge along each valve, with the valves separated by more or less prominent bicarinate keels. In these features, along with the typically persistent fruits, the plants can be distinguished from the similar *Astragalus amphioxys* var. *amphioxys* where their ranges are contiguous, as in northwestern New Mexico and vicinity. Barneby suggested (1947, 1964a) that hybridization between the two species occurs. Variety *amphibolus* occurs in pinyon-juniper and sagebrush communities. It is documented from Arizona by two collections at The Navajo Nation Herbarium (NAVA). Characters distinguishing this variety are questionable.

47b. Astragalus missouriensis Nuttall var. **mimetes** Barneby. Mem. New York Bot. Gard. 13:716. 1964. * [from Greek *mime*, mimic] MIMIC MILKVETCH Native

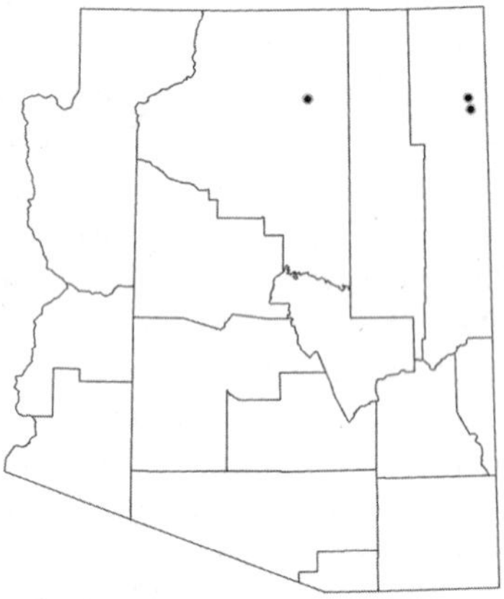

Plants shortly caulescent or, sometimes, subacaulescent. **Stems** to 15 cm. **Inflorescences** with (3–)5–15 flowers. **Flowers**: calyx 5–5.8 mm; tube 4.1–4.8 mm; lobes 0.7–1.1 mm; corolla bright pink-purple or violet-red, wings darker, banner with pale center; banner 9.5–11.8 × 6.6–9 mm; keel 8.9–10.6 mm. **Fruits** ascending, at first nearly cylindrical or somewhat dorsiventrally compressed, straight or nearly so, subsymmetrically oblong-ellipsoid, somewhat laterally dilated and obtuse-angled when mature, 14–24 × 5–7.5(–9) mm, base obtuse or, sometimes, cuneate, apex abruptly contracted into subulate, pungent beak, sutures prominent, strigillose; one-chambered. **Seeds** 36–46.

Flowering Apr–May (Sep). Canyons and slopes on clay soils; [1400–] 1800–2260 m; Great Basin Desertscrub [Chihuahuan Desertscrub]; Apache, Coconino cos.; New Mexico.

Variety *mimetes* is mostly restricted to the Rio Grande Valley in Valencia and Socorro counties, New Mexico, where it is found in creosote bush and juniper communities. Isolated populations have also been found in northeastern and northcentral Arizona. In its small flowers and fruits, it mimics *Astragalus accumbens*, with which it has been confused.

47c. Astragalus missouriensis Nuttall var. **missouriensis** * Native

Plants shortly caulescent or, sometimes, subacaulescent. **Stems** to 15 cm. **Inflorescences** with (3–)5–15 flowers. **Flowers**: calyx 9–12(–14.3) mm; tube 6.3–9(–9.3) mm; lobes 1.4–4.5(–5.3) mm; corolla usually pink-purple, rarely white; banner (14.5–)16–22(–24) × (7.2–)8–12.5 mm; keel (11.5–)12.8–17.3(–18.5) mm. **Fruits** ascending, at first nearly cylindrical or somewhat dorsiventrally compressed, straight or nearly so, subsymmetrically oblong-ellipsoid, somewhat laterally dilated and obtuse-angled when mature, 15–28(–30) × (4–)5–9(–10) mm, base obtuse or sometimes cuneate, apex abruptly contracted into subulate, pungent beak, sutures prominent, strigillose; one-chambered, or with incomplete septum. **Seeds** (33–)40–50(–56). $2n = 22$.

Flowering late Mar–Jul. Found in a wide range of soils but most common on limestone, shale, or sandstone substrates, also on gypsum, on slopes, canyons valleys, plains, erosional breaks, floodplains, arroyos, roadsides, ditches, and other disturbed sites; across its range 300–2400 m; Great Basin Conifer Woodland, [Semidesert Grassland, Plains

Grassland, Great Basin Grassland, Petran Montane Conifer Forest]; Apache Co.; Colorado, Iowa, Kansas, Minnesota, Montana, Nebraska, New Mexico, North Dakota, Oklahoma, South Dakota, Texas, Wyoming; Canada (Alberta, Manitoba, Saskatchewan).

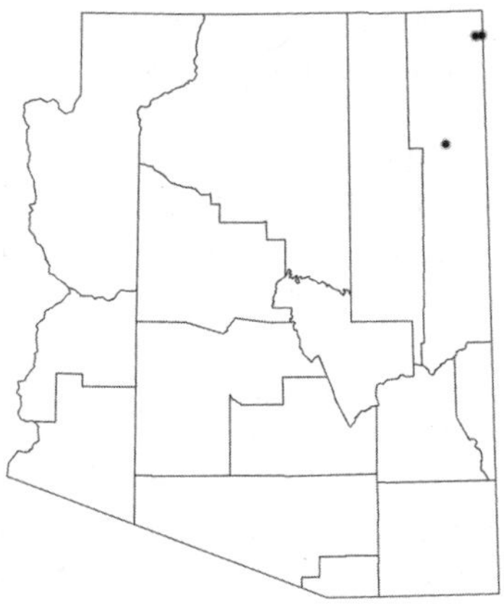

Occurrence of var. *missouriensis* in Arizona is based on only two specimens at the New York Botanical Garden Herbarium (*Barneby 17999* June 1985 and *18244* May 1987) and thus is in need of additional confirmation of its presence. There is an irregular cline in flower size of var. *missouriensis* across its range, smallest in the north, largest in the south; caulescent forms appear to be more common southward (Isely 1998).

48. Astragalus moencoppensis M.E. Jones. Zoë 2:12. 1891. * [of Moenkopi Wash] MOENKOPI MILKVETCH Native

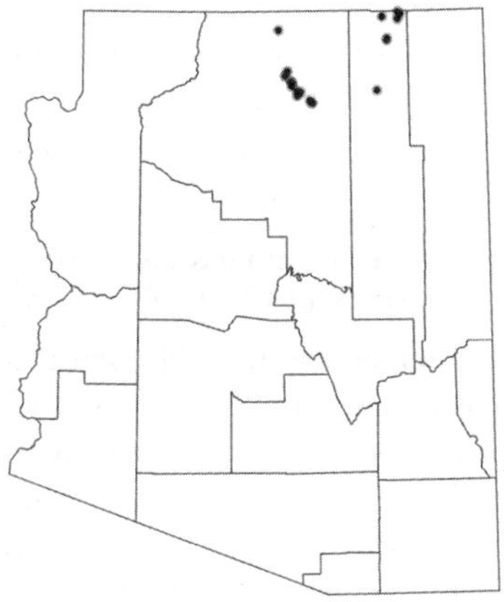

Perennials, slender, often tufted selenophytes, caulescent, 9–25 cm; from superficial or slightly subterranean caudex, branched; strigillose, hairs basifixed. **Stems** numerous to many, erect or ascending, usually shorter than longest peduncles; glabrous to sparsely strigillose. **Leaves** distally reduced or sparsely leafy, to subfiliform phyllodes, 4–17 cm, petiolate; stipules with fused sheathing proximally, fused or distinct distally, lanceolate or deltoid, 1.5–7 mm, scarious. **Leaflets** (1 or) 5–15, involute, mostly strongly folded, often incurved, threadlike to linear or narrowly elliptic, 2–23 × 0.3–2 mm, apex acute to obtuse, surfaces strigose below (abaxially), glabrous above (adaxially); terminal leaflet sometimes decurrent distally, and continuous with the rachis. **Inflorescences** loosely 6–34-flowered, flowers ascending; peduncle erect, 4–25 cm, racemes longer than stem, and much surpassing subtending leaf; axis 3–25 cm in fruit; bracts 1.5–3.5 mm; bracteoles 0 or 1. **Pedicels** ascending, 0.3–2 mm. **Flowers**: calyx campanulate, 5–7.5 mm, white-pilose; tube 3–4 × 2.1–2.8 mm; lobes lanceolate-subulate, 1.8–3.5 mm; corolla drying and persisting, pink-purple, 7–11 mm; banner recurved through 45°, obovate-cuneate or broadly oblanceolate-cuneate, 8.5–11 mm; wings 6.9–9 mm; keel 5.7–7 mm, apex obtuse; ovules 4–6. **Fruits** ascending-spreading; sessile; slightly incurved, ovoid to ellipsoid, dorsiventrally compressed, 5.5–8 × 2.3–3.4 mm, straw-colored, thinly fleshy becoming stiffly papery, strigillose; one-chambered. **Seeds** 4–6.

Flowering May–Jul. Rock outcrops, cliffs, slopes, and roadsides, usually in saline, silty or clay, seleniferous soils; (1120–)1300–1900[–2200] m; Great Basin Desertscrub, Great Basin Conifer Woodland; Apache, Coconino, Navajo cos.; Utah.

Astragalus moencoppensis is marked by its greenish, semi-rushlike appearance and slender racemes bearing small flowers with hirsute-villous calyces which, with the dried, persistent petals, largely enclose the tiny fruits. It is a selenium indicator restricted to southeastern Utah and northeastern Arizona. The species occurs in salt desert shrub, mixed desert shrub, and pinyon-juniper communities.

49. Astragalus mokiacensis A. Gray. Proc. Amer. Acad. Arts 13:367. 1878. * [of Mokiak Pass] MOKIAK MILKVETCH Native

Astragalus lentiginosus Douglas var. *mokiacensis* (A. Gray) M.E. Jones; *Astragalus lentiginosus* Douglas var. *trumbullensis* S.L. Welsh & N.D. Atwood

Perennials, robust, caulescent, 20–60 cm; from a superficial root crown, herbage green or subglabrate, strigillose, hairs basifixed. **Stems** few to several, erect and ascending in clumps. **Leaves** 3–13 cm; shortly petiolate; stipules broadly subherbaceous becoming papery, triangular-acuminate, 3–7 mm, mostly reflexed, almost glabrous to canescent, free. **Leaflets** 7–21, flat, broadly obovate-obcordate, lanceolate, elliptic, or nearly circular-obcordate, 5–13(–19) mm, apex emarginate or retuse to truncate, upper (adaxial) surface glabrous. **Inflorescences** loosely 10–25-flowered; peduncle stout, erect, 4–11(–14) cm; axis elongating, 3.5–18(–22) cm in fruit; bracts pallid or purplish, 1.5–4 mm; bracteoles 0. **Pedicels** 1–3 mm. **Flowers**: calyx 5–8(–9.2) mm; tube (3.5–)4–6.5 × 2.8–3.8 mm; lobes 1–2(–2.7) mm, becoming

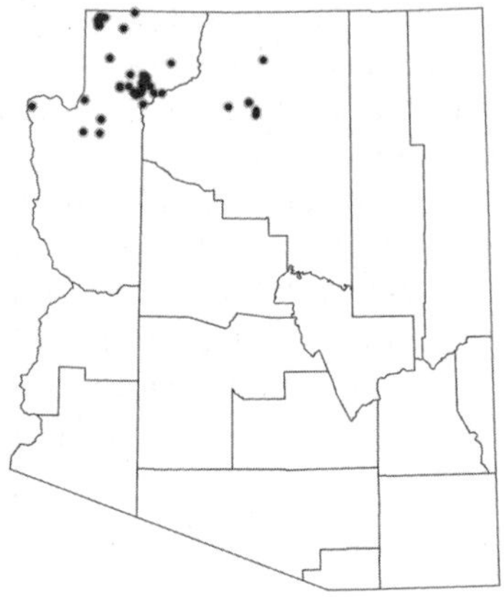

papery, ruptured; corolla pink- to red-purple, usually with pale or white wing tips, drying violet, (10–)14–18(–19) mm; banner 16.3–18.2 mm, recurved ± 50°; wings 15–16.6 mm; keel 14.4–15.7 mm; ovules 22–36. **Fruits** usually ascending-erect to ascending, rarely spreading; subsessile, the stipe 0.5–1 mm; usually nearly straight to incurved, rarely decurved, oblong-ellipsoid, not or scarcely inflated, slightly turgid, dehiscent on plant, 14–28(–32) × 4.5–6.5 mm, somewhat fleshy, green becoming straw-colored, leathery or stiffly papery, glabrous or minutely pubescent; ± two-chambered; beak 3–4 mm, one-chambered. **Seeds** 22–36. $2n = 22$.

Flowering Mar–Jun. Limestone outcrops and gravel, on basaltic or granitic gravels and/or outcrops, dry washes, and outwash fans; 900–1800 m; Mohave Desertscrub, Great Basin Conifer Woodland; Coconino, Mohave cos.; Nevada, Utah.

Barneby (1964a) noted that the Mokiak milkvetch is poorly known. He had earlier (Barneby 1945) regarded this taxon as *Astragalus lentiginosus* var. *mokiacensis*, and it has been consistently confused with *A. lentiginosus* var. *palans*. Barneby later reconsidered its status. Alexander (2005) considers the taxon to be best recognized at the species level, as *A. mokiacensis*. It has persistent fruit otherwise unknown in *A. lentiginosus*. Alexander (2005) also presents a key to similar taxa and to minor variants within *A. mokiacensis*. *Astragalus mokiacensis* is found in Washington County, southwestern Utah; northwestern Mohave County, Arizona; and eastern Clark County, Nevada.

50. Astragalus mollissimus Torrey. Ann. Lyceum Nat. Hist. New York 2:178. 1827. * [from Latin *molli*, soft and *issimus*, very soft] WOOLLY LOCOWEED

Perennials, low, relatively robust, densely or loosely tuft-forming, acaulescent to shortly caulescent, (6–)10–30(–45) cm; from superficial root-crown or caudex; gray- or white-villous-tomentose, hairs basifixed. **Stems** usually few to many, sometimes single, decumbent to ascending; when developed, with several, short internodes to 2 cm. **Leaves** (2–)5–32 cm, petiolate; stipules distinct throughout, lanceolate to deltoid, (3–)4–20 mm, papery. **Leaflets** (9–)11–35, flat or slightly folded, elliptic, nearly circular, ovate, obovate, rhombic-elliptic, rhombic-obovate, rhombic-ovate, oblong-elliptic, or oblanceolate, (2–)3–30(–45) × 2–12 mm, apex acute, obtuse, retuse, or subacute, surfaces villous-tomentose, sometimes densely so. **Inflorescences** with (5–)7–45 flowers; peduncle ascending, becoming scapiform, elongating and nearly always prostrate or reclined in fruit, (1.5–)2.5–24 cm; axis (0.5–)1–18 cm in fruit; bracts 2.5–10(–12) mm; bracteoles 0(–2). **Pedicels** ascending, 0.5–3 mm. **Flowers** calyx cylindric to deeply campanulate, 6.8–15.5 mm, villous-tomentose; tube (4.5–)5–13 × 2–5.2 mm; lobes lanceolate to subulate or subulate bristlelike, 1.6–5.5(–6.8) mm; corolla pink-purple, pale purple, pinkish, cream, or yellowish suffused with dull lavender, or tipped and suffused or margined with dull purple, 11.8–25 mm; banner recurved through 30°, broadly oblanceolate or spatulate, 11.8–25 mm; wings 11.8–25 mm; keel 12.5–20.5 mm, apex round or triangular; ovules 12–38[–41]. **Fruits** spreading or ascending, commonly humistrate; sessile; ± straight to incurved through 0.25+ spiral, oblong-ellipsoid, lanceolate-ellipsoid, ovoid, or ovoid-ellipsoid, sometimes decidedly inflated (not bladdery), cylindrical when narrow, dorsiventrally compressed when broad, (6–)9–24 × (3–)4–13 mm, fleshy, straw-colored, becoming stiffly papery, leathery, or semi-woody, glabrous, puberulent, villosulous, hispidulous, villous-hirsute, villous-tomentulose, or villous-tomentose; two-chambered, with beak two-chambered or one-chambered. **Seeds** 12–38[–41]. $2n = 22$.

Varieties 10 (4 native in the flora): w United States, especially east and west sides of Rocky Mountains, Mexico.

The *Astragalus mollissimus* complex has been variously interpreted as comprising several species, or it has been combined in varietal status under one or more species, or it has been split into species each with segregate varieties and with peripheral taxa still represented at specific rank (Isely 1998). Isely (1983) recognized three species within the complex, restricting *A. mollissimus* to those phases with glabrous or sparsely puberulent fruits, *A. bigelovii* to the bulk of the remainder with shaggy-hairy fruits, and with *A. thompsoniae* (whose fruit has a one-chambered beak) standing by itself. The present treatment follows Barneby (1964a, 1989). Some of the varieties, possibly all, are notorious locoweeds, poisonous to livestock (Ritter 1917, James and Welsh 1992, Ralphs et al. 2002), especially to horses, but also to other domestic livestock.

1. Fruits ovoid, turgid, 6–11 mm wide, beak one-chambered ______ **50c. Astragalus mollissimus** var. **thompsoniae**
1′ Fruits ovoid, lunately ellipsoid or lanceolate-ellipsoid, usually solid, rarely slightly turgid, (4–)4.5–8 mm wide, beak two-chambered.
   2. Plants usually robust, shortly caulescent, stems 3–17 cm; leaves 9–26 cm, leaflets 6–25 mm; keel 13.5–18.5 mm; fruits with hairs to 1–1.6 mm; southeast Arizona ______ **50a. Astragalus mollissimus** var. **bigelovii**
   2′ Plants usually dwarf, acaulescent or nearly so, leaves (3–)5–16 cm, leaflets 3–13 mm; keel 12–18 mm; fruits with hairs 1–2.6 mm, peduncles scapiform; mountains and plateaus of northern and eastern Arizona and northwestern New Mexico.
      3. Flowers 7–12, keel 14.2–18 mm; peduncles 2–8 cm, fruits broadly ovoid, gently to abruptly incurved; mountains of east-central Arizona (Apache Co.) to northwestern New Mexico (McKinley and Sandoval cos.) ______ **50b. Astragalus mollissimus** var. **matthewsii**
      3′ Flowers 12–32, keel 12.3–14.7 mm; peduncles 4–10 cm, fruits narrowly ovoid or lunately ellipsoid, straight to gently incurved; Coconino Plateau and southeast along Mogollon Rim into Catron Co., New Mexico ______ **50c. Astragalus mollissimus** var. **mogollonicus**

50a. Astragalus mollissimus Torrey var. **bigelovii** (A. Gray) Barneby ex B.L. Turner. Legum. Texas 191. 1959. * [for John M. Bigelow, 1804–1878, American botanist] BIGELOW'S WOOLLY LOCOWEED Native

Astragalus bigelovii A. Gray, Smithsonian Contr. Knowl. 5(6):42. 1853

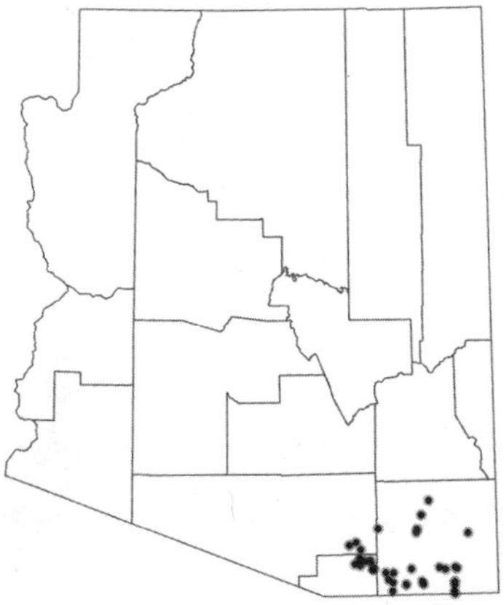

Plants shortly caulescent, robust. **Stems** 3–17 cm. **Leaves** 9–26 cm; stipules 6–20 mm. **Leaflets** (13–)19–27, ovate, obovate, oval, or broadly elliptic, 6–25 mm, apex obtuse or subacute. **Inflorescences** somewhat densely flowered, with (15–)20–45 flowers, flowers subcontiguous or interrupted proximally; peduncle (5–)8–22 cm; axis (4–)5–11 cm in fruit. **Flowers:** calyx 10.5–13.5 mm; tube (8–)8.3–10.3 × (3.2–)4–5.2 mm; lobes (1.7–)2.6–4.4 mm; corolla pink-purple; banner 17–22.5 mm; keel 13.5–18.5 mm. **Fruits** gently incurved or nearly straight, ovoid-acuminate or lanceolate-ellipsoid, usually solid, sometimes slightly turgid, 10–15 × (4–)4.5–8 mm, stiffly papery or leathery, beak two-chambered, densely villous-tomentulose, hairs to 1–1.6 mm. **Seeds** 20–31. $2n = 22$.

Flowering (Jan–)Mar–Jun. Dry plains and foothills, on calcareous soils, sandy loams, basalt gravel, overgrazed and badly eroded cattle ranges, roadsides, and alluvial soils in riparian habitats; 1200–1900(–2300) m; Semidesert Grassland, Great Plains Grassland, Madrean Evergreen Woodland (Petran Montane Conifer Forest); Cochise, Graham, Santa Cruz cos.; New Mexico, Texas; Mexico (Sonora).

Variety *bigelovii* is found in desert- or mesquite-grasslands, short grass prairie, and oak-pinyon-juniper woodland.

50b. Astragalus mollissimus Torrey var. **matthewsii** (S. Watson) Barneby, Mem. New York Bot. Gard. 13:746. 1964 * [for Washington Matthews, 1843–1905, American physician and archeologist] MATTHEWS WOOLLY LOCOWEED Native

Astragalus matthewsii S. Watson, Proc. Amer. Acad. Arts 18:192. 1883; *A. bigelovii* A. Gray var. *matthewsii* (S. Watson) M.E. Jones

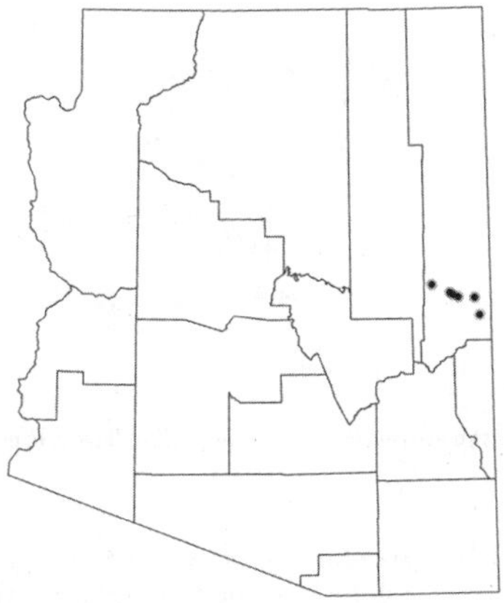

Plants acaulescent or subacaulescent. **Stems** to 1.5 cm, obscured by imbricate stipules. **Leaves** (3–)5–12 cm; stipules (3–)4–8 mm. **Leaflets** 11–23, blades obovate, 3–12 mm. **Inflorescences** densely (5–)7–12-flowered; peduncles scapiform, (1.5–)2.5–8 cm axis (0.5–)1–4.5 cm in fruit, not or scarcely surpassing the foliage. **Flowers:** calyx 10–13 mm; tube 7–8.6 × 3.4–4.7 mm; lobes 2.4–5.2 mm; corolla pale purple; banner 18.5–22.5 mm; keel 14.2–18 mm. **Fruits** gently or abruptly incurved distally, broadly ovoid, turgid, 12–18 × 7–13 mm, widest near obtuse or truncate base, densely villous-tomentose; beak two-chambered. **Seeds** 24–31.

Flowering Apr–Jun. Open slopes and hilltops, mostly in pine forests, along canyons to juniper-pinyon belt, on light, sandy or gravelly, sedimentary, granitic, or volcanic soil; 1900–2600 m; Great Basin Conifer Woodland, Petran Montane Conifer Forest; Apache Co., New Mexico.

Variety *matthewsii* is scattered and uncommon in the mountains of east-central Arizona to northwestern and north-central New Mexico.

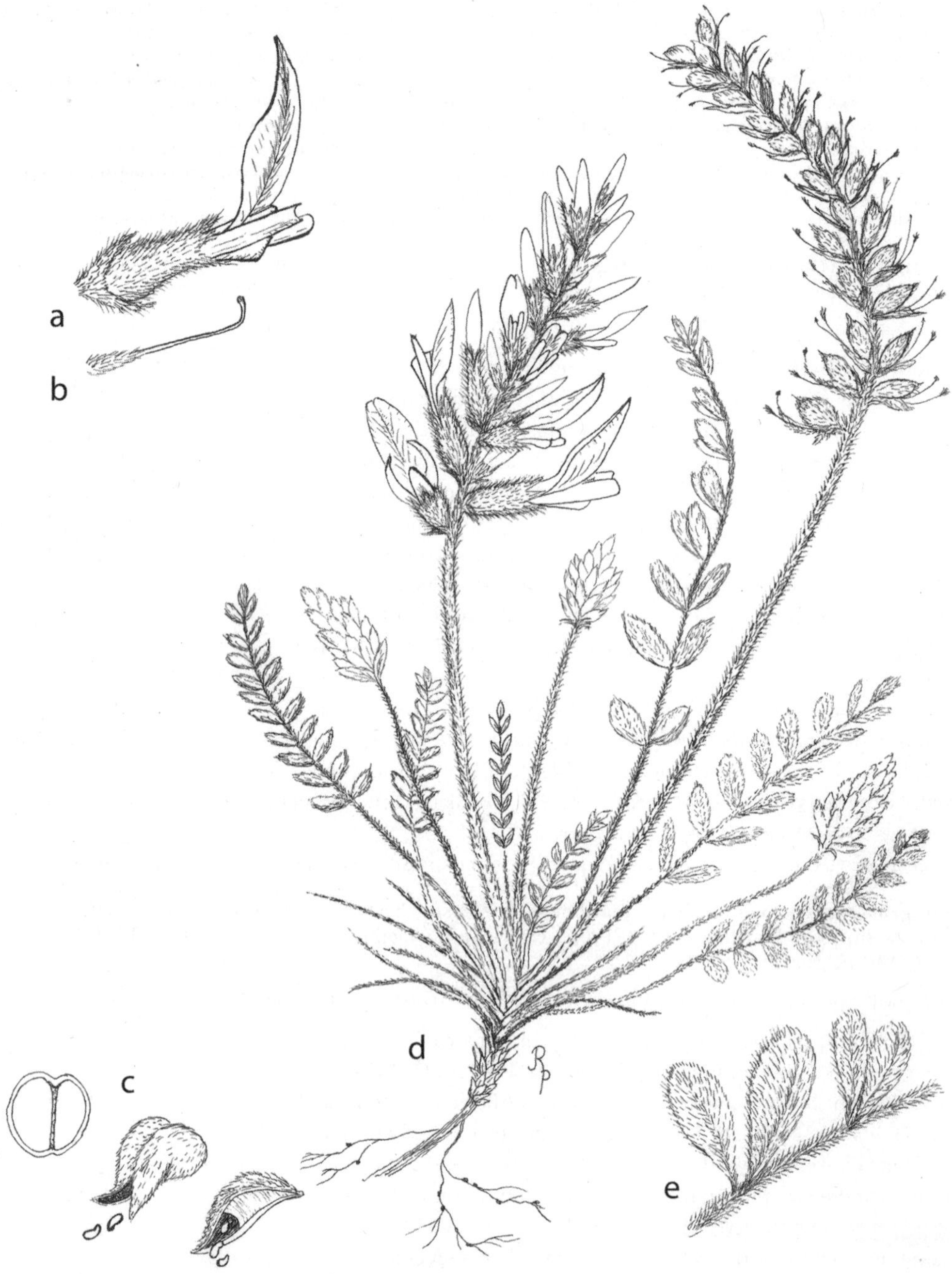

***Astragalus mollissimus* var. *bigelovii*: a. flower; b. pistil; c. fruit detail with cross-section and seeds; d. habit; e. leaf detail. Illustrator: Joyce Riney Peters.**

50c. Astragalus mollissimus Torrey var. **mogollonicus** (Greene) Barneby. Mem. New York Bot. Gard. 13:745. 1964. * [of the Mogollon Mountains] MOGOLLON WOOLLY LOCOWEED Native

Astragalus mogollonicus Greene, Bull. Torrey Bot. Club 8:97. 1881; *Astragalus bigelovii* A. Gray var. *mogollonicus* (Greene) Barneby

Plants acaulescent, usually dwarf. **Stems** reduced to thick crowns, closely invested with stipules. **Leaves** (4–)6–16 cm; stipules 5–12 mm. **Leaflets** (9–)13–23, broadly elliptic to nearly circular, 3–13 mm. **Inflorescences** densely 12–32-flowered; peduncle scapiform, 4–10 cm; axis 1.5–6 cm in fruit. **Flowers**: calyx 10–15.3 mm; tube 6.6–8.5 × 3.2–5 mm; lobes 3.3–6.8 mm; corolla pink-purple; banner 16–21.5 mm; keel 12.5–14.5 mm. **Fruits** nearly straight to gently incurved, narrowly ovoid or lunately ellipsoid, solid, 9–13 × 4.5–6 mm, beak two-chambered, densely villous-tomentose, hairs to 1.6–2.6 mm. **Seeds** 19–28.

Flowering Apr–Jul. Stony flats and hilltops, on pebbly, volcanic soils, on limestone substrates, rocky knolls and disturbed sites; 1800–2300 m; Great Basin Conifer Woodland, Petran Montane Conifer Forest; Apache, Coconino, Navajo cos.; New Mexico.

Astragalus mollissimus var. *mogollonicus* occurs from the Kaibab Plateau south of the Grand Canyon in Arizona to the southeast along the Mogollon Escarpment to western New Mexico in pinyon-juniper woodland and ponderosa pine forests.

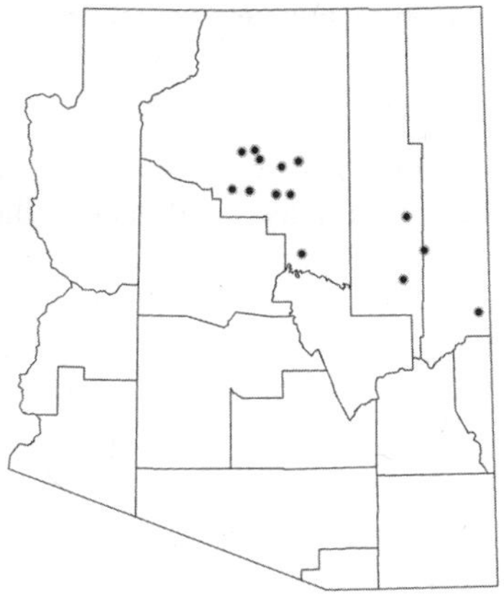

50d. Astragalus mollissimus Torrey var. **thompsoniae** (S. Watson) Barneby. Mem. New York Bot. Gard. 13:746–747. 1964 (as thompsonae) * [for Mrs. Ellen P. Thompson, sister of John Wesley Powell, who collected the type specimen with F. M. Bishop] THOMPSON'S WOOLLY LOCOWEED Native

Astragalus thompsoniae S. Watson, Proc. Amer. Acad. Arts 10:345 1875 (as *Astragalus Thompsonae)*

Plants acaulescent or nearly so, from caudex. **Stems** 0–3 cm, mostly obscured by stipules; densely villous-tomentose. **Leaves** 2–28 cm; stipules 4–13 mm. **Leaflets** 15–35, obovate to nearly circular or elliptic, 2–18 mm, apex obtuse to retuse or acute. **Inflorescences** 7–25-flowered; peduncle 2.5–24 cm; axis 1.5–18 cm in fruit. **Flowers:** calyx 11–15.5 mm; tube 7.7–13 × 3.7–4.6 mm; lobes 2–4.2 mm; corolla pink-purple; banner 18–25 mm; keel 15–18.5(–20.5) mm. **Fruits** curved, ovoid, turgid, 11–23 × 6–11 mm, beak one-chambered, densely villous-tomentose. **Seeds** 28–38.

Flowering Mar–Jun (Oct). Typically on sandy substrates including dunes, also on ridges, rocky slopes, and roadsides; (700–)1100–2400 m; Great Basin Desertscrub, Great Basin Grassland, Great Basin Conifer Woodland; Apache, Coconino, Navajo, Yavapai cos.; Colorado, Idaho, Nevada, New Mexico, Utah.

Variety *thompsoniae* is transitional with var. *matthewsii* in northwestern New Mexico. It occurs in salt desert shrub, mixed desert shrub, grasslands, and pinyon-juniper communities.

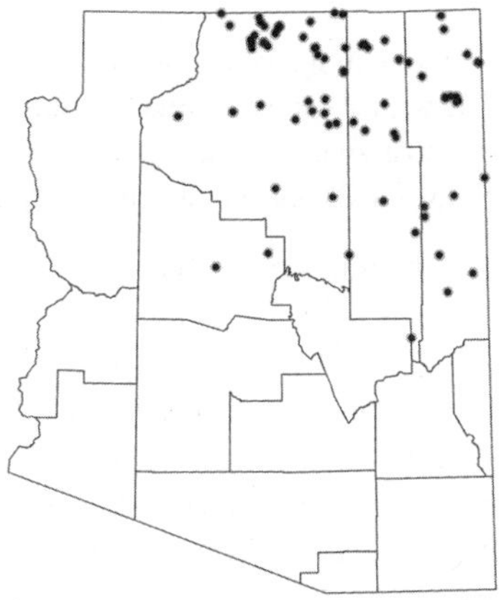

51. Astragalus monumentalis Barneby. Leafl. W. Bot. 7:35. 1953. * [of Monument Valley] MONUMENT VALLEY MILKVETCH

Perennials, dwarf, tuft-forming, acaulescent or subacaulescent, 1.2–18 cm; from caudex, branched, branches often with thatch of dried, persistent leaf bases; strigillose, hairs basifixed or shortly medifixed. **Stems** obsolete or several, ascending, 0–6 cm, internodes usually concealed by stipules; strigillose. **Leaves** 1.2–8(–11) cm, petiole subpersistent, slender, wiry; stipules distinct throughout, broadly ovate or triangular, 2–6 mm, membranous becoming papery, straw-colored. **Leaflets** (5–)9–19(–21), folded or mostly folded, oval to obovate or oblanceolate, 2–9 × 1–4.2 mm, apex obtuse, acute, or emarginate, surfaces strigose to strigillose below (abaxially), strigose or glabrous or glabrate above (adaxially). **Inflorescences** with 3–9(–12) flowers; peduncle erect or ascending, 0.7–12 cm; axis 0.5–7 cm in fruit; bracts 1.5–5 mm; bracteoles 0–2. **Pedicels** ascending, 0.8–2.2 mm. **Flowers**: calyx purplish, campanulate or cylindric, 3.6–8 mm, strigose or strigillose; tube 3–6.7 × 2–3.1 mm; lobes subulate, 0.5–2 mm; corolla pink-purple or bicolored, 7–17 mm; banner recurved through 45°, ovate, 7–16 mm; wings 8–16 mm; keel 7.8–12.4 mm, apex round to narrowly deltoid; ovules 16–30. **Fruits** ascending, spreading-descending; sessile or subsessile; straight or incurved, narrowly oblong to lanceolate, 3-sided compressed, 12–27(–37) × 2–4(–5) mm, thinly fleshy, green, often purple-mottled, becoming straw-colored, papery, strigose; completely or almost completely two-chambered. **Seeds** 16–30.

Varieties 2 (1 native in the flora). Variety *monumentalis* is restricted to Garfield and San Juan cos., Utah.

51a. Astragalus monumentalis Barneby var. **cottamii** (S.L. Welsh) Isely. Syst. Bot. 8(4):423. 1983. * [for Walter P. Cottam, 1894–1988, American botanist and ecologist] COTTAM'S MILKVETCH Native

Astragalus cottamii S.L. Welsh, Rhodora 72:189. 1970

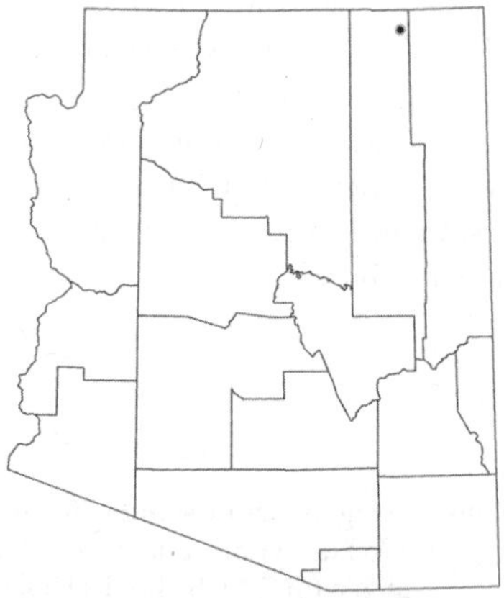

Perennials, sometimes flowering first year, 1.2–15 cm; strigillose. **Leaves** 1.2–8 cm. **Leaflets** mostly folded, elliptic to oval or oblanceolate, apex acute to obtuse, surfaces strigose below (abaxially), strigose or glabrous above (adaxially). **Inflorescences** peduncle erect, 0.7–9(–10) cm; axis 0.5–2 cm in fruit; bracts broadly ovate or lanceolate. **Pedicels** straight, ascending, 0.8–1.2 mm. **Flowers**: calyx purplish, cylindric, 6.2–8 mm, strigillose; tube 4.8–6.7 × 2–3 mm; lobes subulate, 1.2–2 mm; corolla pink-purple or bicolored, 11–17 mm; banner recurved through 45°, ovate, 11–16 mm; keel 9–11.5 mm, apex round to narrowly deltoid; ovules 16–28. **Fruits** spreading-descending; sessile; incurved, oblong to oblong-lanceolate, trigonally compressed, 15–27(–37) × 3–4(–5) mm. $2n = 22$.

Flowering Apr–May. Sandstone rimrock, hogbacks, ledges and in sandy canyons; 1300–1900 m; Great Basin Desertscrub, Great Basin Conifer Woodland; Coconino, Navajo cos.; Colorado, New Mexico, Utah.

Astragalus monumentalis and *A. deterior* are denizens of weathered depressions and crevices in sandstone rimrock. They are allopatric. Variety *cottamii* is contiguous with var. *monumentalis* but its distribution is to the east and southeast in the Four Corners area in Montezuma and La Plata counties, Colorado, San Juan County, Utah, San Juan County, New Mexico, and Coconino and Navajo counties, Arizona where it is known from only two collections. It occurs in shadscale, blackbrush, and pinyon-juniper communities on Kayenta, Cedar Mesa, Entrada, Mesa Verde and Navajo sandstone (rarely other) formations.

52. Astragalus newberryi A. Gray. Proc. Amer. Acad. Arts 12:55. 1876. * [for John S. Newberry, 1822–1892, American geologist and naturalist] NEWBERRY'S MILKVETCH

Perennials, dwarf or low, acaulescent, 2–12 cm; from superficial caudex, branched, branches often with thatch of overlapping stipules and persistent leaf bases; silky-pilose or strigose, hairs basifixed. **Stems** single to several, reduced to thick crowns. **Leaves** 1.5–14(–15) cm, petiolate; stipules distinct throughout, ovate or lanceolate, (2.5–)4–11 mm, submembranous becoming scarious or papery. **Leaflets** 1–13(–15), flat, obovate to elliptic, oblanceolate, or circular, 3–20 × 2–14 mm, apex acute to obtuse or retuse, surfaces villous-tomentulose. **Inflorescences** loosely 2–8-flowered; peduncle ascending, 0.5–11 cm; axis 0.2–2.7 cm in fruit; bracts 3.5–10 mm; bracteoles 0–2. **Pedicels** ascending, 1.4–5 mm. **Flowers**: calyx cylindric or campanulate, 7–20 mm, villous; tube 5–16 × (2.5–)3–4 mm; lobes subulate to lanceolate, 1–4 mm; corolla pink-purple to pale pink, or white tipped and margined with lilac or purple, 14–32 mm; banner recurved through 90°, broadly oblanceolate or rhombic-spatulate, 14–32 mm; wings 13–27 mm; keel 12–26 mm, apex obtuse; ovules 27–40(–46). **Fruits** ascending (humistrate); sessile or subsessile; incurved, obliquely ovoid, dorsiventrally compressed, (13–)18–28(–36) × 7–13(–17) mm, fleshy, green becoming brown-straw-colored, becoming leathery, both villous-tomentose with short curly hairs and densely villous-hirsute with longer, lustrous hairs that conceal pod surface; one-chambered. **Seeds** 27–40(–46). $2n = 22$.

Varieties 5 (4 native in the flora): w United States.

As reviewed by Barneby (1964a), *Astragalus newberryi* is one of several intermountain astragali with fruits so densely villous and tomentulose as to resemble pellets of cotton, a striking feature that tends to overshadow other differences. Except for the uncommon *A. eurekensis* of west-central Utah, this is the only *Astragalus* in the region to combine such a fruit with acaulescence, a thatched base, and a vesture of stiff, shining hairs.

1. Leaflets 1–3; plants of Aquarius Mountains, Mohave County, Arizona ____ **52a. Astragalus newberryi** var. **aquarii**
1′ Leaflets 3–13(–15); plants of Arizona outside of the Aquarius Mountains.
  2. Leaflets (1–)3–5; north, central Arizona ____ **52d. Astragalus newberryi** var. **newberryi**
  2′ Leaflets 7–13; northwest Arizona.
    3. Calyx mostly 12–20 mm; banner 21–32 mm; northwest Arizona ____ **52c. Astragalus newberryi** var. **castoreus**
    3′ Calyx 8–11.5 mm; banner 14–21(–22) mm; west Mohave County, Arizona, south of Colorado River ____ **52b. Astragalus newberryi** var. **blyae**

52a. Astragalus newberryi A. Gray var. **aquarii** Isely. Syst. Bot. 8:423. 1983. * [of the Aquarius Mountains] AQUARIUS MOUNTAIN MILKVETCH Native

Leaflets 1–3, obovate. **Flowers:** calyx deeply campanulate, 7–8 mm; tube 5–6 mm; lobes 1–2 mm; corolla**:** banner 15 mm.

Flowering Feb–Apr. Gravelly slopes on Tertiary lake deposits of soft limestone and volcanic ash deposits; 600–790 m; Sonoran Desertscrub; Mohave Co., endemic.

As presently known, var. *aquarii* is a highly local taxon from Burro Creek area, Mohave County. The soils where it is found do not support many typical Sonoran Desert species that occur on adjacent sites. The taxon is of conservation concern.

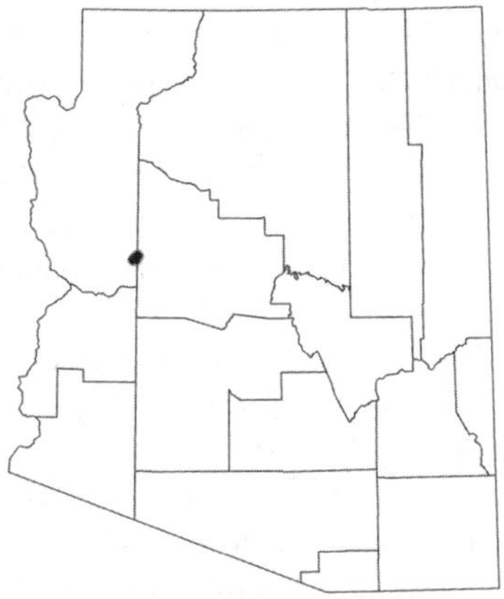

52b. Astragalus newberryi A. Gray var. **blyae** (Rose ex Rydberg) Barneby. Amer. Midl. Naturalist 37:480. 1947. * [for Mrs. Charles Bly, who collected the plant] BLY MILKVETCH Native

Xylophacos blyae Rose ex Rydberg in N.L. Britton et al., N. Amer. Fl. 24:303. 1929

Leaflets (7–)9–13, ovate or elliptic-obovate. **Flowers:** calyx deeply campanulate, 8–11.5 mm; tube (5–)6–8.5 mm; lobes 2–3 mm; corolla**:** banner 14–21 (–22) mm.

Flowering late Feb–May. Dry, rocky hillsides and canyons, on limestone or basaltic gravels; 1000–1400 m; Mohave Desertscrub, Semidesert Grassland; Mohave, Yavapai cos.; endemic.

Variety *blyae*, known from western Mohave County, and possibly Yavapai County, is distinguished from var. *newberryi* by its smaller flowers, more numerous leaflets, and its occurrence within, or at the northern edge of, the creosote bush zone. It is found in a transitional zone between creosote bush desert and desert grassland.

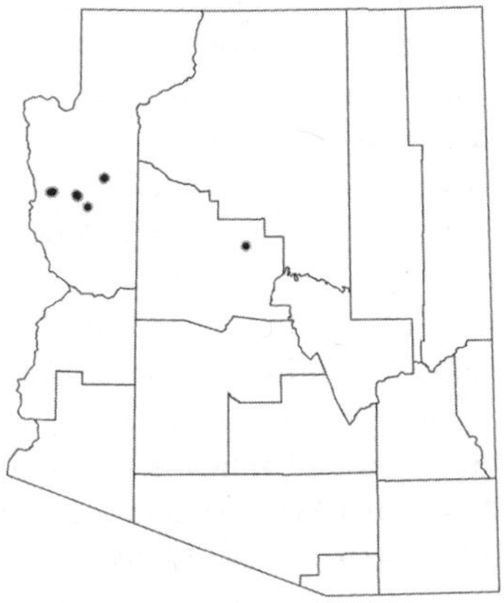

52c. Astragalus newberryi A. Gray var. **castoreus** M.E. Jones. Proc. Calif. Acad. Sci., ser. 2, 5:658. 1895. * [of the Beaver Dam Mountains] BEAVER DAM MOUNTAINS MILKVETCH Native

Leaflets 7–11(–13), ovate or elliptic-obovate. **Flowers**: calyx cylindric, 12–20 mm; tube 10–16 mm; lobes 1.5–4 mm; corolla: banner 21–32 mm.

Flowering Apr–Jun. Slopes and canyons on alluvium consisting of gravel, sand, or silt; (900–)1000–1680[–2400(–3100)] m; Great Basin Desertscrub, Interior Chaparral, Great Basin Conifer Woodland; Coconino, Mohave cos.; California, Idaho, Nevada, Oregon, Utah.

Of the varieties of *Astragalus newberryi*, var. *castoreus* is the most geographically wide-ranging. It is characteristic of sagebrush, pinyon-juniper, and mixed desert shrub communities.

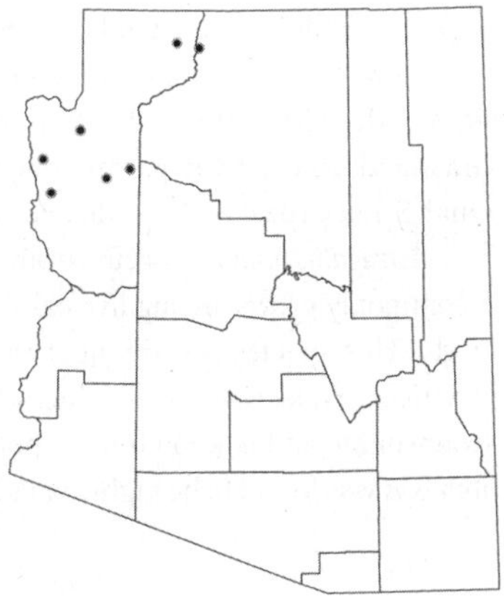

52d. Astragalus newberryi A. Gray var. **newberryi** * Native

Astragalus newberryi var. *escalantinus* Barneby

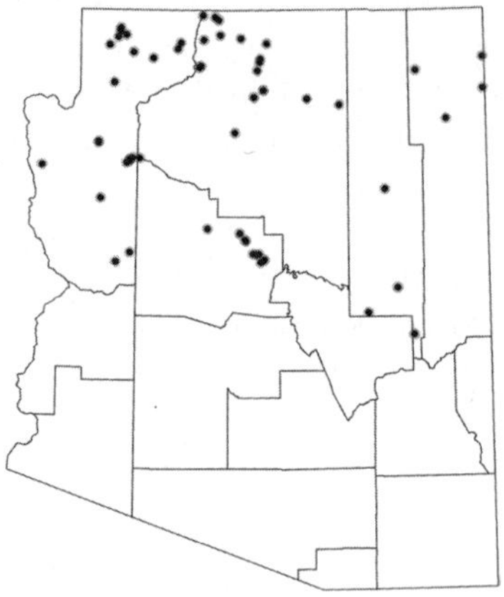

Leaflets (1–)3–5, ovate or elliptic-obovate. **Flowers**: calyx cylindric, 11–14.5 mm; tube 9.5–11.5 mm; lobes 1.5–4 mm; corolla: banner 21.5–32 mm. $2n = 22$.

Flowering Apr–Jun. Ridges, slopes, and canyons in various soil types; (600–)1300–2000 m; Sonoran Desertscrub, Great Basin Desertscrub, Semidesert Grassland, Great Basin Grassland, Interior Chaparral, Great Basin Conifer Woodland; Apache, Coconino, Gila, Mohave, Navajo, Yavapai cos.; California, Colorado, New Mexico, Utah.

Plants described as var. *escalantinus* typically have narrowly elliptic (not ovate to elliptic-obovate) leaflets, differing in no other way from var. *newberryi* (Barneby 1989). Occasional intermediates are known. The range of var. *escalantinus* with narrow leaflets is Kane County in southern Utah, and is contiguous, and partially sympatric, with that of var. *newberryi*. Variety *newberryi* is found in sagebrush, rabbitbrush, matchweed (*Gutierrezia* sp.), pinyon-juniper, canotia, shadscale, and galleta communities.

53. Astragalus nothoxys A. Gray. Proc. Amer. Acad. Arts 6:232. 1864. * [from Greek *notho*, false and *oxys*, sharp, referring to the keel petals resembling those of *Oxytropis*] BEAKED MILKVETCH Native

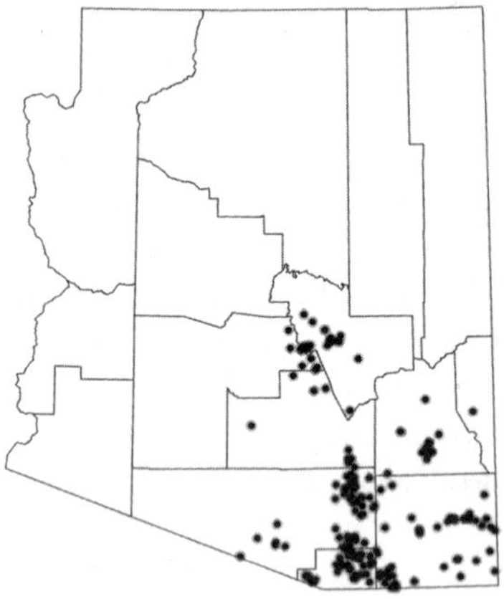

Perennials, short-lived, sometimes flowering first year, caulescent, slender, diffuse, (1.5–)2.5–35(–45) cm; from superficial caudex; strigillose, hairs basifixed. **Stems** several to numerous, decumbent or prostrate with ascending tips. **Leaves** (2–)3–9(–11.5) cm, petiole slender; stipules distinct throughout, ovate-triangular or lanceolate, 2–5 mm, submembranous. **Leaflets** (7–)9–21, flat or loosely folded, obovate, oblong-obovate, broadly oblanceolate, oval, or oval-nearly circular, 2–12(–15) × (0.8–)1–5(–6) mm, apex obtuse or emarginate, surfaces strigillose below (abaxially), glabrous or glabrescent above (adaxially). **Inflorescences** loosely flowered, with (4–)6–20(–25) flowers; peduncle incurved-ascending, (3–)4–14(–15) cm; axis (1–)1.5–9 cm in fruit; bracts 0.8–3(–3.5) mm; bracteoles 0 or 1. **Pedicels** ascending to spreading, 0.5–3.2 mm. **Flowers:** calyx campanulate, 5–7 mm, strigillose; tube 3.6–4.9 × 2.1–2.7(–2.9) mm; lobes lanceolate-subulate, 0.5–1.3 mm; corolla pink-purple, banner and wing tips often whitish, keel mottled, 8.5–12 mm; banner recurved through 45°, ovate- or obovate-cuneate, 8.5–12 mm; wings 8.6–11 mm; keel 6.5–8.4 mm, apex bluntly deltoid, beak narrowly triangular or subulate; ovules (17–)18–26. **Fruits** ascending; sessile; slightly incurved, linear-lanceolate to linear, 3-sided compressed, (13–)15–22 × 2.3–3.5(–4) mm, green or purplish becoming straw-colored or brownish, papery, strigillose; two-chambered. **Seeds** (17–)18–26. $2n = 28$.

Flowering (Jan–)Mar–Jun (summer). In various soils on ridges, slopes, canyons, arroyos, floodplains, and roadsides; (600–)800–2000(--2350) m; (Sonoran Desertscrub, Chihuahuan Desertscrub), Semidesert Grassland, Plains Grassland, Interior Chaparral, Madrean Evergreen Woodland; Cochise, Gila, Graham, Greenlee, Maricopa, Pima, Pinal, Santa Cruz cos.; New Mexico; Mexico (Chihuahua, Sonora).

Astragalus nothoxys occurs from east-central Arizona to southwestern New Mexico, and into northwestern Mexico. It commonly grows among live oaks, mesquite, cholla, and juniper, and is common on caliche soils in live oak woodlands. This species is infrequent in desertscrub. The beak of the keel has been compared, irrelevantly, to that of *Oxytropis* (Watson 1871), in which the cusp is terminal, not laterally as in this species. *Astragalus nothoxys* contains swainsonine and is known to be poisonous to livestock (James and Welsh 1992). In spite of its toxic properties, *A. nothoxys* was found to be highly palatable to white-tail and mule deer in experiments in Arizona (Nichol 1938).

54. Astragalus nutriosensis M.J. Sanderson. Madroño 35:325, fig. 1. 1988. * [of Nutrioso, Arizona] NUTRIOSO MILKVETCH Native

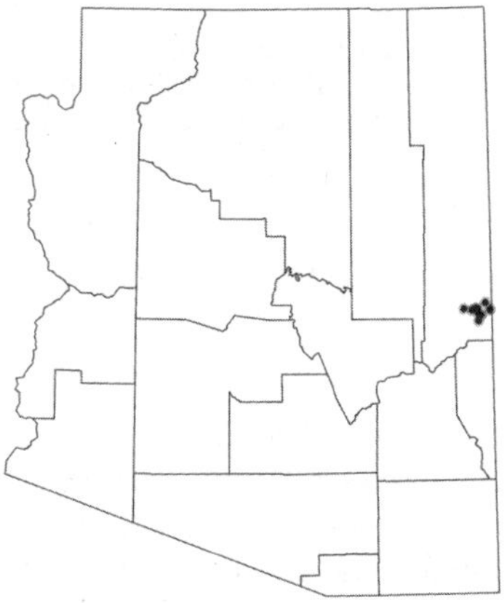

Perennials, dwarf, clump-forming, acaulescent, 3–11 cm; from superficial caudex; cinereous, pilose to strigillose, hairs basifixed. **Stems** absent. **Leaves** 3–11 cm, petiolate; stipules distinct throughout, triangular, 3–9 mm, papery. **Leaflets** 11–19, folded, obovate or elliptic, 4–8 × 1.5–4 mm, apex acute or obtuse, upper surface (adaxial) glabrescent-green, lower surface (abaxial) cinereous with spreading hairs to 2 mm. **Inflorescences** with (1–)2–5(–7) flowers; peduncle ascending, 0–2(–5) cm, almost always much shorter than leaves; axis to 0.6 cm in fruit; bracts 2–5 mm; bracteoles 0. **Pedicels** 2–4 mm. **Flowers**: calyx narrowly cylindric, 11–14 × 2–3 mm, pilose; corolla white to light lavender-pink or veined pink-purple, keel tip lavender-pink; banner recurved through less than 30°, oblanceolate, 20–23 mm × 5–6 mm ; keel 14–17 mm, significantly shorter than wings, apex round or triangular; ovules ca. 20. **Fruits** ascending; sessile on very short peduncles, readily deciduous; nearly straight, semi-ellipsoid, nearly cylindrical to slightly dorsiventrally compressed, 8–10 × 6–8 mm, brownish gray, thin and papery, pilose; two-chambered. **Seeds** ca. 20.

Flowering (Apr–)May–Jun. Volcanic silt-clay soils on gently sloping hillsides; 2140–2400 m; Plains Grassland, Great Basin Conifer Woodland; Apache Co.; New Mexico.

Astragalus nutriosensis is known from the Nutrioso Creek region of eastern Arizona (Apache Co.) and adjacent New Mexico (Catron Co.). It is still relatively poorly known and tentatively placed in section *Mollissimi*. It grows with juniper, blue grama, matchweed (*Gutierrezia* sp.), *Oxytropis lambertii* Greene, and *Astragalus* species *A. brandegeei*, *A. humistratus*, and *A. mollissimus*. This species is of conservation concern.

55. Astragalus nuttallianus de Candolle in A.P. de Candolle and A.L.P.P. de Candolle, Prodr. 2:289. 1825. * [for Thomas Nuttall, 1786–1859, English botanist and zoologist who worked in North America] SMALL-FLOWERED MILKVETCH

Astragalus micranthus Nuttall, not *Astragalus micranthus* Desvaux

Annuals or winter annuals, caulescent, usually slender, often diminutive, (1–)2–30(–45) cm; taproot slender; from superficial root-crown; nearly glabrous to strigose, strigillose, pilosulose, or hirsutulous, hairs basifixed. **Stems** single or few to several (rarely more than 15), erect, ascending, prostrate, or decumbent. **Leaves** 1–6.5(–9.5) cm, petiolate or subsessile; stipules distinct throughout, ovate-triangular, triangular, lanceolate, or linear, (1–)1.5–6(–9) mm, papery proximally, herbaceous distally. **Leaflets** (5–)7–19(–23), flat, elliptic, linear-elliptic, linear-oblong, oblong-cuneate, ovate-cuneate, oblanceolate, obovate, or obcordate, 2–14(–17) × 0.6–5 mm, apex acute, subacute, retuse, obtuse, round, emarginate, or truncate-emarginate, surfaces pubescent below (abaxially), glabrous or pubescent above (adaxially). **Inflorescences** loosely flowered, with (1–)2–10(–27) flowers; peduncle erect or incurved-ascending, (0.2–)0.5–9(–10) cm; axis 0–2(–3) cm in fruit; bracts 0.5–2.5 mm; bracteoles 0 or 1. **Pedicels** ascending to declined, 0.4–1.6 mm. **Flowers**: calyx campanulate, (1.7–)3–5.6(–5.7) mm, nearly glabrous to strigillose, strigillose-pilosulose, hirsutulous, or pilose; tube (1–)1.3–3.1 × 1.2–1.9 mm; lobes subulate to narrowly lanceolate or linear bristle-like, (0.7–)1–3.2 mm; corolla whitish, lilac, or pink-purple, sometimes becoming bluish to violet when dry, 3.7–13 mm; banner recurved through 40–45°, 3.7–13 mm; wings 3.7–8.6(–10.7) mm; keel 3.7–6.8(–9.3) mm, apex triangular, triangular-acute, triangular-subacute, sharply deltoid, or obtusely round, beak-like or not; ovules 10–22. **Fruits** ascending, spreading, or declined; sessile or substipitate, stipe to 0.5–0.9 mm; nearly straight or incurved through 0.3–0.5 spiral, linear or linear-oblanceolate, 3-sided compressed, dehiscent on plant, (7–)10–26 × (1.6–)2.1–3.5 mm, thin, green or purple-tinged becoming straw-colored or brownish then black, papery, glabrous, strigillose, or villosulous; two-chambered. **Seeds** 10–22.

Varieties 9 (4 native in the flora): w United States, Mexico.

The *Astragalus nuttallianus* complex comprises a series of distinct taxa found in a wide region of the south-central and southwestern United States and adjacent Mexico. There are two groups discernable: a western group wherein varietal differentiation is apparently ecologically and geographically controlled, and there is intergradation where one form meets another; and the central and eastern Texas group where several varieties are sympatric, but apparently do not form intermediates. *Astragalus nuttallianus* is often confused with *A. emoryanus* (see discussion under *A. emoryanus*).

1. Keel petals obtusely rounded at tip; axes of inflorescences elongating (0–)0.5–2 cm in fruit; fruits glabrous or minutely strigillose ____ **55d. Astragalus nuttallianus** var. **micranthiformis**
1′ Keel triangular-acute or –subacute or deltoid at tip (sometimes porrect); axes of inflorescences 0–2 cm in fruit; fruits glabrous, strigillose, or hirsutulous.
  2. Calyces pilose, lobes 1.8–2.1 mm ____ **55a. Astragalus nuttallianus** var. **austrinus**
  2′ Calyces strigillose or hirsutulous, lobes (0.7–)1–1.7(–2) mm.
    3. Leaflets monomorphic, all elliptic or oval with the apex acute, obtuse, or rounded; calyx tube 1.9–2.3 mm ____ **55c. Astragalus nuttallianus** var. **imperfectus**
    3′ Leaflets dimorphic, leaflets of proximal leaves mostly cuneate or obovate with the apex retuse or emarginated, those of the distal leaves elliptic with the apex acute; calyx tube (1–)1.4–1.7 mm ____ **55b. Astragalus nuttallianus** var. **cedrosensis**

55a. Astragalus nuttallianus de Candolle var. **austrinus** (Small) Barneby in F. Shreve and I.L. Wiggins, Veg. Fl. Sonoran Des. 709. 1964. * [southern] RÍO FRONTERAS MILKVETCH Native

Hamosa austrina Small, Fl. S.E. U.S. 618, 1332. 1903; *Astragalus austrinus* (Small) E.D. Schulz

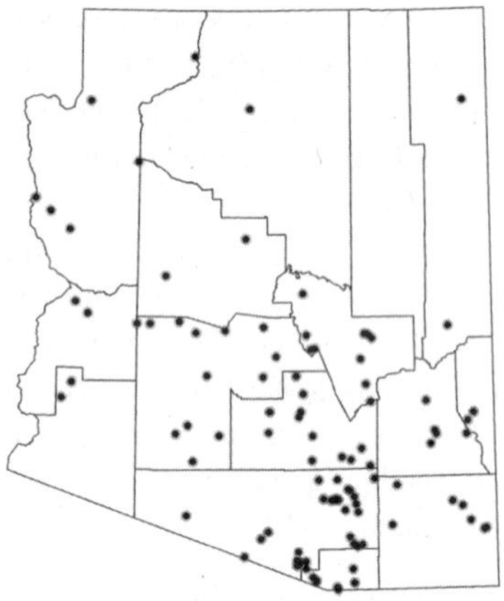

Plants strigillose or hirsutulous, hairs to 0.5 mm. **Stems** (1–)3–25(–35) cm. **Leaves** 1–6.5 cm. **Leaflets** 7–11(–17), often narrowly elliptic, sometimes broader proximally, apex not truncate-emarginate, rounded or acute. **Inflorescences** with (1–)2–5(–8) flowers; peduncle 3–10 cm; axis 0–0.5 cm in fruit. **Flowers**: calyx 3.7–5.4 mm, silvery pilose; tube (1.5–)2–3 mm; lobes 1.8–2.1 mm; corolla white or tipped pink or purple; banner (4–)5.5–7 mm; keel apex triangular-acute or sharply deltoid, usually porrect. **Fruits** 13–24 × 1.8–2.8(–3) mm, glabrous or strigillose. **Seeds** 14–18. 2*n* = 24.

Flowering Feb–May (summer–fall). Occurring on a wide range of substrates, on slopes, plains, canyons, arroyos, floodplains, roadsides, lawns, and other disturbed sites in urban areas; (250–)600–1850[–2200] m; Sonoran Desertscrub, Chihuahuan Desertscrub, Semidesert Grassland, Madrean Evergreen Woodland, (Great Basin Conifer Woodland); Apache, Cochise, Coconino, Gila, Graham, Greenlee, La Paz, Maricopa, Mohave, Pima, Pinal, Santa Cruz, Yavapai, Yuma cos.; New Mexico, Oklahoma, Texas; Mexico (Baja California, Baja California Sur, Chihuahua, Durango, Hidalgo, Nuevo León, Sonora).

Variety *austrinus* occurs from Oklahoma to the southern tip of Texas and to northern Mexico, and westward to south-central Arizona, where it passes clinally into var. *imperfectus* (Isely 1998). The seeds are eaten by Gambel's quail, and the seeds and fruits have been found in the cheek pouches of the hispid pocket mouse (Graham 1941).

55b. Astragalus nuttallianus de Candolle var. **cedrosensis** M.E. Jones. Rev. N.-Amer. Astragalus 270. 1923. * [of Cedros Island] CEDROS ISLAND MILKVETCH Native

Astragalus pertenuis Greene; *Hamosa pertenuis* (Greene) Rydberg

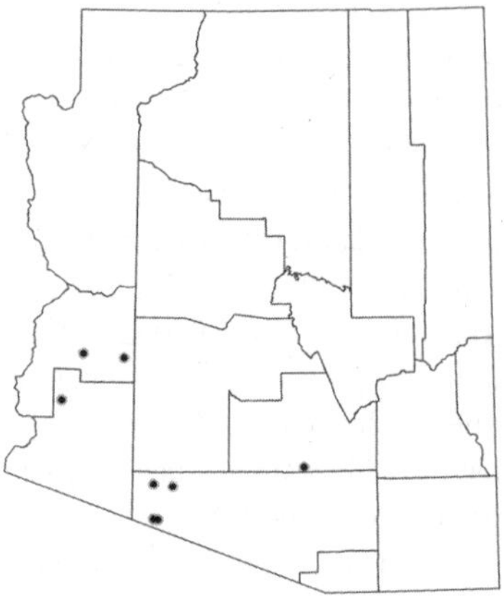

Plants strigillose, hairs to 0.8 mm. **Stems** very slender, 2–30 cm. **Leaves** 1.5–4.5 cm. **Leaflets** (5–)7–11, dimorphic, leaflets of proximal leaves mostly cuneate or obovate with the apex retuse or emarginated, those of the distal leaves elliptic with the apex acute. **Inflorescences** 1–3-flowered; peduncle (0.8–)1.5–8.5 cm; axis 0–2 cm in fruit. **Flowers**: calyx 1.7–3 mm, strigillose; tube (1–)1.4–1.7 mm; lobes (0.7–)1–1.6 mm; corolla whitish or faintly lilac-tinged; banner 3.7–6 mm; keel apex triangular-subacute. **Fruits** (7–)10–20 × 1.6–2 mm, completely or incompletely two-chambered, minutely strigillose. **Seeds** 12–16. 2*n* = 22.

Flowering Dec–Apr. Generally in sandy soil of plains and arroyos; [100–]350–550(–1050) m; Sonoran Desertscrub, (Semidesert Grassland); La Paz, Pima, Pinal, Yuma cos.; California; Mexico (Baja California, Baja California Sur, Sonora).

Variety *cedrosensis* occurs in the Sonoran Desert in southern California and southwestern Arizona, south into Baja California and Sonora, Mexico. The variety rests in large part on the dimorphic proximal leaves with retuse leaflets.

55c. Astragalus nuttallianus de Candolle var. **imperfectus** (Rydberg) Barneby. Leafl. W. Bot. 3:109. 1942. * [from Latin *imperfectus*, imperfect] IMPERFECT MILKVETCH Native

Hamosa imperfecta Rydberg, Bull. Torrey Bot. Club 54:329. 1927

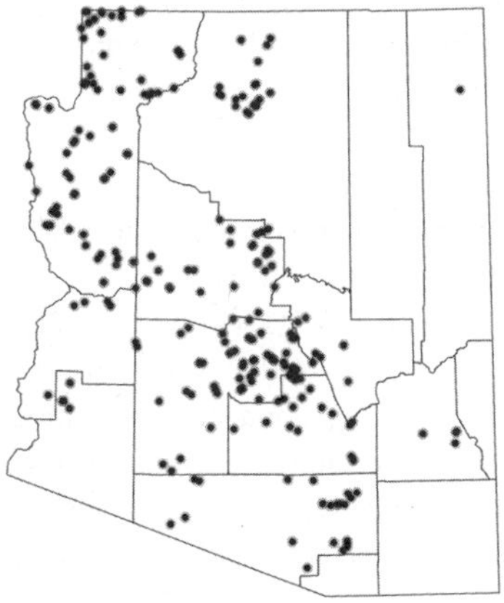

Plants strigillose or hirsutulous, hairs to 1 mm. **Stems** 2–30(–45) cm. **Leaves** 1.5–4.5(–5.5) cm. **Leaflets** 7–11(–13), monomorphic, all elliptic or oval with the apex acute, obtuse, or rounded. **Inflorescences** 1–4-flowered; peduncle (1–)2–8(–9.5) cm; axis (0–)0.2–1 cm in fruit. **Flowers**: calyx 3.2–4.5 mm, strigillose or hirsutulous; tube 1.9–2.8 mm; lobes 1–1.7(–2) mm; corolla whitish or faintly lilac-tinged; banner 4.1–6.5(–7.3) mm; keel apex triangular-subacute. **Fruits** (10–)12–21 × 1.9–3.1 mm, two-chambered or ± one-chambered, glabrous or strigillose. **Seeds** 12–16. $2n = 22$.

Flowering (Jan–)Feb–May(–Jun). Occurring on a wide variety of substrates, on slopes, plains, canyons, arroyos, floodplains, roadsides, lawns, agricultural fields, and other disturbed sites; 200–1600(–1850)[–2000] m; Sonoran Desertscrub, Mohave Desertscrub, Semidesert Grassland, Interior Chaparral, Madrean Evergreen Woodland, (Great Basin Conifer Woodland); Apache, Coconino, Graham, Gila, La Paz, Maricopa, Mohave, Pima, Pinal, Santa Cruz, Yavapai, Yuma cos.; California, Nevada, New Mexico, Utah; Mexico (Baja California, Sonora).

Variety *imperfectus* is found mainly in desertscrub communities. As Isely (1998) noted, var. *imperfectus* was circumscribed to include forms of the species with imperfectly two-chambered fruits but, as characterized here, fruits range from two-chambered to nearly one-chambered. The variety is somewhat distinct in northwestern Arizona, southwestern Utah, southern Nevada, and southeastern California, but merges with var. *austrinus* to the southeast and to the south and west it is confluent with var. *cedrosensis*. Isely (1998), and also Barneby (1964a), suggested vars. *imperfectus* and *cedrosensis* might be merged.

55d. Astragalus nuttallianus de Candolle var. **micranthiformis** Barneby. Mem. New York Bot. Gard. 13:1064. 1964. * [resembling *Astragalus micranthus* (small-flowered). *Astragalus micranthus* is a synonym of *A. nuttallianus* var. *nuttallianus*] MONTEZUMA MILKVETCH Native

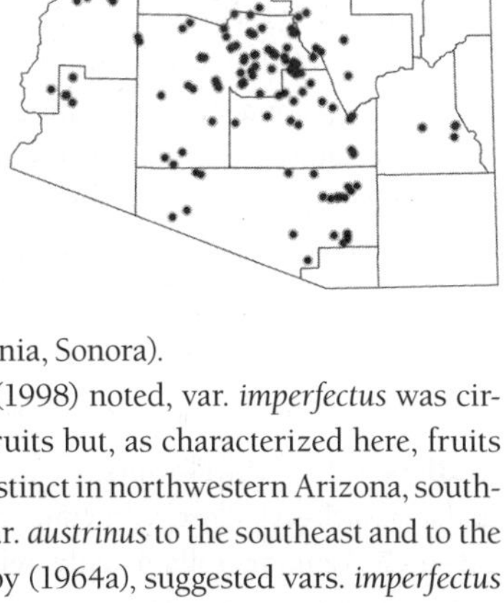

Plants sparsely pilosulose, hairs to 0.7(–0.8) mm. **Stems** 3–25(–39) cm. **Leaves** 1.5–4.5(–6.5) cm. **Leaflets** (7–)9–15(–17), proximally obovate, oblong, or obcordate, distally oblanceolate, elliptic, or linear-oblong, apex proximally retuse or emarginate, distally obtuse to subacute, terminal one sometimes emarginate. **Inflorescences** with (1–)3–7 flowers; peduncle (1.5–)2.5–5.5 (–6.5) cm; axis (0–)0.5–2 cm in fruit, elongating. **Flowers**: calyx 3.4–4.7 mm, loosely strigillose; tube 2–2.8 mm; lobes (1.2–)1.5–2.2(–2.5) mm; corolla bright purple; banner 6.3–7.6(–9.2) mm; keel apex obtusely rounded. **Fruits** gently incurved, (12–)14–20 × 2.1–3.3 mm, glabrous or minutely strigillose. **Seeds** 13–17.

Flowering Mar–May(–Jun). Various substrates including sand and clay, on slopes, canyons, arroyos, and roadsides; (580–)1000–1960 m; (Sonoran Desertscrub), Mohave Desertscrub, Great Basin Desertscrub, Interior Chaparral, Great Basin Conifer Woodland; Apache, Coconino, Gila, Greenlee, Mohave, Navajo, Pima, Yavapai cos.; Colorado, New Mexico, Utah.

The variety *micranthiformis* occurs in approximately one-quarter of each state around the Four Corners area (southeastern Utah, southwestern Colorado, northwestern New Mexico, northeastern Arizona). The record for Pima County (J. F. Wiens 2010-144) in Sonoran Desertscrub is an unusual range extension. Variety *micranthiformis* is found in mixed salt-desert and desert shrub, sagebrush, and pinyon-juniper communities.

56. Astragalus oophorus S. Watson. Botany (Fortieth Parallel) 73. 1871. * [from Greek *oophoros*, egg-bearing] EGG MILKVETCH

Perennials, low, robust or slightly slender, caulescent or shortly so, 15–30(–40) cm; from superficial caudex; glabrous or sparsely hairy, hairs basifixed. **Stems** several or numerous, decumbent to ascending, (2–)15–25 cm, 1 to several internodes elongated. **Leaves** 3–21 cm, petiolate; stipules distinct throughout, deltoid, 1.5–7 mm, thinly herbaceous becoming papery. **Leaflets** 9–21(–25), flat, oval to obovate or circular, 3–20 × 2–11 mm, margins ciliate, apex obtuse to retuse or mucronate, surfaces glabrous. **Inflorescences** 3–14-flowered; peduncle ascending or incurved, 4–13 cm; axis 1–8 cm in fruit; bracts 1.5–5 mm; bracteoles 0–2. **Pedicels** spreading, 2–6 mm. **Flowers**: calyx cylindric or campanulate, 6–12 mm, glabrous or sparsely strigillose; tube 4–8.5 × 2.6–5.5 mm; lobes subulate, 2–6(–7.7) mm; corolla ochroleucous (uniformly colored), or pink-purple and wing tips white (bicolored), 11–24 mm; banner recurved through 85°, oblanceolate to rhombic-elliptic or obovate-cuneate, 11–23 mm; wings 11–21 mm; keel 9.5–16 mm, apex obtuse; ovules 28–54. **Fruits** spreading to pendulous; sessile on and disjointing from a linear gynophore 3.5–11 mm; straight or slightly incurved, ellipsoid, semi-ellipsoid, or semi-ovoid, bladdery-inflated, (20–)25–55(–63) × 10–30 mm, thin, often red-mottled, becoming papery, glabrous; one-chambered, funicular flange narrow or obsolete, 0.2–0.7 mm wide. **Seeds** 28–54.

Varieties 6 (1 native in the flora): w United States.

Barneby (1989) notes that a large, consistently glabrous, bladdery, one-chambered pod of papery texture raised on a gynophore is the distinguishing characteristic of *Astragalus oophorus*, which appears closely related to *A. megacarpus* and *A. beckwithii*. *Astragalus oophorus* differs from *A. megacarpus* by its caulescent habit and usually smaller, more strongly stipitate fruits, while *A. beckwithii* is often distinguished by range and by its smaller, not inflated fruit that is fleshy and becomes leathery or almost woody.

56a. Astragalus oophorus S. Watson var. **caulescens** (M.E. Jones) M.E. Jones. Rev. N.-Amer. Astragalus 121 [line 38]. 1923. * [from Greek *kaulos*, stem, with prominent stems] PALLID EGG MILKVETCH Native

Astragalus megacarpus (Nuttall) A. Gray var. *caulescens* M.E. Jones, Proc. Calif. Acad. Sci., ser. 2, 5:643. 1895

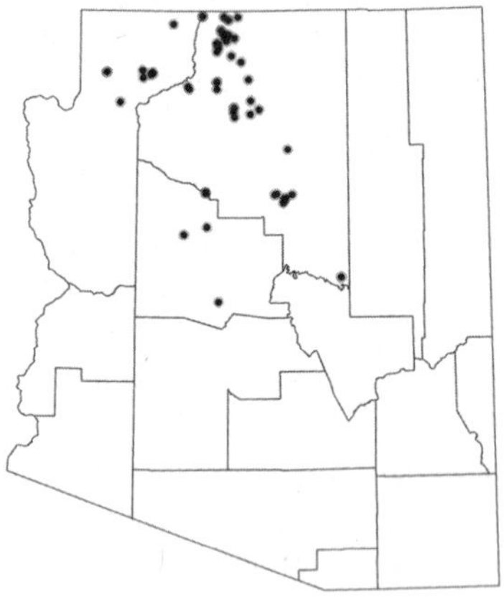

Leaflets 9–19(–25). **Flowers**: calyx campanulate or short-cylindric, often sparsely black-strigillose, sometimes glabrous; tube 4.5–6 × 3.3–5.5 mm; lobes 2.3–6(–7.7) mm; corolla usually ochroleucous (uniformly colored), rarely banner and wings suffused with purple, or purple throughout, 17–21 mm. **Fruits** asymmetric, semi-ovoid or semi-ellipsoid, (25–)30–40(–45)mm, upper surface ± flattened or gently concave, lower surface strongly convex; gynophore 5–9 mm.

Flowering Apr–Jul. Slopes, canyons, roadsides, and other disturbed sites, on various soils including sand and clay, volcanic gravel, limestone, and shale; 1300–2500 m; Great Basin Conifer Woodland, Petran Montane Conifer Forest; Apache, Coconino, Mohave, Yavapai cos.; Colorado, Nevada, Utah.

As noted by Barneby (1964a) plants with bicolored (purple and white) flowers occur occasionally within the range of var. *caulescens* in Mohave County, Arizona. Near the southern limit of the variety in Arizona, plants sometimes have petals suffused or lined with lilac. Plants from the Arizona plateaus also tend to have more numerous leaflets (to 21–25) than found elsewhere within the variety, and the fruit is generally less oblique in outline and with a shorter, less well-defined beak. Variety *caulescens* occurs in sagebrush, pinyon-juniper, ponderosa pine forest, and mountain brush communities.

57. Astragalus pattersonii A. Gray. Bull. U.S. Geol. Geogr. Surv. Territ. 2:235. 1876 (as *Astragalus Pattersoni*) * [for Henry N. Patterson, 1853–1919, American plant collector and printer] PATTERSON'S MILKVETCH Native

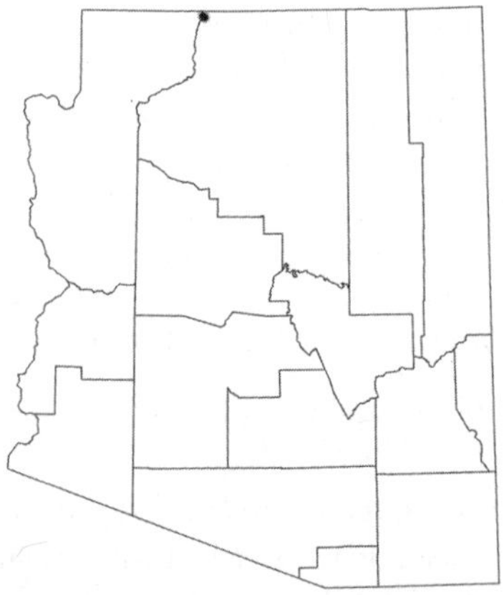

Perennials, robust, clump-forming, caulescent, 20–45(–50) cm; from caudex; glabrous or strigillose; hairs basifixed. **Stems** several to numerous, decumbent to ascending or erect; glabrous proximally, sparsely strigillose distally, usually tinged red. **Leaves** 5–13 cm, shortly petiolate or subsessile; stipules distinct throughout, triangular-lanceolate, 3–8 mm, firm becoming papery, sometimes membranous distally or throughout. **Leaflets** 7–15+, folded, elliptic to lanceolate, oblanceolate, or obovate, 6–38 × 3–16 mm, apex obtuse to acute, retuse, or mucronate, surfaces strigose or glabrous. **Inflorescences** 6–24-flowered, rather dense early but axis soon elongating; peduncle erect, 3–18 cm, axis 2–15 cm in fruit; bracts 2–8 mm; bracteoles 2. **Pedicels** declined-nodding,

1–4.5 mm. **Flowers:** calyx tan or whitish, cylindric, enlarged on one side or base strongly oblique, 8.8–14.2 mm, sparsely pale-strigillose; tube 6–8.8 × 3.5–4.8 mm; lobes spreading, subulate, 2.3–6.5 mm; corolla usually white, uniformly colored, rarely keel tip faintly purplish, 14–22(–25) mm; banner recurved through 50–90°, oval or ovate, 14.6–21.5 mm; wings 14.8–20 mm; keel 12.4–15 mm, apex sharply triangular; ovules 22–38. **Fruits** strictly erect; sessile; straight or slightly incurved, cylindroid to ellipsoid or ovoid, nearly rounded to strongly dorsiventrally compressed, 17–35 × 6–10 mm, green or reddish brown or straw-colored, succulent becoming stiffly leathery, glabrous or puberulent, long persisting; one-chambered. **Seeds** 22–38. 2*n* = 24.

Flowering May–Jul. Slopes on clay and gypsum soils; 1400–1650[–2700] m; Great Basin Desertscrub, Great Basin Conifer Woodland; Coconino cos.; Colorado, New Mexico, Utah.

Astragalus pattersonii fruits are extremely variable, helping to confuse it with *A. praelongus*, but floral characters such as ovule number, shape of the calyx, flower color, and shape of the keel petals provide the distinction. *Astragalus pattersonii* is malodorous (selenium-scented) and is reported to be toxic (Barneby 1964a). The Arizona population (near Fredonia) is disjunct, some 200 miles west of the main distribution of the species and thus, may be introduced (Barneby 1964a). It has also been reported in Arizona from Navajo Co. and other locations in the state but it has not been possible to confirm the identity of these specimens. Across its range, the species is found in pinyon-juniper and mixed desert shrub communities, often on fine textured, seleniferous substrates.

58. Astragalus pinonis M.E. Jones. Contr. W. Bot. 8:14. 1898. * [a pine, the pinyon] PINYON MILKVETCH

Perennials, caulescent, slender, weak, 10–55(–60) cm; from superficial caudex; gray-strigillose, hairs basifixed. **Stems** few (1–5), erect or reclining. **Leaves** sparse, 0.6–11 cm, usually sessile or subsessile; stipules distinct throughout, dimorphic, 1.5–6 mm, the lowest ovate, obtuse, papery, reddish-purple, the upper herbaceous, deltoid or triangular. **Leaflets** (5–)9–19, folded, linear to oblong, 1.5–19 × 0.7–4 mm, apex obtuse to retuse, surfaces strigose; terminal leaflet jointed to or continuous with the rachis. **Inflorescences** 2–20-flowered; peduncle widely diverging or widely incurved-ascending, 1.5–8 cm; axis 2–7(–11) cm in fruit; bracts 1–2 mm; bracteoles 0 or 1. **Pedicels** ascending-spreading, 1–3 mm. **Flowers:** calyx campanulate, 4.2–5.6(–7) mm, strigose; tube 2.3–3.8(–4.5) × 2.5–3.3 mm; lobes subulate, (0.8–)1–2.4 mm; corolla greenish to ochroleucous, suffused with purple, or white and purple-veined, 8.1–10.3 mm; banner recurved through 90°, nearly circular-cuneate, 8.2–10.3 mm; wings 8–9.8 mm; keel 7.5–9.7 mm, apex slightly porrect-beaked; ovules 32–42. **Fruits** spreading or declined; subsessile or elevated on an obscure, glabrous, stipe-like neck to 1 mm; straight to slightly curved, narrowly linear or oblong-ellipsoid, cylindrical or nearly so, 20–35 × (3–)3.5–8.5 mm, brownish or purplish tinged becoming brownish straw-colored, stiffly-papery, strigose; one-chambered. **Seeds** 18–42.

Varieties 2 (1 native in the flora): ec Nevada to s Utah and n Arizona.

58a. Astragalus pinonis M.E. Jones var. **atwoodii** (S.L. Welsh & K.H. Thorne) Barneby & Isely. Intermountain Fl. 3B:67. 1989. * Native

Astragalus atwoodii S.L. Welsh & K.H. Thorne, Great Basin Naturalist 37:103. 1977

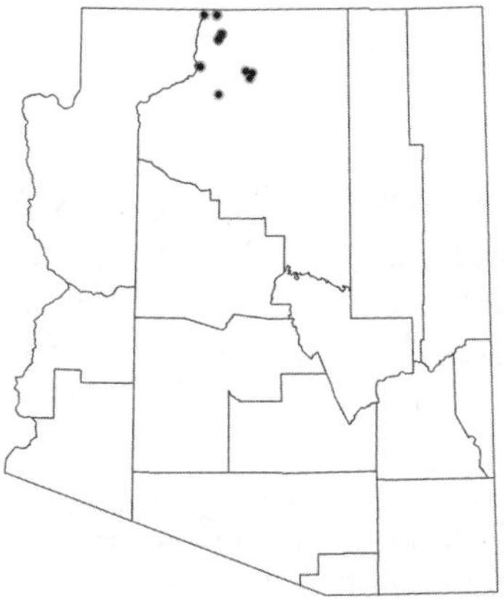

Stems single or few, erect; 2–6 cm underground. **Leaves** 0.6–9 cm. **Leaflets** (7–)9–17, folded, linear to narrowly oblong, 1.5–12.5 × 0.7–2 mm, terminal leaflets of some leaves continuous with the rachis. **Inflorescences** 2–15-flowered. **Fruits** narrowly linear, 3–4.5 mm diameter. **Seeds** 18–25.

Flowering Apr–Jun. Limestone ridges and slopes, rarely in valleys, Kaibab Plateau and House Rock Valley; (1370–)1800–2100 m; Great Basin Conifer Woodland; Coconino Co.; endemic.

Astragalus pinonis var. *atwoodii*, known from Coconino County, is geographically disjunct from *A. pinonis* var. *pinonis*. Variety *pinonis* is widespread across central and southern Nevada east to the south and east margins of the Bonneville Basin in Iron and Juab counties of Utah. Variety *atwoodii* grows in shrubs, displaying leaves and flowers above the shrubs. Found in pinyon-juniper, cliffrose, and sagebrush communities. This variety is of conservation concern.

59. Astragalus praelongus E. Sheldon. Minnesota Bot. Stud. 1:23. 1894. * [very tall] STINKING MILKVETCH

Astragalus procerus A. Gray, not *Astragalus procerus* Boissier & Haussknecht

Perennials, forming moderate to very large, bushy clumps, caulescent, malodorous (selenium-scented), robust to very robust, 10–120 cm; from a superficial root crown or subterranean caudex, branched; strigillose, hairs basifixed. **Stems** few to several, erect or ascending; glabrate, sometimes sparsely strigose distally. **Leaves** 3–22 cm, petiolate or subsessile; stipules distinct throughout [occasionally not distinct], deltoid, 2–7(–9) mm, thinly herbaceous becoming papery-scarious. **Leaflets** 7–33, folded, obovate, elliptic, oblong, lanceolate, or oblanceolate, 3–50 × 2–24 mm, apex obtuse or retuse to acute, surfaces sparsely strigose or glabrescent below (abaxially), glabrous above (adaxially). **Inflorescences** 8–33-flowered; peduncle erect, 4–26 cm; axis 3–16 cm in fruit; bracts 1–7 mm; bracteoles 2. **Pedicels** deflexed, 1–7 mm. **Flowers**: calyx green or yellowish, cylindric, enlarged on one side, 5.8–14 mm, glabrous or sparsely strigose; tube 4.4–7.5 × 3.5–5.4(–6) mm; lobes subulate, deltoid, lanceolate-subulate, lanceolate-attenuate, or triangular-subulate, 0.3–6 mm; corolla ochroleucous, keel often faintly to definitely spotted (purplish-tipped) or not spotted, (11–)15–24 mm; banner recurved through 45°, oval or rhombic-ovate, 15–23.5 mm; wings 15–22 mm; keel 11.5–17 mm, apex bluntly rectangular; ovules 40–75(–84). **Fruits** erect to spreading or declined; sessile, subsessile, or stipitate, stipe 0–8 mm; usually straight, narrowly elliptic to broadly ovate, or nearly globose, turgid or inflated, 18–42 × 5–25 mm, fleshy, green or purplish speckled becoming brown or straw-colored, leathery-woody, glabrous to puberulent or thinly strigillose; almost one-chambered, septum 0.8–2.3 mm wide. **Seeds** 40–75(–84). $2n = 12$.

Varieties 4 (2 native in the flora): sw United States.

Astragalus praelongus is malodorous, smelling sufficiently strong so that, to some, the odor is nauseating.

1. Fruits ellipsoid, oblong-ellipsoid, ellipsoid-cylindroid, or narrowly clavate-ellipsoid; stipes 4.5–8 mm ____ **59a. Astragalus praelongus** var. **lonchopus**

1′ Fruits broadly oblong, ellipsoid, sessile or shortly stipitate; stipes 0–2.5 mm ____ **59b. Astragalus praelongus** var. **praelongus**

59a. Astragalus praelongus E. Sheldon var. **lonchopus** Barneby. Leafl. W. Bot. 9:90. 1960. * [with a long stipe] LONGSTIPE STINKING MILKVETCH Native

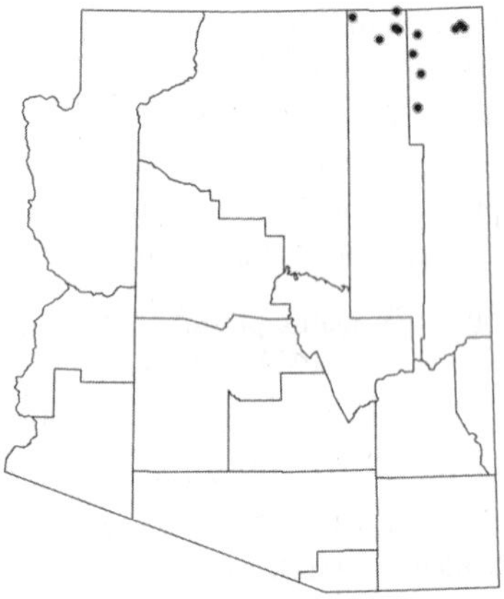

Flowers: calyx with lobes subulate or lanceolate-attenuate, 3–6.5 mm; corolla ochroleucous, keel petals often faintly to definitely spotted, 15–24 mm **Fruits:** stipe 4.5–8 mm; ellipsoid, oblong-ellipsoid, ellipsoid-cylindroid, or narrowly clavate-ellipsoid, 20–30 × 6–9 mm.

Flowering late Apr–Jul. Seleniferous fine-textured, less commonly sandy, substrates of valleys, canyon bottoms, and roadsides; [1200–]1450–1830 [–2100] m; Great Basin Desertscrub, Great Basin Grassland, Great Basin Conifer Woodland; Apache, Navajo cos.; Colorado, New Mexico, Utah.

Variety *lonchopus* is centered in the Four Corners region. It is found in blackbrush, mixed desert shrub, sagebrush, and pinyon-juniper communities.

59b. Astragalus praelongus E. Sheldon var. **praelongus** * Native

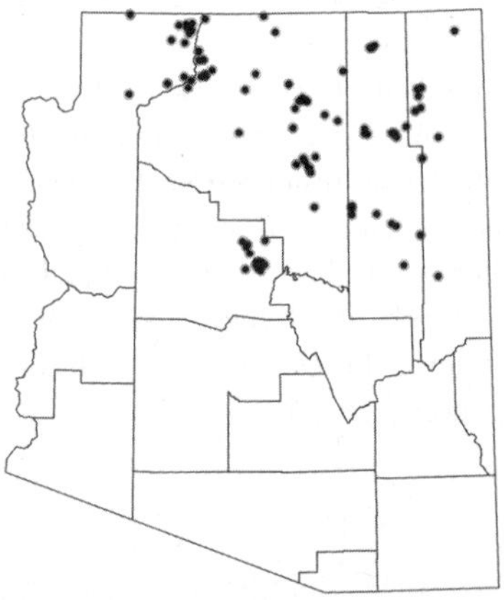

Flowers: calyx with lobes deltoid to lanceolate-subulate, 0.3–4.7 mm; corolla ochroleucous, keel petals often faintly to definitely spotted, 15–24 mm. **Fruits:** stipe obconic when present, 0–2.5 mm; broadly oblong to ellipsoid, 20–38 (–42) × (9–)10–15(–25) mm. $2n = 22, 24$.

Flowering Apr–Jul. Clay and silt of the Cretaceous Mancos and Tropic shales, and Triassic Moenkopi, and Chinle formations, other seleniferous soils of slopes, valleys, arroyos, canyon bottoms, floodplains, and roadsides; 600–2200(–2650) m; Mohave Desertscrub, Great Basin Desertscrub, Semidesert Grassland, Great Basin Grassland, Great Basin Conifer Woodland; Apache, Coconino, Mohave, Navajo, Yavapai cos.; Colorado, Nevada, New Mexico, Utah.

Variety *praelongus* occurs in salt desert shrub, shrubby grassland, pinyon-juniper woodland, and riparian communities. The plants are highly toxic but are seldom grazed by healthy animals and, even then, only as a last resort. Its green foliage in a landscape turns brown from lack of rain. Fox et al. (2010) reported that plants also contained swainsonine. Reportedly, the Hopi used the plant, under the name *siskinga*, in treatment of bladder problems.

60. Astragalus preussii A. Gray. Proc. Amer. Acad. Arts 6:222. 1864. * [for Charles Preuss, 1803–1854, Prussian-born cartographer and surveyor] PREUSS' MILKVETCH

Perennials, short-lived, sometimes flowering first year, forming bushy clumps, caulescent, malodorous (selenium-scented), moderate to robust, (10–)12–53 cm; from superficial woody caudex; glabrous or glabrate appressed-hairy, hairs basifixed. **Stems** few to many, erect or incurved-ascending, diffuse. **Leaves** 3.5–15 cm, shortly petiolate or subsessile; stipules distinct throughout, ovate, deltoid, or lanceolate, 2–7 mm, papery proximally, herbaceous distally, apex sometimes obtuse. **Leaflets** (7–)9–25, folded, obovate, nearly circular, obcordate to oblong, narrowly elliptic, lanceolate, or linear, (1.5–)3–28 × 1–11(–13) mm, apex retuse-emarginate to rounded, obtuse, acute, or truncate, sometimes mucronulate, surfaces glabrous, hairs sometimes on margins or midrib. **Inflorescences** with 3–24 flowers; peduncle erect-ascending, 2–15 cm; axis 1–20 cm in fruit; bracts 1.5–4 mm; bracteoles 2. **Pedicels** ascending, 1–5.5 mm. **Flowers**: calyx purple, cylindric, 6.4–12.3 mm, sparsely strigose; tube 5.1–9.7 × (2.7–)3–4.8 mm; lobes subulate, 1.3–2.6 mm; corolla pink-purple (to bluish when dry), sometimes whitish but distally suffused with lilac-purple, 14–24 mm; banner recurved through 40–45°, rhombic-obovate or rhombic-oblanceolate, (14–)16.5–24 mm; wings (12.6–)14.3–20.5 mm; keel (11.1–)11.5–19 mm, apex blunt; ovules 20–44. **Fruits** persistent, dehiscent on plant, erect to ascending; subsessile or stipitate, stipe 2–7 mm; straight or slightly incurved, cylindroid, ellipsoid, ovoid, or oblong-ellipsoid, inflated but not bladdery, 12–34 × 6–13 mm, green or red-dotted becoming straw-colored, stiffly papery to leathery, glabrous or puberulent; one-chambered. **Seeds** 20–44. $2n = 24$.

Varieties 3 (2 native in the flora): sw United States.

1. Fruits sessile or subsessile; inflorescence axis 4–20 cm in fruit; calyces 6.4–8.5(–9.4) mm, northwest Arizona ________ **60a. Astragalus preussii** var. **laxiflorus**

1′ Fruits stipitate, stipes 2–5+ mm; inflorescences axis 1–7(–9) cm in fruit; calyces 8.5–11 mm, northern Arizona ________ **60b. Astragalus preussii** var. **preussii**

60a. Astragalus preussii A. Gray var. **laxiflorus** A. Gray. Proc. Amer. Acad. Arts 13:369. 1878. * [from Latin *laxi*, loose and *flora*, flower] Native

Astragalus crotalariae (Bentham) A. Gray var. *davidsonii* (Rydberg) Munz & McBurney

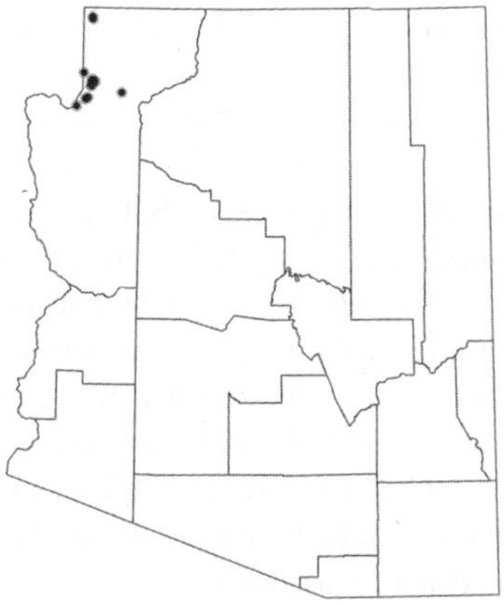

Plants sometimes flowering first year, (10–)30–53 cm. **Leaves** 4–15 cm. **Leaflets** (7–)9–15(–17), broadly obovate to nearly circular, 6–18(–23) × 3–11 (–13) mm, apex obtuse or truncate and mucronulate. **Inflorescences** lax, with (8–)15–24 flowers; peduncle (6–)8–15 cm; axis 4–20 cm, slightly elongated in fruit. **Flowers**: calyx 6.4–8.5(–9.4) mm; tube 5.5–7 mm, color strongly contrasting with corolla; lobes 1.5–1.9 mm; banner 14–17.5 mm. **Fruits:** sessile or subsessile; ovoid to ellipsoid, length usually less than 3 times width, 17–20(–25) × 7–10(–11) mm, abruptly contracted at base.

Flowering late Feb–May. Rocky or gravelly slopes and outcrops on gypsum clay soils; 500–670[–760] m; Mohave Desertscrub; Mohave Co.; California, Nevada, Utah.

Variety *laxiflorus* is restricted to the vicinity of the juncture of Arizona, Nevada, and Utah and remotely disjunct in the southwest Mojave Desert. The variety is of conservation concern. Barneby (1964a) suggests that the widely disjunct station on the western edge of the Mojave Desert near Lancaster, California, may represent an independent origin because of the small flowers and fewer ovules. There are no recent collections and that population is likely extinct. The variety is often found in creosote bush or saltbush communities.

60b. Astragalus preussii A. Gray var. **preussii** * Native

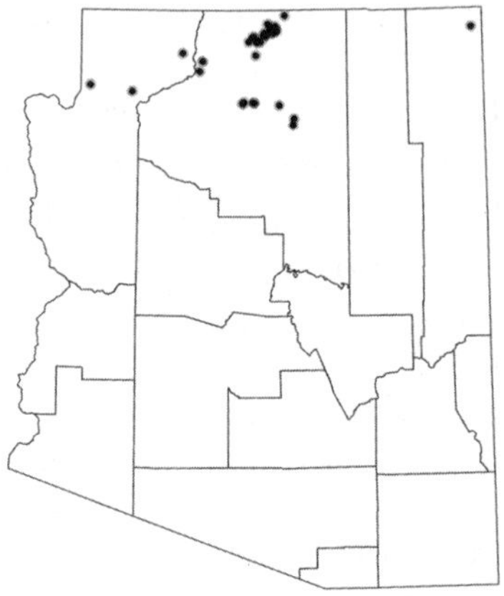

Plants 20–35(–40) cm. **Leaves** 4–8 cm. **Leaflets** usually (7–)11–17, broadly obovate, 3–14 × (2–)5–10 mm, apex retuse-emarginate to truncate, sometimes mucronulate. **Inflorescences** with (8–)10–19 flowers; peduncle (4.5–)8–11.5 cm; axis 1–7(–9) cm, slightly elongated in fruit. **Flowers**: calyx 8.5–11 mm; tube 8–9 mm, color not strongly contrasting with corolla; lobes 1.5–1.9 mm; banner 17.4–24 mm. **Fruits:** stipe 2–5+ mm; cylindric to narrowly ellipsoid, length often more than 3 times width, 24–33 × 6–10 mm, tapering at base. $2n$ = 24.

Flowering late Mar–May. Slopes, valley floors, arroyos, floodplains, and roadsides on sandstone, limestone, or shale, sand, seleniferous and alkaline clays and silts; 560–1880 m; Mohave Desertscrub, Great Basin Desertscrub; Apache, Coconino, Mohave cos.; California, Nevada, Utah.

Variety *preussii* was named for Charles Preuss, the topographer with J. C. Frémont on his second expedition in 1843–44, when the type was taken. The variety is restricted to northern and northwestern Arizona, southwest Utah, the southern tip of Nevada and adjacent San Bernardino County, California where it grows in blackbrush and mixed desert shrub communities. Isely (1998) provides a key to eastern and western phases of this variety.

61. Astragalus recurvus Greene. Bull. Calif. Acad. Sci. 1:155. 1885. * [bent backward] RECURVED MILKVETCH Native

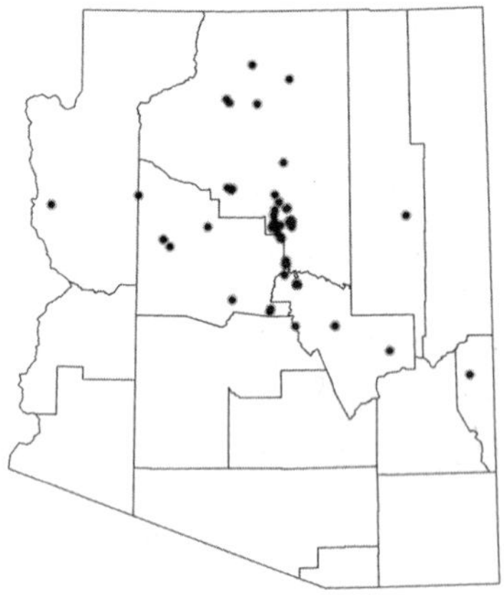

Perennials, caulescent, low, slender, 10–28 cm; from shallow, subterranean caudex; strigillose, hairs basifixed. **Stems** several to numerous, decumbent. **Leaves** (1.5–)3–7 cm, petiolate; stipules fused proximally, fused to 1/2 their length distally, mostly triangular, 1.5–4 mm, membranous becoming papery, apex bidentate or subtruncate. **Leaflets** 9–19, flat or loosely folded, oblong-obovate, elliptic, or linear-oblong, 2–10 × 1.5–6 mm, apex obtuse or emarginate, surfaces pubescent below (abaxially), glabrate above (adaxially). **Inflorescences** with (5–)7–25 flowers; peduncle incurved-ascending, 3–10 cm; axis 1.5–9 cm in fruit; bracts 0.5–1.5 mm; bracteoles 0. **Pedicels** ascending becoming declined, 0.5–1.5 mm. **Flowers:** calyx drying, persistent, campanulate, 3.5–4.7 mm, strigillose, hairs white, sometimes also black; tube 2–3.2 × 1.8–2 mm; lobes subulate, 1–1.6 mm; corolla whitish tinged with dingy lilac, 5.2–7.7 mm; banner recurved through 65–90°, ovate-cuneate, 5.2–7.7 mm; wings 5.1–8 mm; keel incurved through 50–95°, crescent-like, 5–6.5 mm, apex sharply triangular; ovules 12–18. **Fruits** pendulous, sessile, subsessile, or stipitate, stipe 0–0.7 mm; strongly curved downward, lower (abaxial) face concave, lateral faces convex, linear-oblong or linear-ellipsoid, 3-sided compressed, 13–25 × 2.3–3 mm, green, sometimes also purplish, becoming straw-colored, papery, base cuneate or acuminate, apex cuneate or acuminate, surface cross-reticulate, finely strigillose; incompletely two-chambered, inflexed as nearly complete septum 0.7–1.5 mm wide. **Seeds** 12–18. $2n$ = 22.

Flowering Apr–early Jul. Open sites in dry volcanic soils; 1400–2200 m; Madrean Evergreen Woodland, Great Basin Conifer Woodland, Petran Montane Conifer Forest; Coconino, Gila, Greenlee, Mojave, Navajo, Yavapai Cos; endemic to Arizona.

Without fruits, *Astragalus recurvus*, which is restricted to the Coconino Plateau and Mogollon Rim, may be distinguished from *A. rusbyi*, also of basaltic soils of central Arizona, by the vigorous leaves with more numerous leaflets (17–25 versus 9–19) and later flowering period (mid June to September versus April to early July). *Astragalus recurvus* is found in ponderosa pine, mixed oak and pine or juniper woodlands.

62. Astragalus rusbyi Green. Bull. Calif. Acad. Sci. 1:8. 1884. * [for Henry H. Rusby, 1855–1940, American botanist and physician] RUSBY'S MILKVETCH Native

Perennials, caulescent, slender, 15–35(–40) cm; from superficial caudex; thinly strigillose, hairs basifixed. **Stems** several, erect and incurved-ascending; glabrous or nearly so at base. **Leaves** (2.5–)3–8 cm, mostly subsessile; stipules fused proximally, fused or distinct distally, 1.5–5.5 mm, papery-membranous, apex subentire or slightly bidentate. **Leaflets**

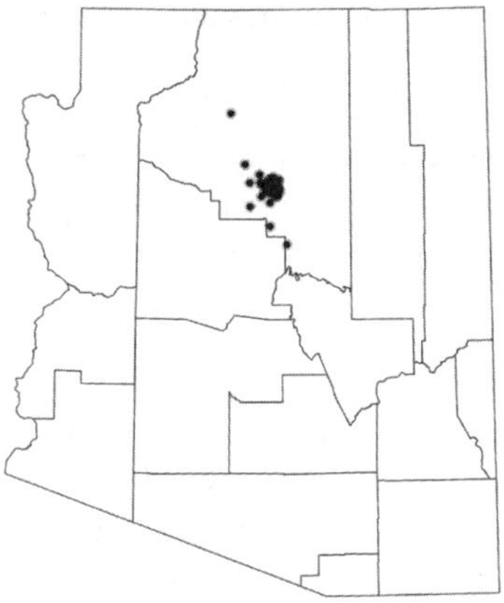

(11–)17–25, flat or loosely folded, oval-oblong, oblanceolate, narrowly obovate, ovate, or narrowly oblong-elliptic, (3–)4–11 × 2–4.5 mm, apex obtuse, truncate, or retuse, surfaces strigillose below (abaxially), glabrous above (adaxially). **Inflorescences** loosely or remotely (10–)15–25(–30)-flowered; peduncle erect or incurved, (2–)4–11 cm; with raceme, usually shorter than stem; axis (2.5–)6–18(–22) cm in fruit; bracts 1–2.5 mm; bracteoles 0. **Pedicels** deflexed, 0.8–1.8 mm, (1.5–)2–3.2 mm in fruit, slightly thickened. **Flowers**: calyx drying, persistent, campanulate, 3.2–4.1 mm, strigillose, hairs white, black, or grayish brown; tube 2.1–2.6 × 1.8–2.3 mm; lobes subulate or triangular-subulate, 1–1.6 mm; corolla whitish, sometimes tinged with lavender, 6.3–7.2 mm; banner abruptly recurved through 100°, ovate-cuneate, 6.3–7.2 mm; wings 6.9–8 mm; keel incurved through 100°, nearly round or obtusely triangular, 4.6–5.5 mm, apex blunt; ovules 7–13. **Fruits** pendulous, stipitate, stipe 2.2–5 mm; pale green becoming straw-colored, straight or nearly so, lower (abaxial) face flat or concave, lateral faces convex, linear-oblong or ellipsoid, obscurely 3-sided compressed, 13–22 × 3.2–4.1 mm, papery, base cuneate or attenuate, apex shortly cuspidate, black-strigillose; incompletely two-chambered, valves inflexed as nearly complete septum 0.8–1.5 mm wide. **Seeds** 7–13. $2n = 22$.

Flowering Jun–Sep. Slopes on basaltic substrates; 1600–2500 m; Petran Montane Conifer Forest; Coconino, Navajo cos.; endemic.

Astragalus rusbyi is restricted to the San Francisco Peaks and the vicinity of Oak Creek in Coconino County. It grows in meadows in ponderosa pine forest and in thickets or aspen groves. The species is similar to the geographically separate *A. egglestonii* (Rydberg) Kearney & Peebles and the northern Mexican *A. longissimus* (M.E. Jones) Barneby, to the point that Isely (1998) suggests that distinction at the species level is dubious, although Barneby (1964a) notes that intermediate populations do not exist. The records for Apache and Navajo counties are each represented by a single specimen.

63. Astragalus sabulonum A. Gray. Proc. Amer. Acad. Arts 13:368. 1878. * [from Latin *sabulum*, of sand and gravel] GRAVEL MILKVETCH Native

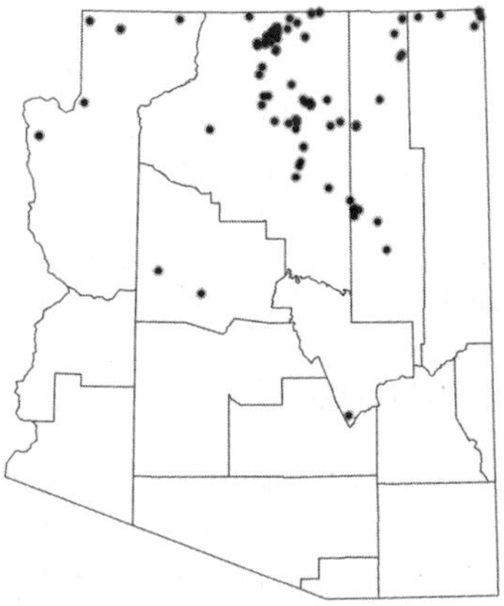

Annuals, winter annuals, or biennials, caulescent, coarse, 4–35 cm; from root-crown; villous-hirsutulous, hairs basifixed. **Stems** single or few, usually decumbent to ascending, rarely erect. **Leaves** 1.5–7 cm, petiolate; stipules distinct throughout, deltoid, 1–4 mm, subherbaceous becoming papery. **Leaflets** (5–)9–15, flat or folded, oblanceolate to oblong or obovate, 2–13 × 1–5 mm, apex retuse to truncate or obtuse, surfaces loosely villous, sometimes glabrate above (adaxially). **Inflorescences** 2–7-flowered; peduncle incurved-ascending, 0.5–4 cm; axis 0.3–2.5 cm in fruit; bracts 1–2.5 mm; bracteoles 0. **Pedicels** ascending-spreading, 0.8–2 mm. **Flowers** calyx campanulate, 3.3–6.2 mm, hirsutulous; tube 1.8–2.5 × (1.8–)2–2.7 mm; lobes subulate, 1.8–3.5 mm; corolla usually ochroleucous, tinged or tipped with purple, 5.2–8 mm; banner recurved 50–70°, ovate, obovate, or nearly circular, 5.2–7.2 mm; wings 5.7–6.7 mm; keel (5–)5.4–6.5 mm, apex narrowly triangular, sometimes slightly beak-like; ovules 10–19. **Fruits** spreading-declined; sessile; lunately incurved or hooked, obliquely ovoid to ovoid-ellipsoid, turgid or somewhat inflated, 9–17(–20) × (4–)5–8(–11) mm, thinly fleshy, green or purple-cheeked or -dotted, becoming brownish or straw-colored, leathery or stiffly papery, white-villous or hirsute, hairs 0.7–2.2 mm; one-chambered. **Seeds** 10–19. $2n = 24$.

Flowering Nov–Jul. Ridges, slopes, valleys, arroyos and floodplains, and roadsides, on sandy substrates or occasionally on sandstone or shale; [(100–)]500–1800[–2200] m; (Sonoran Desertscrub, Mohave Desertscrub), Great Basin Desertscrub, Great Basin Conifer Woodland; Apache, Coconino, Gila, Mohave, Navajo, Yavapai cos.; California, Nevada, New Mexico, Utah; Mexico (Baja California, Sonora).

Though *Astragalus sabulonum* has a wide ecological and altitudinal amplitude, from warm-desert shrubland to pinyon-juniper woodland, it is reasonably uniform morphologically. In Arizona, most populations occur on the Colorado Plateau but descend to the Colorado River in the Grand Canyon and sparingly at lower elevation sites in the Sonoran and Mojave Deserts. Isely (1998) distinguishes it in a key from its closest relatives, *A. pubentissimus* and *A. pardalinus*.

64. Astragalus scopulorum Porter in T.C. Porter and J.M. Coulter, Syn. Fl. Colorado 24. 1874. * [of craggy rocks, referring to the Rocky Mountains] ROCKY MOUNTAIN MILKVETCH Native

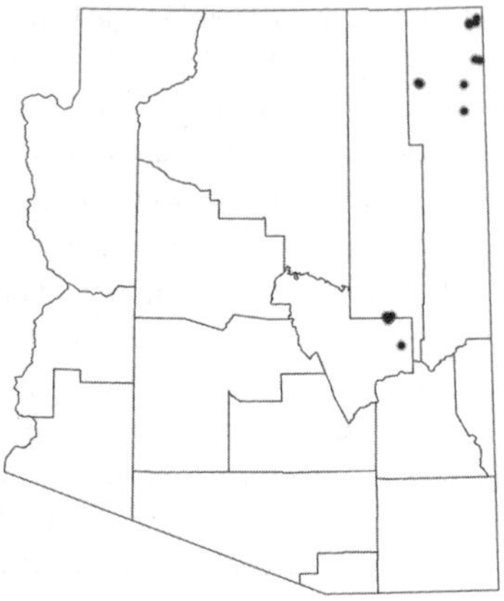

Perennials, caulescent, 15–48 cm; from subterranean caudex; strigillose, hairs basifixed. **Stems** several, decumbent to ascending, 15–48(–65) cm, 3–13 cm underground. **Leaves** 1.5–8.5 cm, shortly petiolate proximally, subsessile distally; stipules with fused sheathing proximally, fused or distinct distally, deltoid or lanceolate, 3–9 mm, stiff-papery proximally. **Leaflets** (13–)15–29(–35), flat or loosely folded, oblong to elliptic or oblanceolate, some narrowly so, 2–19 × 1–8 mm, apex acute to obtuse or mucronate, surfaces sparsely strigose or glabrous below (abaxially), glabrous above (adaxially). **Inflorescences** 4–22-flowered; peduncle erect or incurved-ascending, 2–14 cm, shorter or much longer than leaf; axis 1–7 cm in fruit; bracts 1.5–7 mm; bracteoles 0–2. **Pedicels** declined to nodding, 1–4 mm. **Flowers**: calyx subcylindric, 9–11.5 (–14) mm, strigillose; tube 6.5–8.5 × 3.2–4.1 mm; lobes subulate, 1.5–4(–6) mm; corolla ochroleucous, uniformly colored or keel faintly purplish, 16–24 mm; banner recurved through ±45°, lanceolate-oblong or rhombic-oblanceolate, 18.3–23.5 mm; wings 16.4–19 mm; keel 13.6–15.4 mm, apex obtuse; ovules 18–25. **Fruits** pendulous; stipe 4–9 mm; straight or lunately incurved, oblong, 3-sided compressed, 18–35 × 3–6.5 mm, green, sometimes also purple, becoming brown, papery, glabrous; two-chambered, valves inflexed as complete septum 2–4 mm wide. **Seeds** 18–25. $2n = 22$.

Flowering May–Aug. Valleys, slopes, canyons, road cuts, and disturbed sites in a variety of soil types; 1450–2400[–3300] m; Great Basin Conifer Woodland, Petran Montane Conifer Forest; Apache, Gila, Navajo cos.; Colorado, New Mexico, Utah.

Astragalus scopulorum is found in thickets and on open hillsides of mountain brush, sagebrush, ponderosa pine, pinyon-juniper, and aspen-white fir communities across its range.

65. Astragalus sesquiflorus S. Watson. Proc. Amer. Acad. Arts 10:346. 1875. * [from Latin *sesqui*, one and one-half (one to two) flowered] SANDSTONE MILKVETCH Native

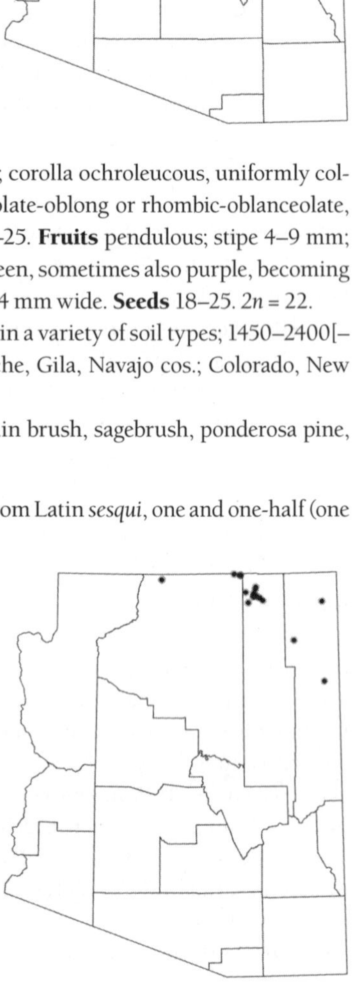

Perennials, densely mat- or cushion-forming, suffrutescent basally, caulescent, 5–28(–37+) cm; from caudex, or superficial root-crown; strigillose-canescent, hairs medifixed, sometimes twisted, especially on stems. **Stems** numerous, prostrate. **Leaves** 1–4(–6) cm, petioles subfiliform, wiry; stipules with fused sheathing throughout, broadly deltoid to ovate or lanceolate, 1.5–4.5 mm, subherbaceous becoming papery. **Leaflets** 7–11(–17), involute, elliptic to obovate, 1.5–10 × 0.6–2.6 mm, apex acute to obtuse, surfaces strigose. **Inflorescences** 1–4-flowered; peduncle widely diverging or incurved-ascending, 0.8–4.5 cm; axis 0.1–0.8 cm in fruit; bracts 1.2–3 mm; bracteoles 0–2. **Pedicels** ascending, 0.7–4 mm. **Flowers:** calyx campanulate, 3.7–5.5 mm, strigillose; tube 1.5–2.8 × 1.8–2.1 mm; lobes subulate, 1.9–3 mm; corollas pink-purple, 6–8 mm; banner strongly recurved through 80°, broadly ovate-cuneate to nearly circular-cuneate, 6–8 mm; wings 6.2–8.5 mm; keel 6–7.4 mm, apex porrectly acuminate or obtuse; ovules 7–10. **Fruits** spreading-ascending; sessile, sometimes attached to a gynophore 0–0.8 mm; incurved, obliquely oblong, 3-sided and laterally compressed, 8–10 × 3–4 mm, purple-mottled becoming straw-colored, papery, strigillose, hairs often contorted; one-chambered. **Seeds** 7–10.

Flowering Apr–Aug. Slopes and canyons on sandstone or sandy substrates; [1100–]1620–3100 m; [Great Basin Desertscrub], Great Basin Conifer Woodland, Petran Montane Conifer Forest, Petran Subalpine Conifer Forest; Apache, Coconino, Navajo cos.; Colorado, Utah.

Astragalus sesquiflorus is a striking little astragalus that forms distinctive mats or cushions on sand below escarpments in northeastern Arizona, southwestern Colorado, and southeastern Utah. With its wide elevation range, this species is found in mixed desert shrub, pinyon-juniper, and ponderosa pine or aspen to spruce-fir communities.

66. Astragalus sophoroides M.E. Jones. Zöe 2:12. 1891. * [resembling *Sophora nuttalliana* B.L. Turner] PAINTED DESERT MILKVETCH Native

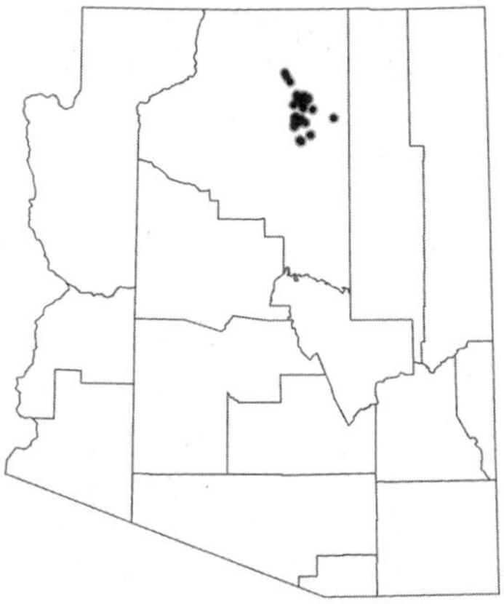

Perennials, low, loosely tufted, shortly caulescent, 2.5–20 cm; from slender woody taproot and forking caudex; densely strigose-strigillose, hairs medifixed, straight, appressed and subappressed, somewhat flattened. **Stems** several, slender, erect or decumbent-ascending. **Leaves** 3–8(–12) cm, petiolate; stipules clasping, with fused sheathing proximally or throughout, triangular, deltoid, or triangular-acuminate, (1–)2–5(–7) mm, scarious. **Leaflets** 7–13, flat or loosely folded, rather distant, narrowly oblong or linear-elliptic, 3–12(–21) × 2–4(–7.4) mm, apex subacute to obtuse, surfaces strigose. **Inflorescences** densely 7–17-flowered; peduncle narrowly ascending, (0.7–)1–3(–4) cm, shorter than the subtending leaf; axis (0.7–)1–4 cm in fruit; bracts 2–5 mm; bracteoles 0. **Pedicels** narrowly ascending, 0.5–0.7 mm. **Flowers:** calyx campanulate or turbinate-campanulate, 5–6.3 mm, white-strigillose; tube 3–3.6 × 2–2.6 mm; lobes lance-subulate, 2–3 mm; corolla reddish lilac, persistent around fruit, 8.5–10 mm; banner recurved through 45°, ovate-cuneate, 8.5–10 mm; wings 7.6–8.8 mm; keel 5.7–6.8 mm, apex bluntly deltoid; ovules 4–6. **Fruits** erect or narrowly ascending; sessile; straight or slightly incurved, oblong-ellipsoid, dorsiventrally compressed, 6.5–8 × 3–4 mm, straw-colored, faintly cross-reticulate becoming stiffly papery, strigillose; one-chambered. **Seeds** 4–6.

Flowering May–Jun. Slopes, cliffs, flats, arroyos, and floodplains on Moenkopi Sandstone or sandy substrates; 1280–1530 m; Great Basin Desertscrub; Coconino Co.; endemic.

Astragalus sophoroides occurs in localized colonies in a small area of east central Coconino County (Navajo Nation). This species appears to be closely related to *A. flavus* var. *candicans* (Barneby 1964a). It is of conservation concern.

67. Astragalus straturensis M.E. Jones. Contr. W. Bot. 8:19. 1898. * [from *stratura*, pavement, in allusion to the lode at Silver Reef mining area] PINE VALLEY MILKVETCH Native

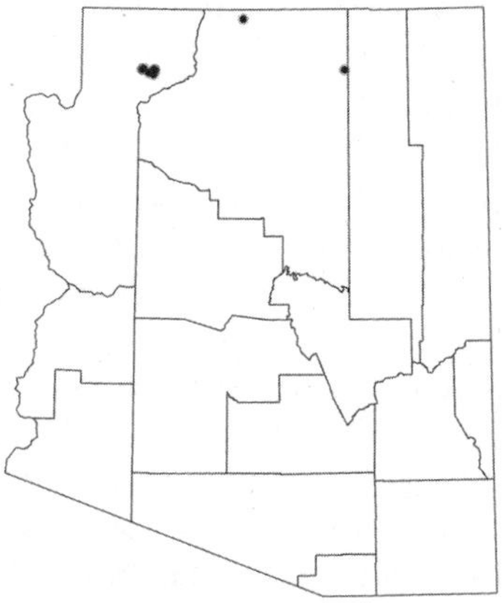

Perennials, slender, wiry, diffuse, caulescent, 13–36(–48) cm; from superficial or shallow, subterranean caudex; strigillose, hairs basifixed. **Stems** several, decumbent to ascending or erect. **Leaves** 3.5–10 cm, shortly petiolate or subsessile; stipules distinct throughout, deltoid or triangular, 1–2.5 mm, thinly herbaceous becoming papery. **Leaflets** (3–)9–19, usually folded, oblong to linear or oval, 3–13 × 1–5 mm, apex obtuse to retuse, somewhat bi-colored with surfaces strigose below (abaxially), glabrous above (adaxially). **Inflorescences** 9–25-flowered; peduncle ascending, 1.5–7.5 cm; axis 1.5–14(–17) cm in fruit; bracts 0.8–1.5 mm; bracteoles 0. **Pedicels** ascending to declined, 0.8–1.5 mm. **Flowers**: calyx campanulate, 3.5–4.2 mm, strigose; tube 2.5–3.5 × 1.9–2.2 mm; lobes triangular, 0.7–1.1 mm; corolla bright purple with white wing tips, 6.5–8.5 mm; banner recurved through 50–90°, ovate or nearly circular-cuneate, 6.3–8.5 mm; wings 6.9–8.4 mm; keel 5–6 mm, apex obtuse to sharply acute; ovules 10–13. **Fruits** pendulous, stipe 1.4–2.5 mm; incurved or straight, oblong, 3-sided compressed, 10–15(–17) × 2.2–3 mm, thinly fleshy, green becoming straw-colored, papery, strigose; ± two-chambered, septum 0.8–1 mm wide. **Seeds** 10–13. $2n = 22$.

Flowering late Apr–Jun. Rocky slopes, sometimes in gypsum soils; [1500–]1950–2020[–2500] m; [Great Basin Desertscrub], Great Basin Conifer Woodland, Petran Montane Conifer Forest; Mohave, Coconino cos.; Nevada, Utah.

The relationships of *Astragalus straturensis* lie with species in the section *Miselli* whose distributions extend southward into the Sierra Madre Occidental, Mexico. It is the only *Astragalus* in its region that is perennial, with pendulous, stipitate, dorsally-grooved fruits and purple flowers. It occurs in sagebrush, pinyon-juniper, ponderosa pine, and mountain brush communities.

68. Astragalus striatiflorus M.E. Jones. Proc. Calif. Acad. Sci., ser. 2, 5:643. 1895. * [with streaked or lined flowers]
ESCARPMENT MILKVETCH Native

Perennials, dwarf, loosely tuft-forming, subacaulescent to shortly caulescent, 1.5–6 cm; taproot slender becoming stout and deep set; from subterranean caudex (branched); densely canescent-hirsutulous, hairs basifixed. **Stems** few, erect or ascending, 0–5 cm, 2–30 cm underground, only tips produced aboveground, often with dried, persistent leaf rachises. **Leaves** 1–4 cm, shortly petiolate; stipules with fused sheathing throughout, ± obscuring stem, ovate, 2–4 mm, papery-scarious. **Leaflets** 5–13, loosely folded, ovate to obovate or oblanceolate, 1–7 × 0.8–2.5 mm, apex obtuse, mucronate, or emarginate, surfaces pilosulose. **Inflorescences** shortly 2–5-flowered; peduncle ascending, 1–3 cm; axis 0.2–1 cm in fruit; bracts 1.5–2.5 mm; bracteoles 0. **Pedicels** ascending, 1–1.5 mm. **Flowers:** calyx campanulate, 5.5–7 mm, hirsutulous; tube 3–4 × 2.8–3.4 mm; lobes subulate, 1.8–3 mm; corolla whitish and, usually, suffused or veined with lilac-purple, keel tip purple, 9–12 mm; banner recurved through 45°, nearly circular, 9.5–12 mm; wings 8.9–11 mm; keel 9–11 mm, apex long-attenuate and beak-like, with spherical stigma protruding; ovules 20–28. **Fruits** spreading; sessile; straight, ellipsoid to subglobose, bladdery-inflated, 12–18 × 8–15 mm, mottled and becoming straw-colored, papery, hirsute-hirsutulous with spreading hairs; two-chambered. **Seeds** 20–28.

Flowering Apr–Jul. Interdune valleys, sandy depressions on ledges, bars, and terraces in stream channels; 1500–2170 m; Great Basin Desertscrub, Great Basin Conifer Woodland; Coconino Co.; Utah.

The distribution of *Astragalus striatiflorus* is almost exclusively limited to the narrow escarpment between the Virgin and Paria rivers in southern Kane and eastern Washington counties, Utah, and the Paria Plateau in adjacent Coconino County, Arizona, in substrates derived from the Vermillion and White Cliffs. It occurs in pinyon-juniper, ponderosa pine, and sandy desert shrub communities.

69. Astragalus subcinereus A. Gray. Proc. Amer. Acad. Arts 13:366. 1878. * [from Latin *cineraceus*, grayish or ashy]
SILER'S MILKVETCH Native

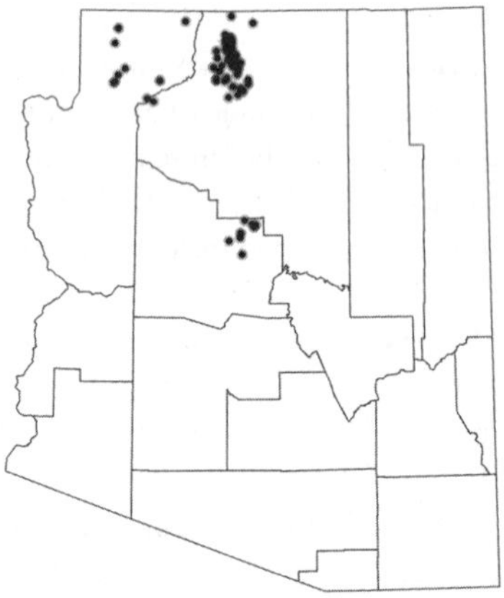

Perennials, diffuse or prostrate, slender or coarse and leafy, strongly caulescent, 14–70(–90) cm; from deeply subterranean branched caudex, or root-crown; strigillose-villosulous or hirsutulous, hairs stiff, straight, spreading or subretrorse, hairs basifixed. **Stems** few to several, prostrate to weakly ascending, often flexuous, (1–)2–10(–15) cm underground. **Leaves** 1.5–8.5 cm, mostly shortly petiolate, sometimes subsessile distally; stipules with fused sheathing proximally, obscurely fused or distinct distally, broadly lanceolate or deltoid, 1.5–6.5 mm, papery proximally, herbaceous or leaf-like distally. **Leaflets** 9–23, flat or folded, oblong to oblanceolate or obovate, 2–16 × 1–8.5(–10) mm, apex obtuse, emarginate, or retuse, surfaces villosulous below (abaxially), villosulous or glabrate above (adaxially). **Inflorescences** (5–)10–35-flowered; peduncle incurved-ascending or widely diverging, ± elongating 1.5–10 cm; axis 1–7 cm in fruit; bracts 1–3 mm; bracteoles 0 or 1. **Pedicels** ascending to declined, 0.5–2.5 mm. **Flowers**: calyx campanulate, 3.4–6.3 mm, villosulous; tube 2.3–3.6 × 1.9–3.2 mm; lobes subulate, 0.9–2.9 mm; corollas white to yellowish, and commonly suffused with lilac or purple, 6–11 mm; banner recurved through 45°, obovate-cuneate, rhombic-obovate, or subflabellate, (6.2–)6.6–9 mm; wings 6.1–8.7 mm; keel (5.7–)6–7.7 mm, apex broadly deltoid; ovules 10–20. **Fruits** spreading to declined; subsessile, with stipe-like neck 0.3–1 mm; straight, ovoid, ovoid-ellipsoid, narrowly to broadly ellipsoid, bladdery-inflated or turgid, nearly cylindrical to dorsiventrally compressed, 12–27 × 3.5–13 mm, green or purplish, brightly red-mottled, becoming papery or papery-membranous, sparsely villosulous; one-chambered. **Seeds** 10–20.

Flowering May–Sep. Level terrain and slopes in various soils, open meadows and beneath trees in ponderosa pine, white fir, pinyon-juniper, aspen, and sagebrush communities; (1400–)1600–2690 m; Great Basin Desertscrub, Great Basin Conifer Woodland, (Madrean Evergreen Woodland), Petran Montane Conifer Forest; Coconino, Mohave, Yavapai cos.; Nevada, Utah.

Astragalus subcinereus is variable throughout its range. As Barneby (1964a) explained, there is no question that the type of *A. subcinereus* represents the Arizona phase of what was described as *Astragalus sileranus* by M.E. Jones, and not the taxon now recognized as *Astragalus fucatus* Barneby. Part of the confusion probably originated from an abnormality in the type specimen; the enlarged calyces of later flowers apparently resulted from infection by a rust fungus. Much of the material in southern Utah differs from Coconino and Mohave County, Arizona, plants in having broader leaflets, longer stems and more firmly-walled fruits.

70. Astragalus tephrodes A. Gray. Smithsonian Contr. Knowl. 5(6):45. 1853. * [from Greek *tephros*, ash-colored, gray] ASHEN MILKVETCH

Perennials, low, loosely or densely tuft-forming, acaulescent or shortly caulescent, 10–48 cm; from caudex, branched, branches often with conspicuous thatch of dried, persistent leaf rachises; pubescence extremely variable, strigillose, pilosulose, or villosulous, hairs basifixed. **Stems** obsolete or several, prostrate, 0–15(–25) cm, internodes to 2(–2.5) cm. **Leaves** 4–24 cm, petiolate; stipules distinct throughout or obscurely fused proximally, deltoid, ovate, or lanceolate, 2–15 mm, submembranous becoming papery-scarious. **Leaflets** 11–27(–31), flat or loosely folded, oblanceolate, obovate, obovate-cuneate, rhombic, rhombic-ovate, or rhombic-elliptic, rarely nearly circular, 2–27 × 1.3–7.5(–13) mm, apex usually obtuse, acute, or subacute, rarely, emarginate, surfaces pubescent or, sometimes, glabrescent or glabrous above (adaxially). **Inflorescences** loosely flowered, with (9–)10–25(–35) flowers; peduncle incurved-ascending or erect, 4–40 cm; axis (1.5–)2–20 cm in fruit; bracts 1.5–11 mm; bracteoles usually 0(–2). **Pedicels** ascending, 0.6–3.4 mm. **Flowers**: calyx cylindric to deeply campanulate, (5–)6.4–14 mm, loosely strigillose, pilosulose, or villous; tube (3.5–)4.5–10.2 × 3.2–5.2 mm; lobes subulate or triangular-subulate, 1.2–3.8 mm; corolla usually pink-purple to dull lilac, rarely dingy white, 11.8–24 mm; banner recurved through 45°, obovate-cuneate, broadly oblanceolate, rhombic-elliptic, or spatulate, 11.8–24 mm; wings (10.8–)11.8–22.7 mm; keel 10.2–20.1 mm, apex obtuse; ovules 24–36. **Fruits** ascending; sessile; lunately incurved or straight proximally, beak incurved, obliquely ovoid-acuminate, lanceolate-, oblong-, or ovoid-ellipsoid, not inflexed, or if so then septum narrow and obscure, not over 1 mm wide, strongly dorsiventrally compressed to flattened, 13–34 × 5–10(–12) mm, thinly to thickly fleshy, green or purplish becoming straw-colored or brownish, thinly to stiffly leathery, or semi-woody, usually densely to sparsely strigillose or pilosulose, rarely glabrous; ± one-chambered. **Seeds** 24–36.

Varieties 3 (2 native in the flora): sw United States, Mexico.

Astragalus tephrodes consists of two rather widely distributed and intergrading varieties (*brachylobus*, *tephrodes*) and one localized to northwest Arizona (*chloridae*). Barneby (1947, 1964a) discussed the taxonomic complexity of the species and generally revisited the situation in 1964, describing a more or less uninterrupted clinal variation across Arizona to western areas of distribution in southwestern Utah, southern Nevada, and southeastern California, resulting in a classification that is necessarily arbitrary.

Barneby discusses reports of var. *tephrodes* in Arizona but indicates they are probably just seasonal variants of var. *brachylobus*. One specimen identified as var. *tephrodes*, from Maricopa County, cites GPS location that is inconsistent with collection locality.

1. Peduncle (4–)5–15(–21) cm; axes of inflorescences (1.5–)2–8(–11) cm in fruit; plants broadly distributed from northwest to southeast Arizona ____ **70a. Astragalus tephrodes** var. **brachylobus**

1′ Peduncle 13–40 cm (stout); axes of inflorescences (4–)7–20 cm in fruit; plants of northwest Arizona ____ **70b. Astragalus tephrodes** var. **chloridae**

70a. Astragalus tephrodes A. Gray var. **brachylobus** (A. Gray) Barneby. Amer. Midl. Naturalist 37:466. 1947. * [from Greek *brachy*, short and *lobus*, lobed] Native

Astragalus shortianus Nuttall var. *brachylobus* A. Gray, Proc. Amer. Acad. Arts 13:367. 1878

Stems 0–8 cm. **Leaves** 4–16 cm; stipules 2.5–11 mm. **Leaflets** 11–27(–31), mostly flat, usually obovate-cuneate, oblanceolate, or rhombic-elliptic, rarely nearly circular, (3–)4–17 mm, apex obtuse, acute, or emarginate, surfaces villosulous, sparsely strigillose, or glabrate below (abaxially). **Inflorescences** with 10–25(–35) flowers; peduncle (4–)5–15(–21) cm; axis (1.5–)2–8(–11) cm in fruit. **Flowers**: calyx 8.8–12.7(–14) mm, loosely strigillose to pilosulose or, sometimes, villous; tube 7.1–10 × 3.4–4.8 mm; lobes 1.7–2.8 mm; corolla (14–)18–24 mm; banner (14–)18–24 mm; keel 14.7–20.1 mm. **Fruits** oblong-ellipsoid to lanceolate-ellipsoid or, sometimes, ovoid-acuminate, 17–30 × 6–10 mm, beak 0.5–2 mm, somewhat fleshy becoming stiffly leathery or semi-woody, not rugulose, usually strigillose or pilosulose, sometimes glabrous. **Seeds** 24–35.

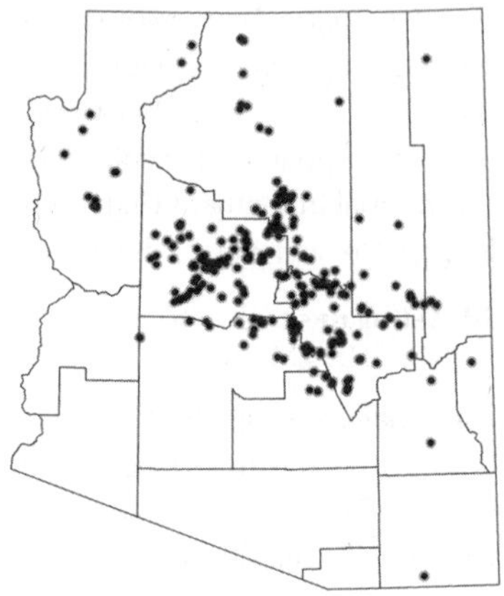

Flowering Mar–Jun. Ridges, slopes, canyons, and valleys on soils derived from volcanic, granitic, or sedimentary bedrock (including limestone); (400–)1000–2300(–2590) m; (Sonoran Desertscrub), Semidesert Grassland, Plains Grassland, Interior Chaparral, Madrean Evergreen Woodland, Great Basin Conifer Woodland, Petran Montane Conifer Forest; Apache, Cochise, Coconino, Gila, Graham, Greenlee, Maricopa, Mohave, Navajo, Pinal, Yavapai cos.; California, Nevada, New Mexico; Mexico (Sonora and presumably Chihuahua).

Variety *brachylobus* is locally plentiful along the Mogollon Rim and in the central mountains south to the Gila River basin. The variety passes gradually into var. *tephrodes* in west-central New Mexico.

70b. Astragalus tephrodes A. Gray var. **chloridae** (M.E. Jones) Barneby. Mem. New York Bot. Gard. 13:642. 1964. * [of Chloride, Arizona] CHLORIDE MILKVETCH Native

Astragalus remulcus M.E. Jones var. *chloridae* M.E. Jones, Rev. N.-Amer. Astragalus, 210. 1923

Stems 2–15 cm. **Leaves** 10–24 cm; stipules 2–15 mm. **Leaflets** 19–25, flat, rhombic, rhombic-ovate, obovate, or oblanceolate, 6–26 mm, apex usually acute, sometimes obtuse, surfaces sparsely strigillose or glabrate below (abaxially). **Inflorescences** 13–24-flowered; peduncle stout, erect at anthesis, 13–40 cm, becoming weighed down by fruits; axis (4–)7–20 cm in fruit. **Flowers:** calyx 11.5–14 mm, pilosulose or loosely strigillose; tube 8.3–10.2 × 3.8–5.2 mm; lobes 2.1–3.8 mm; corolla 16.8–20.4 mm; banner 16.8–20.4 mm; keel 15.2–18.2 mm. **Fruits** lanceolate- or oblong-ellipsoid, 22–34 × 6–10(–12) mm, beak 3–6 mm, stiffly leathery or semi-woody, ± rugulose, strigillose-pilosulose. **Seeds** 28–36. $2n = 22$.

Flowering Mar–May. Valleys and slopes; 900–1400 m; Mohave Desertscrub, Great Basin Conifer Woodland; Mohave Co.; Nevada.

As strictly perceived, var. *chloridae* is restricted to the Cerbat Mountains, Mohave County, Arizona, and the Newberry Mountains, Clark County, Nevada. Isely (1998) allowed a somewhat broader range in Mohave County. A specimen from Yavapai County (Welsh 2007) may belong with this variety.

71. Astragalus tetrapterus A. Gray. Proc. Amer. Acad. Arts 13:369. 1878. * [from Greek *tetra*, four and *pteron*, wing] FOUR-WING MILKVETCH Native

Perennials, caulescent, relatively robust, 10–35 cm; from subterranean caudex; strigillose with basifixed hairs, varying from silvery or cinereous, to dark green and glabrous. **Stems** several, erect, ascending, to becoming diffuse or decumbent, 2–8 cm underground. **Leaves** 1.5–8.5 cm, shortly petiolate; stipules distinct throughout, triangular, 2–5.5 mm, herbaceous. **Leaflets** 9–23, flat or folded, linear, narrowly oblong, or elliptic, 1–33 × 0.3–3.2 mm, apex obtuse to acute, surfaces strigose or glabrous; terminal leaflet sometimes decurrent distally, continuous with rachis. **Inflorescences** 6–15-flowered; peduncle ascending, 1–6.5 cm; axis 1–4 cm in fruit; bracts 1.5–3.5 mm; bracteoles 0–2. **Pedicels** ascending, 1.4–4.3 mm. **Flowers**: calyx cylindric, 5.5–8.7 mm, strigose; tube 4.7–7 × 2.3–3.1 mm; lobes subulate, 0.8–2.8 mm; corolla white, lilac or faintly pink-tinged, sometimes lightly suffused with pink-purple upon drying, keel tip faintly to darkly purple, 15–19 mm; banner recurved through 45°, oblanceolate or rhombic-elliptic, (13.5–)15–19 mm; wings 11.6–16.5 mm; keel 10.2–13 mm, apex obtuse; ovules 28–38. **Fruits** pendulous; sessile; strongly incurved or coiled, obliquely oblong, sharply 4-sided compressed, 20–40 × 6–10 mm, fleshy green or purple-mottled becoming straw-colored or brownish, stiffly papery, smooth and lustrous, glabrous or strigose; one-chambered. **Seeds** 28–38.

Flowering Apr–Jul. Slopes, valleys, and gullies, often on sandy alkaline soil, clays and limestone; [1000–]1350–1700[–2200] m; Great Basin Desertscrub, Great Basin Conifer Woodland; Mohave Co.; Idaho, Nevada, Oregon, Utah.

Astragalus tetrapterus is characterized by its unique, pendulous pod, fleshy when fresh, but collapsing inward along the sutures and becoming four-winged upon ripening (drying). The species contains nitrotoxins that are poisonous to livestock (James and Welsh 1992), but it is seldom sufficiently abundant as to result in large-scale loss. The young fruit is fleshy, with the dorsal suture essentially flat and the ventral one raised in a low ridge; it becomes four-winged upon drying. It is found on gullied, barren knolls and valley floors in sagebrush and pinyon-juniper communities.

72. Astragalus thurberi A. Gray. Pl. Nov. Thurb. 312. 1854. * [for George Thurber, 1821–1890, American botanist] THURBER'S MILKVETCH Native

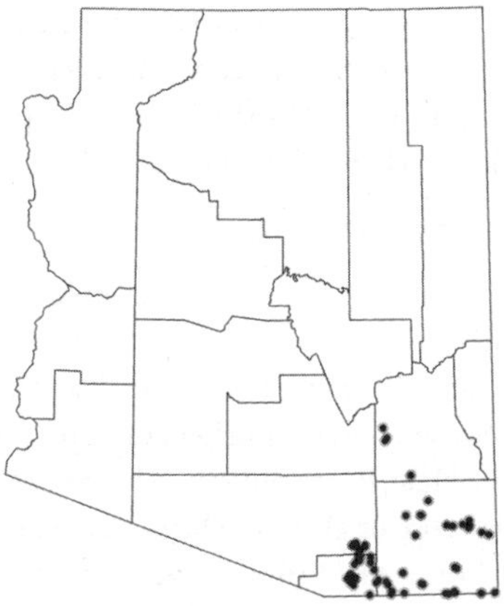

Perennials, short-lived, often flowering first year, forming low, diffuse bushy clumps, caulescent, somewhat coarse, (7–)15–40 cm; from superficial root-crown; strigillose, hairs basifixed. **Stems** several or numerous, decumbent to ascending. **Leaves** (2–)4–10(–11.5) cm, petiolate; stipules distinct throughout, deltoid-ovate or triangular, (2–)3–8 mm, papery proximally, submembranous distally. **Leaflets** (9–)13–21, flat or loosely folded, oblong-elliptic, oblanceolate, or linear-elliptic, 5–15(–18) × 1.2–4(–6) mm, apex obtuse, truncate and mucronulate, or shallowly retuse, surfaces strigillose below (abaxially), glabrous above (adaxially). **Inflorescences** with (7–)10–25(–32) flowers, becoming loosely spreading in age; peduncle erect or incurved-ascending, 0.3–0.9 cm; axis (2–)3–7(–11) cm in fruit; bracts 1–2.4 mm; bracteoles 0–2. **Pedicels** spreading, 0.5–1.8 mm. **Flowers:** calyx campanulate, 3.3–5 mm, strigillose; tube 2.2–2.9 × 1.7–2.2(–2.5) mm; lobes subulate or lanceolate-subulate, (1–)1.3–2.3 mm; corolla reddish lilac or reddish purple, 5.7–7.1 mm; banner recurved through 50°, broadly ovate or nearly circular, 5.7–7.1 mm; wings 5.1–6.7 mm; keel 4.7–5.6 mm, apex obtusely deltoid, sometimes obscurely beak-like; ovules 8–11. **Fruits** spreading or declined; sessile; straight or slightly incurved, broadly and plumply ovoid- or obovoid-ellipsoid, obovoid, or nearly globose, bladdery-inflated, 6–13 × 6–10 mm, thin, green or minutely purple-dotted becoming straw-colored, papery, sparsely strigillose; one-chambered. **Seeds** 8–11. $2n = 22$.

Flowering Mar–May(–Jun). Slopes, plains, valleys, floodplains, roadsides, and other disturbed sites on dry sandy or gravelly soils and on semi-stabilized dunes; 900–1900 m; Chihuahuan Desertscrub, Semidesert Grassland, Plains Grassland, Madrean Evergreen Woodland; Cochise, Graham, Pima, Santa Cruz cos.; New Mexico; Mexico (Chihuahua, Sonora).

With its compact, small, rounded, inflated fruits and small flowers, *Astragalus thurberi* is one of the more distinctive species of the large and difficult section *Inflati*. It may be readily told apart from the similar *A. allochrous* and *A. wootonii* with which it sometimes grows by its much smaller, rounded fruits. It contains swainsonine and is considered poisonous (James and Welsh 1992). *Astragalus thurberi* occurs in creosote bush and mesquite, oak brush and juniper, and in mesquite-grasslands.

73. Astragalus titanophilus Barneby. Amer. Midl. Naturalist 55:477. 1956. * [chalk- or lime-loving] LIMESTONE MILKVETCH Native

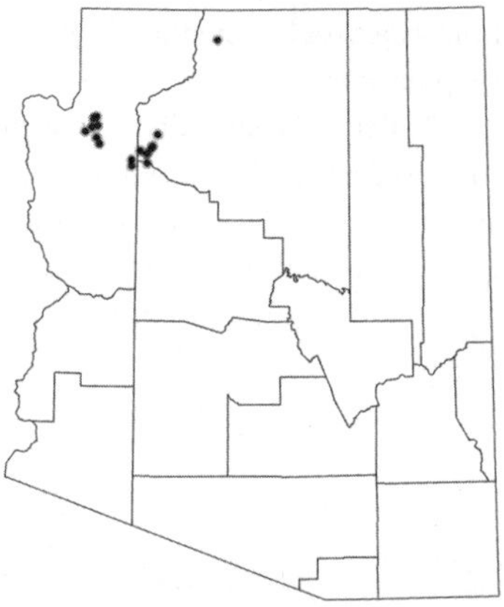

Astragalus convallarius Greene var. *foliolatus* Barneby

Perennials, caulescent, slender, wiry, 15–30 cm; from shallow, subterranean caudex; gray-strigillose, hairs fine, basifixed, sometimes flattened and scale-like. **Stems** few, usually 3–6, prostrate to ascending; 1–7 cm underground. **Leaves** 2.5–9 cm, petiolate; stipules distinct throughout, oblong-obovate, deltoid, or triangular, 2–8 mm, papery proximally, firm membranous distally, apex obtuse or acuminate. **Leaflets** 7–15, jointed to rachis throughout, flat or involute, narrowly linear to linear-oblong, 2–12(–16) mm, apex obtuse to emarginate, surfaces strigillose. **Inflorescences** loosely (3–)7–20-flowered; peduncle ascending or widely diverging and incurved, 3.5–9.5 cm, surpassing leaves; axis 2–14 cm in fruit; bracts 0.8–2 mm; bracteoles usually 0. **Pedicels** spreading or ± decurved, 2–6 mm. **Flowers**: calyx campanulate, 4–5.9 mm, strigose; tube 3.2–4.5 × 2.6–3.2 mm; lobes subulate or triangular-subulate, 0.5–1.4 mm; corolla ochroleucous tinged or veined with lilac, or pale pinkish white, 7.3–8.5 mm; banner strongly recurved through 100–130°, broadly ovate-cuneate or nearly circular-cuneate, 7.3–8.5 mm; wings 7–8.5 mm; keel 7–8 mm, apex bluntly triangular; ovules (16–)20–24. **Fruits** declined or pendulous; subsessile, base contracted into stout, stipe-like neck 1–1.5 mm; straight or gently incurved or decurved, narrowly oblong to ellipsoid, strongly laterally compressed, (18–) 22–35 × 3.2–4.6(–5) mm, green or purplish becoming brown-straw-colored, stiffly papery, strigillose; one-chambered. **Seeds** (16–)20–24.

Flowering Apr–Jul. Sandy or gravelly flats, low hills with fine-textured soils derived from decomposing limestone or rarely basalt; (1300–)1470–1980 m; (Great Basin Desertscrub), Great Basin Conifer Woodland; Coconino, Mohave cos.

Astragalus titanophilus superficially resembles *A. convallarius* but is distinguished by its free lower stipules and typically odd-pinnate leaves. *Astragalus titanophilus* is considered closely related to *A. xiphoides* but additional investigation is warranted. The species is found primarily in pinyon-juniper woodland in a restricted area of northwestern Arizona but can be locally common.

74. Astragalus toanus M.E. Jones. Zoë 3:296. 1893. * [of the Toano Mountains, Nevada] TOANO MILKVETCH

Perennials, selenophyte, rush-like, sparsely leafy, caulescent, robust, 15–50 cm; from shallow, subterranean or superficial caudex; usually strigillose, rarely glabrous, hairs basifixed. **Stems** numerous, erect or ascending, forming bushy clumps. **Leaves** unifoliolate distally, 2–10 cm, petiole stiff, channeled; stipules with fused sheathing proximally, distinct and ovate or deltoid distally, 1.5–6.5 mm, papery-scarious proximally, herbaceous distally. **Leaflets** (1 or) 3–13, involute, linear-filiform to linear-oblong, 3–30 × 1.4–2.5 mm, apex obtuse to acute, surfaces strigose below (abaxially), strigose or glabrous above (adaxially); terminal leaflet of compound leaves sometimes decurrent distally, continuous with rachis. **Inflorescences** loosely 7–35-flowered; peduncle strict, 6–25 cm; axis 3–30 cm in fruit; bracts 1–3 mm; bracteoles 0–2. **Pedicels** ascending, 0.8–3.5 mm. **Flowers:** calyx short-cylindric, 4.6–8 mm, strigose; tube 4.1–6.4 × (2.6–)3–3.9 mm; lobes subulate, 0.5–2 mm; corolla bright pink-purple and wing tips paler or white, or greenish yellow to whitish, (13–)14–20 mm; banner recurved through 40–45°, rhombic-ovate, (13–)15–20 mm; wings (11.6–)13.5–17 mm; keel (9.5–)10.5–13.5 mm, apex obtuse; ovules 14–26. **Fruits** erect; sessile; straight or slightly incurved, oblong, slightly compressed laterally, 13–25 × 3.7–9 mm, green sometimes purple-mottled becoming straw-colored, stiffly leathery to somewhat woody, glabrous or strigose, persistent and dehiscent on plant; one-chambered. **Seeds** 14–26. $2n = 22$.

Varieties 2 (1 native in the flora): w United States.

Astragalus toanus forms broom-like, apparently leafless tufts of stiffly clustered stems, the fruits remaining into late summer, the valves bent outward in a widely gaping dehiscence. The species is attractive in flower, bearing loose spikes of relatively large, brightly colored flowers. The plants are malodorous (selenium-scented).

74a. Astragalus toanus M.E. Jones var. **scidulus** S.L. Welsh & N.D. Atwood. Rhodora 95:403, fig. 11. 1993. * DIAMOND BUTTE MILKVETCH Native

Leaflets (1 or) 11–13. **Flowers:** corolla greenish yellow or whitish, 14–16.5 mm. **Fruits** 6–9 mm wide.

Flowering Apr–May. Flats in selenium-rich red sandy clay soils derived from Moenkopi Formation; 1490–1600 m; Great Basin Desertscrub; Mohave Co.; endemic.

Variety *scidulus* is distantly isolated from var. *toanus* in a tiny area of the Arizona Strip, Mohave County, Arizona. As far as known, it is confined to an area along a single wash at the base of a small mountain. It occurs in mixed desert shrub and grass communities. This taxon is of conservation concern.

75. Astragalus troglodytus S. Watson. Proc. Amer. Acad. Arts 20:362. 1885. * [from Greek *trogle*, hole and *dyein*, go in, dwelling in caves] CLIFF DWELLER'S MILKVETCH Native

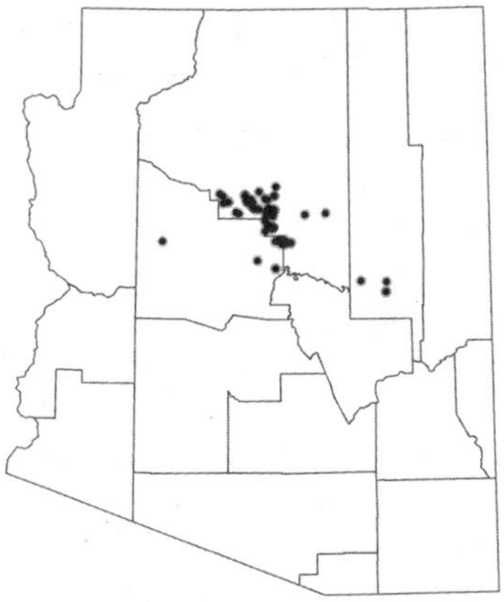

Perennials, tuft-forming, acaulescent or subacaulescent, 6–25 cm; from superficial caudex or knotty root-crown; green or gray-green below silky-pilose pubescence, hairs basifixed. **Stems** obscured by stipules. **Leaves** 3.5–11.5 cm, petiole slender; stipules distinct throughout, ovate, triangular, or lanceolate, 3–11 mm, papery-membranous. **Leaflets** (7–)11–19, flat or loosely folded, obovate, obovate-cuneate, rhombic, or broadly oblanceolate, 4–13 × 2–7 mm, apex subacute, obtuse and mucronulate, or obtuse, surfaces pilose below (abaxially), glabrous above (adaxially). **Inflorescences** densely 10–30-flowered, forming compact globose or oval clusters; peduncle scape-like, incurved-ascending, 4.5–15(–20) cm; axis 0.5–2 cm, not elongating in fruit; bracts 3–8.5 mm; bracteoles 0–2. **Pedicels** ascending to spreading becoming declined proximally, 0.4–1 mm. **Flowers:** calyx campanulate, 5.9–8.2 mm, sparsely hirsute; tube 3.4–5.2 × 2.3–2.9 mm; lobes lanceolate-linear to bristle-like, 1.9–3 mm; corolla dull reddish-purple, keel spotted, 9.5–11.4 mm; banner recurved through 45°, oval-ovate, 9.5–11.4 mm; wings 9.2–11.2 mm; keel 7.9–9.1 mm, apex blunt; ovules 8–12. **Fruits** in oblong or globose clusters, spreading and ascending; sessile; straight, plumply ovoid or nearly spherical, cylindrical or somewhat laterally compressed, 5–7 × 4–4.5 mm, thinly fleshy, green or purple-tinged becoming straw-colored or brownish, stiffly papery, sparsely long-hirsute; one-chambered. **Seeds** 8–12.

Flowering Apr–Jun(–Aug). Dry slopes and flats in ponderosa pine forests, often on basalt rocks or in clay soil; 1580–2260 m; Great Basin Conifer Woodland, Petran Montane Conifer Forest; Coconino, Navajo, Yavapai Cos; endemic.

Astragalus troglodytus is restricted to the Flagstaff Plateau and adjacent Mogollon Rim. This species is endemic to Arizona and is of conservation concern.

76. Astragalus vaccarum A. Gray. Smithsonian Contr. Knowl. 5(6):43. 1853. * [From Latin *vacca*, a cow, for Cow Springs, New Mexico] COW SPRING MILKVETCH Native

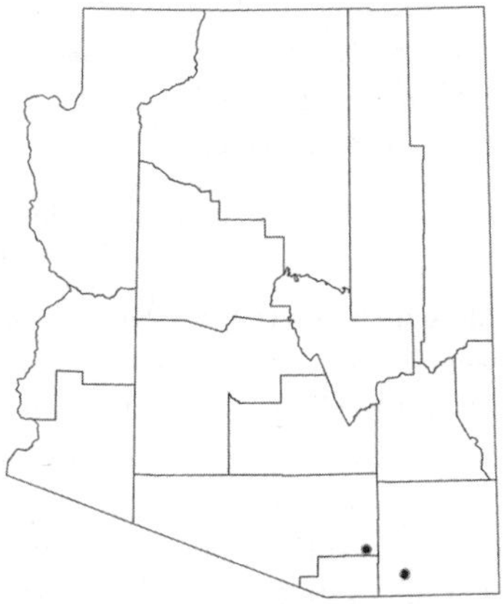

Perennials, caulescent, slender, erect or diffuse, (10–)15–45 cm; from superficial root-crown; strigillose, hairs basifixed. **Stems** few, prostrate to decumbent or ascending. **Leaves** 4–12 cm, shortly petiolate or subsessile, petiole slender; stipules distinct throughout, lanceolate or deltoid, 2–8 mm, subherbaceous or papery. **Leaflets** (9–)11–21(–23), flat or loosely folded, linear-oblong to narrowly elliptic, (3–)5–20(–24) × 1.5–3.2 mm, apex acute to obtuse, surfaces strigillose below (abaxially), glabrous or glabrate above (adaxially). **Inflorescences** with (10–)15–50 flowers, compact early, elongating with age; peduncle erect or incurved-ascending, (2–)5–12 cm; axis (1.5–)3–8 cm in fruit; bracts 1–3 mm; bracteoles 0–2. **Pedicels** arched downward, 0.3–1.2 mm. **Flowers:** calyx campanulate, 2.5–3.6 mm, strigillose; tube 1.7–2 × 1.6–2 mm; lobes subulate, 0.8–1.8 mm; corolla purple or purple-tinged, becoming pale or appearing ochroleucous when dry, to whitish, 4.2–6.2 mm; banner recurved through 45–90°, usually ovate-cuneate, 4.2–6.6 mm; wings 4.9–6.4(–7) mm; keel 3.7–5 mm, apex round; ovules 6–10. **Fruits** deflexed; sessile; incurved, narrowly lanceolate or linear-ellipsoid, 3-sided compressed, 6–12 × 1.3–2 mm, thin, pale green becoming straw-colored, papery, strigillose; two-chambered or incompletely two-chambered. **Seeds** 6–10. $2n = 28$.

Flowering (Apr–)Jun–Oct. Dry floodplains, benches in oak-pine forests, and disturbed sites; 1200[–2500] m; Semidesert Grassland; Cochise, Pima cos.; New Mexico; Mexico (Chihuahua, Coahuila, Durango, Sonora).

Astragalus vaccarum is a northern member of the primarily Mexican section *Micranthi*, long known from only a few old collections from southeastern Arizona and southwestern New Mexico. Contrary to reports of its probable extinction north of the border (Isely 1998, Welsh 2007), it was rediscovered in Hidalgo County, New Mexico, in the 1980s. It is fairly common in northern Mexico where is occurs in oak-pine woodlands. A recent Arizona collection (*E. Makings*

1646, 24 Sep 2003, ASU) was made along the San Pedro River in Cochise County, and a collection (*D. Wolkis 515*, 13 Sep 2009, ASU) was made in the Las Cienegas National Conservation Area in Pima County.

77. Astragalus wingatanus S. Watson. Proc. Amer. Acad. Arts 18:192. 1883. * [of Fort Wingate, New Mexico] FORT WINGATE MILKVETCH Native

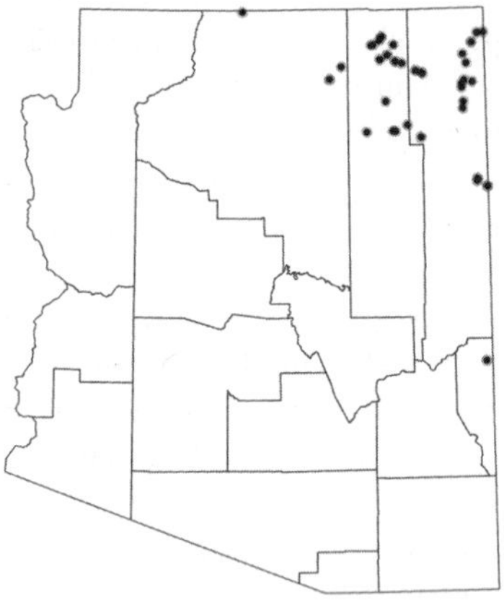

Perennials, usually slender and wiry, often sparsely leafy, clump-forming, caulescent, (10–)15–45(–60) cm; from subterranean caudex, branched, sometimes branches subrhizomatous; strigillose, hairs basifixed. **Stems** several or numerous, spreading-ascending, often distinctly flexuous, (0–)1.5–15 cm underground. **Leaves** 1.5–6.5 cm, shortly petiolate or subsessile distally; stipules with fused sheathing proximally, fused or distinct distally, lanceolate, 1.5–5 mm, scarious proximally, subherbaceous distally. **Leaflets** (3–)7–15(–17), flat or involute, linear to narrowly oblong, elliptic, or oblanceolate, 3–18 × 0.4–3.6 mm, apex acute, obtuse, or retuse, surfaces strigose or glabrous below (abaxially), glabrous above (adaxially); margins often ciliate; terminal leaflet sometimes decurrent distally, continuous with rachis. **Inflorescences** very loosely or remotely 7–35-flowered; peduncle erect or incurved-ascending, stiff, 2–14 cm; axis 3–8 cm in fruit; bracts 0.5–2 mm; bracteoles 0–2. **Pedicels** ascending, 0.8–3 mm. **Flowers**: calyx campanulate, 2.5–3.7 mm, black-strigose; tube 1.5–2.6 × 1.2–1.8(–2) mm; lobes triangular-subulate, 0.4–1.4 mm; corolla lilac or pink-purple, with white wing tips, or whitish with banner and wing tips purple, 5.5–8 mm; banner recurved through 50°, broadly rhombic-obovate or nearly circular, 5.6–8 mm; wings 4.3–7.1 mm; keel 3.5–5.4 mm, apex obtuse or acute-triangular; ovules 4–8. **Fruits** deflexed; subsessile to shortly stipitate, stipe 0–1.7 mm; straight or slightly curved, ellipsoid to oblong, strongly laterally compressed, 2-sided, 9–15 × 3–4.5 mm, thinly fleshy, greenish and purple-speckled becoming straw-colored, papery, glabrous; one-chambered. **Seeds** 4–8. $2n = 22$.

Flowering mid Apr–Jul. Barren clay slopes, valleys, and canyons in various soils; 1500–2150[–2400] m; Great Basin Desertscrub, Great Basin Conifer Woodland, occasional in Petran Montane Conifer Forest; Apache, Coconino, Greenlee, Navajo cos.; Colorado, New Mexico, Utah.

Astragalus wingatanus resembles *A. tenellus* in its small flowers and laterally flattened fruits. The subterranean caudex and elongated fruiting axis are diagnostic for *A. wingatanus*, and closely resemble those of *A. flexuosus* var. *diehlii*. *Astragalus wingatanus* is found in pinyon-juniper woodland, mixed desert shrub, salt desert shrub, and less commonly in mountain brush communities.

78. Astragalus wootonii E. Sheldon. Minnesota Bot. Stud. 1:138. 1894 (as *Astragalus Wootoni*) * [for Elmer O. Wooton, 1865–1945, American botanist] WOOTON'S MILKVETCH

Annuals (winter) or biennials, low, often coarse, caulescent (10–)15–50 cm; from superficial root-crown; strigillose, hairs basifixed. **Stems** sometimes solitary, commonly decumbent to incurved-ascending or almost prostrate. **Leaves** (2–)4–10(–12) cm, shortly petiolate or subsessile; stipules usually distinct, very rarely fused proximally, triangular, (1.5–)2.5–7(–10) mm, submembranous becoming papery. **Leaflets** (7–)11–19[–23], usually folded, rarely flat, narrowly oblanceolate, linear-oblong, or oblong-obovate, 5–20 × 1–3(–5.4) mm, apex retuse-truncate or obtuse, often callous-mucronulate, surfaces strigillose below (abaxially), strigillose or glabrous above (adaxially). **Inflorescences** loosely flowered, with [2–](3–)5–10(–15) flowers; peduncle incurved-ascending, (0.5–)1.5–5.5(–7) cm; axis (0.5–)1–4(–5) cm in fruit; bracts 1–3.2 mm; bracteoles 2. **Pedicels** ascending to spreading or declined, 1–3.5 mm. **Flowers**: calyx campanulate or turbinate-campanulate, 4.3–6.4 mm, strigillose or villosulous; tube 2.1–2.9(–3.2) × 1.8–2.3 mm; lobes lanceolate-subulate, 2–3.5 mm; corolla whitish, sometimes tinged pink or lavender, or pale to vivid reddish-lilac, 4.6–7.5 mm; banner recurved through 50°, ovate-cuneate, obovate-cuneate, or flabellate, 4.6–7.5 mm; wings (4.1–)4.9–7.5 mm; keel (4.1–)4.4–6.4 mm, apex usually broad and blunt, rarely triangular and subacute, sometimes obscurely beak-like; ovules (10–)13–21. **Fruits** spreading or declined, usually humistrate; sessile; straight or slightly incurved, broadly and subsymmetrically, or somewhat obliquely, ovoid, ovoid-ellipsoid, ellipsoid, or nearly circular, bladdery-inflated, sutures subequally convex as seen in profile,(10–)15–37(–43) × (8–)12–20 mm, beak short and obscure, not well differentiated from body, thin, green or purplish-tinged, rarely lightly mottled, becoming straw-colored, papery, surface lustrous, sparsely strigillose, (subvillosulous, or glabrate); one-chambered. **Seeds** (10–)13–21. $2n = 22$.

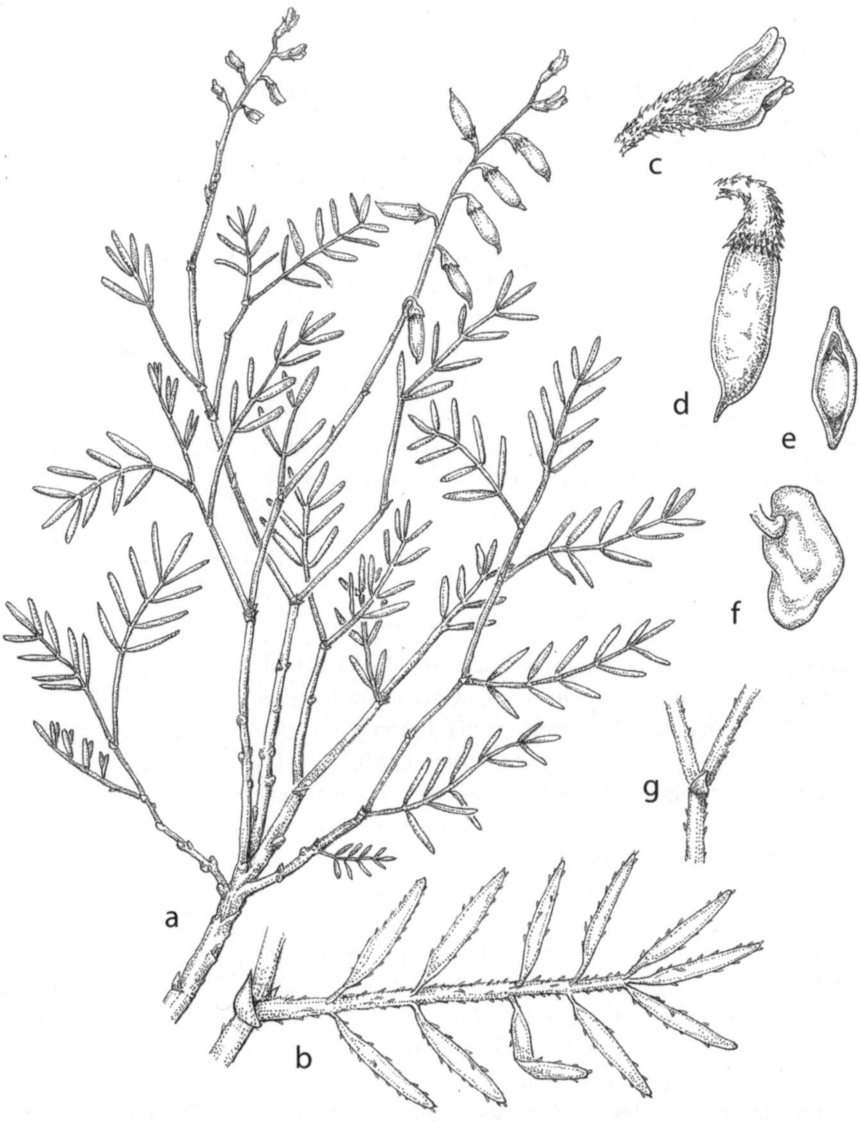

***Astragalus wingatanus*: a. habit; b. leaf; c. flower; d. fruit; e. fruit cross-section; f. seed; g. stem node. Illustrator: Jo Ann Loza.**

Varieties 2 (1 native in the flora): sw United States, Mexico.

Barneby (1964a) provides a discussion of the variation within *Astragalus wootonii*, a broadly distributed taxon, and its morphological similarity to *A. allochrous*. *Astragalus wootonii* differs from *A. allochrous* by its usually less-robust habit, flower color, racemes with rarely more than 10 flowers, and fruits which are subsymmetrically elliptical to globose and beakless or nearly so. Isely (1983) points to the sympatry of *Astragalus wootonii* and *A. allochrous*. He regarded *A. wootonii* as intergradient with *A. allochrous* (with which it shares most of its range) and included it within that species as var. *playanus* (M.E. Jones) Isely. Variety *wootonii* is widespread in Arizona, while variety *candollianus* (Kunth) Barneby is disjunct in central and southeastern Mexico.

78a. Astragalus wootonii E. Sheldon var. **wootonii** * Native

Astragalus allochrous var. *playanus* (M.E. Jones) Isely

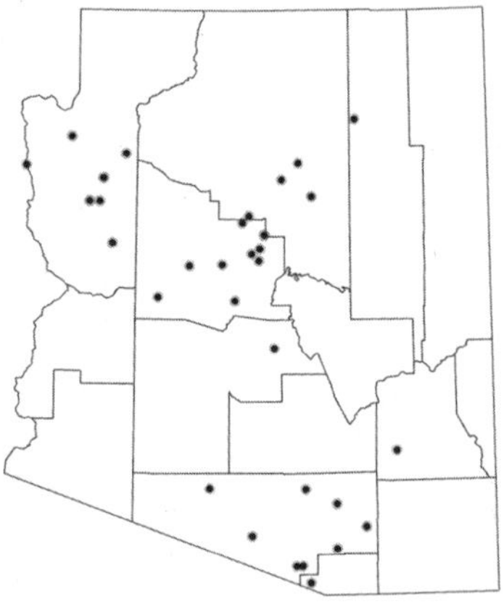

Leaflets (7–)11–19. **Inflorescences** with (3–)5–10(–15) flowers. **Fruits** sparsely strigillose.

Flowering Mar–Jul. Slopes, valleys, arroyos, riparian floodplains, roadsides, and occasionally on dunes, in various soils but especially common on volcanic and granitic soils; (220–)600–2300 m; Sonoran Desertscrub, Chihuahuan Desertscrub, Semidesert Grassland, Plains Grassland, Interior Chaparral, (Madrean Evergreen Woodland), Great Basin Conifer Woodland, (Petran Montane Conifer Forest); Cochise, Coconino, Gila, Graham, La Paz, Maricopa, Mohave, Navajo, Pima, Pinal, Santa Cruz, Yavapai cos.; California, Colorado, New Mexico, Texas; Mexico (Chihuahua, Durango, Sonora).

Variety *wootonii* is widespread in Arizona in many plant communities, particularly in central and southeastern parts of the state. The plants contain the indolizidine alkaloid swainsonine and are potentially poisonous to livestock (James and Welsh 1992).

79. Astragalus xiphoides (Barneby) Barneby. Amer. Midl. Naturalist 55:477. 1956. * [from Greek *xiphos*, sword-like] GLADIATOR MILKVETCH Native

Astragalus convallarius Greene var. *xiphoides* Barneby, Leafl. W. Bot. 5:30. 1947

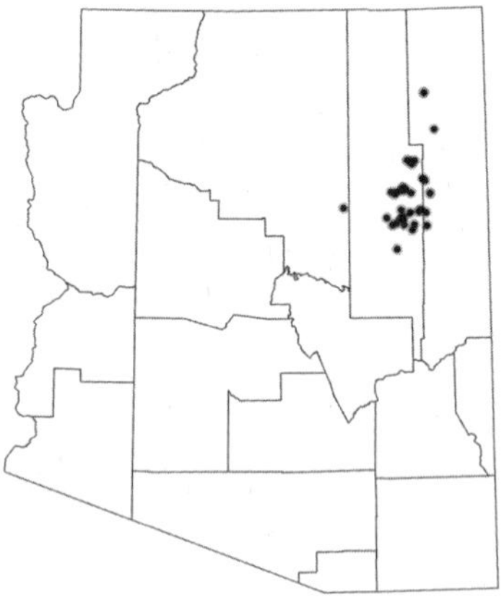

Perennials, caulescent, sparsely leafy or apparently leafless, 30–45 cm; from shallow, subterranean caudex; strigillose-cinereous, hairs basifixed. **Stems** few, diffuse and weakly ascending; 2–7 cm underground. **Leaves** 2.5–13 cm, petiolate, reduced to a linear-filiform or terminally expanded rachis; stipules distinct throughout, ovate, triangular, or broadly lanceolate, 2–7 mm, papery proximally, firm and herbaceous distally, apex obtuse. **Leaflets** (1 or) 3–7, linear-filiform or linear-involute, continuous with rachis, 3–17 mm, apex acute, surfaces strigillose. **Inflorescences** loosely or remotely flowered, with (3–)7–20(–35) flowers; peduncle ascending, 3.5–7.5 cm, early ones stout, slender or thread-like distally; axis 2–18(–33) cm in fruit; bracts 1.2–2.6 mm; bracteoles 0 or 2. **Pedicels** ascending, arched outward, 2–8 mm. **Flowers**: calyx campanulate, 4.6–6 mm, strigillose; tube 3.8–4.1 × 2.8 mm; lobes subulate or triangular-subulate, 0.8–1.9 mm; corolla ochroleucous, 8–9 mm; banner recurved through 100–130°, ovate-cuneate, 9 mm; wings 8.9–9.3 mm; keel 8 mm, apex bluntly triangular; ovules 14–18. **Fruits** declined or pendulous; sessile; straight or nearly so, narrowly lanceolate-oblong, strongly compressed laterally, 25–40 × 4–5.5 mm, thin, pale green becoming straw-colored, papery, strigillose; one-chambered. **Seeds** 14–18.

Flowering Apr–early Jun. Sandstone and clay bluffs, floodplains and arroyos; Great Basin Desertscrub, Great Basin Grassland; 1500–1830 m; Apache, Coconino, Navajo cos.; endemic.

Astragalus xiphoides is confined to a restricted area along the drainage of the Little Colorado River, rare but locally common near Holbrook. The plants are inconspicuous with their often leafless, rush-like stems and small, pale flowers. This taxon is of conservation concern.

80. Astragalus zionis M.E. Jones. Proc. Calif. Acad. Sci., ser. 2, 5:652. 1895. * [of Zion Canyon] ZION MILKVETCH

Perennials, loosely or densely tufted, subacaulescent or shortly caulescent, 3–25 cm; from caudex, branches sometimes with persistent thatch of petioles (sometimes buried in sand); densely villous-tomentulose or strigillose, hairs basifixed. **Stems** several or numerous, prostrate to ascending, 0–11 cm, branched near the base, internodes often concealed by white-hairy stipules and dense white hairs. **Leaves** 2–15 cm, petiole slender, wiry; stipules mostly distinct throughout, sometimes with shortly fused sheathing proximally, broadly ovate or lanceolate, 1.5–5.5 mm, submembranous becoming papery-scarious. **Leaflets** 13–25, flat or loosely folded, elliptic or ovate, 2–16 × 1–6 mm, apex usually acute, sometimes obtuse, surfaces silvery-villous. **Inflorescences** (1–)2–11(–20)-flowered; peduncle ascending in flower, procumbent in fruit, 0.5–15 cm; axis 0.3–6 cm in fruit; bracts 2–5 mm; bracteoles 0–2. **Pedicels** ascending, 1–3 mm. **Flowers:** calyx cylindric, 8.3–18 mm, villous; tube 6.5–13.5 × 3.6–4.7 mm; lobes subulate, 1.5–5.7 mm; corolla often pink-purple, sometimes pale, 18–26 mm; banner recurved through 40°, broadly oblanceolate or spatulate, 18.5–22.5(–26) mm; wings 17–20.5 mm; keel, concealing style, 14.6–18.6(–19) mm, apex obtuse; ovules 24–30. **Fruits** ascending; sessile; brightly purple-mottled, usually curved, obliquely ovoid-oblong, dorsiventrally compressed, 15–35 × 5.5–12 mm, at first green, fleshy, and brightly purple- or red-mottled, becoming somewhat woody, strigose or villosulous; one-chambered. **Seeds** 24–30.

Varieties 2 (1 native in the flora): Arizona, Nevada, New Mexico, Utah.

Astragalus zionis is a variable species, consisting of many isolated populations scattered throughout the deeply dissected sandstones of the Colorado Plateau. The fused stipules are anomalous among its close relatives. It differs consistently from the related *A. argophyllus* in the brightly mottled pod and its distribution on sandstone formations and associated dune habitats across the canyonlands of the southern Colorado Plateau.

80a. Astragalus zionis M.E. Jones var. **zionis** * Native

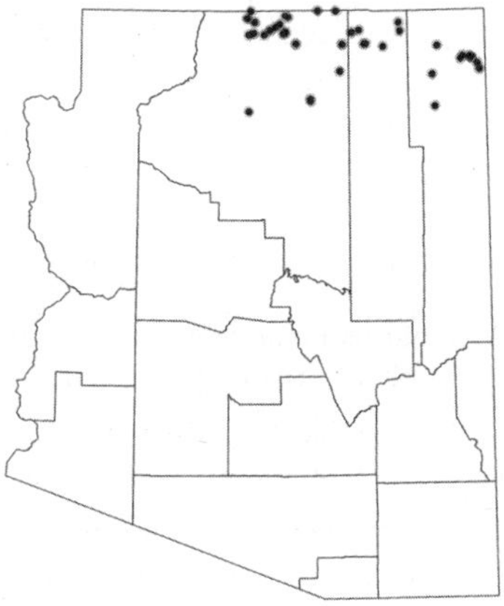

Plants mat- or tuft-forming, usually to 25 cm wide. **Fruits** 15–25(–28) × 5.5–9 mm.

Flowering (Mar–)Apr–Jun. On sandstone and on sandy and gravelly soils, sometimes on sandy floodplains; (860–)1200–2200[–2500] m; Great Basin Desertscrub, Great Basin Conifer Woodland, Petran Montane Conifer Forest; Apache, Coconino, Navajo cos.; Nevada, New Mexico, Utah.

The caudex of var. *zionis* is often slightly subterranean, being buried in duff or in sand, and the stipules are conspicuously white or white-pilose, mostly concealing the internodes. Variety *zionis* occurs in widespread, isolated populations in far northern Arizona in blackbrush, sagebrush, Mormon tea, other mixed desert shrub (sometimes salt desert shrub), mountain brush, ponderosa pine, and riparian communities.

BAUHINIA Linnaeus. Sp. Pl. 1:374. 1753. * [named for Caspar Bauhin, 1560–1624, Swiss botanist and physician, and his elder brother, Jean Bauhin, 1541–1612, Swiss botanist; the bilobed leaves symbolizing the brotherly relationship]

Richard P. Wunderlin

Casparia Kunth

Shrubs or trees. Leaves alternate; 1-foliolate and 2-lobed or 2-foliolate, palmately-veined, margin entire; petiole with a basal pulvinus and apical secondary and tertiary pulvini; stipules small, deciduous. **Inflorescences** terminal or axillary and subterminal racemes or panicles. **Flowers** subcaesalpinoid; calyx 5-lobed, spathe-like at anthesis; corolla bilaterally symmetric with 5 petals, uppermost (banner) innermost, blades clawed; stamens 1 or 5, free or short fused at the base, staminodes sometimes present; ovary 1, 1-chambered, style and stigma simple. **Fruits** dehiscent along lengthwise dorsal and ventral sutures, flattened, thick-walled, elongate. **Seeds** flattened, oblong to nearly circular. $x = 14$.

Species ca. 150 (4 cultivated in flora): ca. 75 in the American tropics, 35 in Africa and Madagascar, ca. 50 in Asia and Malesia.

Bauhinia has been treated in the broad sense as consisting of about 350 species, but recently it is divided into several genera (Lewis et al. 2005, Wunderlin 2010). All of the species treated here are placed in *Bauhinia* in the narrow sense.

Although most commonly planted as ornamentals, the young leaves, flowers, and fruit of several species have been used as food and fodder. The bark is sometimes removed and used for cordage. The wood of some species is useful, but the trees are generally too small to be of importance for timber. Many folk medicinal uses of leaves, bark, flowers, and fruit are described for various species, including treatments for respiratory problems, skin diseases, hemorrhoids, malaria, worms, and diabetes. A complex of chemicals (lectins, alkaloids, flavonoids, sesquiterpenes, and glycosides) are produced, some of which are shown in modern studies to have medicinal potential.

SELECTED REFRENCES: Britton and Rose 1930, Lewis et al. 2005, Wunderlin 1983, 2010.

1. Stamens 1.
   2. Leaves 2-foliolate or 1-foliolate and 2-lobed 3/4 to the base ____ **3. Bauhinia lunarioides**
   2′ Leaves 1-foliolate, 2-lobed 1/4–1/2 to the base ____ **2. Bauhinia divaricata**

1′ Stamens 5.
   3. Flower buds spindle-shaped, tapering toward the apex; stamens subequal; plant producing fruits ____ **4. Bauhinia variegata**
   3′ Flower buds narrowly club-shaped, 5-angled, slightly winged toward the apex; 3 stamens longer than the other 2; plant not (or rarely) producing fruits ____ **1. Bauhinia ×blakeana**

1. Bauhinia ×blakeana Dunn. J. Bot. 46:325. 1908. * [named for Sir Henry Blake, Governor of Hong Kong 1898–1903] HONG KONG ORCHID TREE Cultivated

Trees or shrubs to 6 (10) m. **Stems** slender; branches sparsely pubescent when young, becoming glabrous. **Leaves** with petiole 3–4 cm. **Leaflets** ovate to nearly circular, 7–12(–20) × 7–12(–20) cm, bilobed 1/4–1/3 their length, lobes parallel or widely diverging, base cordate, apex lobes rounded, margin smooth, upper (adaxial) surface glabrous, lower (abaxial) surface puberulent on the veins, 9–13-nerved. **Inflorescences** 3–8-flowered. **Flowers:** buds club-shaped, 5-angled or slightly 5-winged toward the apex; calyx 25–30 mm; petals narrowly obovate or lanceolate, 6–7(–8) cm, rose-lavender with darker veins; stamens 5, arched, 3 slightly longer than the other 2; staminodes 3–5, very short; ovary subequaling the stamens, slightly pubescent. **Fruits** not seen, reportedly rarely formed.

Flowering Oct–Mar. Asia.

Bauhinia ×blakeana, a sterile hybrid of *B. purpurea* Linnaeus and *B. variegata* Linnaeus, has flowers intermediate between the two parent species (Lau et al. 2005). First found in a garden in Hong Kong, it has become a popular cultivated *Bauhinia* in tropical and subtropical areas because of its attractive flowers blooming from fall to spring.

Propagated vegetatively by grafting on *Bauhinia purpurea.*

2. Bauhinia divaricata Linnaeus. Sp. Pl. 1:374. 1753. * [divergent, referring to the slightly spreading leaf lobes] COW HOOF, BULL HOOF, MEXICAN ORCHID TREE; *PIE DE VACA, PATA DE VACA* Cultivated

Bauhinia mexicana Vogel; *Casparia divaricata* (Linnaeus) Kunth ex Britton & Rose; *Casparia mexicana* (Vogel) Britton & Rose; *Casparia spathacea* (de Candolle) Britton & Rose; *Casparia versicolor* (Bertoloni) Britton & Rose

Shrubs or trees to 8 m. **Stems** slender; branches strigose or tomentulose when young, becoming glabrous. **Leaves** with petiole 0.5–4 cm. **Leaflets** nearly circular to ovate to lanceolate, 2–12 × 2–10 cm, emarginate to bilobate about 1/4 their length, lobes parallel or widely diverging, base rounded to cordate, apex lobes acuminate to rounded, margin smooth to slightly crisped, upper surface puberulent or glabrous, lower surface puberulent, 5–9-nerved. **Inflorescences** (10–)20–50-flowered. **Flowers:** buds linear, 1.2–2 cm, strigose or tomentulose, free tips 2 mm; calyx 10–18 mm; petals elliptic-lanceolate, 1.5–2.5 cm, white, turning pink in age; stamen 1, subequaling to 2 times the petal length, slightly arched, glabrous; staminodes white, petal-like, fused for 3/4 or more their length; ovary equaling or slightly exceeding the stamen, pubescent. **Fruit** oblong to linear, 8–15 × 0.8–1.5 cm, puberulent to glabrate. **Seeds** oblong to nearly circular, 8–10 × 6–8 mm.

Flowering, fruiting Jun–Nov or nearly throughout the year with sufficient moisture and warm temperatures. Mexico to Costa Rica, Greater Antilles.

Easily propagated by seeds. The flowers attract a variety of pollinators including bees, butterflies, and hummingbirds.

3. Bauhinia lunarioides A. Gray ex S. Watson. Bibliogr. Index N. Amer. Bot. 205. 1878. * [alluding to the half-moon shaped leaf lobes] TEXAS PLUME, ANACACHO ORCHID TREE Cultivated

Bauhinia congesta (Britton & Rose) Lundell; *Bauhinia jermyana* (Britton) Lundell; *Casparia congesta* Britton & Rose; *Casparia jermyana* Britton

Shrubs or trees to 4 m. **Stems** slender; branches loosely strigose when young, becoming glabrate. **Leaves** with petiole 0.5–1(–1.5) cm. **Leaflets** broadly ovate to nearly circular, (1–)1.5–2.5(–3) × (1.5–)2–2.5(–3) cm, 2-lobate 3/4 or more their length or 2-foliolate, lobes nearly circular, slightly divergent, inner margins straight to slightly convex, base cordate or rarely truncate, apex lobes rounded, margin slightly to moderately crisped, upper surface glabrous, lower surface strigose to glabrate, (5–)11-nerved. **Inflorescences** 2–6(–12)-flowered. **Flowers:** buds narrowly lanceolate, 1–2 cm, loosely white-strigose, occasionally purplish-tinged, free tips 2 mm; calyx 1–2 mm; petals elliptic-lanceolate, 1.5–2.5 cm, subequal, white or occasionally pale pink; stamen 1, subequaling the petals, slightly arched; staminodes 9, fused 1/2 their length. **Fruits** linear, 5–8 × 1–2 cm, glabrate. **Seeds** oblong, 0.7–1 × 0.6–0.8 cm.

Flowering Mar–Jun, reportedly all year. Texas; Mexico.

Bauhinia lunarioides is found on rocky ridges, slopes, in canyons, and along arroyos, often on limestone. It occurs

Bauhinia lunarioides: **a.** habit; **b.** fruits; **c.** leaf. Illustrator: Randall C. Raak.

in Tamaulipan Thornscrub, in the eastern margin of the Chihuahuan Desert and oak and pine-oak woodland in the Mexican states of Coahuila and Nuevo León, and in Kinney, Maverick, and Val Verde counties, Texas. The species is found at elevations from 300–1900 m. Annual rainfall across much of its distribution varies from 375–750 mm.

Anacacho orchid tree makes an ideal small patio tree if pruned up and can be planted as an informal screen. The attractive butterfly-like foliage is deciduous in very cold winters or with prolonged drought. Both white- and pink-flowered plants are available. Though it grows on rocky limestone soils in habitat, this species seems to adapt to a variety of soils in cultivation.

Plants are drought tolerant but performance and appearance are considerably improved with regular irrigation. It can be grown in full sun or partial shade. Anacacho orchid tree is hardy to a least -12°C and perhaps lower. Litter is minimal. Easily propagated by seeds.

4. Bauhinia variegata Linnaeus. Sp. Pl. 1:375. 1753. * [alluding to the various flower colors] ORCHID TREE, MOUNTAIN EBONY Cultivated

Bauhinia candida Aiton; *Bauhinia variegata* Linnaeus var. *alboflava* de Wit; *Bauhinia variegata* Linnaeus var. *candida* (Aiton) Buchanan-Hamilton

Tree to 15 m. **Stems** slender; branches pubescent when young, later glabrous. **Leaves** with petiole 3–4 cm. **Leaflets** broadly ovate to nearly circular, 6–20 × 6–20 cm, bilobate 1/4–1/3 their length, base cordate, apex lobes rounded, upper surface glaucous, lower surface sparsely puberulous on the nerves, 11–13-nerved. **Inflorescences** 3–8-flowered. **Flowers:** buds spindle-shaped, 3–4 cm; calyx 15–20 mm; petals obovate, 4–5.5 cm, pink-, red-, blue-purple, frequently mottled or streaked, or white, somewhat unequal; fertile stamens 5, somewhat unequal; staminodes 5; ovary 2 cm, pubescent mainly on the sutures. **Fruits** linear, 20–30 × 2–2.5 cm, obliquely striate. **Seeds** nearly circular, 1–1.5 × 1–1.5 cm.

Flowering (Dec–)Jan–Apr(–Aug). Asia.

A variety of medicinal uses are widely reported for *Bauhinia variegata*. Santanu Saha et al. (2011) report on studies that support the efficacy of extracts of *B. variegata* in treating inflammatory conditions and indicate that the species is widely used in traditional medicine in India for treating inflammation, skin diseases, leprosy, and in promoting the healing of wounds.

Bauhinia variegata is native to the southern Himalayan region where it occurs in subtropical forests. It is a spectacular flowering tree. In warm-winter areas of Arizona it can be grown for shade and as a street tree. It tolerates a variety of soils so long as drainage is adequate (Jones and Sacamano 2000). This species requires regular irrigation during the growing season in Arizona horticulture. Watering may be reduced during the winter months. Orchid trees can be damaged at temperature below -4°C. Cultivated trees in Tucson have been severely damaged at -7°C but have recovered. The pods and leaves can create litter. Pruning may be needed to control shape and remove freeze-damaged stems, but otherwise, little maintenance is required. Easily propagated by seeds.

Taxa with potential for cultivation

Bauhinia macranthera Bentham ex Hemsley. Diag. Pl. Nov. Mexic. 3:49. 1880

Bauhinia macranthera is grown on the University of Arizona campus and is listed by one wholesale nursery as an option, but is not grown regularly. It resembles *Bauhinia lunarioides* but differs in having larger leaves (5–7 cm vs 1.5–2 cm long) that are bilobate 1/2 to 3/4 their length (rarely bifoliolate) and pink petals. *Bauhinia macranthera* is native to the mountains of northeastern and central Mexico (Wunderlin 1983). It can attain a height and spread of up to 6 m and is an ideal patio tree. Periodic deep irrigation during warm weather maintains a more lush appearance and improves the growth rate, though the plants are drought tolerant once established. Cultivated plants in Tucson have survived temperatures to -9°C.

Bauhinia ramosissima Bentham ex Hemsley. Diag. Pl. Nov. Mexic. 3:49. 1880

Bauhinia ramosissima is grown on the University of Arizona and the Glendale Community College campuses. It is on several "low water-use plants" lists for Arizona, but there currently are no commercial sources. It resembles *Bauhinia lunarioides* in its small, bifoliolate leaves but differs in having purple petals. This slow-growing species can attain a height of 3 m and has survived temperatures of at least -7°C (Starr 1998).

CAESALPINIA Linnaeus. Sp. Pl. 1:380–381. 1753, descr. emended E. Gagnon & G.P. Lewis * [for Andrea Caesalpino, 1519–1603, Italian physician and naturalist]

Matthew B. Johnson

Shrubs or trees. Stems unarmed or with prickles. **Herbage** glabrous or variously pubescent, without glands. **Leaves** alternate, even bipinnate; stipules various. **Leaflets** opposite or alternate, margins entire. **Inflorescences** axillary or terminal racemes or panicles; bracts deciduous. **Pedicels** jointed. **Flowers** caesalpinoid, often showy, bisexual; calyx short with 5 overlapping, unequal lobes, the lowest lobe usually partially enclosing the others, the lobes deciduous leaving a persistent hypanthium; corolla with 5 petals, overlapping, unequal, white, yellow, orange, red, pink, or green; stamens 10, free, pilose at base, declined, exserted; ovary sessile, style and stigma simple. **Fruits** elastically dehiscent with the valves twisting, stiffly papery or leathery; style persistent and forming an acute apex. **Seeds** flattened.

Species ca. 9 (1 cultivated in the flora): Mexico, Central America, and the Caribbean to northern South America.

Gagnon et al. (2016) reclassified some species that are cultivated in Arizona and were once placed in *Caesalpinia* to several genera: *Coulteria, Erythrostemon, Libidibia,* and *Tara*. These include popular landscape plants in Arizona such as *Caesalpinia gilliesii* (now *Erythrostemon gilliesii*), *Caesalpinia mexicana* (now *Erythrostemon mexicanus*), and *Caesalpinia cacalaco* (now *Tara cacalaco*).

Several *Caesalpinia* species are grown as ornamentals. Reported uses for species in the genus include medicinal uses, tannins, dyes, and honey.

SELECTED REFERENCE: Gagnon et al. 2016.

1. Caesalpinia pulcherrima (Linnaeus) Swartz. Observ. Bot. 166. 1791. * [from Latin *pulcherrima*, very beautiful] RED BIRD OF PARADISE, DWARF POINCIANA, PRIDE OF BARBADOS Cultivated

Poinciana pulcherrima Linnaeus, Sp. Pl. 1:380. 1753

Shrubs to 2 (5) m. **Stems** becoming semi-woody with age, unarmed or with scattered bristle-like prickles to 5 mm, or with occasional stout, laterally compressed, recurved prickles to 6 mm; glabrous. **Leaves** 15–30(–40) cm with 4–9 pairs of pinnae, pinnae 4.5–7.4 cm; stipules linear, 0.9–1.2 mm, quickly deciduous; petioles 1.7–4.4 cm; rachis 7–18.5(–25) cm. **Leaflets** opposite, (5–)6–10(–12) pairs per pinna, 6–12(–25) × 4–6.5(–13) mm, oblong or elliptic, the base rounded or oblique, the apex rounded or emarginate. **Inflorescences** axillary or terminal racemes, 15–50 cm, often with 15–60 flowers; bracts 2–2.5 mm, quickly deciduous. **Pedicels** 4–5.5 cm. **Flowers:** calyx 16–18 mm, orange-red, the lobes boat-shaped, at anthesis the lower (abaxial) lobe 14–15 mm, the other lobes 7–10 mm; corolla orange-yellow and red, fading to mostly red, or in some forms yellow or pink only, ± 4 cm across, the petals long-clawed, the banner petal distinctly smaller than the others; stamens with filaments red, long-exserted, 31–56 mm, with rows of erect, white hairs on the basal portion, the anthers brown and yellow; pistil 5.4–6 cm, the ovary green basally and red distally, the style and stigma red. **Fruits** drying brown, elastically dehiscent, 7.3–12 × 1.5–2.2 cm, oblong, laterally compressed, stiffly papery, not constricted between seeds, stipe to 4 mm, apex with a beak to 5 mm. **Seeds** olive green or brown, ± disk-shaped to oblong, compressed, 8–11 × 5–8(–10) mm, dull.

Flowering May–Nov. Presumably Mexico.

Caesalpinia pulcherrima is widely cultivated and also naturalized in tropical and warm subtropical regions around the world. Its original distribution is unknown but is presumed to be Mexico. The species is not documented to have occurred spontaneously in natural environments in Arizona but is widely established in thornscrub and tropical deciduous forest habitats in eastern Sonora. In Mexico, it is reported to occur from sea level to 1200 m elevation (McVaugh 1987).

The seeds of *Caesalpinia pulcherrima* contain tannins and have been used in tanning leather (Standley 1922). Yellow and black dyes are produced from the fruits, and a red dye is obtained from the roots (Standley 1922). The charred stems have been used to make ink (Standley 1922). Good quality honey is reported to be produced from the nectar of *C. pulcherrima* (Standley 1922, Turner et al. 1995). The cooked green seeds are reported to be edible (Standley 1922); however, Lewis and Elvin-Lewis (1977) indicate that the ripe seeds have been used as a fish poison in Central America and to poison criminals, and Standley (1922) mentions the use of the leaves to stupefy fish. Gentry (1942) reports that a wash is made from the roots to treat rattlesnake bites and insect stings. Standley (1922) reports many uses in folk medicine including infusions for treating colds, fever, skin problems and to induce abortion. An infusion of the bark is used as a mouth wash and a decoction of the leaves for ulcers of the mouth and for liver problems (Standley 1922). The flowers have been employed to treat fever, skin and eye inflammation, and as a purgative and to promote menstrual discharge (Standley 1922). Research by Chiang et al. (2003) has found that extracts of *C. pulcherrima* have antiviral activity. Extracts of *C. pulcherrima* have been shown to have anti-inflammatory and antiulcer properties in laboratory research (Sharma and Rajani 2011). Ragasa et al. (2002) found strong antibacterial and antifungal activity in extracts of this species.

Red bird-of-paradise is widely grown at lower elevations in Arizona. Its spectacular flowers provide a tropical appearance and attract butterflies and hummingbirds. A summary of flower pollination by butterflies and hummingbirds can be found in Turner et al. (1995) and also Gentry (1942). Jones and Sacamano (2000) recommend it for mass plantings, as an informal border, and for planting against walls. In tropical regions, *Caesalpinia pulcherrima* may flower nearly throughout the year (McVaugh 1987). The plants should be sited in sunny locations for optimum flowering and a more compact appearance.

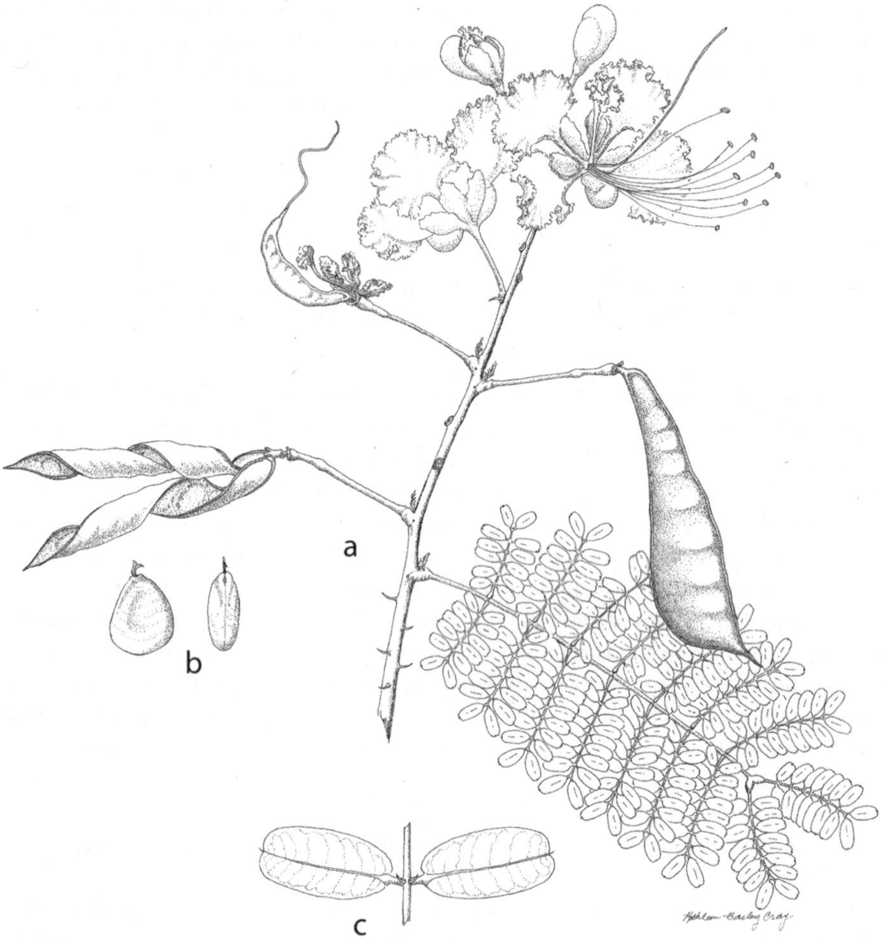

Caesalpinia pulcherrima: **a.** habit; **b.** seeds; **c.** leaflet detail. Illustrator: Kathleen Bailey Cray.

In areas with freezing temperatures in winter, red bird-of-paradise typically becomes deciduous and often suffers stem damage, though it can remain evergreen in warm-winter areas. Stems freeze back below -4°C. The crown is hardy to at least -10°C for brief periods. Jones and Sacamano (2000) recommend mulching the crown of the plant in cold-winter areas. If frozen, red bird-of-paradise should be trimmed back to near the crown in the spring. The plants may also be trimmed back to maintain a more compact habit (Jones and Sacamano 2000). With ample irrigation, established plants can grow up to 2 m in one season. This species requires periodic irrigation for best growth and appearance. Plants occasionally reseed in irrigated landscape settings but are not at risk of becoming naturalized in native habitats in Arizona. This species produces minimal litter. Both the red/orange flowered form and a yellow flowered form are available in Arizona. Selections with pink flowers and cream-colored flowers are known but these have not been available commercially in Arizona so far. Red bird-of-paradise is readily propagated from scarified seeds. It may also be propagated from softwood cuttings (Vines 1960).

CALLIANDRA Bentham. J. Bot. (Hooker) 2:138. 1840, name conserved. * [Greek *calli-*, beautiful, *-andros*, male, alluding the showy stamen filaments] FAIRY DUSTER, FALSE MESQUITE, MESQUITE WEED

Héctor M. Hernández

Anneslia Salisbury, name rejected

Shrubs, perennial herbs [trees], unarmed [armed]. **Stems** erect, ascending or prostrate, glabrous to densely hairy with simple, white trichomes [sometimes intermixed with multicellular, glandular trichomes]. **Leaves** bipinnate; stipules present, leaf-like [spinescent], usually persistent; petiole without glands; 1–7[–26] pairs of pinnae, opposite, sessile. **Leaflets** [1–]4–20[–100] pairs per pinna, thin, leathery, glabrous or hairy. **Inflorescences**: peduncles solitary or clustered at distant nodes, sometimes absent, axillary or terminal, frequently arising from brachyblasts (short, densely branched lateral shoots); hemispherical or obconical flower heads [umbels or pseudoracemes formed through the suppression of subtending leaves], (1–)2–12[–many] flowers; bracts small. **Pedicels** present or absent. **Flowers** mimosoid, homomorphic [heteromorphic], diurnal or nocturnal; calyx bell-shaped, membranous [leathery], usually 5-lobed; corolla bell-shaped or funnel-shaped, membranous [leathery]; stamens [14–]17–30[–many], united proximally into a relatively short tube, [sometimes the tube extremely long in central flowers], white, pink or red, or white proximally and pink or red distally; anthers attached dorsally, with two anther cells, without glands, each anther with 8 polyads; polyads 8-grained, obovate, with a mucilaginous structure on the basal cell; style thread-like; stigma usually hemispherical or disk-shaped. **Fruits** flattened, linear, lanceolate or oblanceolate, usually mucronate, dehiscing longitudinally and elastically along both margins, papery [leathery or woody], margins thickened. **Seeds** obovate or rhomboid, with pleurogram.

Species 136 (3 in the flora; 1 native, 1 native and cultivated, 1 cultivated): s United States, Mexico, West Indies, Central America, and South America.

Calliandra was conceived by Bentham (1875) as a genus distributed in both the New and Old Worlds. However, over the last three decades this concept has been modified. Bentham's *Calliandra* series *Laetevirentes*, together with other species, were segregated into the Neotropical genus *Zapoteca* (Hernández 1986, 1989). On the other hand, the species from Africa, Madagascar, and India have been transferred to three different genera: *Afrocalliandra* (Souza et al. 2013), *Viguieranthus* (Villiers 2002), and *Sanjappa* (Souza et al. 2016), respectively. Thus, in its current concept, *Calliandra* is a Neotropical genus comprising about 136 species grouped into five sections (Barneby 1998).

Numerous species of *Calliandra* are grown as ornamentals for their showy flowers. Some species are suitable as livestock forage and others are used for green manure, fuelwood, pulp for the manufacture of paper, and for erosion control, reforestation, as shade trees, a nectar source for honey production, a source of tannins, and in folk medicine (Hernández 1991, Lewis et al. 2005). Reported folk medicine uses include the treatment of diarrhea, indigestion, and eye problems, and as a substitute for quinine (Allen and Allen 1981, Standley 1922). The wood is hard and strong but uses are limited to carpentry and small wooden articles (Allen and Allen 1981).

SELECTED REFERENCES: Barneby 1998, Bentham 1875, Guinet and Hernández 1989, Hernández 1986,1989, Isely 1972, 1973, Villiers 2002.

1. Stems herbaceous, arising from a woody rhizome, to 32 cm; leaves with (1–)2–7(–12) pairs of pinnae; flowers sessile or pedicellate; filaments to 13 mm when fully erect _____ **3. Calliandra humilis**
1′ Stems woody, to 200 cm; leaves with 1–3(–5) pairs of pinnae; flowers sessile; filaments 15–27 mm when fully erect.
   2. Leaflets pilosulose or strigose on both surfaces, on lower (abaxial) surface only, or entirely glabrous; filaments bright red; fruits to 5.5 cm _____ **1. Calliandra californica**
   2′ Leaflets with upper (adaxial) surface always glabrous; filaments white, pink, or white/pink; fruits to 9 cm _____ **2. Calliandra eriophylla**

1. Calliandra californica Bentham. Bot. Voy. Sulphur 14, pl. 11. 1844. * [of California] BAJA CALIFORNIA FAIRY DUSTER, FAIRY DUSTER, RED FAIRY DUSTER; *CABELLO DE ÁNGEL, TABARDILLO, ZAPOTILLO* Cultivated

Anneslia mixta Britton & Rose; *Anneslia mucronulata* Britton & Rose

Shrubs profusely branched. **Stems** woody, erect or ascending, to 2 m. **Leaves**: stipules linear-lanceolate, 4–6 × 0.5–1 mm; 1–3(–4) pairs of pinna. **Leaflets** 4–12(–15) pairs per pinnae, narrowly oblong to ovate oblong, pilosulose or strigose on both surfaces, on lower (abaxial) surface only, or entirely glabrous, 6–11 × 2–4(–6) mm, smaller proximally, overlapping. **Inflorescences** axillary; heads hemispherical, with 5–11(–13) flowers; peduncles usually clustered, 0.6–1.5(–3.4) cm. **Flowers** sessile; calyx strigose, pilosulose, or glabrate (or glabrous), 1–2 mm; corolla strigose, pilosulose (or glabrous), 5–6(–11) mm; filaments 17–26(–30) per flower, 19–27 mm when fully erect, bright red. **Fruits** brown or light brown, papery, not translucent, densely puberulent, rarely glabrous, to 5.5 × 1.1 cm. **Seeds** 2–7(–8); elliptic or obovate, somewhat flattened, 5–8.5 × 3–5 mm.

Flowering almost all year around in its natural habitat. w Mexico (Baja California, Baja California Sur, Sonora).

Calliandra californica appears to be relatively common in the Peninsula of Baja California (Baja California and Baja California Sur), as well as in a few localities of mainland Sonora (near San Carlos) and in Isla Dátil (Isla Turner). It has been found growing in typical Sonoran Desertscrub and thornscrub, along arroyos, in canyons, and on slopes from near sea level to 550 m altitude.

In its native range *Calliandra californica* is known to be excellent forage and to have medicinal properties. Laboratory studies have shown that compounds extracted from *Calliandra californica* have antimicrobial properties

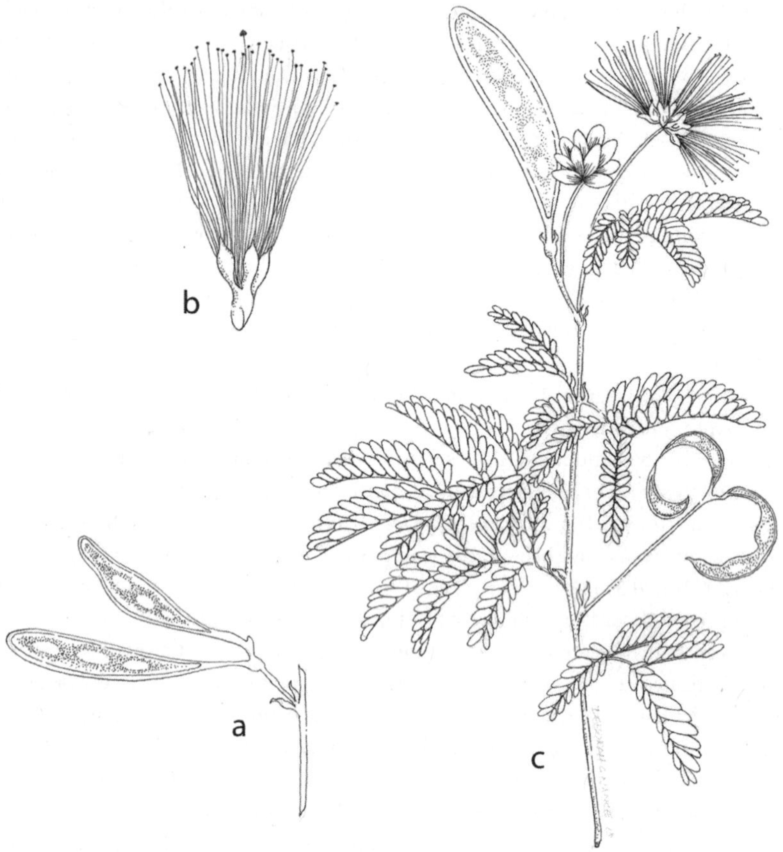

***Calliandra californica*: a. fruits; b. flower; c. habit. Illustrator: Deborah C. Nance.**

(Encarnación-Dimayuga et al. 1994). Further research has found that root extracts have activity against tuberculosis and human cancer cells (Encarnación-Dimayuga et al. 2006).

Calliandra californica is used as an ornamental in warm regions of Arizona, California, Texas, and elsewhere, where it blooms nearly year-round if temperatures do not drop below freezing. Contrasting with the nocturnal flowering behavior of many *Calliandra* species, the flowers of *C. californica* are functional during the day, attracting hummingbirds and butterflies, which very likely act as effective pollinators. Bees, on the other hand, are nectar robbers. Baja California fairy duster is drought tolerant once established, although it requires watering every 1–2 weeks in summer to maintain optimum appearance and flowering and withstands low temperatures to -4 °C. Lower temperatures can result in extensive damage to stems. The foliage is evergreen unless killed by freezing or during extended drought. Little maintenance is required beyond removal of freeze damaged stems or pruning to control size. Scarified seeds germinate readily. Baja California fairy duster may also be propagated by cuttings.

2. Calliandra eriophylla Bentham. London J. Bot. 3:105. 1844. * [Greek *erion*, wool and *phyllon*, leaf] FAIRY DUSTER, FALSE MESQUITE Native, Cultivated

Calliandra chamaedrys Engelmann; *Calliandra conferta* Bentham; *Calliandra eriophylla* Bentham var. *chamaedrys* Isely

Shrubs profusely branched, sometimes without leaves or nearly so at anthesis. **Stems** woody, erect or ascending, to 100 (–120) cm. **Leaves** frequently arising from brachyblasts; stipules linear-lanceolate to narrowly subulate, to 5 mm; 1–3(–4) pairs of pinnae. **Leaflets** 8–10(–14) pairs per pinna, oblong to widely linear, strigose to glabrate on the lower (abaxial) surface, glabrous on the upper (adaxial) surface, 2–5 × 0.8–1.4 mm, overlapping. **Inflorescences** axillary or produced in brachyblasts; peduncles solitary or clustered, 0.4–1.2(–1.8) cm at anthesis; heads hemispherical, 6–8-flowered. **Flowers** sessile; calyx strigose, [0.5–]1–2[–2.5] mm; corolla strigose, rarely glabrate, [3–]4–6 mm; filaments 19–23 per flower, 15–25 mm when fully erect, [white], pink to red-purple, [white in the proximal half and red or pink in the distal half].

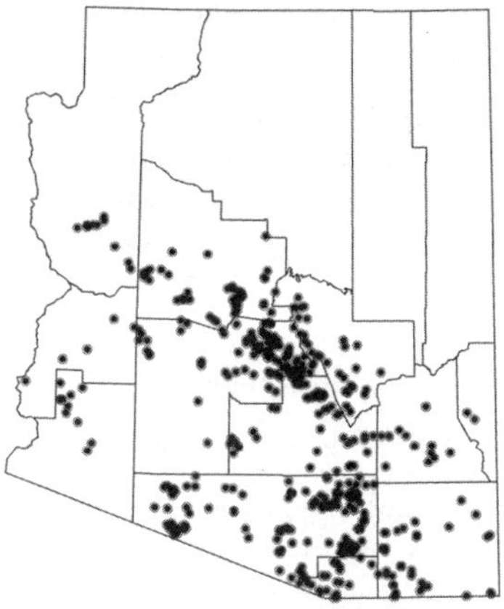

Fruits brown or light brown, linear-lanceolate or oblanceolate, thickly papery, not translucent, pubescent to densely pubescent with short, white trichomes, to 9[–10] × 0.7[–1] cm. **Seeds** 1–6; obovate, flattened, 5–6 × 3–4.5 mm. $2n = 16$.

Flowering Feb–Jun; fruiting Apr–Jun. Dry, rocky hillsides, canyons, and small drainages, limestone, igneous, or granitic soils; [0–]180–1830[–2000] m; Sonoran Desertscrub, Chihuahuan Desertscrub, Mohave Desertscrub, Semidesert Grassland, and rarely in Interior Chaparral and Madrean Evergreen Woodland; Cochise, Coconino, Gila, Graham, La Paz, Maricopa, Mohave, Pima, Pinal, Santa Cruz, Yavapai, Yuma cos.; California, New Mexico, Nevada, Texas; Mexico (Aguascalientes, Baja California Sur, Chiapas, Chihuahua, Coahuila, Durango, Guanajuato, Guerrero, Hidalgo, Jalisco, Michoacán, Morelos, Nayarit, Nuevo León, Oaxaca, Puebla, Querétaro, San Luis Potosí, Sinaloa, Sonora, Tamaulipas, Zacatecas).

Calliandra eriophylla is relatively common in the U.S. portions of the Chihuahuan and Sonoran Deserts, and in parts of the Mohave Desert. It is also widespread throughout the desert areas of northern and central Mexico, extending its range as far south as Guerrero, Oaxaca, and Chiapas, in thornscrub and tropical deciduous forest areas. It is found in desert, mesquite, and palo verde-saguaro scrubland, rarely in oak-juniper woodland. At higher elevations, especially in grassland habitats on exposed slopes, *C. eriophylla* often forms a low, mat-like plant of many short stems spreading from rhizomes. In southern Sonora, *C. eriophylla* grows in sandy coastal dunes (Martin et al. 1998).

The morphological variation of *Calliandra eriophylla* is substantial throughout its range, especially in leaf (number of pairs of pinnae), inflorescence (peduncle length) and flower characters (calyx and corolla length). However, in Arizona its morphology appears to be relatively constant. *Calliandra eriophylla* can be easily distinguished from *C. humilis* by its woody and taller stems, by the hemispherical heads and longer flower filaments, and by the usually larger fruits. In both varieties of *C. humilis* found in Arizona, the stems are herbaceous and shorter, the heads have the form of an inverted cone, the flower filaments are shorter, and the fruits are smaller.

Calliandra eriophylla is an important browse plant for livestock and deer (Allen and Allen 1981, Benson and Darrow 1981), and the seeds are sought by birds (Allen and Allen 1981). The species is useful for erosion control (Benson and Darrow 1981). Allen and Allen (1981) report that nodulation was not observed in *C. eriophylla*.

Fairy duster is commonly grown in low water-use landscapes in Arizona. The species makes an attractive small shrub and is much appreciated for its showy pink or white/pink flowers which attract hummingbirds, as well as butterflies and bees. It can be planted as a low shrub among larger plants, or with succulents or wildflowers. It can be massed for an informal border. Fairy duster tolerates reflected heat. It should have at least a half day of full sun to maintain a more compact appearance. Established plants are drought tolerant, but appearance and flowering are much improved with occasional irrigation. The leaves are deciduous with drought or freezing. Fairy duster is hardy to ca. -12°C. Maintenance is minimal. Fairy duster is readily grown from scarified seeds and may also be propagated from cuttings.

3. Calliandra humilis Bentham. London J. Bot. 5:103. 1846. * [Latin *humili*, low] FALSE MESQUITE, MESQUITE WEED, DWARF STICKPEA

Perennial herbs, profusely branched from the base. **Stems** herbaceous, prostrate or ascending, to 23(–32) cm, arising from a perennial woody rhizome. **Leaves:** stipules linear-lanceolate or elliptic-ovate, to 5 [–7] mm; (1–)2–7[–12] pairs of pinnae. **Leaflets** 5–20[–27] pairs per pinna, widely elliptic, elliptic-ovate, lanceolate or linear-lanceolate, strigose to glabrous on the lower (abaxial) side, glabrous on the upper (adaxial) side, the largest ones 3.5–14 × 1–6 mm. **Inflorescences** axillary, solitary; peduncles solitary or clustered, 0–2.9(–4) cm at anthesis; heads obconic, 5–12-flowered, densely arranged. **Pedicels** absent or present, 0–2 mm. **Flowers:** calyx with few hairs near the apex, (1–)2–3 mm; corolla with few hairs near the apex, 4–5[–7.5] mm; filaments 19–28 per flower, 9–13 mm, white, drying pink. **Fruits** light brown, linear-lanceolate, oblanceolate, papery, translucent, glabrous to sparsely strigose with minute, short, white hairs, to 5.8 × 0.6 cm. **Seeds** 1–6; obovate, flattened, 5–6 × 3–4 mm.

Varieties 3 (2 native in the flora).

The distribution range of *Calliandra humilis* is quite extensive, ranging from central Mexico, as far south as Puebla, to southwestern United States, in Arizona, New Mexico and Texas; however, at the local level, it appears to be ecologically uncommon.

Calliandra humilis can be readily distinguished from the Central and North American members of the genus by its unusual herbaceous habit. The only other herbaceous *Calliandra* from this region is the Texas-Tamaulipas endemic, *C. biflora*, which is characterized by being a taller herb (to 60 cm) with greatly reduced (2-flowered) inflorescences.

Calliandra humilis can be distinguished from *C. eriophylla*, the other native Arizona species, by its herbaceous habit, its shorter stems, its leaves with usually more pairs of pinnae (var. *humilis*), and its obconic flower heads bearing flowers with shorter filaments. In addition, *C. humilis* usually inhabits mesic environments, such as pine and oak forests, and grasslands, whereas *C. eriophylla* typically occurs in desert habitats. Powdered roots of *C. humilis* are reported to have been used by the Zuni to treat skin rash (Camazine and Bye 1980). Nodulation has not been observed in *C. humilis* (Allen and Allen 1981).

1. Leaves with (2–)3–7[–12] pairs of pinnae; leaflets 9–20[–27] pairs per pinna, 3.5–6 × 1–1.8 mm; leaflet venation perceptible only under magnification ______ **3a. Calliandra humilis** var. **humilis**
1′ Leaves with 1–3 pairs of pinnae; leaflets 5–11 pairs per pinna, 6–14 × 2–6 mm; leaflet venation perceptible without magnification ______ **3b. Calliandra humilis** var. **reticulata**

3a. Calliandra humilis Bentham var. **humilis** Native

Calliandra herbacea Engelmann

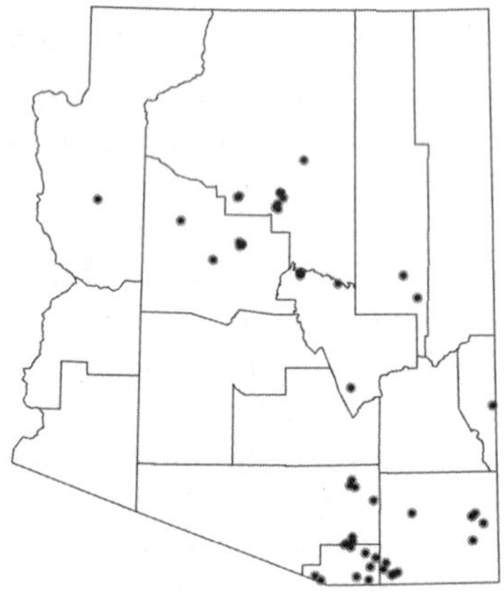

Stems to 23(–32) cm. **Leaves** with (2–)3–7[–12] pairs of pinnae. **Leaflets** 9–20[–27] pairs per pinna, 3.5–6 × 1–1.8 mm, overlapping, leaflet venation perceptible only under magnification.

Flowering [Feb–]May–Jul; fruiting Jul–Oct. Slopes, canyons, and valleys, igneous, rocky sites, in grasslands or open grassy areas [and occasional in tropical deciduous forest]; [1370–]1620–2200[–2440] m; Plains Grassland, Madrean Evergreen Woodland, Great Basin Conifer Woodland, Petran Montane Conifer Forest; Cochise, Coconino, Gila, Greenlee, Navajo, Santa Cruz, Yavapai cos.; New Mexico, Texas; Mexico (Aguascalientes, Coahuila, Chihuahua, Durango, Jalisco, Sonora, Zacatecas).

Calliandra humilis var. *humilis* can be readily distinguished from var. *reticulata* by its usually longer stems, leaves with more pairs of pinnae, and smaller and more numerous leaflets. The leaflets in this variety overlap and have inconspicuous venation. However, there are numerous instances where morphological characters distinguishing these varieties intergrade making identification difficult.

The range of the two varieties overlap in different portions of their extensive distribution area, including in Cochise and Santa Cruz counties, but they are not known to grow side by side. In Arizona, *Calliandra humilis* var. *humilis* occurs in scattered populations in grasslands or open grassy areas in oak woodland, juniper-pinyon or juniper-ponderosa pine forests, and in riparian habitats. There are two low-elevation records from tropical deciduous forest in southeastern Sonora: 350 m (*T. R. & R. K. Van Devender 92-227,* 05 Feb 1992) and 480 m (*Martin s.n.*, 14 Mar 1991). Both are flowering specimens.

3b. Calliandra humilis Bentham var. **reticulata** (A. Gray) L.D. Benson. Amer. J. Bot. 30:630. 1943. * [Latin *reticulon*, net-like] Native

Calliandra reticulata A. Gray, Pl. Wright. 2:53. 1853

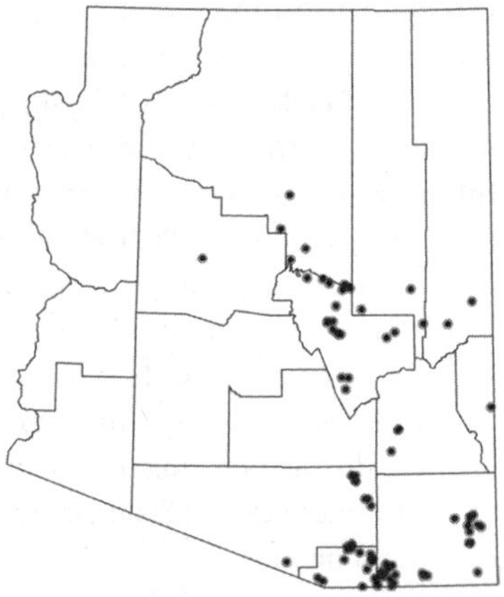

Stems to 15 cm. **Leaves** with 1–3 pairs of pinnae. **Leaflets** 5–11 pairs per pinna, 6–14 × 2–6 mm, commonly well separated, leaflet venation perceptible without magnification.

Flowering Jun–Aug; fruiting Aug–Oct. Rocky slopes, canyons, and open areas; (1070–)1200–2400 m; Semidesert Grassland, Plains Grassland, Madrean Evergreen Woodland, Petran Montane Conifer Forest; Cochise, Coconino, Gila, Graham, Greenlee, Pima, Navajo, Santa Cruz, Yavapai cos.; New Mexico; Mexico (Chihuahua, Durango, Guanajuato, Hidalgo, Jalisco, México, Oaxaca, Puebla, Querétaro, Sonora, Zacatecas).

Calliandra humilis var. *reticulata* can be distinguished from the typical variety by its usually shorter stems, the leaves with fewer pairs of pinnae, and the fewer and larger leaflets, which usually are well spaced and have a conspicuous venation. *Calliandra humilis* var. *reticulata* may be locally common in grasslands, oak woodland, and ponderosa pine forest. The plants grow in the open but can also grow as understory plants in shady forested locations.

CERATONIA Linnaeus. Sp. Pl. 2:1026. 1753. * [Greek, *keras*, horn, for the thick, curved, horn-like fruit]
Douglas H. Goldman

Trees or shrubs, 1–10 m, occasionally dioecious. **Stems** short-appressed pubescent, unarmed. **Leaves** alternate, (1.5–)5–30 cm, odd-pinnate. **Leaflets** 2–12, opposite to alternate, ovate to elliptic or obovate, leathery, entire; stipules minute; stipels absent. **Inflorescences** axillary, spike-like racemes, often emerging from trunks and older branches; bracts 0.5–1 mm, deltoid, often deciduous. **Flowers** caesalpinoid, unisexual (bisexual), small, often with unpleasant odor; calyx puberulent, not persistent, 5-lobed, lobes small, much longer than rudimentary tube; petals absent; stamens (2–)5(–8), free; style short, ovary puberulent, sessile or nearly so, rudimentary in male flowers, with a large, fleshy green disk at base in both male and female flowers; stigma bilobed; ovules generally 12–20. **Fruits** generally pendulous, 1.5–30 cm, indehiscent, compressed, linear to oblong, straight to sickle-shaped, with a sweet, pulpy substance between seeds. **Seeds** 1–17; brown, ovoid to oblong, compressed. $x = 12$.

Species 2 (1 cultivated in the flora): naturalized in California; native to eastern Mediterranean, Arabia, northern Africa.

Ceratonia may be closely related to *Gleditsia* and *Gymnocladus*, which also bear large legumes (Bruneau et al. 2008). *Ceratonia oreothauma*, the only other species in the genus, is native to the Arabian Peninsula and Somalia.

SELECTED REFERENCES: Meikle 1977, Ortiz 1999, Thulin 1993.

1. Ceratonia siliqua Linnaeus. Sp. Pl. 2:1026. 1753. * [Latin, *siliqua*, pod or capsule] CAROB, ST. JOHN'S BREAD
Cultivated

Trees or shrubs 1–10 m. **Stems** stout with rough, brown bark. **Leaves** (1.5–)5–30 cm. **Leaflets** (2)3–10(–12), ovate to elliptic or obovate, 2–6(–8) cm × 1–4(–5) cm, tip blunt or rounded to retuse. **Inflorescences** 1.5–10(–25) cm. **Flowers** with calyx lobes 0.4–2 mm, acute to obtuse. **Fruits** 5–30 cm. **Seeds** 1–17, ovoid to oblong. $2n = 24$.

Flowering summer–fall (winter). Native to the eastern Mediterranean, Arabia; introduced and occasionally naturalized world-wide in areas with Mediterranean and dry-subtropical climates; naturalized in California.

Ceratonia siliqua has a long history of human usage. The ancient Greeks introduced it to Greece and Italy, and the Arabs spread it across the southern Mediterranean to the Iberian Peninsula (Batlle and Tous 1997). Fruit were used in ancient Egypt as a remedy for diarrhea, intestinal worms, and eye ailments, and in ancient Greece it was noted for treating digestive problems; paradoxically, this species has been used not only to treat diarrhea but also as a laxative (Chevallier 1996). The pulp was, and still is, used as a sweetener and for alcoholic beverages. Seeds apparently were the historical source of the carat-standard of jewelers, and in biblical times the fruit was used as forage for pigs (Musselman 2007). More recently it has been used as feed for geese (Sahle et al. 1992) and ruminants (Louca and Papas 1973, Batlle and Tous 1997). Because of the density of its wood, *C. siliqua* has been used for utensils and fuel (Batlle and Tous 1997). The fruit and seeds have several industrial uses, including food additives, cosmetics, and leather tanning (Batlle and Tous 1997), and also are used as a coffee substitute (Lewis 2005). Currently carob is best known as a caffeine-free and low-calorie substitute for chocolate (Whiteside 1981, Craig and Nguyen 1984, Lewis 2005). For a detailed summary of this species see Batlle and Tous (1997).

Horticultural discussion by Matthew B. Johnson

Ceratonia siliqua is an important crop in the Mediterranean Region and is also valued for livestock forage (NRC 1979, Allen and Allen 1981). Numerous selections with faster growth, high yields, and high fruit quality are known (NRC 1979). Trees may remain commercially productive for over 100 years (NRC 1979). In areas that receive less than 550 mm of annual rainfall, some irrigation is necessary for acceptable pod production (NRC 1979). *Ceratonia siliqua* is reported to nodulate (Missbah El Idrissi et al. 1996).

Carob is occasionally cultivated at lower elevations in Arizona as an evergreen shade tree or planted as a screen. The trees develop a thick trunk and a rounded canopy but are slow growing. They have an extensive and deeply penetrating root system that can damage sidewalks and hardscaping (Allen and Allen 1981). Carob is drought resistant, requiring infrequent deep irrigation. Growth can be increased somewhat with generous irrigation. The trees should be planted in well-drained soil and do well in rocky soils. They do not tolerate waterlogging (NRC 1979). Established older plants

Ceratonia siliqua: **a.** habit (pistillate inflorescences); **b.** habit (staminate inflorescence**.**; **c.** pistillate flower; **d.** fruit; **e., f., g.** seeds; **h.** staminate flower; **i.** hermaphroditic flower. Illustrator: Joyce Riney Peters

are hardy to ca. -6° or -7°C. Young plants are frost sensitive (NRC 1979, Jones and Sacamano 2000). Carob produces significant litter from the leaves, flowers, and fruits (Jones and Sacamano 2000), so trees should be planted where this won't be a problem. The species is sensitive to several diseases and pests including Texas root rot or cotton root rot (*Phymatotrichopsis omnivora*), verticilium wilt (*Verticilium dahliae*), sooty canker, and nematodes (Jones and Sacamano 2000). Flowers of male trees have an unpleasant scent, so only female trees should be planted near patios and high traffic areas. Propagation is by scarified seeds. Plants can also be propagated by cuttings, suckers, bud grafting, and air-layering (NRC 1979).

CERCIS Linnaeus. Sp. Pl. 1:374. 1753. * [the ancient Greek name *Kerkis*, referring to a weaver's shuttle, was given by the philosopher Theophrastus as a description of the woody, flattened fruit; *Kerkis* was also applied to poplars] REDBUD
Camille F. Nowell and Peter W. Fritsch

Shrubs or trees, winter-deciduous. **Branchlets** cylindrical, slender, straight to strongly zigzag; new bark reddish becoming gray; buds with many scales. **Leaves** alternate, simple, petiolate; stipules large, membranous, quickly deciduous; petioles glabrous or rarely pubescent, pulvinate at each end. **Leaf blade** circular-reniform to cordate, veins 5–9, palmate, prominent, basal sinus shallow and narrow to deep and broad, or absent, margin entire, flat or rarely undulate, apex rounded to acuminate, leathery (ours) to papery, lower surface glabrous to pubescent, upper surface glabrous, rarely sparsely pubescent, dull, glossy, or glaucous. **Inflorescences** often bearing flowers on trunk, arising from spurs on wood one year or older, fasciculate (ours) or occasionally racemose, few-flowered. **Pedicellate. Flowers** subcaesalpinoid, appearing prior to or during the emergence of the leaves, 5-merous, bilaterally symmetrical; calyx magenta, fused tubular-turbinate-campanulate, wider than long, shallowly 5-lobed; corolla pale to dark pink with magenta nectar guides, rarely white, irregular, petals ovate to elliptic-ovate, clawed, banner in bud inner to two lateral petals, keel larger than banner, comprising two petals locked along their lower margins, forming a pocket; stamens ten, curved and included within keel, distinct, anthers versatile, 2-locular; style curvilinear, glabrous or occasionally pubescent, stigma triangular, terminal, capitellate. **Fruits** turning from green to deep reddish purple at maturity, often persistent on plant over winter, pendulous, indehiscent or tardily dehiscent, one-chambered, usually with prominent wing along dorsal suture (ours), or wingless, flattened, narrowly linear-elliptic to oblong, thin, stiffly papery to leathery, usually with cross-striate or net-like veins, glabrous or pubescent, surface dull to shiny or occasionally glaucous. **Seeds** 2 to many; reddish brown, broadly ovate, longer than wide. $x = 7$.

Species 10 (1 native and cultivated in the flora, 1 cultivated only): North America, n Mediterranean Basin, sw and c Asia, China.

The distribution of *Cercis* is disjunct among the warm-temperate regions of the Northern Hemisphere. In mesic habitats *Cercis* is commonly found in deep woods, flood plains, and river thickets; in semi-xeric habitats it occurs in mid-elevation woodlands, protected canyon corridors, and in dry, boulder-strewn washes.

Cercis is placed in the subfamily Cercidoideae, which forms a clade divergent near the base of the Leguminosae (LPWG 2017). It is the only member of the tribe that retains the base chromosome number $x = 7$ ($2n = 14$), corroborating its early-diverging nature (Polhill and Raven 1981). Several fossil species are known, the earliest from the Late Eocene (Jia and Manchester 2014).

The species of *Cercis* in North America form a clade. Hopkins (1942) established a two-species system for North America which is still widely recognized. Alternatively, based on an exploratory morphometric analysis, Isely (1975) inferred up to six separate entities ("phases"). Barneby (1989) recognized only one continental species and treated all of western North American *Cercis* as *C. canadensis* var. *orbiculata*, but justification was cursory and not definitive.

In contrast to morphology, molecular phylogenetic analyses recover three geographically well-defined clades within North America, with California *Cercis* forming a clade that is sister to a clade formed by Colorado Plateau and eastern North American clades (Fritsch et al. 2018). Molecular dating suggests a divergence time among these three clades of at least 12 million years. These clades were also inferred from a distance-based analysis of Cercis in the United States with isozyme data as reported in an unpublished Ph.D. dissertation (Ballenger 1992). On the basis of these studies, *Cercis* is treated here as comprising three species, with the Colorado Plateau and all Arizona specimens recognized as *C. orbiculata*, distinct from *C. occidentalis* from California and *C. canadensis* from eastern North America. This delimitation of species will also be employed for the treatment of the genus for *Flora of North America* (Ballenger and Vincent, in preparation).

The leaves of *Cercis* appear to have evolved in response to contrasting precipitation regimes. In mesic environments leaves are thin and cordate with an acuminate apex, whereas those of species in semi-arid environments are thick and often reniform with a rounded apex. These morphological differences are thought to be genetically based (Raulston 1990, Davis et al. 2002, Fritsch et al. 2009), and largely transcend species boundaries defined on the basis of the molecular-based data, such that populations appear to be able to easily adapt morphologically to local soils and climate regimes.

In Arizona, *Cercis* flowers appear in early spring before the leaves, producing a bright, showy display. Several species are cultivated as ornamentals. *Cercis* species have been used in folk medicine, and the leaves, flowers, and developing fruits are edible (Vines 1960, Lewis et al. 2005). Native Americans in California managed *Cercis* through coppicing and burning to produce straight young branches well suited for basket making (Anderson 1991). Pollination is performed by bees; fruit is wind-dispersed (Polhill and Raven 1981). The seeds are eaten by wildlife, and the foliage is browsed by deer (Vines 1960). *Cercis* species are suitable for honey production (Vines 1960, Lewis et al. 2005). Allen and Allen (1981) report that nodulation has not been confirmed for *Cercis*. The wood has been used in turnery and cabinet making (Allen and Allen 1981).

SELECTED REFERENCES: Davis et al. 2002, Fritsch and Cruz 2012, Fritsch et al. 2018, Hopkins 1942, Isely 1975.

1. Leaf outline cordate to circular; apices acute to rounded; flowers from base of calyx to end of longest petal 8.5–13 mm; cultivated in the flora ____ **1. Cercis canadensis**
1′ Leaf outline circular-cordate to subreniform; apices rounded-emarginate, rarely acute; flowers from base of calyx to end of longest petal 13–16.5 mm; native in the flora ____ **2. Cercis orbiculata**

1. Cercis canadensis Linnaeus. Sp. Pl. 1:374. 1753. * [of Canada] EASTERN REDBUD

Trees, occasionally shrubs, erect, to 10 m. **Branchlets** glabrous to pubescent; mature bark grayish brown, smooth. **Leaves:** petioles glabrous to densely pubescent. **Leaf blade** light to rich green to bluish green, cordate to broadly ovate to circular, 3.7–12.8 × 4.6–13 cm, soft to somewhat leathery, basal sinus deep and narrow, broadly v-shaped, or absent (leaf blade truncate), margins flat, rarely undulate, apex acute to rounded, lower surface glabrous to pubescent, upper surface glabrous, rarely sparsely pubescent. **Pedicels** glabrous to pubescent. **Flowers** in fascicles of 4–9, base of calyx to end of longest petal 8.5–13 mm. **Fruit** narrowly oblong, tapering at ends, 5–10.2 × 1–1.9 cm, glabrous to sparsely pubescent; wing 1.2–2.5 mm wide. $2n = 14$.

Varieties 3 (2 cultivated in the flora): North America.

Two varieties are cultivated in the flora; the third, *Cercis canadensis* var. *canadensis*, is native and widespread in mesophytic forests of the eastern U.S. It is typically an understory tree in rich woods and bottomlands.

1. Leaf blades densely pubescent beneath, generally 4.6–6.2 cm wide; branchlets pubescent; leaf margin undulate ____ **1a. Cercis canadensis** var. **mexicana**
1′ Leaf blades glabrous to sparsely pubescent beneath, generally 5.7–11 cm wide; branchlets glabrous; leaf margin undulate or rarely flat ____ **1b. Cercis canadensis** var. **texensis**

1a. Cercis canadensis Linnaeus var. **mexicana** (Rose) M. Hopkins. Rhodora 44(522):208. 1942. * [of Mexico] MEXICAN REDBUD Cultivated

Cercis mexicana Rose, N. Amer. Fl. 23(4):202. 1930; *Cercis canadensis* Linnaeus subsp. *mexicana* (Rose) A.E. Murray

Shrubs to small trees, erect, to 5 m, shrubs composed of a number of clustered stems forming a clump. **Branchlets** shortly pubescent. **Leaves:** petioles densely wooly-tomentose; pubescence pale brown to rust-colored. **Leaf blade** deep green to rich bluish green, broadly ovate to circular, 3.7–8.2 × 4.6–6.2(–8) cm, leathery, basal sinus shallowly to deeply v-shaped, margins undulate; apex rounded to subacute, lower surface densely pubescent, upper surface glaucous becoming glossy at maturity, sparsely pubescent to glabrous. **Pedicels** sparsely to densely pubescent. **Flowers:** base of calyx to end of longest petal 10–13 mm. **Fruit** 5–8.8 × 1.2–1.7 cm, sparsely pubescent; wing 1.4–2 mm wide.

Flowering Feb–Mar(–Apr). Texas; Mexico.

Cercis canadensis var. *mexicana* is distributed in parts of southwestern Texas and northeastern Mexico. It is found in the protected limestone canyons and arroyos of the western Edwards Plateau, and is common in the Trans-Pecos and Big Bend regions of Texas. In the mountains of the Sierra Madre Oriental of northeastern Mexico, this variety extends into pine-oak woodland and cloud forest habitats.

Morphometric studies on North American *Cercis* have shown that morphological characters traditionally used to define varieties of *C. canadensis* vary continuously (Fritsch et al. 2009, 2018). An exception is the discontinuous character branchlet pubescence, which is generally absent in the north and present in the south. However, this character correlates with no others for delimiting taxa. The varietal taxonomy has yet to be revised. In the cultivated flora the varieties remain distinct because of the selections used; thus, for the purpose of this treatment, we retain the traditional infraspecific taxonomy of Hopkins (1942) for *C. canadensis*.

Mexican redbud seeds and flowers are utilized for food by wildlife (Powell 1988). Pickled flowers are added to salads and, in Mexico, the flowers are fried (Powell 1988). An extract from the bark has been used for treating diarrhea (Vines 1960, Powell 1988). Horticultural characteristics are similar to those of *Cercis orbiculata*. There are several horticultural selections of *C. canadensis* var. *mexicana*.

1b. Cercis canadensis Linnaeus var. **texensis** (S. Watson) M. Hopkins. Rhodora 44(522):203–207. 1942. * [of Texas] TEXAS REDBUD Cultivated

Cercis occidentalis Torrey ex A. Gray var. *texensis* S. Watson, Bibliogr. Index N. Amer. Bot. 209. 1878; *Cercis canadensis* Linnaeus subsp. *texensis* (S. Watson) A.E. Murray; *Cercis reniformis* Engelmann ex A. Gray; *Cercis texensis* (S. Watson) Sargent

Shrubs to small trees, erect, to 6 m. **Branchlets** glabrous. **Leaves:** petioles glabrous. **Leaf blade** dark to rich green, cordate to circular, 3.7–10.5 × (4.5–)5.7–11 cm, thick, leathery to somewhat leathery, basal sinus shallowly v-shaped, to deep and narrow, margins undulate, rarely flat; apex acute to rounded, lower surface lighter green, glabrous to sparsely pubescent, upper surface subglaucous to strongly glossy and reflective, glabrous to sparsely pubescent. **Pedicels** glabrous. **Flowers:** base of calyx to end of longest petal 8.5–12.5 mm. **Fruit** 5.2–10 × 1.1–1.9 cm, glabrous to sparsely pubescent; wing 1.4–2.5 mm wide.

Flowering Feb–Mar(–Apr). Oklahoma, Texas; Mexico.

Cercis canadensis var. *texensis* occurs naturally in the Arbuckle Mountains of south-central Oklahoma, extending into northern and south-central Texas. In Mexico it occurs in the Sierra Madre Oriental, primarily in the states of Nuevo León and Tamaulipas. This variety is typically found on exposed, Paleozoic limestone outcrops (Hopkins 1942).

The morphological description provided is for the typical cultivated variety of Texas redbud; in the wild, there are multiple morphologically variable ecotypes. Additionally, intermediate forms between *Cercis canadensis* vars. *canadensis* and *texensis* occur at the border of the native habitat. This has been viewed as either a consequence of long continuous "vigorous" introgression (Anderson 1953) or incomplete morphological differentiation of a derivative type (Hopkins 1942).

In Texas and Oklahoma, the variety typically exists with larger leaves. To the south in the Mexican state of Nuevo León, the variety has smaller branches and can have more pubescence on the branchlets and the lower surface of the leaves.

Horticultural characteristics are similar to those of *Cercis orbiculata*, although Texas redbud may have higher irrigation requirements. Several horticultural selections have been made, including one with white flowers.

2. Cercis orbiculata Greene. Repert. Spec. Nov. Regni Veg. 11:111. 1912. * [circular in outline] INTERMOUNTAIN REDBUD Native, Cultivated

Cercis canadensis Linnaeus var. *orbiculata* (Greene) Barneby; *Cercis occidentalis* Torrey ex A. Gray var. *orbiculata* (Greene) Tidestrom

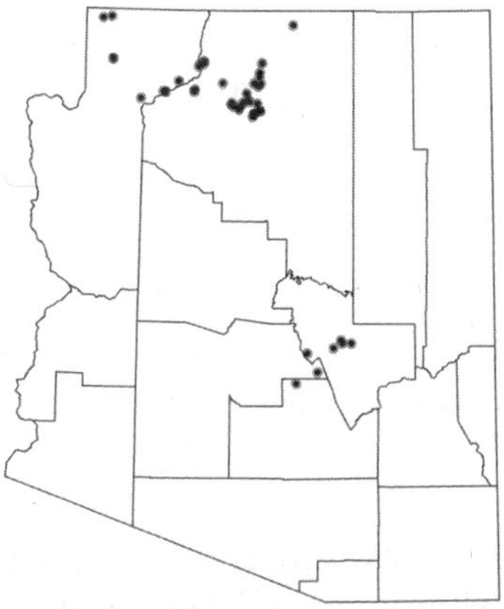

Shrubs to small trees, erect, to 6 m. **Branchlets** glabrous or rarely (Sierra Ancha) sparsely pubescent; new bark reddish, becoming grayish brown, smooth. **Leaves:** petioles glabrous. **Leaf blade** light to rich bluish green, circular-cordate to subreniform, 3.5–6.6 × 5–9.5 cm, leathery to somewhat leathery, basal sinus broadly v-shaped, to deep and narrow, margins flat; apex rounded-emarginate, rarely acute, lower surface glabrous, or if pubescent restricted to vein axils, upper surface glabrous. **Pedicels** glabrous. **Flowers** in fascicles of 3–8, base of calyx to end of longest petal 13–16.5 mm. **Fruit** widely oblong, tapering unequally at ends, 5.9–9 × 1.2–1.8 cm, glabrous; wing 1.7–2.6 mm wide.

Flowering Mar–late May; fruiting May–Aug. Canyons, ledges, and rocky arroyos in granitic, sandy, and limestone soils; 600–2200 m; Sonoran Desertscrub, Mohave Desertscrub, Great Basin Desertscrub, Interior Chaparral, Great Basin Conifer Woodland, (Petran Montane Conifer Forest); Coconino, Gila, Maricopa, Mohave cos.; s Nevada, sw Utah.

Cercis orbiculata occurs only in the Colorado River drainage basin of the U.S. Southwest (see comments under genus discussion for species recognition). In Arizona, it is found in small populations in canyons of the Virgin and Colorado Rivers and also those of montane central Arizona. The southernmost populations occur in the Superstition and Sierra Ancha Mountains, east of Phoenix. The plants of this species typically grow in protected canyons but can be found anchored among rocks in desert washes, and in moist soils of riparian communities. In the Sierra Ancha of central Arizona several specimens have been collected that deviate from the typical intermountain plants in their sparsely pubescent branchlets (versus glabrous) and thinner and broader leaves with slightly acuminate apices (versus rounded).

Because intermountain redbud has only recently been recognized as distinct from western redbud (*C. occidentalis*), the following discussion applies generally to both species unless otherwise specified. Native Americans of the

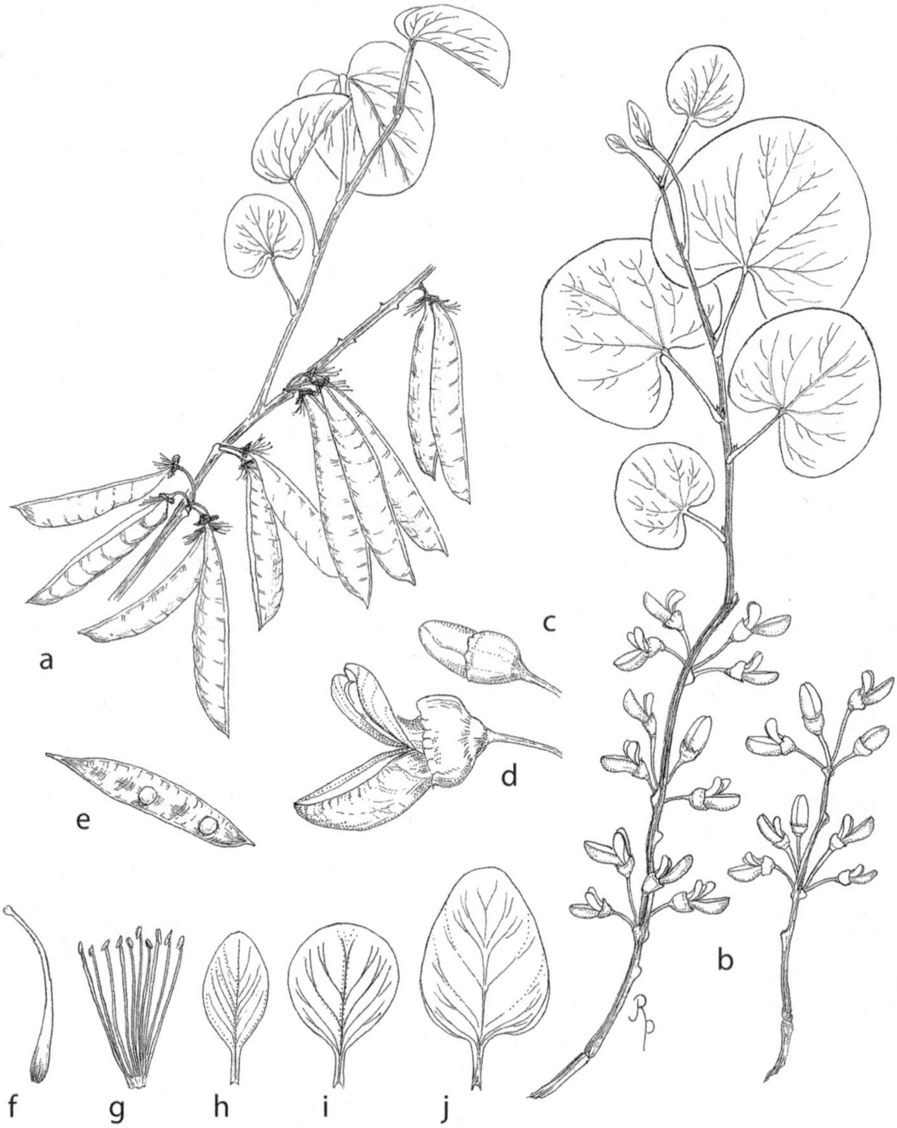

***Cercis orbiculata*: a. fruiting stem; b. habit with flowers; c. flower bud; d. flower; e. fruit valve with seeds; f. pistil; g. stamens; h., i., j. petals. Illustrator: Joyce Riney Peters.**

[Colorado River] canyonlands fashioned bows from the close-grained resilient wood (Barneby 1989). The wood is hard, yellowish brown, and suitable for cabinetry (Vines 1960). Hummingbirds are attracted to the flowers, although pollination is commonly performed by multiple species of bees (Kremen et al. 2002). The foliage is sometimes utilized by leaf-cutter bees (*Megachile* spp.) but this does not harm the plant.

The plants are occasionally grown in Arizona as a flowering shrub or small tree. It is ideal for patios and other areas with limited space because of its small stature. The species can be planted as a focal point, massed for floral display, or planted as an informal screen. Plants typically develop several trunks but can be pruned to a single trunk, and the smooth, gray bark is attractive. Showy, dark rose-pink or magenta flowers are produced in abundance on healthy plants, and a white-flowered selection is known. Some winter chilling may be required for heavy flowering. Plants are best adapted to mid-elevation areas of the state but can be grown at lower elevations with adequate irrigation. Established plants are moderately drought tolerant but periodic irrigation during the growing season is essential to maintain appearance and growth; otherwise they require little maintenance. The plants are adapted to various soils but require adequate drainage and are hardy to at least -18°C. Propagation is by seeds that have been scarified and then stratified. Seeds should be stratified for 30 to 60 days at 5°C (Nokes 1986). Seeds have been observed to germinate without stratification, so this may not be necessary for material originating in warmer locations. *Cercis* may also be propagated by air layering and from softwood cuttings (Vines 1960).

CHAMAECRISTA Moench. Methodus 272. 1794. * [Latin, *chama* and *crista*, crested bivalve shellfish, perhaps referring to the shape of the two-foliolate leaves of some species] SENSITIVE PEA

Brigitte Marazzi

Cassia Linnaeus subg. *Absus* (de Candolle) Symon; *Cassia* Linnaeus subg. *Lasiorhegma* Vogel ex Bentham

Herbs, shrubs and trees to 2 [20] m. **Roots** with bacterial nodules, vertical, rarely horizontal and rhizome-like. **Stems** erect, procumbent, prostrate, rarely from woody rootstock, branches straight or sometimes in zigzag pattern, unarmed. **Leaves** alternate, even pinnately compound, petiolate; stipules [deciduous] or persistent; extrafloral nectaries or glandular hairs present or absent, extrafloral nectaries on petioles, usually cup-shaped, sessile or stipitate. **Leaflets** 1–many pairs, from linear, oblong, elliptic, obovate, ovate, oblanceolate to lanceolate, margins usually entire. **Inflorescences** usually 1- to many-flowered racemes, erect; two-bracteolate, bracteoles deciduous or persistent at anthesis. **Pedicellate. Flowers** caesalpinoid, asymmetric (gynoecium deflected to one side of floral axis); calyx 5-merous, persistent; corolla 5-merous, usually yellow, petal blades narrowed to a claw, claw sometimes red-spotted or reddish orange; androecium (2–)5–10-merous, sometimes stamens of different sizes; anthers attached basally, shorter than filaments, dehiscing by apical pores or short slits, lateral sutures ciliolate; gynoecium shortly stipitate, linear, incurved, few to multiovulate; ovary often hairy, rarely almost glabrous; style usually terminating in a minute stigmatic cavity. **Fruits** elastically dehiscent, stipitate, flat, rhomboid in outline, straight, curved or slightly curved in profile. **Seeds** obovoid-ellipsoid to rhomboid or trapezoid. $2n$ = 12, 16, 28.

Species 330 (3 in the flora): United States; Mexico, West Indies, Central America, South America; Asia, Indian Ocean Islands (Madagascar, Seychelles), Australia.

Chamaecrista was segregated from the large genus *Cassia* Linnaeus and placed in subtribe Cassiinae, together with *Cassia* in the strict sense and *Senna* Miller (Irwin and Barneby 1981, 1982). In molecular phylogenetic studies *Cassia* and *Senna* appear as sister taxa (Bruneau et al. 2001, 2008; Marazzi, Endress, et al. 2006), whereas *Chamaecrista* appears more distantly related, though the three are part of a clade including other caesalpinioid genera and named "cassia clade" (Bruneau et al. 2008, LPWG 2013). Flowers of *Chamaecrista* are asymmetric and specialized in relation to buzz pollination: pollen-collecting bees vibrate the flowers to release pollen from the anthers (Gottsberger and Silberbauer-Gottsberger 1988).

Of the six sections recognized in the current classification of *Chamaecrista* (Irwin and Barneby 1982), only sections *Apoucouita* and *Xerocalyx* appear as monophyletic, whereas the monotypic section *Grimaldia* is embedded within section *Absus*, and sections *Caliciopsis* and *Xerocalyx* are nested within section *Chamaecrista* (de Souza Conceição et al. 2009). These molecular phylogenetic analyses suggest that a shift from a relatively species-poor group of rainforest trees to a more diverse and species-rich clade of savannah shrubs occurred early in the diversification history of *Chamaecrista*. The latter clade is composed of two ecologically and morphologically distinct main subclades (de Souza Conceição et al. 2009). Almost 81% of *Chamaecrista* species occur in the Americas (Lewis et al. 2005).

In North America, *Chamaecrista* is represented by species of sections *Caliciopsis*, *Chamaecrista*, and *Grimaldia*. All but *C. greggii*, a microphyllous shrub or small tree, are annual or perennial herbaceous plants. *Chamaecrista fasciculata* is the only endemic, and *C. pilosa* is the only species that is known to be introduced in the U.S. None seem to be cultivated as ornamentals. The United States represents the northern limit of the geographic distribution of *Chamaecrista* species in the Americas.

Some species of *Chamaecrista* are used in traditional medicine as purgatives and for treating sores, and the seeds are used in the treatment of cataracts (Lewis et al. 2005). Several species are reported as being toxic to livestock (Kingsbury 1964). *Chamaecrista mimosoides* is used as a substitute for tea in east Asia and to treat scorpion envenomation in Tanzania (Lewis et al. 2005).

1. Leaflets 2 pairs ____ **1. Chamaecrista absus**
1′ Leaflets variably 6 or more pairs.
  2. Plants perennial; leaflets [4–](6–)7–10(–12) pairs; seeds lineate and pitted ____ **3. Chamaecrista serpens**
  2′ Plants annual, sometimes biannual; leaflets [6–](7–)10–22(–25)[–40] pairs; seeds rugose ____ **2. Chamaecrista nictitans**

1. Chamaecrista absus (Linnaeus) H.S. Irwin & Barneby. Mem. New York Bot. Gard. 35:664. 1982. * [from Latin, *absus*, a wad (of wool)] TROPICAL SENSITIVE PEA

Cassia absus Linnaeus, Sp. Pl. 1:376. 1753

Herbs or shrubs, erect to 1–2 m. **Leaves** (1.5–)2–9(–11.5) cm; stipules persistent; petiole (7–)10–50 mm; extrafloral nectary 0, sticky glandular hairs present. **Leaflets** 2 pairs, obovate and obtuse to rhombic-elliptic or -ovate and acute, (5–)8–46 × (3–)5–28 mm. **Inflorescences** terminal, usually with 8–20 flowers; bracteoles below calyx, persistent or late deciduous. **Pedicel** 1.5–4 mm. **Flowers**: calyx pale green or sometimes red-tinged, sepal venation reticulate; corolla yellow to reddish orange, fading brownish orange, petals to 4.3–7.2 mm; stamens (2–)4–6(–7), sometimes 1–2 staminodes; anthers 1.1–2.8 mm; ovary usually hairy. **Fruits** flat, straight, or slightly curved, 22–50 × 5–8 mm. **Seeds** 3.4–5.1 × 2.4–4.1 mm; strongly or faintly pitted in lines. $2n = 28$.

Varieties 2 (1 in the flora): introduced, Arizona; s Asia (India); introduced also in w Mexico, Central America.

Chamaecrista absus appears to have been known since 1640 from a plant in Egypt (probably cultivated), called the Absus of Prosper Alpinus. The status of the species is unclear, as some authors consider it likely to be native in the Old World and introduced in the New World, where its occurrence is scattered and discontinuous (Irwin and Barneby 1978, 1982), whereas others argue that the status depends on the variety considered (McVaugh 1987).

Chamaecrista absus is reported to be used in Senegalese traditional medicine to treat hemorrhoids and syphilis (Lewis et al. 2005). Other reported medicinal uses include the treatment of kidney and liver diseases, stomach aches, and to induce blood clotting.

1a. Chamaecrista absus (Linnaeus) H.S. Irwin & Barneby var. **meonandra** (H.S. Irwin & Barneby) H.S. Irwin & Barneby. Mem. New York Bot. Gard. 35:664. 1982. * Apparently Naturalized

Cassia absus Linnaeus var. *meonandra* H.S. Irwin & Barneby, Mem. New York Bot. Gard. 30:282. 1978

Flowers: Stamens (2–)3–4, sometimes 1–2 staminodes.

Flowering Sep–Nov [late summer–early spring]. Canyons; [5–]1280 [–1500] m; Madrean Evergreen Woodland; Pima, Santa Cruz cos.; w Mexico; Central America (Guatemala, Honduras, Costa Rica).

Chamaecrista absus var. *meonandra* may be native to the Americas (McVaugh 1987), but phylogeographic analyses are needed to infer its origin and test, for instance, whether its ancestor colonized North America through long-distance dispersal from Africa, where the var. *absus* is considered to be native. Variety *absus* is characterized by a 5-merous androecium, in which 1–3 (fertile) stamens are often smaller in size (Irwin and Barneby 1978, 1982).

In Arizona, *Chamaecrista absus* var. *meonandra* is found only in the Baboquivari Mountains in Pima County and in Sycamore Canyon, in Santa Cruz County, where it occurs in canyons in oak woodland. In Sonora, it is found in tropical deciduous forest and oak woodland from 750–1500 m elevation (Martin et al. 1998). Across its range, the variety is found in grasslands, disturbed savannas, open places in deciduous forests, and arid to semiarid vegetation in stony or sandy habitats. The Arizona plants appear to be annuals, and the populations are small and highly localized.

2. Chamaecrista nictitans (Linnaeus) Moench. Methodus 272. 1794. * [from Latin *nictitat*, blinking, in reference to the leaf movements] SENSITIVE PARTRIDGE PEA, WILD SENSITIVE PEA

Cassia nictitans Linnaeus, Sp. Pl. 1:380. 1753

Herbs, annual, rarely over-wintering, to 0.8 (1.2) m. **Leaves** 1.5–15(–21) cm; stipules usually persistent; petiole (1.5–)2–6(–7) mm; extrafloral nectary 1(–2), near mid-length of petiole, stipitate. **Leaflets** (6–)8–28(–40) pairs, linear or narrowly oblong or oblong-elliptic, straight or sometimes falcate, (3–)4–26 × 1–3 mm. **Inflorescences** fused to stem, 1(–2)-flowered; bracteoles around mid-length of pedicel. **Pedicel** 0.5–6[–16] mm. **Flowers**: calyx greenish, sepal venation reticulate; corolla yellow, sometimes fading pinkish, longest petal 3.5–9[–16] mm; stamens (2–)4–9(–10); anthers of different sizes, the longest (1.4–)1.6–4.8[–9.5] mm; ovary usually hairy throughout, rarely glabrate. **Fruits** linear-oblong, (14–)17–65(–78) × 2.4–5.5(–5.8) mm. **Seeds** 1.9–3.5(–3.7) mm; rugose [or pitted].

Varieties 13 (1 native in the flora): United States; Mexico, West Indies, Central America; South America (Argentina, Brazil, Paraguay, Peru).

Chamaecrista nictitans is distinguished from the similar *C. fasciculata* (and *C. deeringiana*) by its globose-ovoid floral buds, which are ovoid-acuminate in the other two species. The three varieties found in the United States belong to subsp. *nictitans* and are characterized by 2–9 fertile stamens. All other varieties have ten fertile stamens. Those varieties belong to subspecies *brachypoda* (Bentham) H.S. Irwin & Barneby, subsp. *disadena* (Steudel) H.S. Irwin & Barneby, and subsp. *patellaria* (Colladon) H.S. Irwin & Barneby (Irwin and Barneby 1982).

Burlage (1968) indicates that the leaves of *Chamaecrista nictitans* have cathartic properties. Birds are reported to feed on the seeds of this species in parts of the U.S. (Graham 1941).

2a. Chamaecrista nictitans (Linnaeus) Moench var. **mensalis** (Greenman) H.S. Irwin & Barneby. Phytologia 44(7):501. 1979. * Native

Cassia leptadenia Greenman var. *mensalis* Greenman, Proc. Amer. Acad. Arts 41:238. 1905; *Chamaecrista leptadenia* (Greenman) Cockerell

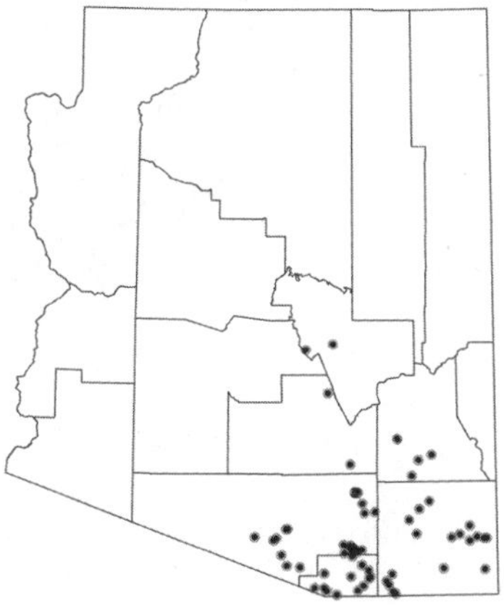

Herbs to 0.6 (0.8) m. **Leaves** (2–)2.5–8(–9) cm; stipules usually persistent; petiole (1.5–)2–5(–6) mm. **Leaflets** (7–)10–22(–25) pairs, (3–)4–14 × 1–3 mm. **Pedicel** 0.5–4 mm. **Flowers**: corolla yellow, often fading pinkish, longest petal 3.5–7(–7.5) mm; stamens (4–)5–8; anthers, the longest 2–3 mm; ovary hairy throughout. **Fruits** (17–)20–48(–56) × 2.5–3.8(–4) mm. **Seeds** 2.5–3.2 mm; rugose.

Flowering (Jul–)Aug–Oct. Slopes, valleys, canyons, arroyos, floodplains, roadsides, and disturbed sites; [10–](850–)1050–1830[–2000] m; Semidesert Grassland, Plains Grassland, (Interior Chaparral), Madrean Evergreen Woodland; Cochise, Gila, Graham, Pima, Pinal, Santa Cruz, Yavapai cos.; New Mexico, Texas; Mexico (Baja California Sur, Chihuahua, Jalisco, Nayarit, Sinaloa, Sonora).

Chamaecrista nictitans var. *mensalis* is widely scattered across parts of southeastern and central Arizona, growing and flowering in response to summer rainfall. It occurs on a variety of soil types, including rocky soil, but is apparently absent from calcareous substrates. The populations can be localized but plants may be locally abundant. Across its range, it is found in arid grassland, desert slopes and washes, thorn-forest, tropical deciduous forest, and open spaces in pine and oak forests.

3. Chamaecrista serpens (Linnaeus) Greene. Pittonia 4:29. 1899. * [from Latin *serpens*, serpent, snake]

Cassia serpens Linnaeus, Syst. Nat. (ed. 10) 2:1018. 1759

Herbs, perennial, erect, rarely procumbent, to 0.4 m. **Leaves** 1–3.5(–4) cm; stipules persistent; petiole 2–4 mm; extrafloral nectary 1(–2), near middle of petiole, rarely a second gland on rachis, sessile. **Leaflets** (4–)5–9(–12) pairs, oblong to oblong-oblanceolate, narrow obovate or obliquely obovate, obtuse or deltately acute, 3–12 × 0.4–6.3 mm. **Inflorescences** axillary, 1(–2)-flowered; bracteoles above mid length of pedicel. **Pedicels** 12–40(–45) mm. **Flowers**: calyx yellowish sometimes also partly reddish, sepal venation reticulate; corolla yellow, longest petal 5–19(–20) mm; stamens 10; anthers the longest 2.5–9.5 mm; ovary loosely hairy. **Fruits** linear-oblong, 14–40(–45) × 2.5–5 mm. **Seeds** 2.3–3(–3.3) mm; lineate and pitted.

Varieties 7 (1 native in the flora): sw United States; Mexico, Central America, South America.

Of the seven varieties found in the United States, variety *wrightii* is the only native one to Arizona. Variety *serpens* is introduced in the U.S. and is the only small-flowered variety, with the longest petal to 7(–8) mm. Among the large-flowered varieties, var. *delicata* (Britton & Rose) H.S. Irwin & Barneby is locally sympatric with var. *wrightii* in Mexico and is distinguished from the latter by its slightly smaller flowers (the longest petal to 12 mm and the longest anther to 5.5 mm). In contrast to both vars. *delicata* and *wrightii*, in the remaining large-flowered varieties (var. *grandiflora* (Bentham) H.S. Irwin & Barneby, var. *isthmogenes* H.S. Irwin & Barneby, var. *mensarum* (Ant. Molina) H.S. Irwin & Barneby, and var. *oaxacana* H.S. Irwin & Barneby), leaves commonly have only 3–7 pairs of leaflets.

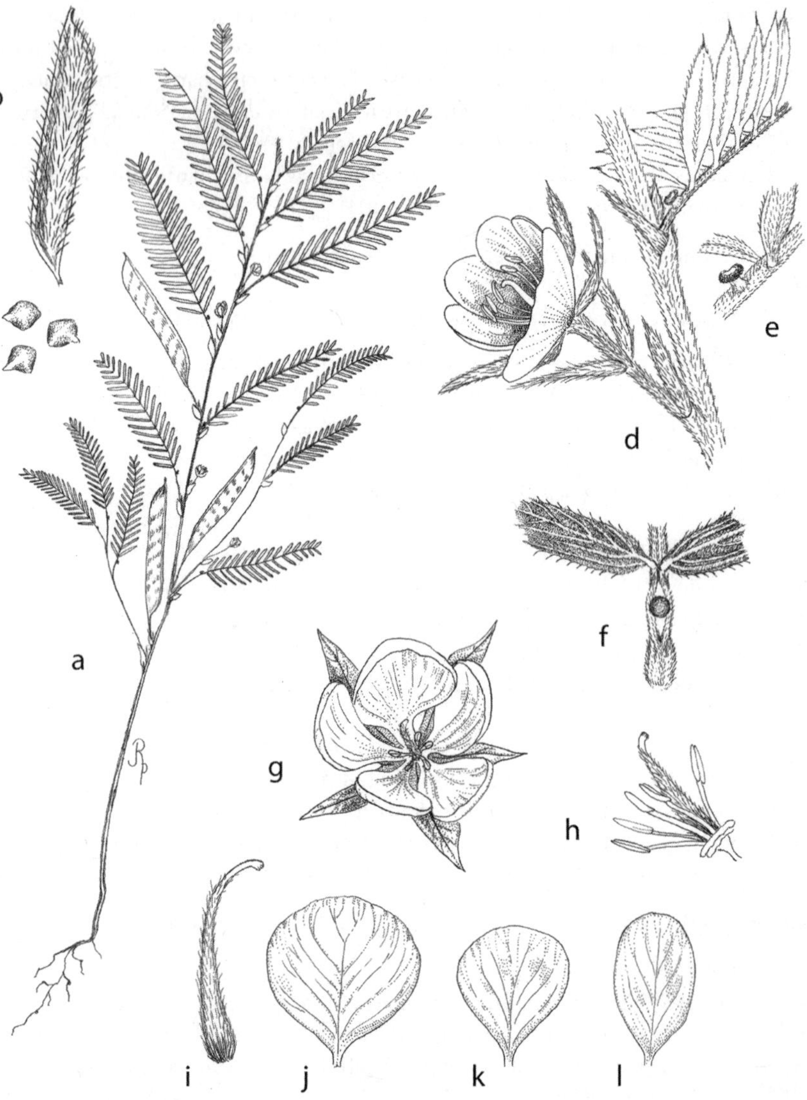

Chamaecrista nictitans var. _mensalis_: a. habit; b. fruit; c. seeds; d. stem node with flower and leaf; e, f. petiole gland detail; g. flower; h. flower with corolla and calyx removed; i. pistil; j., k., l. petals. Illustrator: Joyce Riney Peters.

3a. Chamaecrista serpens (Linnaeus) Green var. **wrightii** (A. Gray) H.S. Irwin & Barneby. Mem. New York Bot. Gard. 35:714. 1982. * [for Charles Wright, 1811–1885, American botanist] WRIGHT'S SENSITIVE PEA Native

Cassia wrightii A. Gray, Smithsonian Contr. Knowl. 5(6):50. 1853; *Cassia palmeri* S. Watson; *Chamaecrista alamosensis* Britton & Rose; *Chamaecrista palmeri* (S. Watson) Greene; *Chamaecrista wrightii* (A. Gray) Wooton & Standley

Leaves 2–3.5(–4) cm. **Leaflets** (6–)7–10(–12) pairs. **Flowers**: longest petal 13–19(–20) mm; longest anthers 2.5–9.5 mm. **Fruits** 14–40(–45) × 2.5–3.5(–4) mm.

Flowering Jul–Oct [late spring or early summer–early winter]. Rocky soil of ridges, slopes, and canyons; [10–]1070–1800[–2100] m; Madrean Evergreen Woodland; Santa Cruz Co.; New Mexico; Mexico (Chiapas, Chihuahua, Durango, Guerrero, Jalisco, México, Nayarit, Sinaloa, Sonora, Zacatecas).

In Arizona, *Chamaecrista serpens* var. *wrightii* is restricted to the mountains of southern Santa Cruz County. Populations are localized but the plants may be common. It prefers sunny locations in rocky soil. Across its range, var. *wrightii* is found on dry hillsides and grassy places, and from coastal plains to tropical deciduous forest and pine-oak forests. Gambel's quail are reported to feed on the seeds (Graham 1941).

CICER Linnaeus. Sp. Pl. 2:738. 1753. * [ancient name in Latin for chickpea]
L.J.G. van der Maesen

Herbs, annual [perennial]; indumentum (hairs) with or without glands, except corolla which is glabrous. **Stems** produced each season from woody roots, erect to prostrate, straight or flexuous, 10–100 cm. **Leaves** alternate, odd-pinnate, (even-pinnate, pinnately 3-foliolate, mutants with phyllodes or unifoliolate leaves are known), rachis ending in a leaflet, a tendril, or a spine; stipules small or somewhat larger than lowest leaflets, toothed or spiny. **Leaflets** 3–36, blade margins partly or almost entirely toothed. **Inflorescences** axillary (pseudo-)racemes. **Flowers** 1–5, papilionoid; calyx subregular or dorsally gibbous, 5 teeth; corolla white to pink or purple, or blue, 5–29 mm, veined; stamens 10, 9 united and 1 free, distal portion of free stamen upturned; ovary ovate, pubescent, style glabrous, stigma small. **Fruits** inflated, elliptic to obovate-rhomboid, to 3 cm. **Seeds** 1 or 2(–10); brown, cream, green, or black, bilobular to globular, to 10 × 8 mm, smooth, wrinkled, or with prickles, beak conspicuous. $x = 7, 8$.

Species 44 (1 cultivated in the flora): s Europe (Mediterranean region including Greece, Turkey, Morocco), c Asia (Afghanistan, Iran, former USSR, Indian subcontinent), Atlantic islands (Canary Islands); introduced to the United States and elsewhere in the Americas.

Cicer is the only genus in tribe Cicereae.

SELECTED REFERENCES: Davies et al. 2007, van der Maesen 1972, 1987, van der Maesen et al. 2007.

1. Cicer arietinum Linnaeus. Sp. Pl. 2:738. 1753. * CHICKPEA, GARBANZO BEAN, BENGAL GRAM Cultivated

Plants annual, deep rooted. **Stems** semi-erect, erect, or almost prostrate, more or less branched from base, (20–)25–60(–100) cm. **Leaves** odd pinnate; stipules ovate to oblique-triangular, 3–5(–11) × (1–)2–4(–6) mm, incised, teeth 2–4(–6); petiole short, 5–10 mm; rachis 25–60(–75) mm, grooved above, middle green above, or somewhat olive, or dark green, with or without anthocyanin color (purple), in extreme cases entirely purple. **Leaflets** 11–13(–17), lowest (first) leaves with fewer leaflets, opposite or not, variously crowded, subsessile, blade obovate-oblong to elliptic, (6–)10–15(–20) × (3–)4–12(–14) mm, base cuneate to rounded, apex rounded to acuminate, margins toothed except near base, teeth to 1.5(–2) mm, sometimes curved; lower (abaxial) surface more prominently ribbed and more glandular pubescent than upper (adaxial) surface. **Inflorescences** with 1 (rarely 2) flowers, axillary (pseudo-)racemes; peduncles (6–)13–17(–30) mm, ending in a small bud scale or bristle of 0.2–4 mm; bracts small, triangular, or tripartite perules, to 1.5 mm. **Pedicels** 6–13 mm, straight when flowering, recurved in fruit. **Flowers:** calyx dorsally gibbous at base, tube 3–4 mm, teeth lanceolate, 5–6 mm, midrib prominent; corolla pinkish, purplish, or red (fading into blue), or white; stamens 9 + 1, ventral filament free, 6–8 mm; ovary ovoid, 2–3 mm; style 3–4 mm; ovules 2 (rarely 3 or 4). **Fruits** late dehiscent, rhomboid-ellipsoid, 14–25(–29) × 8–15(–20) mm, pubescence dense, mainly glandular. **Seeds** ovoid-globular or angular, (4–)7–10(–11) × 5–8 mm; seed coat colorless, white, cream, brown, black, or dull green, rough or smooth (sometimes minute black dots or speckles are present), hilum deep, grayish.

Flowering spring–summer. Asia (India, Mediterranean region), Africa (North Africa and Ethiopia); introduced in Canada, United States (Arizona, California, Idaho, North Dakota, Washington), Mexico, Argentina, s Brazil, Chile.

Cicer arietinum is the third most important pulse crop in the world, after beans and peas, and was introduced from the Mediterranean region and India. It is widely cultivated as a cool-season crop in semi-arid and tropical areas of the world or in summers in temperate zones. In the United States it is cultivated in California, Idaho (particularly the Palouse region), North Dakota, Arizona, and Washington. It is also cultivated in Canada in Saskatchewan. The species occasionally escapes from cultivation but does not appear to be invasive. It is grown at elevations from 0 to 900 meters and occasionally up to 2,400 meters.

In the United States, chickpeas are often canned; large- and cream-seeded Kabuli cultivars are used in salad bars. Hummus, mashed chickpeas with oil and spices, is a popular appetizer in the Near East and is becoming popular in the West too. In fact, Hummus is the Arabic name for chickpea. The largest producer of *Cicer arietinum* is the Indian subcontinent, where it contributes protein to the vegetarian diet in a wide range of dishes for the rich and the poor alike.

In addition to its use as a food crop, *Cicer arietinum* has been reported to have a variety of other uses. Duke (1981) reported that leaves and green pods may be eaten as vegetables. However, Allen and Allen (1981) and Suttie (2000)

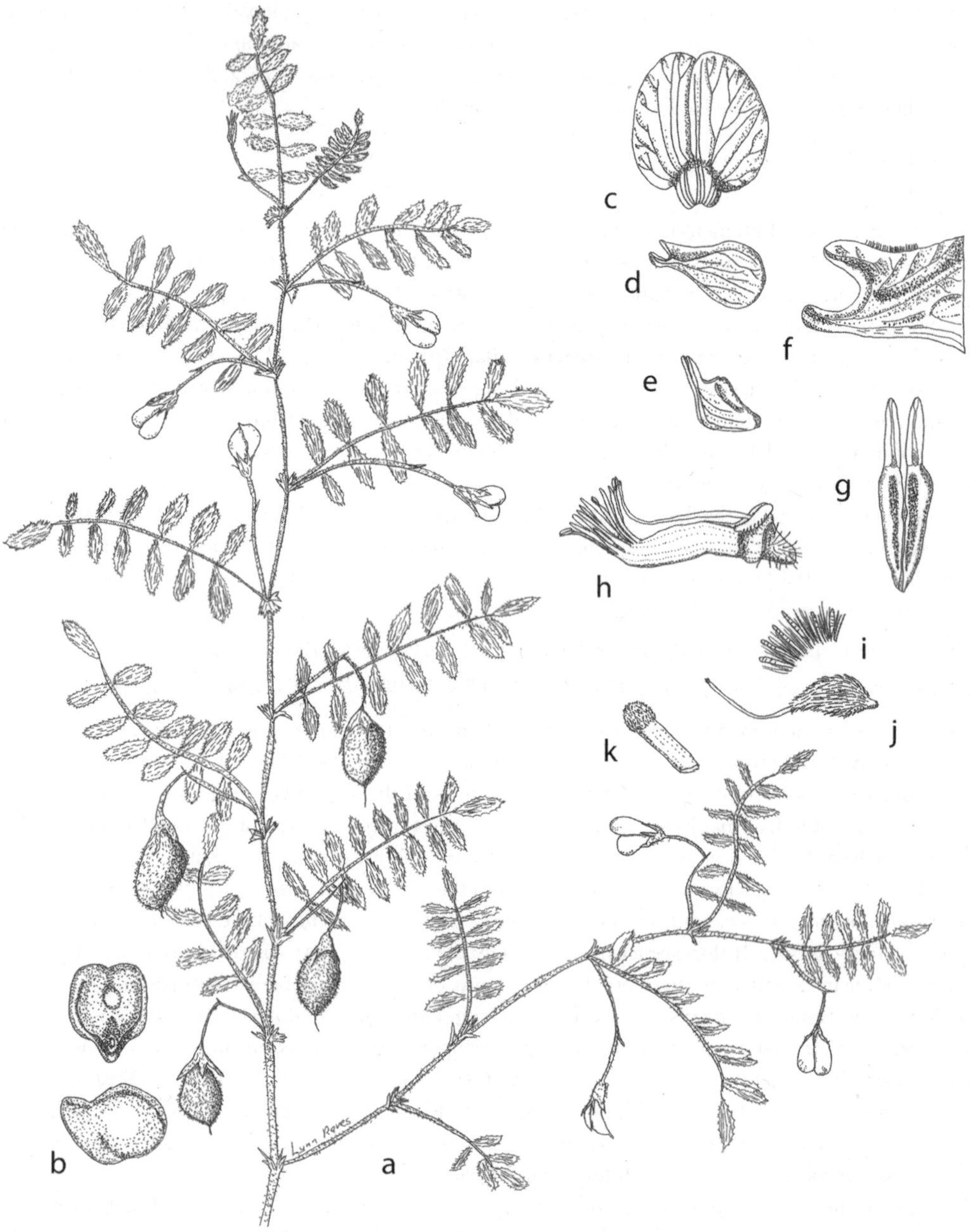

Cicer arietinum: **a.** habit; **b.** seeds; **c., d., e.,** petals; **f.** wing petal detail; **g.** fused keel petals viewed from above; **h.** stamens; **i.** stigma detail; **j.** pistil; **k.** stigma. Illustrator: Lynn Reves.

indicate that the herbage is a poor forage due to high concentrations of oxalic acid. Malic acid is also present in high amounts and contributes to a sour taste (Duke 1981). *Cicer arietinum* is reported to be a source of starch used in textile manufacturing and to yield an adhesive and indigo colored ink (Duke 1981). This species has many reported uses in folk medicine including lowering cholesterol levels and treating bronchitis, catarrh, cutamenia, cholera, constipation, diarrhea, dysentery, dyspepsia, snakebite, sunstroke, warts (with the glandular foliage), and as an aphrodisiac (van der Maesen 1972, Duke 1981). Toxicity of *Cicer* hay in horses has been reported (Pammel 1911).

Cicer arietinum is propagated by seed and is widely reported to nodulate (Allen and Allen 1981). Some cultivars are reported to survive brief exposure to temperatures to -9.5°C when young (Duke 1981). A soil pH of 6–9 is favorable for chickpea (van der Maesen 1972). Though formerly more widely planted in Arizona, only 2 hectares of *C. arietinum* were grown commercially in the state in 2011–2012.

CLITORIA Linnaeus. Sp. Pl. 2:753. 1753. * [from Greek, *kleitoris*, the clitoris, for the superficial similarity of the corolla to the external genitals of the human female] BUTTERFLY PEA, PIGEON-WINGS

Matthew B. Johnson

Herbs, perennial, lianas, shrubs or rarely trees. **Stems** erect, ascending, procumbent or scandent; herbage often variously pubescent. **Leaves** alternate, pinnately 3-foliolate, 1-foliolate, or 5, 7, 9 or 11-foliolate; stipules striate, free, persistent; petiolate; with stipels. **Leaflets** ovate or oblong-lanceolate, margins entire. **Inflorescences** few-flowered axillary pseudoracemes or flowers solitary or paired in leaf axils; bracts 2–6, opposite in pairs, ovate-lanceolate, stipule-like, some or all persistent; bracteoles 2, similar to bracts and stipules, striate, appressed to base of calyx, persistent. **Pedicellate. Flowers** papilionoid, all chasmogamous, or chasmogamous and cleistogamous, resupinate, conspicuous and showy; calyx tubular, two-lipped, the upper (adaxial) 2 lobes partially fused and with 2 teeth, the lower (abaxial) 3 lobes with 3 teeth, the middle lobe longest; corolla purple, blue, lavender, pink or white, banner much longer than the other petals, erect, obovate or rounded, narrowed at base, apex notched, wings curved-oblong, narrowed at base, adherent to middle of the keel, keel shorter than wings, incurved, narrowed at base, apex acute; stamens 10, 9 filaments united and 1 free, anthers all alike; ovary stipitate, with 2–many ovules, style elongate, incurved, bearded along inner (adaxial) side, stigma simple. **Fruits** stipitate, compressed or cylindrical, beaked, 2-valved, the valves twisting at dehiscence. **Seeds** 2–many; spherical or compressed, smooth or sticky. $x = 8, 11, 12$.

Species ca. 60 (1 native in the flora): primarily subtropical and tropical regions of North America, South America, Africa, Asia and Australia. Diversity is highest in northern South America.

The salacious appellative *Clitoria* was considered offensive by some botanists in earlier times and efforts were made to change it to a less indelicate name. Fantz (1977) includes an extensive discussion of the controversy that resulted from the naming of the genus. The use of *Clitoria* as an aphrodisiac in some cultures (discussed below) was perhaps suggested by the appearance of the corolla and derived from a belief in the Doctrine of Signatures where a plant part that resembles an organ or part of the human body would be efficacious in treating a malady affecting that organ or body part (Fantz 1977, Diggs et al. 1999).

Clitoria species have a number of uses. Among these, *C. ternatea* is the most important. It has been used for food, forage, green manure, as a cover crop, a source of blue dye, for medicine, and as an ornamental plant (Fantz 1977). Roots of *C. ternatea* have been widely used in folk medicine in Asia as a laxative, purgative, diuretic, for treating fevers, lung infections, chronic bronchitis, as an antidote for snake bites and scorpion stings, for promoting menstrual discharge, and as a toxic narcotic (Fantz 1977). Research has shown the treatment for snake and scorpion envenomation to be ineffective (Fantz 1977). Reported uses for the leaves include treating earaches, swollen joints, gonorrhea, eczema, and impetigo (Fantz 1977). Fantz (1977) reports that the seeds are used as a laxative and purgative, and to eliminate tapeworms. Several species may have potential as an insect and vertebrate poison (Fantz 1977). The use of *C. ternatea* and several other species as an aphrodisiac for humans and livestock in different cultures is well-documented. A blue dye extracted from the flowers of *C. ternatea* has similar properties to litmus in detecting pH (Allen and Allen 1981) and is used as a dye for food and beverages.

Clitoria laurifolia is planted as a hedge to prevent soil erosion in tea plantations and as a cover crop in teak groves (Allen and Allen 1981). Several species are cultivated in tropical regions for their showy flowers. Wood from the arborescent species of *Clitoria* is stringy, splits easily and has no commercial value (Allen and Allen 1981). *Clitoria* and *Centrosema* are the only native genera of Fabaceae in the U.S. with inverted corollas. Nodulation is reported for several species in the genus (Allen and Allen 1981).

SELECTED REFERENCE: Fantz 1977.

1. Clitoria mariana Linnaeus. Sp. Pl. 2:753. 1753. * [of Maryland] BUTTERFLY PEA, ATLANTIC PIGEON-WINGS, MARYLAND CLITORIA, SPOONFLOWER

Herbs, from a semi-woody rootstock. **Stems** solitary or few to several, erect, ascending, diffuse or procumbent, distal portions sometimes vine-like and twining, to 1.5 m, sparsely to moderately pubescent with minute hooked hairs. **Leaves** 3-foliolate; stipules 4–10 mm, lanceolate or ovate-lanceolate; petioles 1.5–10 cm; stipels 2.5–8 mm, awl-shaped or linear. **Leaflets** ovate, ovate-elliptic or oblong-lanceolate, 2–9(–11.5) × 1–4(–5.5) cm, base rounded to somewhat heart-shaped, apex acute to obtuse, upper (adaxial) surface dark green, glabrous or with a few minute straight hairs, lower (abaxial) surface pale green, sparsely to moderately pubescent with erect to spreading straight hairs, especially along midrib and veins. **Inflorescences** solitary racemes, 0.3–5.5(–9) cm, 1 or 2(–8) flowered; bracts 6, inner pair deciduous, middle pair persistent, outer pair tardily deciduous; bracteoles lanceolate to ovate-lanceolate. **Pedicels** 3–7 mm. **Flowers:** calyx tube (4–)10–14 mm, lobes (2–)6–9(–10) mm, tube and lobes with sparse, straight hairs; corolla pale lavender or bluish purple with a white spot and dark purple veins at base of the banner, banner 40–68 mm, wings 24–30 mm and extending beyond the keel 3–5 mm, or 31–40 mm and extending beyond the keel 7–12 mm, keel 22–35 mm, chasmogamous flowers with staminal tube 21–30 mm, cleistogamous flowers with staminal tube subsessile. **Fruits** elastically dehiscent, 2.5–5.5(–7) × 0.6–0.9 cm, cylindrical with slight constrictions between the seeds, sparsely pubescent with minute hooked hairs, or glabrous; stipe 5–9 or (6–)10–17 mm; beak 2–5 mm. **Seeds** (1–)2–10; dark olive green with small black flecks, brown or black, 4–10 × 3.5–8 mm, spherical or slightly compressed rectangular-spherical, sticky.

Varieties 3 (1 native in the flora): one of the varieties, var. *orientalis* Fantz, is native from eastern India to southern China. Cleistogamy is absent in this variety but present in both New World varieties.

1a. Clitoria mariana Linnaeus var. **mariana** Native

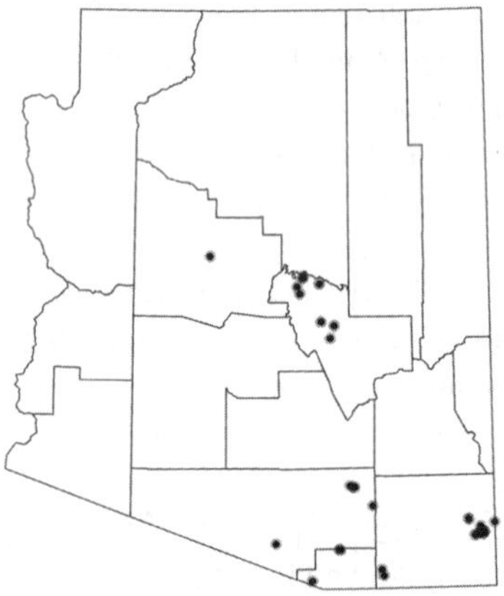

Stems to 1 m. **Leaves:** stipules 4–8 mm, petioles 1.5–6[–10] cm; stipels 2.5–5 [–8] mm. **Leaflets** 2–7(–8) × 1–3.5(–4) cm. **Inflorescences** stout, 0.3–1.5[–9] cm, 1 or 2(–4) flowered. **Flowers:** banner 40–60[–68] mm, wings 31–40 mm and extending beyond the keel by 7–12 mm, keel 22–35 mm. **Fruits:** stipe (6–)10–17 mm. **Seeds** (1–)2–6[–10], dark olive green with small black flecks, or brown, 4–6 × 3.5–5 mm.

Flowering [May–]Jul–Aug(–Sep). Slopes, canyons and stream banks, often in rocky soils; [near sea level–]1280–1700(–2100) m; Madrean Evergreen Woodland; Cochise, Gila, Pima, Santa Cruz, Yavapai cos.; Alabama, Arkansas, Delaware, District of Columbia, Florida, Georgia, Illinois, Indiana, Iowa, Kansas, Kentucky, Louisiana, Maryland, Mississippi, Missouri, New Jersey, New Mexico, New York, North Carolina, Ohio, Oklahoma, Pennsylvania, South Carolina, Tennessee, Texas, Virginia, West Virginia; Mexico (Nuevo León, Sonora, Tamaulipas).

Clitoria mariana var. *mariana* is locally common in mid-elevation montane canyons in southeastern Arizona. It can form dense colonies extending over several hectares in the understory of pine-oak and riparian woodlands. In other locations it occurs as solitary plants or in small groups. The distribution of *C. mariana* var. *mariana* is unusual. Populations in Arizona and the one New Mexico record, as well as the northeastern Mexico populations, are widely disjunct from the main area of distribution for the variety which extends from eastern Texas through the southeastern U.S. The single collections for New Mexico and for Yavapai County are disjunct from the main area of distribution in Arizona. The species is abundant in mountain ranges in southeastern Arizona just north of the International Border, yet collections in Sonora are lacking except for a single collection from near Alamos, in southeastern Sonora. Variety *mariana* occurs in various plant communities and habitats in the eastern U.S. The sticky seed coat likely aids in dispersal of the seeds. The inverted, large, showy and delicately fragrant flowers are unique among Arizona Fabaceae. Nodulation is reported for *Clitoria mariana* (Allen and Allen 1981).

Butterfly pea is occasionally available from specialty native plant nurseries in Arizona. It is worthy of wider cultivation. The plants do well in shaded areas in patios and mini-oasis plantings where the beautiful flowers can be appreciated. Regular irrigation is necessary during the growing season for cultivated plants in desert locations. The stems die back to the rootstock with freezing temperatures and new stems are produced the following season. Butterfly pea is easily propagated by seeds.

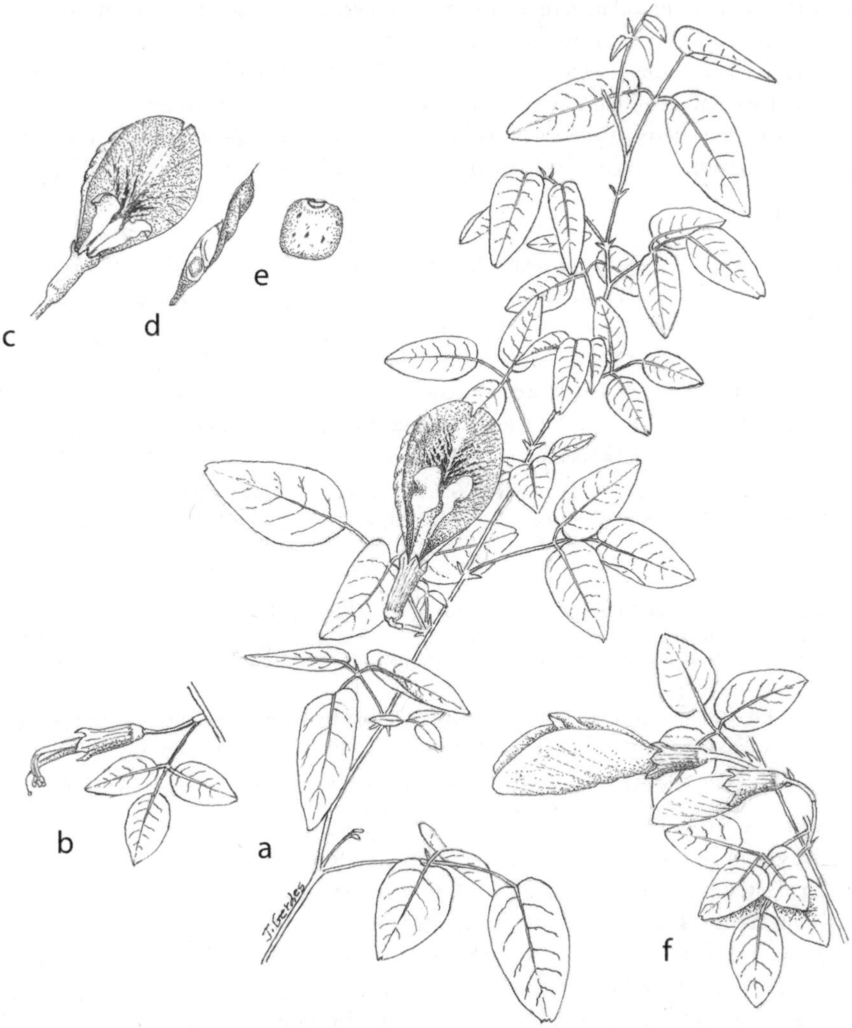

Clitoria mariana var. *mariana*: **a.** habit; **b.** node showing flower with corolla removed; **c.** flower; **d.** dehisced fruit; **e.** seed; **f.** flowers in natural resupinate (upside down) orientation. Illustrator: John H. Gerdes.

COLOGANIA Kunth. Mimos. 205, pl. 57. 1824. * [named for the Cologan family, naturalists from Tenerife, Spain] COLOGANIA

Gabriel Flores-Franco and Alfonso Delgado-Salinas

Herbs perennial, twining or prostrate, rarely erect. **Stems** arising from subterranean lignescent to woody taproots, densely strigose or hirsute, glabrate. **Leaves** alternate, usually pinnately 3-foliolate; stipules persistent; stipels persistent or rarely deciduous. **Leaflets** [1 or] 3 (4–7), linear, lanceolate, ovate to nearly circular, entire, mucronate. **Inflorescences** axillary, lateral, racemose, or in fascicles, usually with both chasmogamous and cleistogamous flowers, or solitary flowers; bracts persistent, usually small; bracteoles small, paired or alternate below calyx, persistent. **Pedicellate. Flowers** papilionoid; calyx tubular [cylindric], upper side gibbous at base, 5-lobed or upper pair fused; corolla purple to magenta or red, glabrous, banner obovate, often emarginate at apex, with small auricles at base, short-clawed, wing petals longer than keel, long-clawed, auriculate, distally spreading, keel slightly incurved, long-clawed; stamens 10, 1 free and 9 united into a tube; pistil with ovary usually pubescent and stipitate, style thread-like, incurved and glabrous. **Fruit** dehiscent, compressed or slightly turgid, linear or falcate, cleistogamous fruits smaller. **Seeds** 2–18, compressed, circular or subquadrate, seed coat smooth, with a conspicuous rim-aril. $x = 11$.

Species ca. 15 (3 native in the flora): Arizona, Nevada, New Mexico, Texas, Mexico, Central America, South

America (Venezuela, Colombia, Ecuador, Perú, Bolivia and northern Argentina); mostly in montane temperate areas. Mexico is the center of diversity for the genus (Flores-Franco 2013).

Cologania species can be distinguished by the combination of the following features: woody taproots; commonly trifoliolate leaves; and few-flowered inflorescences, these with tubular calyxes and brightly colored petals at anthesis (chasmogamous) or with funnelform calyxes and petals that do not fully open (cleistogamous). The cleistogamous flowers may be produced on an inflorescence mixed with chasmogamous flowers or on short separate inflorescences. These flowers are often smaller in size than the chasmogamous flowers, and the number of flower parts may be reduced also. The androecium can be reduced to one or two stamens, and the style may be shorter and reflexed towards the stamens. This often results in two distinctive fruit types, with the cleistogamous fruit shorter and often broader, with fewer seeds (2–6). Due to the polyploid nature, suspected interspecies hybridization, and polymorphic leaves, the taxonomy of this genus is quite difficult (Fearing 1959, McVaugh 1987, Turner 1992).

Plants of *Cologania pulchella*, classified under *C. broussonetii* by Fearing (1959) and Turner (1992), have been reported to occur within the Flora's region; however, according to recent studies, *C. pulchella* is distributed only from northern Mexico to Panama (Flores-Franco 2013) and thus is not treated here. The flowers of *Cologania* species are often attractive and the plants have potential as perennial wildflowers.

1. Plants prostrate; petioles 0.2–0.5 cm; leaflets almost as wide as long, obovate, oblong, obovate-elliptic, obovate-oblong, rarely elliptic or orbicular, mostly obtuse at apex; stipels usually deciduous; inflorescences usually pedunculate ______ **2. Cologania obovata**

1′ Plants twining, sometimes erect, rarely prostrate; petioles 0.8–8 cm; leaflets mostly 2–10 times as long as wide, linear-oblong, oblong, oblong-elliptic, elliptic, lanceolate, or ovate, acute to obtuse at apex; stipels persistent; inflorescences usually sessile, sometimes pedunculate.

  2. Leaves green; calyx tube purple-colored; fruits of chasmogamous flowers 5 × 0.4 cm, linear to slightly falcate, sessile, valves densely appressed pilose or hirsute ______ **1. Cologania angustifolia**

  2′ Leaves pallid-green; calyx tube not purple-colored; fruits of chasmogamous flowers 4.6–6.8 × 0.3–0.4 cm, falcate, stipe 1.6–3.2 mm, valves strigose ______ **3. Cologania pallida**

1. Cologania angustifolia Kunth. Mimos. 209, pl. 58. 1824. * [from Latin, *angustus*, narrow, and *folium*, leaf, referring to the narrow leaflets of some plants] LONGLEAF COLOGANIA, NARROWLEAF COLOGANIA Native

Amphicarpaea angustifolia (Kunth) Taubert in Engler & Prantl; *Cologania longifolia* A. Gray; *Cologania longifolia* A. Gray var. *stricta* M.E. Jones

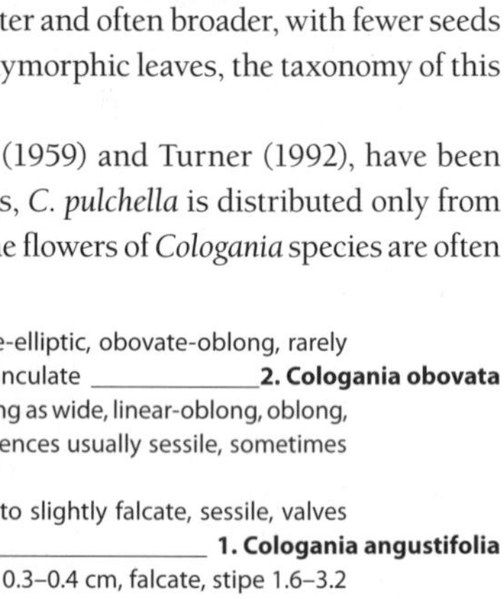

Herbs twining, erect, or prostrate, 10–150 cm. **Stems** canaliculate-striate, strigose, pilose, hirsute to glabrate, with hairs mostly white, rarely yellow. **Leaves** 1.2–22.2 cm; stipules 1.1–6.7 × (0.2–)0.3–2.8 mm, oblong-ovate, oblong, linear-oblong, linear-lanceolate, lanceolate, triangular, striate, hirsutulous, pilose, strigose, or glabrate; petioles 1–8 cm, canaliculate, striate, hirsute, strigose to glabrate; rachis 7–17 mm, canaliculate, corrugate, hirsute or strigose; stipels 0.3–5 mm, linear, strigose, glabrate, or glabrous. **Leaflets** 3 (4–7), 4–17.2 × 0.1–1.6 cm, linear-oblong, oblong, oblong-elliptic, lanceolate, ovate, elliptic, ovate-oblong, rarely circular or obovate, base cuneate, obtuse, rounded, attenuate, sometimes subcordate, acute at apex, obtuse, rarely acuminate, mucronate, venation conspicuous or slightly conspicuous above and conspicuous below, strigose above, slightly to densely strigose below, glabrate, green. **Inflorescences** fasciculate, sessile or pedunculate, with 1–3 (or 4) chasmogamous flowers, and usually with 1 or 2 cleistogamous flowers, 1.8–4.9 cm; peduncle 0.5–4.8(–18) mm, strigose, pilose, hirsutulous, pilosulose, or strigillose; bracts of peduncle 2–7.8 × 0.4–1.3 mm, lanceolate, linear-lanceolate, linear-oblong, oblong, striate, strigose or densely hirsute, glabrate; bracts of pedicel 1.3–4.5 mm, linear, striate, strigose, strigillose, or hirsute, glabrate; bracteoles (0.7–)1.2–3.3(–4.6) mm, linear, strigose, strigillose or hirsutulous, glabrate. **Pedicels** 2–22 mm, densely to slightly strigose or pilose. **Flowers** with calyx 7.5–16 × 2.1–5.9 mm, purple, strigose, hirsute or pilose, glabrate, lobes (1.6–)2.2–4.5(–5.2) × 1.1–2.3(–3.1) mm, triangular, strigose, upper lobe 2.5–4.8 × 2–3.6 mm, deltate, rarely divided; corolla purple, magenta, blue, pink-purple, pink, violet, lavender, deep-lilac; banner 17–33 × 7.6–19 mm, obovate, obcordate at apex, rarely emarginated; wing petals 14–30 × 2.5–7 mm, obovate-oblong or obovate; keel 11–18 × 2.1–3.4 mm, oblong or obovate, slightly curved; style 1.5–3.8 mm, slightly to strongly curved, glabrous. **Fruits** (chasmogamous) 5 × 0.4 cm, linear to slightly falcate, sessile, valves densely appressed-pilose or hirsute. **Seeds** (chasmogamous) ca. 10, ca. 5 mm, brown. **Fruits** (cleistogamous) (1.4–)1.8–4.6 × 0.2–0.4 cm, straight to slightly or strongly curved, sessile or stipitate, stipe 0.8–1 mm, valves densely strigose or

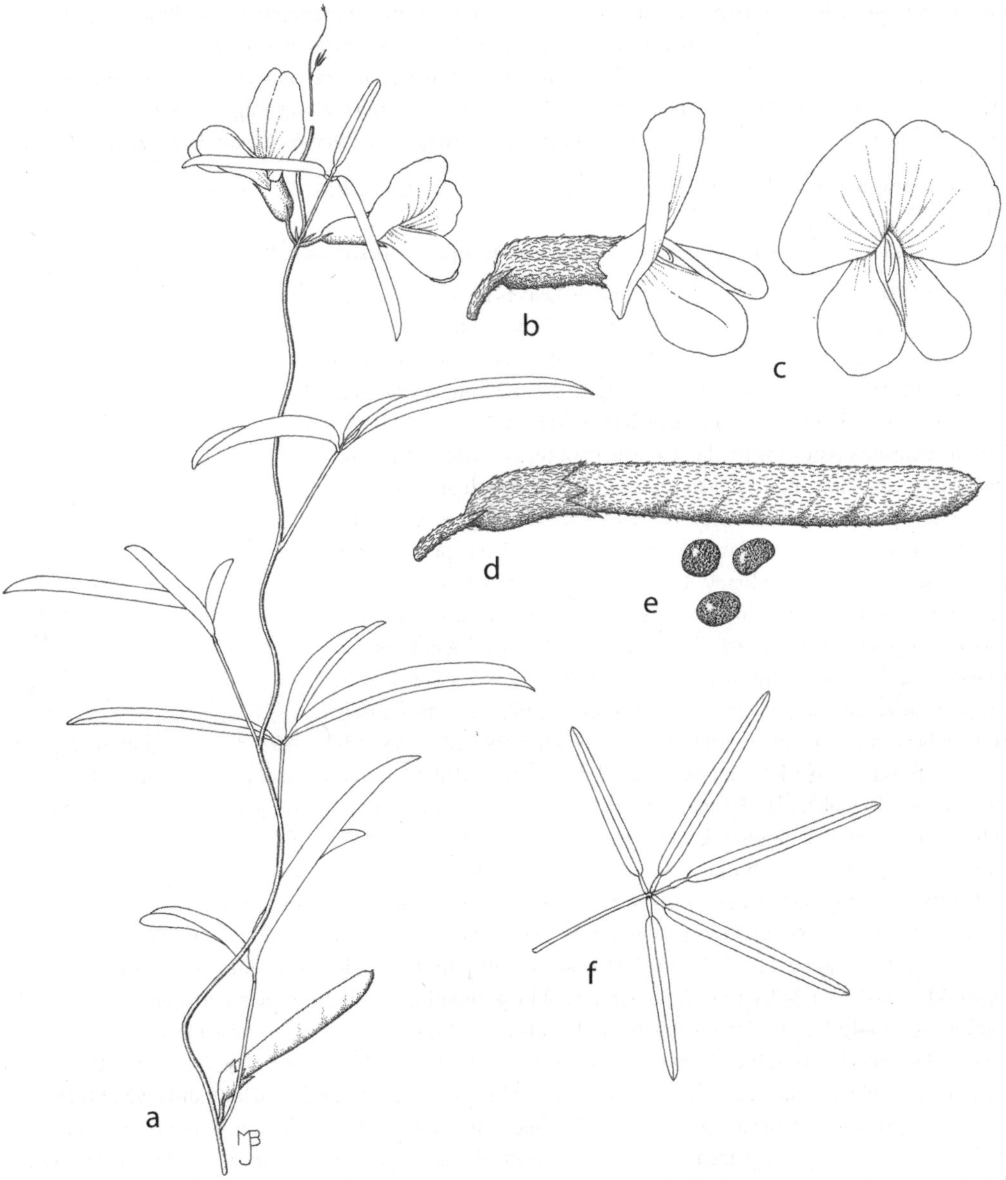

Cologania angustifolia: **a.** habit; **b., c.** flower (side and front views); **d.** fruits; **e.** seeds; **f.** leaf. Illustrator: Matthew B. Johnson.

pilose. **Seeds** (cleistogamous) (2–)4–10, 2–4.8 mm × 1.9–3.1 mm, oblong, circular, subquadrate, brown with darker patterns. $2n = 44$.

Flowering [Mar–]Jun–Sep[–Oct]. Ridges, slopes, canyons, meadows, streambanks, and roadsides, often on rocky soil derived from sandstone, limestone, or igneous parent rocks; [850–]1100–2860 m; (Semidesert Grassland), Madrean Evergreen Woodland, Petran Montane Conifer Forest; Apache, Cochise, Coconino, Gila, Graham, Greenlee, Navajo, Pima, Santa Cruz cos.; New Mexico, Texas; Mexico (Aguascalientes, Chiapas, Chihuahua, Durango, Guanajuato, Hidalgo, Jalisco, México, Michoacán, Morelos, Nuevo León, Oaxaca, Puebla, Querétaro, San Luis Potosí, Sinaloa, Sonora, Tamaulipas, Tlaxcala, Veracruz, Zacatecas).

Cologania angustifolia is characterized by its high variability, not only in leaflet and fruit morphology, but also in its habit. Several different forms throughout its distribution have received species names. These are not recognized in this treatment. Hybridization with other species also has been proposed (Fearing 1959, Turner 1992).

Cologania angustifolia is widespread and often locally abundant in mountainous areas of the state. It is found in various forest and woodland habitats, and occasionally in grasslands, as well as in disturbed sites. The plants grow in the shade beneath trees or in full sun and can be highly variable in appearance. Some plants may have stems that are trailing while others are erect. The size and shape of the leaflets, flowers, and fruits are highly variable. Awareness of this variability may aid in proper identification. *Cologania angustifolia* has been observed to nodulate (Allen and Allen 1981).

2. Cologania obovata Schlechtendal. Linnaea 12:287. 1838. * [from Latin, *ob*, reversed, and *ovat*, egg-shaped] LEMMON'S COLOGANIA Native

Cologania houghii Rose; *Cologania humifusa* Hemsley; *Cologania lemmonii* A. Gray; *Cologania pringlei* S. Watson

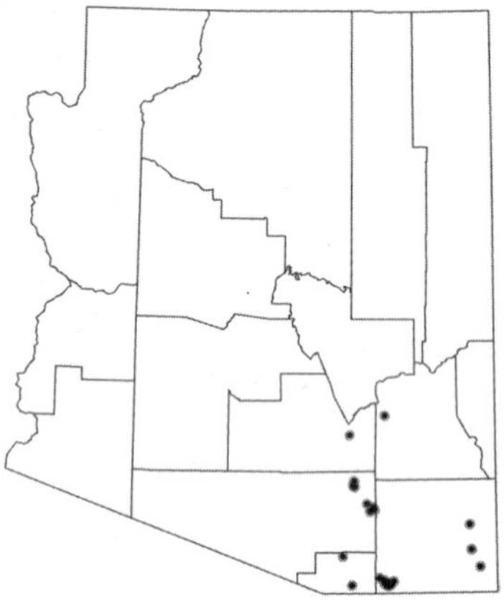

Herbs prostrate. **Stems** canaliculate, densely hirsute. **Leaves** 0.7–5.4(–6.3) cm; stipules 1.8–6.5(–7) × 0.4–3.2(–5.8) mm, oblong to broadly oblong, lanceolate, striate, slightly pubescent; petioles 0.2–0.5 cm, hirsutulous, sometimes with retrorse hairs; rachis (0.7–)1–5.1 mm, hirsute; stipels 0.8–1.2 mm, needle-like, densely pubescent, usually deciduous. **Leaflets** 3, (0.8–)1–5.4 × (0.6–)1–3.6 cm, obovate, obovate-elliptic to circular, cuneate to obtuse at base, obtuse and mucronate at apex, pubescent to densely pubescent, venation slightly conspicuous above, conspicuous below. **Inflorescences** fasciculate, pedunculate, with 1 or 2(–4) flowers, 2.1–10 cm; peduncle (0.3–)2–35 mm, sulcate, pubescent to slightly pubescent; bracts of peduncle 1.8–6.2 × 0.5–3.3 mm, oblong, linear-oblong, ovate, obovate-oblong, rarely lanceolate, acute or obtuse at apex, striate, densely pubescent, glabrate; bracts of pedicel 1.2–5 × 0.1–1.4 mm, linear, linear-lanceolate, lanceolate, linear-oblong; bracteoles 1.2–7.2 × 0.2–5(–5.5) mm, linear to lanceolate, slightly pubescent. **Pedicels** 5–67(–77) mm, densely to slightly pubescent, sometimes with retrorse hairs. **Flowers** with calyx (10–)13–19 × 3–6 mm, pubescent, lobes 2.4–7.5 × 1.7–3.6(–5) mm, upper lobe 2.3–9.5 × 1.5–2.8(–5.2) mm, deltate, never divided, pubescent; corolla purple, blue, magenta, red-purple, violet, lilac, lavender; banner 22–35 × 11–29 mm, broadly obovate to obovate, circular, rounded at apex, obcordate to rarely emarginate; wing petals 22–28.5 × 3–10.5 mm, obovate, slightly obovate to oblong, claws 11–15 mm; keel 15–20 × 3.2–4.5 mm, obovate-oblong, obtuse at apex, claws 9.5–15 mm; style 1.9–3.5 mm, slightly curved. **Fruits** (chasmogamous) 3–4.4 × 0.4–0.5 cm, slightly curved, densely pubescent. **Seeds** (chasmogamous) 7–10, 2.1–2.8 mm × 2.1–3.2 mm, commonly circular, rarely oblong, rhombic, or elliptic, light to dark brown with darker patterns. **Fruits** (cleistogamous) (1.5–)2–3.8 × 0.4–0.5 cm, densely pubescent, stipe ca. 1.5 mm long. **Seeds** (cleistogamous) (1 or) 2–8(–11), 1.8–3.3 × 1.8–3.1 mm, circular, rarely oblong, rhombic, or elliptic, brown with darker patterns. $2n = 44$.

Flowering [Jan–]Jul–Sep. Slopes, canyons, and roadsides, on rocky, sandy, or clayish soils derived from limestone or igneous substrates; [1120–]1490–2800 m; (Semidesert Grassland), Madrean Evergreen Woodland, Petran Montane Conifer Forest; Cochise, Pima, Santa Cruz cos.; Nevada; Mexico (Aguascalientes, Chihuahua, Coahuila, Durango, Hidalgo, México, Michoacán, Nayarit, Oaxaca, Puebla, Querétaro, San Luis Potosí, Sinaloa, Sonora, Tlaxcala, Zacatecas).

Cologania obovata is characterized by its prostrate habit, almost sessile leaves on short petioles and by having fruits either short and wide, or long and wide. Due to this variability Fearing (1959) and Turner (1992) proposed that this species possibly hybridized with *C. angustifolia*.

In Arizona, *Cologania obovata* is found in scattered populations in mountains in the southeastern part of the state. The plants may be locally common. They occur in open pine forests, pine-oak woodland, and occasionally in semidesert grassland and mixed conifer forest. The trailing stems can grow to 1 m in length.

3. Cologania pallida Rose. Contr. U.S. Natl. Herb. 8:38. 1903. * [from Latin, *pallid*, pale] PALE COLOGANIA Native

Herbs, twining, 20–50 cm long. **Stems** canaliculate, corrugate, hirsute to strigose, with pubescence densest basally. **Leaves** 2.9–13.9 cm, 3-foliolate; stipules 1.7–6 × (0.3–)0.8–1.8(–3) mm, lanceolate, oblong, linear-lanceolate, striate, strigose, glabrate or glabrous; petioles 0.9–6 cm long, canaliculate, corrugate, strigose to slightly strigose; rachis 1–19 mm, sulcate, corrugate, strigose; stipels 0.8–2.5 mm, glabrate. **Leaflets** 3, 0.6–5.8 × (0.2–)0.3–3.2 cm, elliptic, oblong-elliptic, ovate, ovate-elliptic, rarely oblong or oblong-obovate, obtuse, cuneate, rounded or rarely subcordate at base, acute, obtuse, or rarely acuminate at apex, mucronate, venation slightly conspicuous above, conspicuous to prominent below, strigose both above and below or sometimes glabrous above, pallid-green. **Inflorescences** fasciculate, sessile, with 1 or 2 chasmogamous flowers, occasionally mixed with 1 or 2 cleistogamous flowers, 2–3.8 cm; bracts of peduncle

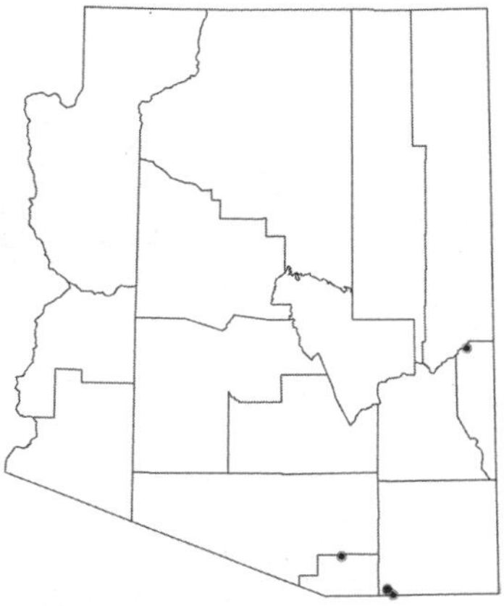

3.3–5.1 × 0.3–0.8 mm, linear-lanceolate, lanceolate, strigose to glabrate; bracts of pedicel 1.5–2.2 mm, linear, strigose; bracteoles linear, ca. 1 mm, strigillose. **Pedicels** 4–18 mm, strigillose or pilosulose. **Flowers** with calyx 8.5–13 × 2.8–4.1 mm, hirsute, rarely hirsutulous, strigillose or strigose, lobes 1.5–3.4(–5) × 1–2.2 mm, upper lobe 2.1–3 × 2.1–3.7 mm, not divided, not purple; corolla purple, purplish blue; banner (16–)21–24.5 × 7.4–14 mm, obovate, obcordate at apex; wing petals 12–21.5 × 2.6–4.8(–5) mm wide, obovate-oblong, claws 7–10 mm; keel 9–15.5 × 2–3.6 mm, obovate, slightly curved, claws 7–10 mm; style 2.2–2.8 mm, slightly to strongly curved, glabrous. **Fruits** (chasmogamous) 4.6–6.8 × 0.3–0.4 cm, slightly to strongly falcate, stipe 1.6–3.2 mm, glabrous, acute at apex, valves strigose, rarely pilose-hirsute. **Seeds** (chasmogamous) 5–11, 3.3–3.5 × 2.2–2.3 mm, oblong, brown. **Fruits** (cleistogamous) 1.9–5.2 × 0.3–0.4 cm, strongly curved, stipe ca. 0.5 mm, acute at apex, valves strigose, strigose-hirsutulous, with white trichomes, rarely glabrate. **Seeds** (cleistogamous) 2–6. $2n = 44$.

Flowering May–Sep. Slopes and canyons on rocky soils derived from limestone or igneous substrates; [1300–] 1740–2650 m; Madrean Evergreen Woodland, Petran Montane Conifer Forest; Cochise, Greenlee, Santa Cruz cos.; New Mexico, Texas; Mexico (Chihuahua, Coahuila, Sonora).

This species, still poorly known, is distinguished by its fascicles of 1 or 2 chasmogamous flowers rarely mixed with 1 or 2 cleistogamous flowers, and by its pallid-green leaves with usually elliptic to ovate-elliptic, strigose or glabrous leaflets. The fruits are often strongly curved.

Cologania pallida is highly localized in Arizona though plants may be locally common. It is documented from the Huachuca, Santa Rita, and White Mountains in pine forest, and in oak, juniper-pine, or pine-oak woodlands. Outside of Arizona it is documented to occur in grasslands.

COULTERIA Kunth. Nov. Gen. Sp. (quarto ed.) 6:328. 1824, descr. emended E. Gagnon, Sotuyo, and G.P. Lewis * [for Thomas Coulter, 1793–1843, Irish botanist who collected in Mexico]

Matthew B. Johnson

Shrubs and trees. Herbage densely pubescent at least when young. **Leaves** alternate, even bipinnate; stipules deciduous. **Leaflets** opposite, margins entire. **Inflorescences** axillary and terminal racemes, bracts at base of primary axis. **Pedicels** jointed. **Flowers** caesalpinoid, unisexual; calyx tube short, the upper 4 lobes nearly equal, the lower (abaxial) lobe elongated, hooded, margin glandular-fringed; corolla with 5 petals, yellow, nearly equal, tapering to a claw; stamens 10, free, somewhat longer than petals; ovary stipitate, style and stigma simple. **Fruits** indehiscent, compressed, papery to stiffly papery; stipe short; apex blunt or with a short point or beak. **Seeds** disk-shaped to oblong, compressed.

Species ca. 7 (1 cultivated in the flora): Mexico and Central America to Colombia, Venezuela, and the Caribbean.

Coulteria differs from the closely related *Tara* in having unisexual rather than bisexual flowers and in fruit and seed characters (Gagnon et al. 2016). Species of *Coulteria* are utilized for timber, woodworking, fuelwood, and as ornamentals.

SELECTED REFERENCE: Gagnon et al. 2016

1. Coulteria platyloba (S. Watson) N. Zamora. Monogr. Syst. Bot. Missouri Bot. Gard. 119:485. 2010. * [from Greek *platy*, broad and *lobus*, lobed] *PALO COLORADO* Cultivated

Caesalpinia platyloba S. Watson, Proc. Amer. Acad. Arts 21:425–426. 1886

Trees to 7 (12+) m. **Stems** with smooth, dark gray bark with conspicuous white lenticels, with age flaking off in irregular plates. **Young herbage** and inflorescences densely pubescent with short, brown velvety hairs; mature herbage glabrate. **Leaves** 27–36 cm with 2 or 3(4) pairs of pinnae, pinnae 10–18(–25) cm; stipules 7–8 mm, linear, quickly deciduous; petioles 1.5–6 cm; rachis 3–12 cm. **Leaflets** 5–7 pairs per pinna, 3–10 × 1–2.3(–3.1) cm, broadly elliptic to ovate. **Inflorescences** axillary or terminal on short axillary branches, 6.5–13 cm, often with 18–30(–40) flowers. **Pedicels** 8–12 mm. **Flowers:** calyx 3–4 mm except lower lobe, hairy, lobes yellow, lower lobe 9–10 mm, about 3 × longer than the others, with a glandular-fringed margin and cupping the stamens; corolla yellow, wing and keel petals 8–10 mm, banner 4–5 mm across; stamens with filaments yellow, 7–9 mm. **Fruits** red-brown or tan, 6–10.5 × 2.7–3.6 cm, oval with a conspicuous rim, persistent; stipe 2–5(–8) mm; apex blunt or with a short point to 2 mm. **Seeds** tan or olive green with margin darker, 9–13 × 7–11 mm, dull.

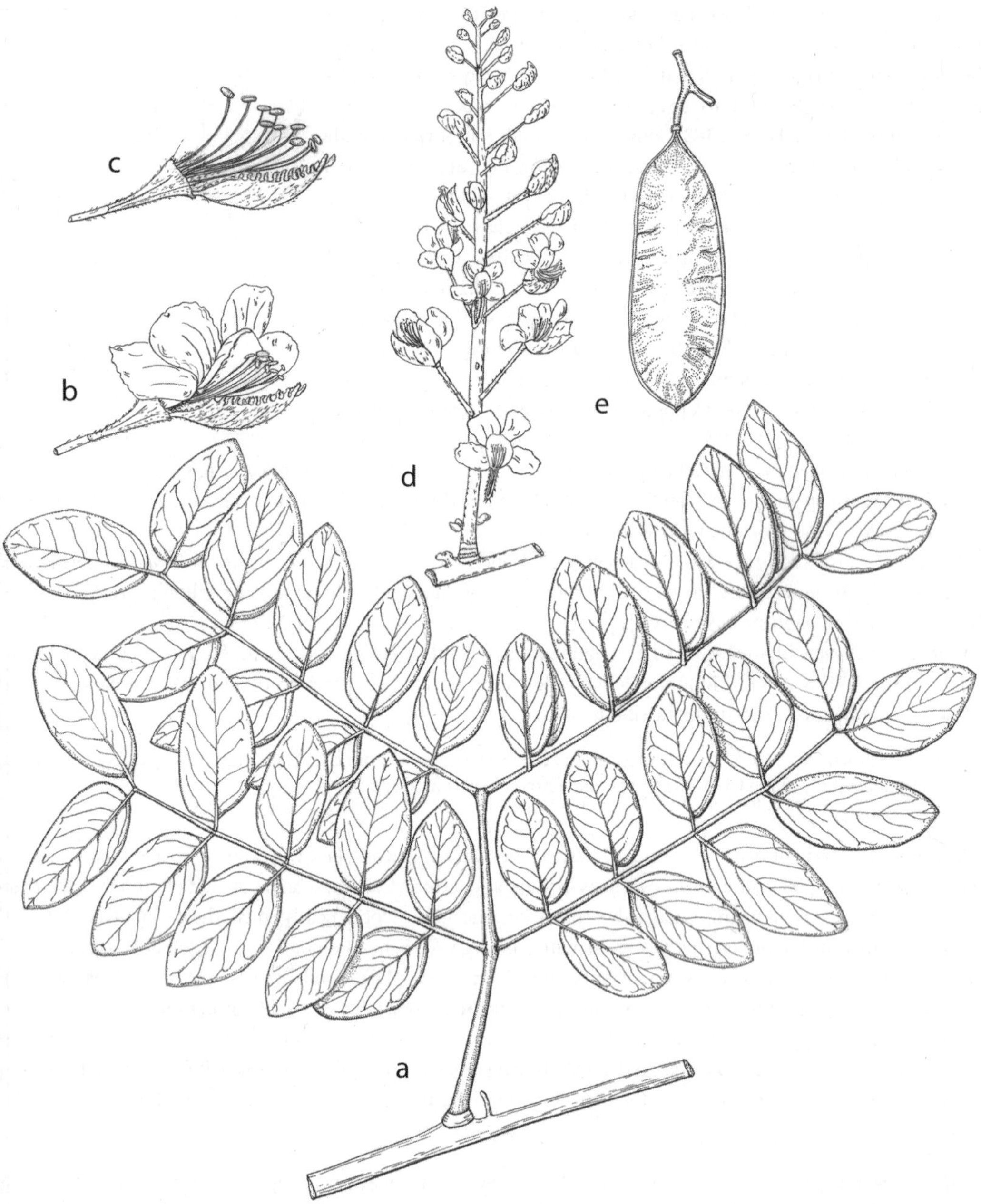

***Coulteria platyloba*: a. habit; b. flower; c. flower cut-away; d. inflorescence; e. fruit. Illustrator: Marsha Bennett.**

Flowering Jun–Jul. Mexico.

Coulteria platyloba is found from southeastern Sonora and Chihuahua to Oaxaca and Tamaulipas and is disjunct in Yucatán. In northwestern Mexico it is scattered on slopes, valleys, and in canyons in thornscrub and tropical deciduous forest. The common name "*palo colorado*" is derived from the red coloration of the leaves in the autumn before the trees become deciduous (Martin et al. 1998).

The hard, durable wood of *Coulteria platyloba* is valued for construction. Posts are said to last 100 years and can root and develop into trees, forming a living fence (Felger et al. 2001). The wood is known commercially as *chakte viga* and is also used for furniture, turnery, flooring, and other purposes. Sapwood is pale yellow and the heartwood varies from yellow orange to reddish orange. The trees have been seriously overharvested in many areas. It has potential for forestry as a timber species in seasonally dry tropical regions.

Palo colorado is rarely available from Arizona nurseries. It is suitable as a patio tree for warm-winter locations. The trees may be planted in narrow spaces due to their upright growth habit and relatively compact canopy. The flowers are attractive and are followed by clusters of persistent seed pods. Leaves are deciduous with drought and turn an attractive reddish color before falling. Typically deciduous in autumn, with irrigation the trees may remain evergreen unless the leaves are killed by freezing. *Palo colorado* requires periodic deep irrigation in warm weather and to maintain an acceptable appearance but is quite drought tolerant once established. A moderate growth rate can be achieved with adequate irrigation. *Palo colorado* suffers damage below -4°C and is severely damaged below -6°C. The large leaflets can produce litter when they fall. Scarified seeds readily germinate and larger stems can be rooted.

Taxa with potential for cultivation

Coulteria pumila (Britton & Rose) Sotuyo & G.P. Lewis. Phytotaxa 29(1):40. 2017
Palo Piojo is an open shrub which grows to 3 m tall with gray green bipinnate leaves and yellow flowers. The plants are deciduous with drought and cold. Hardy to ca. -5°C.

COURSETIA de Candolle. Ann. Sci. Nat. (Paris) 4:92. 1825. * [for George Dumont de Courset, author of *Botaniste Cultivateur*, Paris, 1802; his botanical garden displayed plants of central and southern Europe] BABYBONNETS
Matt Lavin, Kirsten N. Lake, Matthew B. Johnson

Trees, shrubs, subshrubs, and perennial herbs, usually with erect to ascending stems, rarely scandent. **Stems** with young growth sericeous, sometimes sparsely so. **Leaves** alternate, odd- or even-pinnate, [or rarely unifoliate]; stipules ephemeral to persistent, rarely spinescent; stipels absent or present. **Leaflets** [1–]3–27[–46], opposite, rarely alternate, narrowly to widely elliptic, entire. **Inflorescences** racemes [rarely panicles]; bracts quickly deciduous, rarely persistent; bracteoles absent. **Flowers** papilionoid; calyx bilaterally symmetrical, campanulate, 5-lobed; corolla whitish, reddish, purplish, to yellow, usually glabrous; stamens 10, 9 fused and 1 free or partially fused, anthers small, attached basally, dehiscing longitudinally; pistil**:** style glabrous but with a pollen brush tufted to loosely scattered along one side of the distal end; stigma rounded, ciliate, terminal; ovules 8–12. **Fruits** elastically dehiscent, laterally compressed, linear. **Seeds** 1–30, lenticular with an apical hilum. $x = 8$.

Species ca. 40 (3 native in the flora, 1 of those cultivated): throughout the neotropics primarily in grass-poor seasonally dry woodlands, secondarily in savanna woodlands and pine-oak forests; United States (Arizona, Texas, Puerto Rico, and the Virgin Islands), Mexico, Central America, South America.

Coursetia is readily distinguished from other North American legume genera by the combination of racemes with acropetal maturation of flowers where typically one flower per rachis reaches anthesis at a time (in contrast to *Robinia* and *Olneya*), style with a pollen brush born laterally (latrorse), to the inside (introrse), or along the outer side (extrorse), and legumes with seed compartments indicated by lateral or dorsal constrictions in the pod margins or walls between seeds. The few herbaceous *Coursetia* are distinguished from other herbaceous legume genera, which are usually temperate in distribution, by pulvinate leaves. Otherwise, *Coursetia* is highly variable morphologically and not as readily distinguished as related genera. Ecologically, *Coursetia* is centered in grass-poor seasonally dry tropical woodlands often rich in cacti and other succulent plants. This is perhaps why in the U.S. the genus is found mainly in the Sonoran Desert of Arizona and in the Tamaulipan thornscrub of south Texas.

Coursetia species are suitable for livestock browse and some species are useful for soil stabilization. The wood of larger species is strong and is used for construction. Several species are ornamental with attractive flowers.

SELECTED REFERENCES: Lavin 1988, Lavin and Sousa 1995, Lavin et al. 2003.

1. Shrubs to occasionally small trees, usually with multiple woody branches each over 1 m tall; leaves even-pinnate; legume with sinuous margins and without linear transverse impressions on the pod walls that delimit seed compartments ________ **3. Coursetia glandulosa**

1′ Herbs to subshrubs, at most one or two weakly woody branches from base and usually less than 1 m tall; leaves odd-pinnate; legume with straight margins and with linear transverse impressions on the pod walls that delimit seed compartments.

  2. Mostly herbs with prostrate to decumbent stems; roots bearing fusiform tubers; petals predominantly yellow; terminal leaflet no larger than laterals; lower (abaxial) surfaces of leaflets with reddish brown tannin deposits strictly along veins especially at the margins; ovary villous ________ **2. Coursetia glabella**

  2′ Mostly subshrubs with erect to ascending stems; roots usually not bearing fusiform tubers; petals predominantly whitish with a pinkish, yellowish, or reddish tinge; terminal leaflet distinctly larger than laterals; lower (abaxial) surfaces of leaflets without reddish brown tannin deposits or, if deposits present, then deposited in the central portion of the lamina and not extending much to margins; ovary with scattered small glands ________ **1. Coursetia caribaea**

1. Coursetia caribaea (Jacquin) Lavin. Adv. Legume Syst. 3:63. 1987. * [of the Caribbean] *ANIL FALSO*

Galega caribaea Jacquin, Select. Stirp. Amer. Hist. 212, pl. 125. 1763; *Cracca caribaea* (Jacquin) Bentham

Subshrubs and perennial herbs, [shrubs], to 0.6 [3] m tall. **Roots** usually not bearing fusiform tubers. **Stems** erect to ascending; at most one or two weakly woody branches from base. **Leaves** odd-pinnate, 2.5–15 cm; stipules persistent, awl-shaped, 3–12 × 1 mm; stipels minute or up to 2.5 mm; petioles 0.5–3 cm, sericeous; rachis with longitudinal grooves; petiolules 1 mm, glabrate. **Leaflets** 3–27, narrowly to widely elliptic, (6–)10–100 × 3–50 mm, terminal leaflet distinctly larger than laterals, base rounded, apex rounded, glabrate to densely sericeous, lower (abaxial) surfaces of leaflets without reddish brown tannin deposits or, if deposits present, then deposited in the central portion of the lamina and not extending much to margins. **Inflorescences** from leaf axils on new stems, often with 1–30 nodes mostly more than 3 mm apart, axes usually sericeous; bracts linear-lanceolate, 2–10 × 1 mm. **Pedicels** 1.5–8 mm. **Flowers:** calyx tube 2–3 mm, sericeous to hispid, rarely glandular, calyx lobes 2–11 mm, lower-most longer than the two lateral lobes, two upper (adaxial) lobes more fused than laterals; corolla predominantly whitish, sometimes tinged pinkish, yellowish, or reddish, 5–12 mm; filaments subequal; pistil glabrous; ovary with scattered small glands. **Fruits** light to dark brownish, sessile, linear with straight margins, with linear impressions on the pod walls, constricted between seeds, 3–9.5 × 0.3–0.7 cm, base blunt, apex acute and often terminating in persistent style base, glabrous to woolly, rarely glandular. **Seeds** 5–30. $2n = 16$.

Varieties 8 (1 native in the flora): open to closed understory of pine and oak forests, savanna woodlands, and grass-poor seasonally dry tropical forests and woodlands, 0–2000 m; southern Arizona, Puerto Rico, the Virgin Islands, and the Lesser Antilles, and throughout Mexico and Mesoamerica to coastal northern South America.

Flowering variably throughout the year especially at elevations below 1000 m; at higher elevations flowering appears to be concentrated during the wet season (e.g., mid-summer). Nodulation is reported for *Coursetia caribaea* (Dubey et al. 1972).

1a. Coursetia caribaea (Jacquin) Lavin var. **sericea** (A. Gray) Lavin. Syst. Bot. Monogr. 21:122. 1988. * [Latin *serica*, silky] *ANIL FALSO* Native

Cracca edwardsii A. Gray var. *sericea* A. Gray, Proc. Amer. Acad. Arts 17:201. 1882

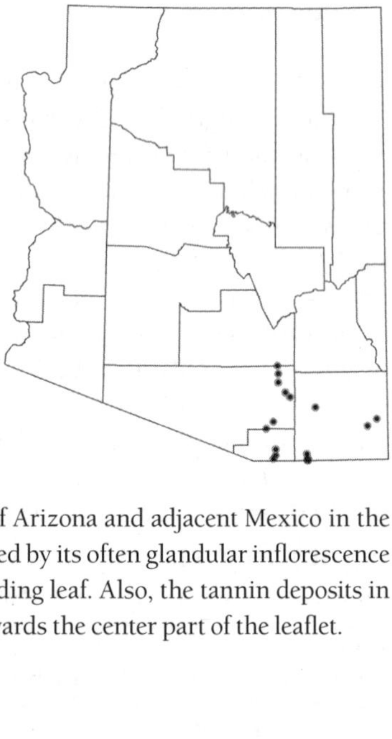

Herbs or subshrubs to 0.6 m. **Leaflets** 7–13, 9–28 × 4–14 mm, strigose to sericeous above, usually densely sericeous beneath, lower (abaxial) surface with tannin deposits present as dark purple reticulations concentrated towards the center of the leaflet. **Inflorescences**: rachis glandular with up to 30 nodes, 1.5–3.5 times the length of the subtending leaf.

Flowering May–Jun, Aug–Oct. Slopes, canyons, and roadsides, rocky soil, often in open to closed understory; [150–]1050–2000 m; occasional at upper elevations in Semidesert Grassland, Madrean Evergreen Woodland; Cochise, Graham, Pima, Pinal, Santa Cruz cos.; Mexico (Chihuahua, Durango, Sinaloa, Sonora).

Coursetia caribaea var. *sericea* is one of the forms of *Coursetia caribaea* that has more than seven leaflets and that also forms distinctive reticulate tannin deposits on the leaflet undersurfaces. In this context, var. *sericea* is distinguished geographically from var. *caribaea* by being the northernmost variety confined to upper elevations in the understory of pine and oak woodlands of Arizona and adjacent Mexico in the Sierra Madre Occidental (Lavin 1988). Morphologically var. *sericea* is distinguished by its often glandular inflorescence rachis with often up to 30 nodes and a length of 1.5–3.5 times that of the subtending leaf. Also, the tannin deposits in this variety are consistently present as dark purple reticulations concentrated towards the center part of the leaflet.

2. Coursetia glabella (A. Gray) Lavin. Adv. Legume Syst. 3:63. 1987. * [Latin *glaber*, hairless] SMOOTH BABYBONNETS Native

Cracca edwardsii A. Gray var. *glabella* A. Gray, Proc. Amer. Acad. Arts 17:201. 1882

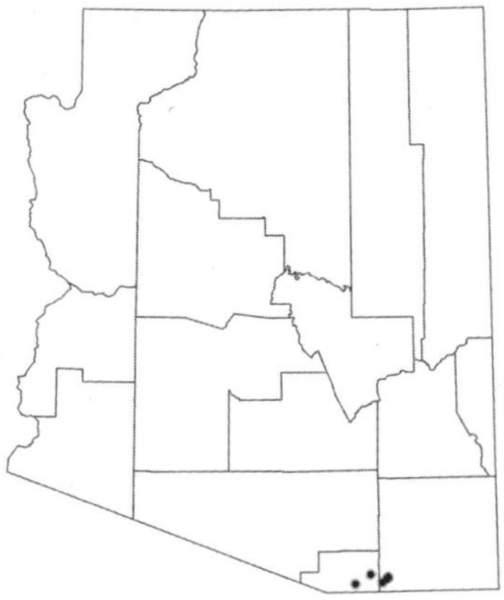

Perennial herbs, to 0.3 m. **Roots** bearing fusiform tubers. **Stems** prostrate to decumbent; at most one or two weakly woody branches from base. **Leaves** odd-pinnate, 3–7.5 cm; stipules persistent, awl-shaped, 3.5–5 × 1 mm; stipels minute; petioles 0.5–1 cm, glabrate; rachis with longitudinal grooves; petiolules 1 mm, glabrate. **Leaflets** 11–21, elliptic, 5–15 × 2–11 mm, terminal leaflet no larger than laterals, base rounded, apex rounded, glabrous above and sericeous beneath, lower (abaxial) surface with reddish brown tannin deposits strictly along veins especially on margins. **Inflorescences** from leaf axils on new stems, often with 5–15 nodes mostly more than 3 mm apart, axes sparsely pilose; bracts lanceolate, 2.5–4.5 × 1 mm. **Pedicels** 1.8–2.2 mm. **Flowers:** calyx tube 2–3 mm, pilose, calyx lobes 3–5 mm, lower-most longer than the two lateral lobes, two upper (adaxial) lobes more fused than laterals; corolla predominantly yellow, 9–12 mm; filaments subequal; pistil glabrous; ovary villous. **Fruits** light to dark brownish, sessile, linear with straight margins, with linear impressions on the pod walls, constricted between seeds, 3.5–6 × 0.3–0.4 cm, base blunt, apex acute and often terminating in persistent style base, pilose to strigose. **Seeds** 7–15. $2n = 16$.

Flowering during the summer. Open to closed understory of pine and oak forests; 1700–2200 m; Madrean Evergreen Woodland; Cochise, Pima, Santa Cruz cos.; nw Mexico (Chihuahua, Sonora, Sinaloa).

Coursetia glabella is the northernmost species of a small group (the *Coursetia pumila* group, Lavin 1988) of four mainly Mexican species that are set off from the rest of the genus by an herbaceous habit, decumbent growth form, and predilection for pine and oak forests. The yellowish petals that contrast with the large reddish calyx lobes readily distinguish *Coursetia glabella* from some of its close relatives to the south. The preference of pine-oak understory may include open grass-poor undisturbed settings. This contrasts with *Coursetia caribaea*, which grows in the vicinity of *Coursetia glabella* and appears to inhabit similar but more disturbed or grassy settings.

In Arizona, *Coursetia glabella* has a restricted distribution. It has been documented as occurring in the Huachuca and Patagonia Mountains and the Canelo Hills. Due to its limited distribution, it is listed as a Species of Concern by the U.S. Fish and Wildlife Service and a Sensitive Species by the U.S. Forest Service. Martin et al. (1998) report flowering for *C. glabella* from July to September in the Río Mayo Region of Sonora and Chihuahua, where it is found from 1540 to 1800 m elevation.

3. Coursetia glandulosa A. Gray. Proc. Amer. Acad. Arts 5:156. 1861. * [Latin, glandular] ROSARY BABYBONNETS; *SAMO PRIETO* Native, Cultivated

Coursetia microphylla A. Gray; *Coursetia seleri* Harms

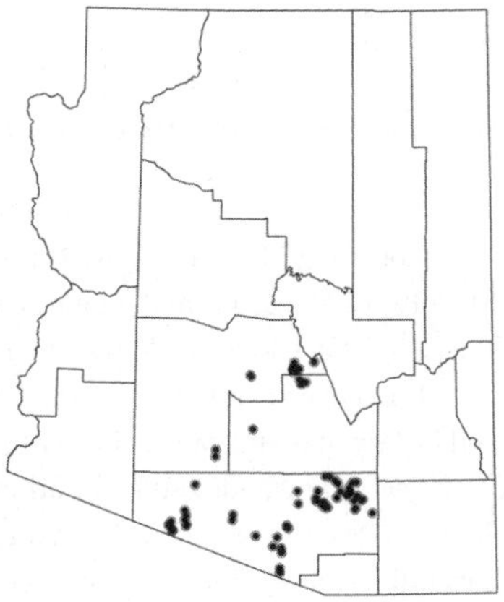

Trees and shrubs, to 3 [10] m. **Roots** diffuse. **Stems** erect; usually with multiple woody branches; bark smooth, gray. **Leaves** even-pinnate, 4.5–15 cm; stipules usually persistent, awl-shaped, 5–8 × 1 mm; stipels absent; petioles 0.5–2 cm, strigose; rachis with longitudinal grooves, extending past distal leaflet pair into a conspicuous mucro; petiolules 1 mm, glabrate. **Leaflets** 8–22, narrowly to widely elliptic, 9–50 × 3–22 mm, base rounded, apex rounded, glabrate to strigose. **Inflorescences** often from short shoots of older twigs, often with 20–40 nodes mostly 1–5 mm apart, axes sparsely to densely stipitate glandular; bracts ovate to lanceolate, 1–3.5 × 1 mm. **Pedicels** 3–10(–12) mm. **Flowers**: calyx tube (2.5–)3–4 mm, stipitate glandular, calyx lobes 2–5 mm, lower-most longer than the two lateral lobes, two upper (adaxial) lobes more fused than laterals; corolla predominantly whitish, 12–15 mm; filaments subequal; pistil glabrous; ovary densely stipitate-glandular. **Fruits** light to dark brownish, sessile, linear but with sinuous margins, without linear transverse impressions on the pod walls that delimit seed compartments (constricted between seeds), 2.5–11 × 0.5–0.7 cm, base blunt, apex acute and often terminating in persistent style base, glandular. **Seeds** 1–13. $2n = 16$.

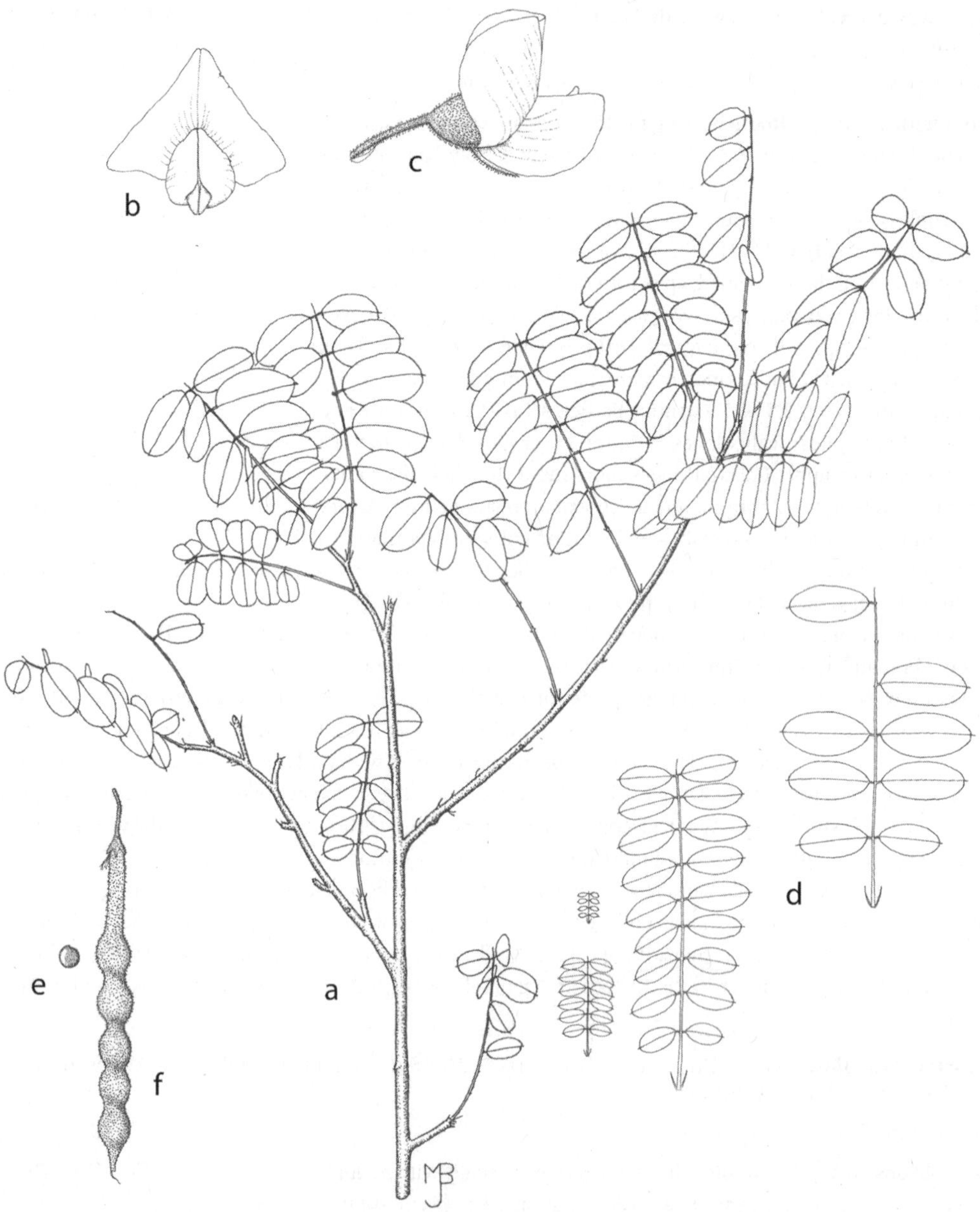

***Coursetia glandulosa*: a. habit; b. flower (front view); c. flower (side view); d. leaves showing size variability; e. seed; f. fruit. Illustrator: Matthew B. Johnson.**

Flowering Mar–May (Jun, Nov–Dec). Rocky slopes, among boulders, canyons, arroyos, and rarely in deep sand; [0–]240–1220[–1800] m; Sonoran Desertscrub, rare in Mohave Desertscrub, and rare at lower elevations in Interior Chaparral; Gila, Mohave, Maricopa, Pima, Pinal cos.; w Mexico.

The range of variation reported in the description tends toward the larger end in southern Mexico and the smaller end in Arizona. This range of variation especially in leaflet size and plant height is clinal.

Coursetia glandulosa is found in open dry shrubby vegetation and seasonally dry woodlands centered in the Sonoran Desert where it is abundant in desert mountain ranges of south central Arizona. It extends from Sonora and Baja California Sur into the Pacific coastal dry forest of Mexico as far south as Oaxaca. A small, disjunct population of this species is documented as growing in deep sand in Mohave Desertscrub, in Mohave County near the Arizona/Nevada border (*P. J. Leary 4315,* 17 May 1995). The plants typically flower in dry seasons, but mass flowering can occur following rains (Felger et al. 2001).

The long stems of *Coursetia glandulosa* can bend to small angles without breaking and have thus been used by the Seri Indians for harpoon shafts, bows, and other tools (Felger and Moser 1985). In southern Sonora, the slender stems are woven into furniture (Martin et al. 1998). The conspicuous, dark reddish, gummy resin found on the stems of this species from throughout its range is the result of a symbiotic relationship between ants and the larva of a scale insect, *Tachardiella* (Hemiptera: Kerriidae). This lac exudate is known as *goma Sonora* or *gomilla de Sonora* in Mexico (Felger et al. 2001) and is highly prized by the Seri Indians for many purposes (Felger and Moser 1985). *Coursetia glandulosa* has been used in folk medicine to treat tuberculosis. The gum from the lac scale is dissolved with sugar in water and taken to prevent chills and fever (Martínez 1959). The gum is also used to make a cough syrup and as an ingredient in varnish (Allen and Allen 1981).

Rosary babybonnets is infrequently cultivated as a landscape plant in Arizona. It can add interest to a desert garden when grouped among boulders with succulents to provide contrast. The small flowers are attractive. The numerous stems and potential size make it suitable as an informal screen. Established plants are drought and heat tolerant but require regular irrigation to maintain an attractive appearance and acceptable growth rate. The foliage is gradually deciduous with drought or freezing temperatures. With adequate irrigation, rosary babybonnets can grow up to 1 m per year. The species can sustain freeze damage below -5°C and is severely damaged at -7°C but recovery is rapid the following season if sufficient irrigation is provided. Pruning is usually necessary only to remove freeze-damaged stems or if a tree-like form is desired. The species is readily propagated from scarified seeds.

CROTALARIA Linnaeus. Sp. Pl. 2:714. 1753, name conserved. * [Greek, *krotalon*, castanet, alluding to the sound of the ripe seeds rattling in the dry pod] RATTLEPOD, RABBITBELLS

Guy L. Nesom

Herbs or shrubs. Stems glabrous to strigose or hirsute. **Leaves** alternate, 1-foliolate, or digitately 3(–7) foliolate; usually petiolate; stipules thread-like to leaf-like or lacking, sometimes decurrent, usually persistent; stipels absent. **Leaflets** lanceolate or ovate to obovate, elliptic, oblong, or linear, entire. **Inflorescences** usually terminal, leaf-opposed, or less commonly axillary racemes or heads, occasionally of solitary or tightly clustered flowers; floral bracts persistent or quickly deciduous, 1–8 mm; bracteoles small, paired below calyx. **Pedicellate. Flowers** papilionoid; calyx bilaterally symmetrical, campanulate to cylindric, 5-lobed; corolla usually yellow, variously marked, less often white or blue, glabrous or hairy outside; stamens 10, filaments partially or mostly united into a single tube, anthers alternately long, attached at base, and small, attached dorsally, dehiscing longitudinally; pistil: style with 1–2 lines of hairs on the upper part; stigma small, terminal, usually bilobed; ovules 1–70. **Fruits** dehiscent, though often tardily so, usually inflated, globose or ovoid to ellipsoid or cylindric, subsessile to long-stipitate. **Seeds** 1–70; oblique-cordiform to oblong-reniform, with a definite hilar sinus, sometimes with a conspicuous aril. $x = 7, 8$.

Species ca. 600 (2 native in the flora): throughout the tropics and extending into subtropics of the Southern Hemisphere, most numerous in eastern and southern tropical Africa.

Several species of *Crotalaria* are of economic importance. These are used for fiber, fodder, green manure, cover crops, erosion control, medicine, and human food (Allen and Allen 1981, Duke 1981, Lewis et al. 2005). *Crotalaria juncea* is especially valued as a source of strong fiber for cordage, fishing nets, and high quality paper, and was used for making camouflage nets during World War II (Allen and Allen 1981, Duke 1981, Lewis et al. 2005). The Guarajío people of Sonora formerly ate the seeds of *C. incana* (Martin et al. 1998). Duke (1981) reports that the seeds of *C. pallida* have been used as a coffee substitute. The seeds of *C. juncea* have been used in folk medicine to treat impetigo and psoriasis, to promote menstrual discharge, and as a blood purifier, and the seeds of *C. spectabilis* are used to treat impetigo and scabies (Duke 1981). Alkaloids found in numerous species, including *C. sagittalis* and *C. spectabilis*, cause crotalism poisoning in livestock, poultry, and humans (Allen and Allen 1981, Duke 1981, Lewis et al. 2005). Certain species are tolerant of soils that contain heavy metals such as cobalt, copper, and manganese, and have been used as indicator plants in mineral prospecting (Allen and Allen 1981). These species may be useful in bioremediation. Nodulation is widely reported for *Crotalaria* (Allen and Allen 1981).

SELECTED REFERENCES: Miller 1967, Windler 1971a, 1971b, 1974a.

1. Leaves trifoliolate; stipules bristle-like, not decurrent, quickly deciduous; stems mostly decumbent to prostrate, minutely and very sparsely strigillose _____ **1. Crotalaria pumila**

1′ Leaves unifoliolate; stipules of at least the middle and upper leaves lanceolate-auriculate, decurrent on the stem, persistent; stems erect to decumbent, hirsute-pilose _____ **2. Crotalaria sagittalis**

1. Crotalaria pumila Ortega. Nov. Pl. Descr. Dec. 2:23. 1797. * [Latin, dwarf] LOW RATTLEBOX; *CASCABEL DE VIBORA* Native

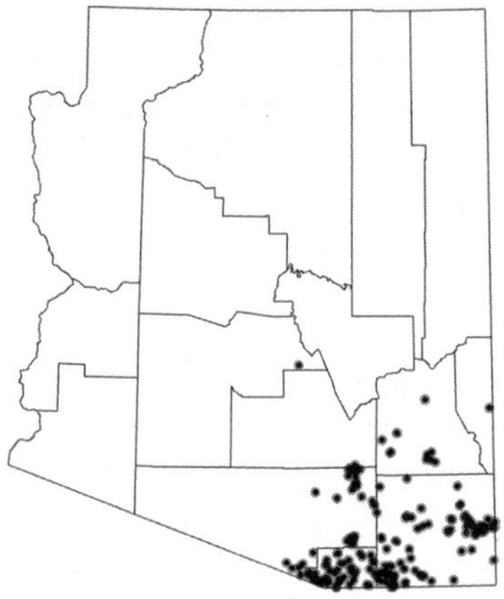

Herbs, annual to perennial. **Stems** mostly decumbent to prostrate, 15–100 cm, sometimes basally woody, minutely and very sparsely strigillose. **Leaves:** stipules bristle-like, quickly deciduous. **Leaflets** 3, narrowly obovate to oblong or elliptic-oblong, 7–15(–35) mm, 1.5–3.5(–6) times longer than wide, strigose below (abaxially), glabrous above (adaxially). **Inflorescences** with 1 or (2–)4–8 flowers in terminal to subterminal or lateral racemes 1–6(–10) cm; floral bracts subulate, quickly deciduous. **Flowers:** calyx 3–5.5 mm, campanulate, strigose, lobes triangular; corollas 7–11 mm, yellow, often red- or orange-tinged or red-lined. **Fruits** 12–20 × 4–8 mm, minutely strigillose, yellowish.

Flowering Dec–May, (Jul–)Aug–Oct in sw United States. Gravelly slopes, low areas among rocks, along streams, alluvial soil; [1–10]1040–2030[–2150] m; Semidesert Grassland, Plains Grassland, Madrean Evergreen Woodland, Petran Montane Conifer Forest, and rarely at upper elevations in riparian habitats in Chihuahuan Desertscrub and Sonoran Desertscrub; Cochise, Graham, Greenlee, Maricopa, Pima, Pinal, Santa Cruz cos.; Florida, Maryland, New Mexico, Oklahoma, Texas; Mexico, West Indies, Central America, South America.

Plants of *Crotalaria pumila* from the sw United States tend to have more elongate leaflets than those from Florida.

In the southeastern U.S., *Crotalaria pumila* occurs in sandy wastes, dunes and sand ridges, sandy pine woods, dune thickets, and hammock margins. In the southwestern U.S., it is found in grassy hillsides, creek bottoms, and alluvium. It grows in the open or as an understory plant beneath trees. In Sonora, *C. pumila* is found in thornscrub and tropical deciduous forest and grows to 2150 m elevation in the Sierra Madre Occidental in Chihuahua (Martin et al. 1998). Further south it is found in a variety of other plant communities.

2. Crotalaria sagittalis Linnaeus. Sp. Pl. 2:714. 1753. * [Latin *sagitta*, arrow-like] ARROW-HEAD RATTLEBOX, COMMON RATTLEBOX

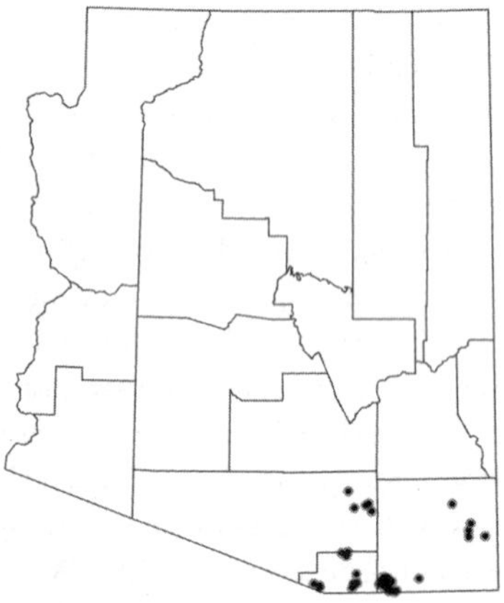

Herbs, annual. **Stems** erect to decumbent, 4–40 cm, hirsute-pilose. **Leaves:** stipules of distal nodes lanceolate-auriculate, decurrent on stems, with 2 free, triangular, distal lobes, persistent, often absent in var. *blumeriana*. **Leaflets** 1, elliptic to oblong or linear-lanceolate, 10–80 mm, 4–8 times longer than wide, subsessile, strigose on both surfaces. **Inflorescences** with 1 or 2–5 flowers in terminal to subterminal or lateral racemes, 2–6 cm; floral bracts linear-triangular, persistent. **Flowers:** calyx 5–8 mm, broadly cylindric, glabrous to slightly puberulous, lobes triangular-lanceolate; corolla 9–11 mm, yellow. **Fruits** 7–35(–40) × 5–11 mm, glabrous. $2n = 32$.

Varieties 2 (2 native in the flora): e United States, se Arizona; Mexico, West Indies, Central America.

1. Stems 4–10 cm, erect to decumbent; stipules reduced or absent; fruits 7–15 mm ________ **2a. Crotalaria sagittalis** var. **blumeriana**
1′ Stems 10–40 cm, erect; stipules prominent; fruits 10–27(–36) mm ________ **2b. Crotalaria sagittalis** var. **sagittalis**

2a. Crotalaria sagittalis Linnaeus var. **blumeriana** H. Senn. Rhodora 41:339. 1939. * Native

Crotalaria pringlei A. Gray

Stems erect to decumbent, 4–10 cm. **Stipules** reduced or absent. **Fruits** 7–15 mm.

Flowering Aug–Oct. Rocky ridges, sandy alluvium; 1830–2400 m; Madrean Evergreen Woodland; Cochise, Pima cos.; Mexico (Chihuahua, Sonora).

Numerous varieties have been described among populations of the widespread and "erratically variable" *Crotalaria sagittalis*, but neither Isely (1998) nor Windler (1974b) recognized varieties, noting that further study was necessary to understand the variation patterns, especially among the Mexican populations. Isely (1998), however, observed that "the isolated Arizona populations (var. *blumeri[ana]* Senn) have small pods (0.7–1.5 cm), and the often tiny plants may be either erect or decumbent. They are not the same as those of the eastern United States." In contrast, garden plants grown from seeds of representative plants of var. *blumeriana* "were as large as any of the plants grown from field seed of typical Mexican *C. sagittalis*" (Windler 1973, p. 301). The type of *C. pringlei* is more like typical *C. sagittalis* and is unusual among the Arizona populations.

Crotalaria sagittalis var. *blumeriana* is formally treated here, recognizing that it is far out of range, geographically and ecologically, from typical *Crotalaria sagittalis* of the eastern United States. Its evolutionary affinities perhaps are with Mexican populations, but almost all of the species in Mexico occurs in the southern half of the country, disjunct from both the Arizona and eastern North American plants.

2b. Crotalaria sagittalis Linnaeus var. **sagittalis** * Native

Crotalaria sagittalis Linnaeus var. *fruticosa* (Miller) Fawcett & Rendle; *Crotalaria sagittalis* Linnaeus var. *oblonga* Michaux

Stems erect, 10–40 cm. **Stipules** prominent. **Fruits** 10–27(–36) mm.

Flowering [Jun–]Aug–Oct(–Nov). Along streams in sandy alluvial soil, often in canyons, and on rocky slopes and ridges; [5–600]880–2130 m; Semidesert Grassland, Madrean Evergreen Woodland, and in riparian habitats at upper elevations in Sonoran Desertscrub; Cochise, Pima, Santa Cruz cos.; Alabama, Arkansas, Connecticut, Delaware, District of Columbia, Georgia, Illinois, Indiana, Iowa, Kansas, Kentucky, Louisiana, Maryland, Michigan, Minnesota, Mississippi, Missouri, Nebraska, New Hampshire, New Jersey, New York, North Carolina, Ohio, Oklahoma, Pennsylvania, Rhode Island, South Carolina, South Dakota, Tennessee, Texas, Vermont, Virginia, West Virginia, Wisconsin; Mexico, West Indies, Central America.

In Arizona, *Crotalaria sagittalis* var. *sagittalis* occurs in scattered populations. It may be locally abundant in the open or in shady locations. It is often found along streams. In the Río Mayo Region of southern Sonora and adjacent Chihuahua, it has been documented in tropical deciduous forest and pine-oak woodland, from 240–2000 m elevation (Martin et al. 1998). The species is widely distributed. Across its range it occupies a variety of habitats including pastures, sandy fields and prairies, roadsides, disturbed sites, shale banks, rocky limestone hills, serpentine barrens, longleaf pine savannas, pine flatwoods, pine barrens, and oak-hickory woods.

Crotalaria sagittalis var. *sagittalis* is known to cause crotalism in cattle, horses, and sheep (Allen and Allen 1981). The animals suffer liver damage and fatalities are documented. This condition was formerly known as Missouri River bottom disease in the central U.S. (Allen and Allen 1981). Nodulation is reported for *C. sagittalis* (Allen and Allen 1981).

DALBERGIA Linnaeus f. Suppl. Pl. 52, 316. 1782. * [for brothers Carl Gustav Dalberg, ca. 1720/1721–1781, and Nils Ericsson Dalberg, 1736–1820, Swedish botanists]

Matthew B. Johnson

Woody lianas, shrubs, or trees, armed or unarmed. **Leaves** alternate; odd or even-pinnate or rarely unifoliolate; stipules variable in size, deciduous or persistent; stipels absent. **Leaflets** alternate or sometimes appearing opposite, margins entire. **Inflorescences** racemes, panicles, or cymes, axillary or terminal, few- to many-flowered; flowers subtended by a pair of bracteoles; bract small, broad, semi-persistent. **Pedicels** present or absent. **Flowers** papilionoid, bilaterally symmetric, small, 5-merous; calyx bell-shaped or tubular, lobes unequal; corolla with petals clawed, banner rounded or ovate, wings oblong, keel oblong, shorter than the wings, united at tip; stamens enclosed by keel, filaments 9 or 10, united into 1 or 2 bundles, or with a single free stamen; ovary 1-chambered, stigma and style simple. **Fruits** indehiscent, either samara-like or drupe-like, linear or oblong, stipitate, margins not thickened at suture. **Seeds** flattened and somewhat kidney-shaped. $x = 10$.

Species ca. 250 (1 cultivated in the flora): pantropical in parts of Mexico, the Caribbean, South America, Africa, and Asia to Australia.

The timber of some species of *Dalbergia* is important for construction, cabinet making, furniture, wood inlay, musical instruments, tool handles, turnery, and carving (Lewis et al. 2005). Various species are used for fodder, fiber, fuel, and have medicinal applications (Lewis et al. 2005). The wood of some species can cause contact dermatitis (Lewis et al. 2005).

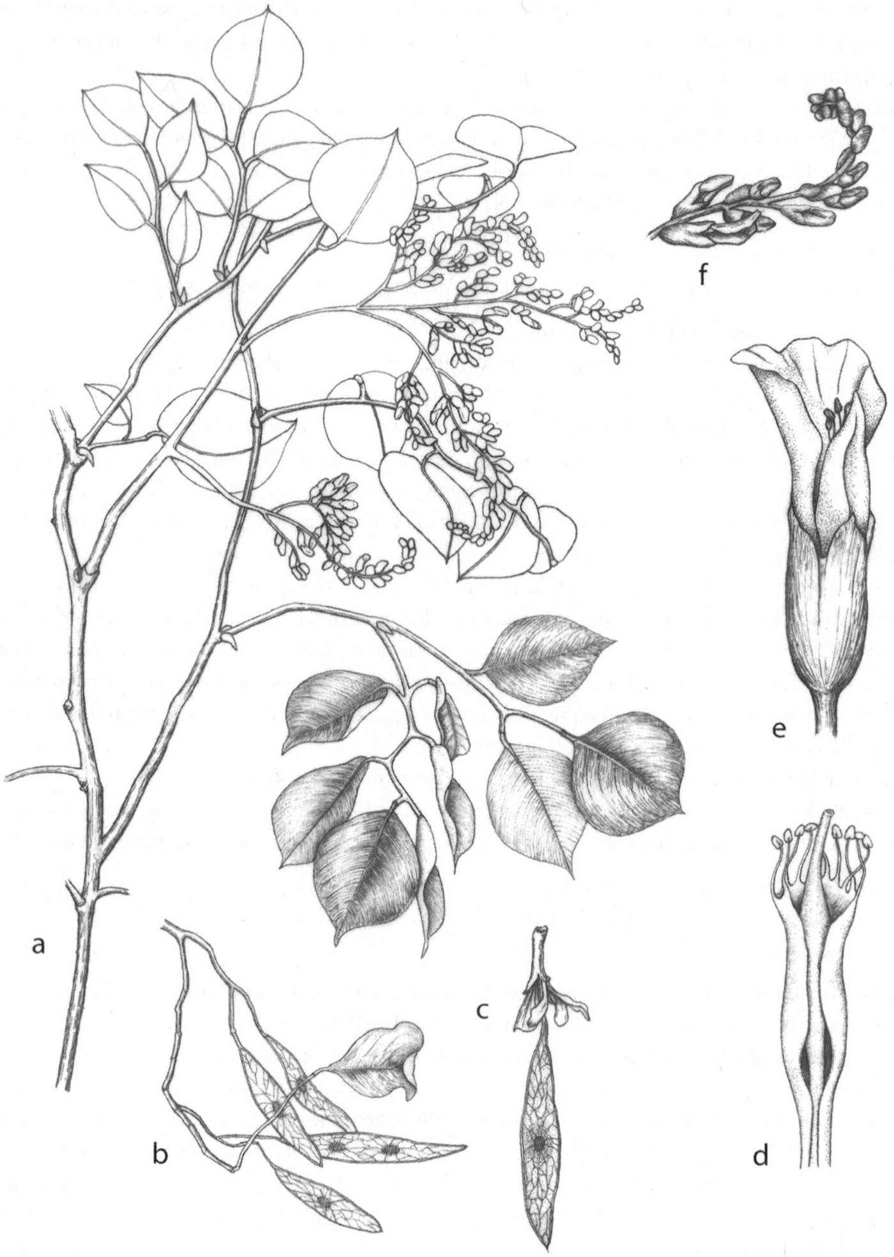

Dalbergia sissoo: **a.** habit; **b.** fruits; **c.** fruit; **d.** stamens and pistil; **e.** flower; **f.** young inflorescence. Illustrator: Chris Bondante.

1. Dalbergia sissoo Roxburgh ex de Candolle. Prodr. 2:416. 1825. * [from a common name for the tree] INDIAN ROSEWOOD, SISSOO, SISSOO TREE, SHISHAM Cultivated

Trees 8–15(–30) m. **Stems:** trunk to 1 m in diameter; bark pale tan and smooth, becoming rough, fissured, and gray; unarmed. **Herbage** with fine white hairs on young twigs, petioles, rachises, petiolules, and inflorescence stalks. **Leaves** odd-pinnate, 7–17 cm; stipules 6–9(–14) mm, lanceolate, quickly deciduous; petioles 1.5–3.5 cm, without glands; rachis 2.7–5.8 cm; petiolules 5–8 mm. **Leaflets** 3–5, alternate, broadly ovate to nearly rounded, 2.5–7.5 × 2–6 cm, base cuneate, apex acuminate, glabrate. **Inflorescences** 30–60 flowered, axillary panicles. **Pedicels** to 1 mm or flowers appearing sessile. **Flowers** appearing tubular; calyx green, 5–7 mm, tubular; corolla white or yellowish white, green basally, 8–9 × 3–3.5 mm, banner 8–9 mm, wings 6.5–7.5 mm, keel 6–7 mm; stamens 9, united into a bundle that is split above (adaxially). **Fruits** tan, conspicuously net-veined, 3.7–10 × 0.7–1.3 cm, samara-like, flattened and thin, oblong to elliptic, stipe 4–5 mm, apex rounded or obtuse. **Seeds** 1–4; brown, 7–9 mm.

Flowering Jan–Feb(–Apr). India, Nepal, Pakistan.

Dalbergia sissoo commonly grows along rivers in the Himalayan foothills at elevations up to 1500 m. This habitat has a 3–4 month dry period and a summer monsoon that provides most of the 500–2000 mm average yearly rainfall (Sheikh 1989). *Dalbergia sissoo* grows in a variety of soils from river alluvium to sand and gravel and is tolerant of some salinity (Sheikh 1989). Nodulation is reported for *D. sissoo* (Allen and Allen 1981).

Dalbergia sissoo is considered to be an excellent quality timber for making cabinets, furniture, and wood veneers (Sheikh 1989). The sapwood is white or pale tan and the heartwood is yellow brown to dark brown. The sapwood is susceptible to fungi and boring insects but the heartwood is extremely durable and resistant to termites (Sheikh 1989). The heartwood has a specific gravity of 0.7–0.8 (Sheikh 1989). Wood of *D. sissoo* is valued for fuel and for making charcoal (Lewis et al. 2005). The wood has a calorific value of 4908–5281 kcal/kg (Sheikh 1989). The trees coppice readily following cutting (NRC 1980). The extensive root system and ability to produce root suckers can be useful for erosion control (Sheikh 1989). The foliage and young stems of *D. sissoo* is browsed by wildlife and used for livestock fodder, especially in the winter and spring months (Sheikh 1989). Use of the leaves as a green vegetable for human consumption has been reported (Sheikh 1989). Sheikh (1989) also indicates that there are numerous medicinal uses employing the leaves, bark, and wood shavings. The tree has been used in folk medicine as an abortifacient, aphrodisiac, expectorant, antihelminthic, antipyretic, and for treating skin diseases and digestive problems (Hajare et al. 2000, Asif and Kumar 2009), and the oil of *D. sissoo* has been used to repel mosquitoes (Asif and Kumar 2009). Niranjan et al. (2010) indicate that leaf extracts of *D. sissoo* have potential in treating diabetes. They mention other folk remedies employing *D. sissoo* including treatment for gonorrhea, syphilis, eye infections, blood diseases, anal disorders, urinary tract infections, scabies, and leprosy. Root extracts of *D. sissoo* are considered a potentially useful antioxidant due to effectiveness in scavenging free radicals (Pooja et al. 2010).

Sissoo is widely planted in lower elevation desert communities of Arizona as a street tree and for shade in residential yards and urban spaces. In appearance, it somewhat resembles cottonwood (*Populus* spp.) and commonly develops a trunk 3–6 dm in diameter. The inconspicuous flowers are fragrant. Growth rate is moderate to fast with sufficient irrigation. Periodic deep irrigation is necessary during the growing season. The foliage is typically evergreen to partially deciduous unless killed by freezing, but the trees can become deciduous with extended drought. Plants sustain twig damage at ca. -6°C and are severely damaged at lower temperatures. Sissoo exhibits rapid recovery following freezing. Some pruning is necessary to shape when young and to remove freeze-damaged stems. Sissoo has generally been free of maintenance problems in Arizona landscapes. However, in some situations the shallow roots have damaged hardscaping, so the trees should be carefully sited. Sissoo is very difficult to eradicate once established as the extensive root system can produce vigorous root sprouts. Propagation is by seeds and cuttings. Seeds do not require scarification but may be soaked in water for 48 hours prior to sowing (Sheikh 1989).

DALEA Linnaeus. Opera Var. 244. 1758, name conserved. * [for Samuel Dale, 1659–1739, English botanist and physician] DALEA, PRAIRIE CLOVER

Michelle M. McMahon, Mary Welch-Keesey, and Matthew B. Johnson

Annual or perennial herbs, subshrubs or shrubs (to small trees in one cultivated species); perennial species often with a well-developed taproot and sometimes a woody caudex. **Herbage** and usually flowers with blister-like glands with aromatic oils; herbage usually pubescent. **Stems** solitary or few to numerous, erect, ascending or prostrate, densely branching in a few shrubby members. **Leaves** alternate, odd-pinnate (rachis in most extended beyond the final pair but if not, and only 3 leaflets, appearing palmate in one species, even-pinnate in one Mexican species); stipules herbaceous or subglandular; leaves usually with petioles (although very short to nearly absent in some); stipels absent. **Leaflets** 3 to several dozen, terminal usually pinnate, some palmate with distal pair. **Inflorescences** spikes (racemes), terminal to primary branches, in some on axillary short shoots, moderately loose to usually dense, then the axis not visible between the crowded persistent calyces, subcapitate to elongated; bracts generally conspicuous, persistent or deciduous, variably glandular and/or pubescent. **Pedicels** very short (≤ 1 mm) to most commonly absent. **Flowers** papilionoid or nonpapilionoid (see corolla, below); calyx tube 10-ribbed, generally glandular in intervals, often asymmetric at the orifice, with a recessed sinus behind the banner, lobes deltate to long bristle-like, greater than or less than tube, often plumose; corolla white, yellow, yellow drying brown to dull purple, magenta, blue-purple, or purple, often bicolored with white or partially white banner, banner arising from the hypanthium rim, the other petals epistemonous, i.e. arising laterally or terminally from the column of fused stamen filaments, easily disjointing and deciduous, when papilionoid, the keel petals well differentiated from the wings, marginally adherent, generally enclosing the androecium, when nonpapilionoid, wings and keel petals very similar, not enclosing the exposed androecium; stamens 5 or 10, rarely some non-functional, monadelphous, i.e., all fused into a single sheath, open towards the banner, when only 5, the wings and keel petals attached in the sinuses between the free portion of the stamens, anthers yellow, often gland-tipped; style slender, or somewhat dilated distally, stigma minute or rarely subcapitate; ovules 2, collateral, only 1 fertile. **Fruits** indehiscent, obliquely obovate to deltate, usually included within calyx, some partially exserted. **Seeds** 1, moderately compressed, smooth. $x = 7(8)$.

Species ca. 165 (28 native in the flora, 2 of those also cultivated, 4 additional non-native cultivated species, and 1 non-native adventive species): North and South America from Canada to Argentina and the Caribbean, with the highest diversity in semi-arid regions of Mexico and the Andes Mountains.

Dalea and its sister genus *Marina* are small-flowered, glandular legumes, favoring dry habitats throughout the Americas. Although it can be difficult to see with a hand lens, the flowers in these two genera are unique among angiosperms: the wings and keel petals appear to arise from the fused stamen filaments. This has led earlier authors to segregate species with only five stamens and petals attaching at the highest points as the true "prairie clovers," in the genus *Petalostemon*, in which the very identity of the petals was questioned because they look like sterile stamens. Barneby (1977) returned these species to *Dalea*, and subsequent evolutionary analysis (McMahon and Hufford 2004) confirmed that they are a clade (i.e., all the prairie clovers are each other's closest relatives), nested well within *Dalea*. Another odd characteristic in many members of *Dalea* is a lack of the usual differentiation among keel and wing petals. In several species, these four petals do not enclose the stamens and the flower does not resemble the standard "pea" flower much at all. This characteristic seems to have arisen multiple times within the genus (McMahon and Hufford 2004) and is notably variable among other lineages in this part of the legume phylogeny (Cardoso et al. 2013).

We refer to these floral forms as papilionoid, resembling the standard "pea" flower, and nonpapilionoid, with undifferentiated wing and keel petals that do not enclose the 5 stamens (in prairie clovers) to 10 stamens (in miscellaneous other daleas). Isely (1998) used the equivalent terms papilionate and nonpapilionate for these general morphologies.

Various species of *Dalea* are used for soil stabilization, green manure, forage, medicine, ornamentals, and to make brooms and baskets (Allen and Allen 1981, Lewis et al. 2005). Allen and Allen (1981) state that *Dalea* species are suitable for livestock forage but are not often utilized due to the resins found in the glands. No species are reported to be toxic to livestock (Allen and Allen 1981). Several species are suitable for landscape use as groundcovers, as flowering shrubs, or in wildflower mixes. The plants may be propagated by seeds and also stem cuttings. Allen and Allen (1981) report nodulation for several species in the genus.

SELECTED REFERENCES: Barneby 1977, Felger and Rutman 2015.

1. Plants shrubs or subshrubs.
  2. Corolla white (sometimes turning ochroleucous on drying) or yellow (in age red- or orange-brown).
    3. Flowers papilionoid, corolla yellow, in age red- or orange-brown; cultivated only ____ **6. Dalea capitata**
    3′ Flowers nonpapilionoid, corolla white (sometimes turning ochroleucous on drying); native and rarely cultivated ____ **1. Dalea albiflora** (in part)
  2′ Corolla bicolored, the banner white or pale yellow and often becoming dull red, the other petals pink, pale lavender, pink-purple, rose-purple, magenta-purple, purplish-violet, or blue.
    4. Plants with stems trailing and forming low, matted thickets; cultivated only ____ **14. Dalea greggii**
    4′ Plants shrubby and not matted with at least some stems erect or ascending; native and cultivated.
      5. Calyx (7.5–)8.5–13.5(–16.2) mm; leaves 0.25–1.1(–1.4) cm ____ **11. Dalea formosa**
      5′ Calyx 2.5–8.5 mm; leaves (0.6–)1–6 cm.
        6. Calyx lobes shorter than tube, 0.5–1.9(–2.2) mm; cultivated only.
          7. Herbage loosely pilosulose to nearly glabrous; bracts puberulent to densely pilose; calyx intervals with 2–3(–4) glands, the lobes 1–1.9(–2.2) mm ____ **3. Dalea bicolor**
          7′ Herbage glabrous; bracts mostly glabrous; calyx intervals with 3–5(–7) glands, the lobes 0.5–1.2 mm ____ **12. Dalea frutescens**
        6′ Calyx lobes equal to or usually longer than tube, 2–5 mm; native and some cultivated.
          8. Leaflets 19–35; bracts persistent ____ **30. Dalea tentaculoides**
          8′ Leaflets 5–19(–21); bracts deciduous.
            9. Leaflets 5–7(–9), densely silvery-pilosulose with clear to reddish glands beneath ____ **27. Dalea pulchra**
            9′ Leaflets 5–19(–21), glabrous to pilosulose, but not silvery, with conspicuous black glands beneath ____ **32. Dalea versicolor**
1′ Plants herbs.
  10. Plants annual.
    11. Flowers papilionoid.
      12. Plants with stems erect or ascending; herbage glabrous, sparsely glandular.

13. Inflorescences 0.8–1.1(–1.2) cm wide (not including petals), bracts persistent; calyx (3.5–)4.4–5.8(–6) mm; corolla yellow or pale yellow, often drying dull orange-red to pink- or purple-brown ________ **4. Dalea brachystachys** (in part)
13′ Inflorescences 0.5–0.7 cm wide (not including petals), bracts deciduous; calyx (2.5–)2.8–3.5 mm; wing and keel petals pink to reddish-purple ________ **9. Dalea filiformis**
12′ Plants with stems prostrate; herbage pilose-pilosulose, prominently glandular.
14. Inflorescences 0.8–1.2 cm wide (not including petals); calyx tube 1.6–2.5 mm ________ **20. Dalea mollis**
14′ Inflorescences (1.2–)1.3–1.6(–1.7) cm wide (not including petals); calyx tube (2.5–)2.7–3.3 mm ________ **21. Dalea mollissima**
11′ Flowers nonpapilionoid.
15. Flowers with 3 petals ________ **31. Dalea urceolata**
15′ Flowers with 5 petals.
16. Leaves with (9–)17–37 leaflets; stamens (8–)10.
17. Stamens (7.5–)8–11 mm ________ **8. Dalea exserta**
17′ Stamens 5–6.8 mm ________ **18. Dalea leporina**
16′ Leaves with (1–)3–7(–9) leaflets; stamens 4–9.
18. Leaflets (0.5–)1–3.7(–4.7) mm ________ **7. Dalea exigua**
18′ Leaflets 5–12(–15) mm ________ **25. Dalea polygonoides**
10′ Plants perennial.
19. Flowers papilionoid.
20. Herbage glabrous.
21. Corolla yellow or pale yellow, often drying dull orange-red to pink- or purple-brown; calyx tube (1.5–)1.6–2.1 mm ________ **4. Dalea brachystachys** (in part)
21′ Corolla pale or vivid violet-purple; calyx tube 2.6–3.4 mm ________ **24. Dalea pogonathera**
20′ Herbage variously pubescent.
22. Corolla pale pink, pale blue, blue to blue-purple, some with white on banner, not yellow fading towards brown.
23. Herbage inconspicuously pubescent; leaves 3–8(–9.5) cm; corolla blue ________ **16. Dalea lachnostachys**
23′ Herbage densely silky-pilosulose; leaves 0.8–3 cm; corolla bicolored ________ **23. Dalea neomexicana**
22′ Corolla yellow, often fading in age and drying dull orange-red to pink- or purple-brown.
24. Calyx 4.2–7.5 mm.
25. Spikes 1.4–2.1 cm wide (not including petals), bracts persistent; banner 6.3–8.6 mm ________ **2. Dalea aurea**
25′ Spikes 0.7–1.3(–1.5) cm wide (not including petals), bracts deciduous by anthesis; banner (4–)4.4–5.5 mm ________ **22. Dalea nana**
24′ Calyx 8.3–13.5 mm.
26. Leaflets 3; banner 5.2–6.6 mm ________ **15. Dalea jamesii**
26′ Leaflets 5; banner 7.8–9.5 mm ________ **33. Dalea wrightii**
19′ Flowers nonpapilionoid.
27. Stamens 5.
28. Leaflets 0.2–0.8(–1.3) mm wide, linear-oblanceolate, linear, or linear-elliptic ________ **28. Dalea purpurea**
28′ Leaflets 0.5–3.1(–5) mm wide, obovate, oblong-elliptic, or linear-oblong.
29. Flowers with corolla rose-purple ________ **29. Dalea searlsiae**
29′ Flowers with corolla white sometimes drying ochroleucous.
30. Inflorescence axis at least partly visible at maturity ________ **5. Dalea candida**
30′ Inflorescence dense, flowers obscuring axis ________ **10. Dalea flavescens**
27′ Stamens (8–)10.
31. Corolla purple, bright red-violet, or magenta-purple.
32. Stems prostrate; calyx lobes (1.3–)1.5–2.3 mm ________ **17. Dalea lanata**
32′ Stems erect; calyx lobes 2.4–4 mm ________ **26. Dalea pringlei**
31′ Corolla white or bicolored white and blue.
33. Calyx intervals without glands ________ **19. Dalea lumholtzii**
33′ Calyx intervals with 3–6 glands.
34. Herbage silky villosulous; wings and keel petals (3.4–)3.8–5.8 mm ________ **1. Dalea albiflora** (in part)
34′ Herbage glabrous; wings and keel petals 2.4–3.6 mm ________ **13. Dalea grayi**

1. Dalea albiflora A. Gray. Smithsonian Contr. Knowl. 5(6):38. 1853. * [with white flowers] WHITE-FLOWER DALEA Native

Dalea gentryi Standley; *Dalea ordiae* A. Gray

Perennial herbs or **subshrubs**, 2–7 dm. **Herbage** silky villosulous, sparsely glandular. **Stems** erect to spreading, branched. **Leaves** 1–4 cm; stipules 1–4.5 mm, linear to awl-shaped; petioles nearly absent or very short. **Leaflets** 13–35, (1.5–)2–7(–10) × 0.5–2.4(–3.9) mm, oblong-elliptic to linear-elliptic. **Inflorescences** dense spikes, (1–)1.5–6(–8) × 0.6–1 cm (not including petals), ovoid or narrowly cylindrical, axis villosulous; bracts persistent, narrowly triangular to triangular-acuminate, pubescent dorsally (or lower glabrate). **Flowers** nonpapilionoid; calyx (2.8–)3.2–5.1 mm, pilosulose externally, tube (1.9–)2–3(–3.3) mm, deeply cleft behind banner, slender ribs becoming prominent, intervals membranous and with 3–6 yellow glands, lobes (0.9–)1.1–1.8(–2.1) mm; corolla white, the banner (4.4–)5–6.4 mm, wings and keel petals similar to one another, (3.4–)3.8–5.8 mm, attached to androecium shortly below filament separation, neither adherent nor enclosing the androecium; stamens 10, (5–)6–8 mm, filaments free for 3.5–4.5 mm. **Fruits** 2.2–2.8 × 1.7–2.1 mm, broadly and obliquely obovoid. **Seeds** 1, 1.6–2 mm.

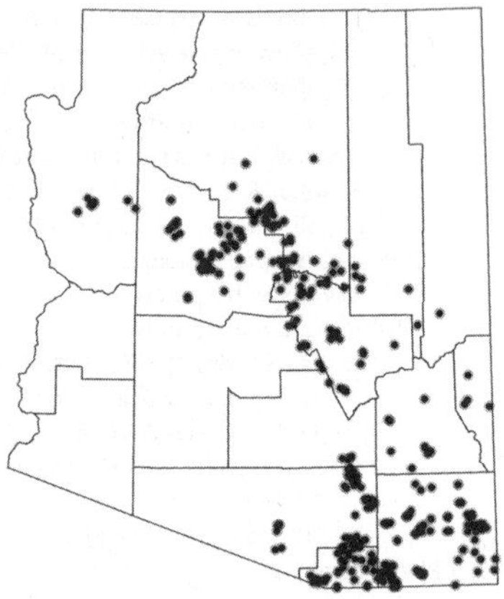

Flowering (May–)Jul–Oct(–Nov). Slopes, canyons, arroyos, floodplains, stream banks, valleys, and roadsides on a wide range of soils; (760–)1100–2200(–2480) m; (riparian habitats in upper elevation Sonoran Desertscrub), Semidesert Grassland, Interior Chaparral, Madrean Evergreen Woodland, Petran Montane Conifer Forest; Apache, Cochise, Coconino, Graham, Greenlee, Gila, Maricopa, Mohave, Pima, Pinal, Navajo, Santa Cruz, Yavapai cos.; New Mexico; Mexico (Chihuahua, Durango, Sinaloa, Sonora).

Dalea albiflora is one of the most widespread and conspicuous members of the genus at mid-elevations in much of southeastern and central Arizona. The northwestern limit of its range is the Hualapai Mountains in Mohave County. It occupies diverse habitats from dry to moist and is found in the open and in shade. The flower of *D. albiflora*, as is the case with many other daleas, is nearly unrecognizable as a papilionoid legume. The nearly identical white wing and keel petals are spread open at anthesis, exposing the 10 stamens, almost giving the tight inflorescence the appearance of an overgrown *Plantago*. *Dalea candida* is easily distinguished from *D. albiflora*: *D. candida* has fewer glands on the calyx; fewer, larger leaflets; and only 5 stamens (a characteristic of the "prairie clovers").

Dalea albiflora varies considerably in stature depending on growing conditions. With its long flowering period, *D. albiflora* is suitable for use as a flowering perennial. The plants require regular irrigation and will flower during their first year. This species is hardy to at least -12°C and probably lower. Old inflorescences may be pruned to encourage flowering and improve appearance. Under cultivation in Tucson, individual plants die out after three to five years, but will reseed in favorable sites in the garden.

2. Dalea aurea Nuttall ex Pursh. Fl. Amer. Sept. 2:740–741. 1814. * [golden, referring to the corolla] GOLDEN DALEA, SILK-TOP DALEA Native

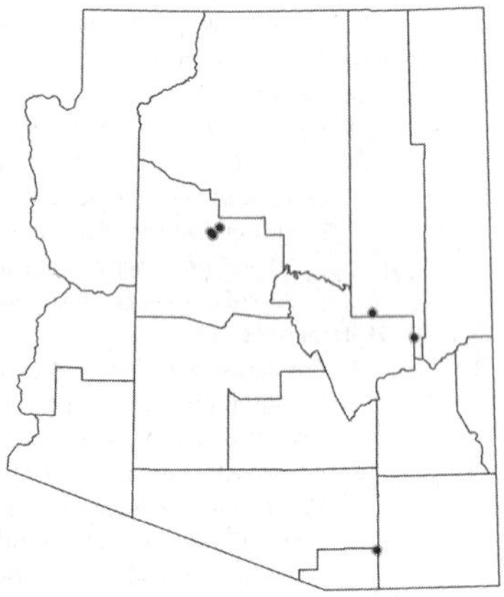

Perennial herbs, 2–8 dm. **Herbage** silky-pilose, inconspicuously glandular although fragrant when crushed. **Stems** stiffly erect, usually unbranched above the base. **Leaves** 1–3.5(–4) cm, scattered, becoming reduced upward; stipules (0.8–)1.2–3 mm, narrowly awl-shaped, papery. **Leaflets** 3–7, always at least 5 on some leaves, (3–)4–16(–20) × 2–4.2 mm, obovate to oblanceolate, (sub-)acute, densely pubescent beneath, sparsely pubescent and darker green above. **Inflorescences** terminal spikes, (1–)1.5–6 × (1.2–)1.4–2.1 cm (not including petals), the axis densely pilosulose; bracts persistent, 2.5–5.5 mm, ovate to lanceolate, acuminate, densely silky-pilose, some finely punctate. **Flowers** papilionoid; calyx 6.1–7.4 mm, densely silky-pilose, tube 2.2–2.8 mm, ribs prominent, intervals with 3–4 small glands, lobes 3.5–5 mm, subequal, longer than tube; corolla pale yellow to yellow, banner 6.3–8.6 mm, other petals attached near middle of androecium, wings 5.2–6 mm, keel petals 5.7–8.5 mm, margins adherent and enclosing androecium; stamens 10, 10–12.5 mm, free for distal fourth. **Fruits** 3–3.5 mm. **Seeds** 1, 2–2.4 mm.

Flowering [Jun–]Aug–Sep. Ridges, slopes, and valleys in various soils, especially on limestone; [10–]1370–1460 [–1950] m; Semidesert Grassland; Gila, Navajo, Santa Cruz, Yavapai cos.; Colorado, Kansas, Nebraska, New Mexico, Oklahoma, South Dakota, Texas; Mexico (Chihuahua, Coahuila).

Populations of *Dalea aurea* are widely scattered and the plants are rarely encountered in Arizona. It is primarily a species of the western Great Plains, extending to the Gulf Coastal Prairie of Texas and just into northeastern Mexico south of the Texas Big Bend region. Across its extensive range it is commonly found on prairies and in open woodlands and scrub. *Dalea aurea* is browsed by livestock and wildlife, and the seeds are eaten by rodents and birds (Stubbendieck and Conard 1989). The leaves were employed by the Lakota to treat colic and dysentery (Stubbendieck and Conard 1989). With its showy yellow flowers, *D. aurea* has landscape potential (Stubbendieck and Conard 1989).

3. Dalea bicolor Humboldt & Bonpland ex Willdenow. Enum. Pl. 2:787. 1809. * [of two colors]

Small trees, shrubs or subshrubs, (2–)4–20(–35) dm. **Herbage** thinly to densely pilosulose or tomentose to nearly glabrous, prominently glandular. **Stems** densely branching, becoming fissured in age. **Leaves** (0.6–)1–3.5(–4.5) cm, the primary cauline ones longer; stipules (0.5–)1–4.5 mm, narrowly triangular; petioles short. **Leaflets** (3–)5–21(–23), (1.5–)2–9.5(–10) × 1–5 mm, obovate, oblong-elliptic, oblanceolate, obtuse or bluntly gland-mucronate, dorsally punctate, the terminal leaflet sessile. **Inflorescences** spikes, (0.5–)1–20(–23) × (0.5–)0.6–1.5 cm (not including petals), calyces nearly touching to well separated, axis villosulous; bracts persistent or only tardily deciduous, 1–5.5(–6) mm, broadly ovate to oblanceolate, dorsally puberulent to densely pilose. **Flowers** papilionoid; calyx 2.5–5.6(–6.8) mm, variably pubescent, the tube 1.6–3.2 mm, with prominent ribs, intervals with 1 row of 2–3(–4) glands, lobes 0.5–2.9 (–3.8) mm, unequal, deltate-ovate to triangular-acuminate; petals bicolored, the banner 3.2–7.5(–8) mm opening white or creamy yellow, quickly becoming reddish, the others rose-purple to vivid blue, attached well below middle of androecium, wings 3.4–7.5 mm, keel 5–10 mm; stamens 10, 5.6–10.5(–12) mm, the filaments free for (1.7–)2–4 mm. **Fruits** 2.2–2.7 mm, obliquely obovate to triangular. **Seeds** 1.

Varieties 5 (1 cultivated in the flora): Mexico.

3a. Dalea bicolor Humboldt & Bonpland ex Willdenow var. **bicolor** * BLUE DALEA Cultivated

Herbage loosely pilosulose to nearly glabrous. **Leaves** (0.7–)1–2.5(–3) cm. **Leaflets** (3–)7–13(–19), 2–7(–8) mm. **Inflorescences** spikes, elongating and usually loose from early anthesis; bracts 1.5–3(–4) mm. **Flowers:** calyx tube 1.6–3.2 mm, lobes 1–1.9(–2.2) mm; wings and keel petals pale to bright rose-purple, violet, or vivid blue.

Flowering (Jun–)Aug–Dec(–Jan). Mexico (Aguascalientes, Chihuahua, Coahuila, Guanajuato, Hidalgo, Jalisco, México, Michoacán, Morelos, Nuevo León, Oaxaca, Puebla, Querétaro, San Luis Potosí, Tamaulipas, Vera Cruz, Zacatecas).

Dalea bicolor var. *bicolor* is widespread across Mexico. It is found in rocky soil on ridges, slopes, canyons, arroyos, and ledges in the Chihuahuan Desert, thornscrub, oak woodland, and field margins at elevations of (500–)600–2700 m elevation (Barneby, 1977). The plants may be locally abundant.

Dalea bicolor var. *bicolor* is grown in Arizona for landscape use. It is suitable as a background plant and as a low, informal hedge. The plants have a rounded form and the green foliage contrasts with the bluish purple flowers. The foliage is deciduous with winter freezing. Established plants have low irrigation requirements but benefit from some moisture to maintain appearance during hot, dry weather. Growth is fast with adequate irrigation. *Dalea bicolor* should be planted in full sun in well-drained soil. The plants are hardy to about -12°C. Older plants may be pruned back to promote fresh growth.

4. Dalea brachystachys A. Gray. Smithsonian Contr. Knowl. 5(6):39. 1853. * [with short spikes] FORT BOWIE DALEA Native

Dalea lemmoni Parry ex A. Gray

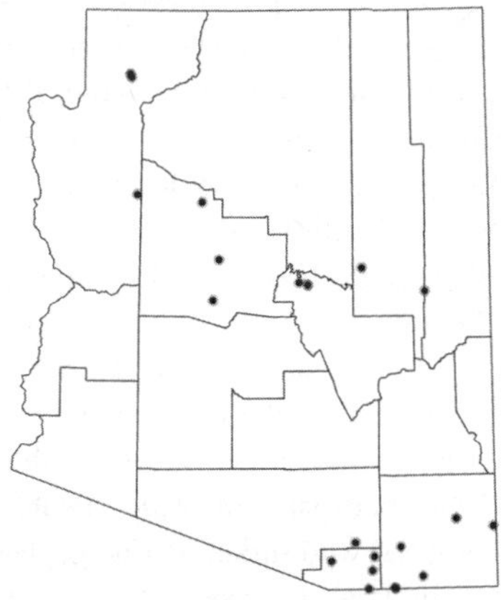

Annual (or biennial) herbs, (0.5–)1–3.5(–4) dm. **Herbage** glabrous to inflorescence, sparsely glandular. **Stems** erect or ascending, slender, diffuse, sparingly branched. **Leaves** 0.8–2.5(–3) cm; stipules 0.8–2.5 mm, narrow. **Leaflets** (3–)5–11, 2–14(–16) × 0.6–1.7 mm, terminal leaflet longest, oblanceolate to linear-oblanceolate, loosely folded, often with reddish margins, prominently glandular beneath. **Inflorescences** spikes, to 1.6 × 0.8–1.1(–1.2) cm (not including petals), ovate or subglobose, axis villosulous; bracts persistent, 2.5–6 mm, rhombic-ovate to lance-acuminate or -caudate, prominently glandular, inconspicuously pubescent. **Flowers** papilionoid; calyx (3.5–)4.4–5.8 (–6) mm, symmetrical, densely pilose, tube (1.5–)1.6–2.1 mm, ribs brown, intervals with 3 (or 0) small glands, lobes (1.5–)2.4–4 mm, usually longer than the tube, spreading when in fruit; corolla pale yellow but drying dull orange-red to pink- or purple-brown, the banner 2–3.4(–6) mm, other petals attached near middle of androecium, wings 2.1–3.4(–4.8) mm, keel petals (2.6–)3–4.3(–6.2) mm, marginally adherent and enclosing androecium; stamens (9–)10, 3.5–5.5(–6.8) mm, free for distal fourth. **Fruits** 2.3–2.6 mm, pilosulose. **Seeds** 1, 1.4–1.7 mm.

Flowering [Apr–Jun], Aug–Oct. Slopes, valleys, and floodplains on sandy to rocky soil and on limestone;

1160–1700(–2040)[–2430] m; Semidesert Grassland, Madrean Evergreen Woodland, Great Basin Conifer Woodland; Cochise, Gila, Mohave, Navajo, Pima, Santa Cruz, Yavapai cos.; New Mexico, Texas; Mexico (Aguascalientes, Chihuahua, Coahuila, Durango, Hidalgo, Puebla, Querétaro, San Luis Potosí, Sonora, Tamaulipas, Zacatecas).

In Arizona, *Dalea brachystachys* principally occurs in the southeastern part of the state at elevations above the deserts. It is also documented sporadically along the Mogollon rim in southern Navajo County and through Yavapai County. Eastward and in Mexico, *D. brachystachys* is widespread in the Chihuahuan Desert and in semidesert and plains grasslands. It occurs southward in thornscrub habitats.

Note that the flowers are yellow when fresh but often dull orange-red in age, and can dry as dark as purplish-brown.

5. Dalea candida Willdenow. Sp. Pl. 3(2):1337. 1802. * [white, referring to the corolla] SLENDER WHITE PRAIRIE CLOVER

Petalostemon candidus (Willdenow) Michaux

Perennial herbs, to 10 dm. **Herbage** glabrous, glandular. **Stems** erect or diffuse, distally branching; prominently ribbed, glandless or nearly so. **Leaves** 1.5–6 cm; stipules 1.5–5.5 mm, triangular or narrowly lanceolate; petioles short. **Leaflets** 5–9, (6–)7–25(–35) × 0.5–3.1(–4.5) mm, obovate to linear-oblong, variable, terminal leaflet longer, acute to emarginate, glabrous above, punctate glandular below. **Inflorescences** spikes, cone-like to cylindroid with axis partly visible at maturity, 0.8–6(–8.5) × 0.6–1 cm (not including petals); bracts green, gland sprinkled, narrowly obovate, shortly acuminate, of two types, lower persistent, reflexed, upper deciduous, erect, and more narrow. **Flowers** nonpapilionoid; calyx (2.9–)3–4.2(–4.4) mm, nearly symmetric, tube 1.9–2.7 mm, glabrous or nearly so, ribs prominent but slender, intervals with 1(–3) small clear or yellow glands above, lobes unequal, 1–1.8 mm; petals white, glandless, banner (4–)4.2–5.7 mm, wings and keel similar, 3–5 mm; stamens 5, as long as and not enclosed by corolla, fused for less than half their lengths, anthers yellow, gland-tipped. **Fruits** (2.6–)2.8–4(–4.5) mm, usually partially exserted. **Seeds** 1, 1.7–2.3 mm, oblong-elliptic.

Varieties 2 (1 in the flora): interior w and c United States, c Canada, south to Durango, Mexico.

5a. Dalea candida Willdenow var. **oligophylla** (Torrey) Shinners. Spring Fl. Dallas-Ft. Worth 409. 1958. * [with few leaflets] Native

Petalostemon gracilis Nuttall var. *oligophyllus* Torrey, Not. Milit. Reconn. 139. 1848; *Dalea oligophylla* (Torrey) Shinners; *Kuhnistera occidentalis* A. Heller ex Britton & Kearney; *Petalostemon occidentalis* (A. Heller ex Britton & Kearney) Fernald

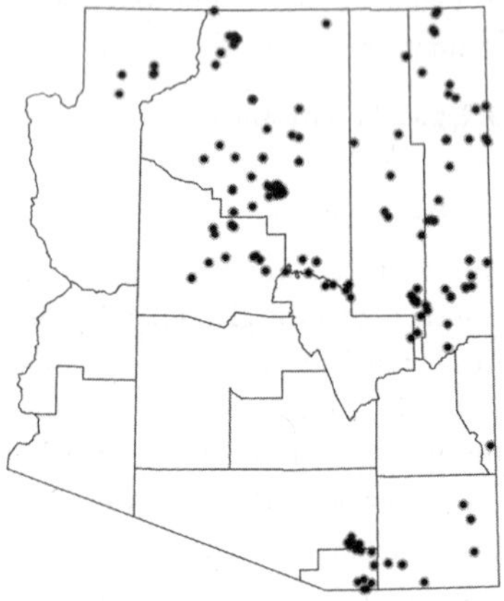

Stems more often diffusely branching than in the typical variety. **Leaflets** relatively small, (6–)9–20(–24) mm. **Inflorescences:** spikes becoming lax by full anthesis, with axis partly visible. **Flowers:** calyx sometimes pilosulose.

Flowering May–Oct. Slopes, canyons, arroyos, valleys, roadsides, floodplains, stream banks, and springs on many soil types; [250–]880–2480 (–2530) m; (upper elevations in Great Basin Desertscrub), Semidesert Grassland, Plains and Great Basin Grassland, Madrean Evergreen Woodland, Great Basin Conifer Woodland, Petran Montane Conifer Forest; Apache, Cochise, Coconino, Gila, Greenlee, Mohave, Navajo, Pima, Santa Cruz, Yavapai cos.; Colorado, Iowa, Kansas, Minnesota, Montana, Nebraska, New Mexico, North Dakota, Oklahoma, South Dakota, Texas, Utah, Wisconsin, Wyoming; Mexico (Chihuahua, Durango, Sonora); Canada (Alberta, Manitoba).

In Arizona, *Dalea candida* var. *oligophylla* is widespread in central and eastern areas of the state where it can be found in a wide range of habitats growing in the open and in partial shade. Across its vast range, var. *oligophylla* occupies diverse habitats on a variety of substrates. Stubbendieck and Conard (1989) state that *D. candida* is suitable as forage for all types of livestock, and the foliage is browsed by deer, elk, pronghorn, and wild turkey. Pocket gophers eat the roots and various birds and rodents eat the seeds (Stubbendieck and Conard 1989). Kearney and Peebles (1942) report that the Hopi have used the plant as an emetic. The Lakota chewed the pleasantly flavored roots as a treat and used the leaves to make a tea while other Plains tribes reportedly used the plant to treat fresh wounds (Stubbendieck and Conard 1989). Stubbendieck and Conard (1989) report that stem tip cuttings will root under mist. The plants have potential for use in soil stabilization (Stubbendieck and Conard 1989) and are an excellent addition to a wildflower garden.

6. Dalea capitata S. Watson. Proc. Amer. Acad. Arts 25:146. 1890. * [head-like] LEMON DALEA

Dalea lloydii (Rydberg) Gentry; *Parosela lloydii* Rydberg

Shrubs to 4 (5) dm; adventitiously rooting. **Herbage** glabrous to the inflorescence or thinly puberulent distally, variably glandular, foliage bright green or glaucescent. **Stems** erect to tortuously diffuse, if procumbent, becoming grayish-brown or whitish and furrowed. **Leaves** 0.6–1.6 cm; stipules 0.4–1.3 mm, narrowly subulate; petioles short. **Leaflets** (5–)7–11(–13), 1.5–5.5 × 0.6–2 mm, obovate to oblanceolate, emarginate, distal leaves smaller with fewer leaflets. **Inflorescences** terminal spikes, oblong to head-like, (0.3–)0.6–1.5(–2.2) × 0.6–1 cm (not including petals), axis pilosulose; bracts persistent or very tardily deciduous, (2.5–)3–5 mm, broadly ovate, acute to short-acuminate, clasping the calyx, firm, reddish or purplish-glaucescent, at least the lowest glabrous dorsally. **Flowers** papilionoid**;** calyx (2.3–)3.3–6.3 mm, the tube (2–)2.2–4 mm, externally either silky-pilosulose or rarely glabrous, the ribs slender, intervals with 1 row of 3–5 small glands, lobes (1–)1.4–2.5(–2.9) mm, unequal, subulate, densely pilosulose at least internally, the ventral pair sometimes united behind banner; petals pale yellow, in age red- or orange-brown, banner 3.5–6.2(–7) mm, wings 3.1–6 mm, keel (3.7–)5.5–7.8(–8.1) mm; stamens 10, (5–)5.5–9.4 mm, the longer filaments free for 1.8–3.5 mm. **Fruits** 2.6–3 mm, pilosulose, gland-sprinkled. **Seeds** 1.

Varieties 4 (1 cultivated in the flora): c and e Mexico.

6a. Dalea capitata S. Watson var. **capitata** * Cultivated

Inflorescences: longest oblong; bracts 2.5–4 mm. **Flowers:** calyx 2.3–3.1 mm.

Flowering (Mar–)Apr–Oct. Mexico (Coahuila, Nuevo León, San Luis Potosí, Zacatecas).

Dalea capitata var. *capitata* is found in Mexico at elevations of 1680–2350 m. Populations are localized at upper elevations in the Chihuahuan Desert and in semidesert grassland and lower elevation pinyon-juniper woodland, often on calcareous substrates (Barneby 1977).

Dalea capitata is grown in Arizona as a groundcover. With its small mature size, the plants are ideal for planting in confined spaces, and it is well-suited for rock gardens. The yellow flowers and green foliage are attractive. *Dalea capitata* will become partially to mostly deciduous with repeated winter freezing or with prolonged drought. Established plants have low to moderate irrigation needs. The plants should be planted in full sun and in well-drained soil. This species appears to be hardy to -18°C. The plants are unappealing to rabbits due to the aromatic foliage. Appearance of the plants may benefit from an occasional hard pruning to remove old stems.

7. Dalea exigua Barneby. Mem. New York Bot. Gard. 27:199–200. 1977. * [frail] CHIHUAHUAN PRAIRIE CLOVER Native

Petalostemon exilis A. Gray

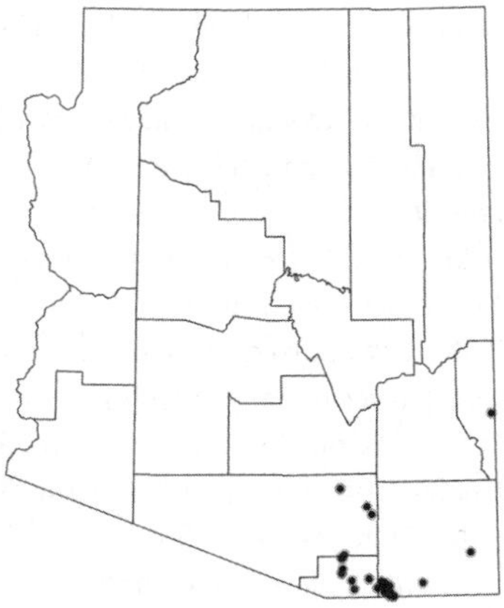

Annual herbs, 1–4(–4.5) dm. **Herbage** glabrate to inflorescence, obscurely glandular. **Stems** stiff, sparsely branching below or near the middle, erect or incurved. **Leaves** 1.5–5.5 cm; stipules 1.5–3(–4) mm, awl-shaped. **Leaflets** 3 or 5, upper leaves often only 1, (0.5–)1–3.7(–4.7) × 0.7–3 mm, linear-oblanceolate to narrowly oblong, mucronate, the terminal one longest, bright green above, paler and glandular below. **Inflorescences** dense spikes, (0.8–)1.3–4.5(–6) × 0.5–0.8 cm (not including petals), axis pilosulose; bracts persistent, the lowest 2.5–3.3 mm, broadly deltate-ovate or rhombic, shortly apiculate, glabrous or nearly so, upper bracts narrower, shortly acuminate, pilosulose. **Flowers** nonpapilionoid; calyx 3.8–4.3 mm, silky-pilose, the tube 1.9–2.2 mm, ribs slender, intervals without glands, lobes 1.2–2.3 mm triangular to triangular-aristate; petals pink-purple, deciduous, banner 3.4–4.3 mm, wings 1.7–2.3 mm, keel petals 2–2.5 mm, wings and keel petals attached just below separation of the filaments; stamens 4–7, 3.2–4 mm, with only 3–6 fertile stamens, the longer filaments free for 1 mm. **Fruits** 2.5 mm, obovate. **Seeds** 1.

Flowering Aug–Oct. Slopes, canyons, roadsides, and occasionally along streams on sandy or rocky soil and rock outcrops; 1460–1850(–2200) m; Madrean Evergreen Woodland; Cochise, Greenlee, Pima, Santa Cruz cos.; New Mexico; Mexico (Chihuahua, Sonora).

Dalea exigua is of scattered occurrence in Arizona. The plants are found in the open and beneath the canopies of trees. It is more widespread in the Sierra Madre Occidental region to the south where it occurs in oak and pine-oak woodlands (Barneby 1977).

8. Dalea exserta (Rydberg) Gentry. Publ. Carnegie Inst. Wash. 527:138. 1942. * [protruding, referring to the androecium] MEXICAN DALEA Native

Parosela exserta Rydberg, N. Amer. Fl. 24(2):73. 1920

Annual herbs, robust, (1.5–)5–10(–12) dm. **Herbage** glabrous to the inflorescence, distally glandular. **Stems** simple or branched from below or near the middle, erect. **Leaves** 3–10 cm; stipules 1–1.5 mm, triangular-subulate. **Leaflets** 17–37, 2–10 × 0.9–3.7 mm, obovate to oblong-oblanceolate, fewer and smaller on upper leaves, smooth above, glandular beneath. **Inflorescences** terminal spikes, (1.5–)3–10 × 1–1.6 cm (not including petals); bracts deciduous, lowest ones broadly lance-caudate, upper more narrow, membranous-margined, glabrate, glandular. **Flowers** nonpapilionoid; calyx 4.8–7 mm, tube (2.5–)2.7–3.8 mm, more deeply cleft behind banner, becoming membranous, ribs slender, intervals with 1–2 rows of glands, lobes (2–)2.3–3.8 mm, aristate, about as long as the tube, glabrous; corolla blue or banner white fading blue-lavender, banner 4.9–7.3 mm, wings 3.6–5 mm, keel petals 3.9–5 mm, not adherent; stamens (8–)10, (7.5–)8–11 mm, well exserted. **Fruits** 2.2–2.8 mm, obovoid to nearly circular. **Seeds** 1.

Flowering [Jan–Mar]Aug–Oct[–Dec]. Slopes and canyons in rocky soil; [350–]1070–1130[–2100] m; Madrean Evergreen Woodland; Santa Cruz Co.; Mexico (Chihuahua, Colima, Guerrero, Jalisco, México, Michoacán, Morelos, Oaxaca, Sonora, Vera Cruz); Costa Rica, Guatemala, Honduras.

Dalea exserta is documented in the U.S. only from Sycamore Canyon along the International Boundary west of Nogales. It is widespread in Mexico and occurs in various habitats, including tropical deciduous forest and oak woodland, and grows on a variety of soils.

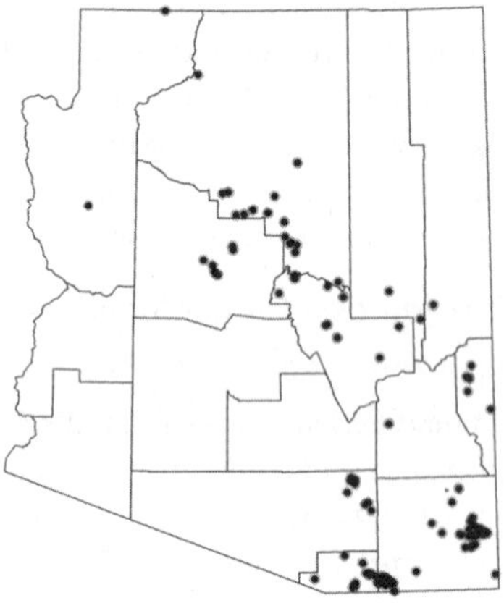

9. Dalea filiformis A. Gray. Smithsonian Contr. Knowl. 5(6):39–40. 1853. * [thread-like, referring to the stems and leaves] SONORAN DALEA Native

Annual herbs, (0.4–)0.7–3(–3.5) dm. **Herbage** glabrous to the inflorescence, inconspicuously glandular. **Stems** slender, branching from base upward or some towards middle, erect. **Leaves** 1–3.5(–4) cm, sparse, green, punctate beneath; stipules 0.5–1.3 mm, narrowly awl-shaped. **Leaflets** 3–5, 5–23(–30) × 0.4–0.7(–2) mm, linear, terminal three palmate or nearly so. **Inflorescences** spikes, often few-flowered, (0.2–)0.3–1.5(–2) × 0.5–0.7 cm (not including petals), axis villosulous; bracts deciduous, 1.5–2.8 mm, rhombic-elliptic to ovate-acuminate, glandular, the lowest glabrous, the upper pilosulose. **Flowers** papilionoid; calyx (2.5–)2.8–3.5 mm, pilosulose, tube (1.3–)1.5–1.8 mm, ribs slender, the intervals glandless or with 1 (2) minute glands, lobes 1.2–1.8 mm, triangular-acuminate, glandular; corolla pink to reddish-purple, or the banner pale at anthesis and becoming dull red, wings and keel petals attached well below middle of androecium, banner 2.6–3.5 mm, wings 1.8–2.4 mm, keel petals 2.4–3.2 mm; stamens 10, but alternating stamens often sterile and reduced in size, the longer ones 2.5–3.5 mm, free for 0.6–0.9 mm, gland-tipped. **Fruits** 2–2.4 mm, obovoid, partially exserted from calyx tube. **Seeds** 1, 1.5–1.9 mm.

Flowering Aug–Oct[–Nov]. Ridges, slopes, valleys, roadsides, and stream banks, on a wide range of soils; (1300–)1650–2440 m; Madrean Evergreen Woodland, Great Basin Conifer Woodland, Petran Montane Conifer Forest; Apache, Cochise, Coconino, Gila, Graham, Greenlee, Mohave, Navajo, Pima, Santa Cruz, Yavapai cos.; New Mexico; Mexico (Chihuahua, Durango, Sonora).

Dalea filiformis is an inconspicuous plant but may be locally abundant in open areas and forest edges. In Arizona it is primarily found in woodland and forest habitats but extends into oak grassland and juniper grassland. In the southeastern part of its range in Mexico, it is also found in grassland habitats (Barneby 1977).

10. Dalea flavescens (S. Watson) S.L. Welsh ex Barneby. Mem. New York Bot. Gard. 27:231. 1977. * [yellowing, referring to the white petals that turn yellow upon drying] CANYONLANDS PRAIRIE CLOVER Native

Petalostemon flavescens S. Watson, Amer. Naturalist 7(5):299–300. 1873

Perennial herbs, (2.5–)3–4.5 dm. **Herbage** pilosulose throughout, glabrous to the spikes, or glabrous throughout, sparsely but prominently glandular. **Stems** erect, often simple, incurved-ascending or diffuse. **Leaves** cauline

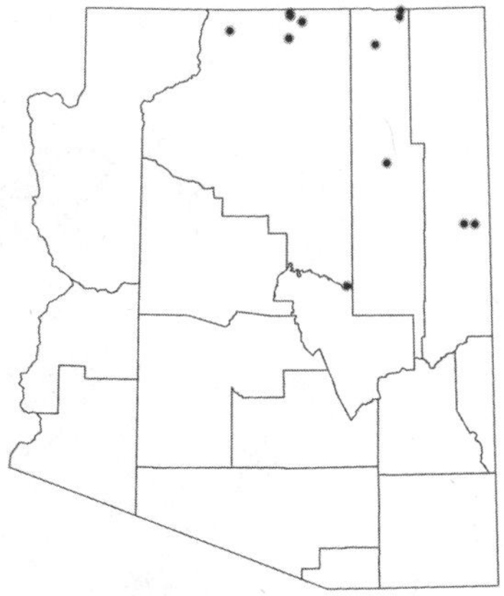

(2–)2.5–4(–4.5) cm, others 1–2 cm; stipules linear-lanceolate or narrowly linear. **Leaflets** 5–7, 7–17(–19) × 1–4.5 mm, oblong-obovate to narrowly oblong-elliptic, obtuse, emarginate, or obtuse and gland-mucronate, usually ± folded and backwardly arched, the terminal often a little longer than the last pair. **Inflorescences** dense spikes, (1–)1.5–7(–22) × 0.85–1.2(–1.3) cm (not including petals), long-peduncled, ovoid-oblong to cylindroid, axis pilosulose but not visible between flowers; bracts 2.5–5.5(–6.5) mm, tardily deciduous, densely pilosulose at least to the tip. **Flowers** nonpapilionoid; calyx (3–)3.3–4.7(–5.2) mm, tube (2–)2.2–2.8 mm, deeply recessed behind the banner, ribs slender, panels between not or obscurely glandular, unequal lobes to (1–)1.2–2.6(–2.9) mm, pilosulose (especially on and within the lobes) with long spreading hairs; corolla white (sometimes drying ochroleucous), banner 5–7.4(–8.4) mm, wings and keel petals 3.8–4.6(–6.3) mm, similar to each other, not adherent, inserted on the androecium at the separation of the filaments; stamens 5, (5.5–)6.2–10(–12.7) mm, free 3.4–6.2(–8.7) mm. **Fruits** 3.1–3.7 mm, obliquely obovoid, minutely gland-sprinkled, densely pilosulose. **Seeds** 1, 2.1–2.4 mm.

Flowering [May–]Jun–Jul[–Sep]. Rimrock, canyons, slopes, roadsides, on sandstone or sandy soil, and occasionally on sand dunes; [1000–]1200–1900(–2000) m; Great Basin Desertscrub; Apache, Coconino, Gila, Navajo cos.; Utah.

Dalea flavescens, although restricted to the Colorado Plateau in northern Arizona and Utah, is highly variable in pubescence, spike dimensions, bract size, and stamen length. Populations near Lake Powell with larger spikes have been recognized as *Dalea flavescens* var. *epica* (S.L. Welsh) S.L. Welsh & Chatterley. *Dalea albiflora* can be distinguished by more leaflets and more stamens; *D. candida* by its less dense spikes, the axis at least partly visible at maturity.

11. Dalea formosa Torrey. Ann. Lyceum Nat. Hist. New York 2:177–178. 1827. * [handsome] FEATHERPLUME DALEA Native

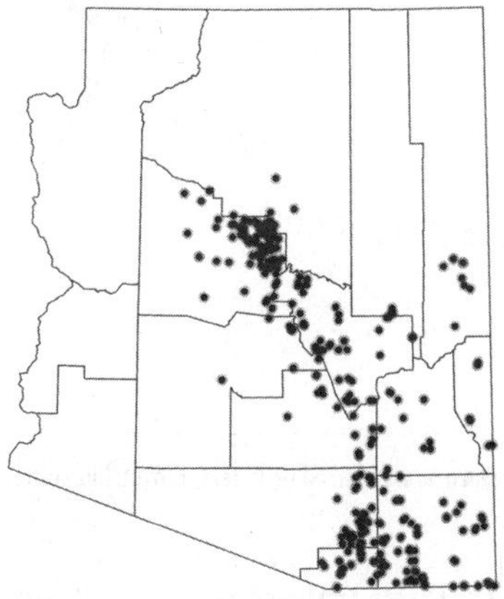

Subshrubs, 1.5–9 dm. **Herbage** glabrous below the inflorescence or nearly so, glandular. **Stems** becoming woody, intricately branching. **Leaves** primary cauline 0.25–1.1(–1.4) cm, deciduous, those on secondary short shoots 0.2–0.5 cm, more persistent; stipules 0.5–0.8(–2.2) mm, awl-shaped. **Leaflets** (5–)7–13(–15), (0.6–)1–6(–7) × 0.3–0.8 mm, obovate-cuneate to narrowly obovate, mostly folded. **Inflorescences** spikes terminal to branches and short shoots, 0.2–0.8(–4) × 1–2 cm (not including petals), 2–9(–18) flowers, subcapitate; bracts early deciduous, 3–6 mm, broadly ovate, tip somewhat recurved, ciliolate, dorsally glabrous, glandular. **Flowers** papilionoid; calyx (7.5–)8.5–13.5(–16.2) mm, plumosely pilose, tube (3–)3.5–5(–5.2) mm, strongly oblique, ribs becoming prominent, intervals with 3–4(–5) reddish glands, longest lobe (0.6–)1–6(–7) mm, bristle-like and arched outward in age; corolla bicolored, banner pale yellow but early becoming dull red, other petals rose- or magenta-purple, attached low on the androecium, banner (6.6–)7–8.8 mm, wings 8–10.4 mm, keel petals 8.2–11.4 mm; stamens 10, 9–12.5 mm, free for distal third. **Fruits** 3–3.5 mm. **Seeds** 1.

Flowering Feb–Oct. Ridges, slopes, canyons, arroyos, and roadsides, in gravelly or rocky, often calcareous soils; [570–]650–1830[–2040] m; Sonoran Desertscrub, Chihuahuan Desertscrub, Semidesert Grassland, Plains Grassland, Interior Chaparral, Madrean Evergreen Woodland, Great Basin Conifer Woodland; Apache, Cochise, Coconino, Gila, Graham, Greenlee, Maricopa, Pima, Pinal, Navajo, Santa Cruz, Yavapai cos.; Colorado, New Mexico, Oklahoma, Texas; Mexico (Chihuahua, Coahuila, Sonora).

Dalea formosa is an unmistakable species that is widespread and often abundant across its range in Arizona. The plants are frequently found on limestone or caliche, and less commonly on sandstone and gypsum. It also occurs on other rock substrates. This species is readily browsed by livestock and deer (Allen and Allen 1981, Stubbendieck and Conard 1989). With its showy purple and yellow flowers and compact, intricately branched form, featherplume dalea would make an attractive addition to rock gardens and desert plantings but is seldom available from nurseries. The foliage is deciduous with drought and freezing temperatures. *Dalea formosa* is drought tolerant and is hardy to at least -20°C.

Dalea formosa: **a.** habit; **b.** leaf; **c.** fruit. Illustrator: Marjorie Leggitt.

12. Dalea frutescens A. Gray. Boston J. Nat. Hist. 6(2):175–176. 1850. * [shrubby] BLACK DALEA Cultivated

Subshrubs or shrubs 3–8(–12) dm. **Herbage** glabrous up to the thinly pilosulose inflorescence axis, glandular. **Stems** greenish, then pale brown and furrowed in age, slender, erect or decumbent and ascending, intricately branching, forming thickets. **Leaves** 1–2 cm on main stems, shorter on distal and shorter branches, brighter green above, paler and punctate beneath; stipules subulate to narrowly triangular-acuminate. **Leaflets** 9–19(–21), 1.5–3.5(–5) × 0.5–1.8 mm, smaller and fewer on distal leaves, obovate to broadly oblanceolate to obcordate. **Inflorescences** spikes, 0.2–1.8(–3.5) × 0.6–1 cm (not including petals), on peduncles 0.2–4.5(–5) cm, subcapitate to oblong, dense or rarely loose, the axis pilosulose or rarely glabrous; bracts 2.2–3.4(–4.5) mm, deciduous, broadly ovate and short- to long-acuminate, mostly glabrous and glandular. **Flowers** papilionoid; calyx 3.8–4.2 mm, tube 2.6–3.2(–3.5) mm, ribs becoming prominent, intervals with 3–5(–7) red or orange glands, lobes deltate-acuminate to broadly triangular, often gland-spurred, the dorsal one 0.5–1.2 mm, the ventral pair broader and shorter; corolla bicolored (exceptionally all white), the banner 5.1–6.2 mm, whitish with green-yellow eye but early becoming dull red, the others vivid pink-purple, wings 5.5–6 mm, keel petals 6.5–8 mm, attached at or below middle of the stamens; stamens 10, 6.8–8.2 mm, free for 1.8–2.7 mm. **Fruits** 2.8–3.5 mm, obliquely triangular, minutely gland-sprinkled, glabrous or thinly pilosulose. **Seeds** 1, brown, smooth.

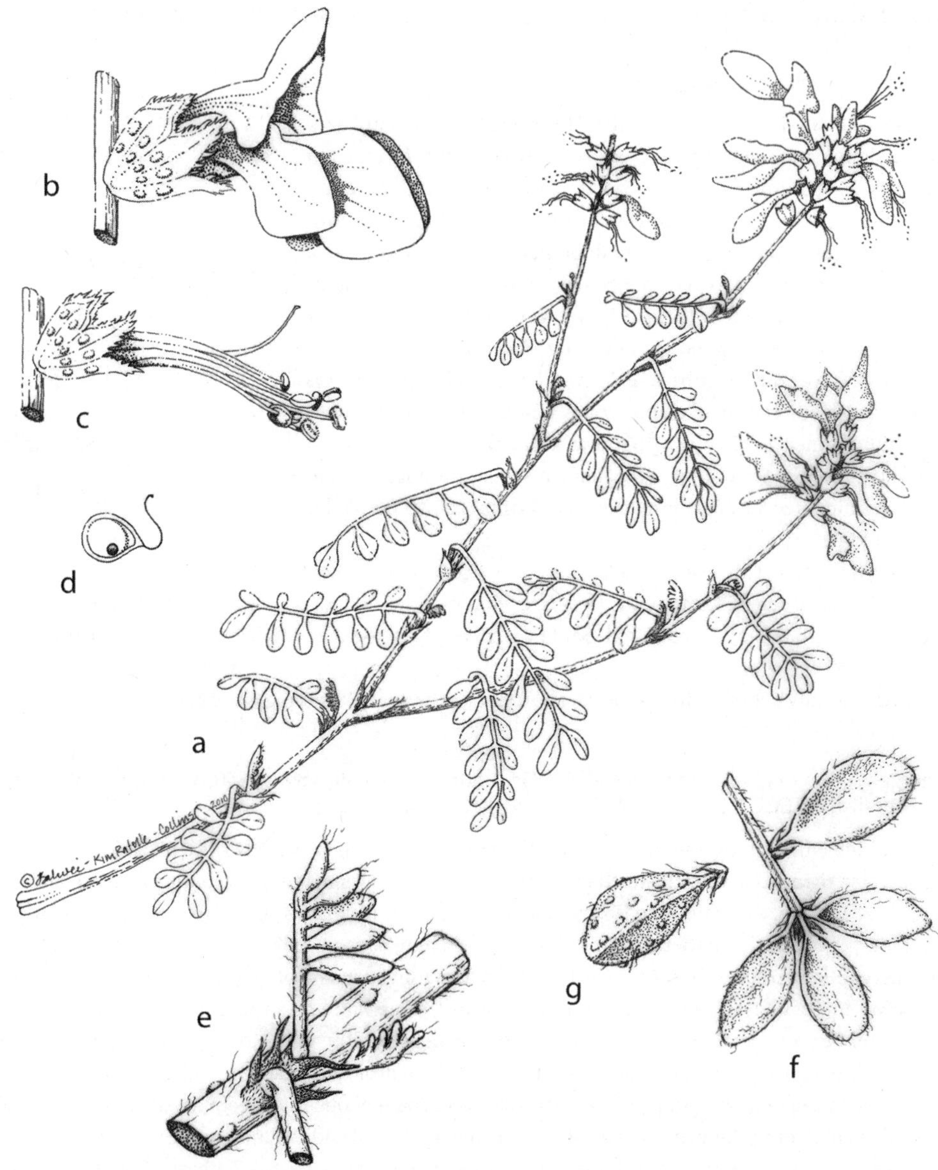

***Dalea frutescens*: a. habit; b. flower; c. flower with corolla removed; d. fruit cross-section; e. stem node; f. leaf; g. leaflet detail (lower/abaxial surface). Illustrator: Katherine Salwei & Kim Ratelle-Collins.**

Flowering (Apr–)Jul–Nov. c Texas, New Mexico, sw Oklahoma; Mexico (Chihuahua, Coahuila, Nuevo León, Zacatecas).

Dalea frutescens grows at elevations of 190–2240 m. It occurs on rocky slopes, arroyos, and valleys in the Chihuahuan Desert and in oak-juniper-mesquite woodland, Tamaulipan thornscrub, and various grassland habitats, generally, but not exclusively, on limestone (Barneby 1977).

Dalea frutescens is grown in Arizona as an attractive low to medium-sized autumn-flowering shrub or tall groundcover noted for its masses of purple flowers. The plants are evergreen or become partially deciduous with winter freezing. Established plants have low irrigation requirements. Growth rate is moderate to fast with irrigation. Black dalea should be planted in well-drained soil in a position where it will receive full sun. This species is hardy to about -18°C. The plants can be pruned to promote fresh growth. Stems that lie on the ground can root and form low thickets (Barneby 1977).

13. Dalea grayi (Vail) L. O. Williams. Ann. Missouri Bot. Gard. 23(3):450. 1936. * [for Asa Gray, 1810–1888, American botanist] GRAY'S DALEA Native

Parosela grayi Vail, Bull. Torrey Bot. Club 24(1):14. 1897

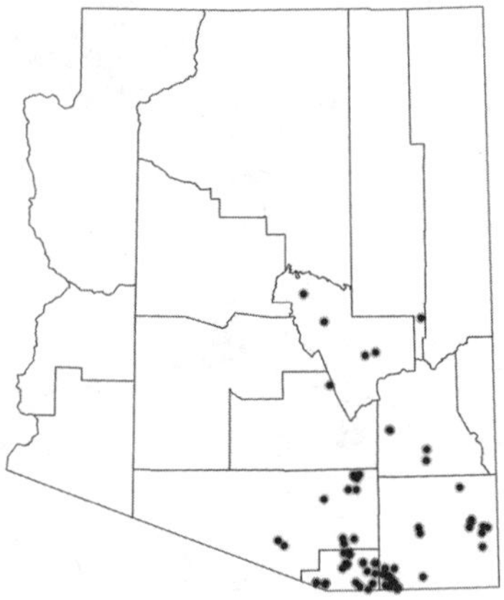

Perennial herbs (3–)3.5–7(–9) dm, wand-like. **Herbage** glabrous to inflorescence, scarcely glandular. **Stems** angled, erect, scarcely branched. **Leaves** 1–5 cm, few; stipules 0.4–1.3 mm, narrowly subulate. **Leaflets** (11–)17–43, 2–5 × 0.5–1.1 mm, elliptic to oblong, often loosely folded, glandular beneath. **Inflorescences** terminal spikes, usually on all branches, (0.4–)0.7–5.5(–7) × 0.7–1 cm (not including petals), ovoid to cylindric, axis densely villosulous; peduncles 0.5–5(–10) cm; bracts (2.2–)2.5–4.5 mm, persistent, ovate or lance-acuminate, all but lowest pubescent dorsally, glandular. **Flowers** non-papilionoid; calyx 3.6–4.8 mm, tube 2–2.7 mm, densely silky-pilose, ribs prominent, intervals with 3–6 small glands, lobes 1.3–2.4(–3.2) mm, triangular; corolla white, banner 4–5.7 mm, wings and keel petals similar, 2.4–3.6 mm, attached at or near the separation of the filaments, margins not adherent; stamens 10, 5.5–7 mm, free for 2.7–4 mm, exserted. **Fruits** 2.2–2.5 mm, pilosulose, finely glandular. **Seeds** 1.

Flowering (May–)Aug–Oct[–Dec]. Rock outcrops, ridges, slopes, and roadsides, often in gravelly or rocky soil; [780–]1160–1770[–2160] m; (Chihuahuan Desertscrub), Semidesert Grassland, Madrean Evergreen Woodland; Apache, Cochise, Gila, Graham, Pima, Pinal, Santa Cruz cos.; New Mexico; Mexico (Chihuahua, Durango, Sinaloa, Sonora).

Dalea grayi is locally common in southeastern Arizona. The species is typically found in rocky soils in exposed habitats.

14. Dalea greggii A. Gray. Pl. Nov. Thurb. 314–315. 1854. * [for Josiah Gregg, 1806–1850, American explorer, merchant, and naturalist] GREGG'S DALEA, TRAILING INDIGO BUSH Cultivated

Subshrubs 3(–5) dm tall, forming low thickets, spreading to 20 dm. **Herbage** densely silky-pilosulose to inflorescence, obscurely glandular. **Stems** 3–8 dm, clustered, numerous, often rooting, spreading, becoming blackish and glabrate in age. **Leaves** 0.6–2.5(–3) cm, the main stem leaves subtending fascicles of smaller ones; stipules 1–2.5(–3) mm, subulate, lanceolate, or linear, pubescent. **Leaflets** 5–9, 2–7(–9) × 0.9–4.6 mm, obovate to broadly oblanceolate, commonly emarginate. **Inflorescences** spikes, 0.4–3(–5.5) × 0.8–1.1 cm (not including petals), subglobose to shortly oblong-cylindroid, axis pilosulose; peduncles 0–4(–6) cm; bracts 2–4 mm, persistent, ovate, lance-acuminate, or linear-elliptic, glandular and densely pubescent. **Flowers** papilionoid; calyx (4–)4.2–6 mm, densely silky-pilosulose with spreading hairs, tube (2–)2.2–2.8(–3) mm, ribs slender, prominent, intervals with 3–8(–10) small clear or golden glands, lobes unequal, the longest 1.7–3(–3.2) mm, deltate-acuminate to lanceolate; petals bicolored, banner opening cream or pale yellow, early becoming dull red, the others pink or pink-purple, attached much below middle of androecium, banner 4.4–6.2 mm, wings 4.4–6.6 mm, keel 5.5–7.2 mm; stamens 10, 5.5–7.5 mm, the longer filaments free for 1.7–2.4 mm, gland-tipped. **Fruits** 2.1–2.8 mm, obliquely obovate, pilosulose from middle upward and sparsely gland-sprinkled. **Seeds** 1, 1.6–2.2 mm.

Flowering Feb–Apr, Jul–Dec. w Texas; Mexico (Chihuahua, Coahuila, south to Oaxaca).

Dalea greggii has been reported to occur in Arizona; however, this was due to a misapplication of the name to *D. pulchra* (e.g., Kearney and Peebles 1960). Both species share silvery foliage, few leaflets, and subcapitate spikes, but *D. pulchra* is an upright shrub. *Dalea greggii* is native to the Trans-Pecos region of Texas and is widespread in Mexico in Chihuahua, Coahuila, Durango, Hidalgo, Nuevo León, Oaxaca, Puebla, Querétaro, San Luis Potosí, and Zacatecas where it occurs in the Chihuahuan Desert, grasslands, coastal savannah, thornscrub and other plant communities from near sea level to 2250 m (Barneby 1977). It is frequently found on calcareous substrates. Allen and Allen (1981) report that *D. greggii* is browsed by deer.

Dalea greggii is a popular, evergreen ground cover in low water-use landscapes in Arizona. The silvery green foliage and small purple flowers are attractive. With some supplemental irrigation it forms a dense, low, spreading mass of stems that root at the nodes. It is well-suited for covering banks or slopes. Established plants have low irrigation requirements but need some irrigation to maintain an acceptable appearance. Growth rate is moderate to fast with irrigation. Trailing indigo bush should have well-drained soil and full sun and is tolerant of heat and reflected light. This species is hardy to at least -12°C.

15. Dalea jamesii (Torrey) Torrey & A. Gray. Fl. N. Amer. 1(2):308. 1838. * [for Edwin James, 1797–1861, American physician, botanist, and geologist] JAMES' DALEA Native

Psoralea jamesii Torrey, Ann. Lyceum Nat. Hist. New York 2:175–176. 1827

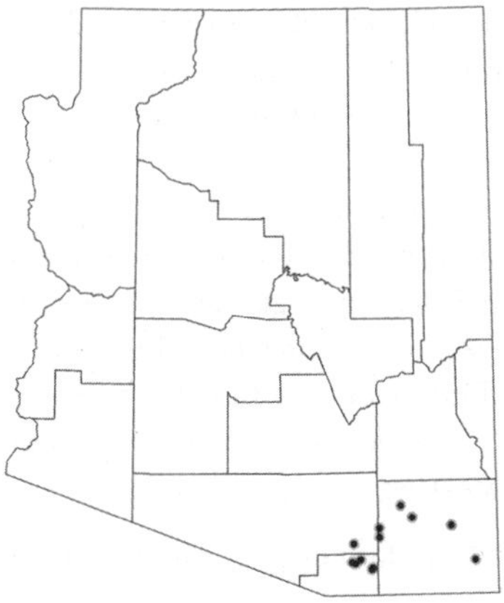

Perennial herbs, 0.1–1(–1.3) dm, low, loosely tufted. **Herbage** densely silky-pilose, obscurely glandular. **Stems** erect when short or decumbent and incurved when longer, rarely branched. **Leaves** (0.6–)1–3(–4) cm, palmate; stipules narrow (1–)2–6 mm, thinly pilosulose. **Leaflets** 3, (3–)5–18 × 2.5–8 mm, obovate to broadly oblanceolate, the terminal often longer than others. **Inflorescences** sessile spikes, 1.5–6.5(–8) × (1.6–)1.8–2.3(–2.5) cm (not including petals), axis pilosulose; bracts deciduous, (4–)5–8(–9) mm, narrowly obovate, enclosing the calyx at the base, thinly pubescent and without glands. **Flowers** papilionoid; calyx (8.3–)8.5–12(–13.3) mm, with silvery pubescence often drying golden, tube (2.4–)2.8–3.5 mm, more recessed behind banner, ribs becoming prominent, intervals with 3–4 tiny glands, lobes aristate, (5–)5.5–9(–10) mm; corolla clear yellow fading orange-brown, banner 5.2–6.6 mm, others attached below middle of stamens, wings 5.3–6.8 mm, keel petals (6.3–)6.5–8.6 mm, marginally adherent and enclosing androecium; stamens 10, (9.5–)10–11.7 mm, filaments free for 1.7–3 mm. **Fruits** 3.5–4 mm, obliquely obovoid. **Seeds** 1, 2.3–3 mm.

Flowering Apr–May[–Aug]. Slopes, valleys, and roadsides in various soils; [920–]1230–1520[–2100] m; Semidesert Grassland; Cochise, Pima, Santa Cruz cos.; Colorado, Kansas, New Mexico, Oklahoma, Texas; Mexico (Chihuahua).

Dalea jamesii is uncommon and localized in Arizona, with only a few records from the southeastern part of the state. The flowering period in Arizona is likely longer than indicated. To be expected from northeastern Sonora (Barneby 1977).

16. Dalea lachnostachys A. Gray. Smithsonian Contr. Knowl. 3(5):46–47. 1852. * [with shaggy inflorescences] GLAND-LEAF DALEA Native

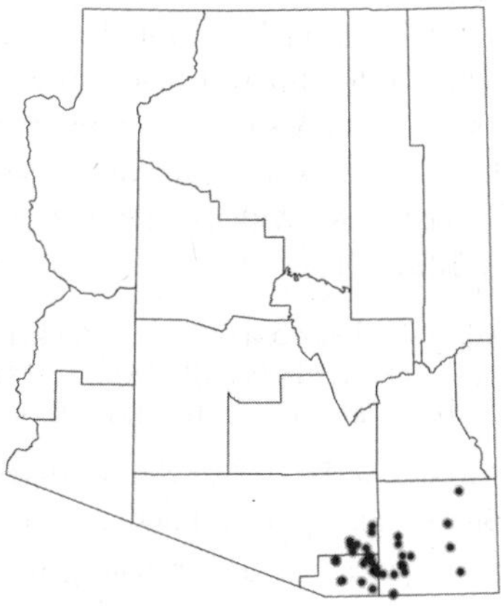

Perennial herbs (1–)1.5–4 dm. **Herbage** inconspicuously pubescent to inflorescence, densely and prominently glandular. **Stems** few, erect, simple or sparingly branched. **Leaves** 3–8(–9.5) cm, golden-green above, glaucescent and prominently glandular beneath; stipules subulate, (1.5–)2–3.5 mm. **Leaflets** (3–)7–9(–11), (5–)7–19 × (2–)3–7.5 mm, obovate or oblong-obovate, obtuse or emarginate. **Inflorescences** racemes, 2–5(–8) × (1.5–)1.6–1.9(–2) cm (not including petals), oblong-cylindroid, the flowers early spreading and the lower ones declined, the axis pilose, pedicels obscure; peduncles 1–8(–11) cm; bracts deciduous, 7–11 mm, ovate-acuminate or -caudate, pilosulose and gland-sprinkled, often pink- or purple-tinged. **Flowers** papilionoid; calyx (4.3–)5.8–7.7 mm, pilose (especially along the prominent, cord-like ribs) with stiff straight spreading spiral hairs up to 3–4.5 mm, tube 2.6–3.1 mm, the intervals with 1 row of small yellowish glands, lobes (2.5–)3.1–4.5 mm, deltate-aristate, plumose; corolla blue (drying violet), banner 5.7–7.6 mm, the others attached low on the stamens, wings 5.2–6.6 mm, keel 5.4–7.1 mm; stamens 10, 6.6–8.7 mm, the longer filaments free for 2.3–2.9 mm. **Fruits** (3–)3.3–4 mm, obliquely obovoid, distally pilosulose. **Seeds** 1, 1.8–2.4 mm, brown, smooth.

Flowering (May–)Jun–Sep(–Oct). Slopes, rock outcrops, and roadsides on various substrates; [1050–]1130–1670 [–1770] m; (Chihuahuan Desertscrub), Semidesert Grassland, Madrean Evergreen Woodland; Cochise, Pima, Santa Cruz cos.; New Mexico, Texas; Mexico (Chihuahua, Coahuila, Sonora).

Dalea lachnostachys is localized and seldom common in Arizona. It is found on various rocky substrates including limestone, granite, and volcanic rocks and grows in sunny, exposed situations.

17. Dalea lanata Sprengel. Syst. Veg. 3:327. 1826. * [woolly] WOOLLY DALEA

Perennial herbs, (2.5–)3–7 dm, prostrate, often mat-forming from an orange-yellow taproot. **Herbage** cinereous-puberulent to glabrate, glandular. **Stems** (2.5–)3–7 dm, with widely diverging branching, sparsely glandular, from densely pilosulose to glabrous. **Leaves** 1–3 cm; stipules narrow; petioles short to nearly absent. **Leaflets** (in main cauline leaves) 9–15, 3–10(–12) × 0.8–5 mm, obovate to broadly oblanceolate, mostly emarginate, prominently glandular beneath. **Inflorescences** spikes, loose and narrow, (1–)1.5–7.5(–9) × 0.6–0.8 cm (not including petals); peduncles 0.4–4 cm; bracts 2.5–3.5 mm, persistent, herbaceous, broadly ovate-acuminate, with prominent glands. **Flowers** nonpapilionoid; calyx 3.3–4.5(–4.8) mm, villosulous throughout or only on the lobes (rarely glabrous), tube 1.8–2.5 mm, ribs slender and inconspicuous, glands small, lobes 1.5–2.3 mm, lanceolate to ovate- or deltate-acuminate; corolla bright red-violet or magenta-purple, banner 2.8–4.3 mm, wings and keel petals 2.2–4.3 mm, similar to each other, not adherent, not enclosing the androecium; stamens 8–10, sometimes only 5 with anthers, 6.5–8.5 mm, filaments free for distal third. **Fruits** 2.5–3.1 mm, obliquely deltate-obovate. **Seeds** 1, 1.8–2.3 mm.

Varieties 2 (1 native in the flora): the species is distributed from Nevada to Arkansas and s Texas and into ne Mexico.

17a. Dalea lanata Sprengel var. **terminalis** (M.E. Jones) Barneby. Mem. New York Bot. Gard. 27:283. 1977. * Native

Dalea terminalis M.E. Jones, Contr. W. Bot. 12:8–9. 1908

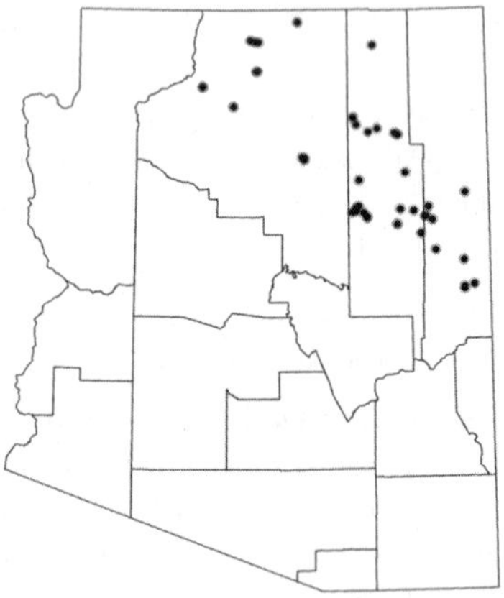

Flowers: calyx tube glabrous or nearly so, lobes villosulous (rarely glabrous).

Flowering Jun–Oct. Slopes, valleys, arroyos, floodplains, roadsides, and sand dunes, usually on sandy soil or occasionally sandstone; 560–1890 m; (Mohave Desertscrub), Great Basin Desertscrub, Great Basin Grassland, Great Basin Conifer Woodland; Apache, Coconino, Navajo cos.; Colorado, New Mexico, Texas, Utah; Mexico (Chihuahua, Coahuila).

Dalea lanata var. *terminalis* is locally common in northern Arizona where it characteristically occupies sandy habitats in desert shrub, grassland, and juniper communities. The low elevation records (below 1200 m) are from along the Colorado River in the Grand Canyon. *Dalea lanata* var. *terminalis* is distinguished from the more eastern var. *lanata* by having the calyx tube glabrous or nearly so (Kearney and Peebles 1942). *Dalea lanata* is browsed by livestock and wildlife, and rodents and birds eat the seeds (Stubbendieck and Conard 1989). With its spreading habit, the plants may have potential for soil stabilization.

18. Dalea leporina (Aiton) Bullock. Bull. Misc. Inform. Kew 1939(4):136. 1939. * [of hares, referring to the soft inflorescences] FOXTAIL DALEA, HARE'S-FOOT DALEA

Psoralea leporina Aiton, Hort. Kew. 3:81–82. 1789; *Dalea lagopus* (Cavanilles) Willdenow

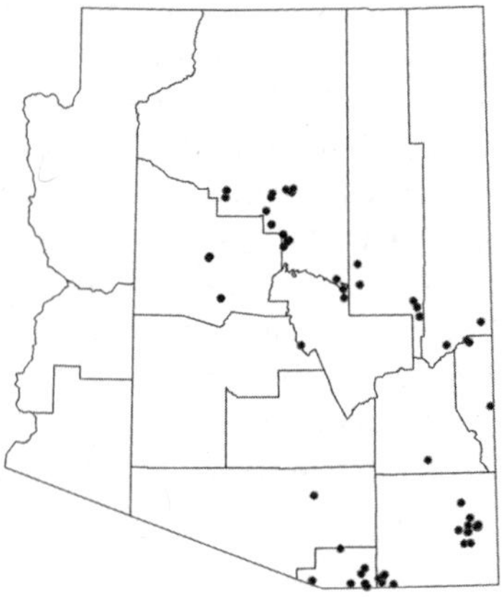

Annual herbs, 1.5–12 dm. **Herbage** glabrous to inflorescence, glandular. **Stems** simple or branching, erect, often incurved-ascending. **Leaves** (1–)2–10 cm; stipules 1–3 mm, subulate; petioles very short. **Leaflets** (9–)17–35, 3–10 × 1–3.5 mm, fewer on upper leaves or on dwarf plants, narrowly oblong. **Inflorescences** terminal spikes, usually on all branches, 1.5–7 × 0.8–1.2(–1.5) cm (not including petals), narrowly ovoid; peduncles 1–12 cm; bracts 4–5.5 mm, tardily deciduous, ovate to lance-acuminate, all but lowest pubescent dorsally, glandular. **Flowers** nonpapilionoid; calyx 3–5.2(–6.5) mm, tube (1.7–)2–2.5(–2.8) mm, more deeply recessed behind banner, ribs slender, intervals with (1–)2 rows of orange glands, lobes 1–2.8(–4) mm, broadly lanceolate, pilose; corolla white, or bicolored with banner white and others pale purple to nearly blue, banner (3.4–)4.4–6 mm, other petals attached shortly below filament separation, wings 2–3 mm, keel petals 2–3 mm, margins adherent; stamens (9–)10, 5–6.8 mm, free for 0.6–1 mm, well exserted. **Fruits** 2.4–3 mm, pilosulose, sometimes dehiscent. **Seeds** 1, 1.7–2.4 mm.

Flowering [Feb–May], Aug–Oct[–Dec]. Ridges, slopes, canyons, valleys, roadsides, disturbed areas, stream banks, and meadows in a variety of soils and often in moist sites; 1370–2520[–2850] m; Madrean Evergreen Woodland, Great Basin Conifer Woodland, Petran Montane Conifer Forest; Apache, Cochise, Coconino, Gila, Graham, Greenlee, Navajo, Pima, Santa Cruz, Yavapai cos.; New Mexico, Texas, widespread north and eastward to North Dakota, Minnesota, Indiana, Missouri; Mexico (widespread); Central America; disjunct in South America in Argentina, Bolivia and Chile.

Dalea leporina has the most extensive distribution of any species in the genus (Barneby 1977) and occurs in a broad range of habitats and plant communities. In Arizona, the species is widespread and may be locally abundant. *Dalea leporina* is occasionally grazed by livestock and has been planted as a green manure (Stubbendieck and Conard 1989).

19. Dalea lumholtzii B. L. Robinson & Fernald. Proc. Amer. Acad. Arts 30:115–116. 1894. * [for Carl S. Lumholtz, 1851–1922, Norwegian explorer and ethnographer] LUMHOLTZ'S DALEA Native

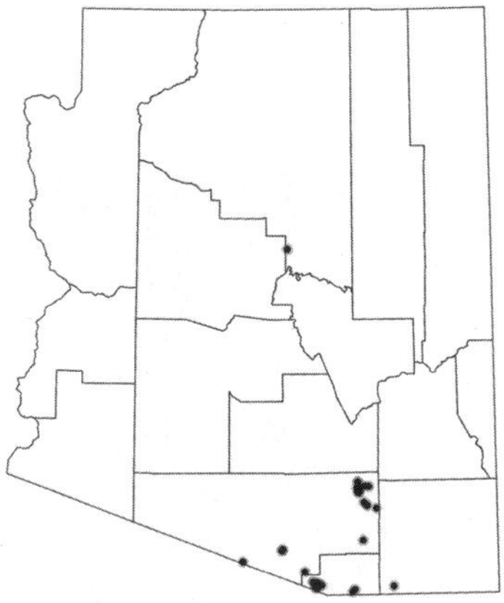

Perennial herbs, 2–5 dm. **Herbage** glabrate to inflorescence or villosulous; glandular. **Stems** erect or spreading, clustered, simple or branched. **Leaves** (2–)2.5–5 cm; stipules 1–2.8 mm, narrowly subulate; petioles short to nearly absent. **Leaflets** 21–35, 2–9(–10) × 0.3–1(–1.7) mm on main leaves, 9–21, 1.5–3 × 0.2–0.4 mm on short shoots, narrowly oblong, usually folded, linear to linear-elliptic, obtuse or retuse. **Inflorescences** terminal spikes, 0.6–2(–3) × 0.6–0.9 cm (not including petals), subspheroid, becoming cylindric; peduncles 2.5–10 cm; bracts persistent, (1.5–)2–4 mm, rhombic-ovate, pubescent dorsally below the middle or throughout, with distal glands. **Flowers** nonpapilionoid; calyx 2.7–4(–4.3) mm, subglabrous, the tube 2.1–2.8(–3) mm, recessed behind banner, ribs slender, intervals without glands, lobes unequal, 0.4–1.3 (–1.5) mm, lanceolate, glabrous or slightly puberulent; corolla white or bicolored with banner white and others blue, banner 3–5.2(–5.9) mm, wing and keel petals similar, 2.6–5.5 mm, margins not adherent; stamens 10, 5–7.7 mm, free for 2.6–3.7 mm, exserted. **Fruits** obliquely obovoid, without glands. **Seeds** 1, ± 1.7 mm.

Flowering Aug–Oct. Slopes and canyons on rocky soil or bedrock; 1070–2000[–2500] m; Semidesert Grassland, Madrean Evergreen Woodland; Cochise, Coconino, Pima, Santa Cruz cos.; Mexico (Chihuahua, Sonora).

Dalea lumholtzii is restricted to a limited area of southern Arizona, and the populations are widely scattered. In Mexico it occurs in the mountains of northeastern Sonora and southward in the Sierra Madre Occidental. The report of this species in Coconino County is based on a single collection (*M. Baker 12273,* 11 Sep 1996) from northeast of Camp Verde.

20. Dalea mollis Bentham. Pl. Hartw. 306. 1848[1849]. * [soft] HAIRY DALEA Native

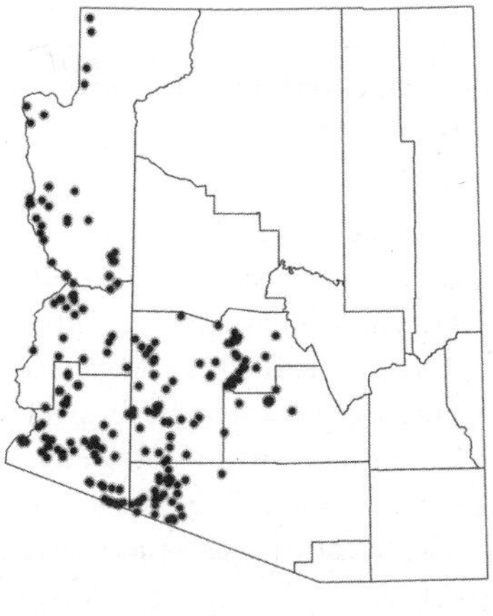

Annual herbs, 0.3–1 × 1–3.5(–6.5) dm, radiating, forming loose mats. **Herbage** pilose-pilosulose throughout or nearly so with fine, divergent or loosely ascending hairs, prominently glandular. **Stems** simple or branched, becoming distally incurved, prostrate. **Leaves** (0.5–)1–3.5 cm; stipules lanceolate, (0.6–) 1–2.5 mm, pubescent dorsally and with one or more large glands. **Leaflets** (7–)9–13(–15), (0.5–)1–7 × 0.5–3.5 mm, obovate, obcordate, cuneate-obcordate, nearly circular, or oblong-obovate, deeply notched to almost emarginate, loosely folded, the leaflets pubescent or glabrate above, prominently glandular, margins flat. **Inflorescences** racemes, shortly pedicellate, (0.5–)1–2(–3.5) × 0.8–1.2 cm (not including petals), ovate becoming oblong, flowers spreading then declining, axis pilosulose; peduncles 0.2–2.5(–4.5) cm; bracts deciduous, (3–)4.5–8 mm, lance- or elliptic-acuminate, glandular and pilose dorsally. **Flowers** papilionoid; calyx 3.3–6.6 mm, silky-pilose, tube 1.6–2.5 mm, ribs slender, intervals with one row of (1–)2–6 small orange glands, lobes deltateor triangular-aristate, of about equal length, (2.5–)2.7–4 mm, nearly always gland-spurred; corolla whitish, but the keel and sometimes the banner tinged with lilac or dull crimson, banner 3–4.2 mm, other petals attached low on androecium, wings 2.9–4.3 mm, usually notched and with a gland, keel 3.4–4.5 mm; stamens 10, 3.5–5.7 mm, free for 1.2–2 mm, gland-tipped. **Fruits** 2.2–2.7 mm, obliquely obovoid, minutely gland-sprinkled and pilosulose. **Seeds** 1, 1.8–2.2 mm, pale-brown, lustrous.

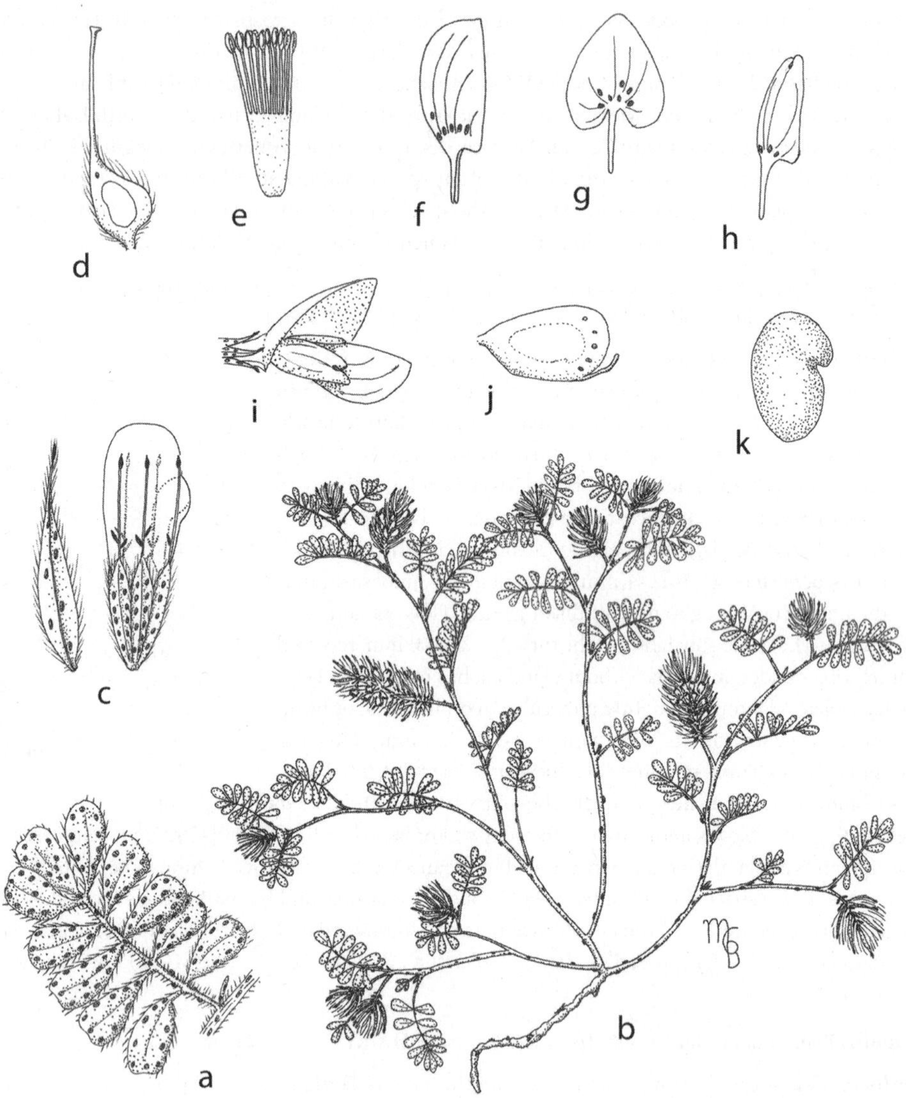

Dalea mollis: **a.** leaf; **b.** habit; **c.** calyx and bract; **d.** pistil; **e.** stamens; **f.** keel petals; **g.** banner petal; **h.** wing petal; **i.** flower; **j.** fruit; **k.** seed. Illustrator: Marceline VandeWater.

Flowering Jan–May(–Jun), (Nov–Dec). Sandy, gravelly, or occasionally rocky soils of slopes, valleys, arroyos, roadsides, and on dunes; [-50–]60–780[–1200] m; Sonoran Desertscrub, Mohave Desertscrub; La Paz, Maricopa, Mohave, Pima, Pinal, Yuma cos.; California, Nevada; Mexico (Baja California, Baja California Sur, Sonora).

Dalea mollis is characteristic of the most arid regions of the Sonoran Desert. In southwestern Sonora *D. mollis* occurs in coastal thornscrub, and it is documented for several islands in the Gulf of California. The plants germinate following autumn or winter rains. *Dalea mollis* differs from the similar and related *D. mollissima* in having the calyx generally < 6 mm long, which is usually surpassed by the corolla, and with wing petals bearing a gland in an apical notch. See also *Dalea neomexicana* var. *neomexicana* for further discussion on distinguishing these closely related species.

21. Dalea mollissima (Rydberg) Munz. Aliso 4(1):93. 1958. * [very soft] SOFT DALEA Native

Annual herbs, 0.3–1 × 1–3.5(–6.5) dm, radiating, prostrate. **Herbage** pilose-pilosulose throughout or nearly so with fine, divergent or loosely ascending hairs, commonly turning golden when dry, prominently glandular. **Stems** simple or branched, becoming distally incurved, prostrate. **Leaves** 1–3(–4) cm; stipules subulate, (1–)1.5–3 mm. **Leaflets** 9–15,

3–9(–10) × 1–4.5 mm, the margins strongly undulate-crenate. **Inflorescences** racemes, shortly pedicellate, 1.5–7.5 × (1.2–)1.3–1.6(–1.7) cm (not including petals); bracts 4.5–7 mm, glandular dorsally and gland-tipped; glands on pedicels prominent, blackish-orange or black. **Flowers** papilionoid; calyx (5.8–)6.1–8.2 mm, long-pilose, tube (2.5–)2.7–3.3 mm, intervals with one row of (3–)4–7 small orange glands, lobes (3.1–)3.4–5.2 mm; corolla whitish, but the keel and sometimes the banner tinged with lilac or dull crimson, banner 2.8–3.7 mm, wings 2.4–3.1 mm, entire and glandless, keel 3.5–4 mm; stamens 10, 4.3–5 mm, free for distal fourth of the filaments. **Fruits** included within the calyx. **Seeds** 1.

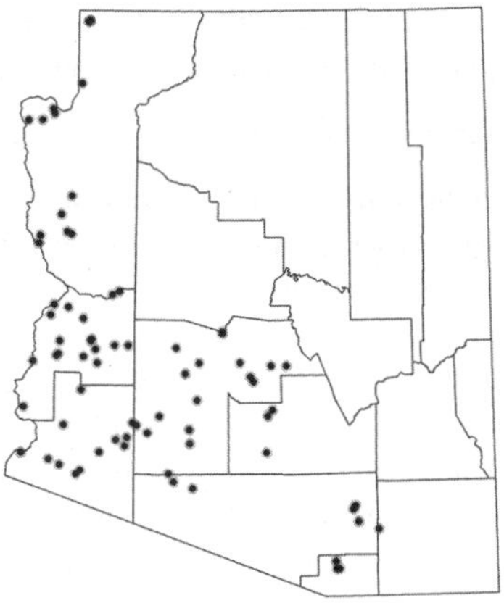

Flowering Feb–Sep[–Nov]. Sandy plains, arroyos, and roadsides, less commonly on loam or gravel and rarely on slopes; [10–]100–730[–1170] m; Sonoran Desertscrub, Mohave Desertscrub; La Paz, Maricopa, Mohave, Pima, Pinal, Santa Cruz, Yuma cos.; California, Nevada; Mexico (Baja California, Sonora).

Dalea mollissima occurs in scattered populations across southwestern Arizona. The plants characteristically form mats on sandy soils but can rarely occur on gravelly or rocky slopes. Flowering can occur nearly year-round. *Dalea mollissima* differs from *D. mollis* by having undulate leaflet margins, the calyx generally ≥ 6 mm and usually surpassing the corolla, and with entire wing petals without glands. See also *Dalea neomexicana* var. *neomexicana* for further discussion on distinguishing these closely related species.

22. Dalea nana Torrey ex A. Gray. Mem. Amer. Acad. Arts, n.s. 4(1):31. 1849. * [dwarf] DWARF DALEA

Perennial herbs, (0.5–)1–3.5 dm, prostrate to decumbent, ascending or erect. **Herbage** silky pilose, silvery-gray drying greenish-golden, glands inconspicuous. **Stems** slender, simple or sparsely branched, prostrate to erect. **Leaves** 1–2.5(–3) cm; stipules 0.5–2 mm. **Leaflets** usually 5 (upper or lowermost sometimes 3 or 7), 3–15(–19) × 1–5 mm, obovate-elliptic to oblanceolate. **Inflorescences** terminal spikes, variable in density, 0.5–4(–5) × 0.7–1.3(–1.5) cm (not including petals); bracts deciduous late in anthesis or later, herbaceous above base, 2.5–5.5 mm, dorsally pilosulose and often glandular. **Flowers** papilionoid; calyx (4.2–)4.5–6.5(–7.5) mm, tube 1.9–2.7 mm, ribs becoming prominent, intervals with 3–4 small glands, lobes 2.2–4.2(–4.7) mm triangular to triangular-aristate, plumose; corolla initially clear yellow, fading brownish or drying pinkish or purplish, banner (4–)4.4–5.5 mm, other petals attached near middle of androecium, wings 2.6–4.3 mm, keel petals 3.8–6.1 mm, marginally adherent and enclosing androecium; stamens 10, (6–)6.4–10 mm, free for distal fourth. **Fruits** 2.5–3 mm. **Seeds** 1, 1.8–2.2 mm.

Varieties 2 (2 native in the flora): sw to sc United States, n Mexico.

1. Spikes permanently dense and cone-like, with no visible axis between flowers; bracts 1.2–2 mm wide, narrowly ovate to lanceolate, acuminate ________ **22a. Dalea nana** var. **carnescens**

1′ Spikes relatively loose in age, with visible axis between flowers; bracts 2–4 mm wide, broadly ovate to elliptic, very shortly acuminate ________ **22b. Dalea nana** var. **nana**

22a. Dalea nana Torrey ex A. Gray var. **carnescens** (Rydberg) Kearney & Peebles. J. Wash. Acad. Sci. 29(11):483. 1939. * [turning flesh-pink, referring to the dried corolla] Native

Dalea rubescens S. Watson, Proc. Amer. Acad. Arts 17:369. 1882

Stems commonly diffuse, ascending, or erect. **Leaflets** 5 in most leaves, sometimes 3 or 7 in lowest and uppermost. **Inflorescences** dense, cone-like spikes, with axis not visible; bracts 1.5–2 mm wide, narrowly ovate to lanceolate, acuminate. **Flowers** with calyx vase-shaped, 1.6–2.4 mm wide.

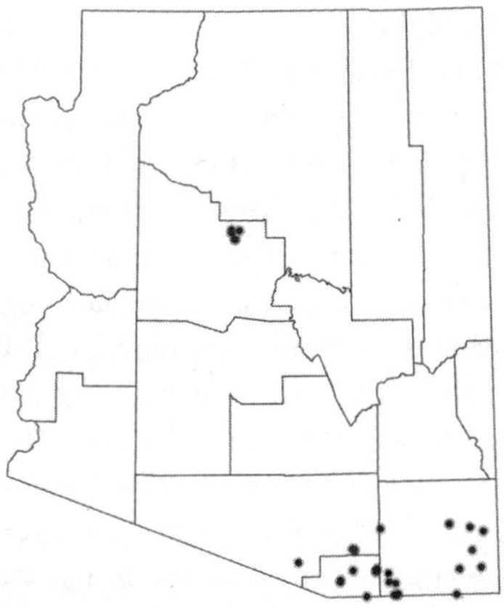

Flowering Apr–Sep(–Oct). Slopes, canyons, and roadsides, usually on gravelly and often calcareous soils; [near sea level–]1040–1900[–1980] m; Semidesert Grassland, Madrean Evergreen Woodland; Cochise, Pima, Santa Cruz, Yavapai cos.; New Mexico, Texas; Mexico (Chihuahua, Coahuila, Durango, Nuevo León, Sonora, Tamaulipas).

Of the two varieties, var. *carnescens* is more widely distributed in Arizona. According to Barneby (1977), this variety is usually found on calcareous substrates while var. *nana* is usually found on neutral, often sandy soils. Across its range, var. *carnescens* occurs in a variety of plant communities including grasslands, Tamaulipan thornscrub, mesquite-oak-juniper woodlands, oak-juniper-pine woodlands, and occasionally in Chihuahuan desertscrub.

22b. Dalea nana Torrey ex A. Gray var. **nana** * Native

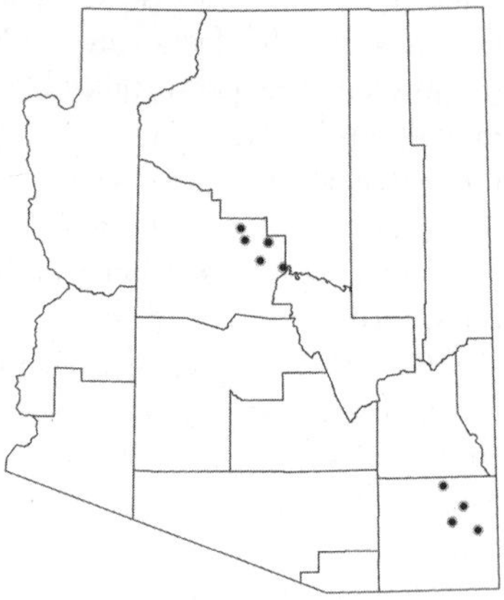

Stems diffuse or prostrate. **Leaflets** 5 in almost all leaves; **Inflorescences:** spikes relatively loose, axis visible with at least 1 mm between mature flowers; bracts 2–4 mm wide, broadly ovate to elliptic, very shortly acuminate. **Flowers:** calyx usually campanulate, 2.2–3 mm wide.

Flowering Apr–Sep. Slopes, valleys, plains, and roadsides, typically on sandy soils; [10–]940–1830 m; (Sonoran Desertscrub), Mohave Desertscrub, Semidesert Grassland; Cochise, Yavapai cos.; Colorado, Kansas, New Mexico, Oklahoma, Texas; Mexico (Chihuahua, Nuevo León, Tamaulipas).

Dalea nana var. *nana* is rather rare in Arizona. Across its range, it is found in a variety of prairie, scrub, and woodland habitats.

23. Dalea neomexicana (A. Gray) Cory. Rhodora 38(455):406. 1936. * [of New Mexico] DOWNY DALEA, NEW MEXICO DALEA

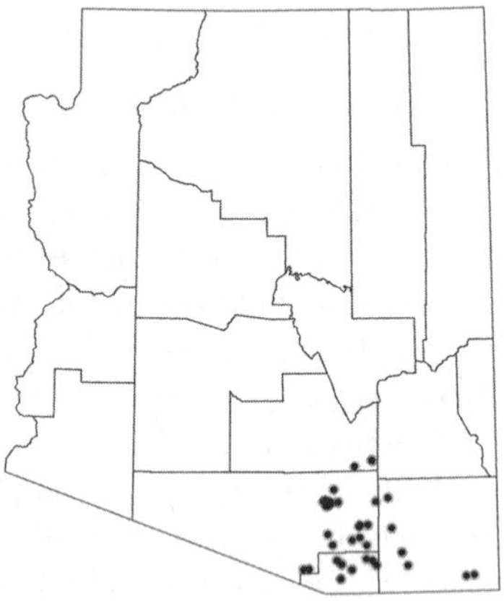

Perennial herbs, diffuse, tufted or matted. **Herbage** densely silky-pilosulose, sparsely or densely orange-glandular. **Stems** (0.2–)0.3–3.5 dm, prostrate, becoming distally incurved. **Leaves** 0.8–4.5 cm; stipules 0.6–1.8(–2.4) mm, narrow; petioles short. **Leaflets** (7–)9–15, 2.5–8 × 1–4 mm, obovate to cuneate, obtuse, margins flat or strongly undulate, prominently glandular below. **Inflorescences** racemes, 1–6 × 1.3–1.6(–1.8) cm (not including petals), ovoid becoming oblong, the open flowers spreading, at least the lowest nodding in age; bracts (3–)3.5–6 mm, lance-acuminate or -caudate, early deciduous. **Flowers** papilionoid; calyx 4.7–8(–8.8) mm, densely pilose, tube 2.1–3.1 mm, ribs slender, intervals with 1 row of 5–7(–9) small glands, lobes 2.5–5.7 mm, triangular-aristate, plumose; corolla usually bicolored, banner whitish with green (turning dull crimson), wings whitish and variably pink-tipped, keel often darker pink to purple, banner 4–5.6 mm, wings 3.3–4.2 mm, keel 4.5–7.1(–7.5) mm (including proximal portion, which is fused to stamens); stamens 10, 5–6.7(–7.2) mm, free for 1.3–2.4 mm. **Fruits** 2.3–3 mm, obliquely obovate. **Seeds** 1, brown, 1.7–2.4 mm.

Varieties 3 (1 native in the flora): sw United States and Mexico.

23a. Dalea neomexicana (A. Gray) Cory var. **neomexicana** * Native

Leaves 0.8–3 cm. **Leaflets** obcordate or truncate, some or all crenate and with conspicuous glands below, pubescent on both sides. **Pedicels** with small but obvious glands.

Flowering Apr–Jun, Jul–Oct[–Nov]. Slopes, valleys, plains, arroyos, and roadsides in various soils; 120–1600 [–2100] m; Sonoran Desertscrub, Chihuahuan Desertscrub, (Mohave Desertscrub), Semidesert Grassland; Cochise, Pima, Pinal, Santa Cruz cos.; New Mexico, Texas; Mexico (Chihuahua, Coahuila, Sonora).

Dalea neomexicana var. *neomexicana* is widely distributed across the arid, southern and central portions of Arizona. Populations are scattered but the plants may be common in some locations. The other varieties have larger pedicel glands or leaflet margins without undulations and are found in northeastern Mexico with one extending into Texas. Two other mat-forming species in southern Arizona can be difficult to distinguish from *D. neomexicana* var. *neomexicana* and from each other, leading to many misidentified specimens in collections. This is particularly the case if the sample in hand is not clearly perennial, e.g., if flowering during the first year. *Dalea mollis* has wing petals almost always with a gland in the notched apex, keel petals visible and exserted just beyond the the calyx teeth and deciduous with the other petals, and plane leaflet margins. *Dalea mollissima* shares with *D. neomexicana* var. *neomexicana* undulate leaflet margins and petals that are difficult to see, surpassed by the plumose calyx teeth. However, *D. mollissima* has larger prominent glands on the pedicels, visible between the flowers and especially after the flowers fall off the stem. *Dalea neomexicana* var. *neomexicana* may also be in Maricopa Co.; this awaits confirmation with accurately identified material.

24. Dalea pogonathera A. Gray. Mem. Amer. Acad. Arts, n.s. 4(1):31–32. 1849. * [with bearded awns, referring to the calyx lobes] BEARDED DALEA

Perennial herbs, to 3 dm. **Herbage** glabrous, sparsely glandular. **Stems** (0.4–)0.7–3.5 dm, decumbent, radiating or ascending, simple or branched. **Leaves** (0.4–)0.8–2(–2.5) cm; stipules (0.7–)1–2.6 mm, narrowly triangular-acuminate; petioles short. **Leaflets** (3–)5–7(–9), (1.5–)2–10 × 0.5–2 mm, narrowly oblanceolate to linear-oblong, emarginate. **Inflorescences** spikes, (0.5–)1–8 × 0.9–1.8(–2) cm (not including petals); bracts 3–6 mm, persistent, pale, broadly ovate and embracing the calyx, short-acuminate. **Flowers** papilionoid; calyx (5–)5.2–10.5 mm, tube 2.6–3.4 mm, densely villosulous, ribs becoming prominent, intervals with 1 row of 4–7 small glands, lobes 1.9–7.2 mm, triangular-aristate, plumose; corolla pale or vivid violet-purple, banner (4.5–)4.8–7.8 mm, wings 2.4–4.2 mm, keel 4.2–7 mm; stamens 9 or 10, (6.5–)7.5–10.5 mm, free for 1.7–3 mm. **Fruits** 2.8–3.5 mm. **Seeds** 1, 1.9–2.4 mm, castaneous, smooth.

Varieties 2 (1 native in the flora): sw United States and Mexico.

24a. Dalea pogonathera A. Gray var. **pogonathera** * Native

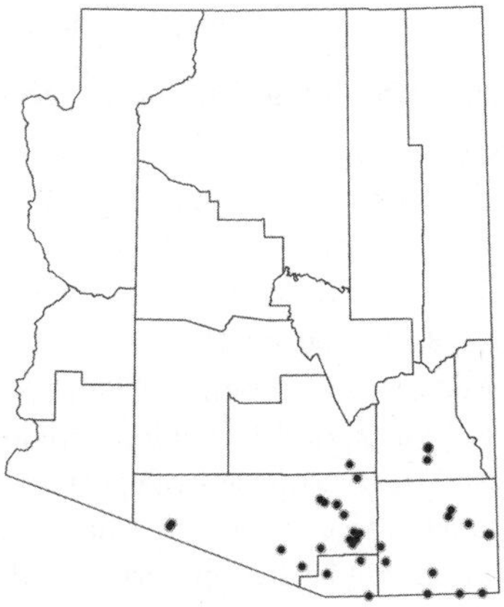

Inflorescences spikes, generally 1.3–1.8(–2) cm wide (not including petals). **Flowers** with longest calyx lobe 4 mm or longer.

Flowering Apr–May, Jul–Sep(–Dec). Slopes, canyons, valleys, and roadsides on various substrates; (610–)910–1680(–2130) m; (Sonoran Desertscrub), Chihuahuan Desertscrub, Semidesert Grassland, Plains Grassland, Madrean Evergreen Woodland; Cochise, Graham, Pima, Pinal, Santa Cruz cos.; New Mexico, Texas; Mexico (Chihuahua, Coahuila, Durango, Nuevo León, San Luis Potosí, Sonora, Zacatecas).

Dalea pogonathera var. *pogonathera* is infrequent to common and widespread across southeastern Arizona in a variety of habitats. The second variety, var. *walkerae*, occurs in southern Texas and northeastern Mexico and differs in having narrower inflorescences, shorter calyx teeth, and shorter hairs on the calyx.

25. Dalea polygonoides A. Gray. Smithsonian Contr. Knowl. 5(6):39. 1853. * [resembling an annual knotgrass] SIXWEEKS DALEA Native

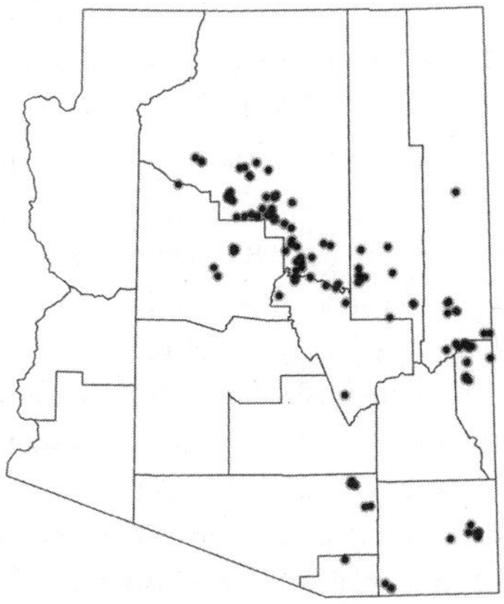

Annual herbs, 0.5–3 dm. **Herbage** glabrous, inconspicuously glandular. **Stems** slender, ascending, simple or branched. **Leaves** (0.7–)1–3 cm; stipules 0.6–2 mm, subulate to subulate-caudate; petioles slender. **Leaflets** (3–)5–7(–9), 5–12(–15) × 0.9–3 mm, emarginate, thick, green above, pale and punctate beneath, oblong-oblanceolate, terminal longest and palmate with the last pair. **Inflorescences** spikes, (0.5–)0.8–2.5(–3) × 0.5–0.8(–0.9) cm (not including petals), ovate becoming cylindric; peduncles (0–)1.5–7(–15) cm; bracts 2–4 mm, persistent, lowest broadly obovate, short-acuminate, the rest narrower, glabrous dorsally and ciliolate, with few large glands. **Flowers** nonpapilionoid; calyx 2.8–4 mm, slightly expanded in fruit, glabrous to thinly pilose, tube 1.6–2.1 mm, recessed behind the banner, ribs prominent, intervals with 1(–2) large orange glands, lobes 1.3–2 mm, aristate; corolla pale rose-purple, banner 2.7–3.8 mm, others early deciduous, attached well below separation of the filaments, wings 2–2.5 mm, keel (2–)2.4–2.9 mm; stamens (6–)7–9, often with only 5 fertile, 2.2–3.5 mm. **Fruits** 2.5–3 mm, obliquely obovoid, exserted from the calyx-tube. **Seeds** 1, 1.5–1.9 mm.

Flowering Aug–Oct. Slopes, valleys, along streams and roadsides in various soils; (1400–)1690–2930 m; Madrean Evergreen Woodland, Great Basin Conifer Woodland, Petran Montane Conifer Forest; Apache, Cochise, Coconino, Gila, Greenlee, Navajo, Pima, Santa Cruz, Yavapai cos.; New Mexico, Texas; Mexico (Chihuahua, Durango).

Dalea polygonoides is commonly found in forest and meadow habitats at higher elevations in eastern parts of the state. The plants can be locally abundant.

26. Dalea pringlei A. Gray. Proc. Amer. Acad. Arts 17:201. 1882. * [for Cyrus G. Pringle, 1838–1911, American botanist] PRINGLE'S DALEA

Perennial herbs, 2–10(–25) dm, becoming suffrutescent in age. **Herbage** glabrous to inflorescence, minutely glandular. **Stems** erect, single to several, ascending, branched distally, ribbed, pale green (to purplish). **Leaves** 1.5–6[–8] cm; stipules narrow, rigid, 0.5–2 mm; petioles short. **Leaflets** 7–45[–67], 1–9[–15] × 0.5–2.5 mm, obovate to elliptic, smooth above, glandular below. **Inflorescences** terminal spikes, (1–)2–10(–14) × 0.8–1.2 cm (not including petals), shortly oblong to cylindric, becoming loose; bracts persistent, 2.5–5.5 mm, the lowest ovate, the rest narrowly ovate- to lance-acuminate, densely ciliolate, distally glandular. **Flowers** nonpapilionoid; calyx (3.5–)4.3–6.1(–6.5) mm, tube 1.7–2.5(–2.7) mm, subsymmetrical, ribs prominent, intervals with one row of 3–6(–7) small pale glands, densely silky-pilose, lobes 2.4–4 mm, triangular-acuminate or shortly aristate, plumose; corolla usually bicolored, banner 3–5.1 mm, white or whitish, or lilac-tipped, becoming dull red, wings and keel petals similar, (1.8–)2–3.1 mm, pink-purple or vivid magenta (rarely white), attached at or near filament separation; stamens 10, 5.5–7 mm, free for distal half. **Fruits** 2.5–3 mm, villosulous with small glands. **Seeds** 1, 1.7–2.2 mm.

Varieties 3 (1 native in the flora): se Arizona and nw Mexico (Sonora, Chihuahua, Sinaloa).

26a. Dalea pringlei A. Gray var. **pringlei** * Native

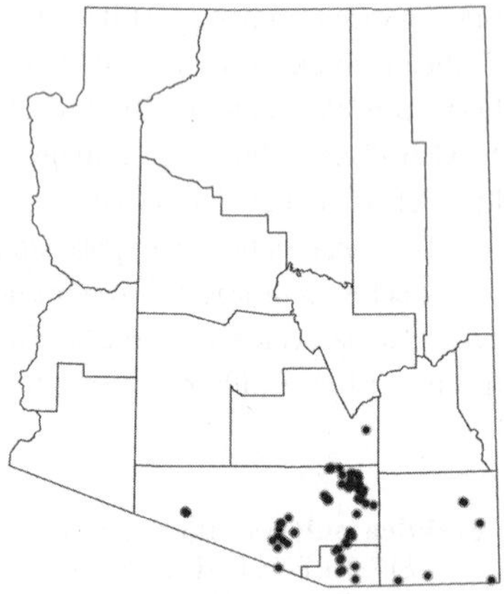

Stems (1.5–)2–6(–8) dm. **Inflorescences** spikes (1–)1.5–6.5 cm; bracts villosulous throughout. **Flowers:** corolla purple.

Flowering Feb–May(–Jul), Aug–Nov[–Dec]. Ridges, slopes, valleys, canyons, arroyos, and roadsides, often in rocky or gravelly soil; [50–]700–1530 (–1780) m; Sonoran Desertscrub, Semidesert Grassland, (Madrean Evergreen Woodland); Cochise, Pima, Pinal, Santa Cruz cos.; Mexico (Sonora).

Dalea pringlei var. *pringlei* is found in southeastern Arizona and in northern Sonora, with outlier populations in the Guaymas region. The other two varieties occur southward in northwestern Mexico. In Arizona, *D. pringlei* var. *pringlei* can be distinguished from the otherwise similar *D. grayi* by its bicolored or bright purple corolla, longer calyx lobes, and looser spikes at maturity (Barneby 1977).

27. Dalea pulchra Gentry. Madroño 10(8):227–230, f. 1 (map), pl. 14. 1950. * [beautiful] SANTA CATALINA DALEA Native, Cultivated

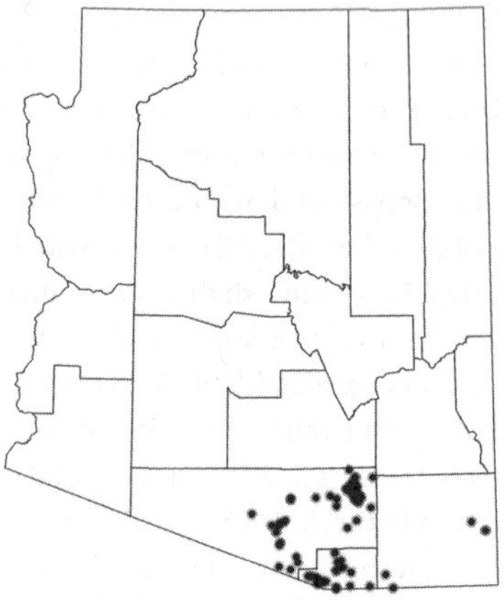

Shrubs to 10 dm. **Herbage** silvery-pilosulose and glandular. **Stems** stiffly and moderately densely branched, in age becoming fissured, glabrate. **Leaves** to 1.5 cm; stipules 0.8–1.5 mm, triangular, livid; petioles short. **Leaflets** 5–7(–9), fewer on short shoots, 1.5–5 × 0.7–2.6 mm, obovate to oblanceolate, mostly folded, generally pubescent both sides, prominently glandular beneath with clear to reddish glands. **Inflorescences** spikes terminal or on axillary short shoots, 0.4–1.7(–2) × 1.1–1.6 cm (not including petals), densely capitate, subglobose or ovate-oblong; bracts deciduous, without glands, lowest 2–3 ranks ovate, overlapping, with only vestigial buds, forming an involucre, others linear-lanceolate, pilose. **Flowers** papilionoid; calyx (4.5–)4.8–7.2 mm, densely pilose, tube 2.4–3.2 mm, deeply recessed behind the banner, ribs becoming prominent, intervals with one row of minute clear glands, lobes 2–4.2 mm, triangular-aristate, plumose, spreading in age; corolla bicolored, banner at first cream-colored, early becoming dull red, other petals pink-purple, attached low on the androecium, banner 6–7.5 mm, wings 6–8 mm, keel petals 7.8–10.2 mm; stamens 10, 7.5–9 mm, free for distal fourth. **Fruits** 2.5 mm, obliquely obovoid, dense line of hairs apically. **Seeds** 1.

Flowering (Jan–)Feb–May(–Nov). Ridges, slopes, canyons, arroyos, and roadsides in rocky or gravelly soil; [50–]730–1530[–1670] m; Sonoran Desertscrub, Semidesert Grassland, lower elevations in Madrean Evergreen Woodland; Cochise, Pima, Santa Cruz cos.; Mexico (Sonora).

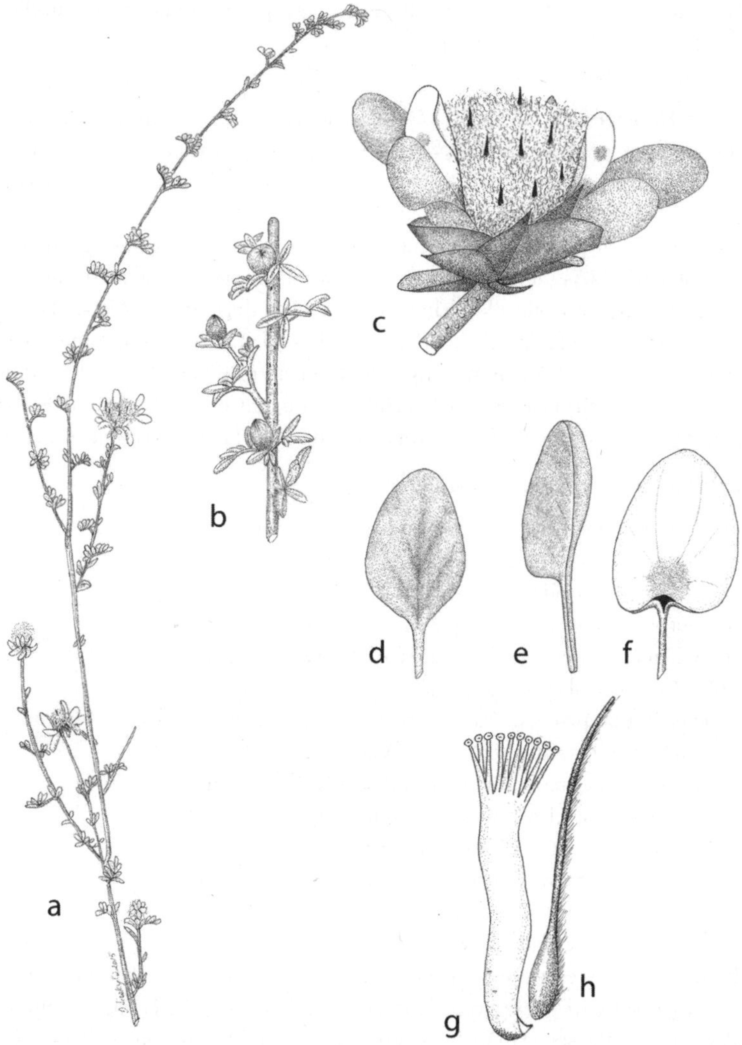

Dalea pulchra: **a.** habit; **b.** stem detail; **c.** inflorescence; **d., e., f.** petals; **g.** stamens; **h.** pistil. Illustrator: Daniela Siroky.

Within its limited area of distribution in Arizona, *Dalea pulchra* is often locally abundant. It occurs in Sonora as far south as the Guaymas region and is also found in thornscrub and Chihuahuan Desert habitats in the state. The common name refers to its type locality in the Santa Catalina Mountains, near Tucson. The names of *Dalea pulchra* and *D. greggii* were formerly confused, e.g., in the Arizona Flora (Kearney and Peebles, 1960). *Dalea greggii* was misapplied to the similarly silvery-foliaged *D. pulchra* plants in southern Arizona, but the true *D. greggii* is easily distinguished by its low, densely spreading habit, with branches that root on contact with the ground.

Santa Catalina dalea is grown in Arizona for its attractive, rounded form and silvery foliage. The flowers are attractive but are not produced in large numbers. It is suitable for low water-use plantings and as an informal, low hedge or background planting. The plants will become partially deciduous with extended drought or winter freezing. Established plants have low irrigation requirements. Growth rate is moderate to fast with irrigation. Santa Catalina dalea should be planted in well-drained soil and receive direct sunlight for most of the day. This species appears to be hardy to at least -12°C. In favorable garden situations, *Dalea pulchra* will often reseed.

28. Dalea purpurea Ventenat. Descr. Pl. Nouv., pl. 40. 1801. * [purple, referring to the corolla] PURPLE PRAIRIE CLOVER, RED TASSELFLOWER

Petalostemon purpureus (Ventenat) Rydberg

Perennial herbs, 2–9 dm. **Herbage** glabrous, thinly pilosulose to the spikes, or densely pilosulose throughout, glands none or minute. **Stems** simple or few-branched distally, erect or spreading, densely leafy from short shoots. **Leaves** primary cauline 1.7–4(–4.5) cm, smaller on the short shoots; stipules (1.5–)2–6(–7) mm, awl-shaped. **Leaflets** (3–)5(–7), (7–)10–24(–28) × 0.2–0.8[–1.7] mm, linear-oblanceolate, linear, or linear-elliptic, acute, often rolled. **Inflorescences** spikes, (1–)1.5–7 × 0.7–1.2(–1.3) cm (not including petals), dense, ovoid becoming oblong-cylindrical; peduncles 0–9 (–15) cm; bracts persistent, lowest 2.5–7 mm, obovate-acuminate, others shorter and more narrow. **Flowers** nonpapilionoid; calyx 3.2–4.5(–5) mm, densely pilosulose (hairs golden when dry), tube (1.7–)2–2.8(–2.9) mm, not recessed behind banner, ribs not prominent, intervals without glands, lobes 1.1–1.9(–2.1) mm; corolla vivid rose- or magenta-purple (to white), banner (4.3–)4.1–6.1(–7.2) mm, wings and keel petals similar, (3–)3.2–4.5(–5) mm; stamens 5, 5.6–8.5(–9) mm, filaments free and purplish for distal half. **Fruits** 2.1–2.6 mm, obliquely obovoid. **Seeds** 1, 1.6–2.1 mm.

Varieties 2 (1 in the flora): Arizona to Georgia, throughout the Great Plains, north into Canada.

28a. Dalea purpurea Ventenat var. **purpurea** * Adventive

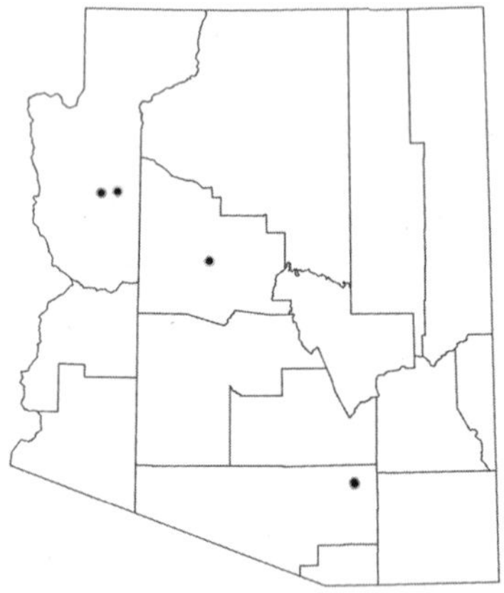

Herbage slightly to clearly pubescent. **Inflorescences** spikes, 0.9–1.2 cm wide (not including petals); peduncles 0–5(–8) cm.

Flowering May–Jun[–Aug]. Roadsides; 1220–1610[–2240] m; Semidesert Grassland, Madrean Evergreen Woodland; Mohave, Pima, Yavapai cos.; Colorado, New Mexico, Oklahoma, Texas, widespread to Alabama, Kentucky, Illinois, and northward to Canada (Alberta, Manitoba).

Barneby (1977) indicates that plants of *Dalea purpurea* var. *purpurea* in Arizona are roadside immigrants. It is native and widespread across the Great Plains states and parts of southern Canada, and eastward. The foliage is considered to be excellent forage for livestock and wildlife. It is used in hay and has a high protein content but may cause bloat (Stubbendieck and Conard 1989). The seeds of *D. purpurea* are eaten by wild turkeys and have been found in pocket gopher caches (Graham 1941). Native Americans ate the leaves of *D. purpurea* fresh and boiled (Stubbendieck and Conard 1989). The dried leaves of *D. purpurea* have been used to make a tea and the roots are pleasant tasting when chewed (Stubbendieck and Conard 1989, Lewis et al. 2005). Crushed leaves were steeped in water and this was applied to open wounds (Stubbendieck and Conard 1989). The Pawnee used the stems to make brooms (Stubbendieck and Conard 1989). Purple prairie clover is grown as a wildflower and used in revegetation in parts of the U.S. but is seldom grown in Arizona. With its beautiful, showy flowers, *D. purpurea* is an excellent addition to wildflower gardens.

29. Dalea searlsiae (A. Gray) Barneby. Mem. New York Bot. Gard. 27:234. 1977. * [for Fanny Searls, 1851–1939, American plant collector who first collected this plant] SEARLS' PRAIRIE CLOVER Native

Petalostemon searlsiae A. Gray, Proc. Amer. Acad. 8:380. 1872

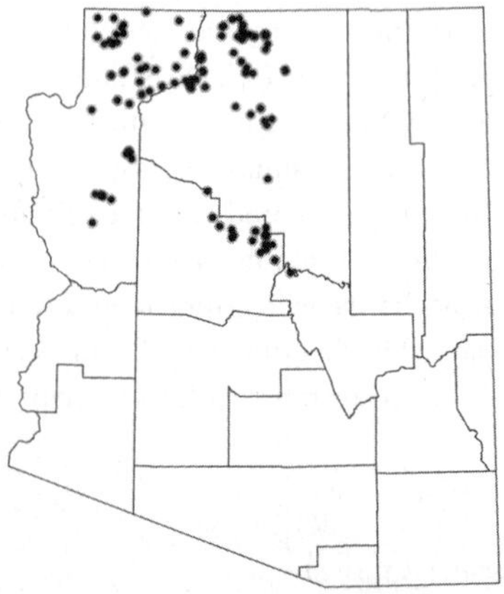

Perennial herbs, (2.5–)3–5.5 dm. **Herbage** commonly glabrous to spikes, rarely villosulous, conspicuously glandular. **Stems** simple or proximally branched, incurved or ascending, leafy in lower half, with short shoots. **Leaves** 2–5.5 cm; stipules 1–2.5(–3) mm, narrowly subulate or linear; petioles short. **Leaflets** 5–7(–9), 7–17(–20) × 0.6–2.5(–5) mm, obovate to oblong-elliptic, emarginate or obtuse, commonly folded, green above, glandular below. **Inflorescences** spikes, (1.5–)2–9(–14) × (0.8–)0.9–1.1 cm (not including petals), initially dense but axis becoming partly visible; peduncles (2.5–)4–16 (–20) cm; bracts late deciduous, 3–5(–6) mm, lanceolate, pubescence variable. **Flowers** nonpapilionoid; calyx (3.2–)3.5–4.6(–4.8) mm, tube 2–2.8 mm, deeply cleft behind banner, glabrous to densely pilosulose, ribs not prominent, intervals with 0 or few small glands, lobes (0.8–)1–2.2 mm, pilose; corolla rose-purple, banner 5.3–7.2 mm, wings and keel similar, (3.6–)3.8–4.8 mm; stamens 5, 5.5–8.5 mm, filaments free and pink in distal half. **Fruits** 3.2–4 mm, half-obovate. **Seeds** 1, 2–2.8 mm.

Flowering (Apr–)May–Jul, Sep–Oct. Slopes, canyons, arroyos, riparian communities, and roadsides on a wide range of substrates; (380–)1060–2290 m; Mohave Desertscrub, Great Basin Desertscrub, Semidesert Grassland, Great Basin Grassland, Great Basin Conifer Woodland, (Petran Montane Conifer Forest); Coconino, Mohave, Yavapai cos.; California, Nevada, Utah.

Dalea searlsiae is widespread and locally abundant across northwestern Arizona. It is recorded from volcanic cinders, shale, sandstone, limestone, gypsum, sand, loam, and clay soils. The magenta flowers are showy and it would make an excellent addition to a perennial wildflower garden or rock garden.

30. Dalea tentaculoides Gentry. Madroño 10(8):228–240, f. 2 (map), pl. 18–19. 1950. * [with long gland-prickles on the bracts and calyx lobes] GENTRY'S INDIGOBUSH Native

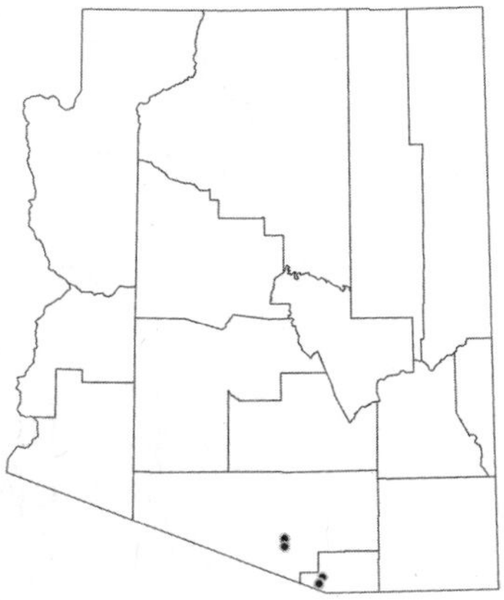

Shrubs or subshrubs, 3–6 dm. **Herbage** glabrous to spikes or thinly pilosulose, glandular. **Stems** with axillary short shoots, branching near base, green and densely glandular, becoming brown and furrowed. **Leaves** primary cauline 3–6 cm, early deciduous; stipules 0.4–2 mm subulate, deciduous. **Leaflets** primary 19–35, 2.5–8 × 1–2.5(–3.5) mm, fewer and smaller on short shoots, oblong-elliptic or obovate, smooth above, prominently glandular beneath. **Inflorescences** spikes terminal to leafy spurs and branchlets, 1–2(–2.5) × 1.1–1.5 cm (not including petals), subsessile, subcapitate becoming shortly oblong; bracts only tardily deciduous, (2.5–)3–5.5 mm, narrowly elliptic-oblanceolate, several prickle-shaped glands on back and margins, glabrous or proximally pilosulose. **Flowers** papilionoid; calyx 5.2–7.8 mm, pilosulose, tube 2.5–3.1 mm, deeply recessed behind banner, ribs scarcely prominent, interval glands 0 or minute, lobes 2.7–4.7 mm, prickle-shaped glands on back and margins; corolla bicolored, banner opening whitish, early becoming dull red, others rose-purple, attaching low on the filament column, banner 6.2–6.7 mm, wings 6.4–7.8 mm, keel 7.7–9.5 mm; stamens 10, 8–9.5 mm, filaments free for distal third. **Fruits** 3 mm. **Seeds** 1, 1.5 mm.

Flowering Apr–Jun, Sep–Oct. Canyons and drainages in sandy or gravelly soil; 1100–1300 m; Madrean Evergreen Woodland; Pima, Santa Cruz cos.; Mexico (Sonora).

Dalea tentaculoides, a species of conservation concern, is rare and highly localized in Arizona where it is known from the Pajarito and Baboquivari Mountains. The species is also known from northern Sonora. The plants grow in sunny locations as well as in shade.

31. Dalea urceolata Greene. Leafl. Bot. Observ. Crit. 1(14):199. 1906. * [urn-shaped, referring to the inflated calyx of the fruit] PINE FOREST DALEA

Annual herbs, to 2.5 dm. **Herbage** glabrous to inflorescences. **Stems** slender, erect or ascending, sparsely branched, scarcely glandular. **Leaves** 1.5–7 cm, green above, pale and glandular below; stipules 0.7–1.7(–2) mm, awl-shaped. **Leaflets** (7–)11–25, 2–10 × 1–2.7 mm, elliptic to oblong, often emarginate. **Inflorescences** terminal spikes, 0.8–4(–5) × 0.9–1 cm (not including petals), ovoid to oblong becoming cylindroid; peduncles 0.5–7.5(–12) cm; bracts 2–5 mm, deciduous except larger lowest 2 or 3, broadly ovate-acuminate, glabrous, some with small glands. **Flowers** nonpapilionoid; calyx 3.5–6 mm, glabrous or sparsely puberulent, tube 2.5–3.5(–4) mm, deeply recessed behind banner, becoming urn-shaped in fruit, ribs prominent, intervals with one row of large orange glands, lobes 0.7–2.4 mm, deltate, ciliate; petals 1–3, white or light blue, banner 3.8–5.5 mm, inner petals (when present) 1.4–2.2(–2.6) mm and early deciduous, not marginally adherent; stamens 10 or only 5 functional, exserted, apically tinged with blue. **Fruits** 2.5 mm. **Seeds** 1, 1.5 mm.

Varieties 3 (1 native in the flora): c Arizona sporadically south to sc Mexico.

Dalea tentaculoides: **a.** habit; **b.** flower; **c.** calyx; **d.** stamens; **e.** leaflet detail showing glands; **f.** leaf detail; **g.** leaf. Illustrator: Jo Ann Loza.

31a. Dalea urceolata Greene var. **urceolata** * Native

Flowers with 3 petals; 7–10 stamens.

Flowering Aug–Oct. Slopes, valleys, and along drainages, often on volcanic soils; 1940–2520[–2550] m; Great Basin Conifer Woodland, Petran Montane Conifer Forest; Apache, Coconino cos.; New Mexico; Mexico (Chihuahua).

Dalea urceolata var. *urceolata* is a high elevation plant that is apparently rare and localized across its range. In Arizona it is found from the White Mountains and along the Mogollon Rim to the vicinity of Flagstaff. The other two varieties occur south of var. *urceolata* in Mexico.

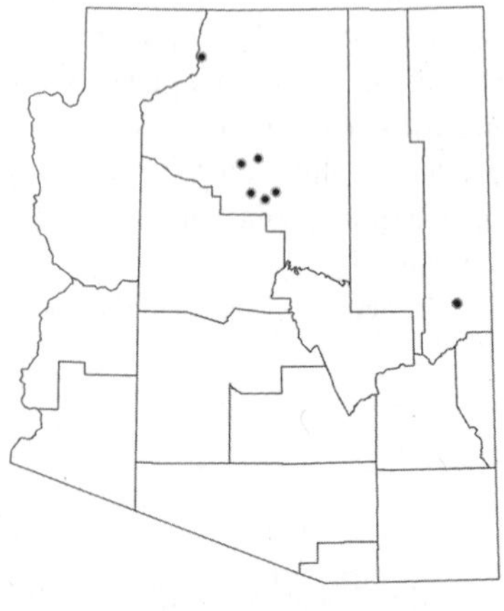

32. Dalea versicolor Zuccarini. Abh. Math.-Phys. Cl. Königl. Bayer. Akad. Wiss. 1:342–343. 1832. * [changeable in color] OAKWOODS DALEA

Subshrubs or shrubs, 2–20 dm. **Herbage** glabrous to pilosulose, variably glandular. **Stems** wiry, erect, ascending, or bending, distally branching. **Leaves** main cauline 1–4 cm, smaller on axillary short shoots; stipules 1–6 mm, awl-shaped, becoming fragile; petioles very short. **Leaflets** (5–)9–35(–37), fewer on axillary leaves, 1–7.5(–9) × 0.5–2(–3) mm, oblanceolate to obovate or linear-oblong, green to silvery, glandular below, rarely above. **Inflorescences** terminal spikes, 0.5–11 × 1–1.5 cm (not including petals), oblong-cylindric to subglobose; peduncles up to 7 (10) cm; bracts deciduous, (2.5–)3–7.5 mm, lance-acuminate to linear-caudate, densely pilose with large glands. **Flowers** papilionoid; calyx 4.5–8.5 mm, densely pilose, tube (2.1–)2.4–3.5 mm, ribs becoming prominent, intervals with one row of 3–6 small to large pale, orange, or black glands, lobes (2.3–)2.6–5 mm, triangular-aristate, usually much longer than tube; corolla bicolored, the banner pale yellow with greenish eye-spot, early becoming dull red, the others (very) pale lavender to bright rosy-purple or dull purplish-violet, attaching much below middle of androecium, banner (4.8–)5–8.2 mm, wings (4.8–)5–9.2 mm, keel (6.4–)6.8–11.5 mm; stamens 10, 7–10.5(–12) mm, free for 1.8–3.2 mm. **Fruits** 2.4–2.9 mm. **Seeds** 1, (1.4–)1.6–2 mm.

Varieties 7 (1 native and cultivated in the flora; Barneby recognized two subspecies with the variety present in Arizona placed under subsp. *versicolor*): se Arizona and adjoining New Mexico through w Mexico to Guatemala.

32a. Dalea versicolor Zuccarini var. **sessilis** (A. Gray) Barneby. Phytologia 26(1):1. 1973. * [stemless, referring to drought-inhibited inflorescences] Native, Cultivated

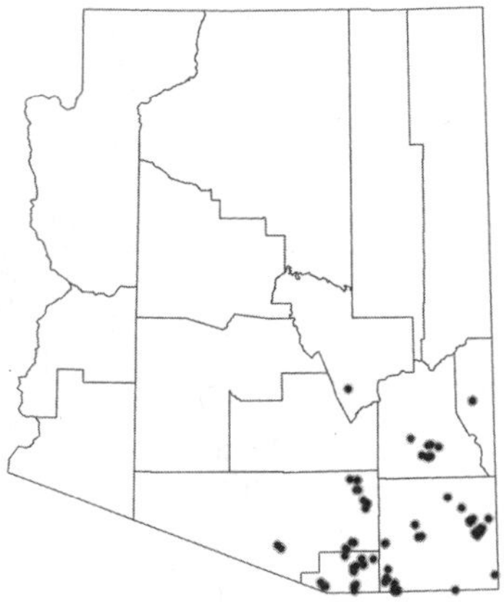

Dalea wislizeni A. Gray var. *sessilis* A. Gray, Proc. Amer. Acad. Arts 16:105. 1880

Leaflets of primary cauline leaves 5–19(–21), with scattered short hairs, not densely hairy and silvery, larger leaflets with 3–6(–8) large black glands below.

Flowering (Mar–)Apr–Jun, Aug–Oct. Rocky ridges, slopes, canyons, arroyos, roadsides, and occasionally above mountain streams; [900–](1070–)1280–2250 m; Semidesert Grassland, Interior Chaparral, Madrean Evergreen Woodland, (Petran Montane Conifer Forest); Cochise, Gila, Graham, Greenlee, Pima, Pinal, Santa Cruz cos.; New Mexico; Mexico (Chihuahua, Sonora).

Dalea versicolor var. *sessilis* is characteristically found in mountainous areas where it occurs in scattered populations. The plants grow in the open as well as beneath the canopies of trees.

Oakwoods prairie clover is occasionally grown as a landscape plant in Arizona. The plants form open, spreading shrubs. In cultivation in warm locations, the plants can continue flowering into the winter months. The foliage is evergreen or can become partially deciduous with winter freezing. Established plants have moderate irrigation requirements. Growth rate is moderate to fast with irrigation. As with other species of *Dalea*, *D. versicolor* var. *sessilis* should be planted in well-drained soil. It is more tolerant of shade than many other species but should receive full sun for at least half a day for best appearance. This species appears to be hardy to about -18°C.

33. Dalea wrightii A. Gray. Smithsonian Contr. Knowl. 3(5):49. 1852. * [for Charles Wright, 1811–1885, American botanist and explorer] WRIGHT'S DALEA Native

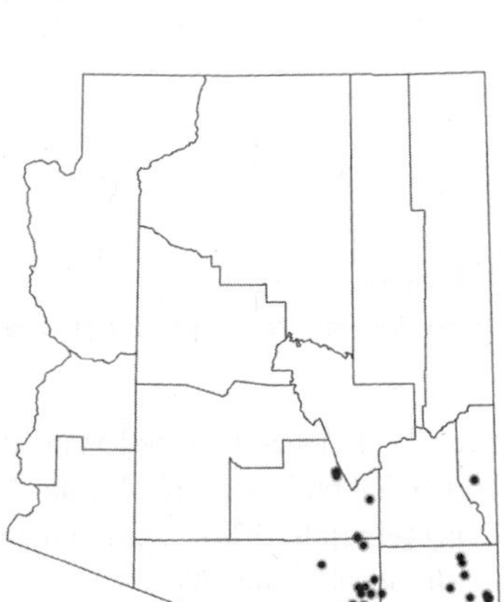

Perennial herbs, 0.3–3 dm, low, branched, from woody caudex in maturity. **Herbage** silky-pilose to pilosulose, inconspicuously glandular. **Stems** few, shortest erect and longer decumbent. **Leaves** 1–3.5(–4) cm; stipules awl-shaped, becoming stiff. **Leaflets** 5, 4–15(–20) × 2–5 mm, elliptic or lanceolate, acuminate, pubescent both sides, terminal largest, terminal three pinnate. **Inflorescences** spikes, (1.5–)2–6 × 1.7–2.3 cm (not including petals), dense, sessile; bracts 6–12 mm, deciduous, lowest obovate-acuminate and densely pubescent, the rest oblong-elliptic, acuminate, nearly glabrous, basally enclosing the calyx. **Flowers** papilionoid; calyx (8.5–)9–12.3 mm, densely pilose, tube (2.5–)2.7–3.5 mm, ribs becoming prominent, intervals with 3–4 small

***Dalea versicolor* var. *sessilis*: a. habit; b. fruits; c. leaves; d. inflorescence. Illustrator: Randall C. Raak.**

clear glands, lobes (6–)6.5–9.3 mm, triangular-aristate, spreading in age, plumose; corolla yellow fading orange-brown, drying dull pink or purple, banner 7.8–9.5 mm, other petals attached at or above middle of androecium, wings 3.3–4 mm, keel petals (3.3–)3.5–5.2 mm, marginally adherent and enclosing androecium; stamens 10, 10–13 mm, filaments free for about 1.5 mm. **Fruits** 2.8–3.5 mm. **Seeds** 1, 2–2.8 mm.

Flowering Apr–Jun, Aug–Nov. Rocky outcrops, ridges, slopes, canyons, roadsides, [and plains], often on limestone substrates; [530–](670–)1100–1590(–2040) m; (Sonoran Desertscrub), Chihuahuan Desertscrub, Semidesert Grassland, lower elevations in Madrean Evergreen Woodland; Cochise, Greenlee, Pima, Pinal, Santa Cruz cos.; New Mexico, Texas; Mexico (Chihuahua, Coahuila, Sonora).

Dalea wrightii is widespread but uncommon across much of its range. It is typically found in open, rocky habitats.

DERMATOPHYLLUM Scheele. Linnaea 21(4):458. 1848. * [derived from the Greek words *dermato-* (meaning skin) and *phyllum* (meaning leaf) in reference to the leathery texture of the leaflets]
Benjamin Brandt

Woody shrubs or small trees, moderately to heavily branched, 1–5 m. **Stems** angular to rounded, unarmed, densely velutinous in new growth becoming glabrescent to glabrous with age. **Leaves** evergreen, alternate; odd-pinnate; with or without stipules; petiole and rachis with upper channel, petiole 0.5–3.5 cm, rachis 1–9.5 cm; pulvinus present at the base of the petiole; leaflets sessile with pulvinus at the base. **Leaflets** 7–15 per pinna, opposite or irregularly arranged, 1.5–4 × 0.5–3 cm, leathery, margins entire. **Inflorescences** terminal, erect or pendulous, racemes, 0.7–15 cm, wooly, flowers 2–many, densely arranged; bracts 2–15 × 1–2 mm, densely sericeous; bracteoles 1–1.5 mm, densely sericeous. **Pedicels** 5–15 mm, densely sericeous. **Flowers** papilionoid, 5-merous, bilaterally symmetric; calyx cone-shaped, 8–13 mm, with 5 triangular lobes 1–3 mm; corolla lilac to violet, violet blue, 1–3 cm, glabrous; stamens 9, filaments free; ovary densely sericeous, ovules 5–10, style glabrous. **Fruits** indehiscent, pendulous, often with sterile intervals between segments and/or at the apical or basal ends, 5–12 × 0.5–1.5 cm, grey–light brown, brown; stipe 5–12 mm. **Seeds** reniform to ovoid, red, brown, orange, yellow, or buff, 7–13 × 6–10 mm.

Species 4 (2 in the flora, 1 native, both cultivated): sw United States, Mexico.

The species now included in *Dermatophyllum* were all formerly treated within *Sophora* but were long believed to be better treated in a separate genus due to molecular and morphological evidence. Yakovlev was the first to formally separate these species into the genus *Calia* and did so based on morphological characteristics (Yakovlev 1967). Additional analysis by Sousa and Rudd (1993), Palomino et al. (1993), Peña et al.(2000), Heenan et al. (2004), and Wojciechowski et al. (2004) largely upheld Yakovlev's treatment while providing further evidence for its accuracy. Issues with the validity of the genus name *Calia* were raised, though, due to its similarity in spelling to the genus *Calea* in the Asteraceae, and therefore a potential parahomonym. The issue was brought to the Nomenclature Committee for Vascular Plants, and in 2011 the committee ruled that *Calia* was an invalid genus name. In efforts to maintain the taxonomic distinctions that had already been supported, Gandhi et al. (2011) combined these species into the genus *Dermatophyllum*, which had been recognized as a synonym of *Calia*.

Species of *Dermatophyllum*, especially *D. secundiflorum*, are cultivated as landscape plants. The seeds and foliage of *D. secundiflorum* are toxic if eaten and have resulted in livestock losses. This is likely true for the entire genus.

SELECTED REFERENCES: Gandhi et al. 2011, Wojciechowski et al. 2004.

1. Leaflets 1.5–3.5 × 0.5–1.2 cm, sericeous on lower (abaxial) surface; fruits laterally compressed ______ **1. Dermatophyllum arizonicum**
1′ Leaflets 2–4 × 1–3 cm, glabrescent to glabrous and shiny on lower (abaxial) surface; fruits turgid ______ **2. Dermatophyllum secundiflorum**

1. Dermatophyllum arizonicum (S. Watson) Vincent. Phytoneuron 2011-57:2. 2011. * [of Arizona] ARIZONA NECKLACEPOD, ARIZONA MESCALBEAN Native, Cultivated

Sophora arizonica S. Watson, Proc. Amer. Acad. Arts 11:135. 1876; *Calia arizonica* (S. Watson) Yakovlev; *Sophora formosa* Kearney & Peebles

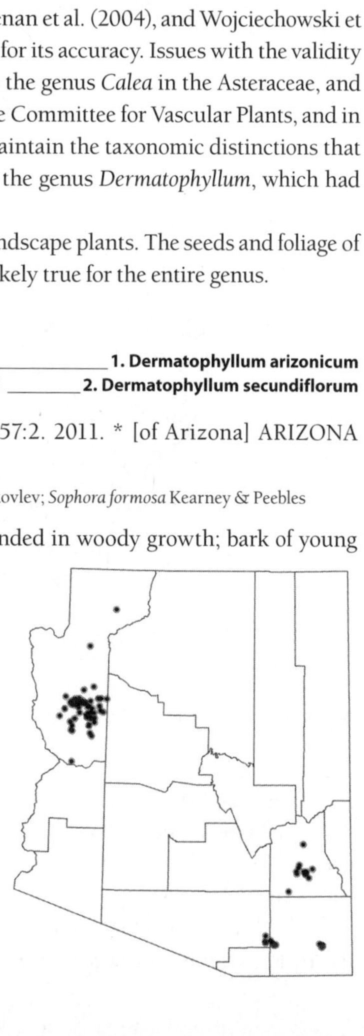

Woody shrubs, heavily branched, 1–4 m. **Stems** angular in new growth, rounded in woody growth; bark of young stems smooth, buff colored, becoming gray on older stems. **Leaves:** stipules 1–5 mm, acuminate; petiole and rachis round to somewhat angular, petiole 0.5–1.7 cm, rachis 1–6 cm, both velutinous. **Leaflets** 7–15 per pinna, elliptic to obovate to oblanceolate, rounded to obtuse to mucronate to mucronulate at apex, rounded to wedge-shaped at base, 1.5–3.5 × 0.5–1.2 cm, sericeous to glabrescent to glabrous and shiny on upper (adaxial) surface, sericeous on lower (abaxial) surface. **Inflorescences** erect, 0.7–5 cm, flowers 2–many; bracts lanceolate to elliptic to oblong, 3–15 × 1–2 mm. **Pedicels** 5–10 mm. **Flowers** calyx 10–13 mm, sericeous; corolla lilac to violet, 2.5–3 cm; ovary oblong, ovules 9–10, style 3–4 mm. **Fruits** light brown, laterally compressed, torulose, often twisted or recurved, 1–10 segments, 5–12 × 0.5–1.5 cm, papery when mature, sparsely sericeous to glabrescent; stipe 6–9 mm. **Seeds** 7–11 × 6–8 mm, dull red to reddish brown, orange, yellow, or buff.

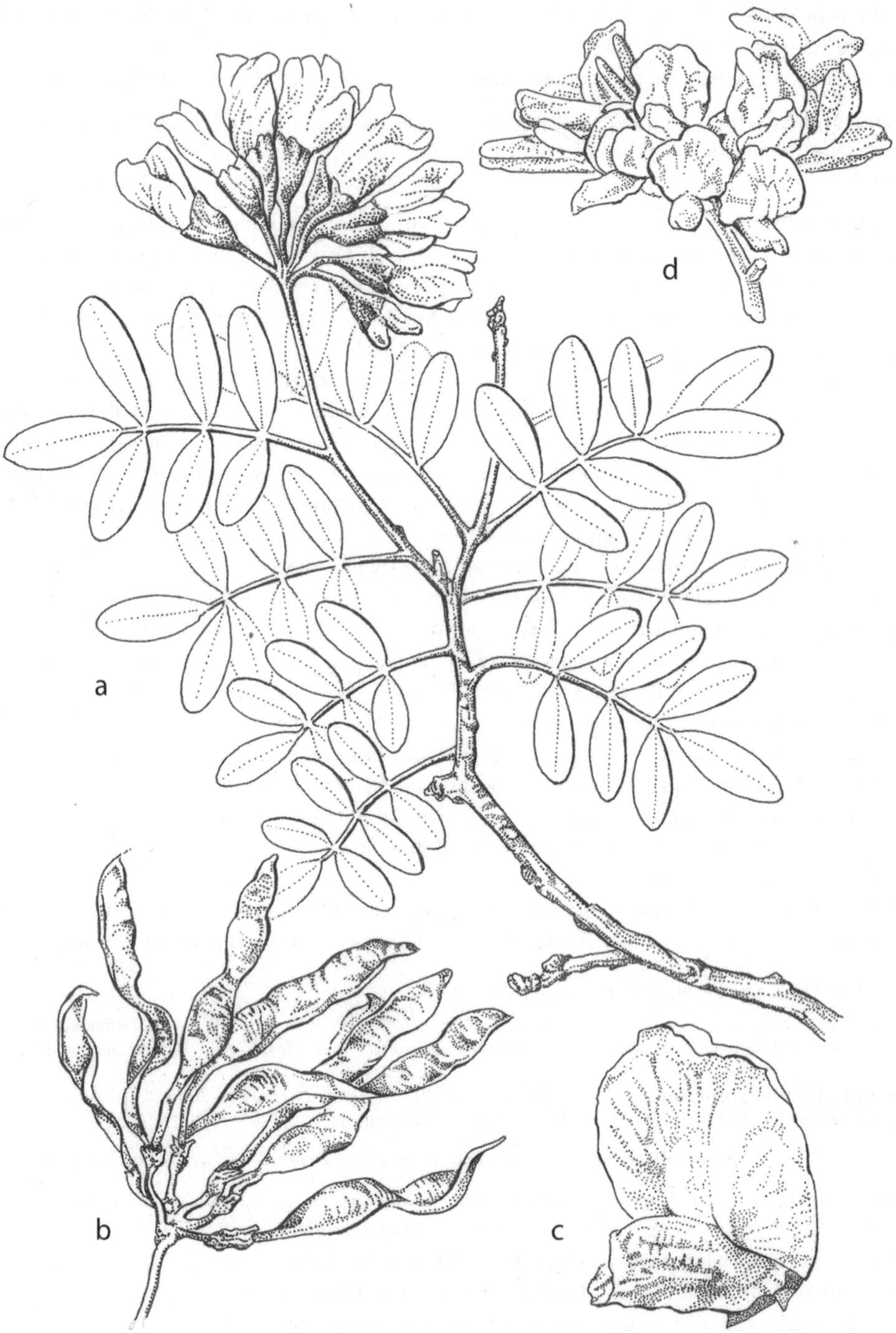

***Dermatophyllum arizonicum*: a. habit; b. fruits; c. flower; d. inflorescence. Illustrator: Randall C. Raak.**

Flowering Mar–Apr(–May). Rocky or gravelly terrain, on hillsides and bajadas, along canyon walls and arroyos, disturbed lands; 600–1585 m; Chihuahuan Desertscrub, Mohave Desertscrub, Semidesert Grassland, Interior Chaparral (often found in localized communities); endemic; Cochise, Graham, Mohave, Pima cos.; also documented in Navajo and Yuma cos.

The distribution of *Dermatophyllum arizonicum* appears to be relictual. In southeastern Arizona, populations are mostly found on limestone substrates. Seeds are often found among the leaf litter beneath the plants, suggesting that they are not utilized by wildlife. The seeds and foliage of closely related taxa contain toxic compounds, and it is likely that those of *D. arizonicum* are poisonous as well. Nodulation has been reported for this taxon (Burton et al. 1974).

Dermatophyllum arizonicum is rarely grown as a landscape plant in the state. These shrubs have attractive blue-green foliage, a compact form, and produce minimal litter. They may be planted as an informal hedge, screen, or windbreak, though growth is slow. Flowers are fragrant, though not as strongly scented as those of *D. secundiflorum*. The plants require well-drained soil and, for best appearance, a sunny location. Established plants are drought tolerant and survive overnight temperatures to at least -18°C. In cultivation, the leaves are sometimes attacked by pyralid moth caterpillars (Lepidoptera: Pyralidae) that can cause extensive damage, though the plants typically recover. Propagation is by scarified seeds with seedlings taking 1–3 weeks to emerge. Seedlings are susceptible to rotting. Careful watering and the addition of sand or other materials to the growing medium to improve drainage and aeration can minimize losses. Use of a fungicide may also help.

2. Dermatophyllum secundiflorum (Ortega) Gandhi & Reveal. Phytoneuron 2011-57:2. 2011. * [with flowers along one side of an inflorescence] TEXAS MOUNTAIN LAUREL, MESCALBEAN; *FRIJOLITO* Cultivated

Broussonetia secundiflora Ortega, Nov. Pl. Descr. Dec. 61, pl. 7. 1798; Agastianis secundiflora (Ortega) Rafinesque; Calia erythrosperma Terán & Berlandier; Calia secundiflora (Ortega) Yakovlev; Cladrastis secundiflora Rafinesque; Dermatophyllum speciosum Scheele; Sophora secundiflora (Ortega) Lagasca ex de Candolle; Sophora sempervirens Engelmann ex A. Gray; Sophora speciosa (Scheele) Bentham ex A. Gray; Virgilia secundiflora (Ortega) Cavanilles

Woody shrubs or small trees, moderately to heavily branched, 1–5 m. **Stems** rounded, velvety tomentose, becoming glabrous; bark smooth, green to brown, becoming dark gray or black, fissured and scaly on older stems. **Leaves** without stipules; petiole and rachis round, petiole 1.5–3.5 cm, rachis 4.5–9.5 cm, both sericeous. **Leaflets** 7–11 per pinna, elliptic to obelliptic to obovate, rounded to retuse to broadly retuse at apex, rounded to wedge-shaped at base, 2–4 × 1–3 cm, glabrescent to glabrous and shiny on both surfaces. **Inflorescences** erect or pendulous, 10–15 cm, flowers 5–many; bracts lanceolate to linear, 2–8 × 1–1.5 mm. **Pedicels** 7–15 mm. **Flowers:** calyx 8–12 mm, densely sericeous; corolla violet-blue, 1–2 cm; ovary cylindrical, ovules 5–6. **Fruits** grey–light brown, turgid, torose to torulose, 1–4 segments, 2–10 × 1–1.5 cm, woody, tomentose when young becoming glabrous with age; stipe 5–12 mm. **Seeds** 11–13 × 8–10 mm, bright red to yellow. $2n = 18$.

Flowering Mar–Apr. Texas and New Mexico; Mexico (Chihuahua, Coahuila, Hidalgo, Nuevo León, Puebla, Queretaro, San Luis Potosí, Tamaulipas).

Dermatophyllum secundiflorum is found in a wide variety of habitats from alluvial plains to slopes and canyons, and occurs in thornscrub, woodland, and upper elevations of the Chihuahuan Desert. It grows on sedimentary and igneous substrates as well as alluvial soil from near sea level to at least 2200 m elevation. The plants are common in some areas.

The seeds and foliage of *Dermatophyllum secundiflorum* are widely reported to be toxic. Schmutz and Hamilton (1979) report that one seed could be fatal to a child, though the hard seed coat would need to be broken for poisoning to occur. In Texas, livestock poisoning has occurred when animals have browsed plants of *D. secundiflorum* growing on rangeland. Sheep and goats usually recover rapidly, but cattle have died from eating as little as 0.25% of their body weight. The leaves were determined to be most toxic in the autumn and winter months (Boughton and Hardy 1935). Allen and Allen (1981) indicate that nodulation has been observed on *D. secundiflorum*.

In regard to the traditional uses of *Dermatophyllum secundiflorum*, its red seeds are most commonly noted as having been used as a ritual hallucinogen by several of the Great Plains and Southern Plains tribes including the Apache, Comanche, Delaware, Iowa, Kansa, Omaha, Oto, Osage, Pawnee, Ponca, Tonkawa, and Wichita. The ritual use has been referred to by a number of names such as Deer Dance, Red Bean Dance, Red Paint Dance, Red Medicine, Wichita Dance, and Whistle Dance (Howard 1957). These rituals, though no longer practiced, have largely been regarded as the possible precursor to rituals involving the peyote cactus (*Lophophora williamsii*) more recently practiced in parts of the region (Howard 1957, La Barre 1957, Campbell 1958). Schultes and Hofmann (1980) noted that, even recently, leaders of peyote ceremonies in the Kiowa and Comanche tribes still wore necklaces made of *D. secundiflorum* seeds as part of their traditional attire. The use of *D. secundiflorum* seeds seems to date back several millennia as evidenced in many archaeological sites throughout western Texas and northern Coahuila where the seeds have been recovered. In the oldest of these sites, Bonfire Shelter in western Texas, with occupied strata dating to as early as 8440 to 8120 B.C. and as late as A.D. 420 to 1040, *D. secundiflorum* seeds were recovered from every stratum. Throughout these sites there has been a strong correlation in the presence of *D. secundiflorum* seeds along with those of the Mexican Buckeye (*Ungnadia speciosa*), and to some degree the remains of peyote (Adovasio and Fry 1976). Despite this correlation there has been no documented use of *U. speciosa* as a hallucinogen. The effects of *D. secundiflorum* have largely been attributed to the presence of cytisine, which is highly toxic and the exposure to which can result in death (Schultes and Hofmann 1980).

In addition to its use as a hallucinogen, Moerman (1998) noted that a decoction prepared from the ground seeds of *Dermatophyllum secundiflorum* had been used topically to treat earaches by the Comanche.

Today *Dermatophyllum secundiflorum* is a popular landscape plant grown throughout the Southern Plains and Desert Southwest. It has proven to be exceptionally drought tolerant, needing little supplemental irrigation. Its lush, evergreen foliage provides shade and an attractive natural barrier all year long. Throughout March and April the plant is covered with large, sweet-smelling blooms.

The plants have a shrubby growth habit but can be pruned to develop a tree form. Growth rates are slow. Plants thrive in full sun or partial shade, though shaded plants are more open. The aroma of the flowers is often compared to that of grape bubblegum or grape soft drinks. The foliage is sometimes attacked by larvae of the pyralid moth as described under *Dermatophyllum arizonicum*. Propagation is by scarified seeds. The seedlings are much more resistant to rotting than those of *D. arizonicum*.

DESMANTHUS Willdenow. Sp. Pl. 4(2):888, 1044–1049. 1806, name conserved. * [from Greek, *desme*, bundle, and *anthos*, flower, referring to the dense flower heads] BUNDLEFLOWER

Matthew B. Johnson

Acuan Medikus, Theodora 62. 1786

Perennial herbs, subshrubs, or shrubs, often with a woody taproot. **Stems** several to numerous, basal, decumbent, ascending or erect, often dying back during dry seasons or with winter freezing. **Leaves** alternate, even bipinnate; stipules awl-shaped or bristle-like, persistent or tardily deciduous; petiolate, with a gland between or below the proximal pair of pinnae and occasionally between other pairs of pinnae; with stipels; nyctinastic movements common. **Leaflets** opposite, small, narrow, numerous, asymmetric, margins ciliate. **Inflorescences** heads axillary, solitary or paired, rarely forming pseudoracemes; bracts persistent or deciduous; bracteoles absent. **Pedicels** absent. **Flowers** mimosoid, sessile, 5-merous, white or pink, perfect or lowermost flowers sterile or functionally male, and some species produce all male flowers in some inflorescences; calyx obconic, teeth short; corolla petals fused below, becoming free; stamens 10 (or 5), filaments free, anthers without glands; ovary sessile, with numerous ovules; style thick; stigma spherical. **Fruits** dehiscent and often twisting (or indehiscent), sessile, linear or oblong, straight or falcate, flattened or somewhat rounded, valves stiffly papery or leathery (or woody), often beaked. **Seeds** flattened, ovate (or rhomboid). $x = 14$.

Species ca. 25 (3 native in the flora); warm temperate and tropical North America, the Caribbean, and South America.

Several species of *Desmanthus* are grown for livestock forage, erosion control, and as ornamentals (Lewis et al. 2005). *Desmanthus virgatus* is reported to be a high-quality forage plant that tolerates heavy grazing but can become weedy in agricultural fields (NRC 1979, Allen and Allen 1981). One species, *D. pernambucanus*, is a pantropical weed (Lewis et al. 2005). *Desmanthus illinoensis* is considered an excellent browse plant for wildlife and livestock (Ajilvsgi 1984). *Desmanthus illinoensis* is utilized for human food (leaves and cooked seeds), for medicine, and has potential as a pulse crop (Lewis et al. 2005). It is also used for revegetation and erosion control (Stubbendieck and Conard 1989). The Pawnee are reported to have made a tea from the leaves of *D. illinoensis* to treat itching (Kindscher 1992), and several reports indicate the Hopi treated conjunctivitis by placing seeds in the eyes. Allen and Allen (1981) report nodulation for species of *Desmanthus*.

Luckow (1993) discusses reports of *Desmanthus glandulosus* in Arizona, which are in part attributed to specimens from the Guadalupe Mountains in New Mexico. Arizona specimens labeled as *D. virgatus* (= *D. glandulosus* in part) were examined and found to be misidentified specimens of *D. cooleyi*. *Desmanthus glandulosus* does not appear to be present in the state. The three species of *Desmanthus* found in Arizona can have both solitary and paired inflorescences with 20 or more flowers per inflorescence, while *D. glandulosus* has only solitary inflorescences with 20 or fewer flowers. It is further distinguished by often having one or more additional glands present along the rachis in addition to the gland at the base of the proximal pair of pinnae.

SELECTED REFERENCE: Luckow 1993.

1. Herbs to 0.5 m; stems ascending or prostrate, herbaceous; widespread ________ **2. Desmanthus cooleyi**
1′ Shrubs or shrubby herbs 0.5–2 m; stems erect, woody at least basally; localized in Pima and Santa Cruz cos.
  2. Leaves 5–12(–18) cm with 6–14(–17) pairs of pinnae ________ **1. Desmanthus bicornutus**
  2′ Leaves 2–5 cm with (1) 2 or 3 (4) pairs of pinnae ________ **3. Desmanthus covillei**

1. Desmanthus bicornutus S. Watson. Proc. Amer. Acad. Arts 21:426. 1886. * [with two horns] RUBY BUNDLEFLOWER, TWO-HORN BUNDLEFLOWER Native

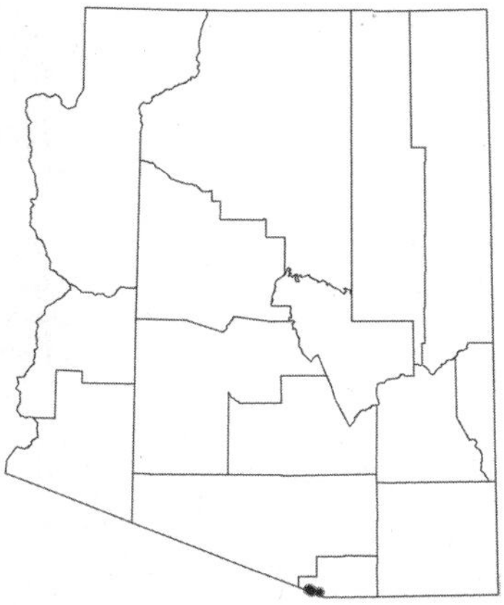

Shrubby herbs to 1.5 [–3] m. **Stems** erect, suffrutescent or becoming woody basally, sparingly branched at the base or unbranched, sparsely strigose to glabrate. **Leaves** 5–12(–18) cm, 6–14(–17) pairs of pinnae; stipules (2–)4–8 mm, narrowly awl-shaped, persistent or tardily deciduous; petiole 0.3–2 cm, sparsely puberulent, with a sessile, ovate to elliptic crater-like gland 0.6–1.5 mm, usually located between the proximal pair of pinnae; rachis 3.4–10 cm; pinnae 1.5–4.6 cm. **Leaflets** 10–35[–45] pairs per pinna, (2–)3–5[–6] × 0.7–1 (–1.3) mm, short petiolate, linear or slightly falcate, base rounded or squared, apex acute to obtuse. **Inflorescences:** heads white, 0.8–1.7 cm, 20–60 flowers; peduncles 1.5–3.7 cm; bracts 1–2 mm, ovate or lanceolate, deciduous. **Flowers:** sterile flowers with calyx 0.5–1.5 mm; corolla 1.5–2.5 mm; staminodes 10, 2.5–3.5 mm; male and perfect flowers with calyx 1.7–3.2 mm; corolla 2.5–4.5 mm; stamens 10, 4–8 mm. **Fruits** 2–12(–19) per peduncle, 4.5–10 × 0.2–0.4 cm, linear, straight or slightly curved, stiffly papery, beak 2–14 mm. **Seeds** (6–)10–22, 2.5–3.5 mm.

Flowering Aug–Sep[–Nov]. Rocky canyon bottoms; [10–]1130–1240[–1500] m; Madrean Evergreen Woodland (often in riparian vegetation); Santa Cruz Co.; Mexico (Baja California Sur, Chihuahua, Colima, Guerrero, Jalisco, México, Michoacán, Nayarit, Sinaloa, Sonora).

Luckow (1993) emphasized differences in stem branching between *Desmanthus bicornutus* and *D. covillei*. This character is not always obvious on herbarium specimens and young stems of either species may be unbranched. *Desmanthus bicornutus* is known from only a few locations in Arizona in the mountains of southern Santa Cruz County. The plants occur in canyons and are localized. In western Mexico, *D. bicornutus* is widespread and occurs in Sonoran Desertscrub, thornscrub, and tropical deciduous forest as well as in oak woodland. Across its range, it is found on plains, slopes, and along arroyos and intermittent and perennial watercourses. Plants of *D. bicornutus* produce some functionally male inflorescences. The percentage of male inflorescences on cultivated plants is highest late in the growing season (Luckow 1993). Martin et al. (1998) report that the stems are bundled together and used as brooms in Sonora.

2. Desmanthus cooleyi (Eaton) Trelease in Branner and Coville, Rep. (Annual) Arkansas Geol. Surv. 1888(4):178. 1891. * [apparently for Dennis Cooley, 1787–1860, American physician and botanist] COOLEY'S BUNDLEFLOWER Native

Acacia cooleyi Eaton, Man. Bot. (ed. 5):89. 1829

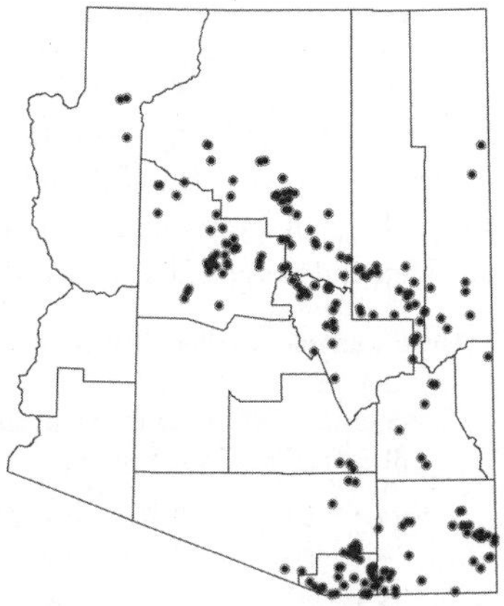

Herbs to 0.5 m. **Stems** ascending or prostrate, herbaceous, much-branched at the base, sparsely pubescent to glabrate. **Leaves** 2–5(–7) cm, (2 or)3–5(–9) pairs of pinnae; stipules 1–2.5 mm, bristle-like, deciduous; petiole 0.3–0.7 cm, puberulent, with a sessile, elliptic or rounded crater-like gland 0.5–1.6 mm, at base of proximal pair of pinnae; rachis 0.6–4 cm; pinnae 1–2.7 cm. **Leaflets** 8–16 pairs per pinna, 2–5 × 0.8–1.5 mm, subsessile, elliptic-oblong, base rounded, apex rounded to acute. **Inflorescences:** heads white, 1–1.7 cm, 25–45 flowers; peduncles 0.5–3(–4) cm; bracts 1–1.5 mm, triangular, persistent. **Flowers:** sterile flowers with calyx 1–1.5 mm; corolla 2–3 mm; staminodes 10, 5–10.5 mm; male and perfect flowers with calyx 2–3 mm; corolla 3.5–5 mm; stamens 10, 6–9 mm. **Fruits** 1–6 per peduncle, 4–9 × 0.3–0.4 cm, linear, straight or slightly curved, leathery, beak 1–6 mm. **Seeds** 6–22, 3.5–4.5 mm.

Flowering Jun–Aug(–Sep). Slopes, canyons, plains, arroyos, pastures, roadsides, and other disturbed sites in various soils; [1030–]1070–2230 [–2590] m; Semidesert Grassland, Plains Grassland, Great Basin Grassland, Interior Chaparral, Madrean Evergreen Woodland, Great Basin Conifer Woodland, Petran Montane Conifer Forest; Apache, Cochise, Coconino, Gila, Graham, Greenlee, Mohave, Navajo, Pima, Pinal, Santa Cruz, Yavapai cos.; Colorado, Kansas, New Mexico, Oklahoma, Texas; Mexico (Chihuahua, Durango, Sonora).

Desmanthus cooleyi is widespread across much of Arizona above the deserts. Populations are scattered though the

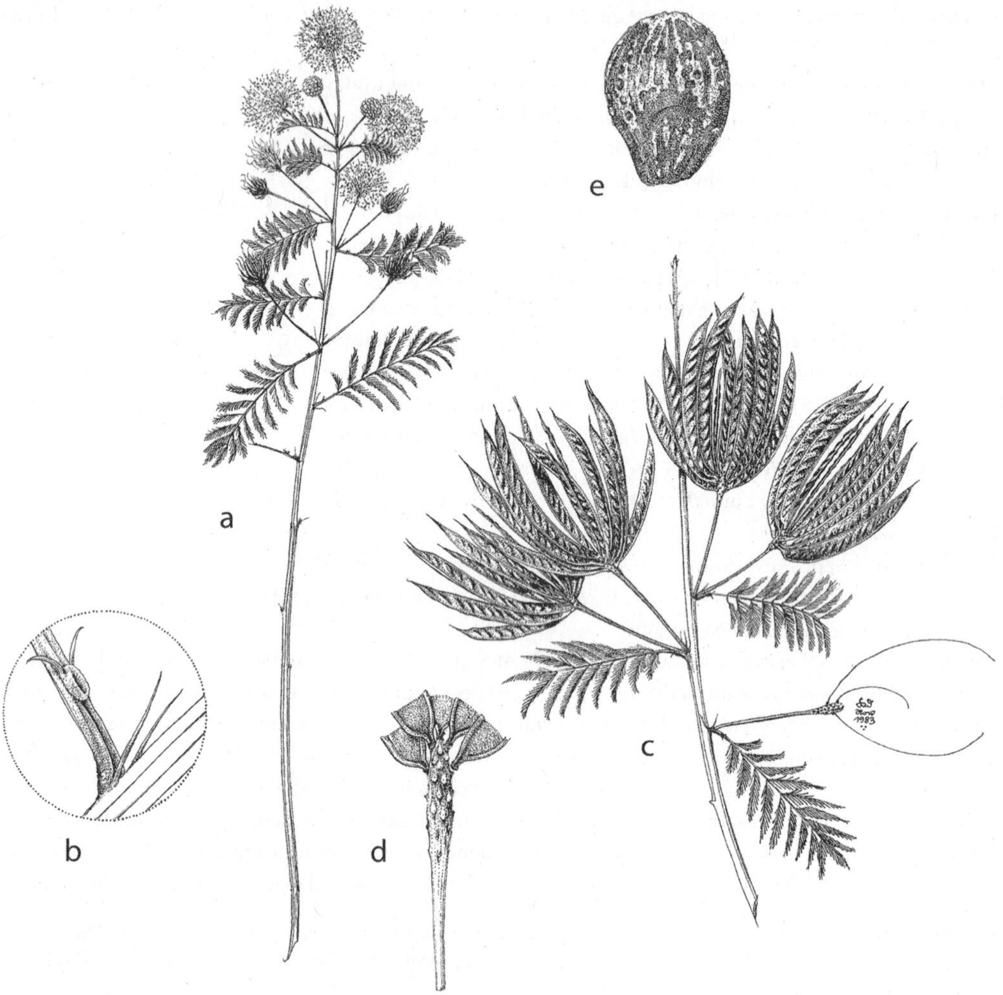

***Desmanthus bicornutus*: a. flowering stem; b. stem node with petiolar nectary and stipules; c. fruiting stem; d. inflorescence axis with a few fruits persisting; e. seed. Drawn by Karen Douthit and reproduced from *Flora Novo-Galiciana*, vol. 5, fig. 21, with permission of the University of Michigan Herbarium.**

plants are often locally abundant. In New Mexico it has been found in upper elevations in the Chihuahuan Desert. This species is browsed by livestock and Luckow (1993) reports that it is good browse for antelope. Seeds of *D. cooleyi* are utilized by quail species (Graham 1941). Graham (1941) indicates this species will colonize disturbed areas such as road shoulders and has potential for use in erosion control.

3. Desmanthus covillei (Britton & Rose) Wiggins ex B.L. Turner. Field and Lab. 18(3):128. 1950. * [for Frederick V. Coville, 1867–1937, American botanist and first director of the U.S. National Arboretum] COVILLE'S BUNDLEFLOWER Native

Acuan covillei Britton & Rose, N. Amer. Fl. 23(2):135. 1928; *Desmanthus covillei* (Britton & Rose) Wiggins ex B.L. Turner var. *arizonicus* B.L. Turner

Shrubs 0.5–2[–2.5] m. **Stems** erect, woody, often much-branched at the base, glabrous. **Leaves** 2–5 cm, (1) 2 or 3 (4) pairs of pinnae; stipules 1.5–4(–6) mm, bristle-like, persistent; petiole 0.5–1.5 cm, sparsely puberulent, with a sessile or short-stipitate, rounded or ovate crater-like gland 0.5–1.5 mm, located between the proximal pair of pinnae; rachis 0.6–2.5 cm; pinnae 1.4–4.1 cm. **Leaflets** 7–17 pairs per pinna, 4–7(–9) × 1–2.5 mm, short petiolate, oblong to obovate, base oblique or squared, apex obtuse or rounded. **Inflorescences:** heads white or cream colored, 1–2.3 cm, 23–43 flowers; peduncles 1–2.3 cm; bracts 1–1.5 mm, obovate or oblanceolate, persistent. **Flowers:** sterile flowers with calyx 1–1.5 mm; corolla 2–3 mm; staminodes 10, 7.5–17 mm; male and perfect flowers with calyx 2.5–4 mm; corolla 4–5.5 mm, stamens 10, 6.5–11.5 mm. **Fruits** 1–4(–6) per peduncle, 5–10(–13) × 0.3–0.4 cm, linear, straight or slightly curved, leathery, apex acute or rarely with a beak 2–5 mm. **Seeds** (7–)10–20[–32], 4–5 mm.

Flowering [Jan, Mar]Jul–Sep[–Oct]. Rocky soil of slopes and canyons; [near sea level–]850–1190 m; Sonoran Desertscrub; Pima Co.; Mexico (Baja California Sur, Sinaloa, Sonora).

Desmanthus covillei is rare and highly localized in Arizona where it is known from a few localities in the Baboquivari, Tucson, and Rincon mountains. It is widespread in the Sonoran Desert and adjacent thornscrub habitats in Sonora and adjacent northern Sinaloa, with a few collections from the Baja California peninsula and on Isla Tiburón in the Gulf of California. Bundles of stems of *D. covillei* are used for brooms in Sonora (Martin et al. 1998).

DESMODIUM Desvaux. J. Bot. Agric. 1(3):122, pl. 5, fig. 15. 1813, name conserved. * [from Greek, *desmo*, chain, and *odion*, like or similar, referring to the chain-like fruits] TICK TREFOIL, TICK-CLOVER, BEGGAR'S TICKS, BEGGAR'S LICE, STICK-TIGHT

Matthew B. Johnson

Meibomia Heister ex Fabricius, Enum. 168. 1759

Herbs, [subshrubs, shrubs or small trees], annual or perennial, usually with a well-developed taproot, perennial species often with a semi-woody, often subterranean crown; herbage and fruits often with hooked hairs. **Stems** solitary or few to numerous, procumbent, decumbent, diffuse, ascending or erect. **Leaves** alternate; pinnately 3-foliolate, rarely 1-foliolate, occasionally 5-foliolate; stipules dry and membranous, free or united, persistent or deciduous; petiolate; with stipels. **Leaflets** elliptic, lanceolate, linear, obovate or ovate, often slightly to markedly paler on lower (abaxial) surface, margins entire. **Inflorescences** pseudoracemes and/or panicles, terminal and/or axillary; bracts striate, awl-shaped, membranous, persistent or deciduous; bracteoles persistent, conspicuous or minute, or absent. **Pedicellate. Flowers** papilionoid, 2–several at each axis node, small or conspicuous and showy; calyx top- or bell-shaped, tube short, two-lipped, the upper (adaxial) 2 lobes fused and 2-toothed, the lower (abaxial) 3 lobes fused and 3-toothed, the middle lobe longest; corolla pink, lavender, purple, white, or yellow, often drying blue or bluish green; banner oblong, ovate or rounded, with a narrow base, typically recurved through 90°; wings obliquely oblong, narrowed at the base; keel straight or incurved, apex obtuse, partially fused; stamens 10, 9 filaments united and 1 free or all united, anthers all alike; ovary with 2–many ovules; style curved or bent inward, deciduous, not elongated in fruit; stigma simple. **Fruits:** loments, compressed or flattened, leathery, with (1 or) 2–many 1-seeded segments, terminal segment usually similar to the others or rarely distinctly different in size and/or shape, separating into individual segments; (sessile), subsessile or stipitate. **Seeds** 1 per fruit segment, compressed, rounded or kidney-shaped. $x = 10$.

Species ca. 275 (14 native in the flora): warm temperate, subtropical, and tropical regions of North and South America, Africa, Asia, and Australia. Diversity is highest in Mexico, Brazil, and eastern Asia.

Desmodium species are valued as cover crops, for erosion control, and as livestock forage in tropical regions. *Desmodium aparines*, *D. intortum*, and *D. uncinatum* are the most important commercially grown species. Allen and Allen (1981) caution that some species may be potentially toxic. The seeds are eaten by birds and other wildlife. *Desmodium* is sometimes intercropped as an insect repellant (Lewis et al. 2005). Numerous uses in folk medicine are reported, including treatments for fever, dysentery, and liver problems. Other medicinal uses include treatments for acne, abscesses, eye problems, ulcers, and catarrh (Uphof 1968). The leaves of *D. oldhami* are made into a tea in Japan (Allen and Allen 1981). Dyes are made from some species of *Desmodium,* and the stems have been a source of fiber. A few species are grown as ornamentals for their showy flowers. The arborescent species are used locally for fuel. Many of the perennial species have tough, semi-woody taproots and crowns, and several can become weedy. The common names refer to the fruit segments which readily cling to clothing and to animal fur, as do the stems of some species, due to the hooked hairs. Nodulation is widely reported for *Desmodium*.

Identification is based primarily on fruit characters; however, with field experience it is possible to identify many of the Arizona species based on vegetative characters alone. Seedlings of most 3-foliolate Arizona species appear to be 1-foliolate initially. In the key, leaflet number is indicated for plants that are reproductive. However, first year flowering stems of 3-foliolate perennial species may have 1-foliolate leaves at proximal nodes.

The widespread tropical species, *Desmodium tortuosum* (Swartz) de Candolle, is reported as occurring in Arizona (McVaugh 1987). No herbarium specimens or other documentation confirming its presence in the state have been located, and this species is not included in the present treatment. It is well-documented for Chihuahua and Sonora, including a collection from south of Nogales, and should be looked for in southeastern Arizona.

1. Leaves all 1-foliolate.
   2. Stems glabrous; leaflets linear or oblong-elliptic, 0.15–0.6(–1.7) cm wide; fruit margins nearly equally notched above and below ____ **1. Desmodium angustifolium**
   2′ Stems minutely hispidulous with hooked hairs; leaflets ovate, 1.1–3.5 cm wide; fruit margins less deeply notched above than below ____ **11. Desmodium psilophyllum**

1′ Leaves 1-foliolate proximally and 3-foliolate distally, or all 3-foliolate.
   3. Plants annual.
      4. Fruit segments 6.5–7.5 mm ____ **14. Desmodium scopulorum**
      4′ Fruit segments 2.5–4.2[–7] mm.
         5. Fruit segments not normally contorted, with straight hairs ____ **13. Desmodium rosei**
         5′ Fruit segments moderately to strongly contorted, with hooked hairs.
            6. Proximal leaflets longer than broad; pedicels 6–8 mm; fruits sessile or subsessile, segments oval to rhomboid, moderately contorted ____ **8. Desmodium neomexicanum**
            6′ Proximal leaflets broader than long; pedicels 7–13 mm; fruits (subsessile or) stipitate, the stipe to 3.5 mm, segments rhomboid, triangular (or oval), usually strongly contorted ____ **9. Desmodium procumbens**
   3′ Plants perennial.
      7. Fruits with margins nearly equally notched above and below, hairs straight.
         8. Stems hispid with hooked hairs; fruit segments 7–13 × 4.5–8 mm, glabrous or sparsely hispidulous with a few straight hairs along the margins ____ **10. Desmodium psilocarpum**
         8′ Stems minutely puberulent with sparse, straight hairs; fruit segments 2.5–5 × 1.8–2.5 mm, finely puberulent with straight hairs ____ **12. Desmodium retinens**
      7′ Fruits with margins less deeply notched above than below, hairs straight or hooked.
         9. Petioles 1.5–4(–5) mm, fruits subsessile or with a stipe to 1 mm ____ **2. Desmodium arizonicum**
         9′ Petioles (4–)7–50 mm; fruits with a stipe 1–5 mm.
            10. Stems and fruit segments with straight hairs ____ **4. Desmodium cinerascens**
            10′ Stems and fruit segments with hooked hairs.
               11. Leaflets lanceolate, ovate-lanceolate, oblong-lanceolate or elliptic.
                  12. Leaflets with upper (adaxial) surface medium green and often with a pale green or whitish green area along midrib, lower (abaxial) surface uniformly pale glaucous green; banner 9–12 mm; fruit segments 2–3 mm wide; stipe 2–3(–5) mm ____ **3. Desmodium batocaulon**
                  12′ Leaflets with upper (adaxial) surface medium green, lower (abaxial) surface slightly paler; banner 5.5–6.5 mm; fruit segments 3.2–4 mm wide; stipe 1–2 mm ____ **7. Desmodium metcalfei**
               11′ Leaflets ovate.
                  13. Leaflets with straight hairs; bracts not overlapping; fruit segments 6–8 × 3.5–5 mm, oval to elliptic-triangular ____ **5. Desmodium grahamii**
                  13′ Leaflets with hooked hairs; bracts overlapping; fruit segments 3–4 × 2–2.5 mm, rounded to somewhat rhomboid ____ **6. Desmodium intortum**

1. Desmodium angustifolium (Kunth) de Candolle. Prodr. 2:328. 1825. * [with narrow leaves] NARROWLEAF TICK TREFOIL

Hedysarum angustifolium Kunth, Nov. Gen. Sp. (quarto ed.) 6:517–518. 1823[1824]

Herbs, perennial, with a semi-woody crown. **Stems** solitary to several, erect, to 1 (1.5) m, glabrous. **Leaves** all 1-foliolate; stipules 0.7–1.2 mm, ovate-lanceolate, free, persistent; petioles 1–30 mm; stipels 0.5–2.5 mm, awl-shaped. **Leaflets** linear or oblong-elliptic, 3–10.5(–15) × 0.15–0.6(–1.7) cm, upper (adaxial) surface medium green, lower (abaxial) surface slightly paler, minutely strigillose with appressed straight hairs on both surfaces or glabrate. **Inflorescences** 15–30 cm in fruit; bracts awl-shaped, overlapping, deciduous; bracteoles absent. **Pedicels** 3.5–16 mm. **Flowers** 2 or 3 at each axis node; calyx: tube 0.8–1.2 mm, lobes 1–1.6 mm; corolla pink or lavender with a 2-lobed green and white spot at base of the banner, drying blue; banner 4.5–6 mm, ovate; wings 4–5.5 mm; keel 3.5–5 mm. **Fruits:** segments (2 or) 3–7, 2.5–4 × 1.7–2 mm, elliptical to rounded or somewhat rhomboid, not normally contorted, terminal segment similar to others, margin nearly equally notched above and below, minutely hispidulous with hooked hairs; subsessile or with a stipe to 2 mm.

Varieties 2 (1 native in the flora). The second variety, var. *angustifolium,* is widespread from Chihuahua, Sonora, and Tamaulipas to northern South America.

1a. Desmodium angustifolium (Kunth) de Candolle var. **gramineum** (A. Gray) B. G. Schubert. Contr. Gray Herb. 129:28. 1940. * [resembling grass, referring to the narrow leaflets] GRASSLEAF TICK TREFOIL Native

Desmodium gramineum A. Gray, Smithsonian Contr. Knowl. 5(6):46. 1853

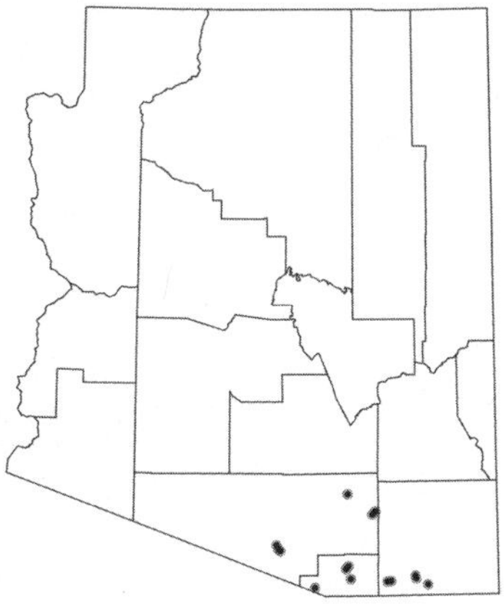

Petioles (10–)17–30 mm. **Leaflets** 0.15–0.25(–0.42) cm wide. **Pedicels** 11–16 mm.

Flowering Jul–Sep. Slopes, canyons, and roadsides; [1050–]1100–1710[–1800] m; Semidesert Grassland, Madrean Evergreen Woodland; Cochise, Pima, Pinal, Santa Cruz cos.; Mexico (Chihuahua, Sonora).

Variety *gramineum* is found in Arizona in an area extending from the Mule and Huachuca Mountains to the Rincon, Santa Catalina, and Baboquivari Mountains, southward into Mexico. The plants are infrequent and characteristically grow on open, often rocky slopes. It is readily distinguished from other *Desmodium* species in the state by the very narrow leaflets. Variety *gramineum* is weakly differentiated from var. *angustifolium* which has shorter petioles to 14 mm long, broader leaflets to 1.7 cm wide and shorter pedicels 3.5–6.5(–10) mm long.

2. Desmodium arizonicum S. Watson. Proc. Amer. Acad. Arts 20:363. 1885. * [of Arizona] ARIZONA TICK TREFOIL Native

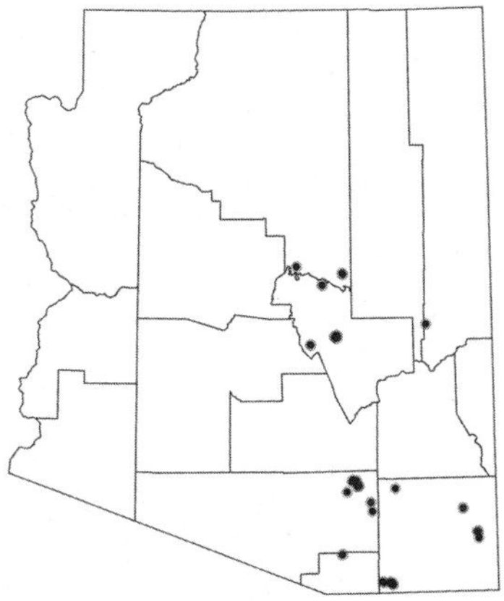

Herbs, perennial, crown not semi-woody. **Stems** solitary or few (or several), erect or ascending, to 0.6 m, sparsely pubescent with short, appressed hairs. **Leaves** all 3-foliolate; stipules 4.6–9 mm, ovate-lanceolate, free, persistent; petioles 1.5–4(–5) mm; stipels 1.5–3 mm, awl-shaped. **Leaflets** lanceolate, narrowly elliptic (or obovate), 2–8 × 0.2–0.9(–1.2) cm, upper (adaxial) surface medium green, lower (abaxial) surface slightly paler, pubescent with appressed straight hairs on both surfaces. **Inflorescences** 26–36 cm in fruit; bracts lanceolate or ovate-lanceolate, overlapping, quickly deciduous; bracteoles absent. **Pedicels** 10–13 mm. **Flowers** 2 at each axis node; calyx: tube 1–1.5 mm, lobes 1.2–3.5 mm; corolla pink or lavender with a non-lobed pale green spot a base of the banner, drying blue; banner 5.5–10 mm, ovate; wings 5–8 mm; keel 4–7.5 mm. **Fruits:** segments 2–4 (or 5), 4.5–6 × 3–4.5 mm, oval to somewhat rhomboid, not normally contorted, terminal segment similar to others, margin less deeply notched above than below, hispidulous with hooked hairs; subsessile or with a stipe to 1 mm.

Flowering Jul–Sep. Slopes, canyons, and roadsides, often in riparian habitats; (1220–)1500–2560[–2620] m; Semidesert Grassland, Madrean Evergreen Woodland, Petran Montane Conifer Forest; Apache, Cochise, Coconino, Gila, Pima, Santa Cruz cos.; New Mexico; Mexico (Chihuahua, Sinaloa, Sonora).

Desmodium arizonicum is a locally common understory plant in upland and riparian forests and woodlands.

3. Desmodium batocaulon A. Gray. Smithsonian Contr. Knowl. 5(6):47. 1853. * [with bramble-like stems, referring to the clinging hairs] SAN PEDRO TICK TREFOIL Native

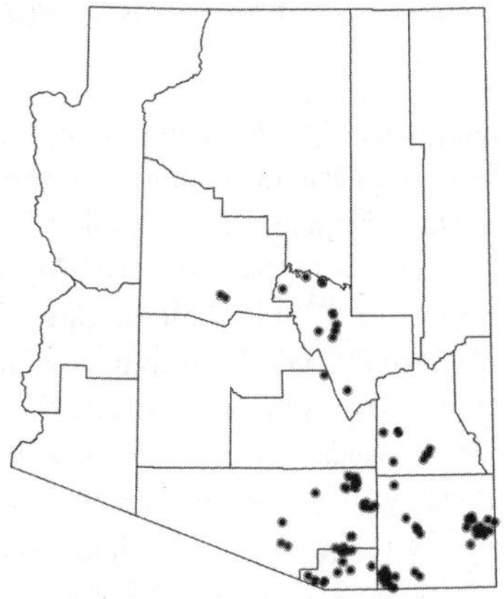

Herbs, perennial, crown not semi-woody. **Stems** several to numerous, ascending, decumbent or procumbent, to 1 m, hispid with hooked hairs. **Leaves** all 3-foliolate; stipules 2.5–4 mm, ovate-lanceolate, free, persistent; petioles (8–)10–30(–50) mm; stipels 0.8–2.2 mm, awl-shaped. **Leaflets** ovate-lanceolate, oblong-lanceolate or elliptic, 1.3–7 × 0.5–2.7 cm, upper (adaxial) surface medium green and often with a pale green or whitish green area along midrib, lower (abaxial) surface uniformly pale glaucous green, sparsely strigose with straight hairs on both surfaces. **Inflorescences** 15–35 cm in fruit; bracts ovate, overlapping, quickly deciduous; bracteoles absent. **Pedicels** (4–)6–8(–11) mm. **Flowers** 2 at each axis node; calyx: tube 1.2–2 mm, lobes 1–4 mm; corolla bright pink or lavender with a 2-lobed pale green and white spot at base of the banner, drying blue or blue-green; banner 9–12

***Desmodium batocaulon*: a. habit; b. fruit; c. fruit segment; d. seed; e. inflorescence; f. stem tip detail; g. leaf. Illustrator: Kathleen Bailey Cray.**

mm, broadly ovate; wings 9–11 mm; keel 7–10 mm. **Fruits:** segments 4–8, 3–5.5 × 2–3 mm, oval or rounded to rhomboid, not normally contorted, terminal segment similar to others, margin less deeply notched above than below, minutely hispidulous with hooked hairs; stipe 2–3(–5) mm.

Flowering (May) Jul–Sep. Slopes, canyons, stream banks, and roadsides, often in rocky or gravelly soil; [230–] (975–)1200–2130[–2300] m; (rare in riparian habitats in Sonoran Desertscrub and Semidesert Grassland), Madrean Evergreen Woodland, Petran Montane Conifer Forest; Cochise, Gila, Graham, Pima, Pinal, Santa Cruz, Yavapai cos.; New Mexico; Mexico (Baja California Sur, Chihuahua, Sinaloa, Sonora; one specimen reported from Colima).

Desmodium batocaulon is widespread and often abundant in dry or moist soils, growing in shaded locations as well as sunny openings in mid-elevation forests and woodlands. It often grows twining among other vegetation. The stems and fruit segments readily cling to shoes and clothing. This species is easily identified in the field by the pale whitish green area along the midrib on each leaflet, when this is present. Southward in Mexico, *D. batocaulon* sometimes occurs in tropical deciduous forest. The species is attractive in flower and suitable for mini-oases and other garden sites where sufficient irrigation is available.

4. Desmodium cinerascens A. Gray. Smithsonian Contr. Knowl. 5(6):48–49. 1853. * [becoming ashy-gray] SPIKED TICK TREFOIL, SHRUBBY TICK-CLOVER Native

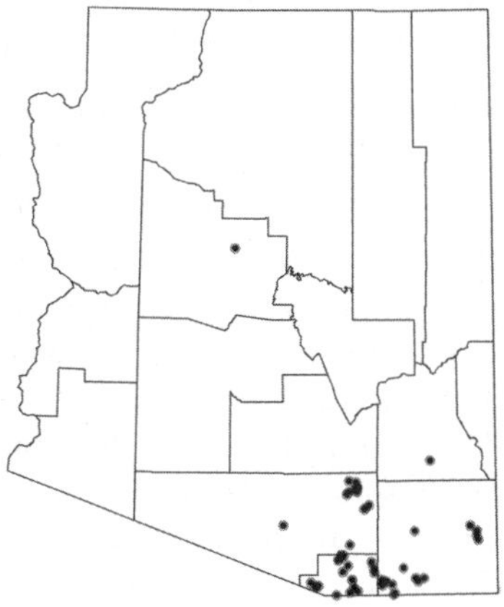

Herbs, perennial, with a robust, semi-woody crown. **Stems** several to numerous, erect or ascending, to 1.5 m, sparsely hispidulous with straight hairs. **Leaves** all 3-foliolate; stipules 2.5–4 mm, lanceolate or ovate lanceolate, free, deciduous; petioles (4–)10–35 mm; stipels 3–4 mm, awl-shaped. **Leaflets** oblong, ovate or obovate, 2.5–7 × 0.8–3.5 cm, upper (adaxial) surface dark green, lower (abaxial) surface slightly paler, sparsely sericeous with appressed, straight hairs on both surfaces. **Inflorescences** 20–35 cm in fruit; bracts lanceolate, not overlapping, quickly deciduous; bracteoles absent. **Pedicels** 4–8 mm. **Flowers** 2(–4) at each axis node; calyx: tube 1.3–1.7 mm, lobes 1.5–2.3 mm; corolla pink or lavender with a 2-lobed pale green spot at the base of the banner and a pale keel, drying dark blue; banner 9–10 mm, broadly ovate; wings 9.5–10 mm; keel 8–8.5 mm. **Fruits:** segments (2 or) 3–7, 4–6.5 × 3.5–4.5 mm, rounded or oval to elliptic-triangular, not normally contorted, terminal segment similar to others, margin less deeply notched above than below, sericeous with short, straight hairs; stipe (2–)2.5–4(–5) mm.

Flowering (May–Jun) Aug–Oct. Rocky slopes, canyons, arroyos, and roadsides; [480–](880–)1200–2300(–2750) m; (rare in riparian canyons in Sonoran Desertscrub), Madrean Evergreen Woodland, (Petran Montane Conifer Forest); Cochise, Graham, Pima, Santa Cruz, Yavapai cos.; Mexico (Chihuahua, Durango, Sinaloa, Sonora).

Desmodium cinerascens is the most robust species of the genus in Arizona. The suffrutescent stems grow to 1.5 m tall and with inflorescences, the plants can exceed 2 m in height. Across much of its range in Arizona, *D. cinerascens* is infrequent and occurs in small, scattered populations. The species is most often found on sunny slopes. The single collection of *D. cinerascens* from Yavapai County (*N.D. Atwood 16991*) is considerably disjunct from the other Arizona populations. In Mexico, the species occurs in tropical deciduous forest. In warm-winter areas of Mexico, flowering has been documented in February and March. With its robust size, showy flowers, and preference for more xeric sites than other native species of the genus, *D. cinerascens* is well-suited as a perennial flowering plant in Arizona gardens. The stems die back to the crown with freezing temperatures in winter. The dead stems may be cut back as new growth appears in spring. With ample irrigation, the plants grow rapidly and flower over several months. Including the inflorescences, they can grow to 2.5 m tall. Cultivated plants of *D. cinerascens* in Tucson have lived for up to 10 years. Plants are easily grown from seeds. These do not require scarification.

5. Desmodium grahamii A. Gray. Smithsonian Contr. Knowl. 5(6):48. 1853. * [for James D. Graham, 1799–1865, American soldier and military topographical engineer] GRAHAM'S TICK TREFOIL Native

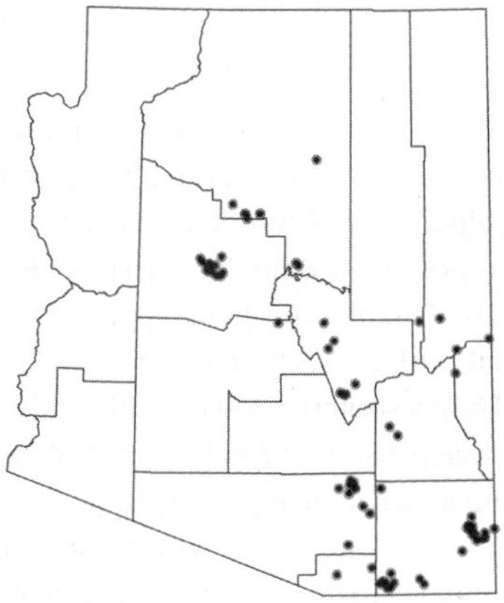

Herbs, perennial, rhizomatous, crown not semi-woody. **Stems** few to several, erect, ascending, diffuse or decumbent, to 0.35 [0.9] m, hispid with hooked hairs. **Leaves** all 3-foliolate; stipules (2–)3–5(–7) mm, lanceolate, free, tardily deciduous; petioles 7–27 mm; stipels 1.6–2.5 mm, awl-shaped. **Leaflets** broadly ovate, 1–5.5 × 0.5–4 cm, upper (adaxial) surface medium green, lower (abaxial) surface pale green, strigose with straight hairs on both surfaces. **Inflorescences** (5–)15–35(–45) cm in fruit; bracts ovate-lanceolate, not overlapping, deciduous; bracteoles absent. **Pedicels** 6–15 mm. **Flowers** 2 at each axis node; calyx: tube 1.2–1.5 mm, lobes 1–3 mm; corolla pink, pale pink, or greenish white with a 2-lobed greenish spot at base of the banner, drying bluish green; banner 9–11 mm, broadly ovate; wings 8–9 mm; keel 8–9 mm. **Fruits:** segments (2 or) 3–5 (or 6), 6–8 × 3.5–5 mm, oval to elliptic-triangular, not normally contorted, terminal segment similar to others, margin less deeply notched above than below, hispid with hooked hairs; stipe 1.5–3 mm.

Flowering (Jun–)Jul–Sep(–Oct). Ridges, slopes, canyons, stream banks, and roadsides; 1580–2300(–2520)[–2700] m; (Interior Chaparral), Madrean Evergreen Woodland, Petran Montane Conifer Forest; Apache, Coconino, Cochise, Gila, Graham, Greenlee, Pima, Santa Cruz, Yavapai cos.; New Mexico, Texas; Mexico (Chihuahua, Coahuila, Durango, Nuevo León, Sonora, Tamaulipas, southward on the Mexican Plateau to Puebla and Vera Cruz).

Desmodium grahamii is locally common as an understory plant in pine-oak woodland where it sometimes forms a low groundcover. It is suitable for cultivation as an informal groundcover but requires regular irrigation during the growing season.

6. Desmodium intortum (Miller) Urban. Symb. Antill. 8(1):292. 1920. * [not twisting, referring to the fruits] GREENLEAF TICK TREFOIL Native

Hedysarum intortum Miller, Gard. Dict. (ed. 8) *Hedysarum* no. 11. 1768; *Desmodium sonorae* A. Gray

Herbs, perennial, crown not semi-woody. **Stems** few to several, erect, ascending or procumbent, to 1 [3] m, distinctly 3-angled, densely to sparsely hispid with hooked hairs (or nearly glabrous). **Leaves** all 3-foliolate; stipules (0.8–) 3–7[–13] mm, narrowly triangular or awl-shaped, free, persistent or tardily deciduous; petioles 8–21[–80] mm; stipels 1.5–2[–3.5] mm, awl-shaped. **Leaflets** ovate, 1–3[–13] × 0.4–1.1[–6] cm, upper (adaxial) surface dark green, lower (abaxial) surface slightly paler, sparsely hispid with hooked hairs on both surfaces. **Inflorescences** 10–35 cm in fruit; bracts ovate, overlapping, deciduous; bracteoles absent. **Pedicels** 6.5–8 mm. **Flowers** 2 at each axis node; calyx: tube 1–1.2 mm, lobes 1.1–1.9 mm; corolla greenish white (pink or lavender) with a 2-lobed whitish spot at base of the banner, drying bluish green; banner 7–9 mm, rounded; wings 6.5–8.5 mm; keel 6–8 mm. **Fruits:** segments 5–10, 3–4 × 2–2.5 mm, rounded to somewhat rhomboid, not normally contorted, terminal segment similar to others, margin less deeply notched above than below, hispid with hooked hairs; stipe 2.5–3 mm.

Flowering Jul–Sep. Slopes and canyons; [400–]1700–1800 m; Madrean Evergreen Woodland; Cochise Co.; Mexico (Sonora, Jalisco, Michoacán, Guerrero, Morelos, Chiapas, Vera Cruz); Central America, Caribbean, South America. Introduced and sometimes escaping elsewhere in tropical regions.

Desmodium intortum is a widespread tropical species that is documented for Arizona only from the Huachuca Mountains where it is rare on open, grassy slopes in oak woodland. It has a more extensive distribution to the south in Sonora where it is found in oak woodland and tropical deciduous forest. It has been introduced into several African countries as a cover crop and for livestock fodder (Lock 1989) as well as Australia. Duke (1981) provides extensive information on cultivation of *D. intortum* as a crop. The plants are tolerant of sun as well as shaded locations. *Desmodium intortum* has the potential to escape from cultivation and become weedy. The stems will root at the nodes and plants can be propagated from stem cuttings (Duke 1981).

7. Desmodium metcalfei (Rose & Painter) Kearney & Peebles. J. Wash. Acad. Sci. 29(11):485. 1939. * [for Orrick B. Metcalfe, 1879–1936, American botanist] METCALFE'S TICK TREFOIL Native

Meibomia metcalfei Rose & Painter, Bot. Gaz. 40(2):144. 1905

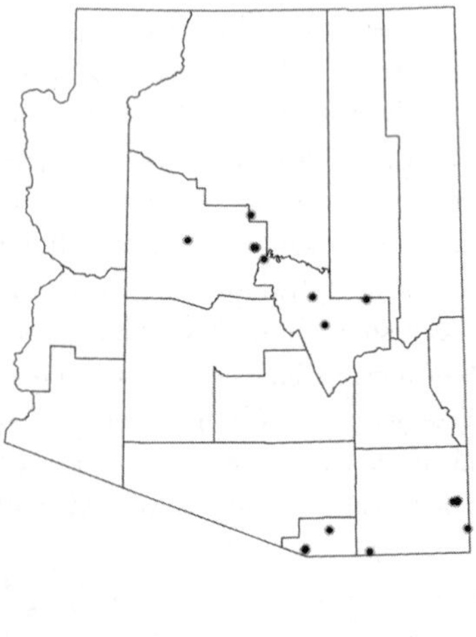

Herbs, perennial, crown not semi-woody. **Stems** few to several, erect or ascending, to 0.3 (0.9) m, with sparse, hooked hairs. **Leaves** all 3-foliolate; stipules 1.8–3 mm, awl-shaped, free, deciduous; petioles (6–)10–30 mm; stipels 0.6–1.8 mm, awl-shaped. **Leaflets** lanceolate or oblong-lanceolate, 1.4–4.3(–6) × 0.4–1 cm, upper (adaxial) surface medium green, lower (abaxial) surface slightly paler, sparsely strigose with straight hairs on both surfaces. **Inflorescences** 5–30 cm in fruit; bracts awl-shaped, not imbricate, quickly deciduous; bracteoles absent. **Pedicels** 5–6 mm. **Flowers** 2 or 3 at each axis node; calyx: tube 1–1.3 mm, lobes 1.6–2 mm; corolla pink or lavender, with a 2-lobed white spot at base of the banner, drying bluish; banner 5.5–6.5 mm, rounded to broadly ovate; wings 5–6 mm; keel 4.5–5.5 mm. **Fruits:** segments (1 or) 2–5, (4–)5–6 × 3.2–4 mm, elliptic or rhomboid, not normally contorted, terminal segment similar to others, margin less deeply notched above than below, hispidulous with hooked hairs; stipe 1–2 mm.

Flowering Jul–Sep. Slopes, canyons, and arroyos; 1220–2010 m; Semidesert Grassland, Madrean Evergreen Woodland, Petran Montane Conifer Forest; Cochise, Coconino, Gila, Santa Cruz, Yavapai cos.; New Mexico; Mexico (Sinaloa).

Desmodium metcalfei is widely distributed in Arizona but is of conservation concern due to the relatively few, localized populations. The specimen from southern Sinaloa (*O. F. Clarke 690901-44*) is widely disjunct from the U.S. populations. *Desmodium metcalfei* is to be expected in northeastern Sonora. Unambiguous herbarium material is scant and specimens should be checked carefully to be sure that they are correctly identified.

8. Desmodium neomexicanum A. Gray. Smithsonian Contr. Knowl. 3(5):53. 1852. * [of New Mexico] NEW MEXICO TICK TREFOIL Native

Desmodium procumbens (Miller) A. S. Hitchcock var. *neomexicanum* (A. Gray) H. Ohashi

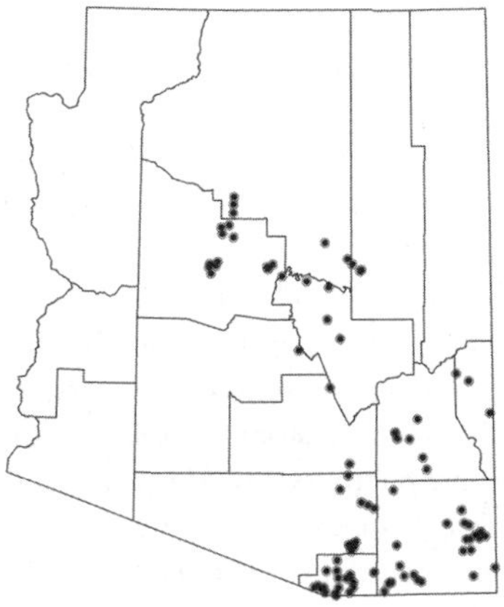

Herbs, annual. **Stems** solitary or with a few branches above, erect, ascending [or procumbent], to 0.3 (0.5) m, hispidulous with hooked hairs. **Leaves** often 1-foliolate proximally (not always apparent on late-season plants) and 3-foliolate distally; stipules 1–3 mm, awl-shaped, free, persistent; petioles 4–11 mm; stipels 0.6–1.7 mm, awl-shaped. **Leaflets** broadest at proximal nodes, lanceolate, ovate (or obovate), (0.5–)1–4(–5) × 0.2–1.8(–2.5) cm, upper (adaxial) surface medium green, lower (abaxial) surface slightly paler, sparsely strigose with straight hairs on both surfaces. **Inflorescences** 7–25 cm in fruit; bracts awl-shaped, overlapping, persistent; bracteoles absent. **Pedicels** 6–8 mm. **Flowers** 2 or 3 at each axis node; calyx: tube 0.4–0.5 mm, lobes 0.6–1 mm; corolla greenish white, white, white and pink, pink or lavender with a 2-lobed green and white spot at base of the banner, drying white or yellow; banner 3.4–4.2 mm, ovate; wings 2.5–3 mm; keel 2.1–2.7 mm. **Fruits:** segments 2–4 (or 5), 3.2–4.2 × 2.5–3 mm, oval to rhomboid, moderately contorted, terminal segment similar to others, margin nearly equally notched above and below, minutely hispidulous with hooked hairs; sessile or subsessile.

Flowering Aug–Oct. Slopes, canyons, arroyos, floodplains, stream banks, roadsides; 1000–2010[–2500] m; Semidesert Grassland, Interior Chaparral, Madrean Evergreen Woodland, (Great Basin Conifer Woodland); Cochise, Coconino, Gila, Graham, Greenlee, Maricopa, Navajo, Pima, Pinal, Santa Cruz, Yavapai cos.; New Mexico, Texas; Mexico (Aguascalientes, Chihuahua, Coahuila, Durango, Guanajuato, Jalisco, San Luis Potosí, Sonora, Zacatecas).

Desmodium neomexicanum is widespread and can be locally common across its range in Arizona. The plants grow in sunny locations and in the shade of other plants. It differs from the similar *D. rosei* by having much broader leaflets and contorted fruits with hooked hairs. The two species sometimes grow together. Herbarium specimens of *D. neomexicanum* often do not show the contorted fruits; however, this character is apparent on live material, at least in Arizona. Plants may or may not have 1-foliolate leaves present proximally. *Desmodium neomexicanum* appears to be closely related to *D. procumbens*, with which it shares many morphological similarities. Ohashi (2013) has combined *D. neomexicanum* as a variety of *D. procumbens*. In the present treatment, it is maintained as a distinct species. In Arizona, *D. neomexicanum* is distinguished from *D. procumbens* by proximal leaflets longer than broad, generally shorter pedicels, and sessile or subsessile fruits.

9. Desmodium procumbens (Miller) A. S. Hitchcock. Rep. (Annual) Missouri Bot. Gard. 4:76. 1893. * [prostrate, referring to the trailing stems] TRAILING TICK TREFOIL

Hedysarum procumbens Miller, Gard. Dict. (ed. 8) *Hedysarum* no. 10. 1768

Herbs, annual. **Stems** solitary or with a few branches above, erect, ascending, diffuse, or procumbent, to 1 [1.5] m, minutely hispidulous with both straight and hooked hairs. **Leaves** typically 1-foliolate proximally (not always apparent on late-season plants) and 3-foliolate distally; stipules 2–5 mm, awl-shaped, free, persistent; petioles 5–30 mm; stipels 0.5–1 mm, awl-shaped. **Leaflets:** proximally ovate, elliptic, rhomboid, fan-shaped, broadly triangular or transversely elliptic, 0.5–2.8 × 0.8–2.2 cm and distally lanceolate or linear-lanceolate, 0.6–5.5(–8) × 0.2–0.7(–1.1) cm, glabrous or sparsely strigose with straight hairs and sometimes also hooked and gland-tipped hairs on both surfaces. **Inflorescences**

5–40 cm in fruit; bracts awl-shaped, overlapping, persistent; bracteoles absent. **Pedicels** 7–13 mm. **Flowers** 2 or 3 at each axis node; calyx: tube 0.7–0.8 mm, lobes 0.7–0.9 mm; corolla white, green and white, pale pink or lavender with a two-lobed green and white spot at base of the banner, drying white or yellow; banner 2.5–3.5 mm, ovate; wings 2–2.6 mm; keel 1.8–2.5 mm. **Fruits:** segments 2–6, 3.5–7 × 2.5–4.5 mm, rhomboid, triangular (or oval), strongly contorted with fruits appearing spiral, terminal segment similar to others, margin nearly equally notched above and below, minutely hispidulous with hooked hairs; subsessile or with a stipe to 3.5 mm.

Varieties 4 (1 native in the flora). All four varieties, vars. *exiguum*, *longipes*, *procumbens*, and *transversum*, are documented for Sonora. These varieties are weakly and not always satisfactorily separable.

9a. Desmodium procumbens (Miller) A. S. Hitchcock var. **exiguum** (A. Gray) B. G. Schubert. Contr. Gray Herb. 129:12. 1940. * [short, small] WESTERN TRAILING TICK TREFOIL Native

Desmodium exiguum A. Gray, Smithsonian Contr. Knowl. 5(6):46. 1853

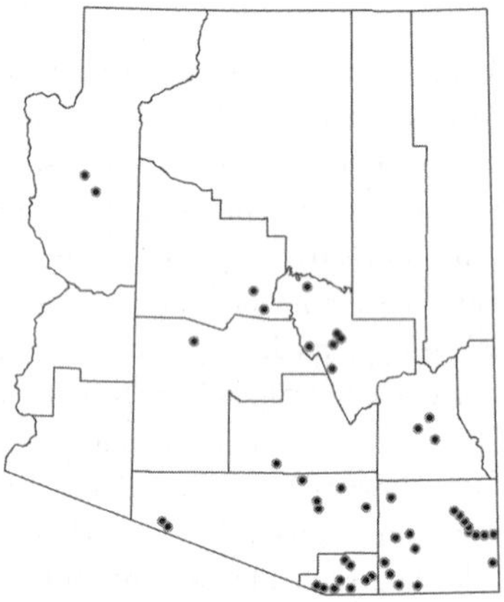

Stems erect or ascending, to 0.3 m. **Leaves** with petioles 5–23 mm. **Leaflets:** proximally ovate, fan-shaped or broadly triangular, 0.5–1.7 × 0.8–2 cm. **Fruits:** segments 2 or 3, 3.5–4 × 2.5–2.8 mm; (subsessile) or with a stipe (1–)2.5–3.5 mm.

Flowering Jul–Oct. Slopes, outcrops, canyons, arroyos, floodplains, stream banks, and roadsides in various soils; [250–](840–)1000–2010 m; (Sonoran Desertscrub), Semidesert Grassland, Interior Chaparral, Madrean Evergreen Woodland; Cochise, Gila, Graham, Maricopa, Mohave, Pima, Pinal, Santa Cruz, Yavapai cos.; New Mexico; Mexico (Baja California Sur, Chihuahua, Guerrero, Jalisco, Oaxaca, Sonora); Guatemala; South America.

Desmodium procumbens var. *exiguum* is widespread and sometimes locally common in Arizona. The plants grow in the open and in the shade of trees or shrubs. In Sonora it occurs in tropical deciduous forest. It is reported to be introduced into the Cape Verde Islands, Malawi, Nigeria, Sierra Leone, and Zimbabwe (Lock 1989). Variety *exiguum* has recently been combined as a synonym of *D. procumbens* var. *procumbens* (Ohashi 2013). This combination has merit. In the present treatment, var. *exiguum* is maintained as distinct. The proximal leaflets are often broader than long. Occasional plants have 1-foliolate leaves throughout, or may have a few 1-foliolate leaves at the distal-most nodes. *Desmodium procumbens* var. *exiguum* is distinguished from the related *D. neomexicanum* by proximal leaflets broader than long, generally longer pedicels, and usually stipitate fruits. The contorted fruits are not always apparent on herbarium specimens.

10. Desmodium psilocarpum A. Gray. Smithsonian Contr. Knowl. 5(6):48. 1853. * [with naked fruits] SANTA CRUZ ISLAND TICK TREFOIL Native

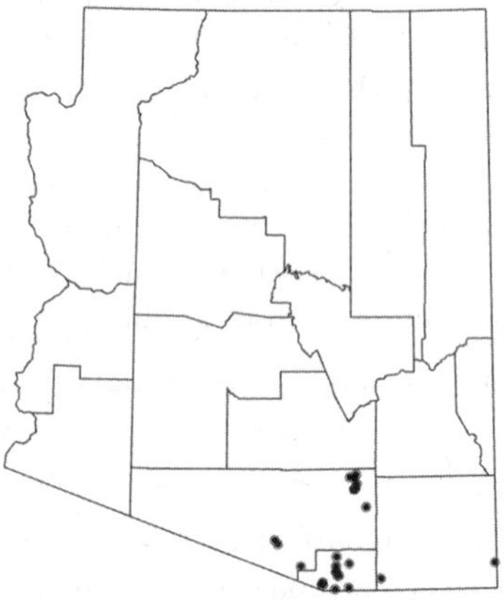

Herbs, perennial, crown semi-woody. **Stems** solitary or few to several, erect or ascending, to 0.6 m, hispid with hooked hairs. **Leaves** 1-foliolate proximally in first year, 3-foliolate distally and throughout in second and subsequent years; stipules 1.5–5 mm, ovate-lanceolate, free, persistent; petioles 8–43 mm; stipels 2–5 mm, awl-shaped. **Leaflets** ovate, 1–6.5[–9] × 0.5–3.7[–5.8] cm, upper (adaxial) surface medium green and at least sometimes with an irregular paler area on either side of the midrib near the widest point of the blade, lower (abaxial) surface slightly paler, sparsely hispid with straight hairs on both surfaces. **Inflorescences** 8–75 cm in fruit; bracts lanceolate, overlapping, deciduous; bracteoles absent. **Pedicels** 10–13 mm. **Flowers** 2 at each axis node; calyx: tube 0.7–0.8 mm, lobes 1.1–1.5 mm; corolla lavender, purple, or whitish, drying blue or pale yellow; banner 8–9 mm, ovate; wings 8–9 mm; keel 7–8 mm. **Fruits:** segments 1–5, 7–13 × 4.5–8 mm, oval to elliptic-rhomboid, not normally contorted, terminal segment similar to others, margin nearly equally notched above and below, glabrous or sparsely hispidulous with a few straight hairs along the margins; stipe 1–3 mm.

Flowering Jul–Sep. Ridges, slopes, canyons, stream banks, and roadsides; [200–]1030–1900[–1980] m; (Semidesert Grassland), Madrean Evergreen Woodland, (Petran Montane Conifer Forest); Cochise, Pima, Santa Cruz cos.; New Mexico; Mexico (Chihuahua, Sonora, Nuevo León, Tamaulipas).

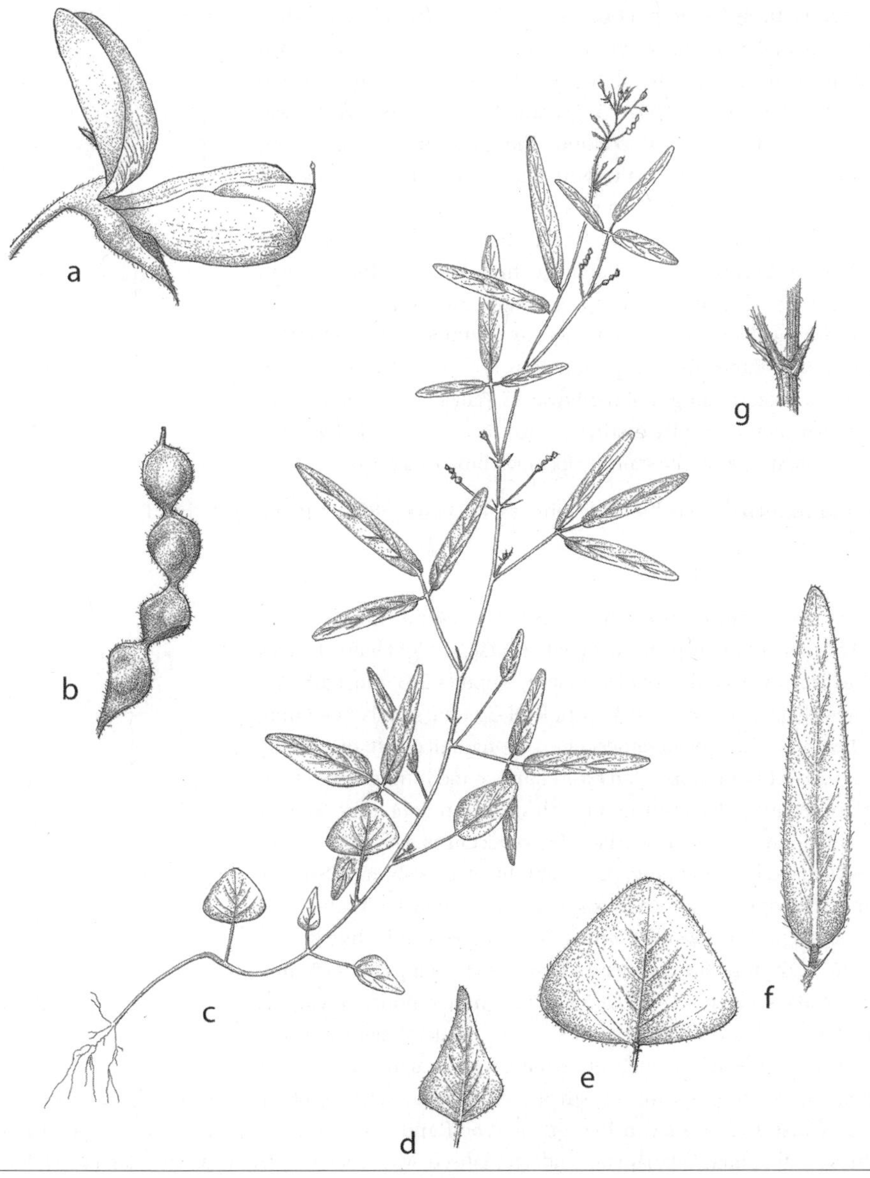

***Desmodium procumbens* var. *exiguum*: a. flower; b. fruit; c. habit; d., e., f., leaflets showing variability in shape; g. stem node with stipules. Illustrator: Sandy Turico.**

Desmodium psilocarpum occurs in scattered populations on rocky slopes and in riparian canyon habitats in Arizona. It has also been reported from Greenlee and Pinal counties. In Sonora, the plants extend into tropical deciduous forest. *Desmodium psilocarpum* and *D. scopulorum* are reported to intergrade in Sonora (Martin et al. 1998). The species is introduced in Zimbabwe (Lock 1989).

11. Desmodium psilophyllum Schlechtendal. Linnaea 12(3):310. 1838. * [with naked leaves] SIMPLELEAF TICK TREFOIL, WRIGHT'S TICK TREFOIL Native

Desmodium wrightii A. Gray

Herbs, perennial, crown semi-woody. **Stems** several to numerous, erect or ascending, to 0.5(–0.9) m, minutely hispidulous with hooked hairs. **Leaves** all 1-foliolate; stipules 2–4 mm, ovate-lanceolate, free, persistent; petioles 4–11 mm; stipels 1–2 mm, awl-shaped. **Leaflets** ovate, 2–7[–9] × 1.1–3.5 cm, upper (adaxial) surface medium green, lower (abaxial) surface slightly paler, glabrous. **Inflorescences** 6–20 cm in fruit; bracts awl-shaped, not overlapping, deciduous;

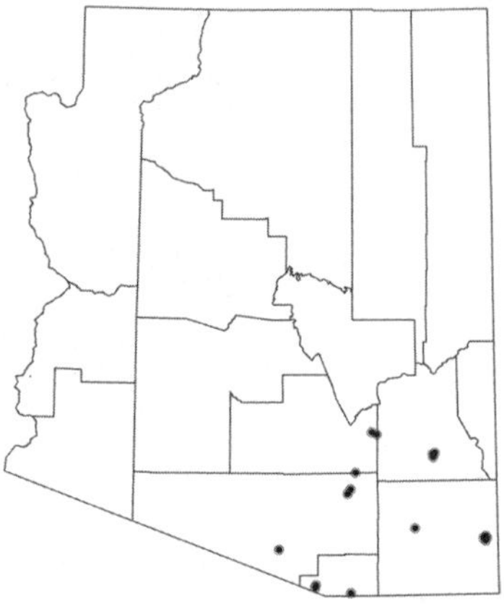

bracteoles absent. **Pedicels** 5–9 mm. **Flowers** 2 (or 3) at each axis node; calyx: tube 1–1.3 mm, lobes 0.8–1.4 mm; corolla pink or lavender with a 2-lobed white or green and white spot at base of the banner, drying blue; banner 4.5–6 mm, broadly ovate to rounded; wings 4.5–5.5 mm; keel 4–5 mm. **Fruits:** segments 2–5, 4.8–6 × 3–3.5 mm, oval to elliptic-rhomboid, not normally contorted, terminal segment similar to others, margin less deeply notched above than below, hispidulous with hooked hairs; stipe 1.5–3 mm.

Flowering Jul–Sep. Canyons and stream banks; [200–]1060–1800 [–2260] m; Madrean Evergreen Woodland; Cochise, Graham, Pima, Pinal, Santa Cruz cos.; Texas; Mexico (Chihuahua, Coahuila, Tamaulipas).

In Arizona, *Desmodium psilophyllum* is locally common as an understory plant in riparian woodlands but the populations are localized. In Texas it extends to near the eastern margin of the Edwards Plateau at low elevations. *Desmodium psilophyllum* is readily distinguished from *D. angustifolium*, the only other 1-foliolate species in the state, by its much broader leaflets.

12. Desmodium retinens Schlechtendal. Linnaea 12(3):311. 1838. * [tenacious] SANTA RITA MOUNTAIN TICK TREFOIL Native

Desmodium wislizeni Engelmann ex A. Gray

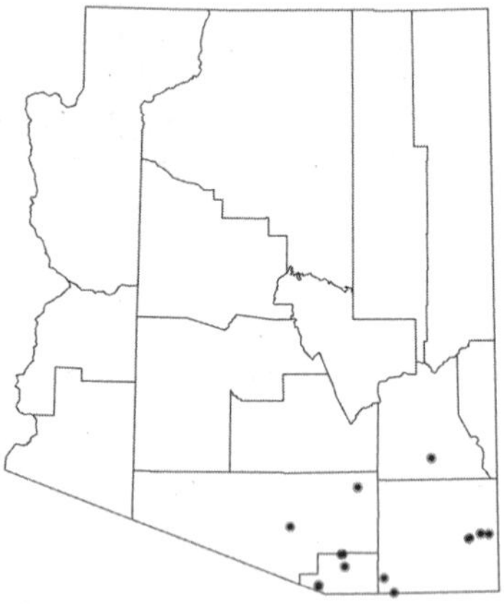

Herbs, perennial, crown not semi-woody. **Stems** few to several, diffuse or procumbent, to 0.6 [1] m, minutely puberulent with sparse, straight hairs. **Leaves** often 1-foliolate proximally and 3-foliolate distally; stipules 2–3 mm, narrowly triangular or awl-shaped, free, persistent; petioles 4–12 mm; stipels 1.6–2 mm, awl-shaped. **Leaflets** ovate, ovate-lanceolate, oblong-lanceolate or elliptic, 0.5–2.2 × 0.2–0.9(–1.6) cm, upper (adaxial) surface medium green, lower (abaxial) surface slightly paler, glabrous or with sparse, straight hairs on margins and midrib of lower (abaxial) surface. **Inflorescences** 5–15(–25) cm in fruit; bracts awl-shaped, overlapping, persistent; bracteoles absent. **Pedicels** 10–15(–23) mm. **Flowers** 2 or 3 at each axis node; calyx: tube 0.9–1 mm, lobes 0.5–1.1 mm; corolla pink or lavender with a 2-lobed green and white spot at base of the banner, drying blue; banner 4–5 mm, ovate; wings 3.5–4.5 mm; keel 3–4 mm. **Fruits:** segments 2–4, 2.5–5 × 1.8–2.5 mm, rounded or elliptic, not normally contorted, terminal segment similar to others, margin nearly equally notched above and below, finely puberulent with straight hairs; stipe 1–1.5 mm.

Flowering Jul–Oct. Slopes, canyons, stream banks, and roadsides, often in rocky soil; [1200–]1430–2140[–2500] m; Semidesert Grassland, Madrean Evergreen Woodland; Cochise, Graham, Pima, Santa Cruz cos.; Mexico (Aguascalientes, Chihuahua, Guanajuato, Hidalgo, Jalisco, Michoacán, Queretaro, San Luis Potosí, Sinaloa, Sonora, Tamaulipas, Zacatecas); Guatemala.

Desmodium retinens is widely distributed in Mexico. In Arizona, the populations are localized. Distal nodes of the inflorescences occasionally have 1-foliolate leaves.

13. Desmodium rosei B. G. Schubert. Contr. Gray Herb. 129:22–25, pl. 1, fig. A, map 6. 1940. * [for Joseph N. Rose, 1862–1928, American botanist] ROSE'S TICK TREFOIL Native

Herbs, annual. **Stems** solitary or with a few branches above, erect, to 0.3 (0.45) m, sparsely and minutely hispidulous with hooked hairs. **Leaves** 1-foliolate proximally (not always apparent on late-season plants), 3-foliolate distally; stipules 2–3 mm, awl-shaped, free, persistent; petioles 5–20 mm; stipels 0.7–1.2 mm, awl-shaped. **Leaflets** linear or linear-lanceolate distally, ovate or elliptic proximally, 1.5–6 × 0.1–0.5 cm, upper (adaxial) surface medium green, lower (abaxial) surface slightly paler, sparsely strigose with straight hairs on margins and along midrib of lower (abaxial) surface. **Inflorescences** 4–25 cm in fruit; bracts awl-shaped, not overlapping, persistent; bracteoles absent. **Pedicels** 10–13 mm. **Flowers** 2 at each axis node; calyx: tube 0.5–0.6 mm, lobes 0.7–1 mm; corolla white, greenish white, pink or lavender, drying bluish green; banner 4–4.5 mm, broadly ovate; wings 3–3.5 mm; keel 2.6–3.2 mm. **Fruits:** segments 2 or 3 (or 4), 3.5–4 × 2.5–3 mm, rounded, not normally contorted, terminal segment similar to others, margin nearly equally notched above and below, minutely hispidulous with straight hairs; subsessile or with a stipe to 1.5 mm.

Flowering Aug–Sep. Slopes, canyons, arroyos, stream banks, and roadsides, in various soils; (950–)1100–2300 m; Semidesert Grassland, Madrean Evergreen Woodland, (Great Basin Conifer Woodland), Petran Montane Conifer Forest; Cochise, Coconino, Gila, Graham, Greenlee, Pima, Santa Cruz, Yavapai cos.; New Mexico; Mexico (Chihuahua, Sonora).

Desmodium rosei is widespread and locally common in Arizona in a variety of habitats. The plants grow in the open and in the understory beneath trees. Lower elevation populations are typically found on riparian floodplains. *Desmodium rosei* is distinguished from the similar *D. neomexicanum* with which it sometimes grows by the much narrower leaflets and non-contorted fruits with straight hairs.

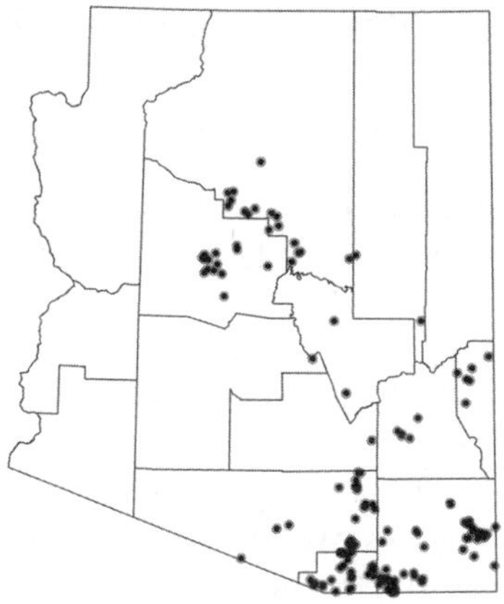

14. Desmodium scopulorum S. Watson. Proc. Amer. Acad. Arts 24:47. 1889. * [of cliffs or rock outcrops] SONORAN TICK TREFOIL Native

Herbs, annual. **Stems** solitary or few-branched above, erect or ascending, to 1 m, minutely hispidulous with hooked hairs. **Leaves** 1-foliolate proximally (not always apparent on late-season plants) and 3-foliolate distally; stipules 2–6.5 mm, awl-shaped, free, tardily deciduous; petioles 8–30[–60] mm; stipels 1.2–3.5 mm, awl-shaped. **Leaflets** ovate on proximal (1-foliolate) leaves and narrowly to broadly lanceolate on distal (3-foliolate) leaves, 1.8–6(–10) × 0.3–2(–3) cm, upper (adaxial) surface medium or dark green, lower (abaxial) surface slightly paler, puberulent with straight hairs on both surfaces. **Inflorescences** 11–40 cm in fruit; bracts awl-shaped, not overlapping, deciduous; bracteoles absent. **Pedicels** 8–13 mm. **Flowers** 2 at each axis node; calyx: tube 0.9–1.1 mm, lobes 1–1.8 mm; corolla white, drying yellow; banner 4.5–5 mm, ovate; wings 4.5–5 mm; keel 4–4.5 mm. **Fruits:** segments 2–4 (or 5), 6.5–7.5 × 3–3.5 mm, rhombic and contorted except terminal segment, terminal segment triangular, not contorted and distinctly larger than the others, 10–12 × 4.5–5 mm, margin nearly equally notched above and below, minutely hispidulous throughout with hooked hairs, terminal segment glabrous except margins with hooked hairs; sessile or with a stipe to 1 mm.

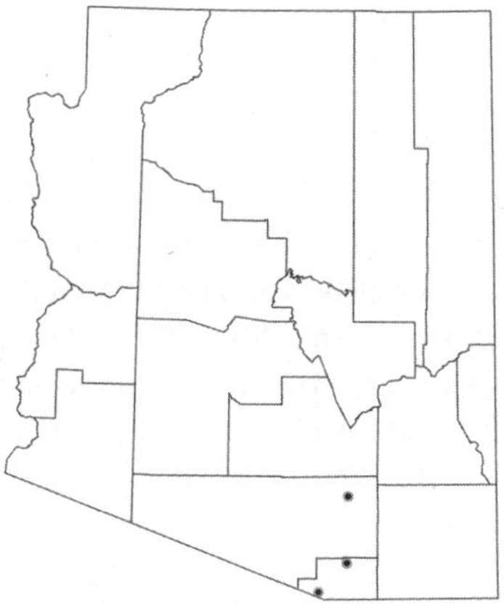

Flowering Aug–Sep(–Oct). Canyons and stream banks; [65–]880–1720 m; (rare in riparian habitats in Sonoran Desertscrub), Madrean Evergreen Woodland; Pima, Santa Cruz cos.; Mexico (Baja California Sur, Sonora).

Desmodium scopulorum is rare and localized in Arizona. It is more extensively distributed in eastern Sonora where it extends into thornscrub and tropical deciduous forest. It is also documented for the Cape Region of Baja California Sur. Martin et al. (1998) report that this species intergrades with *D. psilocarpum* in Sonora.

EBENOPSIS Britton & Rose. N. Amer. Fl. 23(1):33. 1928. * [from a common name for *Ebenopsis ebano*]
Matthew B. Johnson

Shrubs or trees. **Leaves** alternate or fascicled on short shoots; bipinnate; stipules persistent, straight, rigid spines with enlarged bases, paired at nodes, and herbaceous on short shoots; pinnae 1–4 pairs per leaf; glands short-stalked. **Leaflets** opposite, margin entire. **Inflorescences** spikes or heads, arising from axillary short shoots. **Pedicels** absent. **Flowers** mimosoid, radially symmetric, 5-merous; stamens numerous, exserted, filaments fused proximally, free distally; ovary 1-chambered, style and stigma simple. **Fruits** tardily dehiscent along both sutures, thick, woody, somewhat rounded in cross-section, persistent, the seeds separated by conspicuous septa. **Seeds** reddish brown, irregularly oblong and plump.

Species 3 (1 cultivated in the flora): s Texas and Mexico.

Ebenopsis, with its small leaves, conspicuous, thick, persistent woody fruits, and reddish brown seeds is distinctive. The wood is used for fuel, posts, and other purposes.

SELECTED REFERENCE: Barneby and Grimes 1996.

1. Ebenopsis ebano (Berlandier) Barneby & J. W. Grimes. Mem. New York Bot. Gard. 74(1):175. 1996. * [from a common name for the tree] TEXAS EBONY, EBONY BLACKBEAD, EBONY APES-EARRING; *ÉBANO* Cultivated

Mimosa ebano Berlandier, Mosaico Mexicano 4:418. 1840; *Pithecellobium ebano* (Berlandier) C. H. Muller; *Pithecellobium flexicaule* (Bentham) J. M. Coulter

Shrubs or trees 3–8(–20) m with a dense, rounded canopy. **Stems** stout but flexible and often conspicuously zigzag; trunk to 1 m in diameter; bark pale gray and smooth, becoming dark gray, fissured and scaly on trunk and larger limbs. **Leaves** 2.3–7.1 cm; stipules of first leaf at each node spines, 2–10 mm; stipules of subsequent leaves on short shoots, 0.3–0.5 mm; petioles 3–17 mm; rachis 4–11 mm; petiole and rachis with short sparse hairs; 1–3 pairs of pinnae; shallowly-cupped thick-stalked gland 0.3–0.6 × 0.5–1 mm between the first pair of pinnae and usually between subsequent pinnae. **Leaflets** (2 or) 3–6 pairs per pinna, obovate or oval, 4–17 × 2–10 mm, base asymmetrically heart-shaped, apex obtuse or rounded, upper (adaxial) surface lustrous dark green, lower (abaxial) surface dull pale green. **Inflorescences** cylindrical spikes 2–4.5(–5) cm with dense yellowish pubescence, the flowers crowded or interrupted. **Flowers** pale yellow green; calyx 0.8–1.2 mm; corolla 3.3–4.5 mm, funnel-shaped; filaments 7–19 mm, white fading to pale yellow. **Fruits** dark brown or nearly black, (5.5–)9–15.4 × (1.5–)2–2.7 cm, oblong or elliptic and straight or curving, rough-textured, fissured with age, sessile or nearly so, apex blunt with a short point. **Seeds** (5–)7–11; dark reddish brown, 9–15 × 6–10 mm.

Flowering (Jun–)Jul–Aug(–Oct). s Texas; Mexico (Nuevo Léon, San Luis Potosí, Tamaulipas, Vera Cruz, Yucatán).

Ebenopsis ebano is found from near sea level to ca. 500 m elevation. It grows in a variety of habitats including coastal areas, deep alluvial soils of riparian habitats, along arroyos, shallow caliche soils of hills, and rocky soils of mountain slopes and canyons in thornscrub, riparian woodland and forest, and seasonally dry tropical forest. *Ebenopsis ebano* is conspicuous and often abundant where it is found. The trees remain small in xeric sites and reach their best development in alluvial soils in riparian forest. Exceptional trees to 20 m tall have been reported (Isely 1973, Barneby and Grimes 1996). The tallest known *E. ebano* in the U.S. (measured May 2017) is a 14 m tree in Hidalgo County, Texas (Texas Big Tree Registry n.d.). Along the Texas Gulf coast, plants grow on low rises above saline flats and develop a windswept form due to the prevailing strong onshore winds. Bees frequent the flowers, which are also visited by butterflies and other insects. Nodulation is reported for *E. ebano* (Sprent 2009).

The seeds of *Ebenopsis ebano*, which are eaten in Mexico, are boiled when green or roasted when ripe, and have also been used as a coffee substitute (Vines 1960). The sapwood is yellow and the heartwood ranges from dark red or purple to brown. Wood of *E. ebano* is close-grained, durable, hard and heavy with a specific gravity of 1.04 (Vines 1960). The wood is used for posts, in cabinetry, to make furniture and utensils, for fuel, and to make charcoal (Vines 1960, Correll and Johnston 1970). Barneby and Grimes (1996) report that charcoal made from the wood of *E. ebano* is of high quality. Heartwood of *E. ebano* is used for carvings and the seeds are strung as necklaces (Correll and Johnston 1970). The town of Los Ebanos, in Hidalgo County, Texas, is named for the trees of *E. ebano* that grow there.

Texas ebony is commonly grown as a landscape tree in desert regions of southern and central Arizona. It is ideal as a specimen or patio tree and is suited for planting in street medians. With time it can grow large enough to make an effective shade tree due to the dense canopy. Texas ebony can be planted as a screen or barrier hedge (Jones and Sacamano 2000). With regular pruning, the plants can be kept small and compact. The dark green foliage and pleasantly fragrant flowers are attractive. The trees are typically evergreen but can become partially to mostly deciduous in severe winters or following prolonged drought. Growth rate is slow to moderate with adequate irrigation, and eventual size is determined by moisture availability. Texas ebony does well in a wide range of soils and is tolerant of reflected heat. Established plants are hardy to at least -9°C though damage has been observed on young, cultivated plants at ca. -7°C. Due to the spiny stems, Texas ebony should be sited carefully. The plants produce minimal litter except for the woody fruits which persist for long periods of time unless removed. Scarified seeds germinate readily. Texas ebony is successfully grown as a bonsai specimen.

ERRAZURIZIA Philippi. Anales Univ. Chile 41(11):688–689. 1872. * [named in honor of the Errázuriz family of Chile] DUNEBROOM

Michelle M. McMahon and Matthew B. Johnson

Psorobatus Rydberg

Shrubs to 1 m. **Stems and herbage** pubescent and glandular. **Leaves** alternate, odd-pinnate; stipules small, deciduous; petiolate. **Leaflets** obovate to nearly circular. **Inflorescences** spikes; bracts narrow, deciduous [persistent]. **Pedicels** absent. **Flowers** nonpapilionoid; calyx campanulate with 5 lobes; corolla with either 0 or rarely 1 petal, or 5 petals, the

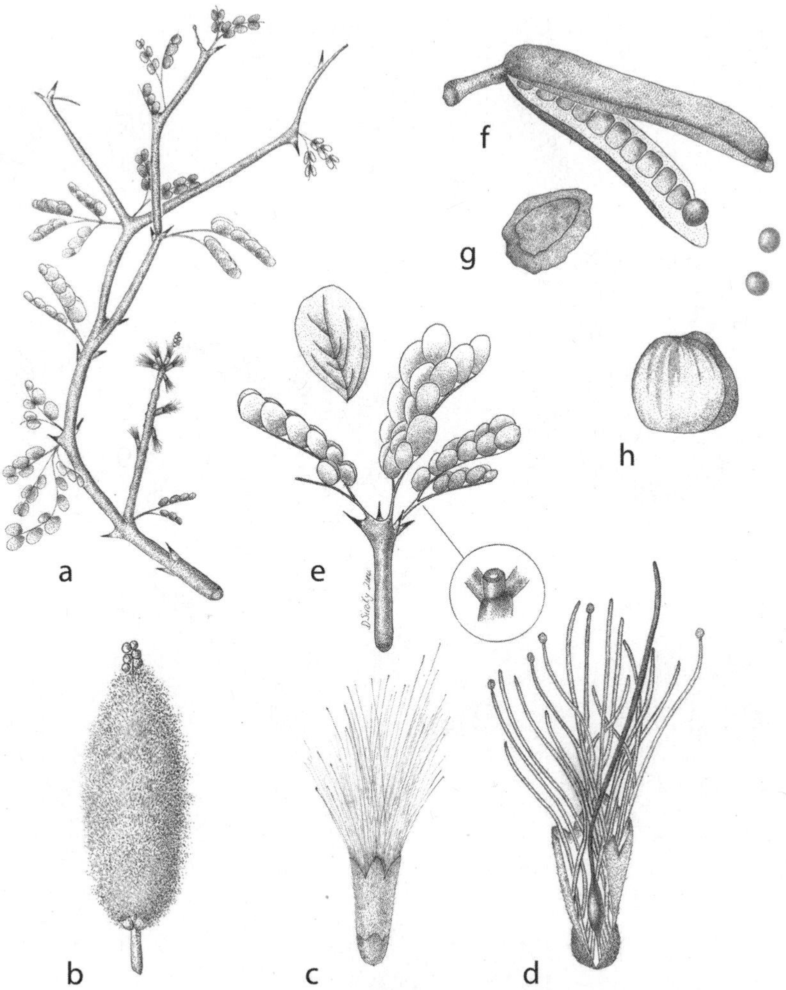

***Ebenopsis ebano*: a. habit; b. inflorescence; c., d. flower detail; e. leaves with detail; f. fruit with seeds; g., h. seeds. Illustrator: Daniela Siroky.**

petals not distinctly clawed, wings and keel petals, if present, not strongly differentiated; stamens partially or mostly united into a sheathing tube, 9- or 10-merous, filaments exserted; ovary with 2 ovules; style slender or if stout, with enlarged stigma and prominent gland near stigma. **Fruits** obovoid, subcompressed, glandular, indehiscent. **Seeds** 1, (3–)4–7.5 mm.

Species 4 (1 native in the flora): arid regions of the sw United States, nw Mexico, and n Chile.

SELECTED REFERENCES: Barneby 1977, Rhodes et al. 2011.

1. Errazurizia rotundata (Wooton) Barneby. Leafl. W. Bot. 9(13–14):210. 1962. * [rounded, referring to the leaflets] ROUND-LEAF DUNEBROOM Native

Parryella rotundata Wooton, Bull. Torrey Bot. Club 25(8):457. 1898

Shrubs, spreading, to 35 cm tall and 1 m in diam. **Herbage:** numerous orange or purple glands on stems, leaf stalks, and lower surfaces of leaflets; pubescence dense with appressed to ascending straight hairs. **Stems** aromatically resinous, stiff and much branched, bark grayish-brown. **Leaves** subsessile, 3–14 cm, often recurving; stipules subtending developed leaves, 0.5–2.5 mm, narrowly triangular or awl-shaped, deciduous; petioles 1–6 mm. **Leaflets** 23–61, occasionally alternate or nearly so, 1–8 × 1–6 mm, oval, obovate, or nearly circular, apices variable, gradually diminishing in size distally along rachis. **Inflorescences** 6–15-flowered, axis 4–15 mm; bracts 1.6–2.5 mm, narrowly lanceolate or awl-shaped, deciduous. **Flowers:** calyx 5–6.5 mm, lobe length less than half the length of tube; corolla absent or rarely with

***Errazurizia rotundata*: a. habit; b. inflorescence; c. fruit; d. seed; e. leaflet; f. stem details; g. stem node; h. stem node. Illustrator: Gillian Rice.**

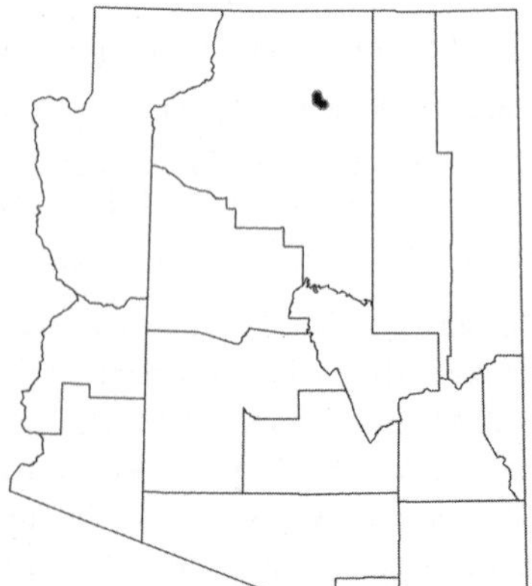

banner only, pale yellow fading reddish, 5 × 2 mm; stamens 10, (7–)8.5–12 mm, filaments basally united into a tubular sheath, free for half their length, greenish-yellow, anthers bright yellow; style slender and glandless. **Fruits** 9–11 × 3–5 mm, ellipsoid or obovoid, beaked, densely pubescent and with numerous small, red or purple glands. **Seeds** 4.6–7 × 2.7–3.7 mm, glossy, brown.

Flowering Apr–May. Open sites in sandy pockets on sandstone ledges and outcrops, gravelly soil on calcareous substrates, and in alluvial volcanic cinders; 1350–1560 m; Great Basin Desertscrub; Coconino, Navajo cos.; endemic.

Errazurizia rotundata is closely related to *Parryella* and may be better placed there or in *Amorpha* (Barneby 1977, McMahon and Hufford 2004, Straub and Doyle 2014). See page xi for updated taxonomy.

Errazurizia rotundata is known from only a few scattered populations in the drainage of the Little Colorado River. It is a species of conservation concern and is designated as Endangered by the Navajo Nation Department of Fish and Wildlife.

ERYTHRINA Linnaeus. Sp. Pl. 2:706–707. 1753. * [Greek *erythro*, red, for the flower color of some species]
Matthew B. Johnson

Perennial herbs, shrubs, and trees. **Stems** often with prickles. **Leaves** alternate, often with prickles; pinnately 3-foliolate; stipules small, persistent or deciduous; stipels gland-like. **Leaflets** of various shapes, broad, lateral leaflets asymmetric, terminal leaflet symmetric, margins usually entire. **Inflorescences** terminal or axillary racemes; bracts densely hairy, deciduous; bracteoles narrow, small, deciduous or absent. **Pedicellate. Flowers** papilionoid; single, paired, or fascicled; calyx tubular or bell-shaped, obliquely truncate, 5-lobed, split on the lower side or rarely 5-toothed; corolla with 5 petals, banner much longer than the others; stamens 10, filaments united at base or one stamen free; ovary 1-chambered, hairy, style without hairs, stigma head-like. **Fruits** stalked, opening along both sutures or the upper suture only, thick-walled, linear or cylindrical, usually constricted between seeds. **Seeds** ovoid, of various colors. $x = 21$.

Species ca. 110–120 (1 native in flora): widespread in tropics and subtropics of North America, South America, Africa, Asia, and Australia.

Erythrina species are widely reported to be poisonous, especially the seeds. The hard seed coat would need to be broken prior to ingestion for poisoning to occur. Crushed seeds of *Erythrina* species are reportedly used as rodent poison. *Erythrina* is used to stupefy fish (Allen and Allen 1981). Numerous studies have examined the alkaloids found in *Erythrina*. Compounds found in several tropical species have bactericidal and fungicidal properties for treating human infection (Mitscher et al. 1987). The bark of *Erythrina* species is reported to be used for treating fever, dysentery, and eye inflammation, and the seeds are employed to treat cancer (Schmutz and Hamilton 1979). Stem sap has been used to treat scorpion stings (Standley 1922). *Erythrina* seeds have an effect similar to curare and have been used to induce hypnosis (Matschat 1935, Allen and Allen 1981). The wood of some species has been used to make surfboards and canoes (Allen and Allen 1981) as well as for carving. Standley (1922) indicates that a yellow dye is obtained from the bark. Fast-growing tropical species of *Erythrina* are planted to provide shade for crops such as cocoa, coffee, pepper, and vanilla. Cut limbs are planted to produce living fences (Allen and Allen 1981). Several species of *Erythrina* are cultivated in warm climates for their attractive flowers and as shade trees. The flowers are adapted for bird pollination.

SELECTED REFERENCE: Krukoff and Barneby 1974.

1. Erythrina flabelliformis Kearney. Trans. New York Acad. Sci. 14(2):32. 1894. * [Latin *flabelliform*, fan-shaped, referring to the leaflets] SOUTHWEST CORAL BEAN; *CHILICOTE* Native, Cultivated

Erythrina coralloides A. Gray, Mem. Amer. Acad. II. 5[Pl. Nov. Thurb.]:301. 1854

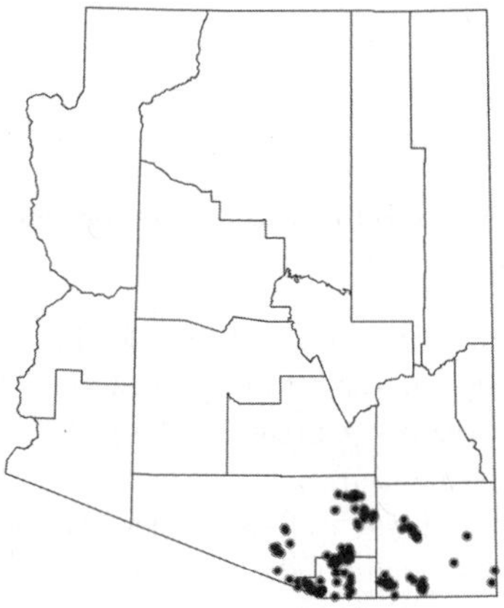

Shrubs or rarely trees to 5 [10] m; young herbage, inflorescence, calyx, ovary, and young pods with dense wooly hairs, becoming glabrate. **Stems** erect or ascending, sparingly branched, often with short, recurved internodal prickles, on larger stems these sometimes developing into blunt protuberances; bark of young stems smooth, light brown to pale gray, becoming dark gray to brown with irregular vertical white stripes; wood soft, pale yellow. **Leaves** 15–25 cm; stipules quickly deciduous; petioles 5–14 cm; rachis 2.5–6.5 cm; petiole and rachis occasionally with small prickles and with nectary glands between the leaflets; petiolules 4–8 mm. **Leaflets** bright green, broadly ovate to deltoid, 3.5–7.5 × (2.8–)5.8–11 cm. **Inflorescences** axillary or sometimes appearing terminal, 7.5–21.5 cm, several–many-flowered; bracteoles quickly deciduous. **Pedicels** 3–5 mm. **Flowers** fascicled or sometimes single or paired, tubular; calyx 8–12 mm, green, with a conspicuous green nectary gland that turns dark after the flower opens; corolla scarlet red, banner to 5.5 cm, narrow, wings and keel reduced; stamens of different lengths, united for about half the length of the longest filaments, 3.2–5.3 cm; pistil with thread-like style, 3.7–4.1 cm, persistent, the stigma scarcely head-like. **Fruits** brown, 13–26.5 cm, opening along upper suture, constricted between the seeds, persistent for several months, somewhat woody. **Seeds** 1–12; bright red or sometimes yellow, orange, or tan, 15 × 10 mm.

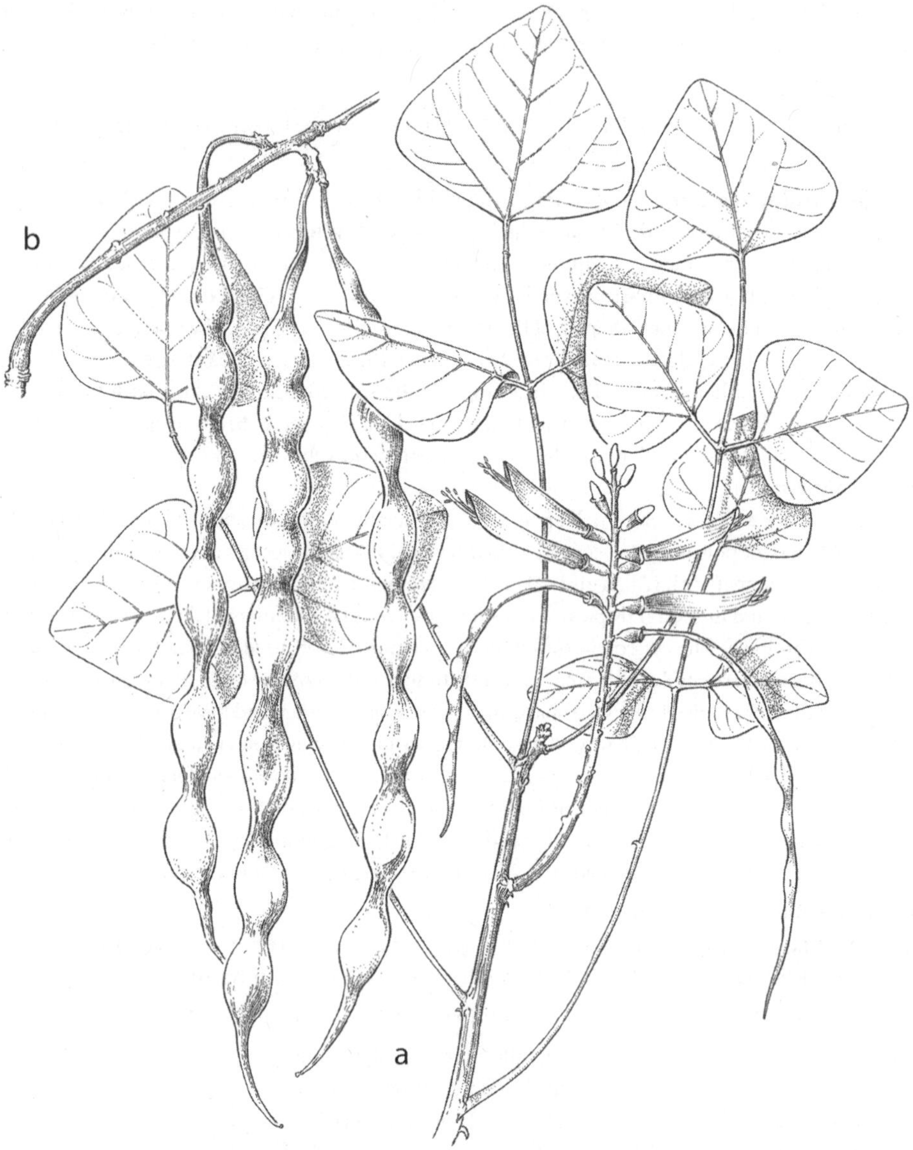

***Erythrina flabelliformis*: a. habit with inflorescence and developing fruits; b. mature fruits. Illustrator: Marjorie Leggitt.**

Flowering May–July. Rocky slopes, ridges, cliffs, canyons [and arroyos] in mountains; upper edge of Sonoran Desertscrub, Semidesert Grassland, Madrean Evergreen Woodland; [60–]900–1860[–2000] m; Cochise, Pima, Santa Cruz cos.; extreme southwest New Mexico; western Mexico (from Sonora, Chihuahua, and Baja California Sur to Michoacán).

Erythrina flabelliformis is a tropical species that reaches the northern limit of its distribution in Arizona. It is present in most of the mountain ranges of southeastern Arizona north to the Santa Catalina Mountains and west to the Baboquivari Mountains. Populations are scattered but plants may be locally abundant. Its preference for rocky sites appears to be related to the importance of higher soil moisture for seedling establishment and survival (Conn 1976). The rocks also retain heat that may ameliorate temperatures on cold winter nights. In Arizona, plants are typically multiple-stemmed shrubs 1–2 m tall due to periodic damage from winter freezing. In thornscrub and tropical deciduous forest habitats from central Sonora southward, *E. flabelliformis* develops into narrow trees to 10 m tall with a well-developed trunk to 1 m in diameter. In essentially frost-free regions the plants are not restricted to rocky habitats.

Hummingbirds visit the nectar-rich flowers of *Erythrina flabelliformis* and are presumably the pollinators. Leafcutter bees (*Megachile* sp.) harvest circular sections of the leaflets for their nests. Ants visit the extrafloral nectaries of the calyces and leaves, and appear to play a role in protecting the plants from other insects (Sherbrooke and Scheerens 1979). Bruchid beetles do not utilize the seeds of *E. flabelliformis* (Kingsolver and Decelle 1979); however, Sherbrooke and Scheerens (1979) have observed larvae of a moth that destroys the seeds. Nodulation is documented for many species of *Erythrina*, but has not been reported for *E. flabelliformis* (Allen and Allen 1981).

A common name for *Erythrina flabelliformis* in parts of Mexico is *mata raton*, as the seeds and bark are crushed and used as a rat poison (Mason and Mason 1987). The leaves are occasionally browsed, but stems and seeds of southwestern coral bean are reported to be poisonous to livestock (Benson and Darrow 1981). Vines (1960) reports that all parts of *E. flabelliformis* have poisonous properties. The seeds have been used in Durango, Mexico to treat toothache (Standley 1922, Usher 1974). The wood of *E. flabelliformis* is soft and easy to carve. It has been used for bottle and gourd stoppers and to make toys and ceremonial dance masks (Felger et al. 2001). The colorful seeds of *E. flabelliformis* are commonly strung to make necklaces. Young green pods are cooked and eaten by Guarajio and Mayo people in Sonora (Felger et al. 2001). Schmutz and Hamilton (1979) report that roasted or boiled seeds and cooked leaves and flowers are edible. However, the seeds are known to be highly toxic and should not be eaten as cooking may not render them safe.

Southwestern coral bean is occasionally cultivated in Arizona gardens for its showy flowers and lush foliage. The flowers emerge before or with the new leaves in late spring. The leaves turn bright yellow and quickly fall in September or October. Plants remain dormant until the following May. Coral bean is suitable for planting with succulents and other xerophytes, and in mini-oases. The plants provide a tropical appearance and are ideal for small spaces. Due to the long dormant period, they are often planted with other plants that can provide color and foliage. In warm-winter areas, coral bean can develop into a spectacular specimen tree. Coral bean requires a moderate amount of moisture during the growing season but is drought resistant at other times. It prefers well-drained soil. The stems are damaged at temperatures below ca. -4°C. In sites with damaging winter freezes, plants produce numerous stems from a large base. In locations where plants are protected from freezing, they can develop into small, upright trees with one or more trunks and a compact canopy. Recovery following freezing is rapid the following season; however, flowers are produced from the previous years' stems and plants that freeze each year are unlikely to flower. Propagation is by seeds and by stem cuttings made during the growing season. Despite the hard seed coat, fresh, untreated seeds have been reported to germinate. Older seeds generally require scarification.

ERYTHROSTEMON Klotzsch in Link, Klotzsch, and Otto, Icon. Pl. Rar. 2:97, pl. 39. 1844, descr. emended E. Gagnon and G.P. Lewis * [from Greek, *erythro*, red and *stemon*, stamen]
Matthew B. Johnson

Poincianella Britton & Rose, in part

Shrubs and trees (perennial herbs), rarely armed. **Stems** often with pustulate lenticels, without glands or glandular throughout. **Herbage** glabrous or pubescent. **Leaves** alternate; usually odd bipinnate; stipules deciduous or persistent. **Leaflets** opposite, margins entire. **Inflorescences** axillary or terminal racemes, or in panicle-like clusters; bracts large, serrate, deciduous. **Pedicels** jointed, elongated. **Flowers** caesalpinoid, bisexual (or rarely staminate); calyx tube small, turban-shaped, oblique, the lobes nearly equal, the margins serrate, deciduous leaving a persistent hypanthium; corolla with 5 petals, pale to bright yellow, orange-pink or reddish pink, often with orange markings on the banner, nearly equal, clawed; stamens 10, free, longer than petals, hairy below; ovary sessile, style and stigma simple. **Fruits** elastically dehiscent with valves twisting, stiffly papery, leathery, or slightly woody, with or without glands, the style persistent as a short beak. **Seeds** somewhat disk-shaped to oblong or irregular, compressed. $x = 12$.

Species 31 (3 cultivated and 1 of these also naturalized in the flora): United States, Mexico, Central America, the Caribbean, and South America.

Erythrostemon is closely related to *Pomaria* (Gagnon et al. 2016). The genus includes species formerly placed in *Poincianella* (Gagnon et al. 2016). Reported uses include planting as ornamentals and for revegetation (Lewis et al. 2005). Some species have been used in folk medicine. One species, *Erythrostemon gilliesii*, has become widely naturalized as an escapee from horticulture outside of its natural range.

SELECTED REFERENCE: Gagnon et al. 2016.

1. Leaves with 6–13 pairs of pinnae plus a terminal pinna; filaments 70–100 mm, bright red ____ **1. Erythrostemon gilliesii**
1′ Leaves with 2–5 pairs of pinnae plus a terminal pinna; filaments 8–12 mm, green or yellow-green.
  2. Young stems smooth; leaves (5–)14–25(–33) cm; calyx green, 11–13 mm; fruits 6.2–7.6 cm ____ **2. Erythrostemon mexicanus**
  2′ Young stems with conspicuous raised lenticels; leaves 3.5–14 cm; calyx maroon, 8–10 mm; fruits 2.8–4.8 cm ____ **3. Erythrostemon palmeri**

1. Erythrostemon gilliesii (Wallich ex Hooker) Klotzsch in Link, Klotzsch, and Otto, Icon. Pl. Rar. 2(3):97, pl. 39. 1844. * [for John Gillies, 1792–1834, Scottish botanist and physician] YELLOW BIRD-OF-PARADISE; *BARBA DE CHIVO* Naturalized, Cultivated

Poinciana gilliesii Wallich ex Hooker, Bot. Misc. 1:129–131. 1830; *Caesalpinia gilliesii* (Wallich ex Hooker) D. Dietrich

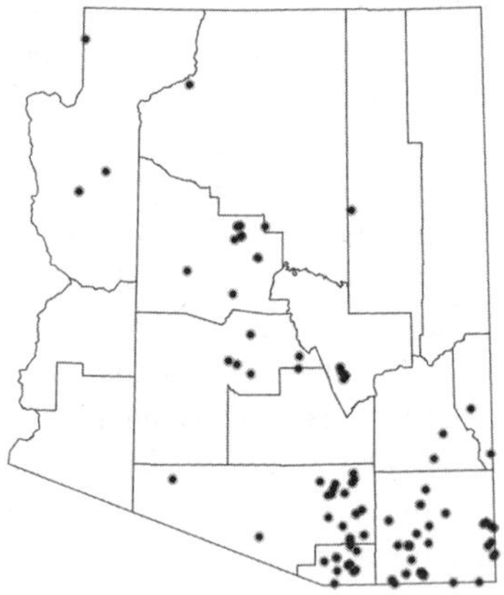

Shrubs to 2.5 (3) m, all parts of the plant have an unpleasant odor. **Stems** with bark smooth, green with raised white lenticels, becoming gray with age. **Herbage** glabrous. **Leaves** 6–18 cm with 6–13 pairs of pinnae and a terminal pinna, pinnae 1.1–3.4 cm; stipules ovate with margins densely fringed, 2.5–4 mm, becoming dry and stiff, persistent; petioles 0.9–3 cm. **Leaflets** 5–9 pairs per pinna, 2–6(–7) × 1–3 mm, oblong or oblong-elliptic. **Inflorescences** terminal racemes, 6–16 cm, commonly with 10–40 flowers; inflorescence axis, pedicels and outer surface of calyx with abundant, stalked, red glands; bracts membranous, fimbriate, ovate, 10–16 mm. **Pedicels** 12–25 mm. **Flowers**: calyx green, 16–18 mm, the lobes somewhat cupped; corolla pale yellow, 23–25 mm across, the petals 20–25 mm, the banner similar to the other petals; stamens with filaments 70–100 mm, long-exserted, bright red and showy; ovary green with dense white hairs on the margins and with stalked glands, style 70–100 mm, red with sparse white hairs, or pistil absent in some flowers. **Fruits** straw colored, villosulous and with abundant stalked red glands, 7–9 × 1.6–1.9 cm, sessile, stiffly papery, somewhat rhomboid with an acute apex. **Seeds** tan, brown, or olive green with dark flecks, 9–11 × 6–9 mm, lustrous.

Flowering Apr–Oct. Disturbed sites such as vacant lots, drainage ditches, roadsides, along arroyos and floodplains, and less commonly in undisturbed habitats; 340–1650 m; Sonoran Desertscrub, Chihuahuan Desertscrub, Mohave Desertscrub, Semidesert Grassland; naturalized in Cochise, Coconino, Gila, Graham, Greenlee, Maricopa, Mohave, Navajo, Pima, Santa Cruz, Yavapai cos.; naturalized elsewhere including California, Georgia, Nevada, New Mexico, Oklahoma, Texas, Utah; naturalized in Mexico and Australia; native to South America (Argentina, Uruguay).

Erythrostemon gilliesii is native to central and northeastern Argentina and Uruguay. In Argentina it grows in savannahs and subtropical woodlands in the Gran Chaco. The plants tolerate a wide range of soil types and can naturalize in areas that receive 250 mm or more of yearly rainfall. Some plants of *E. gilliesii* can have both staminate and perfect flowers (Calviño and Galetto 2010). A study by Moré et al. (2006) examined pollination in native populations of *E. gilliesii* in Argentina and determined that several species of hawkmoths are the primary pollinators.

Extracts from the seeds of *Erythrostemon gilliesii* have shown antitumor activity (Ulubelen et al. 1967). Anticarcinogenic activity of caesalin from *E. gilliesii* is selective in which cancers it shows activity against (Cheatham et al. 2000). The green fruits and seeds of *E. gilliesii* are widely reported to be toxic and can cause vomiting and diarrhea if eaten (Kingsbury 1964, Schmutz and Hamilton 1979, Tull 1987, Cheatham et al. 2000).

Yellow bird-of-paradise has long been grown in Arizona landscapes and gardens for its showy flowers. This distinctive, rugged plant can provide summer color and interest in a variety of landscape settings. The canopy is open. Hummingbirds and butterflies are attracted to the flowers. This species should be placed in a sunny location for best appearance. The plants become deciduous with extended drought or freezing. The species can grow 1 m or more per year with adequate irrigation. Yellow bird-of-paradise is drought tolerant but growth and flowering are greatly improved with periodic irrigation during warm weather. The plants are hardy to at least -12°C and probably lower. Stems on apparently healthy plants occasionally die back, requiring pruning to maintain an attractive appearance. The stems are brittle. At higher elevation sites, plants may freeze back in cold winters, requiring pruning of the damaged stems. Recovery is rapid during the following season. Yellow bird-of-paradise frequently reseeds in the landscape and has become naturalized at numerous locations near settlements in the southwestern U.S. Propagation is by scarified seeds.

2. Erythrostemon mexicanus (A. Gray) E. Gagnon & G.P. Lewis. PhytoKeys 71:124. 2016. * [of Mexico] MEXICAN BIRD-OF-PARADISE, MEXICAN POINCIANA Cultivated

Caesalpinia mexicana A. Gray, Proc. Amer. Acad. Arts 5:157. 1861; *Poincianella mexicana* (A. Gray) Britton & Rose

Shrubs to 3 m or trees to 5 (8) m. **Stems** with smooth green bark, ageing to gray and becoming slightly rough. **Herbage** glabrous or nearly so. **Leaves** (5–)14–25(–33) cm with 3–5 pairs of pinnae and a terminal pinna, pinnae (2–)3.5–8.5 cm; stipules ovate and somewhat cupped with fimbriate margins, 3–4 mm, quickly deciduous; petioles (2–)3.5–4.5(–5.3) cm. **Leaflets** (2)3 or 4(5) pairs per pinna, (6–)10–25 × (3–)5–15(–18) mm, obovate, broadly elliptic or oblong.

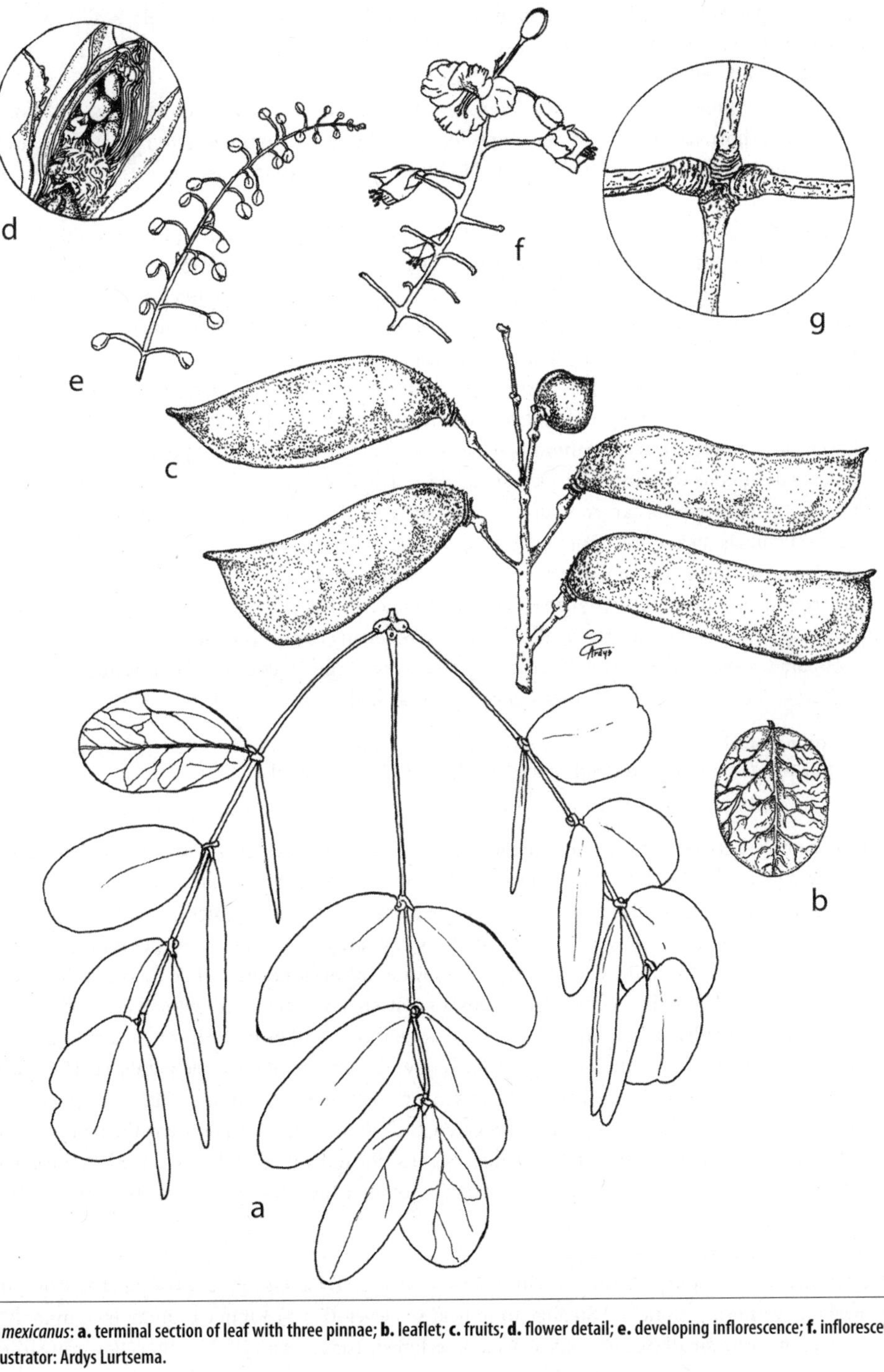

***Erythrostemon mexicanus*: a. terminal section of leaf with three pinnae; b. leaflet; c. fruits; d. flower detail; e. developing inflorescence; f. inflorescence; g. rachis node detail. Illustrator: Ardys Lurtsema.**

Inflorescences axillary and terminal racemes and in terminal panicle-like clusters, (5–)8–30 cm, with 10–60 flowers; stalks, pedicels, and calyces glabrous; bracts membranous, fimbriate, ovate, 1.5–2 mm. **Pedicels** 12–25 mm. **Flowers**: calyx green, 11–13 mm, the lower (abaxial) lobe 3-parted and about half again as long as the other lobes; corolla yellow, 16–22 mm across, the petals 9–13 mm, the banner reflexed and slightly smaller than the other petals with faint minute red spots in the mid-region; stamens with filaments green, 10–12 mm; ovary moderately covered with brown glands and white hairs, style 8–10 mm, with sparse white hairs on basal portion. **Fruits** straw colored, glabrous and with scattered stalked red glands most abundant near the margins, 6.2–7.6 × 1.7–2.2 cm, sessile, stiffly papery, somewhat rhomboid with an acute apex. **Seeds** olive green, 7–11 × 7–11 mm, lustrous.

Flowering Apr–Sep. Texas; Mexico.

Across its range, *Erythrostemon mexicanus* is found on slopes, canyons, valleys, floodplains, and along roadsides and disturbed areas in thornscrub, tropical deciduous forest, and tropical semi-evergreen forest from near sea level–450(–1700) m (McVaugh 1987). It is widely distributed in Mexico along both the east and west coasts and extends into extreme southern Texas.

The seeds of *Erythrostemon mexicanus* are reported to be poisonous (Jones and Sacamano 2000). Cheatham et al. (2000) report that the dried and powdered roots were used by the Tarahumara people of Mexico to treat cuts, sores, and ulcers. The Maya people prepared a mixture of cooked pods and salt to treat stomach infections (Cheatham et al. 2000).

Mexican bird-of-paradise is widely planted at lower elevations in Arizona as a flowering landscape shrub. It may be grown as a shrub or small tree in a patio, planted as an informal hedge or screen, and planted in street medians. Plants tend to be taller than wide and can be sited in narrow spaces. The showy yellow flowers, long blooming period, and fast growth are all positive attributes. This species is reported to flower year-round in frost-free regions (Jones and Sacamano 2000). The flowers attract butterflies and hummingbirds (Cheatham et al. 2000). The plants remain evergreen in mild winters but become deciduous with freezing. Occasional deep irrigation is desirable to maintain appearance and flowering. With ample water, the plants can grow up to 1.5 m per year. Mexican bird-of-paradise tolerates a wide range of soils. Plants sustain damage below -6°C but usually recover rapidly the following season. This species requires some pruning to develop a tree form. The spent inflorescences remain on the plant and are considered unsightly by some people. Fallen leaves, fruit valves, and seeds can create litter problems. The explosively dehiscent fruits can propel seeds some distance and these can germinate in irrigated situations, requiring removal. Though it can reseed in the landscape, Mexican bird-of-paradise poses little threat of becoming invasive in natural habitats in Arizona due to aridity and/or low winter temperatures. The species is readily propagated from scarified seeds.

3. Erythrostemon palmeri (S. Watson) E. Gagnon & G.P. Lewis. PhytoKeys 71:125. 2016. * [for Edward Palmer, 1831–1911, English-American explorer and collector] *PALO PIOJO* Cultivated

Caesalpinia palmeri S. Watson, Proc. Amer. Acad. Arts 24:47. 1889; *Poincianella palmeri* (S. Watson) Britton & Rose

Shrubs to 4 m or rarely slender trees to 8 m. **Stems** with reddish twigs; bark gray with conspicuous raised lenticels. **Herbage** glabrous or young leaves often with white hairs. **Leaves** 3.5–14 cm with (2 or) 3–5 pairs of pinnae and a terminal pinna, pinnae (0.5–)1–3 cm; stipules ovate with fimbriate margins, 1–1.5 mm, quickly deciduous; petioles 1–2.2 cm; rachises often with a stalked gland between the pairs of pinnae. **Leaflets** (1 or) 2–4 pairs per pinna, (2.5–)6–10(–14) × (1–)2–8(–11) mm, elliptic, oval, oblong, oblanceolate, or obovate. **Inflorescences** axillary and terminal racemes, 3–18 cm, with (2–)5–40 flowers, the stalks, pedicels, and calyces densely pubescent with white hairs; bracts membranous, fimbriate, ovate, 2–3 mm. **Pedicels** 7–14 mm. **Flowers**: calyx maroon, 8–10 mm, the lower (abaxial) lobe 3-parted and sometimes slightly larger than the other lobes; corolla yellow, 17–22 mm across, the petals 11–15 mm, the banner reflexed and about half as long as the other petals and with few to numerous, minute, bright red spots forming an arc across the center; stamens with filaments pale yellow green, 8–11 mm; ovary with dense white hairs, style 6–8.5 mm, with dense white hairs. **Fruits** tan or straw-colored, finely pubescent and dotted with stalked red brown glands, 2.8–4.8 × 1–1.4 cm, sessile, stiffly papery, somewhat sickle-shaped with an acute apex. **Seeds** pale tan, straw colored, or pale yellow green, 6–8 × 5–7 mm, lustrous.

Flowering (Dec–)Mar–May(–Jul). Mexico.

Erythrostemon palmeri is found on hills and in valleys in Sonoran Desertscrub, thornscrub, and tropical deciduous forest in central and western Sonora and Sinaloa, from near sea level–625(–730) m elevation. It commonly grows as a shrub but can develop into a small tree in tropical deciduous forest. Turner et al. (1995) report flowering from February through May and in October in response to rainfall.

The wood of *Erythrostemon palmeri* is hard and flexible. In Sonora, the Seri use the wood to make shafts for fishing harpoons, and it is a preferred wood to make bows and cradleboard frames (Felger and Moser 1985). The red sap is used as paint for arrows and decorating other objects (Felger and Moser 1985). The common name "*palo piojo*" refers to the white lenticels on the stems which resemble lice (Martin et al. 1998).

Palo piojo is occasionally planted in Arizona landscapes. It grows as a multiple-stemmed shrub but can be trained into a single-trunked tree. It is suitable in a patio and may be used to provide filtered shade for smaller flowering perennials and succulents. The bark is interesting, and the small but attractive flowers are scattered through the canopy and are produced over several months. The foliage is deciduous with drought or freezing temperatures. Moderate growth is achieved with sufficient irrigation, though established plants tolerate extended drought. This species appears to be reliably hardy to at least -6°C. During the severe freeze of February 2011, cultivated plants in Tucson suffered only minor stem damage at -7°C and up to 18 hours of subfreezing temperatures. The small leaflets produce minimal litter. Numerous plants have been observed to volunteer from seeds and become established in an irrigated field in Tucson. This species is unlikely to become invasive outside of well-irrigated settings in Arizona. Propagation is easy using scarified seeds.

EYSENHARDTIA Kunth in A. von Humboldt et al., Nov. Gen. Sp. 6(fol.):382; 6(qto.):489, pl. 592. 1824, name conserved. * [for Carl Wilhelm Eysenhardt, 1794–1825, German physician, botanist, zoologist, professor at Königsberg (Kaliningrad)] KIDNEYWOOD

Douglas H. Goldman

Shrubs or small trees, unarmed, 1–5[–25] m, generally glandular punctate throughout newer growth and reproductive material. **Stems** brown- to gray-pubescent in youth, often becoming glabrescent, without glands or glandular punctate. **Leaves** scented when crushed, alternate; odd-pinnate, 0.5–19 cm; stipules small, subulate; stipels generally present, small, subulate. **Leaflets** (5–)7–47[–61], opposite, nearly circular to oblong-elliptic or obovate, margins entire. **Inflorescences** spike-like racemes, terminal or rarely axillary; bracts early-deciduous to persistent, minute, generally lanceolate; bracteoles minute to absent. **Flowers** non-papilionoid, fragrant; calyx persistent, funnel-shaped, pubescent; lobes 5, shorter than tube; corolla white to yellow-white [pink], sometimes aging to purple, slightly irregular, without glands, glabrous; stamens 10, 9 fused and 1 free, fused filaments united for about 1/2 their length; ovary sessile or nearly so, glabrous; style variously bent apically, often with a large gland at bend or near tip; stigma capitate; ovules 2(–4). **Fruits:** indehiscent, ascending to deflexed, small, laterally compressed, ovoid to oblong-lanceolate, occasionally weakly sickle-shaped. **Seeds** 1(–2), near fruit apex, ellipsoid, obovoid, oblanceolate, to oblong-reniform or sickle-shaped, variously compressed. $x = 10$.

Species 10–15 (1 cultivated in the flora, 1 native and cultivated): sw, sc United States, Mexico, Central America (El Salvador, Guatemala).

Eysenhardtia is a small genus, often with vague species boundaries, especially in Mexico. It is closely related to *Amorpha*, *Apoplanesia*, *Errazurizia*, and *Parryella*, all members of the papilionoid tribe Amorpheae (McMahon and Hufford 2004).

Eysenhardtia polystachya, a Mexican species, was used to treat kidney ailments as a New World source of the diuretic called *lignum nephriticum* (Lang and Isely 1982). Wood chips from this plant in water produce a solution that is blue in bright light. The tropical Asian legume tree *Pterocarpus indicus* has also been called *lignum nephriticum*. The chemical properties of the fluorescent solutions of *lignum nephriticum* are described in Muyskens (2006).

SELECTED REFERENCES: Lang and Isely 1982, McVaugh 1987.

1. Largest lower (abaxial) surface leaflet glands scattered; fruit reflexed, seed filling about 1/2 of fruit, whole fruit appearing flattened; seed strongly compressed, edges sharp; Arizona, New Mexico ____ **1. Eysenhardtia orthocarpa**

1′ Largest lower (abaxial) surface leaflet glands restricted to midvein and margins; fruit ascending, seed filling about 2/3 of fruit, remainder flattened; seed slightly compressed, edges blunt; Texas ____ **2. Eysenhardtia texana**

1. Eysenhardtia orthocarpa (A. Gray) S. Watson. Proc. Amer. Acad. Arts 17:339. 1882. * [Greek *ortho*, straight, and *carpus*, fruit] KIDNEYWOOD; *PALO DULCE* Native, Cultivated

Eysenhardtia amorphoides Kunth var. *orthocarpa* A. Gray, Smithsonian Contr. Knowl. 5(6):37. 1853

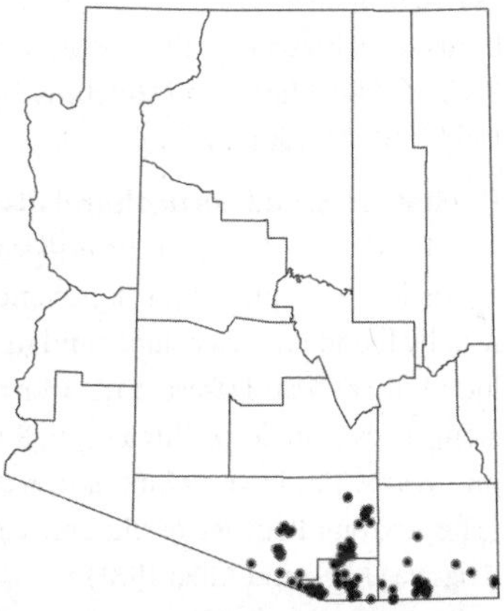

Shrubs 1.5–5 m. **Leaves** (3–)5–14 cm. **Leaflets** (13–)17–35(–47), oblong to narrow-oblong, 5–20 mm, some lower (abaxial) surface glands much larger than others, largest randomly scattered. **Inflorescences** (3–)5–10(–16) cm. **Flowers:** calyx tube 2–3.5 mm; lobes acute, 0.3–0.8 mm; corolla white, 6–10 mm; style with gland near tip. **Fruits** reflexed, oblong-lanceolate, sickle-shaped to straight, 10–16 × 2.5–4 mm, whole fruit appearing flattened. **Seeds** filling about 1/2 of fruit, obovoid to oblanceolate, strongly compressed, edges sharp.

Flowering May–Aug; fruiting Jul–Oct[–Dec]. Generally rocky soil; [100–] 1100–1800[–2000] m; shrubby Semidesert Grassland and the transition to

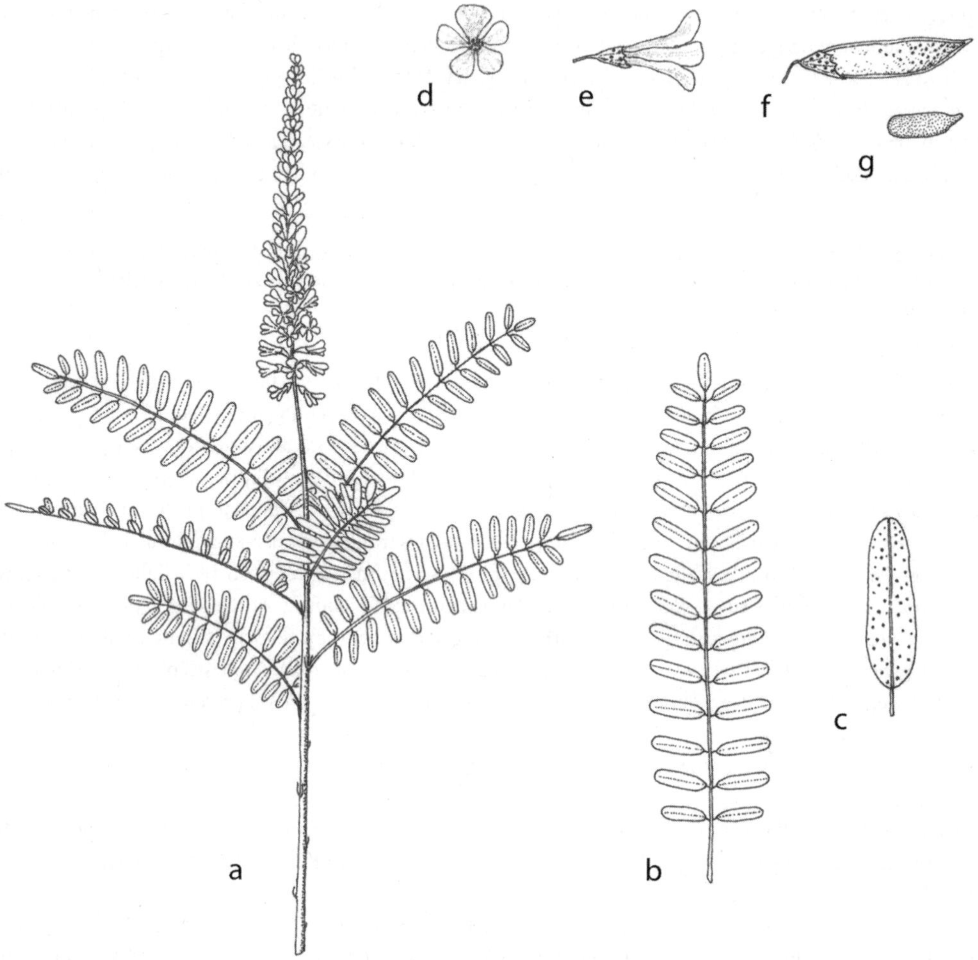

Eysenhardtia orthocarpa: **a.** habit; **b.** leaf; **c.** leaflet; **d.**, **e.** flower (front and side views); **f.** fruit; **g.** seed. Illustrator: Matthew B. Johnson.

open Madrean Evergreen Woodland; Cochise, Graham (rare), Pima, Pinal (rare), Santa Cruz cos.; New Mexico; Mexico (Chihuahua, Jalisco, Sinaloa, Sonora).

Eysenhardtia orthocarpa is found at higher elevations in the northern versus southern part of its range (Turner et al. 1995), generally above 1100 m in Arizona. In the United States only the widespread, typical variety of *E. orthocarpa* is found, whereas *E. orthocarpa* var. *tenuifolia* Lang is native only to Mexico (Lang and Isely 1982, McVaugh 1987; treated as *R. reticulata* by Shreve and Wiggins 1964). The former variety has minutely hairy leaflets whereas the latter has glabrous to glabrescent leaflets. Plants of *E. orthocarpa* in the United States have been called *E. polystachya* (Kearney and Peebles 1960, Martin and Hutchins 1980), whereas that species is native only to central and southern Mexico (Lang and Isely 1982, McVaugh 1987).

Habitat, uses, and horticultural discussion by Matthew B. Johnson

In Arizona, *Eysenhardtia orthocarpa* is rarely found in upper elevation Sonoran Desertscrub (Rincon Mountains) and is locally common on slopes and other warm habitats in grassland and oak woodland. Arizona populations are usually found on rocky slopes and in canyons, and the plants are locally abundant. Southward in Mexico the species is found in Sonoran Desertscrub, thornscrub, and tropical deciduous forest where the plants also occur in valleys and along arroyos in deep alluvial soil. Rainfall across much of the Arizona distribution of *E. orthocarpa* averages 375–500 mm yearly. The largest plant of *E. orthocarpa* in Arizona, found in the Coyote Mountains in Pima County, is 6.4 m tall (The Arizona Register of Big Trees 2005). Nodulation was not observed in experiments involving *E. orthocarpa* in Arizona (Allen and Allen 1981).

Eysenhardtia orthocarpa is an important browse plant for white-tail deer and mule deer (Anthony and Smith 1977). Bees frequent the flowers, and other species of *Eysenhardtia* are considered good honey plants. The wood has a variety of uses. Gentry (1942, p. 136) states that "The wood is used for building purposes, tool handles, sugarcane mills, etc., and is reported very hard and durable, the strongest available."

Kidneywood makes an excellent patio tree due to its upright growth form and small stature. Its showy and delightfully fragrant flowers attract butterflies and other insects. It can also be planted as an informal screen or barrier (Jones and Sacamano 2000). The plants typically develop several trunks but can be pruned to a single trunk. Growth rate is rapid with adequate irrigation. The plants are drought tolerant once established but require supplemental irrigation to develop a tree form in most landscape settings and to maintain an acceptable appearance. Kidneywood is deciduous following freezing temperatures but may remain semi-evergreen in mild-winter areas. The leaves are gradually shed during extended drought. The bark on older trunks breaks into strips that curl away from the stem, giving the trunks an interesting shaggy appearance. The foliage is aromatic when crushed. The plants are not particular as to soil so long as it is reasonably well-drained. Cultivated plants in Tucson have been observed to flower from April into November. Kidneywood is hardy to ca. -9 C. Propagation is similar to that described for *Eysenhardtia texana*. The seeds will germinate without scarification. Semi-hardwood cuttings taken in warm weather will root readily.

2. Eysenhardtia texana Scheele. Linnaea 21:462. 1848. * [of Texas] TEXAS KIDNEYWOOD; *VARA DULCE* Cultivated

Shrubs 1–3(–5) m. **Leaves** (1–)2–7(–10) cm. **Leaflets** (7–)13–35(–47), elliptic to oblong, 3–13 mm, some lower (abaxial) surface glands much larger than others, largest along midvein and margins. **Inflorescences** 3–11(–15) cm. **Flowers:** calyx tube 2–4 mm; lobes obtuse to acute, 0.3–0.7 mm; corolla white, 5–8 mm; style with a gland near tip. **Fruits** ascending, oblong, sickle-shaped, (straight), 6–13 × 2–3 mm, flattened near base. **Seeds** filling about 2/3 of fruit, oblong-subreniform to sickle-shaped-ellipsoid, slightly compressed, edges blunt. $2n = 20$.

Flowering Mar–Nov. Texas; Mexico (Chihuahua, Coahuila, Nuevo León, San Luis Potosí, Tamaulipas, Veracruz).

Eysenhardtia texana is a widespread, common species in parts of central, southern, and western Texas, particularly in calcareous soil; stems and leaves of this species apparently have antibacterial and antifungal chemical properties (Wächter et al. 1999). It is found in grassland, open woodland, shrubland, semi-desert, and mildly disturbed areas. Across its range *E. texana* is found from 10 to 2500 meters. In Arizona it is cultivated up to 1600 meters.

Uses and horticultural discussion by Matthew B. Johnson

Lewis et al. (2005) list numerous uses for *Eysenhardtia texana* including medicine, livestock fodder, forage for wildlife, honey production, erosion control, reforestation, and a source of a yellow-brown dye. *Eysenhardtia texana* is considered a good honey plant (Nokes 1986). Whisenant and Ueckert (1982) consider *E. texana* to have good potential as a range forage shrub for livestock and wildlife. Species of *Eysenhardtia* readily resprout following severe browsing (Allen and Allen 1981).

Texas kidneywood is ideally suited for planting as a screen or informal hedge and can be trained as a small patio tree. The flowers, which are produced over several months, are attractive and perfume the air with a delicate sweet fragrance. These attract butterflies and other insects. The growth rate for *Eysenhardtia texana* is slower than for *E. orthocarpa*. Established plants are drought tolerant but require occasional irrigation during the growing season to maintain acceptable appearance. Texas kidneywood is deciduous with freezing or extended drought. Crushed leaves are aromatic. The plants are not particular about soil but should have adequate drainage. Flowering has been observed on cultivated plants in Tucson from April into November. Texas kidneywood is hardy to at least -12°C. A plant of Texas kidneywood cultivated at Boyce Thompson Arboretum is 8.2 m tall (The Arizona Register of Big Trees 2005). Untreated seeds of *E. texana* germinate easily (Nokes 1986). Whisenant and Ueckert (1982) report no difference in germination between treated and untreated seeds of *E. texana*. These authors report an optimum temperature of 30°C for *E. texana* with 75% germination, while only 2% germination was observed at 15°C and 35% germination was observed at 40°C. Nokes (1986) reports that Texas kidneywood may be grown from softwood and semi-hardwood cutting which root easily during summer and early autumn when treated with rooting hormone and kept under intermittent mist. Annual rainfall across much of its range averages 375–800 mm. Allen and Allen (1981) report nodulation on *E. texana* in Texas.

GALACTIA P. Browne. Civ. Nat. Hist. Jamaica 298. 1756. * [Greek, *galaktos*, milk, alluding to the "milky" branches of the original species, *Galactia pendula* Persoon]

Guy Nesom

Heterocarpaea Scheele; *Odonia* Bertoloni

Perennial herbs, subshrubs, rarely shrubs, from rhizomes, an elongate taproot, or a fusiform taproot. **Stems** procumbent or twining and climbing. **Leaves** mostly 1- or 3-foliolate, less commonly 5-foliolate or 7(–9)-foliolate; stipules small, deciduous or persistent; stipels persistent. **Inflorescences** pseudoracemes, axillary [terminal]; pedunculate or without an axis and flowers in axillary fascicles, sometimes axillary and of a single flower; rachis with slightly swollen nodes; bracts small, setaceous; bracteoles minute. **Pedicellate. Flowers** papilionoid; solitary, paired, or in fascicles of 2–3 at rachis nodes, axillary when peduncle and rachis absent; each flower with 1 bract and 2 bracteoles, deltoid, lanceolate or subulate; calyx campanulate, lobes 5 appearing as 4 (upper 2 completely fused), linear to lanceolate, longer than the tube, lateral slightly smaller, lowest often longest; corolla purplish to bluish or pink [red]; petals subequal in length; banner circular to ovate or obovate-circular, margin slightly inflexed or appendaged, apex rounded; wings narrow or obovate, adherent to keel; keel obtuse and almost straight, subequal to or longer than wings; stamens 10, 9 united and 1 free or united at the middle (pseudomonadelphous); vexillary stamen free or fused from middle downward; ovary subsessile; style thread-like, glabrous; stigma small, rounded, terminal; nectary at ovary base. **Fruits** linear, laterally compressed, straight or weakly to strongly falcate, with false septa between seeds, elastically dehiscent. **Seeds** 1–12. $x = 10$.

Species ca. 111 worldwide, 21 in the United States (1 native in the flora): tropical and subtropical regions, native to the United States, Mexico, Central America, West Indies, South America, Africa, Asia, Australia.

Galactia P. Browne (Phaseoleae, Diocleinae) is estimated here to include at least 111 species, most native to the Americas, distributed among South America, Central America, West Indies, Mexico, and the southeastern and south-central U.S., with the greatest concentrations in South America and the U.S. Diversity in the U.S. is concentrated in Florida and in Texas. Four *Galactia* species are native to Asia, Australia, and Africa, including *G. tenuiflora* (Willdenow) Wight & Arnott, which apparently occurs natively in all three areas. Chen et al. (2010) described the range of *G. tenuiflora* as southern China, Taiwan, India, Malaysia, Philippines, Sri Lanka, Thailand, Vietnam, and Africa. *Galactia tashiroi* Maximowicz is endemic to Taiwan and the Ryukyu Islands of Japan.

Taxonomic studies of *Galactia* have been completed relatively recently for Argentina and neighboring countries (Burkart 1971), Venezuela (Torres et al. 1983), Brazil (Ceolin 2011, Ceolin and Miotto 2013), and Cuba (Beyra Matos et al. 2005), and various regional floras have included substantial numbers of species. In the U.S., the first study devoted solely to *Galactia* was by Vail (1895), who included 14 species. Small (1933) included 14 species for the southeastern U.S.; Duncan (1979) recognized 8 for the same area. Ward and Hall (2004) recognized 11 species for Florida, and Isely's overview (1998) for the U.S. included 17. Significant disagreement over species concepts and nomenclature has characterized the U.S. studies—Isely in 1998 (p. 563) noted that "Herbarium annotations remain author-idiosyncratic and chaotic."

Galactia encompasses a wide diversity of habit, root and rhizome structure, leaf venation, inflorescence structure, and floral morphology.

Various species of *Galactia* serve as host plants for butterflies (Lewis et al. 2005), and the seeds are eaten by some birds (Graham 1941). Some species may be useful in erosion control (Graham 1941). Nodulation is reported for some species of *Galactia* (Allen and Allen 1981).

1. Galactia wrightii A. Gray. Smithsonian Contr. Knowl. 3(5):44. 1852. * [for Charles Wright, 1811–1885, American botanist] WRIGHT'S MILKPEA Native

Galactia tephrodes A. Gray; *Galactia wrightii* A. Gray var. *mollissima* Kearney & Peebles

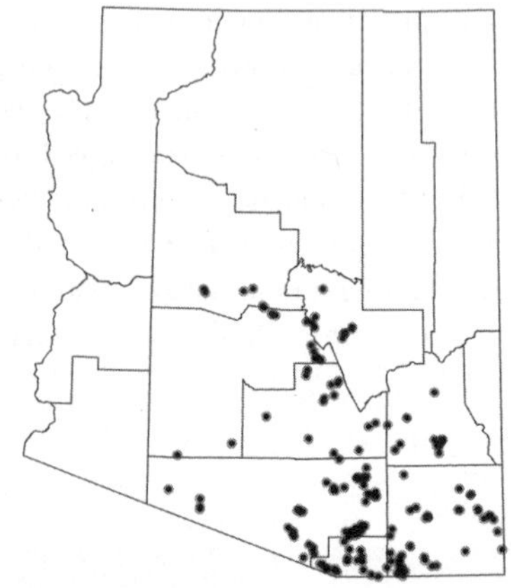

Perennial herbs, arising from a woody taproot. **Stems** first procumbent, then weakly twining distally, strongly lignescent, densely to sparsely strigose with retrorsely appressed to slightly spreading hairs. **Leaflets** 3, blades oblong-elliptic to elliptic-ovate, 15–48 × 5–18 mm, herbaceous, upper side (adaxially) sparsely to densely short strigose-sericeous with closely appressed hairs to pubescent with soft, spreading-erect hairs, moderately to densely strigose-sericeous and glaucous below (abaxially), sometimes sparsely strigose to glabrous on upper side (adaxially), veins not raised, apices rounded to obtuse or

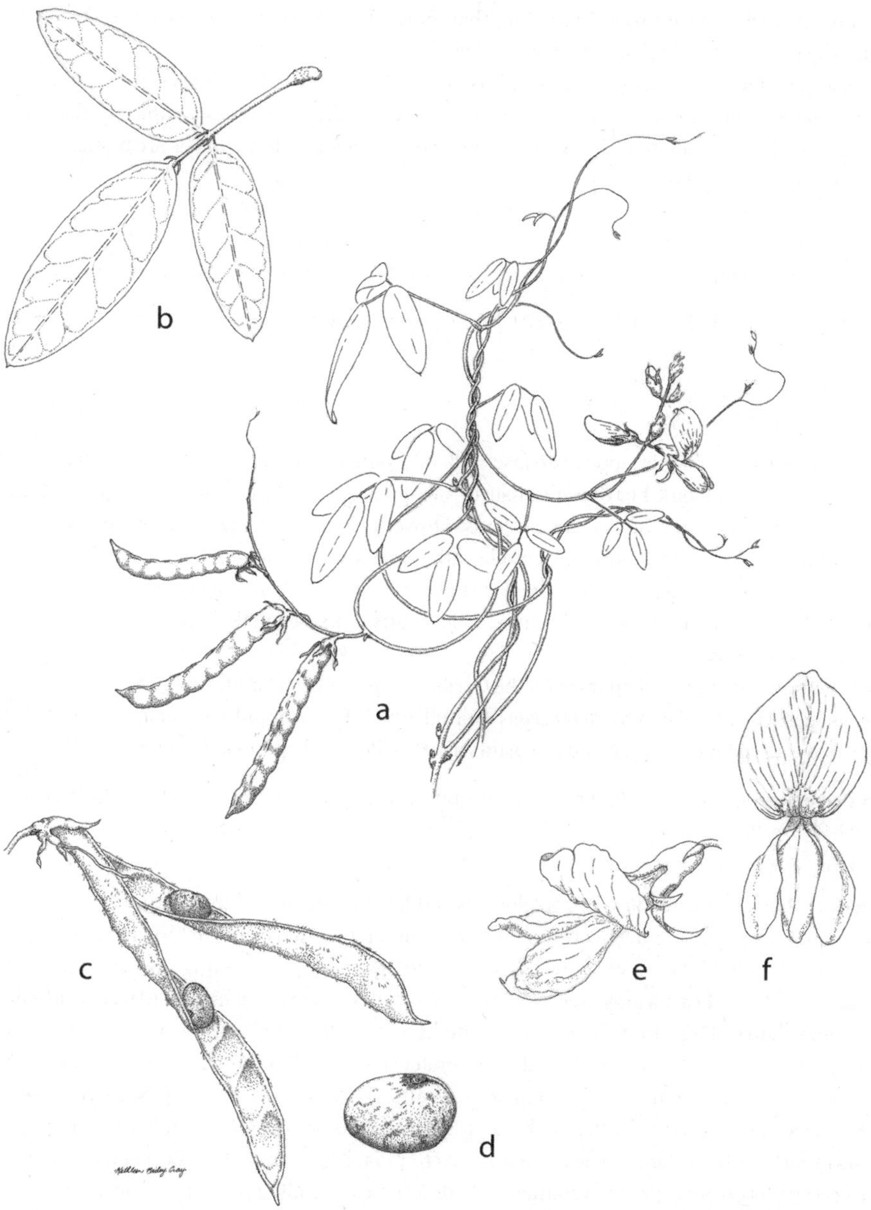

***Galactia wrightii*: a. habit; b. leaf; c. dehisced fruit; d. seed; e., f. flower. Illustrator: Kathleen Bailey Cray.**

shallowly retuse. **Inflorescence** axis 35–200(–250) mm, flowers (2–)5–18(–30) in a pseudoraceme, commonly fasciculate on the raceme, axis rarely 10 mm and 2–3 flowered. **Flowers**: calyx 5–7 mm, loosely short-strigose to hirsute-pubescent; corolla 11–12 mm, pink to purple-rose to lavender. **Fruits** 25–50 × 4–5 mm, straight, moderately to densely strigose to strigose-sericeous. **Seeds** (3–)5–9.

Flowering May–Sep(–Oct). Ridges, slopes, canyons, rocky outcrops, arroyos, and roadcuts, usually in rocky soil including granite crevices, igneous talus, among boulders and gravelly alluvium; [350–](680–)800–2100 m; upper elevations in Sonoran and Chihuahuan Desertscrub, Semidesert Grassland, Interior Chaparral, Madrean Evergreen Woodland; Cochise, Coconino, Gila, Graham, Maricopa, Pima, Pinal, Santa Cruz, Yavapai cos.; New Mexico, Texas; Mexico (Chihuahua, Sonora).

Galactia wrightii is characterized by twining stems, mostly oblong-elliptic leaflets with glaucous abaxial surfaces, long inflorescences with numerous flowers and relatively large corollas, and strigose-sericeous fruits. The type of var.

mollissima is a plant with vestiture more spreading than normal–such variants are scattered through the range in Arizona, as cited by Kearney and Peebles (1942), as well as in Texas.

Plants in southwestern New Mexico and southeastern Arizona (especially the Huachuca Mountains but not the Chiricahua Mountains) tend to have reduced vestiture—the leaves are sometimes completely glabrous. Such plants extend southward in Mexico along the Chihuahua-Sonora border region. Intermediates are numerous, however, especially in Arizona, and no distinct boundary seems evident.

Galactia wrightii is locally common at lower elevations in the mountains of Arizona where it trails on the ground or climbs on other vegetation. It is found in tropical deciduous forest in Mexico. The seeds of this species are utilized by several species of quail (Graham 1941). Nodulation has not been observed on *G. wrightii* (Allen and Allen 1981).

GEOFFROEA Jacquin. Enum. Syst. Pl. 7, 28. 1760. * [for Claude Joseph Geoffroy (1685–1752), French pharmacist and botanist]

Matthew B. Johnson

Gourliea Gillies ex Hooker & Arnott

Shrubs or trees. **Leaves** alternate or appearing fascicled; odd-pinnate; stipules small, awl-shaped, deciduous; short-petioluled; stipels small, deciduous. **Leaflets** alternate or subopposite, oblong, obtuse or emarginate. **Inflorescences** of axillary racemes or flowers clustered near tips of stems. **Flowers** papilionoid, bilaterally symmetric; calyx 5-lobed, 2 upper (adaxial) lobes united or all lobes united; corolla with 5 petals, free, blades clawed, uppermost (adaxial) (banner) petal outermost, keel petals overlapping and joined at the base; stamens 10, in two groups or sometimes united into a single column; ovary 1-chambered, style and stigma simple. **Fruits** indehiscent, fleshy, drupe-like, ovoid, with thin pulp over a woody endocarp, 1-seeded.

Species 2 (1 cultivated in flora): Panama and the Caribbean to southern South America.

Geoffroea is most closely related to *Pterocarpus* (Polhill 1981). The fruit pulp is edible and utilized by people and animals. The wood is used for carpentry, and the plants have medicinal properties (Lewis et al. 2005).

1. Geoffroea decorticans (Gillies ex Hooker & Arnott) Burkart. Darwiniana 9(1):19. 1949. * [referring to the exfoliating or decorticating bark] CHILEAN PALO VERDE; *CHAÑAR* Cultivated

Gourliea decorticans Gillies ex Hooker & Arnott, Bot. Misc. 3:208. 1833; *Geoffroea spinosa* Jacquin; *Gourliea chilensis* Clos

Shrubs or trees to 8 m, often spreading by root sprouts and forming thickets. **Stems:** trunks often with buttress-like longitudinal ridges; bark green or gray green and relatively smooth textured or tan to brown and shredding into strips, exfoliating in thick strips and flakes; wood whitish yellow; twigs rigid, often ending in a sharp point. **Herbage** with short, white appressed hairs. **Leaves** gray green, 2.2–4.1 cm; petioles 5–9 mm; rachis 10–20 mm. **Leaflets** (3)5–11, 3–16 × 2–6 mm. **Inflorescences** 8–20 mm, with short white hairs, 2–5 flowered or flowers solitary. **Pedicels** 2–5 mm. **Flowers:** calyx green, 3–4 × 3–4 mm, bell-shaped, lobes united; corolla yellow with faint orange lines on banner, 7–10 × 7–10 mm, banner cleft at apex, wings longer than keel. **Fruits** orange-brown when ripe, 18–24 × 15–20 mm, with a smooth surface. **Seeds** tan, ± elliptical, often with transverse wrinkles or ridges, 8–12 mm × 4–6 mm.

Flowering in early spring before new leaves appear. Argentina, Bolivia, Brazil, Chile, Paraguay, Peru, and Uruguay.

Geoffroea decorticans is widespread in southern South America including the Gran Chaco, Pampas, Monte Desert, and Atacama Desert. It is found in a variety of habitats including periodically inundated sites. Becker (1983) reports that the plants are tolerant of saline soils. In favorable habitats, *G. decorticans* can form dense, nearly impenetrable thickets from root sprouts. These thickets can be over 30 m across. At least two varieties of *G. decorticans* have been described (Burkart 1949). Bark characters vary from plant to plant. The bark of some plants is a uniform green or gray green and mostly smooth-textured while that of other plants is tan to brown and shreds into strips with little if any green bark visible beneath. Nodulation has been observed on *G. decorticans* seedlings in the laboratory when inoculated with *Rhizobium* species from *Prosopis glandulosa* and from *Olneya tesota* (Becker 1983).

Plants are occasionally planted as living fences in South America and are useful in combating erosion and desertification, but can become weedy, particularly in overgrazed rangeland. New plants can emerge from the roots up to 10 m from the parent plant (Becker 1983). Eradication is difficult. Feldman and Feldman (1987) report 8,500–30,000 *chañar* plants per hectare from root sprouts in pastures in Santa Fe Province, Argentina, two years after plowing.

The wood of *Geoffroea decorticans* is occasionally used in light carpentry, for tool handles, and for fuelwood and charcoal. The wood weighs 585–600 kg/m^3 (López et al. 1987). The bark has been used to dye textiles (Becker 1983). Nectar from the flowers is considered good bee forage and yields high quality honey. The fruit pulp is sweet with a flavor

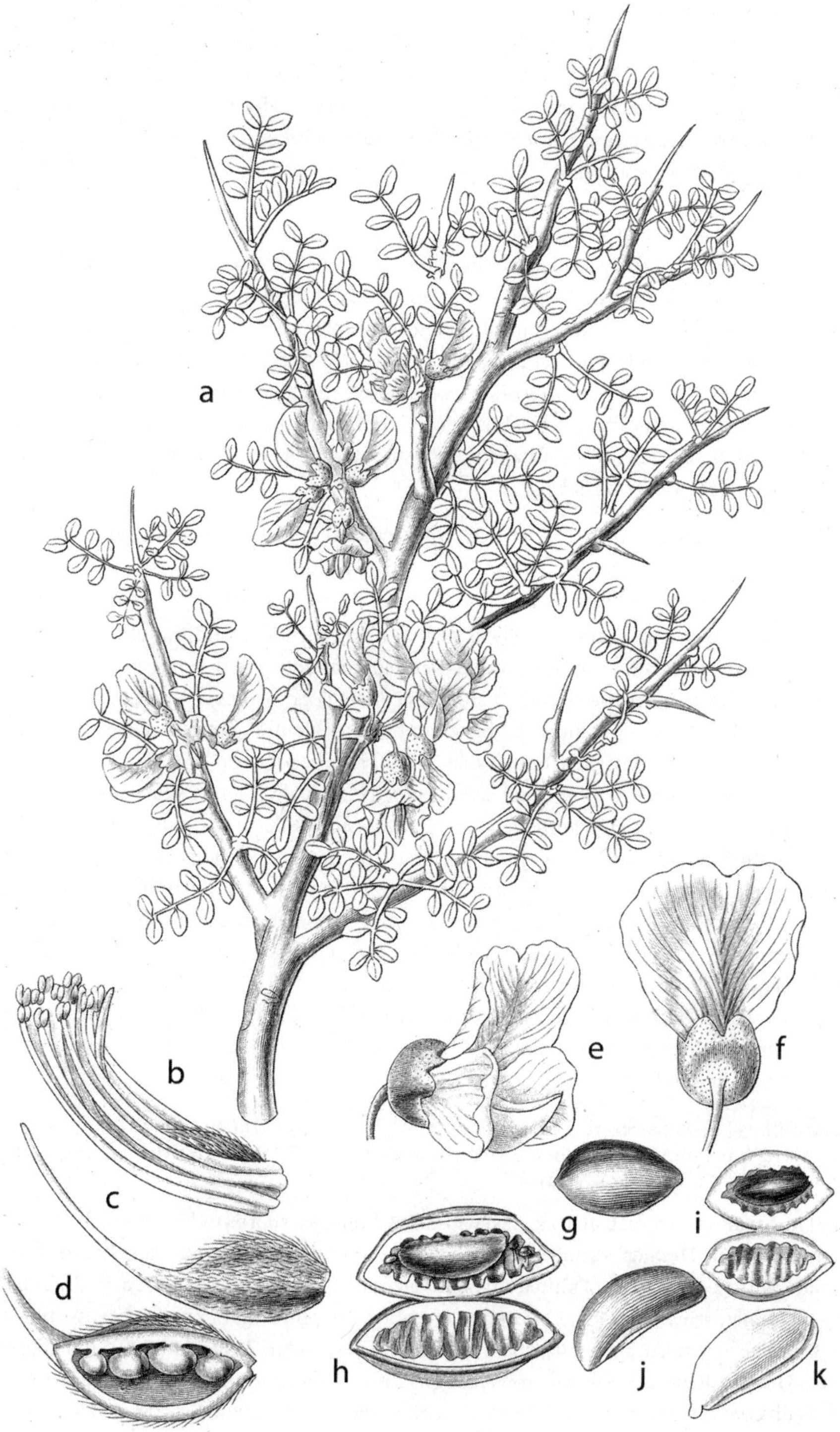

Geoffroea decorticans: **a.** habit; **b.** stamens; **c.** pistil; **d.** ovary (longitudinal section); **e., f.** flowers; **g.** fruit; **h., i.** fruit (longitudinal section showing seed); **j., k.** seeds. Illustration from Hooker (1833).

similar to that of gingerbread, and is eaten fresh or dried by people, livestock, and wildlife. The pericarp of the fruit is high in sugar and the seeds contain high levels of protein and fat, but also contain low levels of cyanide (Becker 1983). The fruits are not suitable for animal feed when eaten exclusively (Becker 1983). Aronson (1990) reports that goats thrive on *chañar* in Chile and that the trees are valued for their long period of fruit-bearing. A fermented beverage called *aloja* is produced from the fruits. *Aloja* can be distilled into a sweet liquor called *arrope*. *Arrope* is used in herbal medicine to treat asthma (Becker 1983, Aronson 1990), and López et al. (1987) report that a decoction of the leaves and bark is used to treat convulsions.

Chañar is occasionally cultivated as a landscape tree at lower elevations in Arizona. The green bark, sculptural trunks, and showy masses of flowers are attractive. The plants are drought tolerant and do best in deep soils. Foliage is drought and cold deciduous. Flowering in Arizona is typically in March and April. Root sprouts are a problem in landscapes where moisture is available. These can come up among other plants and between pavers and present an ongoing maintenance issue. *Chañar* is cold tolerant to at least -12°C. Cultivated plants growing at Boyce Thompson Arboretum since the late 1920's have not been observed to sustain freeze damage. *Chañar* may be propagated from seeds that have been removed from the woody endocarp. Seed viability and germination rates are typically high although some people in Arizona have reported difficulty with germination. Root sprouts have been successfully side-boxed. Becker (1983) indicates that plants can be rooted from cuttings.

GLEDITSIA J. Clayton in Linnaeus, Sp. Pl. 2:1056. 1753. * [for Johann Gottlieb Gleditsch, 1714–1786, German botanist] HONEY LOCUST

Matthew B. Johnson and Kirsten N. Lake

Trees, shrubs (1 species), to 30+ m. **Stems** usually armed with conspicuous, rigid, simple or branched thorns; bark becoming furrowed. **Leaves** alternate, even-pinnate and even-bipinnate, often both types on the same plant, deciduous; stipules inconspicuous, deciduous; petiolate; stipels absent. **Leaflets** opposite or subopposite, margins shallowly crenate. **Inflorescences** racemes, axillary, simple or clustered; bracts scale-like, minute, deciduous; bracteoles absent. **Pedicellate**. **Flowers** caesalpinoid, unisexual or bisexual, small; calyx bell-shaped, with 3–5 subequal lobes, distinct or overlapping at anthesis; corolla with 3–5 subequal, overlapping petals, pale green or yellow green; stamens 6–10, free, inserted on calyx disk, exserted, straight; anthers uniform; ovary subsessile, with 2–many ovules; style simple, short; stigma simple. **Fruits** stipitate, dehiscent or indehiscent, elongated, compressed and often curved or twisted, leathery or somewhat woody, sometimes with pulpy mesocarp. **Seeds** brown, compressed, ovoid or nearly circular, smooth.

Species ca. 13–16 (1 cultivated and adventive in the flora): temperate and subtropical regions of Asia and the Americas. The center of diversity is in central, eastern, and southeastern Asia to the central Indo-Pacific region with 1 species in South America and 2 or 3 species in eastern North America.

Fruits of various species have been used as a source of soap (Allen and Allen 1981). Fruits have also been used in folk medicine as an emetic and expectorant (Allen and Allen 1981; Lewis et al. 2005). The roots of *Gleditsia* species are not known to nodulate (Allen and Allen 1981).

1. Gleditsia triacanthos Linnaeus. Sp. Pl. 2:1056–1057. 1753. * [from Greek *tria*, three, and *acanthos*, spine; referring to the thorns which commonly have three principal branches] COMMON HONEY LOCUST, HONEYSHUCK, THORNY LOCUST Cultivated, Adventive

Trees to 20 [30+] m. **Stems** and trunk usually armed with stout, branched thorns to 15 (30) cm; bark dark gray, smooth, becoming fissured with age. **Herbage** variably pubescent when young, becoming glabrate. **Leaves** both even-pinnate and even-bipinnate, with (2–)4–8 pairs of pinnae on bipinnate leaves; petioles 1.5–4.5; rachis 5–28.5 cm. **Leaflets** 9–18 pairs of leaflets on pinnate leaves and (8–)10–14 pairs of leaflets per pinna on bipinnate leaves, oblong to lanceolate, 2.8–5.5 × 0.5–1.9 cm, base rounded, apex mucronate. **Inflorescences** 3–7 cm. **Pedicels** 1.5–2.5 mm. **Flowers** with calyx 5–6.5 mm, tube 3–3.5 mm, lobes 2.5–3 mm; corolla pale greenish, 4–5 mm; stamens 4–5 mm; ovary densely pubescent along margins. **Fruits** dark brown or reddish brown, indehiscent, broadly linear, 15–30[–45] × 1.5–3[–3.8] cm, with pulpy mesocarp, stipe 1–2[–3] cm, apex with or without a persistent style. **Seeds** 9–13 × 4–8 mm.

Flowering May–Jun. Floodplains, arroyos, stream banks, meadows, pastures, roadsides; 370–1790 m; Sonoran Desertscrub, Chihuahuan Desertscrub, Semidesert Grassland, Madrean Evergreen Woodland; Apache, Cochise, Coconino, Maricopa, Santa Cruz, Yavapai cos.; native to the eastern U.S. from Texas to South Dakota, Pennsylvania and Mississippi, and widely but sporadically naturalized in the U.S. outside of its native distribution; naturalized in Mexico (Sonora); Canada.

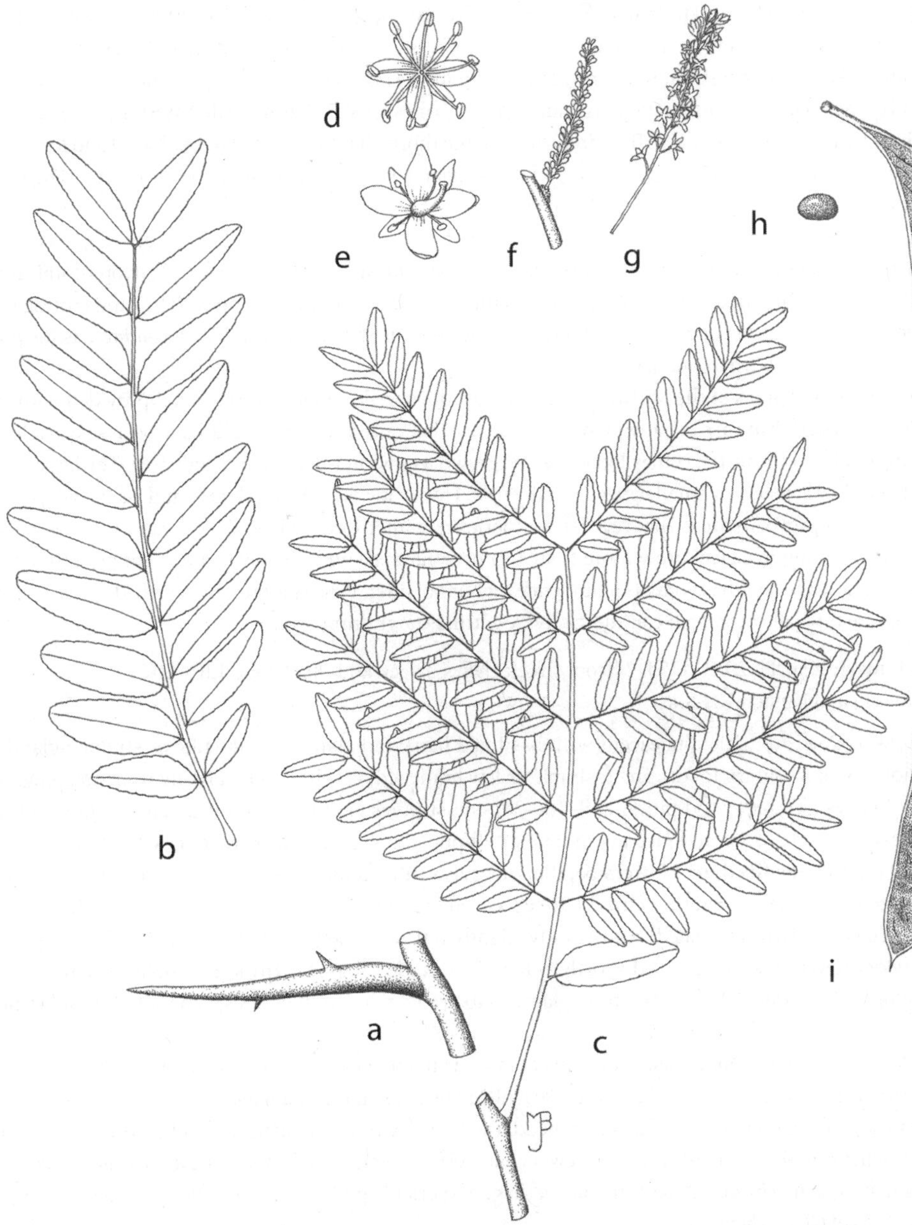

***Gleditsia triacanthos*: a. thorn; b. pinna; c. leaf; d, e. flowers; f., g. inflorescences; h. seed; i. fruit. Illustrator: Matthew B. Johnson.**

Gleditsia triacanthos is occasionally cultivated in Arizona as a landscape and shade tree. It persists around abandoned settlements and is adventive in the state at a few locations with sufficient moisture. The species is native to the Ohio and Mississippi River drainages where it favors moist soils. A number of varieties and forms have been described. There seems to be little taxonomic significance to these.

Gleditsia triacanthos is widely planted in the Midwest for erosion control and for windbreaks (Graham 1941, Allen and Allen 1981). Livestock and a variety of wildlife species eat the fruits (Graham 1941, Allen and Allen 1981, Diggs et al. 1999, Lewis et al. 2005) as well as the foliage (NRC 1979). Productive trees can produce 450 kg of fruits per year (NRC 1979). The fruits contain 12–14% sugar (Allen and Allen 1981) though others report sugar content of 30% or higher from fruits of selected trees. Crosswhite (1981) states that pioneers in Texas ate the sweet, young fruits of *G. triacanthos*, while avoiding older fruits which were bitter. The fruits have been used in flavoring beer (Lewis et al. 2005). Brown and Brown (1972) report that the fruits have been used to make an alcoholic beverage and that the seeds have been used as a coffee substitute. Graham (1941) states that the flowers are a source of honey. The wood is hard, strong and heavy, and is used for fence posts and occasionally for construction and railroad ties (Vines 1960, Allen and Allen 1981, Lewis et al. 2005). It is also an excellent fuelwood. Lewis et al. (2005) indicate that the thorns have been used to card wool and for hunting frogs. The thorns are also reported to have been used by native people for tipping spears used for fishing (Diggs et al. 1999). *Gleditsia triacanthos* has been employed medicinally by various native peoples. Uses have included treatments for indigestion, measles, whooping cough, colds, fever, and smallpox (Duke and Wain 1981). The trees are adapted to a wide range of soil types and tolerate temperature extremes, making them ideal for agroforestry and silvopastoral uses in many regions (NRC 1979). This species can produce abundant root suckers and will coppice following cutting (NRC 1979). Introductions into regions beyond its natural range should be done with caution as *G. triacanthos* has the potential to become invasive in areas with a suitable climate.

Though native to humid areas of the Midwest, honey locust does well in much of Arizona if provided with sufficient irrigation. Thornless cultivars are usually planted. There are also various cultivars that have been selected for leaf color, growth form, and lack of fruiting. Honey locust can be grown as a winter-deciduous shade tree. The large mature size of some forms limits their use in smaller residential yards. Honey locust is heat tolerant and seems adaptable to various soils. The species grows in areas that regularly experience winter lows to below -20°C and may tolerate low temperatures to -40°C. Young trees can grow more than 1 m per year and typically have an upright growth habit, so pruning is minimal. The leaves and fruits create a moderate amount of litter. Propagation is by scarified seeds. Cultivars with desirable traits are typically produced commercially by cuttings or grafting.

GLYCYRRHIZA Linnaeus. Sp. Pl. 2:741. 1753. * [from Greek *glykys*, sweet, and *rhiza*, root] LICORICE
Martin Wojciechowski and Kirsten N. Lake

Herbs, rhizomatous, coarse, aromatic; pubescent with basifixed hairs and glandular-punctate or stipitate-glandular, sometimes glutinous. **Stems** solitary to few, erect, simple or few branched, 0.3–1(–1.2) m, arising from creeping, often woody rhizomes. **Leaves** alternate; odd pinnate; 8–18(–24) cm; stipules cauline, free, ovate, elliptic to acuminate. **Leaflets** 9–19, narrowly ovate, acute to lance-acuminate or elliptic, 2.5–5.5 cm. **Inflorescences** 10–many flowered, congested; axillary racemes. **Pedicellate. Flowers** papilionoid; calyx tube campanulate, teeth subequal but the 2 upper (adaxial) teeth united about half of length behind banner; corolla dingy yellow, whitish, faint blue or dull purplish tinged or tipped, petals regularly graduated, banner only slightly reflexed; stamens 10, 9 united and 1 free, generally monomorphic; style glabrous, gently incurved shortly below the rounded stigma. **Fruits** essentially indehiscent, narrowly oblong to ellipsoid or ovate, 10–30 × 4–6 mm; glabrous to armed with stout bristles, generally glandular. **Seeds** 2–several. $x = 8$.

Species ca. 20 (1 native in the flora): distributed in cool to warm temperate regions of Asia, Australia, Europe, and the Americas. North American species can become an agricultural weed in moist habitats.

The European *Glycyrrhiza glabra* is used as a commercial source of licorice flavoring for food products and tobacco. It is also used in medicine and as a foaming agent (Lewis et al. 2005). Species of *Glycyrrhiza* can be mistaken for large *Astragalus* but are distinguished by glandular-punctate herbage and prickly pods. Allen and Allen (1981) report nodulation from some species of *Glycyrrhiza*.

SELECTED REFERENCE: Barneby 1989.

1. Glycyrrhiza lepidota Pursh. Fl. Amer. Sept. 2:480. 1814. * [from Greek, *lepidotos*, scaly, in reference to the young leaves] AMERICAN LICORICE, WILD LICORICE, NUTTALL'S LICORICE Native

Glycyrrhiza glutinosa Nuttall ex Torrey & Gray

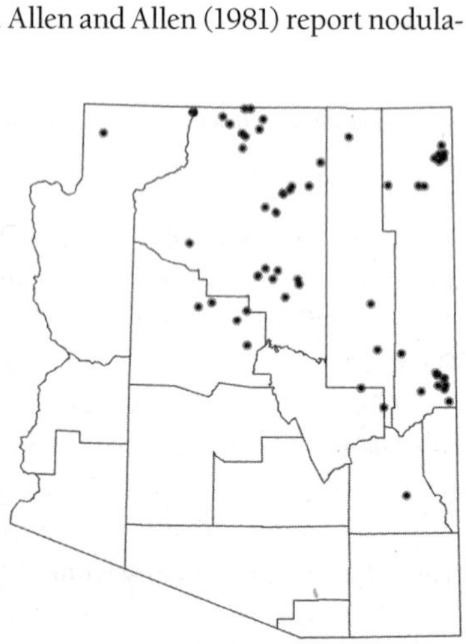

Perennial herbs, thinly pubescent with both fine basifixed hairs and foliage and inflorescences with immersed, lens-like, sessile and globose or stalked glands, often glutinous to touch, the density of hairs and glands variable but the leaflets always glandular-punctate below. **Stems** 0.3–1 m. **Leaves:** stipules ovate, acute or lance-acuminate, 3–10 mm, deciduous. **Leaflets** 9–19; (2–)2.5–4 cm, narrowly ovate, acute, lance-acuminate, or elliptic, mucronulate. **Inflorescences** ascending racemes, compact early on but then elongating;

Glycyrrhiza lepidota: **a.** habit (including rhizome); **b.** leaflet (upper/adaxial surface); **c.** leaflet (lower/abaxial surface); **d.** fruit; **e.** distal portion of inflorescence; **f.** flower; **g.** pistil; **h.** stamens; **i., j., k.** petals. Illustration: Joyce Riney Peters.

peduncles 2–7(–10) cm; floral bracts lanceolate, 3–8 mm. **Pedicels** 0.5–1 mm. **Flowers** yellowish or whitish, faintly purple-tinged or –tipped flowers; calyx 5–7.5 mm, tube (2–)2.5–3 mm, teeth 2–4.5 mm, banner erect, lanceolate, acute, 11–15 mm, wings and subacute keel progressively shorter, anthers dimorphic. **Fruits** ellipsoid or oblong-ellipsoid, 10–20 × 4–6 mm, stiffly chartaceous valves armed with stout, erect, straw-colored, brown or purplish, apically hooked bristles up to 2.5–3 mm. **Seeds** 2–4; purplish-brown, 2.5–3 mm.

Flowering late May–Aug(–Oct). Growing generally in open sites in moist soils, along rivers and streams (riparian communities), roadsides, agricultural areas, sometimes in dry sagebrush areas; (880–)1200–2300 m; (Sonoran

Desertscrub), Great Basin Desertscrub, Great Basin Conifer Woodland, Petran Montane Conifer Forest; Apache, Coconino, Graham, Greenlee, Mohave, Navajo, Yavapai cos.; Alaska, Washington south to California, east to Wisconsin, Virginia, and Texas; Mexico (Baja California); Canada (British Columbia to Ontario).

Glycyrrhiza lepidota, easily distinguished by its glandular-punctate leaves and bur-like fruits, is widespread in western U.S. and adjacent areas of Canada and Mexico; its range east of Mississippi River is sporadic.

The roots and rhizomes of *Glycyrrhiza lepidota* are used as food and medicine by Native Americans. The species was used to treat fever, sore throat, earache, and toothache (Stubbendieck and Conard 1989, Wiener 1990). It has also been used as a remedy for diarrhea, stomach ache, coughs, and chest pains (Kindscher 1987).

American licorice can be used as a soil binder (Graham 1941). A variety of wildlife species have been reported to feed on the leaves, stems, seeds, and roots (Graham 1941, Stubbendieck and Conard 1989). The bur-like fruits cling to the fur of animals and help to distribute the seeds (Stubbendieck and Conard 1989). The plants may be propagated by rhizome cuttings and by stem cuttings, which will root if placed under mist (Stubbendieck and Conard 1989). Nodulation is documented for *Glycyrrhiza lepidota* (Allen and Allen 1981).

HAEMATOXYLUM Linnaeus. Sp. Pl. 1:384. 1753. * [*Haematoxylon* Linnaeus (1764) is an orthographic variation. Greek *haima*, blood, and *xylon*, wood, alluding to the red heartwood]
Matthew B. Johnson

Shrubs or trees. Leaves alternate, sometimes fascicled; even pinnate (rarely bipinnate); stipules small, bristle-like, soon deciduous. **Leaflets** opposite, with closely-spaced parallel veins, margin entire. **Inflorescences** axillary racemes, bracts minute. **Flowers** caesalpinoid; calyx 5-lobed, overlapping, lobes deciduous soon after the flowers open, tube persistent; corolla slightly bilaterally symmetric with 5 petals, uppermost (banner) petal innermost, blades clawed; stamens 10, filaments free; ovary with a short stipe, 1 chambered, style and stigma simple. **Fruits** dehiscing along a lengthwise suture in mid area of each valve, flattened, thin-walled, oval or elliptic. **Seeds** flattened, oblong, two-lobed. $x = 12$.

Species 5 (1 cultivated in flora): 4 (1 undescribed) in the American tropics and subtropics, 1 in Africa.

Haematoxylum species were used by indigenous peoples long before European conquest of the New World. The wood has a density of 960–1040 kg/m^3 (Allen and Allen 1981). Logwood or Campeche wood, *H. campechianum*, is the source of haematoxylin, a high quality black or bluish dye that is used as a stain in cytology and histology due to its lasting properties and staining characteristics. Haematoxylin is also used in inks and to dye cloth. More than three dozen formulae of haematoxylin are known (Allen and Allen 1981). Logwood and brazilwood are widely reported to have been major exports from regions where they grow and have continued to be important dye sources since the development of synthetic dyes. Logwood was sought by British privateers who seized Spanish ships for their cargoes of the wood (Everett 1968). *Haematoxylum campechianum* is used in making furniture (Usher 1974). The wood has a tannin content of 10% (Standley 1922). The heartwood of *H. campechianum* is astringent and is used to treat dysentery (Standley 1922, Usher 1974). *Haematoxylum* honey is reported to be of high quality (Lewis et al. 2005).

1. Haematoxylum brasiletto H. Karsten. Fl. Columb. 2:27–28, pl. 114. 1862. * [brasil, referring to several other legume dyewoods, or brazilwoods, and the basis for the name of the country of Brazil] BRAZILWOOD; *BRASIL*, *PALO BRASIL* Cultivated

Haematoxylon boreale S. Watson

Shrubs or trees to 10 (12) m with multiple trunks. **Stems:** trunks and limbs developing conspicuous fluted ridges and furrows; bark gray or red-brown, becoming moderately rough-textured; heartwood blood red, sapwood pale yellow; twigs slightly zig-zag, with a stout, straight axillary spine 4–20 mm at most nodes or some stems unarmed. **Herbage** with short, appressed, yellowish worm-like hairs, becoming hairless with age. **Leaves** 2.5–8.3 cm; petioles 6–19 mm; rachis 6–50 mm; petiolules to 1 mm or leaflets sessile. **Leaflets** bright green above (adaxial), pale green beneath (abaxial), 2–4 pairs per leaf, blades usually broadest above middle, cuneate or obovate, or rarely ovate, 7–28 × 5–21 mm, base cuneate or rounded, apex truncate, emarginate, obcordate or rarely rounded. **Inflorescences** 3–11 cm, sparsely to densely covered with short erect hairs, 10–35 flowered. **Pedicels** 1.1–2.1 cm. **Flowers:** calyx tube green or red, 13 × 4–5 mm, becoming laterally compressed, unequal, boat-shaped, lobes yellow, 3–4 mm; corolla 8–20 mm across, bright yellow and red-lined at base of banner, (5–)6–10 mm, anthers with filaments 7–8 mm, pilose at base; ovary sparsely pilose, stipe 0.3–0.5 mm. **Fruits** tan or straw-colored, tardily dehiscent, 1.8–5 × 1.1–1.4 cm, base cuneate or rounded, apex rounded or mucronate. **Seeds** 1–4; tan, 7–13 × 3–5 mm.

Flowering nearly year-round with sufficient moisture and warm temperatures. w Mexico (widespread); Central America; South America (Colombia, Venezuela).

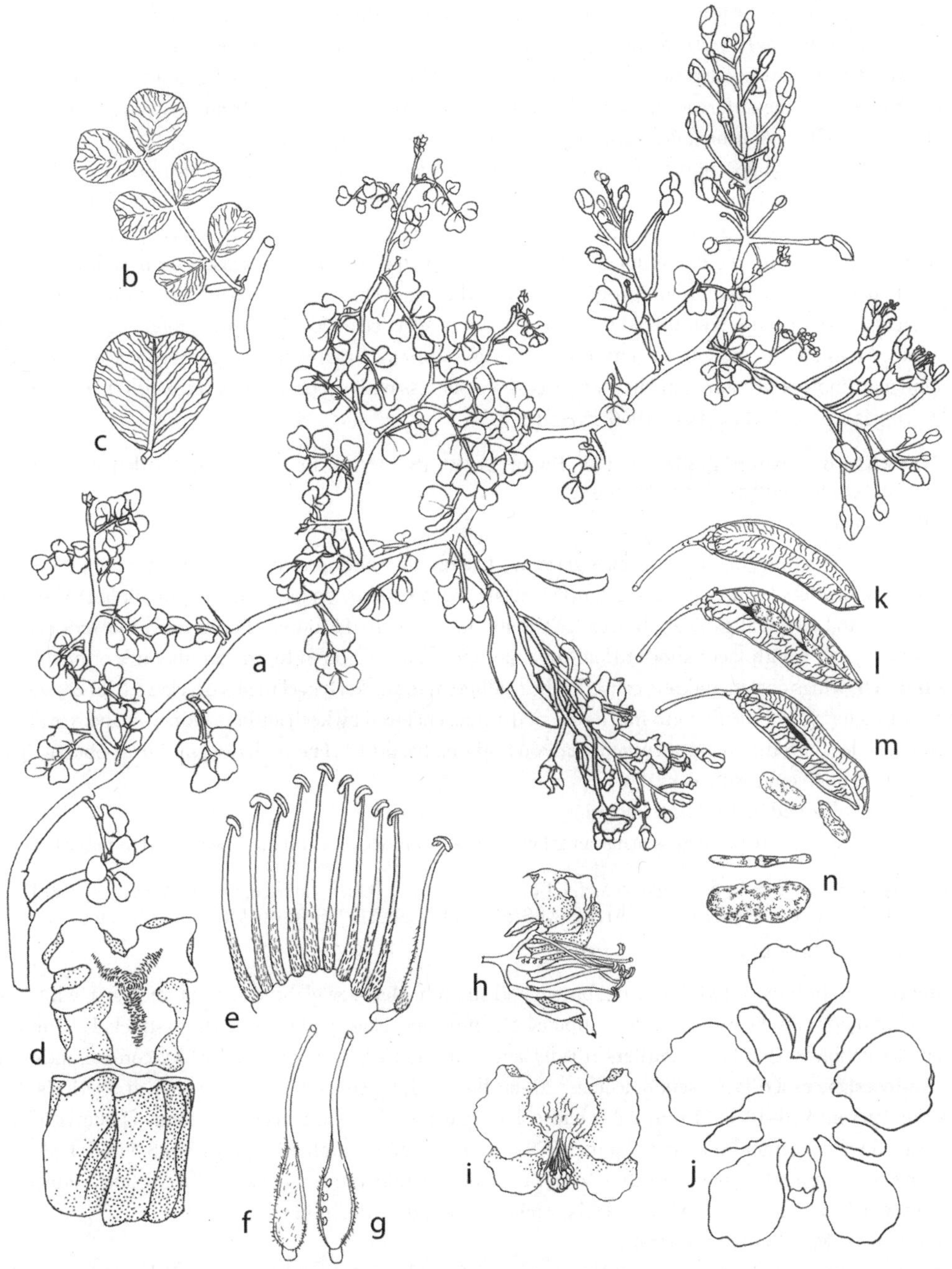

***Haematoxylum brasiletto*: a.** habit; **b.** leaf; **c.** leaflet; **d.** stem cross-sections; **e.** stamens; f, **g.** pistil; **h.** flower cross-section; **i.** flower; **j.** petals and sepals; **k.** fruit; **l., m.** fruits showing dehiscence; **n.** seeds. Illustrator: Margaret Pope.

Haematoxylum brasiletto is common in various habitats in thornscrub and tropical deciduous forest in western Mexico, reaching the southern margin of the Sonoran Desert in central Sonora and southern Baja California Sur. *Haematoxylum brasiletto* does not nodulate (Hughes and Styles 1984) and nodulation has not been observed on *H. campechianum* (Allen and Allen 1981).

The heartwood of *Haematoxylum brasiletto* has a long history of use in dyeing. The bright red dye, brazilein, is used as a stain in cytology and histology. The wood is used for a variety of purposes including fuelwood, carpentry, and light construction. "The wood is excellent fuel and burns with a greenish flame. It is also used in the construction of houses, corrals, etc. A light-red dye or ink and also rosary crosses are made from the deep-red heartwood" (Gentry 1942). Martínez (1959) reports that *H. brasiletto* has been used to tint wine and to make toothpicks. The species has been used in folk-medicine for a variety of purposes. According to Standley (1922) *H. brasiletto* "is refrigerant, febrifuge, astringent, and corroborative." Sánchez-Marroquín et al. (1958) found that sawdust extracts from the wood had antibacterial properties and were effective against *Brucella*, *Shigella*, and *Staphylococcus* species.

Brazilwood is occasionally cultivated as a landscape plant at lower elevations in Arizona. Brazilwood is a beautiful shrub or small tree with unique sculptural trunks, attractive heart-shaped leaflets, and a free-flowering habit. It is ideal for patios and other small spaces, and it is effective when planted as a security hedge or screen. Established plants are drought tolerant but require regular supplemental irrigation during the growing season for best development and appearance. In habitat they are eventually drought deciduous but may remain evergreen in cultivation, particularly with irrigation. Stem damage occurs at temperatures below ca. -4°C (Felger et al. 2001). Recovery following freeze damage is usually rapid the following season. Leaf and pod litter is minimal. Brazilwood is easily propagated from seeds. Scarification may improve germination but is not necessary. Success with propagation of *Haematoxylum* using cuttings and air layering is reported (Chittenden 1956, Everett 1981).

HARDENBERGIA Bentham in Endlicher et al., Enum. Pl. 40. 1837. * [for Franziska von Hardenberg, 19th century Austrian countess and patron of botany]

Matthew B. Johnson

Vines, prostrate or climbing, or shrubs. **Stems** glabrous. **Leaves** alternate, odd-pinnate; petiolate; stipules deciduous; stipels persistent. **Leaflets** 1 or 3 (or 5), dark green, lanceolate or ovate, veins conspicuous, margins entire. **Inflorescences** axillary racemes and terminal panicles; bracteoles absent; bracts small, deciduous. **Pedicellate. Flowers** papilionoid, small but showy; calyx with teeth shorter than tube, 2 upper (adaxial) teeth fused; corolla pink or purple, banner rounded and cleft, wings obovate-falcate, clawed, lobed, adherent to the keel, keel incurved, clawed, lobed; stamens 10, 9 filaments fused for part of their length into a tube and 1 free, enclosed by keel, anthers kidney-shaped; ovary sessile, 1-chambered, style short and thick, incurved, glabrous, stigma head-like. **Fruits** dehiscent, linear-oblong, flattened, rigid. **Seeds** ovoid or oblong, arillate. $x = 11$.

Species 3 (1 cultivated in the flora): Australia.

Hardenbergia species are planted as ornamental vines and groundcovers and have been used to make beverages.

1. Hardenbergia violacea (Schneevoogt) Stearn. J. Bot. 78:70. 1940. * [Latin *violacea*, violet-colored] PURPLE CORAL PEA, LILAC VINE, FALSE LILAC, FALSE SARSAPARILLA, NATIVE SARSAPARILLA Cultivated

Glycine violacea Schneevoogt, Icon. Pl. Rar. 29. 1781–1793

Vines. Stems 2+ m, trailing or twining, wiry, brown and smooth. **Herbage** glabrous except for sparse white hairs on inflorescence and pedicels. **Leaves** 1-foliolate; stipules 2–3 mm, lanceolate; petioles 6–28 mm; stipels 1–1.7 mm, lanceolate or oblong; petiolules 1–2.3 mm. **Leaflets** broadly lanceolate or ovate, 3–10 × 0.5–6 cm, base cordate, apex rounded, leathery. **Inflorescences** axillary pseudoracemes, 20–60-flowered; bracts 2–2.5 mm, awl-shaped. **Pedicels** 2–5 mm. **Flowers** 1 or 2 per node, 6–10 × 5–7 mm; calyx blackish purple, tube 3–3.5 mm, teeth 1–1.2 mm, ciliate; corolla purple with a greenish yellow center on banner, banner 8–9 mm, recurved through 90°, wings 7.5–8.5 mm, keel 5–6 mm; androecium 4.5–5.5 mm, 9 filaments fused for 2/3 of their length. **Fruits** brown, 4–5 × 0.7–0.8 cm, apex rounded. **Seeds** 6–8, dark brown, 3–5.5 mm, with a white aril (elaiosome) surrounding the hilum.

Flowering Jan–Mar. Native to Australia.

Hardenbergia violacea is native to coastal regions and some inland areas of New South Wales, Queensland, South Australia, Victoria, and Tasmania. It grows in a variety of habitats from coastal scrub to open woodland and forest in mountains (ANPSA 2015) at elevations from near sea level to 1250 m. It is less common in inland areas. The elaiosome attracts ants which disperse the seeds (Hocking and Kortt 1987). The roots and also the boiled leaves have been used to make a sweet beverage (Cunningham et al. 1992). Cunningham et al. (1992) mention that *H. violacea* does well in areas of Australia with dense populations of feral rabbits, indicating it is not heavily utilized by these animals. A colic-like condition has been reported in horses that have eaten *H. violacea* (Cunningham et al. 1992). Allen and Allen (1981) and Davidson and Davidson (1993) indicate that *H. violacea* nodulates.

Purple coral pea is grown in Arizona and elsewhere as a twining vine or trailing groundcover. Shrubby forms are known (ANPSA 2015). It makes an effective screen when grown on a trellis. The large, coarse leaves contrast with the attractive purple flowers. Cultivars with pink and white flowers are available (ANPSA 2015). Purple coral pea produces

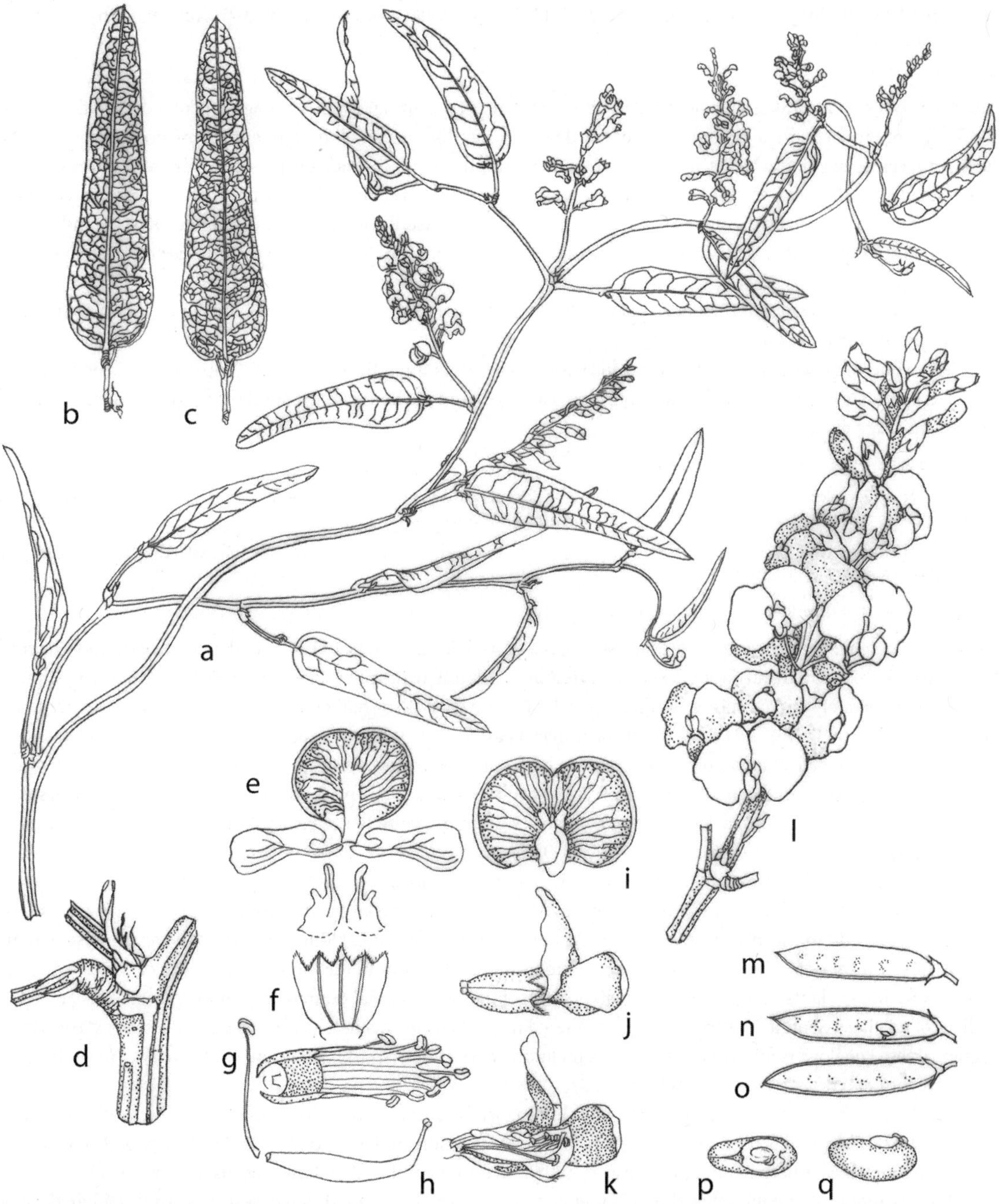

***Hardenbergia violacea*: a. habit; b., c. leaves; d. stem node; e. petals; f. calyx; g. stamens; h. pistil; i., j., k. flowers; l. inflorescence; m., n., o. fruits; p. seeds. Illustrator: Margaret Pope.**

showy flowers during the winter and spring months. It prefers full sun or partial shade. Plants that are excessively shaded will grow but flower poorly. Purple coral pea is hardy to ca. -4°C. Freeze damaged plants usually recover the following growing season. Purple coral pea should be grown in well-drained soil. Propagation is by scarified seeds and by cuttings of mature stems from the current growing season (ANPSA 2015).

HAVARDIA Small. Bull. New York Bot. Gard. 2:91. 1901. * [for Valéry Havard, 1846–1927, American physician and botanist]

Matthew B. Johnson

Shrubs or trees. Leaves alternate; bipinnate; stipules mostly persistent rigid spines, straight or curved, paired at nodes; 1–13 pairs of pinnae per leaf; glands sessile. **Leaflets** opposite [or alternate], margin entire. **Inflorescences** globose heads or racemes of globose heads, axillary, terminal, or on spur branches. **Pedicels** present or absent. **Flowers** mimosoid, radially symmetric, 5-merous; calyx shallowly to deeply bell-shaped; corolla bell-shaped, lobes erect or recurved; stamens numerous, exerted, filaments fused proximally and free distally, white; ovary 1-chambered, style and stigma simple. **Fruits** tardily dehiscent along both sutures, flattened, broadly linear, stiffly papery. **Seeds** disk-shaped to oblong-elliptic.

Species 5 (2 cultivated in the flora): seasonally dry tropical and subtropical regions of Mexico, Central America, and extreme southern Texas.

The wood of *Havardia* species is used for various purposes including fuel, posts, and utensils. The foliage is browsed by livestock, and the flowers are considered a good source of honey. Some species are utilized in revegetation and as landscape plants.

SELECTED REFERENCE: Barneby and Grimes 1996.

1. Trees with a well-developed trunk and high crown; leaflets 5–11 pairs per pinna; flowers short-pediceled ____ **1. Havardia mexicana**
1′ Shrubs or trees with multiple trunks; leaflets (10–)12–21 pairs per pinna; flowers sessile ____ **2. Havardia pallens**

1. Havardia mexicana (Rose) Britton & Rose. N. Amer. Fl. 23:41. 1928. * [of Mexico] MEXICAN EBONY; *CHINO, PALO CHINO* Cultivated

Pithecellobium mexicanum Rose, Contr. U.S. Natl. Herb. 1:100. 1891

Trees with a high, open crown to 12 (15) m. **Stems:** trunk well developed, to 1 (1.5) m in diameter, bark smooth, gray, becoming rough, fissured and dark gray; armed. **Herbage** with fine white pubescence. **Leaves** 2.5–6(–8.5) cm; stipules straight to moderately curved spines, 1–4 mm; petioles 0.4–1.6 cm with a shallow-cupped gland below first pair of pinnae, 0.3–0.5 mm; rachis 0.4–3.2 cm; 2–6 pairs of pinnae. **Leaflets** 5–11 pairs per pinna, narrowly oblong or obovate, 5–7 × 1.5–3 mm, base lobed, apex obtuse, gray green on both surfaces. **Inflorescences** with heads 1.5 cm, greenish white, axillary and solitary or in terminal racemes of 5–30 heads with 1–3 heads per node. **Pedicels** 1–3 mm. **Flowers** pale green; calyx 1.3–1.8 mm; corolla 3.2–4.5 mm; filaments 9–11 mm. **Fruits** tan or brown, 5.5–14 × 1.3–2.5 cm, stipe 2–5 mm, apex obtuse. **Seeds** 3–9; dark brown, 6–10 × 4–8 mm.

Flowering Mar–May. Mexico (Baja California Sur, Sonora, Sinaloa).

Havardia mexicana is found in alluvial soils on floodplains and along larger arroyos near the eastern margin of the Sonoran Desert and in thornscrub and tropical deciduous forest from 60–1000 m elevation. It is rare or absent along small arroyos or on slopes away from the valley bottoms and is absent from the more arid portions of western Sonora. The northernmost plants occur several kilometers south of the International Boundary in the Río Magdalena and Altar Valleys of northern Sonora. The trees grow as scattered individuals or in small groves, mixed with other xeroriparian legume trees. The flowers attract various insects including bees and wasps. Fruits ripen in 4–6 weeks. *Havardia mexicana* is reported to nodulate (Sprent 2009).

The bark of *Havardia mexicana* is utilized for tanning leather, and the trees have been heavily exploited in Sonora (Gentry 1942). The wood is used to make furniture, for construction, and is used in stills. The heartwood is red and is used for carving in southern Sonora (Martin et al. 1998). Livestock eat the fallen flower heads (Gentry 1942).

Mexican ebony is suitable for parks and other large spaces. The high canopy and fine leaves produce filtered shade that breaks the heat of the desert sun yet allows sufficient light for the cultivation of many different plants beneath the tree. This species has an upright growth form when young, minimizing the need for pruning to control its shape. The flowers are fragrant and are produced in profusion. Young trees grow rapidly with adequate irrigation. Deep soil is necessary for optimum performance and the trees do not thrive in rocky upland soils. Mexican ebony needs supplemental irrigation in desert regions. The trees are deciduous with freezing and are hardy to at least -10°C. The spines are a consideration when selecting a planting site. These are absorbed as the trunk increases in diameter, and with the high-branched canopy, are not a problem after a few years. Mexican ebony produces minimal litter. The plants are easily propagated from scarified seeds.

1. *Acmispon brachycarpus* / JC; **2.** *Acmispon greenei* / JC; **3.** *Acmispon greenei* flowers / JC; **4.** *Acmispon wrightii* / CK; **5.** *Alhagi maurorum* / CK; **6.** *Amorpha fruticosa* / JC; **7.** *Astragalus allochrous* / JC; **8.** *Astragalus allochrous* fruits / JC

Jillian Cowles = JC; Chris Kline = CK; Martin Wojciechowski = MW; Chris Bondante = CB; Joseph Charboneau = JCh

1. *Astragalus arizonicus* / JC; **2.** *Astragalus arizonicus* flowers / JC; **3.** *Astragalus amphioxys* / MW; **4.** *Astragalus ceramicus* / MW; **5.** *Astragalus hallii* var. *fallax* / JCh; **6.** *Astragalus lentiginosus* var. *diphysus* / CK; **7.** *Astragalus newberryi* var. *newberryi* / MW

1. *Astragalus nothoxys* / MW; **2.** *Astragalus nothoxys* flowers / JC; **3.** *Astragalus nutriosensis* / MW; **4.** *Astragalus thurberi* / CK; **5.** *Calliandra eriophylla* / JC; **6.** *Calliandra eriophylla* flowers / JC; **7.** *Chamaecrista nictitans* var. *mensalis* / JC; **8.** *Chamaecrista nictitans* var. *mensalis* flower / JC; **9.** *Clitoria mariana* var. *mariana* / CK

1. *Cologania angustifolia* / JC; **2.** *Coursetia caribaea* var. *sericea* / JC; **3.** *Crotalaria pumila* / JC; **4.** *Cologania angustifolia* flower / JC; **5.** *Coursetia caribaea* var. *sericea* flower / JC; **6.** *Crotalaria pumila* flower / JC; **7.** *Crotalaria sagittalis* flower / JC; **8.** *Coursetia glandulosa* / CK; **9.** *Crotalaria sagittalis* / JC; **10.** *Dalea albiflora* / JC; **11.** *Dalea albiflora* flowers / JC

1. *Dalea formosa* / JC; **2.** *Dalea jamesii* / JC; **3.** *Dalea jamesii* flowers / JC; **4.** *Dalea lachnostachys* flowers / JC; **5.** *Dalea neomexicana* flowers / JC; **6.** *Dalea pringlei* flowers / JC; **7.** *Dalea lachnostachys* / JC; **8.** *Dalea pringlei* / JC; **9.** *Dalea neomexicana* / JC

1. *Dalea pulchra* / CK; **2.** *Dalea pulchra* flowers / CK; **3.** *Desmanthus cooleyi* / CK; **4.** *Desmodium angustifolium* var. *gramineum* / JC; **5.** *Desmodium batocaulon* / JC; **6.** *Desmodium batocaulon* flowers / JC; **7.** *Desmodium grahamii* fruits / JC; **8.** *Desmodium grahamii* flower / JC; **9.** *Desmodium angustifolium* var. *gramineum* fruit / JC; **10.** *Desmodium angustifolium* var. *gramineum* flower / JC; **11.** *Desmodium grahamii* / JC

1. *Errazurizia rotundata* / MW; **2.** *Erythrina flabelliformis* fruit / CK; **3.** *Indigofera sphaerocarpa* / JC; **4.** *Indigofera sphaerocarpa* flowers / JC; **5.** *Erythrina flabelliformis* flowers / CK; **6.** *Lathyrus graminifolius* fruits / CK; **7.** *Lathyrus graminifolius* / CK; **8.** *Lathyrus lanszwertii* / CK

1. *Lotus corniculatus* / CK; **2.** *Lupinus argenteus* / CK; **3.** *Lupinus arizonicus* / CK; **4.** *Lupinus brevicaulis* / JC; **5.** *Lupinus brevicaulis* flowers / JC; **6.** *Lupinus concinnus* flowers / JC; **7.** *Lupinus pusillus* / CK; **8.** *Lupinus concinnus* / JC

1. *Lupinus sparsiflorus* / CK; **2.** *Lupinus kingii* / MW; **3.** *Macroptilium gibbosifolium* / JC; **4.** *Marina parryi* / JC; **5.** *Medicago lupulina* / JC; **6.** *Macroptilium gibbosifolium* flowers / JC; **7.** *Medicago lupulina* flowers / JC; **8.** *Medicago sativa* / CK; **9.** *Melilotus albus* / JC; **10.** *Melilotus albus* flowers / JC

1. *Melilotus indicus* / JC; **2.** *Mimosa biuncifera* flowers / JC; **3.** *Mimosa biuncifera* / JC; **4.** *Mimosa dysocarpa* flowers / CK; **5.** *Mimosa grahamii* var. *grahamii* flowers / JC; **6.** *Olneya tesota* flowers / CK; **7.** *Oxytropis lambertii* flowers / CK

1. 

2. 

3. 

4. 

5. 

1. *Parkinsonia microphylla* / JC; **2.** *Parkinsonia microphylla* flowers / JC; **3.** *Parkinsonia florida* flowers / JC; **4.** *Phaseolus acutifolius* var. *tenuifolius* flower / JC; **5.** *Phaseolus acutifolius* var. *tenuifolius* / JC

1. *Phaseolus maculatus* / JC; **2.** *Phaseolus maculatus* flower / JC; **3.** *Phaseolus maculatus* fruit / JC; **4.** *Phaseolus ritensis* / JC; **5.** *Phaseolus ritensis* flowers / JC; **6.** *Phaseolus scabrellus* / CK

1. *Pomaria jamesii* / CK; **2.** *Prosopis velutina* / CK; **3.** *Robinia neomexicana* / JC; **4.** *Robinia neomexicana* flowers / MW; **5.** *Rhynchosia texana* / JC; **6.** *Rhynchosia texana* flower / JC

1. *Senegalia greggii* / JC; **2.** *Senegalia greggii* fruits / JC; **3.** *Senna covesii* / JC; **4.** *Senna covesii* flowers / CK; **5.** *Senna hirsuta* var. *glaberrima* flowers / JC; **6.** *Senna hirsuta* var. *glaberrima* / JC

1. *Senna wislizeni* / JC; **2.** *Trifolium sonorense* / JC; **3.** *Trifolium repens* / CK; **4.** *Trifolium repens* flowers / CK; **5.** *Trifolium sonorense* flowers / JC; 6. *Vachellia constricta* with flowers / JC; **7.** *Vachellia constricta* with fruits / JC

1. *Vicia americana* / CK; **2.** *Vicia pulchella* / CK; **3.** *Vicia ludoviciana* var. *ludoviciana* / CK; **4.** *Vicia ludoviciana* var. *ludoviciana* flower / JC; **5.** *Vicia pulchella* flowers / JC; **6.** *Zapoteca formosa* subsp. *schottii* / CB

2. Havardia pallens (Bentham) Britton & Rose. N. Amer. Fl. 23:42. 1928. * [Latin *pallens*, pale] *HUAJILLO, TENAZA* Cultivated

Calliandra pallens Bentham, London J. Bot. 5:102. 1846; *Pithecellobium pallens* (Bentham) Standley

Shrubs or trees 5–8(–12) m. **Stems:** multiple trunks; bark smooth, gray; armed. **Herbage** with fine white pubescence, becoming glabrate. **Leaves** 5–12 cm; stipules straight spines, 1–14(–16) mm, the larger spines laterally flattened; petioles 1.2–3.7 cm with a shallow-cupped gland at or below midpoint of petiole and a second gland near the first pair of pinnae, 0.5–1.6 mm; rachis 1.7–4.8 cm; 2–6 pairs of pinnae. **Leaflets** (10–)12–21 pairs per pinna, linear or narrowly oblong, 5–10 × (1–)2–3(–4.5) mm, base lobed, apex mucronate, upper (adaxial) surface dark green, lower (abaxial) surface paler. **Inflorescences** with heads 1.5–2 cm, greenish white, axillary and solitary or 2 or 3 per node, or in terminal racemes of 7–40 heads with 1–3 heads per node. **Pedicels** absent. **Flowers** pale green; calyx 1–1.7 mm; corolla 3.5–6 mm; filaments 9–14 mm. **Fruits** dull brown, 8–12.5 × 1.4–2 cm, stipe 6–16 mm, apex acuminate with the prominent style 1–4 mm. **Seeds** 6–13; dark brown, 5–7.5 × 3.5–6 mm.

Flowering Jun–Sep. Extreme s Texas; Mexico (Coahuila, Durango, Hidalgo, Nuevo León, Querétaro, San Luis Potosi, Tamaulipas, Vera Cruz, disjunct in Chiapas and Oaxaca).

Havardia pallens is found in alluvial soils along watercourses, as well as in canyons and on slopes in thornscrub and tropical deciduous forest, and at a few locations in mountains in the Chihuahuan Desert, from near sea level–1100 m elevation. The plants are a major component of some thornscrub communities in Tamaulipas and Nuevo León. At the northern limit of its range in southern Texas the plants are localized at low elevations. Nodulation is reported for *H. pallens* (Sprent 2009).

Havardia pallens is considered good bee forage for honey production, and the leaves are occasionally browsed by goats and sheep (Vines 1960). The wood is reddish brown, hard and heavy (Vines 1960). It has been used for fence posts and to make tool handles and other items.

Tenaza is occasionally grown in Arizona landscapes. It makes an attractive patio tree and can be planted to form an informal screen or a barrier. It has a multiple-trunked, upright growth form and is suitable for narrow spaces. Masses of fragrant flowers are produced at intervals in the summer in response to adequate rainfall or irrigation. The flowers attract butterflies which feed on the nectar. Bees also frequent the flowers. *Tenaza* is typically evergreen but the leaves can be killed by significant freezing. The stems are hardy to -7°C. Recovery following freezing is rapid. The plants produce minimal litter. *Tenaza* has persistent, painfully sharp spines that are a consideration when selecting planting locations. Propagation is by scarified seeds which germinate readily in warm weather. The largest *Havardia pallens* in the U.S. is a cultivated tree on The University of Arizona campus, in Tucson. This tree is 12.5 m tall (The Arizona Register of Big Trees 2005).

HEDYSARUM Linnaeus. Sp. Pl. 2:745. 1753. * [from Greek, *hedys*, sweet and *saron*, broom] SWEETVETCH
Matthew B. Johnson and Kirsten N. Lake

Herbs, annual and perennial, or rarely shrubs. **Stems** diverse. **Herbage** pubescent with simple hairs. **Leaves** alternate, odd-pinnate or 1-foliolate; stipules becoming dry and membranous, free or fused and forming a sheath, persistent; petiolate; stipels absent. **Leaflets** opposite, sessile, margins entire. **Inflorescences** racemes, axillary; bracts bristle-like to lanceolate; bracteoles 2, awl-like to bristle-like. **Pedicellate. Flowers** papilionoid, showy; calyx bell-shaped with 5 subequal, awl-shaped lobes; corolla pink, purple, red, white, or yellow; banner ovate or obcordate, reflexed; wings oblong; keel oblong or sometimes curved, with a blunt apex; stamens 10, filaments 9 fused and 1 free, anthers uniform; ovary stipitate, with 3–10 ovules; style thread-like; stigma minute. **Fruits** loments, stipitate, separating into indehiscent segments, compressed, with (1) 2–several segments, reticulate or with bristles, usually constricted between the segments. **Seeds** 1 per fruit segment, brown, compressed, kidney-shaped.

Species 140–180 (2 native in the flora): widespread across much of Asia and Europe to North Africa. Four species in North America.

The foliage of *Hedysarum* species is palatable to livestock (Allen and Allen 1981). Several species are used in traditional Chinese medicine to treat diabetes and hypertension and have anti-tumor and anti-aging applications (Dong et al. 2013). The Old World species *H. coronarium* contains compounds that may be suitable for use in skin care products (Burlando et al. 2017). Nodulation is widely reported for the genus (Allen and Allen 1981).

1. Stipules 3–5 mm; ovary puberulent; fruits with a stipe 1–4.5 mm; widespread in Apache, Coconino, Mohave and Navajo cos. ________ **1. Hedysarum boreale**

1′ Stipules 10–20 mm; ovary strigillose or glabrous; fruits with a stipe 2–10 mm; rare in northern Apache Co. ________ **2. Hedysarum occidentale**

Havardia pallens: **a.** habit; **b.** fruits; **c.** seeds; **d.** terminal pair of leaf pinnae. Illustrator: Ardys Lurtsema.

1. Hedysarum boreale Nuttall. Gen. N. Am. Pl. 2:110. 1818. * [northern] NORTHERN SWEETVETCH, UTAH SWEETVETCH, BOREAL SWEETVETCH Native

Herbs to 50 (60) cm, perennial. **Stems** erect, ascending or diffuse, simple or with several branches. **Herbage** pubescence with, appressed fine, short, white hairs; some plants appear green while others have a silvery appearance. **Leaves** odd-pinnate, 5–12 cm; stipules fused and forming a bidentate sheath, 3–5 mm; petioles 0.7–2.2(–4.2) cm; rachis 2–6 cm. **Leaflets** (5–)7–15, 0.6–2.6 × 0.2–1.2 cm, narrowly elliptic to broadly ovate. **Inflorescences** 7.5–26[–29] cm, (5–)10–45-flowered; bracts 1.5–7.5 mm membranous, lanceolate, persistent; bracteoles 1–4[–7.5] mm, awl-like, persistent. **Pedicels** 1–2 mm. **Flowers** with calyx 5–7[–8] mm, tube 2–3[–4] mm, lobes 3–4.5[–5.5] mm; corolla dark pink,

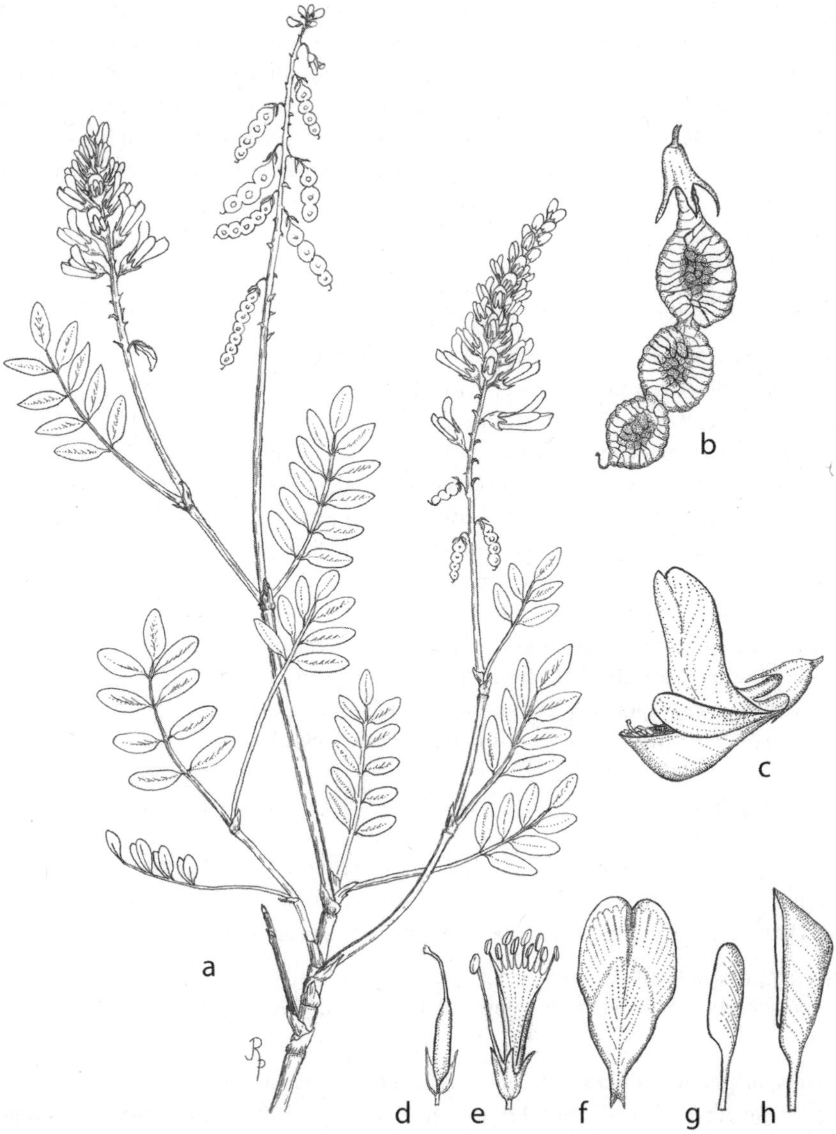

***Hedysarum boreale*: a. habit; b. fruit; c. flower; d. pistil; e. stamens; f., g., h. petals. Illustrator: Joyce Riney Peters.**

magenta, reddish purple (or white), banner 10–13[–16] mm, wings 9–11 mm, keel 12–15 mm; stamens 12–15 mm; ovary puberulent with 4–6(–8) ovules. **Fruits** with a stipe 1–4.5 mm, with 1–5 segments, segments broadly elliptic to circular, greenish straw-colored, 5–10 mm, puberulent or glabrate, constricted between seeds, conspicuously reticulate veined, the veins occasionally with tubercles or bristle-like hairs to 3 mm. **Seeds** ± 3.5 × 2.5 mm.

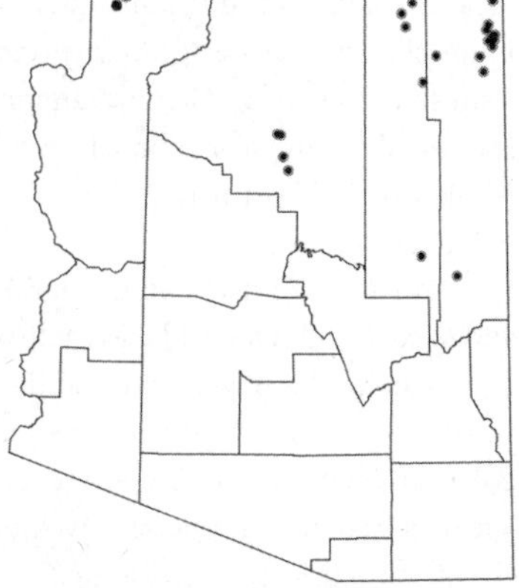

Flowering May–Aug. Slopes, canyons, arroyos, and roadsides on various soil and rock types; 1250–2300(–2900) m; Great Basin Desertscrub, Great Basin Conifer Woodland, (Petran Montane Conifer Forest); Apache, Coconino, Mohave, Navajo cos.; Alaska, Colorado, Idaho, Montana, Nevada, New Mexico, North Dakota, Oklahoma, Oregon, South Dakota, Texas, Utah, Montana, Wyoming; Canada; Russia.

Three weakly differentiated and intergrading subspecies or varieties of *Hedysarum boreale* are sometimes recognized. These are not treated here due to a lack of clear distinctions between them. The Arizona plants fall under subsp. *borelae* which is widespread in the southwest. This taxon varies greatly in the amount of leaf pubescence. This is primarily a species of pinyon-juniper and sagebrush communities. It is widespread but the populations are scattered.

Kearney and Peebles (1942) state that *Hedysarum boreale* provides excellent forage but is not sufficiently common in Arizona to be of significance. The roots are reported to have been used as a food source by some native peoples in North America (Rosales 2014). Rosales (2014) also states that *H. boreale* is suitable for soil stabilization. With its showy, fragrant flowers, *H. boreale* would make an excellent addition to a wildflower garden. Nodulation is reported for *H. boreale* (Allen and Allen 1981).

2. Hedysarum occidentale Greene. Pittonia 3(13):19. 1896. * [western] WESTERN SWEETVETCH Native

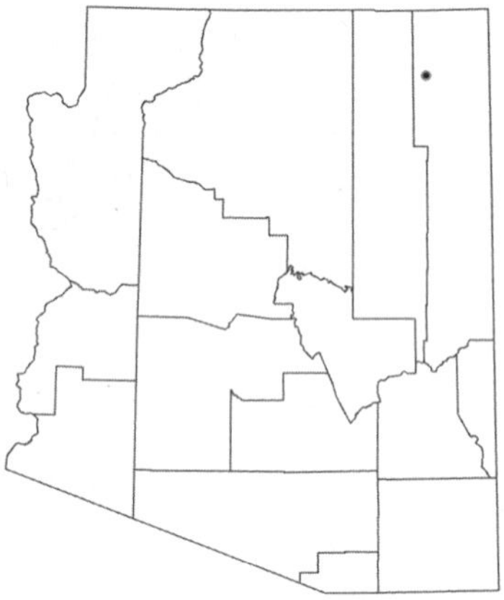

Herbs to 85 cm, perennial. **Stems** erect, ascending or diffuse. **Herbage** variously pubescent with short, white hairs, but with an overall greenish appearance. **Leaves** odd-pinnate, 8–14[–17] cm; stipules fused and forming a bidentate sheath, 10–20 mm; petioles 1.3–3.4 cm; rachis 4–11(–15.5) cm. **Leaflets** (7–)9–17(–21), 1.3–2.5(–4) × 0.4–1.4 cm, lanceolate to elliptic or ovate. **Inflorescences** 12–30 cm, 10–40-flowered; bracts 5–10 mm, lanceolate, persistent; bracteoles 2–4.5 mm, awl-like, persistent. **Pedicels** 1.5–3.5[–4.5] mm. **Flowers** with calyx 4.5–7 mm, the tube 2.5–3.5 mm, the lobes 1–3.5 mm; corolla dark pink to magenta purple, the banner 9–14 mm, the wings 10–15.5 mm, the keel 12–18 mm; stamens 12–14 mm; ovary strigillose or glabrous with 3–5(–6) ovules. **Fruits** with a stipe 2–10 mm, with (1)2–4 segments, the segments elliptic-obovate, straw-colored, 8–14 mm, pubescent, becoming glabrate at maturity, constricted between seeds, faintly reticulate veined. **Seeds** ± 3 × 2 mm.

Flowering Jun–Aug. Slopes, arroyos; [1700–]2000–2100[–3200] m; Great Basin Conifer Woodland; Apache Co.; Colorado, Idaho, Montana, Utah, Washington, Wyoming; Canada.

Hedysarum occidentale is known from only a few records in northern Apache County where it occurs in pinyon-juniper woodland. Elsewhere, it extends into a variety of plant communities including montane scrub, pine forest, and spruce-fir forest. The roots are widely reported to be edible and have a licorice-like flavor.

HOFFMANNSEGGIA Cavanilles. Icon. 4:63, pl. 392. 1798. * [for Johann Centurius von Hoffmannsegg, 1766–1849, a German botanist who explored France, Hungary, Portugal, and Spain] RUSH-PEA

Beryl B. Simpson

Shrubs, subshrubs, or perennial herbs, unarmed, arising from taproots and/or thick caudexes, sometimes with spreading, tuber-forming underground stems. **Leaves** odd-bipinnate; stipules present, persistent, ovate to lanceolate, entire. **Leaflets** oblong to irregularly ovate or obovate. **Inflorescences** of terminal or lateral racemes with 3–27 flowers. **Pedicels** present. **Flowers** caesalpinoid, slightly to strongly bilaterally symmetrical; sepals 5, slightly overlapping or valvate; petals 5, distinct, clawed, spatulate, yellow, yellow-orange, or pink, usually with the claw of the banner petal concave and bearing variable amounts of trichomes in the fold and sometimes with glandular trichomes on the lower (abaxial) surfaces and margins; stamens 10, fused at the base, subequal, about as long as the petals, filaments with clear, multicellular processes primarily on the basal portions; anthers attached dorsally, dehiscing inward; gynoecium consisting of an elongate, laterally compressed ovary on a short stipe with style as long as or longer than the ovary, stigma impressed, terminal or obliquely terminal. **Fruits** dehiscent or indehiscent multi-seeded legumes, compressed laterally. **Seeds** 1–11[–15] per pod; 2–7 × 1.5–6 mm, compressed, generally ovoid or oval in outline with the funiculus attached obliquely to the seed creating a "shoulder" at the apex, green to dark brown. $x = 12$.

Species 25 (3 native in the flora): occurring amphitropically in the arid and semiarid regions of the United States and Mexico and semi-arid and Andean areas of South America from Peru to sc Argentina.

A molecular phylogenetic study of the genus (Simpson et al. 2004a) has shown that *Hoffmannseggia* consists of a woody clade and an herbaceous clade with two members of the herbaceous clade occurring in the flora area. Biogeographic studies (Simpson et al. 2004b) indicate that there have been four independent colonizations of North America from South America, two in each clade. While species of the genus often resemble those of *Caesalpinia* (*sensu lato)* and *Pomaria*, phylogenetic studies have shown that the closest relative of *Hoffmannseggia* is the South American monotypic *Stenodrepanum* with this pair sister to *Zuccagnia* and *Balsamocarpon*.

SELECTED REFERENCES: Fisher 1892, Gagnon et al. 2016, Nores et al. 2012, Simpson 1999, Simpson and Ulibarri 2006, Simpson et al. 2004a, 2004b.

1. Plants woody, upright shrubs; fruits sickle-shaped to crescent-shaped, broad in the center tapering to acute ends, dehiscent ____ **3. Hoffmannseggia microphylla**
1′ Plants perennial herbs; fruits rectangular to arched or sickle-shaped sometimes almost forming a complete circle, margins more or less parallel, indehiscent.
   2. Petal claws, sepals, and pedicels variously vestitured but lacking glandular trichomes; fruits sickle-shaped, apex round, valves puberulent ____ **1. Hoffmannseggia drepanocarpa**
   2′ Petal claws, sepals and pedicels with conspicuous multicellular glandular trichomes; fruits rectangular to slightly arched, obtuse at apex, valves sparsely tomentose ____ **2. Hoffmannseggia glauca**

1. Hoffmannseggia drepanocarpa A. Gray. Pl. Wright. 1:58. 1852. * [Greek *drepane*, sickle, and *karpos*, fruit] SICKLE-POD, RUSH-PEA Native

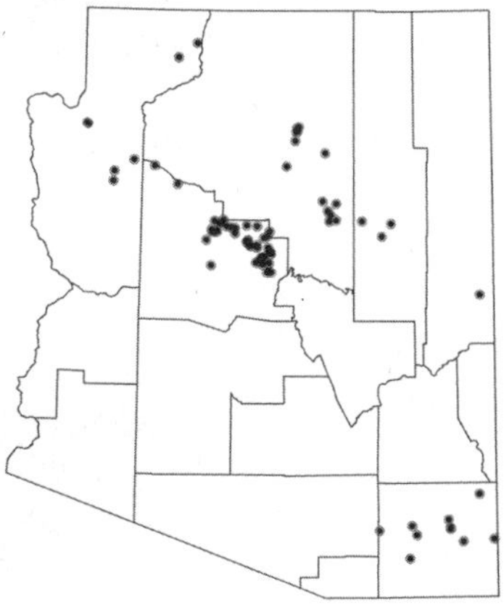

Perennial herbs with very short internodes producing an almost clustered appearance, 8–30 cm, arising from a taproot. **Leaves** with 4–9 pairs of lateral pinnae and a terminal pinna, 3–10 (including a petiole) × 1.3–3 cm; stipules lanceolate, entire, 2–3 × 1–1.3 mm. **Leaflets** 6–9 per lateral pinna, ovate-obtuse in outline, 1–5.5 × 1–2.3 mm, glabrous to slightly pubescent above (adaxially) and glabrous to slightly strigose below (abaxially). **Inflorescences** terminal, 6–27 cm, 3–11 flowered, the rachis and pedicels strigose. **Flowers** top-shaped in side view, 6–9 × 3.5–5 mm; free portion of sepals 3–5 × 1–2 mm, below (abaxially) pubescent to strigose; banner yellow, sometimes tinged with red, 5–8 × 2.5–4.5 mm, bearing a small tuft of trichomes at the base of the inside of the claw and glabrous on the lower (abaxial) surface; lateral petals yellow sometimes tinged with red, 6.5–8 × 2–5 mm, glabrous below (abaxially). **Fruits** indehiscent, sickle-shaped (sometimes almost making a full circle) with parallel margins, reticulately veined, pale brown or reddish, 23–40 (across from tip to tip) × 5–8 mm; apex round; valves puberulent, brown or red; margins pronounced, puberulent. **Seeds** 6–11. $2n = 24$ (Texas).

Flowering and fruiting [Apr–]May–Aug[–Oct]. Sandy or clay limestone soils of slopes, valleys, floodplains and riparian habitats; [900–]1070–1560[–2000] m; Chihuahuan Desertscrub, Mohave Desertscrub, Great Basin Desertscrub, Semidesert Grassland, Plains and Great Basin Grasslands, Great Basin Conifer Woodland; Apache, Cochise, Coconino, Graham, Mojave, Navajo, Yavapai cos.; Kansas, Colorado, New Mexico, Texas; Mexico (Chihuahua).

This small, inconspicuous species is one of the few *Hoffmannseggia* species that lack glandular trichomes. The seeds are utilized as food by quail (Allen and Allen 1981).

2. Hoffmannseggia glauca (Ortega) Eifert. Sida 5(1):43. 1972. * [Latin, *glaucus*, with a pale coating or bloom] HOG POTATO, INDIAN RUSH-PEA Native

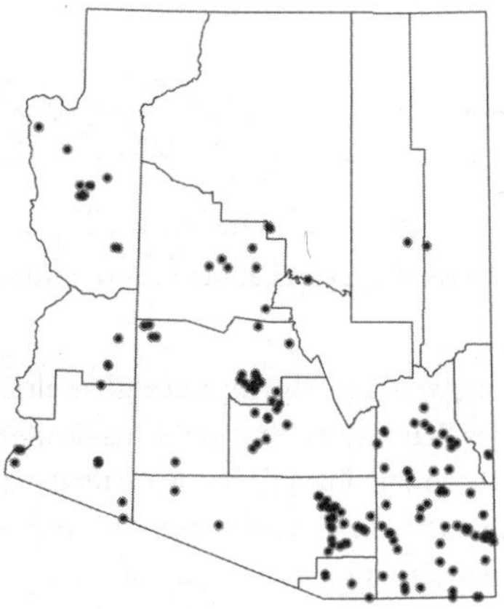

Larrea glauca Ortega, Nov. Pl. Descr. Dec. 15, pl. 2. 1797; *Caesalpinia falcaria* var. *capitata* (Fisher) Fisher; *Caesalpinia falcaria* var. *pringlei* (Fisher) Fisher; *Caesalpinia falcaria* var. *rusbyi* (Fisher) Fisher; *Hoffmannseggia densiflora* Bentham in A. Gray; *Hoffmannseggia falcaria* Cavanilles; *Hoffmannseggia falcaria* var. *capitata* Fisher; *Hoffmannseggia falcaria* var. *pringlei* Fisher; *Hoffmannseggia falcaria* var. *rusbyi* Fisher; *Hoffmannseggia stricta* Bentham in A. Gray; *Hoffmannseggia stricta* var. *demissa* A. Gray

Perennial herbs, 5–30(–50) cm with a deep taproot and underground stems producing small (to 2 cm) tuber-shaped roots. **Leaves** with 3–13 pairs of lateral pinnae and a single terminal pinna, 3.8–15 (including petiole) × 1.3–4.2 cm; stipules ovate, ciliate, 1.5–4 × 1.5–3 mm. **Leaflets** 4–13 per lateral pinna, ovate-obtuse in outline, 2–6 × 1–4.5 mm, glabrous above (adaxially) and strigose below (abaxially). **Inflorescences** terminal, 5–23 cm, 4–15 flowered, the rachis and pedicels puberulent to strigose mixed with multicellular glandular trichomes. **Flowers** broadly flaring, bending down with age, 10–16 × 10–18 mm; free portion of sepal 7–9.5 × 2–3 mm, lower surface (abaxially) pubescent

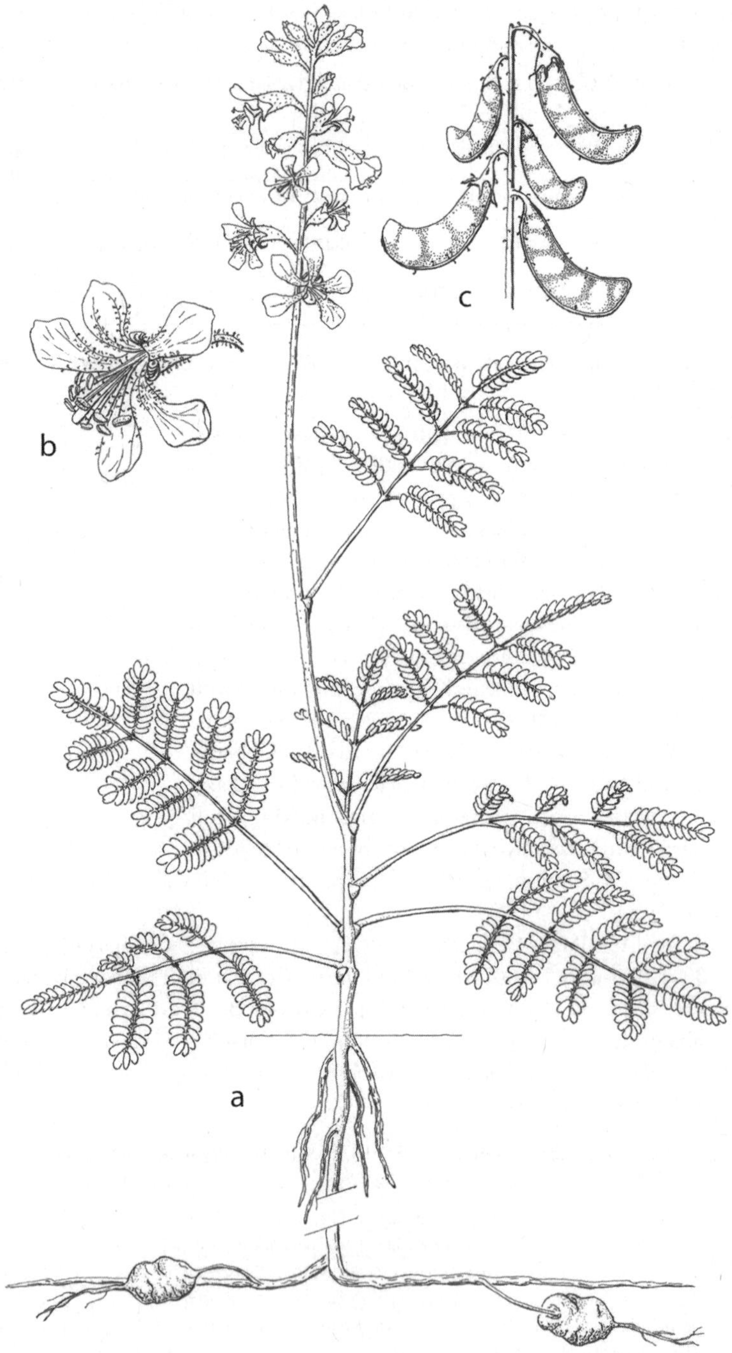

***Hoffmannseggia glauca*: a. habit with roots; b. flower; c. fruits. Illustrator: Rebecca Hall.**

and very densely covered with multicellular glandular trichomes; banner yellow with red markings, 5–14 × 5 mm, with few hairs at the base of the inside of the folded claw and conspicuous multicellular glandular trichomes on claw and on the lower (abaxial) surface; lateral petals bright yellow, 13 × 6 mm, with multicellular glandular trichomes on claw margins and base of the lower (abaxial) surface. **Fruits** indehiscent, rectangular to arched or even expanded near the tip, margins more or less parallel, light brown to tan, 20–40 × 5–8 mm; apex obtuse; valves lightly tomentose with a few scattered multicellular glandular trichomes, tan; margins obscure. **Seeds** up to 10. $2n$ = 24 (Texas and Argentina).

Flowering (Feb–)Apr–Oct. Primarily in disturbed areas, especially along roads, also slopes, valleys, arroyos, floodplains, dry playas, and riparian areas; [near sea level–]150–1560[–1950]; widespread in Sonoran Desertscrub, Chihuahuan Desertscrub, Mohave Desertscrub, Great Basin Desertscrub, Semidesert Grassland, Plains and Great Basin Grassland; Apache, Cochise, Coconino, Graham, Greenlee, La Paz, Maricopa, Mohave, Navajo, Pima, Pinal, Santa Cruz, Yavapai, Yuma cos.; Kansas, Oklahoma, Texas, New Mexico, California; Mexico (Baja California, Chihuahua, Coahuila, Durango, Nuevo León, San Luis Potosí, Sonora, Zacatecas); Peru, Bolivia, Chile, Argentina.

Hoffmannseggia glauca is considered a weedy pest in agricultural and pasture lands of the U.S. Mid- and Southwest. It spreads aggressively by underground tuber-shaped roots. While it is possible that *H. glauca* was introduced into North America by humans, the historical use of the tubers by indigenous people in the American Southwest indicates a long association and perhaps natural long-distance dispersal from South America.

In Mexico, *Hoffmannseggia glauca* is known as *camote de ratón, coco, cocos, coquito,* or *papilla*. According to Moerman (1998) the Apache, Pima, and Pueblo ate the swollen tuber-like roots either cooked or raw. Castetter (1935) commented that the roots tasted like Irish potatoes when roasted. However, in the U.S., the species is now mostly known as an invasive weed along roadsides and vacant lots where it spreads readily via rhizomes.

3. Hoffmannseggia microphylla Torrey. Rep. U.S. Mex. Bound. 2:58. 1859. * [Greek *micro*, small, and *phyll*, leaf] WAND HOLDBACK Native

Caesalpinia virgata Fisher, Bot. Gaz. 18:123. 1893; *Larrea microphylla* (Torrey) Britton

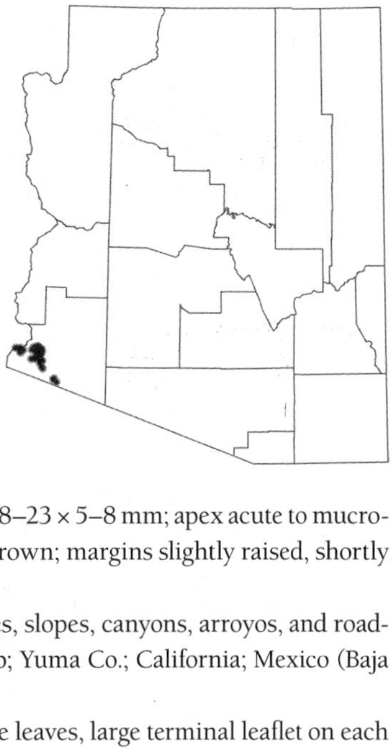

Shrubs 0.5–2.5 m. **Stems** upright, almost leafless. **Leaves**, where present, consisting of one pair of pinnae and a conspicuously longer terminal pinna, 2–5 (including a petiole) × 0.9–2.6 cm; stipules linear-lanceolate, fringed, 0.5–1.5 × 0.2–0.5 mm. **Leaflets** of lateral pinnae 3–7 per pinna, oblong-obtuse in outline, 1–3.5 × 1–2 mm, glabrous to slightly villous above (adaxially) and sparsely to densely villous below (abaxially). **Inflorescence** a terminal or lateral raceme, 9–16 cm, 10–27 flowered, the rachis and pedicels strigose. **Flowers** openly flared in side view, 12–15 × 6–12 mm; free portion of sepals 4–12 × 5–8 mm, below (abaxially) strigose or strigose and with multicellular glandular trichomes; banner yellow, with red markings, 5–10 × 4–7 mm, bearing a small tuft of trichomes at the base of the inside of the claw and with short glandular trichomes on the lower (abaxial) surface; lateral petals yellow, 5–10 × 3–5 mm, with short glandular trichomes below (abaxially). **Fruits** dehiscent with the valves flaring and often each twisting around itself, sickle-shaped to crescent-shaped in outline, bearing a jagged ring where the sepals have dehisced, brown, 18–23 × 5–8 mm; apex acute to mucronate; valves puberulent with scattered multicellular glandular trichomes, light brown; margins slightly raised, shortly villous. **Seeds** 3–6. $2n = 24$ (California).

Flowering [Jan–]Apr–Oct[–Nov]; sporadic flowering in Jul, Sep–Dec. Ridges, slopes, canyons, arroyos, and roadsides in sandy or rocky soils; [sea level–]40–300[–1000 m]; Sonoran Desertscrub; Yuma Co.; California; Mexico (Baja California, Sonora).

Hoffmannseggia microphylla is easily recognized by its upright stature, sparse leaves, large terminal leaflet on each pinna, and the jagged ring around the base of the mature fruit where the sepals break off tardily and unevenly as the fruit matures.

HOSACKIA Douglas ex Bentham. Edwards's Bot. Reg. 15:pl. 1257. 1829. * [in honor of David Hosack, 1769–1835, American physician and botanist] LOTUS, DEER VETCH

Ries Lindley

Perennial herbs, unarmed, glabrous or sparsely pubescent with simple hairs. **Leaves** alternate, odd-pinnate; stipules scarious or leaf-like, distinct; stipels absent. **Leaflets** 3–15, often ± opposite, margins entire. **Inflorescences** umbels, 1–20 flowered, axillary; bracts often present; peduncles generally present. **Flowers** papilionoid; corolla yellow, white, pink, red, or green; filaments 10, 9 fused, 1 free. **Fruits** dehiscent, exserted from calyx, linear to oblong, ± beaked. **Seeds** few–many.

Species ca. 11 (2 native in the flora): sw Canada, w United States, Mexico, Guatemala.

Hosackia comprises some of the New World taxa formerly placed in *Lotus*. Other New World taxa formerly in *Lotus*

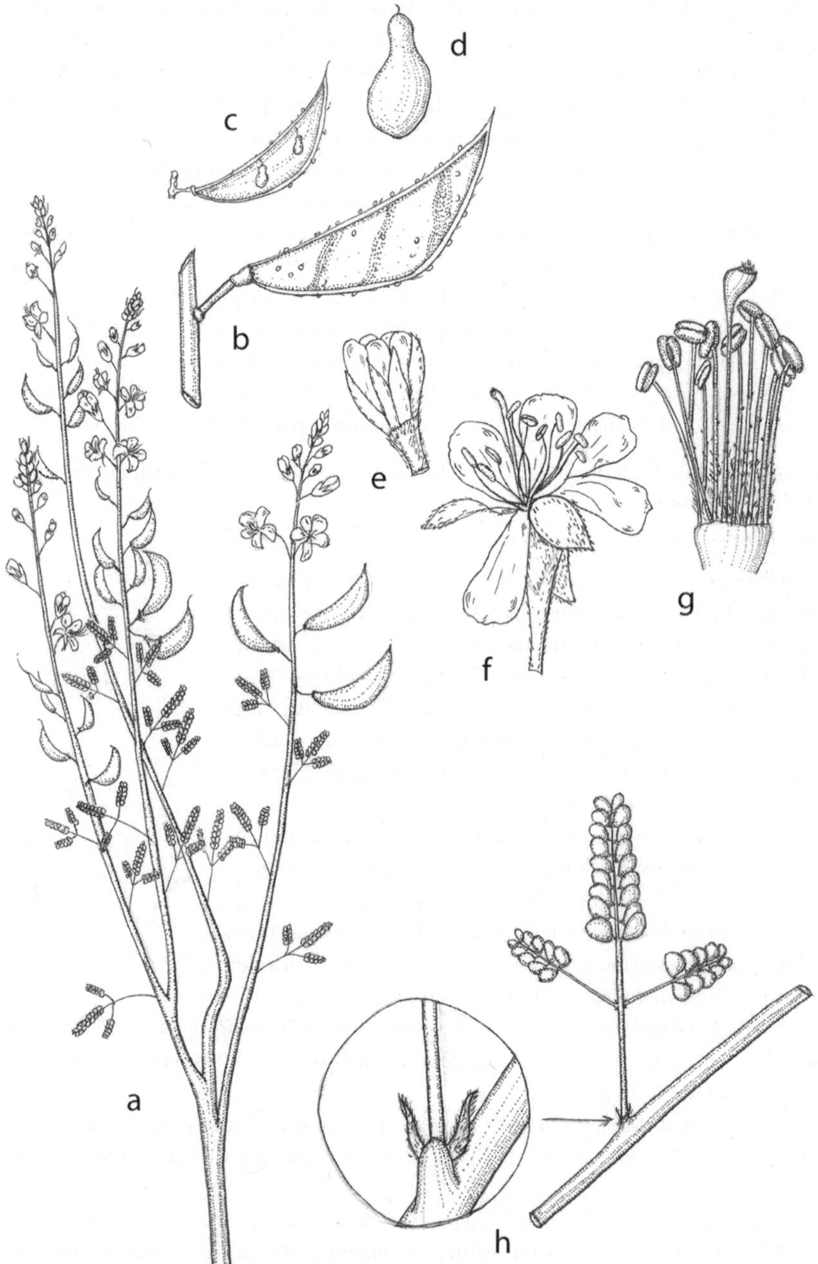

***Hoffmannseggia microphylla*: a.** habit; **b.** fruit; **c.** fruit cut-away; **d.** seed; **e.** flower bud; **f.** flower; **g.** reproductive structures; **h.** leaflet with stem node detail. Illustrator: Marsha Bennett.

are treated under *Acmispon*. Most *Hosackia* species occur in mesic or moist sites. *Hosackia gracilis* is occasionally cultivated for its showy, bicolored flowers (Isely 1981).

1. Leaflets 3–5, obovate ____ 1. **Hosackia alamosana**
1′ Leaflets 7–13, elliptic to oblong ____ 2. **Hosackia oblongifolia**

1. Hosackia alamosana Rose. Contr. U.S. Natl. Herb. 1(4):96. 1891. * [of the Sierra de Alamos, Sonora, Mexico] SONORAN BIRD'S-FOOT TREFOIL, ALAMOS LOTUS Native

Lotus alamosanus (Rose) Gentry

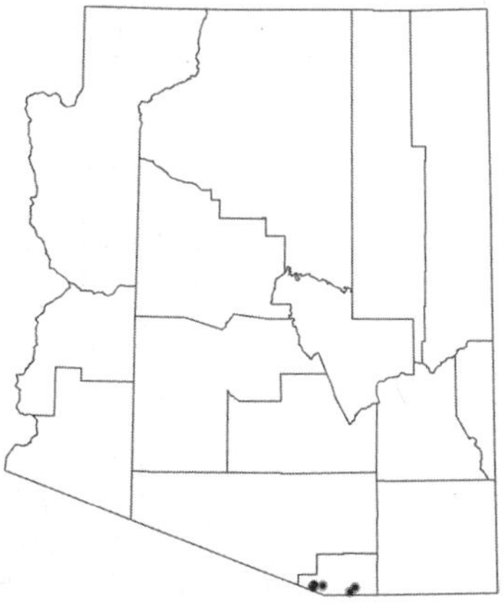

Plants mat-forming; glabrous except new growth. **Stems** to 45 cm, prostrate, sometimes rooting at the nodes. **Leaves**: stipules membranous, drying to a pale straw color; petiole 2–17 mm; axis 0–10 mm. **Leaflets** 3–5; obovate, 4–12 × 2–6.5 mm. **Inflorescences** (1–)2–3[–4] flowered; bracts absent; peduncles 7–20 cm. **Pedicels** 0.5–1 mm. **Flowers** 5.5–9 mm; calyx 2.5–4 mm; corolla yellow and white. **Fruits** erect, linear, 25–32 × 1.1–2 mm, thinly coriaceous, glabrate. **Seeds** many.

Flowering (Jan–)Feb–May(–Jul), possibly year-round. Sandy or rocky soil of canyon bottoms, along streams and seeps; [900–]1070–1520[–2400] m; Madrean Evergreen Woodland; Santa Cruz Co.; Mexico (Chihuahua, Durango, Sinaloa, Sonora).

In Arizona, *Hosackia alamosana* is found only in Sycamore Canyon, in the Pajarito Mountains, and Miller Spring in the Patagonia Mountains. It is more widespread in northwestern Mexico. The plants are restricted to moist sites.

2. Hosackia oblongifolia Bentham. Pl. Hartw. 305. 1848. * [from Latin *oblongus*, oblong, and *folia*, leaf] STREAMBANK BIRD'S-FOOT TREFOIL, NARROW-LEAF LOTUS

Lotus oblongifolius (Bentham) Greene

Plants with numerous stems; hairy or glabrous. **Stems** to 60(–140) cm, spreading or ascending, rhizomatous. **Leaves**: stipules membraneous, drying to a pale straw color; petiole 2–20(–30) mm; axis (0–)10–42 mm. **Leaflets** 3–11 (–13); elliptic to oblong, (6–)10–25(–32) × 1–12 mm. **Inflorescences** 2–6(–8)-flowered; bracts present just below flowers or absent, peduncles 0.7–12 cm. **Pedicels** 0.5–1 mm. **Flowers** 8–13 mm; calyx 4–6.5 mm; corolla bicolored yellow and white. **Fruits** erect, oblong, 25–50(–70) × 1.5–3 mm, thinly coriaceous. **Seeds** few.

Varieties 2 (1 native in the flora): w US from s Oregon to n Baja California, e to Nevada and disjunct in Arizona and Chihuahua. Variety *cuprea* is endemic to a limited area of the Sierra Nevada Mountains in eastern California.

2a. Hosackia oblongifolia Bentham var. **oblongifolia** * Native

Leaflets 7–13. **Flowers** 9–13 mm.

Flowering May–Sep. Seeps, springs, streams, moist meadows; [200–]ca. 1,890[–2,600] m elevation; Madrean Evergreen Woodland. Cochise, Greenlee, Santa Cruz cos.; California, Oregon, Nevada; Mexico (Baja California, Chihuahua).

Hosackia oblongifolia is primarily a species of far western North America. Variety *oblongifolia* is found throughout the range of the species. The populations in Arizona are highly localized. There are three historic collections of *H. oblongifolia* var. *oblongifolia* from Arizona from the late 19th and early 20th centuries. Two recent collections (*M. Pelto 6.5,* 22 July 2006) from Greenlee County (specimen not examined) and (*J. Verrier 1603,* 03 July 2020) from the east slope of the Santa Rita Mountains in Santa Cruz County confirm that this taxon is present in Arizona. Both collections are from mid-elevation sites in Madrean Evergreen Woodland. The population in the Santa Rita Mountains consists of numerous plants growing at a spring (J. Verrier, personal communication). Plants from this population have leaves with 7–13 leaflets that are 4–12 × 2–7 mm with a 1 mm apical point. Leaflets have fine, appressed pubescence along the margin and on both surfaces. The stems are up to 140 cm long, distinctly angled or ridged, and are rhizomatous (J. Verrier, personal communication). The flowers are showy. Nodulation is reported for *H. oblongifolia* (Allen and Allen 1981).

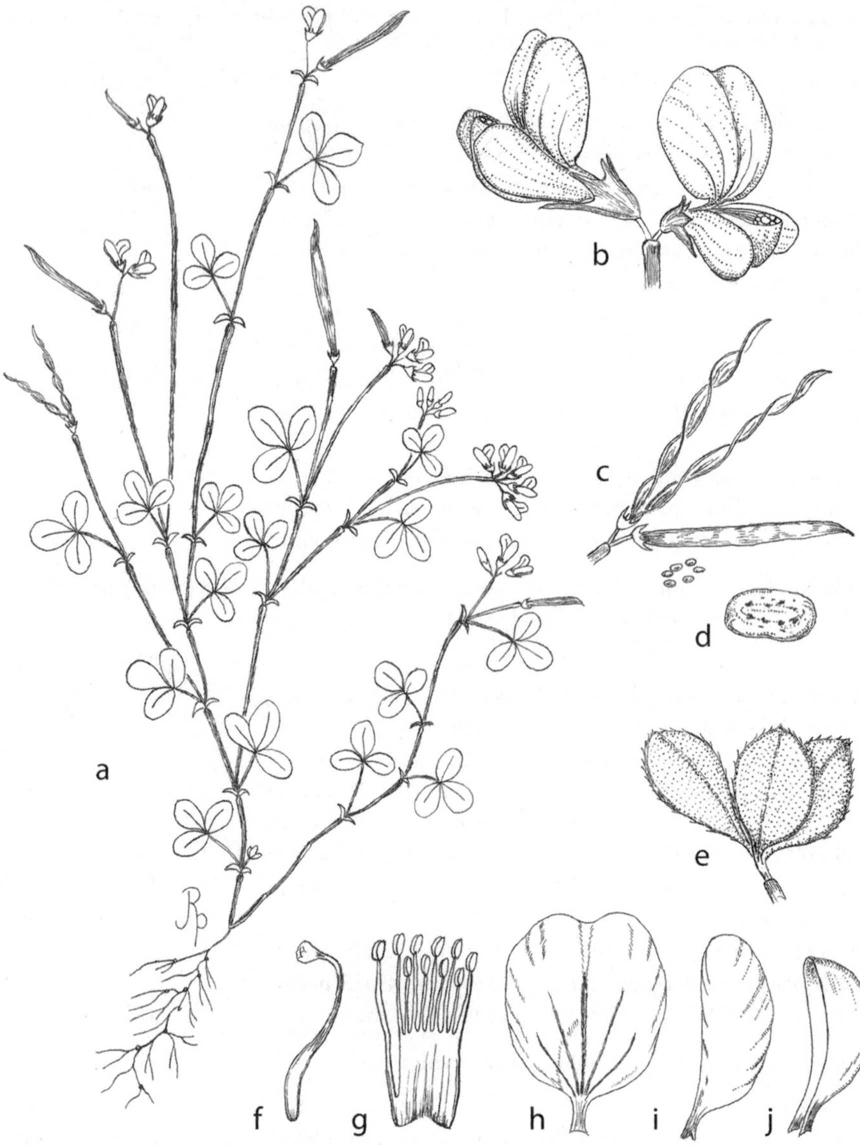

Hosackia alamosana: **a.** habit; **b.** flowers; **c.** fruits (showing dehiscence.; **d.** seed; **e.** leaf; **f.** pistl; **g.** stamens; **h.**, **i.**, **j.** petals. Illustrator: Joyce Riney Peters.

INDIGOFERA Linnaeus. Sp. Pl. 2:751. 1753. * [from Latin, *indigus*, indigo, and *fero*, to bear or produce, in reference to the species which are the source of indigo dye] INDIGO

Matthew B. Johnson

Herbs, annual or perennial, shrubs or rarely trees. **Herbage** usually with appressed medifixed hairs. **Leaves** alternate, odd-pinnate, pinnately or palmately 3-foliolate, or 1-foliolate; stipules small, awl- or bristle-shaped, partially adhering to the petiole, persistent; petiolate; stipels present or absent. **Leaflets** 1–many, linear, oblong or oblanceolate, margins entire. **Inflorescences** racemes or spikes, axillary, solitary or paired; bracts small, deciduous; bracteoles absent. **Pedicellate** or not. **Flowers** papilionoid, small; calyx bell-shaped, often short, lobes 5, narrow, acute, subequal or the lowest lobe longest; corolla orange, pink, red, purple or whitish green with orange or red markings, banner rounded, sessile or narrowed at the base, apex rounded or obtuse, sometimes with hairs on the back, wings oblique-obovate, oblong, oblanceolate or linear, loosely adherent to the keel, keel oblique-lanceolate, gradually narrowed at the base, apex obtuse, rarely beaked, with a short spur on each side; stamens 10, 9 united and 1 free, anthers all alike; ovary subsessile, with 1–many ovules, usually strigose, style incurved, often with a tuft of hairs, stigma simple. **Fruits** linear, oblong (or spherical), straight or curved, cylindrical, angled or compressed, 2-valved; stipitate. **Seeds** 1–many; spherical or nearly square.

Species ca. 700 (1 native in the flora): tropical, subtropical, and warm temperate regions. Widespread except for Europe, with the highest diversity in Africa and Madagascar.

Indigofera has been valued as the source of indigo dye since ancient times. *Indigofera arrecta, I. articulata, I. longeracemosa, I. suffruticosa*, and *I. tinctoria* were the principal species of commerce (Allen and Allen 1981, Lewis et al. 2005). Indigo is extracted from the leaves. Duke (1981) provides information on cultivation, harvesting, and yields of several *Indigofera* species. Allen and Allen (1981) present a brief account of the complex historical importance of indigo that includes the founding of the Dutch East India Company to facilitate a reliable supply of indigo dye. Fear of competition with woad dye, which is produced from the woad plant (*Isatis tinctoria*), led to a ban on importation of indigo in parts of Europe for a time. Violators were subject to execution (Allen and Allen 1981). Natural indigo production declined greatly by the end of the 19th century with the introduction of synthetic dyes. Additionally, indigo production posed health risks for those involved in extracting and handling the dye (McVaugh 1987). Today, indigo dye is produced on a much smaller scale than in the past.

Though its use as a dye source is diminished and only a few species are grown ornamentally for their colorful flowers, *Indigofera* species have other potential uses. *Indigofera glandulosa* appears to have potential for use in human nutrition. A number of species, including *I. glandulosa*, are widely grown in tropical regions as fodder for livestock (Allen and Allen 1981, Duke 1981, Ghane et al. 2010). Several *Indigofera* species including the dye species *I. arrecta, I. suffruticosa*, and *I. tinctoria* are grown as cover crops and green manure (Allen and Allen 1981, Duke 1981). Various species are also useful for erosion control. Allen and Allen (1981) state that *I. teysmannii* can grow up to 7 m in one year and is planted in Sri Lanka for shade and as a windbreak in cacao, coffee, and tea plantations. Native peoples in Africa obtain an arrow poison from the roots of *I. simplicifolia* (Uphof 1968). The crushed roots of *I. lespedezioides* and *I. suffruticosa* are used to stupefy fish in Brazil (Blohm 1962). Though several species are toxic to livestock, others have insecticidal properties and are applied externally on livestock to repel parasites (Allen and Allen 1981). *Indigofera* species have many reported uses in folk medicine. These plants are used as antiseptics and astringents, as well as for treating bladder infections, bronchitis, diarrhea, epilepsy, fever, hemorrhoids, insect bites, heart problems, hepatitis, nervous disorders, skin diseases, snakebite, spleen problems, and ulcers (Allen and Allen 1981, Duke 1981).

Nodulation is widely reported for many *Indigofera* species. *Indigofera* is the third largest genus in the Fabaceae.

1. Indigofera sphaerocarpa A. Gray. Smithsonian Contr. Knowl. 5(6):37. 1853. * [with spherical fruits] SONORAN INDIGO, PILLPOD INDIGO Native

Shrubs to 1.2 m. **Stems** moderately pubescent with medifixed hairs. **Leaves** odd-pinnate; stipules 1.2–4 mm, awl-shaped; petioles 8–17 mm; stipels minute. **Leaflets** 7–19, 8–20(–25) × 2–7(–9) mm, lateral leaflets opposite, narrowly oblanceolate or oblong-oblanceolate, base wedge-shaped, apex mucronate, upper (adaxial) surface glabrous, lower (abaxial) surface moderately pubescent with medifixed hairs. **Inflorescences** racemes, 4.5–10(–16) cm in fruit, often with 30–50 flowers, solitary; bracts 1.4–2 mm. **Pedicels** 0.9–1.2 mm. **Flowers:** calyx: tube 0.5–0.6 mm, lobes 0.4–0.8 mm; corolla greenish white proximally and pale to dark coral or salmon-colored distally, banner 6–7 mm, with straight hairs on the back, wings 6–6.5 mm, keel 4–5 mm. **Fruits** 4–5 × 2.1–2.9 mm including beak, brown, nearly spherical to tear-drop shaped, with medifixed hairs, tardily dehiscent; downcurved stipe 1.3–1.5 mm; beak 0.5–1.2 mm. **Seeds** 1; spherical, tan to brown, 2–2.3 mm.

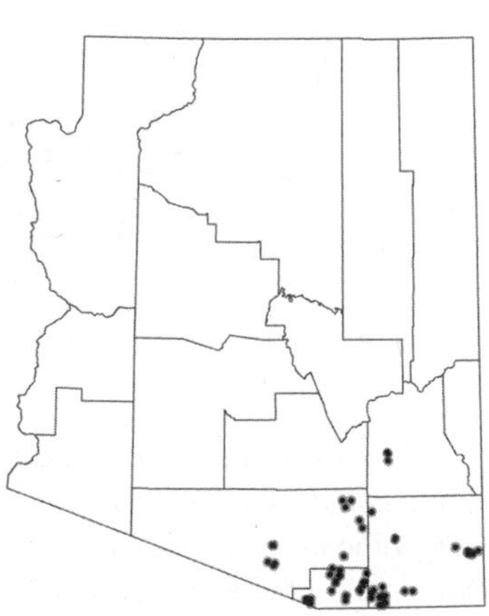

Flowering Jul–Sep. Slopes, canyons, arroyos, and roadsides, in rocky, sandy, or clay soils; 1070–1800(–2040) m; Madrean Evergreen Woodland; Cochise, Graham, Pima, Santa Cruz cos.; New Mexico; Mexico (Chihuahua, Sonora).

In southeastern Arizona, *Indigofera sphaerocarpa* occurs as scattered plants or in localized populations, often on dry, rocky slopes in full sun. It is found in oak savannah and oak and pine-oak woodland. In most areas the plants are infrequently encountered. Its range extends into extreme southwestern New Mexico and across northeastern Sonora and adjacent Chihuahua. *Indigofera sphaerocarpa* is reported to be a suitable browse plant and to have potential for use in erosion control (Graham 1941, Kearney and Peebles 1942).

Indigofera sphaerocarpa: **a.** habit; **b.** flower with corolla removed; **c.**, **d.** flower; **e.** fruit cut-away; **f.** seed. Illustrator: John H. Gerdes.

Indigofera sphaerocarpa is rarely available from specialty native plant nurseries in Arizona. Supplemental irrigation is necessary for cultivation at lower elevations. Cultivated plants in Tucson have survived overnight temperatures to -8°C without damage. At least in cultivation, the plants appear to be short-lived; declining after four to six years. Propagation is by seeds. Scarification greatly facilitates germination.

LADEANIA A. N. Egan & Reveal. Novon 19(3):310–314. 2009. * [named for LaDean Egan, b. 1949, mother of and long-time field assistant for botanist Ashley N. Egan, b. 1977] SCURFPEA

Ashley N. Egan

Rhynchosia sect. *Arcyphyllum* (Elliott) Torrey & A. Gray, Fl. N. Amer. 1:284. 1838; = non *Psoralea* sect. *Psoralidium* H. E. Baillon; *Psoralidium* Rydberg

Plants, herbaceous or shrub-like, perennial, usually much-branched and glandular. **Roots** deep and woody, or rhizomatous. **Stems** usually much-branched basally but sparingly above, glabrate or sericeous to densely pubescent throughout or at least basally, occasionally becoming glabrate with age, green, smooth or lightly striate, occasionally distinctly ribbed apically, lowermost portions often with thin, tan to brown cataphylls. **Leaves** palmately compound or rarely pseudopalmate, (1–)3–5-folioliate; stipules quickly to tardily deciduous, usually fused to petiole, rarely fused behind petiole, often modified to form cataphylls at base of stem; petioles not jointed to stem. **Leaflets** linear to oblanceolate, persistent or quickly deciduous and sometimes lacking at maturity, stipels lacking. **Inflorescence** an indeterminate pseudoraceme with 3–22 nodes, 1–3(–4) flowers per node; bracts quickly deciduous, lanceolate to rhombic. **Flowers** papilionoid; calyx bell-shaped with pale green triangular teeth or green throughout, glandular, pubescent though sometimes glabrate to sericeous; teeth 5, usually less than half as long as calyx tube, mostly equal in length, the lower (abaxial) tooth sometimes slightly longer, flaring back and tearing along a lateral sinus in fruit; corolla to 7 mm, violet to blue-purple, white or pale yellow; stamens 10 with 9 filaments united and 1 mostly free, the distinct portion of filaments thread-like; ovary 1-ovulate. **Fruit** an indehiscent, 1-seeded pod, globose, pilose, deciduous above receptacle. **Seed** round to elliptic, somewhat compressed, brown, smooth. $x = 11$.

Species 2 (2 native in the flora): western half of the United States and up into Canada.

Historically, *Ladeania* has been described as having as many as 14 species (Rydberg 1919) based on morphometric characters. The widespread species *Ladeania lanceolata* is morphologically variable in raceme length, flower color, and leaflet shape. However, most specimens are intermediate with each other, leaving no clear break in morphological characters as to delineate separate species. Therefore, I followed the advice of Isely (1998) and Grimes (1990) in recognizing a single widespread species, *L. lanceolata*, along with the narrowly endemic *L. juncea*.

SELECTED REFERENCE: Grimes 1990.

1. Mature stems rush-like, leafless or with only a few basal leaves; peduncle 15–40 cm or more; calyx 2.5–4 mm; fruit eglandular ____ **1. Ladeania juncea**

1′ Mature stems not rush-like, with numerous leaves; peduncle to 12 cm; calyx 1.5–2.5 mm; fruit glandular ____ **2. Ladeania lanceolata**

1. Ladeania juncea (Eastwood) A. N. Egan & Reveal. Novon 19(3):310–314. 2009. * [named for habit which superficially resembles that of *Juncus*] RUSH LEMONWEED, RUSH SCURFPEA Native

Psoralea juncea Eastwood, Proc. Calif. Acad. Sci., ser. 2, 6:286. 1896 [1897]; *Psoralidium junceum* (Eastwood) Rydberg

Plants perennial, herbaceous or shrub-like. **Stems** much-branched below but sparingly above, rush-like in appearance. **Leaves** palmate, (1–)3(–5) folioliate, present on young plants but often deciduous by anthesis except at the base; petiole 1–7 cm. **Leaflets** oblanceolate to elliptic, strigose. **Inflorescence** with peduncle 15–40 cm or more; rachis 5–22 cm, internodes 0.5–1 cm but elongating up to 4.5 cm in fruit. **Flowers** with calyx 2.5–4 mm; corolla deep purple, 4–6.5 mm. **Fruit** densely pilose, without glands. $2n = 22$.

Flowering Apr–May; fruiting Jun–Aug. Open sand dunes, semistabilized sands, or rocky slopes; 1000–1750 m; Great Basin Desertscrub; Coconino Co. (along Colorado River and Lake Powell); Utah (localized near Paria, San Juan, and Colorado Rivers in Kane, San Juan, Garfield cos.).

Although *Ladeania juncea* is locally abundant, its geographic distribution is endemic and narrowly restricted mainly to Kane and San Juan counties in Utah along the Paria, San Juan, and Colorado Rivers where it is often dominant in the community. It is apparently disjunct and rare in Garfield County, Utah and Coconino County, Arizona. In addition, habitat in a portion of its range is in decline due to grazing and recreational use. These characteristics coupled with infrequent seed set (Isely 1998) suggest that *L. juncea* may be of conservation concern.

Phytochemical studies have discovered antimutagenic (Menon et al. 1999, Turchi et al. 2009) and antibacterial (Schmitt et al. 1991) properties stemming from the compound plicatin B.

2. Ladeania lanceolata (Pursh) A. N. Egan & Reveal. Novon 19(3):310–314. 2009. * [Latin *lancea*, lance-shaped] LEMON SCURFPEA, LANCELEAF SCURFPEA Native

Psoralea lanceolata Pursh, Fl. Amer. Sept. 2:475. 1814 [actual date, Dec. 1813]; *Lotodes ellipticum* (Pursh) Kuntze; *Psoralea elliptica* Pursh; *Psoralea laxiflora* Nuttall ex Torrey & A. Gray; *Psoralea micrantha* A. Gray; *Psoralea scabra* Nuttall ex Torrey & A. Gray; *Psoralea stenophylla* Rydberg; *Psoralea stenostachys* Rydberg; *Psoralidium lanceolatum* (Pursh) Rydberg

Plants perennial, herbaceous. **Stems** branching below and above, not rush-like. **Leaves** palmate or seldom pseudopalmate, 3(–5) folioliate, persistent through flowering; petiole (0.5–)0.9–2.9 cm. **Leaflets** obovate or oblanceolate to linear, glabrate to sparsely pubescent. **Inflorescence** with peduncle 0.7–12 cm; rachis 0.5–14 cm, internodes to 1.5 cm,

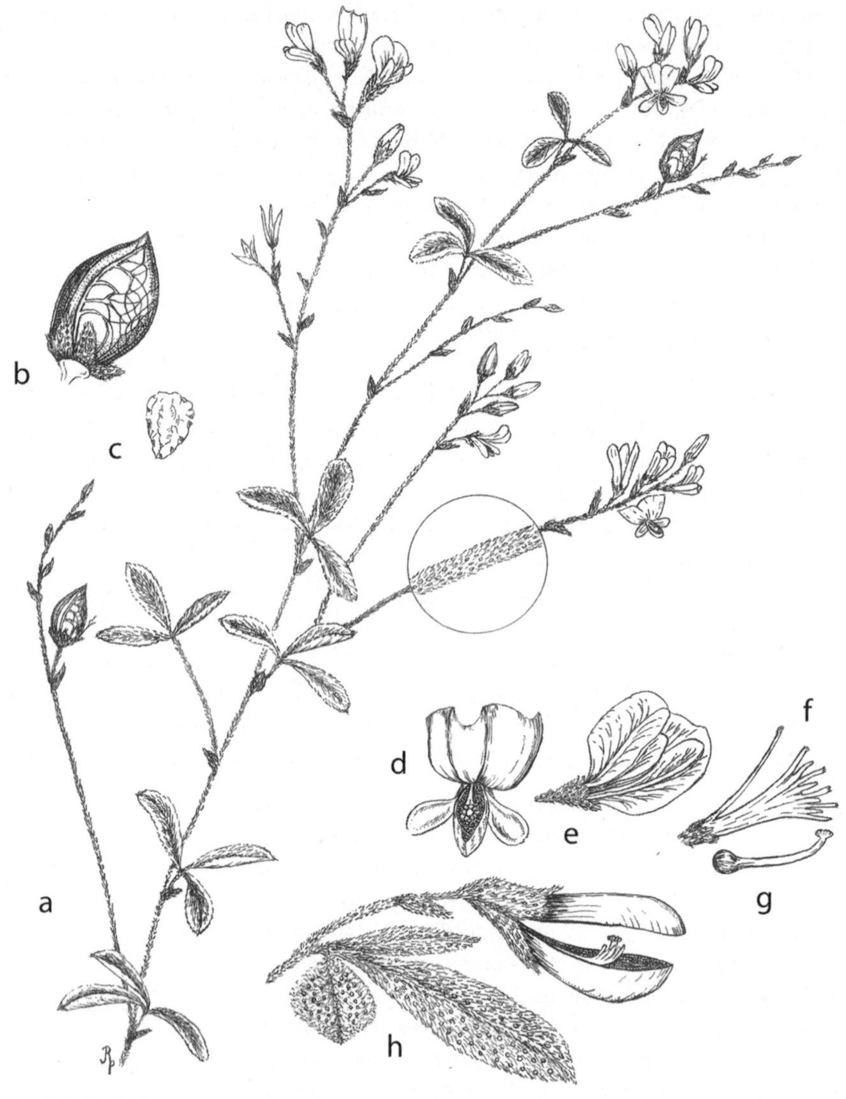

***Ladeania lanceolata*: a. habit; b. fruit; c. seed; d., e. flower; f. stamens; g. pistil; h. flower and leaf detail. Illustrator: Joyce Riney Peters.**

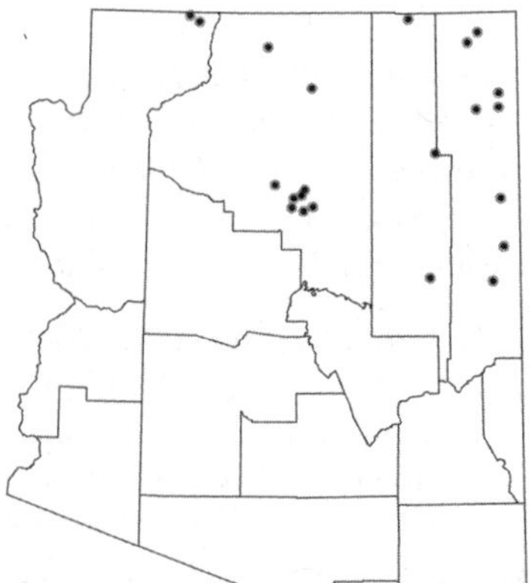

elongating or not in fruit. **Flowers** with calyx 1.5–2.5 mm; corolla blue to purple or white or bicolored, 5–7 mm. **Fruit** sparingly strigose to glabrate, conspicuously glandular. $2n = 22$.

Flowering Apr–Jul; fruiting Jul–Sep. Open areas, sand or occasionally in gravel or volcanic cinders in valleys, slopes, canyons, alluvial soils along watercourses, and roadsides; [200–]840–2500 m; Great Basin Conifer Woodland, Petran Montane Conifer Forest, and in riparian habitats at upper elevations of Great Basin Desertscrub and Semidesert Grassland; Apache, Coconino, Mohave, Navajo, Yavapai cos.; California, Colorado, Idaho, Iowa, Kansas, Montana, Nebraska, Nevada, New Mexico, North Dakota, Oklahoma, Oregon, South Dakota, Texas, Utah, Washington, Wyoming; Canada (se Alberta, s Saskatchewan).

Ladeania lanceolata is a widespread species found largely in deep sand deposits. It often invades disturbed areas and is a pioneer of sand dunes, playing an important role in soil stabilization and plant succession of dunes (Chadwick and Dalke 1965). The species can become a problem in agricultural fields and its presence on rangeland can indicate overgrazing (Kearney and Peebles 1942). *Ladeania lanceolata* is highly variable in morphology, which has resulted in the recognition of up to five taxa (Rydberg 1919). The length of the inflorescence, flower color, pod vestiture, and leaflet shape all vary along geographic trendlines throughout its range. Toft and Welsh (1972) recognized three varieties. However, in my estimation, morphometric variation presents no distinct groups, but seems to grade from one trend to another, making species distinctions based on morphology alone untenable.

LATHYRUS Linnaeus. Sp. Pl. 2:729. 1753. * [Greek name for pea or pulse] SWEET PEA, VETCHLING
Steven L. Broich

Herbs, annual or perennial, glabrous or pubescent, trichomes unicellular and without glands, and/or multicellular and glandular; from rhizomes, rootstock, or taproots. **Stems** erect, sprawling, or climbing, angled, sometimes winged, glabrous or pubescent. **Leaves** alternate, even-pinnate; stipules present, leaf-like, base usually semisagittate, rarely sagittate or hastate; petiolate; rachis sometimes winged, usually terminated in simple or branched, prehensile tendril, sometimes reduced to bristle or absent. **Leaflets** (0 or) 2–16 (or 18), paired or unpaired, usually decreasing in size distally, stipels absent, blade margins entire, surfaces glabrous or sparsely to densely pubescent. **Inflorescences** 1–20-flowered, axillary racemes; bracts absent. **Flowers** papilionoid; calyx symmetric or obliquely campanulate, lobes 5, equal or unequal, when unequal, lower lobes short and triangular, lateral and keel lobes longer and lanceolate or linear; corolla white, cream, yellow, pink, lavender, lilac, rose, red, reddish orange, crimson, blue to purple, violet, magenta, or bicolor, wings and keel often lighter than banner, petals clawed, distinct from blade; stamens 10, 9 united and vexillary stamen free or partially united with others, anthers attached dorsally, uniform; style compressed perpendicular to the floral axis, sometimes twisted, pubescent on upper side. **Fruits** usually sessile, rarely stipitate, flattened or cylindrical, linear, dehiscent, usually with obscure margins, splitting along both margins, glabrous or sparsely pubescent; valves twisted after dehiscence. **Seeds** 3–15, usually dull brown, black, tan, or dark green, often mottled, spherical, sometimes angular. $x = 7$ (polyploidy rare).

Species 140 (5 native in the flora and 2 cultivated and rarely escaping from cultivation): North America, South America, Europe, Asia, n Africa; introduced widely in temperate regions.

Lathyrus species are widely grown as cover crops, for forage, for erosion control, and as garden plants for their showy flowers (Allen and Allen 1981, Lewis et al. 2005). *Lathyrus sativus* is grown for human consumption. The tubers of *L. montanus* are eaten for food, used to make an alcoholic beverage, and have been employed to treat diarrhea (Allen and Allen 1981). Some species contain toxic compounds that cause lathyrism in humans and livestock. Horses and pigs are reported to be particularly sensitive to lathyrism (Allen and Allen 1981). Nodulation is widely reported for the genus.

1. Leaflets 2.
  2. Perennial; inflorescences 5–15-flowered ____ **6. Lathyrus latifolius**
  2′ Annual; inflorescences 2–4-flowered ____ **7. Lathyrus odoratus**
1′ Leaflets greater than 2.
  3. Tendrils absent or reduced to bristles, not well developed.
    4. Corollas white or cream ____ **5a. Lathyrus lanszwertii** var. **leucanthus**
    4′ Corollas usually blue-purple, sometimes violet to magenta, lavender, or pinkish ____ **1a. Lathyrus brachycalyx** var. **zionis** (in part)
  3′ Tendrils well developed, often branched.
    5. Leaflets very often paired ____ **4. Lathyrus laetivirens**
    5′ Leaflets scattered.
      6. Leaflets usually linear; racemes generally with more than 5 flowers ____ **3. Lathyrus graminifolius**
      6′ Leaflets elliptic or ovate to lanceolate; racemes with generally with less than 5 flowers.
        7. Leaves 5–7 cm; flowers 13–16 mm ____ **1a. Lathyrus brachycalyx** var. **zionis** (in part)
        7′ Leaves 2–5 cm; flowers > 16 mm ____ **2. Lathyrus eucosmus**

1. Lathyrus brachycalyx Rydberg. Bull. Torrey Bot. Club 34:425. 1907. * [from Greek, *brachy*, short, and *calyx*, the calyx] BUSH PEA

Herbs perennial, glabrous or villous; from rhizomes. **Stems** angled, erect, sprawling, or climbing, basally branched 0–3 times, 1–5 dm. **Leaves** 2–9 cm; stipules lanceolate, 5–12 × 1–3 mm, much smaller than leaflets; tendrils usually well developed, sometimes reduced to a single bristle. **Leaflets** 6–12, scattered, elliptic to narrowly so, 8–45 × 3–8 mm, surfaces glabrous or villous. **Inflorescences** 2–5-flowered, 5–15 cm. **Flowers** 13–20 mm; calyx lobes unequal, lateral lobes deltate, shorter than tube; corolla pinkish to deep blue-purple, banner erect, blade nearly 2 times as long as claw, wings equal to keel; ovary glabrous. **Fruits** 30–40 × 6–8 mm.

Varieties 2 (1 native in the flora): w United States.

1a. Lathyrus brachycalyx Rydberg var. **zionis** (C.L. Hitchcock) S.L. Welsh. Great Basin Naturalist 38:317. 1978. * Native

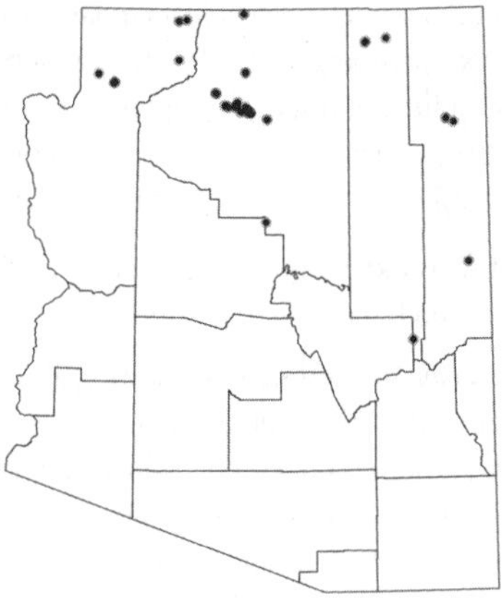

Herbs glabrous or sparsely villous. **Stems** sprawling or climbing, 3–5 dm. **Leaves** 5–7 cm. **Leaflets** 6–10, 18–45(–70) × 3–8 mm. **Flowers** 13–16 mm, banner deeply cordate. $2n = 14$.

Flowering Apr–Jun(–Aug). Slopes, canyons, valleys, arroyos, stream banks, roadsides, disturbed areas; (1040–)1300–2160(–2300) m; (Mohave Desertscrub), Great Basin Desertscrub, Plains and Great Basin Grassland, Interior Chaparral, Great Basin Conifer Woodland; Apache, Coconino, Mohave, Navajo, Yavapai cos.; Colorado, New Mexico, Utah.

Lathyrus brachycalyx var. *zionis* is found principally on the Colorado Plateau where it often occurs in drier sites and at lower elevations than other species of *Lathyrus* in Arizona. Populations are scattered though plants may be locally common.

2. Lathyrus eucosmus Butters & H. St. John. Rhodora 19:160. 1917. * [from Greek, *eu*, good, and *cosm*, the world, the universe] BUSH VETCHLING Native

Lathyrus brachycalyx Rydberg subsp. *eucosmus* (Butters & H. St. John) S.L. Welsh; *Lathyrus brachycalyx* Rydberg var. *eucosmus* (Butters & H. St. John) S.L. Welsh

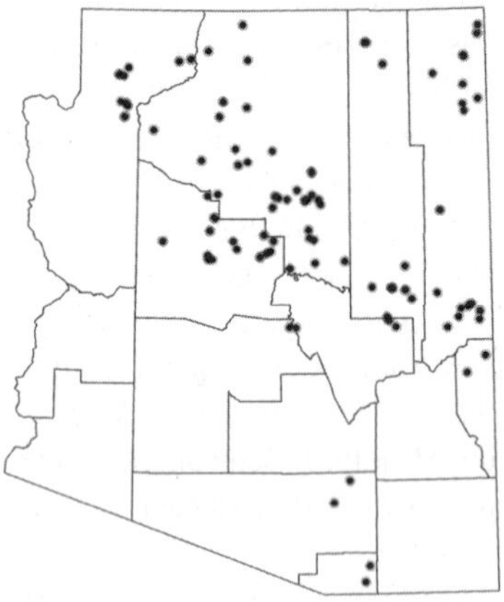

Herbs perennial, glabrous or puberulent; from rhizomes. **Stems** angled, erect, basally branched 0–2 times, 2–5 dm. **Leaves** 2–4 cm; stipules linear, 5–15 × 1–2 mm, much smaller than leaflets; tendrils simple or branched, sometimes prehensile. **Leaflets** 6 or 8, scattered, ovate to lanceolate, 15–45 × 4–10 mm, surfaces glabrous throughout or puberulent on the lower surface, often somewhat coriaceous. **Inflorescences** 2–4-flowered, 7–12 cm. **Flowers** 18–20 mm; calyx lobes unequal, lateral lobes deltate, shorter than tube; corolla blue-purple, banner erect, blade longer than claw, wings equal to keel (held above keel); ovary glabrous. **Fruits** 40–50 × 8–10 mm, short-stipitate. $2n = 14$.

Flowering Apr–Sep. Slopes, canyons, arroyos, stream banks, roadsides in various soils including sand and clay; (915–)1200–2500 m; (Great Basin Desertscrub, Semidesert Grassland), Interior Chaparral, (Madrean Evergreen Woodland), Great Basin Conifer Woodland, Petran Montane Conifer Forest; Apache, Coconino, Gila, Greenlee, Mohave, Navajo, Pima, Santa Cruz, Yavapai cos.; Colorado, New Mexico, Texas, Utah, Wyoming; Mexico (Coahuila).

Lathyrus eucosmus is locally common at mid-elevations across northern Arizona. It is apparently rare in the southern part of the state. The flowers are showy. With its spreading growth habit, it has potential for erosion control (Kearney and Peebles 1942). Nodulation is reported for *L. eucosmus* (Allen and Allen 1981).

3. Lathyrus graminifolius (S. Watson) T. G. White. Bull. Torrey Bot. Club 21:454. 1894. * [from Latin, *gramin*, grass, and *folia*, leaf] GRASSLEAF PEA Native

Lathyrus palustris Linnaeus var. *graminifolius* S. Watson, Proc. Amer. Acad. Arts 23:263. 1888

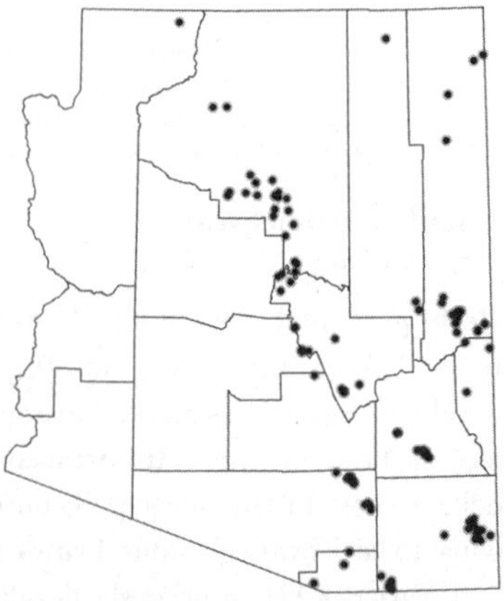

Herbs perennial, glabrous or sparsely pubescent at base of leaflets; from rhizomes or woody rootstock. **Stems** angled, sprawling or climbing, sometimes branched at flowering nodes, 2–6 dm. **Leaves** 5–9 cm; stipules lanceolate to linear, 8–12 × 1–5 mm, much smaller than leaflets; tendrils usually well developed. **Leaflets** 4–8, scattered, usually linear, rarely lanceolate, 40–80 × 1–10 mm, surfaces glabrous. **Inflorescences** 5–8-flowered, 10–18 cm. **Flowers** 8–15 mm; calyx lobes subequal, lateral lobes deltate, shorter than tube; corolla white to lavender, banner erect, blade equal to claw, wings equal to keel; ovary glabrous. **Fruits** 30–50 × 4–8 mm. $2n = 14$.

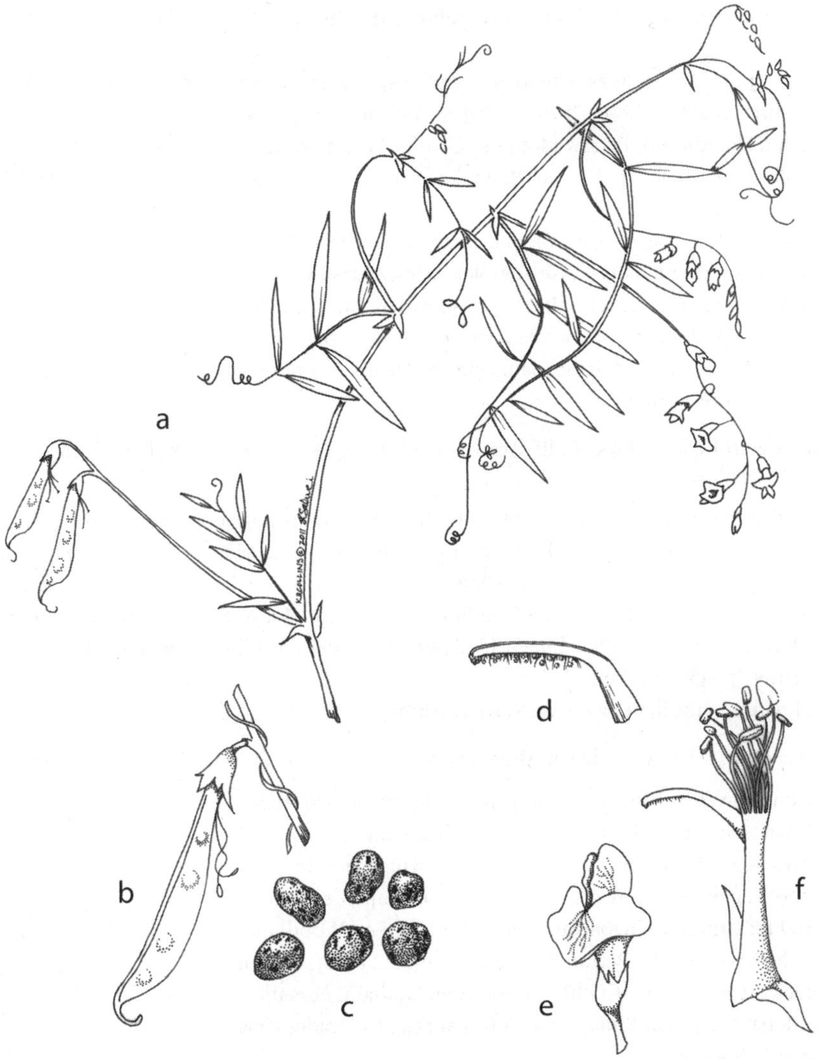

***Lathyrus graminifolius*: a. habit; b. fruit; c. seeds; d. style; e. flower; f. reproductive structures. Illustrator: Katherine Salwei & Kim Ratelle-Collins.**

Flowering Apr–Aug(–Oct). Ridges, slopes, canyons, meadows, stream banks, roadsides, disturbed areas in various soils; [1000–]1160–2800(–3170) m; (Semidesert Grassland, Interior Chaparral) Madrean Evergreen Woodland, (Great Basin Conifer Woodland), Petran Montane Conifer Forest, (Petran Subalpine Conifer Forest); Apache, Cochise, Coconino, Gila, Graham, Greenlee, Maricopa, Mohave, Navajo, Pima, Pinal, Santa Cruz, Yavapai cos.; New Mexico, Texas; Mexico (Chihuahua, Hidalgo, San Luis Potosí, Sonora).

Lathyrus graminifolius is widespread across Arizona at mid- and higher elevations. The plants may be locally common. Allen and Allen (1981) suggest that *L. graminifolius* has potential for erosion control.

4. Lathyrus laetivirens Greene ex Rydberg. Fl. Colorado 217. 1906. * [from Latin, *laet*, pleasing, and *virens*, green] PLATEAU VETCHLING, ASPEN PEA Native

Lathyrus lanszwertii Kellogg var. *laetivirens* (Greene ex Rydberg) S.L. Welsh; *Lathyrus leucanthus* Rydberg var. *laetivirens* (Greene ex Rydberg) C.L. Hitchcock

Herbs perennial, glabrous; from rhizomes. **Stems** angled, sprawling to erect, basally branched 0–3 times, 2–8 dm. **Leaves** 3–8 cm; stipules lanceolate, 8–12 × 2–5 mm, much smaller than leaflets; tendrils well developed. **Leaflets** 6–10, often paired, ovate, 20–40 × 10–20 mm, surfaces glabrous. **Inflorescences** 2–5-flowered, 4–7 cm. **Flowers** 15–20 mm; calyx lobes unequal, lateral lobes deltate, shorter than tube; corolla white to slightly pink or lavender, banner erect,

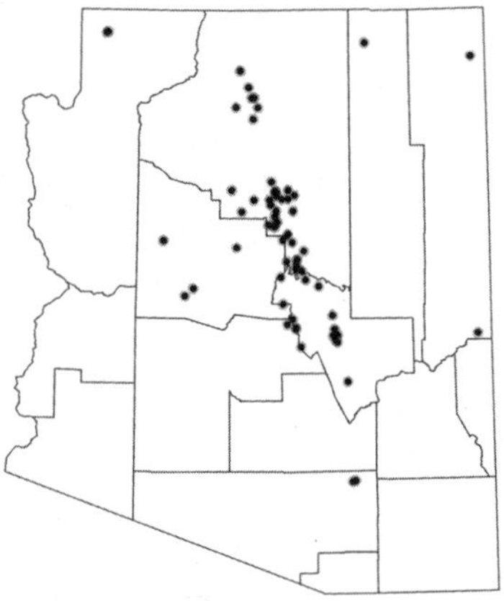

blade longer than claw, wings equal to keel; ovary glabrous. **Fruits** 30–50 × 5–15 mm.

Flowering Apr–Jul(–Oct). Slopes, canyons, meadows, stream banks, roadsides, and disturbed areas; 1200–2870[–3200] m; Interior Chaparral, (Madrean Evergreen Woodland), Petran Montane Conifer Forest; Apache, Coconino, Gila, Maricopa, Mohave, Navajo, Pima, Yavapai cos.; Colorado, Nevada, New Mexico, Utah.

Hitchcock (1952), Welsh et al. (1987), and Isely (1998) have placed *Lathyrus laetivirens* within the *L. lanszwertii* complex, while Barneby (1989) treated *L. laetivirens* as a separate species. Its final disposition awaits further study. In Arizona, this species is most abundant in the northern part of the state with scattered populations elsewhere. Plants may be locally common in the open or beneath the forest canopy.

5. Lathyrus lanszwertii Kellogg. Proc. Calif. Acad. Sci. 2:150, fig. 44. 1863. * [from Greek, *leuco*, white, and *anthus*, flower] NEVADA PEA

Herbs perennial, glabrous to sparsely pubescent; from rhizomes. **Stems** angled, erect, sprawling, or climbing, basally branched 0–4 times, 1–8 dm. **Leaves** (0.5–)1–8 cm; stipules linear to lanceolate, 5–25 × 1–6 mm, much smaller than leaflets; tendrils mucronate to well developed. **Leaflets** 2–10, scattered, lanceolate or linear, 20–70 × 2–20 mm, surfaces glabrous or sparsely pubescent. **Inflorescences** 2–6-flowered, 2–10 cm. **Flowers** 7–15 mm; calyx lobes unequal, lateral lobes deltate, shorter than tube; corolla white or blue-purple, banner erect, blade equal to claw, wings equal to keel; ovary glabrous. **Fruits** 25–45 × 6–10 mm.

Varieties 5 (1 native in the flora): western North America.

5a. Lathyrus lanszwertii Kellogg var. **leucanthus** (Rydberg) Dorn. Vasc. Pl. Wyoming 171, 297. 1988. * Native

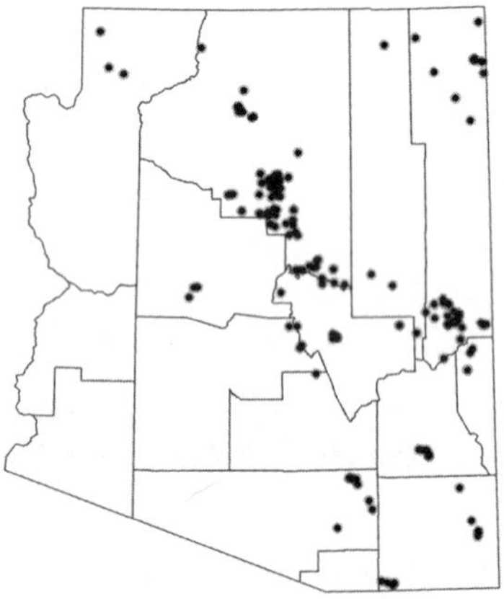

Stems erect, 1–3 dm. **Leaves** 2–3 cm, terminating in simple bristle, less than 1 cm. **Leaflets** 4–6, lanceolate and 25–50 × 5–15 mm, or linear and 35–70 × 2–5 mm. **Inflorescences** 3–5 cm. **Flowers** 10–15 mm; corolla white. $2n = 28$.

Flowering Apr–Aug. Slopes, canyons, meadows, stream banks, roadsides; (1430–)2000–3450 m; (Interior Chaparral, Madrean Evergreen Woodland, Great Basin Conifer Woodland), Petran Montane Conifer Forest, Petran Subalpine Conifer Forest; Apache, Cochise, Coconino, Graham, Greenlee, Gila, Maricopa, Mohave, Navajo, Pima, Pinal, Yavapai cos.; Colorado, New Mexico, Utah; Mexico (Sonora).

Lathyrus lanszwertii consists of five varieties distributed throughout the western states (Broich 2007). Variety *leucanthus* includes two distinct sympatric races: one with lanceolate leaflets (= *L. leucanthus*), and one with linear leaflets (= *L. arizonicus* Britton). This taxon is widely distributed across higher elevation areas in Arizona.

6. Lathyrus latifolius Linnaeus. Sp. Pl. 2:733. 1753. * [from Latin, *lati*, broad, and *folia*, leaf] PERENNIAL PEA, BROAD-LEAF EVERLASTING PEA Cultivated, Adventive

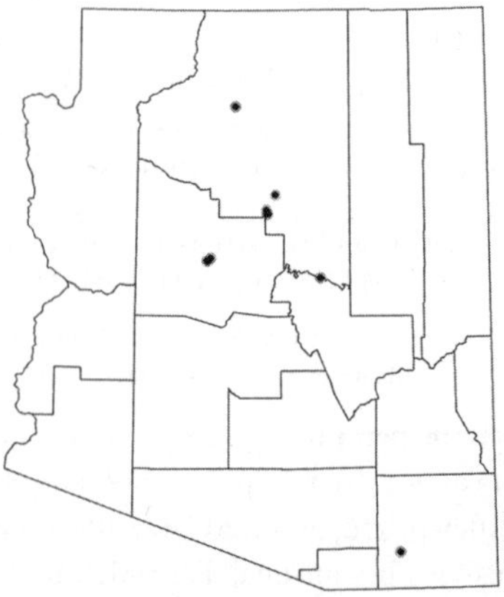

Herbs perennial, glabrous; from rhizomes. **Stems** broadly winged, sprawling or climbing, basally branched 1–5 times, 8–20 dm. **Leaves** 2–5 cm; stipules lanceolate, 30–40 × 7–15 mm, at least 1/2 stem width, much smaller than leaflets; rachises broadly winged, tendrils well developed, branched. **Leaflets** 2, lanceolate-elliptic, 30–120 × 7–50 mm, surfaces glabrous. **Inflorescences** 5–15-flowered, 10–33 cm. **Flowers** 15–23 mm; calyx lobes unequal, lateral lobes linear-triangular, shorter than tube; corolla purple, magenta, pink, or white, banner erect, blade longer than claw, wings equal to keel; ovary densely glandular-pubescent, style rotated 90° from ovary axis. **Fruits** 50–110 × 5–10 mm. $2n = 14$.

Flowering May–Sep. Canyons, stream banks, roadsides, disturbed areas; [50–]1180–2130 m; (Chihuahuan Desertscrub), Semidesert Grassland, Great Basin Conifer Woodland, Petran Montane Conifer Forest; Apache, Cochise, Coconino, Gila, Yavapai cos.; cultivated and adventive. Native to Europe.

Lathyrus latifolius is widely naturalized in North America as an escapee from gardens. It is weakly adventive in Arizona. The plants are occasionally cultivated in the state for their attractive flowers and interesting, winged stems. This species is widely reported to nodulate.

7. Lathyrus odoratus Linnaeus. Sp. Pl. 2:732. 1753. * [from Latin, *odor*, smell, for the fragrant flowers] SWEET PEA Cultivated, rarely escaping

Herbs annual, glabrous to sparsely pubescent. **Stems** broadly winged, climbing, basally branched 1–3 times, 8–30 dm. **Leaves** 2–6 cm; stipules lanceolate, 10–25 × 2–6 mm, smaller than leaflets; rachises winged, tendrils well developed. **Leaflets** 2, ovate or obovate, 15–50 × 10–40 mm, surfaces glabrous or sparsely pustulate-hirsute. **Inflorescences** 2–4-flowered, 10–20 cm. **Flowers** 15–25 mm; calyx lobes equal, lateral lobes linear-triangular to lanceolate, equal to tube; corolla white, pink, purple, violet, or blue, banner erect, blade much longer than claw, wings longer than keel; ovary densely pustulate-hirsute, style rotated 90° from ovary axis. **Fruits** 50–70 × 8–15 mm. $2n = 14$.

Flowering Apr–May. Disturbed areas; [50–400] m; Sonoran Desertscrub; Pima Co.; widely cultivated and rarely escaping. Native to the central Mediterranean region of Europe.

Lathyrus odoratus is cultivated as an ornamental and is an occasional escapee, not persisting in Arizona. There is a single record (*C.T. Mason 3172*, 05 Apr 1976) of a spontaneously growing plant in the state. *Lathyrus odoratus* is prized for its showy, fragrant flowers which are popular with gardeners and florists. The plants grow well during the cool months at lower elevations in Arizona, and during the warm months at higher elevations. Nodulation in *L. odoratus* is widely reported.

LEUCAENA Bentham. J. Bot. (Hooker) 4:416. 1842. * [Greek *leukos*, white, referring to the flowers]
Colin Hughes

Rhyncholeucaena Britton & Rose [=*Leucaena greggii* S. Watson]; *Caudoleucaena* Britton & Rose [=*Leucaena retusa* Bentham]

Small or medium-sized trees or large shrubs, unarmed. **Leaves** alternate, bipinnate; stipules ovate, awl-shaped or lanceolate with small asymmetric striately-nerved basal wings, persistent or tardily deciduous; rachis with a petiolar nectary immediately below first pair of pinnae; (2–)4–many pairs of pinnae. **Leaflets** few to many, opposite, short petiolate or nearly sessile, linear-oblong, elliptic, obovate or elliptic-lanceolate, always strongly oblique at base. **Inflorescences** globose heads, one to several in the axils of developing leaves, always with more than 30 flowers per head; the flower heads subtended by a ring of fused bracts at the base or just below the flower head. **Flowers** mimosoid, radially symmetric; floral bracts peltate; calyx tubular or bell-shaped, united for two-thirds of its length, with five blunt teeth; petals 5, 1-nerved, usually free or partially united; stamens 10, free to the base, usually white, but sometimes yellow or pink; anthers usually sparsely or densely pilose (not *L. retusa*) and this unique within mimosoid legumes. **Fruits** pendulous, dehiscent along one or both sutures, linear or linear-oblong, more or less flattened pods, the valves stiffly papery, in some species leathery, never truly woody. **Seeds** ovoid, laterally compressed, the seed coat glossy rich chestnut-brown, U-shaped pleurogram. $x = 26$ or 28.

Species 25 (2 cultivated in flora, 1 of these also adventive): in the American tropics from southern United States (Texas, New Mexico), Mexico, Central America, South America (to Peru). *Leucaena leucocephala* widely cultivated and naturalized pantropically.

Species of *Leucaena* are grown for timber, fuel, charcoal, livestock fodder, green manure, shade, soil improvement, and reforestation. The unripe fruits and seeds are harvested for human consumption (Allen and Allen 1981, Brewbaker 1987, Hughes 1998b).

SELECTED REFERENCES: Hughes 1998a, Govindarajulu, Hughes, and Bailey 2011.

1. Leaflets 13–21 pairs per pinna, 9–20 × 2–4.5 mm; flowers white ____ **1. Leucaena leucocephala**
1′ Leaflets (4–)5–6(–8) pairs per pinna, 20–26 × 10–12 mm; flowers yellow ____ **2. Leucaena retusa**

1. Leucaena leucocephala (Lamarck) de Wit. Taxon 10:54. 1961. * [Greek *leukos*, white, and *kephale*, head, referring to the white globose flower heads] LEAD TREE

Mimosa leucocephala Lamarck, Encycl. 1:12. 1783; *Acacia leucocephala* (Lamarck) Link

Small tree 2–10(–15) m, with a short clear trunk, 10–30(–50) cm in diameter, and upright angular branching. **Leaves** 12–15(–25) × (5–)7–16 cm; stipules 3.5–4 mm, ovate or lanceolate, persistent, but shrinking and inconspicuous on older

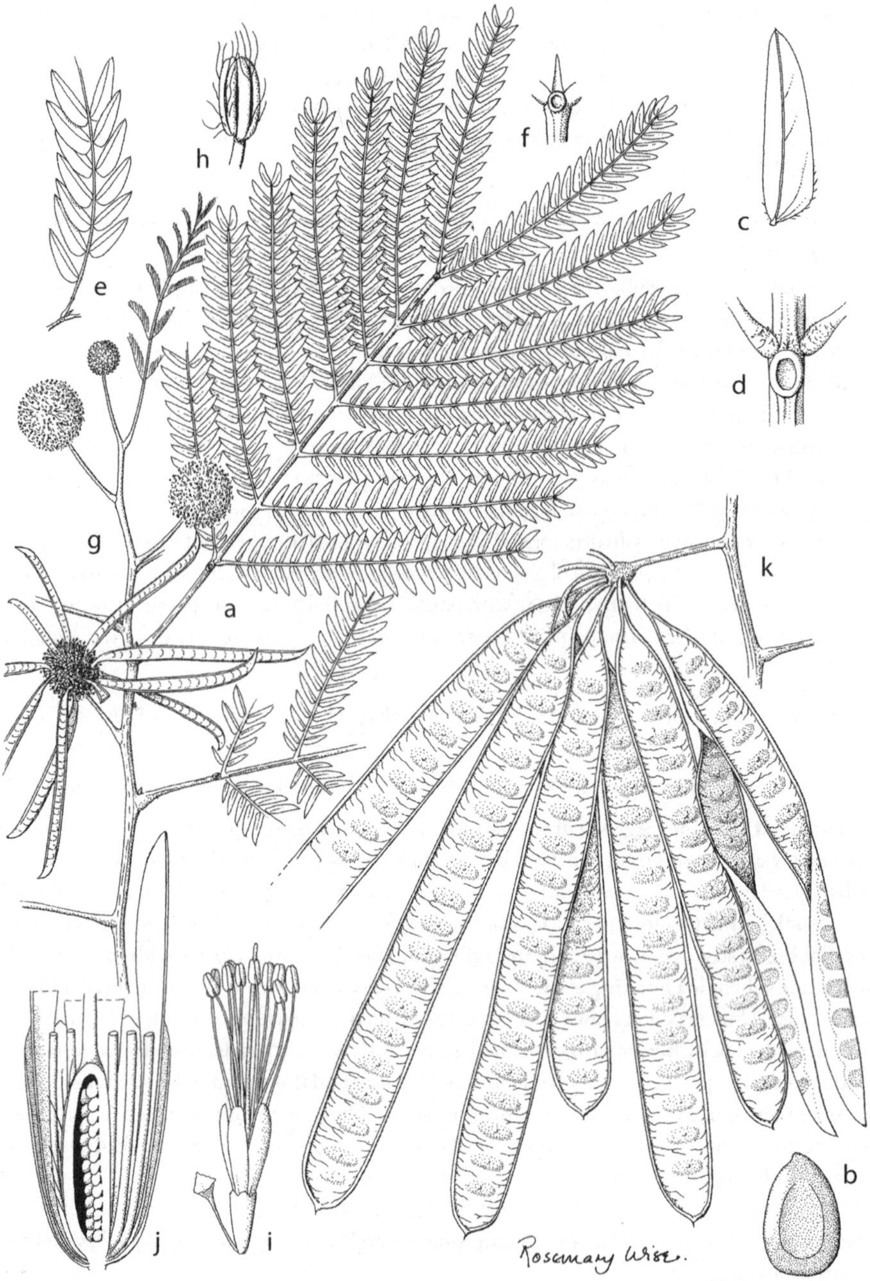

***Leucaena leucocephala*: a. leaf; b. seed; c. leaf; d. petiolar nectary; e. inset pinna; f. rachis nectary; g. flowering shoot; h. anther, i. flower; j. longitudinal section of flower; k. fruits. Illustrator: Rosemary Wise, reproduced from *Systematic Botany Monographs*, vol. 55, fig. 40, with permission of The American Society of Plant Taxonomists.**

shoots, glabrous or short white pubescent; petioles 13–34 mm with a single sessile, elliptic, concave extrafloral nectary at the base of the first pair of pinnae; pinnae (4–)6–8(–9) pairs, pinnular rachis 5–10 cm. **Leaflets** 13–21 pairs per pinna, 9–20 × 2–4.5 mm, linear-oblong, strongly asymmetric about mid vein, glabrous, mid-green and slightly discolorous. **Inflorescences** globose heads in fascicles of 2–6 in leaf axils on actively growing shoots, each flower head densely flowered with 100–180 flowers; peduncles 15–20 mm. **Flowers** subtended by 2–3 mm peltate glabrous bracts; calyx 2.3–3.8 mm, pubescent on lobes, pale greenish-cream; petals 4–5.6 mm, free at base, free or partially united above, pale green; stamen filaments 6–11 mm, white, anthers hairy, especially at distal end, pale cream-white; ovary 1.8–2.7 mm with

18–28 ovules, style 8–12 mm, white. **Fruits** 5–20(–45) per flower head; 11–19 × 1.5–2.1 cm, pendulous on 4–14 mm stipes, dehiscent along both sutures, linear-oblong, sometimes with a short curled beak, narrowly compressed, valves mid- to orange-brown, stiffly papery, the margins slightly thickened. **Seeds** 8–18; 6.7–9.6 × 4–6.3 mm, ovoid, aligned transversely in pod, pleurogram visible, deeply U-shaped. $2n = 4x = 104$, self-compatible tetraploid.

Subspecies 3 (1 cultivated in the flora and adventive).

1a. Leucaena leucocephala (Lamarck) de Wit subsp. **glabrata** (Rose) S. Zárate. Phytologia 63(4):305. 1987. * [becoming hairless with age] WHITE LEAD TREE; *GUAJE, GUAJE VERDE* Cultivated, Adventive

Leucaena glabrata Rose, Contr. U.S. Natl. Herb. 5(3):140. 1897

Small trees 4–15 m, 10–35 cm trunk diameter, arborescent with a short clear trunk to 2–3 m and an open irregular crown, dying back after a harsh frost, sometimes to the ground, but resprouting to form a multi-stemmed shrub; leafy shoots largely glabrous. **Leaves** (17–)19–25 × 12–16 cm; petioles 24–34 mm, glabrous; petiolar nectary 2–3 mm; leaf rachis 11–15.5 cm; pinnae (4–)6–8(–9) pairs, pinnular rachis 8–10 cm. **Leaflets** 15–21 pairs per pinna, 11–16 × 3–4.5 mm. **Inflorescences** 18–21 mm in diameter when flowers open; 120–180 flowers per flower head; peduncles 17–20 mm. **Flowers** bracts 2.4–3 mm; calyx 2.3–3.2 mm, pubescent on lobes, pale greenish-cream; petals 4–5.6 mm, free, pale green; ovary 2–2.7 mm with 24–28 ovules, style 10–12 mm. **Fruits** 6–20(–45) per flower head; 12–19 × 1.8–2.1 cm, rounded apically with a short curled beak, glabrous, slightly lustrous. **Seeds** 13–18; 7.6–8.2 × 5.5 mm.

Flowering and fruiting throughout the year as moisture permits. Riparian habitats, roadsides and other disturbed sites; 50–570 m; Sonoran Desertscrub; Maricopa, Pinal, Yuma cos.; Texas, New Mexico, s California; Mexico, Central America, n South America and widely cultivated throughout the tropics and subtropics.

The true natural distribution of *Leucaena leucocephala* subsp. *glabrata* remains unknown. Within Mexico it is extremely widely distributed as a cultivated tree, but no verifiably natural populations of this subspecies are known to date. In Arizona, it is cultivated and has not established reproducing populations but could become established in the future.

Leucaena leucocephala is an allopolyploid species with *L. pulverulenta* (Schlechtendal) Bentham and *L. cruziana* Britton & Rose as the likely diploid parents (Govindarajulu, Hughes, Alexander, and Bailey 2011). Hughes et al. (2007) speculate that *L. leucocephala* arose via spontaneous hybridization potentially facilitated by pre-Colombian cultivation of *Leucaena* as a minor food plant.

Leucaena leucocephala subsp. *glabrata* is a very common backyard, street, and orchard tree throughout Mexico, where it is cultivated for the production of unripe pods and seeds, which are consumed for food and widely marketed. Within Mexico it is known as the *guaje verde* in reference to its pods which are bright green when unripe.

While *Leucaena leucocephala* subsp. *leucocephala* was introduced pantropically by the late 19th century and is now widely naturalized and a well-known invasive weed in some areas, subspecies *glabrata* was introduced only from the 1950s onwards, but is now one of the most widely cultivated tropical trees. It is grown mainly for its leaves which have a high crude protein content and are a source of high quality forage for livestock (*the 'alfalfa of the tropics'*) or green manure in cropping. The wood is used for firewood and poles, and more generally, this subspecies is planted as a fast-growing multipurpose nitrogen-fixing tree in agroforestry and subsistence agriculture in many tropical countries (Brewbaker 1987, Hughes 1998b). In tropical agriculture this subspecies is sometimes known as the Salvador type following introduction of superior strains from El Salvador. The seeds are sometimes used to make necklaces and yield a gum that has commercial possibilities (NRC 1977). The plant has been used in folk medicine to treat stomach problems and as a contraceptive and abortifacient (NRC 1977). The fruits and wood yield dyes (NRC 1977). Nodulation in *L. leucocephala* is widely reported (Allen and Allen 1981).

While not generally available from Arizona nurseries, white lead tree is sporadically cultivated for shade in gardens at lower elevations in the state. It is apparently most frequent in Maricopa County (Tempe and Phoenix), but also found in Pinal County. Growth is very rapid, often 5 m or more per year with ample irrigation. Regular irrigation is necessary during warm months to maintain an acceptable appearance and promote growth. The plants remain evergreen in mild-winter locations. White lead tree suffers stem damage at temperatures of -4°C and is severely damaged below -6°C. Recovery is extremely rapid the following season, as trees coppice or resprout readily following cutting or frost damage. The abundant fruits create considerable litter, and cold-damaged stems require pruning. The tree readily reseeds wherever moisture is available and is occasionally adventive and weedy on disturbed ruderal sites. The seedlings develop a deep taproot and are difficult to remove if allowed to remain for 2 years or longer. The plants grow readily from scarified seeds and can be propagated from stem cuttings.

2. Leucaena retusa Bentham in Gray, Plantae Wrightianae 1:64. 1852. * [*retuse*, with a shallowly notched, rounded apex of leaflets] LITTLE-LEAF LEAD TREE, GOLDEN-BALL LEAD TREE, LEAD TREE [lead tree referring to the dense wood and golden-ball to the bright yellow globose inflorescences], WOOHOO; *HUISACHE* (Texas) Cultivated

Acacia sabeana Buckley; *Caudoleucaena retusa* (Bentham) Britton & Rose

Large shrubs and small trees (1.5–)2–5(–8) m, 10–15 cm trunk diameter, sometimes multi-stemmed from base and forming a small round crown. **Leaves** 18–25 × 15–22 cm; stipules 7–7.5 mm, ovate, long-pointed or subulate at apex with asymmetric membranous wings at base, striate, midrib prominent, persistent, glabrous; petioles 40–50 mm; in addition to petiolar nectary, rachis with a series of cylindrical columnar peg-shaped extrafloral nectaries, 1–3 mm tall on ventral side of petiole and rachis at base of each pair of pinnae; pinnae (2–)3–4(–5) pairs, pinnular rachis 9–11 cm. **Leaflets** (4–)5–6(–8) pairs per pinna, 20–26 × 10–12 mm, weakly asymmetric, obliquely obovate or elliptic lanceolate, rounded and apiculate, rarely slightly retuse apiculate, entire, villous pubescent on young leaflets soon glabrous at maturity, thin, blue-green. **Inflorescences** globose heads in fascicles of 2–4 in leaf axils on actively growing shoots, each flower head with 150–190 flowers, the unopened flower heads 'cone-like' in appearance; peduncles 25–90 mm with a pair of caudate bracts at distal end. **Flowers** subtended by peltate 5–7 mm caudate villous floral bracts conspicuously exserted in bud; calyx 4–4.5 mm, glabrous, pale yellow-green; petals 5.4–6.6 mm, free, pale yellow-green; stamen filaments 12–13.5 mm, bright yellow, anthers glabrous (unlike most species of *Leucaena*), the connective with a small apiculum, yellow; ovary 2 mm with 16–28 ovules, style 12.5–14 mm, bright yellow, narrowing to a tubular stigma held above the anthers. **Fruits** 1–2(–5) per flower head; (12–)16–20(–25) × 1.1–1.3 cm pods, pendulous on sturdy 4–5 mm stipes, tardily dehiscent along both sutures, linear or somewhat falcate, terminating in a thickened persistent style, compressed, valves green and slightly fleshy when unripe, turning mid-brown and leathery when ripe, the margins thickened. **Seeds** 13–15; 7–8 × 4.5–6.5 mm, rhomboidal, aligned obliquely in pods, pleurogram deeply U-shaped. $x = 28$.

Flowering April–June; fruiting August–November; partially or completely leafless during the cold winter season from December to February. Texas, New Mexico; Mexico (Chihuahua, Coahuila).

Leucaena retusa is the most northerly species of *Leucaena,* occurring naturally from 26°43'N to 31°52'N and extending into mountains as high as 1900 m, and is thus the most cold-tolerant species in the genus. Across its range, *L. retusa* has been found from 360–1920 m elevation. It occurs in the north of Mexico, mainly in Coahuila and the southern U.S., mainly in south-central and western Texas. Sporadic collections from New Mexico and Chihuahua, Mexico are clearly outliers, and the species is apparently very rare in those states.

Leucaena retusa grows in moist to semi-arid montane and canyon sites in mixed pine-oak forest with *Pinus ponderosa*; scrub oak-juniper forest with *Quercus fusiformis*, *Q. grisea*, *Q. intricata* and *Juniperus ashei*, *Fraxinus*, *Eysenhardtia*, *Rhus*, *Berberis*, *Pistacia,* and *Cercis*; and in subtropical deciduous forest and semi-arid thorn scrub with *Senegalia berlandieri*, *Mimosa*, *Prosopis*, *Opuntia*, *Agave,* and *Yucca*. In the drier portions of its range *L. retusa* is often confined to canyons and along rocky arroyo courses where it forms small, open groves. The species occurs primarily on limestone substrates but is also found in igneous soils. The foliage is browsed by cattle and deer (Vines 1960, Powell 1988).

The relatively small mature size, unarmed stems, lush foliage, and attractive yellow pom-pom-like flower heads make golden-ball lead tree suitable as a patio or garden tree. It is occasionally cultivated as an ornamental at lower elevations in Arizona, for example at both the Glendale Xeriscape Botanical Garden and the Desert Arboretum Park, Maricopa County, and at Boyce Thompson Arboretum, Pinal County. It is also suited for transition areas and provides a tropical appearance in the landscape. It can be especially effective when several trees are planted to form a small grove, providing shade for succulents and flowering perennials. Golden-ball lead tree is adapted to a variety of soils including rocky soils. This species is typically slow-growing, but has a moderate growth rate with sufficient irrigation. Regular irrigation during hot weather is necessary to maintain an attractive appearance. The leaves are winter-deciduous. Golden-ball lead tree appears to be hardy to at least -12°C. The fallen fruits and leaves produce some litter. Individual limbs on some cultivated plants periodically die and the branches can be brittle. This may be the result of insufficient irrigation. Propagation is by scarified seeds.

LIBIDIBIA (de Candolle) Schlechtendal. Linnaea 5:192. 1830, descr. emended E. Gagnon and G.P. Lewis * [from *"dividivi"* or *"libi-dibi,"* local names for some species]

Matthew B. Johnson

Shrubs and trees. **Stems** with bark smooth and exfoliating in thin plates and having a mottled appearance, or fissured and peeling into vertical strips. **Herbage** glabrous or puberulent. **Leaves** odd bipinnate [or once pinnate]; stipules quickly deciduous. **Leaflets** opposite, margins entire. **Inflorescences** axillary and terminal racemes or panicles; bracts absent or present. **Pedicels** jointed. **Flowers** caesalpinoid, mostly with slender papillae; calyx with a top-shaped tube and 5 nearly equal or unequal lobes, the lobes deciduous leaving a persistent hypanthium; corolla with 5 petals, yellow or white, about as long as the calyx, 1 petal reduced or not; stamens 10, free, somewhat longer than the petals, exserted, anthers yellow or red; ovary stipitate, style and stigma simple. **Fruits** brown or black (or red), indehiscent, thick, leathery or woody, compressed. **Seeds** oblong, compressed, smooth.

Species 7+ (1 cultivated in the flora): Mexico, Caribbean, and South America.

Libidibia is in need of revision and there are likely unnamed species (Gagnon et al. 2016). The wood of *Libidibia* species is valued for timber, fuelwood, and for making charcoal. It is also used in turnery and for making musical instruments. The fruits contain tannins and have been used commercially in leather production. Fruits are also used for fodder, to make ink, and in folk medicine. Species of *Libidibia* are planted as landscape trees and have potential for use in agroforestry.

SELECTED REFERENCE: Gagnon et al. 2016.

1. Libidibia paraguariensis (D. Parodi) G.P. Lewis. Pl.-Book (ed. 3) 1021. 2008. * [of Paraguay] *GUAYACÁN* Cultivated

Acacia paraguariensis D. Parodi, Revista Farm. 3:7. 1862; *Caesalpinia paraguariensis* (D. Parodi) Burkart

Trees 8–15(–18) m. **Stems** with bark mottled with various shades of green, gray, and tan or predominantly of a single color, exfoliating in thin irregular plates. **Herbage** glabrous. **Leaves** 5.5–8 cm with 2–4 pairs of pinnae plus a terminal pinna, pinnae 2.1–5 cm; stipules 0.5–1 mm, membranous, linear; petioles 0.7–1.3 cm; rachis 1–3.2 cm. **Leaflets** 4–10 pairs per pinna, 3–9 × 1.5–4 mm, oblong. **Inflorescences** axillary racemes, 1.5–4 cm, with 8–15 flowers; ovate bracts 0.5–1 mm, quickly deciduous. **Pedicels** 3–4 mm. **Flowers:** calyx 8–10 mm, yellow green with an orange blush towards the base at anthesis, the lobes 5–5.5 mm, the lower (abaxial) lobe wider than the others; corolla yellow, 10–12 mm across, the banner wider than the other petals and often with a faint orange area in the center; stamens with filaments yellow, 12–13 mm, with sparse, minute, stalked glands along most of their length, the anthers yellow green; pistil orange with the distal-most portion of the style and the stigma yellow green. **Fruits** brown or blackish, 1.7–3.8 × 1.5–2 cm, nearly disk-shaped to oblong with a somewhat irregular, lustrous surface; woody; stipe 4–10 mm; apex blunt with a mucro 0.5–1.5 mm. **Seeds** brown or greenish brown, 6–9 × 4.5–6 mm, lustrous.

Flowering May–Aug. South America (Argentina, Bolivia, Brazil, Paraguay).

Libidibia paraguariensis is widespread in the Gran Chaco region of South America where it is a characteristic plant of semiarid and subhumid forests on plains, slopes, and along watercourses. It grows as scattered individuals or in small groves and is seldom abundant in drier habitats. In more humid regions in the central and eastern Chaco, it can be a common overstory species (López et al. 1987). However, in the southernmost portion of its range in Argentina, it has been practically eliminated due to overharvesting (Aronson and Saravia Toledo 1992). It occurs at elevations from 50–1150 m (Aronson and Saravia Toledo 1992). Annual rainfall across the range of *L. paraguariensis* varies from 375–1350 mm, and the trees experience temperature extremes of 45°C in summer and -7.5°C in winter (Aronson and Saravia Toledo 1992). Mature size appears to be correlated with moisture availability. Near the dry limits of its distribution, trees are often only 4 m tall or less, while in wet areas the trees can reach at least 18 m tall with a high canopy and a trunk to 1 m in diameter (Aronson and Saravia Toledo 1992). *Libidibia paraguariensis* is estimated to reach an age of 700 years in habitat (Saravia Toledo and Del Castillo 1986). *Libidibia paraguariensis* is found in a wide range of soils from sandy to clay with a pH range of 6–7.5 but appears to be intolerant of alkaline or saline soils (Aronson and Saravia Toledo 1992). The species has an extensive root system and withstands high winds as well as inundation for periods of up to 150 days (Aronson and Saravia Toledo 1992). Aronson and Saravia Toledo (1992) indicate that seeds of *L. paraguariensis* are dispersed by a variety of wildlife as well as livestock.

The wood of *Libidibia paraguariensis,* known as brown ebony in commerce, is highly valued and is extensively harvested from wild populations. The trees coppice well following cutting but the new stems must be protected from livestock (Aronson and Saravia Toledo 1992). Leonardis (1975) indicates that the texture of the wood is fine and homogenous with an interlocking grain that makes it difficult to work. The sapwood is pinkish white, and the heartwood is dull brown or purplish to almost black (Leonardis 1975). The wood is heavy and hard with a specific gravity of 1.180 kg/dm^3 and a calorific value of 4200 kcal/kg (Leonardis 1975). Large trees are difficult to find near settlements as it is the preferred species for making charcoal across much of its range. The wood is highly resistant to decay in the ground as well as in water and has been used in construction (for posts, gates, railroad ties, bridge frames, and pilings) and for making furniture (Leonardis 1975). The wood is also used in the manufacture of musical instruments including tuning

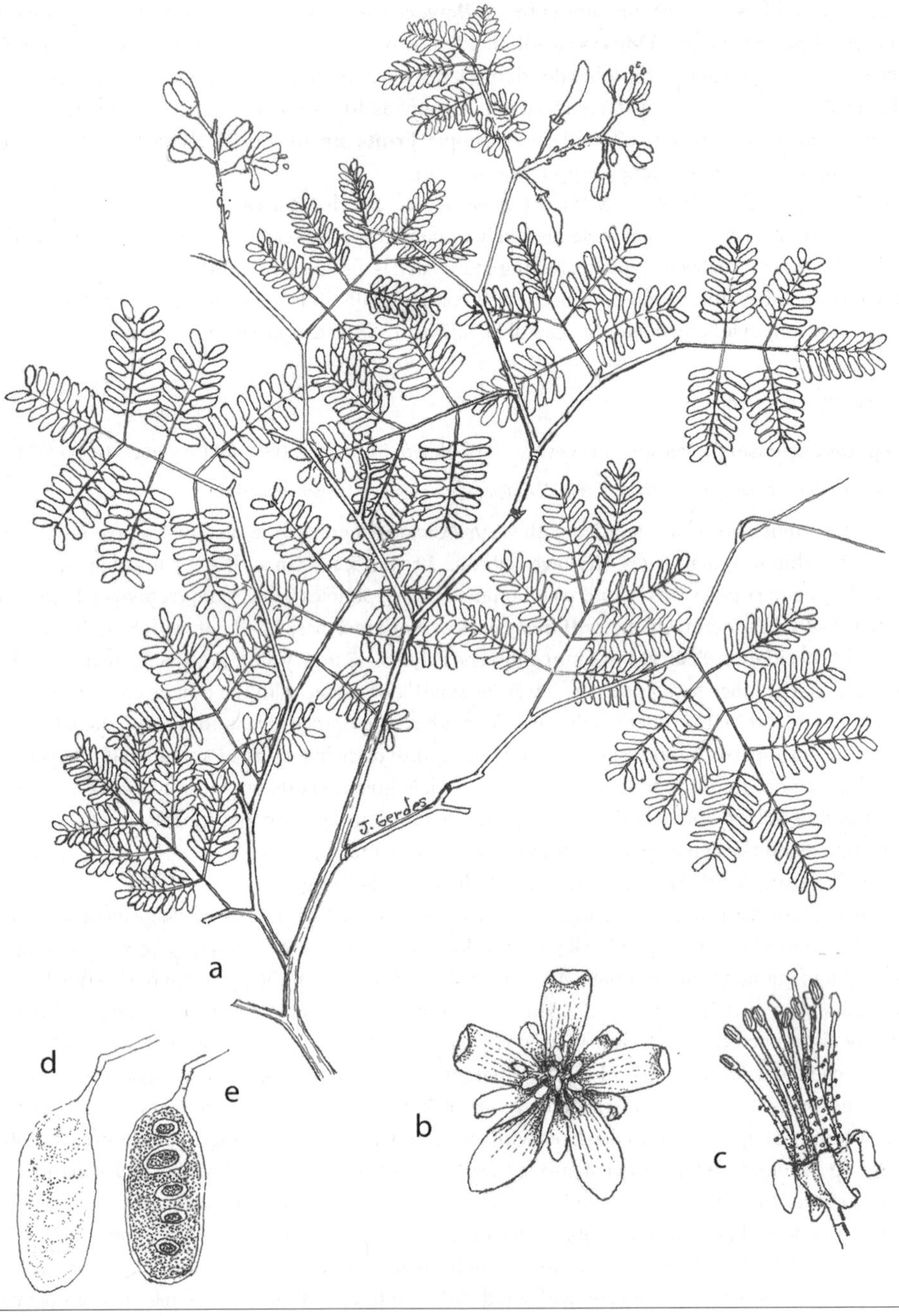

***Libidibia paraguariensis*: a. habit; b. flower; c. stamen and pistil; d. fruit; e. longitudinal section of fruit. Illustrator: John H. Gerdes.**

forks and bridges and pegs for guitars and violins (Leonardis 1975). Aronson and Saravia Toledo (1992) present an excellent overview of *L. paraguariensis* which discusses its potential as a superior forage tree for silvopastoral systems, for wood production, and for other uses. They indicate that the trees produce quantities of fruits reliably each year, even during drought, and that the tannins in the fruits are not a problem for livestock. Both the foliage and pods are considered valuable livestock forage. A particular advantage of *L. paraguariensis* is that the fruits gradually fall over a period of 7–9 months, providing forage over much of the year (Aronson and Saravia Toledo 1992). *Libidibia paraguariensis* was formerly used to a limited extent for tanning leather. The fruits contain 15–25% tannin, and the wood contains 8% tannin (Leonardis 1975). Indigenous peoples in the Chaco utilize the fruits to produce a bluish black dye (López et al. 1987,

Aronson and Saravia Toledo 1992). An extract from the bark has been used in folk medicine as a remedy for dysentery and for treating skin problems and syphilis (López et al. 1987). Extracts of *L. paraguariensis* have shown strong activity against a variety of bacteria including antibiotic resistant strains in laboratory studies (Woldemichael et al. 2003). Research by Sgariglia et al. (2011) suggests that bark infusions of *L. paraguariensis* may be suitable as a natural food preservative. The flowers are frequented by various bees which produce honey of excellent quality (López et al. 1987, Aronson and Saravia Toledo 1992).

Guayacán is a recent introduction to the Arizona landscape palette. It has excellent potential as a specimen and large patio tree. It is suitable for planting in parks and as a street tree, though its apparent slow growth rate in this region is a limitation. It is likely adaptable for use as a lawn tree. The bark is particularly attractive and is reminiscent of some species of *Eucalyptus*. The canopy produces light filtered shade ideal for many plants that do poorly in full sunlight. Small yellow flowers are scattered through the canopy during the summer, followed by the persistent fruits that remain on the leafless branches through the winter months to provide interest. The leaves are deciduous with freezing. Emerging leaves in the spring are a beautiful rust color, turning to green as they develop. *Guayacán* may grow 0.6 m per year with irrigation but is relatively slow to develop into a tree in Arizona horticulture. Though it can reach large sizes in wet areas, it is unlikely to become nearly so large in cultivation in Arizona and other dry regions. Periodic deep irrigation will maintain appearance and growth though established plants are drought resistant. Plants in Arizona horticulture appear to be hardy to at least -6.5°C. Two cultivated plants in Tucson had the outer twigs killed at -8°C but recovery was rapid the following season. This species develops a low, horizontal branching habit when young and may require considerable pruning and staking to develop a tree form. Litter is minimal. Propagation is by scarified seeds. Aronson and Saravia Toledo (1992) recommend some shading for *guayacán* seedlings during their first several months to avoid heat stress and suggest the species as an excellent bonsai subject.

LOTUS Linnaeus. Sp. Pl. 2:773. 1753. * [from the Greek, *lotus*, which has been applied to various plants] TREFOIL
Matthew B. Johnson and Kirsten N. Lake

Annual and perennial herbs, [subshrubs, and shrubs]. Leaves alternate, odd pinnate (palmate or simple); stipules gland-like; petiole present or absent; stipels absent. **Leaflets** with margins entire. **Inflorescences** umbels or flowers solitary, axillary; bracteate; bracteoles usually absent. **Pedicellate. Flowers** papilionoid; calyx cylindrical or bell-shaped with 5 free lobes or 2 upper lobes sometimes partially united; corolla yellow (white or pink); stamens with 9 filaments fused and 1 free, these often 5 longer and 5 shorter, anthers uniform; ovary sessile, style thread-like; stigma terminal or lateral, often head-like. **Fruits** cylindrical, oblong, or ovoid, exserted or included within calyx, usually prominently beaked, dehiscent or indehiscent. **Seeds** 2–many. $x = 12, 24$.

Species ca. 125 (1 naturalized in the flora): mostly temperate areas of Europe to east Asia and the Mediterranean Region, with several species in Africa, 2 species in Australia, and 1 in the tropical western Pacific. Forage species have been widely introduced and have sometimes become naturalized.

As treated here, *Lotus* comprises taxa native to the Old World. New World taxa that are found in Arizona and were formerly treated as *Lotus* are placed in *Acmispon* and *Hosackia*. *Lotus corniculatus* and *L. pedunculatus* are widely grown for livestock forage and for soil improvement, as cover crops and green manure (Allen and Allen 1981, Lewis et al. 2005). Other species have been used in cosmetics (Lewis et al. 2005), and a few are grown for their showy flowers. Nodulation is widely reported for the genus (Allen and Allen 1981).

1. Lotus corniculatus Linnaeus. Sp. Pl. 2:775–776. 1753. * [from Latin, *corniculatus*, horned] BIRD'S FOOT TREFOIL, COMMON BIRD'S FOOT TREFOIL Naturalized

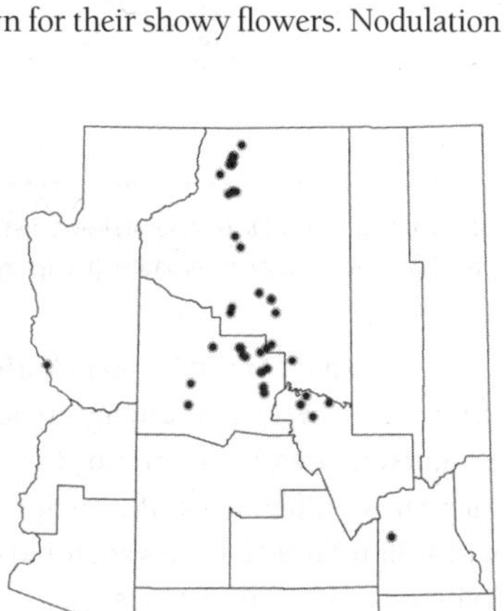

Perennial, mat-forming herbs to 30 (60) cm high. **Stems** decumbent or ascending, to 50 (60) cm, glabrous or strigose. **Leaves** pinnate, with the proximal pair of leaflets stipular in position and the distal 3 leaflets appearing palmate; stipules sometimes not apparent; petiole absent; rachis 2–6 mm. **Leaflets** 5, 5–20 × 1–4.5 mm, linear, lanceolate, elliptic, obovate, or ovate. **Inflorescences** umbels with 3–8 flowers, peduncles 5–10 cm; bracts 3–5 mm, lanceolate; bracteoles absent. **Pedicels** 1–1.5 mm. **Flowers** with calyx 2–3.5 mm, lobes as long as the tube; corolla yellow, 8–14 mm; stamens with 5 long and 5 short filaments. **Fruits** 15–35 × 2–3 mm; narrowly cylindrical; dehiscent with valves twisting. **Seeds** brown to blackish, nearly spherical, 1.1–1.4 mm.

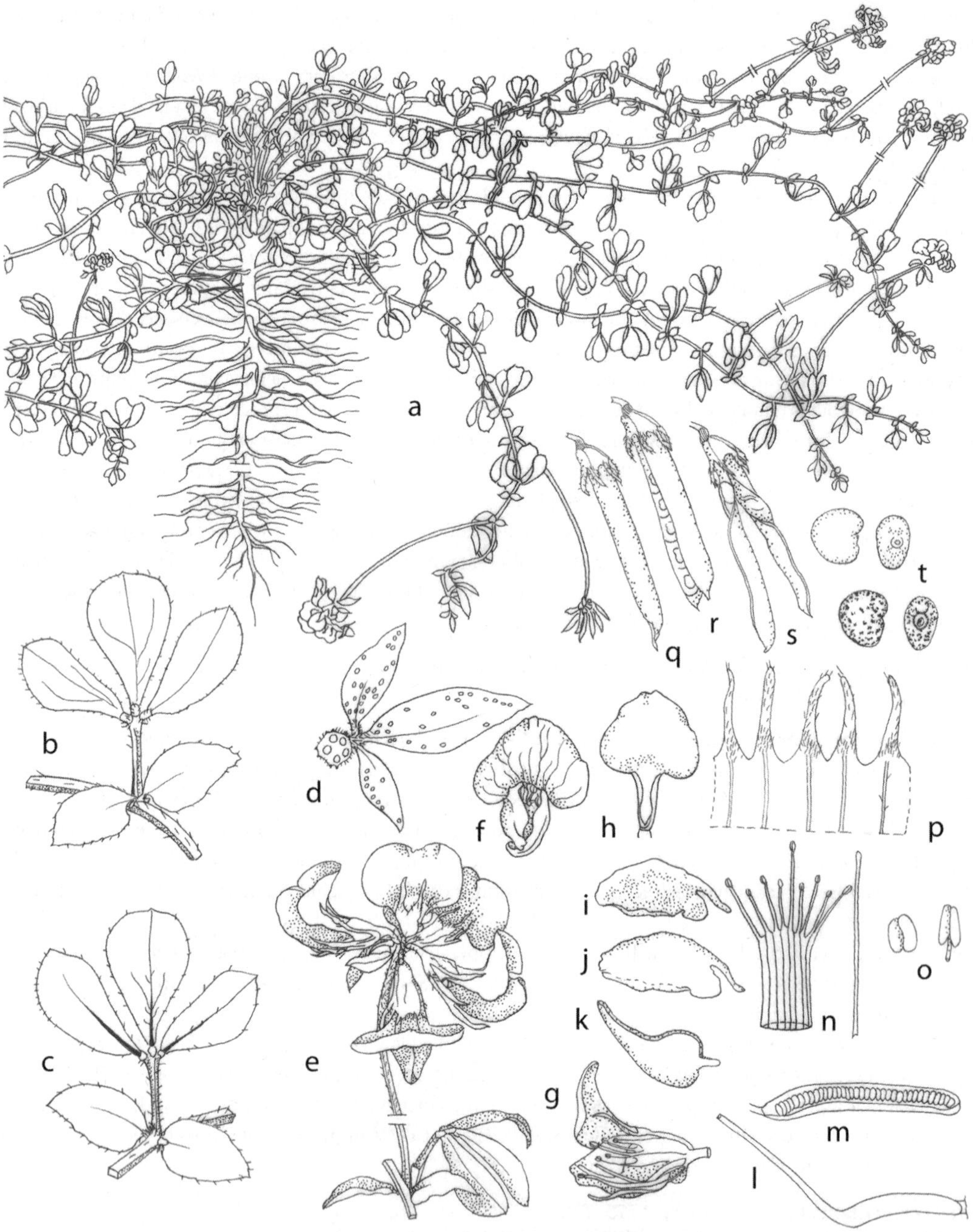

***Lotus corniculatus*: a. habit; b., c. sessile leaves, five leaflets; d. leaf, three leaflets, associated with inflorescence; e. inflorescence; f., g. flowers; h., i., j., k. petals, l. pistil; m. ovary; n. stamens; o. anther; p. calyx; q, r, s. fruits showing dehiscence; t. seeds. Illustrator: Margaret Pope.**

Flowering (Apr–)May–Sep. Various soils of slopes, meadows, floodplains, arroyo banks, marshy areas, and pastures, fields, irrigation ditches, roadsides, parking lots, and other disturbed areas; [near sea level–]910–2920 m; Semidesert Grassland, Interior Chaparral, Madrean Evergreen Woodland, Great Basin Conifer Woodland, Petran Montane Conifer Forest, Petran Subalpine Conifer Forest; Coconino, Gila, Graham, Pima, Santa Cruz, Yavapai cos.; widely introduced for cultivation and extensively naturalized in the United States and elsewhere; native and widespread in Eurasia and parts of Africa.

Lotus corniculatus is rather widely naturalized in Arizona, usually in areas of human disturbance. It is also found in more remote locations, especially along riparian corridors. At lower elevations in the state it is narrowly confined to sites with moist soil. Healthy plants produce an abundance of seeds. This species does not appear to pose any environmental risks in Arizona though it may locally compete with native plants. It is considered a problem species in some states in the U.S.

Lotus corniculatus is widely cultivated for forage, green manure, as a cover crop, and for erosion control. Allen and Allen (1981) state that it is long-lived, deep-rooted, and drought resistant. Under certain conditions, the plants produce cyanogenic compounds (Mabberley 1987) but do not appear to cause problems with livestock. Cured hay of *L. corniculatus* is comparable to that of other high-quality legumes and is non-bloating (Duke 1981). It is especially favored by horses and cattle (Duke 1981). Gambel's quail have been found to eat the seeds of *L. corniculatus* (Graham 1941). The species is reported to nodulate (Allen and Allen 1981). Pollination is primarily by bees (Duke 1981). The common name, bird's foot trefoil, is from the appearance of the clusters of fruit.

Lotus corniculatus is readily grown from scarified seeds. Seedlings are initially weak and slow to establish. It may also be propagated from stem cuttings (Duke 1981). A sunny location is necessary for optimum performance. The plants will flower in their first year. This species is tolerant of a wide range of soil conditions including those that are infertile, poorly drained, acidic, and mildly alkaline (Duke 1981). Is also grows on coastal dunes. There are numerous cultivars of *L. corniculatus* (Duke 1981).

LUPINUS Linnaeus. Sp. Pl. 2:721. 1753. * [derivation uncertain; Latin *lupinus*, wolf, or *lupe*, sadness, and *inus*, possession, perhaps alluding either to plants supposedly overrunning the ground as an animal might or to harsh taste of seeds causing facial contortion] LUPINE, BLUEBONNET; *LUPINO*

Teresa Sholars

Herbs, annual, biennial, or perennial, [shrubs, or subshrubs], unarmed; usually from taproots or woody crowns, rarely rhizomes. **Cotyledons** usually deciduous, usually petiolate but can be sessile fused into a cup or disk. **Stems** erect to decumbent or prostrate, usually pubescent, sometimes glabrous. **Leaves** alternate, usually palmately compound, [rarely 3-foliolate or unifoliolate], usually cauline, sometimes crowded near base, sometimes basal; stipules present, bristle-like to leaf-like, fused to petiole; petiolate. **Leaflets** [(1 or 3)] 5–11(–17), stipels absent, blade margins entire, surfaces glabrous or pubescent. **Inflorescences** 1–100±flowered, terminal and axillary racemes, erect, pedunculate, axillary rarely reduced to 1 or 2 flowers; bracts present, persistent or deciduous. **Flowers** papilionoid, chasmogamous, scattered or verticillate; calyx two-lipped, lobes united, appearing as lower and upper lip, these entire or toothed, usually with appendages (often inconspicuous) between lobes; corolla usually blue to purple, sometimes white, yellow, pink, or rose; banner with central groove, sides reflexed; wings touching at tips, corrugated; keel usually attenuate; stamens 10, united; anthers basifixed, dimorphic, alternately long on short filaments, short on long filaments; style brushy. **Fruits** legumes, sessile, straight or curved, usually oblong, papery, splitting along both margins, valves usually twisted after dehiscence, usually pubescent, rarely glabrous. **Seeds** (1 or) 2–12, usually smooth, rarely ridged or tuberculate, spheric, lentiform, or angular. $x = 6$.

Species 287 (19 in the flora, 17 native, 2 naturalized and one of these also cultivated): North America, South America, Europe (Mediterranean), Africa; introduced widely in s Africa, Pacific Islands (New Zealand), Australia.

Most species of *Lupinus* occur in western North America and western South America.

The taxonomy of *Lupinus* is complicated. The Integrated Taxonomic Information System has 1,010 names for Lupines; circumscription is difficult, made problematic by the vast number of species recognized then lumped and split in various ways by different taxonomists. Lupines exhibit some of the highest known rates of net diversification in plants (Drummond et al. 2012). Many papers show gene flow and introgression through outcrossing in perennials. Self-pollination is predominant in annual *Lupinus*, and that characteristic has resulted in the establishment of localized variants that have been recognized as distinct species. For recent phylogenetic work on *Lupinus* see Drummond et al. (2012).

Some species of *Lupinus* contain alkaloids, especially in their seeds, fruits, and young leaves, that are toxic to humans and livestock, especially sheep. Species of "sweet" lupines (e.g. *L. albus* from Europe and *L. mutabilis* from the Andes Mountains of South America) produce edible seeds that are grown for human consumption as a grain and as a vegetable. Several species have been grown for cover crops and green manure though they are now seldom utilized for this purpose (Allen and Allen 1981). Most lupines produce showy flowers and some are grown as wildflowers or garden plants. Cultivars and hybrids with a range of flower colors have been developed from species such as *L. polyphyllus* and *L. texensis*. The Texas bluebonnet, *Lupinus texensis*, is the state flower of Texas. Nodulation is widely reported for the genus.

1. Herbs annual.
  2. Cotyledons sessile, fused into a persistent disk or cup (if deciduous, leaving a circular scar); seeds 1 or 2(–6), usually tuberculate or rugose.
    3. Flowers in crowded or widely spaced whorls; bracts reflexed; upper keel margin ciliate near claw ____ **11. Lupinus microcarpus**
    3′ Flowers usually scattered or solitary; bracts straight; upper keel margin glabrous.
      4. Leaves basal; herbage usually glabrous, sometimes sparsely pubescent when young, rarely at anthesis; pedicels 3–5 mm ____ **13. Lupinus odoratus**
      4′ Leaves basal and/or cauline (often crowded near base); herbage sparsely pubescent to canescent or pilose; pedicels 0.5–3 (–4) mm.
        5. Herbage canescent, hairs 0.6–1 mm; fruits undulate, sides with short inflated hairs becoming scaly when dry ____ **17. Lupinus shockleyi**
        5′ Herbage hairs appressed or spreading, more than 1 mm; fruits not obviously undulate, thinly pilose to coarsely hirsute.
          6. Upper calyx lobe more than 1/2 as long as lower lobe ____ **8. Lupinus kingii**
          6′ Upper calyx lobe less than 1/2 as long as lower lobe.
            7. Racemes head-like; seeds smooth ____ **4. Lupinus brevicaulis**
            7′ Racemes usually elongate; seeds wrinkled or ridged.
              8. Leaves crowded near base; racemes exceeding foliage; peduncles 3–5 cm; fruits 0.5–1 × ± 0.5 cm, upper margin not constricted between seeds ____ **6. Lupinus flavoculatus**
              8′ Leaves well distributed along stems; racemes shorter than or slightly exceeding foliage; peduncles 0–1 cm; fruits 1.5–2 × 0.55–0.7(–0.8) cm, upper margin constricted between seeds ____ **15. Lupinus pusillus**
  2′ Cotyledons petiolate, usually withering and deciduous; seeds usually more than 2, usually smooth.
    9. Lower, and often upper, margin of keel petals ciliate near claw, glabrous near apex.
      10. Flowers distinctly whorled; corollas usually blue-purple, rarely white, lavender, or pink ____ **19. Lupinus succulentus**
      10′ Flowers scattered (if appearing ± whorled, pedicel bases clearly spiraled, not whorled); corollas yellow, white, pink to magenta, or blue to purple.
        11. Leaflet blades 5–10 mm wide, surfaces glabrous above (adaxially); corollas dark pink to magenta ____ **2. Lupinus arizonicus**
        11′ Leaflet blades 2–4 mm wide, surfaces pubescent above (adaxially) at least near margins; corollas usually blue, rarely pinkish ____ **18. Lupinus sparsiflorus**
    9′ Lower and upper margin of keel petals glabrous near claw, upper margin ciliate near apex, or glabrous throughout.
      12. Flowers scattered; bracts usually persistent; keel margins glabrous ____ **5. Lupinus concinnus**
      12′ Flowers whorled at some or all nodes; bracts deciduous; upper keel margin usually ciliate near apex ____ **3. Lupinus bicolor**
1′ Herbs perennial.
  13. Upper keel margin ciliate near claw or only from claw to middle (glabrous middle to tip) ____ **9. Lupinus latifolius**
  13′ Upper keel margin usually glabrous or ciliate throughout, or ciliate from middle to tip.
    14. Calyx spur 1–3 mm ____ **1c. Lupinus argenteus** var. **argophyllus**
    14′ Calyx bulge or spur 0–1 mm.
      15. Banner back hairy.
        16. Flower 6–8 mm; banner back thinly hairy ____ **1e. Lupinus argenteus** var. **hillii**
        16′ Flower 8–14(–18) mm; banner back densely hairy ____ **16. Lupinus sericeus**
      15′ Banner back glabrous.
        17. Racemes dense; bracts persistent; stems very short or absent, densely tufted ____ **10. Lupinus lepidus**
        17′ Racemes ± open (scattered or whorled); bracts usually ± deciduous; stems present.
          18. Herbs 0.1–3.5(–5) dm; stems usually ± prostrate to decumbent, rarely ascending ____ **7. Lupinus huachucanus**
          18′ Herbs (1–)2–15 dm; stems usually erect, ascending, or spreading, rarely decumbent.
            19. Longest leaflet blades 20–40 mm; fruits 2 cm wide ____ **12. Lupinus neomexicanus**
            19′ Longest leaflet blades (20–)30–95 mm; fruits 0.8–1 cm wide.
              20. Basal leaves not present at anthesis, if present then petioles less than 3 times as long as leaflets ____ **1. Lupinus argenteus** (in part)
              20′ Basal leaves present, long-petiolate, more than 3 times the leaflets at anthesis ____ **14. Lupinus polyphyllus**

1. Lupinus argenteus Pursh. Fl. Amer. Sept. 2:468. 1814 * [from Latin, *argent*, silver, silvery] SILVER LUPINE

Herbs, perennial, 1–15 dm, green and glabrous or silvery-hairy; from superficial or shallowly buried root crown. **Cotyledons** deciduous, petiolate. **Stems** erect or ascending, green or purplish, clustered, branched or unbranched. **Leaves** basal and/or cauline; stipules 2–12 mm; petiole 1–15 cm. **Leaflets** (5 or) 6–10, blades 10–60 × 4–10 mm, lower (abaxial) surface hairy, upper (adaxial) surface glabrous or hairy. **Inflorescences** loose racemes, 5–16(–25) cm, flowers whorled or not; peduncles (1–)4–25(–30) cm; bracts usually deciduous, 3–4 mm. **Pedicels** (1–)2–5(–7) mm. **Flowers** 5–15 mm; calyx 4–8(–10) mm, bulge or spur 0–3 mm, may be variable on plant, lower lobe entire or 3-toothed, upper lobe entire or 2-toothed; corolla usually blue, sometimes purple, violet, pink, lilac, or white, banner patch yellowish to whitish, blue, brown, or absent; banner not much reflexed-recurved beyond midpoint, this less than 3 mm proximal to apex, banner back glabrous to hairy; wings glabrous; lower keel margins glabrous, upper margin ciliate. **Fruits** (1–)2–3 × 0.6–0.9 cm, hairy or silky. **Seeds** (2 or) 3–5 (or 6), tan, brown, or red.

Varieties 14 (9 native in the flora): w North America. Variety *utahensis* is reported from northern Arizona; however, all specimens examined have been misidentified. It is not included in the present treatment. Within the 14 varieties in North America 109 names and combinations have been proposed. Most varieties intergrade and overlap in characters.

Lupinus argenteus is reported to be a serious cause of livestock poisoning in the western U.S. (Parker 1972). The seeds are particularly toxic. Sheep are especially susceptible to poisoning. Crooked calf disease can result from pregnant cows feeding on lupines, though wild ruminants that browse on lupines are seldom affected (Stubbendieck and Conard 1989). Stubbendieck and Conard (1989) state that children have been poisoned by eating small quantities of seeds of *L. argenteus*.

1. Calyx spur 1–3 mm ____ **1c. Lupinus argenteus** var. **argophyllus**
1′ Calyx might bulge 0–1 mm, but not elongated into a spur.
  2. Leaflet blades oblanceolate, plane, green, surfaces glabrous or pubescent above (adaxially); plants of cool, moist mountain meadows, stream banks, lakeshores.
    3. Corollas pale blue with brown banner patch; pedicels 1–2.5 mm; flower 6–8 mm long ____ **1d. Lupinus argenteus** var. **fulvomaculatus**
    3′ Corollas usually blue with yellow banner patch or none; pedicels usually 3–4 mm; flower 5–11 mm long.
      4. Stems unbranched and racemes solitary, terminal ____ **1i. Lupinus argenteus** var. **rubricaulis**
      4′ Stems normally branched, giving rise to several racemes.
        5. Flowers large, wings (7.5–)8–10 mm ____ **1b. Lupinus argenteus** var. **argentatus**
        5′ Flowers small, wings 5–7.5 mm ____ **1h. Lupinus argenteus** var. **parviflorus**
  2′ Leaflet blades oblanceolate or elliptic-oblanceolate, narrow and often folded, surfaces gray or silver-pubescent; plants of dry open areas, foothills and mountains.
    6. Stem hairs descending in backward direction or widely spreading ____ **1g. Lupinus argenteus** var. **palmeri**
    6′ Stem hairs forwardly appressed.
      7. Flowers 5–7(–9) mm ____ **1e. Lupinus argenteus** var. **hillii**
      7′ Flowers (7–)8–15 mm.
        8. Flowers (7–)8–12 mm ____ **1a. Lupinus argenteus** var. **argenteus**
        8′ Flowers 12–15 mm ____ **1f. Lupinus argenteus** var. **moabensis**

1a. Lupinus argenteus Pursh var. **argenteus** * SILVER LUPINE Native

Lupinus argenteus Pursh var. *lemmonii* (C.P. Smith) Isely; *Lupinus lemmonii* C.P. Smith; *Lupinus sitgreavesii* S. Watson

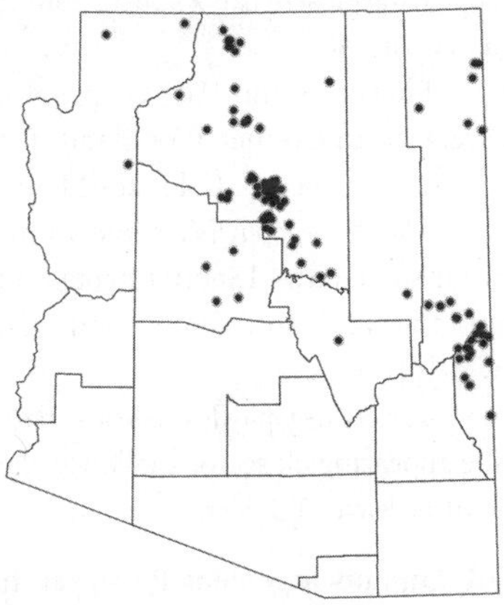

Herbs 2–15 dm, hairs forwardly appressed. **Stems** branched. **Leaves** cauline; petiole 1–5 cm. **Leaflet** blades oblanceolate or elliptic-oblanceolate, narrow and often folded, surfaces gray or silver-pubescent. **Pedicels** (1–)2–5(–6) mm. **Flowers** (7–)8–12 mm; calyx bulge less than 1 mm; corolla blue, purple, pink, or white, banner back ± hairy. $2n = 48$.

Flowering (May–)Jun–Oct. Slopes, canyons, valleys, meadows, stream banks, and roadsides in a variety of soils; (1000–)1600–3540 m; (Madrean Evergreen Woodland), Great Basin Conifer Woodland, Petran Montane Conifer Forest, Petran Subalpine Conifer Forest; Apache, Coconino, Gila, Greenlee, Mohave, Navajo, Yavapai cos.; California, Idaho, Nevada, New Mexico, North Dakota, Oregon, South Dakota, Utah, Wyoming; Canada (Alberta, British Columbia, Saskatchewan).

Variety *argenteus* is known from the Panamint Mountains of eastern California to southern Canada, western North Dakota, western South Dakota, and central New Mexico. It is widespread in the intermountain region in Arizona, southern Idaho, Nevada, eastern Oregon, Utah, and southwestern Wyoming. Variety *argenteus* is rather widespread at higher elevations in moist and relatively dry sites in the northern half of Arizona. The plants grow in the open as well as in the forest understory.

Lupinus lemmonii C.P. Smith belongs here since Isely (1998) and Kearney and Peebles (1960) stated that the only diagnostic character that differentiates this species from var. *argenteus* is its southern location.

1b. Lupinus argenteus Pursh var. **argentatus** (Rydberg) Barneby in A. Cronquist et al., Intermount. Fl. 3(B):246. 1989. * [from Latin, *argentat*, silver, silvery] Native

Lupinus decumbens Torrey var. *argentatus* Rydberg, Fl. Colorado 197. 1906

Herbs 2–5.5 dm, glabrous or sparsely pubescent. **Stems** normally branched. **Leaves** mostly cauline, basal usually absent at flowering. **Leaflet** blades oblanceolate, plane, surfaces glabrous or pubescent above (adaxially). **Inflorescences** several racemes. **Pedicels** 3–4 mm. **Flowers** (7.5–)8–10 mm; calyx often distinctly gibbous on lower surface at base; corolla blue, large, banner spot blue or yellow, back glabrous to hairy, wings (7.5–)8–10 mm. $2n = 48$.

Flowering Jun–Sep. Cool, moist mountain meadows, stream banks, lakeshores, high mountain elevations, sometimes to or above timberline; [1700–]2100–2350[–3400] m; Petran Montane Conifer Forest; Coconino, Mohave cos.; Colorado, Idaho, Montana, New Mexico, Utah.

Variety *argentatus* is known from the Rocky Mountains in Colorado to Idaho and Montana, south to New Mexico and Arizona, and in the Wasatch Plateau and adjacent areas of Utah, from Salt Lake to eastern Iron County, Utah. In Arizona, it is apparently restricted to the Kaibab Plateau where it is rare.

Variety *argentatus* can be separated from the other varieties by the absence of a spur and its abaxially gibbous calyx. The herbs are mesomorphic plants of cool, moist or wet mountain meadows; the leaflets are green; flowers are relatively large; and the stems are branched.

1c. Lupinus argenteus Pursh var. **argophyllus** (A. Gray) S. Watson. Proc. Amer. Acad. Arts 8:541. 1873. * [from Greek, *argo*, shining and *phyll*, leaf] Native

Lupinus decumbens Torrey var. *argophyllus* A. Gray, Mem. Amer. Acad. Arts, n. s. 4:37. 1849; *Lupinus aduncus* Greene; *Lupinus argophyllus* (A. Gray) Cockerell; *Lupinus caudatus* Kellogg subsp. *argophyllus* (A. Gray) L. Ll. Phillips; *Lupinus caudatus* Kellogg var. *argophyllus* (A. Gray) S.L. Welsh

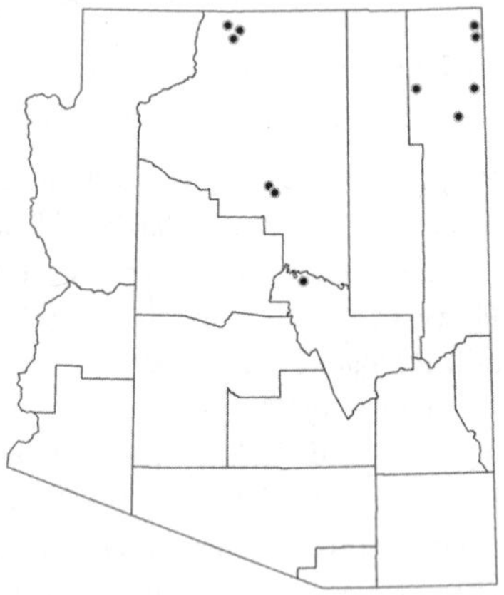

Herbs 2–8 dm, pubescent. **Stems** branched. **Leaves** cauline; petiole of proximal cauline leaves 1.5–3 cm. **Leaflet** blade surfaces silvery-pubescent. **Pedicels** 2–4 mm in flower, to 8 mm in fruit. **Flowers** 8–12 mm, in profile appearing closed; calyx spur 1–3 mm; corolla light blue, banner back generally hairy. $2n = 48$.

Flowering Jun–Aug. Slopes, valleys, meadows; [1200–]1830–2500 m; Great Basin Conifer Woodland, Petran Montane Conifer Forest; Apache, Coconino, Gila cos.; Colorado, Montana, New Mexico, South Dakota.

Variety *argophyllus* is known from the foothills of the Rocky Mountains from Montana and South Dakota south to northern New Mexico and Arizona and the San Juan Basin in southwestern Colorado. Arizona populations are widely scattered.

Variety *argophyllus* can be separated by the spurred calyx, flowers in profile appearing closed or shallowly gaping, and petioles of the lowest or largest cauline leaves 1.5–3 cm.

1d. Lupinus argenteus Pursh var. **fulvomaculatus** (Payson) Barneby. Great Basin Naturalist 46:257. 1986. * [from Latin, *fulv*, reddish yellow and *maculat*, spotted] Native

Lupinus fulvomaculatus Payson, Bot. Gaz. 60:376. 1915

Herbs 1–10 dm, glabrous or pubescent. **Stems** branched. **Leaves** usually cauline, basal leaves usually absent at flowering, if present then petioles less than 3 times as long as leaflets. **Leaflet** blades oblanceolate, plane, surfaces glabrous or pubescent above (adaxially). **Pedicels** 1–2.5 mm. **Flowers** 6–8 mm; calyx bulge 0–1 mm; corolla pale blue with brown banner patch, banner back glabrous to hairy. $2n = 48$.

Flowering Jun–Aug. Cool, moist mountain meadows, stream banks, lakeshores; [2000–]2470[–3600] m; Petran Montane Conifer Forest; Apache Co.; Colorado, New Mexico, Utah.

Variety *fulvomaculatus* is known from the Abajo and La Sal Mountains in southeastern Utah and mountainous Colorado, in Santa Fe National Forest and vicinity in New Mexico, and in Arizona, where it is rare in northeastern Apache County.

1e. Lupinus argenteus Pursh var. **hillii** (Greene) Barneby in A. Cronquist et al., Intermount. Fl. 3(B):246. 1989. * HILL'S LUPINE Native

Lupinus hillii Greene, Leafl. Bot. Observ. Crit. 2:236. 1912; *Lupinus hallii* Abrams var. *osterhoutianus* (C.P. Smith) Harmon; *Lupinus ingratus* Greene var. *arizonicus* C.P. Smith; *Lupinus osterhoutianus* C.P. Smith

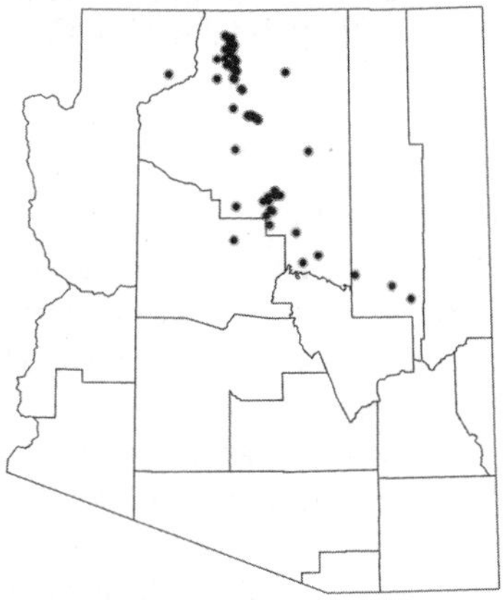

Herbs to 3 dm, hair inconspicuous and forwardly appressed to spreading. **Stems** branched. **Leaves** usually cauline, basal leaves usually absent at flowering, if present then petioles less than 3 times as long as leaflets. **Leaflet** blades oblanceolate or elliptic-oblanceolate, and folded, surfaces gray or silvery-pubescent. **Pedicels** (1–)2–5(–6) mm. **Flowers** 6–8 mm; calyx bulge 0–1 mm; corolla blue-purple, banner back usually thinly strigillose, wings 5.5–7 mm. $2n = 48$.

Flowering (May–)Jun–Sep. Ridges, slopes, valleys, arroyos, and meadows in various soils; (1630–)2000–2740[–2800] m; Great Basin Conifer Woodland, Petran Montane Conifer Forest, Petran Subalpine Conifer Forest; Coconino, Mohave, Navajo, Yavapai cos.; Colorado, Nevada, New Mexico.

Variety *hillii* is known from northern and central Arizona where it is uncommon. It is the small-flowered form that occurs in the Southwest in which the flowers are bud-like and scarcely gaping.

1f. Lupinus argenteus Pursh var. **moabensis** S.L. Welsh. Great Basin Naturalist 46:262. 1986. * [of Moab, Utah] Native

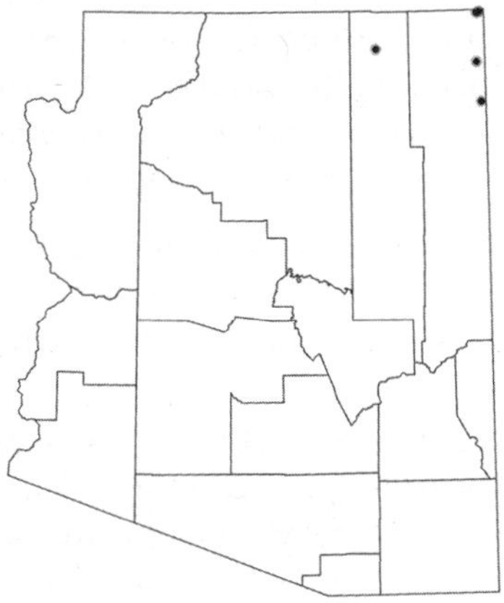

Herbs 2–9 dm, hairs forwardly appressed. **Stems** branched. **Leaves** usually cauline, basal usually absent at flowering, if present then petioles less than 3 times as long as leaflets. **Leaflet** blades oblanceolate or elliptic-oblanceolate, narrow and folded, surfaces gray or silvery-pubescent with appressed hairs. **Pedicels** 5–7 mm. **Flowers** 12–15 mm; calyx bulge 0–1 mm; corolla purple, banner back ± hairy, wings 12–14 mm. $2n = 48$.

Flowering Apr–Jul. Ridges, badlands, sandy washes, and roadsides; 1500–2410[–2500] m; Great Basin Desertscrub, Great Basin Conifer Woodland, Petran Montane Conifer Forest; Apache, Navajo cos.; Colorado, Utah.

Variety *moabensis* is known from the badlands of the Colorado Plateau in Utah and adjacent Colorado. It has been found in only a few sites in Arizona though it may be locally common. This variety occurs in unusually dry habitats for the species. It is known for its very large, showy flowers.

1g. Lupinus argenteus Pursh var. **palmeri** (S. Watson) Barneby. Great Basin Naturalist 46:257. 1986. * [for Edward Palmer, 1829–1911, English born American botanist] PALMER'S LUPINE Native

Lupinus palmeri S. Watson, Proc. Amer. Acad. Arts 8:530. 1873

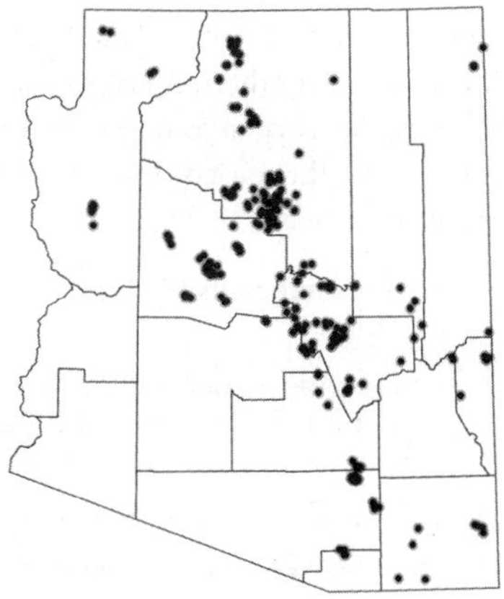

Herbs 3–6 dm, hairs descending in backward direction or widely spreading. **Stems** branched. **Leaves** cauline; petiole 4–10 cm. **Leaflet** blades oblanceolate or elliptic-oblanceolate, narrow, folded, surfaces densely gray spreading-hairy and silvery-silky. **Pedicels** 2–7 mm. **Flowers** 8–10(–12) mm; calyx bulge or spur less than 1 mm; corolla blue, banner back hairy. $2n = 48$.

Flowering (Apr–)May–Jun(–Nov). Ridges, slopes, canyons, meadows, stream banks, roadsides; (1310–)1700–2700[–3100] m; Interior Chaparral, Madrean Evergreen Woodland, (Great Basin Conifer Woodland), Petran Montane Conifer Forest; Apache, Cochise, Coconino, Gila, Graham, Greenlee, Maricopa, Mohave, Navajo, Pima, Pinal, Santa Cruz, Yavapai cos.; California, Nevada, New Mexico, Utah, Washington.

Variety *palmeri* is widespread over central and northwestern Arizona and adjacent New Mexico, and through high elevations in California, Nevada, Utah, and Washington. It is the most frequently encountered variety of *Lupinus argenteus* in central and southern areas of Arizona. Plants may be locally abundant and are often found in comparatively dry sites. Peak flowering is in May and June, however, plants can continue to flower into autumn.

Variety *palmeri* is similar to variety *argenteus* except stem pubescence is of mixed long-spreading and short hairs.

1h. Lupinus argenteus Pursh var. **parviflorus** (Nuttall) C.L. Hitchcock in C.L. Hitchcock et al., Vasc. Pl. Pacif. N.W. 3:302. 1961. * [from Latin, *parvi*, small and *flor*, flower] LODGEPOLE LUPINE Native

Lupinus parviflorus Nuttall in W.J. Hooker and G. A. W. Arnott, Bot. Beechey Voy. 336. 1838; *Lupinus argenteus* Pursh var. *myrianthus* (Greene) Isely; *Lupinus argenteus* Pursh subsp. *parviflorus* (Nuttall) L. Ll. Phillips; *Lupinus myrianthus* Greene; *Lupinus parviflorus* Nuttall subsp. *myrianthus* (Greene) Harmon

Herbs 2–5(–8) dm, mostly glabrous. **Stems** branched. **Leaves** cauline. **Leaflet** blades green, oblanceolate, plane, surfaces glabrous or glabrate. **Inflorescences** several racemes. **Pedicels** 3–4 mm. **Flowers** 5–7.5 mm; calyx bulge 0–1 mm, but not elongated into a spur; corolla blue, banner spot yellow, back glabrous to hairy, wings 5–7.5 mm. $2n = 48$.

Flowering Jun–Sep. Meadows, wet places; [1500–]1800[–2800] m; Petran Montane Conifer Forest; Mohave Co.; Idaho, Montana, South Dakota, Wyoming.

Variety *parviflorus* is known from southwestern Idaho, western Montana, northern Wasatch Mountains and mountainous Owyhee County in Idaho, the Black Hills in South Dakota, and western Wyoming, where it often occurs in spruce-aspen or lodgepole pine forests. In Arizona, it is known from two specimens (*USGS Brooks Crew,* 05 Aug 2006) in ponderosa pine-sagebrush woodland on the Shivwits Plateau. This represents a substantial range extension.

The combination of small flowers, villous pedicels, and broad, green, glabrous (or nearly so) upper (adaxial) leaflet surfaces separates this variety from others of the species.

1i. Lupinus argenteus Pursh var. **rubricaulis** (Greene) S.L. Welsh. Great Basin Naturalist 38:326. 1978. * [from Latin, *rubri*, red and *caulis*, stem] RED-STEMMED SILVER LUPINE Native

Lupinus rubricaulis Greene, Pl. Baker. 3:35. 1901; *Lupinus argenteus* Pursh var. *depressus* (Rydberg) C.L. Hitchcock; *Lupinus caudatus* Kellogg var. *rubricaulis* (Greene) C.P. Smith; *Lupinus depressus* Rydberg

Herbs 2–5.5 dm, glabrous or pubescent. **Stems** unbranched. **Leaves** cauline. **Leaflet** blades green, oblanceolate, plane, surfaces glabrous or pubescent above (adaxially). **Inflorescences** solitary, terminal racemes. **Pedicels** 3–4 mm. **Flowers** 8–11 mm; calyx bulge 0–1 mm, but not elongated into a spur; corolla blue, banner spot yellow, back glabrous to hairy. $2n = 48$.

Flowering Jun–Sep. Slopes, springs, moist places in forests; 1820–[2200–3400] m; Great Basin Conifer Woodland; Apache Co.; Colorado, Idaho, Montana, Nevada, Utah.

Variety *rubricaulis* is known from the mountains in northeastern Nevada and along the crests of the Uinta and Wasatch Mountains in Utah. In Arizona it is found in northern Apache County with a single record from Coconino County. The Arizona plants are unusual in occurring at lower elevations than is typical for this variety. It is very similar to var. *argenteus* but found in more moist or wet (mesic) habitats.

2. Lupinus arizonicus (S. Watson) S. Watson. Proc. Amer. Acad. Arts 12:250. 1877. * [of Arizona] ARIZONA LUPINE Native

Lupinus concinnus J. Agardh var. *arizonicus* S. Watson, Proc. Amer. Acad. Arts 8:537. 1873; *Lupinus arizonicus* (S. Watson) S. Watson var. *barbatulus* (Thornberry ex C.P. Smith) I. M. Johnston; *Lupinus sparsiflorus* Bentham var. *arizonicus* (S. Watson) C.P. Smith; *Lupinus sparsiflorus* Bentham var. *barbatulus* Thornberry ex C.P. Smith

Herbs, annual, 1–5 dm, with short-appressed and long, spreading hairs. **Cotyledons** deciduous, petiolate. **Stems** erect, branched. **Leaves** cauline; stipules 2–4 mm; petiole 2.5–7 cm. **Leaflets** 6–10, blades 10–40 × 5–10 mm, upper (adaxial) surface glabrous. **Inflorescences** racemes 6–30 cm, flowers scattered (or if appearing ± whorled, pedicel bases clearly

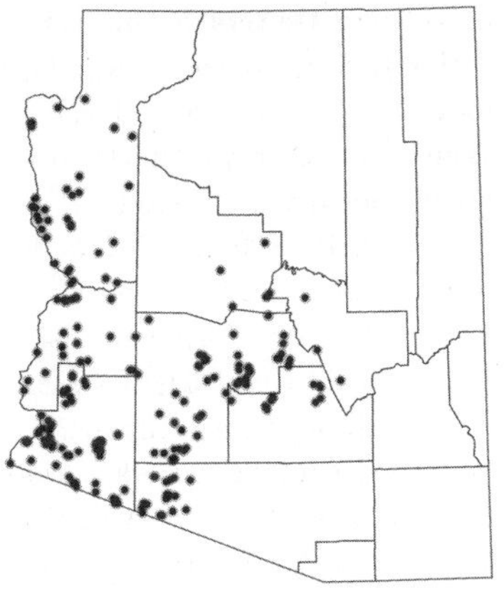

spiraled, not whorled); peduncles 2–6 cm; bracts usually persistent, 4–8 mm. **Pedicels** 2–4 mm. **Flowers** 7–10 mm; calyx 3–6 mm, lobes ± equal, lower lobe entire, upper lobe deeply cleft; corolla banner and wings dark pink to magenta, drying blue-purple or whitish, banner spot yellowish, becoming darker magenta; lower margin of keel ciliate near claw, upper margin glabrous. **Fruits** often secund, 1–2 × to ± 0.5 cm, coarsely pubescent. **Seeds** 4–6, smooth. 2*n* = 48.

Flowering Jan–May. Sandy washes, dunes, valleys, slopes, canyons, roadsides, open areas in sandy, gravelly, or rocky soils; [near sea level–]40–1100 m; Sonoran Desertscrub, Mohave Desertscrub, (Interior Chaparral); La Paz, Maricopa, Mohave, Pima, Pinal, Yavapai, Yuma cos., California, Nevada; Mexico (Baja California, Baja California Sur, Sonora).

Lupinus arizonicus occurs in the eastern Mohave and Sonoran Deserts of southeastern California, southwestern Arizona, and southern Nevada, plus adjacent areas in northern Mexico. Arizona lupine is often widespread and abundant in years with adequate rainfall. In some places the plants may form pure stands. Robust plants have been named var. *barbatulus*. Graham (1941) reports that scaled quail and Gambel's quail eat the seeds of *L. arizonicus*. Arizona lupine is occasionally used in wildflower and revegetation mixes.

3. Lupinus bicolor Lindley. Bot. Reg. 13:pl. 1109. 1827. * [from Latin, *bi*, two and *color*, color] MINIATURE LUPINE Naturalized

Lupinus congdonii (C.P. Smith) D.B. Dunn; *Lupinus polycarpus* Greene

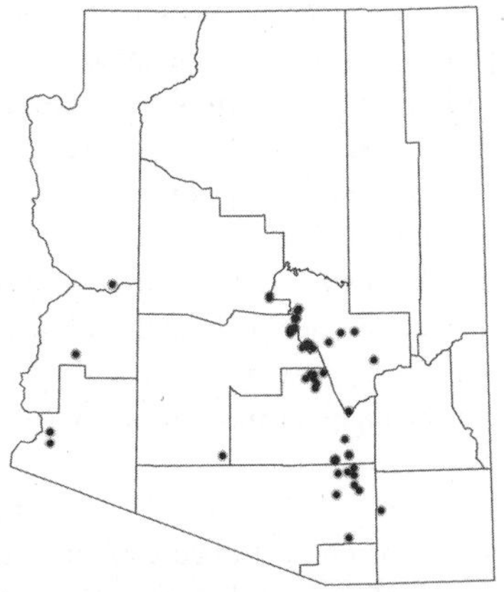

Herbs, annual, 1–4 dm, pubescent. **Cotyledons** deciduous, petiolate. **Stems** ascending or erect, branched. **Leaves** cauline; stipules 4–8 mm; petiole 1–7 cm. **Leaflets** 5–7, blades 10–40 × 1–5 mm, upper (adaxial) surface glabrous or sparsely pubescent. **Inflorescences** racemes, 4–20 cm, flowers usually in fewer than 5 whorls, sometimes scattered at nodes; peduncles 3–10 cm; bracts deciduous, 4–6 mm. **Pedicels** 1–3.5 mm. **Flowers** 4–10 mm; lower calyx lobe entire, 4–6 mm, upper lobe deeply cleft, 2–4 mm; corolla usually blue, rarely light blue, pink, or white, banner spot white, becoming magenta; banner length greater than width; upper margin of keel usually ciliate near apex, rarely glabrous, sometimes blunt. **Fruits** 1–3 × 0.3–0.6 cm wide, pubescent. **Seeds** 5–8, smooth. 2*n* = 48.

Flowering Feb–May. Ridges, slopes, flats, dunes, arroyos, stream banks, roadsides, open or disturbed areas in a variety of soils; [near sea level–](60–)600–1710 m; Sonoran Desertscrub, (Chihuahuan Desertscrub, Mohave Desertscrub), Semidesert Grassland, Interior Chaparral; Cochise, Gila, La Paz, Maricopa, Mohave, Pima, Pinal, Yavapai, Yuma cos.; California, Oregon, Washington; Mexico (Baja California, Baja California Sur, Sonora); Canada (British Columbia).

Lupinus bicolor is naturalized in Arizona. It can be locally abundant in wet years. The named subspecies and varieties do not conform to consistently recognizable geographical or morphological entities. Vigorous plants with larger flowers may be confused with *L. nanus*. In California, plants on the outer North Coast Range may persist for two growing seasons. *Lupinus bicolor* (as *L. polycarpus*) was recently reported as a weed in Alabama (Diamond 2016). In California, the seeds are eaten by quail and the leaves are browsed by mule deer (Graham 1941). Allen and Allen (1981) report nodulation for *L. bicolor*.

Lupinus micranthus Douglas (1829, not Gussone 1828) is an illegitimate name that pertains here.

4. Lupinus brevicaulis S. Watson. Botany (Fortieth Parallel) 53, pl. 7, figs. 1–4. 1871. * [from Latin *brevi*, short and *caulis*, stem] SAND LUPINE, SHORT-STEMMED LUPINE Native

Herbs, annual, less than 1 dm, pubescent, hairs more than 1 mm. **Cotyledons** persistent, disc-like, sessile. **Stems** very short, tufted, branched. **Leaves** basal and cauline, crowded near base; stipules 4–6 mm; petiole 2–5 cm. **Leaflets** 6–8, blades 10–15 × 3–6 mm, upper (adaxial) surface glabrous. **Inflorescences** head-like racemes, 3+-flowered, 3–8 cm, longer than leaves, flowers scattered (spiraled), dense; peduncles 2–5 cm; bracts persistent, straight, 2–3 mm. **Pedicels**

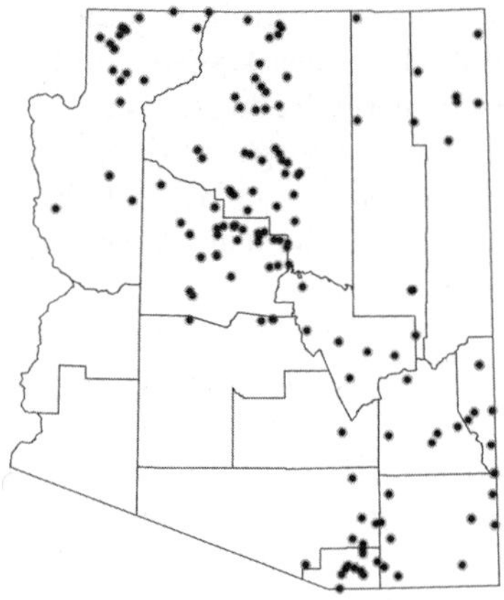

0.5–1.5 mm. **Flowers** 6–8 mm; calyx appendages absent, lower lobe entire or shallowly cleft, ± 6 mm, upper lobe 2-toothed, 3 mm, less than 1/2 as long as lower lobe; corolla bright blue, banner spot white or yellow; keel glabrous. **Fruits** not obviously undulate, to +- 1 × to ± 0.5 cm, thinly pilose to coarsely hirsute. **Seeds** 1 or 2, smooth.

Flowering Mar–Jun. Ridges, slopes, valleys, canyons, floodplains, sandy washes, dunes, roadsides, open areas on a wide range of substrates; [300–]600–2210[–2300] m; Sonoran Desertscrub, Chihuahuan Desertscrub, Mohave Desertscrub, Great Basin Desertscrub, Semidesert Grassland, Plains Grassland, Great Basin Grassland, Interior Chaparral, Madrean Evergreen Woodland, Great Basin Conifer Woodland; Apache, Cochise, Coconino, Gila, Graham, Greenlee, Maricopa, Mohave, Navajo, Pima, Pinal, Santa Cruz, Yavapai cos.; California, Colorado, Nevada, New Mexico, Oregon; Mexico (Chihuahua, Sonora).

Lupinus brevicaulis is widely distributed in diverse habitats at middle elevations across Arizona. The plants sometimes form pure stands in favorable locations. *Lupinus brevicaulis* resembles *L. flavoculatus* but has smaller flowers.

5. Lupinus concinnus J. Agardh. Syn. Gen. Lupini 6, pl. 1, fig. 1. 1835. * [from Greek, *con*, with and Latin *cinnus*, curled hair] BAJADA LUPINE Native

Lupinus concinnus J. Agardh var. *desertorum* (A. Heller) C.P. Smith; *Lupinus concinnus* J. Agardh subsp. *orcuttii* (S. Watson) D.B. Dunn; *Lupinus concinnus* J. Agardh var. *orcuttii* (S. Watson) C.P. Smith

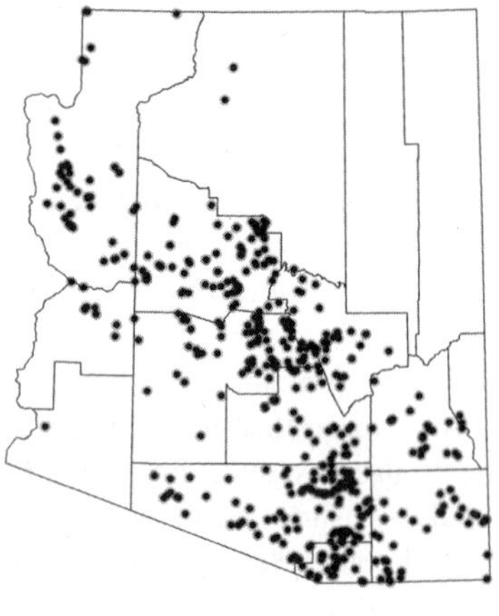

Herbs, annual, 1–3 dm, spreading-pubescent. **Cotyledons** deciduous, petiolate. **Stems** erect or decumbent, branched. **Leaves** cauline; stipules 2–8 mm; petiole 2–7 cm, spreading-pubescent. **Leaflets** 5–9, blades 10–30 × 1.5–8 mm, surfaces pubescent. **Inflorescences** terminal racemes, 1–18 cm, rarely axillary, then usually reduced to 1 flower in proximal branches, flowers loosely to densely scattered; primary peduncles erect, lateral peduncles sometimes decumbent, 0–8 cm; bracts persistent, straight, 2.5–4 mm. **Pedicels** 0.7–2 mm. **Flowers** 5–12 mm; calyx 3–5 mm, lobes ± equal, lower lobe entire, upper lobe deeply cleft; corolla usually pink to purple, rarely white, banner spot white or yellowish; keel usually glabrous, rarely with few, minute cilia on lower margin. **Fruits** 1–1.5 × 0.3–0.5 cm, pubescent. **Seeds** 3–5, smooth. $2n$ = 48.

Flowering Feb–May. Slopes, canyons, valleys, floodplains, arroyos, roadsides, open or disturbed areas in various soils; [near sea level–]60–1700 m; Sonoran Desertscrub, Chihuahuan Desertscrub, Mohave Desertscrub, Great Basin Desertscrub, Semidesert Grassland, Interior Chaparral, Madrean Evergreen Woodland, Great Basin Conifer Woodland; Cochise, Coconino, Gila, Graham, Greenlee, La Paz, Maricopa, Mohave, Pima, Pinal, Santa Cruz, Yavapai, Yuma cos.; California, New Mexico, Texas, Utah; Mexico (Baja California, Baja California Sur, Sonora).

Lupinus concinnus is widely distributed across Arizona below the Colorado Plateau and Mogollon Rim. It is often abundant in years with ample rainfall. In Texas, *L. concinnus* is known from the Trans-Pecos region; in California it is known primarily in central and southern areas. Graham (1941) reports the plants are browsed by mule deer in California.

Lupinus concinnus is a highly variable, predominantly self-pollinated complex, and the named varieties cannot be consistently segregated. Desert plants with linear, coarsely hairy leaflets and few, minute cilia on lower edges of keel (at times recognized as var. *desertorum*) may be confused with *L. sparsiflorus*.

6. Lupinus flavoculatus A. Heller. Muhlenbergia 5:149, pl. 5. 1909. * [from Latin, *flav*, yellow and *cula*, little] YELLOW-EYED LUPINE Native

Lupinus rubens Rydberg var. *flavoculatus* (A. Heller) C.P. Smith

Herbs, annual, 0.5–2 dm, pubescent, hairs more than 1 mm. **Cotyledons** persistent, disc-like, sessile. **Stems** relatively short, decumbent or erect, branched. **Leaves** basal and cauline, crowded near base; stipules 3–8 mm; petiole 3–6 cm. **Leaflets** 7–9, blades 10–20 × 5–8 mm, upper (adaxial) surface glabrous. **Inflorescences** elongated racemes, 3+-flowered,

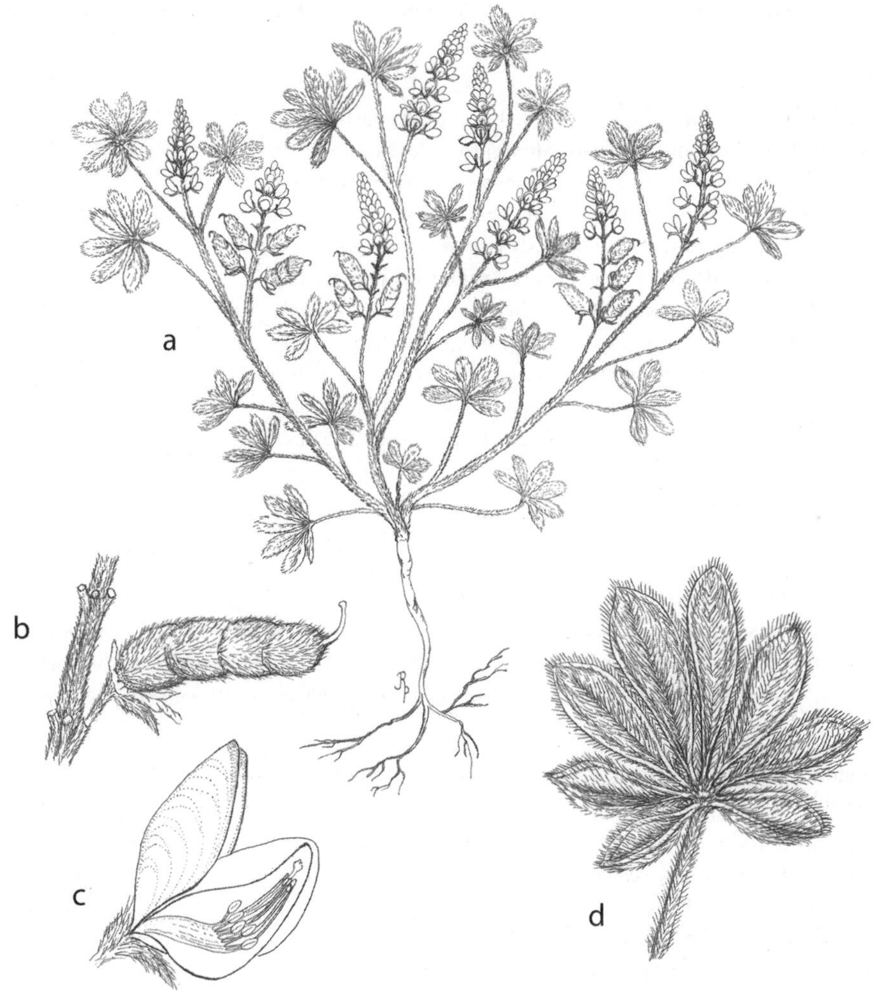

***Lupinus concinnus*: a. habit; b. fruit; c. flower detail; d. leaf. Illustrator: Joyce Riney Peters.**

6–12 cm, exceeding leaves, flowers spiraled, dense; peduncles 3–5(–10 in fruit) cm; bracts persistent, straight, 2–3 mm. **Pedicels** 2–3 mm. **Flowers** 7–10 mm; calyx appendages absent, lower lobe shallowly cleft, 4–5 mm, upper lobe deeply cleft, 1–3 mm, less than 1/2 as long as lower lobe; corolla bright blue, banner spot yellow; keel blunt, glabrous. **Fruits** 0.5–1 × ± 0.5 cm, not obviously undulate, often secund, upper margin not constricted between seeds, thinly pilose to coarsely hirsute. **Seeds** 1 or 2, ridged.

Flowering Mar–May. Slopes, valleys, arroyos, roadsides in sandy or gravelly soils; [600–]850–1840[–2300] m; Mohave Desertscrub, Great Basin Desertscrub, Great Basin Conifer Woodland; Coconino, Mohave cos.; California, Nevada, Utah.

Lupinus flavoculatus is uncommon in Arizona and the populations are widely scattered. The species resembles a hairy form of *L. odoratus*. Nodulation is reported for *L. flavoculatus* (Allen and Allen 1981).

7. Lupinus huachucanus M.E. Jones. Contr. W. Bot. 12:10. 1908. * [of the Huachuca Mountains, Arizona] HUACHUCA MOUNTAIN LUPINE Native

Lupinus platanophilus M.E. Jones

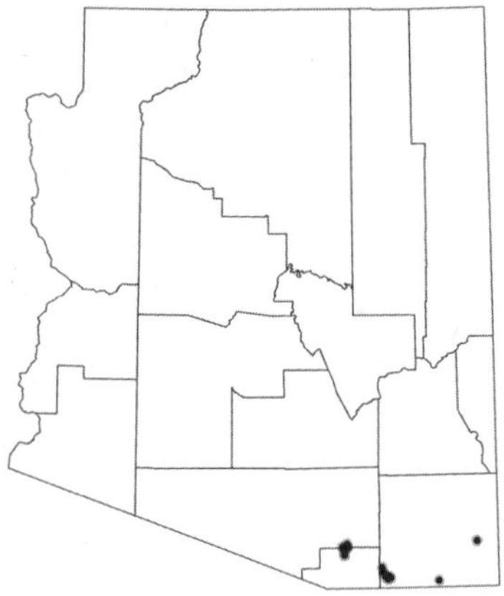

Herbs, perennial, (often with annual aspect), 0.1–2 dm, conspicuously pilose, from taproot. **Cotyledons** deciduous, petiolate. **Stems** prostrate to decumbent, clustered, acaulescent or short-spreading and unbranched. **Leaves** mostly near base, in a rosette; stipules 4–10 mm; petiole 1–9 cm. **Leaflets** 5–7 (or 8), blades 10–55 × 4–12 mm, greenish, both surfaces copiously villous-hirsute with long, spreading hairs, lower (abaxial) surface more dense. **Inflorescences** racemes, 6–23 cm, flowers scattered; peduncles 3–4.5 cm; bracts deciduous, 4–8 mm. **Pedicels** 1–4 mm. **Flowers** 7–13 mm; calyx bulge or spur 0–1 mm, lobes entire, 5–7.5 mm; corolla violet-blue, banner yellow toward center, tip of keel purple, wings yellow toward center; banner glabrous on outer surface; upper keel margin ciliate to densely ciliate. **Fruits** 1.5–2 cm, hispid. **Seeds** 3–5, dark with light spots.

Flowering Mar–May. Mountain slopes and canyons; of conservation concern; 1500–2000(–2100) m; Madrean Evergreen Woodland; Cochise, Pima, Santa Cruz cos.; Mexico (Chihuahua, Durango, Sonora).

Lupinus huachucanus occurs in the Santa Rita Mountains in Santa Cruz and Pima counties and in the Chiricahua and Huachuca Mountains in Cochise County. Populations are scattered though the plants may be locally common. *Lupinus huachucanus* somewhat resembles *L. concinnus* but is readily distinguished by its spreading habit, racemes surpassing the foliage, violet-blue corollas, and ciliate keel. *Lupinus concinnus* is an annual with a more erect habit, pink corollas, and a non-ciliate keel.

8. Lupinus kingii S. Watson. Proc. Amer. Acad. Arts 8:534. 1873. * [for Clarence R. King, 1842–1901, American geologist and explorer, first director of the U.S. Geological Survey] KING'S LUPINE Native

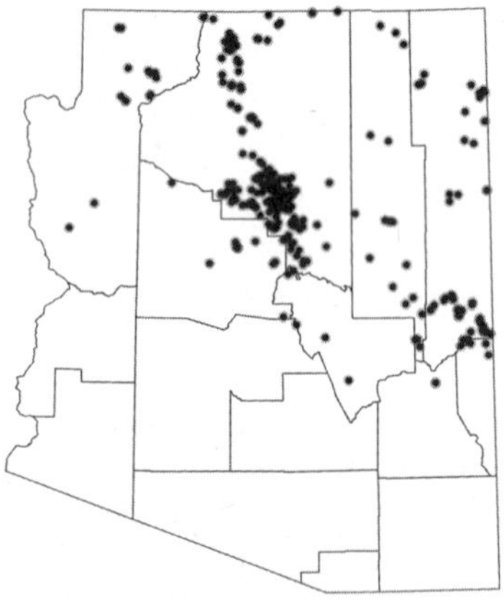

Herbs, annual, 1–2.5(–4) dm, pubescent, hairs soft, flexuous, more than 1 mm. **Cotyledons** early deciduous, sessile, fused into a membranous cup. **Stems** erect and few-branched from beyond middle, or diffusely many-branched from base or near it, softly pilose with fine silvery, yellowish hairs. **Leaves** cauline; stipules 3–12 mm; petiole 1.3–3.3 cm. **Leaflets** (3 or) 4–7, blades 7–20(–24) × 3–5 mm, upper (adaxial) surface pubescent. **Inflorescences** head-like racemes, 3+-flowered, 0.3–2(–3) cm, flowers scattered; peduncles (0–)1–8(–10) cm; bracts ± persistent, straight, 3–4 mm. **Pedicels** 0.8–2 mm. **Flowers** 5–9 mm; calyx 5.5–7.5 mm, densely or thinly pilose, tube 1.5–2 mm, appendages present at sinuses, bidentate, lower lobe 3-denticulate, 4–5.5 mm, upper lobe cleft, 3–4 mm, more than 1/2 as long as lower lobe; corolla usually blue with pale banner patch, sometimes entirely white; banner little recurved, (5–)5.5–8 mm; keel 9–10 mm, glabrous; style abruptly recurved, 4–5 mm. **Fruits** not obviously undulate, 0.9–1.3 × 0.45–0.55 cm, sparsely or densely pilose. **Seeds** 2.

Flowering (Apr–)May–Sep. Ridges, slopes, canyons, valleys, meadows, stream banks, roadsides, often in sandy or igneous soils; (940–)1160–2750[–3000] m; Great Basin Desertscrub, (Madrean Evergreen Woodland), Great Basin Conifer Woodland, Petran Montane Conifer Forest, Petran Subalpine Conifer Forest; Apache, Coconino, Gila, Graham, Greenlee, Mohave, Navajo, Yavapai cos.; Colorado, Nevada, New Mexico, Utah.

Lupinus kingii is widespread and often abundant at higher elevations in the northern part of Arizona. It grows in the open as well as in the forest understory in moist or dry soils. Parker (1972) indicates that *L. kingii* contains toxic alkaloids but cases of livestock poisoning have not been reported. Seeds are eaten by scaled quail and Gambel's quail (Graham 1941). Allen and Allen (1981) report nodulation for *L. kingii*.

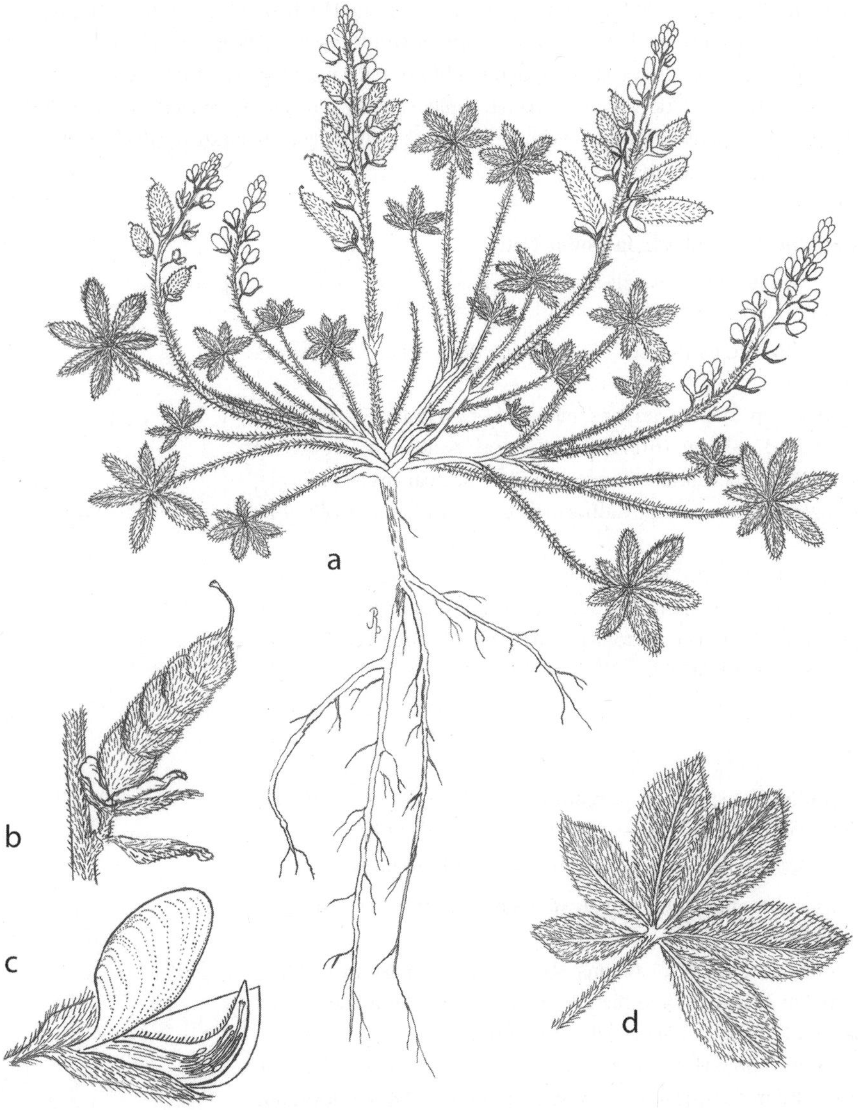

***Lupinus huachucanus*: a. habit; b. fruit; c. flower detail; d. leaf. Illustrator: Joyce Riney Peters.**

9. Lupinus latifolius J. Agardh. Syn. Gen. Lupini 18. 1835. * [from Latin, *lati*, wide and *folium*, leaf] BROAD-LEAVED LUPINE

Lupinus rivularis Douglas ex Lindley var. *latifolius* (J. Agardh) S. Watson

Herbs, perennial, 3–24 dm, green, glabrous or hairy. **Cotyledons** deciduous, petiolate. **Stems** erect or spreading, branched or unbranched. **Leaves** cauline, basal when present withered by anthesis; stipules 5–10 mm; petiole 4–20 cm. **Leaflets** 5–11, blades 40–100 × 6–24 mm, lower (abaxial) surface ± hairy, upper (adaxial) surface glabrous or hairy. **Inflorescences** racemes, 16–60 cm, flowers whorled or not, open; peduncles 8–20 cm; bracts deciduous, 8–12 mm. **Pedicels** 2–12 mm. **Flowers** 8–18 mm; lower calyx lobe entire or notched, 4–8 mm, upper lobe entire or 2-toothed, 5–10 mm; corolla blue or purple to white, banner patch usually white to yellowish turning purple; banner glabrous on outer surface; lower keel margins usually ciliate, upper margin ciliate from claw to middle. **Fruits** 2–4.5 cm, ± densely hairy. **Seeds** 6–10, mottled dark brown, 3–4 mm.

Varieties 5 (2 native in the flora): w North America.

Various authors have differed in their circumscriptions of *Lupinus latifolius*. For example, Vaughn and Dunn (1977)

recognized three varieties, Isely (1998) recognized six varieties, and Hitchcock et al. (1961) recognized three varieties. None of the varieties completely overlap. The most conservative approach has been taken here by recognizing taxa that have the clearest characteristics, but that approach might not reflect phylogeny. Much study is needed to clarify the varieties and particularly the relationships between *L. latifolius* and *L. arcticus*, *L. perennis*, and *L. polyphyllus*.

Maternal ingestion of *Lupinus latifolius* is known to cause birth defects in livestock (Keeler et al. 1977).

1. Flowers 10–14(–16) mm; stems not hollow ____ **9a. Lupinus latifolius** var. **latifolius**
1′ Flowers 14–18 mm; stems hollow ____ **9b. Lupinus latifolius** var. **parishii**

9a. Lupinus latifolius J. Agardh var. **latifolius** Native

Lupinus latifolius J. Agardh var. *leucanthus* (Rydberg) Isely; *Lupinus latifolius* J. Agardh var. *thompsonianus* (C.P. Smith) C.L. Hitchcock; *Lupinus leucanthus* Rydberg; *Lupinus sericeus* Pursh var. *thompsonianus* C.P. Smith

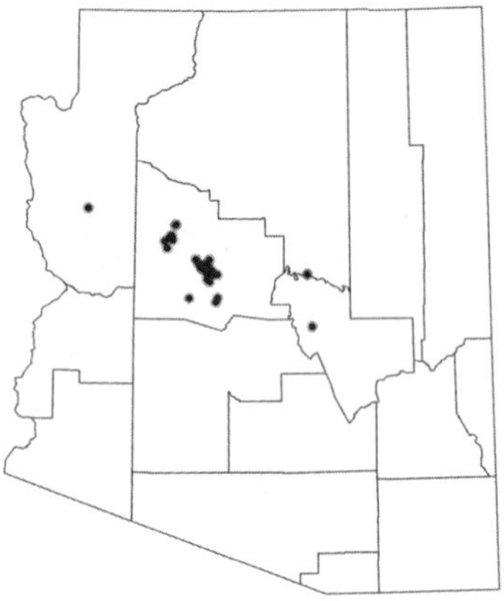

Herbs to 10(–15) dm. **Stems** not hollow, glabrous or puberulent. **Flowers** 10–14(–16) mm.

Flowering Apr–Sep. Slopes, canyons, moist areas, streamsides; [near sea level–]1350–2090[–3500] m; Interior Chaparral, Madrean Evergreen Woodland, Petran Montane Conifer Forest; Mohave, Yavapai cos.; California, Oregon, Washington; Mexico (Baja California); Canada (British Columbia, Yukon).

Variety *latifolius* is known from Arizona to the coast ranges, Cascades, and Sierra Nevada in California, throughout Oregon and Washington, and into British Columbia and Yukon. The Arizona populations are widely scattered but plants may be locally abundant.

9b. Lupinus latifolius J. Agardh var. **parishii** C.P. Smith in W.L. Jepson, Man. Fl. Pl. Calif. 530. 1925 (as var. *parishi*). * [for Samuel B. Parish, 1838–1928, American botanist] Native

Lupinus latifolius J. Agardh subsp. *parishii* (C.P. Smith) Kenney & D.B. Dunn; *Lupinus parishii* Eastwood

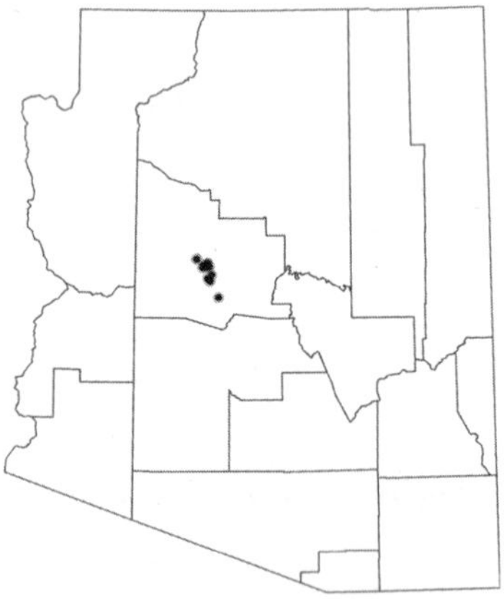

Herbs 5–20 dm, caudex often unbranched. **Stems** hollow, glabrate to sparsely strigose. **Flowers** 14–18 mm.

Flowering May–Aug. Slopes, canyons, stream banks, moist areas; [50–]1370–1950[–3500] m; Madrean Evergreen Woodland, Petran Montane Conifer Forest; Yavapai Co.; California; Mexico (Baja California).

Variety *parishii* is known from Arizona to central and southern California, Santa Barbara and Tulare to San Diego counties. It is found at higher elevations in the central and southern Sierra Nevadas and at lower elevations along the south coast and inland ranges. In Arizona, var. *parishii* occurs in mid-elevation montane habitats. Most collections in the state are from Yavapai County.

10. Lupinus lepidus Douglas ex Lindley. Bot. Reg. 14:pl. 1149. 1828. * [from Greek, *lepidos*, a scale] DWARF LUPINE

Herbs, perennial, less than 6 dm, matted, hairy. **Cotyledons** deciduous, petiolate. **Stems** acualescent to prostrate, ± unbranched or branched. **Leaves** usually basal, sometimes cauline present; stipules 3–25 mm; petiole 2–10 cm. **Leaflets** 5–8, blades 5–40 × 3–7 mm, surfaces pubescent. **Inflorescences** racemes, (2–)3–30 cm, less than to more than leaves, flowers usually dense; peduncles (0–)1–14 cm; bracts usually persistent, 4–15 mm. **Pedicels** 0.4–3 mm. **Flowers** 6–13 mm; calyx bulge or spur 0–1 mm, lower lobe entire or 2-toothed, 4–7 mm; corolla pink, violet, or blue; banner glabrous on outer surface; lower keel margins glabrous, upper margin ciliate. **Fruits** 0.9–1.4 × 0.3–0.55 cm, hairy. **Seeds** 2–6, ± mottled tan to green or brown, 2–4 mm.

Varieties 10 (1 native in the flora): w North America.

Dwarf perennial lupines are usually characterized by a dwarf, caespitose habit, persistent inflorescence bracts, banners that are glabrous on outer surface, and ciliate keel petals. Lupines have a long history of lumping and splitting with various authors. The history of the taxonomy of this species was discussed in detail by Cox (1972), Barneby (1989), and Weitemier (2010). Barneby (1989) gave justification for his conservative treatment of this taxon, which is generally followed here. Genetic analysis in lupines has shown little separation, according to Weitemier (2010), but he suggested that there is good evidence to maintain these variations as varieties, following Barneby (1989), rather than elevating them to species level.

10a. Lupinus lepidus Douglas ex Lindley var. **utahensis** (S. Watson) C.L. Hitchcock in C.L. Hitchcock et al., Vasc. Pl. Pacif. N.W. 3:315. 1961. * [of Utah] Native

Lupinus aridus Douglas ex Lindley var. *utahensis* S. Watson, Proc. Amer. Acad. Arts 8:534. 1873.

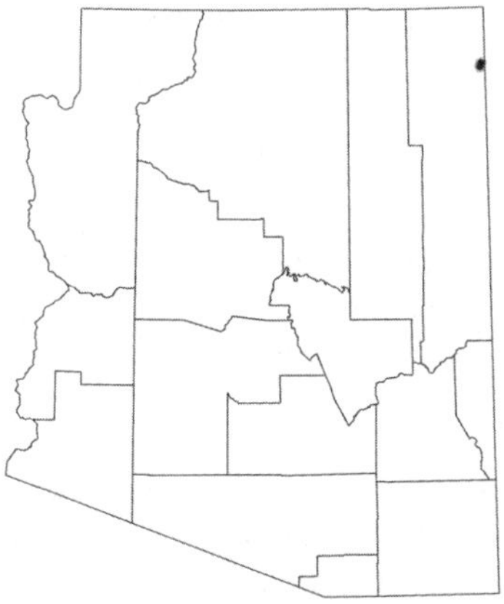

Herbs short-lived, 1–2.5 dm, matted, densely hairy. **Cotyledons** deciduous, petiolate. **Stems** very short or absent, densely tufted. **Leaves** basal. **Leaflet** blades 8–20 mm, surfaces densely hairy. **Inflorescences** racemes, (2–)3–6 cm, less than leaves, flowers dense, in 2 or 3 whorls; peduncles 0–1.4(–2) cm; bracts 8–15 mm. **Pedicels** 0.4–1.5 mm. **Flowers** 7–10 mm; corolla purple or violet, banner glabrous on outer surface, banner patch white, scarcely recurved, 2.5–3 mm wide; lower keel margins glabrous, upper keel margins ciliate near tip. 2*n* = 48.

Flowering Jun–Aug. Mountain meadows, gravel bars, streambeds, lakeshores; (1500–)2610–2780(–3500) m; Petran Montane Conifer Forest; Apache Co.; California, Colorado, Idaho, Montana, Nevada, New Mexico, Oregon, Utah, Wyoming.

Variety *utahensis* is widespread in the Rocky Mountains from central Idaho and western Montana, Colorado, eastern Oregon, northwestern New Mexico, and Arizona, west to eastern Lassen County in California, and disjunct in the White Mountains on the California and Nevada border. In Arizona, var. *utahensis* is known from the Chuska Mountains in northeastern Apache County. Across its range, the plants are found in sagebrush communities, lodgepole pine forests, ponderosa pine forests, spruce-fir forests, and in areas above the timberline.

11. Lupinus microcarpus Sims. Bot. Mag. 50, pl. 2413. 1823. * [from Greek, *micro*, small and *carpus*, fruit] CHICK LUPINE

Herbs, annual, 1–8 dm, sparsely to densely pubescent. **Cotyledons** persistent or deciduous, leaving a circular scar, disc-like, sessile. **Stems** erect or ascending, branched mostly from base, hollow, at least near base. **Leaves** cauline; stipules 6–17 mm; petiole 3–15 cm. **Leaflets** 5–9(–11), blades 10–50 × 2–12 mm, upper (adaxial) surface glabrous. **Inflorescences** racemes, 4–60 cm, flowers in crowded to widely spaced whorls; peduncles 2–30 cm; bracts persistent, reflexed, 3.5–12 mm. **Pedicels** 0.5–5 mm. **Flowers** 8–18 mm; calyx appendages usually absent, sometimes present, lower lobe 5–10 mm, upper lobe 2–6 mm; corolla white to dark yellow, pink to dark rose, or lavender to purple; lower margin of wings sometimes ciliate, upper margin usually ciliate near claw; lower margin of keel usually ciliate near claw, upper margin sometimes ciliate but not as dense. **Fruits** 1–1.5 × to 1 cm, pubescent. **Seeds** 2, tan to brown, usually mottled, ridged or smooth.

Varieties 3 (2 native in the flora): w North America, South America.

All Arizona records for this species are from Gila County and are widely disjunct from populations in California. *Lupinus microcarpus* is highly variable and with varieties intergrading. The species is reported to nodulate (Allen and Allen 1981).

1. Calyx and flower bracts with long shaggy hairs; fruits usually erect or spreading ______ **11b. Lupinus microcarpus** var. **microcarpus**
1′ Calyx and flower bracts with few short and appressed, or long and spreading hairs; fruits usually secund, sometimes spreading ______ **11a. Lupinus microcarpus** var. **densiflorus**

11a. Lupinus microcarpus Sims var. **densiflorus** (Bentham) Jepson. Fl. Calif. 2:279. 1936. * [from Latin, *densi*, thick and *florus*, flower] Native

Lupinus densiflorus Bentham, Trans. Hort. Soc. London, ser. 2, 1:410. 1835

Bracts with short-appressed to long-spreading hairs. **Calyces** with sparsely appressed to spreading hairs; appendages usually absent. **Corolla** usually white to yellow, sometimes tinged pink or lavender, rarely rose or purple; wings withering, oblanceolate, lower margin of wings rarely ciliate near claw, upper margin usually ciliate near claw; lower margin of keel sometimes sparsely ciliate near claw. **Fruits** usually secund, sometimes spreading. 2*n* = 48.

Flowering Apr–May. Slopes, meadows, floodplains, and roadsides in open or disturbed areas; [near sea level–]1040–1540(–2190) m; Semidesert Grassland, Interior Chaparral, Madrean Evergreen Woodland, (Petran Montane Conifer Forest); Gila Co.; California; Mexico (Baja California).

In Arizona, variety *densiflorus* is apparently localized in mountainous areas in Gila County. It occurs widely in western and central California. Graham (1941) indicates that valley quail consume the seeds in California. Variety *densiflorus* has potential for erosion control on roadsides (Graham 1941).

11b. Lupinus microcarpus Sims var. **microcarpus** Native

Lupinus densiflorus Bentham subsp. *austrocollium* (C.P. Smith) D.B. Dunn; *Lupinus densiflorus* Bentham var. *austrocollium* C.P. Smith; *Lupinus densiflorus* Bentham var. *crinitus* Eastwood ex C.P. Smith; *Lupinus densiflorus* Bentham var. *palustris* (Kellogg) C.P. Smith; *Lupinus densiflorus* Bentham var. *persecundus* C.P. Smith; *Lupinus microcarpus* Sims var. *ruber* (A. Heller) C.P. Smith; *Lupinus ruber* A. Heller; *Lupinus subvexus* C.P. Smith

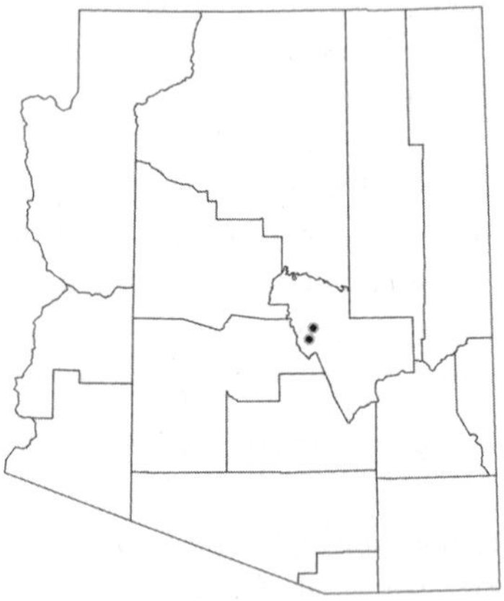

Bracts with long, shaggy hairs. **Calyces** with long, shaggy hairs; appendages usually absent. **Corolla** pink to rose, or lavender to purple, rarely yellowish or white; wings withering, linear to lanceolate, lower margin of wings rarely ciliate, upper margin ciliate near claw; lower margin of keel sometimes ciliate near claw. **Fruits** usually erect or spreading. $2n = 48$.

Flowering May–Jun. Roadsides and floodplains in open or disturbed areas; [near sea level–]1220[–1600] m; Interior Chaparral; Gila Co.; California, Nevada, Oregon, Washington; Mexico (Baja California); Canada (British Columbia); South America (Argentina, Chile).

There are only two records of var. *microcarpus* in Arizona (*R. G. Engard 529*, 26 May 1975 and *W. Hodgson 5030*, 01 June 1988). Both of these records are from Gila County. The plants are reported to be locally abundant and may be more widely distributed than is indicated by these collections.

12. Lupinus neomexicanus Greene. Pittonia 4:133. 1900 (as *Lupinus Neo-Mexicanus*) * [of New Mexico] NEW MEXICO LUPINE Native

Lupinus blumeri Greene

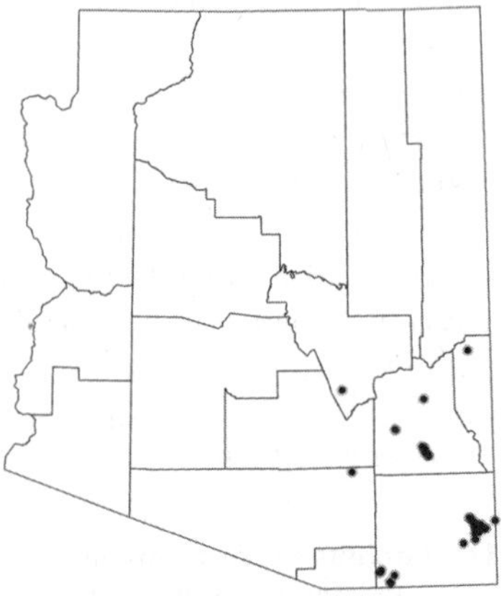

Herbs, perennial, (3–)4–10 dm, hirsutulous or shortly pilose (green and inconspicuously hairy). **Cotyledons** deciduous, petiolate. **Stems** erect or ascending, few-clustered, unbranched or branched. **Leaves** cauline, (basal not present at anthesis); stipules 4–13 mm; petiole 2.5–6 cm. **Leaflets** 5–8, blades 20–40(–50) × 4–11 mm, broadly oblanceolate, lower (abaxial) surface appressed-hairy, upper (adaxial) surface glabrate. **Inflorescences** racemes, 4–15 cm, flowers scattered in subwhorls; peduncles 5–12 cm; bracts deciduous, 4–6 mm. **Pedicels** 5–8 mm. **Flowers** (10–)12–14 mm; calyx lobes entire, 6 mm; corolla pale lavender to reddish purple, banner spot yellow or white, banner back glabrous; keel distally ciliolate. **Fruits** 4 × 2 cm, villous. **Seeds** 3–6, smooth.

Flowering May–Sep. Mountain slopes, meadows, canyons, stream banks, roadsides on various substrates; (1190–)1500–2710 m; (Interior Chaparral), Madrean Evergreen Woodland, Petran Montane Conifer Forest; Cochise, Gila, Graham, Greenlee, Pima cos.; New Mexico; Mexico (Chihuahua, Durango, Oaxaca, Sonora).

Lupinus neomexicanus is known from southeastern Arizona where it is locally common in montane habitats, to Rio Arriba County in New Mexico and in adjacent Chihuahua and Sonora to Durango. A disjunct population has been recorded from the Oaxaca area in Mexico. Chipmunks have been observed to eat seeds of *L. neomexicanus* in Arizona (Graham 1941).

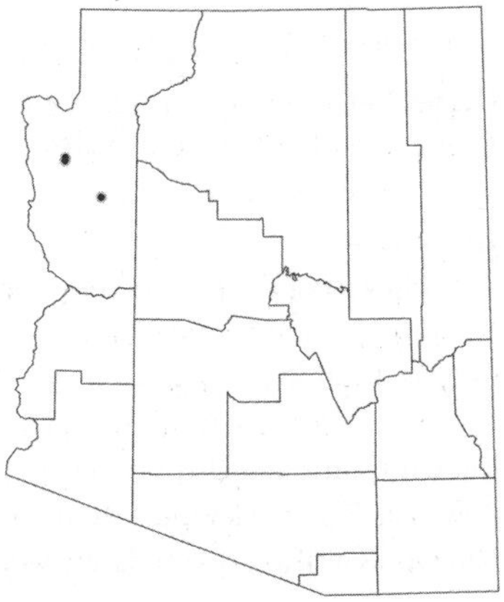

13. Lupinus odoratus A. Heller. Muhlenbergia 2:71. 1905. * [from Latin, *odori*, smell, scent] MOHAVE LUPINE Native

Lupinus odoratus A. Heller var. *pilosellus* C.P. Smith

Herbs, annual, 1–3 dm, usually glabrous, sometimes sparsely pubescent when young, rarely at anthesis, hairs less than 0.5 mm. **Cotyledons** persistent, disc-like, sessile. **Stems** very short, tufted, branched. **Leaves** basal; stipules 5–7 mm; petiole 2–12 cm. **Leaflets** 7–9, blades bright green, 10–20 × 3–8 mm, upper (adaxial) surface glabrous. **Inflorescences** racemes, 4–25 cm, flowers scattered; peduncles hollow, 6–15 cm; bracts persistent, straight, 2–4 mm, tips sparsely ciliate. **Pedicels** 3–5 mm. **Flowers** 7–10 mm; calyx lobes sometimes ciliate at tips, lower lobe entire, 4–5 mm, upper lobe rounded or shallowly 2-toothed, 3–3.5 mm; corolla deep blue-purple, banner spot white or yellow becoming magenta; keel glabrous. **Fruits** 1.5–2.5 × ± 0.8 cm, upper suture wavy and ciliate with long dense hairs, sides with a few short hairs becoming scaly on drying. **Seeds** 2–6, ridged.

Flowering Mar–May. Sandy desert flats, open areas; 500–1600 m; Mohave Desertscrub, (Semidesert Grassland); Mohave Co.; California, Nevada.

Lupinus odoratus occurs in the Mohave Desert region of California, northward to Inyo and Mono counties, and eastward into southern Nevada and northwestern Arizona. It has rarely been collected in Arizona. The fresh flowers of *L. odoratus* smell like violets. Pilose plants can be confused with *L. flavoculatus*.

14. Lupinus polyphyllus Lindley. Bot. Reg. 13:pl. 1096. 1827. * [from Greek, poly, many and phyll, leaf] MEADOW LUPINE

Herbs, perennial, 1–15 dm, green, glabrous to pubescent; caudex subterranean and divisions rhizome-like, or superficial and divisions closely tufted. **Cotyledons** deciduous, petiolate. **Stems** erect, tufted, unbranched, stout, usually hollow. **Leaves** basal and cauline; stipules not leaf-like, green to silvery, 5–30 mm; petiole 3–45 cm, basal leaves long-petiolate and forming a basal tuft, cauline petioles shorter distally. **Leaflets** (5 or) 6–12(–17), blades (25–)30–150 × 5–25 mm, upper (adaxial) surface hairy or glabrous. **Inflorescences** open racemes, 6–40 cm, flowers ± whorled; peduncles 3–13 cm; bracts deciduous, 7–11 mm. **Pedicels** (2–)3–15 mm. **Flowers** 9–15 mm; calyx 4–8 mm, bulge or spur 0–1 mm, lobes entire; corolla violet to lavender, pink, or white, banner patch yellow to white, sometimes turning red-purple; banner glabrous on outer surface; keel upcurved, glabrous, ciliate, or ciliolate. **Fruits** 2.5–5 × 0.7–1 cm, hairy. **Seeds** 3–9.

Varieties 6 (2 native in the flora): w North America; introduced in South America (Argentina, Chile), Eurasia, Pacific Islands (New Zealand), Australia.

Lupinus polyphyllus was hybridized with *L. arboreus* and other species to create the Russell Lupine garden hybrid, which is invasive and occurs in New Zealand and the United Kingdom.

1. Upper (adaxial) surfaces of leaflets silky strigillose, gray or silver ____ **14b. Lupinus polyphyllus** var. **humicola**
1′ Upper (adaxial) surfaces of leaflets glabrous or ± puberulent, green ____ **14a. Lupinus polyphyllus** var. **ammophilus**

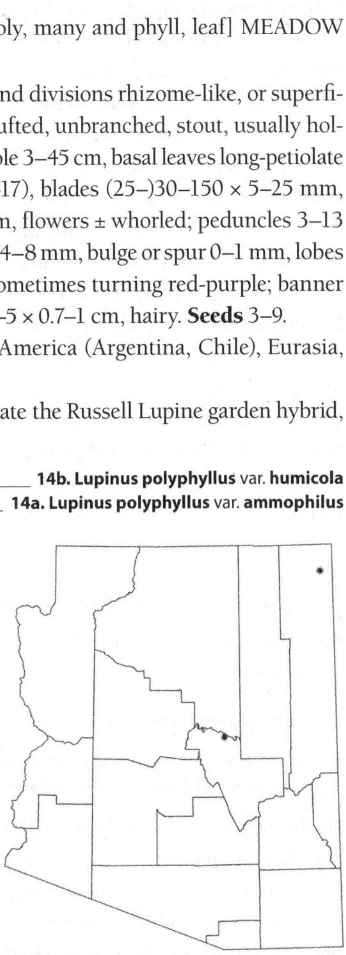

14a. Lupinus polyphyllus Lindley var. **ammophilus** (Greene) Barneby. Great Basin Naturalist 46:257. 1986. * [from Greek, *ammo*, sand and *philo*, loving] Native

Lupinus ammophilus Greene, Pittonia 4:136. 1900

Herbs 1.5–5.5 dm, hirsute, caudex subterranean, divisions rhizome-like. **Stems** diffuse, subhollow. **Leaflets** 6–9 (or 10), largest blade (25–)30–60 mm, upper (adaxial) surfaces of leaflets glabrous or ± puberulent, green, lower (abaxial) surface hirsute. **Flowers:** upper margin of keel ciliate. $2n = 48$.

Flowering Apr–Jul. Barren clay, seeps; 1200–3000 m; Great Basin Conifer Woodland; Apache Co.; Colorado, New Mexico, Utah.

Variety *ammophilus* is rare in northeastern Arizona. A collection (*G. Rink 3364*, 31 May 2004) from the Navajo Reservation at 2111 m elevation is the only recent collection from the state. The variety is reported from Great Basin Desertscrub outside of Arizona.

14b. Lupinus polyphyllus Lindley var. **humicola** (A. Nelson) Barneby. Great Basin Naturalist 46:257. 1986. * [from Latin, *humi*, ground and *cola*, dwell] Native

Lupinus humicola A. Nelson, Bull. Torrey Bot. Club 25:204. 1898

Herbs 2–4.5(–7) dm, caudex above ground. **Stems** ascending, not hollow. **Leaflets** (5–)7–12, blades 30–55(–70) mm, surfaces silky strigillose, gray or silver, often folded, appearing linear. **Flowers:** upper margin of keel usually ciliate. 2*n* = 48.

Flowering (Apr–)May–Jun(–Sep). Slopes, dry washes; 1400–2500 m; Great Basin Desertscrub; Navajo Co.; California, Colorado, Idaho, Nevada, Oregon, Utah, Washington; Canada (British Columbia).

Variety *humicola* is of conservation concern in California. In Arizona, it is known from a single collection (*os 7007,* 11 Jun 2003) from south of Lake Powell at 1739 m elevation. Across its range, var. *humicola* is found in desert habitats as well as grasslands and woodland sites.

15. Lupinus pusillus Pursh. Fl. Amer. Sept. 2:468. 1814. * [from Latin, *pusill*, tiny] RUSTY LUPINE, SMALL LUPINE Native

Lupinus intermontanus A. Heller; *Lupinus odoratus* A. Heller var. *rubens* (Rydberg) Jepson; *Lupinus pusillus* Pursh subsp. *intermontanus* (A. Heller) D.B. Dunn; *Lupinus pusillus* Pursh var. *intermontanus* (A. Heller) C.P. Smith; *Lupinus pusillus* Pursh subsp. *rubens* (Rydberg) D.B. Dunn; *Lupinus pusillus* Pursh var. *rubens* (Rydberg) S.L. Welsh; *Lupinus rubens* Rydberg

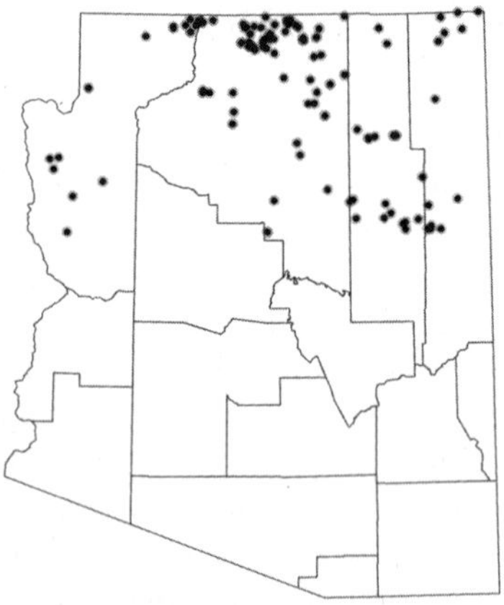

Herbs, annual, less than 1 dm, sparsely pubescent, hairs more than 1 mm. **Cotyledons** persistent, disc-like, sessile. **Stems** short and tufted or erect, branched. **Leaves** basal and cauline, well scattered along stems; stipules 4–11 mm; petiole 3–6 cm. **Leaflets** 5 or 6, blades 10–20 × 5 mm, upper (adaxial) surface glabrous. **Inflorescences** elongated racemes, 3+-flowered, 3 cm, shorter than or slightly exceeding leaves, flowers scattered, dense; peduncles 0–1 cm; bracts persistent, straight, 3 mm. **Pedicels** to 1 mm. **Flowers** 6 mm; lower calyx lobe entire, ± 5 mm, upper lobe cleft, ± 2.5 mm, less than 1/2 as long as lower lobe; corolla white to pale blue becoming blue, banner spot white or yellowish; keel glabrous. **Fruits** not obviously undulate, 1.5–2 × 0.55–0.7 (–0.8) cm, upper margin constricted between seeds, thinly pilose to coarsely hirsute. **Seeds** 2, wrinkled or ridged. 2*n* = 48.

Flowering Apr–May. Ridges, slopes, canyons, valleys, dunes, floodplains, arroyos, roadsides in open, often sandy areas; 500–2100 m; Mohave Desertscrub, Great Basin Desertscrub, Semidesert Grassland, Plains Grassland, (Interior Chaparral), Great Basin Conifer Woodland; Apache, Coconino, Gila, Mohave, Navajo cos.; California, Colorado, Idaho, Kansas, Montana, Nebraska, Nevada, New Mexico, North Dakota, Oklahoma, Oregon, South Dakota, Utah, Washington, Wyoming; Canada (Alberta, Saskatchewan).

Lupinus pusillus is distributed across northern Arizona where it may be locally common. The plants show a marked preference for sandy soils but are also found in clay, gravel, and rocky soils. *Lupinus pusillus* is reported to cause livestock poisoning (Parker 1972). Graham (1941) reports that the species is browsed by elk. Stubbendieck and Conard (1989) state that the plants are occasionally eaten by deer, elk, and pronghorn. Allen and Allen (1981) report that *L. pusillus* nodulates.

16. Lupinus sericeus Pursh. Fl. Amer. Sept. 2:468. 1814. * [from Greek, *serico*, silky] SILKY LUPINE Native

Lupinus barbiger S. Watson; *Lupinus huffmannii* C.P. Smith; *Lupinus jonesii* Rydberg; *Lupinus sericeus* Pursh var. *barbiger* (S. Watson) S.L. Welsh; *Lupinus sericeus* Pursh subsp. *huffmannii* (C.P. Smith) Fleak & D.B. Dunn; *Lupinus sericeus* Pursh var. *jonesii* (Rydberg) Isely

Herbs, perennial, (2–)4–14 dm, silky strigose to finely hirsute (with both short and long hairs). **Cotyledons** deciduous, petiolate. **Stems** erect, ascending, or decumbent, branched distally. **Leaves** mostly cauline, basal normally absent; stipules 5–7 mm; petiole 2–12 cm (longer proximally becoming shorter distally). **Leaflets** 5–9, blades 20–60(–90) × 3–11 mm, surfaces usually silky, rarely thinly silky appearing glabrous below (abaxially). **Inflorescences** racemes, dense in bud, 12–25(–42) cm, flowers in whorls or scattered; peduncles 2–4 cm; bracts subdeciduous, 5–8 mm. **Pedicels** 2–5(–6)

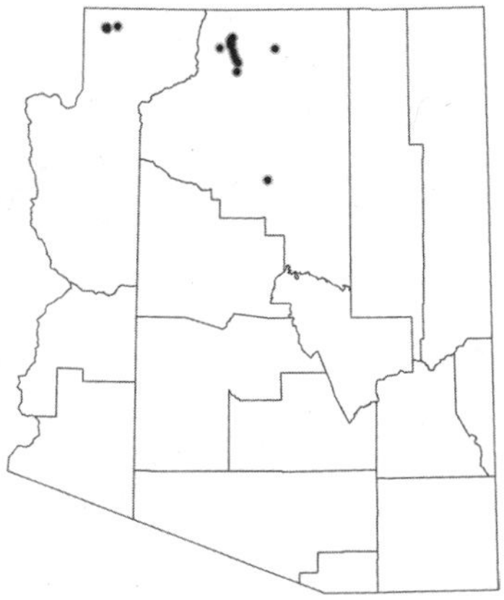

mm. **Flowers** 8–14(–18) mm; calyx sometimes slightly pouch-shaped, lower lobe subentire or entire, 5 mm, upper lobe 2-fid, 6 mm; corolla pale purple to bright blue, sometimes yellowish or whitish, banner spot white to yellow turning brown; banner well reflexed-recurved at or proximal to midpoint, this 3.5–6 mm proximal to apex, banner back densely hairy; keel moderately curved, upper margin of keel ciliate almost full length, not reaching tip. **Fruits** 2–3 cm, densely pilosulose. **Seeds** (2 or) 3–5(–7).

Flowering May–Aug(–Oct). Slopes, meadows, dry banks, openings in conifer forests, roadsides in various soils; [200–](1590–)2000–2750[–3100] m; (Great Basin Conifer Woodland), Petran Montane Conifer Forest, Petran Subalpine Conifer Forest; Apache, Coconino, Graham, Mohave cos.; Idaho, Nevada, New Mexico, Oregon, Utah, Washington, Wyoming; Canada (British Columbia).

Lupinus sericeus is widespread and common from the Rocky Mountains in British Columbia southward to eastern Washington, eastern Oregon to northern New Mexico and northwestern Arizona, northeastern Nevada through Idaho, and Wyoming to the Black Hills in South Dakota. In Arizona, the species is found at higher elevations in the northern part of the state where it often grows in the understory among pines.

17. Lupinus shockleyi S. Watson. Proc. Amer. Acad. Arts 22:470. 1887. * [for William H. Shockley, 1855–1925, American mining engineer and plant collector] DESERT LUPINE, SHOCKLEY'S LUPINE Native

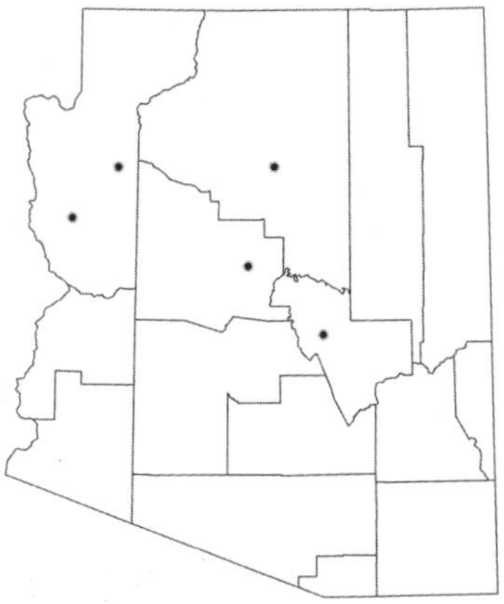

Herbs, annual, 0.5–3 dm, canescent, hairs 0.6–1 mm. **Cotyledons** persistent, disc-like, sessile. **Stems** very short, tufted or spreading, branched. **Leaves** basal and cauline, crowded near base; stipules 3–5 mm; petiole 4–12 cm. **Leaflets** 8–10, blades 10–30 × 4–10 mm, upper (adaxial) surface glabrous. **Inflorescences** racemes, 3+-flowered, 3–14 cm, flowers scattered; peduncles 1–10 cm; bracts persistent, straight, 2–4 mm. **Pedicels** 1–3 mm. **Flowers** 4.5–6 mm; calyx 3–6 mm, lobes ± equal, lower lobe entire, upper lobe cleft; corolla dark blue-purple, banner spot white becoming yellow; keel blunt, glabrous. **Fruits** undulate, 1.5–2 × 0.8–1.2 cm, ciliate with long, dense hairs, sides with short, inflated hairs, becoming scaly on drying. **Seeds** 2, rugose.

Flowering Mar–May(–Jun). Slopes, roadsides; [near sea level–]1220–1720 m; Semidesert Grassland, Plains Grassland, Interior Chaparral, Madrean Evergreen Woodland; Coconino, Mohave, Yavapai cos.; California, Nevada.

Lupinus shockleyi occurs in the desert areas of southern California, adjacent areas of southern Nevada, and northwestern Arizona. In Arizona this species is rare and reported populations are widely scattered at elevations above the deserts.

18. Lupinus sparsiflorus Bentham. Pl. Hartw. 303. 1849. * [from Latin, *sparsi*, sparse and *florus*, flower] COULTER'S LUPINE Native, Cultivated

Lupinus sparsiflorus Bentham subsp. *mohavensis* Dziekanowski & D.B. Dunn; *Lupinus sparsiflorus* Bentham var. *mohavensis* (Dziekanowski & D.B. Dunn) S.L. Welsh

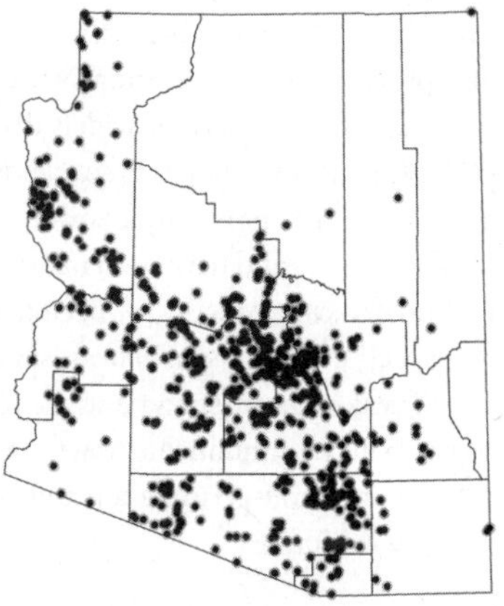

Herbs, annual, 2–4 dm, hairs 0.5–2 mm, shorter appressed, longer usually spreading. **Cotyledons** deciduous, petiolate. **Stems** erect, usually branched from base. **Leaves** cauline; stipules 3–7 mm; petiole 3–4 cm. **Leaflets** 7–11, blades 15–30 × 2–4 mm, upper (adaxial) surface pubescent at least marginally. **Inflorescences** racemes, 15–25 cm, flowers scattered, sometimes appearing ± whorled; peduncles 2–4 cm; bracts usually deciduous, 3–5 mm, shorter than buds. **Pedicels** 2–4 mm. **Flowers** 10–12 mm; calyx 3–6 mm, lobes ± equal, lower lobe entire, upper lobe deeply cleft; corolla usually blue, rarely pinkish, drying darker, banner spot whitish becoming magenta; lower margin of keel ciliate near claw, upper margin often ciliate near claw. **Fruits** 1–2.5 × 0.4–0.5 cm, coarsely pubescent. **Seeds** 4 or 5, smooth.

Flowering Feb–May. Ridges, slopes, canyons, valleys, floodplains, arroyos,

Lupinus sparsiflorus**: a.** **flower; b.** **leaflet detail; c.** **habit; d.** **fruit; e.** **seeds. Illustrator: Joyce Riney Peters.**

roadsides, in various usually sandy, gravelly, or rocky soils; [near sea level–]80–1680 m; Sonoran Desertscrub, (Chihuahuan Desertscrub), Mohave Desertscrub, Semidesert Grassland, Interior Chaparral, (Madrean Evergreen Woodland); Apache, Cochise, Coconino, Gila, Graham, La Paz, Maricopa, Mohave, Navajo, Pima, Pinal, Santa Cruz, Yavapai, Yuma cos.; California, Nevada, Utah; Mexico (Baja California, Baja California Sur, Sonora).

Lupinus sparsiflorus is widely distributed in Arizona south of the Mogollon Rim. There are a few records from north of the Rim. In wet years this species forms extensive and sometimes almost pure populations in valleys, on slopes, and along roadsides, especially at lower elevations in the central part of the state. Plants from the Mohave Desert often have smaller flowers and have been named subsp. *mohavensis*, and those from western San Diego County with pinkish flowers and truncate leaflets have been named var. *inopinatus*. Graham (1941) indicates that California quail eat the seeds of this species. *Lupinus sparsiflorus* is reported to nodulate (Allen and Allen 1981).

The flowers are showy and the plants are often grown as a wildflower and used in revegetation mixes. Seeds should be sown in the autumn after daytime temperatures have moderated in well-drained soil and provided with regular irrigation to promote germination and establishment. In dry years it may be necessary to protect the plants from depredations by wildlife.

19. Lupinus succulentus Douglas ex K. Koch. Wochenschr. Vereines Beförd. Gartenbaues Königl. Preuss. Staaten 4:277. 1861. * [from Latin, *succus*, juicy] ARROYO LUPINE Naturalized, Cultivated

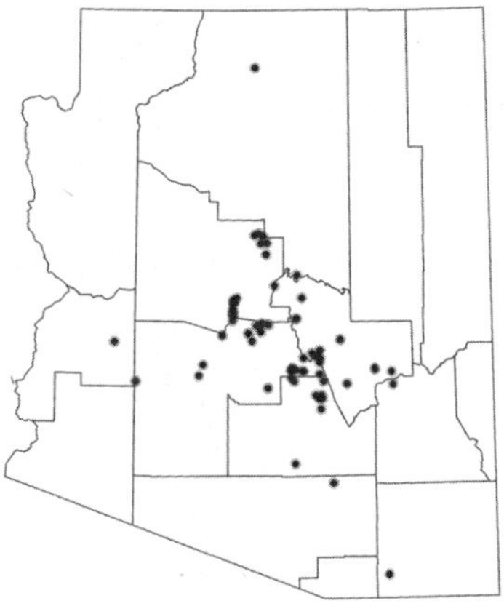

Herbs, annual, 2–10 dm, sparsely pubescent. **Cotyledons** deciduous, petiolate. **Stems** erect, unbranched or branched, usually succulent. **Leaves** cauline (may be crowded at base on new growth); stipules 5–8 mm; petiole 6–15 cm. **Leaflets** 7–9, blades 20–60 × 7–20 mm, upper (adaxial) surface glabrous. **Inflorescences** racemes, 15–25 cm, flowers distinctly whorled; peduncles 5–9 cm; bracts deciduous, 3–5 mm. **Pedicels** 3–7 mm. **Flowers** 12–18 mm; calyx 4–7 mm, lobes ± equal, lower lobe entire, upper lobe cleft; corolla usually blue-purple, rarely white, lavender, or pink, banner spot white, becoming magenta; upper margin of wings ciliate near claw; lower and upper margins of keel ciliate near claw. **Fruits** 3.5–5 cm, coarsely pubescent to tomentose. **Seeds** 6–9. $2n = 48$.

Flowering Feb–May. Ridges, slopes, canyons, floodplains, arroyos, roadsides, and other disturbed areas, often in gravelly or rocky soils; [near sea level–]390–1400 m; Sonoran Desertscrub, Semidesert Grassland, Interior Chaparral; Cochise, Coconino, Gila, Graham, La Paz, Maricopa, Pima, Pinal, Yavapai cos.; California; Mexico (Baja California, Baja California Sur).

Lupinus succulentus occurs widely throughout California except in the Great Basin and desert regions and extends into northern Mexico. It is introduced in Arizona where it is locally common in wet years in central areas of the state. California quail are reported to consume the seeds (Graham 1941). Allen and Allen (1981) report nodulation for *L. succulentus*.

Lupinus succulentus is grown as a wildflower in Arizona. Seeds should be sown in the autumn and provided with regular watering to promote germination. The plants can become robust and produce showy displays of flowers.

LYSILOMA Bentham. London J. Bot. 3:82. 1844. * [Greek *lysi*, loose, and *loma*, fringe, for the detachable rim of the fruits]

Matthew B. Johnson

Shrubs or trees. Leaves alternate; bipinnate; stipules leafy and conspicuous, usually quickly deciduous; 1–30 pairs of pinnae per leaf; glands sessile. **Leaflets** opposite, margin entire. **Inflorescences** globose heads or subglobose spikes or racemes, axillary or appearing terminal, solitary or clustered. **Pedicels** present or absent. **Flowers** mimisoid, radially symmetric, 5-merous; calyx bell-shaped; stamens numerous, exerted, filaments fused proximally and free distally, white or greenish white; ovary 1-chambered, style and stigma simple. **Fruits** tardily dehiscent or indehiscent, flattened, papery, linear-oblong, stipitate, margins forming a thickened rim that separates from the fruit valves in most but not all species, valves not recurving. **Seeds** flattened disk-shaped.

Species 8 or 9 (1 native in the flora and also cultivated): s Arizona, Mexico, Central America, the Caribbean to Florida.

The thickened, detachable rim of the fruits of most *Lysiloma* species is distinctive. The wood of various species is used for posts, construction, railroad ties, furniture and cabinets, utensils, musical instruments, carving, and fuel, and the trees are also utilized for fodder, shade, erosion control, and medicine (Allen and Allen 1981, Lewis et al. 2005). Timber from *L. sabicu* Bentham is durable in water and was used for shipbuilding as well as for bobbins and shuttles (Standley 1922). The bark of several species is used for tanning (Standley 1922, Allen and Allen 1981).

SELECTED REFERENCE: Barneby and Grimes 1996.

1. Lysiloma watsonii Rose. Contr. U.S. Natl. Herb. 1(4):99. 1891. * [for Sereno Watson, 1826–1892, American botanist] FEATHER TREE, FEATHER BUSH; *TEPEGUAJE* Native, Cultivated

Lysiloma microphyllum Bentham var. *thornberi* (Britton & Rose) Isely; *Lysiloma thornberi* Britton & Rose; *Lysiloma watsonii* (Rose) subsp. *thornberi* (Britton & Rose) Felger & C.H. Lowe

Shrubs or trees 1–5[–20+] m with a spreading canopy and a trunk to 30 cm [1+ m] in diameter. **Stems** with conspicuous lenticels, swollen at nodes; bark of trunk and larger limbs dark brown, rough and checkered with large scale-like pieces. Young herbage, petioles, rachises, inflorescences and calyces with short white pubescence, leaflets glabrate or glabrous. **Leaves** 10–15 cm; stipules 2.5–6 × 1.2–2 mm, broadly lanceolate, quickly deciduous; petioles 4–15(–22) mm with a

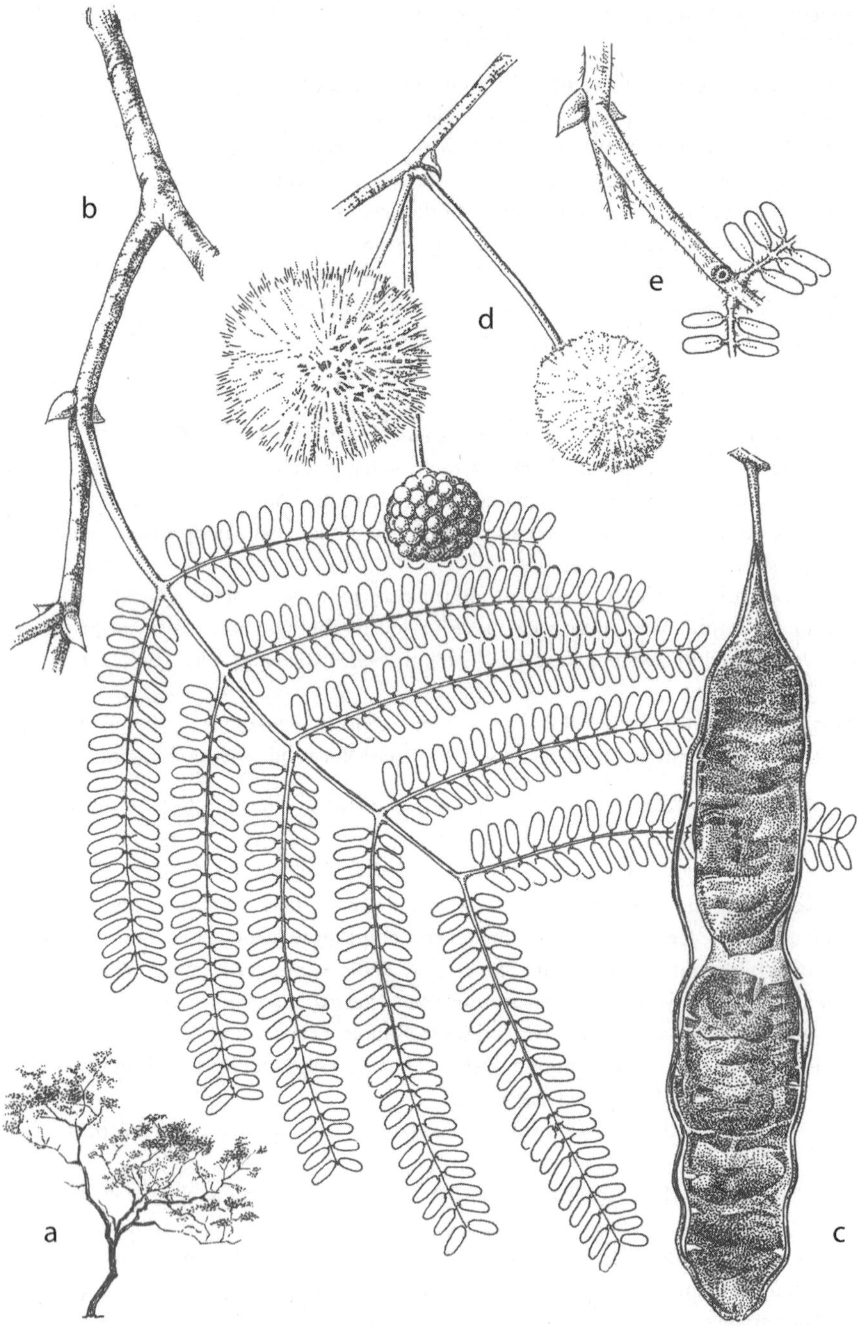

Lysiloma watsonii: **a.** habit; **b.** leaf; **c.** fruit; **d.** inflorescences; **e.** petiole gland detail. Illustrator: Randall C. Raak.

raised gland 0.4–1.9 × 0.3–0.8 mm and 0.2–1.2 mm high between the petiole midpoint and the first pair of pinnae, and sometimes with additional glands along the rachis between or below subsequent pairs of pinnae; rachis (2–)4–9.5 cm; 5–10(–11) pairs of pinnae. **Leaflets** 20–76 pairs per pinna, oblong, 3–9 × 1.2–2.7 mm, somewhat glaucous. **Inflorescences** rounded subglobose racemes ± 1.5 cm, axillary and solitary or 2 or 3 fascicled or rarely appearing terminal. **Pedicels** 0.2–0.3 mm. **Flowers** pale green; calyx 2–2.5 mm; corolla 3.5–4 mm; filaments 6–9 mm, white. **Fruits** dark brown or nearly black, 12–24.5 × 1.5–2.7[–4.5] cm, stipe 5–14 mm, apex acuminate; rim separating from valves. **Seeds** (2–)5–8(–12); brown, 8–10 mm.

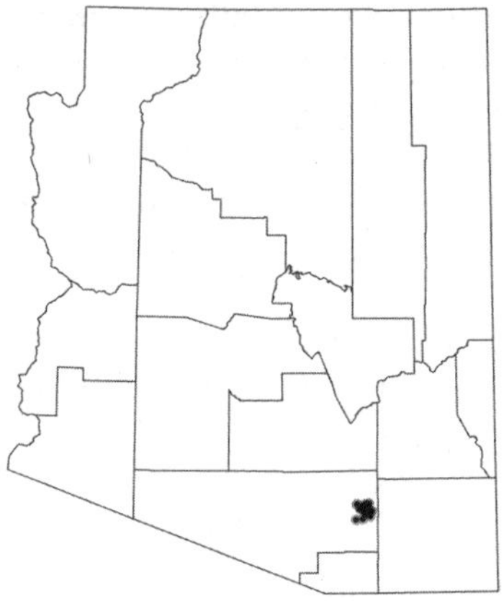

Flowering April–May(Jul–Oct). Canyons and slopes; Sonoran Desertscrub and Semidesert Grassland; [near sea level–]1000–1435 m; Pima Co. (Rincon Mountains); Mexico (Chihuahua, Durango, Sinaloa, Sonora).

In Arizona, *Lysiloma watsonii* is restricted to several drainages along the south side of the Rincon Mountains within Saguaro National Park. The plants are locally common. In Mexico, the species is widespread and locally abundant, especially in the ecotone between thornscrub or tropical deciduous forest and oak woodland. Annual rainfall across much of its range averages 375–750 mm, though it has been found in a few Sonoran localities that receive less than this. At the northern and upper elevational limits of its range, *L. watsonii* typically grows as a multiple-stemmed shrub 1–3 m tall. These plants are periodically frozen back in cold winters. In nearby protected sites such as canyons, they can develop into small trees to 4 or 5 m. Stature increases in warmer-winter regions to the south. In central Sonora, large, isolated trees are sometimes found along smaller arroyos in dry thornscrub. The species is sporadic along the eastern margin of the Sonoran Desert. Felger et al. (2001) report that *L. watsonii* exceptionally attains heights of 20 m in tropical canyons of southeastern Sonora and mention one tree from the upper Río Mayo drainage with a height and spread of ca. 30 m. *Lysiloma watsonii* is nearly evergreen unless the leaves are killed by freezing temperatures. The leaves turn yellow and fall in April, shortly before the new leaves appear. New leaves are bright green and change to a dull glaucous green as they mature. The size and position of the petiole gland is variable, and the gland is generally larger on specimens from Mexico. Specimens from Durango and Sinaloa have pods up to 4.5 cm wide. *Lysiloma watsonii* is reported to nodulate (Allen and Allen 1981, Sprent 2009).

The seeds of *Lysiloma watsonii* are nutritious and have been ground as flour for human consumption (Thorn et al. 1983, Felger et al. 2001). The wood is reddish brown. It is strong and durable, and has been used in construction, for rollers in a sugar cane mill, and for utensils (Gentry 1942). In Sonora, forked limbs of *L. watsonii* are often used for posts for sliding beam gates (Felger et al. 2001). The bark is used for hardening gums, and a mouth rinse is made by boiling the bark (Gentry 1942). Felger et al. (2001) report that the Guarajío people of Sonora use a preparation of the bark for cleaning wounds. In Mexico, the species is used to treat gastrointestinal ailments in traditional medicine, and extracts of *L. watsonii* have been shown to have antibacterial activity against several species of enteropathogenic bacteria (Moreno Salazar et al. 2008).

Feather tree is cultivated in Arizona landscapes for shade and as a patio tree. Its dense, luxuriant, fern-like foliage provides a tropical appearance and is effective for screening. In cold-winter locations where the plant freezes to the base, it can be grown as a shrub for seasonal foliage. The flowers are showy and fragrant, appearing with the new leaves in spring and sporadically later in the summer. Growth rate is moderate with irrigation. Some supplemental irrigation is necessary during warm weather in desert locations in Arizona for acceptable growth and appearance. Excessive irrigation or poorly drained soil can result in chlorosis. The plants require a sunny location for best performance. Feather tree sustains stem damage below ca. -5°C. Recovery is generally rapid the following growing season. Even if regularly frozen to the base, plants will vigorously grow numerous stems, though are unlikely to flower. The trees develop a low, horizontal-branched spreading canopy if not pruned. Freeze-killed wood should be pruned out. The fine leaf litter can form a dense mat beneath the canopy, retarding the growth of smaller plants, and the fallen pods create some litter. The spring leaf-drop can alarm people who are not aware that this is typical. Feather tree is easily propagated from scarified seeds.

MACROPTILIUM (Bentham) Urban. Symb. Antill. 9(4):457. 1928. * [from Greek *makros*, large, and *ptilon*, feather, wing] BUSHBEAN

Leticia Torres-Colín and Alfonso Delgado-Salinas

Vines or sometimes erect, semierect or prostrate herbs, biannual or perennial. **Roots** turnip-shaped, elongated, slender taproots or thick tubers, eventually becoming somewhat woody. **Stems** slender, solid or sometimes with hollow pith, densely canescent, sericeous, strigillose, hirsute or glabrate. **Leaves** alternate, 3-foliolate [very rarely unifoliolate]; stipules persistent, ovate, ovate-lanceolate or widely deltate, spreading or retrorse; stipels present, linear or subulate, often reflexed. **Leaflets** linear, elliptic, circular, ovate, broadly ovate or obovate, membranous, chartaceous or coriaceous, entire or 2-lobed. **Inflorescences** pseudoracemes, axillary (cleistogamous flowers in some species with modified inflorescences); tufts of bracts at base of peduncle either persistent or deciduous; swollen nodes present, node bracts deciduous; bracteoles present, usually deciduous. **Pedicels** subsessile. **Flowers** papilionoid; calyx tubular or campanulate, 5-lobed, free or upper (adaxial) lobes partly fused; corolla pink, salmon-orange, red, purple-black, or yellow, flower colors fading to purple when dry, banner circular, obovate, oblong, emarginate with 2 small basal auricles, mostly pale green, wings larger than the banner, obovate or oblong-ovate, long stipitate, brightly colored, glabrous or pubescent, keel petals distally fused and hook-like, basally fused to staminal tube; stamens 10, 9 fused and 1 free, style with introrse beard, and apically recurved and thickened. **Fruits** dehiscent, valves plane or twisting, linear to oblong-falcate, without septa, spreading or pendent. **Seeds** 2–20, small, oblong or reniform, seed coat brown; short hilum not centric, oblong or ovate. $x = 11$.

Species 18 (3 native in the flora): s United States, Mexico, Central America, Caribbean, South America to n Argentina; some species have been introduced for fodder and have become naturalized in the tropics and subtropics of the world.

Molecular evidence establishes the monophyletic placement of *Macroptilium* amongst the other New World genera of subtribe Phaseolinae (Delgado-Salinas et al. 2011). All *Macroptilium* species are characterized by a diagnostic feature of the flower architecture: the left-hand wing petal is directed upward to adopt the function of the banner while it also becomes a supporting structure of the other wing petal and encloses the distally-hooked keel.

1. Vines; taproot elongated, sometimes tuberous; inflorescence with extrafloral nectaries in conspicuous nodes; tufts of persistent bracts at base of peduncle; flowers more than 1 cm long, corolla purple-black; fruits 7–10 cm long, valves twisting at dehiscence, linear, spreading; seeds 12–13 ____ **1. Macroptilium atropurpureum**
1′ Prostrate herbs; root turnip-shaped, small; inflorescence with extrafloral nectaries in inconspicuous floral nodes; tufts of deciduous bracts at base of peduncle; flowers less than 1 cm long, corolla salmon-orange or red; fruits not more than 2.5 cm long, valves not twisting at dehiscence, oblong-falcate, pendent; seeds 2–5.
    2. Leaflets circular, broadly ovate, obovate, sometimes basally 2-lobed; corolla salmon-orange; ovary with 6 ovules ____ **2. Macroptilium gibbosifolium**
    2′ Leaflets linear, entire; corolla coral-red or red; ovary with 4 ovules ____ **3. Macroptilium supinum**

1. Macroptilium atropurpureum (de Candolle) Urban. Symb. Antill. 9:457. 1928. * [from Latin, *atro*, dark or black, and *purpureum*, purple, referring to the flower color] SIRATRO BEAN, PURPLE BEAN; *CONCHITO* Native

Phaseolus atropurpureus de Candolle, Prodr. 2:395. 1825

Vines, perennial; taproot elongated. **Stems** solid, sericeous. **Leaves** with stipules lanceolate or ovate, 5–6 × 1.8–2 mm, retrorse; petioles 3.5–8 cm, canescent; rachis 5–8 mm; stipels 1.2–2.5 mm, linear. **Leaflets** chartaceous, elliptical or ovate, basally cuneate or obtuse and apically acute, upper surface strigose and often canescent below, terminal leaflets 4.5–4.7 × 3–4.3 cm, often hastate, lateral leaflets 3.3–4 × 2.7–3.3 cm, 2-lobed. **Inflorescences** axillary, 26–36 cm, axes pilose; tufts of bracts persistent, 5.5 mm, lanceolate, canescent; conspicuous extrafloral nectaries; bracts and bracteoles lanceolate or subulate, deciduous. **Pedicels** 1–2 mm. **Flowers** 1.7–3 cm; calyx 6.2–8 mm, campanulate, canescent, all calyx lobes shorter than tube; corolla purple-black; banner 13–17 × 9–10.4 mm, oblong; wings 13–24 × 8–14.3 mm, obovate; keel petals 9–16 mm; ovary with nectary disc, lower side longer, ovules 13. **Fruits** 7–10 × 0.3–0.4 cm, valves twisting at dehiscence, linear, spreading. **Seeds** 12–13, 3.5–5.2 × 2–3.2 mm, oblong. $2n = 22$.

Flowering mostly Jun–Sep[–Feb]. In sandy soils of canyons; [near sea level–]1520–1880[–1900] m; Semidesert Grassland, Madrean Evergreen Woodland; Pima Co.; California, Florida, Mississippi, New Mexico, Texas; Mexico (Baja California Sur, Campeche, Chiapas, Chihuahua, Coahuila, Durango, Guanajuato Guerrero, Hidalgo, Jalisco, México, Michoacán, Nuevo León, Oaxaca, Sinaloa, Sonora, Tabasco, Tamaulipas, Veracruz); Caribbean; Central America; South America (southward to n Argentina).

In Arizona, *Macroptilium atropurpureum* is rare and localized in the Baboquivari and Coyote Mountains. Herbarium specimens documenting this species in the state are from the 1930s. The species is more common southward in Sonora and Baja California Sur and into the tropics where it occurs in a wide range of habitats, including coastal dunes and disturbed sites, and in diverse biotic communities.

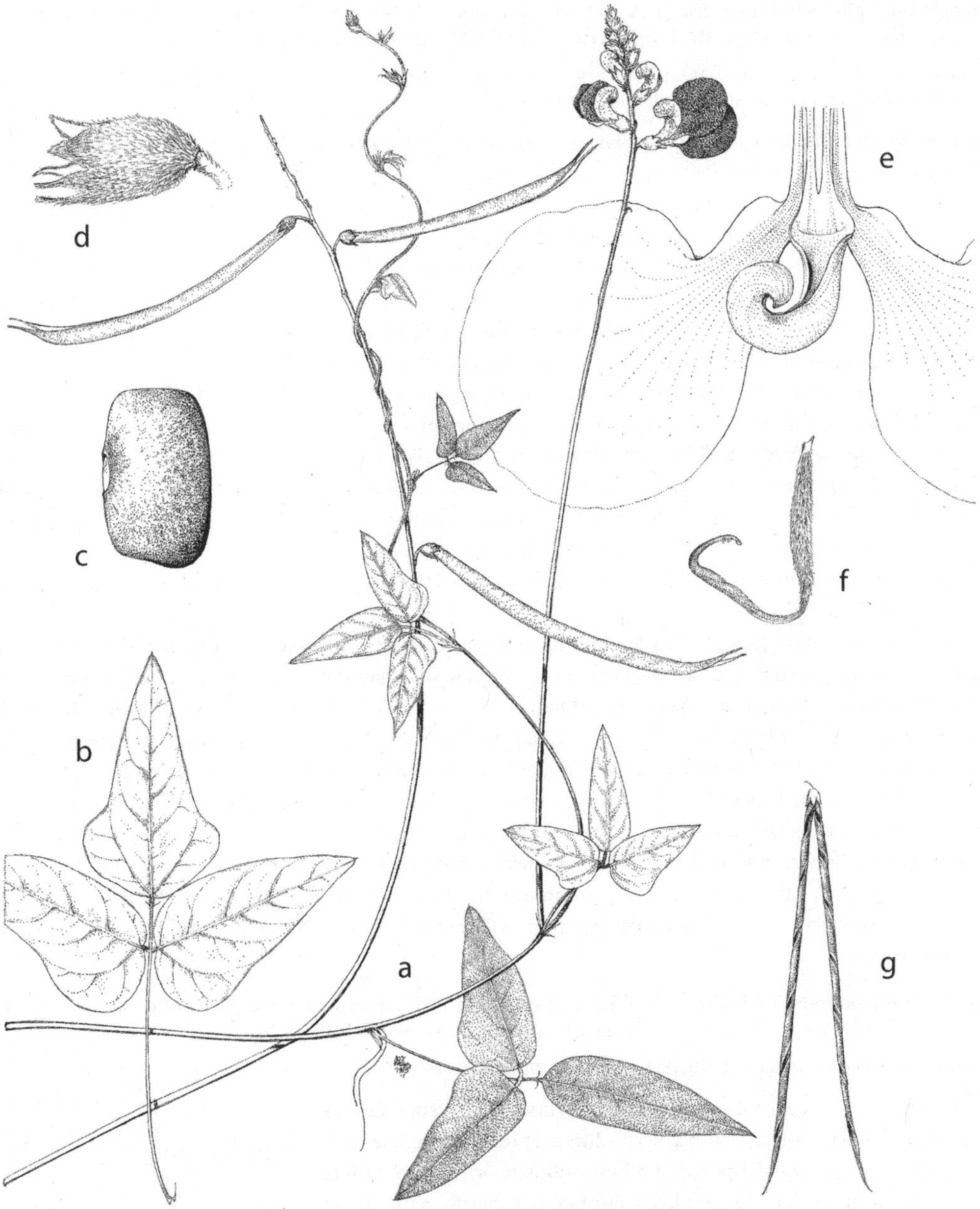

Macroptilium atropurpureum: **a.** flowering and fruiting stems; **b.** leaf; **c.** seed; **d.** calyx; **e.** interior view of keel and adherent wings; **f.** pistil; **g.** fruit after dehiscence. Drawn by Karen Douthit and reproduced from *Flora Novo-Galiciana*, vol 5, fig. 80, with permission of the University of Michigan Herbarium.

Though not grown for this purpose in Arizona, *Macroptilium atropurpureum* is widely cultivated as a forage for livestock, especially cattle, and as a cover crop in tropical regions around the world. The seeds and foliage are also suitable as wildlife forage (Duke 1981). The plants are moderately drought-tolerant due to the taproot but do not tolerate prolonged waterlogging (Duke 1981). It has become naturalized in some regions. *Macroptilium atropurpureum* is known to nodulate (Allen and Allen 1981).

2. Macroptilium gibbosifolium (Ortega) A. Delgado. Syst. Bot. 6(3):294. 1981. * [from Latin, *gibbus*, swollen or hump, and *folium*, leaf, referring to the lobes on the leaflets] VARIABLELEAF BUSHBEAN Native

Phaseolus gibbosifolius Ortega, Nov. Pl. Descr. Dec. 25. 1797; *Macroptilium heterophyllum* (Humboldt & Bonpland ex Willdenow) Maréchal & Baudet; *Phaseolus heterophyllus* Humboldt & Bonpland ex Willdenow

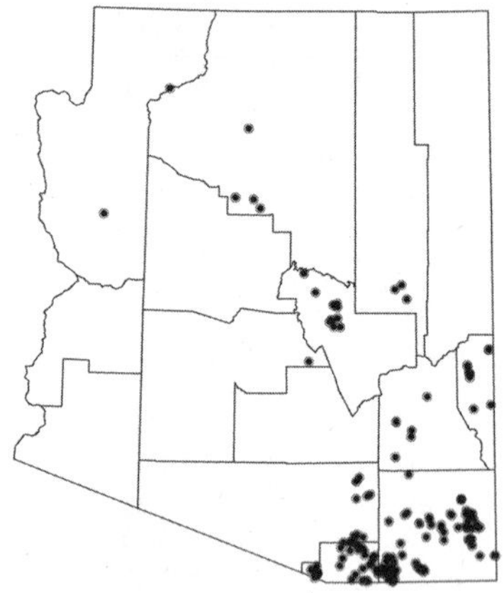

Herbs prostrate, perennial; root turnip-shaped. **Stems** solid, strigillose or hirsute. **Leaves** with stipules widely deltate or ovate, 4.5–8 × 2–2.7 mm, spreading; petioles 1.7–6 cm, short strigose or hirsute; rachis 5–9 mm; stipels 1–5.5 mm, subulate. **Leaflets** chartaceous, circular, ovate, broadly ovate or obovate, basally attenuate or cuneate and apically acute or obtuse, upper surface hirsutulous, lower surface strigose, terminal leaflets 2.3–4.7 × 1.6–2.8 cm, lateral leaflets 2–4.3 × 1.6–2.9 cm, sometimes 2-lobed. **Inflorescences** axillary, 14–25 cm (subterranean cleistogamous inflorescences 1.5–2.2 mm, flowers 3.5 mm, corolla whitish), axes adpressed-pilose; tufts of bracts mostly deciduous; extrafloral nectaries on inconspicuous floral nodes; bracts and bracteoles subulate or lanceolate, deciduous. **Pedicels** 0.7–1 mm. **Flowers** ca. 1 cm; calyx 3.5–4.5 mm, campanulate, hirsutulous, all calyx lobes as long as or longer than the tube; corolla salmon-orange; banner 5.5–7.5 × 4.5–6.5 mm, circular; wings 5.1–10 × 3.3–6.6 mm, oblong-ovate; keel petals 4–8 mm; ovary with nectary disc, entire, ovules 6. **Fruits** 1.6–2.5 × 2.7–3 cm, valves not twisting at dehiscence, oblong-falcate, pendent. **Seeds** 4–5, 2.9–3 × 2–2.2 mm, oblong.

Flowering (Apr–May) Jul–Oct. Rocky summits, slopes, canyons, meadows, and floodplains; 1160–2770 m; Semidesert Grassland, (Plains Grassland), Madrean Evergreen Woodland, Petran Montane Conifer Forest; Cochise, Coconino, Greenlee, Gila, Graham, (Maricopa, Mohave), Navajo, Pima, Santa Cruz cos.; New Mexico, Texas; Mexico (Aguascalientes, Chiapas, Chihuahua, Durango, Guanajuato, Guerrero, Hidalgo, Jalisco, México, Michoacán, Morelos, Nayarit, Oaxaca, Puebla, San Luis Potosí, Sonora, Tamaulipas, Veracruz, Zacatecas); Central America.

Macroptilium gibbosifolium is the only member of the genus that is widely distributed in Arizona. It is locally common in mountains in the southeastern part of the state and is more sporadically distributed elsewhere. With its distinctive salmon-colored flowers, it should be easily distinguished from species of *Phaseolus*. This species and *M. supinum* are unique among Arizona Fabaceae in having subterranean inflorescences with self-pollinating flowers in addition to inflorescences on aerial portions of the plant. The stems will root at the nodes. McVaugh (1987) reports that the tuber-like roots are edible.

3. Macroptilium supinum (Wiggins & Rollins) A. Delgado & L. Torres. Anales Inst. Biol. Univ. Nac. Autón. México, Bot. 66(2):189. 1995. * [from Latin, *supinus*, bent backwards] SUPINEBEAN Native

Phaseolus supinus Wiggins & Rollins, Contr. Dudley Herb. 3(8):270, pl. 60. 1943

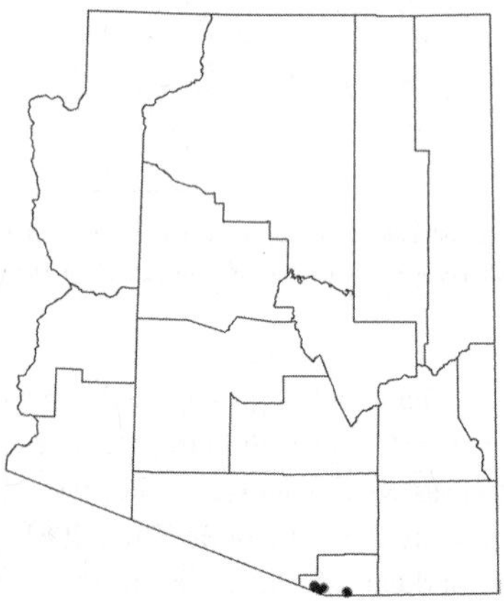

Herbs prostrate, perennial; root turnip-shaped. **Stems** solid, hirsute. **Leaves** with stipules ovate-lanceolate, 2.3–6.2 × 1.5–2 mm, spreading; petioles 2–5 cm, hirsute; rachis 5–7 mm; stipels 1–2.5 mm, subulate or linear. **Leaflets** coriaceous, linear, entire, basally rounded or cuneate and apically acute, upper surface glabrate and densely hispid below, terminal leaflets 2.5–4.2 × 0.5–0.8 cm, lateral leaflets 3.8–4.5 × 0.9–1.1 cm. **Inflorescences** axillary, 11–15 cm (subterranean cleistogamous inflorescence 3.9–6.3 mm, flowers 6 mm; corolla whitish); axes glabrous; tufts of bracts mostly deciduous; extrafloral nectaries on inconspicuous floral nodes; bracts and bracteoles subulate, deciduous. **Pedicels** 0.6 mm. **Flowers** 0.9 cm; calyx 2–2.3 mm, campanulate, hirsute or glabrate, all calyx lobes as long as or longer than the tube; corolla red; banner 5.3 × 5 mm, obovate; wings 8.1–8.9 × 3.2–5.4 mm, oblong-ovate; keel petals 6.2 mm; ovary with nectary disc, entire, ovules 4. **Fruits** 1–1.8 × 0.19–0.25 cm, valves not twisting at dehiscence, oblong-falcate, pendent. **Seeds** 2–3(–4), 1.7–2.3 × 1.2–3.5 mm, reniform.

Flowering Sep–Nov. Ridges and slopes; 1100–1500 m; Semidesert Grassland, Madrean Evergreen Woodland; Santa Cruz Co.; Mexico (Nayarit, Sonora). Of conservation concern.

Macroptilium supinum is found only in the Atascosa, Pajarito, Patagonia, and San Luis Mountains in Arizona. It also occurs in northwestern Mexico (Delgado-Salinas and Torres-Colín 1995). The distribution is patchy and populations are generally small. *Macroptilium supinum* and *M. gibbosifolium* are the only Arizona Fabaceae with subterranean inflorescences. These develop among grasses, leaf litter, and soil. They also bear inflorescences on aerial portions of the plants.

MARINA Liebmann. Vidensk. Meddel. Dansk Naturhist. Foren. Kjφbenhavn 1853(3–4):103–104. 1854. * [named for Doña Marina or La Malinche, the Nahua woman who served as interpreter and companion for Hernán Cortés during the conquest of Mexico] FALSE PRAIRIE CLOVER

Michelle M. McMahon, Mary Welch-Keesey, and Matthew B. Johnson

Annual or perennial herbs, or shrubs. Herbage pubescent, usually glandular. **Leaves** alternate, odd-pinnate, [rarely 1-foliolate]; stipules small, deciduous; petiolate; stipels absent. **Leaflets** glandular, with pallid sinuous diagonal lines, or "lineoles"; terminal leaflet stalked. **Inflorescences** racemes or racemose panicles, terminal or appearing axillary; bracts small, deciduous. **Pedicellate. Flowers** subpapilionoid (petals attached to stamen sheath); calyx campanulate, 5-lobed; corolla indigo-blue to magenta or bicolored, with petals clawed, the wings and keel enclosing the staminal column, keel adherent; stamens partially or mostly united into a sheathing tube, 10 [rarely 9 or 5]; ovary with 1 ovule, style glabrous distally, stigma small. **Fruits** either harp-shaped and compressed, compressed and obliquely obovate, or obovoid and scarcely compressed, glandular, indehiscent. **Seed** 1, small, smooth, glossy. $x = 10$.

Species 38 (3 native in the flora and 1 of these occasionally cultivated): all are native to Mexico with a few species extending into the southwest United States and Guatemala, and one species extending south to Venezuela and the Caribbean.

Stems of *Marina* have been used to make brooms (Martin et al. 1998, Lewis et al. 2005). A yellow dye is produced from the roots of some species (Lewis et al. 2005). Some species of *Marina* have potential as flowering perennials in landscapes and wildflower gardens. Allen and Allen (1981) report nodulation for one species of *Marina* (as *Dalea*).

SELECTED REFERENCES: Barneby 1977, Isely 1998.

1. Calyx not pubescent, intervals between ridges densely glandular; sparsely branching shrub with slender red-purple stems to 3 m; inflorescence loose (internodes much longer than flower length); limited to the Patagonia Mtns ____ **2. Marina diffusa**

1′ Calyx pubescent, intervals between ridges with small glands, usually obscured by hairs; prostrate herbs or subshrubs up to 1 m, not red-purple; inflorescence moderately dense; s AZ.

2. Flowers and fruits spreading to ascending, not declined; stems densely glandular; taprooted ____ **3. Marina parryi**

2′ Flowers and fruits declining at maturity; herbage not or scarcely glandular; branching from subterranean caudex in age ____ **1. Marina calycosa**

1. Marina calycosa (A. Gray) Barneby. Mem. New York Bot. Gard. 27:83. 1977. * [with large calyces] SAN PEDRO FALSE PRAIRIE CLOVER Native

Dalea calycosa A. Gray, Smithsonian Contr. Knowl. 5(6):40. 1853

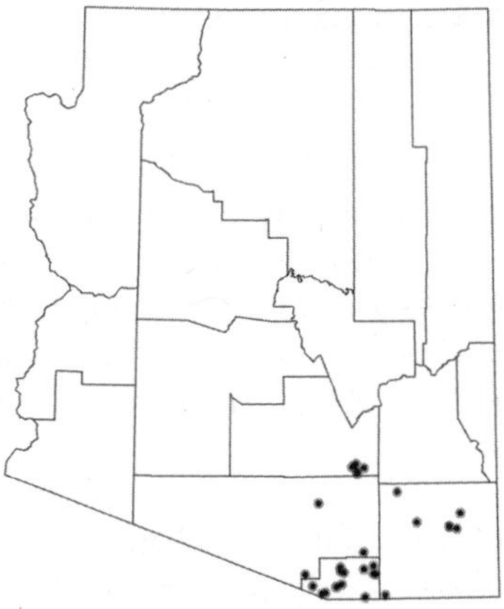

Perennial, tufted decumbent to prostrate herbs from a short caudex, to 30 cm. **Stems and herbage** strigulose with appressed hairs, commonly gray or canescent, obscurely or not glandular. **Leaves** 1–3 cm; stipules (1–)1.5–3.5 mm, lanceolate, papery; petioles shorter to slightly longer than length of lowest leaflets. **Leaflets** (9–)10–25(–35), 1.5–7 × 1–2(–3) mm, obovate to oblong-obovate. **Inflorescences** 10–40-flowered in an initially dense raceme, becoming loose with reflexed flowers, the axis (1–)1.5–6.5 cm; bracts 2.5–4 mm, narrowly lanceolate, quickly deciduous. **Pedicels** 0.3–0.6 mm. **Flowers** declined at maturity; calyx silvery with long ascending hairs, 5.2–7 mm, the tube 1.8–2.5 mm, the lobes 3.2–5.1 mm; corolla bicolored, whitish and dull blue or brownish purple, banner 3.7–5 mm, wings 4.5–6.3 mm, keel 6.3–9.4 mm; stamens 10, 7–8 mm, filaments free distally for 2.5–3 mm, anthers pale to dark blue. **Fruits** 2.6–3 × 2–2.2 mm, obliquely obovate, compressed, valves pilosulous and with glands in two crescent lines. **Seed** 1, 2–2.5 mm, straw-colored.

Flowering Apr–Jun, Aug–Sep(–Nov). Ridges, slopes, canyons, arroyos, valleys, and roadsides in various soils; 940–1650[1770] m; (upper elevations in Sonoran Desertscrub), Semidesert Grassland, Plains Grassland, Madrean Evergreen Woodland; Cochise, Graham, Pima, Pinal, Santa Cruz cos., New Mexico; Mexico (Chihuahua, Sonora).

Marina calycosa is primarily a species of grassland and open oak-grassland, rarely extending into upper elevation

Sonoran Desert habitats. It is found in scattered colonies but may be locally abundant and occurs on a variety of soil types from sandy to gravelly alluvium and rocky slopes.

2. Marina diffusa (Moricand) Barneby. Mem. New York Bot. Gard. 27:126. 1977. * [with a diffuse inflorescence]

Dalea diffusa Moricand, Pl. Nouv. Amer. 8, pl. 6. 1833[1834]

Shrubs 1–2.5 m. **Stems** wand-like and much branched distally, often glabrous, glandular, purple-red. **Leaves** 1.5–3.5(–5) cm, glabrous, green above and pallid beneath, often quickly deciduous; stipules 0.3–1.4 mm, ovate or narrowly awl-shaped, membranous; petioles shorter to slightly longer than length of lowest leaflets. **Leaflets** 11–25(–35), 3–6.5(–11) × 1–2 mm, oblong or oblong-oblanceolate, marginally and sometimes dorsally gland dotted, glabrous, distal-most leaves smaller with 1–7 leaflets, 0.7–5 mm, mostly obovate to broadly oblanceolate. **Inflorescences** loosely 1–25-flowered, the axis (0–)0.1–4.5 cm; bracts 1 mm, obovate-cuneate, submembranous, quickly deciduous. **Pedicels** 0.6–1.5 mm. **Flowers** ascending and spreading; calyx 2.5–3.6 mm, glabrous, the tube 1.5–2.2 mm, the lobes 0.7–1.6 mm; corolla bicolored, the banner mostly white, otherwise rose- or magenta-purple or maroon-violet, banner 2.5–3.5 mm, wings 3.5–4.7 mm, keel 4.4–5.9 mm; stamens 10, (5.5–)6–8 mm, the filaments free distally for 1.5–2.1 mm, the anthers pale blue. **Fruits** 2.7–2.9 × 2–2.5 mm, plumply obovoid, not or scarcely compressed, the valves irregularly glandular. **Seed** 1, 2.1–2.5 mm, dark olive green.

Varieties 2 (1 native in the flora). The second variety, var. *radiolata*, is found in the Mexican states of Jalisco, Nayarit, and Zacatecas.

2a. Marina diffusa (Moricand) Barneby var. **diffusa**. * DIFFUSE FALSE PRAIRIE-CLOVER; *ESCOBILLA*, *ESCOPETA*, *ARENILLA* Native

Leaves of panicles 1–4.5 mm. **Leaflets** 1–5, 0.8–3 mm. **Inflorescences** 1–10 flowered.

Flowering Sep–Dec[–Feb]. Slopes and roadsides; [200–]1530[–1800] m; Madrean Evergreen Woodland; Santa Cruz Co.; Mexico (Chiapas, Chihuahua, Colima, Durango, Guerrero, Jalisco, México, Michoacán, Morelos, Oaxaca, Sinaloa, Sonora); Guatemala.

Marina diffusa var. *diffusa* is known in Arizona only from the Patagonia Mountains where it is highly localized, however it is widespread in western and southern Mexico (except where it is replaced by var. *radiolata*). In southeastern Sonora, *M. diffusa* var. *diffusa* is relatively widespread and occurs on slopes and along arroyos and canyon streams in tropical deciduous forest and pine-oak woodland (Martin et al. 1998). In Mexico, the flexible stems of *M. diffusa* have been bundled and used as brooms (Barneby 1977, Martin et al. 1998).

3. Marina parryi (Torrey and A. Gray) Barneby. Mem. New York Bot. Gard. 27:68. 1977. * [for Charles C. Parry, 1823–1890, English-born American physician and botanist] PARRY'S FALSE PRAIRIE CLOVER, PARRY'S MARINA Native, Cultivated

Dalea parryi Torrey and A. Gray, Proc. Amer. Acad. Arts 7(2):397. 1868

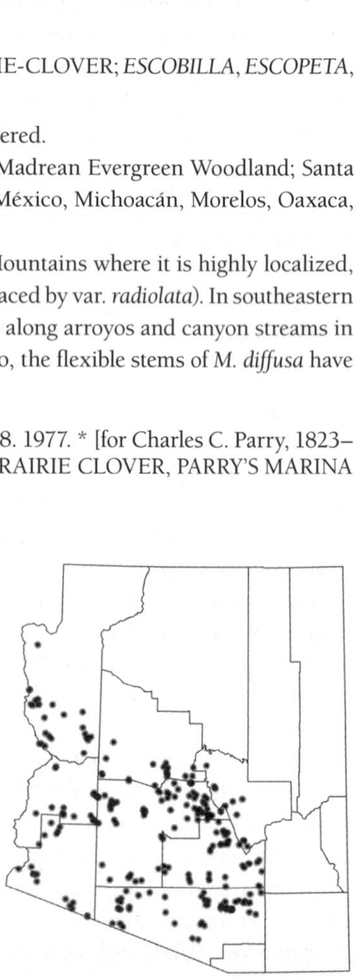

Herbs or suffrutescent subshrubs or shrubs, 15–70(–120) cm; yellowish taproot often visible on herbarium specimens. **Stems** wiry, diffuse, glabrous or glabrate proximally, dense with dark glands. **Herbage** thinly to densely strigulose with appressed hairs, leaflets pubescent on both surfaces or sometimes glabrous above, often densely glandular below. **Leaves** 1.5–4.5(–6) cm; stipules 0.3–1.4 mm, triangular or awl-shaped; petioles 5–8 mm. **Leaflets** (9–)10–23(–29), 0.5–7 × 2–4 mm, orbicular-obcordate to broadly oblong-obovate, thick. **Inflorescences** 10–40-flowered in an elongated, loose raceme, flowers ascending or spreading, the axis 2–10 cm; bracts 1.5–3 mm, lanceolate or elliptic, deciduous, glandular. **Pedicels** 0.3–0.6 mm. **Flowers** with calyx 2.8–4.5 mm, the tube 1.6–2.1 mm, the lobes 1.2–2.6 mm, unequal; corolla usually bicolored, blue and white or yellow, rarely blue-violet throughout, banner 2.8–4 mm, wings 3.3–4.8 mm, keel 5.1–7.1 mm; stamens 10, (5–)5.5–7 mm, filaments free distally for ca. 2 mm, anthers blue. **Fruits** 1.8–2.4 × 1.6–1.8

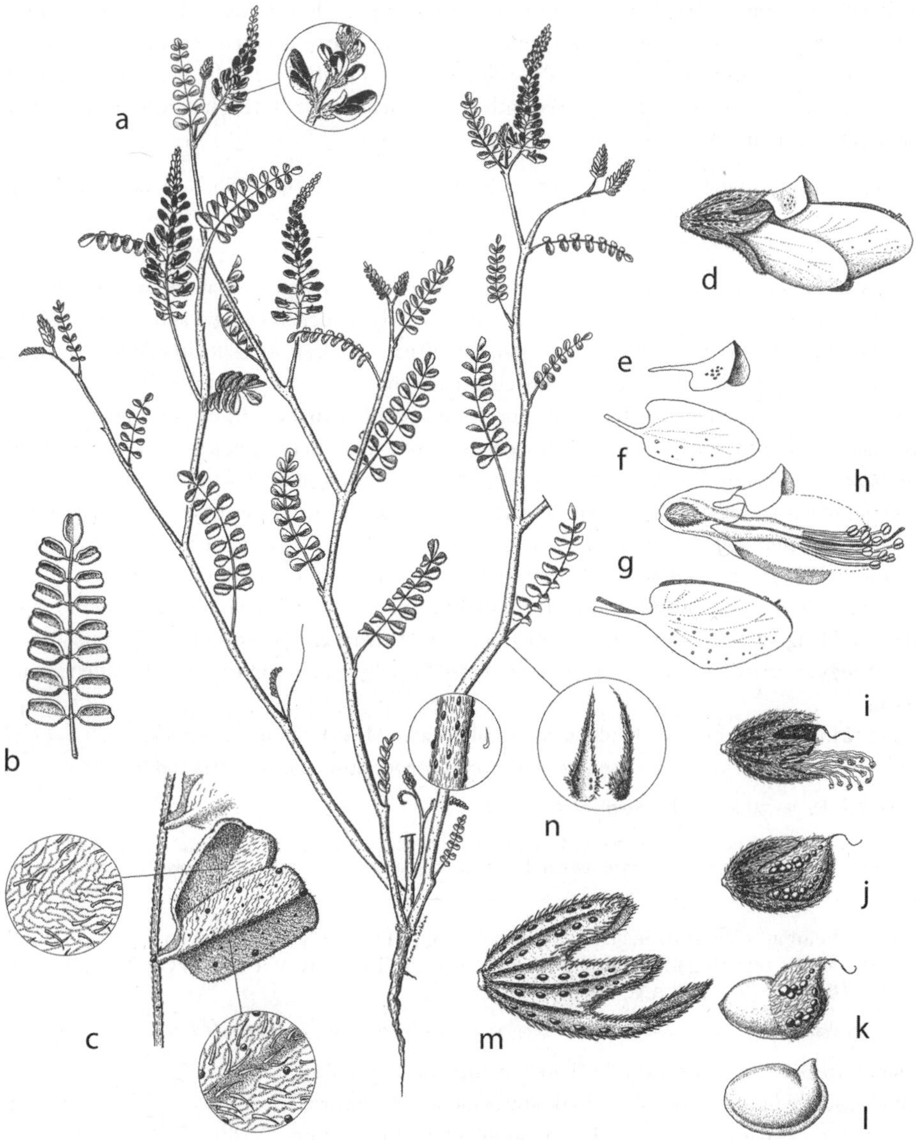

Marina parryi: **a.** habit; **b.** leaf; **c.** leaflet detail; **d.** flower; **e., f., g.** petals; **h.** longitudinal section of flower; **i.** immature fruit; **j.** mature fruit; **k.** fruit dehiscence; **l.** seed; **m.** calyx; **n.** stipules. Illustrator: Susan E. Ashton.

mm, obovate, laterally compressed, the valves puberulent and glandular distally in two short rows. **Seed** 1, 1.6–2 mm, brown.

Flowering opportunistically nearly year-round with peak flowering Feb–Apr. Ridges, slopes, canyons, arroyos, valleys, plains, and roadsides in a wide variety of soil types from sandy to rocky; [near sea level–]60–1130(–1220) m; Sonoran Desertscrub, (Mohave Desertscrub and ecotones with Interior Chaparral and Semidesert Grassland); Gila, La Paz, Maricopa, Mohave, Pima, Pinal, Yavapai, Yuma cos., California, Nevada; Mexico (Baja California, Baja California Sur, Sonora).

Marina parryi is a distinctive and widespread Sonoran Desert plant that may be locally abundant. Its distribution nearly matches the outline of the Sonoran Desert (Barneby 1977). It is found occasionally in the southern Mohave Desert and rarely extends into mixed desertscrub/chaparral or desertscrub/grassland habitats. At the southern limits of its distribution in Sonora, it is found in thornscrub (Martin et al. 1998). This species may flower nearly year-round in response to rainfall.

The dark blue-purple and white flowers of Parry's marina are especially attractive when the plants grow in masses, and they are sometimes cultivated as a flowering perennial. The plants are well-suited for natural desert landscapes and rock gardens. *Marina parryi* requires full sun and a well-drained soil to thrive. Established plants are highly drought tolerant but appearance and flowering are improved with occasional irrigation during hot weather. This species is hardy to at least -10 C. Propagation is by seeds.

MARIOSOUSA Seigler & Ebinger in Seigler, Ebinger & Miller, Novon 16:415. 2006. * [honors Mario Sousa, former Director of the Herbarium of the Instituto de Biología (MEXU), Universidad Autónoma de México]
David S. Seigler and John E. Ebinger

Shrubs or trees, erect. **Stems** lightly to highly branched; twigs rounded or angular, glabrous to pubescent; short shoots usually absent; prickles and stipular spines absent (stipules weakly hardened or rigid and sometimes interpreted as spines in *M. millefolia*). **Leaves** alternate; bipinnate, pinnae 1–50 pairs per leaf, opposite; stipules mostly small, not obvious, persistent; petioles round or flattened, usually channeled above, petiolar gland usually present; rachis usually channeled above, a small gland between the upper 1 or 2 pinna pairs. **Leaflets** 4–65 pairs per pinna, opposite, subsessile, membranous, glabrous to pubescent. **Inflorescences** loosely to densely flowered cylindrical spikes, solitary to clustered in the leaf axis; peduncles short, glabrous to pubescent, not elongating in fruit; floral bracts present. **Pedicels** absent. **Flowers** mimosoid, radially symmetric, 5-merous; calyx cup-shaped, lobes triangular, glabrous to pubescent; corolla cup-shaped, membranous, lobes triangular, glabrous to pubescent, valvate in bud; nectariferous disk absent; stamens 50–100, 5–13 mm, exserted, white or cream; anthers attached dorsally, lacking a gland; ovary 1, stipitate; style thread-like; stigma thread-like. **Fruits** straw-colored in Arizona, erect to pendulous, usually dehiscing along both sutures, linear to oblong, straight, flattened, dry, stiffly papery to leathery, glabrous to pubescent; stipe short; apex sometimes beaked. **Seeds** arranged in a single row, not surrounded by pulp, oval to elliptic, strongly flattened, smooth, pleurogram present.

Species 14 (2 in the flora, 1 native, 1 cultivated): United States, Mexico, Central America.

The wood of some species has been used for construction, fence posts, tool handles and fuel.

SELECTED REFERENCES: Jawad et al. 2000, Seigler et al. 2006b.

1. Pinna pairs mostly 6–14 per leaf; petioles 30–75 mm; bark smooth [to fissured], gray ______ **1. Mariosousa millefolia**
1' Pinna pair 1 (rarely 2 or 3); many petioles more than 100 mm; bark exfoliating and papery, white, orange brown to yellow on inner surface ______ **2. Mariosousa heterophylla**

1. Mariosousa millefolia (S. Watson) Seigler & Ebinger in Seigler, Ebinger & Miller, Novon 16:419. 2006. * [Latin *millefolia*, 1,000 leaves, referring to the numerous leaflets] FERNLEAF ACACIA, MILFOIL WATTLE, SANTA RITA ACACIA; *TEPEMESQUITE BLANCO* Native, Cultivated

Acacia millefolia S. Watson, Proc. Amer. Acad. Arts 21:427. 1886; *Senegalia millefolia* (S. Watson) Britton & Rose

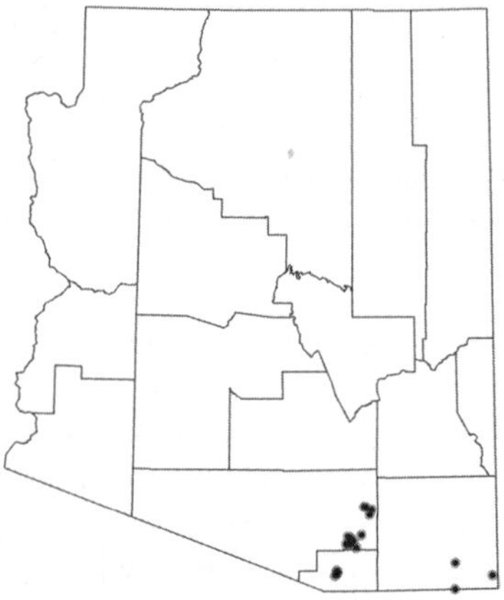

Shrubs or small trees to 3 m. **Stems** slender, lightly branched; twigs not zigzagged, round, usually lightly puberulent; bark gray, smooth [to fissured]. **Leaves:** pinnae (2–)6–14 pairs per leaf, 30–55 mm; stipules linear, to 6.5 mm, glabrous, scarious and weakly hardened or rigid, but sometimes interpreted as spines; petioles round, channeled above, 30–75 mm, glabrous; petiolar gland absent; rachis 50–190 mm. **Leaflets** 20–37 pairs per pinna, oblong, oblique and obtuse at base, acuminate at apex, 2.0–6.5 × 0.7–1.4 mm, glabrous above, lightly pubescent beneath. **Inflorescences** loosely-flowered cylindrical spikes 30–75 mm, solitary (rarely 2–3) in the leaf axil; peduncles 5–15 mm. **Flowers** creamy-white; calyx 1.1–1.6 mm, glabrous; corolla 2.0–2.7 mm, glabrous; filaments 4.5–6.5 mm. **Fruits** 70–170 × 12–21 mm, not constricted between the seeds, stiffly papery, glabrous; stipe 6–12 mm; apex acute to obtuse. **Seeds** dark brown, oval, 6.2–9.5 mm across.

Flowering Jun–Aug. Rocky slopes; [230–]1050–1560[–1620] m; Semidesert Grassland, open Madrean Evergreen Woodlands and occasional at upper edge of Sonoran Desertscrub; Cochise, Pima, Santa Cruz cos.; New Mexico; Mexico (Sonora, w Chihuahua).

Mariosousa millefolia occurs in widely scattered populations in southeastern Arizona but is often abundant where it is found. The plants commonly form open, spreading shrubs with several stems arising from the base. The large, pale fruits are distinctive. A low elevation record of 731 m on an herbarium specimen (*J.J. Thornber 5,* 22 Sep 1903) with the

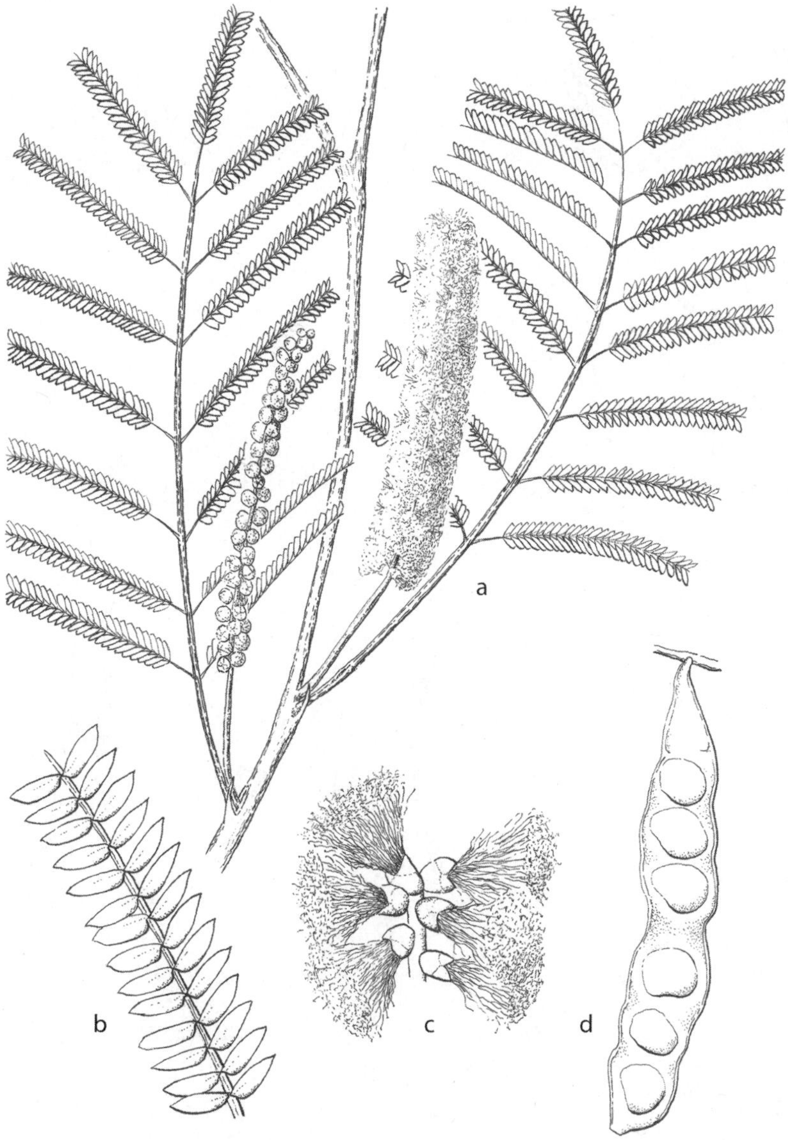

Mariosousa millefolia**: a. habit; b. pinna; c. flowers; d. fruit. Illustrator: Christine Hubbell.**

locality information: "USA, Arizona, Pima, Tucson; Santa Cruz Valley" does not list a specific locality. This elevation would place the plant within the area presently occupied by the city of Tucson. No records of naturally occurring plants of this species below 1000 m have otherwise been documented for Arizona. In Sonora it extends into inland Sinaloan Thornscrub.

Mariosousa millefolia is occasionally grown in Arizona for its lacy foliage and showy flowers. Supplemental irrigation during hot weather helps to maintain an attractive appearance when grown at lower elevation sites. Plants are winter deciduous. Cultivated plants of Santa Rita acacia in Tucson have survived overnight temperatures of -9°C. Individual stems on at least some plants die after several years which can be a maintenance issue in some landscape settings. Plants are readily propagated from scarified seeds.

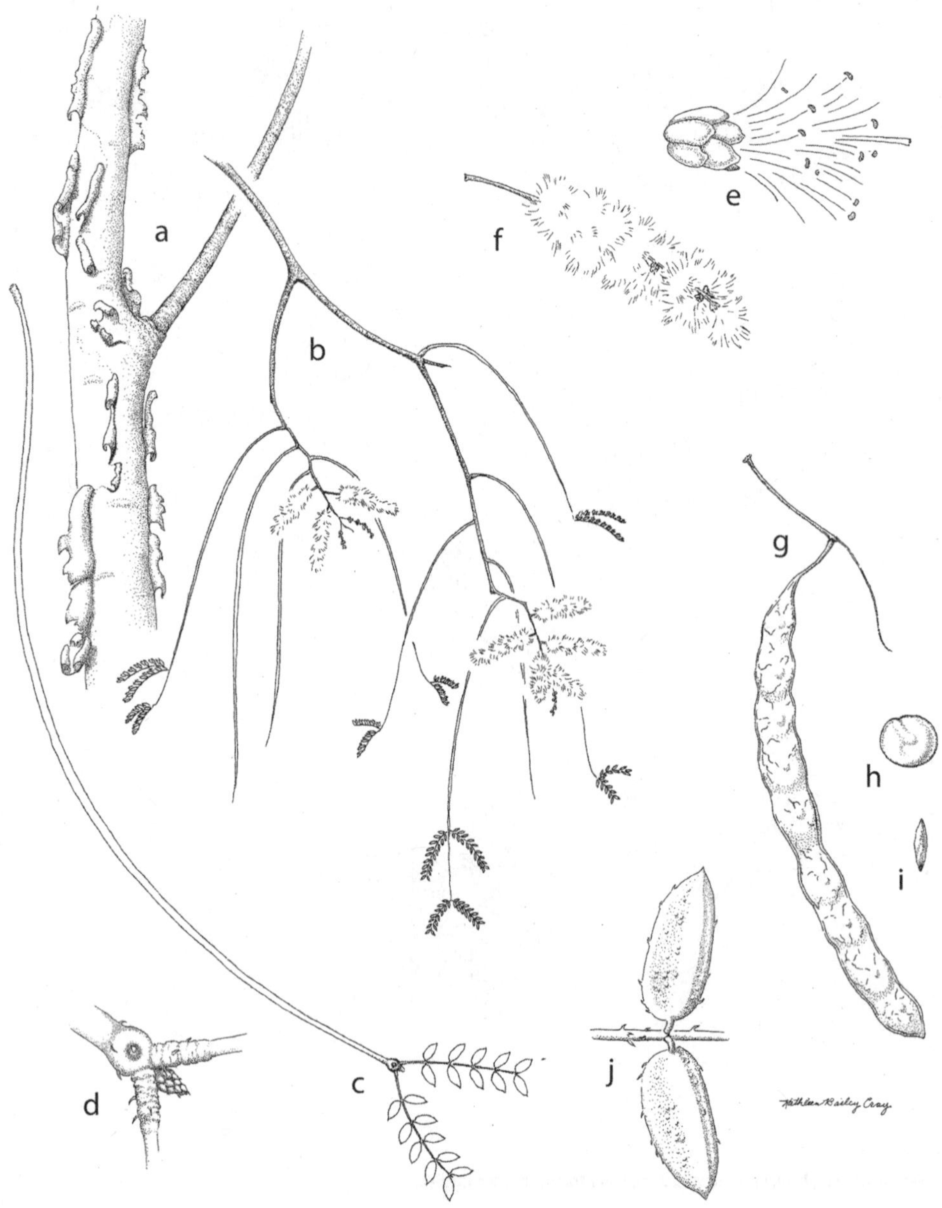

Mariosousa heterophylla: **a.** stem with exfoliating bark; **b.** habit; **c.** leaf; **d.** petiole nectary detail; **e.** flower; **f.** inflorescence; **g.** fruit; **h., i.** seeds; **j.** leaf detail. Illustrator: Kathleen Bailey Cray.

2. Mariosousa heterophylla (Bentham)Seigler & Ebinger, Phytologia 100:257. 1018. * *PALO BLANCO* Cultivated

Prosopis heterophylla Bentham London J. Bot. 5:82. 1846; *Acacia willardiana* Rose in Vasey and Rose; *Senegalia heterophylla* (Bentham) Britton & Rose

Small trees to 10 m. **Stems** slender, lightly branched; twigs not zigzagged, round, glabrous; bark exfoliating and papery, white, orange brown to yellow on inner surface. **Leaves:** 1(–3) pairs of pinnae per leaf, 16–80 mm; stipules linear, to 1.1 mm, glabrous; petioles flattened, not channeled above, 20–400 mm, glabrous; petiolar gland solitary, sessile, oval, 0.3–0.7 mm across; rachis 0–100 mm. **Leaflets** 4–20 pairs per pinna, oblong to elliptic, oblique and obtuse at base, narrowly acute to acuminate at apex, 3.0–7.5(–12.0) × 1.0–2.3 mm, glabrous to rarely appressed hairs on both surfaces. **Inflorescences** loosely-flowered cylindrical spikes 30–90 mm, solitary in the leaf axil or in short raceme-like clusters; peduncles 5–25 mm. **Flowers** creamy-white; calyx 1.3–2.2 mm, glabrous; corolla 2.4–3.6 mm, glabrous; filaments 6–8 mm. **Fruits** 70–180 × 8–22 mm, not constricted between the seeds, stiffly papery, glabrous; stipe 8–14 mm; apex obtuse. **Seeds** dark brown, oval, 6–11 mm across.

Flowering Feb–Jun. Mexico (Sonora, nw Sinaloa).

Mariosousa heterophylla is unmistakable with its slender, wispy appearance, peeling bark, and long narrow petioles commonly remaining attached to the plant after the leaflets and pinnae fall. These remaining petioles resemble those of some Australian species of *Acacia*. In habitat, it is especially distinctive when growing in open groves against dark igneous slopes or silhouetted against the sky along rocky ridgelines. It is found on arid hills, rocky slopes, cliffs and arroyos in Sonoran Desertscrub and Sinaloan Thornscrub from sea level to 500 m.

The wood of *Mariosousa heterophylla* is hard and the stems are flexible. The Seri of western Sonora use the wood in light construction and to make digging sticks, agave pry bars, carrying yokes, bows and foreshafts for turtle harpoons (Felger and Moser 1985). It is also used in Seri basketry (Moerman 1998).

Cultivated in landscapes in warm-winter desert locations in Arizona, palo blanco is a unique small tree for patios and narrow spaces. The open canopy provides sparse shade. The pinnae are soon deciduous, leaving the petiole to carry out photosynthesis. Palo blanco is extremely tolerant of drought and reflected heat once established. The stems are damaged by overnight temperatures below ca. -6°C. Propagation is by scarified seeds.

MEDICAGO Linnaeus. Sp. Pl. 2:778. 1753. * [from Greek *medike*, from Medea (Media), ancient empire where alfalfa was supposedly derived] ALFALFA, LUCERNE, MEDIC

Ernest Small

Herbs [shrubs], annual, biennial, or perennial, unarmed, sometimes base woody; roots fibrous or branched taproots; rhizomes sometimes present. **Stems** prostrate, procumbent, decumbent, ascending, or erect, branched or unbranched, especially when small or crowded by vegetation, usually slightly to moderately pubescent, hairs eglandular or sometimes gland-tipped, sometimes with a septum. **Leaves** alternate, odd-pinnate; stipules present, margins entire, toothed, or lobed, glabrous or pubescent, hairs eglandular and/or gland-tipped; epulvinate; petiolate. **Leaflets** 3, pulvinate, blade margins partly serrate, sometimes laciniate or incised, surfaces glabrous or pubescent, hairs eglandular and/or gland-tipped, especially below (abaxially). **Inflorescences** 1–50-flowered, axillary or terminal, usually racemes or heads, sometimes umbels or flowers subsolitary; bracts 0–10 cm. **Flowers** papilionoid; calyx funnelform to campanulate, lobes 5, triangular to lanceolate, equal or subequal, upper (adaxial) pair longer; corolla usually yellow or orange-yellow, rarely purple, violet, variegated violet-yellow, or white; banner subequal to or longer than wings and keel; wings strongly adherent to keel by wing spur in keel invagination; keel rounded, not strongly incurved, hooding stamens; stamens 10, 9 united and 1 free; anthers attached dorsally, uniform, free portion of alternating filaments often thick. **Fruits** tan to brown or black, sessile or short-stipitate, cylindrical, compressed, or flat [moniliform, plicate], straight, curved, or coiled, indehiscent or dehiscent at sutures; dorsal sutures spiny, tuberculate, or smooth [winged and fimbriate], faces papery, soft, or extremely hard (sometimes with proliferative alveolar, spongy tissue), glabrous or pubescent, hairs eglandular and/or glandular, intersuture venation obscure to prominent, often anastomosing in species-specific patterns, sometimes a submarginal vein present (parallel to dorsal suture but somewhat remote from it). **Seeds** 1–30, usually reniform, sometimes oval, triangular, or rhomboid-ovoid, smooth, rugose, or tuberculate. $x = 7, 8, 9$.

Species ca. 85 (4 in the flora, 3 naturalized and 1 cultivated and adventive): Europe, Asia, n Africa; introduced also nearly worldwide in temperate and tropical areas.

All of the species treated here have shown weedy tendencies in areas where they are introduced; however, none is significantly troublesome.

Medicago sativa is arguably the most important temperate region forage plant in the world. Several other perennial species, and, more commonly, annual species, which are commonly called medics or medicks, are cultivated for forage, but infrequently in North America. About a dozen species, none of which is treated here, are rare, very poorly studied endemics in need of conservation measures.

Species in sect. *Spirocarpos,* which includes *Medicago minima*, all have coiled fruits and normally have two rows of spines on the dorsal suture. However, in some plants these are reduced to tubercles or the dorsal sutures are without ornamentation. Cultivars have often been selected for lack of fruit spines, as this prevents entanglement in fur and makes the nutritious fruits easier for livestock consumption. Cultivars have been grown in North America, increasing the likelihood that spineless-fruited forms will be encountered. The spines are protective to the fruit, serve to adhere to fur for distribution, and perhaps also to facilitate burial of the indehiscent fruits in soil. In their native area, spiny and spineless forms occur randomly on a geographic basis. The literature is burdened with extensive, misguided, infraspecific classifications of the species on the basis of presence and length of spines.

In the nineteenth and twentieth centuries, *Medicago* was controversially separated from *Trigonella*, less controversially from *Melilotus*, with given species assigned to these and many other genera depending on taxonomic treatment. Recent morphological, chemical, and molecular (DNA) evidence has made it clear that *Medicago* is well demarcated from its nearest allies, including *Trigonella*, *Melilotus*, and *Trifolium*. *Medicago* flowers have an irreversible explosively tripping pollination mechanism, which is degenerate in the annual species but evident by a syndrome of floral characteristics. Although fruits of *Medicago* are usually indispensable for species identification, flowers suffice for distinguishing *Medicago* species from those of allied genera. Contrasting characters possessed by *Medicago* but not related genera include: banner, except when very small, with major basal vein usually with more than three branches near base; horn of wing petal large, about one-third of wing limb; keel very strongly adherent to wing by a wing spur in a keel invagination; androecium with apex of fused portion of anther column arched, alternating free portions of filaments often relatively thick; stigma fungiliform (shaped like a culinary mushroom); style always short.

Three species found in Arizona have unique character states for North American species. *Medicago lupulina* is the only species that has one-seeded fruits. *Medicago sativa* is the only species that can have violet or purple flowers. *Medicago polymorpha* (along with *M. truncatula* and *M. turbinata* outside Arizona) sometimes develops fruits that are extremely hard-walled at maturity, while the remaining species in North America always have fruit walls that are flexible or relatively easily bent.

Several *Medicago* species used for forage have a number of widely used common names, names often more or less confined to Europe, Australia, or different areas of the United States. Only vernacular names that have been used repeatedly in the literature are reported here.

Perennial species of *Medicago* are slightly to substantially outbreeding, while annual species are mostly strongly inbreeding.

SELECTED REFERENCES: Bena 2001, Goertzen and Small 1993, Heyn 1963, Lesins and Lesins 1979, Small 1985, 1987 and 2011, Small and Brookes 1984, Small and Jomphe 1989, Small et al. 1987, Turkington and Cavers 1979.

1. Corollas usually uniformly purple or variegated yellow-violet, sometimes violet or greenish, rarely white ________ **4. Medicago sativa** (in part)
1′ Corollas yellow, rarely orange-yellow.
  2. Fruits lacking spines or tubercles on dorsal suture.
    3. Fruits 1-seeded, nutlet-like ________ **1. Medicago lupulina**
    3′ Fruits 2–12-seeded, a falcate or slightly coiled pod ________ **4. Medicago sativa** (in part)
  2′ Fruits with spines or tubercles on dorsal suture.
    4. Shoot pubescent ________ **2. Medicago minima**
    4′ Shoot glabrescent or glabrous ________ **3. Medicago polymorpha**

1. Medicago lupulina Linnaeus. Sp. Pl. 2:779. 1753. * [from Latin, *lupulus*, the hop plant] BLACK MEDIC, YELLOW TREFOIL Naturalized

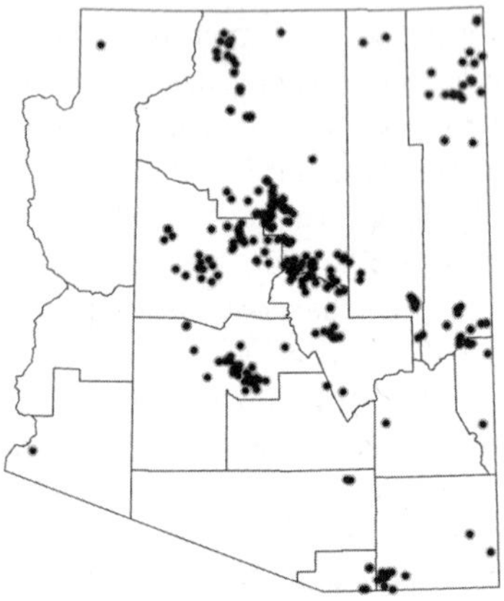

Herbs with glabrescent to densely pubescent vegetative shoots, hairs eglandular, appressed, sometimes glandular. **Stems** prostrate, decumbent, or semi-erect. **Leaves**: stipule margins entire or irregularly toothed. **Leaflets:** blades elliptic, ovate, or obovate, 10–20 × 6–15 mm, distal 1/2 margin serrate. **Inflorescences** (5–)15–50-flowered, cylindrical heads. **Flowers** 2–4 mm; calyx pubescent, hairs eglandular or glandular, lobes equal to tube; corolla yellow, 2 times length of calyx. **Fruits** ± ovoid, 2–3.5 × 1 mm, covered with eglandular (sometimes also gland-tipped) hairs when young; face with somewhat fusing, prominent veins (sometimes appearing as ridges) from ventral suture obliquely to dorsal suture. **Seeds** 1, yellow to olive-green, oval to reniform, 1.5–2 × 1–1.15 mm. $2n = 16, 32$.

Flowering (Mar–)May–Aug(–Oct). Usually in moist sites of slopes, canyon bottoms, meadows, floodplains, streambanks, lakeshores, marshes, springs, roadsides, agricultural fields, lawns, landscaped areas, and other disturbed sites in various soils; [near sea level–]290–2900[–3000] m; (Sonoran Desertscrub, Mohave Desertscrub, Semidesert Grassland, Interior Chaparral), Madrean Evergreen Woodland, Great Basin Conifer Woodland, Petran Montane Conifer Forest, Petran Subalpine Conifer Forest; documented for all counties in Arizona except La Paz Co.; cosmopolitan across the U.S.; Mexico (Baja California and Sonora to Vera Cruz and Zacatecas); Canada; introduced also nearly worldwide in temperate and tropical regions; native to Europe, Asia, and northern Africa.

Medicago lupulina is primarily found at higher elevations in Arizona but also occurs in moist situations in urban areas at low elevations. The species is widely distributed and often abundant in the state and has become widely naturalized worldwide.

Medicago lupulina is a variable species, but the variation is not structured in ways that can reasonably be classified formally. Of the many criteria that have been used to delimit infraspecific groups, presence of gland-tipped hairs and whether annual/biennial or perennial have been most frequently employed.

Medicago lupulina is valued as a pasture plant (there are several cultivars), cover crop, and as a green manure plant; it is typically plowed under in the fall as part of a crop rotation. Its use as a forage plant for livestock in Arizona has not been reported. Although *M. lupulina* is often considered a lawn weed, nitrogen fixation associated with this inconspicuous plant contributes to lawn health.

2. Medicago minima (Linnaeus) Bartalini. Cat. Piante Siena 61. 1776. * [from Latin, *minim*, small] GOLDFIELDS or KAALGOORLIE or LEAST or LITTLE MEDIC, LITTLE BUR or SMALL BUR CLOVER, LESSER BUR or SMALL BUR or SMALL WOOLY BUR, WOOLLY BUR MEDIC Naturalized

Medicago polymorpha Linnaeus var. *minima* Linnaeus, Sp. Pl. 2:780. 1753

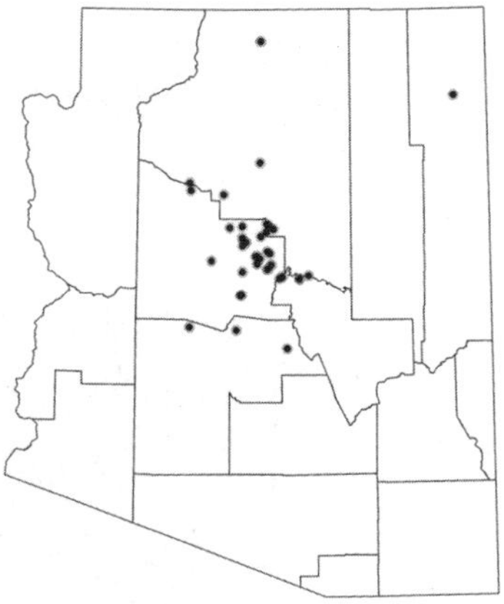

Herbs with pubescent vegetative shoots, hairs eglandular and/or gland-tipped. **Stems** ascending or procumbent. **Leaves:** stipule margins entire or minutely dentate at base. **Leaflets:** blades usually obovate, rarely oblanceolate, 5–8(–12) × 2–7 mm, distal 1/3 margin serrate. **Inflorescences** 2–5(–8)-flowered, racemes. **Flowers** 2–4.5(–6) mm; calyx pubescent, hairs eglandular and/or glandular, lobes equal to tube; corolla yellow, less than 2 times length of calyx. **Fruits** with 3–5 coils, disk-shaped, cylindrical, or ovoid, 3–5 × 2.5–6 mm, hairs glandular and/or eglandular, margin usually spiny, sometimes tuberculate, spines, when present, often relatively thin and flexible, base of spines 2-rooted, 1 root arising in dorsal suture, other in submarginal vein; coil face with strongly curved, unbranched radial veins entering broad, veinless margin that occupies 1/3 outer radius of coil face. **Seeds** 3–10, yellow to light brown, somewhat reniform, 1.5–2.5 × 0.9–1.3 mm. $2n = 16$.

Flowering Mar–May(–Jul). Rocky slopes, canyon bottoms, floodplains, arroyos, roadsides, pastures, vacant lots, and other disturbed sites in various soils; [near sea level–]610–2090 m; (Sonoran Desertscrub), Semidesert Grassland, Madrean Evergreen Woodland, Great Basin Conifer Woodland, (Petran Montane Conifer Forest); Apache, Coconino, Gila, Maricopa, Yavapai cos.; California, Texas, the Pacific coast north to Washington, southcentral U.S., and scattered in the eastern U.S.; introduced also in South America, Hawaii, New Zealand, Australia; native to Eurasia and Africa.

Medicago minima has been only slightly developed to date for agricultural purposes, although in areas where it grows it is considered to be good forage. In Argentina and Australia, *M. minima* is an extremely important component of the rangelands used for livestock grazing. The first cultivar of *M. minima* to be registered ('Devine') was released in 2005 by the Texas Agricultural Experiment Station.

3. Medicago polymorpha Linnaeus. Sp. Pl. 2:779. 1753. * [from Greek, *poly*, many, and *morpha*, form] BUR CLOVER, BUR MEDIC Naturalized

Medicago denticulata Willdenow; *Medicago hispida* Gaertner; *Medicago nigra* (Linnaeus) Krocker

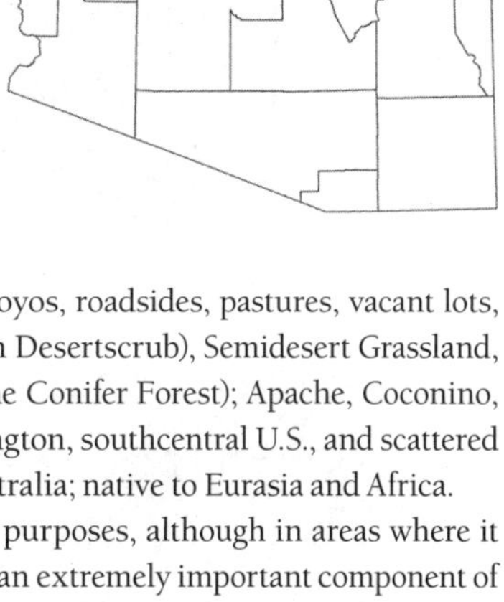

Herbs with glabrescent vegetative shoots, hairs eglandular. **Stems** procumbent, decumbent, or ascending. **Leaves:** stipule margins laciniate. **Leaflets:** blades obovate, obcordate, or cuneate, 8–20 × 7–18(–20) mm, distal 1/3–1/2 margin usually serrate, rarely laciniate. **Inflorescences** (1 or) 2–6(–10)-flowered, racemes. **Flowers** 3.5–6 mm; calyx sparsely pubescent, hairs eglandular, lobes equal to tube; corolla yellow, usually less than 2 times length of calyx. **Fruits** with 1.5–7 coils, disk-shaped, short to long cylindrical, or conical-truncate, 2–12 × 4–10 mm, usually glabrous, margin usually spiny, sometimes spines reduced to tubercules, spines, when present, often relatively thin and flexible, base of spines 2-rooted, 1 root arising in dorsal suture, other in submarginal vein; faces moderately soft, sometimes very hard at maturity,

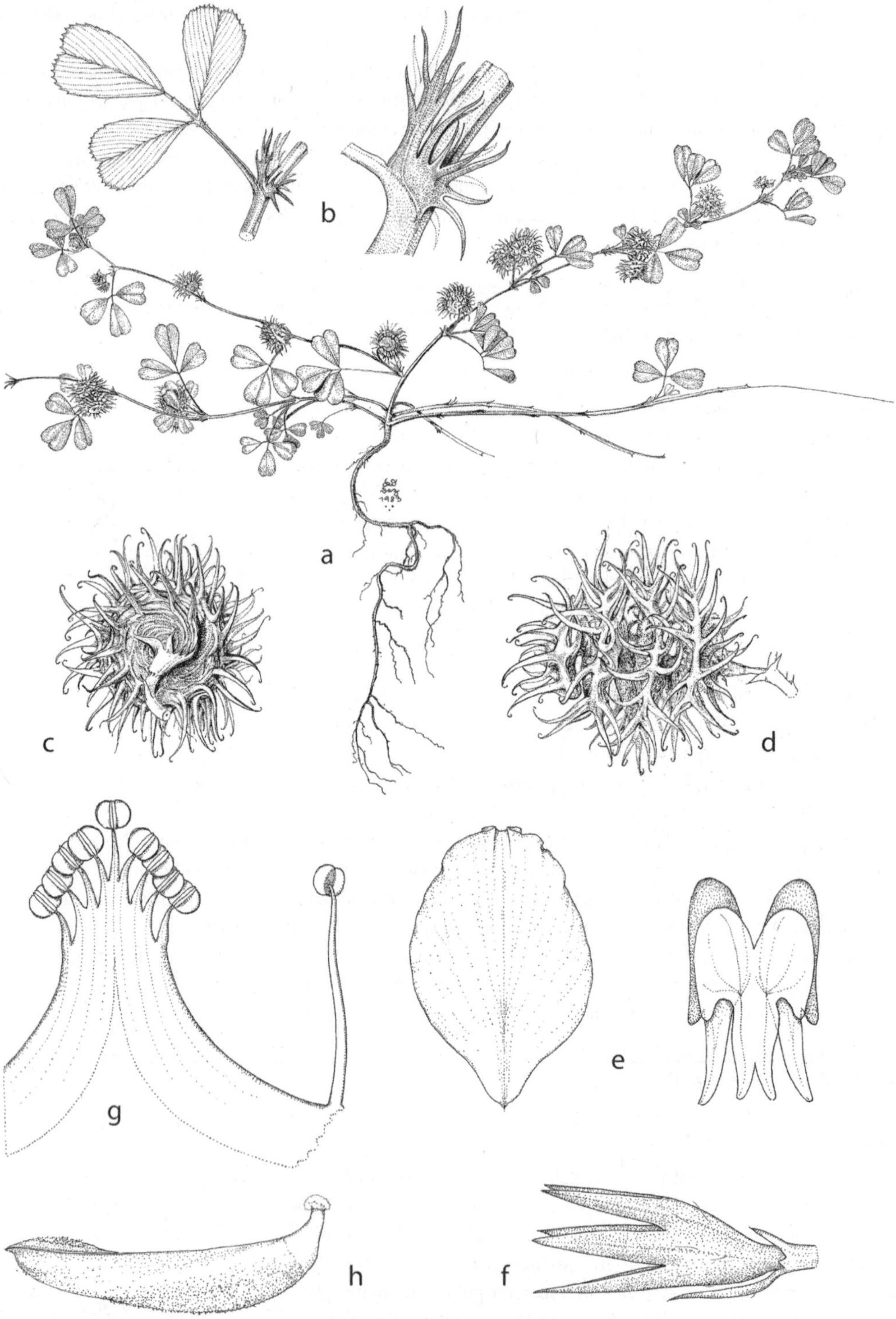

Medicago polymorpha: **a.** habit; **b.** nodes with leaf and stipules; **c.** fruit (basal view); **d.** fruit (side view); **e.** petals; **f.** calyx; **g.** stamens; **h.** pistil. Drawn by Karen Douthit and reproduced from *Flora Novo-Galiciana*, vol. 5, fig. 83, with permission of the University of Michigan Herbarium.

coil face with transverse veins anastomosing in outer part of coil before entering lateral vein near coil edge. **Seeds** 2–12, light yellow to brownish, reniform, 2–4 × 1.5–2.2 mm. $2n$ = 14, 16.

Flowering Feb–Jun(–Jul). Slopes, canyon bottoms, floodplains, arroyos, stream banks, roadsides, vacant lots, lawns, and other disturbed sites in various soils; [near sea level–]260–2900 m; Sonoran Desertscrub, Great Basin Desertscrub, Semidesert Grassland, Interior Chaparral, Madrean Evergreen Woodland, (Great Basin Conifer Woodland, Petran Montane Conifer Forest, Petran Subalpine Conifer Forest); Apache, Cochise, Coconino, Gila, Graham, La Paz, Maricopa, Pima, Pinal, Santa Cruz, Yavapai, Yuma cos.; California, Nevada, New Mexico, Texas, Utah and widespread in much of the U.S except for the upper Great Plains region; Mexico (widespread from Baja California, Chihuahua, and Sonora to Chiapas and Oaxaca); introduced also in Canada, Central America, South America, New Zealand, Australia; native to Eurasia and Africa.

Medicago polymorpha is one of the more important annual medics that have been developed for use as pasture forage for dry, hot environments. About a dozen cultivars have been bred. *Medicago polymorpha* is widely introduced and naturalized outside of its native range. It is most common at lower elevations in Arizona.

Medicago polymorpha is the most likely species of the genus to be confused with other species. The fruits are easily misidentified as one of the hard-fruited *Medicago* (*M. rigidula*, *M. truncatula*, *M. turbinata*), but are fairly similar to those of certain of the soft-fruited taxa, particularly *M. laciniata and M. minima*, two species that are also quite common as weeds. The fruit coil faces of both of the latter species have distal veinless areas. The coil face of *M. polymorpha* has quite reticulate venation, whereas that of *M. laciniata* has notably S-shaped radial veins that anastomose little. *Medicago minima* is a quite hairy plant, whereas *M. polymorpha* is almost always glabrous.

4. Medicago sativa Linnaeus. Sp. Pl. 2:778. 1753. * [from Latin, *sativa*, cultivated] ALFALFA, LUCERNE

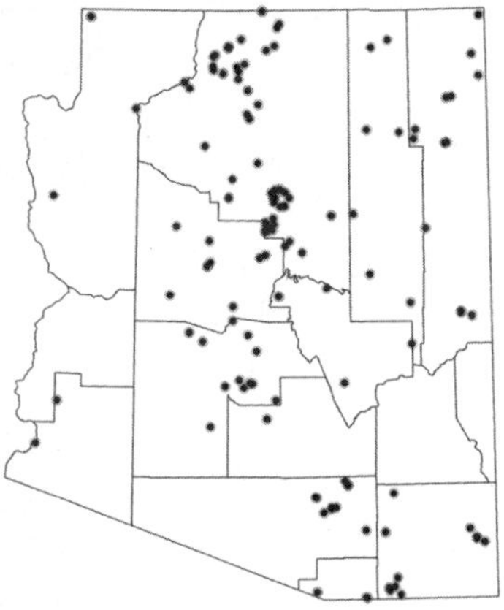

Herbs with glabrescent to pubescent vegetative shoots, hairs eglandular [glandular]. **Stems** prostrate to erect. **Leaves:** stipule margins entire or basally toothed. **Leaflets:** blades obovate to linear or lanceolate, 5–30 × 2–15 mm, distal margin serrate. **Inflorescences** 3–30(–50)-flowered, racemes. **Flowers** 5–15 mm; calyx glabrous or pubescent, hairs eglandular or glandular, lobes equal to tube; corolla usually purple, yellow, or variegated yellow-violet, rarely violet, green, or white, [yellow-orange, pink], 2 times length of calyx. **Fruits** curved or with 1.5–6 coils, falcate when curved, lenticular, ovoid, or cylindrical when coiled, 7–15 × 1.5–3 mm when falcate, 4–14 × 3–9 mm when coiled, glabrescent or pubescent with eglandular and/or glandular hairs; face (coil face when coiled) veins oblique from ventral suture, slightly branched, fusing towards dorsal suture. **Seeds** 2–12, yellow, brownish, greenish yellow, or violet-brown, reniform, 1–2.5 × 1–1.5 mm. $2n$ = 16, 32.

Subspecies 6 (3 cultivated and adventive, including 1 hybrid, in the flora): introduced in U.S. and n Mexico, also in West Indies, Central America, South America (Argentina, Brazil, Chile, Uruguay), Pacific Islands, Australia; native to Eurasia.

Because most collections of *Medicago sativa* from Arizona are not identified to infraspecific taxon, a general distribution statement is included here rather than following each subspecies. All three subspecies are documented for the state.

Flowering Apr–Oct. Slopes, valleys, canyons, floodplains, streambanks, lakeshores, roadsides, agricultural fields, pastures, vacant lots, lawns, and other disturbed areas in various soils; [near sea level–]50–2740[–5000] m; Sonoran Desertscrub, Chihuahuan Desertscrub, Mohave Desertscrub, Great Basin Desertscrub, Semidesert Grassland, Great Basin Grassland, Interior Chaparral, Madrean Evergreen Woodland, Great Basin Conifer Woodland, Petran Montane Conifer Forest; Apache, Cochise, Coconino, Gila, Graham, Maricopa, Mohave, Navajo, Pima, Pinal, Santa Cruz, Yavapai, Yuma cos.; cosmopolitan across the U.S.; widely cultivated and naturalized in Mexico and Canada; introduced also in the West Indies, Central America, South America, Pacific islands, Australia; native to Eurasia.

Medicago sativa is the most widely grown of the temperate forage legumes. Wherever it is cultivated, escapes are likely to be found in the vicinity, and the species has become established in most countries. This polymorphic Old World species is complicated by polyploidy, hybridization, and domestication (Small 2011) and has been divided by some into several species (dozens, by some Russian taxonomists) and innumerable infraspecific taxa. The natural habitats of the wild progenitors of *M. sativa* in Asia (mostly in the former U.S.S.R.) are rapidly being decimated, and there is considerable danger that valuable genetic diversity is being lost.

Medicago sativa: a. habit; b. leaflet; c., d. flower; e. flower with wings and keel petals removed; f. stamens; g. seed; h. fruits. Illustrator: Katherine Salwei & Kim Ratelle-Collins.

1. Flowers usually purple, sometimes violet (not bicolored), very rarely white; fruits with at least 1.5 coils (usually to 6) ______ **4b. Medicago sativa** subsp. **sativa**

1′ Flowers yellow or variegated yellow-violet, rarely green; fruits falcate or with fewer than 1.5 coils.

 2. Flowers yellow; fruits falcate (less than 0.5 coil) ______ **4a. Medicago sativa** subsp. **falcata**

 2′ Flowers usually variegated yellow-violet, sometimes green, yellow, or violet; fruits with 0.8–1.4 coils ______ **4c. Medicago sativa** nothosubsp. ×**varia**

4a. Medicago sativa Linnaeus subsp. **falcata** (Linnaeus) Arcangeli. Comp. Fl. Ital. 160. 1882. * [from Latin, *falci*, a sickle] YELLOW-FLOWERED or SIBERIAN or SICKLE-PODDED ALFALFA, SICKLE-PODDED MEDIC Cultivated, Adventive

Medicago falcata Linnaeus, Sp. Pl. 2:779. 1753

Flowers yellow. **Fruits** falcate. $2n = 16, 32$.

Subspecies *falcata* is very widespread in northern Eurasia. It has given rise to a few domesticated forms, but there are only a few cultivars. Although not often cultivated for forage, subsp. *falcata* has been sown in wild areas of the northern United States and Canada to promote wildlife. It is much used for breeding alfalfa resistant to cold, acid soils, and disease, and it is able to thrive in many colder areas of North America. In Eurasia, it is both diploid ($2n = 16$) and tetraploid ($2n = 32$).

4b. Medicago sativa Linnaeus subsp. **sativa** * PURPLE ALFALFA, PURPLE LUCERNE Cultivated, Adventive

Flowers usually purple, sometimes violet, very rarely white. **Fruits** with 1.5–6 coils. $2n = 32$.

Subspecies *sativa* includes most cultivated forms of alfalfa (including most of the thousands of cultivars) and escapes from cultivation and is the most common subspecies in the flora area. Many of the plants assignable here have been introgressed from subsp. *falcata*, but not sufficiently to warrant recognition as the hybrid taxon subsp. ×*varia*, discussed here. Domesticated forms often escape from cultivation but are not aggressive weeds.

4c. Medicago sativa Linnaeus nothosubsp. ×**varia** (Martyn) Arcangeli. Comp. Fl. Ital. 160. 1882, as subspecies * [from Latin, *varia*, variegated] BASTARD or HYBRID or SAND or VARIEGATED ALFALFA, BASTARD or HYBRID or SAND or VARIEGATED LUCERNE Cultivated, Adventive

Medicago ×*varia* Martyn, Fl. Rust. 3:87. 1793, as species

Flowers usually variegated yellow-violet, sometimes green, yellow, or violet. **Fruits** often with 0.8–1.5 coils. $2n = 32$ [also 16 in Eurasia].

Subspecies ×*varia* is the result of hybridization of subspp. *falcata* and *sativa*. Most plants are not F_1, but the result of local backcrossing that has produced introgressants showing intermediacy in fruit characters (neither notably coiled nor more or less straight) and flower color (neither completely yellow nor completely violet, but usually a mixture or variegation of these colors). F_1 plants are generally recognizable by the presence of greenish flowers, the result of a blending of the yellow pigment of subsp. *falcata* and the violet color pigments of subsp. *sativa*. Since several of the diagnostic characters segregate and combine more or less independently, precise delimitation from the parental species can be difficult. Subspecies ×*varia* is extensively cultivated, especially in more northern areas of the world. It has a greater tendency to weediness than the other subspecies, and so it is also very widely distributed.

MELILOTUS Miller. Gard. Dict. Abr. (ed. 4), 2. 1754. * [from Greek *mel*, honey, alluding to sweet fragrance of foliage or plants being favorite resort of bees, and *lotos*, alluding to similarity to lotus] SWEET-CLOVER, SWEET CLOVER, SWEETCLOVER, MELILOT

Ernest Small

Herbs, annual, biennial, or short-lived perennial, unarmed. **Stems** usually erect or ascending, sometimes decumbent, glabrous or glabrescent. **Leaves** alternate, odd-pinnate; stipules present, fused to petiole; petiolate. **Leaflets** 3, blade margins usually dentate, rarely entire or subentire, surfaces glabrous or pubescent. **Inflorescences** pedunculate, (5–)8–80(–120)-flowered, axillary, racemes, usually elongating in fruit; bracts present or absent. **Pedicellate. Flowers** papilionoid, usually pendulous, sometimes upright; calyx campanulate, lobes 5, ± subequal; corolla yellow or white; stamens 10, 9 united, 1 free; anthers basifixed. **Fruits** stipitate, usually ± compressed, usually globose, subglobose, obovoid, or ovoid, sometimes lanceolate-rhomboid or elongated-elliptic, indehiscent or tardily dehiscent, reticulately or concentrically patterned with raised venation, usually glabrous (pubescent in *M. altissimus*). **Seeds** 1 or 2 (or 3)[–5], usually ovoid, obovoid, oblong to subglobose, or ellipsoid, smooth or slightly tuberculate. $x = 8$.

Species ca. 18 (3 naturalized in the flora): Europe, Asia, n Africa; introduced in North America, also in temperate areas worldwide.

Trigonella and *Melilotus* may merit amalgamation (Coulot and Rabaute 2013), and *Melilotus* has been subdivided in various ways (reviewed by Stevenson 1969). Most species of *Melilotus* have a characteristic vanilla-like odor and a bitter taste, due to the presence of coumarins.

Melilotus officinalis and *M. albus* are sometimes merged, the latter treated as a synonym of the former. However, there are very strong barriers to interbreeding between the two (Webster 1955, Maekawa et al. 1991), in addition to geographical and ecological differences, which justify their continued recognition as separate species (Darbyshire and Small 2018).

Some species of *Melilotus* are cultivated as forage crops, for soil improvement, erosion control, revegetation, and reclamation plantings, and as honey plants. Some are significant weeds, including the cultivated species. Nodulation is widely reported for the genus (Allen and Allen 1981).

SELECTED REFERENCES: Isely 1954, Schulz 1901, Stevenson 1969, Turkington et al. 1978.

1. Corollas white ____ **1. Melilotus albus**
1′ Corollas yellow.
   2. Flowers 1.5–3 mm ____ **2. Melilotus indicus**
   2′ Flowers 4–7 mm ____ **3. Melilotus officinalis**

1. Melilotus albus Medikus. Vorles. Churpfälz. Phys.-Öcon. Ges. 2:382. 1787 (as *Melilotus alba*) * [from Latin, *albi*, white] WHITE SWEET-CLOVER, WHITE MELILOT, HONEY-CLOVER; *MÉLILOT BLANC* Naturalized

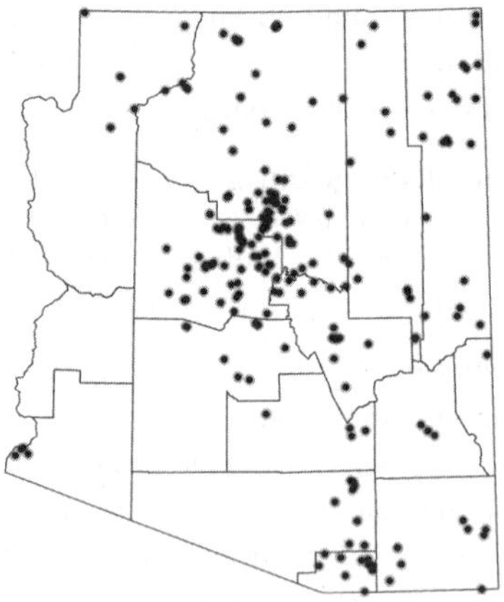

Herbs usually biennial, rarely annual, 30–150(–260) cm. **Stems** erect or ascending. **Leaves:** stipules lanceolate-subulate to bristle-like, 4–6(–10) mm, margins usually entire, sometimes toothed at base. **Leaflets** narrowly oblong-obovate to suborbicular-oblong, 10–24(–50) × 5–12(–15) mm, margins dentate. **Inflorescences** racemes 40–80(–120)-flowered. **Pedicels** 1–1.5(–2) mm. **Flowers** 3.5–5(–6) mm; corolla white; ovary glabrous. **Fruits** obovoid, subglobose, or globose-ovoid, 3–5 mm, reticulate-veined, glabrous. **Seeds** 1 or 2 (or 3), ovoid, (1.7–)2–2.5(–3.5) mm. 2*n* = 16.

Flowering (Mar–)Apr–Oct. Slopes, streams, floodplains, arroyos, lakeshores, springs, meadows, roadsides, irrigation canals, pastures, agricultural fields, urban locations, and other disturbed habitats in various soils; [near sea level–]60–2560[–2700] m; suitable habitats in essentially all biotic communities in the state; statewide except for La Paz County; widely naturalized across North America including Mexico; also naturalized in West Indies, South America, Australia; native to Eurasia.

Most botanical literature incorrectly lists the authority for *Melilotus albus* as Desrousseaux in Lamarck instead of Medikus. Coulot and Rabaute (2013) included *Melilotus albus* in *Trigonella sulcata* (Desfontaines) Coulot & Rabaute.

Melilotus albus is extremely variable in Eurasia and has been divided into many infraspecific taxa. It is the most important economic species in *Melilotus*, often grown as a crop, green manure, and honey plant, and posing a widespread weed problem. Cattle that eat the plants can develop sweet clover poisoning (Diggs et al. 1999).

Essentially all white-petaled *Melilotus* plants growing outside of cultivation in North America are *M. albus*; herbarium specimens often do not show the color well, sometimes resulting in misidentification, particularly between *M. albus* and *M. officinalis*. Aside from petal color, the species are quite similar. The venation areolae on the mature pods tend to differ: the raised venation ridges tend to form an irregular reticulation on the mature fruits of *M. albus*; they tend to form transverse areolae on the fruits of *M. officinalis*. This species is widely reported to nodulate.

2. Melilotus indicus (Linnaeus) Allioni. Fl. Pedem. 1:308. 1785 (as *Melilotus indica*) * [of India] ANNUAL YELLOW or INDIAN or SMALL-FLOWERED SWEET CLOVER, SOUR-CLOVER; *ALFALFILLA* Naturalized

Trifolium indicum Linnaeus, Sp. Pl. 2:765. 1753 (as *Trifolium indica*); *Trigonella smallii* Coulot & Rabaute

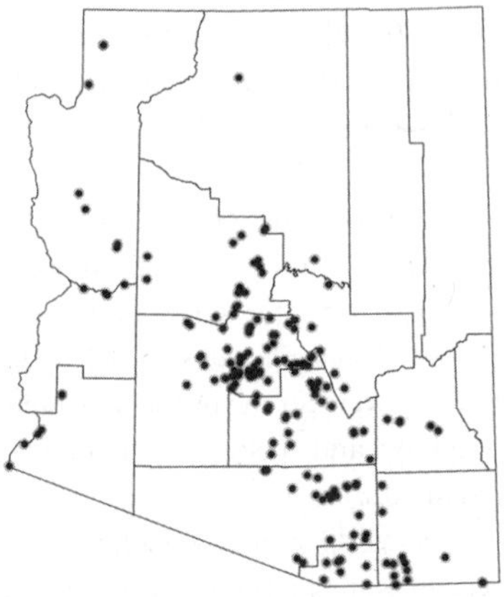

Herbs annual, (10–)15–60 cm. **Stems** erect or ascending. **Leaves**: stipules lanceolate-subulate to bristle-like, (3–)4–6(–8) mm, margins subentire (entire or toothed at base). **Leaflets** lanceolate-oblong, oblanceolate, or obovate, (8–)12–25 × (2–)7–10 mm, margins dentate. **Inflorescences** racemes 10–50-flowered. **Pedicels** 1 mm. **Flowers** 1.5–3 mm; corolla yellow; ovary glabrous. **Fruits** subglobose, 1.5–3(–4) mm, strongly reticulate-veined, glabrous. **Seeds** 1 (or 2), ovoid, 1.8–2.3 mm. 2*n* = 16 (counts from flora area, Eurasia; a count of 18 has been recorded from Asia).

Flowering Feb–Sep. Slopes, floodplains, streams, arroyos, lakeshores, springs, meadows, roadsides, pastures, agricultural fields, urban locations, and other disturbed sites in various soils; [near sea level–]40–2070(–2710) m; Sonoran Desertscrub, (Chihuahuan Desertscrub, Mohave Desertscrub), Semidesert Grassland, Great Plains Grassland, Interior Chaparral, (Madrean Evergreen Woodland, Great Basin Conifer Woodland, Petran Montane Conifer Forest); Cochise, Coconino, Gila, Graham, La Paz, Maricopa, Mohave, Pima, Pinal, Santa Cruz, Yavapai, Yuma counties; widely naturalized in the U.S., Canada, and Mexico; introduced also in Central America, South America, and New Zealand; native to Europe, Asia, Africa.

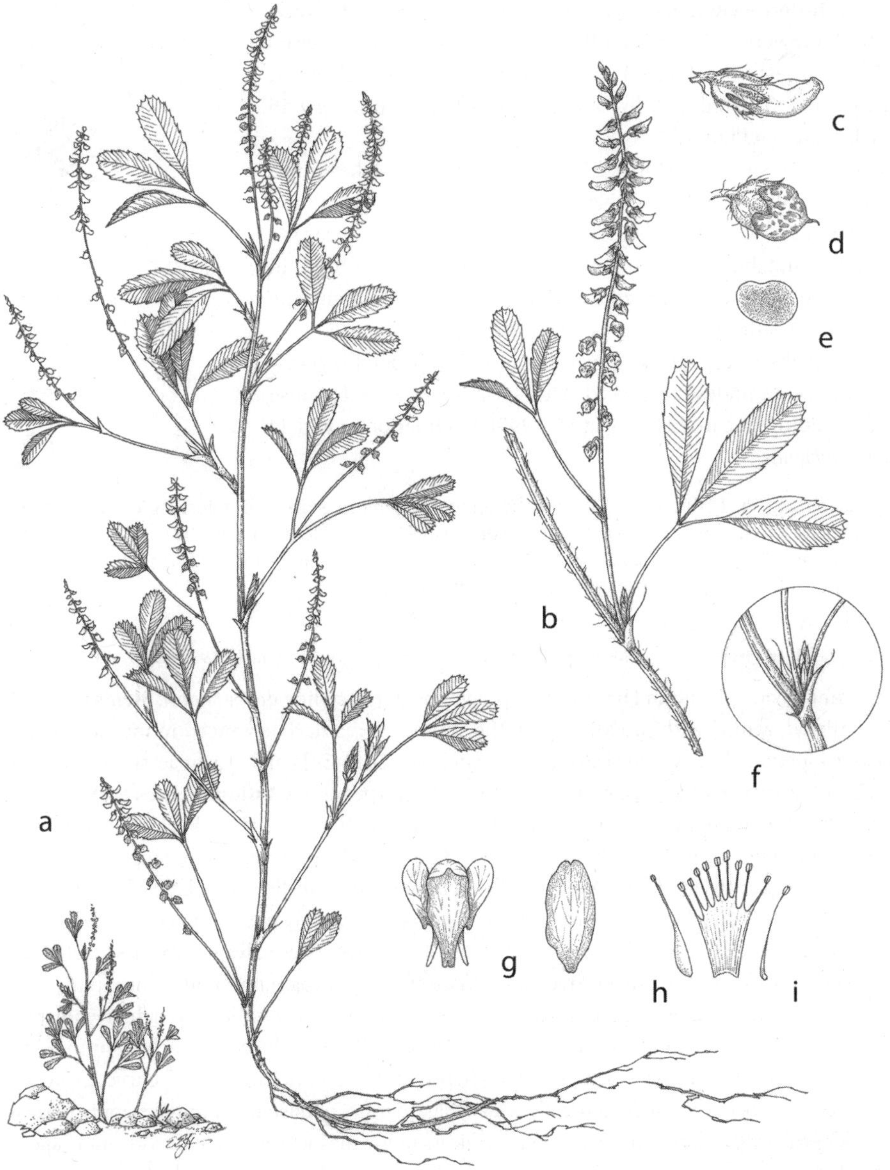

***Melilotus indicus*: a. habit; b. stem section enlargement; c. flower (side view); d. fruit; e. seed; f. stem node; g. petals; h. pistil; i. stamens. Illustrator: Elaine Hultgren.**

Melilotus indicus occasionally has been grown as a forage crop in the southern United States; it is generally considered to be an undesirable weed.

Melilotus indicus is widely naturalized across the southern half of Arizona mostly below the Mogollon Rim and downstream from the Grand Canyon. Flowering progresses to higher elevations later in the year. The species is widely reported to nodulate.

3. Melilotus officinalis (Linnaeus) Lamarck. Fl. Franç. 2:594. 1779. * [from medieval Latin, referring to an officina, a medieval storeroom for medicines] YELLOW SWEET-CLOVER, FIELD or RIBBED or YELLOW MELILOT Naturalized

Trifolium officinale Linnaeus, Sp. Pl. 2:765. 1753 (as *Trifolium officinalis*); *Trigonella officinalis* (Linnaeus) Coulot & Rabaute

Herbs usually biennial, sometimes annual, (15–)30–280 cm. **Stems** decumbent to erect. **Leaves**: stipules lanceolate or subulate, 3–6(–12) mm, margins entire. **Leaflets** obovate, obovate-oblong, ovate, or ovate-lanceolate, 8–25 × 4–15 mm, margins dentate. **Inflorescences** racemes with 30–70(–80) flowers. **Pedicels** (1.5–)2–2.5 mm. **Flowers** 4–7 mm; corolla yellow; ovary glabrous. **Fruits** ovoid, 2.5–5 mm, transversely rugose, areoles notably elongated, glabrous. **Seeds** 1 (or 2), ovoid, obovoid, or ellipsoid, (1.5–)1.8–2(–2.5) mm. 2*n* = 16 (counts from flora area and Eurasia).

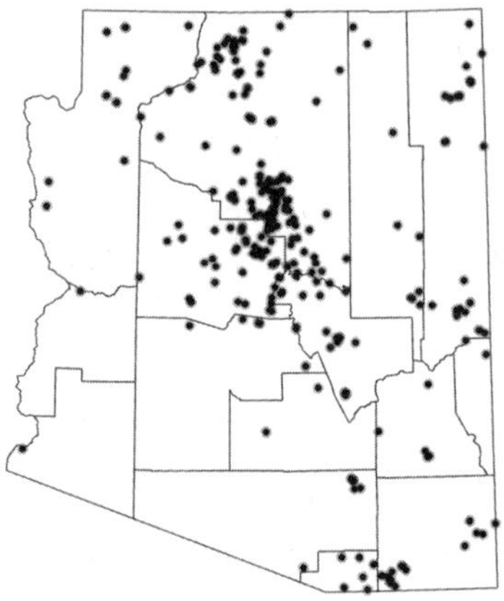

Flowering (Jan–)Apr–Oct. Slopes, floodplains, streams, lakeshores, springs, marshes, meadows, roadsides, irrigation canals, pastures, agricultural fields, urban locations, and other disturbed sites in various soils; [near sea level–]170–3200 m; suitable habitats in essentially all biotic communities in the state; statewide; widely naturalized across North America including Canada and Mexico, South America, Africa, Australia; native to Eurasia.

Melilotus officinalis is occasionally grown as a forage crop, but it is generally considered to be an undesirable weed. This species is known to cause sweet clover poisoning in cattle (Diggs et al. 1999). Nodulation is widely reported for *M. officinalis.*

MIMOSA Linnaeus. Sp. Pl. 1:516. 1753. * [from the Spanish, *mimoso*, sensitive, referring to the thigmotactic foliage (leaves closing when touched) of some species; it was first applied to a plant in the vernacular form *yerva mimosa* by Portuguese botanist Cristóbal Acosta, and later by English botanist John Parkinson, as a more formal Latin generic name, Latin *mimus*, mime or mimic, perhaps an error]

Rosaura Grether

Leptoglottis de Candolle ex Torrey & A. Gray; *Mimosopsis* Britton & Rose; *Morongia* Britton; *Schrankia* Willdenow

Shrubs [trees, semi-woody plants, and herbs] erect, procumbent, [sprawling or scandent]. **Stems** rounded in cross section, striate or ribbed; armed with prickles below the stipules, irregularly arranged in internodes, [along ribs], or unarmed. **Leaves** bipinnate [rarely some leaves unipinnate], [0–]1–11[–33] pairs of pinnae; stipules 1–8[–13] mm; petiole without a gland [rarely glandular]. **Leaflets** [1–]2–16[–95] pairs per pinna. **Inflorescences** of globose [to subglobose] heads or of cylindrical spikes or racemes with pedicellate flowers, axillary or disposed in lateral or terminal pseudo-racemes; peduncles unarmed [or prickly]; floral bracts [1/7–]1/5–1/2[–3/4 or as long as or longer than] the length of the corolla. **Pedicels** present or absent. **Flowers** mimosoid, all bisexual or male flowers at base of inflorescence; calyx with fused sepals, [1/10–]1/5–3/4 the length of the corolla, bell-shaped [or deeply cut into narrow lobes or reduced to a few bristles], 4–5-lobed, lobes valvate; corolla with fused petals, 4–5-lobed, lobes valvate, [1/5–]1/4–1/2[–2/3] the length of the corolla; stamens 8–10 [or 4–5], filaments free, rarely fused at bases, surpassing corolla, white, pink, [or lilac], anthers not glandular, attached dorsally, opening inward; ovary sessile or with stipe; style longer than stamens, stigma pore-shaped [cup-shaped, tubular, or obliquely funnel-shaped]. **Fruits** valvately dehiscent detaching from the rim or separating in segments that detach from rim, linear to oblong [tetragonal or elliptic], straight or curved; valves segmented or entire, constricted or not between seeds, generally wider than margin; persistent margin generally narrow [sometimes as wide as or wider than valves], prickly or unarmed; lacking pulpy mesocarp; sessile or with stipe, apex acute, with a mucro, an apiculus, a beak, or obtuse. **Seeds** brown, reddish-brown, dark brown to black [or olive-ochre], lens-shaped, more or less spherical, elliptic, oblong or rhomboid, seed coat smooth or porous, pleurogram horseshoe-shaped. $x = 13$.

Species ca. 530 (4 native in the flora, 1 of those cultivated): occurring mostly in south-central United States, Mexico, Central America, South America, and Antilles, some species growing in Africa, Madagascar, and Asia. Common in tropical, arid, and semiarid regions, 0–2250 m; some species grow in temperate areas within the tropics and subtropics, 2000–3500 m.

Mimosa is divided into five sections. Sections *Batocaulon* and *Habbasia* are represented in the United States; only the first one is found in Arizona. Members of sections *Mimadenia*, *Habbasia*, *Mimosa*, and *Calothamnos* are not found in the flora.

Mimosa diplotricha C. Wright ex Sauvalle, *M. pigra* Linnaeus, and *M. pudica* Linnaeus are pantropical weeds. *Mimosa diplotricha* and *M. pigra* are noxious weeds mainly in Australia. *Mimosa diplotricha* is found in Hawaii and Puerto Rico, but not in continental United States. *Mimosa pigra* is a prohibited noxious weed in the U.S. *Mimosa pudica* is cultivated in the United States as an ornamental plant. It is also used for experimental projects due to its sensitive leaves, but it is not well established or naturalized in Arizona. Other species of *Mimosa* are widely used for honey production and as medicinal plants, forage for goats, living fences, firewood sources, and reforestation by rural communities in tropical, arid, and semiarid regions, mainly in Latin America.

SELECTED REFERENCES: Barneby 1991, Dahmer et al. 2011, Grether et al. 2007, Martínez-Bernal and Grether 2006, Martínez-Bernal et al. 2008, Simon et al. 2011, Turner 1959.

1. Inflorescences of spikes or racemes with short pedicellate flowers; fruits with segmented valves.
   2. Leaves with 5–11 pairs of pinnae, 6–14 pairs of leaflets per pinna; inflorescences of spikes; flowers with corolla densely silky, with long appressed hairs; fruits with short matted soft hairs ___ **3. Mimosa dysocarpa**
   2′ Leaves with 2–4(–5) pairs of pinnae, 2–6(–7) pairs of leaflets per pinna; inflorescences of racemes with pedicellate flowers [or spikes]; flowers with corolla hairless or puberulous with fine, short hairs; fruits hairless or with fine, short hairs and firm, stiff hairs ___ **2. Mimosa distachya**

1′ Inflorescences of globose heads; fruits with entire valves.
   3. Prickles paired, recurved; heads 8–12 mm diam., 17–40 flowers each; corolla with short, matted, soft hairs; fruits linear, 0.3–0.4 cm wide ___ **1. Mimosa biuncifera**
   3′ Prickles in groups of 3, 2 straight and 1 recurved; heads 13–20 mm diam., 50–120 flowers each; corolla with minute, straight, soft hairs on lobes or hairless; fruits oblong [to linear-oblong], (0.5–)0.6–0.8 cm wide ___ **4. Mimosa grahamii**

1. Mimosa biuncifera Bentham. Pl. Hartw. 12. 1839. * [from the Latin *uncus* meaning hook, with two hooked or recurved prickles per node] WAIT-A-MINUTE BUSH, WAIT-A-BIT, CATCLAW MIMOSA Native

Mimosa aculeaticarpa Ortega var. *biuncifera* (Bentham) Barneby; *Mimosa lindheimeri* A. Gray; *Mimosa warnockii* B.L. Turner

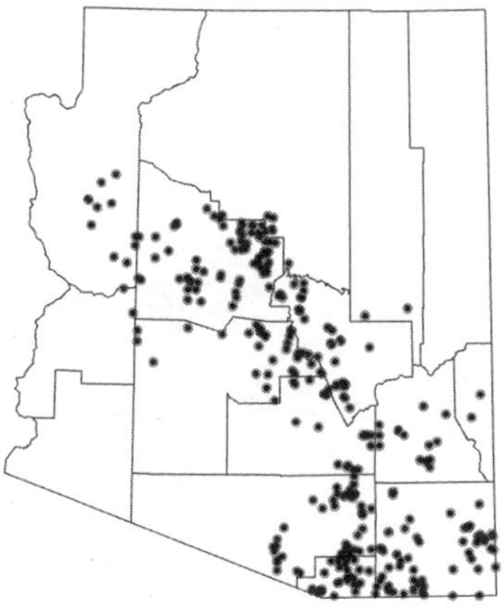

Shrubs erect, 1–2 m. **Stems** striate, with recurved prickles in pairs below stipules, and short, matted, soft hairs to hairless. **Leaves** 2–8 pairs of pinnae; stipules linear to awl-shaped, 2–4 mm, puberulent; petiole 0.1–0.7 cm; primary rachis 0.3–2 cm. **Leaflets** 5–9(–12) pairs per pinna, oblong, 2–4 × 0.4–1.3 mm, base oblique, margins ciliate, apex acute to rounded, lower surface puberulent or hairless, upper surface hairless, reticulate veins slightly visible on the lower surface. **Inflorescences** of globose heads, axillary, solitary or in clusters of 2–3, 0.8–1.2 cm diam., 17–40 flowers; peduncle 0.7–1.5 cm; bracts spatula-shaped, 1/4–1/2 of corolla length. **Pedicels** absent. **Flowers** bisexual; calyx (1/3–)1/2–3/4 the length of the corolla, 5-lobed; corolla 5-lobed, lobes 1/4–1/3 the length of the corolla, with short, matted, soft hairs, white or purplish-pink on lobes; stamens 10, filaments free to bases, white; ovary with stipe, hairless or pubescent, style attenuate at the apex, stigma pore-shaped. **Fruits** linear, 2.5–4.5 × 0.3–0.4 cm, straight to curved; valves hairless, constricted between seeds, entire; sessile or with stipe 0.5–1.5 mm, apex acute to acuminate; margin prickly on the external suture or unarmed. **Seeds** 3–8; dark brown, oblong, 3.5–6 × 2–2.7 × 1–1.8 mm, seed coat smooth or porous, pleurogram horseshoe-shaped, 40–50%. $2n = 52$ (Texas).

Flowering May–Aug; fruiting Sep–Dec. Slopes, canyons, valleys, arroyos, floodplains in a wide range of soils; [200–](480–)900–1700(–1860)[–2000] m; Sonoran Desertscrub, Chihuahuan Desertscrub, Semidesert Grassland, Interior Chaparral, Madrean Evergreen Woodland, (Great Basin Conifer Woodland); Cochise, Coconino, Gila, Graham, Greenlee, La Paz, Maricopa, Mohave, Navajo, Pima, Pinal, Santa Cruz, Yavapai cos.; New Mexico, Texas; Mexico (Chihuahua, Coahuila, Durango, Guanajuato, Hidalgo, Jalisco, México, Morelos, Nuevo León, Puebla, Querétaro, San Luis Potosí, Sinaloa, Sonora, Tamaulipas, Tlaxcala, Veracruz).

Mimosa biuncifera has been considered as a variety of *Mimosa aculeaticarpa* Ortega (Barneby 1991); however, I distinguish it at the specific level, as it frequently occurs in arid or semiarid regions, up to 2000 m elevation, while *M. aculeaticarpa*, endemic to Mexico, can grow at the edges of desertscrub, ca. 2000 m, but it is generally found in temperate forests at higher elevations, up to 2,850 m in central Mexico. *Mimosa biuncifera* has been determined to be tetraploid, particularly in populations from Arizona.

Mimosa biuncifera is widely distributed and often abundant at mid-elevations across much of central and southeastern Arizona. It is occasionally found in pine-oak woodland at the highest elevations. At low elevations in desertscrub, it is confined to floodplains and other low-lying areas where additional moisture is available. *Mimosa biuncifera* occurs as low as 200 m elevation in thornscrub in Sonora (Martin et al. 1998). In some locations it forms impenetrable thickets in canyons and on slopes. The sharp prickles readily snag and tear clothing and skin, hence the common names. These thickets offer protection to small birds and other wildlife. The seeds of *M. biuncifera* are an important food source for Gambel's quail and scaled quail and are highly palatable for white-tailed deer (Graham 1941). Kearny and Peebles (1942) report that it is browsed by livestock when other forage is scarce. The plants are reported to be suitable for honey production (Kearney and Peebles 1942). Due to its abundance, this species is useful in erosion control.

With its dense, prickly growth habit, *Mimosa biuncifera* is suitable for barrier plantings. *Mimosa biuncifera* is rarely

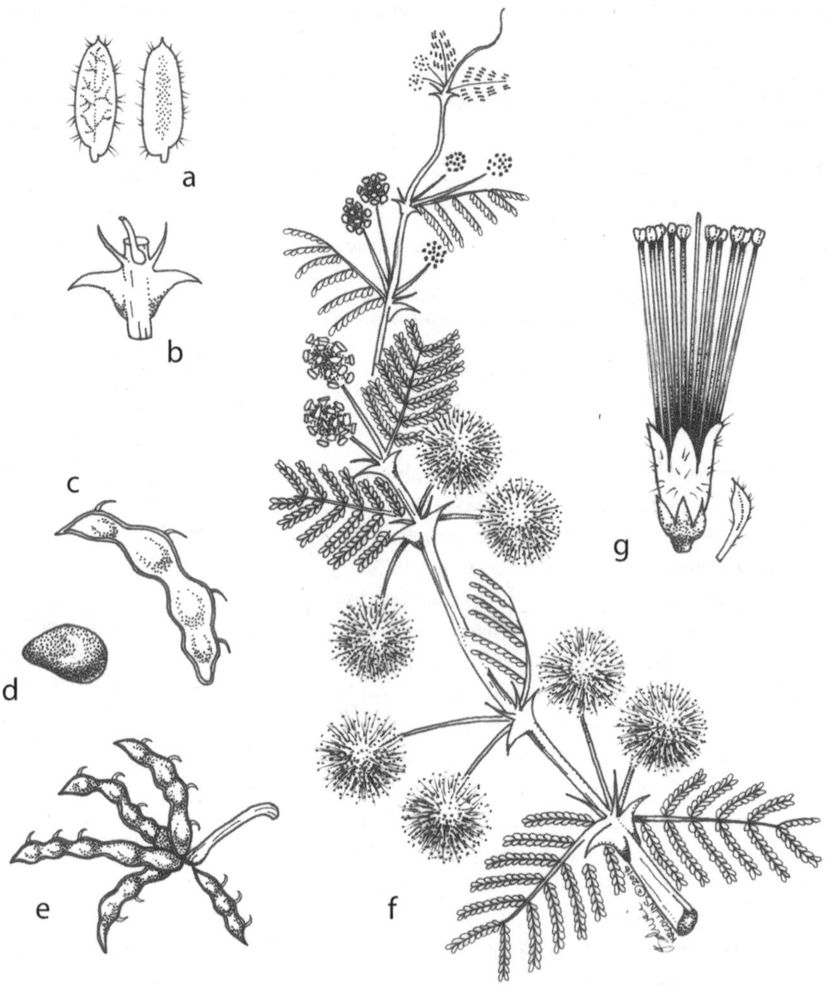

***Mimosa biuncifera*: a. leaflet (lower and upper surfaces with detail); b. stem node with stipules and prickles; c. fruit; d. seed; e. fruit cluster; f. habit; g. flower and bract. Illustrator: Katherine Salwei & Kim Ratelle-Collins.**

cultivated due to its unpleasant prickles and unruly, sprawling growth form. The plants are deciduous with freezing and are able to survive temperatures to at least -20°C. Propagation is by scarified seeds.

2. Mimosa distachya Cavanilles. Icon. 3:48, pl. 295. 1795. * [from the Latin *stachys* meaning spike, with spikes in clusters of two]

Mimosa brandegei B. L. Robinson

Shrubs erect or trees, 1–4[–6] m. **Stems** slightly ribbed, brown, puberulent when young; striate, with recurved prickles, solitary or paired at nodes, irregularly arranged along internodes or unarmed, grayish, hairless, when mature. **Leaves** (1–)2–4(–5) pairs of pinnae; stipules awl-shaped, 1–5 mm, with fine, short hairs; petiole (0.5–)1–3.5 cm; primary rachis (0.5–)3–5(–7) cm. **Leaflets** 2–6(–7) pairs per pinna, widely oblong or elliptic to ovate or obovate, (3.5–)5–12(–17) × (1.5–)3–8(–11) mm, base oblique, margins entire or ciliate, plane or revolute, apex rounded or acute, with a mucro or notch, hairless or puberulent on both surfaces or the lower (abaxial) surface puberulent, bluish green, and with reticulate veins visible. **Inflorescences** of spikes or racemes with short pedicellate flowers, axillary, solitary or in clusters of 2–4, 1.8–4.5 cm, 20–130 flowers; peduncle 0.5–1.5 cm; bracts spatula-shaped, 1/5–1/4 of corolla length. **Pedicels** 0.2–0.8 mm or absent. **Flowers** bisexual; calyx 1/5–1/2 the length of the corolla, 4–5-lobed, with fine, short hairs or hairless; corolla 4–5-lobed, lobes 1/3–1/2 the length of the corolla, purple pink or white, with fine, short hairs or hairless; stamens 8–10, filaments free or fused at bases, pink or white; ovary with a stipe or sessile, with short, matted, soft hairs or hairless, style attenuate at the apex, stigma pore-shaped. **Fruits** linear to oblong, 2–7.5 × 0.5–1.5 cm, straight or curved; valves hairless or with fine, short hairs and firm, stiff hairs, constricted between seeds, segmented; with a stipe 2–8 mm or sessile, apex with cusp, or beak; margin prickly, with firm, stiff hairs or unarmed. **Seeds** (2–)3–9(–11); dark brown to black, lens-shaped, 3–5 × 2.5–4 × 1.6–2.1 mm, seed coat shiny, smooth or porous, pleurogram horseshoe-shaped, 25%.

Varieties 4 (1 native in the flora): sw United States, Mexico, West Indies (Cuba), South America (Colombia, Venezuela).

2a. Mimosa distachya Cavanilles var. **laxiflora** (Bentham) Barneby. Mem. New York Bot. Gard. 65:83. 1991. * [from the Latin *laxus* meaning loose, and *floris* meaning flower, with flowers in a loose arrangement] MEXICAN MIMOSA; *GATUÑA, UÑA DE GATO* Native

Mimosa laxiflora Bentham, London J. Bot. 5:93. 1846

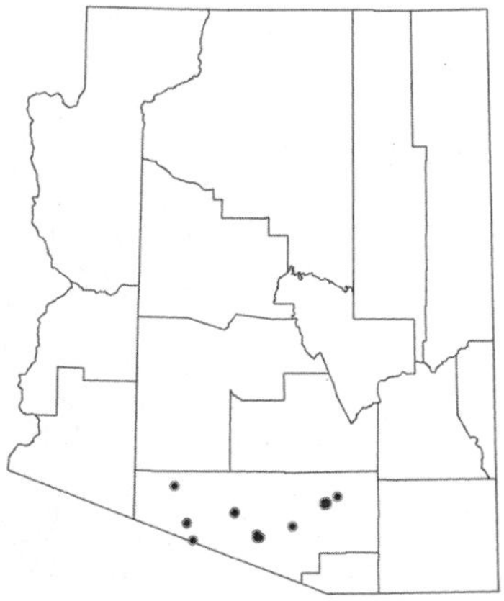

Shrubs 1–2.5 m. **Leaflets** 2–4(–5) pairs per pinna, (3.5–)5–11 × (1.5–)3.5–8 mm, margins ciliate, plane. **Inflorescences** of lax racemes, solitary or in clusters of 2, 20–55 flowers. **Pedicels** 0.2–0.5 mm. **Flowers** with hairless calyx; corolla lobes 1/2 the length of the corolla, purple pink, hairless; stamens with filaments fused at bases; ovary with a stipe, hairless. **Fruits** 2.5–4.5 × 0.5–0.75 cm, valves hairless, margin unarmed; stipitate. **Seeds** 4–9.

Flowering Apr–Aug; fruiting Aug–Oct. Dry stony hillsides and along desert washes, among rocks on exposed ridge tops; [near sea level–]500–1040 m; Sonoran Desertscrub; Pima Co.; Mexico (Chihuahua, Sinaloa, Sonora).

In Arizona, *Mimosa distachya* var. *laxiflora* is restricted to a few locations in Organ Pipe Cactus National Monument, near Sells, and in the Tucson Mts. The plants are limited by aridity and freezing temperatures in winter. *Mimosa distachya* var. *laxiflora* is widespread and abundant in thornscrub and tropical deciduous forest habitats in Sonora where it may rarely become tree-sized. The showy pink or white flowers are attractive. Plants are rarely available from specialty nurseries in Arizona. The plants are deciduous with drought and freezing. Stem damage occurs below -5°C, but recovery is usually rapid the following season. Supplemental irrigation is necessary to maintain appearance and growth. Propagation is by scarified seeds.

3. Mimosa dysocarpa Bentham in A. Gray, Pl. Wright. 1:62. 1852. * [from the Greek *dyso* referring to the soft pubescence, and from the Latin *carpium* meaning fruit] VELVET-POD MIMOSA; *GATUÑO* Native, Cultivated

Mimosa wrightii A. Gray

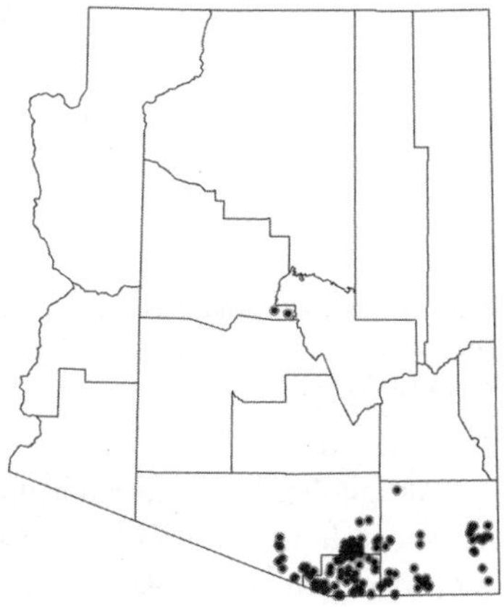

Shrubs erect, 0.7–1.5 m. **Stems** ribbed, with recurved to straight prickles irregularly arranged along internodes; with short, matted, soft hairs, becoming hairless. **Leaves** 5–11 pairs of pinnae; stipules linear to awl-shaped, 4–7 mm, with short, matted, soft hairs; petiole 0.6–1 cm; primary rachis 2.5–7.5 cm. **Leaflets** 6–14 pairs per pinna, linear-lanceolate or lanceolate-oblong, 3–6 × 1–2.5 mm, base oblique, margins densely silky with long, appressed hairs, apex acute, with mucro, lower surface densely silky with long, appressed hairs, light green, with 1–2 prominent eccentric veins, upper surface silky with sparse, long, appressed hairs or hairless, dark bluish-green. **Inflorescences** of dense spikes, axillary, solitary, in clusters of 2 or 3 or in pseudo-racemes, 2–5 (–8) cm, 50–150 flowers; peduncle 1–1.5(–2) cm; bracts linear to spatula-shaped, 1/3–1/2 of corolla length. **Pedicels** absent. **Flowers** bisexual and male; calyx 1/2 the length of the corolla, 4–5-lobed; corolla 4–5-lobed, lobes 1/3–1/2 the length of the corolla, densely silky, with long appressed hairs, purple pink; stamens 8–10, filaments free to bases, pink; ovary sessile, with short, matted, soft hairs, style attenuate at the apex, stigma pore-shaped. **Fruits** linear-oblong, 4–5 × 0.5–0.7 cm, straight or curved; valves with short, matted, soft hairs, constricted between seeds, segmented; sessile, apex with beak; margin unarmed or sparsely prickly. **Seeds** (2–)4–9; reddish brown, lens-shaped, 5 × 4 × 3 mm, seed coat smooth, pleurogram horseshoe-shaped, 50%. 2*n* = 26.

Flowering Jun–Sep; fruiting Sep. Slopes, canyons, valleys, arroyos, floodplains, and roadsides, often in rocky soil; [300–](980–)1160–2100 m; (Sonoran Desertscrub), Semidesert Grassland, Madrean Evergreen Woodland; Cochise, Pima, Pinal, Santa Cruz, Yavapai cos.; New Mexico, Texas; Mexico (Chihuahua, Durango, Sonora, Zacatecas).

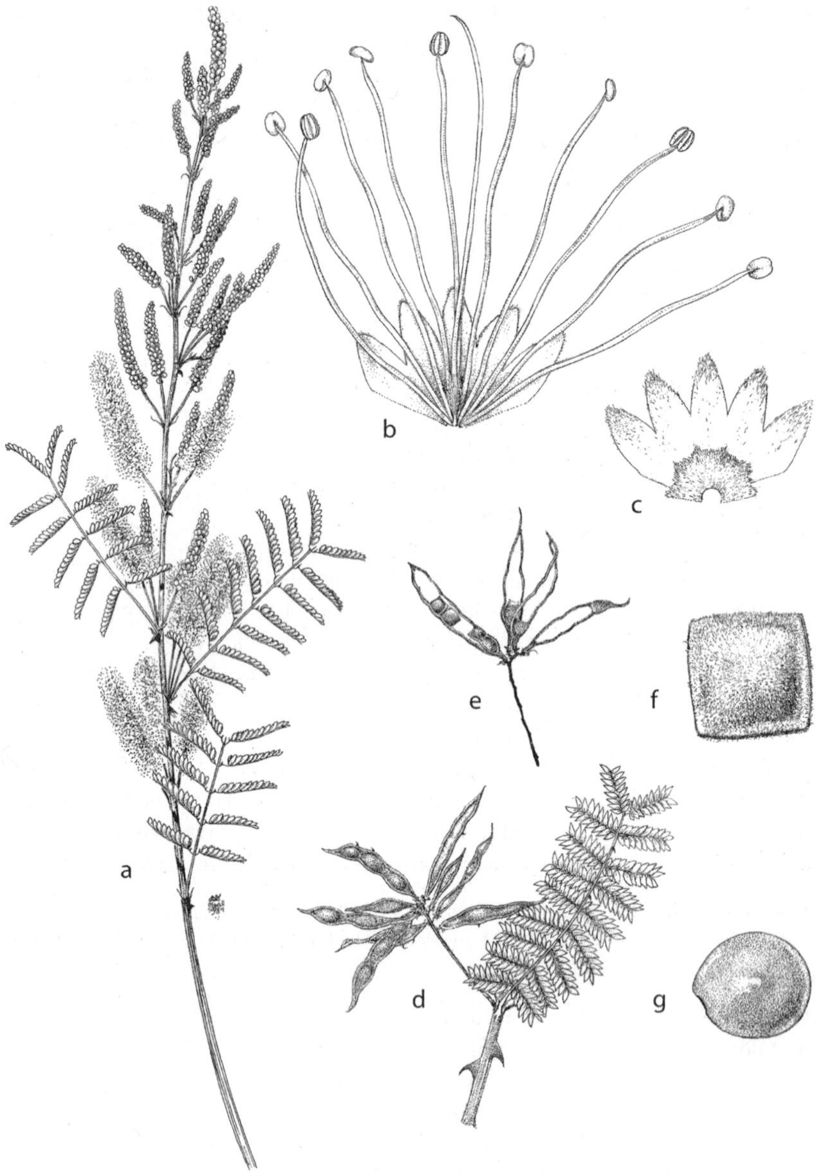

Mimosa dysocarpa: **a.** flowering branch; **b.** flower laid open to show insertion of stamens; **c.** calyx and corolla; **d.** leafy twig with fruits; **e.** fruit disarticulation from persistent margin; **f.** fruit segment; **g.** seed. Drawn by Karen Douthit and reproduced from *Flora Novo-Galiciana*, vol. 5, fig. 28, with permission of the University of Michigan Herbarium.

Mimosa dysocarpa is locally abundant in mesquite and oak-juniper grassland habitats in southeastern Arizona. It is also known from one location in Yavapai County and one location in Pinal County. It barely enters the Sonoran Desert in riparian habitats. The plants are conspicuous in mid- and late summer with their showy pink flowers. In Sonora, *M. dysocarpa* is found as low as 300 m elevation in tropical deciduous forest (Martin et al. 1998).

Velvet pod mimosa is occasionally grown in Arizona for its ornamental value. The growth form is often spreading with multiple stems. Some supplemental irrigation in the growing season is necessary to maintain appearance at lower elevation locations. The foliage is deciduous with freezing. The stems appear to be hardy to at least -12°C. Due to its prickly stems and rangy growth habit, the plants are best sited away from traffic areas, and are suitable for planting on slopes or banks. Mix with succulents or evergreen shrubs to provide interest and winter color when *Mimosa dysocarpa* is dormant and leafless. This species is readily grown from scarified seeds.

4. Mimosa grahamii A. Gray. Pl. Wright. 2:52. 1853. * [the epithet honors Colonel Graham, of the Topographical Engineers, under whose auspice Charles Wright's collection of plants was made in 1851 while attached to the scientific corps of the U.S. Boundary Commission, then under Colonel Graham's charge] GRAHAM'S MIMOSA

Shrubs erect or procumbent 0.6–1(–2) m. **Stems** striate to ribbed, with 2 prickles paired, straight, below stipules, and 1 recurved prickle in the internodes; pilose, with soft, erect-ascending hairs or hairless. **Leaves** (4–)5–8(–12) pairs of pinnae; stipules strap-shaped to awl-shaped, 2–5(–12) mm, pilose, with soft, erect-ascending hairs to hairless; petiole 0.3–2 cm; primary rachis (2–)3–9 cm. **Leaflets** 7–16(–20) pairs per pinna, oblong to lanceolate-oblong or elliptic, (2.5–)3–6(–6.5) × 1–2.5 mm, base oblique, margins ciliate or entire, apex acute, with mucro, pilose, with soft, erect-ascending hairs or hairless on both surfaces, bluish green, with reticulate veins prominent on the lower (abaxial) surface. **Inflorescences** of globose heads, axillary, solitary or in clusters of 2 or 3, 13–20 cm diam., 50–120 flowers; peduncle (1–)1.5–3.5 cm; bracts spatula-shaped, 1/3–1/2 of corolla length. **Pedicels** 0.2–0.3 mm or absent. **Flowers** bisexual; calyx 1/5–1/3 the length of the corolla, 5-lobed; corolla 5-lobed, lobes 1/4–1/3 the length of the corolla, with minute, straight, soft hairs on lobes or hairless, purplish-pink; stamens 10, filaments free to bases, white; ovary with a stipe, with short, matted, soft hairs, style attenuate at the apex, stigma pore-shaped. **Fruits** oblong [to linear-oblong], (1.5–)2.5–4.5(–6) × (0.5–)0.6–0.8 cm, straight or slightly curved; valves with short, matted, soft hairs, or hairless, not constricted or slightly constricted between seeds, entire; with stipe 1.5–2.5 mm or sessile, apex acute, with mucro or beak 3–4 mm; margin prickly. **Seeds** 3–8, reddish brown, lens-shaped, oblong or rhomboid, 5–6 × 4.1–4.6 × 1.8–2 mm, seed coat porous, pleurogram horseshoe-shaped, 50%.

Varieties 2 (1 native in the flora): sw United States, Mexico (Chihuahua, Durango, Sonora).

4a. Mimosa grahamii A. Gray var. **grahamii** Native

Mimosa grahamii A. Gray var. *lemmonii* (A. Gray) Kearney & Peebles; *Mimosa lemmonii* A. Gray

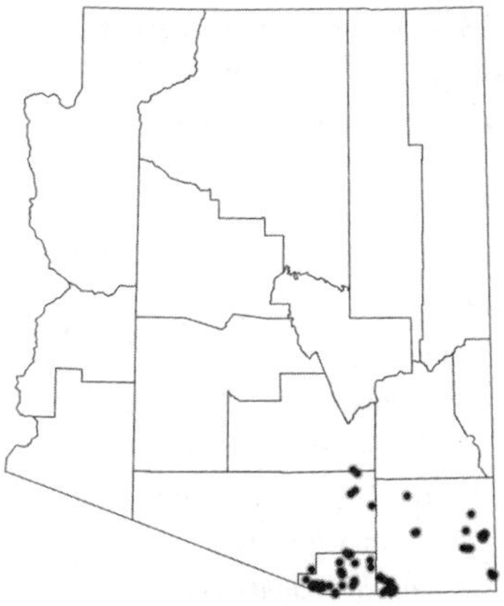

Shrubs 0.6–1 m. **Leaflets** 7–16 pairs per pinna. **Fruits** oblong, 2.5–4.5(–6) × (0.5–)0.6–0.8 cm, not constricted between seeds.

Flowering Apr–May(–Oct); fruiting Aug–Sep. Slopes, canyons, and arroyos, often in rocky soils; 1150–1830(–2000) m; Semidesert Grassland, Madrean Evergreen Woodland; Cochise, Pima, Pinal, Santa Cruz cos.; New Mexico; Mexico (Chihuahua, Durango, Sonora).

Some populations of *Mimosa grahamii* occurring in Santa Cruz and Cochise counties are densely hairy, while those from Pinal County are less hairy on branchlets, leaves, and flowers; fruits of these latter are slightly curved, longer and narrower (6 × 0.5 cm) than those of typical variety, and they could correspond to *M. grahamii* var. *prolifica* (S. Watson) Barneby, described from canyons and rocky hills of the Sierra Madre Occidental in central Chihuahua, Mexico. The study of material from the whole range of distribution of this species is needed in order to distinguish varieties.

Mimosa grahamii occurs in scattered populations at mid-elevations in southeastern Arizona; reports from Yavapai Co. exist but should be confirmed. The plants may be locally common. They achieve their best development in semi-shaded riparian woodland habitats, but can be found on dry, sunny slopes. Peak flowering is in spring with sporadic flowering into early autumn. *Mimosa grahamii* tolerates temperatures well below freezing. As with the other *Mimosa* species, plants are easily grown from scarified seeds.

NISSOLIA Jacquin. Enum. Syst. Pl. 7, 27. 1760. * [for William Nissole, 1647–1735, French physician and botanist] YELLOWHOOD

Matthew B. Johnson

Perennial twining herbaceous vines, shrubs, (or prostrate herbs). **Stems** herbaceous, becoming woody in most species, from a woody rootstock. **Herbage** glabrous or variously pubescent. **Leaves** alternate, odd-pinnate; stipules joined at base, triangular or lanceolate, entire or with small gland-tipped teeth; petiolate; stipels absent. **Leaflets** 5 (or 7), elliptic, rounded, obovate or obcordate, mucronate, margins entire. **Inflorescences** of several to numerous flowers in axillary clusters or less commonly racemes or paniculate; bracts awl-shaped, resembling stipules; bracteoles absent (or if present subtending pedicel joint at base of calyx). **Pedicellate**. **Flowers** papilionoid, small; calyx bell-shaped, abruptly narrowed at base and jointed at pedicel, tube with 5 teeth, or truncate with remote linear or awl-shaped teeth; corolla yellow; banner slightly longer than others, spatulate to nearly rounded, straight or reflexed, pubescent on outer (abaxial) surface, short-clawed; wings falcate-oblong, clawed; keel obtuse, incurved; stamens 10, 9 with filaments united below the middle, 1 stamen free at base and united at middle, tube splitting along vexillary side as fruit develops, anthers all alike; ovary: style glabrous; stigma simple, head-like. **Fruits** indehiscent; samara-like and lomentaceous, compressed, linear or spatulate, jointed, terminal segment sterile, flattened, wing-like. **Seeds** kidney-shaped, compressed.

Species 13 (2 native in the flora): dry tropical and warm-temperate regions of Mexico and the Caribbean with 1 species extending south into Argentina, 2 species extending into Arizona, and 1 species extending into Texas.

Nissolia fruticosa is reported to be used to stupefy fish and in folk medicine to treat snakebite (Allen and Allen 1981, Lewis et al. 2005, Rudd 1956). Nodulation apparently does not occur in *Nissolia* (Sprent 2009).

1. Calyx tube glabrous, equaling or shorter than the teeth; fruit flattened, scarcely constricted between seeds, falcate; widespread (Maricopa, Pima, Pinal, Santa Cruz cos.) ____ **1. Nissolia schottii**
1′ Calyx tube pubescent, longer than the teeth; fruit turgid, deeply constricted between seeds, straight; localized (Cochise Co.) ____ **2. Nissolia wislizeni**

1. Nissolia schottii (Torrey) A. Gray. J. Proc. Linn. Soc., Bot. 5:26. 1861. * [for Arthur Carl Victor Schott, 1814–1875, German-American artist, cartographer, and botanist] SCHOTT'S YELLOWHOOD Native

Chaetocalyx schottii Torrey ex Emory, Rep. U.S. Mex. Bound. 2(1):56, pl. 18. 1859

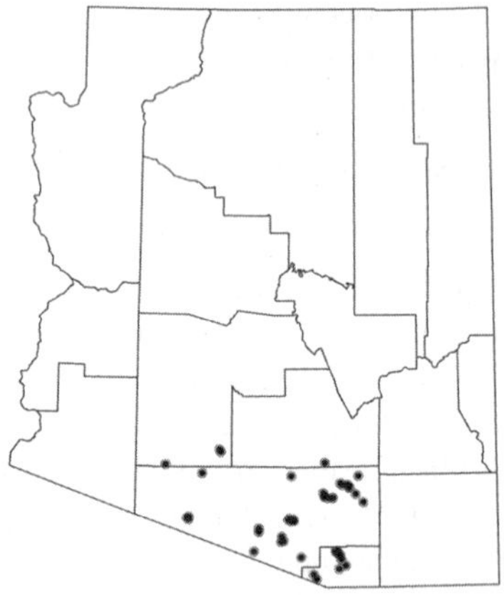

Twining vine. Stems herbaceous or becoming semi-woody at base, from a subterranean crown, to 2 m long, sparsely pubescent or subglabrous and occasionally with soft yellow bristles. **Leaves** 3–10[–13.5] cm; stipules 3–5 mm, lanceolate, apex acute, margins entire, herbaceous, persistent; petiole 1–5.5 cm. **Leaflets** 5, 0.4–3[–4] × 0.2–2[–2.5] cm, elliptic or subrhombic, base obtuse, apex acute or obtuse, mucronulate, subglabrous on both surfaces. **Inflorescences** sessile clusters, 1–8-flowered; a single bract 0.5–1 mm, awl-shaped; bracteoles absent. **Pedicels** 6–10 mm. **Flowers** 8–10[–12] mm; calyx 5–7 mm, tube glabrous, (2–)3–4 mm, teeth 2–4 mm, awl-shaped; corolla yellow, banner 8–10[–12] mm, somewhat reflexed, wings 8–9[–11] mm, keel 7–8[–10] mm; stamens 6–7 mm. **Fruits:** stipe 1–2 mm; 2–3.5[–4] cm, with 2–4 segments, flattened, scarcely constricted between seeds, falcate, subglabrous or sparsely pubescent, fertile segments 4–6 × 4–5 mm, the terminal segment flattened and broadly wing-like with conspicuous net-like veins, 10–15 × 6–11 mm. **Seeds** 3–3.5 mm.

Flowering (Jun–)Jul–Sep(–Nov). Slopes, canyons, and along arroyos, usually in rocky soil; [50–]730–1430 m; Sonoran Desertscrub and lower elevations in Semidesert Grassland; Maricopa, Pima, Pinal, Santa Cruz cos.; Mexico (Chihuahua, Sonora).

Nissolia schottii occurs in scattered populations on rocky slopes and canyons of mountains in the southern part of the state. The plants may be locally abundant with stems that twine through and over shrubs or trees. Occasional winter flowering is reported for *N. schottii* in southern Sonora by Martin et al. (1998) who also report that it is found on the Baja California peninsula. The Baja California plants represent *N. setosa*, which is usually recognized as a distinct species. In Sonora, *N. schottii* grows in thornscrub and tropical deciduous forest. The herbage is browsed by quail, white-tail deer, and other wildlife. The stems die back with freezing temperatures. Schott's yellowhood is easily grown from seeds but is only rarely available from specialty nurseries.

2. Nissolia wislizeni (A. Gray) A. Gray. J. Proc. Linn. Soc., Bot. 5:25. 1861. * [for Friedrich Adolph Wislizenus, 1810–1889, German-American physician and botanist] ARIZONA YELLOWHOOD Native

Chaetocalyx wislizeni A. Gray, Smithsonian Contr. Knowl. 3(5):51. 1852

Prostrate herbs. Stems herbaceous from a subterranean crown, trailing, to at least 1 m long, with sparse white pubescence and soft yellowish bristles. **Leaves** (0.5–)2–5.5 cm; stipules 2–7 mm, lanceolate to triangular or oblanceolate, apex acute or acuminate, margins glandular-toothed, herbaceous, persistent; petiole 0.9–3 cm. **Leaflets** 5, 0.5–1.3[–2] × 0.4–1[–1.8] cm, rounded or elliptic-obovate, the base obtuse or subcordate, the apex obtuse to emarginated, mucronulate, glabrous on upper (adaxial) surface, subglabrous or sparsely pilose on lower (abaxial) surface. **Inflorescences** of sessile or short-pedunculate clusters, 1–5-flowered; a single bract 1.5–4 mm, awl-shaped; bracteoles narrowly awl-shaped, 1–1.5 mm. **Pedicels** 3–10(–15) mm. **Flowers** 10–15 mm; calyx 3.5–5.5[–6] mm, tube 2–4 mm, pubescent and with soft

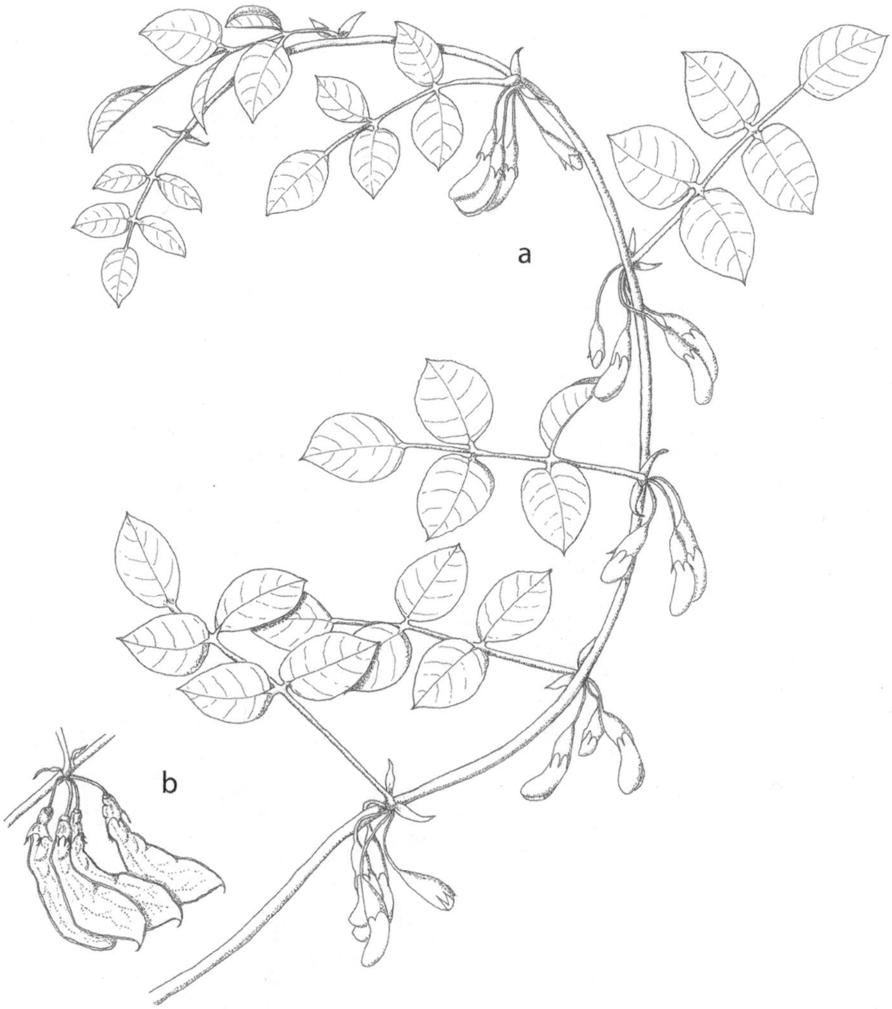

Nissolia schottii: **a.** habit; **b.** fruits. Illustrator: Barbara Vucich.

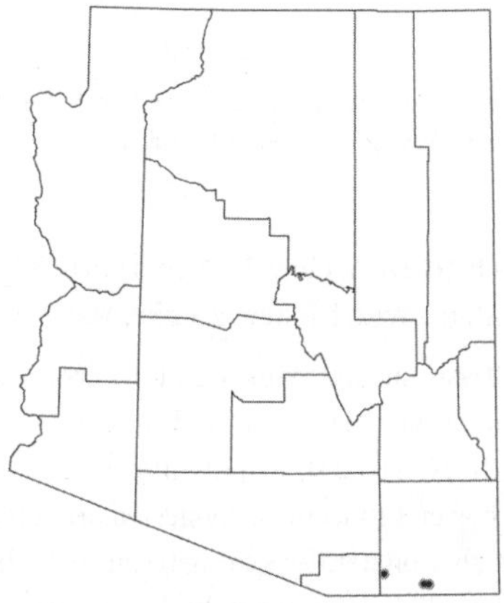

bristles from glandular bases, teeth 1–2[–3] mm, narrowly triangular to awl-shaped; corolla yellow or yellow marked with pink, banner 8–11[–13] mm, straight or slightly reflexed, wings 7–9[–11] mm, keel 6–7[–10] mm; stamens 6–7[–10] mm. **Fruits:** stipe 1–2 mm; 1.5–3.5 cm, with 2–5 segments, turgid, deeply constricted between seeds, straight, sparsely pubescent or subglabrous, fertile segments 4–7 × 3.5–5 mm, the terminal segment flat or concave, wing-like, smaller than to as long as the fertile segments. **Seeds** 4–5 mm.

Flowering Jul–Aug. Slopes, canyons, and floodplains; [320–]1460–1520[–2440] m; Semidesert Grassland; Cochise Co.; Mexico (Aguascalientes, Chihuahua, Durango, Guanajuato, Hidalgo, Jalisco, Queretaro, San Luis Potosí, Sonora, Zacatecas).

Nissolia wislizeni is documented for Arizona from two localities just north of the International Boundary. It has an extensive distribution in Mexico, however populations are scattered. In Mexico it occurs in grassland as well as desertscrub and montane woodland communities. The slight development of the sterile, terminal fruit segment of *N. wislizeni* contrasts markedly with the broad, wing-like terminal segment of *N. schottii*. The Arizona material of *N. wislizeni* closely matches specimens from further south in Mexico except in having somewhat smaller measurements.

Nissolia wislizeni: **a.** habit with flowers; **b.** leaf with stipules; **c.** young flower (lateral view); **d.** stem section with fruits. Drawn by Karen Douthit and reproduced from *Flora Novo-Galiciana*, vol. 5, vol. 88, with permission of the University of Michigan Herbarium.

OLNEYA A. Gray. Pl. Nov. Thurb. 328. 1854. * [for Stephen Thayer Olney, 1812–1878, Rhode Island botanist]
Matt Lavin, Kirsten N. Lake, Matthew B. Johnson

Trees, armed. **Stems** usually erect to ascending, rarely scandent; young growth densely sericeous to woolly. **Leaves** alternate; even- and odd-pinnate; stipules present, quickly deciduous or persistent; petiolate; stipels absent. **Leaflets** (8–)9–21(–24), usually alternate, rarely opposite, blade margins entire, surfaces sericeous. **Inflorescences** 3–29-flowered; racemes, congested and often appearing as panicles; bracts present; bracteoles absent. **Flowers** papilionoid; calyx bilaterally symmetrical, bell-shaped, lobes 5; corolla whitish to pinkish or purplish, wing petals and keel tips usually purple; stamens 10, 9 united and 1 free; anthers attached basally; style glabrous proximally, with pollen brush surrounding distal 1/2; stigma terminal, rounded, ciliate; ovules 10–14(–17). **Fruits** pedicellate, elastically dehiscent, cylindrical, spindle-shaped, sparsely to densely covered with stipitate glands. **Seeds** 1–2(–5), spheroidal; hilum subapical. $x = 9$.

Species 1 (1 native and cultivated in flora): sw United States, nw Mexico.

Olneya is native to the Sonoran Desert in Arizona and adjacent California, and to northwestern Mexico.

Olneya is readily distinguished from other woody papilionoid genera by the combination of broadly triangular calyx lobes, tomentose dorsal surfaces of the standard and wing petals, leaf rachises terminated by a spinose bristle but that also bear usually an odd number of leaflets in alternate arrangement, and spinescent stipules that are especially prominent on the younger twigs.

1. Olneya tesota A. Gray. Pl. Nov. Thurb. 328. 1854. * [a local Native American name] DESERT IRONWOOD; *PALO FIERRO* Native, Cultivated

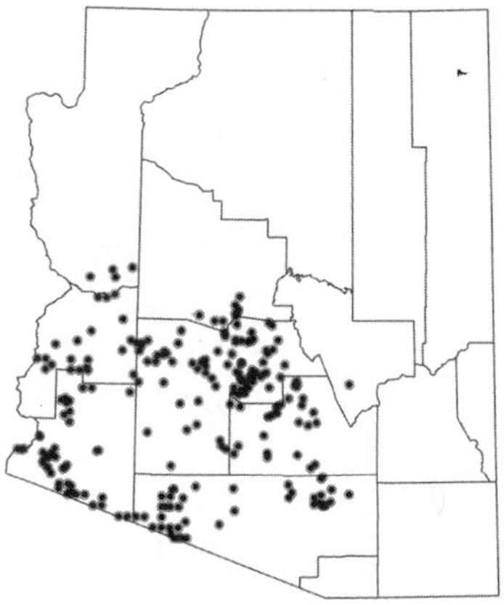

Trees to 10 m. **Stems** with bark smooth and gray, becoming fissured and eventually shredding on trunk and large limbs. **Leaves** 2–10 cm; stipules awl-shaped, 4–10 × 1 mm, spinescent; petiole 0.5–1 cm, strigose; rachis with longitudinal grooves; petiolules 1 mm, nearly glabrescent. **Leaflets** narrowly obovate to elliptic, (5–)7–21 × 2.5–7(–9) mm, base attenuate to rounded, apex rounded. **Inflorescences:** rachises with fewer than 30 nodes, mostly 0.5–6 mm apart, axis tomentose to woolly, stipitate glands absent or sparse; bracts persistent to quickly deciduous, 1 × 1 mm, blunt. **Pedicels** 3–4 mm. **Flowers:** calyx tube (1.5–)2–3 mm, sericeous to tomentose, lobes equal, broadly triangular, 7–11 mm; corolla 10–14 mm, glabrous or sericeous; filaments subequal; anthers relatively small, dehiscing longitudinally; ovary with stipitate glands, densely sericeous. **Fruits** light to dark grayish or brownish, 2–5(–6) × 1.1–1.3 cm, base blunt, apex acute, often terminating in persistent style base, usually sparsely covered with stipitate glands, slightly constricted between seeds. $2n = 18$.

Flowering early summer (Apr–)May(–Jun); flowering does not occur each year. Open, dry shrubby vegetation of slopes, valleys, plains, and arroyos; [near sea level–]50–890[–1300] m; Sonoran Desertscrub; Gila, La Paz, Maricopa, Mohave, Pima, Pinal, Yavapai, Yuma cos.; California; Mexico (Baja California, Baja California Sur, Sonora).

Olneya tesota is one of the keystone tree species of the Sonoran Desert. The canopy provides shaded microhabitats for the establishment of several species of columnar cacti and many other plants, as well as shelter and feeding grounds for numerous animals. Dimmitt (2000) reports that, in the most arid portions of the Sonoran Desert, 165 species of plants have been found to use *O. tesota* as a nurse plant. The flowers attract a variety of bees as well as hummingbirds (Turner et al. 1995). In 2000, Ironwood National Monument was established in south central Arizona to protect extensive populations of *O. tesota* and many other plants and animals characteristic of the Sonoran Desert. *Olneya tesota* occasionally occurs in Sinaloan Thornscrub in Sonora.

Annual rainfall across the range of *Olneya tesota* averages 75–450 mm. In the driest parts of its range, *O. tesota* is mostly restricted to drainages where additional moisture is available. In upland sites in wetter areas of the Sonoran Desert, it can form open to relatively dense woodlands where it is dominant or mixed with other woody Fabaceae.

The largest *Olneya tesota* in the state has a height of 13.7 m (The Arizona Register of Big Trees 2005). *Olneya tesota* is long-lived with some trees estimated to exceed 300 years in age though they may be significantly older (Dimmitt 2000). The wood is extremely durable, and dead trees may remain standing for 1000 years (Dimmitt 2000). Nodulation is reported for *O. tesota* (Allen and Allen 1981, Felker and Clark 1981).

Olneya tesota is an economically important species to the Seri Indians of the Sonoran Desert (Felger and Moser 1985). It provides nutrition (the seeds and unripe pods) and medicine. The wood is used to produce tools, weapons, musical instruments, and artistic sculptures. Tinctures were made from the green wood of *O. tesota* and used by the Seris in vision quests, which they believed gave them power.

The wood makes excellent fuelwood and charcoal, and in Sonora it is being seriously overharvested despite legal protection. In Arizona, *Olneya tesota* is a protected species and harvesting is regulated by the Arizona Department of Agriculture. The wood is brittle, hard, and dense. It has a specific gravity of 1.14 (Peattie 1953) and will not float in water. The foliage is browsed to some extent by livestock (Kearney and Peebles 1942) as well as wildlife.

Desert ironwood is frequently planted in hot desert regions of Arizona. It is suitable for street medians and larger spaces where it can serve as a specimen tree. It can also be planted as a screen or barrier (Jones and Sacamano 2000). The dusty pink flowers are produced in dense masses in some years, giving this species a unique appearance among landscape trees grown in Arizona. The leaves are often evergreen but may become gradually deciduous during extended drought or if killed by freezing temperatures. The trees also become deciduous prior to flowering. Growth is rather slow

Olneya tesota: **a.** habit; **b.** fruits; **c.** fruit detail; **d.** seed; **e.** leaflet; **f.** flowers; **g.** pistil; **h.** stamens, **i.**, **j.**, **k.** petals. Illustrator: Joyce Riney Peters.

with rainfall alone but can be increased significantly with adequate irrigation. Established trees of desert ironwood are drought tolerant, but may require some irrigation during extreme drought events. Occasional irrigation improves performance. Established trees are hardy to about -7°C. Due to the spiny twigs, this species should be sited away from traffic areas. The trees produce fine leaf litter. Large desert ironwood trees are boxed and moved with a high success rate. Seeds will germinate without scarification when fresh. Scarification enhances germination of older seeds.

OXYTROPIS de Candolle. Astragalogia (qto.), 24–28, 66–98, plates 2–8. 1802, name conserved. * [from Greek *oxy*, sharp, and *tropis*, keel] LOCOWEED, CRAZYWEED, OXYTROPE

Matthew B. Johnson

Herbs or subshrubs, perennial. **Stems** absent or forming cushions of short stems from a rootstock. **Pubescence** often silky, basifixed (or medifixed), the hairs sometimes glandular. **Leaves** alternate or whorled, odd-pinnate; stipules free, partially fused, or fused; petioles deciduous [or persistent, sometimes stiff and spine-like]; stipels absent. **Leaflets** sessile or shortly petiolulate, the margins entire. **Inflorescences** axillary racemes or spikes, lateral or scapose; bracts membranous, small or absent; bracteoles small or absent. **Pedicels** present or absent. **Flowers** papilionoid; calyx tubular or bell-shaped, lobes 5, unequal; corolla white, yellow, violet or purple, keel apex porrect; stamens 10, the filaments 9 fused and 1 free, slender; ovary sessile or stipitate. **Fruits** sessile or stipitate, membranous, papery or leathery, appearing partially or completely 2-chambered due to the funicular flange. **Seeds** several to many; compressed, mitten- or kidney-shaped, smooth. $x = 8$.

Species 300–400 (2 native in the flora): cold temperate and boreal regions of Eurasia and North America. The genus has its highest diversity in the mountains and subarctic steppes of central Asia.

Several species of *Oxytropis* are responsible for livestock poisoning known as locoism. Other species are utilized for forage. Lewis et al. (2005) report that some species are used medicinally, as food ingredients, and are grown as ornamentals. Polyploidy in *Oxytropis* is frequent (Barneby 1989). Nodulation is widely reported for the genus (Allen and Allen 1981). In the field, *Oxytropis* might be confused with *Astragalus* as the two genera share a number of similarities.

1. Inflorescences 20–42 cm; keel 13–18 mm; fruits stiffly papery, turgid but not bladdery inflated, 0.3–0.5 cm wide; widespread ______ **1. Oxytropis lambertii**

1′ Inflorescences 6–16.5 cm; keel 5.5–12.5 mm; fruits thinly papery, bladdery inflated, 0.4–1.2 cm wide; localized in Apache and Coconino cos. ______ **2. Oxytropis oreophila**

1. Oxytropis lambertii Pursh. Fl. Amer. Sept. 2:740. 1814[1813]. * [for Aylmer Lambert, 1761–1842, English botanist] LAMBERT'S LOCOWEED, PURPLE LOCOWEED, CRAZYWEED, LOCO, WHITE LOCO, POINT LOCO Native

Oxytropis lambertii Pursh var. *bigelovii* A. Gray

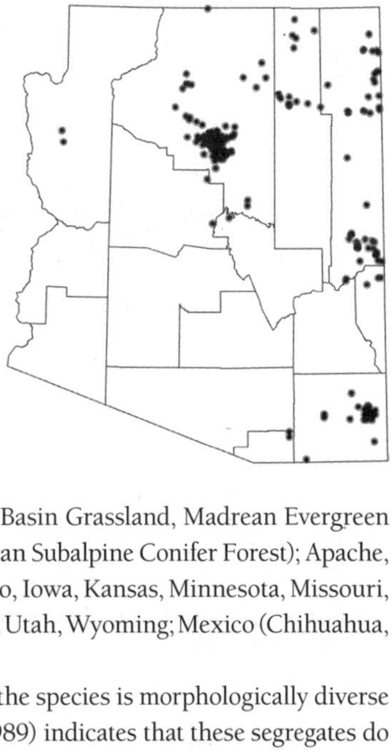

Herbs, clustering. **Stems** absent. **Pubescence** basifixed and also sometimes medifixed, thinly to densely pilose or villous. **Leaves** 3–20 cm, basal; stipules triangular, partially fused, free portion 2–9 mm; petioles (2–)4–9 cm; rachis (2.5–)4–12 cm. **Leaflets** (7–)9–15(–19), 5–40 × 2–7 mm, linear, lanceolate, elliptic, or ovate. **Inflorescences** scapose spike-like racemes, 6–25-flowered, 20–42 cm, spreading to erect; bracteoles 3–12 mm, awl-shaped, persistent. **Pedicels** to 1 mm or flowers subsessile. **Flowers** with calyx bell-shaped, 7–12 mm, tube 6–8.5 mm, lobes 1–3.5 mm; corolla pink-purple or occasionally white to pink or purple, banner 15–25 mm, wings 12–23 mm, keel 13–18 mm. **Fruits** sessile or subsessile, erect or ascending, 1.5–2.8 × 0.3–0.5 cm, turgid but not bladdery inflated, oblong to lance-ellipsoid with a conspicuous beak, stiffly papery, funicular flange forming a nearly complete septum.

Flowering Apr–Oct. Various soils including clay, sand, and rocky sites of ridges, slopes, canyons, stream terraces and banks, meadows and roadsides; (1350–)1700–2700(–3050)[–3150] m; (Semidesert Grassland), Plains and Great Basin Grassland, Madrean Evergreen Woodland, Great Basin Conifer Woodland, Petran Montane Conifer Forest, (Petran Subalpine Conifer Forest); Apache, Cochise, Coconino, Greenlee, Mohave, Navajo, Santa Cruz, Yavapai cos.; Colorado, Iowa, Kansas, Minnesota, Missouri, Montana, Nebraska, New Mexico, North Dakota, Oklahoma, South Dakota, Texas, Utah, Wyoming; Mexico (Chihuahua, Sonora, Zacatecas); Canada.

Numerous segregates of *Oxytropis lambertii* have been proposed; however, the species is morphologically diverse with a lack of distinctive characters that can be used for separation. Barneby (1989) indicates that these segregates do not appear to be worthy of taxonomic recognition. Arizona populations have been recognized as var. *bigelovii* A. Gray but are not treated separately here.

Oxytropis lambertii is widely distributed at mid- and higher elevations in Arizona and may be locally common. The plants are most commonly found within the ponderosa pine zone. They occur in open areas and sometimes beneath trees. *Oxytropis lambertii* is highly toxic if eaten and causes locoism in livestock including cattle, sheep, goats, and especially horses (Stubbendieck and Conard 1989). Animals become habituated to the plants but must consume a significant portion of their body weight for lethal poisoning to occur. Symptoms of locoism include trembling, erratic behavior, clumsiness, lethargy, and paralysis (Stubbendieck and Conard 1989). Eradication of the plants from rangeland is difficult (Allen and Allen 1981). *Oxytropis lambertii* is reported to nodulate (Allen and Allen 1981). The brightly colored flowers and contrasting gray green foliage are attractive, and the species is suitable as a perennial wildflower in the garden.

Oxytropis lambertii: **a.** habit; **b.** fruit; **c.** seeds; **d.** fruit cross-section; **e.** leaflet; **f.** pistil; **g.** stamens; **h., i., j.** petals; **k.** flower (front view); **l.** flower (lateral view). Illustrator: Joyce Riney Peters.

2. Oxytropis oreophila A. Gray. Proc. Amer. Acad. Arts 20:3. 1884. * [from Greek, *oreo*, mountain, and *phila*, loving] MOUNTAIN OXYTROPE, JONES' OXYTROPE

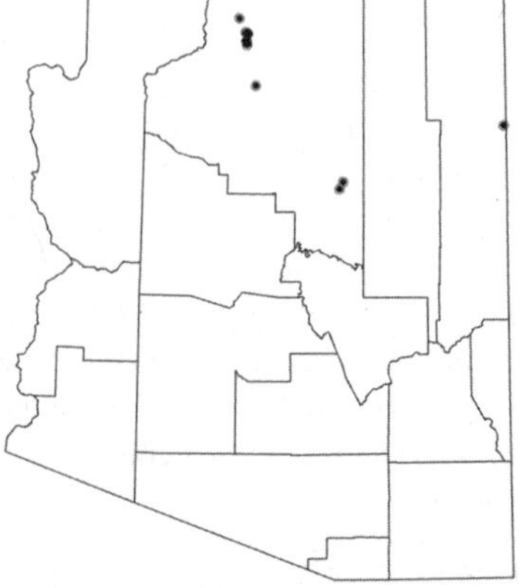

Herbs, clustering or forming cushions. **Stems** absent. **Pubescence** basifixed, silky pilose. **Leaves** 0.5–15 cm, basal; stipules triangular, partially fused, free portion 1–5 mm; petioles 1.5–3.5 cm; rachis 1.7–5.5 cm. **Leaflets** (1–)3–15 (–17), 4–12 × 1–5 mm, elliptic to ovate. **Inflorescences** scapose spike-like racemes, (1 or) 2–15-flowered, 6–16.5 cm, ascending to erect; bracteoles 3–5 mm, awl-shaped, persistent. **Pedicels** to 1 mm or flowers subsessile. **Flowers** with calyx bell-shaped, 4.5–9 mm, tube 3.5–6 mm, lobes 1–3.5 mm; corolla pink-purple (white), banner 8–16.5 mm, wings 5–15 mm, keel 5.5–12.5 mm. **Fruits** sessile, spreading, 0.9–2.5 × 0.4–1.2 cm, ovoid and bladdery-inflated with a beak, thinly papery, funicular flange forming only a partial septum.

Varieties 3 (2 native in the flora): a third variety, var. *jonesii*, is found in parts of central Utah.

1. Plants cushion-forming; leaves 0.5–3 cm; leaflets (1–)3–11; inflorescences 1–3(–5)-flowered, the peduncle shorter than foliage ______ **2a. Oxytropis oreophila** var. **juniperina**

1′ Plants clustering but not cushion-forming; leaves 2–15 cm; leaflets 9–15(–17); inflorescence (3–)4–12(–15)-flowered, the peduncle tmostly equaling or longer than foliage ______ **2b. Oxytropis oreophila** var. **oreophila**

2a. Oxytropis oreophila A. Gray var. **juniperina** S.L. Welsh. Great Basin Naturalist 38(3):339. 1978. * [for the occurrence of the plant in pinyon-juniper woodland] Native

Plants cushion-forming. **Leaves** 0.5–3 cm. **Leaflets** (1–)3–11. **Inflorescences** 1–3(–5)-flowered, the peduncle shorter than foliage.

Flowering Apr–Sep. Rocky or clay soils of ridges and slopes; 1750 [1890–2570] m; Great Basin Desertscrub, [Great Basin Conifer Woodland]; Coconino Co.; Nevada, Utah.

Oxytropis oreophila var. *juniperina* is documented for Arizona only from Meteor Crater in Coconino County. Its primary distribution lies in eastern Nevada and central Utah.

2b. Oxytropis oreophila A. Gray var. **oreophila** Native

Plants clustering but not cushion-forming. **Leaves** 2–15 cm. **Leaflets** 9–15(–17). **Inflorescences** (3–)4–12(–15)-flowered, the peduncle mostly equaling or longer than foliage.

Flowering Apr–Sep. Rocky outcrops and slopes of limestone and igneous substrates, and meadows; 2100–2700 [–3400] m; Great Basin Conifer Woodland, Petran Montane Conifer Forest, Petran Subalpine Conifer Forest; Apache, Coconino cos.; California, Nevada, Utah.

In Arizona, *Oxytropis oreophila* var. *oreophila* is apparently restricted to Coconino County except for a single record from Window Rock, in Apache County. Populations are isolated, but the plants are often locally abundant. They prefer open, dry sites within pinyon-juniper woodland and in ponderosa pine, mixed conifer and spruce-fir forests. Allen and Allen (1981) report nodulation for *O. oreophila*.

PARASENEGALIA Seigler & Ebinger. Novon 25(2):181–183. 2017. * [similar to *Senegalia*]
David S. Seigler and John E. Ebinger

Trees, shrubs, or lianas, some to 25 m, mostly erect to spreading. **Stems** lightly to highly branched; twigs rounded or angular, glabrous to pubescent; short-shoots absent; prickles and spines absent. **Leaves** alternate, bipinnate; pinnae 3–20 pairs/leaf; opposite; stipules small, not obvious, tardily deciduous to persistent, rarely absent; petioles round, channeled above; petiole gland present; rachis channeled above, sessile to stalked glands at various positions. **Leaflets** 1–numerous pairs per pinna, opposite, subsessile, membranous, glabrous to pubescent. **Inflorescences** loosely- to densely-flowered globose head to cylindrical spike, solitary to clustered in the leaf axil, mostly in terminal or axillary raceme-like or paniculate clusters; peduncles short, glabrous to pubescent, rarely elongating in fruit; bracts present. **Pedicels** absent to short. **Flowers** mimosoid, radially symmetric, 5-merous; calyx cup-shaped, lobes triangular, glabrous to pubescent; corolla cup-shaped, membranous, lobes triangular, glabrous to pubescent, valvate in bud, nectiferous disc mostly present; stamens 40–140, 5–13 mm, exserted, white, cream, or yellow; anthers attached dorsally, sometimes with a small gland; ovary 1, stipitate; style thread-like; stigma thread-like. **Fruits** erect to pendulous, usually dehiscing along both sutures (sometimes tardily so), linear to oblong, straight to slightly curved, mostly flattened, mostly dry, leathery, glabrous to pubescent; stipe short; apex sometimes beaked. **Seeds** arranged in a single row, not surrounded by pulp, circular to ellipsoid to oblong, flattened, smooth; pleurogram present. $x = 12$.

Species ca. 11 in the New World (1 cultivated in the flora): native in Central America, Caribbean, and South America (Argentina). Not known from the Old World.

Seigler and Ebinger (Seigler et al. 2017) described the genus *Parasenegalia* when they removed seven species from the genus *Senegalia* based mostly on the absence of prickles on the twigs, petiole, and rachis in these species. Later, they added four mostly recently described species of *Senegalia* to this genus (Seigler and Ebinger 2018).

1. Parasenegalia visco (Lorentz ex Grisebach) Seigler & Ebinger. Novon 25(2):194–197. 2017. * [*visco*, from the common name] *VISCO, VISCOTE* Cultivated

Acacia visco Lorentz ex Grisebach, Abh. Königl. Ges. Wiss. Göttingen 19:135, 279. 1874 (as *Acacia visite*), emended page 279 (as *Acacia visco*); *Acacia concinna* F. Philippi (Willdenow) de Candolle; *Acacia platensis* Manganaro; *Acacia polyphylla* Clos; *Acacia visite* Grisebach; *Manganaroa platensis* (Manganaro) Spegazzini; *Manganaroa subsericea* Spegazzini; *Senegalia visco* (Lorentz ex Grisebach) Seigler & Ebinger

Trees, erect, to 20 m. **Stems** slender, lightly branched; twigs not zigzagged, round, glabrous to lightly puberulent; short shoots absent; prickles absent; bark brown, fissured. **Leaves:** pinnae 4–10 pairs per leaf, 40–70 mm; stipules, linear, to 5 mm; petiole channeled above, 25–45 mm, lightly puberulent; petiolar gland solitary, sessile, oblong, 0.7–2.7 mm; rachis 35–130 mm. **Leaflets** 25–40 pairs per pinna, oblong, oblique and truncate on one side at base, acute to acuminate apex, 3.5–7.5 × 0.9–2.1 mm, appressed pubescent on both surfaces. **Inflorescences** densely-flowered globose heads 16–23 mm across, 1–3 in the leaf axils; peduncles 15–40 mm. **Pedicels** absent. **Flowers** white; calyx 1.7–2.8 mm, puberulent; corolla 3.2–4.3 mm, lightly puberulent; filaments 9–11 mm. **Fruits** 80–150 × 18–30 mm, straight, flattened, not constricted between the seeds, stiffly papery, lightly puberulent; stipe 4–10 mm; apex obtuse, beak to 10 mm. **Seeds** light brown, oval to oblong, strongly flattened, 9–13 × 7–10 mm.

Flowering Oct–Jan. South America (n Argentina, Bolivia, n Chile); commonly cultivated in Peru as well as where native (Rico-Arce 2007).

Parasenegalia visco is found on slopes, valleys, and arroyos in seasonally wet tropical forests, deciduous forests, riparian forests, yungas, the Monte Desert, and disturbed second growth forests and thickets. It has a wide tolerance for environmental conditions, even extending into the arid Monte Desert along larger arroyos where additional moisture is available. Elevations range from 1000 to 3000 m.

The wood is hard and heavy, and is used for fuel, posts, and in light carpentry (Leonardis 1975).

Visco can develop into an impressive tree with a high, open canopy of lacy foliage that has a tropical effect and provides filtered light. It should be positioned in the landscape with its potential mature size in mind. This species was first cultivated in Arizona at Boyce Thompson Arboretum in the late 1920's and several of the original trees continue to thrive. In recent years, young landscape plants of this species in Arizona have suffered fatal infestations of a wood-boring insect. Cultivated plants have been undamaged by temperature down to -9°C. The foliage and pods produce modest litter. Scarified seeds germinate readily.

PARKINSONIA Linnaeus. Sp. Pl. 1:375. 1753. * [for John Parkinson, 1567–1650, apothecary to King James I] PALO VERDE

Alexander Krings

Cercidiopsis Britton & Rose; *Cercidium* Tulasne; *Peltophoropsis* Chiovenda

Trees or shrubs, armed or unarmed. **Stems:** bark green or yellowish green; glabrous or strigillose becoming glabrescent. **Leaves** from spurs, bipinnate or appearing once pinnate from obsolete petiole, drought deciduous; stipules minute, quickly deciduous, or spinescent; pinnae 2–4 per leaf; stipels absent; glands linear, minute, in sparse to dense patches between leaflets or in leaflet axils. **Leaflets** [2–]4–76 per pinna, alternate, subopposite, or opposite, blade oblong, oblanceolate, obovate, elliptic, or nearly circular, margins entire. **Inflorescences** 1–15-flowered, axillary, racemes, [sometimes corymbiform], sometimes terminally congested; bracteoles persistent, tardily deciduous, or quickly deciduous. **Flowers:** caesalpinoid; calyx radially symmetric, sepals nearly free, lobes bent backward, oblong, subequal; corolla yellow, glabrous except claw pubescent, one petal dissimilar from the others; filaments nearly equal or distinctly unequal; stamens free or united into a single tube, anthers attached basally, dehiscing apically; ovary sessile, style glabrous, stigma terminal; ovules 1–5. **Fruits** indehiscent, flattened, or almost round in cross section, irregularly and, sometimes, indistinctly torulose, oblong to linear. **Seeds** 1–5, oblong to nearly circular in outline; hilum apical or subapical. $x = 14$.

Species 11 or 12 (In the flora, 2 native, 1 naturalized, 1 adventive, all cultivated.): s, w United States, Mexico, West Indies, Central America, South America, Asia, Africa, Australia (introduced).

Taxa referable here have usually been treated as in either of two distinct genera, *Cercidium* and *Parkinsonia* (e.g., Sargent 1889, Johnston 1924, Benson 1940, Carter 1974a), or within a broad *Parkinsonia* (e.g., Watson 1876, Brenan 1963, Isely 1975, Polhill and Vidal 1981, Hawkins 1996). Recent molecular evidence from the chloroplast genome indicates that, together, *Cercidium* and *Parkinsonia* form a strongly supported monophyletic group (Haston et al. 2005), and a broad concept is adopted here.

Several species of *Parkinsonia* are grown as landscape plants in dry regions and are valued for their showy masses of yellow flowers. The plants provide food, cover, and nesting sites for birds and other wildlife. The flowers are a source of nectar for honey. Fruits of some species are browsed by livestock when other forage is scarce. Seeds and green pods have been utilized by indigenous people. The wood is sometimes used for small articles but is not suitable as timber. The common name *palo verde*, applied to several species in the U.S. Southwest and northern Mexico, is from Spanish meaning *green stick*.

1. Secondary rachises (100–)150–600 mm, winged; leaflets 40–76 per pinna, alternate, subopposite, and opposite, usually all on the same rachis; inflorescence axes 8–18 cm ____ **1. Parkinsonia aculeata**

1′ Secondary rachises 1.5–40 mm, not winged; leaflets 4–16(–34) per pinna, opposite; inflorescence axes 0.4–3.8 cm.

2. Branch tips spinescent, nodal spines absent; leaves appearing once pinnate due to obsolete petiole, leaflets (6–)12–16 per pinna, rachises with glandular patches surrounding leaflet pulvini or, at least, on opposing sides; calyx upper (adaxial) surface pubescent; petals light yellow, except the upper one white or cream-yellow, not orange-dotted; ovaries densely sericeous; fruits constricted between the seeds ____ **3. Parkinsonia microphylla**

2′ Branch tips not spinescent, nodal spines present; leaves clearly bipinnate with petiole well-developed, leaflets 4–)6–14(–34) per pinna, secondary rachises with glandular patches on upper (adaxial) side, not surrounding leaflet pulvini; calyx upper (adaxial) surface glabrous (or if pubescent [*P. praecox*] then ovaries glabrate); petals deep yellow, the upper usually orange-dotted basally; ovaries glabrate; fruits not constricted between the seeds.

3. Pinnae 2, leaflets (4–)6(–8) per pinna; inflorescence axis strigillose (hairs appressed); upper (adaxial) calyx surface glabrous ____ **2. Parkinsonia florida**

3′ Pinnae 2–4, leaflets (6–)10–14(–34) per pinna; inflorescence axis villous (hairs spreading); upper (adaxial) calyx surface pubescent, especially apically ____ **4. Parkinsonia praecox**

1. Parkinsonia aculeata Linnaeus. Sp. Pl. 1:375. 1753. * [from Latin, *aculeus*, thorn] JERUSALEM THORN, MEXICAN PALO VERDE, PALO VERDE; *RETAMA* Naturalized, Cultivated

Parkinsonia thornberi M.E. Jones

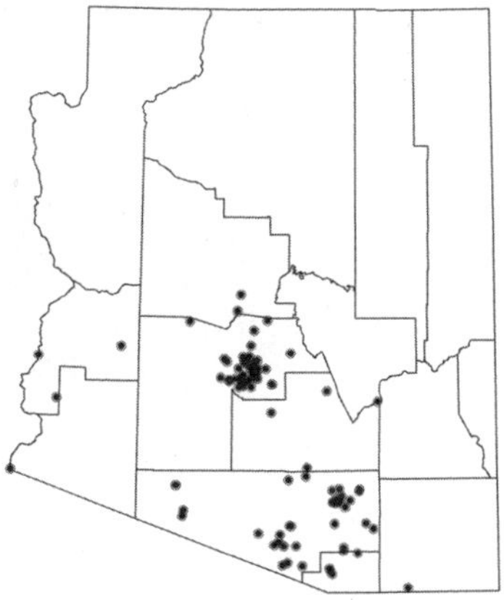

Trees to 10 m, armed. **Stems** with yellowish green bark, becoming gray, rough and shallowly fissured with age; twigs glabrous or sparsely pubescent (at least around nodes), hairs upward appressed, ca. 0.2 mm. **Leaves** bipinnate, appearing once pinnate due to obsolete petiole; stipules spinescent, spines to 2 cm; petiole obsolete; pinnae 2 (or 3) pairs; petiolules 0.2–0.3 mm, sparsely pubescent, hairs to 0.2 mm; secondary rachises (100–)150–600 mm, narrowly winged, glandular patches in leaflet axils (sparse). **Leaflets** 40–76 per pinna, alternate, subopposite, or opposite, blade oblanceolate to oblong, 2–8 × 0.9–3 mm, base attenuate, apex obtuse, apiculate, lower (abaxial) surface sparsely pubescent, upper (adaxial) surface glabrate. **Inflorescences** 2–15-flowered; axis 8–18 cm, strigillose to glabrate; bracteoles quickly deciduous or subpersistent, linear, 1.7–1.8 × 0.1–0.2 mm, apex acute. **Pedicels** 5.4–13.8 mm, glabrous or sparsely pubescent, hairs upward-appressed or directed upward, joints 1.5–3.5 mm from flower, not bearded, nearer flower than pedicel base. **Flowers:** calyx lobes (typically deciduous), 6–7 mm, lower (abaxial) surface sparsely pubescent, upper (adaxial) surface pubescent; corolla 13–20 mm diam., petals light to deep yellow, the upper usually orange-dotted basally, erose-margined, pubescent basally; filaments nearly equal, pubescent basally; ovary sericeous, hairs directed upward. **Fruits** almost round in cross section, constricted between seeds, 2–12 × 0.5–0.8 cm, flat-beaked, glabrous. **Seeds** (1–)2–5. $2n = 28$.

Flowering Mar–Sep; fruiting May–Jan. Low, poorly drained areas, washes, open woods, roadsides, disturbed places, in cultivation; 0–1300 m; Sonoran Desertscrub, Chihuahuan Desertscrub, Semidesert Grassland, and rarely at lower elevations in Madrean Evergreen Woodland; to be expected statewide, but not yet documented for Apache, Greenlee, Mohave, and Navajo cos.; Alabama, California, Florida, Georgia, Louisiana, Mississippi, Nevada, New Mexico, South Carolina, Texas, Utah; Mexico; West Indies; Central America; South America; Asia; Africa; Australia.

The original distribution of *Parkinsonia aculeata* is likely Central America and southern Mexico, but the species is now extensively naturalized in the tropics and subtropics following escape from cultivation (Felger et al. 2001).

Hybrids between *Parkinsonia aculeata* and *P. microphylla* are known from Arizona. Like the parent species, hybrids between *P. aculeata* and *P. microphylla* generally exhibit pubescent upper (adaxial) calyx lobes and sericeous ovaries. However, individuals of *P. aculeata* × *P. microphylla* can be distinguished by the combination of secondary rachises of intermediate length (usually 2–6 times as long as in *P. microphylla*) and glandular patches completely or nearly completely surrounding the pulvini.

A noteworthy, unarmed, cutting-propagated cultivar derived from a cross between *Parkinsonia aculeata* × *P. microphylla* and *P. florida* is *Parkinsonia* 'Desert Museum' (Dimmitt 1987). This cultivar usually exhibits four pinnae per leaf, as well as well-developed petioles and rachises. *Parkinsonia aculeata* × *P. microphylla*, in contrast, usually exhibits two pinnae per leaf, although, due to the extreme reduction of the petiole, it may appear as one pinna emerging from the twig. Glandular patches are limited to one side of the leaflet pulvini in *P.* "Desert Museum.". Both *P.* "Desert Museum" and *P. aculeata* × *P. microphylla* exhibit inflorescences with more or less upward-appressed hairs.

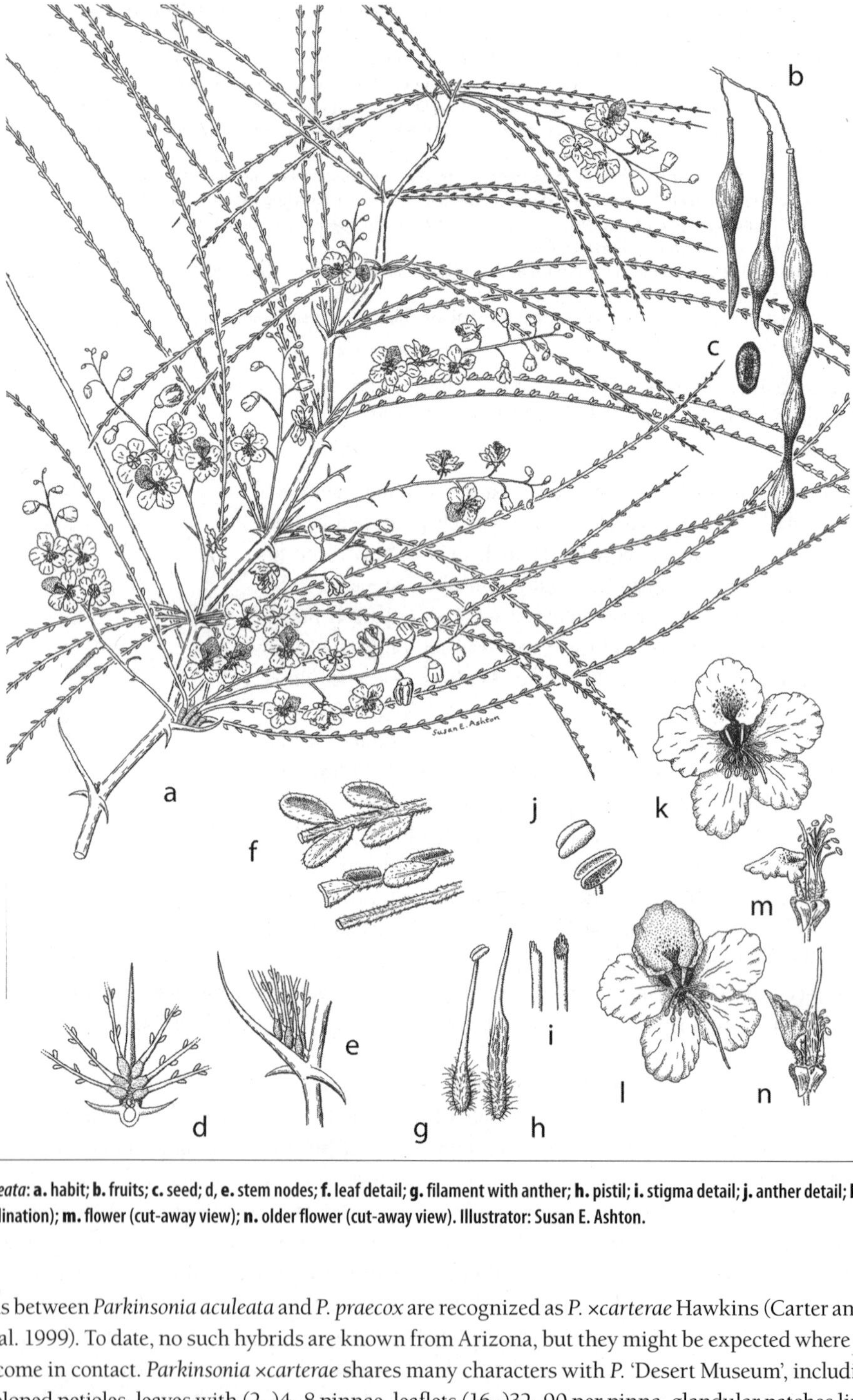

***Parkinsonia aculeata*: a.** habit; **b.** fruits; **c.** seed; d, **e.** stem nodes; **f.** leaf detail; **g.** filament with anther; **h.** pistil; **i.** stigma detail; **j.** anther detail; **k.** flower; **l.** older flower (post-pollination); **m.** flower (cut-away view); **n.** older flower (cut-away view). Illustrator: Susan E. Ashton.

Hybrids between *Parkinsonia aculeata* and *P. praecox* are recognized as *P.* ×*carterae* Hawkins (Carter and Rem 1974, Hawkins et al. 1999). To date, no such hybrids are known from Arizona, but they might be expected where the two parent species come in contact. *Parkinsonia* ×*carterae* shares many characters with *P.* 'Desert Museum', including reduced, though developed petioles, leaves with (2–)4–8 pinnae, leaflets (16–)32–90 per pinna, glandular patches limited to one side of the leaflet pulvini, trichomes of the inflorescence axes and pedicels upward-appressed or directed upward, upper (adaxial) calyx lobes pubescent, ovaries very sparsely pubescent to glabrous, and fruits flattened, not or only slightly constricted between seeds (see also Hawkins et al. 1999). Under ultra-violet (UV) light, all petals of *P.* ×*carterae* appear strongly absorptive, whereas in *P.* 'Desert Museum' only the upper petal is absorptive (dark) and the remaining petals reflective (light). However, these observations are based on only three specimens (*P.* ×*carterae: Hinton 9968,* 25 Mar 1937, ARIZ; *P.* 'Desert Museum': *Mason 05-31,* 21 Apr 2005, *Johnson s.n.*, ARIZ). The extent of natural variation in UV reflectance is not known.

Parkinsonia aculeata is widely naturalized in urban areas in desert regions of Arizona. It commonly forms thickets along arroyos, vacant lots, alleys, and other disturbed sites. It is also found in rural areas and extends above the desert into grassland and lower oak woodland, occasionally up to 1370 m elevation. The species is widely naturalized in tropical and subtropical regions around the world and occurs in many diverse habitats including thornscrub, grassland, and forests. Along the Texas Gulf coast, it is found in periodically waterlogged soils and on shell reef islands. The trees are tolerant of high salinity. Rainfall in many areas where *P. aculeata* has become naturalized ranges from 200–1000 mm per year (NRC 1980).

The wood of *Parkinsonia aculeata* is hard and brittle with a specific gravity of 0.6 and is used in some countries as fuel and to make charcoal (NRC 1980). Standley (1922) mentioned the use of this species in papermaking. Young branches and fruits are sometimes used for livestock fodder (Standley 1922, Vines 1960, NRC 1980). The twigs were utilized by the Seri to make a tea for relief from stomach pain (Felger and Moser 1985). Felger et al. (2001) stated that the Guarijío people of Sonora use *P. aculeata* to make a tea to treat cough and flu. In the American tropics infusions of the leaves have been employed in folk medicine to treat fever, diabetes, epilepsy, to produce sweating, and to induce abortion (Standley 1922). Vines (1960) reported that the seeds were ground into flour by native peoples. Lewis et al. (2005) stated that the flowers of *P. aculeata* are eaten in Senegal and that the species has medicinal uses. NRC (1980) suggested that this species can be planted to combat erosion.

Mexican palo verde is occasionally planted in Arizona landscapes but has fallen out of favor due to its weedy tendencies, abundant leaf litter, and the availability of more attractive alternatives. The canopy has a graceful appearance due to the foliage and provides delicately filtered light. Healthy trees are showy when in flower. With ample moisture, Mexican palo verde can grow up to 2 m per year. Cultivated plants benefit from occasional irrigation to maintain an acceptable appearance but are quite drought tolerant (Jones and Sacamano 2000). Plants are deciduous with extended drought or freezing temperatures. Mexican palo verde is hardy to about -12°C. The leaf rachises produce considerable litter that is especially difficult to remove from cacti, agaves, and other spiny plants. The wood of Mexican palo verde is relatively weak and limbs may break in high winds. The trees should be sited away from traffic areas due to the spiny stems. This species is short-lived, seldom living more than 25 or 30 years. The trees are especially susceptible to the palo verde root borer (*Derobrachus geminatus*). The larvae of this large beetle feed on the roots and often kill the trees. Other species of *Parkinsonia* are less commonly affected. Mexican palo verde readily volunteers from seeds in irrigated landscapes and other sites where extra moisture is available. This species is easily grown from scarified seeds. The hybrid *Parkinsonia* 'Desert Museum' is a superior plant that is widely planted in landscapes, parking lots, and medians in Arizona and elsewhere in the Southwest. It is fast growing, unarmed, has a long flowering period, and is not weedy. It produces considerable litter of fallen flowers but leaf litter is much less than for *P. aculeata*. *Parkinsonia* 'Desert Museum' is propagated by cuttings. There are other hybrid selections of *P. aculeata* with specific desirable horticultural characteristics also available in the nursery trade.

2. Parkinsonia florida (Bentham ex A. Gray) S. Watson. Proc. Amer. Acad. Arts 11:135. 1876. * [from Latin, *florid*, flowery] BLUE PALO VERDE Native, Cultivated

Cercidium floridum Bentham ex A. Gray, Smithsonian Contr. Knowl. 3(5):58. 1852; *Cercidium torreyanum* (S. Watson) Sargent; *Parkinsonia torreyana* S. Watson

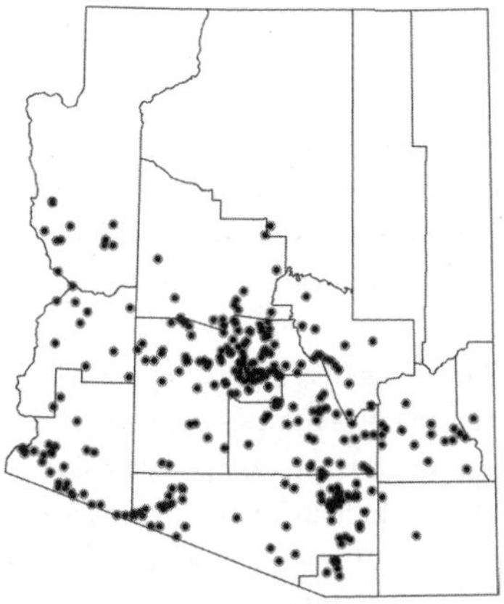

Trees 2.5–8(–12) m, armed. **Stems** with pallid green bark, becoming gray and scaly with age; twigs moderately pubescent, hairs spreading or upward-appressed, ca. 0.2 mm; armed with nodal spines. **Leaves** bipinnate; stipules not spinescent, linear-lanceolate, 0.6 mm; petiole 1.6–3.5 mm, pubescent, hairs 0.2–0.3 mm; pinnae 2; petiolules 0.2 mm, densely pubescent, hairs to 0.2 mm; secondary rachises 3–7 mm, not winged, glandular patches on upper (adaxial) side, not surrounding leaflet pulvini. **Leaflets** (4–)6(–8) per pinna, opposite, blade elliptic to obovate, 3–7 × 1.6–3 mm, base cuneate to rounded, apex truncate to rounded, not apiculate, lower (abaxial) surface sparsely pubescent, upper (adaxial) surface sparsely appressed-pubescent to glabrate. **Inflorescences** 2–10-flowered; axis 1.1–3.5 cm, strigillose to glabrate; bracteoles quickly deciduous. **Pedicels** 10–14 mm, sparsely pubescent or glabrate, hairs upward-appressed or directed upward, joints 1–3 mm from flower, bearded, rarely glabrous, typically nearer flower than pedicel base, rarely nearer pedicel base than flower. **Flowers:** calyx lobes 5–7 mm, lower (abaxial) surface sparsely pubescent, upper (adaxial) surface glabrous; corolla 8–16 mm diam., petals deep yellow, the upper usually orange-dotted basally, erose-margined, pubescent basally; filaments distinctly unequal, pubescent basally; ovary glabrate. **Fruits** flattened, not constricted between seeds, 3.8–12 × 1–1.3 cm, flat-beaked, glabrous in age. **Seeds** 1–2(–4). *2n* = 28.

Flowering Mar–Jun(Aug–Nov); fruiting Apr–Jun(Sep–Dec). Fine soils along washes and flood plains; [-30–]30–1000(–1240) m; Sonoran Desertscrub and occasional at lower elevations in Semidesert Grassland; statewide, except for Apache, Coconino, and Navajo cos.; California, Nevada; Mexico (Sonora, nw Sinaloa).

Parkinsonia florida is currently listed as a Salvaged Assessed Protected Native Plant by the Arizona Department of Agriculture under its basionym. It is the only species of *Parkinsonia* in Arizona that exhibits glabrous upper (adaxial) calyx lobes. *Parkinsonia florida* may hybridize with *P. microphylla* (see discussion under latter species). A closely related subspecies, *Parkinsonia florida* subsp. *peninsulare*, is restricted to Baja California Sur.

Parkinsonia florida is a distinctive component of xeroriparian woodlands along arroyos and floodplains in the Sonoran Desert. It extends into semidesert grassland and is also found in thornscrub in Sonora. It is seldom found in dry upland sites within the Arizona portion of its distribution except near its upper elevational limits, suggesting a requirement for more moisture than *P. microphylla*. It is known from only a single specimen in Nevada. Annual rainfall across the range of *P. florida* varies from 75–500 mm. *Parkinsonia florida* produces masses of bright yellow flowers in the spring. Its flowering period usually begins two or three weeks before that of *P. microphylla*, so there is little overlap in flowering. Occasional flowering occurs in the autumn in response to rainfall. The flowers are visited by several species of bees. *Parkinsonia florida and P. microphylla*, together, are listed as the official state tree of Arizona.

The wood of *Parkinsonia florida* has a density of 500 kg/m^3 and is soft and close-grained (Allen and Allen 1981). It makes low quality fuelwood. The Seri use the wood of this species and *P. microphylla* to make pry bars for harvesting agaves (Felger and Moser 1985). The green seeds are edible, with a flavor similar to green peas (Felger et al. 2001). The Seri people of Sonora ate the flowers, seeds, and green fruits (Felger and Moser 1985). The ripe seeds were toasted and ground to make flour and the seeds were also strung as necklaces (Felger and Moser 1985). Seeds of *P. florida* were used for food by several indigenous groups in Arizona (Castetter and Bell 1951).

Blue palo verde is widely planted in hot desert regions in the Southwest U.S. It is used for shade, as a screen, and is valued for its masses of showy yellow flowers. The trees are leafless much of the year, but the stems provide green color year-round (Jones and Sacamano 2000). Plants will grow 0.8 m per year with adequate irrigation. Established plants are drought resistant, but growth is slow in landscape settings without occasional deep irrigation. Jones and Sacamano (2000) stated that this species is more tolerant of growing in lawns than most other palo verdes. Blue palo verde is hardy to -12°C. The dense canopy offers substantial wind resistance and the weak stems are prone to break during high winds. The canopy of blue palo verde is naturally low-branched. Regular pruning is required to allow headroom beneath the canopy. In cultivation, mite infestations can cause an unsightly "witches broom" growth (Jones and Sacamano 2000). The fruits are produced in abundance and can be a maintenance issue. The small leaflets produce a fine mulch layer beneath the canopy. Blue palo verde is easily propagated from scarified seeds.

3. Parkinsonia microphylla Torrey in War Department [U.S.], Pacif. Railr. Rep. 4(5):82. 1857. * [from Greek, *micro*, small, and *phyll*, a leaf] FOOTHILL PALO VERDE, LITTLELEAF PALO VERDE, YELLOW PALO VERDE Native, Cultivated

Cercidiopsis microphylla (Torrey) Britton & Rose; *Cercidium microphyllum* (Torrey) Rose & I. M. Johnston

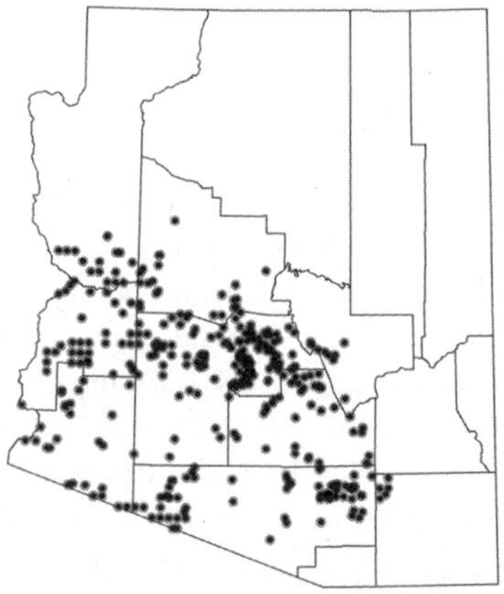

Shrubs or small trees 3–4.6 m, armed. **Stems** with yellowish green bark having a waxy appearance; twigs moderately pubescent, hairs spreading or upward appressed, ca. 0.2 mm; nodal spines absent, but branch tips frequently spinescent. **Leaves** bipinnate but appearing once pinnate due to obsolete petiole; stipules not spinescent, linear-lanceolate, 1.5 mm; petiole obsolete; pinnae 2 (or 3); petiolules 0.2 mm, sparsely pubescent, hairs to 0.12 mm; secondary rachises 10–40 mm, not winged, glandular patches surrounding leaflet pulvini or, at least, on opposing sides. **Leaflets** (6–)12–16 per pinna, opposite, blade elliptic to nearly circular, 2–3 × 0.9–2.9 mm, base rounded, apex obtuse to rounded, not apiculate, lower (abaxial) and upper (adaxial) surfaces sparsely pubescent. **Inflorescences** 2–15-flowered; axis 0.6–3.8 cm, strigillose; brac-

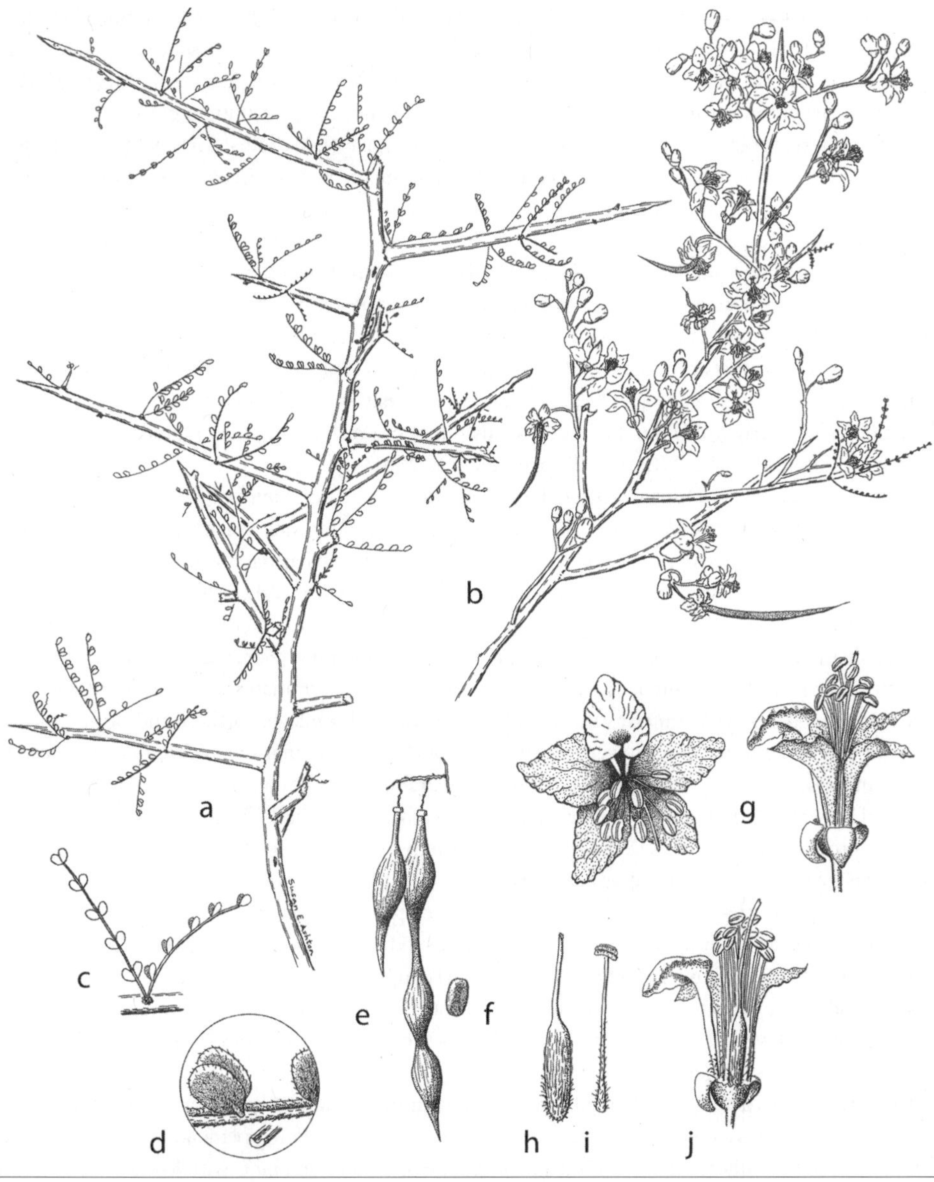

***Parkinsonia microphylla*: a.** habit; **b.** stem with flowers and developing fruits; **c.** leaf; **d.** leaf detail; **e.** fruits; **f.** seed; **g.** flower (front and side views); **h.** pistil; **i.** anther with filament; **j.** flower cross-section. Illustrator: Susan E. Ashton.

teoles persistent, lanceolate, 1.3–1.4 × 0.3–0.35 mm, apex acute. **Pedicels** 4.6–12.1 mm, pubescent, hairs upward-appressed or directed upward, joints 1–3.6 mm from flower, bearded, distinctly nearer flower than pedicel base. **Flowers:** calyx lobes 6–7 mm, lower (abaxial) and upper (adaxial) surfaces pubescent; corolla 8–12 mm diam., petals light yellow, except the upper one white or cream-yellow, not orange-dotted, erose-margined, pubescent basally; filaments distinctly unequal, pubescent basally; ovary sericeous, hairs directed upward. **Fruits** almost round in cross section, constricted between seeds, 3–12 × 0.8–1 cm, flat-beaked, pubescent at least basally in age. **Seeds** 1–4. $2n = 28$.

Flowering Mar–May; fruiting Apr–Jun. Coarse soils of plains and hill slopes; 50–1100(–1220) m; Sonoran Desertscrub; nearly statewide, except for Apache, Coconino, Graham, Greenlee, and Navajo cos. (expected for Santa Cruz Co.); California; Mexico (Baja California, Baja California Sur, Sonora).

Parkinsonia microphylla is currently listed as a Salvaged Assessed Protected Native Plant by the Arizona Department of Agriculture under its synonym in *Cercidium*.

Hybrids between *Parkinsonia microphylla* and *P. florida* can usually be recognized by intermediacy in characters (e.g., leaflets too large for *P. microphylla*, but too numerous for *P. florida*), as well as presence of characters from both parents (e.g., nodal spines, petiolate leaves, glandular patches surrounding pulvini, pubescent ovaries) (Carter 1974a).

Hybrids between *Parkinsonia microphylla* and *P. praecox* occur where the distributions of the two species overlap in Mexico and have been recognized as *P.* ×*sonorae* (Rose & I. M. Johnston ex I. M. Johnston) Hawkins and Felger. A few of these hybrids have been documented from Arizona occurring spontaneously in areas adjacent to where *P. praecox* is cultivated. Individuals of *P.* ×*sonorae* generally exhibit inflorescence axes and pedicels with the spreading pubescence of *P. praecox* and the pulvini-surrounding glandular patches of *P. microphylla* (this latter character state appears to be present in most hybrids involving *P. microphylla* and a single other species in North America). Fruits of *P.* ×*sonorae* are flattened and only slightly constricted. In contrast to *P. microphylla*, petioles in *P.* ×*sonorae*, although reduced, are clearly distinct under magnification and branches are not spinescent at the tip. In addition, leaves of *P.* ×*sonorae* usually comprise two pinnae and leaflets that are about twice the length of those of *P. microphylla*. The hybrid can be further distinguished from *P. praecox* by the bark on older branches inconspicuously longitudinally striate (vs. quadrate-pustulate in *P. praecox*), petioles 0–3 (rarely to 12) mm long (vs. usually 6 mm or greater in *P. praecox*), and the banner petals often white or pale (vs. yellow in *P. praecox*) (Carter 1974b, Felger et al. 2001).

See the discussion under *Parkinsonia aculeata* for hybrids between it and *P. microphylla*.

Parkinsonia microphylla is endemic to the Sonoran Desert. It is one of the most conspicuous and abundant plants in many upland sites in the eastern Sonoran Desert. It is especially abundant on bajadas and lower mountain slopes but can also be common in valley locations. *Parkinsonia microphylla* grows to near sea level in Mexico. The trees form low, open "woodlands" with other woody legumes, cacti, and shrubs to form distinctive and diverse plant communities. In dry habitats, especially in the western Sonoran Desert, *P. microphylla* is sometimes restricted to drainageways. This species is an important nurse plant for columnar cacti. *Parkinsonia microphylla* germinates with summer rains. The western boundary of its range appears to be limited by the lack of summer rainfall, while its northern and altitudinal distribution is likely determined by winter freezing. Flowering begins two or three weeks after that of *P. florida*. The flowers are produced in dense masses, giving the canopies a pale yellow appearance. Where roots of *P. microphylla* are exposed along arroyos and road cuts, they commonly sprout new plants from adventitious buds (Palzkill 1998). This suggests that *P. microphylla* may produce clonal groves (Palzkill 1998). Research by Szarek and Woodhouse (1978) showed that the green bark of *P. microphylla* accounts for approximately 74% of total photosynthesis.

Carter (1974a) reported that the branches of *Parkinsonia microphylla* were cut for livestock fodder and that the seeds are edible. The seeds were an important food resource for various indigenous groups in Arizona (Castetter and Underhill 1935, Castetter and Bell 1942, Castetter and Bell 1951). The Seri people of Sonora eat the flowers, fresh or boiled, and the raw green seeds and green fruits, which are cooked with meat (Felger and Moser 1985). Ripe seeds were toasted and ground to make flour and were also strung to make necklaces (Felger and Moser 1985). The residue from boiled flour was used to waterproof ollas (Felger and Moser 1985).

Due to its slower growth rate, foothill palo verde is less commonly planted in Arizona landscapes than blue palo verde or palo brea. It provides a distinctive appearance, makes an ideal screening plant and is very attractive when in flower. The canopy provides filtered light that is ideal for succulents and other plants with low water requirements that will not tolerate the full intensity of the desert sun. The flowers are produced in masses and are of a soft yellow hue. Foothill palo verde thrives in rocky and caliche soils that are high in clay, provided that it is not subject to waterlogging. Growth rate with irrigation is typically 0.5 m or less per year. The species will survive on rainfall across its range, but plants benefit from infrequent, deep irrigation during severe drought to prevent the possible loss of limbs. Foothill palo verde is hardy to about -10°C. The very small leaves produce fine litter that is generally not a problem. The fruits can produce some litter that may be undesirable in manicured landscapes. Foothill palo verde is naturally low-branched with several trunks arising from near the base. The canopy may be lifted to allow the form of the limbs to be appreciated and to allow headroom beneath the canopy. Plants should be sited away from traffic areas and care should be taken when working around the trees to avoid eye injuries from the sharp-tipped twigs. This and other *Parkinsonia* species are sometimes boxed and moved with a high survival rate; however, the plants are intolerant of being bare-rooted. Plants are easily grown from scarified seeds.

4. Parkinsonia praecox (Ruiz & Pavón ex Hooker) Hawkins. Pl. Syst. Evol. 216:63. 1999. * [from Latin, *praecox*, early, premature] PALO BREA Cultivated, Adventive

Caesalpinia praecox Ruiz & Pavón ex Hooker, Bot. Misc. 3:208. 1833; *Cercidium praecox* (Ruiz & Pavón ex Hooker) Harms

Shrubs or trees 1–9 m, armed. **Stems** dark to olive green, bark of older stems and trunks pale pea green; twigs glabrous to densely pubescent, hairs upward appressed, 0.2 mm; armed with nodal spines. **Leaves** bipinnate; stipules not spine-tipped, lanceolate to elongate deltate, 1.9–2.6 mm; petiole 6–8.5 mm, pubescent, hairs to 0.2 mm; pinnae 2–4; petiolules 3.9–5.2 mm, densely pubescent, hairs to 0.3 mm; secondary rachises 10–25 mm, not winged, glandular patches on upper (adaxial) side, not surrounding leaflet pulvini. **Leaflets** (6–)10–14(–34) per pinna, opposite, blade oblanceolate to oblong, 3.5–8 × 1.9–3.5 mm, base cuneate to rounded, apex truncate or rounded, apiculate or not, lower (abaxial) surface sparsely to moderately pubescent, upper (adaxial) surface sparsely pubescent to glabrate. **Inflorescences** 1–11-flowered; axis 1.1–2.5 cm, moderately to densely villous; bracteoles persistent or quickly deciduous, deltate or oblong, 0.4–1.6 × 0.3–0.5 mm, apex obtuse. **Pedicels** 8–11.5 mm, moderately to densely pubescent, hairs spreading, joints 2.6–3.2 mm from flower, bearded, equidistant between flower and pedicel base or nearer flower than pedicel base. **Flowers:** calyx lobes 4.9–6.5 mm, lower (abaxial) surface moderately pubescent, upper (adaxial) surface sparsely to densely pubescent, especially at the apex; corolla 7.5–15.9 mm diam., petals deep yellow, the upper usually orange-dotted basally, erose-margined, pubescent basally; filaments distinctly unequal, pubescent basally; ovary glabrate. **Fruit** flattened, not constricted between seeds, 2.6–5.6 × 0.6–0.8 cm, glabrous. **Seeds** 1–5. $2n = 28$

Flowering Feb–Jun(–Sep); fruiting Mar–Jul(–Sep). Adventive along roadways and adjacent to landscapes; Sonoran Desertscrub; Maricopa, Pima, Pinal Co.; Mexico; West Indies; South America.

Parkinsonia praecox is known from cultivation in Arizona though the species is now spreading from cultivated plants in the state. The species can be distinguished from all other species of *Parkinsonia* in the flora area by the spreading pubescence of the inflorescence axis and pedicels (appressed or distinctly directed upward in other species). Hybrids between *P. praecox* and *P. aculeata* have been treated as *P.* ×*carterae* Hawkins (see discussion under *P. aculeata*).

Parkinsonia praecox has the most extensive native distribution of any species in the genus. In Mexico, it has been found in northern Sonora within 60 km of the Arizona border and extends southward discontinuously to northern Argentina. *Parkinsonia praecox* occurs on slopes, plains, and along arroyos in desertscrub, thornscrub, tropical deciduous forest, savannah, and occasionally in grassland. In Sonora, it is found from near sea level to 1065 m elevation and in South America it occurs up to 2000 m. Annual rainfall within it range in northwestern Mexico is 200–625 mm. *Parkinsonia praecox* appears to be extending its range in northern Sonora and thrives on overgrazed rangeland (Felger et al. 2001). It is not yet considered naturalized in Arizona; however, it has been observed to reseed in and adjacent to landscaped areas in central and southern Arizona. This species has been observed in urban areas as well as along highways in rural areas remote from cultivated plants. Many of these plants are of reproductive size. A second form, *P. praecox* subsp. *glauca*, occurs in northwestern and west central Argentina. It grows in valleys in the Monte Desert that receive as little as 80–100 mm of rain per year.

Allen and Allen (1981) reported that the hard, brittle wood of *Parkinsonia praecox* is easy to work and finishes smoothly. It is sometimes used as fuelwood. Allen and Allen (1981) report it burns quickly and gives off an unpleasant odor. Stems can exude a yellowish transparent gum that has been used in making soap (Record and Hess 1943). The gum is sweet, sticky, and edible. It is reported to be used medicinally in treating cattle in Bolivia (Killeen et al. 1993). Carter (1974a) states, "The common name, *palo brea*, is derived from the fact that the waxy substance coating the bark, after being scraped from the branches and melted by heat, is used as a 'gum' for gluing together leather objects and furniture; thus, it is used just like *la verdadera brea*." In Sonora, the Seri make a tea from the bark to aid in the expulsion of the placenta (Felger and Moser 1985). The bark is ground with sea turtle oil and salt to make a lotion to treat bruises and sprains (Felger and Moser 1985).

Palo brea is commonly planted in street medians, parking lots, and landscapes in southern Arizona. It makes an outstanding specimen plant or patio tree where it can be pruned so its bark and horizontal branching structure can be appreciated. The unique pale green bark contrasts with the blue green foliage. The spreading canopy produces dappled shade that can protect tender plants from intense sunlight. Palo brea produces a dense covering of bright yellow flowers over several weeks. With ample irrigation, palo brea will grow up to 1 m per year. Established plants require infrequent deep irrigation to maintain an acceptable appearance. The foliage is deciduous during prolonged drought or following freezing temperatures. Palo brea is generally hardy to -6.5°C. Due to the thorny stems, palo brea should be located away from traffic areas. Some pruning is necessary to establish scaffold branches and raise the canopy. Field grown trees are commonly boxed and transplanted. Propagation is by scarified seeds.

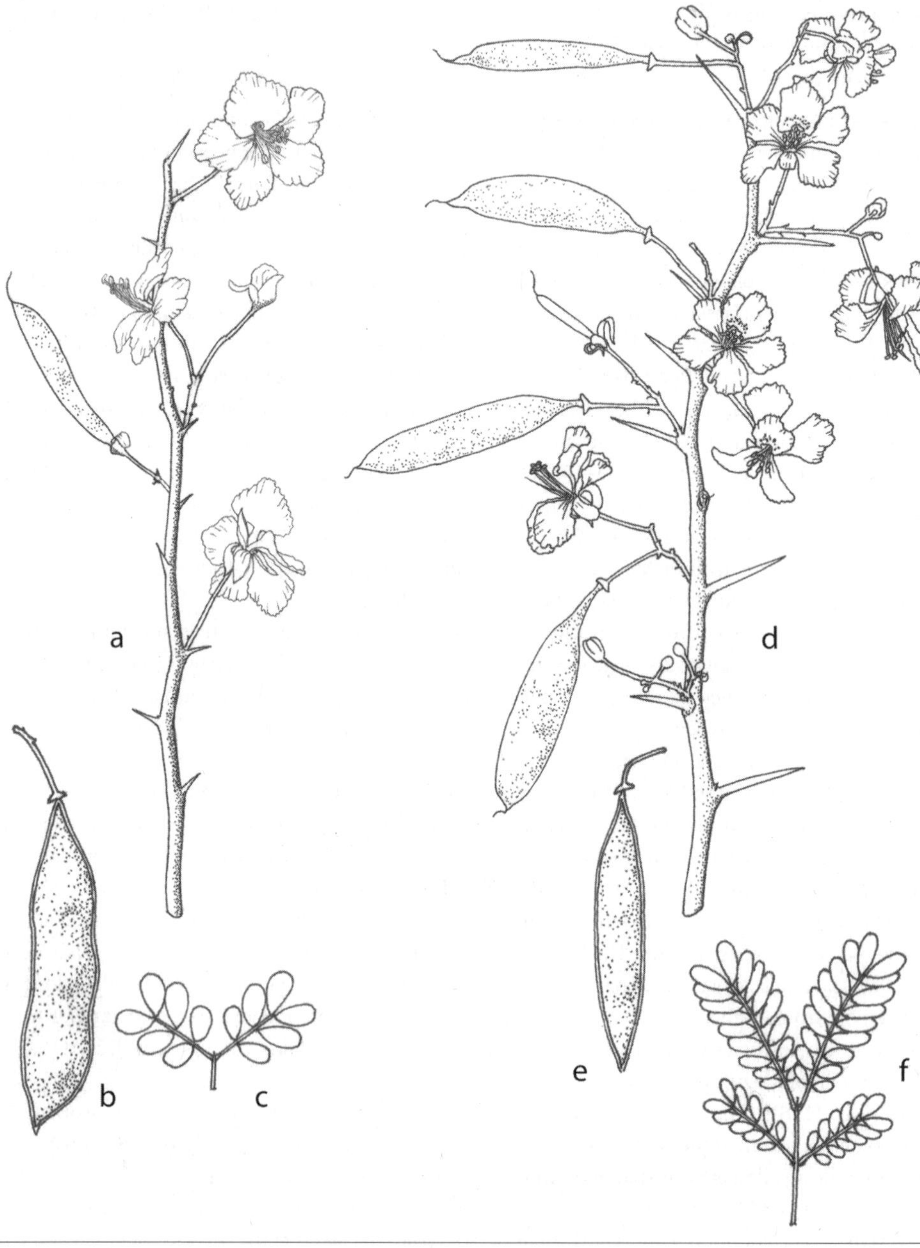

Parkinsonia spp.: **a., b. c.** *P. florida* **a.** flowering branch; **b.** fruit; **c.** leaf; **d., e., f.** *P. praecox* **d.** flowering and fruiting branch; **e.** fruit; **f.** leaf. Illustrator: Matthew B. Johnson.

PARRYELLA Torrey & A. Gray. Proc. Amer. Acad. Arts. 7:397. 1868. * [named for Charles C. Parry, 1823–1890, English-born American botanist] DUNEBROOM

Michelle M. McMahon, Mary Welch-Keesey and Matthew B. Johnson

Shrubs with erect, pliant stems. **Herbage** with lens-like or prickle-like glands. **Leaves** alternate, odd-pinnate, deciduous; stipules awl-shaped, deciduous; petiolate. **Leaflets** numerous, narrow. **Inflorescences** racemes, solitary or in panicles; bracts minute, deciduous; bracteoles absent. **Pedicellate. Flowers** nonpapilionoid; calyx campanulate or vase-shaped, shortly 5-lobed; corolla absent; stamens 10, unequal, filaments free for most of their length, anthers yellow; ovary with 2 ovules; style sinuous, pubescent; stigma small. **Fruits** obliquely narrow-obovoid, slightly compressed, beak persistent, conspicuously glandular-pustulate, indehiscent. **Seeds** 1. $x = 10$.

Species 1 (1 native in the flora): Colorado Plateau and upper Rio Grande Valley in the southwestern U.S.

Parryella is closely related to *Errazurizia rotundata* (McMahon and Hufford 2004). It is also related to *Amorpha* and *Eysenhardtia*, differing in having apetalous flowers, essentially free filaments, and a broom-like growth habit (Barneby 1989).

SELECTED REFERENCES: Barneby 1989, Isely 1998

1. Parryella filifolia Torrey & A. Gray ex A. Gray. Proc. Amer. Acad. Arts. 7:397. 1868. * [with thread-like leaves] DUNEBROOM, COMMON DUNEBROOM Native

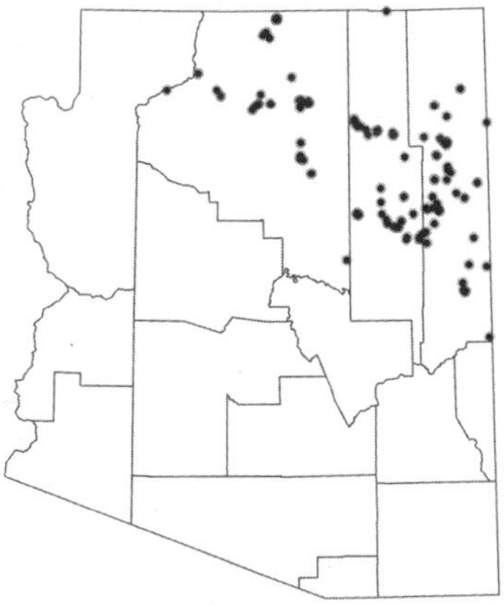

Shrubs to 1.5 m. **Stems** slender, flexible, bark becoming gray. **Leaves** (3–)5–17 cm; sparsely puberulent or glabrate; petioles 4–13 mm. **Leaflets** 17–47, 2–17 (–20) × 0.3–1.5 mm, linear-filiform or oblong-elliptic to oblanceolate or narrowly obovate, revolute to concave. **Inflorescences** subsessile, 10–90-flowered, the axis 4–14 cm. **Pedicels** 0.4–1.2 mm. **Flowers** with calyx 2–3.8 mm, the tube 2–3 mm, the lobes much shorter than the tube, ciliate; stamens 2–3.5 mm. **Fruits** 5–8 × 2.5–3.5 mm, pale with conspicuous red or purple glands. **Seeds** 3.3–5 × 1.5–2.5 mm, elliptic, smooth, glossy, tan or pale brown.

Flowering (Apr–)May–Sep. Sandy, sandy loam, or sometimes clay loam soils of ridges, slopes, valleys, dunes, canyons, arroyos, sandy river beaches and floodplains, sandy pockets on ledges and tallus, and roadsides; (530–) 1100–2040 m; Great Basin Desertscrub, Plains and Great Basin Grassland, (lower elevations in Great Basin Conifer Woodland); Apache, Coconino, Navajo cos.; Colorado, New Mexico, Utah.

Parryella filifolia is a distinctive shrub of northeastern Arizona. The species is typically found on sandy alluvium and on substrates derived from sandstone, but occasionally occurs with basalt or limestone. The many-stemmed shrubs often form hummocks in sandy soils. Records for the species below 1000 m elevation are from along the Colorado River in the Grand Canyon. Reports of *P. filifolia* from Yavapai County (Barneby 1989) have not been verified. The herbage has a strong, citrus-like aroma when crushed. Allen and Allen (1981) report that *P. filifolia* roots have nodule-inducing rhizobia. The Hopi have used the stems of *P. filifolia* in basket making and as brooms, the seeds to treat toothache, and the leaves as an insecticide (Whiting 1939, Allen and Allen 1981, Rhodes et al. 2011). This species is also reported to have been used in Navajo ceremonies as an emetic (Vestal 1952). The plants have potential for erosion control in areas with sandy soil (Allen and Allen 1981, Rhodes et al. 2011). The foliage is not palatable to livestock (Allen and Allen 1981). Dunebroom would make an attractive, drought tolerant landscape plant for areas to which it is adapted, but it does not appear to be grown by the nursery industry in Arizona at present.

PEDIOMELUM Rydberg in Britton, N. Amer. Fl. 24:18. 1919. * [from the Greek *pedio*, plain, and *melo*, apple, in reference to Pomme de Prairie, the French name for the type] INDIAN BREADROOT, SCURFPEA

Ashley N. Egan

Psoralea Linnaeus subg. *Pediomelum* (Rydberg) Ockendon

Herbs, perennial. **Roots** deep, a fusiform or apically-swollen, woody taproot, or more rarely fibrous with pendulant tubers. **Stems** either prostrate-ascending, decumbent, or erect, or much reduced, branching or not, glabrous to densely hirsute. **Leaves** usually palmately (1–)3–7-foliolate, pseudopalmate, or pinnately trifoliolate, with or without glands, margins straight (undulate); stipules deciduous or persistent. **Inflorescences** axillary pseudoracemes, persistent or deciduous; bracts persistent to tardily deciduous. **Pedicellate** to rarely subsessile. **Flowers** papilionoid; calyx campanulate, usually enlarging through fruiting or rarely not enlarging, but flaring backwards and tearing along a lateral sinus [*P. tenuiflorum*], 5-toothed, the lower often enlarged; corolla rarely reddish-brown, sometimes white to ochrolecous, or more commonly purple to blue; stamens 10; pistil 1, style arched to sharply reflexed, 1-ovulate. **Fruit** a beaked pod, persistent on the receptacle. **Seeds** 1, round to elliptic. $x = 11$.

Species 25 (9 native in the flora): xeric to mesic habitats in temperate deserts, rock outcrops, and woodlands of North America, 2 species in Mexico.

Endemism is high in *Pediomelum* with most species having restricted geographical ranges. This is no exception in Arizona. Limited range coupled with habitat degradation from grazing and urbanization has resulted in a number of *Pediomelum* species being listed as rare, threatened, or endangered (NatureServe 2007, Walter and Gillett 1998). The

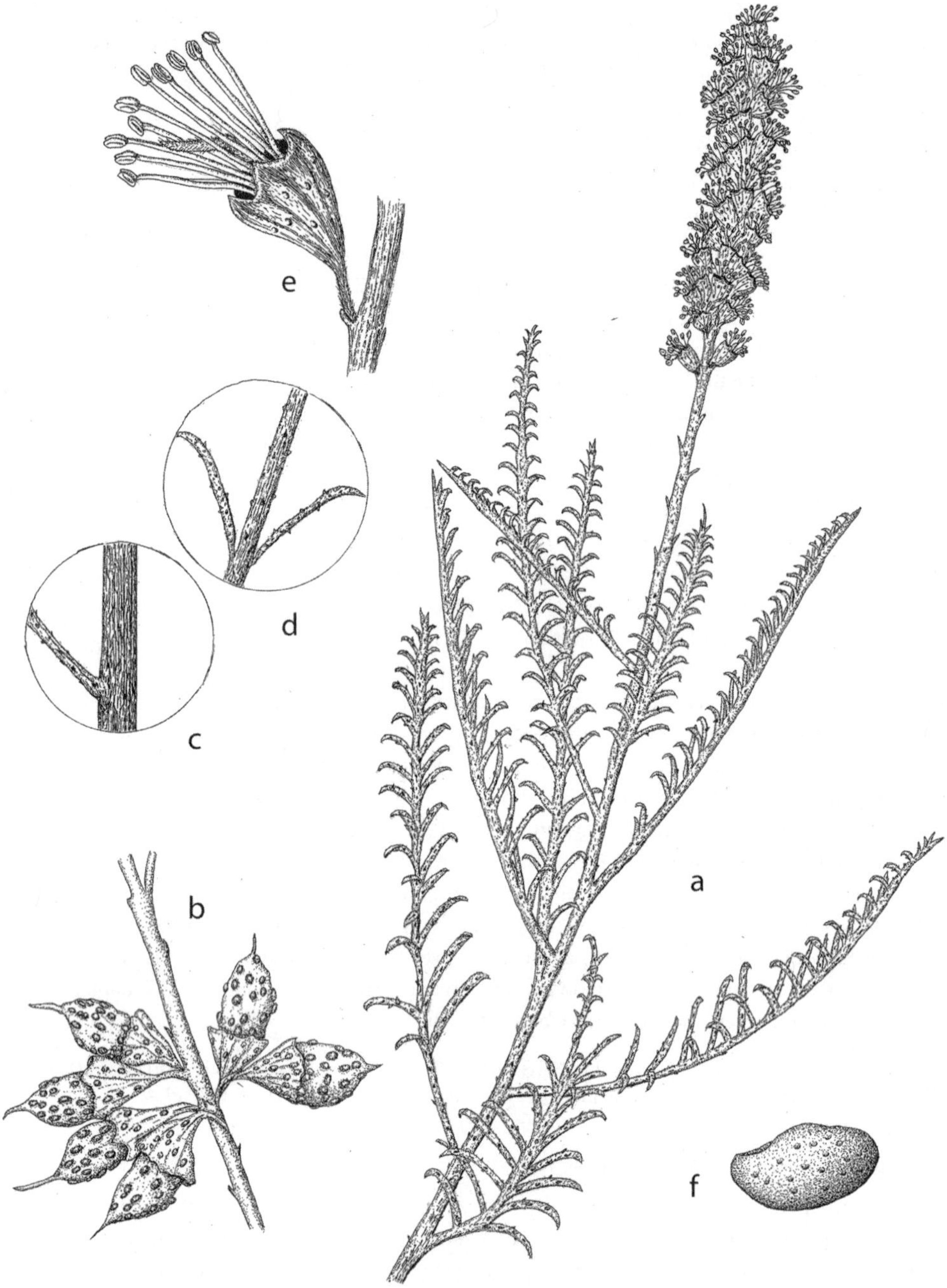

***Parryella filifolia*: a. habit; b. fruits; c. stem node detail with petiole; d. leaflet detail; e. flower; f. seed. Illustrator: Karen Gengle.**

rapid and recent evolutionary diversification of *Pediomelum* may have contributed to the level of endemism within the group (Egan and Crandall 2008a).

Grimes (1990) divided the genus into three subgenera: subg. *Leucocraspedon* to accommodate two prostrate species with salmon, brick-red, or yellowish flowers and a white ridge surrounding the hilum of the seed (*Pediomelum rhombifolium* and *P. palmeri*); subg. *Pediomelum* to accommodate those species that were usually caulescent and had a persistent inflorescence; and subg. *Disarticulatum* whose members were largely acaulescent and whose inflorescence disjointed with age at the base of the peduncle. Molecular phylogenetic studies strongly support subg. *Leucocraspedon*, and somewhat follow membership of the other two subgenera, but not completely. Associations surrounding *P. esculentum*, the type species, and *P. aromaticum* are in particular problematic (Egan and Crandall 2008b).

Several species of *Pediomelum* have reported uses. Perhaps the best-known example is *Pediomelum esculentum* (Pursh) Rydberg, also known as the prairie turnip or Indian breadroot, which was one of the most important sources of food and commerce among Native American tribes of the Great Plains, its root acting as a main vegetable staple. Lewis and Clark, during their expedition to the Pacific Coast, repeatedly documented the use and importance of *Pediomelum* species by various Native American tribes (Lewis et al. 2003). Some have been used medicinally and as a stimulant; and the roots have been utilized for food (Graham 1941, Stubbendieck and Conard 1989). The seeds are eaten by rodents and birds (Graham 1941, Stubbendieck and Conard 1989), but poisoning in humans has been reported (Owensby 1980). Deer and antelope browse the foliage, but the plants are seldom grazed by livestock (Stubbendieck and Conard 1989). Green stems have been used for basketry by Native Americans (Stubbendieck and Conard 1989). Rhizomatous species may have potential for erosion control.

SELECTED REFERENCES: Egan 2014, Egan and Reveal 2009, Gandhi 2015, Grimes 1990, Isely 1998, Kartesz and Gandhi 1992, Ockendon 1965, 1966, Tonne 2000, Welsh and Licher 2010, Welsh et al. 1993.

1. Leaves mostly pseudopalmately 5-foliolate, the terminal petiolule more than twice as long as lateral petiolules; petals ochroleucous throughout or suffused with salmon-pink or reddish-brown, but not purple; seeds with a hilum surrounded by a raised, white ridge ___ **6. Pediomelum palmeri**

1′ Leaves palmately (1–)3–5(–8)-foliolate, the terminal petiolule usually equal in length to lateral petiolules, but if longer, not more than twice as long; petals various shades of blue, violet, or purple, or ochroleucous tinged blue-purple, rarely ochroleucous throughout; seeds without a hilum surrounded by a raised, white ridge.

2. Plants caulescent, often secondarily branched; inflorescence persistent; fruit glandular.

3. Branches mostly arising from pseudoscapes or lowermost nodes; fruit short-strigose; seeds gray-green ___ **1. Pediomelum aromaticum**

3′ Branches mostly arising from middle to upper nodes; fruit glabrous; seeds brown ___ **8. Pediomelum tenuiflorum**

2′ Plants acaulescent or shortly caulescent, if branched, only from basal nodes and subtended by cataphylls; inflorescence disjointing at the base of the peduncle with age; fruit without glands.

4. Calyx tube 6–10 mm.

5. Upper surfaces of leaflets glabrous to pubescent only along base of veins; leaves obviously bicolored, gray-green beneath, yellow-green above ___ **3. Pediomelum epipsilum**

5′ Upper surfaces of leaflets pubescent throughout; leaves mainly greenish to grey-green on both sides, or if bicolored, paler beneath and green above ___ **4. Pediomelum megalanthum**

4′ Calyx tube 2.5–5 mm.

6. Calyx teeth unequal, the lower oblanceolate and often longer and much broader than upper four; seeds rugose.

7. Flowers 9–13 mm; calyx teeth 4–6 mm ___ **2. Pediomelum castoreum**

7′ Flowers (12–)14–18 mm; calyx teeth 7–12 mm ___ **7. Pediomelum pentaphyllum**

6′ Calyx teeth subequal, the lower linear-lanceolate to elliptic; seeds smooth.

8. Bracts (7–)8–12 mm; calyx tube 2.5–4 mm; plants of sw UT, nw AZ (Mohave Co.), se NV ___ **5. Pediomelum mephiticum**

8′ Bracts 4–8(–9) mm; calyx tube (3–)4–5 mm; plants endemic to Mohave and Yavapai cos., AZ ___ **9. Pediomelum verdiense**

1. Pediomelum aromaticum (Payson) W.A. Weber. Phytologia 53:188. 1983. * [from Greek, *aromat*, aromatic, spice, seasoning, in reference to the smell of the plant when crushed] PARADOX BREADROOT Native

Psoralea aromatica Payson, Bot. Gaz. 60(5):379. 1915; *Pediomelum aromaticum* (Payson) W.A. Weber var. *barnebyi* S.L. Welsh; *Pediomelum aromaticum* (Payson) W.A. Weber var. *tuhyi* S.L. Welsh

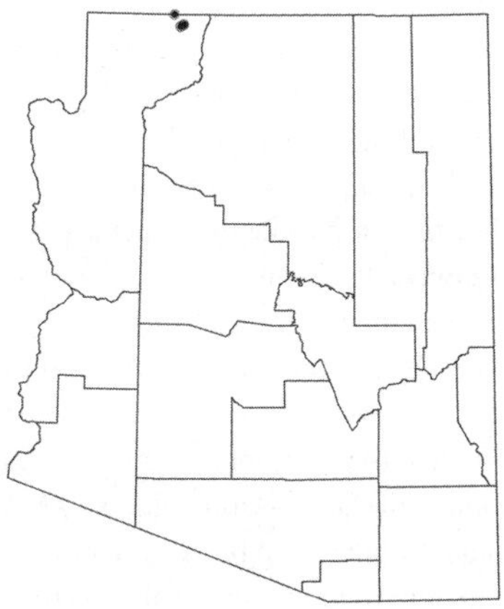

Plants to 20 cm, caulescent, prostrate-ascending, decumbent to erect, spreading by rhizomes and often forming patches, branching, essentially glandular and strigose throughout. **Roots** a branching, rhizome-like caudex atop an apically enlarged, deep, woody root. **Stems** 1–many, 0–15 cm, branched with branches arising mostly from pseudoscapes or lowermost nodes; pseudoscapes 0–few, 0–12 cm; cataphylls 0–7 mm, straw-colored and prominently veined. **Leaves** palmately (3–)5–7-foliolate; petioles 1.2–8 cm, jointed to the stem; stipules triangular, 2–8(–9) × 1–5 mm, fused-free to free, persistent; petiolules 1–3 mm, the terminal petiolule usually equal in length to lateral petiolules, but if longer, not more than twice as long, pubescent. **Leaflets** oblanceolate, ovate, to rhombic, (0.6–)1.2–2.6 × (0.3–)1–2 cm, cuneate basally, rounded to retuse apically, glabrate except along margins or sparingly on veins above, sparsely but uniformly pubescent below. **Inflorescences** persistent, reduced to one- or few-flowered pseudoracemes, 0.5–2 cm, with 1–7 nodes and 3–7 or 10–12 flowers per node; peduncle 0.2–1.3 cm, subtending the leaves; rachis 0–0.8 cm at anthesis; bracts persistent or more rarely tardily deciduous, oblanceolate to short-cupulate, 2–7 × 1–3 mm. **Pedicels** persistent, 1–2.5(–4)

mm at anthesis. **Flowers** (5–)8–11(–13) mm; calyx 5–10(–11) mm, calyx tube 3–4 mm, sparsely pubescent with upright hairs, glandular, upper teeth triangular, 1.5–4 × 1 mm, lower tooth elliptic to oblanceolate, 2.5–7 × 1.5–2 mm, calyx gibbous-campanulate in fruit; petals blue-purple, banner paler in color than others, oblanceolate, 9–11 × 4–5.5 mm with claw 2.5–3 mm, wings 9–11 × 2–3 mm with claw 3.5–4 mm, keel 7–8 × 2 mm with claw 3–4 mm; filaments 6–6.5 mm; anthers ovate, 0.33 mm; ovary glabrous with style pubescent basally. **Fruits** with the body round to ovate, 5–6 × 4 mm, glandular and short-strigose above, beak 5–6 mm, surpassing the calyx teeth. **Seeds** gray-green, 4.5–5 × 3.5–4 mm, oblong-reniform, without a hilum surrounded by a raised, white ridge.

Flowering May–June; fruiting July–Aug. In rocky clay or sandstone soils in barren or open places; [1000–]1500–1670[–2000] m; Great Basin Conifer Woodland; Mohave Co.; s Utah, w Colorado.

In the past, *Pediomelum aromaticum* has had several varieties recognized based on number of flowers per node, flower and peduncle length, and stem robustness. These varieties are largely geographically distinct with *P. aromaticum* var. *barnebyi* comprising those populations from the Canaan Mountain region in Washington County, Utah and adjacent Arizona, an area at the westernmost edge of *P. aromaticum*'s distribution. These plants were said to be more robust than others, but plants from across the range show similar robustness. Var. *tuhyi* is separated from the main *P. aromaticum* distribution on its north and east edge and is said to differ from var. *aromaticum* in having flowers smaller than 7.3 mm and calyces smaller than 5.3 mm coupled with decumbent stems. However, as Grimes (1990) points out, flowers in several collections, including the holotype, have mature flowers longer than 9 mm. Therefore, we side with Grimes (1990) and others and recognize *P. aromaticum* as a single species without subspecies. In Arizona, populations of *P. aromaticum* are restricted to a few localities in northern Mohave County, where they occur in pinyon-juniper woodland.

2. Pediomelum castoreum (S. Watson) Rydberg in Britton, N. Amer. Fl. 24:22. 1919. * [from Greek, *castor*, beaver, for the Beaver Dam Mountains] BEAVER INDIAN BREADROOT Native

Psoralea castorea S. Watson, Amer. Naturalist 12(9):601. 1878

Plants to 16 cm, acaulescent or shortly caulescent, if branched then from lowermost nodes and subtended by cataphylls, glandular on leaves, less so on bracts and stipules, essentially silvery-strigose throughout. **Roots** apically enlarged, deep, woody. **Stems** 0–5 cm (lateral decumbent stems up to 25 cm), unbranched without cluster of leaves or inflorescences at end; pseudoscapes to 10(–14) cm; cataphylls 5–20 mm, glabrous or variously strigose. **Leaves** clustered (alternate), palmately or pseudopalmately (3–)5–6-foliolate; petioles (3–)6.5–15 cm, slightly grooved longitudinally, slightly enlarged and jointed basally; stipules triangular or lanceolate to narrowly oblong, 5–13.5 × 2–5 mm, fused-free below, free above, deciduous or persistent; petiolules 1.5–2.5 mm, the terminal one usually equal in length to lateral petiolules, but if longer, not more than twice as long, sometimes glabrate. **Leaflets** cuneate-obovate to circular, (1.5–)2–4.5 × (1.5–)2.3–4 cm, attenuate basally, broadly acute to rounded or retuse apically, gray-green, pubescent below, yellow-green and sparsely pubescent along veins to pubescent above, not white-veined. **Inflorescences** disjointing with age at the base of the peduncle, ellipsoid to ovoid, 3–10 cm, shorter than subtending petiole, with 3–11 nodes and 2–3-flowers per node; peduncle (1.5–)2.4–7 cm; rachis 1.5–1.6 cm at anthesis, elongating to 2–5.5 cm in fruit; bracts tardily deciduous or persistent, narrowly elliptic or spatulate to oval, 3.5–8 × 3–7 mm. **Pedicels** 0–2 mm at anthesis. **Flowers** 9–13 mm; calyx 9–11 mm to upper tooth, elongating to 14 mm in fruit, 10–12 mm to lower tooth, elongating to 18(–20) mm in fruit, calyx tube 3–4 mm, sometimes setose, usually without glands, upper teeth linear or oblanceolate, 4–5 × 1–1.5 mm, lower tooth broadly oblanceolate and often longer and broader than upper four, 4–6 × 4–4.5 mm, often with 3 prominent veins, calyx gibbous-campanulate in fruit; petals ochroleucous, often purplish-tinged, to purple, banner usually elliptic, 9–13 × 3.5–5 mm with claw 2–4 mm, wings 10–13 × 2 mm with claw 4–4.5 mm, keel 6.5–8 × 2 mm with claw 3–4 mm; filaments 6–7.5 mm; anthers elliptic, 0.35 mm; ovary pubescent apically, style glabrous or occasionally pubescent basally. **Fruits** elliptic to ovoid, body 6–8 × 5–6 mm, strigose, without glands, beak 8–11(–15) mm, equal to or longer than calyx teeth. **Seeds** gray-green to dark brown, 5.5–6 × 3.5–4 mm, plump, narrowly reniform, rugose, without a hilum surrounded by a raised, white ridge.

Flowering Apr–Jun. In sand or sandy soils of valleys, arroyos, and roadsides; [400–]530–730[–1000] m; Mohave Desertscrub; Mohave Co.; California, Nevada.

Pediomelum castoreum is of limited distribution in Arizona. It is found in extreme northwestern Mohave County where it grows in open desertscrub communities.

3. Pediomelum epipsilum (Barneby) S.L. Welsh. Great Basin Naturalist 46:257. 1986. * [from Greek, *epi*, upon, and *psilum*, bare, naked] KANE BREADROOT Native

Psoralea epipsila Barneby, Leafl. W. Bot. 3(9):193. 1943; *Pediomelum megalanthum* (Wooton & Standley) Rydberg var. *epipsilum* (Barneby) J.W. Grimes

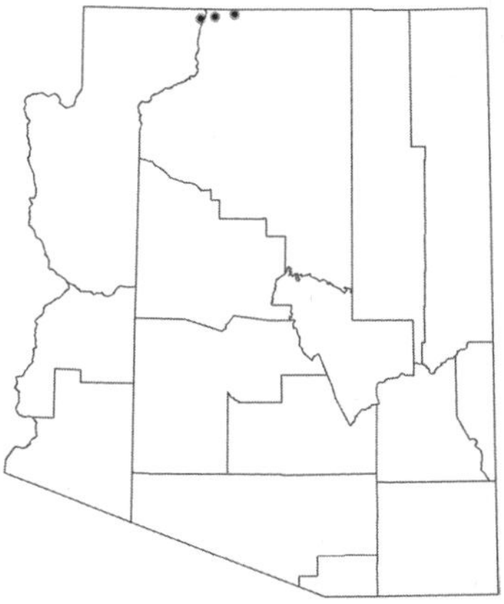

Plants 3–16(–22) cm, clump-forming, acaulescent or shortly caulescent, essentially glandular throughout and pubescent throughout with semi-erect to appressed hairs, except on upper surfaces of leaves. **Roots** apically enlarged, deep, woody. **Stems** 0.5–2 cm (lateral decumbent stems up to 15 (22) cm); pseudoscapes 1–3, up to 4.5 cm; cataphylls 0.5–1.5 mm, conspicuously nerved. **Leaves** palmately 5(–8)-foliolate; petioles (0.8–)5–14 cm, jointed basally; stipules lanceolate to slightly elliptic, 5–11 × 2.5–4.5 mm, fused-free below, free above, conspicuously veined, persistent; petiolules 2–4 mm, the terminal one usually equal in length to lateral petiolules, but if longer, not more than twice as long, pubescent. **Leaflets** ovate, circular, obovate to oblanceolate or slightly rhomboid, (0.8–)1.5–2.5(–4) × (0.5–)1–2.8 cm, cuneate basally, broadly acute to rounded apically, glandular, conspicuously glabrate (to sparingly strigose along veins) and yellow-green above, pubescent and gray-green below. **Inflorescences** disjointing with age at the base of the peduncle, subglobose to elongate, 2–6 cm with 2–10 nodes and (1–)3(–4) flowers per node; peduncle (0.5–)1.4–5 cm, pubescent with erect-ascending hairs; rachis 1.4–4 cm at anthesis; bracts tardily deciduous or persistent, oblanceolate to elliptic and caudate, (10–)13–18(–22) × 6–10 mm. **Pedicels** 3–4 mm at anthesis. **Flowers** (11–)14–19 mm; calyx 11–14(–18) mm, calyx tube 5–6(–9) mm, setose or with appressed hairs, sparingly covered or without minute glands, teeth linear-lanceolate to elliptic, upper teeth 6–9 × 1–1.5 mm, lower tooth 7–10 × 2–3 mm, calyx gibbous-campanulate in fruit; petals purple, banner sometimes paler, oblanceolate, 18–19 × 6–7 mm with claw 6–7 mm, wings 13–18 × 2–3 mm with claw 7–8.5 mm, keel 14–16 × 2–3 mm with claw 8–8.5 mm; filaments 13–16 mm; anthers elliptic, 0.33 mm; ovary pubescent or only apically so, style pubescent basally. **Fruits** ovoid, 6–9 × 4–4.5 mm, without glands, beak 5–8 mm. **Seeds** brown, 4–5 × 3 mm, reniform, without a hilum surrounded by a raised, white ridge.

Flowering May–Jun. In rocky to clay soil of hills, valleys, and ravines; 1600–1700 m; Great Basin Desertscrub, Great Basin Conifer Woodland; Coconino, Mohave cos.; Utah (Kane Co.).

Pediomelum epipsilum has been variously treated at specific and subspecific rank (subsp. of *P. megalanthum*). However, careful comparison of *P. epipsilum* with other *P. megalanthum* varieties reveals sufficient distinguishing characters that delineate *P. epipsilum*, including having leaflets smaller and glabrate above or sparingly strigose along veins (vs. leaflets larger and pubescent above and below in vars. *megalanthum* and *retrorsum*), and bracts larger and caudate (vs. smaller, acuminate, acute, or shortly caudate in the others). Indeed, Grimes' (1990) quantitative character comparison shows non-overlapping ranges for leaflet and bract size, separating *P. epipsilum* from the others. Barneby (1943), in his description of the species as *Psoralea epipsila* Barneby, states that it differs from *Ps. mephitica* by its caulescent nature and conspicuous bicolored leaves and that it is intermediate between *Ps. mephitica* and its variety *Ps. mephitica* var. *retrorsa* (Rydberg) Kearney & Peebles in flower size. He also states that the banner is barely exserted from the calyx, whereas the others are well exserted. This preponderance of both quantitative and qualitative differences, coupled with the fact that *P. epipsilum* is set apart in phylogenetic analyses (Egan and Crandall 2008b), supports recognition of *P. epipsilum* at the species level, as others before have also done (Barneby 1943, Welsh et al. 1993).

Pediomelum epipsilum occurs in pine or juniper woodlands and desert shrub communities. It is otherwise known only from Kane County, Utah.

4. Pediomelum megalanthum (Wooten & Standley) Rydberg in Britton, N. Amer. Fl. 24:22. 1919. * [from Greek, *mega*, large, and *lanthum*, unseen, concealed]

Psoralea megalantha Wooton & Standley, Contr. U.S. Natl. Herb. 16(4):140. 1913

Plants 4–25 cm, mostly acaulescent or shortly caulescent, pubescent, glandular on leaves, sparingly so or without glands on calyx teeth, bracts and stipules. **Roots** a fusiform or apically-swollen woody taproot. **Stems** 1 to few, all short, 2–10(–15) cm, and totally hidden by stipules, erect or often some lateral ones decumbent, with a dense cluster of leaves

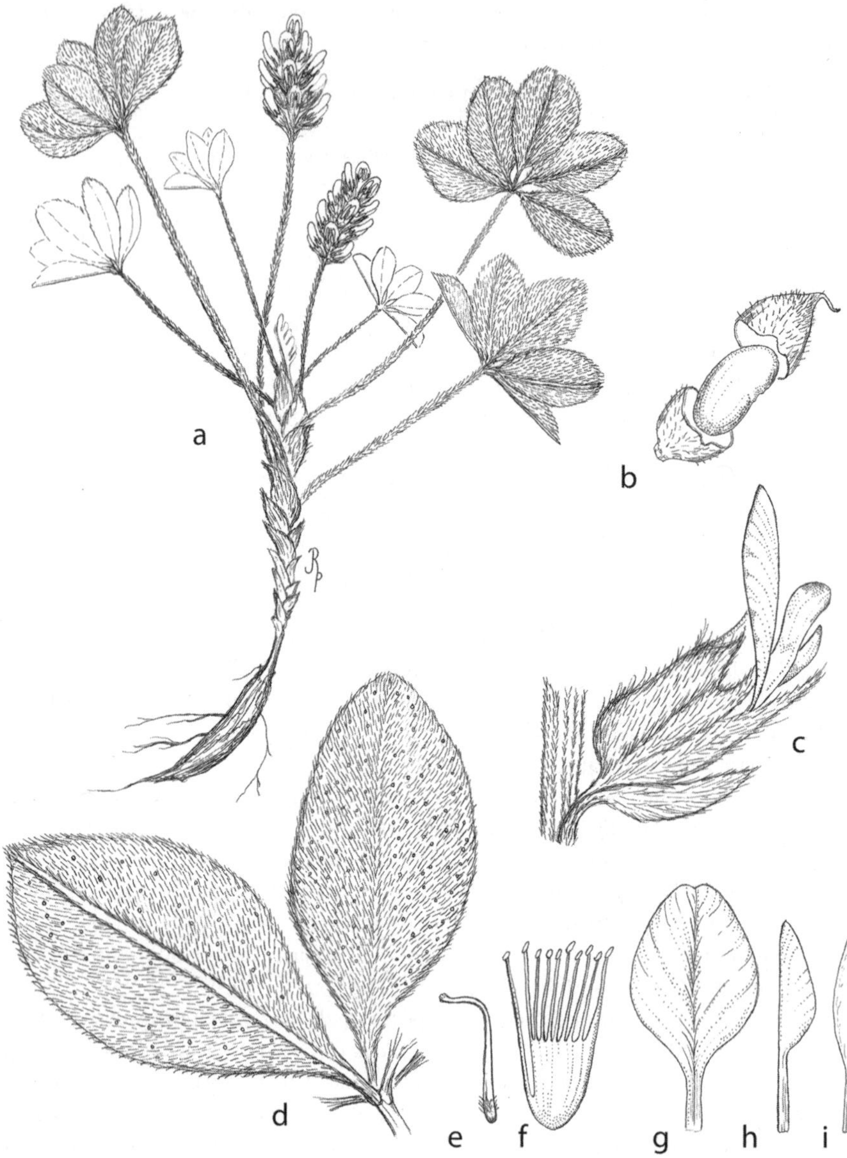

***Pediomelum megalanthum*: a. habit; b. open fruit with seeds; c. flower (side view); d. leaflets; e. pistil; f. stamens; g., h., i. petals. Illustrator: Joyce Riney Peters.**

and inflorescences at the end; pseudoscapes 1–few, to 4.5 cm; cataphylls brown to straw-colored, prominently nerved, to 15 mm. **Leaves** palmately 5(–8)-foliolate, those 6–8-foliolate usually subtend the first inflorescences, glandular above, less so below; petioles 4–15 cm, glabrate to sparsely pubescent, jointed to the leaf spur; stipules lanceolate to elliptic, (2–)5–15 × 2.5–4 mm, lower ones fused-free, becoming free above, veined, persistent; petiolules 1–3(–4) mm, the terminal one usually equal in length to lateral petiolules, but if longer, not more than twice as long. **Leaflets** broadly ovate, circular, oblanceolate or somewhat rhombic, 1–4 × 0.6–2.5 cm, apex rounded to broadly acute, base cuneate or attenuate, pubescent above and below, mainly greenish to grey-green on both sides, or if bicolored, paler beneath and green above. **Inflorescences** disjointing with age at the base of the peduncle, subglobose to elongate, 4–15(–20)cm, with 2–10 nodes and (1–)3(–4) flowers per node; peduncle 2–14 cm; rachis 0.5–5 cm, elongating in fruit, internodes to 9 mm; bracts tardily deciduous, oblanceolate, lanceolate or elliptic, (3–)6–18 × 2.5–10 mm. **Pedicels** persistent, 1.5–6(–8) mm. **Flowers** (13–)15–20(–22) mm; calyx 15–17 mm to upper teeth, 16–20 mm to lower tooth, calyx tube 6–8(–10) mm, setose or with appressed hairs, with or without minute glands, upper four teeth lanceolate to slightly oblanceolate, 6–8

× 1–2 mm, lower tooth lanceolate, elliptic to oblanceolate, 7–10 × 2.5–3 mm, teeth internally glabrate to pubescent, with one prominent vein per tooth, veins to sinuses less apparent, calyx gibbous-campanulate in fruit; petals whitish-blue to purple, banner similar to or more often lighter than wings and keel, obovate-lanceolate, 16–22 × 6–9 mm, the claw 6–7 mm, biauriculate with auricles internally callous, wings 13–20 × 2–3 mm, claw 8–10 mm, auricle to 2 mm, keel petals 14–16 × 2–3 mm, the claw 8–9 mm, the blade with a darker blotch above the middle; filaments 13–17 mm, anthers elliptic, 0.33 mm; ovary pubescent or only on upper half, style pubescent at base. **Fruits** oval-elliptic in profile, 6–9 × 4–5 mm, without glands, erect- to appressed-pubescent on upper portion, the attenuate beak (3–)5–8 mm, at least 2 mm broad at point of attachment, included within the calyx. **Seeds** brown, 4–5 × 3–4 mm, reniform-elliptic, smooth, without a hilum surrounded by a raised, white ridge.

Varieties 2 (2 native in flora): Arizona, Colorado, Nevada, New Mexico, Utah.

Pediomelum megalanthum and its varieties have variably been recognized at specific or varietal levels by different botanists in the past. Grimes (1990) and Isely (1998) included *P. epipsilum* as a variety under *P. megalanthum* along with *P. megalanthum* var. *retrorsum* and var. *megalanthum* while others have recognized these at the specific level (e.g. Welsh et al. 1993). *Pediomelum epipsilum* presents a suite of distinguishing characters supporting specific status (see *P. epipsilum* discussion). However, a recent morphometric analysis based on quantitative characters showed no clear break between vars. *megalanthum* and *retrorsum* (Egan 2014). Nevertheless, in honor of tradition, vars. *megalanthum* and *retrorsum* are detailed here due to the differences in vestiture.

1. Pubescence of peduncle either of sparse, long curly hairs going in all directions, or more commonly of two types: shorter hairs strictly appressed to incurved-ascending hairs, and longer erect ones ________ **4a. Pediomelum megalanthum** var. **megalanthum**
1′ Pubescence of peduncle mostly of long straight erect or reflexed hairs, or rarely with both short and long hairs, but these then simultaneously erect ________ **4b. Pediomelum megalanthum** var. **retrorsum**

4a. Pediomelum megalanthum (Wooton & Standley) Rydberg var. **megalanthum.** N. Amer. Fl. 24:22. 1919. * LARGE-FLOWERED BREADROOT Native

Psoralea megalantha Wooton & Standley

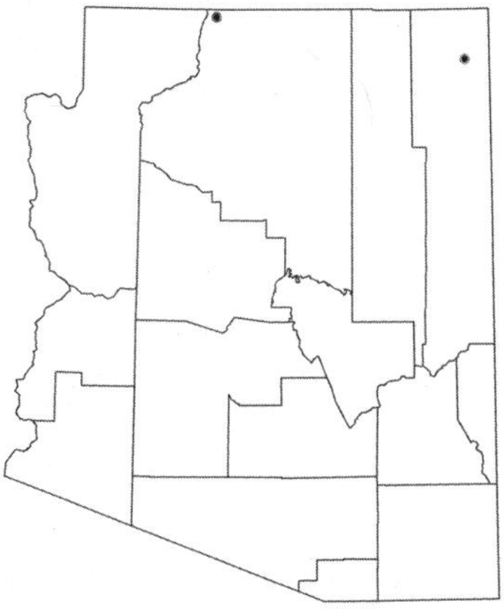

Plants 4–25 cm, erect to decumbent. **Leaves** with petioles 4–10(–12) cm. **Leaflets** 1.5–4 × 1–2.5 cm, usually bicolored with gray-green below, gray-green to yellow-green above. **Inflorescences** with peduncles 2.5–5 cm, pubescent either with sparse, long curly hairs going in all directions, or more commonly of two types: shorter hairs strictly appressed to incurved-ascending hairs, and longer erect ones; rachis 1–4 cm, with 2–9 nodes.

Flowering Apr–Jun. In decaying sandstone and clay soils on rock outcrops; 1400–2000 m; Great Basin Desertscrub, Great Basin Conifer Woodland; Apache, Coconino cos.; Colorado, New Mexico, e Utah.

4b. Pediomelum megalanthum (Wooton & Standley) Rydberg var. **retrorsum** (Rydberg) J.W. Grimes. Brittonia 38:185. 1986. * [from Latin, *retro*, backward, back, in reference to peduncle hairs] PEACH SPRINGS BREADROOT Native

Pediomelum retrorsum Rydberg, N. Amer. Fl. 24(1):22. 1919; *Psoralea mephitica* S. Watson var. *retrorsa* (Rydberg) Kearney & Peebles; *Psoralea retrorsa* (Rydberg) Tidestrom

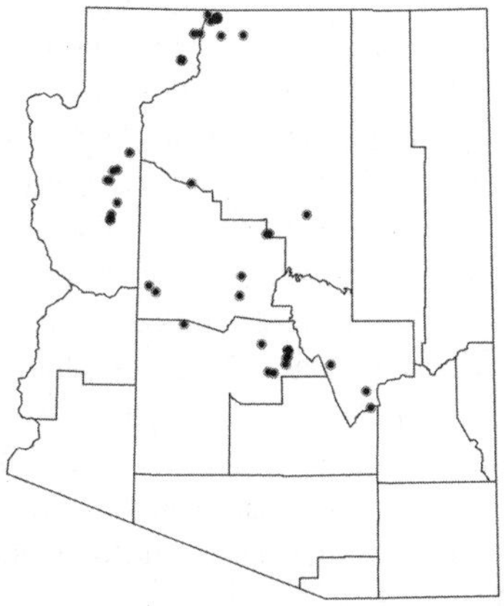

Plants 4–15 cm, erect to decumbent. **Leaves** with petioles 5–13(–15) cm. **Leaflets** 1–3.5 × 0.6–4 cm, sometimes bicolored with gray-green below, gray-green or barely different colored above. **Inflorescences** with peduncles 3–8 cm, pubescent with mostly long straight erect or reflexed hairs, or rarely with both short and long hairs, but these then simultaneously erect; rachis 1.5–7 cm, with 3–10 nodes.

Flowering Mar–Jun. On decaying sandstone and clay soils, and granitic substrates on slopes, valleys, and along drainages; [500–]640–1680 m; Sonoran Desertscrub, Great Basin Desertscrub, Interior Chaparral, Great Basin Conifer Woodland; Coconino, Gila, Graham, Greenlee, Maricopa, Mohave, Yavapai cos.; Nevada.

In Arizona, *Pediomelum megalanthum* var. *retrorsum* is relatively widely distributed. It occurs in several different plant communities and on various substrates. Populations are scattered but the plants may be locally abundant.

5. Pediomelum mephiticum (S. Watson) Rydberg in Britton, N. Amer. Fl. 24:22. 1919. * [from Latin, *mephitis*, a foul odor] SKUNKTOP Native

Psoralea mephitica S. Watson, Amer. Naturalist 12(9):601. 1878

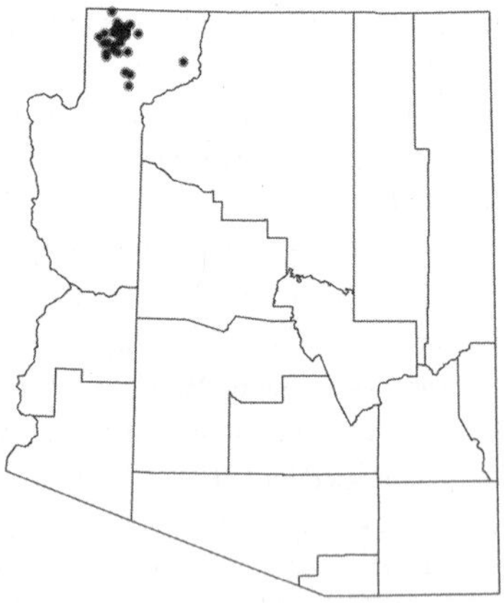

Plants 4–16 cm, acaulescent or subacaulescent, essentially glandular and pubescent throughout. **Roots** apically enlarged, deep, woody. **Stems** 0–4 cm; pseudoscapes 1–3, up to 4 cm, mainly subterranean; cataphylls 0–5 mm, glabrous to pubescent. **Leaves** clustered, palmately 5(–6)-foliolate; petioles (3–)5–12 cm, retrorsely hairy, jointed basally; stipules lanceolate to elliptic, (4–)7–11 × 2.5–8 mm, fused-free below, free above, straw-colored, tardily deciduous to persistent; petiolules 2–3 mm, the terminal one usually equal in length to lateral petiolules, but if longer, not more than twice as long, pubescent. **Leaflets** cuneate-obovate to circular, (1.5–)2–4 × (0.8–)1.2–3.5 cm, cuneate basally, broadly acute to rounded or retuse apically, glandular and pubescent, gray-green below, green to yellow-green above. **Inflorescences** disjointing with age at the base of the peduncle, globose to elongate, 1.5–5(–8) cm, with 4–6(–10) nodes and (2)3 flowers per node; peduncle (2–)4–8(–10) cm, equal to or shorter than the petioles, pilose with hairs generally retrorse; rachis 15–25 mm at anthesis; bracts tardily deciduous to persistent, ovate to elliptic, (7–)8–12 × 3–7 mm. **Pedicels** persistent, 2–4(–6 mm) at anthesis. **Flowers** 10–13 mm; calyx (8–)10–13 mm, calyx tube 2.5–4 mm, setose to pubescent with hairs to 2 mm, glandular, teeth subequal, upper teeth linear or linear-lanceolate to elliptic, 4–8 × 1–1.5 mm, lower tooth linear-lanceolate 7–9 × 2–3 mm, calyx gibbous-campanulate in fruit; petals mostly purple, banner white, elliptic to somewhat oblanceolate, 9–12 × 6 mm with claw 2–4 mm, wings 10–12 × 2–2.5 mm with claw 4–4.5 mm, keel 8–9 × 2–3 mm with claw 3.5–4.5 mm; filaments 7–8.5 mm; anthers elliptic, 0.33 mm; ovary glabrous to pubescent apically, style concomitantly so basally. **Fruits** round to ovoid, body 5–7 × 3.5–5 mm, pubescent, without glands, beak 1–4 mm, not exerted beyond calyx. **Seeds** red- to dark brown or gray-green and black-mottled, 4–5 × 3 mm, oval to reniform, smooth, without a hilum surrounded by a raised, white ridge.

Flowering Apr–Jun. In rocky or sandy soil, often of volcanic or limestone origin, on gentle slopes, valleys, and roadsides; [700–]1000–1770[–2000] m; (Semidesert Grassland), Great Basin Conifer Woodland; Mohave Co.; Nevada, Utah.

The type locality is stated as "Beaver City, Utah" which is largely agreed to be erroneous. The type locality is more likely "Beaver Dam, Arizona" (Grimes 1990). *Pediomelum mephiticum* is primarily a species of pinyon-juniper woodland in northern Mohave County and adjacent Nevada and Utah.

6. Pediomelum palmeri (Ockendon) J.W. Grimes ex Gandhi. Harvard Pap. Bot. 20(2):213. 2015. * [named for Edward Palmer (1831–1911), American botanist] Native

Psoralea palmeri Ockendon, SouthW. Naturalist 11:412. 1966; *Pediomelum palmeri* (Ockendon) J.W. Grimes

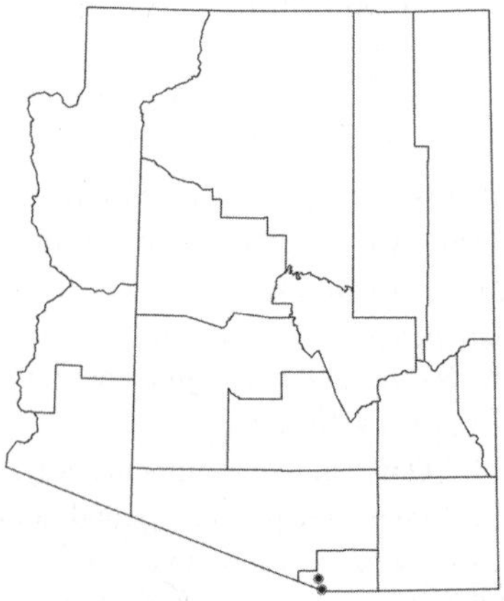

Plants prostrate, to 60 cm, caulescent, glandular on upper leaflet surfaces, sometimes obscurely so below, rarely on calyx teeth, stipules or bracts. **Roots** apically enlarged and with swollen lateral roots, deep, woody. **Stems** 1–many, often branching from the base, to 60 cm; pseudoscapes, when present, to 30 cm; cataphylls to 15 mm, blond to brown, veined. **Leaves** mostly pseudopalmately 5-foliolate, sometimes lowermost leaves pinnately trifoliolate; petioles 0.6–7.5 cm, slightly grooved longitudinally, pubescent, not jointed to the stem; stipules lanceolate to linear-lanceolate, 2–6.5 × 1–2 mm, free, persistent; petiolules 0.75–1 cm, blond-green and usually glabrous, the terminal one more than twice as long as the lateral petiolules. **Leaflets** lanceolate to rhombic or circular, sometimes dimorphically so, 0.7–3.1 × 0.3–1.3 cm, broadly acute to acuminate apically, sometimes obscurely glandular and glabrate to sparsely pubescent above (adaxially), without glands and glabrous or pubescent along veins below (abaxially). **Inflorescences** persistent, subcapitate, ovoid, 1–12 cm, with 1–5 nodes, internodes to 5 mm, 3 flowers per node; peduncle 0.5–10.5 cm; rachis 0.4–1.8 cm at anthesis; bracts persistent, lanceolate to spatulate, 1–3 × 0.5–2 mm. **Pedicels** persistent, 0.5–2 mm. **Flowers** 6–9 mm; calyx 4.5–5.5 mm to upper tooth and 4.5–6 mm to lower tooth in flower, strigose throughout, minutely glandular, 4.5–6 mm to upper teeth, 4.5–6 mm to lower tooth, calyx tube 2–4 mm, upper teeth lanceolate, 2–3 × 1 mm, lower tooth elliptic to slightly oblanceolate, 3.5–4 × 2 mm, enlarging in fruit; petals salmon-pink to reddish-brown at apex, whitish-green to ochroleucous at the base, or ochroleucous throughout, banner oblanceolate-obovate, 5–9 × 2–2.5 mm with claw 1.5–2.5 mm, wings 6–8 × 1.5–2.5 mm with claw 2–4 mm, keel 5–7 × 2–2.5 mm with claw 2–3.5 mm; filaments 4.5–6.5 mm; anthers elliptic, 0.33 mm; ovary sparsely pubescent apically. **Fruits** round-oval in outline, 4.5–5.5 × 4–5 mm, without glands, sparingly strigose on upper half, abruptly narrowing to beak 5.5–9 mm, exerted beyond calyx teeth. **Seeds** light brown or gray-brown, 4.5 × 3–4 mm, lenticular, hilum surrounded by a raised, white ridge.

Flowering Aug–Sep. In open areas of slopes and canyons; [0–]1200–1700[–2000] m; Semidesert Grassland, Madrean Evergreen Woodland; Santa Cruz Co.; Mexico (Chihuahua, Coahuila, Durango, Guanajuato, San Luis Potosí, Sonora).

Pediomelum palmeri has a rather difficult nomenclatural history. The species was first described under the misapplied name of *Psoralea pentaphylla* in the sense of Rydberg (1919). Ockendon (1965) discovered this, and in a footnote, renamed the species *Psoralea palmeri* Ockendon but without a Latin description, then required for valid naming of a new species. Grimes (1990) transferred *Psoralea palmeri* to *Pediomelum*, with the name then purported to be *Pediomelum palmeri* (Ockendon) Grimes, but again without a valid Latin description. Kartesz and Gandhi (1992) recognized this and presented a Latin description with name *Pediomelum ockendonii* Gandhi & Kartesz, but both Kartesz and Gandhi (1992) and Grimes (1990) overlooked the fact that Ockendon (1966) validated the name with a Latin description and holotype designation. This was recently discovered and rectified by Gandhi (2015) wherein *Pediomelum palmeri* was validated.

This species is widespread in Mexico where it is found in a varied habitat including grasslands, pine forest communities, and along the seashore. Two historical collections have been made in southern Arizona near the border with Mexico. Natureserve (2007) lists this species as possibly extirpated in Arizona. However, recent collections have been made from Peck Canyon and vicinity in Santa Cruz, Co. Flowering of plants in Arizona is documented for August and September. The species is of conservation concern in the state.

7. Pediomelum pentaphyllum (Linnaeus) Rydberg in Britton, N. Amer. Fl. 24:23. 1919. * [from Greek, *penta*, five, and *phyllum*, a leaf, referring to the typically 5-foliolate leaves] CHIHUAHUA SCURFPEA Native

Psoralea pentaphylla Linnaeus, Sp. Pl. 2:764. 1753; *Pediomelum trinervatum* Rydberg

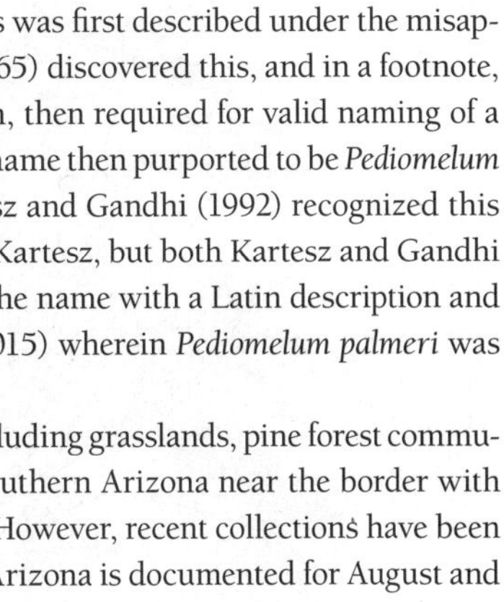

Plants to 30 cm, acaulescent or subacaulescent, essentially glandular and strigose throughout with appressed-spreading hairs, sparingly so on upper leaflet surfaces. **Roots** deep taproots with swollen portions. **Stems** 0–1, 0–4 cm but with leaves clustered at base (rarely with lateral decumbent stems 0–15 cm with leaves and inflorescences clustered at the tips); pseudoscapes 1, 0–4 cm; cataphylls, when present, 7–15 mm, emarginated apically. **Leaves** clustered or contiguous, palmately (pinnately) 5(–6)-foliolate, sometimes with 2 leaflets set above others; petioles 6–15 cm, strigose, jointed basally; stipules lanceolate to linear, 9–15 × 5 mm, somewhat fused-free below, free above, persistent; petiolules 1.5–2 mm, terminal one usually equal in length to lateral petiolules, but if longer, not more than twice as long, pubescent. **Leaflets** lanceolate to obovate, elliptic, or rhombic, 2–5.5 × (1.5–)2–3.2 cm, cuneate basally, mucronate apically, uniformly strigose to pubescent on veins and margins above, more densely so below, margins undulate. **Inflorescences** disjointing with age at the base of the peduncle, globose to long-ovoid, to 6 cm, with 6–9 nodes and 3 flowers per node; peduncle (3–)4–9.5 cm, equal to or shorter than petioles; rachis 2–4.5 cm at anthesis; bracts persistent to tardily deciduous, lanceolate to elliptic, often ensiform (sword-shaped), 5–7 × 2–3 mm. **Pedicels** persistent, 2–3 mm at anthesis. **Flowers** (12–)14–18 mm; calyx 11–12 mm to upper tooth and 14–16 mm to lower tooth in flower, elongating slightly in fruit, calyx tube 4–5 mm, strigose to setose, glandular, upper teeth linear-lanceolate to lanceolate, 7–8 × 1–1.5 mm, lower tooth broadly oblanceolate and often broader and longer than upper four, 10–12 × 3.5–4 mm, gibbous-campanulate in fruit; petals white to purple, banner white to lavender, paler than other petals, lanceolate to elliptic,

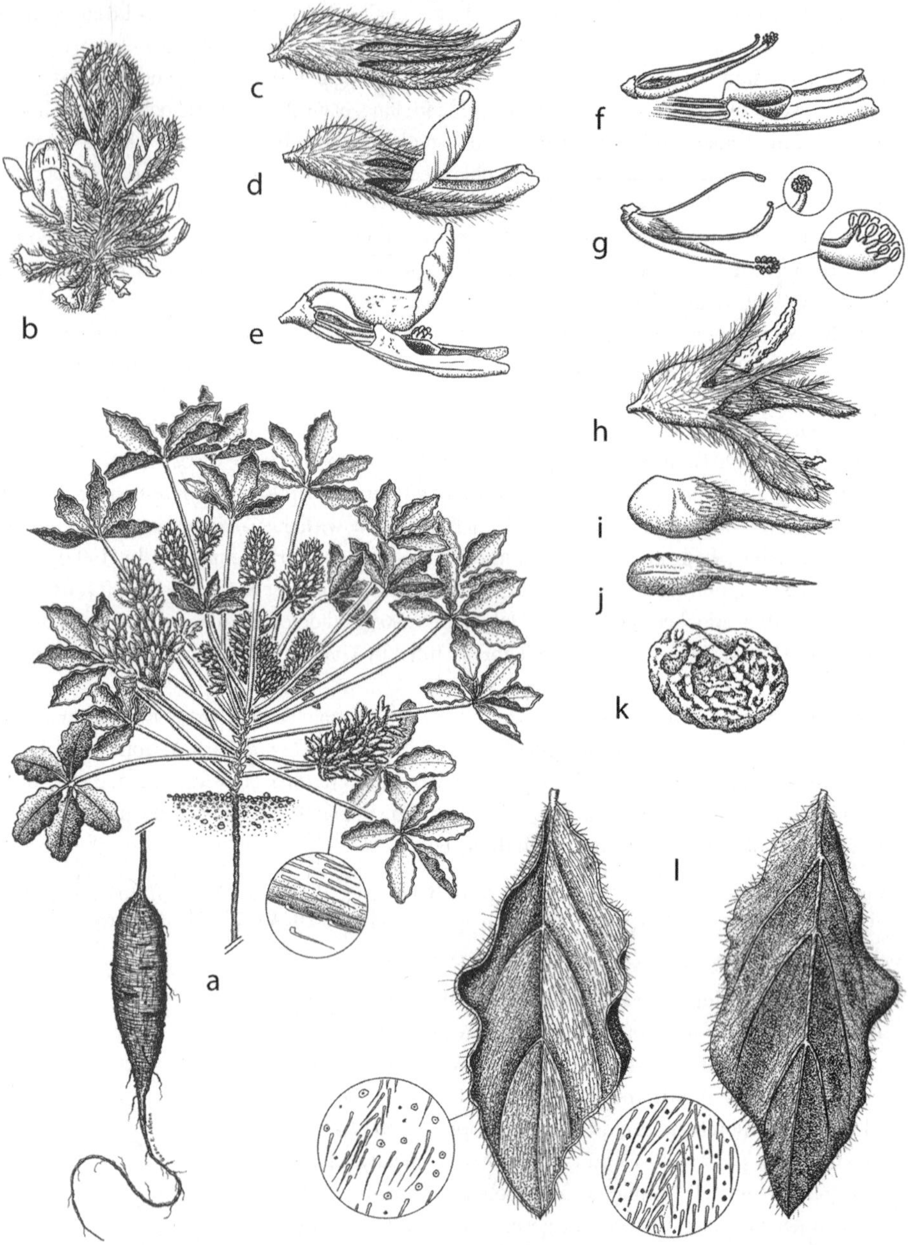

Pediomelum pentaphyllum: **a.** habit (with tuberous root); **b.** inflorescence; **c.** flower bud; **d.** flower (side view); **e.** flower detail; **f.** stamens; **g.** detail of pistil and stamens; **h.** calyx with fruit; **i.** fruit (side view); **j.** fruit (top view); **k.** seed; **l.** leaflet detail. Illustrator: Susan E. Ashton.

15–17 × 6–7 mm with claw 4–5 mm, wings 15–16 × 3–3.5 mm with claw 6 mm, keel 9–12 × 3 mm with claw 6–7 mm; filaments 8.5–10.5 mm; anthers round-elliptic, 0.33 mm; ovary pubescent apically, style pubescent basally. **Fruits** elliptic-oblanceolate, 7–8 × 4–5 mm, without glands, with broad, flat beak 10–15 mm, barely exserted from calyx. **Seeds** gray-green to brown, 5–6 × 3–4 mm, oblong-reniform, rugose, without a hilum surrounded by a raised, white ridge.

Flowering Apr–Jun. Sandy alluvial soils of gentle slopes, valleys, and roadsides; 1000–1320[–2000] m; Chihuahuan Desertscrub, Semidesert Grassland; Cochise, Graham cos., New Mexico, Texas; Mexico (Chihuahua).

Pediomelum pentaphyllum has a rather confusing taxonomic history, with the authority and actual species to which the name refers changing several times since its initial description. Good descriptions of this history are given elsewhere (Tonne 2000, WildEarth Guardians 2008), but it is important to note that the name was misapplied by Rydberg (1919).

The plant that Rydberg (1919) described as *P. pentaphyllum* is actually *Pediomelum palmeri* (Ockendon) J.W. Grimes; the plant denoted herein as *P. pentaphyllum* was called *Pediomelum trinervatum* Rydberg. Unfortunately, J.W. Grimes, before completion of his monograph (1990), erroneously annotated several *P. pentaphyllum* herbarium specimens as *P. trinervatum*.

Pediomelum pentaphyllum is a rare plant, with only a few confirmed populations in Arizona, although it has been historically collected in New Mexico (Hidalgo Co.), Texas (Presidio Co.), and Mexico (Tonne 2000). Because of this rarity, the species is currently under consideration to be listed as an endangered species by the U.S. federal government (WildEarth Guardians 2008). Some have theorized that its decline may have been due to grazing, herbicide use, or over-collecting by Native Americans and indigenous Mexicans, who used it as a fever reducer, the use of which may have contributed to its spread through cultivation, historically (WildEarth Guardians 2008).

8. Pediomelum tenuiflorum (Pursh) A.N. Egan in A.N. Egan & Reveal, Novon 19:311. 2009. * [from Latin, *tenui*, thin, slender, and *florum*, a flower] SLIMFLOWER SCURFPEA Native

Psoralea tenuiflora Pursh, Fl. Amer. Sept. 2:475. 1814[1813]; *Psoralidium tenuiflorum* (Pursh) Rydberg

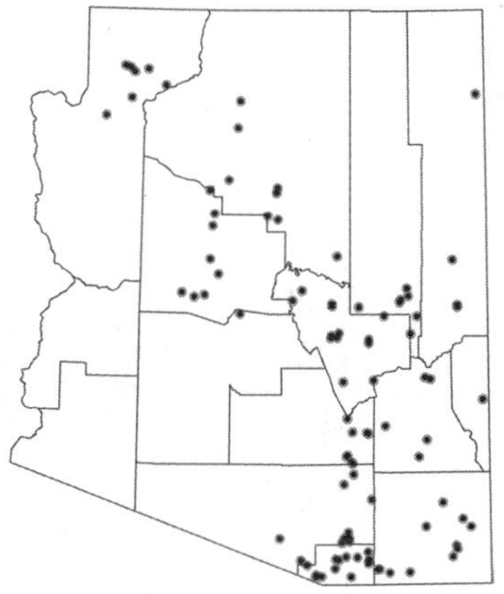

Plants to 130 cm, caulescent, glandular, glabrate to pubescent, arising from a deep, branched woody root. **Roots** woody, branching extensively and forming a diffuse caudex of short rhizomes. **Stems** 1(–3), sparingly to much branched, branches mostly arising from middle to upper nodes; pseudoscapes 0; cataphylls 4–12 mm, glabrous and papery. **Leaves** palmately (1–)3–5-foliolate; petioles (0.1–)1.5–17(–22) mm, glabrate to pubescent, rarely from a swollen pulvinus; stipules linear-lanceolate, (2–)4–7(–13) × 1 mm, free, persistent to tardily deciduous; petiolules 1–2.5 mm, terminal one usually equal in length to lateral petiolules, but if longer, not more than twice as long, glabrate to pubescent. **Leaflets** elliptic to narrowly oblanceolate, (10–)14–40(–50) × (2–)4–8(–15) mm, rounded to attenuate basally, rounded to retuse apically, glabrous above, strigose below. **Inflorescence** persistent, long-ovoid to lanceolate, (1–)2–6 cm, with 3–12 nodes and 1–3 flowers per node, elongating to 11 cm in fruit; peduncle (0.5–)2–9.5 cm; rachis 1.5–6 cm at anthesis, elongating through fruiting; bracts persistent, trullate to lanceolate, 1.5–3(–5) × 0.5–2 mm. **Pedicels** 1.5–3 mm at anthesis. **Flowers** 5–7 mm; calyx 1.5–4 mm, sometimes flaring backwards and tearing along lateral sinus, calyx tube 1.5–2.5(–3) mm, pubescent, densely glandular, teeth triangular, upper teeth 0.5–1 × 0.5–1 mm, lower tooth 1–2 × 1 mm, campanulate, not gibbous in fruit; petals dark blue to purple (white), banner usually more pale, elliptic to obovate, 4.5–6 × 4.5–6 mm with claw 1–2 mm, wings 6 × 1.5–2.5 mm with claw 1.5–3 mm, keel 3.5–4.5 × 1.5–2 mm with claw 1.5–2 mm; filaments 3.5–4 mm; anthers elliptic, 0.25 mm; ovary glabrous to pubescent apically. **Fruits** elliptic in shape, 7–8 × 3–4 mm, glandular, glabrous, with broad short beak to 2 mm long, well exserted from calyx, deciduous with calyx and pedicel. **Seeds** brown, 5–6 × 3–5 mm, reniform, without a hilum surrounded by a raised, white ridge.

Flowering Apr–Sep. Various soils of slopes, valleys, canyons, and roadsides in habitats from desert to mixed conifer forest; [200–](700–)1000–2300(–2490) m; (Sonoran Desertscrub), Semidesert Grassland, Plains Grassland, Interior Chaparral, Madrean Evergreen Woodland; Apache, Cochise, Coconino, Gila, Graham, Greenlee, Mohave, Maricopa, Navajo, Pima, Pinal, Santa Cruz, Yavapai cos.; Colorado, Illinois, Indiana, Kansas, Minnesota, Missouri, Montana, Nebraska, New Mexico, Oklahoma, South Dakota, Texas, Utah, Wisconsin, Wyoming; Mexico (Chihuahua, Sonora).

Pediomelum tenuiflorum was moved from the genus *Psoralidium* based on molecular phylogenetic work in which both nuclear and chloroplast markers grouped this species within *Pediomelum* (Egan and Crandall 2008a, 2008b). The calyx morphology of *P. tenuiflorum* resembles that of *Psoralidium* but differs by the fruit being persistent on the receptacle and falling with the calyx, traits found within *Pediomelum*. This species is widespread with morphological gradations across its distribution, especially in leaf and inflorescence size and shape and in calyx pubescence. Rydberg (1919) recognized as many as four separate species based on these differences, which seem to be environmentally influenced.

Pediomelum tenuiflorum is the most widely distributed and frequently encountered member of the genus in Arizona. It grows on various soils in a wide range of elevations and habitats from the Sonoran Desert to grasslands, chaparral, pinyon-juniper and pine-oak woodlands, and ponderosa pine and pine-fir forests. Populations are scattered.

Hay made from *Pediomelum tenuiflorum* is reported to be eaten by livestock, though it has been rumored to be toxic to horses and cattle (Stubbendieck and Conard 1989), and possibly other mammals as well. Stubbendieck and Conard

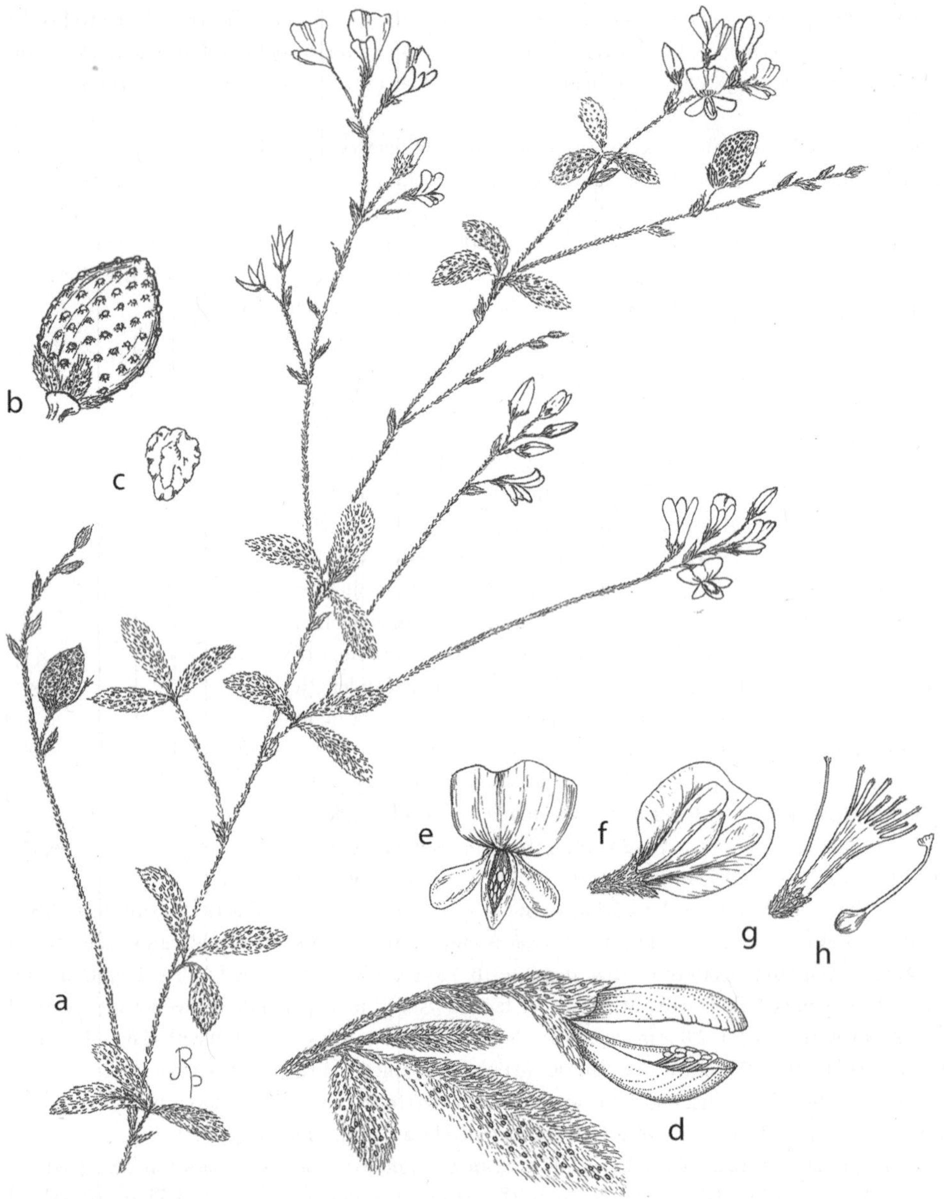

Pediomelum tenuiflorum: **a.** habit; **b.** fruit; **c.** seed; **d.** flower with leaf; **e., f.** flowers; **g.** stamens; **h.** pistil. Illustrator: Joyce Riney Peters.

(1989) report that the plants were used as a fish poison by some Native Americans. Rogers (1980) documents its use by Lakota Native American tribes as a remedy for headaches and as a mosquito repellent. Recent research has shown *P. tenuiflorum* to contain bakuchiol, a chemical with putative antibacterial properties, especially against *Staphylococcus epidermidis* (Hsu et al. 2009), a finding that may scientifically support its documented use as a disinfectant by the Zuni Native Americans (Stevenson 1915).

9. Pediomelum verdiense S.L. Welsh & Licher. W. N. Amer. Naturalist 70:12. 2010. * [for the Verde Limestone Formation] VERDE BREADROOT Native

Pediomelum pauperitense S.L. Welsh, Licher & N.D. Atwood

Plants 4.5–13(–15) cm, acaulescent to short caulescent, essentially glandular and pubescent throughout. **Roots** underground caudex arising from a deep, tuberous root. **Stems** 0–4(–6) cm; pseudoscapes 0–3, up to 6 cm, mainly

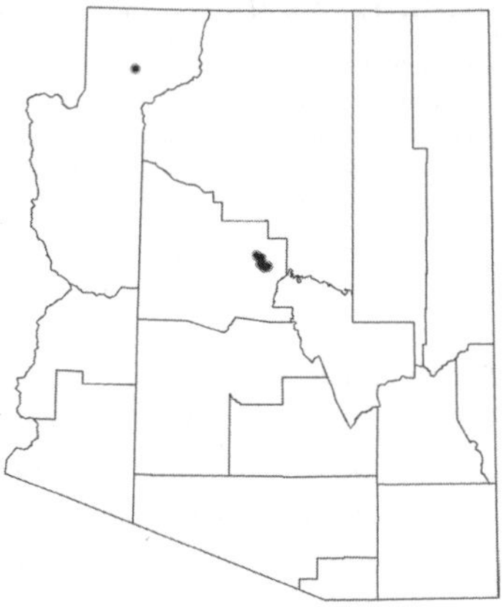

subterranean; cataphylls 0–5 mm, glabrous to pubescent. **Leaves** clustered, palmately (3)5-foliolate; petioles 2–10(–11.5) cm, with hairs appressed-ascending, jointed basally; stipules lanceolate to elliptic, 4–16 × 2–8 mm, fused-free below, free above, scarious, tardily deciduous to persistent; petiolules 2–3 mm, terminal one usually equal in length to lateral petiolules, but if longer, not more than twice as long, pubescent. **Leaflets** cuneate-obovate, (0.8–)1.2–3 × 0.7–1.8(–2.2) cm, cuneate basally, broadly acute to rounded or retuse apically, glandular and pubescent with more hairs along veins above and on lower surface, gray-green below, green to yellow-green above. **Inflorescences** disjointing with age at the base of the peduncle, globose, 1.5–3 cm, with (1–) 2–4(–6) nodes and (2)3 flowers per node; peduncles 0.5–4.5(–6) cm, shorter than the petioles, spreading or spreading-ascending white-hairy, sometimes with longer spreading white hairs; rachis 1–3(–4) cm at anthesis; bracts tardily deciduous to persistent, elliptic, 4–8(–9) × 2–6 mm. **Pedicels** persistent, thread-like, 2.5–4.5(–6) mm. **Flowers** (8–)10–13.5(–15) mm; calyx (7.5–)8.5–12(–13) mm, calyx tube (3–)4–5 mm, setose, glandular, teeth subequal, upper teeth lanceolate to oblong or elliptic, 4–7(–8) × 1–2.5 mm, lower tooth elliptic, (4–)5–9 × (1.5–)2–3.5 mm, gibbous-campanulate in fruit; petals white to purple, the banner white, cream, purple or suffused with pale purple, the wings and keel dark purple, with the wings sometimes lighter in color; banner elliptic to somewhat oblanceolate, 9–12(–14) × 6–8 mm with claw 2–5 mm, wings 10–13 × 2–3 mm with claw 4–5 mm, keel 8–10 × 2–3.5 mm with claw 3–5 mm; filaments 7–8.5 mm; anthers elliptic, 0.33 mm; ovary glabrous to pubescent apically, style concomitantly so basally. **Fruits** round to ovoid, body 5–7 × 3.5–5 mm, pubescent, without glands, beak 1–4 mm, not exerted beyond calyx. **Seeds** oval to reniform, 3.5–5 mm × 2.5–3 mm, olive to gray brown and with or without purple mottling, smooth, without a hilum surrounded by a raised, white ridge.

Flowering Apr–Jun. Limestone soils; 900–1750 m; Yavapai Co.(Verde Valley Formation) and Mohave Co. (near Poverty Mountain); endemic.

This species was described as two separate species by Welsh & Licher (2010): *Pediomelum verdiense* S.L. Welsh & Licher and *P. pauperitense* S.L. Welsh, Licher & N.D. Atwood, with the former endemic to the silty, white limestone soils of the Verde formation, Yavapai Co., AZ and the latter endemic to the pink limestone soils of Poverty Mountain, Mohave Co., AZ. *Pediomelum pauperitense* was said to have smaller flowers, pedicels, bracts, and seeds and to differ in the banner and wing coloration from *P. verdiense*, with *P. pauperitense* having white or cream-colored banner and wings (more reminiscent of the geographically proximate *P. mephiticum*) while *P. verdiense* has purple or purple suffused banner and wings. However, a morphometric analysis based on 10 quantitative floral and vegetative characters found extensive overlap in nearly all characters, especially floral (Egan 2014), supporting the conclusion that these be recognized as a single species. It is possible that *P. pauperitense* represents a hybrid between *P. mephiticum* and *P. verdiense*, as several characters are overlapping or intermediate between the two. However, this hypothesis needs confirmation by molecular means.

PETERIA A. Gray. Pl. Wright. 1:50. 1852. * [for Dr. Robert Peter, 1805–1894, American botanist, active in floral research of Kentucky]

Matt Lavin, Kirsten N. Lake, Matthew B. Johnson

Herbs with usually erect to ascending stems, rarely scandent. **Stems** with young growth sericeous, sometimes sparsely so. **Leaves** alternate, pulvinate, odd-pinnate; stipules persistent and spine-tipped; stipels absent. **Leaflets** 7–31, opposite, rarely alternate, narrowly to broadly elliptic, entire. **Inflorescences** terminal, racemose; bracts persistent, sometimes spine-tipped; bracteoles absent. **Flowers** papilionoid; calyx bilaterally symmetrical, tubular, 5-lobed; corolla whitish with a purplish tinge, glabrous; stamens 10, 9 united and 1 free, anthers small, attached basally, dehiscing longitudinally; pistil: style glabrous but with a pollen brush tufted at the distal end; stigma rounded, ciliate, terminal; ovules 8–12. **Fruits** elastically dehiscent, laterally compressed, elliptic, linear. **Seeds** 2–7, cylindrical with an apical hilum. $x = 8$.

Species 4 (2 native in the flora): Great Basin, Chihuahuan, and Sonoran Deserts, Mexican pine forests; Arizona, California, Colorado, Idaho, Nevada, New Mexico, Utah, Texas, Mexico.

Peteria is readily distinguished from other temperate herbaceous legume genera by the combination of its terminal raceme, spine-tipped stipules, and cylindrical (or at least not strongly laterally compressed) seeds. *Peteria* produces edible tubers borne from rhizomes.

SELECTED REFERENCE: Lavin and Sousa 1995.

1. Leaflets linear to narrowly elliptic, not over 10 mm; flowers at anthesis not congested along the rachis, flowering internodes (3–) 5–35 mm, rarely with more than 1 flower reaching anthesis at a time; banner petals 14–16 mm _____ **1. Peteria scoparia**
1′ Leaflets broadly elliptic to circular, if narrowly elliptic, then leaflets usually well over 12 mm; flowers at anthesis usually congested along the rachis, flowering internodes 0.1–5(–12) mm, usually more than 2 consecutive flowers simultaneously at anthesis; banner petals 16–22 mm _____ **2. Peteria thompsoniae**

1. Peteria scoparia A. Gray. Pl. Wright. 1:50. 1852. * [Latin *scopa*, a broom] RUSH PETERIA; *CAMOTE DEL MONTE*
Native

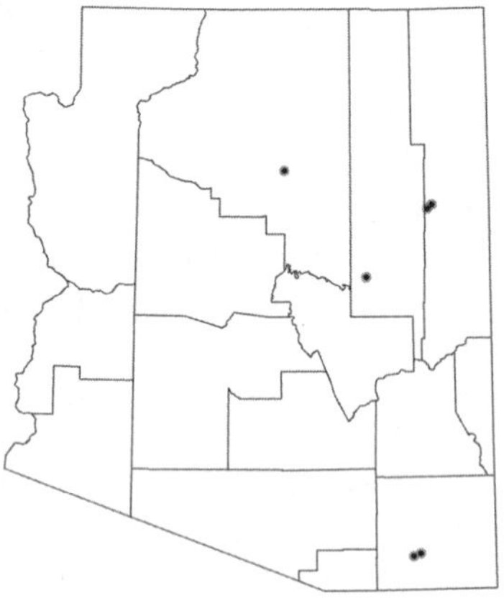

Perennial herbs to 0.8 m. **Leaves** 4–5 cm; stipules awl-shaped, spine-tipped, 3–5 × 1 mm; petioles 0.5–1 cm, glabrate to strigose; rachis with longitudinal grooves; petiolules 1 mm, glabrate. **Leaflets** 11–13 (terminal leaflet often reduced to a spine-like bristle), linear to very narrowly elliptical, 3–6(–9) × 1–2(–9) mm, base rounded, apex acute, glabrate to strigose. **Inflorescences** loose, with 8–20 nodes (3–)5–35 mm apart, rarely with more than 1 flower reaching anthesis at a time, axes glabrate to sparsely strigose to stipitate glandular distally; bracts awl-shaped, 3–5 × 0.5–1 mm, apex often spine tipped. **Pedicels** 3–8 mm. **Flowers:** calyx tube 5.5–7 mm, glabrate to strigose, calyx lobes 2–4 mm, lower-most longer than the two lateral lobes, two upper (adaxial) lobes more fused than laterals; corolla creamy white with a purplish tinge, 14–16 mm; filaments subequal; pistil glabrous. **Fruits** sessile, 3.5–5(–6.2) × 0.5–0.6 cm, base blunt, apex acute and often terminating in persistent style base, light to dark brown, glabrous. **Seeds** 2–5. $2n = 18$.

Flowering during the summer. Open dry shrubby vegetation on slopes and valleys, often in gravelly soils derived from sandstone or limestone; 1000–2100 m; Chihuahuan Desertscrub, Plains and Great Basin Grassland, Great Basin Conifer Woodland, Petran Montane Conifer Forest; Apache, Cochise, Coconino, Navajo cos.; Colorado, New Mexico, Texas; nc Mexico.

Peteria scoparia occasionally has even-pinnate leaves by the reduction of the terminal leaflet to a small spine. Evenly pinnate leaves are also found in related legume genera including *Sesbania*, *Olneya*, and certain species of *Coursetia* and *Poitea* (but no leaf spines are present in these genera). Some populations of *Peteria scoparia* have early-deciduous leaflets but this is not a consistent character for the species. For example, specimens from southwestern Colorado and the northernmost populations in New Mexico and Arizona display more luxuriant foliage that approaches *P. thompsoniae*. Although these northernmost specimens of *P. scoparia* are vegetatively very similar to *P. thompsoniae*, they have inflorescences with widely spaced flowers, standard petals up to 16 mm long, and leaflets up to 4 mm wide (Lavin and Sousa 1995).

Powell (1988) states that the palatability of the tubers of *Peteria scoparia* in Trans-Pecos Texas is unknown. The Navajo people employ preparations of *P. scoparia* to treat influenza and porcupine punctures, and smoke from the dried herbage is used to treat cough in sheep (Moerman 1998).

2. Peteria thompsoniae S. Watson. Amer. Naturalist 7:300. 1873. * [named in honor of Ellen P. Thompson who first collected the plant while on an expedition with her brother, Major John Wesley Powell] THOMPSON'S PETERIA
Native

Peteria nevadensis Tidestrom

Perennial herbs to 1 m. **Leaves** (5.5–)7–15.5 cm; stipules awl-shaped, spine-tipped, 4–6 × 1 mm; petioles 0.5–1.5 cm, sericeous; rachis with longitudinal grooves; petiolules 1 mm, nearly glabrescent. **Leaflets** (9–)15–31, broadly elliptical to circular, if narrowly elliptic, then leaflets usually well over 12 mm, 12–21 × 7–13 mm, base rounded, apex rounded, glabrate to sericeous. **Inflorescences** congested, often with 20–100+ nodes 0.1–5(–12) mm apart, usually more than 2 consecutive flowers simultaneously at anthesis, axes sparsely strigose to stipitate glandular distally; bracts awl-shaped, 5–11 × 0.5–1 mm, apex often spine-tipped. **Pedicels** 2–3 mm. **Flowers:** calyx tube (6–)7–11 mm, strigose to sericeous, calyx lobes 7–11 mm, lower-most longer than the

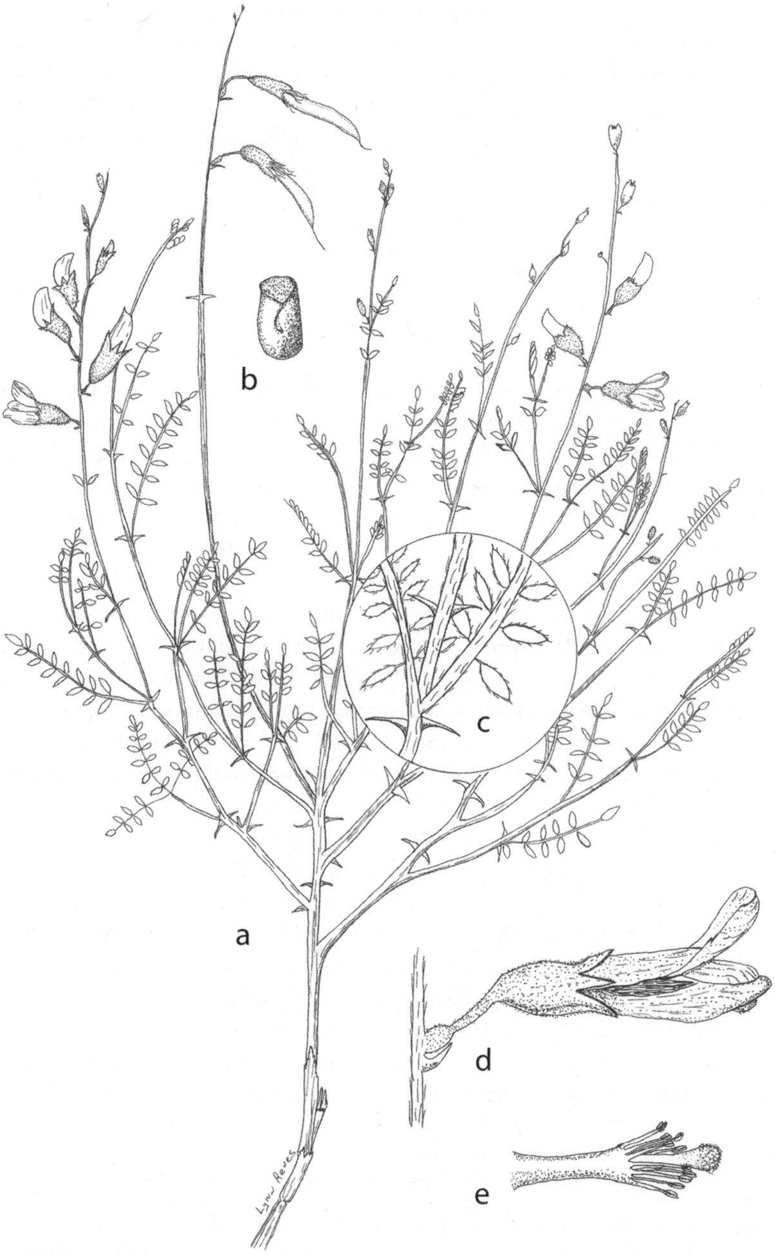

Peteria scoparia: **a.** habit; **b.** seed; **c.** node detail with stipular spines; **d.** flower; **e.** staminal column with style and stigma. Illustrator: Lynn Reves.

two lateral lobes, two upper (adaxial) lobes more fused than laterals; corolla creamy white, 16–22 mm; filaments subequal; pistil glabrous. **Fruits** sessile, 4–6.5 × 0.6–0.7 cm, base blunt, apex acute and often terminating in persistent style base, light to dark brown, glabrous. **Seeds** 2–5. $2n$ = 16, 18.

Flowering usually during early summer. Open dry shrubby vegetation in sandy and clay soils of valleys, slopes, and along arroyos; 830–1800 m; Great Basin Desertscrub; Mohave, Apache (1 record) cos.; Nevada and adjacent California and Idaho, Utah.

In northern Arizona (Mohave County), populations of *Peteria thompsoniae*, compared to those in other locations, are distinctive in having a longer flowering season, the end of April through August; the leaflets are commonly more numerous (25–31) and consistently more glabrate and bear more scattered tannin deposits on the leaflet surfaces.

Furthermore, the flowers in these populations have a generally smaller calyx tube (6–7 mm long) and a greater number of ovules per ovary (10–12). Populations from southern Utah (Kane and Washington counties) have leaves similar to those of the Arizona populations in that they tend to be more distinctly marked with finely scattered tannin deposits and have glabrate to sparsely strigose surfaces, but leaflet number per leaf is not as consistently high as in the Arizona populations. Populations from eastern Utah (Emery, Wayne, and San Juan counties) are distinct in having leaves with narrowly elliptic leaflets that are glabrate at least on the upper surface (Lavin and Sousa 1995). The single record from Apache County (*M. W. Mohlenbrock 2253*) is disjunct from the primary area of distribution in northern Mohave County.

PHASEOLUS Linnaeus. Sp. Pl. 2:723–725. 1753. * [from Latin, *phaseolus,* derived from the Greek, *phaselos,* a light boat or canoe, in reference to the shape of the legume; for extensive discussion see Genaust, 1976] BEAN; *FRIJOL*

Alfonso Delgado-Salinas

Alepidocalyx Piper; *Lipusa* Alefeld; *Minkelersia* M. Martens & Galeotti

Herbaceous vines, prostrate to climbing, annual or perennial; roots tuberous, woody, or fibrous. **Stems** striate, often lignescent, covered with oblique to appressed, retrorse hairs, interspersed with minute hooked hairs. **Leaves** pinnate, usually 3-foliolate; stipules and stipels sometimes covered with glandular hairs, these nectariferous. **Leaflets** (1 or) 3 [or 5], membranous, firm, or leathery, often lobed. **Inflorescences** of pseudoracemes often with lateral branches, 4–60 floral nodes; primary nodal bracts 0.1–1.5 cm, entire or dissected; usually persistent; bracteoles 0.1–1 cm, usually persistent; bracts and bracteoles often with nectariferous glandular hairs. **Pedicels** arched to reflexed in fruit. **Flowers** papilionoid; calyces campanulate or campanulate-tubular, with lobes usually shorter than tube; upper lobes usually entirely fused or almost so, sometimes forming a lip; corollas pink, purple, red, or white; banner in flower recurved ca. 90° with a claw; wings 0.6–2 cm, with claws; keel 1.5–2 times laterally-vertically coiled, with claws; diadelphous with 1 stamen free and 9 stamens united into a tube that is coiled within the keel; ovary with a nectary disk at the base, 1–20 ovules, style thickened distally, bearded introrsely; stigma mostly introrse. **Fruits** usually falcate, rarely oblong-straight or rhombic; valves membranous, stiffly papery, or leathery. **Seeds** 1–20, 0.25–1 cm on wild species. $n = 11, 10$.

Species ca. 70 (7 native in the flora): Canada, United States, Mexico, Caribbean, Central America, South America; tropical to warm temperate areas.

Phaseolus contains five cultivated species. Four of them (*P. acutifolius* (tepary bean), *P. coccineus* (runner bean), *P. lunatus* (lima bean), and *P. vulgaris* (common bean)) are grown in the U.S. At least two of them (*P. lunatus* and *P. vulgaris*) have a multitude of cultivars and are important agricultural crops worldwide. Due to their limited cultivation in Arizona, descriptions are not provided for *P. coccineus*, *P. lunatus*, and *P. vulgaris*. These species have similar cultivation requirements to *P. acutifolius*, though most cultivated selections take longer to mature their fruits. *Phaseolus coccineus* is not adapted to low desert regions. Nodulation is widely reported for *Phaseolus*.

There is single record of *Phaseolus leptostachyus* Bentham for Arizona, collected at Fairbanks, Cochise Co. (*Morris et al. A11584,* 25 Sep 1939, ASU). The species is distributed from Sonora to Central America and should be looked for in Arizona. It is included in the key to facilitate identification, but no description is provided.

1. Leaflets entire, rarely lobed.
  2. Leaflets membranous to slightly stiff and papery, long-petiolate.
    3. Primary nodal bracts 2.5–7 × 0.5–3 mm; pedicels 1–3.5(–5) mm; bracteoles 0.5–1.5 mm; fruits 1.9–3.2 × 0.3–0.7 cm; seeds 4–6 ________ **Phaseolus leptostachyus**
    3′ Primary nodal bracts 1–4 × 0.3–1.2 mm; pedicels 3–9 mm; bracteoles 1.5–1.8 mm; fruits 3.2–7.5 × 0.5–0.9 cm; seeds 5–10 ________ **1a. Phaseolus acutifolius** var. **acutifolius**
  2′ Leaflets thick or leathery, short-petiolate.
    4. Leaflets with obtuse to rounded apices; petioles 1.5–6.5 cm; corollas reddish, violet, bright pink; fruits sessile, falcate to oblong, valves leathery, passively or weakly dehiscent, 3–7.4 × 1.2–1.7 cm ________ **4. Phaseolus maculatus**
    4′ Leaflets with acute apices; petioles up to 3.5 cm; corollas violet to purplish; fruits stipitate, falcate, valves thin to somewhat leathery, elastically dehiscent, 3.5–5.5 × 0.6–1.2 cm ________ **6. Phaseolus ritensis**
1′ Leaflets slightly to distinctly lobed.
  5. Calyces campanulate-tubular; corollas 1.5–2.5 cm; bracteoles ovate or absent, ca. 0.1 mm long; fruits linear-straight to slightly falcate, 3.5–4.5 × 0.3–0.4 cm; seeds 8–11 ________ **5. Phaseolus parvulus**
  5′ Calyces campanulate; corollas 1–1.5 cm; bracteoles ovate to linear-lanceolate, persistent, 1–7 × 1–4 mm; fruits slightly curved to falcate, 2–5.5 × 0.3–1.5 cm; seeds 2–10.
    6. Leaflets broadly ovate, rhomboid, or deltoid, the upper surface often variegated with a pale area along midrib ________ **7. Phaseolus scabrellus**
    6′ Leaflets linear to lanceolate, ovate, or narrowly oblong, the upper surface uniform in color without a pale area along midrib.
      7. Bracteoles 1.5–1.8 mm long; fruits linear-falcate, 3.2–5.4 × 0.5–0.6 cm; seeds 7–10, smooth ________ **1b. Phaseolus acutifolius** var. **tenuifolius**
      7′ Bracteoles ca. 1 mm long; fruits falcate, 2.2–3.5 × 0.4–0.8 cm; seeds 2–6, slightly rugose to tuberculate.

8. Plants from slender taproot; leaflets ovate, entire to distinctly lobate; bracteoles 0.6–1.3 × 3–5 mm; seeds 4–6 ____ **3. Phaseolus filiformis**
8′ Plants with a thickened and elongate woody taproot; leaflets linear, narrowly oblong to lanceolate, often lobed at base; bracteoles 1–1.5 × ca. 0.3 mm; seeds 2–4 ____ **2. Phaseolus angustissimus**

1. **Phaseolus acutifolius** A. Gray. Pl. Wright. 1:43–44. 1852. * [from Latin *a cut*, sharp and *folium*, leaf] TEPARY BEAN; *EJOTILLO, EJOTITO DEL MONTE, FRIJOL CIMARRÓN*

Herbaceous, annual vines to 4 m; weak or vigorous twiners, often trailing from fibrous roots. **Stems** cylindrical, usually slender, branching diffusely from proximal nodes; covered with oblique to appressed, retrorse hairs, interspersed with minute hooked hairs. **Leaves** with stipules triangular to oblong, 1.5–3 × 0.5–2 mm, acute at tip, prominently 3–6-veined, ascending; petioles 2–8 cm; rachises 0.5–2.5 cm; upper stipels linear to lanceolate, often clavate, ascending; lower stipels lanceolate, often clavate, acute at tip, 1-veined, spreading, often nectariferous. **Leaflets** membranous to rigid, linear-lanceolate to lanceolate or ovate to widely ovate, 2.5–11.5 × 0.3–5 cm, acute to attenuate at tip, base rounded, truncate to cuneate, often weakly to conspicuously lobed on 1 or both sides toward the base; upper surfaces sparsely covered with straight and hooked hairs; lower surfaces sparsely to densely covered with minute hooked hairs, with straight hairs mostly along veins and at margins; venation often prominent. **Inflorescences** of pseudoracemes, to 26 cm; peduncles 0.8–11.5 cm; rachises mostly 2–24 cm, with 2–6 usually biflorous nodes; peduncular basal bracts linear to lanceolate; primary nodal bracts ovate to lanceolate, acute at tip, 1–3-veined, usually persistent; secondary nodal bracts triangular to lanceolate, 0.7–1.5 × 2–4 mm, acute at tip, persistent; bracteoles linear to lanceolate, acute at tip, 1- or 2-veined, persistent. **Pedicels** 3–7 mm, covered with minute hooked hairs, thicker and arched at fruit. **Flowers** with calyces campanulate, mostly 3–4 mm, upper lobes usually fused into an apiculum, lateral and lower lobes triangular to lanceolate; corollas pink to lavender, violet, or purple, ca. 1 cm long × 0.7 cm high; banner nearly circular, wider than long, 7–8 × 8–9 mm, emarginate at apex, slightly biauriculate toward base; wings obovate to oblong, mostly 1–1.2 cm; keel ca. 6 × 7 mm, mostly 1 1/2 coiled, coils at anthesis facing forward; staminal tube ca. 1.3 cm; ovules 7–10; style thickened distally, bearded introrsely; stigma introrse. **Fruits** linear-falcate, 3.2–7.5 × 0.5–0.9 cm, pendent, elastically dehiscent, compressed; valves thin to papery, expanding slightly over seeds, sparsely covered with appressed and hooked hairs; beak often 3 mm. **Seeds** 5–10; oblong, 3–6.5(–8) × 2.5–5.5 mm; smooth to slightly rugose, brown to ivory colored, punctuated and mottled with black.

Varieties 3 (2 native in the flora).

Phaseolus acutifolius sensu lato is a widespread polymorphic assemblage that shows a seemingly continuous series of variants, which have developed marked leaflet differences toward the extremes of an almost continuous range. These differences apparently reflect selective forces within microhabitats (Nabhan and Felger 1978), var. *tenuifolius* with narrow to extremely narrow linear-lanceolate leaflets and var. *acutifolius* with ovate-lanceolate leaflets. The extreme forms of var. *acutifolius* with small ovate leaflets are also recognized here as typical var. *acutifolius*, and what has been considered var. *latifolius* with broader leaflets is also here considered var. *acutifolius*.

Floral characters within *Phaseolus acutifolius* have proven to be fairly uniform. Nevertheless, the inflorescences and fruit size of var. *acutifolius* tend to be relatively larger than those of var. *tenuifolius*. However, in the case of fruit length, this seems to be a variable trait, and thus should not be over-emphasized. I have therefore, distinguished two weakly differentiated varieties of *P. acutifolius*. The characters used to distinguish these, mainly leaf shape and habitat preferences (usually minor in other groups), are fairly constant and have been important in past classifications.

1. Leaflets mostly ovate to widely ovate, acute at apex, at base mostly round to cuneate; terminal leaflets 2.5–8.5 × 1–5 cm; fruits 3.6–7 × 0.5–0.9 cm; seeds 4–6 × 3.3–5.5 mm ____ **1a. Phaseolus acutifolius** var. **acutifolius**
1′ Leaflets linear to lanceolate, acute to attenuate at apex, at base mostly truncate to subtruncate, often weakly to conspicuously lobed at base; terminal leaflets 3.5–11.5 × 0.3–2.3 cm; fruits 3.2–5.4 × 0.5–0.6 cm; seeds 3–5 × 2.5–4.5 mm ____ **1b. Phaseolus acutifolius** var. **tenuifolius**

1a. **Phaseolus acutifolius** A. Gray var. **acutifolius** * Native

Phaseolus acutifolius var. *latifolius* Freeman

Leaflets mostly ovate to widely ovate, acute at apex, at base mostly round to cuneate; terminal leaflets 2.5–8.5 × 1–5 cm. **Fruits** 3.6–7 × 0.5–0.9 cm. **Seeds** 4–6 × 3.3–5.5 mm.

Flowering Jun–Nov. Slopes, canyons, valleys, drainages, roadsides on various soils; [200–]850–1610[–1700] m elevation; Semidesert Grassland, Plains Grassland, Madrean Evergreen Woodland; Cochise, Gila, Pima, Santa Cruz cos.; New Mexico, Texas; Mexico (Baja California, Chihuahua, Coahuila, Durango, Jalisco, Nayarit, Querétaro, Sinaloa, Sonora).

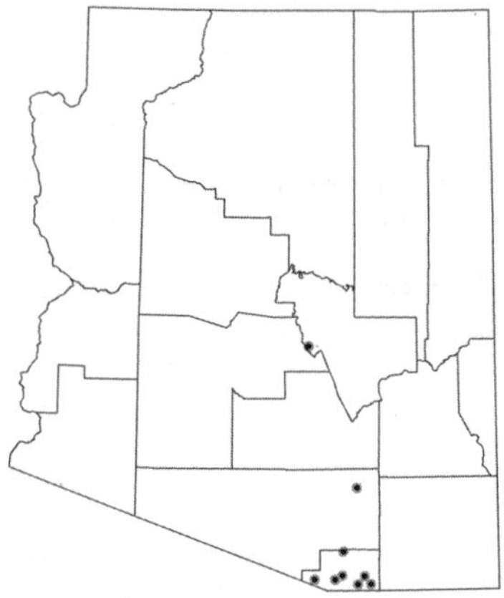

The leaflets of the type collection of *Phaseolus acutifolius*, as well as others from specimens of western Texas, are characterized by having smaller and narrower leaflets than those from populations elsewhere.

In Arizona, var. *acutifolius* is highly localized. It becomes more common southward in Mexico where it occurs in a variety of plant communities including thornscrub, tropical deciduous forest, grasslands, chaparral, and pine-oak woodland.

Tepary bean has a long history of cultivation by native people. It is still cultivated today on a limited basis and has potential as a commercial crop for hot, dry regions. It is able to mature its seeds more rapidly than other cultivated beans. Traditionally, teparies were grown by taking advantage of natural runoff from summer storms. With irrigation, the plants can produce mature fruits within two months (Nabhan 1985). If irrigation is available, seeds can be sown in the late spring after danger of frost is past. Scarification is not necessary. Teparies do not tolerate excessive moisture and require only occasional deep irrigation. With increased interest in conserving crop diversity and growing regionally adapted crops, many varieties of teparies have been made available through the non-profit seed bank Native Seeds/SEARCH in Tucson, Arizona. The plants are suitable as green manure. Nodulation is widely reported.

Tepary bean cultivars produce larger fruits and seeds than are typically found in wild populations with variable seed color. Fruits of cultivars are 5–9 × 0.8–1.3 cm and elastically to weakly dehiscent. Seeds are white, yellow-brown, tan-mottled, and blue-black, and there are 2–7 per fruit (Pratt and Nabhan 1988).

1b. Phaseolus acutifolius A. Gray var. **tenuifolius** A. Gray. Pl. Wright. 2:33. 1853. * [from Latin *tenui*, slender and *folium*, leaf] Native

Phaseolus tenuifolius (A. Gray) Wooton & Standley

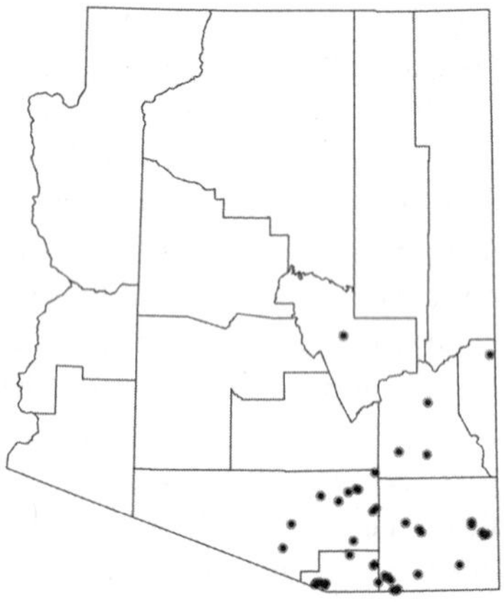

Leaflets linear to lanceolate, acute to attenuate at apex, at base mostly truncate to subtruncate, often weakly to conspicuously lobed; terminal leaflets 3.5–11.5 × 0.3–2.3 cm. **Fruits** 3.2–5.4 × 0.5–0.6 cm. **Seeds** 3–5 × 3.5–4.5 mm.

Flowering Jun–Nov. Slopes, canyons, valleys, drainages, roadsides in various soils; [900–]1070–2010 m elevation; Semidesert Grassland, Plains Grassland, (Interior Chaparral), Madrean Evergreen Woodland; Cochise, Gila, Graham, Greenlee, Pima, Pinal, Santa Cruz cos.; New Mexico; Mexico (Baja California, Chihuahua, Colima, Durango, Jalisco, Sinaloa, Sonora).

According to Johnston (1940) both localities cited in Gray's description of this variety are situated in New Mexico. The first location was in Grant Co. about 13 miles east of Silver City (Santa Rita del Cobre); the other at a locality closer to the Mexican boundary. Both collections apparently were mounted on the same herbarium specimen at Gray Herbarium, Harvard University, and it is impossible to tell which plant was obtained from which site.

Phaseolus acutifolius var. *tenuifolius* is common and widespread during wet summers in southeastern Arizona. The stems trail along the ground and over rocks, or twine into adjacent vegetation. Plants typically grow in full sun but may be found in partial shade. Variety *tenuifolius* is more often associated with open, drier habitats than is var. *acutifolius*. *Phaseolus angustissimus* occurs within the range of *P. acutifolius* and might be confused with it. Plants of var. *tenuifolius* can be distinguished from *P. angustissimus* by their fibrous roots and many-seeded fruits. The seeds of var. *tenuifolius* are edible (Nabhan 1985).

2. Phaseolus angustissimus A. Gray. Pl. Wright. 2:33. 1852. * [from Latin *angusti*, narrow and *-issimus*, a suffix denoting an extreme degree, referring to the narrow leaflets] SLIMLEAF BEAN Native

Phaseolus angustissimus var. *latus* M.E. Jones; *Phaseolus dilatatus* Wooton & Standley

Herbaceous, perennial vines to 1.5 m, climbing or trailing, often forming dense growth; from thick and elongate woody taproots, ca. 20 cm long and 2.5 cm thick. **Stems** cylindrical, striate, usually more than one from root crown, often lowermost branches opposite; stems glabrous or sparsely to densely covered with hooked hairs, purplish pigmented, later

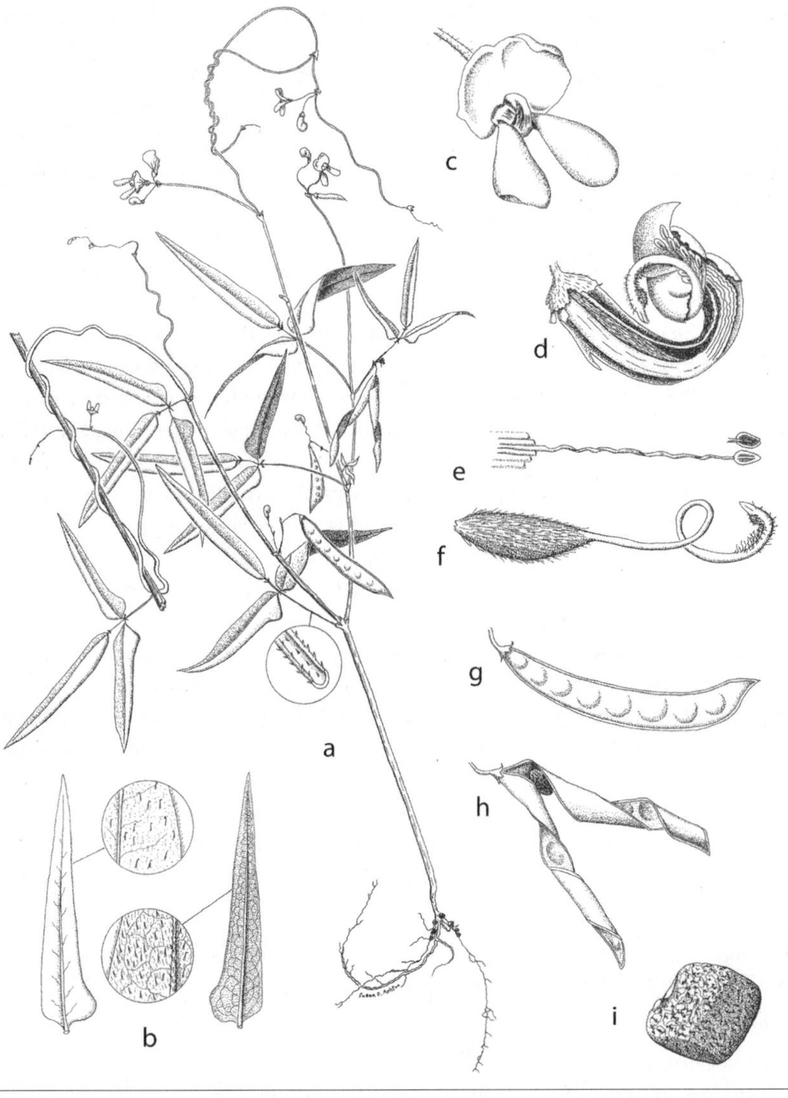

Phaseolus acutifolius var. *tenuifolius*: **a.** habit; **b.** leaflet with surface detail; **c.** flower; **d.** stamens; **e.** filament with anther; **f.** pistil; **g.** fruit; **h.** open fruit showing seeds; **i.** seed. Illustrator: Susan E. Ashton.

woody and persistent. **Leaves** with stipules ovate to triangular, 2.5–4 × 1–2 mm, often faintly lobed at base, 3–5-veined, spreading to reflexed; petioles 2–5 cm; rachises 0.8–1.7 cm; upper stipels linear, ascending or spreading; lower stipels narrowly triangular to lanceolate, ascending or spreading. **Leaflets** thin to somewhat leathery, linear, narrowly oblong to lanceolate or narrowly trullate, 4–5.5 × 0.2–1.7 cm, often at base on each side with an angular corner, or narrow, short, usually obtuse at tip, or long attenuate, truncate to broadly cuneate at base; the blades covered with minute hooked hairs, with straight hairs along the veins beneath and on margins. **Inflorescences** of pseudoracemes, to 20 cm, rarely with axillary flowers at base; peduncles 3–11 cm; rachises 2–9 cm, with 2–6 biflorous nodes; peduncular basal bracts ovate, 0.5–1 × ca. 0.5 mm, more than 2 covering buds; primary nodal bracts ovate, 3–5-veined, acute at tip, persistent; secondary nodal bracts ovate to lanceolate, acute at tip, usually persistent; bracteoles ovate to lanceolate, often oblong,

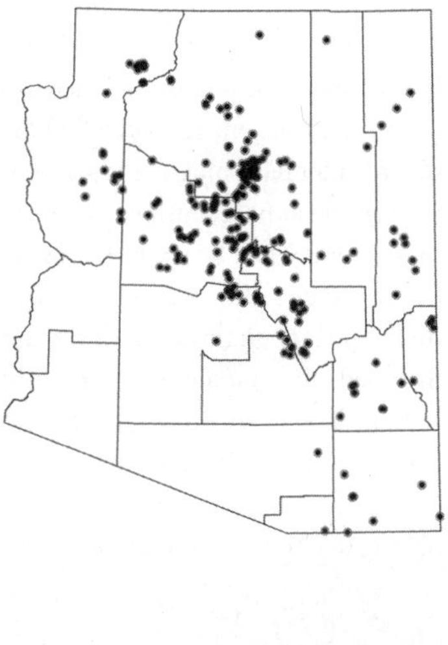

acute at tip, 1–3-veined, often densely covered with straight hairs, often at midpoint of pedicels, persistent. **Pedicels** 3–5 mm, arched at fruit. **Flowers** with calyces campanulate, 2.5–3.5 mm, prominently veined; upper lobes fused, emarginate or cleft at tip; lateral lobes mostly triangular, acute at tip; lower lobes triangular to narrowly so, acute at tip; corollas pink to light purple, ca. 1.2 cm long × 1 cm high; banner oblong to circular, ca. 1 × 1 mm, slightly emarginate at apex, biauriculate toward base; wings obovate, ca. 1.2 × 0.6 cm; keel ca. 5.5 long × 5 mm high, 1 1/4–1 1/2 coiled, coil twisted and facing forward; staminal tube ca. 1.3 cm; ovules 2–4; styles thickened distally, bearded introrsely; stigma introrse, tip laterally introrse. **Fruits** linear-falcate, 2.3–3 × 0.4–0.8 cm wide at widest point (in distal half), pendent, elastically dehiscent, compressed; valves thin, expanding slightly over seeds, sparsely covered with minute appressed hairs, beak ca. 1.5 mm. **Seeds** 3–4; oblong, 3.5–5 × 3–4 mm; rugose, reticulate-areolate patterns with rounded mounds; brown often mottled with black.

Flowering May–Oct. Slopes, canyons, valleys, stream banks and drainages, roadsides, disturbed sites, on various soils; (760–)1050–2560 m; (Sonoran Desertscrub, Chihuahuan Desertscrub, Semidesert Grassland), Interior Chaparral, Great Basin Conifer Woodland, Madrean Evergreen Woodland, Petran Montane Conifer Forest; Apache, Cochise, Coconino, Gila, Graham, Greenlee, Maricopa, Mohave, Navajo, Pima, Pinal, Santa Cruz, Yavapai cos.; New Mexico, Texas; Mexico (Sonora).

Wright's specimen (951) at GH has two intermingled, at least in part, plants (both *Phaseolus angustissimus*). The material mounted in the left-hand side of the sheet (*Wright 500*) is rather scanty, consisting mainly of the basal portion of a plant; the plant on the right-hand side shows flowers and immature pods, features obviously used by Gray in his original description (*Wright 68*). This latter plant was therefore selected to serve as the lectotype. I consider var. *latus* and *P. dilatatus* as merely forms of a variable *P. angustissimus*.

The present species is readily distinguished from others by a combination of characters including habit, linear leaflets, few floral nodes per inflorescence, and falcate fruits with 3–4 rugosely ornamented seeds.

Phaseolus angustissimus is most likely to be confused with *P. filiformis* because of a combination of characters including superficial appearance. Differences in root system, stipule length, ovule number, and fruit width are usually sufficient to separate the two.

In Arizona, *Phaseolus angustissimus* is widely distributed at mid-elevations and can be locally abundant in a variety of habitats. Low elevation plants are confined to riparian habitats. The plants grow in full sun or often in the shade of trees or twining upon shrubs. The leaves, flowers, and crushed roots have been reported by the Zuni to promote vigor in children (Stevenson 1915).

3. Phaseolus filiformis Bentham. Bot. Voy. Sulph.:13. 1844. * [from Latin *fili*, a thread and *forma*, shape] SLIMJIM BEAN, WRIGHT'S BEAN, SLENDER-STEMMED BEAN Native

Phaseolus wrightii A. Gray

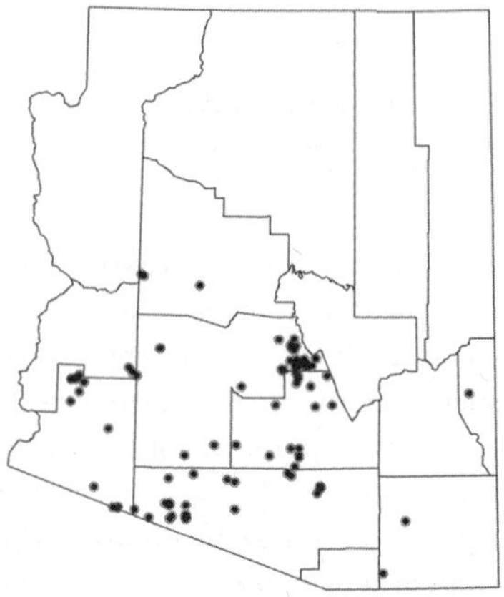

Herbaceous, annual or short-lived perennial vines, to 2.5 m; trailing or climbing, frequently forming matted growth, from slender taproots. **Stems** cylindrical, angulate, branching profusely from the base, often lowermost stems opposite; sparsely to densely cover with oblique to appressed straight hairs, retrorse; interspersed with hooked hairs. **Leaves** with stipules ovate to broadly triangular, 1–2.5 × 0.8–2.4 mm, acute, 3–5-veined, usually ascending; petioles 2–5.5 cm; rachises 0.5–2 cm; upper stipels linear, spreading or ascending; lower stipels linear to ovate-lanceolate. **Leaflets** membranous to rigid, ovate in overall outline, entire to deeply lobed toward base, lobes squarish-obtuse to oblong-attenuate, mostly 1–5 × 1–4.5 cm, obtuse to acute at tip, broadly cuneate to truncate at base; both surfaces covered with hooked hairs, mostly straight hairs along veins beneath. **Inflorescences** of pseudoracemes, mostly 1.2–16 cm; peduncles often angulate, 1–10 cm; rachises 0.2–7.5 cm with 2–6 biflorous nodes; peduncular basal bracts ovate; primary bracts ovate to lanceolate, acute, 3-veined, often pigmented, persistent; secondary bracts ovate to narrowly triangular, 0.5–1 × 2–5 mm, acute, often persistent; bracteoles oblong-ovate to narrowly triangular, acute or obtuse, 1–3-veined, persistent. **Pedicels** 2–5 mm, arched in fruit. **Flowers** with calyces campanulate, 2–3.5 mm, upper lobes fused, emarginate, lateral and lower lobes triangular, acute; corollas pink, lavender, reddish to light purple, often white fading yellowish, ca. 1 × 0.8 cm; banner oblong to circular, ca. 1 × 0.8 cm, emarginate at apex, biauriculate; wings obovate, ca. 1 × 0.45–0.7 cm; keel ca. 6 × 5.5

mm, mostly 1 1/2 coiled; staminal tube ca. 1.3 cm; ovules 3–6; styles introrsely bearded; stigma introrse. **Fruits** linear-falcate, 2.3–3.5 × 0.4–0.5 cm, pendent, elastically dehiscent, compressed; valves membranous, expanding over seeds, sparsely covered with straight and hooked hairs; beak ca. 2 mm. **Seeds** 4–6; oblong to squarish, rarely reniform, 2–4.3 × 2.3–4 mm; rugose, reticulate-alveolate, mounds with spherical tips in magnified view; brown colored.

Flowering nearly year-round with adequate moisture. Ridges, slopes, cliffs, canyons, valleys, lava flows, stabilized dunes, arroyos, roadsides, among rocks, or in sandy soils often derived from gypsum, limestone, and volcanic substrates; [near sea level–]240–1430(–1830) m; Sonoran Desertscrub, (Interior Chaparral, Madrean Evergreen Woodland); Cochise, Greenlee, La Paz, Maricopa, Pima, Pinal, Yavapai, Yuma cos.; California, New Mexico, Texas; Mexico (Baja California, Baja California Sur, Chihuahua, Coahuila, Durango, Sonora).

The nomenclature of *Phaseolus filiformis* seems to be moderately straightforward. However, it is noteworthy to consider that Gray cited in the protologue of *P. wrightii* that the species was very close to *P. filiformis* and that it may be best treated as a variety of it. A year later, in Pl. Wright. 2:33. 1853, Gray modified the concept of *P. wrightii* when he misjudged *C. Wright 952* (*P. grayanus*) identifying it as a better representative of the species type (*C. Wright 108*). This led to the interpretation that the second collection represented a "better type." As a result, Kearney and Peebles (1939) reduced *P. grayanus* to a variety of *P. wrightii*, creating additional confusion.

According to Bentham (1844), the fruits of *Phaseolus filiformis* may contain up to eight seeds, more than are found in the specimens studied. *Phaseolus filiformis* is easily distinguished by the combination of the following features: slender taproot or fibrous root system, conspicuous variation in leaflet lobes, small number of flowers per inflorescence, falcate fruits, and seeds with reticulate-areolate surfaces.

Phaseolus filiformis is fairly common in desert habitats in southwestern and central Arizona during wet periods. It is uncommon and localized in chaparral and oak-juniper-pine communities. In Mexico, it can be found in coastal vegetation and thornscrub. The stems trail on the ground or twine on other plants. Flowers, mature seeds, and immature fruits were eaten by native people in southwestern Arizona (Nabhan 1985; Newton 2013).

4. Phaseolus maculatus Scheele. Linnaea 21:465–466. 1848. * [from Latin *maculat*, spotted] METCALF BEAN; *FRIJOLILLO* Native

Phaseolus metcalfei Wooton & Standley; not *Phaseolus maculatus* Martius in Mart. & Schrank, Hort. Reg. Monac. 185. 1829

Herbaceous, perennial, coarse trailing or climbing vines to 6 m, from large, woody taproots to 30 cm long or longer and to 20 cm broad. **Stems** coarse, angulate, covered with hooked hairs, often with oblique to appressed hairs, retrorse. **Leaves** with stipules ovate to lanceolate, 0.5–1 × 0.1–0.25 cm, acute, often slightly lobed towards the base, 5–6-veined, spreading, later reflexed; petioles and rachises deeply grooved longitudinally; petioles 1.5–6.5 cm; rachises 1.4–3.6 cm; upper stipels ovate, ascending; lower stipels ovate to triangular, 3-veined, spreading or retrorse. **Leaflets** (occasionally the first pair of leaves on new growth can be unifoliolate or bifoliolate), leathery [those of central México undulate when dry], widely ovate to very widely depressed ovate or rhombic, terminal leaflets 3–9.5 × 3–7.4 cm, lateral leaflets 3–7 × 2.5–5 cm, obtuse to rounded, often acute, retuse to emarginate at tip, strongly apiculate, base rounded to subcordate, or subtruncate to cuneate; both surfaces covered with minute hooked hairs, margins and often lower surfaces, along prominent veins with straight hairs. **Inflorescences** of pseudoracemes, to 50 cm; peduncles often with lateral inflorescences, occasionally with a floral node close to the base, to 28 cm; rachises 5–24 cm with 6–20 nodes, often proximal nodes subopposite, mostly biflorous; peduncular basal bracts 2 or more, mostly covering buds, triangular to lanceolate; primary nodal bracts lanceolate, occasionally circular, 3–6-veined, persistent; secondary nodal bracts linear, often deciduous; bracteoles lanceolate to oblanceolate, often club-shaped, 0.8–2 × 0.1–0.3 mm, 1-veined, often persistent. **Pedicels** 0.35–0.7(–1) cm, covered with hooked hairs, often with scattered straight hairs, arched to reflexed in fruit. **Flowers** with calyces campanulate, mostly 4–5.5 mm, upper lobes fused, emarginate (often developing as partially divided, triangular, obtuse to acute), lateral lobes triangular, acute, lower lobes narrowly triangular, acute; corollas reddish, violet, bright pink to purple, 1–1.4 long × ca. 1 cm high; banner oblong to obovate, ca. 1 × 1 cm, emarginate at apex, auricles small; wings obovate to widely obovate, 1–1.5 × 0.6–0.75 cm,; keel ca. 7 × 6 mm, mostly 1 1/2 coiled; staminal tube ca. 1.5 cm; ovules 3–5; style with a short, introrse beard; stigma introrse. **Fruits** oblong, somewhat curved, obovate when 1-seeded, 3–7.4 × 1.2–1.7 cm, sessile or rarely stipitate, compressed, at times with constriction in proximal half, pendent, weakly dehiscent; valves leathery, expanding slightly over seeds, mostly glabrous; beak curved. **Seeds** (1–)3–4 (or 5); nearly circular to nearly square, (0.6–)0.8–1.3 cm broad; smooth, brown, pointed, mottled and often streaked with black.

Flowering Jun–Oct. Ridges, slopes, canyons, valleys, along streams, and roadsides, mostly on well-drained sandy or rocky soils; 1520–2130[–2700] m; Semidesert Grassland, Madrean Evergreen Woodland, (Petran Montane Conifer Forest); Apache, Cochise, Coconino, Gila, Greenlee, Navajo, Pima, Yavapai cos.; New Mexico, Texas; Mexico (Aguascalientes, Chihuahua, Coahuila, Durango, Hidalgo, Jalisco, Puebla, San Luis Potosí, Sonora, Guanajuato, Querétaro, Tlaxcala, Zacatecas).

Phaseolus maculatus appears to be composed of three races. The first is a northern race (Texas, New Mexico, Arizona, Sonora and Chihuahua) which includes the type of *P. maculatus* and is characterized by leaves with rigid margins and mostly large, passively or weakly dehiscent fruits. Curiously, the type specimen of *P. maculatus* seems to be elastically dehiscent. The second race is an interior or more eastern race with mostly rhombic leaflets with undulate margins. The fruits although weakly dehiscent appear shorter and narrower than the northern race. The second race is from scattered localities in central Zacatecas, Aguascalientes, San Luis Potosí, Querétaro, Hidalgo, Tlaxcala, and Puebla; which includes the type of *P. retusus*. The third race is a western race, characterized by its appressed cauline pubescence as well as the rigid leaflet-margins and small, elastically dehiscent fruits with small seeds and circular bracts (ca. 2 mm broad). It is perhaps the most distinctive of the three, ranging from the southern corner of Durango to southwestern Zacatecas, northern Jalisco, western Aguascalientes, and central Guanajuato, thus, including the type of *P. venosus*. The third race perhaps deserves taxonomic recognition, but until more material is collected it seems better to treat these as a single species. Without doubt these three races of *Phaseolus maculatus* are more closely related one to another than to another species.

Phaseolus maculatus is closely related to *P. ritensis*. Both possess deep and enlarged taproots, trailing habits, coriaceous leaves with short petioles and long inflorescences with few to many flowers. Additional comparisons with *P. ritensis* will be found under that species. *P. maculatus* and *P. ritensis* have parallel distributional ranges and it seems reasonable to suppose that their histories have followed parallel pathways, divergent by natural selection and by human intervention (Nabhan et al. 1980). The leaves and seeds of this species are reported to be eaten by masked bobwhite quail (Graham 1941).

In Arizona, *Phaseolus maculatus* occurs in widely scattered and usually small populations. Individual plants can be impressively large with their leafy stems spreading over the ground for several meters beneath trees or on steep, rocky slopes. This species is most common in pine-oak woodland, but also occurs in more xeric plant communities across its range.

5. Phaseolus parvulus Greene. Bot. Gaz. 6:217. 1881. * [from Latin *parvi*, small] PINOS ALTOS MOUNTAIN BEAN Native

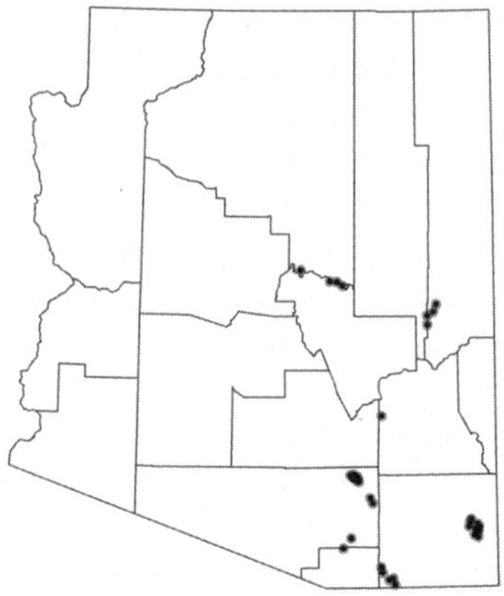

Herbaceous, perennial erect, trailing or twining vines, 5–50 cm, from small, deep-seated oblong or spherical taproot, 1–2 cm broad. **Stems** solitary or branching, cylindrical, mainly glabrous, sometimes sparsely covered with minute hooked hairs. **Leaves** with stipules ovate to ovate-lanceolate, 1.8–6.5 (–8.5) × 0.7–3(–4) mm, larger distally, acute to acuminate at tip, mostly 5–9-veined, sometimes red-purple pigmented, appressed to stem, reflexed or distally spreading; petioles 1.8–3.4 cm; rachises 0.4–1 cm; upper stipels subulate, present or absent; lower stipels ovate to lanceolate, ascending. **Leaflets** thin, linear to lanceolate, sometimes narrowly trullate, terminal leaflets 2.4–5.6 × 0.3–1.2(–2.7) cm, lateral leaflets 2–4.7 × 0.5–1.3(–2.8) cm, occasionally the terminal leaflets with two basal lobes on each side, lateral leaflets often more or less distinctly rounded-lobed at the base on one or both sides, rounded or acute at tip, obscurely apiculate, attenuate at base; with minute, glandular hairs on lower surfaces, upper surfaces glabrous. **Inflorescences** of precociously flowering pseudoracemes, 2–18 cm; peduncles 2–12 cm; rachises 2–6 mm with 1–3 (usually 1-flowered) nodes; peduncular basal bracts absent; primary nodal bracts ovate, 3-veined, often deciduous; secondary nodal bracts deciduous; bracteoles ovate, present or absent. **Pedicels** 2–4.5 mm, straight to slightly arched in fruit, sometimes purple pigmented. **Flowers** with calyces campanulate-tubular, with upper basal portion rounded, papillose, 5–8 mm, upper lobes entirely fused or almost so, acute, lateral lobes triangular, sometimes falcate, acute, lower lobes lanceolate; corollas violet, lavender-pink or purple, 1.5–2.5 cm; banner obovate, spatulate, 1.2–1.6 × 1–1.2 cm, obcordate at apex, biauriculate towards base, at anthesis the margins of the middle portion of the banner involute and clasped together, wrapping around the other petals and forming a tubule; wings obovate, 1.5–2.7 × 0.5–1 cm; keel ca. 1 × 0.4–0.6 cm, almost 2 coiled; staminal tube 1.6–2.5 cm; ovules 10–12; style with sparse, introrse-lateral beard; stigma introrse, laterally twisted by style torsion. **Fruits** linear-straight to slightly falcate, 3.5–4.5 × 0.3–0.4 cm, horizontal to pendent, elastically dehiscent, cylindrical; valves leathery, mostly glabrous, beak ca. 1 mm. **Seeds** 8–11; mostly oblong, reniform, 2–3.7 × 1.7–2.6 mm; smooth, brown to light brown or greyish green, punctuated and mottled black, occasionally light green.

Flowering Jul–Sep. Slopes, canyons, drainages, wet meadows, roadsides, and disturbed sites in deep soils or rocky soils often derived from sandstone or igneous substrates; (1750–)2000–2840[–3000] m; (Madrean Evergreen Woodland), Petran Montane Conifer Forest; Apache, Cochise, Coconino, Graham, Navajo, Pima, Santa Cruz cos.; New Mexico; Mexico (Chihuahua, Durango, Sinaloa, Sonora, Zacatecas).

Phaseolus parvulus is a high elevation species that ranges from the mountains of Arizona and southwestern New Mexico, southward along the axis of the Sierra Madre Occidental in Mexico. In Arizona it is rather localized but can be common in suitable habitats. The plants typically trail along the ground beneath forest trees or in the open.

6. Phaseolus ritensis M.E. Jones. Contr. W. Bot. 12:14. 1908. * [for the Santa Rita Mountains, Arizona, where the species was first collected] SANTA RITA BEAN, EGGLEAF STRING BEAN; *COCOLMECA, FRIJOL DEL MONTE* Native

Phaseolus maculatus Scheele subsp. *ritensis* (M.E. Jones) Freytag; *Phaseolus ovatifolius* Piper

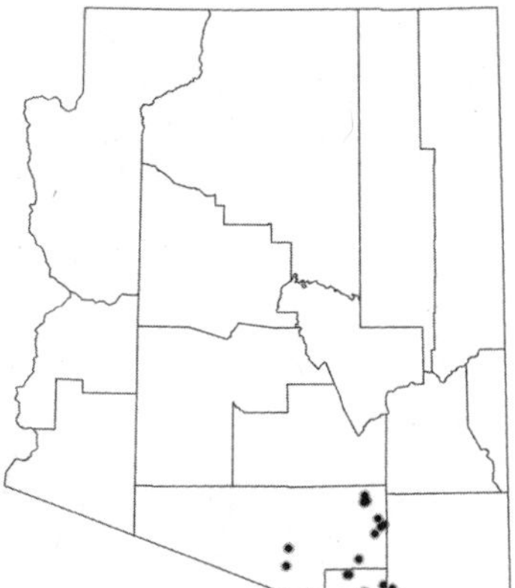

Herbaceous, perennial stiff-branched trailing vines, occasionally climbing, to 3 m, from large taproots, to 40 cm long × 20 cm broad. **Stems** coarse, angulate, scattered with hooked and straight hairs, retrorse, glabrate. **Leaves** with stipules triangular to lanceolate, 2–4.5 × 0.5–2 mm, acute, ca. 5-veined, spreading or reflexed; petioles reduced, shorter than terminal leaflet (often only the pulvinus present) 0.4–3.5 cm, with deep longitudinal grooves; rachises 0.5–2 cm, with deep longitudinal grooves; upper stipels oblong, acute, ascending or spreading; lower stipels ovate to triangular, spreading or retrorse; leaves often unifoliolate on the first through fifth nodes of new growth. **Leaflets** leathery, ovate to widely ovate, or trullate to widely trullate, occasionally lanceolate, terminal leaflets 3–8(–10.5) × 1.4–6(–9.5) cm, lateral leaflets 2.5–7(–8.6) × 1–5(–6.5) cm, often rounded or retuse at tip, strongly apiculate, at base oblique, cuneate or truncate; glabrous to sparsely covered with sraight and hooked hairs along veins and margins on both surfaces. **Inflorescences** mostly of pseudoracemes, to 85 cm; peduncles often with lateral inflorescences, to 40 cm; rachises to 45 cm, with 6–25 biflorous nodes (occasionally proximal nodes subopposite); peduncular basal bracts in groups, ovate to narrowly triangular, acute, covering buds; primary nodal bracts triangular, acute, mostly 3-veined, glabrous to strigillose, usually persistent; secondary nodal bracts linear to narrowly triangular, acute, 1-veined, deciduous; bracteoles triangular to lanceolate, acute, ciliate at tip, often persistent. **Pedicels** 0.3–1 cm, glabrous to sparsely-densely covered with hooked hairs, arched to often reflexed in fruit. **Flowers** with calyces campanulate, 3–4 mm; upper lobes almost fused, emarginate, obtuse, lateral lobes triangular, acute, lower lobes triangular, acute; corolla lavender to purple or violet, ca. 1.3 × 1 cm; banner obovate to oblong, ca. 1 × 1 cm, emarginate at apex, biauriculate towards base; wings obovate to oblong, 1–1.3 × ca. 0.6 cm; keel ca. 6 × 6.5 mm, 1 1/2 to almost 2-coiled; staminal tube ca. 1.3 cm; ovules 4–6; style bearded, mostly introrse; stigma introrse. **Fruit** oblong-falcate, (2.5–)3.5–5.3 × 0.6–1.1 cm, pendent, elastically dehiscent, compressed; valves leathery, mostly glabrous, expanding slightly over seeds; stipe to 5.5 mm; beak ca. 1 mm. **Seeds** 3–6; oblong to transversely oblong, 4–7.5 × 4–5.7 mm; grayish, light brown pointed, mottled and streaked with black.

Flowering Jul–Oct. Ridges, slopes, canyons, drainages, and roadsides in sandy, gravelly, or rocky soils often of igneous or volcanic origin, or ash; [600–]1200–2290[–2700] m; (Semidesert Grassland), Madrean Evergreen Woodland, (Petran Montane Conifer Forest); Cochise, Graham, Pima, Santa Cruz cos.; Texas; Mexico (Chihuahua, Durango, Jalisco, Nayarit, Sinaloa, Sonora).

Phaseolus ritensis occurs in scattered populations in the mountains of southeastern Arizona, extending southward along the Sierra Madre Occidental and outlying ranges in Mexico. It is found in pine-oak woodland and less commonly

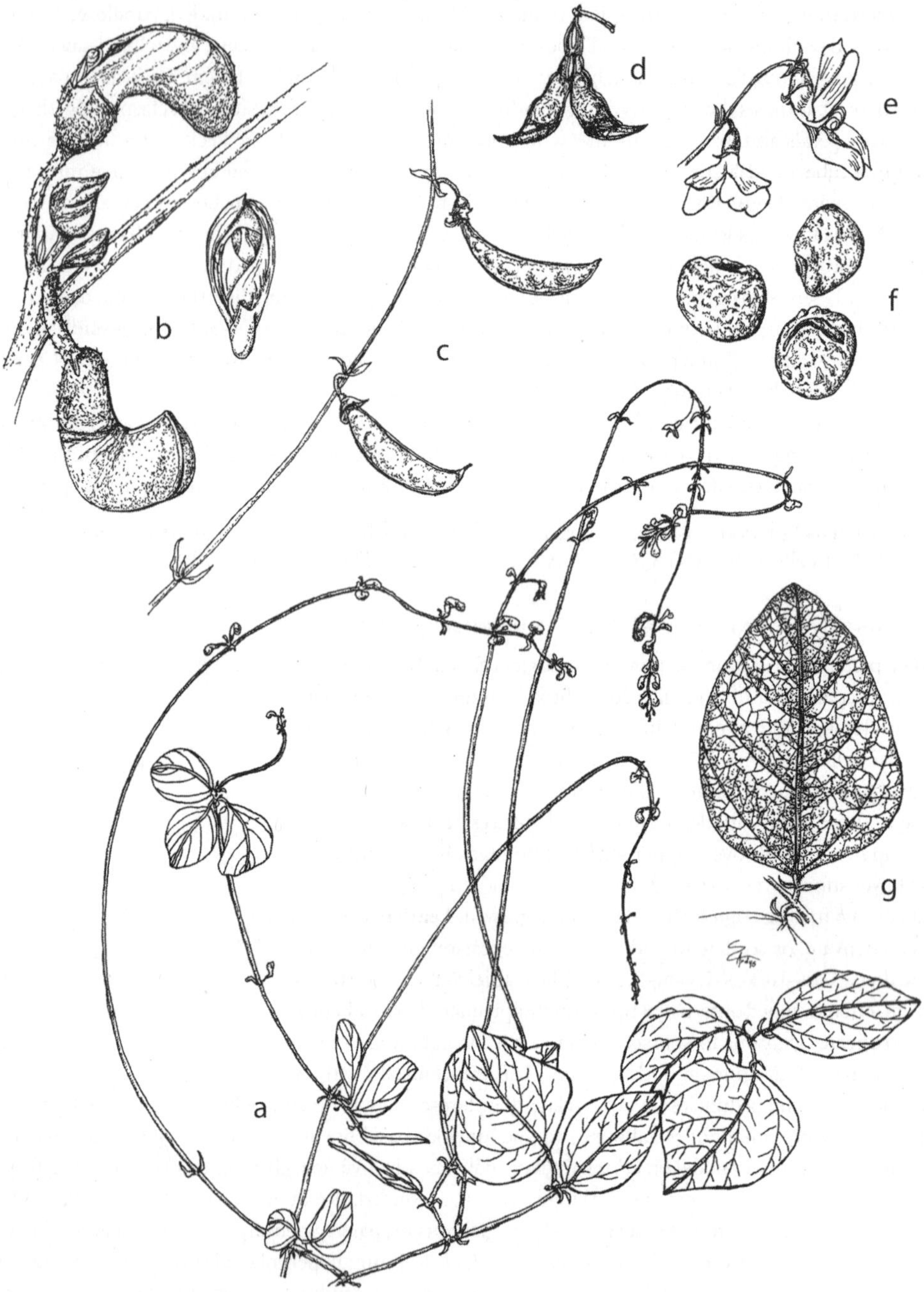

Phaseolus ritensis: **a.** habit; **b.** inflorescence; **c.** fruits; **d.** dehisced fruit; **e.** flowers; **f.** seeds; **g.** leaflet. Illustrator: Ardys Lurtsema.

in open oak woodland in the open or in shade. An unusual, low elevation specimen at 517 m was collected at Canyon Lake, Maricopa County (*A. Nelson 11234a*, 20 Mar 1930).

Roots of *Phaseolus ritensis* are reportedly used as a base for fermenting an alcoholic beverage '*batari*' by the Guarajío people of Mexico (Gentry 1942). The species is also reported to be used in folk medicine as a purgative and stimulant. The foliage has been used for forage. For additional information see Nabhan et al. (1980).

Phaseolus ritensis and *P. maculatus* are very closely related and thus, very similar in appearance. Both species, in flower, have been regularly confused. Jones, describing *P. ritensis*, sensed the similarity with *P. maculatus* (*P. retusus*) and explained that this species rarely has obtuse and never retuse leaflets tips; that the bracts are minute and not conspicuous and persistent as in *P. maculatus*; and the fruits of *P. ritensis* are distinctly falcate and long-stipitate. Jones' diagnosis specified quite neatly the differences between the species. Later, Nabhan et al. (1980) and Buhrow (1983) revealed new distinguishing features between both taxa such as eophyll morphology, number of leaflets on the third node, leaf color, inflorescence length, fruit length, fruit valves and their distinct degrees of dehiscence, and seed dimensions.

It is noteworthy to point out that even though *Phaseolus ritensis* primarily has leaflets with acute apices, plants from Nayarit and Jalisco with broader blades have obtuse to rounded apices.

Isely (1998) commented that, "The evident differences between *Phaseolus ritensis* and *P. maculatus* are primarily of vigor and size of plant parts; the two kinds overlap in such quantitative features. Therefore, their independent specific status has been considered dubious. But the stipitate legumes of *P. ritensis*, much smaller than the sessile ones of *P. maculatus*, are distinctive. This plus the quantitative differences otherwise seems to justify the independent status of the two taxa."

7. Phaseolus scabrellus Bentham ex S. Watson. Proc. Amer. Acad. Arts 17:346–347. 1882. * [from Latin scabr, rough and ellus, small] GRAY'S BEAN, SILVERSPOT LIMA Native

Phaseolus grayanus Wooton & Standley; *Phaseolus pedicellatus* Bentham var. *grayanus* (Wooton & Standley) A. Delgado ex Isely; *Phaseolus wrightii* A. Gray var. *grayanus* (Wooton & Standley) Kearney & Peebles

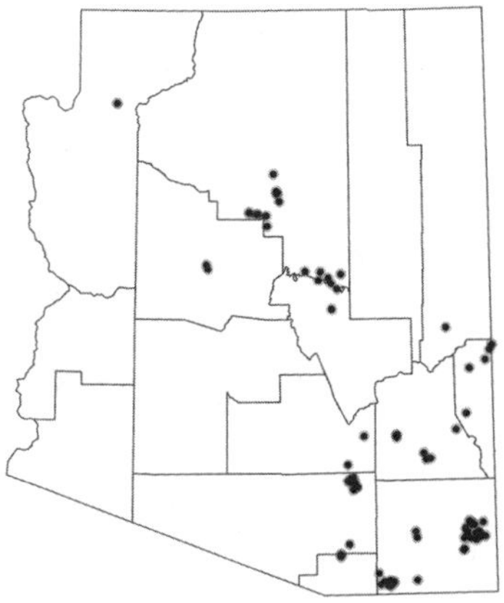

Herbaceous, perennial, usually climbing vines, developing from thick, elongate, tuberous taproots. **Stems** angulate, sometimes with longitudinal grooves, glabrous or sparsely to densely covered with retrorse straight and hooked hairs, often purple pigmented. **Leaves** with stipules oblong to triangular or lanceolate, 3–5.5 × 1–2.5(–3) mm, acute or obtuse at tip, 3–7 veined, ascending, spreading, or reflexed; petioles 1–6.5(–7.3) cm; rachises 0.5–2 cm; upper stipels oblong-subulate, 1–2 veined, spreading. **Leaflets** membranous to firm, broadly ovate, rhomboid, deltoid, [or lanceolate], terminal leaflets (1.5–)2.5–6.5(–7.5) × (0.5–)1.5–4.6(–7.5) cm, lateral leaflets (0.7–)2–5.5(–6.5) × (0.4–)1.3–4(–5.5) cm, faintly or deeply lobed, middle lobe triangular to oblong or linear-lanceolate, basal lobe rounded to squarish, acute at tip, rounded to truncate at base; upper surfaces glabrous to sparsely covered with ascending hairs, mainly along the veins and often variegated with a pale area along midrib (at least in Arizona populations). **Inflorescences** of pseudoracemes, often with lateral floral branches, mostly (1–)5–30(–40) cm; peduncles 1.5–18.5 cm; rachises mainly 2.5–10 cm, with 3–14 often biflorous nodes; peduncular nodal bracts ovate to lanceolate, acute at tip, 1–3-veined; primary nodal bracts ovate to lanceolate or broadly ovate, acute, 3–9 veined, entire or lobed, persistent; secondary nodal bracts linear-ovate to narrowly triangular, acute at tips, 1-veined, usually deciduous; bracteoles ovate to linear-lanceolate, acute or obtuse at tip, 1–3-veined, usually persistent. **Pedicels** 0.3–1 cm, covered with hooked hairs only or with ascending and interspersed hooked hairs. **Flowers** with calyces campanulate, 2–3(–4) mm, upper lobes fused, emarginate, lateral and lower lobes triangular; corolla pink, purple, rarely white, 0.12–0.15 cm long × 0.7–1 cm high; banner oblong to circular, ca. 1 × 1 mm, emarginate at apex, biauriculate toward base; wings obovate, 1.2–1.6 × 0.5–0.8 cm; keel 5.5–9 mm long × 5–7 mm high, 1 1/2–1 3/4 coiled, twisted and facing forward; staminal tube ca. 1.5 cm; ovules 3–6; style bearded introrsely; stigma introrse. **Fruit** linear-falcate, 3–4.5(–5.1) × 0.8–1(–1.5) cm, pendent, elastically dehiscent, compressed; valves thin, expanding slightly over seeds, sparsely covered with ascending hairs. **Seeds** 2–5; oblong to reniform, 4.2–7.5 × 3.5–7 mm; smooth, brown or gray, pointed and mottled with black.

Flowering Jun–Oct. Slopes, valleys, canyons, stream banks, meadows, and roadsides on sandy, gravelly, or rocky soils often derived from limestone or igneous rocks. 1600–2590[–2800] m; Madrean Evergreen Woodland, Petran Montane Conifer Forest; Apache, Cochise, Coconino, Gila, Graham, Greenlee, Mohave, Pima, Pinal, Santa Cruz, Yavapai cos.; New Mexico, Texas; Mexico (Chihuahua, Coahuila, Durango, Querétaro, San Luis Potosí, Sonora, Zacatecas).

Phaseolus scabrellus is mainly restricted to the Chihuahuan Desert Region and adjacent mountains, reaching as far south as Querétaro. In Arizona it is characterized by its usually deeply 3-lobed and often variegated leaflets, rounded to squarish basal lobes, and smaller flowers with globose appendages at the base of the vexillary stamen.

In Arizona, *P. scabrellus* is widespread and locally common in mountainous areas, usually as an understory plant in

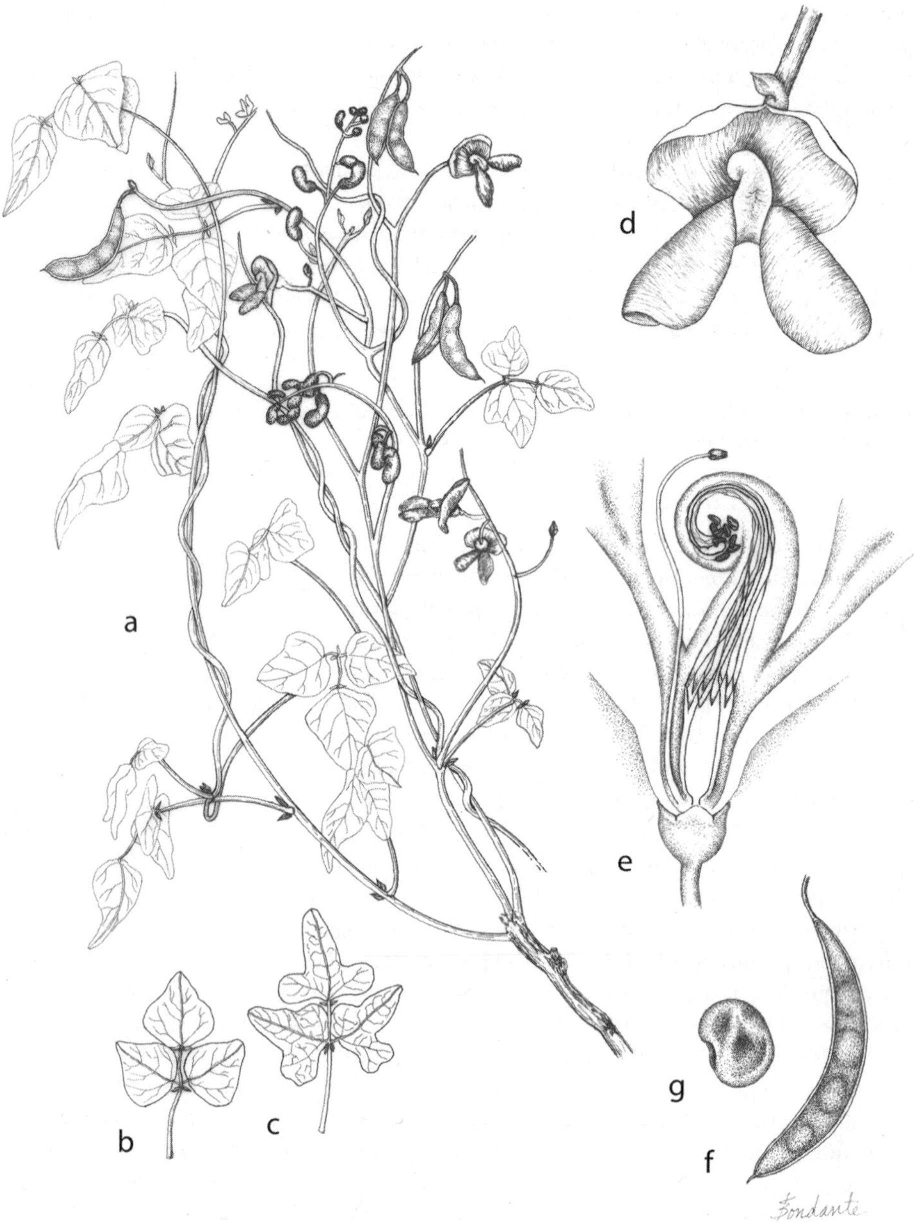

Phaseolus scabrellus: **a.** habit; **b., c.** leaves; **d.** flower; **e.** longitudinal section of flower; **f.** fruit; **g.** seed. Illustrator: Chris Bondante.

open pine forest or pine-oak woodland, and in riparian habitats. It is also found in chaparral and scrub across it range. A low elevation record of 752 m is documented from Aravaipa Canyon, Pinal Co. (*D. Wolgast 197*, 12 Sep 2008). The plants may form mats of trailing stems on the ground or climb upon adjacent vegetation.

PISUM Linnaeus. Sp. Pl. 2:727. 1753. * [Greek *pisos*, the pea plant] PEA
Steven L. Broich

Herbs, annual, mostly glabrous. **Stems** erect or sprawling, branched, ribbed. **Leaves** even-pinnate; rachis winged, usually terminating in a branched tendril; stipules present, foliose, without nectariferous patch; stipels absent. **Leaflets** [2–]4–6[–8], paired, folded in bud, blade ovate to lanceolate, usually decreasing in size from base of rachis, margins entire or dentate. **Inflorescences** racemes, axillary, erect or lax, 1–3[or 4]-flowered; bracts quickly deciduous; bracteoles absent. **Flowers:** papilionoid; calyx base symmetrical, tube bell-shaped, symmetrical, lobes 5, equal, lanceolate, exceeding tube; corolla petals differentiated into claw and blade; banner blade folded perpendicular to axis of flower, claw convolute around other petal claws; wing petals adherent to keel petals by means of a boss and pocket system, usually subequal to keel petal; keel incurved, attenuate, proximal and distal margins fused; stamens 10, 9 united and 1 completely or partially free; anthers uniform; ovary glabrous or pubescent; style dorsally compressed, folded longitudinally, upper (adaxial)surface pubescent. **Fruits** sessile, oblong-linear, cylindrical, without septa, margins usually obscure, splitting; valves twisted after dehiscence. **Seeds** 3–10 per fruit, usually yellow or green, sometimes mottled, spherical or angular, smooth or papillose, sometimes wrinkled. $x = 7$.

Species 3 (1 cultivated in the flora): wild species native to s Europe, sw Asia, n Africa; cultivated forms introduced in North America, n Europe, n, c Asia.

Pisum consists of three species (Kosterin and Bogdanova 2008, Smýkal et al. 2011), two of which are cultivated and one known only in the wild. *Pisum sativum* has been divided into two primary subspecies (Maxted and Ambrose 2001) one of which, subsp. *sativum*, includes both culinary and forage varieties widely cultivated in North America. Recent phylogenetic studies have placed *Pisum* consistently within the Vicieae (Wojciechowski et al. 2004).

Pisum has served as a food source for humans since prehistoric times and is used for livestock feed, as a cover crop, and for green manure (Zohary and Hopf 1973). The Austrian monk Gregor Mendel, known as the father of modern genetics, worked out the basics of genetic inheritance in experiments using *P. sativum*.

1. Pisum sativum Linnaeus. Sp. Pl. 2:727. 1753. * [Latin, cultivated] COMMON PEA, FIELD PEA, GARDEN PEA, SUGAR PEA Cultivated

Stems 1–20 dm, branched basally, 1–4 times. **Leaves** 5–10 cm, flattened, tendrils well developed; stipules ovate to lanceolate, semi-sagittate, 15–80 × 10–40 mm, usually larger than leaflets. **Leaflets:** 20–50 × 15–30 mm. **Inflorescences** 3–10 cm. **Flowers** 15–30 mm; corolla white or bicolored. **Fruits** 50–90 × 15–20 mm. **Seeds** 4–12 mm diam.

Subspecies 2, 1 wild, 1 cultivated, both with multiple varieties/cultivars (1 cultivated in the flora): subsp. *sativum* cultivated and introduced in North America, n Europe, n, c Asia; subsp. *elatius* native to nw Africa, the Middle East, w Asia. Both varieties cultivated in North America belong to subsp. *sativum*.

Nodulation is widely reported for *Pisum sativum*.

1. Corollas bicolored (banner lilac, wings darker reddish purple); seeds 4–8 mm, angular, sometimes mottled ____ **1a. Pisum sativum** subsp. **sativum** var. **arvense**

1′ Corollas white; seeds 6–12 mm, spherical, not mottled ____ **1b. Pisum sativum** subsp. **sativum** var. **sativum**

1a. Pisum sativum Linnaeus subsp. **sativum** var. **arvense** (Linnaeus) Poiret in Lamarck et al., Encycl. 5:456. 1804. * [Latin *arvens*, of fields] FIELD PEA Cultivated

Pisum arvense Linnaeus, Sp. Pl. 2:727. 1753

Flowers: corollas bicolored, banner lilac, wings darker reddish purple. **Seeds** 4–8 mm, angular, sometimes mottled. $2n = 14$.

Flowering dependent on region, a cool season plant. Occasionally escaped, ruderal habitats; introduced Idaho, Minnesota, Montana, North Dakota, South Dakota, Washington, Wisconsin; Canada (Alberta, Manitoba, Saskatchewan); native to nw Africa, the Middle East, c and s Asia.

Variety *arvense* is cultivated for soil improvement or as a seed crop for export.

1b. Pisum sativum Linnaeus subsp. **sativum** var. **sativum** * GARDEN PEA Cultivated

Flowers: corollas white. **Seeds** 6–12 mm, spherical, not mottled. $2n = 14$.

Flowering dependent on region, a cool season plant. Introduced and grown widely in Canada and US; native to nw Africa, the Middle East, c and s Asia.

Variety *sativum* is cultivated as a cool season garden vegetable crop. It is occasionally grown in Arizona gardens.

POMARIA Cavanilles. Icon. 5:2, pl. 402. 1799. * [named for Dominic Pomar, 1598–1621, botanist and physician of Philip III of Spain]

Beryl B. Simpson

Melanosticta de Candolle, Prodr. 2:484. 1825

Subshrubs or perennials herbs, unarmed, with black, clustered roots or thick taproots. **Stems** unarmed; villous, pubescent, pilose, or canescent, often with orange (drying black) multicellular glandular-punctate trichomes (appearing as dots) and complex multicellular projections consisting of a cylindrical stalk bearing plumose and/or apical trichomes. **Leaves** odd-bipinnate, with point of pinna attachment often marked by a ring of orange punctate (black when dry) glandular trichomes and thread-like stipels; stipules persistent or tardily abscissing, fringed, serrate, or laciniate, often villous or pubescent and with glandular-punctate trichomes. **Leaflets** variously vestitured with simple trichomes but always with conspicuous orange (black on drying) glandular-punctate trichomes on the lower (abaxial) surface. **Inflorescence** terminal or axillary racemes. **Pedicels** present. **Flowers** caesalpinoid, pedicellate, bilaterally symmetrical; sepals 5, fused at the base, valvate in bud, entire, lanceolate, acute, unequal in size with the lowermost (abaxial) largest, projecting between the 2 lower lateral petals and with the stamens and gynoecium nestled inside, lower (abaxial) surface vestitured on the margins and veins and with glandular-punctate trichomes between the margins and veins producing a striped appearance, abscissing in fruit with the sepal bases forming a small ring at the base of the legume; petals 5, distinct, more or less spatulate, arranged with the innermost petal upright forming the banner and the others in two lateral pairs, clawed, yellow usually blotched with red, often fading to pink; stamens 10, unequal in size, arched and lying in the lowermost sepal with the two basally abruptly bent, lower (abaxial) filaments densely villous and projecting into the lower opening of the banner petal, variously vestitured along the filaments; anthers 1 mm, attached dorsally, usually red; gynoecium consisting of an ovoid ovary often villous and densely covered with glandular-punctate trichomes mixed with complex multicellular projections, style arched, stigma impressed, lateral. **Fruits** dehiscent along both sutures, usually multiseeded, laterally compressed, mucronate by the persistent style, abruptly narrowing at the base and with the valves and margins variously vestitured with simple trichomes, glandular-punctate trichomes and a variable number of complex multicellular projections consisting of a cylindrical stalk bearing plumose and/or stellate trichomes. **Seeds** 1–4; 3–7 × 2–5 mm, compressed, borne on broad funicles.

Species 15 (1 native in the flora): occurring amphitropically in the subtropical areas of North America (9 species), South America (3 species), and southern Africa (3 species).

As the synonyms of *Pomaria jamesii* indicate, the genus was confounded for years with *Hoffmannseggia* (and also *Caesalpinia*). The confusion with *Hoffmannseggia* occurred because species of both are small shrubs or perennial herbs, while most Caesalpinia species are large shrubs or trees. The confusion with *Caesalpinia* occurred because of *Caesalpinia caudata*, described as a *Hoffmannseggia* and one of the few small shrubby species of *Caesalpinia* in North America (Simpson 1998). Morphologically *Pomaria* is readily distinguished from both by the lateral stigmas and the presence of complex multicellular processes on the fruits. Phylogenetic studies have shown that *Pomaria* is sister to *Erythrostemon*, a segregate of *Caesalpinia* in the broad sense.

SELECTED REFERENCES: Gagnon et al. 2016, Simpson 1998, Simpson and Lewis 2003, Simpson et al. 2006.

1. Pomaria jamesii (Torrey & A. Gray) Walpers. Repert. Bot. Syst. 1:811. 1843. * [for Edwin James, 1797–1861, American botanical explorer] FALSE MESQUITE, RUSH-PEA, JAMES' RUSH PEA, JAMES' HOLDBACK Native

Hoffmannseggia jamesii Torrey & A. Gray, Fl. N. Amer. 1(3):393. 1840; *Caesalpinia jamesii* (Torrey & A. Gray) Fisher; *Hoffmannseggia jamesii* Torrey & A. Gray var. *popinoensis* Fisher; *Larrea jamesii* (Torrey & A. Gray) Britton

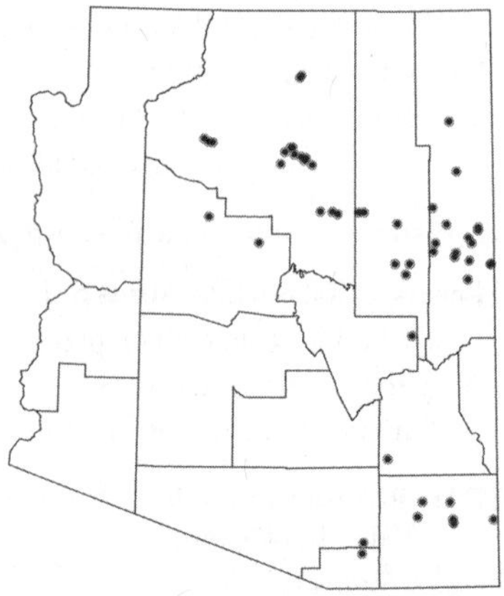

Herbaceous perennials to 50 cm with a thick, spindle-shaped taproot. **Stems**: old stems yellow, striate, puberulent. **Leaves** with 2–3 pairs of lateral pinnae and a terminal one, 42–75 (including petiole) × 25–60 mm; stipules linear to lanceolate, more or less entire, pubescent with a few glandular-punctate trichomes, 4 × 5 mm. **Leaflets** 5–10 pairs per pinna, oblong with a rounded apex, 2.5–7 × 1–2.5 mm with a few scattered trichomes to moderately strigose above (adaxially), densely strigose or covered with curly trichomes mixed with glandular-punctate trichomes, especially marginally, below (abaxially). **Inflorescences** lateral and terminal, 12–15 cm, bearing 16–26 flowers. **Pedicels** simple. **Flowers** generally drooping at anthesis, top-shaped in lateral view, 6–10 × 9–10 mm; sepals villous below (abaxially) or with short, curled trichomes at the base, tips, and margins with scattered glandular-punctate trichomes; lowermost sepal 7–9 × 4 mm; lateral sepals 6–9 × 9–10 mm; banner bright yellow with red markings, 5–9 × 2–2.5 mm, villous inside the fold of the claw and bearing glandular-punctate trichomes below (abaxially) on claw and basal portion of the blade; lateral petals yellow, sometimes red basally, 6–8 × 3–3.5 mm, sometimes with a few glandular-punctate trichomes below (abaxially) at the base and slightly villous on the inner claw surface. **Fruits** decumbent, crescent-shaped in outline, 20–25 × 8–10

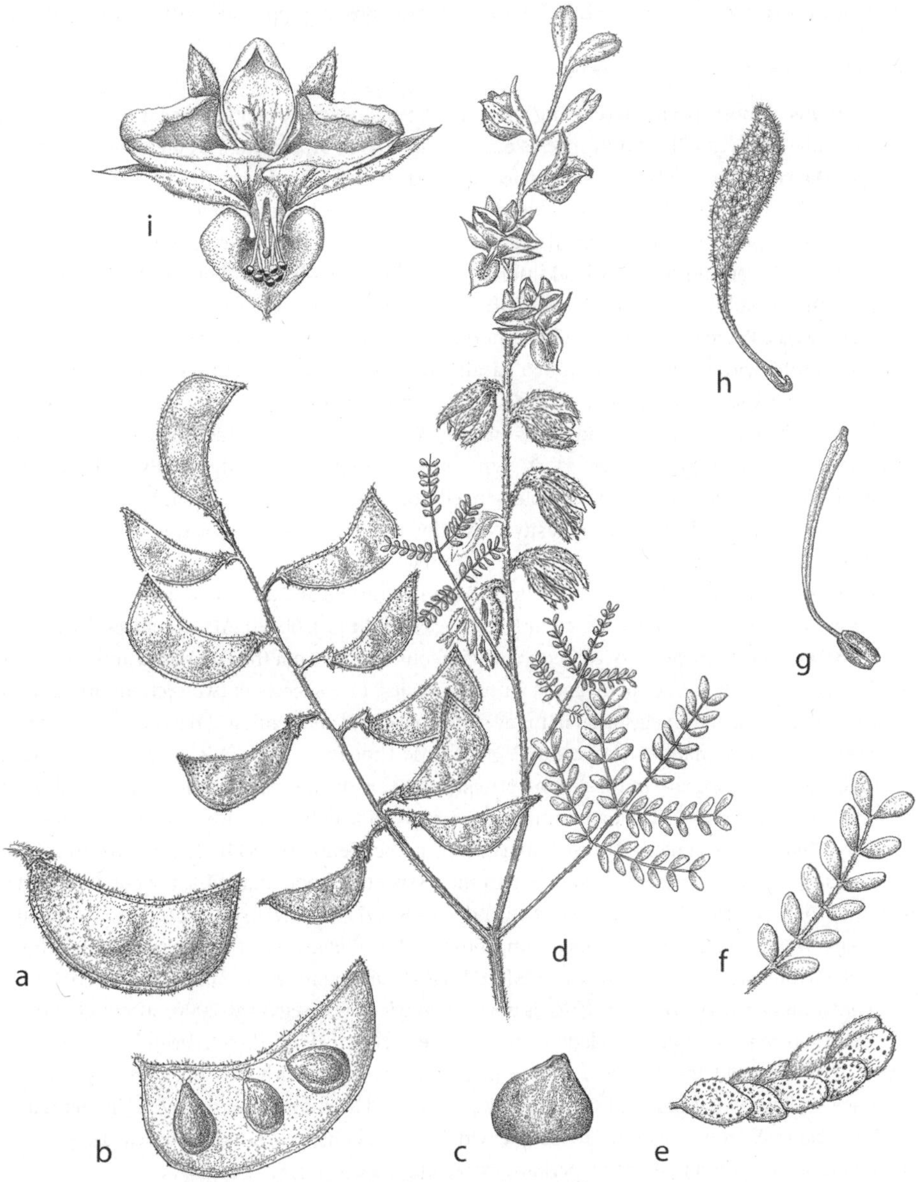

***Pomaria jamesii*: a. fruit; b. fruit valve showing seed arrangement; c. seed; d. habit; e. leaflet details; f. pinna; g. anther; h. pistil; i. flower. Illustrator: Sandy Turico.**

mm; valves with scattered glandular-punctate trichomes and short (to 0.4 mm) multicellular projections, primarily along the upper fruit valve and tip, that appear stellate; margins densely covered with complex multicellular trichomes similar to those of the valves. **Seeds** 2.

Flowering [Mar–]May–Oct. Sandy or clay soils, on slopes and valleys and along arroyos and roadsides; 600–2300 m; Great Basin Desertscrub, Semidesert Grassland, Plains and Great Basin Grassland, occasional in Great Basin Conifer Woodland; Apache, Cochise, Coconino, Gila, Navajo, Santa Cruz, Yavapai cos.; Kansas, Colorado, Oklahoma, Texas, New Mexico; Mexico.

Pomaria jamesii is the most widespread species of the genus in North America, and the only species to reach as far west as Arizona. The Arizona populations are widely scattered and are most commonly found in loose, sandy soils.

PROSOPIS Linnaeus. Syst. Nat. (ed. 12) 2:282, 293. 1767. * [from an obscure application of the Greek name for burdock (*Arctium lappa*)]

Richard Stephen Felger and Matthew B. Johnson

Shrubs or trees. Stems of most species developing short shoots; usually armed with either stipular spines or spines at the nodes or internodal prickles (Old World species except *P. africana*), or unarmed (*P. africana*), and/or twigs sometimes thorn-tipped. **Leaves** alternate and often fascicled on short shoots, bipinnate with 1–4 pairs of pinnae (*P. kuntzei* in South America has minute, quickly deciduous bipinnate leaves and is thus essentially aphyllous, and the related *P. sericantha* is aphyllous); stipules herbaceous and deciduous or spinescent and persistent; petiole often with a gland at or near the lowest pair of pinnae and often with additional glands on rachis below additional pairs of pinnae. **Leaflets** opposite, margins entire; pulvini well developed. **Inflorescences** axillary, densely flowered cylindrical spikes or racemes, or globose heads. **Pedicels:** absent or with a short pedicel. **Flowers** bisexual, mimosoid, small; calyx 5-lobed, shallowly toothed; corolla with petals free or united basally; stamens 10, in two series, filaments free, anthers with a small terminal gland; ovary sessile or stalked, style and stigma simple; flowers usually protogynous (the slender style protrudes from the bud before the stamens and petals emerge). **Fruits** indehiscent, thickened, usually rounded in cross section and moderately to strongly compressed, straight, curved or spirally coiled, usually with carbohydrate-rich mesocarp. **Seeds** pale brown, compressed, ovoid, oblong or elliptic, hard, with each seed encased in a leathery endocarp or bony endocarp, closed or sometimes opening easily; cotyledons green even within the endocarp in the intact pods (at least the sect. *Algarobia* in North America). $x = 14, 28$.

Species 44 (3 taxa native and cultivated in the flora, 2 taxa cultivated and adventive, and 1 taxon cultivated only): widespread from sw and sc U.S. and Mexico to s South America and the Caribbean, Africa, and sw Asia.

Burkart (1976) separates *Prosopis* into five sections, with four species from the Old World and the remainder from the New World. Argentina is the center of diversity with 28 species. Five species in two sections are native to the U.S. *Prosopis* is a major component of the vegetation in many subtropical arid and semi-arid regions worldwide.

Three *Prosopis* species are native in Arizona, *P. glandulosa* (honey mesquite), *P. pubescens* (screwbean) and *P. velutina* (velvet mesquite). Screwbean, in the section *Strombocarpa*, is distinguished by its unique, tightly coiled pods and is relatively localized in the state. The remaining species, native and cultivated, are in section *Algarobia*.

In 1839 George Bentham designated *Algarobia* as a generic name (Bentham 1839). "Because the trees resembled the 'algorroba' (*Ceratonia siliqua*) of the Iberian Peninsula in their aspect and potential uses, the conquistadores gave this Spanish name to various species of *Prosopis*" (D'Antoni and Solbrig 1977) as well to other economically significant New World legume trees. The term *algarroba* has its origin in the Arabic *al-kharrubah* or *al-jarrúba* (al-k̲arrūba; خرنوب), the name for the carob, *Ceratonia siliqua* (Austin 2010, Oxford English Dictionary 2012, Avicenna 1025/1976).

Section *Algarobia* as defined by Burkart (1976) is not monophyletic (Bessega et al. 2006), although those in Arizona, native and cultivated, form a monophyletic clade within a more narrowly defined sect. *Algarobia*. Honey mesquite and velvet mesquite are widespread and common. Honey mesquite has leaves with only one pair of pinnae (jugate) and velvet mesquite leaves have two pairs of pinnae (bijugate; or some leaves sometimes jugate). The remaining *Prosopis* species, sect. *Algarobia* in Arizona are cultivated, some widely so, and native to South America. They are occasionally encountered outside of cultivation in Arizona, perhaps increasingly so.

Hybrids among South American members of section *Algarobia* are suspected among trees cultivated in Arizona. Trees of hybrid origin between South American taxa and native Arizona taxa are suspected by Arizona horticulturalists but have not been confirmed. Accurate identification of some cultivated and feral *Prosopis* of South American origin may be difficult due to significant overlaps in leaf morphology and possible hybridization. For example, a cultivated tree in Yuma was observed to have leaves on some branches that resembled those of *P. alba* in terms of leaflet number, length, and spacing, while leaves on other branches were like those of *P. chilensis*. The extreme forms of these plants, however, should be readily identifiable. In their native habitats in Argentina, plants from sect. *Algarobia* show unexpectedly low genetic differentiation, considerable infra-population variation, and a complicated and apparently low incidence of hybridization (Bessega et al. 2000).

Section *Algarobia* includes about two dozen species in South America and half a dozen species in North America. The North American taxa tend to be allopatric (occupying different geographic ranges) and in general are notorious in having indistinct ("fuzzy") species boundaries with often-broad areas of populations with intermediate morphologies. These distributional patterns probably represent continuous (clinal) variation or even reticulate evolution, and sometimes can confound attempts at realistic taxonomy (e.g., Bessega et al. 2000, Bessega et al. 2006, Burghardt and Espert 2007). Although some intermediate populations might be of hybrid origin rather than continuous (clinal) evolution, our extensive observations indicate clinal variation to be common among native North American sect. *Algarobia* (e.g., Felger 2000).

Palacios (2006) offers an intriguing and radical taxonomy of native sect. *Algarobia* species in Mexico, and by extension the U.S. Mexico borderlands. The taxonomy is based solely on morphology, including leaves. We find this work unusable and unrealistic. Palacios' distribution maps do match reality.

Stored mesquite seeds can remain viable for half a century if dry and enclosed in their endocarp (Martin 1948). The sugary mesocarp and tough endocarp of the pod are adapted for animal dispersal. Germination of the endocarp-enclosed seeds is greatly enhanced after passing through the digestive tract of various animals. Although the major Pleistocene herbivores with which the mesquites likely co-evolved are gone, coyotes, javelina, and even bears and other animals consume large quantities of the fruit. In modern times cattle fill the role of the extinct Pleistocene eaters of mesquite pods and are largely responsible for expanding ecologic and geographic ranges since at least the late 19th century (e.g., Mooney et al. 1977, Turner et al. 1995). In some of the drier habitats, local mesquite populations wax and wane with cycles of wetter or drier decades (Felger 2000), and these fluctuations may even occur in places free from human disturbance and cattle grazing (Turner 1990).

Prosopis species are widely grown as landscape trees in lower elevation communities in the state where they are valued for their shade. Horticultural selections of several species are available from nurseries, and naturally occurring trees are retained in some developments. Native and South American *Prosopis* are propagated by scarified seeds and can also be propagated from semi-hardwood cuttings and by air layering. The bark on trunks and larger limbs often becomes oily black due to oozing bacterial slime flux from wounds beneath the bark; younger stems often exuding light-colored gum. Horticultural information is provided in the discussion section for each species

The last two species listed in the key are unlikely to be found in Arizona and do not receive complete treatments here. However, they are included in the key because they have leaf characters that may overlap with some plants of *Prosopis alba* and *P. chilensis*. Recent work (Hughes et al. 2022) indicates that New World *Prosopis* taxa are better placed in *Neltuma* and *Strombocarpa*. These changes arrived just prior to publication and are listed here as synonyms; this is not a rejection of the new combinations.

SELECTED REFERENCES: Bell and Castetter 1937, Benson 1941, Bessega et al. 2006, Burkart 1976, D'Antoni and Solbrig 1977, Desert Harvesters 2018, Felger 1977, 2007, Felger and Logan 2010, 2018; Felger and Moser 1985, Felger and Rutman 2015, Felker and Clark 1980, Hughes et al. 2022, Johnston 1962, Lancaster 2010, Meyer 1984, Rzedowski 1988, Simpson 1977.

1. Twigs with paired stipular spines; flowers sessile, the petals united below; pods tightly coiled ____ **4. Prosopis pubescens**
1′ Twigs often with 1 or 2 spines at nodes but not stipular spines; flowers on pedicels to 1 mm, the petals free; pods straight or curved (occasionally loosely coiled) but not tightly coiled.
  2. Most or all leaves generally with 1 pair of pinnae (juvenile plants or a minority of leaves on older plant may have 2 pairs of pinnae); leaves generally glabrous or glabrate, the mature leaves bright green; leaflets 6–20 pairs per pinna ____ **3. Prosopis glandulosa**
  2′ Most or all leaves with 2 or 4 pairs of pinnae (a minority leaves occasionally have 1 pair of pinnae); leaves glabrous or hairy, the mature leaves dull, medium green; leaflets 10–56 pairs per pinna.
    3. Leaves conspicuously velvety pubescent at maturity; leaves generally with 2 pairs of pinnae; leaflets 14–24(–32) pairs per pinna; native and widespread, and sometimes cultivated ____ **5. Prosopis velutina**
    3′ Leaves generally glabrous, glabrate, or sparsely hairy; leaves generally with (1 or) 2–4 pairs of pinnae; leaflets 10–56 pairs per pinna; not native, and sometimes weakly and locally adventive, some are widely cultivated.
      4. Leaflets with 29–56 pairs per pinna, the leaflets usually overlapping or separated by a distance of less than the leaflet width ____ **1. Prosopis alba**
      4′ Leaflets with 10–31 pairs per pinna, the leaflets separated by a distance greater than the leaflet width.
        5. Leaflets 11–54 mm, 4–12 mm apart ____ **2. Prosopis chilensis**
        5′ Leaflets 4–15 mm, 1–5 mm apart.
          6. Leaflets 4–15 mm, 3–15 times longer than wide, 2–5 mm apart ____ **Prosopis flexuosa**
          6′ Leaflets 4–7 mm, 2–7 times longer than wide, 1–2 mm apart ____ **Prosopis nigra**

1. Prosopis alba Grisebach. Abh. Königl. Ges. Wiss. Göttingen 19:131. 1874. * [Latin *albi*, white] ARGENTINE MESQUITE; *ALGARROBO BLANCO* Cultivated

Neltuma alba (Grisebach) C.E. Hughes & G.P. Lewis

Trees to 15 (25) m. **Stems** often unarmed or with paired nodal spines to 4 cm long; bark brown or reddish brown and smooth, becoming dark brown or dark gray, rough and longitudinally fissured on larger limbs and trunk. **Herbage** glabrous or essentially so. **Leaves** 5.2–25.9 cm with (1 or) 2–4 pairs of pinnae, pinnae 4.4–19.4 cm; stipules narrowly lanceolate, 1–2 mm; petiole 2.6–5 cm; rachis (0.4–)1.6–5.3 cm; the mature leaves dull, medium green. **Leaflets** 29–56 pairs per pinna, linear, 5–17 × 1–2 mm, 5–8 times longer than wide, usually overlapping or separated by a distance of less than the leaflet width, base rounded, apex acute. **Inflorescences** cylindrical racemes, 7–11 cm. **Pedicel** 0.4–0.7 mm.

Flowers pale yellow or greenish white; calyx 1–1.5 mm, pubescent on outer surfaces; corolla with petals free, 3–3.5 mm; filaments 4–5 mm; ovary stalked, curved, pubescent. **Fruits** straw colored, 12–25 × 1.1–2 cm, curved to ring-shaped; stipe 4–15 mm; apex acute, 1–10 mm. **Seeds** 5–7 × 2.5–4 mm.

Flowering Apr–Jun. Argentina, Bolivia, Paraguay, Peru, Uruguay.

Prosopis alba is widespread in the northern Gran Chaco region of southern South America, extending northward to Peru. It occurs on plains, valleys, and floodplains of intermittent watercourses in thornscrub, dry tropical deciduous forest, and humid savannah habitats. It also occurs in the Atacama Desert of Chile where it was introduced and grown by people in prehistoric times (Aronson 1990). The trees can be abundant and are often a major component of the vegetation. *Prosopis alba* often colonizes saline soils in Paraguay (López et al. 1987). As with other members of the genus, *P. alba* is reported to nodulate (e.g., NRC 1980).

Since the late 20th century plants fitting the description of *Prosopis alba* are sometimes becoming established in landscaped areas and large drainages in southern Arizona such as the Pantano Wash in Tucson. It may have the potential to spread in areas with sufficient moisture but does not appear likely to move out into upland habitats in Arizona due to its higher water requirements. Possible hybridization of this and other South American species in section *Algarobia* with native *P. glandulosa* and *P. velutina* and/or *P. glandulosa* var. *torreyana* is a concern and may be occurring but has not been scientifically verified. For example, nursery stock grown from seeds collected from native Arizona trees in the Tucson region as well other urban areas in southern Arizona often show wide variation suggesting hybridization, whereas ones grown from seeds collected in remote "natural" areas are relatively uniform (Jim Verrier, personal communication, February 2013).

Prosopis alba is often called *el árbol* (the tree) in parts of northern Argentina due to its importance (Burkart 1976). Burkart (1976) and Leonardis (1975) state that *P. alba* is an important timber tree with the wood used for construction, prefabricated houses, doors, furniture, and parquet flooring. Some trees have straight trunks up to 10 m. The trees are being seriously overharvested, and Burkart (1976) and Aronson (1990) discuss negative artificial selection where superior trees are cut first, leaving less economically desirable individuals. Burkart (1976) advocates selection of superior trees for plantation production. Plantations of *P. alba* in Argentina with trees on 2 × 2 m spacing have produced 7 m^3/ha. of wood per year (López et al. 1987). The wood is durable, weighs 760 kg/m^3, and has yellowish white sapwood and purplish brown to brown heartwood (Leonardis 1975). It has a calorific value of 5000 kcal/kg and is highly regarded for fuelwood and charcoal (López et al. 1987). *Prosopis alba,* as well as other species from sect. *Algarobia*, has excellent potential for dryland forestry. Experimental plantations have been established in southern California and southern Texas. The pods are readily consumed by livestock and are used to make flour and to produce a sweet fermented beverage, *aloja de algarrobo* (López et al. 1987).

Argentine mesquite is often planted in Arizona landscapes, and this or related South American taxa are the most commonly planted urban trees in the Sonoran Desert region. It is an excellent shade tree although it becomes too large for many residential lots. The trees should be sited carefully, as they can potentially reach 25 m in height and spread under favorable conditions (Aronson 1990). The trunk can grow to 1.5 m in diameter. The roots can damage nearby pavement and walls and can invade sewer lines. The dense canopy provides considerable wind resistance. The trees are subject to blowing over in high winds, particularly if planted in shallow soils or confined planting spaces, or if improperly irrigated so that strong root development is inhibited. Furthermore, selecting the largest plant in a container may portend future trouble, as the roots of such plants become tightly coiled in the container and seldom develop into strongly root-anchored trees. In the right setting, however, Argentine mesquite makes an attractive tree with a rounded canopy of dark green foliage. Argentine mesquite can grow up to 1.5 m per year when young if adequately irrigated but is tolerant of extended dry conditions once established. Plants in Arizona horticulture tolerate overnight temperatures to -7° or -8°C although at least some individual plants are not as cold resistant as other cultivated South American mesquites. The plants are nearly evergreen in mild winters, with at least some leaves remaining on the trees until the new leaves appear in spring. The leaves and pods can create a litter problem, though the fine leaflets form a good mulch layer where allowed to remain. This mulch layer does not inhibit the growth of other plants. Vegetatively propagated, spineless selections of *Prosopis alba* are available, and most older seed-grown trees are either unarmed or have only small thorns. Scarified seeds germinate readily, but the seed-grown plants can exhibit variable traits, so commercial production often involves vegetative clonal propagation (from cuttings) from selected plants with desirable characteristics.

2. Prosopis chilensis (Molina) Stuntz. U.S.D.A. Bur. Pl. Industr. Invent. Seeds 31:85. 1914. * [of Chile] CHILEAN MESQUITE; *ALGARROBO DE CHILE* Cultivated, Adventive

Ceratonia chilensis Molina, Sag. Stor. Nat. Chili 172. 1782; *Neltuma chilensis* (Molina) C.E. Hughes & G.P. Lewis

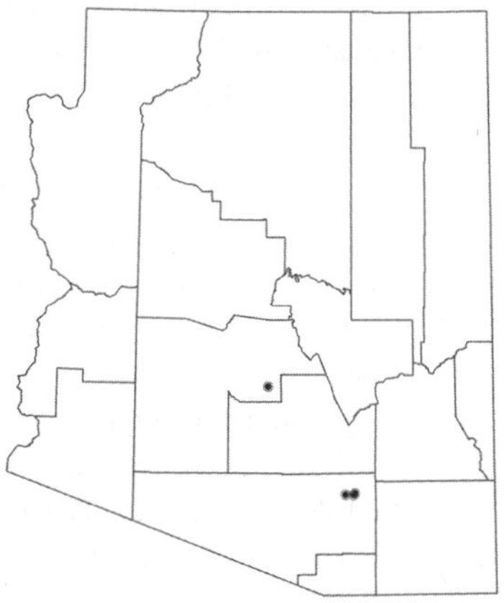

Trees to 12 m. **Stems** unarmed or often with stout nodal spines to 6 cm; bark greenish brown to brown and smooth, becoming dark brown or nearly black, rough and fissured longitudinally on larger limbs and trunk. **Herbage** glabrous. **Leaves** 12–25 cm with (1 or) 2 or 3 pairs of pinnae, pinnae 8–24.5 cm; stipules narrowly lanceolate, ca. 2 mm; petiole 1.6–7.5 cm; rachis when present 2.1–5 cm; the mature leaves dull, medium green. **Leaflets** 10–29 pairs per pinna, linear, 11–54 × 1.1–3 mm, 10–18 times longer than wide, 4–12 mm apart, base rounded, apex acute. **Inflorescences** cylindrical racemes, 7–12 cm. **Pedicel** 0.6–0.8 mm. **Flowers** pale yellow or greenish white; calyx 1 mm; corolla with petals free, 3 mm, pubescent on inner (ventral) surface; filaments 5–6 mm; ovary stalked, straight or curved, pubescent. **Fruits** straw colored or rarely dark purple, 12–18 × 1–1.8 cm, straight to strongly curved; stipe 3–11 mm; apex acute, 1–6 mm. **Seeds** (5.5–)6–7.5 × 3–4(–6) mm.

Flowering Apr–Jun. Argentina, Bolivia, Chile, Peru.

In its native lands *Prosopis chilensis* occupies various habitats including slopes, valleys, and floodplains, usually in areas with deep soil, in desert, thornscrub, savannah, and dry tropical deciduous forest. This species has been found up to 2900 m in the Peruvian Andes (Burkart 1976). *Prosopis chilensis* and putative or suspected hybrids with other South American *Prosopis* have become adventive in some localized areas in Arizona. Pantano Wash and the Rillito River in Tucson support mature, reproducing individuals of *P. chilensis*. Some of these trees, however, have leaves that overlap in measurements for those of *P. flexuosa*, although the pods are typical of those of *P. chilensis*. Chilean mesquite is of concern due to its potential to expand into native habitats in Arizona as well as the possibility of hybridizing with native mesquites. The name *Prosopis chilensis* has been misapplied to *P. glandulosa* var. *torreyana* in some older literature. Nodulation is reported for *P. chilensis* (e.g. Allen and Allen 1981).

In South America *Prosopis chilensis* is important as a shade tree, for timber, fuel, forage, and people and livestock utilize the pods as food (e.g., Burkart 1976, D'Antoni and Solbrig 1977). Burkart suggests that a program to select trees for superior growth habit and pod production among unarmed individuals would benefit forestry programs in dry regions. As with other *Prosopis* in sect. *Algarobia*, the wood is hard, strong, and resistant to decay; it has a specific gravity of 0.80–0.92 (NRC 1980). The sapwood is yellowish, and the heartwood is dark brown. The foliage and pods are eaten by livestock (NRC 1980).

Chilean mesquite has the potential to increase and spread into natural areas. Although cultivation information is provided here, this tree should not be planted in locations adjacent to native vegetation.

Chilean mesquite has similar horticultural characteristics to that of Argentine mesquite, but the foliage is somewhat more open and lacy due to the wider spacing between leaflets. The plants have a similar growth rate under irrigation and develop a spreading canopy. Proper irrigation is essential for the formation of a strong root system to avoid problems with wind throw. Chilean mesquite is an ideal shade tree and is suitable for street plantings, parks, and other larger spaces. Plants of this species in Arizona horticulture appear to be hardy to at least -9°C. The stems often have stout nodal spines, especially on young plants.

3. Prosopis glandulosa Torrey. Ann. Lyceum Nat. Hist. New York 2:192. 1827. * [Latin *gland* and *osa*, having glands] HONEY MESQUITE, MESQUITE

Neltuma glandulosa (Torrey) Britton & Rose

Shrubs or trees to 10 [15] m. **Stems** often armed with solitary or paired nodal spines to 5 cm; bark pale gray or gray green and smooth, becoming dark brown to blackish and longitudinally fissured and shredding on larger limbs and trunk. **Herbage** glabrous or glabrate, or some plants sparsely pubescent with minute hairs. **Leaves** 8–26 cm with 1 (or 2) pairs of pinnae, pinnae 6–17 cm; stipules narrowly lanceolate, 2–3 mm; petiole 0.7–13 cm, with a circular nectar gland at the base of the pinnae; rachis 0.5–4.3 cm; pulvini well developed at base of the pinnae and leaflets; the mature leaves bright green. **Leaflets** 6–20 pairs per pinna, linear or oblong, 10–62 × 1.5–4.5 mm, 5–15 times longer than wide, (1–)2–18 mm apart, base rounded, apex obtuse. **Inflorescences** cylindrical racemes, 5–14 cm. **Pedicel** 0.5–1 mm. **Flowers** pale yellow or greenish white; calyx 1–1.5(–2) mm, shallowly toothed; corolla with petals free, 2.5–3.5 mm, pubescent near tips; filaments 3.5–7 mm; ovary stalked, straight, pubescent. **Fruits** straw colored and often marked with red or purple, (7–)10–23 × 0.7–1.4 cm, straight or moderately curved; stipe 8–17(–23) mm; apex with an acute beak (1–)4–13(–22) mm. **Seeds** 4–6(–7) × (2.5–)3–5.5(–6) mm.

Varieties 2 (1 native and 1 adventive, and both sometimes cultivated): sc and sw U.S. and n Mexico.

Nodulation is reported for *Prosopis glandulosa* (e.g., Allen and Allen 1981, Johnson and Mayeux 1990).

No differences have been found in human usage of the native *Prosopis* species in sect. *Algarobia*, *P. glandulosa* and *P. velutina* (e.g., Felger 2000, 2007). Therefore, the following summary discussion applies to both species in Arizona. Most of the uses would be similar for various other mesquite species in Mexico and even among various South America species of sect. *Algarobia*. It should be noted, however, that there is great variation within species and populations. Two species are notable for not having palatable pods: *P. articulata* S. Watson in Mexico has minimal mesocarp "pulp" and the "tropical" *P. juliflora* Linnaeus *sensu stricto* generally has bitter, inedible pods.

The literature on traditional as well as modern uses and knowledge of mesquite is extensive and only a small fraction of such information is highlighted here, abridged and interpreted from diverse reports, including Bell and Castetter (1937), Castetter (1935), Castetter and Bell (1951), Castetter and Opler (1936), Curtin (1949), Felger (1977, 1979, 1990, 1992, 2000, 2007), Felger and Logan (2010, 2018), Felger and Moser (1985), Gallagher (1976), Gifford (1933), Hodgson (2001), Kearney and Peebles (1960), Kelly (1977), Lancaster (2010), Lumholtz (1912), Moerman (1998), Parker (1982), and Rea (1997).

From early prehistoric until modern times mesquite served native peoples across the arid lands of North and South America as a primary resource (e.g., D'Antoni and Solbrig 1977, Felger 1977, 2007). Mesquite was the single most useful plant in the Sonoran Desert (Bell and Castetter 1937, Felger and Moser 1985, Hodgson 2001, Rea 1997, Betty Melvin in Zepeda 1985, pp. 49, 77).

Every part of the tree was used (e.g., Felger 2007, Rea 1997). Mesquite was the common denominator among the diverse peoples of the region, settled agriculturalists as well as hunters and gatherers. It was part of everyday life from cradle to grave and featured prominently in oral literature. Mesquite pods provided an essential staple in many Native American cultures.

Mesquites thrive under human-disturbed conditions and have long been common landscape features around regional homes and settlements—many a desert home was shaded and sheltered by mesquite trees—among Native Americans as well pioneers, ranchers, and present-day Arizonans.

Mesquite was one of the first food plants recorded by Europeans in the American Southwest, beginning with Álvar Núñez Cabeza de Vaca in 1535 (Adorno and Pautz 1999) and the Francisco Coronado expedition in 1540 (Flint and Flint 2005). Hernando Alarcón, associated with the Coronado expedition, was given "*un pane di Mizquiqui*" (a bread made from mesquite) at the Colorado River delta in August 1540 (Flint and Flint 2005, p. 208). The Nahuatl-derived term *mizquiqui* undoubtedly resulted from Alarcón's knowledge of the tree in southern Mexico.

The largest mesquite groves grew along rivers, major washes, and watercourses, providing ample shade and pleasant places to camp and work. Mesquite pods, being relatively large, are easy to pick, either from the tree when ripe or from the ground soon after they fall. Older, fallen pods, especially if moldy, were assiduously avoided, and present-day harvesters generally only pick ripe pods still on the tree in order to avoid potential contamination. The Cocopahs and others used a pole with a short crosspiece lashed with mesquite bark at an acute angle to serve as a hook for bending down spiny branches to harvest the pods (Gifford 1933, p. 267). The pod is indehiscent, and thus the contents do not fall out, which is a significant harvest feature. People generally knew the location of trees or groves that produced sweeter pods and superior yields, or ones with interesting flavors. For example, the fresh, ripe pods of some velvet mesquites in the Tucson region have an apple-like flavor. Ripe pods would be broken open and tasted for sweetness and ripeness. In addition, mesquite and screwbean (*Prosopis pubescens*) pods sometimes were obtained from packrat nests, which allowed an extension of the harvest season (Felger 2007, Felger and Moser 1985, Kelly 1977). The Seris in Sonora, Mexico, distinguish and name eight different growth stages of mesquite pods (Felger and Moser 1985, Moser and Marlett 2010).

Harvested pods would be sun-dried or fire-parched, making them easier to process, and generally crushed or ground into flour. There were numerous variations in the preparation of the pods and flour, for mesquite "was a versatile staple" (Rea 1997, p. 187). The pods were often crushed and ground in bedrock mortars using a large pestle, which was likely to be fashioned from mesquite wood. The pods also were often crushed in large wooden mortars, such as ones made from a cottonwood (*Populus fremontii*) or mesquite log. Different methods of winnowing would separate flour and chaff. The gyratory crusher/mortar or grinder for processing mesquite pods is discussed below.

Mesquite and screwbean food products generally were prepared without cooking. The flour was most often steeped in water and consumed as *atole de mezquite* or sometimes made into cakes. Pods dry or fresh and even still moist were also crushed or merely twisted and boiled or steeped in water to make beverages. Most often whole mesquite pods were stored. Processed flour was not stored because it is highly hygroscopic and soon becomes hard and undesirable. Mesquite flowers were also sometimes eaten, gently mashed in water and the sweet liquid drunk, but this was not a staple food resource.

Mesquite pods were generally a dependable crop, but in certain areas and times extreme freezing weather or extended drought might reduce or cancel the usually copious crop. Although these were rare events, such years would have severe consequences, especially since other wild crops might likewise fail. For the most part, however, mesquite pods were predictably available for harvest at the height of the hot, early summer dry season, making this a time of plenty, and the pods continued to be available through the summer and fall but in reduced quantities. People often temporarily moved to areas with superior mesquite groves or pods (e.g., Lumholtz 1912, p. 331). For the most part mesquite pods were readily available in quantities far in excess of people's ability or desire to harvest them, which is a likely explanation of why mesquite trees were not cultivated.

Native Americans in the Southwest in pre-contact time did not have cool-season agricultural crops. Their farmed trinity, corn (maize, *Zea*), beans (*Phaseolus*), and squash (*Cucurbita*), were frost-sensitive warm/hot season crops. Introduction of wheat, a cool season (winter-spring) crop harvested in late spring, offered a viable substitute for wild harvested staples, especially mesquite. Spanish colonial era Jesuit missions employed wheat to substitute and discourage mesquite and other pre-summer wild harvests.

The mesocarp (pulp) of mesquite pods provided a major carbohydrate (calorie-rich) component in traditional Native American diets. The pod husk (exocarp) is not digestible, but ground with the mesocarp adds dietary fiber to the flour. The actual seeds, each encased in the tough, leathery-woody pit (endocarp), were apparently not widely used although some people included the protein-rich mesquite seeds in their diet, such as the Seris (Comcaac) in Sonora and likely the Hia C'ed O'odham in southwestern Arizona and nearby Sonora. However, many reports of people eating the seeds should be interpreted with caution—the term "bean" is often loosely and uncritically used for seeds, seeds with endocarps, or the whole pods.

A distinct kind of prehistoric stone mortar was called a "gyratory crusher" by Julian Hayden (1969) because the design and function seemed analogous to that of modern or industrial gyratory crushers. These stone gyratory crusher/grinder mortars, found in various prehistoric sites in southern Arizona and Sonora, were likely used to process mesquite pods including separation of the seeds from the leathery pit (Felger 1977, 2000). These mortars resemble the usual bedrock mortars ("bedrock grinding holes") but differ in being free-standing and having the opening (the bowel or cup) extending through the stone (Adams 2002). These are not *metates* that have been "worn through" since the undersides are worked and smoothened. Toasted, or dried pods, fed into the bowl of the gyratory mortar, are easily ground by turning a large pestle, with surprisingly little effort, and many or most of the endocarps are cracked open and the seeds freed from the endocarp (Felger and Moser 1985). Winnowing can readily separate seeds, mesocarp flour, and chaff. The seeds can then be ground into flour with the same gyratory grinder.

A whitish or amber-colored gum, oozing from wounds on the branches or trunk was eaten as candy, chewed as gum, or mixed with saguaro fruit and eaten like jam.

Nowadays there is substantial renewed interest in adapting traditional harvest and preparation methods for present-day needs, and well-prepared mesquite flour is increasing available (e.g., Desert Harvesters 2018, Lancaster 2010). For present-day acceptance, mesquite flour should be finely ground—coarsely ground mesquite flour is scarcely acceptable. Mesquite flour is gluten-free. Selections and agronomic development of mesquites, especially certain South American species, are becoming increasingly significant. The literature providing information for agronomic development is extensive: key references include those of Becker (1981), Felker et al. (2003), Lee and Felker (1992), Meyer et al. (1982), Saunders et al. (1986), Zolfaghari and Harden (1982), and Zolfaghari et al. (1986). Felger and others predict that these *Prosopis* species in sect. *Algarobia* will become major global food crops, changing agriculture in hot, arid regions (e.g., Felger 1979, 1990; Felger and Logan 2010, 2018).

In modern times people such as the Cocopahs and others also used mesquite pods as fodder or forage for cattle, horses, and pigs. Horses, however, can succumb to colic-impaction due to over-eating the pods.

The black sap or pitch collected from the bark was used as a general-purpose black dye. It was used for decorating pottery but was most commonly boiled to make black hair dye; for example, Cocopahs and Gila River Pimas mixed this black dye with river mud and left it overnight on the hair to keep it black and prevent lice (Gifford 1933, Rea 1997).

Medicinal uses of mesquite were diverse and widespread (e.g., Moerman 1998). For example, a whitish or amber-colored gum, oozing from wounds on the branches or trunk, was extensively employed as medicine, especially for eye ailments. This whitish gum also was used as a mucilage or gargle to soothe a sore throat (Curtin 1949). The black sap or pitch collected from the bark was used for many medicinal purposes including eye ailments. Bark from the roots was used by Pimas to treat wounds, and they ate the "sap" (the whitish gum?) to treat respiratory conditions (Curtin 1947, Kearney and Peebles 1960). The O'odham drank a decoction of the powdered white inner bark as an intestinal antispasmodic. The crushed green leaves were mixed into mother's milk, and the mixture strained into a baby's inflamed eye. For treating older people, the crushed leaves were mixed with water. The "seeds" were eaten as a stomachic, and a decoction of the leaves was drunk for bladder trouble. The Gila River Pimas boiled the root and used the liquid to dress wounds; or they pounded the root, dried it, and then ground it into a fine powder (Rea 1997, Russell 1908). Tea made from the root was also used to treat menstrual problems (Curtin 1949).

Mesquite digging sticks were used for planting, and a similar pole flattened at one end was used for weeding (Kelly 1977). The post-contact plow was made from a single piece of mesquite wood by the O'odham and others. The roots were widely fashioned into multi-purpose cordage (Felger and Moser 1985). For kickball games, played by many traditional people in the Greater Southwest, the ball was made of mesquite wood or a pebble embedded in mesquite gum. Mesquite-wood fishhooks are reported for people living along the Colorado River (Bourke 1889, Heintzelman 1857). The wood was also used for weapons, including arrow points consisting of sharpened foreshafts (MacDougal 1906), clubs, and maces (Heintzelman 1857). Alarcón (Flint and Flint 2005, p. 190) reported that among the people along the lower Colorado River "their weapons were bows and arrows made of hard wood and two or three sorts of clubs made from fire seasoned wood." The hardwood arrows could have been made from various species of trees or shrubs, the bows might have been from screwbean, and the clubs most likely were mesquite.

House posts most often were made of mesquite trunks, and the forked trunks or main branches were especially popular in traditional local architecture. The Cocopahs and others fired pottery with mesquite wood, and pottery designs were made by using a chewed end of a mesquite twig to apply black mesquite pitch (e.g., Gifford 1933). Large basketry granaries, commonplace features on top of traditional houses in the region, were used to store mesquite pods and were often made with arrowweed (*Pluchea sericea*), willow (*Salix gooddingii*), or mesquite stems. Chittenden (1901) reports baskets made by Cocopah women from mesquite-root fiber, but such use has been questioned (Felger 2007).

The mesquites produce high-quality hardwood that has been the preferred cooking fuel in desert regions of southwestern North America from ancient to present-day times. Mesquite continues to be the preferred regional cooking fuel, not only because it is readily available but also because of the superior burning qualities and flavor it imparts to food.

1. Leaflets (25–)30–45(–62) mm, 8–15 times longer than wide ________ **3a. Prosopis glandulosa** var. **glandulosa**
1′ Leaflets (10–)15–25(–30) mm, 5–8 times longer than wide ________ **3b. Prosopis glandulosa** var. **torreyana**

3a. Prosopis glandulosa Torrey var. glandulosa * TEXAS HONEY MESQUITE Cultivated, Adventive

Prosopis juliflora (Swartz) de Candolle var. *glandulosa* (Torrey) Cockerell; *Neltuma glandulosa* (Torrey) Britton & Rose var. *glandulosa*

Shrubs or trees to 10 [15] m. **Stems** with single or occasionally paired spines above leaves in axillary buds (short shoots), often 0.8–4(–5) cm, straight, rigid and sharp, or sometimes unarmed; the spines at first dark colored, becoming whitish with age. **Herbage** glabrous or glabrate. **Leaves** 12.5–26 cm with 1 pair of pinnae (a few leaves may have 2 pairs of pinnae and a rachis 3–4.3 cm), the pinnae 8–17 cm; petioles 3.9–13 cm. **Leaflets** 6–13(–15) pairs per pinna, (25–) 30–45(–62) mm, 8–15 times longer than wide, 5–18 mm apart. **Flowers** numerous, crowded, greenish-white or pale yellowish, 5–7 mm; filaments 4.5–6 mm, longer than the petals; ovary and inner surface of petals densely white hairy. **Fruits** straight to moderately curved, often 10–23 × 0.7–1.4 cm, slightly compressed, slightly to conspicuously constricted between seeds; mesocarp sweet and well developed.

Flowering Apr–May(–Aug). Adventive in Arizona primarily along road shoulders and medians of major highways; 300–1560 m; Sonoran Desertscrub, Chihuahuan Desertscrub, Semidesert Grassland; Cochise, Maricopa, Pima, Pinal cos.; native in Colorado, Kansas, Louisiana, Missouri, New Mexico, Oklahoma, Texas; Mexico (Chihuahua, Coahuila, Nuevo León, San Luis Potosí, Zacatecas). Its presence in much of the northern and eastern geographic area in the United States is the result of modern range expansion, largely due to cattle-related causes.

Prosopis glandulosa var. *glandulosa* is one of the most abundant woody plants in much of southern Texas and parts of northeastern Mexico. It grows in various soils on slopes, canyons, valleys, plains, and on floodplains in desert, thornscrub, grassland, subtropical forest, oak-juniper woodland, along stock ponds, and opportunistically in other habitats such as pastures and along roads. It is found naturally in areas that receive between 250 to 1250 mm of yearly rainfall. This mesquite occurs as a tree in favorable habitats and as a shrub in colder or drier habitats. Established plants are rugged, deep-rooted and drought resistant, and often resprout after cutting or freezing. As with other mesquites, livestock eating the pods facilitates germination on rangeland. Correll and Johnston (1970) report that *P. glandulosa* has increased greatly in abundance but has not substantially expanded its distribution. Variety *glandulosa* has become naturalized in parts of Africa and Australia. In Arizona, it is adventive along major highways in the southern part of the state, particularly along the westbound side of Interstate Highways 8 and 10, where seeds are presumably disseminated from trucks hauling livestock from Texas to California. Many of these trees are mature and appear to be spreading along the roadways. The additional moisture from rainfall running off of the paved road surface aids establishment of the trees and allows them to thrive in areas that would normally be too dry for them. The Texas Big Tree Registry (n.d.) includes a tree of *P. glandulosa* var. *glandulosa* with a height of 17 m and canopy spread of 26.5 m (measured January 2009).

Texas honey mesquite has the potential to increase and spread into natural areas. Although cultivation information is provided here, caution should be used in deciding whether this tree should be planted in locations adjacent to native vegetation.

Texas honey mesquite is increasingly being planted as a landscape tree in warmer parts of Arizona. It is valued as a shade tree and is used for larger yards, streets and open spaces. The lacy, bright green, weeping foliage of this species provides filtered light and is reminiscent of the Peruvian pepper-tree (*Schinus molle*). The leaves are quickly deciduous following freezing. Plants from the northern part of its range are hardy to at least -20°C for brief periods, and in colder locations, the plants grow as shrubs that sometimes freeze to the ground during exceptionally cold winters. With irrigation, honey mesquite can grow 1 m per year. Honey mesquite tolerates a wide range of soils and will thrive in lawns. The spiny stems are a consideration when selecting planting locations, and the pods and leaves create some litter. Plants of var. *glandulosa,* at least from Texas, are considerably more cold-hardy than those of var. *torreyana.*

3b. Prosopis glandulosa Torrey var. **torreyana** (L.D. Benson) M.C. Johnston. Brittonia 14:82. 1962. * [for John Torrey, 1796–1873, American botanist] WESTERN HONEY MESQUITE Native, Cultivated

Prosopis juliflora (Swartz) de Candolle var. *torreyana* L.D. Benson, Amer. J. Bot. 28:751, fig. 4. 1941; *Neltuma odorata* (Torrey & Frémont) C.E. Hughes & G.P. Lewis; *Prosopis odorata* Torrey & Frémont; *Strombocarpa odorata* (Torrey & Frémont) A. Gray

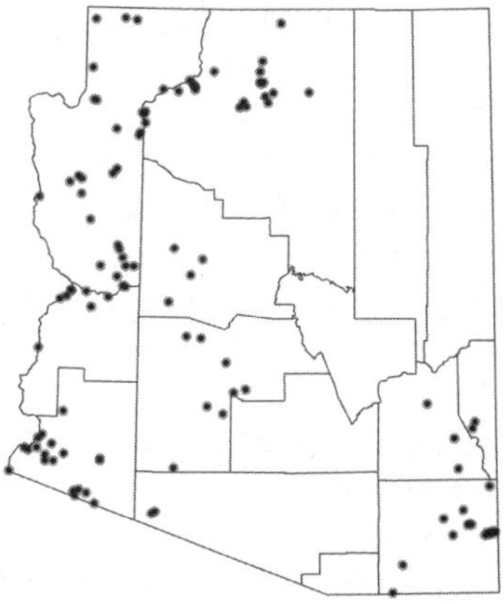

Shrubs or trees to 6 (12) m. **Stems:** paired or sometimes single spines above leaves in axillary buds (short shoots), or sometimes absent or reduced from some or most stems, and especially prominent among juvenile and salt-stressed plants; these spines 0.8–4 cm, straight, rigid, and sharp, at first dark colored, becoming whitish with age. **Herbage** glabrous, glabrate, or sometime sparsely pubescent with minute hairs. **Leaves** 8–21 cm with 1 pair of pinnae (a few leaves may have 2 pairs of pinnae and a rachis 0.5–2.7 cm, the pinnae 6–15.3 cm; petiole (0.7–)1–8(–11) cm. **Leaflets** (8–)10–15(–20) pairs per pinna, (10–)15–25(–30) mm, 5–8 times longer than wide, (1–)2–5 mm apart. **Flowers** numerous, crowded, pale yellowish, 4.5–5 mm; filaments 3.5–5 mm, longer than the petals; ovary and inner surface of petals densely white hairy. **Fruits** straight to moderately curved, often (7–)10–20 × 0.7–1 cm, slightly compressed, slightly to conspicuously constricted between seeds; mesocarp sweet and well developed.

Flowering April–May, and sporadically through early fall; pods ripening in early summer and sporadically through fall. Slopes, valleys, canyons, arroyos, and floodplains; [near sea level–]140–1500(–1840)[–1950] m; Sonoran Desertscrub, Chihuahuan Desertscrub, Mohave Desertscrub, rare in Great Basin Desertscrub, Semidesert Grassland, Madrean Evergreen Woodland; Cochise, Coconino, Graham, Greenlee, La Paz, Maricopa, Mohave, Pima, Yavapai, Yuma cos.; California, Nevada, New Mexico, Texas, Utah; Mexico (Baja California, Baja California Sur, Chihuahua, Coahuila, Sinaloa, Sonora).

Prosopis glandulosa var. *torreyana* occupies areas in Arizona to the east and west of the range of *P. velutina.* Specimens are lacking for Greenlee County where it is to be expected. Felger (2000) and Felger et al. (2001) discuss populations that exhibit leaf and pubescence morphology intermediate between *P. glandulosa* var. *torreyana* and *P. velutina* across rather substantial areas where their distributions meet in southwestern Arizona and adjacent Sonora. Benson and Darrow (1981) also discuss what they call hybridization between these taxa. This taxon is common in thornscrub and tropical deciduous forest in northwestern Mexico.

In favorable sites *Prosopis glandulosa* var. *torreyana* develops into a substantial tree, while in less favorable locations it is a shrub. In sandy valley locations, especially in the Chihuahuan Desert and dunes near the Arizona border in extreme northwestern Sonora, it forms low thickets on domes of wind-blown soil collecting around the crowns of the plants. A tree in Big Bend National Park, Texas, is reported to be 12.8 m tall with a canopy spread of 15.8 m (Powell 1988). Rainfall across the range of *P. glandulosa* var. *torreyana* varies from 75 to 750 mm. In the driest localities, it generally occurs along xeroriparian drainages where extra soil moisture is available.

Populations in westernmost Arizona and adjacent Sonora, such as in the vicinity of the Colorado River are glabrous or glabrate, and tend to have elongate and widely spaced leaflets, falling into the usually accepted concept of *Prosopis glandulosa* var. *torreyana*. Farther eastward, approaching the domain of *P. velutina* and encompassing substantial areas of western Arizona, the mesquites have leaves with one pair of pinnae (jugate) and elongated and often widely spaced leaflets of *P. glandulosa*. These plants also have pubescent leaves and twigs characteristic of *P. velutina*, indicating intermediate populations. There seems to be a tendency for these intermediate populations to have proportionally broader (and shorter) leaflets to the east, where they approach *P. velutina*, so that except for the jugate leaves one could place them with *P. velutina*. Even during favorable times, however, these mesquites produce jugate leaves characteristic of the *glandulosa* taxon. Although there is a continuum in leaflet characters, the jugate versus bijugate condition is closely correlated with the Lower Colorado Valley and the Arizona Upland phytogeographic zones, respectively (Felger 2000).

Western honey mesquite has similar horticultural characteristics to Texas honey mesquite but should do well with less irrigation. The eventual mature size of the former is somewhat smaller than the latter, and the smaller leaflets provide a different overall appearance. In cultivation, germplasm from low-elevation subtropical areas of Mexico is not as cold hardy as plants originating from the Chihuahuan Desert region. Western honey mesquite should do well with less irrigation than Texas honey mesquite and is useful for revegetation within its geographic range.

4. Prosopis pubescens Bentham. London J. Bot. 5:82. 1846. * [Latin *pubesc*, downy] SCREWBEAN; *TORNILLO* Native, Cultivated

Strombocarpa pubescens (Bentham) A. Gray

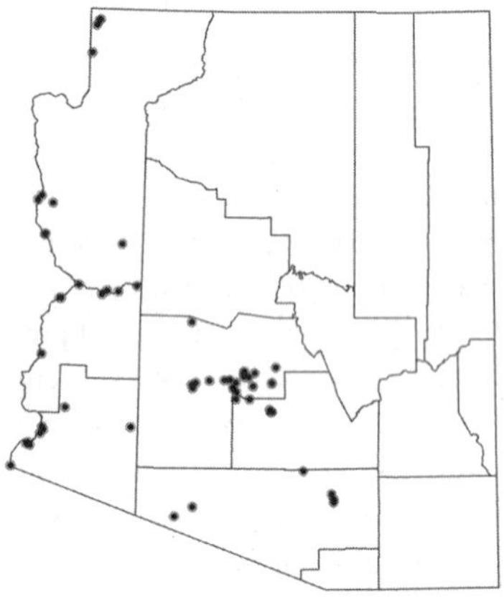

Shrubs or trees to 6 (10) m. **Stems** slender; armed with paired stipular spines; bark pale gray and smooth, becoming dark gray or black, rough and separating into strips on larger limbs and trunk. **Herbage** with minute hairs. **Leaves** 3–6 cm, with 1 pair pinnae (exceptionally 2), the pinnae 1.6–2.4 cm; stipules spinescent, 2–9(–25) mm, straight, stout, rounded, white, paired at nodes; petiole 0.7–1.2 cm. **Leaflets** 4–9 pairs per pinna, oblong or elliptic oblong, 4.5–13 × 2–4 mm, 2–3 times longer than wide, overlapping–2 mm apart, base rounded, apex acute. **Inflorescences** cylindrical spikes, 3–4.5(–7) cm. **Pedicels** absent. **Flowers** bright yellow (calyces, corollas, filaments, anthers, and anther glands); calyx 1 mm, pubescent on outer surfaces; corolla lobes united below, 2–3 mm, pubescent on outer surfaces; stamens well exserted, the filaments 4–5 mm with some reddish flecks; ovary sessile, coiled, densely pubescent (ovary and style obscured by white hairs). **Fruits** tan in color, 2.5–5.5 × 0.4–0.6 cm, forming a tightly spiraled cylinder, usually in clusters of (3–)6–20+, the clusters sessile on their inflorescence axis and often persistent long after ripening; apex blunt. **Seeds** 3–3.5 × 2–2.5 mm.

Flowering May (Jun–Oct); pods ripening mostly during summer months. Generally in alkaline or partially saline alluvial soils of floodplains and arroyos with a shallow water table; [near sea level–]60–730[–1520] m; riparian or semiriparian habitats in Sonoran Desertscrub, [Chihuahuan Desertscrub, Semidesert Grassland]; La Paz, Maricopa, Mohave, Pima, Pinal, Yuma cos.; California, Nevada, New Mexico, Texas, Utah; Mexico (Baja California, Chihuahua, Coahuila, Sonora).

Screwbean in Arizona is found primarily along the Colorado and Gila River drainages and their tributaries, and occasionally at oases in the western part of the state. It typically grows in alkaline or saline soils. *Prosopis pubescens* had a more extensive distribution in Arizona in the recent past, as evidenced by herbarium specimens and historic records, with the loss largely due to declining water tables and subsequent destruction of riparian habitat (Mauz 2011). Screwbean, formerly common along the Santa Cruz River in the vicinity of Tucson, has been virtually extirpated from the region since at least the 1950s (Mauz 2011). Powell (1988) reports a tree measuring 9.1 m tall in Big Bend National Park. The Texas Big Tree Registry (n.d.) lists the largest plant in the state as 8.8 m tall (measured January 2018).

Prosopis pubescens is morphologically and geographically isolated from other members of section *Strombocarpa*,

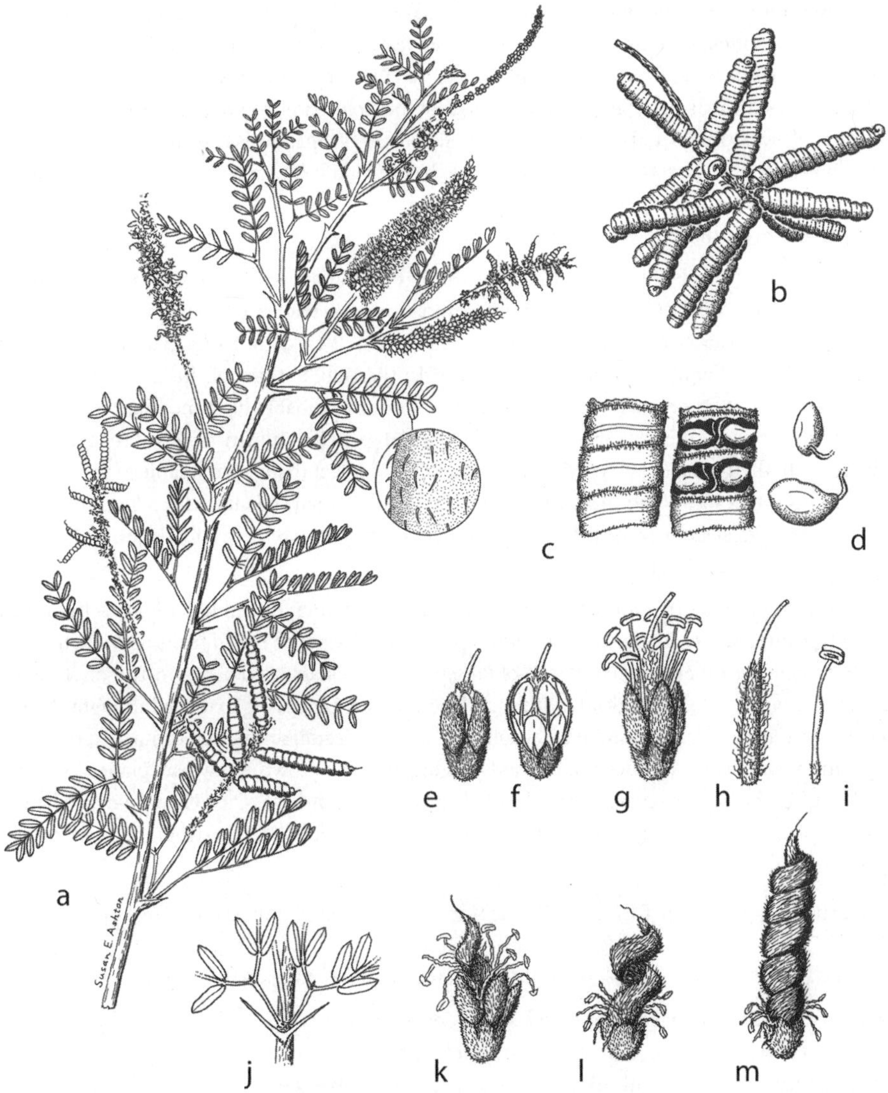

***Prosopis pubescens*: a. habit; b. fruit cluster; c. detail of fruit; d. seeds; e., f., g. detail of flower bud and flower; h. pistil; i. stamen; j. stem node with leaves and spines; k., l., m. detail of fruit development. Illustrator: Susan E. Ashton.**

the nearest of which is *P. palmeri* S. Watson in Baja California Sur. Allen and Allen (1981) indicate that nodulation is not reported for *P. pubescens*; however, other species in the genus are known to nodulate, and this should be further investigated.

Screwbean pods have been an important food resource for wildlife and livestock—the mesocarp of the pods is sweet and a calorie-rich source of carbohydrate. The pods were harvested in early summer as a significant food resource by Native Americans. The pods often were parched prior to preparation and often buried in a pit to "ripen" or partially ferment, and then ground into flour. The flour ("*pinole*") was soaked in water to make a beverage or boiled to make syrup. The pods were also sometimes fermented as a beverage (Bell and Castetter 1937, Castetter and Bell 1951). In 1699 people near present-day Yuma gave Manje "some bread" made from screwbean flour (Manje 1954, p. 114), although Kelly (1977, p. 34) says, "screwbean meal was never made into cakes." Like mesquite (*Prosopis glandulosa*) pods, whole screwbean pods often were stored for extended periods in large, rooftop basketry granaries.

Native Americans made extensive medicinal uses of screwbean bark and its gummy exudate, roots, and pods (Moerman 1998). For example, Gila River Pimas used the root bark as remedies for wounds and for menstrual problems

(Curtin 1949, Russell 1908). The hard wood was used in similar ways to that of mesquite, including for fuel, house construction, posts, and weapons, such as war clubs and lances (e.g., Bell and Castetter 1937, Gifford 1933, p. 274). On January 1, 1828, James Pattie "examined twenty-three bows" of "Pipi warriors" (River Pai Pai) that he and his companions had just killed in a skirmish on the Baja California side of the Colorado River delta. He said the bows were "six feet in length and made of a very tough and elastic kind of wood, which the Spaniards call *Tarnio*" [*tornillo*] (Pattie 1988, p. 111). Hardwood bows seen by Alarcón in 1540 might have been made from screwbean or mesquite, or perhaps more likely catclaw acacia *Senegalia* (*Acacia*) *greggii*.

Mearns (1892–93) gives the following account:

> 'Tornilla.' Screw Bean. Jan. 28, 1894. This tree grows abundantly along the Rio Grande at El Paso and thence down to Fort Hancock, Texas. Going west, on the boundary line, we did not see any more 'Screw-Bean' until we reached the Santa Cruz River. It was quite common at Tucson. West of the Santa Cruz it was not again seen until we reached Quitovaquito, on the Sonoyta River. In the rich 'bottom' of the Colorado River, at Fort Mohave, I saw forests of screw beans. Some of the trees would probably have measured from one to two metres in circumference at the base. The Mojave Indians had gathered immense piles of them for food. They were stacked up in their villages and covered with a thatch to protect them from weather. I was told that they ground them up into a meal, and mixed [the flour] with other materials for use as food. At Tucson and Quitobaquito the trees seldom exceeded 1/3 of a metre in diameter, or 8 metres in height, while the average size was about one-half as much.

Screwbean is an ideal small tree for patios and mini-oases, providing filtered light. It is useful in revegetation of floodplain habitats in areas where it is native. The white spines and distinctive coiled fruits provide interest. Screwbean is not as drought resistant as other Arizona species of *Prosopis*. Large, established plants in Tucson have been observed to suffer considerable stem die back or death with insufficient irrigation during extended drought. Young plants can grow up to 1 m per year under ideal conditions. The foliage is quickly deciduous with freezing temperatures. Screwbean tolerates temperatures to at least -20°C for brief periods. Young, nursery-grown screwbean plants 2 to 3 m tall suffered loss of smaller branches during the severe freeze of early February 2011 at Silver City, New Mexico, when temperatures fell to -23°C. These plants subsequently recovered. Screwbean should be located away from traffic areas due to the spiny stems. Litter is minimal.

5. Prosopis velutina Wooton. Bull. Torr. Bot. Club 25:456. 1898. * [New Latin *velutin*, velvety] VELVET MESQUITE Native, Cultivated

Prosopis juliflora (Swartz) de Candolle var. *velutina* (Wooton) Sargent; *Neltuma velutina* (Wooton) Britton & Rose

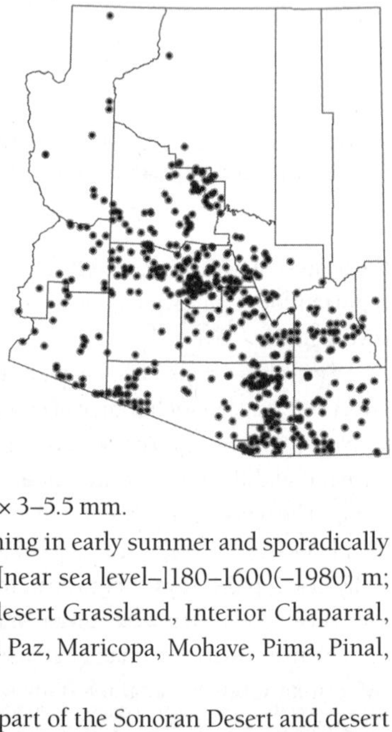

Shrubs or trees to 15 m. **Stems** unarmed or with paired nodal spines to 2 cm; bark dull gray green and smooth, becoming dark gray, rough and longitudinally fissured on larger limbs and trunk. **Herbage** densely velvety pubescent with minute white hairs. **Leaves** 6–14 cm with 2 pairs of pinnae (rarely with a few leaves with 1 pair of pinnae sometimes on new growth in spring and drought-stressed leaves), pinnae 4.7–9 cm; stipules narrowly lanceolate, 2–3 mm; petiole 1–5(–7.3) cm; rachis 1.2–3.3 cm; the mature leaves dull, medium green. **Leaflets** 14–24(–32) pairs per pinna, oblong, 6–13 × 2–3.5 mm, 3–4 times longer than wide, 1–2 mm apart, base rounded, apex obtuse. **Inflorescences** cylindrical racemes, 5–15 cm. **Pedicel** 0.4–0.6 mm. **Flowers** pale yellow or greenish white; calyx ca. 1 mm, pubescent on outer surface; corolla with petals free, 2–3 mm, somewhat pubescent on outer surface; filaments 4–6 mm; ovary stalked, straight or curved, pubescent. **Fruits** straw colored, often with reddish or dark purple markings, 8–16 × 0.6–1 cm, straight or curved; stipe 2–10 mm; apex with an acute beak 2–7(–11) mm. **Seeds** 5–7(–8) × 3–5.5 mm.

Flowering (Mar–)Apr–Jun, and sporadically through fall; pods mostly ripening in early summer and sporadically through summer and fall. Slopes, canyons, valleys, arroyos, and floodplains; [near sea level–]180–1600(–1980) m; Sonoran Desertscrub, Chihuahuan Desertscrub, Mohave Desertscrub, Semidesert Grassland, Interior Chaparral, Madrean Evergreen Woodland; Cochise, Coconino, Gila, Graham, Greenlee, La Paz, Maricopa, Mohave, Pima, Pinal, Santa Cruz, Yavapai, Yuma cos.; Mexico (Sonora).

Prosopis velutina is one of the most common trees or shrubs of the eastern part of the Sonoran Desert and desert

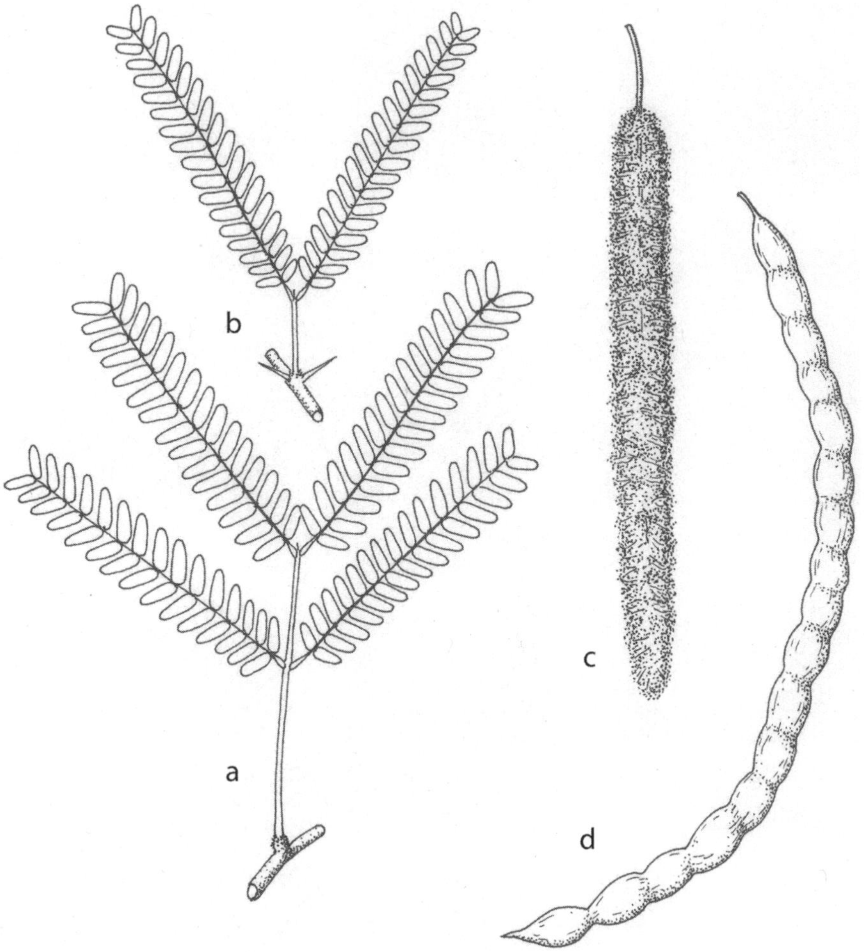

***Prosopis velutina*: a. leaf; b. leaf; c. inflorescence; d. fruit. Illustrator: Matthew B. Johnson.**

grassland habitats in Arizona and adjacent Sonora. It occupies a region more or less between the eastern and western populations of *P. glandulosa* var. *torreyana*, and intermediates occur where their ranges meet. In Sonora *P. velutina* is occasionally found in thornscrub.

Prosopis velutina is similar to *P. glandulosa* var. *torreyana* differing most notably by having (1) the herbage and inflorescences pubescent, (2) 2 pair of pinnae (new growth in spring and drought-stressed leaves sometimes have only 1 pair of pinnae), and (3) the leaflets proportionally broader, not attaining as large a size, usually more closely spaced, and tending to average more leaflets per pinnae.

Extensive forests of velvet mesquite grew on the floodplains of the major rivers in central and southern Arizona, and southward to central Sonora. Cutting for wood and fuel, clearing for agriculture, and declining water tables have eliminated or greatly reduced these mesquite forests. Velvet mesquite has increased greatly in upland areas, particularly in former grassland, since Anglo-American settlement of Arizona due to overgrazing by livestock and a reduction in wildfires that formerly eliminated mesquite seedlings. Livestock readily eat the fruits and disperse the seeds. Annual rainfall across much of the range varies from 150 to 500 mm. In drier sites it is largely restricted to deeper soils where it receives runoff or has access to ground water. Under favorable conditions *Prosopis velutina* can develop into an impressive and deep-rooted tree. Roots of this species have been found at a depth of 80 m south of Tucson (Solbrig and Cantino 1975), among the deepest roots in the world. The largest *P. velutina* in Arizona has a height of 14 m, and a canopy spread of 18 m (The Arizona Register of Big Trees 2005). An excellent ecological summary of *P. velutina* is provided by Turner et al. (1995), and an extensive study is available in Simpson (1977). *Prosopis velutina* is reported to not form nodules (Allen and Allen 1981); however, this needs to be investigated further.

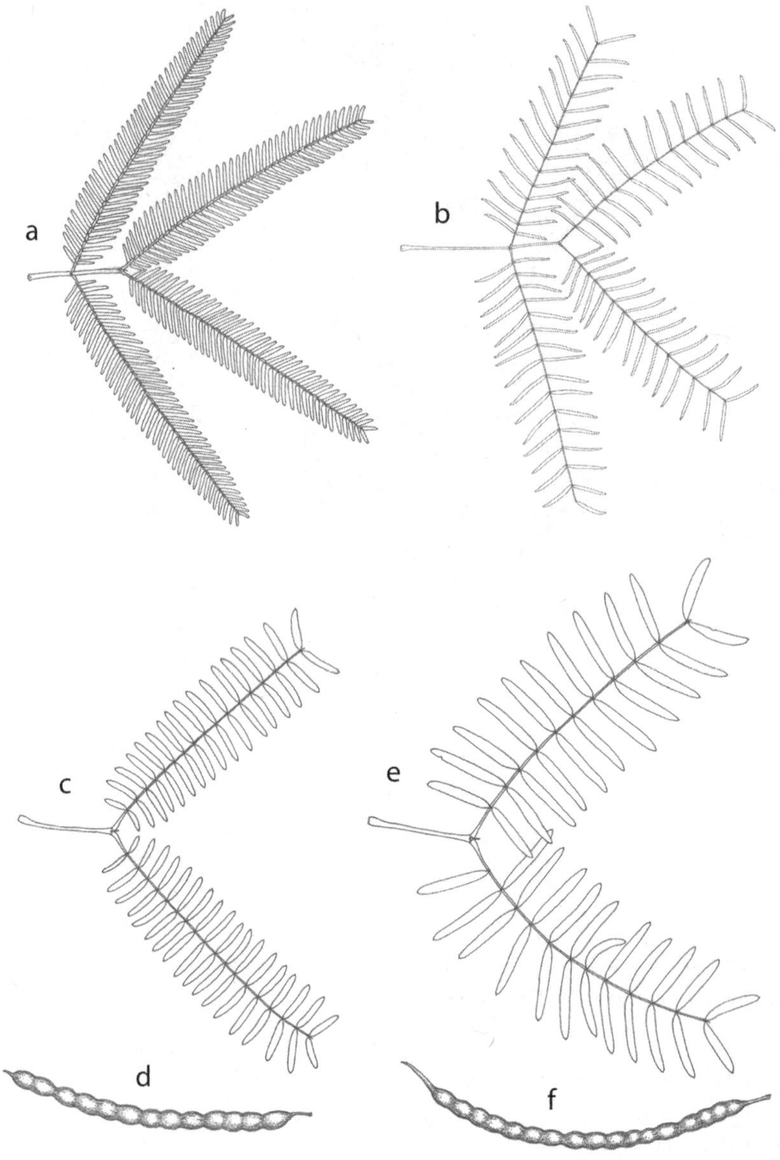

***Prosopis* spp.: a. *P. alba*, leaf; b. *P. chilensis*, leaf; c., d. *P. glandulosa* var. *torreyana*, leaf and fruit; e., f. *P. glandulosa* var. *glandulosa*, leaf and fruit. Illustrator: Matthew B. Johnson.**

The fruits of *Prosopis velutina* are eagerly eaten by a variety of wildlife species and by livestock. Native Americans heavily utilized the fruits to make flour. It is an excellent candidate for development as a dryland tree crop. Sweetness varies considerably from plant to plant and selection of high yielding trees with sweet fruits is desirable. Small scale plantations have been established in Arizona (Moody and Moody 2009). Mesquite honey is a popular food item. Velvet mesquite wood has long been the most important source of fuelwood within its range and is especially popular for barbeques. The species is heavily overexploited for making charcoal in Sonora. The wood is used for fence posts, in construction and to make furniture and for carpentry. Wood of velvet mesquite is hard and heavy. The sapwood is yellow, and the heartwood is dark brown.

For an extensive discussion of the ethnobotanical uses of *Prosopis velutina*, please see the discussion under *Prosopis glandulosa*.

Velvet mesquite is widely planted in warmer regions of the state for landscaping. This species makes a beautiful specimen tree and is suitable for planting along streets and as a large patio tree. Velvet mesquite does well in lawn areas

(Jones and Sacamano 2000). The spreading canopy produces filtered light that is ideal for many smaller plants. The trees are gradually deciduous following freezing temperatures. Young plants receiving ample irrigation can grow up to 1 m per year. This species is hardy to approximately -15°C. During the severe freeze of December 1978, plants of *Prosopis velutina* in southeastern Arizona suffered freeze damage and in some locations were killed to ground level. Native trees often have a picturesque appearance and are successfully boxed and moved on construction sites. Young trees can be spiny but older trees are generally unarmed or have only small spines. The fruits and leaves create some litter. Many of the outer twigs are sometimes killed by the mesquite girdler (*Oncideres rhodosticta*), a longhorn beetle. Such twigs, with their dried leaves, may remain on the trees for several weeks or longer. This is mostly a cosmetic issue and does not seem to affect the overall health of the tree.

Taxa with potential for cultivation

Prosopis flexuosa de Candolle. Prodr. 2:447. 1825

Prosopis flexuosa is not known for certain from Arizona horticulture. Adventive trees that appear to be *P. chilensis* (or hybrids?) sometimes have leaf characters that more closely resemble those of *P. flexuosa* than other South American species of *Prosopis* in sect. *Algarobia*. *Prosopis flexuosa* ranges farther south into colder and more arid habitats in Argentina than do the other South American species included in this treatment.

Shrubs or Trees to 10 m. **Leaves** 4.4–17.6 cm with (1) 2 or 3 pairs of pinnae, the pinnae 3.8–13 cm; stipules narrowly lanceolate, 1–2 mm; petiole 1.2–3.5 cm; rachis 1.1–2 cm. **Leaflets** 12–31 pairs per pinna, linear, 4–15 × 1–2 mm, 3–15 times longer than wide, 2–5 mm apart, base rounded, apex obtuse or acute.

Prosopis nigra (Grisebach) Hieronymous, Bol. Acad. Nac. Ci. 4:283. 1881 [1882]

Prosopis nigra is not presently grown by Arizona nurseries; however, individuals that key to this species have been found in landscapes, presumably from a commercial source. *Prosopis nigra* has excellent potential for use in larger spaces and is similar to *P. alba* and *P. chilensis* in its horticultural characteristics and requirements. The much smaller leaves and leaflets readily distinguish *P. nigra* from those species.

Prosopis nigra (black mesquite) is a common tree in South America in thornscrub and tropical dry forests of the Gran Chaco and in humid savannah habitats, on plains, valleys, and floodplains. It can form extensive nearly pure stands under ideal conditions. The plants often grow in deep clay soils and are tolerant of alkaline conditions, poorly drained soils, and periods of extended inundation (e.g., Burkart 1976, López et al. 1987). Established trees of black mesquite have exhibited considerable drought tolerance. Cultivated plants in Tucson have been undamaged by temperatures of -9°C.

Trees to 10+ m. **Leaves** 3.2–19 cm with 1–3 pairs of pinnae, pinnae 5–10 cm; stipules narrowly lanceolate, 1–2 mm; petiole 0.7–4.3 cm; rachis when present 0.5–2.2 cm. **Leaflets** 11–31 pairs per pinna, oblong-elliptic, 4–7 × 1.2 mm, 2–7 times longer than wide, 1–2 mm apart, base rounded, apex obtuse.

PSOROTHAMNUS Rydberg. N. Amer. Fl. 24(1):45–46. 1919. * [from Greek, *psoro*, scab, and *thamnos*, shrub, referring to the conspicuous glands on the plants] INDIGO BUSH

Michelle M. McMahon, Mary Welch-Keesey, and Matthew B. Johnson

Psorodendron Rydberg

Subshrubs, shrubs, or shrubby trees. **Herbage** variously pubescent or rarely glabrous, with abundant glands especially on stems, lower leaf surfaces, inflorescence axes, calyces, and fruit. **Stems** sometimes thorn-tipped, often intricately branched. **Leaves** alternate and sometimes also fascicled, odd pinnate or 1-foliolate, deciduous; stipules small, deciduous; petiolate. **Leaflets** 1 or 3–17(–19), [in one variety outside AZ, leaflets are confluent with rachis, the leaf then appearing simple and deeply dissected]; plants not with leaves all trifoliolate. **Inflorescences** spikes (in *P. emoryi*) or usually racemes with short to very short pedicels, appearing terminal; bracts small, deciduous; bracteoles variable, minute. **Flowers** papilionoid; calyx campanulate, ribbed, the lobes unequal; corolla pink, reddish purple, purple, violet, or blue, the petals clawed; stamens 10, filaments united for a portion of their length with alternating longer and shorter filaments; ovary with 2 (rarely to 7) ovules. **Fruits** indehiscent. **Seeds** 1 (2). $x = 10$.

Species 9 (7 native in the flora, 1 of these also cultivated): arid regions of sw U.S. and nw Mexico.

A few taxonomic problems remain unresolved in this genus. The four varieties of *Psorothamnus arborescens* may not be independent of the two varieties of *P. fremontii*, themselves poorly distinguished. The isolated *P. arborescens* var. *pubescens*, of northern Arizona and of conservation concern, may well be a separate lineage. Molecular phylogenetic analysis, although incompletely sampled, indicates that *P. spinosus*, *schottii*, *kingii*, *fremontii*, and *arborescens* would be better placed in a resurrected *Psorodendron* Rydberg (McMahon and Hufford 2004).

Psorothamnus species have potential for use in revegetation within their natural distributions. The aromatic resins have potential for use as novel fragrances. The oils contained in the glands appear to deter browsing by livestock and wildlife. Several of the species are ornamental, with interesting forms and showy flowers, and are drought tolerant. These plants deserve wider use in landscaping.

SELECTED REFERENCES: Barneby 1977, Isely 1998, McMahon and Hufford 2004.

1. Leaves 1-foliolate, rarely with a few leaves 3-foliolate.
  2. Shrubs or subshrubs broom-like, stems not thorn-tipped; inflorescences short and dense; Navajo, Coconino, Cochise counties ____ **5. Psorothamnus scoparius**
  2′ Shrubs or shrubby trees, stems thorn-tipped; inflorescences elongate with visible axis; La Paz, Maricopa, Mohave, Pima, Yavapai, Yuma counties.
    3. Shrubby trees with widely diverging branches, pale gray turning tan, leaves quickly deciduous; inflorescences becoming thorn-tipped; widespread and locally common ____ **6. Psorothamnus spinosus**
    3′ Shrubs with spreading branches, green turning straw-colored or whitish, leaves moderately persistent; inflorescences not becoming thorn-tipped; rare and local in Yuma County ____ **4. Psorothamnus schottii**

1′ Leaves 5- or more foliolate.
  4. Inflorescences head-like or subglobose spikes; southwest Arizona ____ **2. Psorothamnus emoryi**
  4′ Inflorescences elongated racemes, with visible axes and short pedicels; northwest to north-central Arizona.
    5. Flowers bright pink to violet-purple, fruit scarcely exserted from calyx; in Arizona, known only from near Wupatki Natl Mon on the Little Colorado River ____ **7. Psorothamnus thompsoniae**
    5′ Flowers reddish purple to indigo-purple, fruit well-exserted from calyx; west and northwest Arizona, north-eastward along Colorado River.
      6. Fruits glabrous with many small, often confluent, glands; widespread within range above ____ **3. Psorothamnus fremontii**
      6′ Fruits pilosulose with large distinct glands; narrowly distributed near Lee's Ferry and Vermillion Cliffs ____ **1. Psorothamnus arborescens** var. **pubescens**

1. Psorothamnus arborescens (Torrey ex A. Gray) Barneby. Mem. New York Bot. Gard. 27:33. 1977. * [tree-like] MOHAVE INDIGO BUSH

Dalea arborescens Torrey ex A. Gray, Pl. Nov. Thurb. 316. 1854; *Psorodendron arborescens* (Torrey ex A. Gray) Rydberg

Shrubs spreading, to 1 m. **Herbage** glabrous to pubescent, pubescence variable; leaves gray tomentulose, silky-canescent, thinly pilosulose, or glabrate; young stems and lower leaf surfaces glandular. **Stems** irregularly branched, sometimes thorn-tipped. **Leaves** 1–3.5(–5.5) cm; stipules 0.7–2 mm, narrowly awl-shaped to linear attenuate; petiolate. **Leaflets** (1 or 3) 5–7(–17), 3–16 × 0.4–4 mm, ovate or lanceolate to linear, terminal leaflet often continuous with rachis. **Inflorescences** sessile racemes, or nearly so, the persistent axis 1–12 cm, 8–48-flowered; bracts 2.5–4 mm, papery, narrowly awl-shaped or linear, deciduous. **Flowers** with calyx 5–9 mm, the tube 3–4.5 mm, the lobes 2–4.5 mm, glabrous to pubescent; corolla indigo-blue or violet-purple, the banner 6.3–10.5 mm, the wings 6.5–11 mm, the keel 8.2–11.2 mm; stamens (5.5–)6.5–9.8 mm. **Fruits** 8–10 × 3.5–6 mm, obliquely ellipsoid or ovoid-ellipsoid, often with persistent style, the valves glabrous to densely pilose, with large, round or ellipsoid blister glands. **Seeds** 1, 4.7–7 × 3–5 mm, whitish or straw-colored.

Varieties 4 (1 native in the flora). The other varieties, var. *arborescens*, var. *minutifolius*, and var. *simplifolius*, are found in California and Nevada, with one disjunct record of var. *arborescens* in northwestern Sonora.

1a. Psorothamnus arborescens (Torrey ex A. Gray) Barneby var. **pubescens** (Parish) Barneby. Mem. New York Bot. Gard. 27:38. 1977. * [loosely hairy compared with *Psorodendron johnsonii*] MARBLE CANYON DALEA Native

Parosela johnsonii (S. Watson) Vail var. *pubescens* Parish, Bot. Gaz. 55(4):308. 1913; *Dalea amoena* S. Watson var. *pubescens* (Parish) Peebles; *Dalea fremontii* Torrey ex A. Gray var. *pubescens* (Parish) L.D. Benson; *Psorodendron pubescens* (Parish) Rydberg

Shrubs to 80 cm. **Leaves** 1.5–3.5 cm. **Leaflets** 5–13, 3–16 × 0.4–1.4 mm, linear or narrowly oblanceolate. **Flowers** with calyx 7.3–9 mm; corolla indigo blue, banner 7 mm, keel 8.2–9.1 mm. **Fruits** with valves usually densely pilose.

Flowering Apr–Jun. Rocky clay soil of low hills and sandstone talus at the base of cliffs; 970–1490 m; Great Basin Desertscrub; Coconino Co.; endemic.

Psorothamnus arborescens var. *pubescens* is confined to Marble Canyon and adjacent areas in the eastern Grand Canyon, near Lee's Ferry and the Vermillion Cliffs. It is of conservation concern due to its small populations and limited distribution.

2. Psorothamnus emoryi (A. Gray) Rydberg. N. Amer. Fl. 24(1):47. 1919. * [for William H. Emory, 1811–1887, American army officer, cartographer, and explorer]

Dalea emoryi A. Gray, Pl. Nov. Thurb. 315. 1854

Shrubs or subshrubs, rounded or spreading, to 1 m. **Herbage** densely and softly tomentulose (rarely stems or stems and leaves glabrescent or glabrous) and with reddish-orange glands. **Stems** suffrutescent. **Leaves** highly variable, main stem leaves 1–5 cm, distal leaves shorter and 3-foliolate or simple, ultimate leaves minute and bract-like; stipules 0.6–1.1 mm, awl-shaped, deciduous; petiolate. **Leaflets** of main stem leaves 5–11(–13), distal leaves 3-foliolate or simple; lateral leaflets of main leaves 2–10 × 0.8–3 mm, terminal leaflet up to three times in length, leaflets on distal leaves progressively smaller, all varying from nearly circular to oblong-oblanceolate, to narrowly elliptic. **Inflorescences** spikes, dense, head-like or subglobose, (or oblong), the axis 0.1–2 cm, 6–25-flowered; bracts 1.5–3 mm, narrowly lance-elliptic or awl-shaped to sublinear, quickly deciduous. **Flowers** with calyx 2.7–7.2 mm, the tube 1.7–3.4 mm, the lobes 1.6–3.8 mm; corolla purple, violet, or bicolored with distal portions pale or whitish, the banner 4.2–6.3 mm, the wings 5.2–7.8 mm, the keel 5.7–8.5 mm; stamens 5.7–7.8 mm. **Fruits** 2.3–2.8 mm, subsymmetrically obovoid-ellipsoid, the valves glabrous proximally, pilose and glandular distally, enclosed within calyx. **Seeds** 1, 2.5–3 × 1.3–1.5 mm, straw-colored to dull brown.

Varieties 2 (1 native in the flora). The second variety, var. *arenarius*, occurs in the Baja California peninsula and adjacent islands.

2a. Psorothamnus emoryi (A. Gray) Rydberg var. **emoryi** * EMORY DALEA, DYEBUSH Native

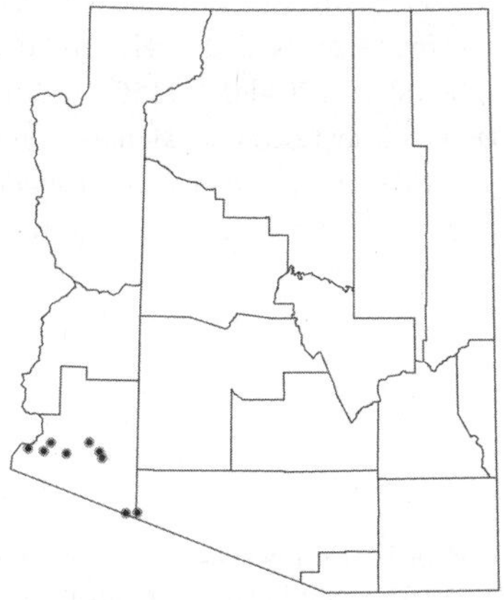

Flowers with calyx 4.1–7.2 mm, the tube 2.2–3.4 mm, the lobes 1.8–3.8 mm; corolla with banner mostly 5.3–6.3 mm, the margins of the blade pink-purple.

Flowering Jan–May, Sep–Dec. Sandy soil of valleys, sand dunes, arroyos, and roadsides; [-35–]40–340[–760] m; Sonoran Desertscrub, [Mohave Desertscrub]; La Paz, Pima, Yuma cos.; California; Mexico (Baja California, Baja California Sur, Sonora).

Psorothamnus emoryi var. *emoryi* occurs in scattered populations in sandy habitats in very arid desert in southwestern Arizona. The plants commonly occur with *Ambrosia dumosa*, *Larrea divaricata*, and *Pleuraphis rigida*, and can be locally abundant. Around the Gulf of California, *P. emoryi* often occurs on coastal dunes (Felger 2000). The plants are sometimes parasitized by *Pilostyles thurberi* (Felger 2000). This species appears to flower in response to rain. The stems have been used in basket making (Allen and Allen 1981). Emory dalea would be an excellent landscape plant for hot, sandy desert regions.

3. Psorothamnus fremontii (Torrey ex A. Gray) Barneby. Mem. New York Bot. Gard. 27:40. 1977. * [for John C. Frémont, 1813–1890, American soldier, explorer, and politician] FREMONT'S DALEA

Dalea fremontii Torrey ex A. Gray, Pl. Nov. Thurb. 316. 1854; *Psorodendron fremontii* (Torrey ex A. Gray) Rydberg

Shrubs, spreading or erect, to 1 m. **Herbage** glandular, sometimes obscurely so, ashen- or silvery-strigillose, leaflets often more densely pubescent above than beneath. **Stems** irregularly branched, stiff but not thorn-tipped or with only an occasional thorn-tipped twig, bark of older stems pale or whitish. **Leaves** solitary or, on older stems, fascicled on short shoots, 1–4 cm; stipules 0.7–1.2 mm, awl-shaped or attenuate awl-shaped; petiolate. **Leaflets** 3–7(–9), 2–25 × 0.7–4 mm, terminal often longest, narrowly obovate to linear. **Inflorescences** racemes, proximal flowers sometimes solitary and subtended by leaf, axis persistent, glandular, 2–18 cm, loosely 5–40(–70)-flowered; bracts 1.2–2 mm, papery, quickly deciduous. **Flowers** with calyx 5–7.7 mm, the tube 2.8–3.9 mm, the lobes 2.2–4.1 mm; corolla purple to reddish purple, the banner 7.3–9.3 mm, the wings 7.5–9.9 mm, the keel 7.8–9.8 mm; stamens 6.6–9.1 mm. **Fruits** 7–11 × 4.2–5.3 mm, obliquely ovoid-ellipsoid, the valves with numerous small often confluent orange glands. **Seeds** 1, 4.4–5.8 × 2.2–3.2 mm, tan to brown.

Varieties 2 (2 native in the flora). Varieties arguably distinct; specimens from near the Colorado River appear intermediate.

1. Leaflets all narrowly linear, under 1 (–1.2) mm wide; mostly on granitic, volcanic, or rarely sandstone substrates ____ **3a. Psorothamnus fremontii** var. **attenuatus**

1′ Leaflets narrowly ovate-elliptic to narrowly obovate, linear-elliptic, or linear-lanceolate, at least some 1.5–4 mm wide; mostly on limestone or sandstone substrates ____ **3b. Psorothamnus fremontii** var. **fremontii**

3a. Psorothamnus fremontii (Torrey ex A. Gray) Barneby var. **attenuatus** Barneby. Mem. New York Bot. Gard. 27:41, pl. 4. 1977. * [narrowed, referring to the leaflets] Native

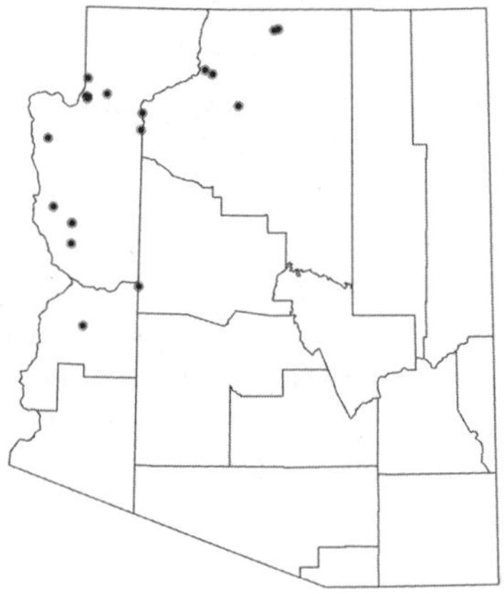

Leaflets 3–25 × less than 1 (–1.2) mm, all linear.

Flowering Apr–May. Rocky or occasionally sandy soils of slopes, canyons, arroyos, and roadsides on granitic, volcanic, or rarely sandstone substrates; [170–]400–1230 m; Sonoran Desertscrub, Mohave Desertscrub, Great Basin Desertscrub; Coconino, La Paz, Mohave, Yavapai cos.; California, Nevada.

Psorothamnus fremontii var. *attenuatus* occurs in scattered populations where it may be locally common. In Arizona, this variety is generally south of the Colorado River; specimens on the river approach the nominate variety. Both varieties of *P. fremontii* would make interesting additions to the landscape palette.

3b. Psorothamnus fremontii (Torrey ex A. Gray) Barneby var. **fremontii** Native

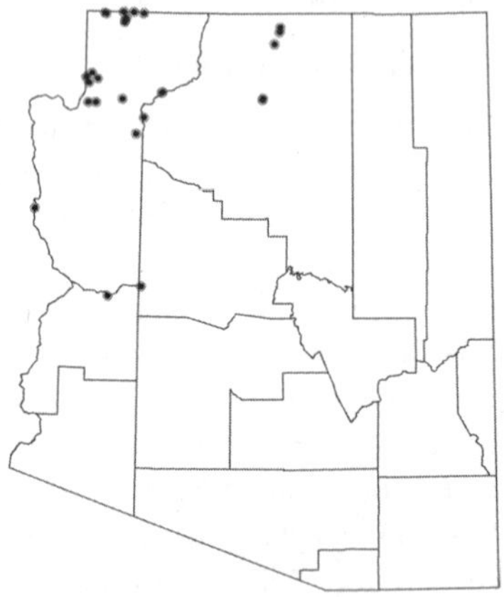

Leaflets 2–20 × 1–4 mm, narrowly ovate-elliptic to narrowly obovate, linear-elliptic, or linear-lanceolate.

Flowering Apr–Jun. Rocky, gravelly, or rarely sandy soils of ridges, slopes, canyons, arroyos, and rarely sand dunes, usually on limestone or sandstone substrates; [270–]430–1150[–1600] m; Mohave Desertscrub, Great Basin Desertscrub; Coconino, Mohave, Navajo cos.; California, Nevada, Utah.

Psorothamnus fremontii var. *fremontii* grows in widely scattered populations in northwestern Arizona, generally north of the Colorado River.

4. Psorothamnus schottii (Torrey) Barneby. Mem. New York Bot. Gard. 27:31. 1977. * [for Arthur C. Schott, 1814–1875, German-born American artist, cartographer, botanist, and geologist] SCHOTT'S INDIGO BUSH Native

Dalea schottii Torrey, Rep. U.S. Mex. Bound. (1):53. 1859; *Psorodendron schottii* (Torrey) Rydberg

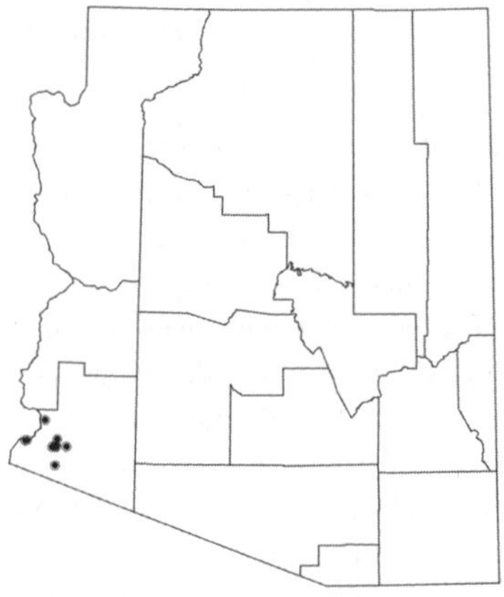

Shrubs, spreading or erect, 1–2.5 m. **Herbage** sparsely to silvery canescent, leaves nearly glabrous below, gland-dotted. **Stems** often spur-branched, young twigs green, eventually thorn-tipped, bark straw-colored or whitish, furrowed. **Leaves** solitary or fascicled, (0.6–)1–2.8 cm; stipules 0.2–0.4 mm, triangular, quickly deciduous; very shortly petiolate. **Leaflets** 1(–3), (6–)10–28 × 0.5–1 mm, linear. **Inflorescences** racemes, the axis 1–9 cm, (3–)5–20-flowered; bracts 0.5–1.3 mm, broadly awl-shaped, deciduous. **Flowers** with conspicuously glandular, moderately strigose calyx 4–6.5 mm, the tube 2.8–4.3 mm, the lobes 1–2 mm; corolla purple to reddish purple, the banner 6.4–10.5 mm, the wings 7–10.2 mm, the keel 8.5–12 mm; stamens 7–11 mm. **Fruits** prominently glandular, 7–10 mm, obovoid-cuneate. **Seeds** 1, 6–8 × 3.3–4 mm, smooth, brown to purplish brown.

Flowering (Nov–)Feb–May. Sandy or gravelly soils of slopes, valleys, arroyos, and roadsides; [5–]60–300[–750] m; Sonoran Desertscrub; Yuma Co.; California; Mexico (Baja California).

Psorothamnus schottii is found in a few locations in arid Sonoran Desert habitats in extreme southwestern Arizona. It is more widely distributed in California and in northern Baja California. The plants sometimes form sand hummocks by arresting wind-blown sand (Barneby 1977). This species has potential for soil stabilization in sandy desert locations though it is probably not well-suited for planting on sand dunes. *Psorothamnus schottii* produces a profusion of showy purple flowers and is suitable for xeric landscapes. It is not currently in the landscape trade in Arizona.

5. Psorothamnus scoparius (A. Gray) Rydberg. N. Amer. Fl. 24(1):48. 1919. * [broom-like] BROOM DALEA Native

Dalea scoparia A. Gray, Mem. Amer. Acad. Arts, n.s. 4(1):32. 1849

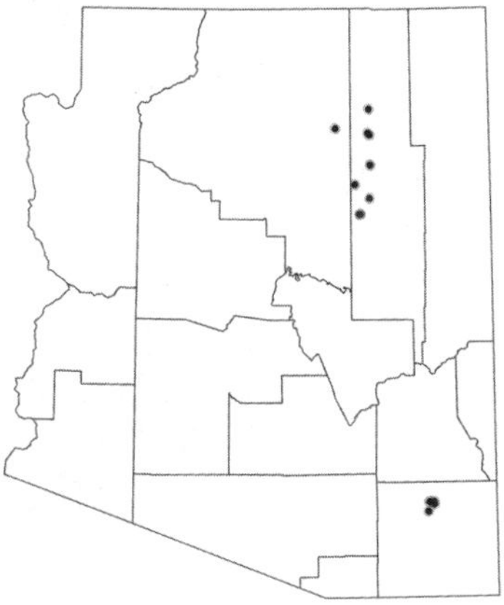

Shrubs or subshrubs, broom-like, 1(–1.5) m. **Herbage** gray-strigillose to canescent, occasionally glabrous, with abundant orange glands. **Stems** green, flexible, forking, erect and incurved-ascending. **Leaves** sparse, early deciduous, 0.5–2 cm; stipules 0.5–0.7 mm, awl-shaped, deciduous; petiolate. **Leaflets** 1(–3), 2–20 × 0.8–1.2 mm, linear-oblanceolate, lateral leaflets, if present, much smaller than terminal leaflet. **Inflorescences** dense head-like racemes, the axis (0.1–)0.3–1.8 cm, 5–15-flowered; bracts 1.5–2.5 mm, ovate or oblanceolate-acuminate, quickly deciduous. **Flowers** with calyx 3.5–4.5 mm, densely pilose, the tube 2.2–3 mm, the lobes 1.3–2 mm; corolla blue-violet, the banner 6.4–9 mm, the wings 7.1–9.1 mm, the keel 6.1–8.3 mm; stamens 5.5–8.5 mm. **Fruits** only slightly exserted from calyx, 4 mm, the valves pilosulose and densely glandular distally. **Seeds** 1, 2.1–2.7 × 1.1–1.5 mm, brown, sometimes with purple speckles.

Flowering Jun–Oct. Sandy soil of valleys, arroyos, roadsides, and on sand dunes; [1100–]1200–1500[–1800] m; Chihuahuan Desertscrub, Great Basin Desertscrub, Semidesert Grassland; Cochise, Coconino, Navajo cos.; New Mexico, Texas; Mexico (Chihuahua, Coahuila).

In Arizona, *Psorothamnus scoparius* occurs in two widely separated areas. The species is found in the vicinity of the Willcox Playa in Cochise County. In northern Arizona it is found along the drainage of the Little Colorado River. These areas are widely disjunct from the main area of distribution which lies along the Río Grande in New Mexico and adjacent Texas and Chihuahua. The plants may be locally common but in Arizona they are of restricted distribution. *Psorothamnus scoparius* often grows with *Atriplex*, *Artemisia*, *Larrea*, *Prosopis*, *Yucca*, and perennial grasses. The numerous stems help to trap blowing sand and often form hummocks (Barneby 1977). Various species of bees serve as pollinators for this species (Rozen, Jr. and Rozen 1986). Nodulation is reported for *P. scoparius* (Allen and Allen 1981). This species has been used for basket weaving, medicine, dyes, and revegetation (Lewis et al. 2005). Broom dalea is an attractive flowering shrub, with fragrant, dark blue flowers. It is ideally suited for the sandy desert garden but is not currently available from Arizona nurseries.

6. Psorothamnus spinosus (A. Gray) Barneby. Mem. New York Bot. Gard. 27:25. 1977. * [spiny] DESERT SMOKE TREE Native, Cultivated

Dalea spinosa A. Gray, Pl. Nov. Thurb. 315–316. 1854; *Psorodendron spinosum* (A. Gray) Rydberg

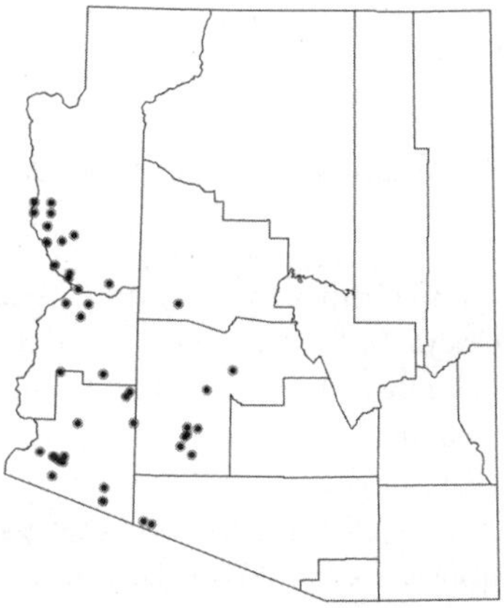

Shrubs or shrubby trees, 6(–10) m, plants appearing leafless or truly so for much of the year. **Herbage** strigillose, prominently glandular. **Stems** with widely diverging branches, twigs thorn-tipped, bark pale gray turning tan in age; pubescence densely silvery-silky, appressed or subappressed, rarely becoming glabrescent, bearing numerous orange glands. **Leaves** 0.3–2 (–3.5) cm; stipules 0.2–1 mm, broadly deltate, quickly deciduous; petiolate. **Leaflets** 1, 3–20(–35) × 1–4(–6) mm, oblanceolate or cuneate-oblanceolate. **Inflorescences** racemes, the thorn-tipped axis 1–4 cm, loosely 5–15-flowered; bracts ovate, 0.5–0.8 mm, deciduous. **Flowers** with calyx 4.5–5.2 mm, the tube 3–3.8 mm, the lobes 1–1.7 mm; corolla indigo blue, the banner 6.4–7.5 mm, the wings 7–8.8 mm, the keel 9.2–11 mm; stamens 9.5–11 mm. **Fruits** 4.2–5.3 mm, compressed-obovoid, the valves pilosulose distally and glandular. **Seeds** 1 or 2, 3.7–4.7 × 2–2.9 mm, smooth, brown with purple speckles.

Flowering (Feb–)May–Jul(Sep–Nov). Sandy or gravelly soils of arroyos on lower bajadas and valleys, and roadsides; [-70–]60–350(–760) m; Sonoran Desertscrub, Mohave Desertscrub; La Paz, Maricopa, Mohave, Pima, Yavapai, Yuma cos.; California, Nevada; Mexico (Baja California, Sonora).

Psorothamnus spinosus is locally abundant in xeroriparian woodland along larger arroyos in the Sonoran Desert

***Psorothamnus spinosus*: a., b. habit; c. twig with glands; d. fruit; e. flower. Illustrator: Randall C. Raak.**

and southern Mohave Desert in Arizona. It frequently grows with *Ambrosia*, *Chilopsis*, *Hymenoclea*, *Condea*, *Larrea*, *Olneya*, *Parkinsonia*, *Prosopis*, and *Senegalia*. This species is a phreatophyte, requiring sites that receive supplemental moisture (Nilsen et al. 1984). A deep root system allows the plants to utilize ground water during periods of drought. Seedlings rapidly develop a deep taproot. A 10 cm tall seedling excavated from an arroyo had a root that was 45 cm long (Johnson et al. 1993). *Psorothamnus spinosus* is reported to nodulate (Allen and Allen 1981, Jenkins et al. 1988).

Psorothamnus spinosus is leafless much of the year, relying on the twigs for photosynthesis. The total leaf area of a plant is less than 0.1% of the total photosynthetic area (Nilsen et al. 1984). Plants of *P. spinosus* appear to flower in as little as 2 or 3 years under ideal conditions, and in the nursery flowering has been observed on plants less than 8 months old (Johnson et al. 1993). The flowers are visited by several species of bees. The fruits fall from the plants soon after ripening. Ants frequently collect the fallen fruits, extracting the seeds from the calyx (Johnson et al. 1993).

The dried wood of *Psorothamnus spinosus* has been used for fuel and produces aromatic smoke (Barneby 1977).

Psorothamnus spinosus has been evaluated for its potential as a fragrance source for use in cosmetics and cleaning products, with trial plantings and extraction of resins from the calyx glands (Johnson et al. 1993). *Psorothamnus spinosus* has potential for planting as a windbreak.

The common name, desert smoke tree, is derived from the color and appearance of the plant when seen at a distance. Desert smoke tree is occasionally planted for landscaping in Arizona. The plants are ideally adapted to hot, arid regions where winters are not too severe. Its distinctive appearance and masses of showy indigo flowers are unique among native landscape plants. It is suitable as a focal point in a xeric landscape, planted singly or in groups. This species can also be planted as an informal hedge. With its spinescent twigs, desert smoke tree can be planted as a barrier if encouraged to develop a shrubby habit. The plants tolerate extreme heat. While they typically occur in sandy soils, the trees will grow well in heavier soils as long a drainage is adequate, though growth may not be as rapid (Johnson et al. 1993). Established plants are drought tolerant but need occasional deep irrigation to maintain an acceptable appearance. Desert smoke tree should be planted in a sunny location. Growth rates of 1 m per year have been observed for young plants growing in sandy soil and receiving regular supplemental irrigation. Cultivated plants have proven to be hardy to -8°C and are likely able to survive lower temperatures for brief periods.

Desert smoke tree seems to suffer periodic die-back of many of the twigs and even major limbs, even when irrigated in cultivation. The cause for this is not clear and does not always appear to be related to insufficient irrigation. Dead stems may be pruned out or left for a natural appearance. Desert smoke tree is reported to be relatively short-lived (Barneby 1977) and this has been confirmed with at least some cultivated specimens dying out after 15 years. However, a cultivated plant in Tucson has survived for 23 years. Desert smoke tree should be sited away from traffic areas due to the spine-tipped twigs; however, these are not particularly hazardous. Propagation is by seeds. Unscarified seeds germinate readily (Johnson et al. 1993). To avoid problems with rotting, seedlings should be grown in a sandy, well-drained soil mix and care should be taken when watering young seedlings, particularly during hot, humid weather (Johnson et al. 1993).

7. Psorothamnus thompsoniae (Vail) S.L. Welsh & N.D. Atwood. Great Basin Naturalist 35(4):354. 1975 [1976]. * [for Ellen P. Thompson, sister of John W. Powell, who first collected this species]

Parosela thompsoniae Vail, Bull. Torrey Bot. Club 24(1):18. 1897; *Dalea thompsoniae* (Vail) L.O. Williams

Subshrubs or shrubs to 0.9 (1.5) m, rounded to spreading. **Herbage** glabrous, puberulent, or densely silky-pilosulose; glands orange-brown, prominent. **Stems** with widely diverging branches, twigs often thorn-tipped. **Leaves** (1–)1.4–3 cm; stipules 0.2–0.9 mm, triangular, deciduous; petiolate. **Leaflets** (5–)7–17(–19), 2–7 × 0.3–2.5 mm, obovate to linear-oblanceolate or linear oblong, thick. **Inflorescences** racemes, the axis 1–9 cm, loosely 6–25-flowered; bracts 0.6–1.5 mm, ovate-elliptic, submembranous, quickly deciduous. **Flowers** with calyx 3.7–5 mm, the tube 2.4–3.5 mm, the lobes 1.1–2.3 mm, glabrous or pilosulose; corolla bright pink or violet-purple, the banner 5.5–8.4 mm, the wings 7.6–8.6 mm, the keel 7.4–9 mm; stamens 6–7.7 mm. **Fruits** 4–4.5 mm, the valves pilosulose and densely glandular, scarcely exserted from calyx. **Seeds** 1, 2.1–2.6 mm.

Varieties 2 (1 native in the flora).

7a. Psorothamnus thompsoniae (Vail) S.L. Welsh & N.D. Atwood var. **whitingii** (Kearney & Peebles) Barneby. Mem. New York Bot. Gard. 27:54. 1977. * [for Alfred F. Whiting, 1912–1978, American botanist and anthropologist who first collected this plant] WHITING'S DALEA Native

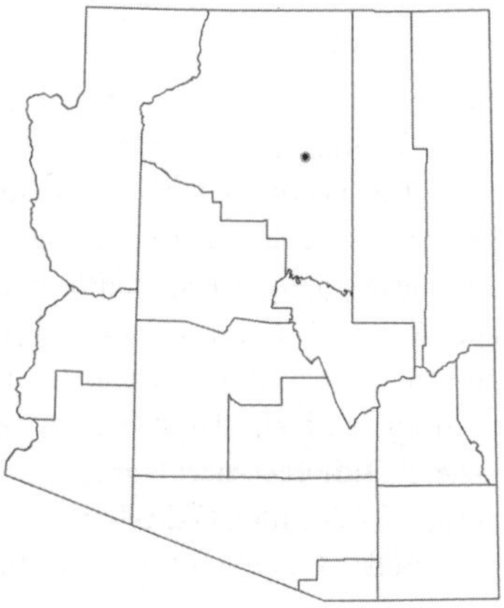

Dalea whitingii Kearney & Peebles, J. Wash. Acad. Sci. 29(11):484. 1939

Leaflets 3–7 × 0.3–0.9 mm, linear-oblanceolate to linear-oblong. **Flowers** with calyx tube pilosulose.

Flowering May–Aug. Sand or clay soils of slopes, valleys, along arroyos, and on sandstone tallus and benches; 1150–1400 m; Great Basin Desertscrub; Coconino Co.; Utah.

Psorothamnus thompsoniae var. *whitingii* is found in a small area of Coconino County along the Little Colorado River drainage, near Wupatki National Monument, and in adjacent San Juan County in southern Utah. Relatively few localized populations are known but the plants may be abundant. Of conservation concern.

RHYNCHOSIA Loureiro. Fl. Cochinch. 2:425, 460. 1790, name conserved. * [Greek *rhynchos*, a beak, in reference to the shape of the keel in the type species] SNOUTBEAN

Michael Woods

Arcyphyllum Elliott; *Austerium* Poiteau; *Chrysonias* Bentham; *Chrysoscias* E. Meyer; *Copisma* E. Meyer; *Cyanospermum* Wight & Arnott; *Hidrosia* E. Meyer; *Hydrosia* A. Jessieu; *Leucopterum* Small; *Leycephyllum* Piper; *Nomismia* Wight & Arnott; *Orthodanum* E. Meyer; *Phaseolus* Linnaeus subg. *Rhynchosia* (Loureiro) Eaton & Wright; *Phyllomatia* Bentham; *Pitcheria* Nuttall; *Polytropia* Presl; *Ptychocentrum* Bentham; *Rhinchosia* Desvaux; *Rynchosia* J. Macfadyen; *Sigmodostyles* Meisner; *Stipellaria* Klotzsch; *Walpersia* Harvey

Herbs, vines or subshrubs, prostrate, procumbent, twining, ascending, or erect. **Stems:** herbaceous, occasionally woody at the base, cylindrical to angular and striate, simple or branched. **Leaves** alternate, subsessile to long-petioles; simple or pinnately 3-foliolate; stipules persistent or deciduous, striate; stipels persistent or deciduous, 0.5–2.5 mm. **Leaflets** lanceolate to round to ovate to reniform or rhombic, resin-dotted on both surfaces or only below. **Inflorescence:** racemes in the leaf axils, flowers spaced or congested, rarely solitary; bracts striate, persistent or deciduous. **Pedicels** absent to 12 mm. **Flowers** papilionoid; calyx bell-shaped with five lanceolate lobes or tube-shaped with five short lobes; corolla typically yellow, often striped or blotched purple or reddish-brown; keel petals fused, sickle-shaped and broader, slightly incurved at apex; wing petals narrowly oblong to ovate; stamens 10, 9 fused and 1 separate; anthers uniform; ovary sessile to subsessile, densely villous, 1–2 ovules; style thread-like, glabrous; stigma rounded. **Fruits** ovate or oblong-elliptic to sickle-shaped, usually compressed, hairy, gland-dotted, beaked. **Seeds** red and brown to black to greenish and mottled or bi-colored red and black, shiny, nearly circular to ovate-reniform; hilum linear to ovate, 0.5–3 mm. $x = 11$.

Species approximately 200 (4 species native in flora): warm temperate and tropical regions in both Eastern and Western Hemispheres.

Various species of *Rhynchosia* have been planted for forage or as cover crops. The brightly colored seeds of several species are strung as necklaces and rosaries, hence a common name of rosary bean. The seeds have been used as weights (Lewis et al. 2005) and those of some species have narcotic properties (Schmutz and Hamilton 1979, Lewis et al. 2005). A yellow dye is obtained from the seeds of *R. corylifolia,* and the roots and leaves of certain species have been used as food by people (Allen and Allen 1981).

SELECTED REFERENCES: Grear 1978, Turner 2011.

1. Stems hairs yellowish; seeds widely ovate to subglobose ____ **1. Rhynchosia edulis**
1′ Stems hairs silverish or grayish or absent; seeds almost circular to ovoid or ovate-reniform.
   2. Flowers 8–10 mm, stamens 8–12 mm; seeds 5–8 mm ____ **3. Rhynchosia precatoria**
   2′ Flowers 4–7 mm, stamens 4–7 mm; seeds 3–4 mm.
      3. Wing petals gland-dotted, pubescent at tip; fruits oblong-ovate; seeds ovate-reniform ____ **2. Rhynchosia minima**
      3′ Wing petals lacking glandular dots, glabrous; fruits oblong; seeds almost circular ____ **4. Rhynchosia texana**

1. Rhynchosia edulis Grisebach. Abh. Königl. Ges. Wiss. Göttingen 19:123. 1874. * CHIHUAHUAN SNOUTBEAN Native

Dolicholus apoloensis Rusby; *Dolicholus ixodes* Standley; *Dolicholus melanostictus* (Grisebach) Kuntze; *Eriosema edule* (Grisebach) Burkhart; *Eriosema nigropunctatum* Brandegee; *Eriosema volubile* Micheli; *Eriosema volubile* Micheli forma *pube densiore* Chodat & Hassler; *Rhynchosia melanosticta* Grisebach; *Rhynchosia nigropunctata* S. Watson; *Rhynchosia pinetorum* Standley; *Rhynchosia rariflora* Standley; *Rhynchosia rupicola* Brandegee

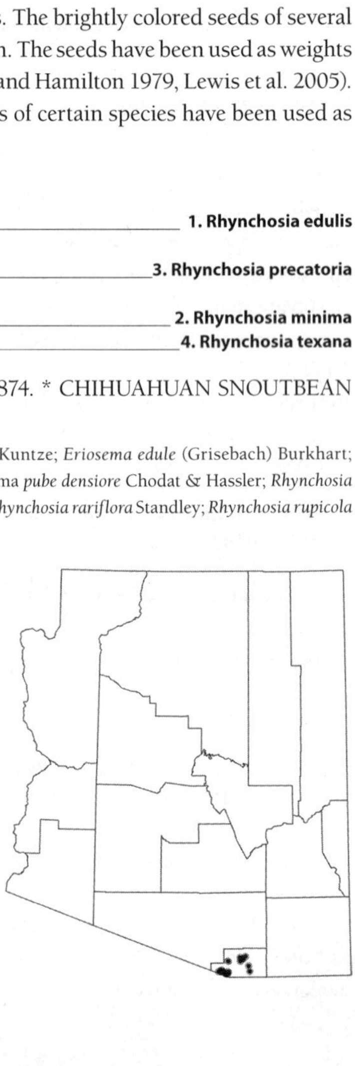

Herbs from woody taproot. **Stems** several, simple or branched, erect or suberect, decumbent, procumbent or twining, viscid, yellowish hirsute. **Leaves** 3-foliolate; petioles 1.5–5.5 cm, hirsute to viscid; stipules persistent, linear to narrowly lanceolate, acute to acuminate, hirsute, 3–9 × 0.5–1.5 mm. **Leaflets** lanceolate to long ovate or rounded, 1.5–7 × 1–6 cm, membranous or leathery, apex acute to acuminate, base obtuse to rounded or truncate, margins ciliate and slightly revolute, hairy above and beneath, especially along the veins, yellow or black gland dotted, often viscid on both surfaces, strongly reticulate beneath. **Inflorescence:** branched, 2.5–20 cm; bracts ovate, acuminate, early deciduous, 2–3 mm. **Pedicels** 1–3 mm. **Flowers** 7–10 mm; calyx shorter than the corolla, 4–7 mm, viscid, tube bell-shaped, 1.5–2 mm, lobes lanceolate, long-acuminate, 2–2.5 mm; corolla yellow, striped reddish-brown or purple, keel glabrous, 6.5–9 × 2–3 mm, wing petals narrowly oblong, gland-dotted and hairy at apex, 6.5–9 × 1.5–2 mm; banner ovate to widely ovate, slightly apiculate, hairy and gland-dotted, 7–10 × 4–5 mm; stamens 6–9 mm. **Fruits** long ovate, compressed, often viscid, long and short hairs intermixed, becoming glabrous with age, 16–26 × 6–8 mm, beak oblique 3–5 mm. **Seeds** greenish-brown or mottled, ovate to subglobose, compressed, 4–5 × 3–4 mm; hilum linear-ovate, 1–2 mm.

Flowering May–Jun; fruiting Jun–Sep. Rocky, sandy, or clay soils of canyons, slopes, ravines, pastures, riverbanks, roadsides; [150–](870–)1160–1460[–2130] m; Semidesert Grassland, Madrean Evergreen Woodland; Santa Cruz Co.; Mexico; Central America; South America.

In Arizona, *Rhynchosia edulis* has been collected from a few locations in the Atascosa, Pajarito, Patagonia, and Tumacacori Mountains in Santa Cruz County. The plants are locally common in some localities. Vouchers are deposited at both The University of Arizona (ARIZ) and Arizona State University (ASU). In Arizona, the stems typically die back to the crown each winter.

2. Rhynchosia minima (Linnaeus) de Candolle. Prodr. 2:385. 1825. * LEAST SNOUTBEAN Native

Dolichos minimus Linnaeus, Sp. Pl. 2:726. 1753; *Dolicholus minimus* (Linnaeus) Medikus; *Glycine lamarckii* Kunth; *Glycine littoralis* Vahl ex de Candolle; *Glycine reflexa* Nuttall; *Phaseolus caribaeus* Eaton & Wright; *Rhynchosia aureo-guttata* Andersson; *Rhynchosia exigua* Andersson; *Rhynchosia mexicana* Hooker & Arnott; *Rhynchosia minima* (Linnaeus) de Candolle var. *diminifolia* Walraven; *Rhynchosia minima* (Linnaeus) de Candolle var. *lutea* Eggers; *Rhynchosia minima* (Linnaeus) de Candolle var. *pauciflora* Kuntze; *Rhynchosia punctata* de Candolle

Herbs or semi-woody shrubs with long branched taproot. **Stems** one to several, branched, prostrate or twining, rarely ascending, glabrous to downy or villous, rarely viscid, hairs spreading or reflexed, silvery or grayish. **Leaves** 3-foliolate; petioles 0.4–10 cm, glabrous, tomentose or villous; stipules persistent, linear to lanceolate, acuminate, puberulent to villous, 1–6 × 1–2 mm. **Leaflets** rhombic-circular, rhombic to subrhombic or ovate, 3–9 × 0.5–7 cm, membranous or leathery, apex acuminate to obtuse, base obtuse to rounded, acute or truncate, glabrous or finely strigose, villous, puberulent above, glabrous to puberulent or sparsely to densely villous beneath, gland-dotted above and beneath. **Inflorescence** 2–8-flowered, unbranched, 2–20 cm; bracts lanceolate, acuminate, early deciduous, 1–2 mm. **Pedicels** 1–5 mm. **Flowers** 4–7 mm; calyx shorter than the corolla, 3–6 mm, puberulent or villous, tube bell-shaped, 1.5–2 mm, lobes lanceolate, attenuate, 1.5–4 mm; corolla yellow, veined purple or brown, keel glabrous, 5–6 × 2–3 mm; wing petals narrowly oblong, gland-dotted and pubescent at apex, 4.5–5 × 1–2 mm; banner circular-ovate emarginate, puberulent and gland-dotted, 4–6 × 3–5 mm; stamens 5–7 mm. **Fruits** sickle-shaped, oblong-ovate, compressed, glabrous to puberulent or villous, 9–20 × 3–5 mm, beak 0.5–1 mm. **Seeds** grayish, brown, black or mottled, ovate-reniform, plump, 3–4 × 2–3 mm; hilum ovate, 0.5–1 mm.

Flowering Apr–Sep; fruiting Aug–Nov. Slopes in arid grasslands; [100–]1280[–2000] m; Semidesert Grassland; Santa Cruz Co.; Alabama, Arkansas, Florida, Louisiana, Mississippi, Missouri, Texas; Central America; South America; West Indies; Bermuda.

In Arizona, *Rhynchosia minima* has been reported from only three locations in Santa Cruz County: Sonoita Creek State Natural Area, San Rafael Ranch State Park, and Salero Ranch.

Whyte et al. (1953) reports that *Rhynchosia minima* is an important pasture plant for livestock in some tropical regions, and Tull (1987) mentions that it is a good native browse plant. Mali and Mahale (2008) determined in laboratory experiments that leaf extracts of *R. minima* were effective in killing parasitic worms. They indicate that this species is used in folk medicine in India to treat wounds and parasitic worms, and as an abortifacient.

3. Rhynchosia precatoria (Humboldt & Bonpland ex Willdenow) de Candolle. Prodr. 2:385. 1825. * PRAYING SNOUTBEAN Native

Glycine precatoria Humboldt & Bonpland ex Willdenow, Enum. Pl. 2:755–756. 1809; *Dolicholus vailiae* Rose; *Rhynchosia bicolor* Micheli

Semi-woody shrubs from woody rootstock. **Stems** one to several, simple or branched, sprawling or twining, villous, hairs silvery or grayish and silvery intermixed, gland dotted, rarely viscid. **Leaves** 3-foliolate; petioles 0.1–0.7 cm, villous, rarely viscid; stipules early deciduous, ovate or lanceolate, acute or acuminate, villous, 4–6 × 1.5–2 mm. **Leaflets** obovate to ovate-lanceolate, ovate-rhomboid to rhomboid, 2–13 × 1.5–10 cm, leathery, apex acute to acuminate, base rounded to obtuse or cuneate, villous above, rarely becoming glabrous, gland-dotted, villous beneath. **Inflorescence** branched, 5–30 cm; bracts linear to narrowly-lanceolate, acuminate, deciduous, 2–4 mm. **Pedicels** 1–2 mm. **Flowers**

8–10 mm; calyx equal to or shorter than the corolla, 4–8 mm, villous, often viscid, tube bell-shaped, 2–2.5 mm, lobes lanceolate, attenuate, 2–3 mm; corolla green-yellow, streaked with purple or brown, keel puberulent at apex, 6–8 × 3–3.5 mm; wing petals oblong, often gland dotted and puberulent at apex, 6–7 × 2–2.5 mm; banner circular, emarginate, villous and gland-dotted, 8–9 × 6–7 mm; stamens 8–12 mm. **Fruits** widely ovate-oblong, subinflated, constricted, densely villosulous to puberulent, rarely viscid, 20–30 × 7–12 mm, beak 1–2 mm. **Seeds** red and black, almost round to ovoid, plump, 5–8 × 4–6 mm wide; hilum ovate, 2–3 mm.

Flowering Feb–Jun; fruiting Jun–Aug. Rocky soils of canyons; [200–]1220[–2000] m; Madrean Evergreen Woodland; Santa Cruz Co.; Mexico; Central America; South America.

In Arizona, *Rhynchosia precatoria* has been collected from only one location, Rock Corral Canyon, in the Tumacacori Mountains in Santa Cruz County. A voucher is deposited at The University of Arizona (ARIZ). Across its range, *R. precatoria* is found in sandy soils or limestone, in open areas, thickets, and forests. In Sonora, some people consider the seeds to bring good luck (Martin et al. 1998).

4. Rhynchosia texana Torrey & A. Gray. Fl. N. Amer. 1:687. 1840. * TEXAS SNOUTBEAN Native

Dolicholus angustifolius (Engelmann ex A. Gray) Kuntze; *Rhynchosia angustifolia* Engelmann ex A. Gray; *Rhynchosia senna* Gillies ex Hooker var. *angustifolia* (A. Gray) Grear; *Rhynchosia senna* Gillies ex Hooker var. *texana* (Torrey & A. Gray) M.C. Johnston; *Rhynchosia texana* Torrey & A. Gray var. *angustifolia* A. Gray

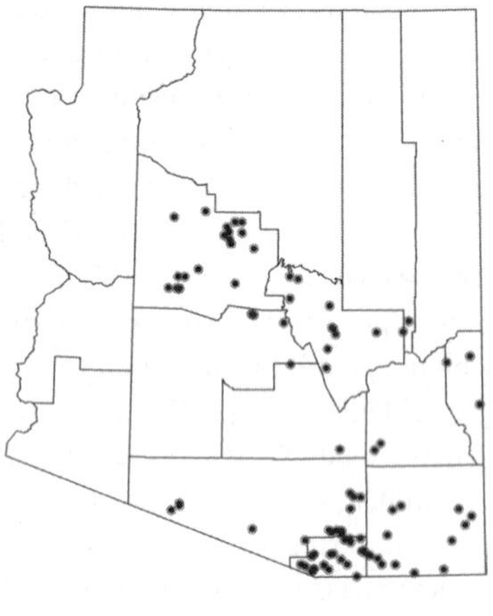

Herb from woody taproots. **Stems** branched, prostrate, twining, rarely sub-ascending, puberulent to strigillose or villosulous with silvery pubescence. **Leaves** 3-foliolate; petioles 0.3–1.5 cm, puberulent, strigillose or villosulous; stipules persistent, linear or linear-lanceolate, 1.5–3 × 0.5 mm. **Leaflets** linear-lanceolate to rhombic-ovate, margins revolute, 0.5–4.5 × 0.2–2.5 cm, leathery, apex acute or obtuse, base obtuse or subcordate, puberulent, strigillose or villosulous above and beneath, gland-dotted above and beneath. **Inflorescence** unbranched, 0.6–1.5 cm, of a single flower, rarely fasciculate with 2–3 per axil; bracts bristle-like or linear, deciduous, 0.5–1 mm. **Pedicels** 1–6 mm. **Flowers** 5–7 mm; calyx shorter than corolla, 3–5.5 mm, villous or villosulous, tube bell-shaped, 1.8–2 mm, lobes subulate, acuminate, 1–3.5 mm; corolla yellow, with brown veins, keel glabrous, 5–7 × 2–2.5 mm, claw 1.5–2 mm; wing petals oblong to ovate, lacking glandular dots, glabrous, 5–6 × 1–2 mm, claw 1.5–2 mm; banner widely ovate, puberulent, 5–7 × 3.5–5 mm, auricles 0.5 mm, claw 1–2 mm; stamens 4–6 mm. **Fruits** oblong, compressed, puberulent-hirsutulous intermixed, 15–25 × 4–6 mm, beak 1 mm. **Seeds** brown, black or mottled, nearly circular, compressed, 3–4 × 3–4 mm; hilum ovate, 0.5–1 mm. $2n = 22$.

Flowering Feb–Jun; fruiting Jun–Sep. Rocky outcrops, slopes, canyons, arroyos, valleys, riparian habitats, and roadsides in various soils including dry limestone, sandy, or igneous soils; [200–]1000–1860[–2700] m; Semidesert Grassland, Plains Grassland, Interior Chaparral, Madrean Evergreen Woodland; Cochise, Gila, Graham, Greenlee, Maricopa, Pima, Navajo, Santa Cruz, Yavapai cos.; New Mexico, Texas; Mexico; South America.

In Arizona, *Rhynchosia texana* is found in a wide variety of habitats and soil types. It is found in the open with the stems trailing along the ground or clambering over other plants in thickets. Allen and Allen (1981) state that *R. texana* (as *R. senna* var. *texana*) has good potential as a groundcover.

ROBINIA Linnaeus. Sp. Pl. 2:722. 1753. * [for Jean Robin, 1550–1629, and his son, Vespasian, 1579–1662, Royal Gardeners in Paris during the reign of Henry IV, who brought *Robinia pseudoacacia* into cultivation in Europe] LOCUST

Matt Lavin, Kirsten N. Lake, Matthew B. Johnson

Trees and shrubs, variably armed. **Stems** erect to usually ascending, often root-sprouting; young growth glabrous to sericeous, tomentose, or hispid, sometimes with stipitate glandular hairs. **Leaves** alternate; odd-pinnate; stipules linear, quickly deciduous or persistent and then becoming spinescent; petiolate; rachis with a longitudinal groove; stipels

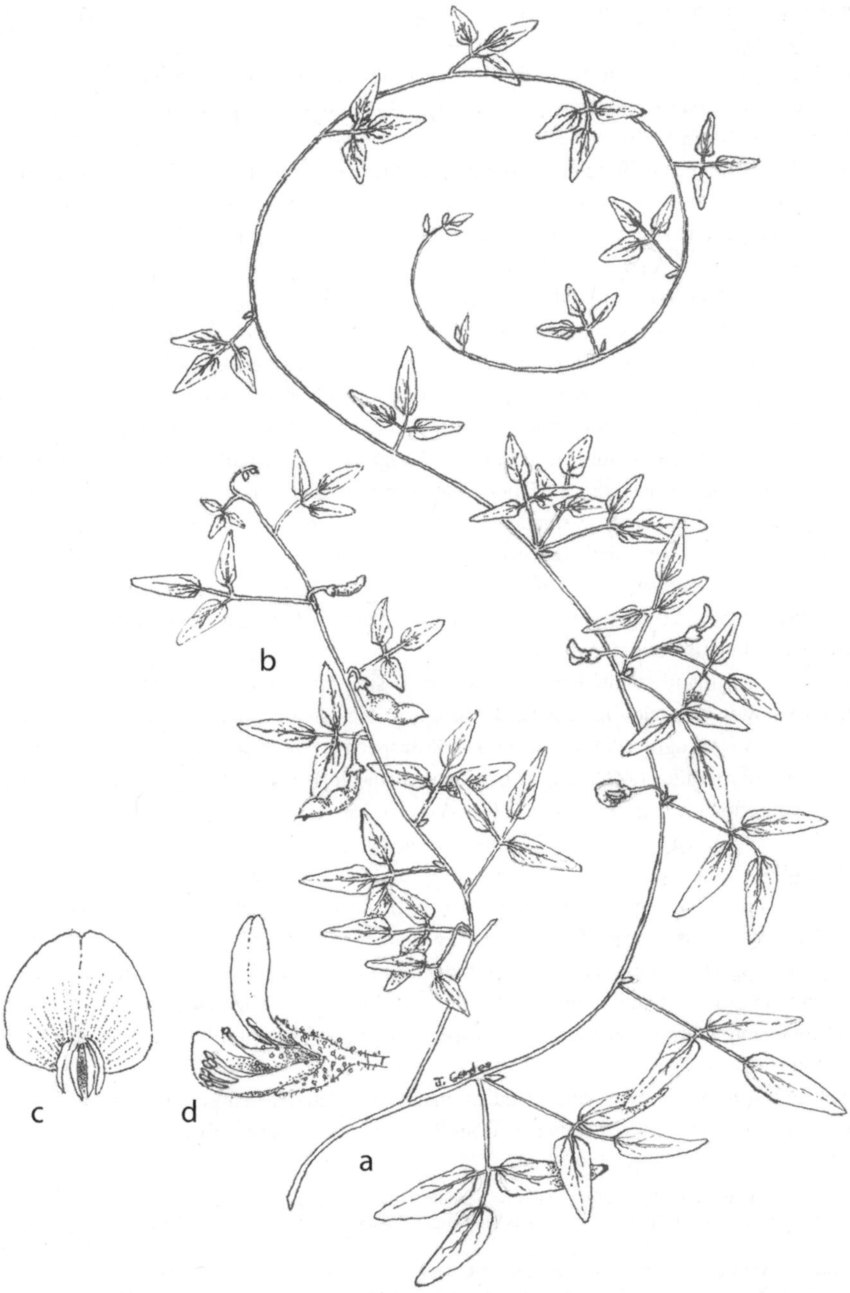

Rhynchosia texana: **a.** habit; **b.** fruiting stem; **c.** flower; **d.** longitudinal section of flower. Illustrator: John H. Gerdes.

present. **Leaflets** 7–25, usually opposite to subopposite, rarely alternate; elliptic, base rounded, apex rounded; margins entire, surfaces glabrate to strigose or sericeous. **Inflorescences** with mostly 4–25 flowers; axillary, racemes; bracts quickly deciduous; bracteoles absent. **Flowers** papilionoid; calyx bilaterally symmetrical, bell-shaped, lobes 5; corolla whitish to mostly pinkish, glabrous; stamens 10, 9 united and 1 free, unequal, anthers attached basally, relatively small, dehiscing longitudinally; style glabrous, with a pollen brush loosely scattered along 1 side distally; stigma terminal, spherical, ciliate; ovules 9–17. **Fruits** sessile, tardily elastically dehiscent, laterally compressed, linear, not compartmented between seeds, base rounded, apex rounded and often terminating in persistent style base, glabrous, sericeous, tomentose, hispid, or glandular. **Seeds** 3–16, lenticular, with a slightly recessed hilum. $x = 10$.

Species 4 (1 native and cultivated in the flora, 1 adventive and cultivated): United States, Mexico (northern Sierra Madre Occidental in Chihuahua and Sonora).

Species of *Robinia* are often colonizers of disturbed settings (episodic or regularly) in temperate deciduous and conifer forests, woodlands, and shrubby vegetation, and they sometimes escape from ornamental plantings (Isely and Peabody 1984, Lavin and Sousa 1995).

Robinia is readily distinguished from other North American legume genera by pods that are tardily dehiscent and not very elastically (or explosively) so. Seeds are transversely arranged (e.g., the hilum faces toward the base of the fruit rather than towards the placental margin). Each seed is born from an elongate funiculus, which presumably accommodates the transverse orientation. Also, *Robinia* racemes are different from those of its closest relatives: they are formed from the leafy axils of current season's growth and produce flowers that reach anthesis simultaneously.

The genus *Robinia* represents one of the very few temperate North American lineages to have evolved from neotropical ancestry (e.g., like *Strophostyles*, Riley-Hulting et al. 2004). Most temperate North American legume groups (e.g., *Astragalus*, *Lupinus*, *Vicia*) stem from Eurasia. Like its neotropical relatives, *Robinia* species are primarily inhabitants of disturbance-prone settings and often root sprout on unstable or exposed soils and thus can form dense colonies that stabilize loose soil. The growth habit ranges from small and large shrubs to small and large trees (*Robinia pseudoacacia* adopts a large-tree-habit most consistently). Variation in such growth form is likely a function of disturbance, with more recent and regular disturbance indicated by smaller stature. Growth of *Robinia* species also appears to be regulated more by temperature than day length (personal observation, M. Lavin), which could be a result of a recent tropical ancestry. Consequently, *Robinia* seems outcompeted in undisturbed settings and prone to die-back related to cold temperatures. Apart from a variably shrubby to tree growth form, this may explain the generally crooked stems and unkempt appearance of many *Robinia* individuals and populations.

Excepting the predominantly white-flowered and outcrossing *Robinia pseudoacacia* (Bongarten 1992), the three predominantly pink-flowered species, *R. hispida*, *R. viscosa*, and *R. neomexicana*, harbor much inter-population variation that is often formally recognized at the species or infraspecific level (e.g., Isely and Peabody 1984). This variation is often the result of hybridization (e.g., *R. viscosa* var. *hartwigii* (Koehne) Ashe), clonal variation due, in part, to reproduction involving triploidy (e.g., segregates *R. hispida*), plasticity of growth form (e.g., *R. hispida* var. *nana* (Elliott) de Candolle), or geographic structuring (e.g., *R. neomexicana* var. *rusbyi* (Wooton & Standley) W. C. Martin & C. R. Hutchins ex Peabody). This historically recognized intraspecific variation is not formally treated here because the genus *Robinia* is first in need of a population genetic analysis that can ultimately differentiate morphological variation marking distinctive evolutionary lineages from morphological variation caused by other factors (e.g., ecological setting, Mendelian sorting, interspecific hybridization, and founder events especially involving asexual forms).

Robinia species are grown as landscape trees for their showy flowers. They are also planted for soil stabilization and timber, fuelwood, and honey production. *Robinia* species have been used for human food, as a dye source, and in folk medicine. Fiber can be made from the bark (Allen and Allen 1981, Lewis et al. 2005). *Robinia* species contain toxic compounds that can cause poisoning in humans and animals if ingested (Allen and Allen 1981, Lewis et al. 2005).

The following treatment is adapted primarily from Radford et al. (1968), Isely and Peabody (1984), and Gleason and Cronquist (1991).

1. Flowers with whitish, rarely pinkish, petals, 15–20 mm; racemes pendant, often with 15–25 flowers; fruit glabrous, the upper margin narrowly winged; inflorescence axis and calyces sericeous, lacking hispid or glandular hairs; length of calyx lobes shorter than that of the calyx tube ____ **2. Robinia pseudoacacia**

1′ Flowers with pinkish, rarely whitish, petals, 20–25 mm; racemes lax to ascending or erect, with mostly 10–20 flowers; fruits conspicuously hispid, the margin not winged; inflorescence axis and calyces with stipitate glands, tomentose to sericeous; length of calyx lobes about equal to that of the calyx tube ____ **1. Robinia neomexicana**

1. Robinia neomexicana A. Gray. Pl. Nov. Thurb. 314. 1854. * [of New Mexico] NEW MEXICO LOCUST; *GARROBO*
Native, Cultivated

Trees to small shrubs, 1–5(–15) m. **Stems** with smooth gray bark, becoming fissured on older branches and trunks. **Leaves** mostly 10–25 cm; stipules 8–11 mm, both deciduous and persistent (spiny); petioles 0.5–1.6 cm, tomentose to sericeous; petiolules 2–3 mm, tomentose to sericeous. **Leaflets** 13–23, mostly 25–35 × 15–20 mm, upper (adaxial) surface strigose to sericeous, lower (abaxial) surface strigose to sericeous. **Inflorescences:** mostly 3–10 cm, lax to erect, with mostly 10–20 flowers, axis tomentose to sericeous, often with stipitate glands; bracts lanceolate, 9–12 × 2–3 mm, margins entire. **Pedicels** mostly 3–5 mm. **Flowers:** calyx tube 6–7 mm, tomentose to sericeous, often with stipitate glands, lobes 5–7 mm, lower (abaxial) longer than laterals, upper (adaxial) united, more so than laterals; corolla pinkish

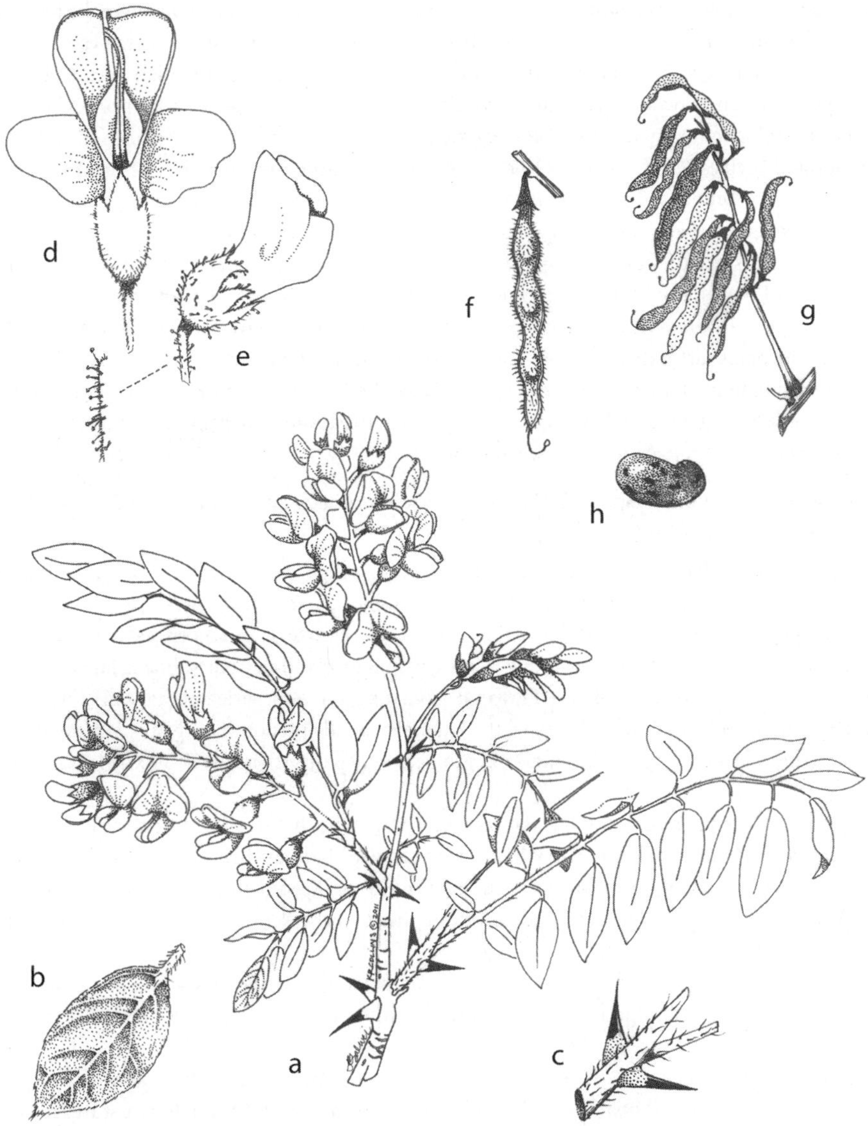

***Robinia neomexicana*: a. habit; b. leaflet; c. stem node; d., e. flowers with pedicel detail; f. fruit; g. fruits; h. seed. Illustrators: Katherine Salwei & Kim Ratelle-Collins.**

(whitish), 20–25 mm. **Fruits** light to dark brown, upper (placental) margin not winged, 4–8 × 0.9–1.1 cm, generally conspicuously hispid. **Seeds** 4–10(–15).

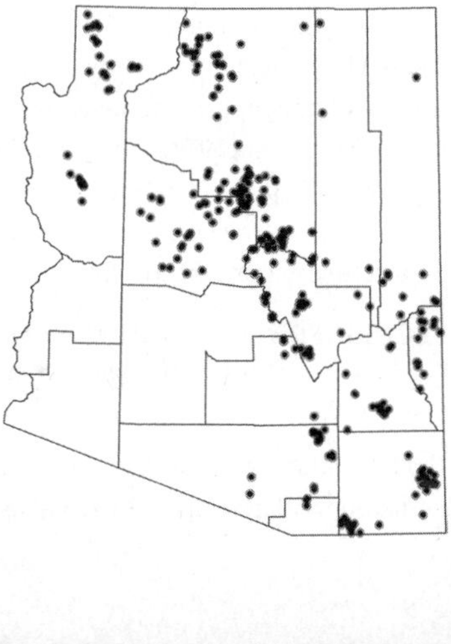

Flowering May–Jul(–Sep). Open disturbed settings, exposed rocky outcrops, steep slopes, and understory of pine forests and woodlands and in riparian habitats; (850–)1200–2900 m; Interior Chaparral, Madrean Evergreen Woodland, Great Basin Conifer Woodland, Petran Montane Conifer Forest; all Arizona counties except La Paz and Yuma cos.; native to mountainous regions of California, Colorado, Nevada, New Mexico, Texas, and Utah; Mexico (the Sierra Madre Occidental of Chihuahua and Sonora).

The tomentose to sericeous leaf rachises and branches, often stipitate glandular inflorescence axis, often conspicuously hispid fruits, and distribution in southwestern United States and adjacent Mexico distinguish *Robinia neomexicana* from the other pink-flowered species of *Robinia*. Forms of *Robinia neomexicana* lacking the hispid and glandular hairs on the inflorescence axis and pod still retain the relatively abundant tomentose to sericeous indumentum on the leaves and branches, which remains fairly evident even in age.

Robinia neomexicana is widespread in mountainous areas of the state, though populations are often scattered and localized. It is especially common along riparian canyons and on steep rocky slopes and road cuts where it forms clonal colonies that may extend over many meters. Cattle eat the flowers of *R. neomexicana*, and the foliage is eaten by cattle and deer (Kearney and Peebles 1942, Allen and Allen 1981). Schmutz and Hamilton (1979) state that *R. neomexicana* is suspected of being toxic to humans. Toxicity of *R. pseudoacacia* in livestock and humans is well documented. The Hopi utilize *R. neomexicana* to treat rheumatism and as an emetic (Kearney and Peebles 1942).

New Mexico locust is occasionally planted in landscapes at mid- and higher-elevation areas of the state where it is appreciated for its flowers. It will do well at lower elevation sites if sufficient irrigation is provided. New Mexico locust is suitable as a patio tree and in a mini oasis. This species can grow rapidly with ample irrigation. The leaves are winter-deciduous. The plants are extremely cold hardy, easily tolerating temperatures below -20°C and many consecutive days of subfreezing temperatures. Due to its spines, New Mexico locust should be sited away from traffic areas. The trees produce some leaf litter when they become deciduous in the autumn. In some situations, the root sprouts may become a maintenance issue. Propagation is by scarified seeds.

2. Robinia pseudoacacia Linnaeus. Sp. Pl. 2:722. 1753. * [Greek *pseudo*, false, and acacia] BLACK LOCUST, FALSE ACACIA Cultivated, Adventive

Trees, (1–)4–25 m. **Stems with** bark grey or light brown, becoming rough and furrowed with age. **Leaves** mostly 20–30 cm; stipules 5–10 mm, both deciduous and persistent (spiny); petioles 0.7–2.5 cm, strigose; petiolules 2–3 mm, glabrate. **Leaflets** (7–)15–19, mostly 25–45(–65) × 10–20(–25) mm, surfaces glabrate. **Inflorescences**: mostly 5–19 cm, pendant, most often with (10–)15–25 flowers, axis sericeous lacking hispid or glandular hairs; bracts lanceolate, 8–10 mm × 1 mm, margins entire. **Pedicels** mostly 7–12 mm. **Flowers:** calyx tube 5–6 mm, sericeous, lobes 1–2 mm, lower (abaxial) longer than laterals, upper (adaxial) united, more so than laterals; corolla predominantly whitish, rarely pinkish, 15–20 mm. **Fruits** light to dark brown, upper (placental) margin narrowly winged, mostly 4–10 × 1–1.5 cm, glabrous. **Seeds** 4–8(–16).

Flowering Mar–Jun. Temperate deciduous forests and woodlands, disturbed settings; [near sea level–]580–1580 [–2000] m; adventive in riparian habitats in Mohave Desertscrub, Madrean Evergreen Woodland, and Great Basin Conifer Woodland; Mohave, Navajo, Yavapai cos.; native at least in the Appalachian and other mountainous regions of eastern North America, widely cultivated and escaped throughout North America and many temperate regions of the world.

Populations growing at latitudinal extremes (e.g., in Canada) are more likely to be shorter-statured (e.g., 3 m or less tall). The winged and glabrous pods and flowers with white petals arranged on pendant racemes set *Robinia pseudoacacia* apart from the other species of *Robinia*.

In Arizona, *Robinia pseudoacacia* has been documented as established in only a few localities in riparian habitats. Cultivated trees can persist around abandoned buildings in situations with adequate moisture. It has potential to become more widely invasive. Nodulation in *R. pseudoacacia* is widely reported.

Flowers of *Robinia pseudoacacia* are highly regarded for honey production (Allen and Allen 1981, Barneby 1989). The wood of *R. pseudoacacia* is valued as timber. It is hard, flexible, highly resistant to decay, and has been widely used for fence posts, in construction, and for making furniture (Allen and Allen 1981, Barneby 1989). It has also been used to make wagon wheel hubs, telephone pole insulator pins, and wooden pins used in the construction of ships (Allen and Allen 1981). The wood is difficult to work and weighs 640–800 kg/m^3 (Allen and Allen 1981). The foliage, seeds, and inner bark are reported to be toxic to humans and livestock if ingested and fatalities have occurred (Kingsbury 1964, Hardin and Arena 1974, Lewis and Elvin-Lewis 1977, Schmutz and Hamilton 1979).

Black locust is infrequently planted in Arizona as a landscape tree. Its horticultural characteristics are similar to those of *Robinia neomexicana*. It is extremely cold tolerant and will grow at lower elevations with sufficient irrigation. Spineless selections are available.

SECURIGERA de Candolle. Fl. Franç. (de Candolle & Lamarck) ed. 3, 4:609. 1805, name conserved. * [Latin *securiger*, axe-bearing, referring to the shape of the fruit of *Securigera securidaca*]

Peter W. Ball

Herbs [annual or] perennial. **Stems** furrowed or ridged. **Leaves** alternate; odd-pinnate; stipules free, membranous [or herbaceous], dark-tipped. **Leaflets** usually opposite, margins entire. **Inflorescences** axillary, pedunculate [umbels or]

heads; bracteoles free, membranous [or herbaceous], dark-tipped. **Flowers**: papilionoid; calyx bell-shaped, 5-toothed, weakly bilaterally symmetrical; corolla papilionoid; stamens 10, upper (adaxial) stamen free, remaining 9 with fused filaments, anthers fixed basally; ovary 1-chambered, style and stigma simple. **Fruits** loments, indehiscent or tardily dehiscent, rounded or angled in cross-section, not or slightly constricted between seeds, with weak longitudinal anastomosing veins. **Seeds** 2–12; oblong, smooth, hilum lateral. $x = 6$.

Species 12 (1 weakly naturalized): Mediterranean region, widely introduced in North America and temperate regions of the world.

The circumscription of *Securigera, Coronilla,* and *Hippocrepis* was revised by Lassen (1989) who transferred a high proportion of the species of *Coronilla* to *Securigera* and *Hippocrepis* based on characters other than fruit structure. The molecular phylogenetic studies of Allen and Porter (2000) confirmed that *Coronilla*, as traditionally defined, is polyphyletic, although they did not examine a wide enough range of species to clearly support Lassen's rearrangement.

SELECTED REFERENCES: Allen and Porter 2000, Lassen 1989.

1. Securigera varia (Linnaeus) Lassen. Svensk Bot. Tidskr. 83:86. 1989. * [Latin *varius*, variegated, multicolored, referring to the flower color] CROWN VETCH, PURPLE CROWN VETCH Weakly Naturalized

Coronilla varia Linnaeus, Sp. Pl. 2:743. 1753

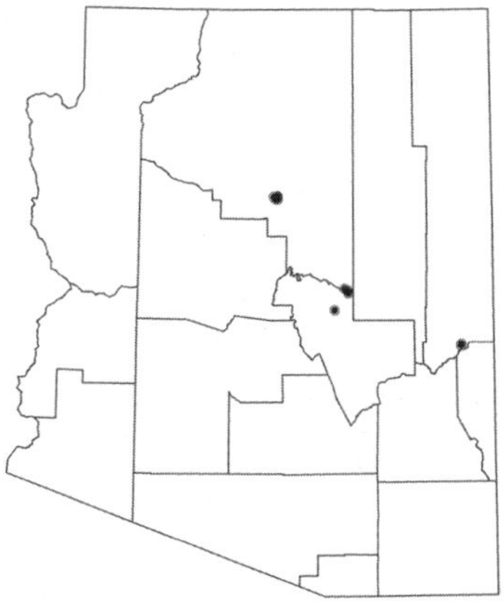

Perennial herbs, rhizomatous. **Stems** erect to procumbent, 20–150 cm, hollow, angular, glabrous. **Leaves**: stipules membranous, linear to oblong, 1–6 mm; rachis 4–15 cm. **Leaflets** 5–12 pairs, linear to oblong, adaxial (upper) surface usually densely covered with minute dark purple dots, abaxial (lower) surface often with a minute purple network, 6–25 × 2.5–12 mm, apex rounded to retuse, mucronate, base cuneate-rounded, margins narrowly scarious. **Inflorescences**: peduncles 4–14 cm; bracteoles linear, 1–2 mm; heads 6–25-flowered, 2–3.5 cm in diameter. **Pedicels** 2–5 mm. **Flowers**: calyx 2.2–4 mm, teeth 0.7–1 mm; corolla white, pink, purple, or bicolored, keel tip usually darker, (8–)10–15 mm. **Fruits** 20–60 × 1.5–2.2 mm; segments 2–10(–12), 4–6 mm, oblong, 4-angled in cross-section, with weakly anastomosing, longitudinal veins on the surfaces. **Seeds** cream to pale brown, oblong, 3.5–4 × 1–2 mm. $2n = 24$.

Flowering, fruiting Jun–Jul, then erratically until Oct. Roadsides, grassy banks, and streamsides; 1700–2400 m; Madrean Evergreen Woodland and Petran Montane Conifer Forest; Apache, Cochise, Coconino, Gila cos.; s Europe and sw Asia, widely introduced in temperate regions of the globe.

Crown vetch was introduced into North America as a forage plant in the 20th century although it was not commonly naturalized. From about 1960 onwards it was much planted along roadsides and on steep banks as a soil stabilizer from which areas it has often spread into waste places, fields, grassland, and various native habitats. Because of this propensity to spread, its planting has now been banned in some areas. In Arizona it has been planted along roadsides on a limited scale in the cooler parts of the state and seems to have persisted or spread along roads, streamsides, and into parking lots to some extent.

Allen and Allen (1981) and Duke (1981) indicate that *Securigera varia* is used as forage. However, it has been reported that, although it is not toxic to ruminants, it can cause poisoning in nonruminants such as horses (Stubbendieck and Conrad 1989, Knight and Walter 2001). Burrows and Tyrl (2013) also report that cardiotoxins and nitrotoxins found in the plant are unlikely to affect ruminants. They state that horses may be more susceptible, but that has not been tested. Reported signs of poisoning in nonruminants include weight loss, poor growth rates, depression, ataxia, and posterior paralysis leading to death (Knight and Walter 2001). The plant has reportedly been employed as a cardiotonic, a diuretic, and a purgative (Duke 1981). *Securigera varia* is widely reported to nodulate.

SENEGALIA Rafinesque. Sylva Tellur. 119. 1838. * [based on *Mimosa senegal* L.]

David S. Seigler and John E. Ebinger

Shrubs, trees or lianas, mostly erect or spreading. **Stems** lightly to highly branched; twigs rounded or angular, glabrous to pubescent; short shoots present or absent; stipular spines absent; prickles present (very rarely reduced to small projections on the stems, petiole, or rachis), straight to recurved, usually scattered along the twig, sometimes on petiole and rachis, rarely paired at the nodes. **Leaves** alternate; bipinnate, pinnae 1–numerous pairs per leaf, mostly opposite; stipules mostly small, not obvious, usually early deciduous; petioles round, usually channeled above, petiolar

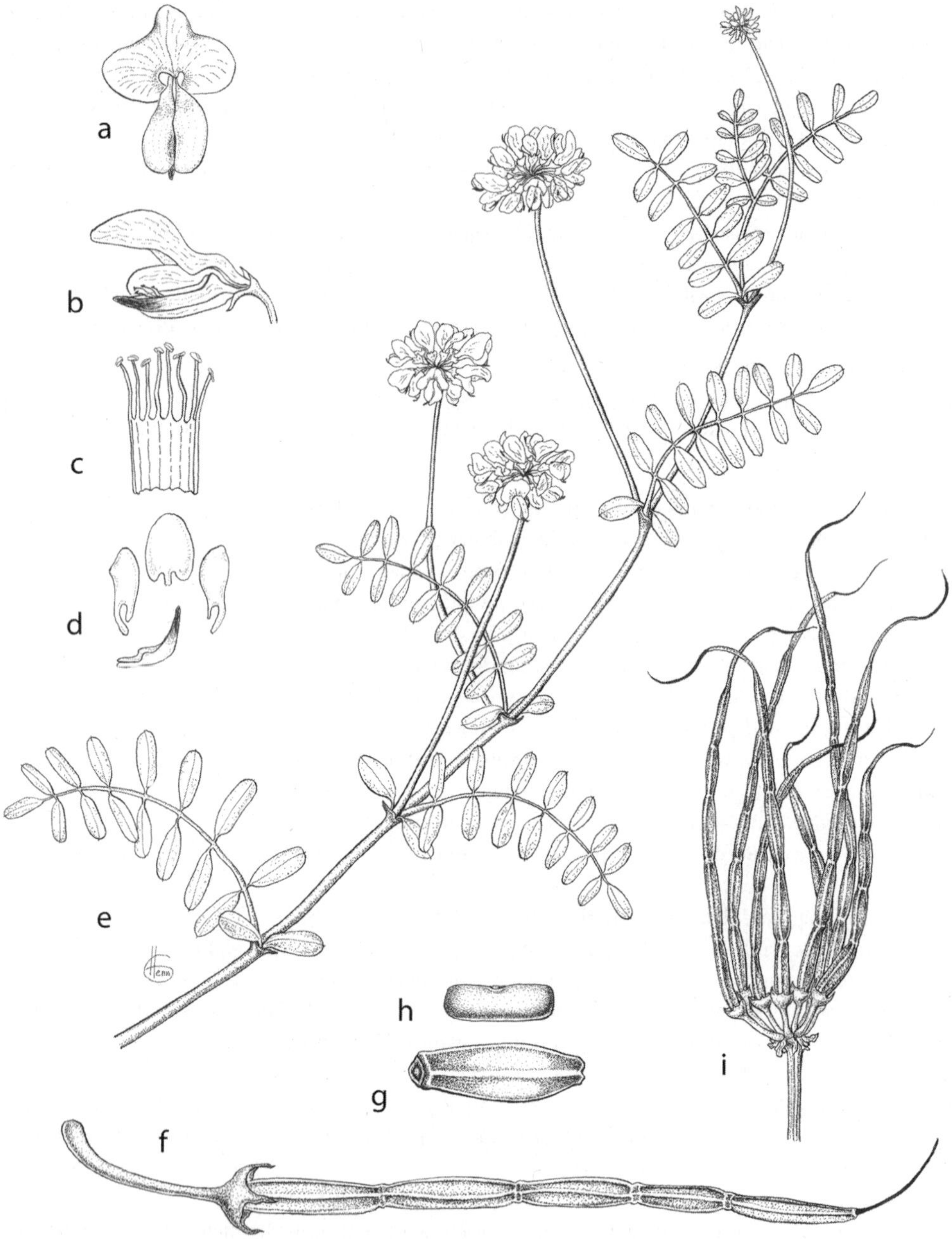

Securigera varia: **a.** flower (front view); **b.** flower cut-away view; **c.** stamens; **d.** petals; **e.** habit; **f.** fruit; **g.** fruit segment; **h.** seed; **i.** fruit cluster. Illustrator: Gabriele Henn.

gland present; rachis channeled above, sessile to stalked glands at various positions. **Leaflets** 1–numerous pairs per pinna, usually opposite, subsessile, membranous, glabrous to pubescent. **Inflorescences** loosely- to densely-flowered globose heads to cylindrical spikes, solitary to clustered in the leaf axil or on spur branches, sometimes in terminal or axillary raceme-like or paniculate clusters; peduncles short, glabrous to pubescent, rarely elongating in fruit; bracts present. **Pedicels** absent to short. **Flowers** mimosoid, radially symmetric, 5-merous; calyx cup-shaped, lobes triangular, glabrous to pubescent; corolla cup-shaped, membranous, lobes triangular, glabrous to pubescent, valvate in bud; nectariferous disk present in all flowers; stamens 30–90, 5–13 mm, exserted, white, cream, or yellow; anthers attached dorsally, sometimes with a small gland; ovary 1, sessile to stipitate; style thread-like; stigma thread-like. **Fruits** erect to pendulous, usually dehiscing along both sutures, linear to oblong, straight to slightly curled, mostly flattened, mostly dry, stiffly papery to leathery, glabrous to pubescent; stipe short; apex sometimes beaked. **Seeds** arranged in a single-row, not surrounded by pulp, circular to oblong, sometimes flattened, usually smooth, pleurogram present. $x = 12$.

Species ca. 100 in the New World (4 in the flora, 2 native, 2 cultivated): Arizona, California, Nevada, New Mexico, Texas, Utah; Mexico, Central America, South America (n Argentina). Also in the Old World.

Senegalia species are widely utilized in regions where they occur. The wood is used for construction, carpentry, utility items, for fuelwood, and to make charcoal. Gum Arabic is obtained from *Senegalia senegal* (Linnaeus) Britton and from several other species and has wide use as a food additive and in many industrial applications. Excellent quality honey is produced by bees that forage on *Senegalia* flowers. Species of *Senegalia* are used for a variety of medical purposes.

SELECTED REFERENCE: Seigler et al. 2006a.

1. Leaflets 25–55 pairs/pinna, pinnae 30–75 mm ____ **1. Senegalia berlandieri**
1′ Leaflet 3–15 pairs/pinna, pinnae < 24 mm.
  2. Inflorescence globose to slightly elongated ____ **3. Senegalia occidentalis**
  2′ Inflorescence cylindrical.
    3. Leaflets 2.8–5.5 mm ____ **2. Senegalia greggii**
    3′ Leaflets 5.5–9.2 mm ____ **4. Senegalia wrightii**

1. Senegalia berlandieri (Bentham) Britton & Rose. N. Amer. Fl. 23:109. 1928. * [for Jean Louis Berlandier, 1805–1851, French botanist who collected in Texas and Mexico] BERLANDIER'S ACACIA; *GUAJILLO* Cultivated

Acacia berlandieri Bentham, London J. Bot. 1:522. 1842; *Acacia tephroloba* A. Gray

Shrubs or small trees, erect, to 6 m. **Stems** slender, lightly branched; twigs not zigzagged, round, puberulent to rarely glabrous; short shoots absent; prickles straight, widely scattered along the twig, commonly absent on petiole and rachis; bark gray, smooth, becoming fissured on oldest stems. **Leaves:** pinnae 7–14(–19) pairs per leaf, 30–75 mm; stipules linear, to 4.5 mm, puberulent; petioles channeled above, 7–28 mm, usually puberulent; petiolar gland solitary, sessile, narrowly elliptic, 1.2–3.5 mm; rachis 25–145 mm. **Leaflets** 25–55 pairs per pinna, linear to oblong, oblique and obtuse at base, acute at apex, 2.8–7.2 × 0.7–1.4(–1.8) mm, glabrous to puberulent on both surfaces. **Inflorescences** densely-flowered globose heads 12–18 mm across, solitary in the leaf axil or in terminal raceme-like to paniculate clusters; peduncles 6–30 mm. **Pedicels** absent. **Flowers** white; calyx 1.2–2.0 mm, puberulent; corolla 2–3 mm, puberulent; filaments 6–8 mm. **Fruits** 70–160 × 15–30 mm, straight, flattened, not constricted between the seeds, leathery, puberulent; stipe 3–10 mm; apex obtuse, short beaked. **Seeds** light brown, oval to circular to oblong, somewhat flattened, 10–13 × 9–11 mm.

Flowering Feb–May. sw Texas and adjacent n Mexico (Chihuahua, Coahuila, Durango, Guanajuato, Hidalgo, Nuevo León, Querétaro, San Luis Potosí, Tamaulipas, Zacatecas).

In Texas and northeastern Mexico, *Senegalia berlandieri* is a dominant component of the Tamaulipan Thornscrub vegetation over extensive areas. It is also found in Semidesert Grassland, occasionally in Chihuahuan Desertscrub, and in disturbed vegetation. Elevations range from 100–2,000 m. Plants are found on dry to relatively moist sites on limestone outcrops and calcareous soil, ridges, slopes, and plains. Near the northern limit of its distribution, it grows as a multiple-stemmed shrub 1–2 m tall. Further south, the plants are often 4 m tall with a spreading canopy.

The wood of *Senegalia berlandieri* is occasionally used to make tool handles and other items, and the plants are valued as a source of high-quality honey (Vines 1960, Powell 1998). The leaves are considered acceptable livestock browse when other forage is available; however, sheep, goats, and possibly cattle can develop a potentially fatal condition called guajillo wobbles or limber leg when a large portion of their diet is composed of *S. berlandieri* over an extended period such as during drought (Tull 1987, Smith and Sherman 1994, Cheatham and Johnston 1995, Hart et al. 2003). Ashes from the roots were used to treat poison ivy and mixed with prickly-pear pulp to make a salve for skin irritation (Cheatham and Johnston 1995).

Berlandier's acacia is planted in Arizona landscapes. Its lacy foliage and showy, fragrant flowers are attractive. The plants are ideal as a screen or informal hedge. Berlandier's acacia is evergreen with adequate irrigation and mild winter temperatures. Cultivated plants in Tucson tolerate overnight temperatures to ca. -9°C. The leaflets produce a fine litter though the pods may be undesirable in more formal landscapes. It is readily grown from scarified seeds.

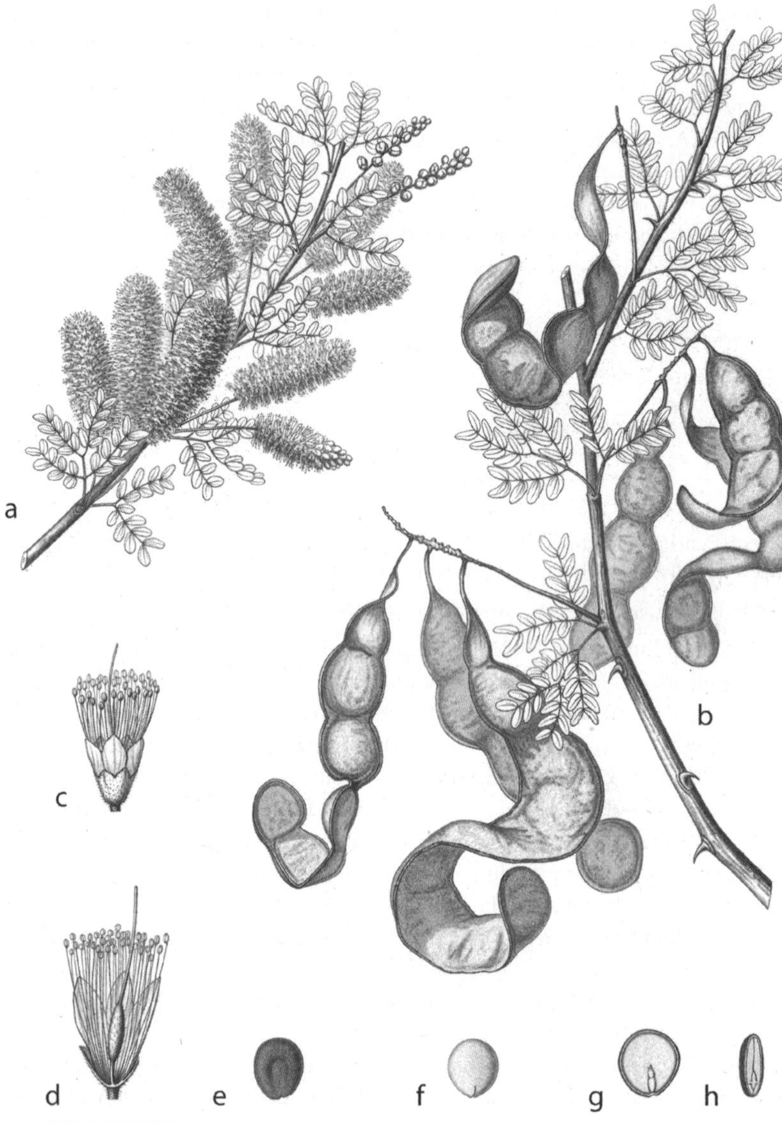

***Senegalia greggii*: a. flowering stem habit; b. fruiting stem habit; c., d. flowers; e., f., g., h. seeds. Illustration from Silva of North America, vol. 3.**

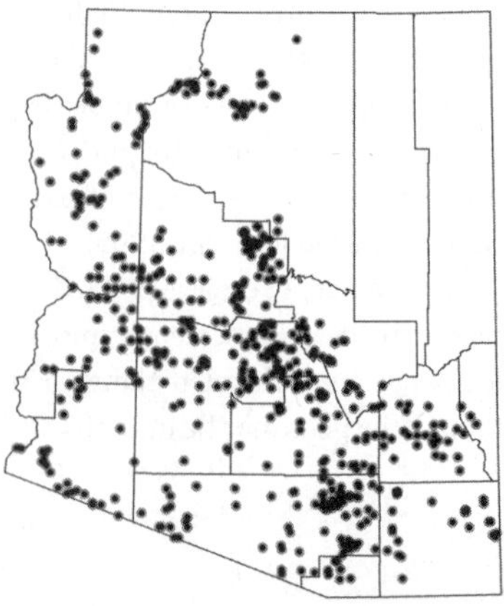

2. Senegalia greggii (A. Gray) Britton & Rose. N. Amer. Fl. 23:110. 1928. * [for Josiah Gregg, 1806–1850, American botanist] CATCLAW ACACIA; *UÑA DE GATO* Native, Cultivated

Acacia greggii A. Gray, Pl. Wright. 1:65. 1852; *Acacia durandiana* Buckley; *Acacia greggii* A. Gray var. *arizonica* Isely

Shrubs or small trees, erect or spreading, to 6 m. **Stems** slender, highly branched; twigs slightly zigzagged, round, glabrous to pubescent; short shoots present; prickles recurved, scattered along the twig; bark gray, fissured and shredding. **Leaves:** pinnae 1–3(–5) pairs per leaf, 5–14 mm; stipules narrowly triangular, to 1.1 mm, glabrous; petioles channeled above, 3–11 mm, glabrous to pubescent; petiole gland solitary, sessile, oval to circular, 0.2–0.5 mm across; rachis 0–12 mm. **Leaflets** 3–6(–8) pairs per pinna, oblong to obovate, the terminal leaflets slightly larger and obovate, oblique and obtuse at base, rounded to acute at apex, 2.8–5.5 × 0.9–3.2 mm, glabrous to pubescent on both surfaces. **Inflorescences** loosely- to densely-flowered cylindrical spikes or spike-like racemes 15–40 mm, 1–4 from short shoots; peduncles 8–30 mm. **Pedicels** absent or short. **Flowers** creamy white to yellowish; calyx 1.3–2.0 mm, glabrous to rarely puberulent; corolla 1.8–3.1 mm, glabrous; filaments 5–7 mm. **Fruits** 40–150 × 13–21 mm, somewhat curled to straight, flattened, constricted between some seeds, stiffly papery, glabrous; stipe 5–12 mm; apex obtuse, not beaked. **Seeds** dark brown, oblong, somewhat flattened, 8–12 × 6–9 mm.

Flowering May–Sep. Slopes, canyons, arroyos, and valleys, disturbed dry areas in calcareous or sandy soils; [near sea level–]100–1500 m; Sonoran Desertscrub, Chihuahuan Desertscrub, Semidesert Grassland; Cochise, Coconino, Gila, Graham, La Paz, Maricopa, Mohave, Pima, Pinal, Santa Cruz, Yavapai, Yuma cos.; California, New Mexico, Nevada, Texas, Utah; Mexico (Baja California, Baja California Sur, Chihuahua, Coahuila, Durango, Nuevo León, Sonora, Tamaulipas).

Isely (1973) recognized var. *arizonica* Isely on the presence of leaflet pubescence. This characteristic is highly variable and cannot be used for consistent separation.

Senegalia greggii is widely distributed across Arizona. It is often abundant, especially along arroyos and in semidesert grasslands. Upland plants frequently form thicket-like shrubs several meters across. Along larger arroyos, the plants develop into trees with trunks 30 cm or more in diameter. Outside of Arizona it is found in Sinaloan Thornscrub and Tamaulipan Thornscrub.

The wood of catclaw acacia is hard and durable. The heartwood is reddish brown, and the sapwood is pale yellow. The wood is used for fuel and to make small articles. The Seri of western Sonora use it for a variety of purposes including digging sticks, agave pry-bars, harpoon points, and violin bows (Felger and Moser 1985). Other native peoples have used the twigs in basketry and the wood for building materials and cradle frames and to clean and prepare animal hides (Cheatham and Johnston 1995, Moerman 1998). The outer bark can be used to dye wool (Cheatham and Johnston 1995). The lac scale (*Kerria lacca*) was sometimes harvested from *Senegalia greggii* and used to repair pottery (Tull 1987, Powell 1998). The foliage may be toxic to livestock (Kingsbury 1964). Tull (1987) indicates that *S. greggii* is a good browse plant but that cattle and sheep have been poisoned by eating large quantities over several months. Cheatham and Johnston (1995) state that the foliage is browsed by livestock and wild ruminants. Quail eat the seeds, and the plants afford excellent cover for wildlife (Powell 1998). Cheatham and Johnston (1995) report that the powdered leaves and fruits are made into an infusion for treating diarrhea, and a poultice is made from the powdered fruits for sprains, bruises, and muscle pain. An excellent honey is produced from catclaw acacia. Various native peoples used the fruits and seeds for food (Cheatham and Johnston 1995, Moerman 1998, Powell 1998).

Catclaw acacia is grown for landscaping and revegetation. It makes an excellent barrier plant due to the prickly stems and affords good cover for wildlife, though should not be planted near traffic areas. The plants may be trimmed to develop a tree form. On older plants, new growth often lacks prickles. Catclaw acacia is deciduous with freezing temperatures. This species appears to be hardy to at least -20°C for brief periods and will tolerate temperatures below freezing for several days. Propagation is by scarified seeds.

3. Senegalia occidentalis (Rose) Britton & Rose. N. Amer. Fl. 23:117. 1928. * [Latin *occidental*, western] TREE CATCLAW; *TÉSOTA* Cultivated

Acacia occidentalis Rose, Contr. U.S. Natl. Herb. 8:32. 1903

Shrubs or trees, erect, to 15 m. **Stems** slender, highly branched; twigs zigzagged, round, usually pubescent; short shoots present; prickles recurved, scattered along the twig; bark gray, fissured and shredding. **Leaves:** pinnae 2–5 pairs per leaf, 7–23 mm; stipules triangular, glabrous, to 0.8 mm; petioles channeled above, 3–11 mm, pubescent; petiolar gland solitary, sessile, circular, 0.4–1.5 mm; rachis 4–30 mm. **Leaflets** 6–15 pairs per pinna, oblong, the terminal leaflets slightly enlarged and obovate, oblique and obtuse at base, acute at apex, 2.8–5.2 × 0.7–1.4 mm, glabrous to lightly appressed pubescent beneath. **Inflorescences** densely-flowered, globose to slightly elongated heads, 15–20 × 13–18 mm, usually 1 to 4 from the leaf axils or short shoots; peduncles 6–16 mm. **Pedicels** 0.2–0.6 mm. **Flowers** creamy white; calyx 1.2–1.8 mm, puberulent; corolla 2.6–3.3 mm, glabrous; filaments 5.5–7.5 mm. **Fruits** 60–165 × 20–25 mm, straight, flattened, not constricted between the seeds, stiffly papery, glabrous; stipe 6–16 mm; apex obtuse, not beaked. **Seeds** dark brown, oval, flattened, 11.0–14.5 × 9.0–11.5 mm.

Flowering Feb–Jun. Mexico (Sinaloa, Sonora).

Senegalia occidentalis appears a short distance south of the International border in Sonora. It is found on gentle slopes, valleys, arroyos, and dry sites on calcareous or sandy soils. It reaches its best development along larger arroyos and river valleys with deep soil profiles and presumably a relatively shallow water table. It is found from sea level to 1200 m. Biotic communities in which it is found include Sinaloan Deciduous Forest, Sinaloan Thornscrub, and Sonoran Desertscrub. Large trees can have a canopy spread of 18 m with a well-developed trunk to at least 1.2 m in diameter. These impressive trees are now rare near settlements but can still be found in more remote areas. On drier upland site *S. occidentalis* is much smaller.

The wood is used to carve bowls and utensils (Gentry 1942).

Tree catclaw is becoming more widely grown as a landscape tree in Arizona. It requires space for best development and would be an ideal shade tree for parks and other large areas; however, growth rates are not fast. The flowers scent the air with a pleasant fragrance. The trees are deciduous with freezing. Cultivated plants in southern Arizona have survived -9°C without damage. Propagation is by scarified seeds.

4. **Senegalia wrightii** (Bentham) Britton & Rose. N. Amer. Fl. 23:110. 1928. * [for Charles Wright, 1811–1885, American botanist] CATCLAW, WRIGHT'S ACACIA; *UÑA DE GATO* Native

Acacia wrightii Bentham in A. Gray, Pl. Wright. 1:64. 1852; *Acacia greggii* A. Gray var. *wrightii* (Bentham) Isely

Shrubs or small trees, erect, to 12 m. **Stems** slender, lightly branched; twigs not zigzagged, round, glabrous; short shoots present; prickles recurved, scattered along the twig; bark brown, fissured. **Leaves:** pinnae 1–3 pairs per leaf, 3–22 mm, stipules narrowly triangular, to 1.1 mm, puberulent; petioles channeled above, 2–20 mm, glabrous to rarely pubescent; petiolar gland solitary, sessile, oval to oblong, 0.3–1.1 mm; rachis 0–24 mm. **Leaflets** 3–7(–9) pairs per pinna, oblong to obovate to oblanceolate, the terminal leaflets slightly larger and obovate, oblique and obtuse at base, obtuse at apex, 5.5–9.2 × 2.2–4.5 mm, appressed pubescence beneath and usually above. **Inflorescences** loosely-flowered cylindrical, spike-like racemes, 20–70 × 11–18 mm, 1–4 from the short shoots and leaf axils; peduncles 10–40 mm. **Pedicels** 0.4–2.1 mm. **Flowers** creamy white to yellow; calyx 1.2–2.0 mm, glabrous to puberulent; corolla 1.8–2.8 mm, glabrous; filaments 5–7 mm. **Fruits** 40–150 × 15–25 mm, straight to somewhat curled, flattened, constricted between some seeds, stiffly papery, glabrous; stipe 4–10 mm; apex obtuse, short beaked. **Seeds** dark brown, circular to oblong, flattened, 8–12 × 6–9 mm.

Flowering Apr–Jun. Valleys, floodplains, arroyos, limestone outcrops, and dry, rocky calcareous soils, often in thickets and disturbed sites; [sea level–]1000–1200 m; Semidesert Grassland; Cochise, Coconino, Gila, Greenlee, Mojave, Maricopa, Pinal cos.; California, Nevada, New Mexico, Texas; Mexico (Baja California Sur, Chihuahua, Coahuila, Nuevo León, Tamaulipas).

Senegalia wrightii appears to be rare in Arizona. We have seen specimens from Gila Co.: US 70, 9 mi E of Globe, alt. 3500 ft (*E. Lehto 1178,* 15 Jul 1962 (ASU)); Greenlee Co.: Eagle Creek, alt. 4000 ft (*W. L. Minckley s.n.*, 28 Jun 1977 (ASU)); Maricopa Co.: Fish Creek Canyon, Tonto National Forest, alt. 3600 ft (*K. C. Rice 828*, 6 Jul 1991 (ASU)); and Pinal Co.: Fish Creek Canyon, Apache Trail (*S. D. McKelvey 728*, 18 Feb 1929 (A)). In Texas and northeastern Mexico, *S. wrightii* occurs in Tamaulipan Thornscrub, Chihuahuan Desertscrub, and oak-juniper-mesquite woodland.

Wright's acacia is considered to be a good source of honey (Vines 1960, Cheatham and Johnston 1995), and the wood is used for fuel and posts (Vines 1960).

Senegalia wrightii is infrequently cultivated in Arizona. The foliage appears more luxuriant than that of *S. greggii* due to the larger leaflets, and the plant more readily develops a tree form, but in other horticultural aspects the two species are similar.

SENNA Miller. Gard. Dict. Abr. (ed. 4), vol. 3. 1754. * [from Arabic *sana*, akin to Aramaic, *sanya*, thorn-bush]
Brigitte Marazzi

Cassia sect. *Senna* de Candolle ex Colladon, Histoire Naturelle et Médicale des Casses:92. 1816; *Cassia* subgen. *Senna* Bentham

Herbs, **shrubs**, [trees to 30 m]. **Roots** without bacterial nodules. **Leaves** alternate, even pinnately compound, petiolate, some sclerophyllous, rarely modified into phyllodes, or absent; stipules deciduous or persistent; extrafloral nectaries 0 or 1–many, on petioles, on rachis between proximal pair of leaflets only, or between proximal and subsequent pairs of leaflets, gland-like and concave, sessile or stipitate. **Leaflets** 1–many pairs, blade from elliptic, obovate, ovate, oblanceolate, lanceolate, to linear, margins usually entire. **Inflorescences:** racemes of 1–many flowers, erect; bracts early or late deciduous, rarely persistent at anthesis; bracteoles absent. **Flowers** caesalpinoid, bilaterally symmetric or asymmetric (gynoecium deflected to one side of the floral axis); calyx 5-merous, early or late deciduous or rarely persistent after anthesis, greenish to yellowish; corolla 5-merous, from yellow to yellow-orange, petal blades narrowed to a claw; androecium 5+5-merous, often heterantherous with one upper (adaxial) set of 3 staminodes, one middle set of 4 short stamens and one lower (abaxial) set of 2–3 long stamens, or not heterantherous with stamens similar in shape and only slightly different in sizes; filaments stiff, arched; anthers stiff, attached below, usually longer than filaments, dehiscing by 1 or 2 apical pores or short slits; gynoecium shortly stipitate, linear, incurved, few to multi-ovulate; ovary hairy, rarely glabrous; style linear or incurved, usually terminating in a minute stigmatic tip. **Fruits** indehiscent or partly dehiscent along suture at edge of valves or from distal end; flat to cylindrical, straight, simply curved to spiral, often corrugated over seeds. **Seeds** ovoid, obovoid, ellipsoid to rhomboid, usually areolate. $2n = 22, 23, 24, 26, 28$.

Species 300–350 (7 in the flora, 3 native, 3 native and cultivated, 1 cultivated and adventive): circumtropical and subtropical regions, few warm temperate regions, especially in the Americas.

Senna was segregated from the large *Cassia* L. and placed in subtribe Cassiinae, together with *Cassia sensu stricta* and *Chamaecrista* Moench (Irwin and Barneby 1981, 1982).

Senna is a remarkable example of floral specialization in relation to buzz pollination (i.e. pollen-collecting bees vibrate flowers; Buchmann 1974). The nectarless flowers of *Senna* display a wide range of traits typical of buzz-pollinated flowers: poricidal anthers, heteranthery, point-tipped stigma, and enantiostyly (deflection of the carpel to the left or right) often accompanied by an asymmetric corolla and androecium (Marazzi et al. 2007, Marazzi and Endress 2008). Another characteristic feature of *Senna* is the conspicuous gland-like extrafloral nectaries (EFNs) on leaves and sometimes also at the base of pedicels in 80% of the species (Irwin and Barneby 1982, Marazzi and Sanderson 2010). Less well-known are recently reported, cryptic EFNs embedded within stipules, bracts, and sepals in species previously thought to lack EFNs (Marazzi et al. 2013). EFNs offer nectar to ants in return for their protection from insect herbivores (Fleet and Young 2000).

Senna is a relatively old genus with several fossil fruits described from the Eocene (Herendeen 1992, Calvillo-Canadell and Cevallos-Ferriz 2005) and probably diverged from its sister genus *Cassia sensu stricta* about 50 million years ago (Marazzi and Sanderson 2010). The current classification of *Senna* (Irwin and Barneby 1982), recognizing 35 series in six sections (*Astroites*, *Chamaefistula*, *Paradictyon*, *Peiranisia*, *Psilorhegma*, *Senna*), has been adopted worldwide but received little support in molecular phylogenetic studies, as only section *Psilorhegma* appears to be monophyletic (Marazzi, Endress, et al. 2006).

In North America, all of the sections are represented except for the monotypic *Astroites* and *Paradictyon* from central Mexico and South America, respectively. North America represents the northern range of geographic distribution of the genus on the American continent. Several of the typical ornamental species are cultivated in warmer areas of North America, especially Florida and California, including *S. bacillaris*, *S. corymbosa*, *S. didymobotrya*, *S. multijuga*, *S. surattensis*, and *S. spectabilis*. Currently, two of them, *S. corymbosa* and *S. surattensis*, are considered as naturalized in North America along with at least six other introduced species. Although a few of the introduced species have regionally been recognized as problematic invasive plants (e.g., *S. pendula* in Florida, *S. multiglandulosa* in California), and two other species, *S. obtusifolia and S. occidentalis,* are considered noxious weeds in many countries worldwide, most of the introduced species are of no particular concern in North America.

A number of *Senna* species have long been used in the European medicinal tradition (as a purge), several other species have been cultivated as ornamentals (Colladon 1816, Luckow 1996). *Senna alexandrina* and *Senna italica* have been used as laxatives since ancient times. *Senna alata* is widely used to treat skin diseases (Lewis et al. 2005). The wood of larger species is occasionally used for construction and to make charcoal (Lewis et al. 2005). Other uses include tanning leather and as a fish poison, and the seeds of some species have been used as a substitute for coffee (Lewis et al. 2005). Allen and Allen (1981) report that the bark of *S. auriculata*, a species that has a variety of medicinal uses, has a tannin content of approximately 20%. Several species of *Senna* are grown as landscape and garden plants for shade or for their showy flowers.

1. Extrafloral nectaries absent from leaves __________ **7. Senna wislizeni**
1′ Extrafloral nectaries present on leaves at the base of the petiole only or between one or more leaflets, or along dorsal (upper) margin of phyllodes, where putative leaflets should be inserted.
  2. Shrubs; leaves sclerophyllous, modified into phyllodes, or rarely absent; leaflets reduced or absent.
    3. Stamens 7, staminodes 3; flowers monosymmetric; leaves variably reduced to absent on the same plant; branch tips often acuminate; inflorescences of 1–2 flowers; fruits erect __________ **1. Senna armata**
    3′ Stamens 10, staminodes 0; flowers slightly asymmetric, style deflected to one side of floral axis; leaves always present; branch tips not acuminate; inflorescences of 2–10 flowers; fruits pendulous __________ **2. Senna artemisioides**
  2′ Herbs; leaves not or only slightly sclerophyllous; leaflets well-developed.
    4. Extrafloral nectary at the base of the petiole; leaves glabrous to nearly glabrous __________ **5. Senna hirsuta**
    4′ Extrafloral nectaries between first pair of leaflets only or also between subsequent leaflet pairs; leaves hairy.
      5. Leaves with 1 pair of leaflets __________ **3. Senna bauhinioides**
      5′ Leaves with 2 or more pairs of leaflets.
        6. Leaves with 2–4(–5) pairs of leaflets; stipules up to 1 mm wide; inflorescence with (2–)4–8 flowers __________ **4. Senna covesii**
        6′ Leaves with (4–)5–8 pairs of leaflets; stipules 1–3 mm wide; inflorescence with (5–)7–25 flowers __________ **6. Senna lindheimeriana**

1. Senna armata (S. Watson) H.S. Irwin & Barneby. Mem. New York Bot. Gard. 35:292. 1982. * [from Latin *armata*, armed] DESERT SENNA, SPINY SENNA Native

Cassia armata S. Watson, Proc. Amer. Acad. Arts 11:136. 1876

Shrubs to 1(–1.5) m, branches chlorophyllous, branch tips often acuminate. **Leaves** sclerophyllous, modified into phyllodes, variably reduced to absent on the same plant, 2–9 cm, sparsely hairy to nearly glabrous; stipules minute, not apparent; extrafloral nectaries (0–)1–2, highly reduced, on rachis. **Leaflets** (0–)2–8(–10) pairs, often irregularly inserted or absent, ovate, apex obtuse or subacute, 0.2–0.9 × 0.1–0.6 cm. **Inflorescences** of 1–2 flowers; bracts deciduous. **Pedicels** 8–21 mm. **Flowers** monosymmetric; calyx yellow; corolla yellow, longest petal 7.5–13 mm; androecium not heterantherous, stamens 7, staminodes 3; anthers 3–4.3 mm, dehiscing by 1 apical pore, apical appendage 0; gynoecium linear, ovules 6–12; ovary hairy; style incurved. **Fruits** dehiscing apically downward, flat, straight, erect, 2–4.5 × 0.5–0.65 cm. **Seeds** ovoid, gray or brown, smooth or wrinkled, dull. $2n$ = unknown.

Flowering (Jan–)Mar–Jun (Aug–Nov). Sandy to gravelly desert washes, alluvial fans, and flood plains; 100–[150–1100(–1420)] m; Sonoran Desertscrub; Mohave Co.; California, Nevada; Mexico (Baja California).

Representative of the Mohave and western Sonoran Deserts, *Senna armata* is the only North American senna displaying a highly xerophytic habit with chlorophyllous, nearly leafless stems (described as "rush-like" in desert floras such as Turner et al. 1995). Otherwise this habit characterizes the unrelated group of a dozen species of *Senna* series *Aphyllae* from the arid lands in southern South America (Irwin and Barneby 1982). Owing to the highly xerophytic habit, *S. armata* was considered taxonomically "obviously isolated" (Irwin and Barneby 1982), but according to molecular phylogenetic analyses (Marazzi, Endress, et al. 2006, Marazzi and Sanderson 2010), it is in fact included in the same clade as species of series *Brachycarpe* (e.g., *S. bauhinioides*, *S. covesii*, and *S. lindheimeriana*, which also occur in Arizona). This result is further supported by a similarity in floral morphology (Marazzi et al. 2007, Marazzi and Endress 2008).

In Arizona, *Senna armata* is known from only a few localities near the Colorado River in western Mohave County. The plants are attractive in flower and would make an interesting addition to the landscape palate for hot, arid desert regions. Flowering is primarily from March into June with occasional flowering at other times in response to adequate rainfall. With their dense, spreading growth habit, the plants afford cover for smaller animals.

2. Senna artemisioides (Gaudichaud ex de Candolle) Randell. J. Adelaide Bot. Gard. 12(2):220. 1989. * [resembling *Artemisia*, for the Greek goddess Artemis] SILVER SENNA Cultivated, Adventive

Cassia artemisioides Gaudichaud ex de Candolle, Prodr. 2:495. 1825

Shrubs to 3 m, branch tips not acuminate. **Leaves** slightly to highly sclerophyllous or reduced to phyllodes, always present, 0.8–5 cm, finely hairy; stipules deciduous; extrafloral nectaries 1–8, sessile, between proximal leaflets or sometimes also one to all subsequent pairs of leaflets. **Leaflets** 0–8 pairs, linear to narrow elliptic, 2–7 × 0.1–0.8 cm. **Inflorescences** of 2–10 flowers; bracts early deciduous. **Pedicels** 4–15 mm. **Flowers** slightly asymmetric; calyx yellowish to greenish; corolla yellow, longest petal 6–10 mm; androecium not heterantherous, stamens 10, staminodes 0; anthers 1.5–5 mm, dehiscing by two apical pores, apical appendage 0; gynoecium incurved, ovules unknown; ovary slightly hairy; style slightly incurved. **Fruits** dehiscing apically downward, flat, straight or slightly curved, corrugated over seeds, pendulous, 2–8 × 0.6–1.2 cm. **Seeds** obovoid, dark brown or black, smooth, glossy or dull. $2n = 28, 56$.

Flowering primarily late Fall to early Spring with some flowering year-round; occasionally adventive on slopes, valleys, arroyos, roadsides, vacant lots and other disturbed areas in various soils; to 810 m; Sonoran Desertscrub; Maricopa, Pima, Pinal, Yuma cos.; California; Australia.

Due to its remarkable morphological and genetic variation (especially in leaflet number and form), *Senna artemisioides* has been characterized by a shifting taxonomic history. The species as circumscribed by Randell (1989) was segregated into several, taxonomically non-accepted "*Senna* taxon forms" by Randell and Barlow (1998). Because these forms have not been adopted by other floristic studies in United States, *S. artemisioides* as presented here encompasses the description of four common taxon forms found in United States: *Senna* form taxon "*artemisioides,*" "*filifolia,*" "*petiolaris,*" and "*sturtii.*" Symon (1998) considered individuals of taxon form '*petiolaris*', characterized by arched phyllodes, as a separate species, *S. phyllodinea*, mainly because of the absence of individuals with an intermediate leaf morphology transitioning to the phyllode shape. Although this argument is reasonable, *S. phyllodinea* has not been accepted and adopted in other floristic studies. It must be noted that phyllodes consist of the lateral compression of both the petiole and the rachis and not only of the petiole, as reported in Randell and Barlow (1998). Along the upper margin of the phyllode there is in fact a reduced extrafloral nectary at each point of insertion of the suppressed leaflets.

Senna artemisioides is widely distributed across inland Australia, where it is often a conspicuous element in arid plant communities. Up to a half dozen of these form taxa can be found growing sympatrically in mixed shrub associations in some areas, such as at Uluru National Park, Northern Territory. Cunningham et al. (1992) state that *S. artemisioides* is sparingly browsed by sheep, goats, and cattle in Australia. Populations of the plants have increased greatly in many grassland and woodland areas in New South Wales since European settlement, perhaps due to overgrazing and a lack of wildfires (Cunningham et al. 1992). The seeds have been utilized as a food resource by Aboriginal people in central Australia (Cleland 1966).

Senna artemisioides has the potential to increase and spread into natural areas. This species has become adventive in Arizona, primarily in urban/suburban areas, where it can become locally abundant. Occasional plants are found in natural habitats at some distance from the nearest cultivated landscape plants. In particular, form taxa '*filifolia*' and '*artemisioides*' have significant potential to reseed and establish in disturbed sites and in natural areas.

Although cultivation information is provided here, *Senna artemisioides* should not be planted, especially not in locations adjacent to native vegetation. The various forms of *Senna artemisioides* are widely grown in landscapes in lower elevation desert communities in Arizona. These plants are popular for their masses of bright yellow flowers that appear during the winter and spring when few other plants are in flower. Sporadic flowering can continue through the year with sufficient irrigation or rain. These plants are effective when massed and can be planted as an informal hedge.

Plants grow best in a sunny position and are tolerant of reflected light and heat. Occasional irrigation is necessary to maintain best appearance though established plants exhibit considerable drought tolerance. Hardiness varies between -5° and -8°C, depending on the form. Older plants that suffer extensive freeze damage often exhibit poor recovery. *Senna artemisioides* does not respond well to shearing and does best when allowed to develop naturally. Excessive pruning or shearing in the fall or winter months removes the flower buds which appear on the previous season's growth. Overall, these are short-lived plants that tend to lose vigor after a decade or so. The plants produce an abundance of seeds. The profusion of seeds, seedlings, and dried fruits can be a maintenance issue in hardscaped areas. In one suburban Tucson location, many thousands of seedlings were observed germinating around three mother plants following heavy summer rains in 2006. These seedlings appear to have been eaten by rabbits or other animals over the following two months. All of the forms of *S. artemisioides* are readily propagated from scarified seeds.

The following form taxa of *Senna artemisioides* are commonly grown in Arizona. These are distinguished primarily by differences in their leaves. Characters are based mostly on cultivated Arizona material and do not necessarily reflect the full range of morphological variation found in native populations in Australia. Because all form taxa have the potential to reseed and spread, this plant is not recommended for planting near natural areas.

Senna form taxon **'artemisioides'** (feathery cassia)

Leaflets (3–)4–6(–8) pairs, (1–)1.5–2.5 × 0.1 cm, linear, rounded or longitudinally channeled above (adaxially) or inrolled, apex acute or blunt, silvery or dull green. Hardy to ca. -7°C. This form is likely of hybrid origin (Randell 1989).

Senna form taxon **'filifolia'** (desert cassia)

Leaflets 1 or 2(–4) pairs, 2–4 × 0.1 cm, linear, rounded, apex acute, green. Hardy to ca. –8°C.

Senna form taxon **'petiolaris'** (silvery cassia)

Leaves modified into flattened phyllodes 2–7.5 × 0.2–0.8 cm, oblanceolate or elliptic and straight or upturned, usually silvery but dull green forms exist. Leaflets usually absent in material from Arizona horticulture but up to 2 (or 3) pairs sometimes present in Australian populations. Hardy to ca. -6°C.

Senna form taxon **'sturtii'** (Sturt's cassia)

Leaflets 2–8 pairs, 1.5–2.5 × 0.2–0.8 cm, elliptic (or linear), margins inrolled, apex acute, dull green. Hardy to ca. -5°C. This form is likely of hybrid origin, apparently involving different form taxa (Randell 1989).

3. Senna bauhinioides (A. Gray) H.S. Irwin & Barneby. Phytologia 44(7):499. 1979. * [resembling *Bauhinia*, in reference to the paired leaflets] TWINLEAF SENNA Native

Cassia bauhinioides A. Gray, Boston J. Nat. Hist. 6(2):180. 1850

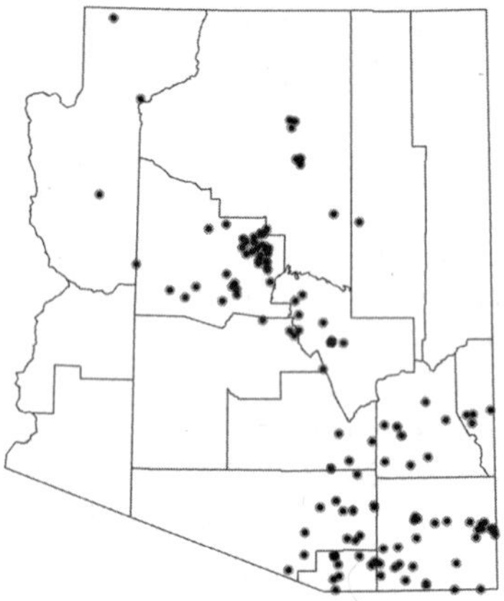

Herbs to 0.35 m, perennial. **Leaves** slightly sclerophyllous, 1.5–5.5 cm, hairy; stipules deciduous; extrafloral nectary 1, between pair of leaflets, stipitate or nearly sessile. **Leaflets** 1 pair, obliquely oblong or ovate-oblong, 0.8–4.6 × 0.5–1.6 cm. **Inflorescences** of 1–3 flowers; bracts deciduous. **Pedicels** 3–11 mm. **Flowers** monosymmetric; calyx yellowish to pale green, deciduous; corolla yellow, longest petal 5–10 mm; androecium not heterantherous, stamens 7, staminodes 3; longest anther 1.8–2.8 mm, dehiscing by 1 apical pore, apical appendage 0; gynoecium nearly linear, ovules 22–36; ovary densely hairy; style incurved. **Fruits** dehiscing apically downward, cylindrical, slightly corrugated over seeds, curved, erect, 1.7–3 × 0.45–0.65 cm. **Seeds** paddle-shaped to pear-shaped, olive green-brown turning gray. $2n$ = unknown.

Flowering Apr–Oct. Ridges, slopes, valleys, floodplains, arroyos, roadsides, and disturbed areas in a wide range of soil types; [80–]760–1680[–1800] m; Sonoran Desertscrub, Chihuahuan Desertscrub, (Mohave Desertscrub, Great Basin Desertscrub), Semidesert Grassland, Interior Chaparral, (Madrean Evergreen Woodland); Apache, Cochise, Coconino, Gila, Graham, Greenlee, Maricopa, Mohave, Navajo, Pima, Pinal, Santa Cruz, Yavapai cos.; New Mexico, Texas; Mexico (Chihuahua, Coahuila, Nuevo Léon, Sonora).

Senna bauhinioides is one of three sennas in the United States (along with *S. pumilio* and *S. ripleyana)* which can form and sprout from a woody taproot (Irwin and Barneby 1982).

Senna bauhinioides is widespread in Arizona though the plants are seldom abundant. It grows in open sites on a wide range of soils, dying back during unfavorable periods and growing anew after sufficient rainfall. This species would make an interesting addition to the xeric wildflower garden.

4. Senna covesii (A. Gray) H.S. Irwin & Barneby. Phytologia 44(7):499. 1979. * [for Elliott L. Coues, 1842–1899, American military surgeon and naturalist] DESERT SENNA, COUES' CASSIA Native, Cultivated

Cassia covesii A. Gray, Proc. Amer. Acad. Arts 7:399. 1868

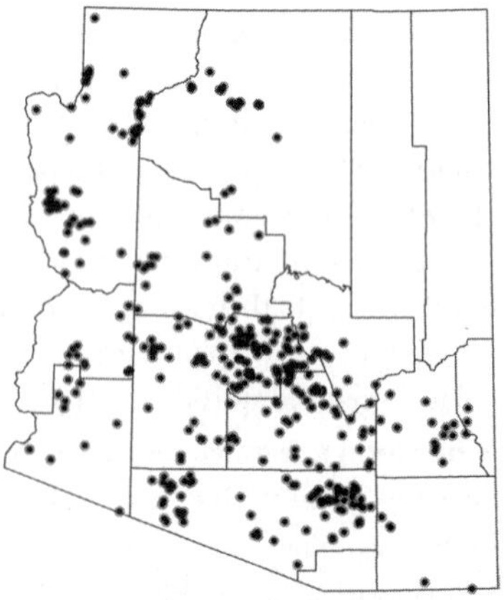

Herbs to 0.7 m, perennial. **Leaves** slightly sclerophyllous, 2–10 cm, hairy; stipules persistent, up to 1 mm wide; extrafloral nectaries between all pairs of leaflets, stipitate. **Leaflets** 2–4(–5) pairs, obovate, elliptic-obovate or oblong-elliptic, 1–3.8 × 0.55–1.9 cm. **Inflorescences** of (2–)4–8 flowers; bracts deciduous. **Pedicels** 8–17 mm. **Flowers** monosymmetric; calyx pale green, pinkish, or yellowish; corolla golden-yellow, longest petal 9–15 mm; androecium not heterantherous, stamens 7, similar in form and size, staminodes 3; anthers 2.5–4.2 mm, dehiscing by 1 apical pore, apical appendage 0; gynoecium linear, slightly incurved, ovules 28–42; ovary hairy; style incurved at ovary, then straight. **Fruits** dehiscing apically and downward, cylindrical, slightly curved, erect; 2–3.8 × 0.5–0.8 cm. **Seeds** rhomboid, brown. $2n$ = unknown.

Flowering Mar–Oct(–Dec). Slopes, canyons, valleys, arroyos, floodplains, roadsides, and disturbed areas in various soils; 220–1200(–1400); Sonoran Desertscrub, Chihuahuan Desertscrub, Mohave Desertscrub, Semidesert Grassland, (Interior Chaparral); Cochise,

Coconino, Gila, Graham, Greenlee, La Paz, Maricopa, Mohave, Pima, Pinal, Yavapai, Yuma cos.; California, Nevada, New Mexico; Mexico (Baja California, Sinaloa, Sonora).

Senna covesii is widespread and often abundant in Arizona. Plants can flower in their first season and typically live for 3 or 4 years. Stems die back after freezing weather or drought, but new stems emerge following sufficient rain. Although this plant has not been collected in Santa Cruz County, it could be there and should be looked for. A disjunct record from Williams, Coconino County (*Demaree 42205-A*) at 2024 m elevation could represent an accidental introduction. Quail are reported to eat the seeds of *S. covesii*. The yellow flowers present a showy display where the plants occur in abundance along roads and on hillsides. In some situations, *S. covesii* can become weedy. *Senna covesii* is used in revegetation and wildflower seed mixes. It is not typically sold as individual plants in nursery containers. Plants can be readily grown from scarified seeds. Unscarified seeds in a seed mix will germinate over time with sufficient moisture. In the wildflower garden, dried stems may be removed to suit individual aesthetics.

5. Senna hirsuta (Linnaeus) H.S. Irwin & Barneby. Phytologia 44(7):499. 1979. * [from Latin *hirsut*, hairy, rough] WOOLLY SENNA

Cassia hirsuta Linnaeus, Sp. Pl. 1:378. 1753

Herbs, to 2 m, perennial. **Leaves** not sclerophyllous, 5.5–28(–33) cm, very finely hairy to nearly glabrous; stipules deciduous; extrafloral nectary 1, sessile, at the base of petiole. **Leaflets** 3–7(–8) pairs, ovate- or lance-acuminate, 3–10.5 × 0.8–4 cm. **Inflorescences** of 2–8 flowers (n of equator), of 15–43 flowers (Argentina and Brazil)**;** bracts deciduous. **Pedicels** 9–25 mm. **Flowers** monosymmetric; calyx greenish or yellowish or brownish; corolla yellow, longest petal 8–15 mm; androecium heterantherous, stamens 7, staminodes 3; anthers of middle short stamens 3.8–5.4 mm, of abaxial long stamens 5–7 mm, dehiscing by two apical pores, apical appendage short, tongue-shaped; gynoecium incurved, ovules 50–108; ovary hairy; style incurved. **Fruits** dehiscing from the tip downward, flat, curved, ascending to deflexed; 10–28 × (0.25–)0.3–0.65(–0.7)cm. **Seeds** obovoid, brown to brown-olive green-brown.

Varieties 7 (1 native in the flora): Arizona, New Mexico; Mexico; worldwide in tropical and subtropical regions.

Molecular phylogenetic analyses (Marazzi, Endress, et al. 2006, Marazzi and Sanderson 2010) including a number of different varieties of *Senna hirsuta* suggest that this species is paraphyletic and may in fact represent more than one species.

5a. Senna hirsuta (Linnaeus) H.S. Irwin & Barneby var. **glaberrima** (M.E. Jones) H.S. Irwin & Barneby. Phytologia 44(7):499. 1979. * [from Latin *glaber*, smooth] SLIMPOD SENNA Native

Cassia leptocarpa Bentham var. *glaberrima* M.E. Jones, Contr. W. Bot. 12:7. 1908; *Cassia leptocarpa* Bentham

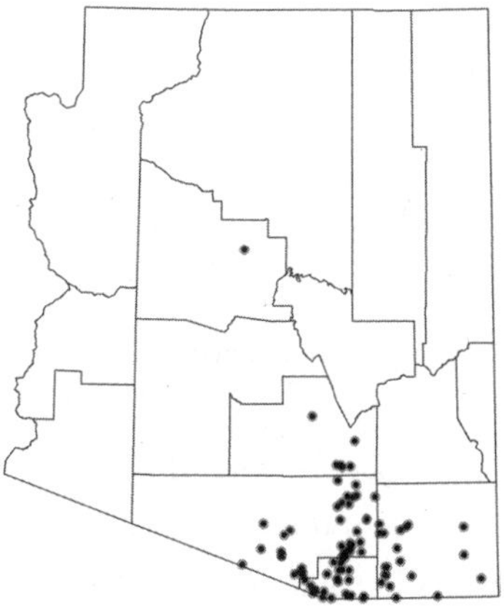

Leaves 5.5–18 cm. **Leaflets** 4–7(–8) pairs, ovate- or lance- acuminate, the longest 3–8 × 0.8–3.2 cm, upper (adaxial) face glabrous or very minutely puberulent, elsewhere thinly puberulent or strigillose. **Flowers**: longest petal 10–15 mm; ovules 60–78. **Fruits**: 12–23 × 0.25–0.45 cm.

Flowering Jul–Sep. Slopes, valleys, canyons, arroyos, floodplains, roadsides, and disturbed areas in various soils; [400–]650–1680[–2100] m; (Sonoran Desertscrub), Semidesert Grassland, Madrean Evergreen Woodland; Cochise, Pima, Pinal, Santa Cruz cos.; New Mexico; Mexico (Aguascalientes, Chihuahua, Guanajuato, Jalisco, Michoacán, Morelos, Oaxaca, Sonora, Tamaulipas, Zacatecas).

Senna hirsuta var. *glaberrima* and var. *leptocarpa* are the only two varieties in which the upper side of the leaflets is glabrous or nearly glabrous (Irwin and Barneby 1982). However, these two varieties have disjunct distributions, with var. *leptocarpa* occurring locally in ne Argentina and adjoining regions of Brazil (Irwin and Barneby 1982) and Paraguay (Marazzi, Fortunato, et al. 2006).

Senna hirsuta var. *glaberrima* is scattered across the southeastern part of Arizona. It is most frequently seen along drainages and roadsides but is seldom abundant. In Sonora, this taxon is also found in tropical deciduous forest (Martin et al. 1998). The plants grow in sandy, silty, gravelly, and rocky soils. New stems grow from the subterranean base each season, dying back in the winter. The plants are short-lived. Gambel's quail eat the seeds of *S. hirsuta* var. *glaberrima* (Graham 1941). The bruised or crushed herbage is strongly malodorous.

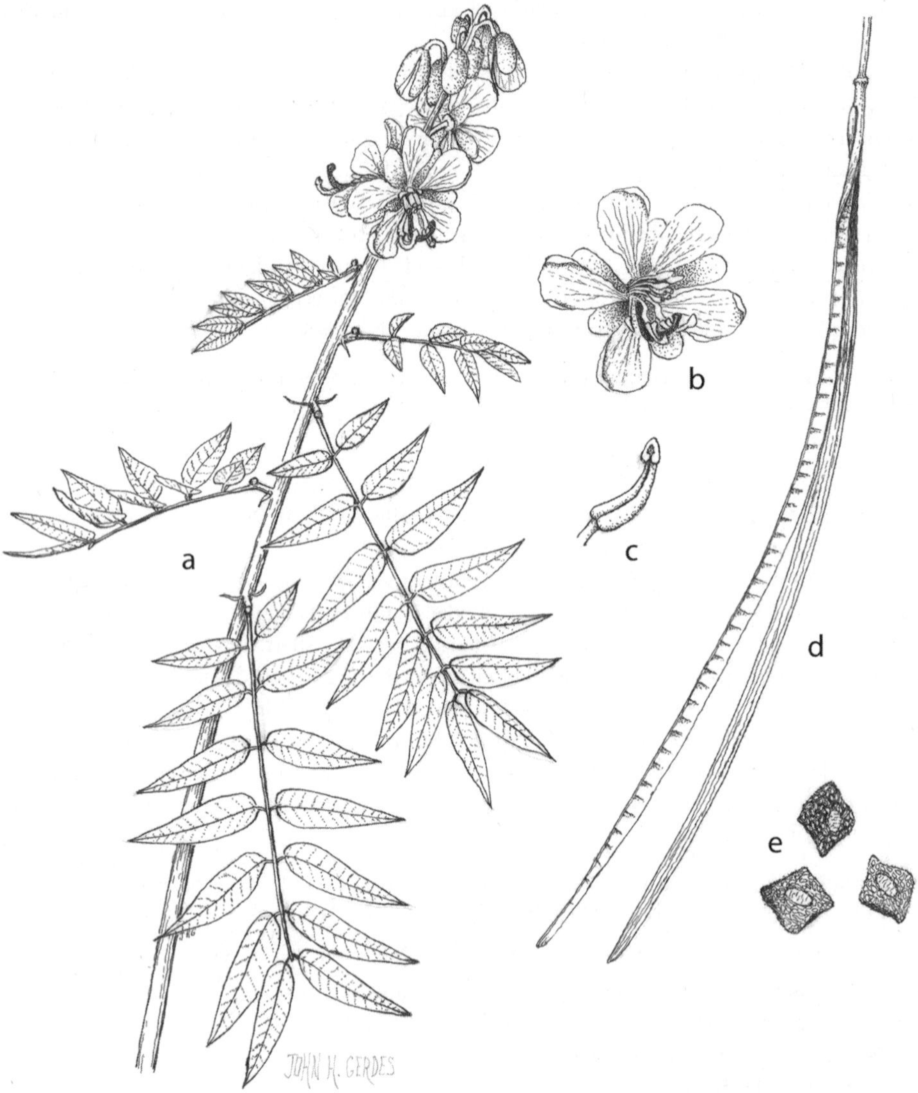

***Senna hirsuta* var. *glaberrima*: a. habit; b. flower; c. anther detail; d. fruit after dehiscence; e. seeds. Illustrator: John H. Gerdes.**

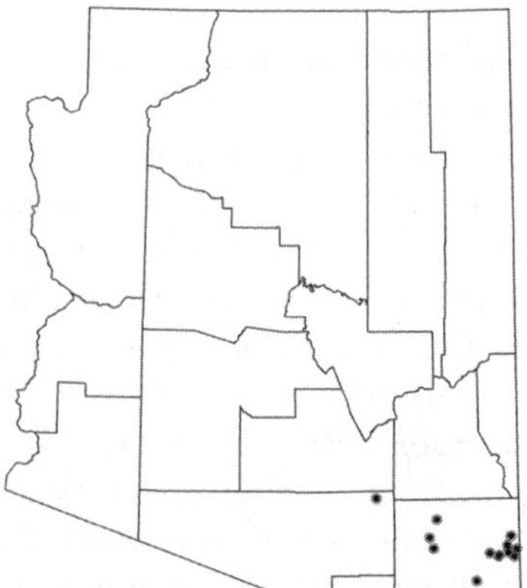

6. Senna lindheimeriana (Scheele) H.S. Irwin & Barneby. Phytologia 44(7): 500. 1979. * [for Ferdinand J. Lindheimer, 1801–1879, German-born American botanist] VELVET LEAF SENNA, LINDHEIMER'S CASSIA Native, Cultivated

Cassia lindheimeriana Scheele, Linnaea 21(4):457–458. 1848

Herbs to 1.5 m, perennial. **Leaves** slightly sclerophyllous, 6–16 cm, densely hairy; stipules deciduous; extrafloral nectaries between all pairs of leaflets, stipitate. **Leaflets** (4–)5–8 pairs, blade oblong-, ovate- to obovate-elliptic, 2–5 × 0.8–2 cm. **Inflorescences** of (5–)7–25 flowers; bracts deciduous. **Pedicels** 6–22 mm. **Flowers** monosymmetric; calyx, pale green, pinkish, or rarely yellow; corolla yellow, longest petal 10.5–16 mm; androecium heterantherous, stamens 7, staminodes 3; anthers 3–5.1 mm, dehiscing by an apical pore, apical appendage 0; gynoecium linear, ovules 20–28; ovary densely hairy; style

Senna lindheimeriana: **a.** habit; **b., c.** leaves; **d.** flower; **e.** anther; **f., g.** immature fruit; **h.** mature fruit. Illustrator: Chris Bondante.

incurved. **Fruits** dehiscing apically and downward, flat, straight, ascending to deflexed; 3–6.5 × 0.5–0.9 cm. **Seeds** obovate or oblong-obovate, brown. $2n = 28$.

Flowering Jun–Sep. Ridges, slopes, valleys, canyons, arroyos, and roadsides, usually on limestone soils; [200–](600–)1370–2200 m; (Sonoran Desertscrub), Chihuahuan Desertscrub, Madrean Evergreen Woodland; Cochise, Pima, Pinal cos.; New Mexico, Texas; Mexico (Chihuahua, Coahuila, Nuevo León, Sonora, Tamaulipas).

Senna lindheimeriana primarily occurs in southern Cochise County in Arizona where it may be locally common. It grows in rocky limestone soils or on soils derived from limestone. There are two disjunct records in the state: J. *Verrier 336*, from Edgar Canyon, in the Santa Catalina Mountains, and *R. G. Walter 738*, from along Hwy. 77, 1 mile north of Winkelman. The later record is reported as being in Pinal County but would be located in Gila County if north of Winkelman.

The leaves of *Senna lindheimeriana* are widely reported as being toxic to cattle. Ajilvsgi (1984) states that the plant has laxative properties. In Arizona, velvet leaf senna is occasionally grown as a flowering perennial for wildflower and low water-use gardens. The plants are attractive in flower and the velvety-soft leaves provide a pleasant tactile experience. Velvet leaf senna thrives in full sun or partial shade. Periodic irrigation maintains vigor and appearance. The plants die back with winter freezing, usually resprouting again in the spring. This species is rather short-lived but reseeds readily in favorable situations. Dead stems may be removed when new growth appears in the spring. The tendency to reseed can become a maintenance issue in some garden settings. Plants are readily propagated by scarified seeds.

7. Senna wislizeni (A. Gray) H.S. Irwin & Barneby. Phytologia 44(7):500. 1979. * [for Friedrich A. Wislizenus, 1810–1889, German-born American physician, explorer, and botanist]

Cassia wislizeni A. Gray, Smithsonian Contr. Knowl. 3(5):60. 1852

Shrubs 1–5(–6) m. **Leaves** xerophyllous, 1.2–3.7 cm, the first pair of leaflets often deciduous, glabrous to strigulose or pilosulous; stipules persistent or deciduous; extrafloral nectaries 0. **Leaflets** 3 pairs, obovate, 0.4–1 × 0.25–0.75 cm. **Inflorescences** of 2–17 flowers; bracts deciduous. **Pedicels** 12–26 mm. **Flowers** asymmetric (gynoecium deflected to one side of floral access); calyx yellowish, greenish, or purplish-brown; corolla bright yellow, with lower petals strongly concave, longest petal 16–28 mm; androecium heterantherous, stamens 7, staminodes 3; anthers of middle stamens 4–6.2 mm, of long stamens 4.8–6.2 mm, dehiscing by 2 apical slits, apical appendage 0; gynoecium incurved, ovules 22–25; ovary glabrous; style incurved. **Fruits** dehiscing apically and downwards, flat, straight to slightly curved, corrugated or not over seeds, ascending, 7–24 × 0.55–0.8 cm. **Seeds** rhomboid to paddle-shaped, brown to pinkish-brown.

Senna wislizeni var. *wislizeni* and var. *villosa* display persistent stipules and fruits corrugated over the seeds, whereas stipules are promptly deciduous and fruits not corrugated in the other two varieties, var. *pringlei* and var. *painteri*. Variety *villosa* and var. *wislizeni* are distinguished by their vestiture: the former is pilosulose, whereas the latter is strigillose (Irwin and Barneby 1982).

Varieties 4 (1 native in flora): Arizona, New Mexico, Texas; Mexico.

7a. Senna wislizeni (A. Gray) H.S. Irwin & Barneby var. **wislizeni**. Phytologia 44(7):500. 1979. * WISLIZENUS' SENNA, SHRUBBY SENNA Native, Cultivated

Shrubs to 1 (2) m, young stems strigillose. **Leaves**: stipules persistent. **Fruits** corrugated over seeds.

Flowering (Apr–)Jun–Oct. Rocky limestone, granitic and volcanic soils of slopes, canyons and arroyos, and roadsides; [850–](910–)1220–1440[–1570] m; (Sonoran Desertscrub), Chihuahuan Desertscrub, (Semidesert Grassland); Cochise, Pima cos.; New Mexico, Texas; Mexico (Chihuahua, Coahuila, Sonora).

Senna wislizeni var. *wislizeni* is found in southern Cochise County in Arizona, primarily on rocky slopes in the Chihuahuan Desert. Populations are widely scattered but the plants may be locally common. The records of *S. wislizeni* for Pima County are all adjacent to the Catalina Highway from Tucson to Mt. Lemmon and date from after the widening of the highway in the 1990s. These plants were almost certainly introduced during revegetation following construction.

Wislizenus' senna is cultivated as a low water-use landscape shrub in Arizona, owing to spectacular masses of showy yellow flowers. It can be planted as an informal hedge or as a background plant. In cultivation, these plants can grow to 2.5 m in height. *Senna wislizeni* should be planted in full sun and does best in well-drained soil. Occasional irrigation during hot, dry weather will maintain appearance but established plants are quite drought-tolerant. *Senna wislizeni* appears to be hardy to at least -18°C. The plants become deciduous with freezing. New leaves appear in mid-spring. The long winter deciduous period is a potential drawback for some garden applications (Jones and Sacamano 2000). Plants are readily grown from scarified seeds.

Taxa with potential for cultivation

Senna corymbosa (Lamarck) H.S. Irwin & Barneby. Mem. New York Bot. Gard. 35:397. 1982
Flowering or autumn senna is a shrub that grows to 3 m tall with evergreen leaves and showy yellow flowers through much of the year. It is hardy to at least -7°C.

Senna didymobotrya (Fresenius) H.S. Irwin & Barneby. Mem. New York Bot. Gard. 35:467. 1982
Candle bush or peanut-butter cassia is an herbaceous or suffrutescent plant growing to 2.5 m tall. The leaves are evergreen unless killed by freezing. Flowers may be produced year round in suitable locations with adequate moisture. The stems are killed by freezing but the crown will survive if the cold is not too severe. New stems will grow the following season.

Senna polyantha (Colladon) H.S. Irwin & Barneby. Mem. New York Bot. Gard. 35:448. 1982
Rompe bota is a multiple-stemmed shrub to 2 or 3 m tall with rather small, deciduous leaves and yellow flowers that appear in the summer. It is hardy to ca. -6°C.

Senna purpusii (Brandegee) H.S. Irwin & Barneby. Mem. New York Bot. Gard. 35:319. 1982
Baja California senna is a stiffly branched shrub to 1.5 m tall with distinctive bluish green leaflets. The leaves are deciduous with extended drought or freezing. Showy yellow flowers appear in the spring. The plants are damaged by temperatures below ca. -4°C.

SESBANIA Scopoli. Intr. Hist. Nat. 308. 1777. * [from *sesban*, an ancient Arabic name for one of species of this genus] RIVERHEMP

Frank T. Farruggia

Aeschynomene sesban Linnaeus, Sp. Pl. 2:714. 1753; *Agati* Adanson; *Darwinia* Rafinesque; *Daubentonia* de Candolle; *Daubentoniopsis* Rydberg; *Emerus* Kuntze; *Glottidium* Desvaux; *Resupinaria* Rafinesque; *Sesban* Adanson

Trees, shrubs, or herbs, annual or perennial. **Stems** unarmed or armed with prickles, young growth glabrous or with long close-pressed to spreading simple hairs; pith solid, spongy, partitioned, or hollow; nitrogen-fixing nodules found on submerged stems in some species. **Leaves** alternate; even-pinnate; forward folding nyctinasty similar to Mimosoideae; stipules deciduous to long persistent, with or without a ridge of laminar tissue along inner surface; petiole base with stipitate multicellular glands; rachis with longitudinal groove; stipels often persistent and glandular. **Leaflets** 12–80, opposite, elliptical-ovate, rectangular to oblong, base acute to obtuse, often unequal, apex mucronate, margins entire. **Inflorescences** axillary, with 1–15 flowers, racemes or panicles, ascending, lax or pendant; bracts and paired bracteoles early deciduous. **Flowers** papilionoid, 5-merous, bilaterally symmetrical; calyx radially symmetrical or bilaterally symmetrical due to distribution of lobes around the rim, bell-shaped; rim of tube with or without spreading hairs and stalked glands between lobes; corolla white, pale yellow, orange to orange-red, or crimson, with or without purple spots on outer surface of banner, glabrous; banner with pair of calli along claw (these highly reduced in some species); stamens 9 united and 1 free, anthers attached dorsally; pistil glabrous or style with spreading hairs; stigma spherical or slightly elongate; ovules 1–51. **Fruits** indehiscent or dehiscent, inflated, winged, quadrangular, torulose or cylindrical. **Seeds** 1–40; reniform-circular or columnar. $x = 6$.

Species ca. 68–85, of which 13 are native or introduced to the New World (1 native and 1 cultivated in the flora): United States, Mexico, Canada, West Indies, Central America, South America, Asia, Africa, Australia, Pacific Islands.

Sesbania species can be aggressive colonizers in agricultural and riparian settings throughout tropical, subtropical, and warm temperate regions of the world.

Various *Sesbania* species are used for food, fiber, medicine, forage, fuelwood, green manure, erosion control, reforestation, and to produce dyes, gum (from seeds), tannins, and pulp (to make paper and construction materials). They may also be grown as ornamental plants (NRC 1979, Allen and Allen 1981, Felger 2000). The wood of arborescent species is soft and not durable. It has been used to make toys and charcoal for gunpowder (Allen and Allen 1981, NRC 1983). Some species are reported to be toxic and have caused poisoning in livestock and humans.

1. Pods cylindrical, without wings, with 29–36 seeds; blade of keel petals bearing a tooth along the middle of the upper (adaxial) margin; banner with obvious calli along claw; seeds columnar ____ **1. Sesbania herbacea**

1′ Pods quadrangular in cross section, 4-winged, with 6–9 seeds; blade of keel petals lacking a tooth along the upper (adaxial) margin; banner with calli reduced to small teeth; seeds globose to reniform-circular ____ **2. Sesbania punicea**

1. Sesbania herbacea (Miller) McVaugh. Fl. Novo-Galiciana 5:695. 1987. * COLORADO RIVERHEMP, BIGPOD SESBANIA, DANGLE POD, HEMP SESBANIA Native

Emerus herbacea Miller, Gard. Dict. (ed. 8), Emerus no. 3. 1768; *Aeschynomene emerus* Aublet; *Aeschynomene picta* Cavanilles; *Coronilla occidentalis* Willdenow; *Darwinia exaltata* Rafinesque; *Sesban sonorae* Rydberg; *Sesbania exaltata* (Rafinesque) Cory; *Sesbania macrocarpa* Muhlenberg

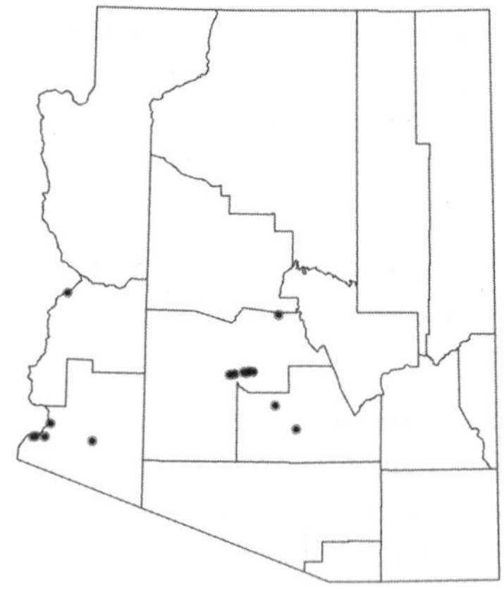

Herbs or small shrubs, 4.5 m, annual, mostly herbaceous, becoming woody at base. **Stems** erect, spreading, glabrous; pith spongy, becoming partitioned. **Leaves** 10–30 cm; stipules 10.3–10.5 mm, with inner fold running for nearly entire length, inner fold and base with stipitate multicellular glands; pulvinus as long as or at least half of the length of the petiole; rachis glabrous, occasionally with prickles, with stipitate glands in canal at base of each leaflet pair; stipels narrow, ± glandular. **Leaflets** 20–80, rectangular to oblong, (9.1–)13.8–16.5(–32.3) × (1.6–)3–3.8(–7.2) mm, the base obtuse, the apex truncate to obtuse, the upper (adaxial) surface glabrous, the lower (abaxial) surface glabrous. **Inflorescences** with 1–several flowers, racemes; peduncle (2.1–)16.2–18.5(–42.4) mm. **Flowers** (10.1–)13.5–14.6(–19.3) mm; calyx (4–)5.4–5.8(–8.2) mm, radially symmetrical; teeth 5, (0.6–)1.4–1.6(–2.6) mm, awl-shaped to narrowly pointed; banner (7.1–)12.2–13.1(–16.8) × (5.8–)12.8–14(–18.4) mm, yellow-orange, with purple-maroon spots on outer surface, ovate to obovate, the base truncate, the apex obtuse-cordate; the pair of calli as shallow ridges to wing-like, spanning length of claw, the apex rounded to truncate, lacking a thickened knob-like region at the base of blade-top of claw; wings (7.3–)11.6–12.5(–14.1) × (2.4–)4.3–4.8(–6.4) mm, yellow-orange, with a blunt and short tooth along upper (adaxial) margin; keel (7.7–)10.7–11.3(–12.6) × (4.9–)7–7.4(–8.3) mm, yellow-orange, purple or maroon at apex, apex rounded-acute, curved upward to inward towards banner, acute tooth along middle of upper (adaxial) margin projecting upward-inward; stamens strongly curved inward within keel; pistil glabrous; style curved back towards banner; ovules (12–)29–36(–51). **Fruits** (82.8–)169.2–193.9(–235.1) × (2.4–)2.7–3.1(–4.1) mm, brown with maroon-red mottling, tardily elastic dehiscent, narrow-elongate, rounded to elliptic in cross-section, straight to sickle-shaped, without wings; stipe (2.3–)4.6–5.2(–7.2) mm in fruit; beak (2.6–)6.3–7.5(–10.7) mm, narrowly tapering. **Seeds** (12–)29–36(–51); 2.7–3.8 × 0.9–2 mm, 1.5–2.5 mm thick, columnar, green-brown to reddish, often with purple-black mottling; hilum 0.4–0.5 mm, recessed, circular in outline.

Flowering, fruiting (Apr–)May–Sep(–Oct). Seasonally wet areas especially riparian, wetland, and agricultural sites; 0–640 m; Sonoran Desertscrub; Gila, La Paz, Maricopa, Pinal, Yuma cos.; throughout much of the Southern United States; Mexico; Caribbean; Central America; South America.

The identification of *Sesbania herbacea* is made easy by unique features of flower and stem structure. The flowers are yellow with purple-maroon mottling, have keel petals which turn upward abruptly and have apices that are purple-maroon, and stems have a septate pith.

In Arizona, *Sesbania herbacea* is found in moist soils of floodplains, ditches, and irrigated areas.

Seeds of *Sesbania herbacea* are reported to be toxic to livestock and poultry (Allen and Allen 1981, Hardin and Brownie 1993), however, Vines (1960) and Duke (1981) report that these are eaten by wild bird species. The plants have been grown as green manure and as a cover crop in southwestern Arizona, but often spread to adjacent agricultural fields (Parker 1972). The strong fibers from the stems have been used to make nets and fishing lines by native peoples along the lower Colorado River (Kearney and Peebles 1942, Parker 1972).

Hemp sesbania is propagated by seeds which do not require scarification. The plants require moist soil and are damaged or killed by freezing temperatures. Nodulation in *Sesbania herbacea* is widely reported.

2. Sesbania punicea (Cavanilles) Bentham in Martius, Fl. Bras. (Martius) 15(1):43. 1859. * [reddish-purple] RATTLE BOX, SCARLET SESBAN Cultivated

Piscidia punicea Cavanilles, Icon. 4(1):8–9, pl. 316. 1797; *Daubentonia punicea* (Cavanilles) de Candolle; *Emerus puniceus* (Cavanilles) Kuntze; *Piscidia ovalifolia* D.A. Larrañaga

Shrubs [or small trees] 4 m, perennial, woody. **Stems** erect, spreading, unarmed and glabrous when mature; pith spongy, soon becoming obsolete. **Leaves** 4.5–30 cm; stipules 4–4.5 mm, with inconspicuous inner fold running for two-thirds length of stipule, inner fold and base often with stipitate multicellular glands and long simple hairs; pulvinus less than half as long as petiole; rachis with ± appressed pilose hairs, with or without stipitate glands in canal, stipitate glands

Sesbania herbacea: **a.** leaf; **b.** flowering stem; **c.** inflorescence with nearly mature fruit; **d.** fruits after dehiscence; **e.** seed. Drawn by Karen Douthit and reproduced from *Flora Novo-Galiciana*, vol. 5, fig. 98, with permission of the University of Michigan Herbarium.

present at petiolule base; stipels present, small and glandular. **Leaflets** 14–40, elliptical to oblong-obovate, (16.9–)18.3–19.7(–29.8) × (4.9–)5.5–6(–10.5) mm, the base acute-obtuse, the apex truncate to obtuse, the upper (adaxial) surface glabrous, the lower (abaxial) surface glabrous or sparingly pubescent. **Inflorescences** with 5–15 flowers, racemes; peduncle (11.1–)21–23.8(–33.3) mm. **Flowers** (17.5–)20.1–22.2(–25.1) mm; calyx (4.7–)5.6–6.3(–7.3) mm, ± bilaterally symmetrical; teeth 5, (0.1–)0.3–0.5(–0.6) mm, deltoid to short-broad; banner (15.7–)18.5–20.5(–22.5) × (18–)19.4–20.5(–21.8) mm, orange-grenadine-red, without purple-maroon mottling on the outer surface, ovate, the base truncate, the apex cordate, the pair of calli as small pair of teeth with acute apex at base of claw; wings 17.1–18.7 × 4.2–5.2 mm, orange-grenadine-red, without tooth along upper (adaxial) margin; keel (14.5–)15–16.4(–17.5) × (9.5–)10.7–12.2(–13.1) mm, orange-grenadine-red, ± same color throughout, the apex rounded-obtuse, curved upward, without tooth along upper (adaxial) margin; stamens curved upward within keel; pistil glabrous; style curved upward with stamen; ovules 6–7(–10). **Fruits** (52.1–)87.9–95.2(–107.1) × (7.6–)8.2–8.4(–8.9) mm, red-brown with or without horizontal mottling, tardily dehiscent, quadrangular in cross-section, ± straight, with pronounced spreading wings from suture margins; stipe (7.8–)11.2–12.3(–16.3) mm in fruit; beak (5.1–)8.9–10.2(–13.6) mm, pyramidal. **Seeds** (1–)6–7(–10); 5.4–5.9 × 3.5–3.9 mm, 3–4 mm thick, globose to reniform-circular, red-brown to grey without mottling; hilum 1–1.8 mm, recessed, oblong in outline.

Flowering, fruiting May–Sep. Native to Brazil, Argentina, Uruguay, and Paraguay. Introduced to U.S. Known as invasive in Africa and U.S. (e.g. California). Occasionally cultivated as a landscape plant at lower elevations in Arizona.

Sesbania punicea with its striking orange-red flowers is used as an ornamental and is commonly sold as *Sesbania tripetii* or "Scarlet Sesban" in North America. This native of central South America can survive short freezes and has escaped and become a noxious weed in wetland locations in California and South Africa. The orange-scarlet color of the flowers is so different from all other New World Sesbania species that identification is simple. The fruits, when mature, have broad wings that obscure the location of the seeds when viewing the sides of the fruit parallel to the sutures.

Seeds of *Sesbania punicea* are reported to be toxic to domestic fowl (Giusti 1938) and may cause fatal poisoning in humans if eaten (Russell et al. 1997). Research involving seed extracts of *S. punicea* suggests that it may be effective in treating certain cancers (Powell et al. 1976).

Sesbania punicea has become invasive in areas of California and the southeastern U.S.; however, it is unlikely to be a problem in Arizona except in moist or irrigated areas. This plant is readily propagated by seeds. Scarification speeds germination. Cultivated plants in Tucson have survived overnight temperatures of -6°C without damage.

SOPHORA Linnaeus. Sp. Pl. 1:373–374. 1753. * [derived from the Arabic word *sufayra*, which means leguminous tree]
Benjamin Brandt

Herbaceous perennials [shrubs or trees to 15 m], heavily branched, erect to ascending, to 50 cm; some herbaceous species with rhizomatous roots. **Stems** slender, angular, unarmed. **Leaves** alternate, odd-pinnate; stipules 4–30 × 1–2 mm; petiole and rachis with upper channel, petiole 1–10 mm, rachis 1–6 cm; petiolulate or leaflets sessile. **Leaflets** 11–17 per pinna, opposite or irregularly arranged, 5–30 × 1–5 mm, margins entire. **Inflorescences** terminal, racemes, 3–20 cm long, flowers 5–many; floral bracts 4–6 mm. **Pedicels** 1–6 mm. **Flowers** papilionoid, 5-merous, bilaterally symmetric; calyx asymmetrically cup-shaped, gibbous on upper (adaxial) side, 5–12 mm, with 5 unequal triangular lobes 1–5 mm; corolla recurved, 12–23 mm, glabrous; stamens 8–17 mm, filaments free; ovary densely sericeous, ovules 4–9, style 3–6 mm, glabrous. **Fruits** indehiscent or tardily dehiscent, erect, torulose, often with sterile intervals between segments and/or at the apical or basal ends, not inflated, segments 1–8, 4–7 × 0.5–0.6 cm, villous; stipe 2–10 mm.

Species 50–60 (2 native in the flora): sw United States, Mexico, Central America, South America, Pacific Islands, New Zealand, Africa, Australia, Asia, Europe.

Sophora has long been recognized as being rife with problems in regard to its taxonomy. In efforts to bring more uniformity to the genus, Yakovlev (1967) and Sousa and Rudd (1993) suggested recombining some of the species into the genera *Calia* and *Styphnolobium*. However, the name *Calia* was subsequently determined to be invalid, and Gandhi et al. (2011) published those species under the name *Dermatophyllum*. The genus *Sophora* as it is treated here is with the exclusion of these recombined species. These recombinations have largely been made based upon morphological characteristics, chromosome numbers, chemical composition, and methods of germination (Yakovlev 1967, Sousa and Rudd 1993, Peña et al. 2000). Recent phylogenetic analysis by Heenan et al. (2004) has provided further support for this treatment of the genus.

Several species of *Sophora* are cultivated for their showy flowers. *Sophora tetraptera,* with beautiful yellow flowers, is the national flower of New Zealand (Allen and Allen 1981). The wood of larger species is valued for use as bearings and in turnery and cabinetry (Lewis et al. 2005). Other species are used medicinally, and some are toxic and have been employed as insecticides (Lewis et al. 2005). The dried roots of *S. flavescens* are used to treat dysentery in Chinese medicine (Allen and Allen 1981).

SELECTED REFERENCES: Heenan et al. 2004, Peña et al. 2000, Sousa and Rudd 1993, Yakovlev 1967.

1. Leaflets obelliptic to oblong to obovate, 5–15 × 3–5 mm ____ **1. Sophora nuttalliana**
1' Leaflets narrowly linear, 15–33 × 1–2 mm ____ **2. Sophora stenophylla**

1. Sophora nuttalliana B.L. Turner. Field & Lab. 24(2):[42]. 1956. * [for Thomas Nuttall, 1786–1859, English botanist who worked extensively in the United States] SILKY SOPHORA, WHITE LOCO Native

Sophora carnosa (Pursh) Yakovlev; *Sophora sericea* Andrews

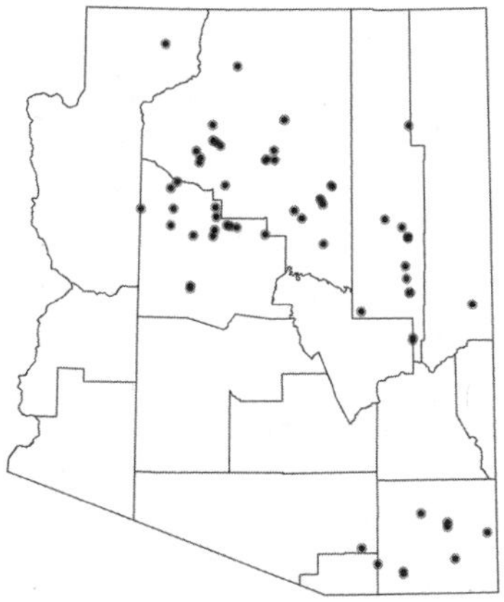

Herbaceous perennials, to 50 cm. **Stems** erect or ascending from rhizomes; sericeous. **Leaves:** stipules heteromorphic, often lacking, elliptic to linear, 4–6 × 1–2 mm, sericeous; petiole and rachis round, petiole 1–5 mm, rachis 1.3–6 cm, both sericeous; petiolules ± 1 mm. **Leaflets** 13–17 per pinna, often folded along midrib, obelliptic to oblong to obovate, rounded to obtuse at apex, rounded to wedge-shaped at base, 5–15 × 3–5 mm, glabrescent to glabrous on upper (adaxial) surface, sericeous on lower (abaxial) surface. **Inflorescences** 3–12 cm, sericeous, flowers 6–many, loosely arranged, floral bracts 4–6 mm. **Pedicels** 1–6 mm. **Flowers:** calyx 5–9 mm, lobes 1–3 mm, blue, sericeous; corolla white to yellowish-white, 12–16 mm; stamens 8–9 mm; ovule 4–8, style 3–4 mm. **Fruits:** segments 1–8, 4–7 × 0.5–0.6 cm; stipe 2–3 mm. **Seeds** 1–8.

Flowering Apr–Jun. Sandy soils, dry or moist areas, along canyon floors, floodplains, marshes, lakesides, valleys, meadows, and riparian areas; 1250–2100 m; Semidesert Grassland, Plains and Great Basin Grassland, Great Basin Conifer Woodland, Montane Conifer Forest; Apache, Cochise, Coconino, Navajo, Pima, Yavapai cos.; Colorado, Kansas, Nebraska, New Mexico, Oklahoma, South Dakota, Texas, Utah, Wyoming; n Mexico (Chihuahua, Sonora, Nuevo León).

The vegetative characteristics of *Sophora nuttalliana* can look quite similar to those of several species of *Astragalus*. The fertile characteristics, specifically its separate stamens and pod shape, make it easily distinguishable. This similarity however perhaps prompted a significant error in what is now understood as the first published description of the species, *Astragalus carnosus* Pursh. In it Pursh lists characteristics of both *S. nuttalliana* as well as what is now known as *A. crassicarpus* Nuttall. Nuttall, in his 1818 description of *S. sericea* Nuttall, noted this error and used the *Astragalus* elements of the description to describe *A. carnosus* Pursh emend Nuttall. In his 1956 publication, Turner suggests the rejection of the species *A. carnosus* citing the documented errors in its original description. In the same article, he also suggests recognizing *S. sericea* Nuttall as *S. nuttalliana*, undoubtedly in honor of the first author to correctly describe the species (Turner 1956). It is important to note though, that there was an unfortunate omission in Turner's original article which was corrected in an errata statement at the beginning of the following journal issue.

Moerman (1998) reports the roots of *Sophora nuttalliana*, which are noted for their sweet flavor, were chewed by members of the Acoma, Western Keres, Laguna, and San Felipe tribes as a delectable treat. In addition, the Navajo were also known to use this plant to treat ailments of the respiratory system, including influenza and tuberculosis (Wyman and Harris 1941). The Navajo also noted this plant as being a sought-after forage by their sheep, who would scratch at the ground beneath the plant to get at its roots (Elmore 1978).

2. Sophora stenophylla A. Gray in Ives, Rep. Colorado R. 4:10. 1860. * [with narrow leaves] FRINGELEAF NECKLACEPOD, SILVERY SOPHORA Native

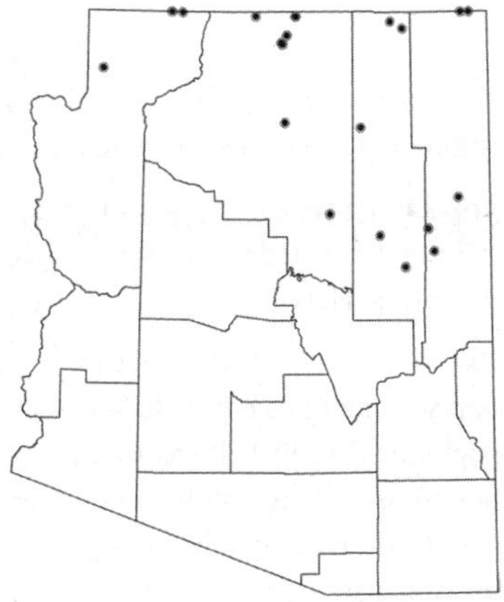

Herbaceous perennials, to 35 cm. **Stems** erect or ascending from spreading rhizomes; densely sericeous giving a silvery appearance. **Leaves:** stipules heteromorphic, narrowly linear, 15–30 × 1–2 mm, densely sericeous; petiole and rachis somewhat flattened, petiole 7–10 mm, rachis 1–4 cm, both densely sericeous; leaflets sessile. **Leaflets** 11–15 per pinna, narrowly linear, slightly rounded to acute at apex, 15–33 × 1–2 mm, densely sericeous on both surfaces. **Inflorescences** 5–20 cm, densely sericeous, flowers 5–many, loosely or densely arranged; floral bracts 4–5 mm. **Pedicels** 2–6 mm. **Flowers:** calyx 9–12 mm, lobes 2–5 mm, densely sericeous, deep blue to violet; corolla deep blue to violet, 20–23 mm; stamens 16–17 mm; ovules 7–9, style 5–6 mm. **Fruits:** segments 1–5, 4–5 × 0.5–0.6 cm; stipe 5–10 mm. **Seeds** 1–5.

Flowering (Apr–)Jun–Jul. Sandy soils, on dunes, atop mesas, in valleys, disturbed areas; 1150–1750 m; Great Basin Desertscrub, Great Basin Grassland, occasionally in Great Basin Conifer Woodland; Apache, Coconino, Mohave, Navajo cos.; Colorado, New Mexico, Utah.

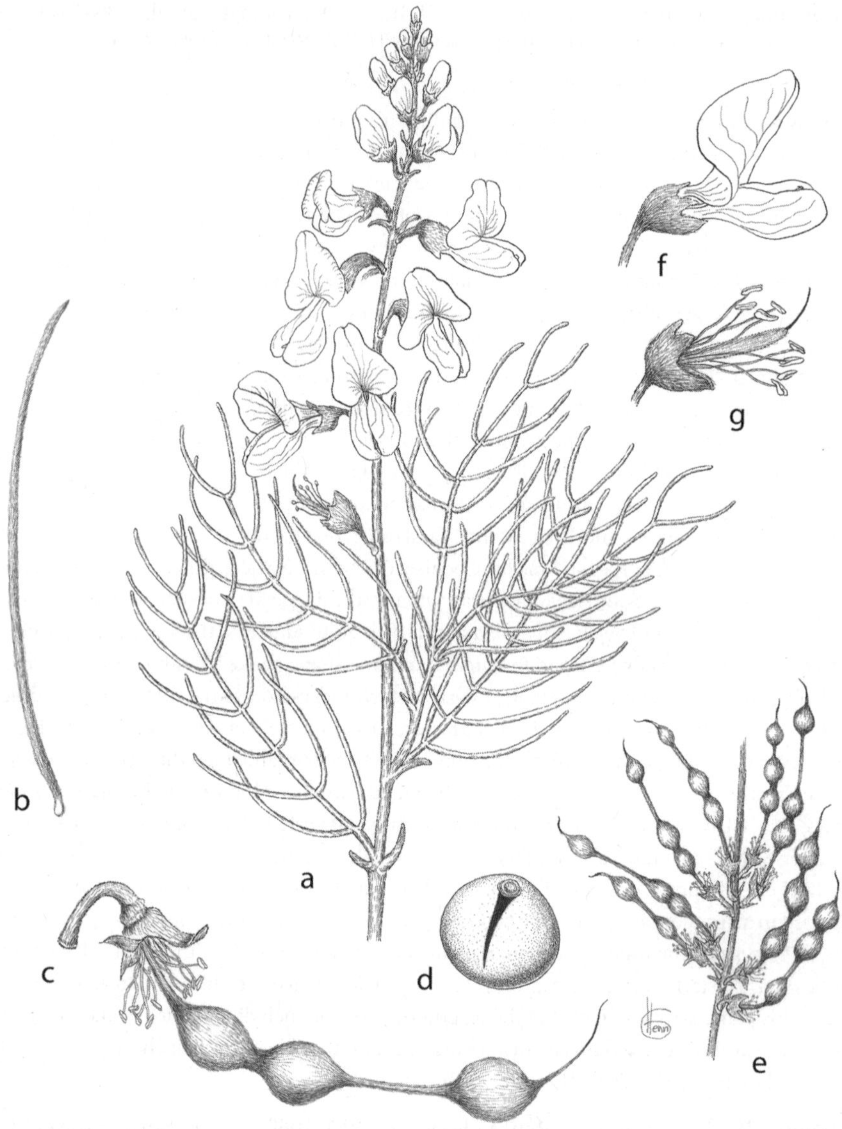

***Sophora stenophylla*: a. habit; b. leaflet; c. fruit; d. seed; e. fruits; f. flower; g. flower with corolla removed. Illustrator: Gabriele Henn.**

Sophora stenophylla is distributed across northern Arizona. The populations are scattered but plants may be locally common. This species would make an attractive addition to a wildflower garden. The showy flowers contrast with foliage.

SPHAEROPHYSA de Candolle. Mem. sur Legum. 288. 1825. * [Greek *sphaire*, sphere, and *physa*, bladder, alluding to the inflated fruit]

Richard R. Halse

Herbs, perennial. **Stems** unarmed, erect, strigillose. **Leaves** alternate; odd-pinnate; stipules small. **Leaflets** (9–)15–25, opposite, margins entire. **Inflorescences** axillary, racemes; bracteoles present, immediately subtending calyx. **Flowers** papilionoid, calyx bell-shaped, lobes subequal, shorter than tube; corolla bilaterally symmetric with 5 petals, dull brick- or orange-red, pinkish-brown, bright red, or reddish, [blue]; stamens 10, 9 fused and 1 free; style sparsely hairy below stigma; ovary 1-chambered, stipitate. **Fruits** indehiscent, inflated, papery, membranous, semi-transparent. **Seeds** ca. 40, reniform. $x = 8$.

Species 2 (1 adventive in the flora): naturalized in North America, western Europe (Turkey), central and eastern Asia.

SELECTED REFERENCES: Barneby 1964b, Isely 1998.

1. **Sphaerophysa salsula** (Pallas) de Candolle. Prodr. 2:271. 1825. * [Latin *sals*, salted, and *ula*, little] AUSTRIAN PEAWEED, AUSTRIAN FIELD PEA, ALKALI SWAINSONPEA, SWAINSONA Adventive

Phaca salsula Pallas, Reise Russ. Reich. 3:747. 1776; *Astragalus iochrous* Barneby; *Astragalus violaceus* H. St. John; *Swainsona salsula* (Pallas) Taubert in Engler & Prantl

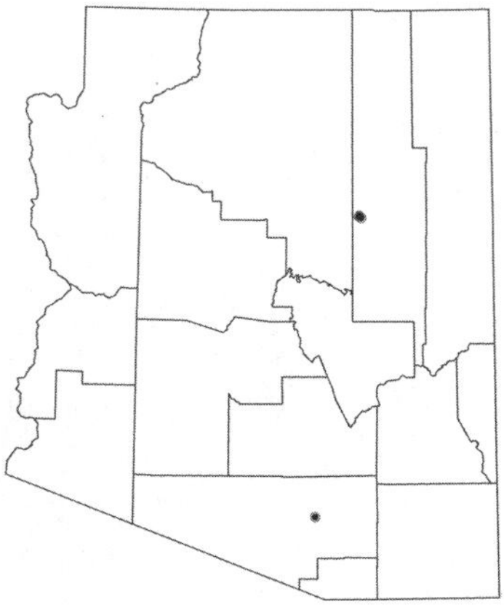

Herbs, perennial. **Roots** woody, creeping, horizontal. **Stems** simple or sparsely branched, 4–10(–15) dm. **Leaves** 3–11 cm; petioles 3–20 mm; stipules 1–4 mm. **Leaflets** oblong-obovate, narrowly oblong or narrowly elliptic, 3–18 × 3–9 mm, apex retuse to obtuse, apiculate or acute, lower (abaxial) side strigillose, upper (adaxial) side glabrous. **Inflorescences** ascending; peduncle 3–9 cm, 5–18 cm in fruit; 5–17-flowered; single bract 1–4 mm, bracteoles 0.5–1 mm. **Pedicels** 2.5–5 mm, ± recurved in fruit, thickened, 3–8 mm. **Flowers:** calyx persistent, 5–6 mm, strigillose, the teeth triangular, 1.2–2 mm, becoming papery; corolla drying lavender or brownish, 10–15 mm; ovary densely strigillose. **Fruits** green to purplish tinged, bladdery-inflated, opening by weathering and decay, ovoid or globose, 1.3–3.5 cm, strigillose to glabrate; stipe 4–8 mm. **Seeds** brown to greenish or reddish-brown, smooth, dull.

Flowering May–Jun, fruiting Jun. Disturbed sites, roadsides; 1480–1520 m; Great Basin Desertscrub; Navajo Co.; California, Colorado, Idaho, Kansas, Montana, Nevada, Oregon, Texas, Utah, Washington, Wyoming; Canada (Saskatchewan).

In Arizona *Sphaerophysa salsula* has only been collected in the 1930's near Winslow. It may no longer be present in the state. Throughout its range in North America it flowers from May to September and fruits from June to October; its elevational range is 100–2320 meters. It is a state listed noxious weed in California, Nevada, Oregon, and Washington.

Sphaerophysa salsula is distributed from Afghanistan and Iran to China, Mongolia, and Siberia. It is found in semi-arid habitats in regions with a temperate continental climate (Lewis et al. 2005). The plants are halophytic and occur in saline soils (Allen and Allen 1981, Lewis et al. 2005). Nodulation has been documented for *S. salsula* (Xu et al. 2009).

Sphaerophysa salsula is considered a threat to grain and alfalfa production due to seed contamination (Washington State NWCB 2008). The seeds of *S. salsula* are similar in size to those of alfalfa and can be introduced into new areas as a contaminant in alfalfa seeds. Eradication is difficult as the plants produce an extensive system of woody rhizomes (Washington State NWCB 2008). *Sphaerophysa salsula* is valued for revegetating alkaline soils in China (Xu et al. 2009). Allen and Allen (1981) suggest that *S. salsula* may be poisonous if eaten in sufficient quantity as the plants contain spherophysine, an alkaloid that blocks nerve ganglia; however, Lewis et al (2005) indicate that the plants are utilized for forage and are also used medicinally. Spherophysine from *S. salsula* is reported to be used to treat uterine problems and hypertension (Shreter 2010). The use of cattle to graze *S. salsula* for reduction of seed production in pastures is suggested (CDFA n.d.).

SPHINCTOSPERMUM Rose. Contr. U.S. Natl. Herb. 10:107. 1906. * [from Greek, *sphinktos*, constricted, and *sperma*, seed, in reference to the seeds with a central constriction] HOURGLASS PEASEED

Matt Lavin, Kirsten N. Lake, Matthew B. Johnson

Annual herbs with erect to ascending stems. **Stems** with young growth strigose. **Leaves** alternate; unifoliolate; stipules persistent; stipels absent. **Leaflet** linear-lanceolate, entire. **Inflorescences** of solitary flowers in axils of mostly distally positioned leaves; bracts early deciduous; bracteoles absent. **Flowers** papilionoid, calyx bilaterally symmetrical, bell-shaped, 5-lobed; corolla pinkish, glabrous; stamens 10, 9 united and 1 free, anthers small, attached basally, dehiscing longitudinally; pistil glabrous with a pollen brush uniformly surrounding the distal end; stigma rounded, ciliate, terminal; ovules 10–12. **Fruits** elastically dehiscent, laterally compressed, linear. **Seeds** 3–12, quadrangular with a hilum recessed in a central constriction that typically surrounds the entire seed. $x = 8$.

Species 1 (1 native in the flora): Sonoran Desert in Arizona and adjacent Mexico and Pacific coastal dry forests in Mexico.

Sphinctospermum is readily distinguished by its annual habit, unifoliolate (appearing simple) linear-lanceolate leaves, and solitary flowers measuring much less than 10 mm long. This annual legume is broadly distributed along Pacific coastal Mexico but is only locally abundant and is not often collected. The unifoliolate leaves and solitary flowers

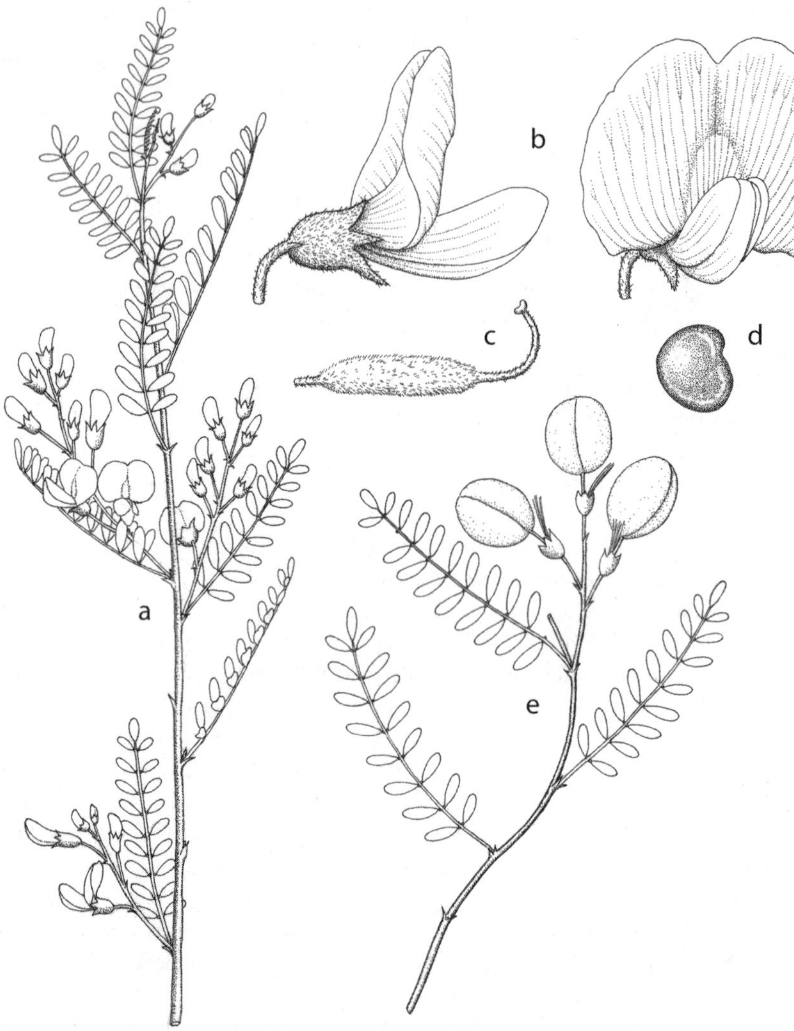

***Sphaerophysa salsula*: a. habit; b. flower (side and front view); c. pistil; d. seed; e. fruiting stem. Illustrators: Gigi Wilson & Matthew B. Johnson.**

are superficially similar to those of *Genistidium*, and results of molecular phylogenetic analysis suggest these two are sister genera (e.g., Duno de Stefano et al. 2010).

SELECTED REFERENCES: Lavin 1990, Lavin and Sousa 1995.

1. Sphinctospermum constrictum (S. Watson) Rose. Contr. U.S. Natl. Herb. 10:107. 1906. * [Latin *constrictum*, constricted or contracted] HOURGLASS PEASEED Native

Annual herbs to 0.8 m. **Leaves** glabrate, unifoliolate (appearing simple), (1–)2–8 (–9) cm; stipules awl-shaped, 1–2 × 0.5 mm; petioles minute, leaves subsessile. **Inflorescences:** stipitate glands absent; bracts linear-lanceolate, 1–2 × 0.5 mm. **Pedicels** 1–2 mm. **Flowers:** calyx tube 1.5–2.5 mm, glabrate, calyx lobes 1.5–2.5 mm, the lowermost the longest, the upper two fused slightly more with each other than with the laterals; corolla pinkish, 5–6 mm; filaments subequal; pistil glabrous. **Fruits** sessile, 2–3 × 0.2–0.3 cm, base blunt, apex acute and often terminating in persistent style base, light brown, glabrous. $2n = 16$.

Flowering Aug–Nov. Rocky slopes and outcrops, alluvial soils of riparian habitats, and disturbed settings; [50–]840–1700 m; Sonoran Desertscrub,

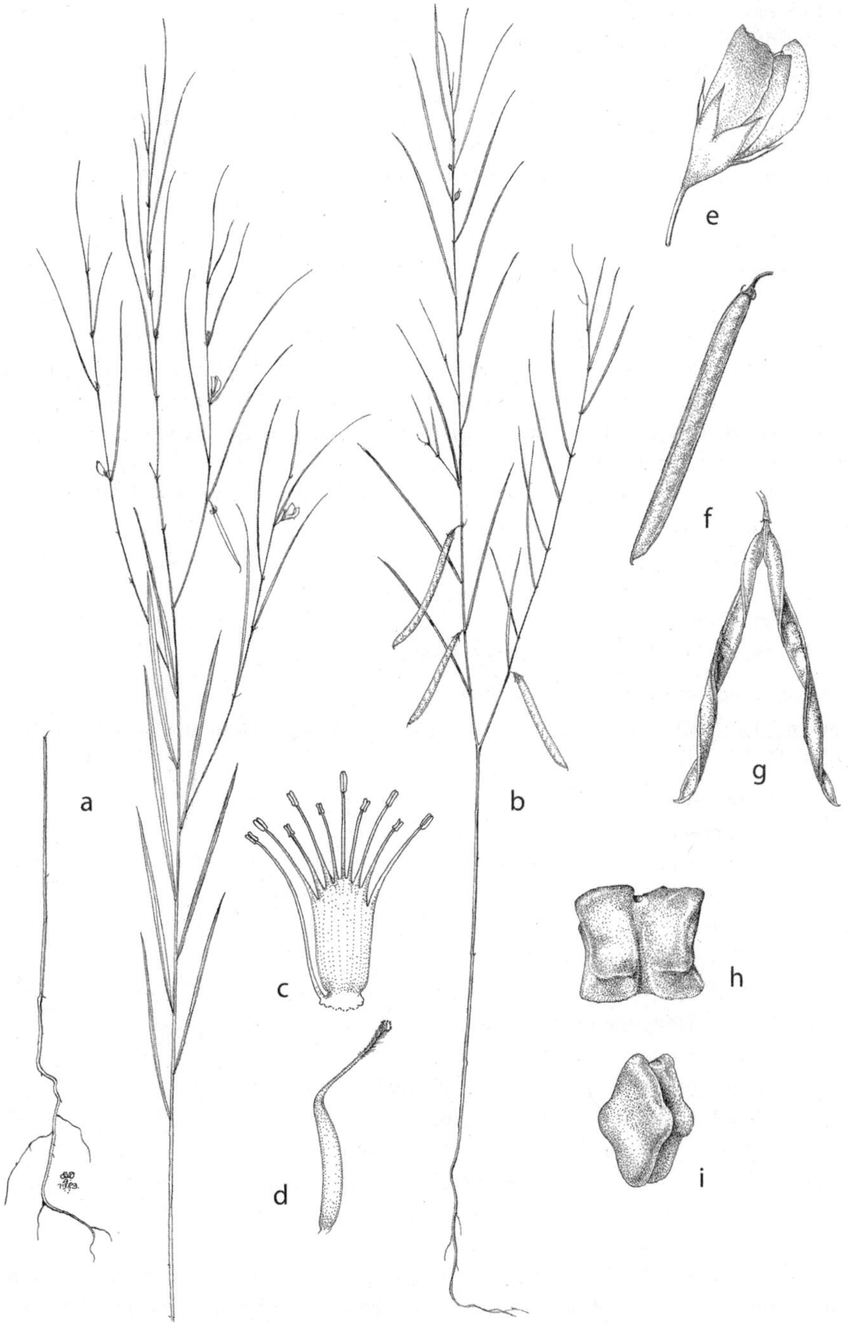

***Sphinctospermum constrictum*: a. habit (flower); b. habit (fruit); c. stamens; d. pistil; e. flower; f. fruit prior to dehiscence; g. fruit after dehiscence; h., i. seeds. Drawn by Karen Douthit and reproduced from *Flora Novo-Galiciana*, vol. 5, fig. 99, with permission of the University of Michigan Herbarium.**

Semidesert Grassland, (Madrean Evergreen Woodland); Cochise, Pima, Pinal, Santa Cruz cos.; Pacific coastal Mexico from Sonora to the southern end of the Baja California Peninsula and to Oaxaca.

The inflorescence of a solitary flower includes a very short rachis and floral bract subtending the floral pedicel. Sometimes a short axillary shoot bearing a flower is born from this same axil. This results in two flowers from a leaf axil along the main stem. In Arizona, *Sphinctospermum constrictum* occurs in widely scattered populations. In Mexico this taxon also occurs in thornscrub and tropical deciduous forest.

STYLOSANTHES Swartz. Prodr. 7, 108. 1788. * [from Greek *stylo*, pillar, and *anthos*, flower, referring to the long, often persistent style] PENCIL-FLOWER

Matthew B. Johnson and Kirsten N. Lake

Herbs or subshrubs to 50 cm, perennial. **Stems** mostly trailing or prostrate, sometimes ascending. **Pubescence** absent or of bristles or viscid hairs. **Leaves** alternate, pinnately 3-foliolate; stipules sheath-like, fused to petiole base or axis, tips awl-shaped, persistent; petiolate; stipels absent. **Leaflets** elliptic, lanceolate, or oblanceolate, sometimes with a sharp-tipped apex, margins entire or with bristle-like hairs. **Inflorescences** spike-like racemes or heads, congested, terminal or axillary; bracts conspicuous, membranous-leafy, persistent; bracteoles persistent. **Pedicellate**. **Flowers** papilionoid, solitary or paired; calyx tubular with 5 unequal lobes, the 4 upper lobes fused, the lowest lobe longer and free; corolla yellow or orange, often with purple stripes, or rarely apetalous, petals short-clawed and often lobed at base; banner orbicular-obovate; wings oblong, free; keel petals incurved; stamens 10, alternating long and short, filaments fused for most of their length, the tube splitting with age, anthers dimorphic; ovary with 2 (or 3) ovules; style long, thread-like, the base persistent and curved; stigma minute. **Fruits** loments, sessile, indehiscent or dehiscent at apex, compressed, with 2 segments, glabrous or pubescent, reticulate, beak often curved, hooked, or coiled. **Seeds** 1 per fruit segment, brown or black, glossy, ovoid, smooth.

Species ca. 25 (1 native in the flora): primarily tropical and subtropical regions of North America and South America, also Africa and southern Asia; naturalized in Australia.

Brazilian lucerne, *Stylosanthes guianensis*, and other species are important fodder plants in tropical and subtropical regions around the world (Allen and Allen 1981, Lewis et al. 2005). Species are planted as cover crops, green manure, and for soil stabilization (Allen and Allen 1981, Lewis et al. 2005). Some species are used medicinally (Lewis et al. 2005). Nodulation is widely reported for the genus (Allen and Allen 1981).

SELECTED REFERENCE: Mohlenbrock 1957.

1. Stylosanthes humilis Kunth in Humboldt, Nov. Gen. Sp. (quarto ed.) 6:506, pl. 594. 1823 [1824]. * [from Latin *humilis*, low] TOWNSVILLE STYLO Native

Herbs to 10 [50] cm. **Stems** ascending or sometimes prostrate, simple or with a few lateral branches. **Pubescence** of scattered bristles on stems, petioles, leaflet margins, and inflorescence bracts. **Leaves** 1.2–2.7 cm; stipules 5–9 mm; petioles 4–10 mm; rachis 0.5–2 mm. **Leaflets** lanceolate to elliptic, 5–17 × 1–3.5 mm, the base rounded to cuneate, the apex acuminate or aristate, the margins with bristle-like hairs. **Inflorescences** 1–3 cm with the flowering portion 0.8–1.2 cm, 3–4(–6)[–10]-flowered; bracts 5–6 mm; bracteoles 2–3.5 mm. **Flowers** with calyx [2–]5.5–7 mm, the tube 4–5 mm, the lobes ca. 1.5 mm; corolla yellow, the banner 3–4 mm, the wings 3–4 mm, the keel 3–3.5 mm. **Fruits** with only the distal segment fertile, this 1.5–2.5 mm, puberulent, with the beak hooked or coiled, 2–4 mm. **Seeds** 1–1.5 mm, black.

Flowering Jul–Sep. Along stream channels and arroyos, and on adjacent slopes in sandy or rocky soil; [10–]1630–1680 m; Madrean Evergreen Woodland; Cochise, Santa Cruz cos.; Mexico (Sonora, Sinaloa, to southern Mexico); Central America; Caribbean.

In Arizona, *Stylosanthes humilis* is rare and highly localized in the Huachuca Mountains and Canelo Hills. It is widely distributed in Mexico to Central America and the Caribbean. The species has been introduced as a pasture plant outside of its native range and has become naturalized in Australia and Malaysia (Mohlenbrock 1957). Plants identified as *S. humilis* in Arizona may represent a different, possibly undescribed taxon (P. Jenkins, personal communication). Nodulation of *S. humilis* is reported from Zimbabwe (Allen and Allen 1981). Of conservation concern in Arizona.

STYPHNOLOBIUM Schott. Wiener Z. Kunst 3:844. 1830. * [derived from the Greek words *styphno-* (meaning sour, astringent) and *lobion* (meaning pod), in reference to the taste of the fleshy portion of the fruits]

Benjamin Brandt

Woody shrubs or small trees, bark smooth or fissured to scaly, or occasionally exfoliating. **Stems** unarmed, rounded. **Leaves** alternate, odd pinnate, deciduous; stipulate; petiolate. **Leaflets** 9–21, opposite proximally and alternate distally or all opposite or alternate; margins entire. **Inflorescences** racemes, axillary, pendulous; bracts 1, bracteoles 2. **Pedicellate. Flowers** papilionoid, 5-merous, bilaterally symmetric; calyx cup-shaped, with 5 shallow rounded lobes

***Stylosanthes humilis*: a. habit with enlargement of roots showing nodules; b., c. petals; d. stamens; e. pistil; f. fruit; g. leaf with stipules; h. leaf detail; i. stem section with fruit; j. inflorescence; k. inflorescence; l. seeds. Illustrator: Elaine Hultgren.**

to truncate, gibbous or not gibbous; corolla white to yellowish-white to pale pink, petals clawed; banner ± round and reflexed; wings symmetric to asymmetric; keel petals free; filaments free. **Fruits** indehiscent, pendulous, torulose and sometimes appearing bead-like, exocarp leathery, mesocarp fleshy, black at maturity. **Seeds** oblong-ellipsoid, smooth, brown to black.

Species 9 (1 cultivated in the flora): s United States, Mexico, Central America, China.

All of the species now treated in *Styphnolobium* were formerly treated within *Sophora*. Schott was the first to separate *Styphnolobium* from *Sophora*, listing *Styphnolobium japonicum* (Linnaeus) Schott as the type species for this new genus. Walpers later added *Styphnolobium affine* (Torrey & A. Gray) Walpers to this genus, listing morphological characteristics as the bases for his recombination (Walpers 1842). Much later Sousa, Rudd, and Medrano added several other species to the genus, basing their recombination on morphological characteristics, chromosome numbers, and methods of germination (Sousa and Rudd 1993).

Styphnolobium japonicum is planted as an ornamental and is used for lumber and medicine. The leaves and flowers are edible (Lewis et al. 2005). The fruits yield a yellow dye which is used to dye silk and in batik (Lewis et al. 2005).

SELECTED REFERENCES: Heenan et al. 2004, Peña et al. 2000, Sousa and Rudd 1993, Walpers 1842.

1. Styphnolobium affine (Torrey & A. Gray) Walpers. Repert. Bot. Syst. 1(5):807. 1843. * EVE'S NECKLACE, EVE'S NECKLACEPOD, TEXAS SOPHORA, PINK SOPHORA, NECKLACE TREE Cultivated

Sophora affinis Torrey & A. Gray, Fl. N. Amer. 1(3):390. 1840

Woody shrubs or small trees, heavily branched, 2–10.5 m. **Stems** slightly tomentose to sericeous when young, becoming glabrate with age; green or brown, bark of older stems gray or brown, scaly. **Leaves:** stipules 2–3 mm, linear, often lacking; petioles 1.2–3.5 cm; rachises 4.5–13.5 cm; petiolules 3–4 mm. **Leaflets** 13–17 per pinna, opposite near the base becoming irregularly alternate towards the apex, elliptic to ovate to obovate, obtuse to acute at apex, acute to cuneate to obtuse at base, 1–4.5 × 0.5–2.5 cm, glabrescent to glabrous on upper (adaxial) surface, sparsely sericeous on lower (abaxial) surface. **Inflorescences** 2–8 cm, flowers 6–15, loosely arranged; bracteoles 0.2–0.5 mm. **Pedicels**: 5–20 mm, sparsely canescent. **Flowers:** calyx 3–5 mm, 5-lobed, not gibbous, sparsely sericeous; corolla white to yellowish-white to rose-pink, 11–14 mm; ovaries oblong, moderately to densely sericeous, ovules 16–20. **Fruits** black, stiffly leathery, cylindrical and constricted at regular intervals, 1–6 segments, 1.5–15 × 0.6–0.9 cm, slightly tomentose when young; stipe 3–5 mm. **Seeds** 6–7 × 4–5 mm, brown to black. $2n = 28$.

Flowering Apr–Jun. Arkansas, Louisiana, Oklahoma, and Texas.

Styphnolobium affine has been used as a beautiful addition to landscapes throughout the Southern Plains and shows considerable promise for Arizona landscapes as well. It is quite adaptable to virtually any well-drained soil type, is drought tolerant, and can do well from part-shade to full sun conditions. The fragrant rosy-pink blossoms begin to develop in early to mid-spring and can remain through early to mid-summer. Once flowering is through, the unique, fleshy, black pods begin to develop, which are sure to attract attention in any landscape (Gilman and Watson 1994). The leaves are winter deciduous. Propagation is by scarified seeds. The seeds are reportedly poisonous (Correll and Johnston 1970).

TARA Molina. Sag. Stor. Chili (ed. 2) 153, 282–283, descr. emended E. Gagnon and G.P. Lewis * [derived from *tara*, a vernacular name in South America]

Matthew B. Johnson

Shrubs or trees. **Stems** armed with prickles. **Herbage** glabrous or puberulent. **Leaves** alternate, even bipinnate; stipules apparently absent. **Leaflets** opposite, margins entire. **Inflorescences** axillary or terminal racemes or panicles; bracts minute, deciduous. **Pedicels** jointed. **Flowers** caesalpinoid, often showy, bisexual; calyx with 5 unequal lobes, the lowermost lobe covering the other 4 lobes in bud, lowermost lobe with margins pectinate, fimbriate or entire, the lobes deciduous leaving a persistent hypanthium; corolla with 5 petals, overlapping, unequal, yellow with the banner sometimes orange or with red markings; stamens 10, free, filaments pubescent at base, declined, exserted; ovary sessile, style and stigma simple. **Fruits** sessile, indehiscent, turgid and somewhat fleshy, leathery. **Seeds** rounded to oblong or ellipsoid.

Species 3 (1 cultivated in the flora): Mexico, Guatemala, Nicaragua, and the Caribbean with one species from Ecuador and Peru.

Tara is closely related to *Coulteria*, differing in having bisexual vs. unisexual flowers and in fruit and seed characters (Gagnon et al. 2016). The fruits are a source of tannins. These plants are sometimes cultivated as ornamentals.

SELECTED REFERENCE: Gagnon et al. 2016.

1. Tara cacalaco (Bonpland) Molinari & Sánchez Ocharan. Weberbauerella 1(8):5. 2016. * *CASCALOTE* Cultivated

Caesalpinia cacalaco Bonpland, Pl. Aequinoct. 2:173–174, pl. 137. 1817

Shrubs or trees to 5 (10) m. **Stems** usually armed with stout, flattened prickles to 1 cm, solitary or paired at the petiole base (appearing stipular); bark brown and smooth, becoming dark brown or blackish, rough and fissured with the prickles becoming enlarged and woody. **Herbage** with dense, short, tan hairs, especially when young; hairs on inflorescence axis, pedicels, and calyx persistent. **Leaves** 11–20 cm with 2–6 pairs of pinnae, pinnae 4.5–9(–10) cm;

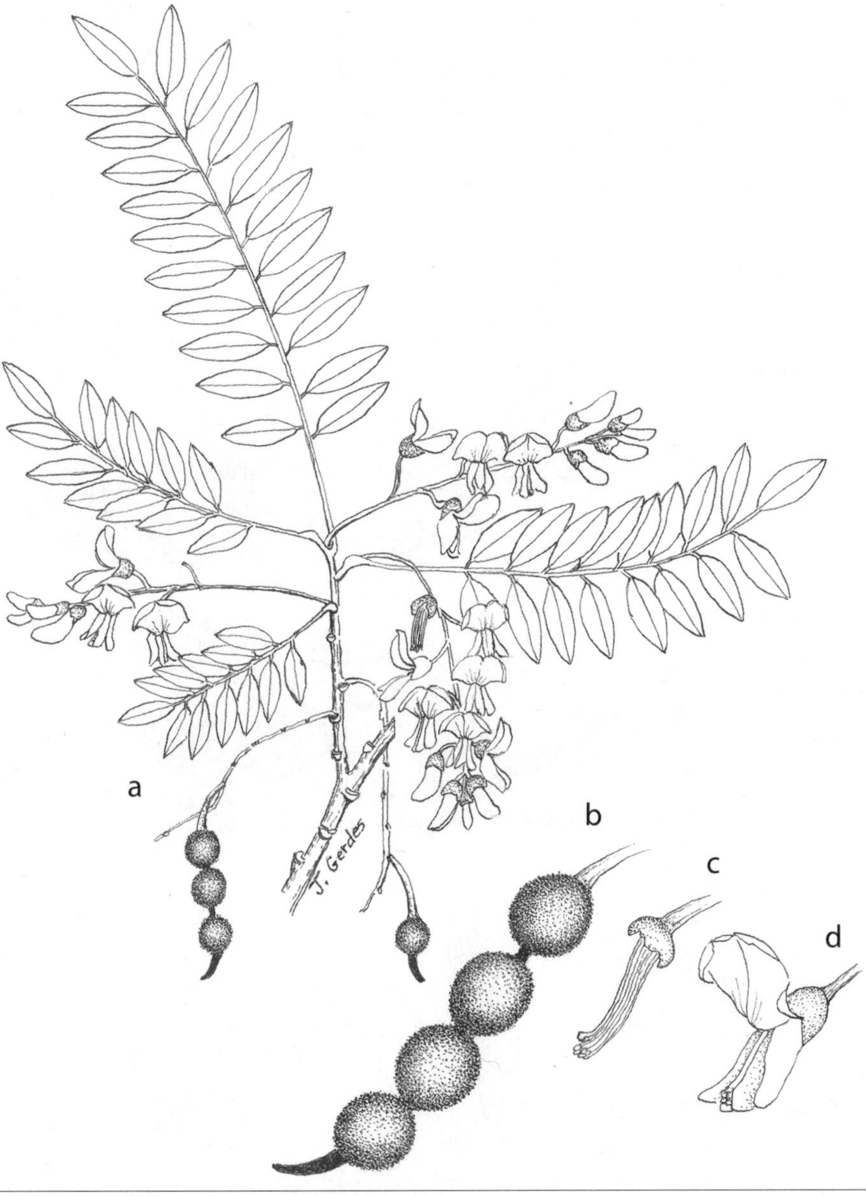

Styphnolobium affine: a. habit; b. fruit; c. stamens; d. flower. Illustrator: John H. Gerdes.

petioles 1.5–4(–5.5) cm; rachis 2–15 cm, occasionally with small prickles. **Leaflets** 2–5 pairs per pinna, 1.5–3 × 1.2–2.1 cm, elliptic or obovate, the base rounded or oblique, the apex rounded, sparsely pubescent. **Inflorescences** axillary and terminal racemes and terminal panicles, 10–30 cm, commonly with 30–50 flowers; bracts not seen. **Pedicels** 3–8 cm. **Flowers:** calyx 1–1.5 cm, green with yellow lobes; lower lobe 1–1.2 cm, cupped, fimbriate with stalked glands on the margin; corolla yellow or the banner petal sometimes orange, 9–12 mm; stamens with filaments pale yellow or greenish yellow, 11–13 mm; pistil 7–9 mm. **Fruits** dull red, drying brown, 5.4–15 × 1–1.3 cm, straight or slightly curved, constricted between seeds, pubescent, apex flattened into a beak 5–10 mm. **Seeds** greenish brown, rounded to oblong and only slightly compressed, 7–12 × 7–10 mm, slightly lustrous.

Flowering Oct–Feb(–May). Mexico.

Tara cacalaco is native to western and southern Mexico. Its distribution extends along the Pacific slope from northern Sinaloa to the Isthmus of Tehuantepec and inland in southern Mexico to México and Puebla. The species grows in thornscrub, tropical dry forest, savannahs, and the margins of mangrove swamps from sea level to 1000 m elevation

Tara cacalaco: **a.** flower; **b.** flower detail; **c.** inflorescence; **d.** inflorescence tip; **e.** stem segment with leaf and fruit; **f.** portion of fruit with seed; **g.** stem prickles. Illustrator: Daniela Siroky.

(McVaugh 1987). Pods of *T. cacalaco* contain tannins, and extracts from this species have shown promise as antioxidants (Veloz-García et al. 2004) as well as having antimicrobial activity (Veloz-García et al. 2010).

Cascalote is gaining popularity as a landscape plant in warm winter locations in Arizona. It is suitable as a patio tree and may be planted against walls and in street medians. The prickles become enlarged and woody on the trunk and older limbs, providing visual interest. The showy yellow flowers appear at a time of year when many other plants are out of flower. The compact canopy provides welcome shade and the trunks are attractive. Cultivated plants are typically evergreen unless the foliage is killed by freezing. Occasional deep irrigation in warm weather is necessary to maintain an acceptable appearance. With adequate irrigation, the plants may grow 1.5 m per year when young. Cascalote thrives in heat and full sun but will tolerate some shade. Established plants appear to be hardy to -6°C. The plants should be located away from traffic areas due to the armed stems. An unarmed form is available. In some situations, the plants produce numerous stems from the base and pruning may be necessary to develop a tree form. Propagation is by scarified seeds.

TEPHROSIA Persoon. Syn. Pl. 2:328. 1807, name conserved. * [Greek, *tephros*, ash-colored, probably alluding to silvery- or gray-sericeous vestiture of many of the species]

Guy L. Nesom

Cracca Linnaeus, name rejected

Perennial herbs or shrubs, rarely annual. **Stems** erect to prostrate, or decumbent; pubescent or glabrous. **Leaves** alternate, odd-pinnate; leaflets opposite; stipules present, stipels absent. **Inflorescence** pseudoracemes, terminal and/or axillary opposite leaves. **Flowers** papilionoid; calyx 5-toothed, two teeth superior, more or less united and much shorter; corolla reddish to purple or white; stamens 10, all united or 9 united and 1 free, banner stamen free or frequently apically fused to staminal tube and basally free; style barbate or glabrous, stigma glabrous or penicillate. **Fruits** oblong, compressed, straight or slightly curved or apically incurved, pubescent or glabrous, dehiscent by 2 sutures. **Seeds** 6–12, cylindrical to nearly circular or reniform. $x = 11$.

Species ca. 400 (3 native in the flora): native to North America, Mexico, West Indies, Central America, South America, Asia, Africa, Pacific Islands; the species are primarily tropical and subtropical.

Many species of *Tephrosia* produce rotenone, which is widely used as an insect and fish poison. The name "*barbasco*" is used for some of the fish-poison species in Mexico and Central America. Other uses for species of *Tephrosia* include cover crops, fodder for livestock, medicines, and as ornamentals (Lewis et al. 2005). *Tephrosia* has reportedly been used as an arrow poison, and some tropical species are planted as hedges, windbreaks, as shade for tea and coffee plantations, and for erosion control (Allen and Allen 1981). Allen and Allen (1981) state that *Tephrosia* is a low-quality fodder. Kearney and Pebbles (1942) indicate the Arizona species apparently contain low concentrations of rotenone but are still suspected of being poisonous. Nodulation is widely reported in the genus (Allen and Allen 1981).

1. Leaflets (3–)5–9(–11); corollas 6–9 mm; styles glabrous ____ **2. Tephrosia tenella**
1′ Leaflets (9–)11–27; corollas 13–18 mm; styles barbellate.
  2. Fruits glabrous; flowers 4–8(–10) in a loose inflorescence 5–8(–15) cm; corollas white, darkening reddish; leaflets 5–7(–9) mm wide ____ **1. Tephrosia leiocarpa**
  2′ Fruits villous to strigose-villous or strigose; flowers 8–14 in a congested inflorescence 2–4 cm; corollas weakly bicolored to yellowish and becoming pink-purple; leaflets (5–)7–13 mm wide ____ **3. Tephrosia thurberi**

1. Tephrosia leiocarpa A. Gray. Smithsonian Contr. Knowl. 5(6):36. 1853. * [Greek *leio*, smooth, *karpos*, fruit] SMOOTH-POD HOARY-PEA

Cracca leiocarpa (A. Gray) Kuntze, Revis. Gen. Pl. 1:175. 1891

Herbs to subshrubs. **Stems** erect, 30–100[–300] cm, minutely strigillose. **Leaves**: petioles 7–20 mm. **Leaflets** 11–23, oblong to elliptic-oblanceolate, (15–)20–40 × 5–7(–9) mm, 3–6.5 times longer than wide, strigillose on lower side (abaxially), glabrate on upper side (adaxially), apices obtuse or rounded to truncate or retuse. **Inflorescences** terminal and axillary loose pseudoracemes 5–8(–15) cm; floral bracts bristle-like, quickly deciduous. **Flowers** 4–8(–10); corollas 13–16 mm, white (pink to deep red or purplish, fide Wiggins 1964), darkening reddish; stamens 9 united, 1 partially free; styles barbellate. **Fruits** 4–6.5 × 0.5–0.6 cm, glabrous.

Varieties 2 (1 native in the flora): sw United States, n Mexico.

Tephrosia leiocarpa var. *costenya* McVaugh occurs only in the Pacific lowlands of Jalisco, Mexico (and noted by McVaugh (1987) to be disjunct in Oaxaca). It is distinct in its leaflets with elongate terminal mucros ((1–)1.5–2.5 mm long), separate geography, and apparently in its later flowering (Sep–Jan). The disjunct population system of *T. leiocarpa* var. *leiocarpa* in Oaxaca, with minutely puberulent stems, perhaps deserves to be formally recognized.

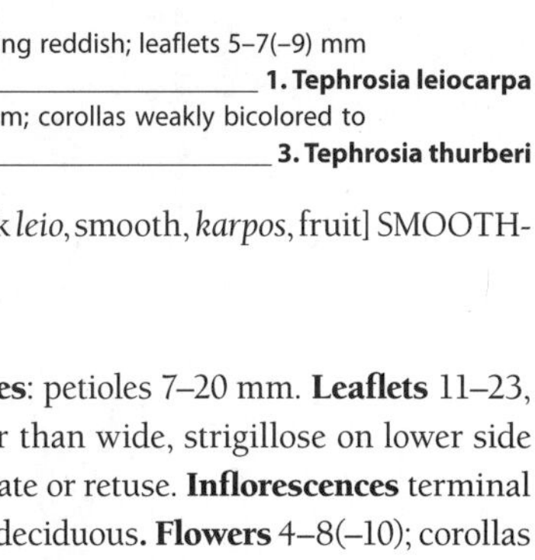

1a. Tephrosia leiocarpa A. Gray var. **leiocarpa** * *PALOMITAS DE LA SIERRA*
Native

Tephrosia affinis S. Watson; *Tephrosia viridis* M.E. Jones

Features as in the species description, described from plants of Arizona and Sonora. **Leaflets** with terminal mucros 0.4–0.8 mm long.

Flowering Jun–Aug(–Sep). Often in rocky soil or among boulders on open slopes and canyons; [200–]1060–1750[–1800] m; occasional in Semidesert Grassland, Madrean Evergreen Woodland; Pima, Santa Cruz cos.; Mexico (Sonora, Chihuahua, Durango, Nayarit, Jalisco, disjunct to Oaxaca).

Tephrosia leiocarpa **var.** ***leiocarpa*****: a. habit; b. flower; c. flower (longitudinal section); d. fruits; e. fruit; f. dehisced fruit; g. seeds. Illustrator: Chris Bondante.**

Tephrosia leiocarpa var. *leiocarpa* is typically found in small, localized populations in the mountains of southern Arizona. In Mexico it is found in tropical deciduous forest and other communities. It makes an attractive summer-flowering perennial in wildflower gardens and mini oases. *Palomitas de la sierra* grows well in full sun or in partial shade. The plants require supplemental irrigation for best appearance and performance at lower elevations in the state. The stems typically die back with freezing temperatures and should be pruned when new growth appears the following season. Plants are readily propagated from scarified seeds.

2. Tephrosia tenella A. Gray. Smithsonian Contr. Knowl. 5(6):36. 1853. * [Latin *tener*, delicate] SLENDER HOARY-PEA Native

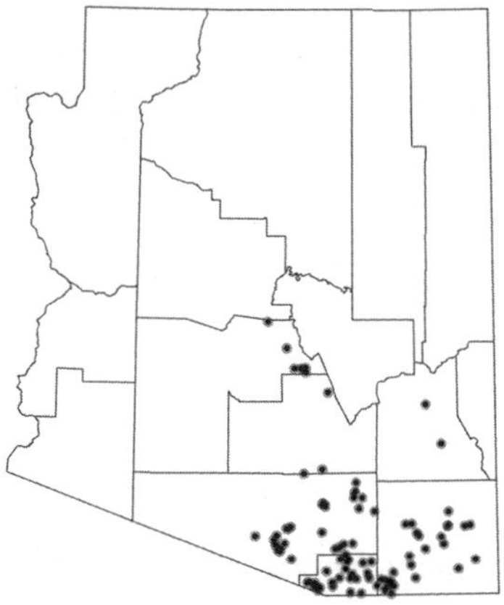

Herbs. **Stems** erect to decumbent or ascending, 10–50(–100) cm, sparsely strigillose to glabrescent. **Leaves**: petioles 10–30 mm. **Leaflets** (3–)5–9 (–11), linear to linear-oblong, narrowly oblong or oblong-elliptic, 15–40 × 2–6 mm, 5–9 times longer than wide, sparsely strigillose on lower side (abaxially), glabrous on upper side (adaxially), apices acute to obtuse or rounded. **Inflorescences** terminal and axillary pseudoracemes, 10–30 cm; floral bracts bristle-like and awl-shaped, subpersistent. **Flowers** (2–)8–17; corollas 6–9 mm, pink to rose-pink or rose-purple, aging darker reddish to purple; stamens 9 united and 1 free; styles glabrous. **Fruits** 3–5 × 0.3–0.4 cm, sparsely strigillose.

Flowering Apr–Oct(–Nov). Open slopes and flats, canyon slopes and bottoms, ledges and crevices, among boulders, riparian habitats, washes, gravelly alluvium, roadsides; [sea level–](670–)800–1900(–2000) m; Sonoran Desertscrub, (Chihuahuan Desertscrub), Semidesert Grassland, Plains Grassland, Madrean Evergreen Woodland; Cochise, Graham, Maricopa, Pima, Pinal, Santa Cruz cos.; New Mexico, Texas; Mexico; Central America.

Isely (1998, p. 860) noted that this species (as *Tephrosia tenella*) seemingly is "a peripheral manifestation of the bihemispheric *T. purpurea* complex." The type of *T. purpurea* (Linnaeus) Persoon is from Africa, and the species in the U.S. has been recorded only from chrome ore piles (originating from Rhodesia) in Baltimore, Maryland (in 1958, 1959, and 1969; MO) and in Newport News, Virginia (in 1959; MO).

Tephrosia purpurea and *T. tenella* are closely similar and in need of taxonomic study. Shreve and Wiggins (1964) identified the species in the Sonoran Desert as *Tephrosia purpurea* and observed that plants in the northern part of its range (*T. tenella*, the type from Sonora) seem to have narrower, more acute leaflets but intergrade completely with the more southern populations.

Tephrosia vicioides Schlechtendal as recently been suggested to be the correct name for these plants (McVaugh 1987). The type of *T. vicioides* is from Veracruz, Mexico. In contrast, Téllez (2001) used *T. tenella* as the name for plants from northern Mexico south to Panama, noting that *T. vicioides* is very similar and also a widely distributed American weed; he treated both species as occurring in Nicaragua, distinguishing *T. vicioides* by its greater number (11–25) of leaflets. In the limited observations here, plants with a consistently greater number of leaflets occur only in Veracruz and the Tehuacán region of Oaxaca.

Most commonly in *Tephrosia tenella* as identified here, the taproot becomes thick and distinctly perennial, and the plants often develop a woody, caudex-like region as well. Some plants, however, including those of the type collection, flower as annuals, especially in Cochise and Santa Cruz counties, Arizona, and northeastern Sonora. These plants arise from a slender taproot, with stems 8–30 cm and (1–)2–5 flowers per raceme. They apparently differ in ecology, characteristically occurring in oak or oak-pine woodland, but there appear to be no other morphological differences (e.g., in vestiture or flower morphology) besides duration, and annual and perennial individuals apparently co-occur. These plants need further taxonomic study; glabrous-styled taxa were not included in Wood's study of *Tephrosia* (1949). Nodulation is reported for *T. tenella* (Allen and Allen 1981).

3. Tephrosia thurberi (Rydberg) C.E. Wood. Rhodora 51:265. 1949. * [for George Thurber, 1821–1890, American botanist] THURBER'S HOARY-PEA Native

Cracca thurberi Rydberg in Britton et al., N. Amer. Fl. 24:165. 1923

Herbs. **Stems** erect, 30–60 cm, hirsute to hirsute-villous. **Leaves**: petioles 10–30(–40) mm. **Leaflets** (9–)11–27, elliptic to narrowly oblong, 15–30 × (5–)7–13 mm, 2.5–3.5(–4) times longer than wide, loosely strigose to sericeous on both surfaces, apices obtuse to rounded. **Inflorescences** terminal and axillary congested pseudoracemes 2–4 cm; floral bracts bristle-like, quickly deciduous. **Flowers** 8–14; corollas 15–18 mm, weakly bicolored with pale pink standard and pink-purple wings or all yellowish, aging pink-purple; stamens 9 united, 1 partially free; styles barbellate. **Fruits** 3–6 × 0.4 cm, villous to strigose-villous or strigose.

Flowering Jun–Sep. Rocky soil of ridges, slopes, roadsides; (1430–)1600–2000(–2130) m; Madrean Evergreen Woodland; Cochise, Pima, Santa Cruz cos.; Mexico (Sonora, Chihuahua, Durango, Sinaloa).

Tephrosia thurberi is very similar to the widespread *T. virginica* Bigelow of the eastern USA but far-disjunct and in a different biome. Plants may be locally abundant but the populations are widely scattered. With its attractive flowers, this species is suitable as a perennial in a summer wildflower garden or rock garden. Its horticultural characteristics and requirements are similar to those of *T. leiocarpa.*

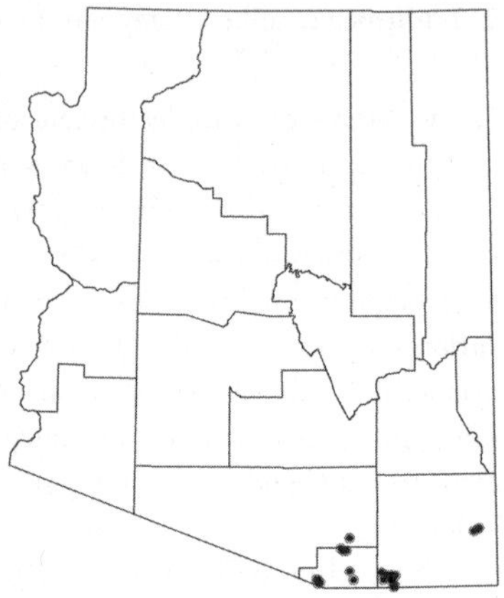

THERMOPSIS R. Brown in W.T. Aiton, Hortus Kew., ed. 2, 3:3. 1811. * [Greek *thermos*, lupine and *opsis*, outward appearance] FALSE LUPIN, GOLDEN PEA

Matthew B. Johnson and Martin Wojciechowski

Herbs, perennial, usually rhizomatous. **Stems** erect or ascending, few-branched to well-branched; branches ascending or spreading; glabrate to variously pubescent; unarmed. **Leaves** palmately 3-foliolate, to 6(–9) × to 3(–6) cm, smaller and narrower on branches; stipules usually persistent, ± leaf-like, dimorphic, most proximal clasping, dry-membraneous, and not blade-bearing, remainder broadly ovate to lanceolate and subtending leaves; stipels absent; subpetiolulate. **Leaflets** narrowly elliptic to obovate, 3–11 × 0.5–6.5 cm, margins entire. **Inflorescences** terminal (lateral) racemes, or commonly solitary, 5–90-flowered, scattered along axis or in whorls; bracts deciduous or persistent, widely ovate to lanceolate, apex acute to acuminate; bracteoles absent. **Pedicels** 2–17 mm. **Flowers** papilionoid; calyx bell-shaped, appearing 4-lobed, subequal, upper (adaxial) lobe double, broader, and slightly longer than others, apex truncate to emarginate, lower (abaxial) lobes deltoid to lanceolate; corolla yellow; banner very widely ovate, shorter than wings and keel, apex cleft (emarginate and short mucronate), claw wedge-shaped to oblong; wings not fused to keel, usually asymmetric, elliptic to oblong-rhombic, auriculate, ± length of keel, claw narrowly oblong; keel asymmetrically oblong-elliptic to keel-shaped, with auricles, posteriorly fused, without a beak, claw narrowly oblong; stamens 10, free, uniform; anthers attached dorsally; pistil short-stipitate; ovary oblong, usually longer than style, velutinous or densely appressed pubescence; ovules [5–]10–16; style glabrous; stigma minute. **Fruits** persistent, erect or divergent; tan to brown, narrowly elliptic, laterally compressed, straight, arched, or rarely ring-shaped, margins straight or, as a consequence of ovule abortion, often wavy and/or constricted between seeds, sometimes resembling loments; pericarp stiffly papery, not elastic, variously pubescent; stipe 2–4 mm; slowly dehiscent, valves separating from apex through both sutures. **Seeds** [1–]6–16 per fruit, tan to brown-black, oblong to reniform, often with a minute, membranous rim-aril. $x = 9$.

Species 23 (1 native in the flora): North America, Asia.

Thermopsis species are useful for erosion control due to their rhizomatous roots (Allen and Allen 1981, Lewis et al. 2005). These deep-rooted plants are drought-resistant (Allen and Allen 1981). In some situations, they are considered weedy. Allen and Allen (1981) report that *T. rhombifolia* (Nutt. ex Pursh) Richardson is toxic to cattle. *Thermopsis lupinoides* (L.) Link is used in folk medicine in the Ukraine as an expectorant (Allen and Allen 1981), and other species are also used in folk medicine. The flowers are showy and several species have ornamental potential. Nodulation has been documented for several members of the genus (Allen and Allen 1981).

SELECTED REFERENCE: Chen et al. 1994.

1. Thermopsis montana Nuttall ex Torrey & A. Gray. Fl. N. Amer. 1:388. 1840. * [of mountains] FALSE LUPIN, GOLDEN PEA, MOUNTAIN THERMOPSIS

Perennial herbs, 20–80 cm, glabrous to appressed pubescent or thinly villous. **Stems** slender, erect, solitary or clustered, moderately or few branched, arising from a woody rootstock or rhizomes; branches ascending at 20°–45°, weakly to moderately zigzag. **Leaves**: stipules persistent, broadly ovate or elliptic, 1.7–4.5 × 0.7–2.5 cm, base oblique to wedge-shaped, apex acuminate or acute; petioles 1–4 cm. **Leaflets** narrowly oblanceolate to elliptic, or obovate, 3.5–8 × 0.7–3 (–5) cm, apex acute or obtuse, sometimes short mucronate or emarginate, sparsely appressed pubescent to glabrous, lateral veins 6–11 pairs, conspicuously net-veined beneath or not. **Inflorescences** terminal racemes, 6–25-flowered, flowers in whorls of 2–3 or scattered, 4–25 cm; peduncles 2–8 cm; bracts semipersistent, narrowly elliptic to widely ovate, 6–10 × 2.5–6 mm. **Pedicels** 3.5–5 mm, villosulous. **Flowers** 1.6–2.2 cm; calyx 9–11 × 5–8 mm at the limb, lobes 3–5 mm, shorter than tube; wing and keel petals asymmetrically oblong-elliptic; ovary velutinous or densely appressed pubescent; ovules 10–16. **Fruits** ascending, straight, 4.5–6.5 × 0.4–0.6 cm, villosulous or appressed pubescent. **Seeds** 6–16, brown-black, oblong with a beak, 3.5–5 × 2.5–3 mm.

Varieties 2 (1 native in the flora): w United States.

1a. Thermopsis montana Nuttall ex Torrey & A. Gray var. **montana** Native

Thermopsis divaricarpa A. Nelson; *Thermopsis rhombifolia* (Nuttall ex Pursh) Richardson var. *montana* (Nuttall ex Torrey & A. Gray) Isely

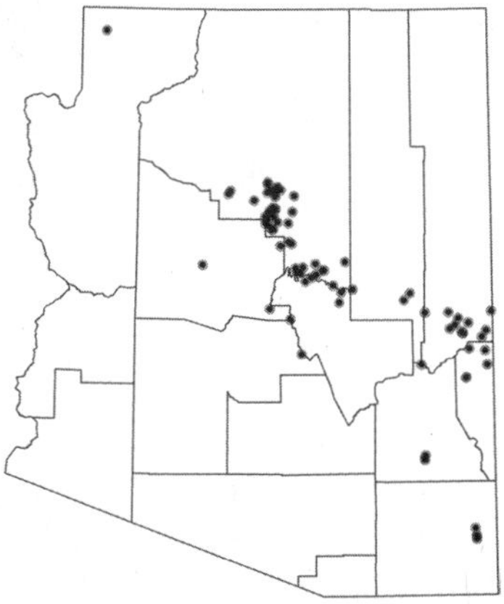

Plants 20–70 cm, appressed pubescent to thinly villous. **Stems** stiffly erect, solitary or clustered, few branched, arising from a woody rootstock or rhizomes; branches moderately zigzag. **Leaves**: petioles 1–2.5 cm. **Leaflets** elliptic to narrowly elliptic or obovate, 3.5–8 × 0.7–2.5 cm, appressed pubescent, not conspicuously net-veined beneath. **Inflorescences** 4–18 cm; bracts elliptic to widely ovate. **Flowers** 1.6–2 cm. **Fruits** villosulous. $2n = 18$.

Flowering May–Jun. Slopes, meadows, canyons, and along streams and roadsides in various soils; 1640–3050[–3200] m; Madrean Evergreen Woodland, Great Basin Conifer Woodland, Petran Montane Conifer Forest, Petran Subalpine Conifer Forest; Apache, Cochise, Coconino, Gila, Graham, Greenlee, Navajo, Yavapai cos.; Colorado, Idaho, Montana, Nevada, New Mexico, Oregon, Utah, Washington, Wyoming.

Thermopsis montana var. *montana* is widespread and highly variable but can usually be distinguished from var. *ovata* by its smaller habit, narrower leaflets and bracts, and greater tendency to spread by rhizomes. In southwestern Oregon the taxa appear to intergrade.

The strictly erect pods distinguish *Thermopsis montana* from *T. gracilis* and from the superficially similar *T. divaricarpa*.

Thermopsis montana var. *montana* is widespread in mountainous areas across the central and eastern portion of Arizona. Populations are scattered, though plants may be locally abundant due to the rhizomatous growth habit. The flowers of *T. montana* are showy and large flowering patches are spectacular. The plants are resistant to drought and trampling and are unpalatable to livestock and wildlife (Craighead et al. 1963). *Thermopsis montana* is documented to be toxic to cattle (Keeler et al. 1986) and humans (McGrath-Hill and Vicas 1997) though fatal poisoning in humans is not reported. This species, as *T. pinetorum* Greene, was reportedly used in traditional Navajo medicine as a fumigant for headache and sore eyes, to make a decoction used as cough medicine, and to make a lotion to provide protection against witches (Vestal 1952). Nodulation is documented for *T. montana* (Allen and Allen 1981).

TIPUANA (Bentham) Bentham. J. Proc. Linn. Soc., Bot. 4 (Suppl.):72. 1859 [1860]. * [for the Tipuani Valley, Bolivia]
Matthew B. Johnson

Machaerium sect. *Tipuana* Bentham, Hooker's J. Bot. Kew Gard. Misc. 5:267. 1853

Trees. Leaves alternate or sometimes subopposite; odd-pinnate; tardily deciduous; stipules minute, deciduous; stipels none. **Leaflets** alternate, opposite or subopposite, elliptic, obovate or oblong, the base rounded to cuneate, the apex emarginate. **Inflorescences** of terminal and axillary panicles; bracts and bracteoles small, deciduous. **Pedicellate. Flowers** papilionoid, bilaterally symmetric; calyx 5-lobed, the 2 upper (adaxial) lobes longer than the lower (abaxial) lobes; corolla with 5 petals, free, blades clawed, uppermost (banner) petal outermost, keel petals overlapping and joined at the base, enclosing the reproductive parts; androecium of 9 sterile staminodes and a single fertile stamen [or with 10 fertile stamens], 9 of the filaments united for two-thirds of their length into a tube surrounding the pistil and 1 filament free nearly to the base; ovary 1-chambered, style and stigma simple. **Fruits** winged, samara-like, stipitate, indehiscent. **Seeds** 1–3. $x = 10$.

Species 1 (cultivated in flora): Argentina, Bolivia, Brazil.

Tipuana is closely related to *Pterocarpus* (Polhill 1981).

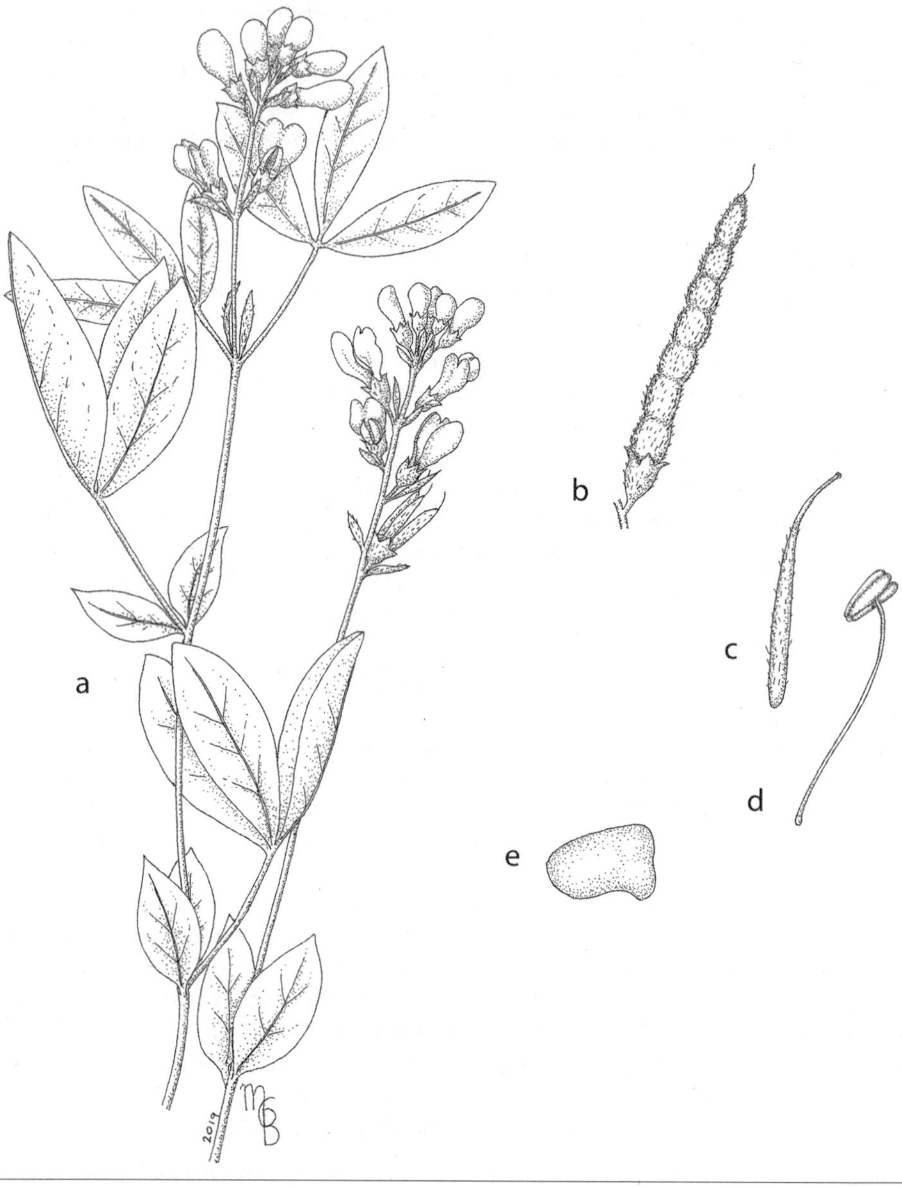

Thermopsis montana var. *montana*: **a.** habit; **b.** fruit; **c.** pistil; **d.** anther; **e.** seed. Illustrator: Marceline VandeWater.

1. Tipuana tipu (Bentham) Kuntze. Revis. Gen. Pl. 3(3):72. 1898. * [*tipu* is a local name for the tree in South America] PRIDE OF BOLIVIA, RACEHORSE TREE, ROSEWOOD; *TIPU BLANCA, TIPU* Cultivated

Machaerium tipu Bentham, Hooker's J. Bot. Kew Gard. Misc. 5:267. 1853

Trees to 12 [35] m. **Stems** with bark of limbs and trunk dark gray and rough with longitudinal fissures; wood whitish yellow; young stems green, with short, fine white pubescence, becoming glabrous. **Leaves** 11.5–33 cm; the petiole, rachis, pedicels, and midribs with short, fine white pubescence; petiole 1.6–3.4 cm; rachis 7–24.5 cm; petiolules 1.5–2 mm, the blades with sparse, white pubescence on the lower (abaxial) surface. **Leaflets** (9–)11–17(–25) per leaf, 2.5–5.4 × 0.8–1.6 cm. **Inflorescences** 5–11 cm, often 6–15 flowered. **Pedicels** 8–10 mm. **Flowers**: calyx pale green, 9–11 mm, with scattered, minute reddish hairs, lobes 1.5–3 mm; corolla yellow, banner 16–24 mm, reflexed 90° with the apex emarginate, the lower central portion of the banner with an elongated pale brown 'corona' around a pale yellow center, wings 15–20 mm, keel 10–13 mm; staminal tube 10–12 mm, filaments pale yellow, distal (free) portions incurved,

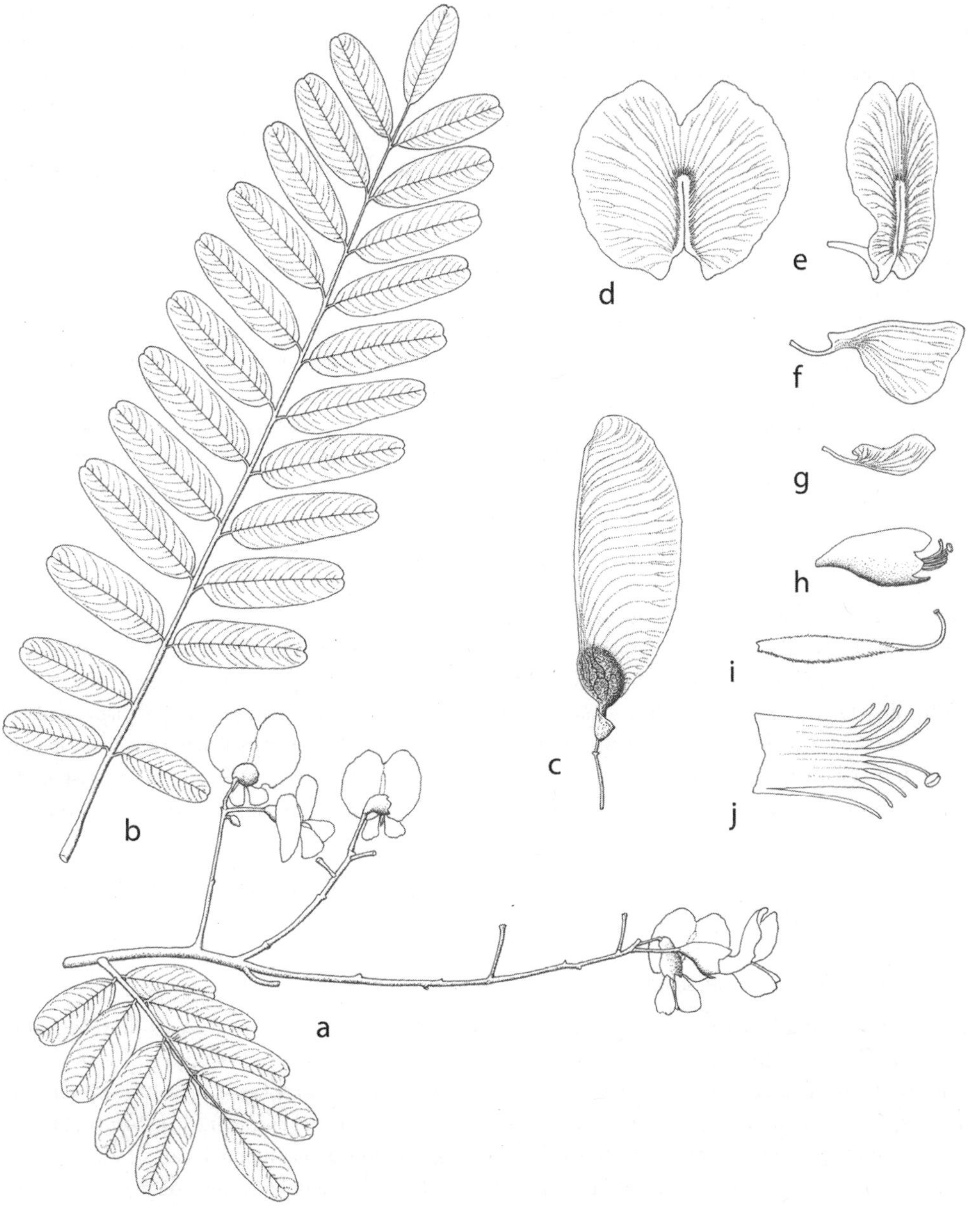

Tipuana tipu: **a.** inflorescence; **b.** leaf; **c.** fruit; **d., e., f., g.** petals; **h.** calyx, **i.** pistil; **j.** stamens. Illustrator: Matthew B. Johnson.

anther pale yellow; pistil pale green, 11–13 mm, ovary densely covered with fine, appressed white pubescence, style glabrous and incurved. **Fruits** orange-brown or straw colored when ripe, 6.7–9 × 1.9–2.8 cm, stipe 4–11 mm. **Seeds** tan, ± elliptical, often with transverse wrinkles or ridges, 8–12 × 4–6 mm.

Flowering Apr–May(–Jun). Argentina, Bolivia, Brazil.

Tipuana tipu is found in humid, forested locations along the mountains in northwestern Argentina and Bolivia to Brazil. In Argentina, it is a component of the Selva Tucumano Boliviana, growing at elevations of 1200–1500 m (Leonardis 1975). The trunks and limbs are frequently covered with epiphytes in habitat (Burkart 1952). *Tipuana tipu* is a large tree in habitat with an open canopy and a trunk diameter of up to 1.75 m; however, cultivated trees outside of its native range seldom grow more than 10 m tall (Allen and Allen 1981). Flowering often occurs in spring before or as the new leaves appear. Published descriptions of *T. tipu* indicate the flowers have 10 stamens (Burkart 1952, Allen and Allen 1981, Killeen et al. 1993); however, from the limited material available from cultivated trees in Arizona, there are 9 staminodes and a single fertile stamen. With this exception, the Arizona material matches the published descriptions and also an illustration found in Burkart (1952). Burkart (1952) mentions that cultivated trees in South America frequently "rain" exudate in the spring from the larvae of an insect (*Cephisus siccifolius*) that feeds upon the tree. This phenomenon, however, has not been observed on cultivated trees in Arizona. Nodulation has been observed on *T. tipu* (Allen and Allen 1981).

The wood, called "tipa" in Argentina, is highly valued. It is resistant to splitting and shrinking as it dries (Leonardis 1975). The wood is used in cabinetry, turnery, for fine furniture, tool handles, and sports equipment, as well as for fuelwood and charcoal (Allen and Allen 1981, Burkart 1952, Lewis et al. 2005). It is also used to make carts and was previously used in automobile and railroad coach interiors (Burkart 1952). The wood has a density of 640–750 kg/m^3 (Record and Hess 1943). In addition to being planted for shade and as an ornamental, *Tipuana tipu* is used in reforestation and for windbreaks and erosion control (Lewis et al. 2005). A red resin obtained from the trunks is used medicinally to treat uterine problems (Burkart 1952) and has been used in tanning and dyeing (Allen and Allen 1981). Leaves and unripe fruits contain saponins and peroxidase (Burkart 1952). The young leaves are desirable forage (Burkart 1952).

Tipu is widely planted along streets and in parks and plazas in northern Argentina and elsewhere in South America and is also cultivated in southern Europe, the Middle East, Australia, and the United States. It is infrequently grown in Arizona. Though native to relatively moist habitats, it has proven remarkably well-adapted to hot desert climates in Arizona. Several trees planted in the late 1920s or 1930s at Boyce Thompson Arboretum, Superior, AZ, continue to thrive and the species grows well in Yuma. This species is surprisingly drought resistant when established though it requires some supplemental irrigation in Arizona. The trees at Boyce Thompson have not received direct irrigation for many years, but receive some indirect irrigation from nearby plants. With adequate irrigation *tipu* can grow rapidly. Mature plants can be transplanted and regrowth following pruning is rapid (Burkart 1952). *Tipu* is cold tolerant to at least -7°C. The canopy provides welcome shade from the sun and the trees are quite showy when in flower. It is tardily deciduous in late winter. The leaves and fruits can be a litter issue. The roots have the potential to heave walkways and invade sewer lines in some situations. In 2008, the *tipu* psyllid (*Platycorypha nigrivirga*) was found on *Tipuana tipu* in southern California with heavy infestations causing extensive damage to the foliage (Nisson n.d.). In eastern Australia, *tipu* has escaped from cultivation and become established in native habitats. Plants may be propagated from seeds that have been removed from the fruit.

TRIFOLIUM Linnaeus. Sp. Pl. 2:764. 1753. * [Latin *tres*, three, and *folium*, leaf, alluding to three leaflets characteristic of many species] CLOVER; *TRÈFLE*

Michael A. Vincent

Amoria C. Presl; *Chrysaspis* Desvaux; *Lupinaster* Fabricius

Herbs, annual or perennial, unarmed. **Stems** prostrate to erect, glabrous or pubescent. **Leaves** alternate, odd-pinnate or palmate; stipules present, persistent, fused to petiole; petiolate or sessile. **Leaflets** 3–9, stipels absent, blade margins usually entire or toothed, rarely lobed, surfaces glabrous or pubescent. **Inflorescences** (1–)25–100+ flowered, racemes, umbelliform, spike-like, or head-like, axillary or terminal; peduncles present or absent; bracts present or absent, united or distinct; bracteoles present or absent. **Pedicels** present or absent. **Flowers** papilionoid; calyx radially or bilaterally symmetrical, tubular or campanulate, sometimes inflated in fruit, lobes 5; corolla white, cream, pink, red, magenta, purple, or yellow, glabrous; stamens 10, 9 fused and 1 free; anthers basifixed; pistil linear to lanceolate, style glabrous or pubescent, stigma terminal. **Fruits** sessile or stipitate, cylindric or laterally compressed, lanceolate to elliptic, dehiscent usually longitudinally, rarely transversely, or indehiscent, membranous or leathery, glabrous or pubescent. **Seeds** 1–4[–9], globose, ellipsoid, ovoid-oblong, mitten-shaped, or reniform, hilum lateral. $x = 5, 6, 7, 8$.

Species ca. 280 (18 in the flora, 13 native, 2 naturalized, 3 adventive): North America, Mexico, Central America, South America, Eurasia, Africa; introduced in West Indies, Pacific Islands, Australia.

In North America, *Trifolium* is superficially similar to *Medicago*, *Melilotus*, and *Trigonella*. *Trifolium* differs from those genera by persistence of the corolla after anthesis, a legume that is included in or only slightly exserted from the persistent corolla, and by distally dilated filaments (Gleason and Cronquist 1991, Wei and Vincent 2010).

Trifolium has three major centers of diversity: the Mediterranean region of southern Europe, western Asia, and northern Africa; the California region in North America; and the eastern African highlands (Zohary and Heller 1984). In the New World, diversity is highest in the California region of the United States, the mountainous regions of western North America, and western South America. In North America, a suite of closely related species, centered in the southeastern United States, represents another locus of diversification.

Trifolium consists of subgenus *Chronosemium* (Seringe) Reichenbach and subg. *Trifolium*; the latter is subdivided into eight sections (Ellison et al. 2006). Subgenus *Chronosemium* is represented in North America by the Old World species *T. aureum*, *T. campestre*, and *T. dubium*. All native New World species of subg. *Trifolium* represent a monophyletic lineage, section *Involucrarium* Hooker. The remaining introduced species of subg. *Trifolium* represent other sections of subg. *Trifolium*: sect. *Lupinaster* (Fabricius) Seringe (*T. lupinaster*), sect. *Trichocephalum* W. D. J. Koch (*T. subterraneum*), sect. *Trifoliastrum* Gray (*T. cernuum*, *T. glomeratum*, *T. nigrescens*, *T. repens*, *T. retusum*, and *T. suffocatum*), sect. *Trifolium* (*T. alexandrinum*, *T. angustifolium*, *T. arvense*, *T. echinatum*, *T. hirtum*, *T. incarnatum*, *T. lappaceum*, *T. medium*, *T. pratense*, and *T. striatum*), and sect. *Vesicastrum* Seringe (*T. fragiferum*, *T. hybridum*, *T. ornithopodioides*, *T. resupinatum*, *T. tomentosum*, and *T. vesiculosum*). This differs dramatically from the phylogenetic arrangement of Zohary and Heller (1984), where the genus is subdivided into eight sections and native New World *Trifolium* species are placed in sects. *Involucrarium* and *Lotoidea* Crantz (in a narrow sense).

Species of *Trifolium* have been cultivated for centuries in parts of Europe as forage crops; some species were introduced into North America for that purpose. Of these, *T. repens* appears to have been the earliest introduction, perhaps as early as the mid-1600s (Mack 2003), and its spread across the continent was very rapid. Other commonly cultivated species include *T. hirtum*, *T. hybridum*, *T. incarnatum*, *T. pratense*, *T. subterraneum*, and *T. vesiculosum*; other species may also be cultivated in North America and may be important regionally (Taylor 1985, Gillett and Taylor 2001). *Trifolium fragiferum*, *T. pratense*, and *T. repens* have been grown on a limited basis as lawns in Arizona. Nodulation is widely reported for *Trifolium*.

SELECTED REFERENCES: Ellison et al. 2006, Gillett 1965, 1969, 1971, 1972, 1980, Gillett and Taylor 2001, McDermott 1910, Zohary and Heller 1984.

1. Inflorescences subtended by 2–3 broad membranous bract-like stipules (pseudoinvolucre), involucre of bracts absent.
   2. Stems growing in dense tufts from a woody caudex, usually forming a mat; leaflet blades oblanceolate, often folded, 0.4–1.5 × 0.2–0.4 cm; calyx 6–9 mm, lobes subequal, subulate, orifice open; corolla 10–15 mm, banner oblong, 9–13 × 3–4 mm, apex rounded or truncate ____ **2. Trifolium andinum**
   2′ Stems not in dense tufts from a woody caudex, not forming a mat; leaflet blades elliptic to obovate, 1.5–5 × 0.7–1.5 cm; calyx 8–11 mm, lobes upper (adaxial) equaling tube, lateral and lower (abaxial) 2 times tube, orifice hairy, slightly closed; corolla 13–18 mm, banner elliptic-ovate, 10–13 × 3–5 mm, apex narrowly rounded ____ **13. Trifolium pratense**
1′ Inflorescences not subtended by pseudoinvolucre, involucre of bracts present or absent.
   3. Inflorescences subtended by a noticeable involucre of bracts.
      4. Calyces becoming inflated, markedly asymmetric-2-lipped, with veins connected by reticulating lateral veins ____ **4. Trifolium fragiferum**
      4′ Calyces not inflated or markedly asymmetric-2-lipped, with veins not connected by reticulating lateral veins.
         5. Plants annual.
            6. Plants densely or rarely sparsely pubescent; involucres flattened or bowl-shaped, when compressed laterally, nearly completely hiding calyces, lobes divided 1/2 their length, lanceolate-ovate, entire or slightly toothed proximally, acuminate ____ **10. Trifolium microcephalum**
            6′ Plants glabrous or sparsely pubescent; involucres rotate, bracts sometimes separate, or involucres flattened to bowl- or vase-shaped, when compressed laterally, not hiding flowers except proximally, lobes more deeply divided.
               7. Calyx tubes not slit between upper (adaxial) lobes, lobes longer than tube, usually entire and without shoulders, rarely with 1–3 lateral teeth; leaflet blades obovate to oblanceolate, obcordate, or elliptic; banner petals 3–10 mm ____ **16. Trifolium variegatum**
               7′ Calyx tubes slit between upper (adaxial) lobes, lobes shorter than or equal to tube, sometimes distally cleft with three teeth or shouldered below subulate apex; leaflet blades linear, lanceolate, or elliptic; banner petals 12–20 mm ____ **17. Trifolium willdenovii**
         5′ Plants perennial.
            8. Lower (abaxial) calyx lobe rarely longer than tube; involucre with lobes sharply and acutely serrate; leaflet blades obovate to elliptic ____ **18. Trifolium wormskioldii**
            8′ Lower (abaxial) calyx lobe longer than tube; involucre with lobes entire, bifurcate, or triaristate to lacerate; leaflet blades oblanceolate, obovate, linear, or elliptic.
               9. Leaflets narrowly linear, elliptic, or obovate; involucral lobes triaristate to lacerate, or irregular to bifurcate; stipule margins toothed or lacerate; roots taproots, slender, rhizomes short or absent ____ **11. Trifolium mucronatum**
               9′ Leaflets oblanceolate or obovate; involucral lobes linear-lanceolate, entire; stipule margins entire; roots stout, branched, rhizomes absent ____ **12. Trifolium pinetorum**
   3′ Involucre reduced or absent.
      10. Plants annual.
         11. Leaflets pubescent; inflorescences dense and head-like; calyces densely pubescent; flowers not evidently reflexed after anthesis; corollas white with purple tips ____ **1. Trifolium albopurpureum**
         11′ Leaflets glabrous or sparsely pubescent; inflorescences umbels or racemes, congested but not dense and head-like; calyces glabrous; flowers reflexed after anthesis; corollas white, pink, purple, or yellow.
            12. Calyx teeth nearly same length as tube or shorter; corollas yellow ____ **3. Trifolium dubium**
            12′ Calyx teeth much longer than tube; corollas white, pink, or purple ____ **5. Trifolium gracilentum**
      10′ Plants perennial, sometimes short-lived.
         13. Involucre present but reduced.

14. Plants from a branched, woody crown; leaflets elliptic or obovate, pubescent at least below; calyces strigose; pedicels erect, or rarely those of proximal flowers reflexed; involucres narrow, membranous, dentate rims, 0.5 mm ____ **6. Trifolium gymnocarpon**
14′ Plants with evident prostrate stems; leaflets obovate, glabrous or nearly so; calyces pubescent; pedicels reflexed in fruit; involucre a narrow rim, or of bracts to 0.5 mm ____ **15. Trifolium sonorense**
13′ Involucre absent.
15. Calyx teeth hairy, 2–3 times as long as tube; peduncles pubescent, especially toward the apex; leaflet blades linear, lanceolate to elliptic, oblong, or ovate ____ **8. Trifolium longipes**
15′ Calyx teeth glabrous or nearly so, less than twice as long as tube; peduncles glabrous or nearly so; leaflet blades obovate, ovate, or lanceolate, rarely rhombic.
16. Stems creeping, rooting at nodes; leaflets obovate, apices emarginate; stipules lanceolate, apices short-subulate ____ **14. Trifolium repens**
16′ Stems erect or ascending, not rooting at nodes; leaflets ovate, obovate, lanceolate, or rhombic, apices broadly acute, acuminate, obtuse, or emarginate; stipules obovate, lanceolate, lanceolate-elliptic, or triangular-ovate, apices acute, acuminate, or long-acuminate.
17. Inflorescence rachises not surpassing flowers; banners ovate-oblong, apex rounded-denticulate ____ **7. Trifolium hybridum**
17′ Inflorescence rachises sometimes surpassing flowers; banners elliptic, obovate, ovate-oblong, or lanceolate-ovate, apex not denticulate ____ **9. Trifolium macilentum**

1. Trifolium albopurpureum Torrey & A. Gray. Fl. N. Amer. 1:313. 1838. * [from Latin, *alba*, white, and *purpure*, purple] RANCHERIA CLOVER Native

Trifolium albopurpureum Torrey & A. Gray var. *neolagopus* (Lojacono) McDermott; *Trifolium helleri* P.B. Kennedy; *Trifolium insularum* P.B. Kennedy; *Trifolium macraei* Hooker & Arnott var. *albopurpureum* (Torrey & A. Gray) Greene; *Trifolium neolagopus* Lojacono; *Trifolium pseudoalbopurpureum* P.B. Kennedy

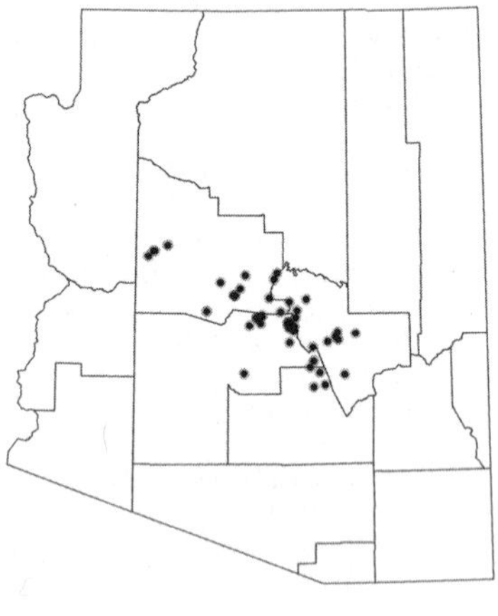

Plants annual, erect, branched from base, 5–40 cm, pubescent. **Leaves** palmate; stipules elliptic-ovate, 0.6–1 cm, margins entire, toothed, or remotely lobed, ciliate, apices acuminate; petiole 0.5–7 cm; petiolules 1 mm. **Leaflets** 3, blades obovate or elliptic, 0.5–2 × 0.2–1 cm, base cuneate, veins obscure or slightly thickened, margins serrate, apex acute or obtuse, often retuse, surfaces pubescent. **Inflorescences** axillary or terminal, dense and head-like, 25–50-flowered, ovoid-ellipsoid, 1–1.8 × 0.8–1.5 cm; peduncles 0.5–10 cm; involucres absent; bracteoles absent. **Pedicels** absent. **Flowers** 6–9 mm, not evidently reflexed after anthesis; calyx campanulate, 6–8 mm, densely pubescent, veins 10 (additional faint veins sometimes present), tube 1–2 mm, lobes unequal, narrowly triangular, subulate, sinuses rounded, lobes divergent, orifice open; corolla white with purple tips, 4–6 mm, banner narrowly elliptic, 6–7 × 2 mm, apex acute, slightly erose. **Fruits** broadly ellipsoid, 2.8–3.2 mm. **Seeds** 1 (or 2), yellow or brown, often red-mottled, ellipsoid, 2–2.6 mm, surface smooth, glossy. $2n = 16$.

Flowering Apr–May. Grassy foothills and valleys, streambanks, and roadsides; [0–]690–1600[–2100] m; Sonoran Desertscrub, Semidesert Grassland, Interior Chaparral, Madrean Evergreen Woodland; Gila, Maricopa, Pinal, Yavapai cos.; California, Oregon, Washington; Mexico (Baja California).

Trifolium albopurpureum is a widespread member of a closely allied group of species, including *T. amoenum*, *T. dichotomum*, and *T. olivaceum*, which intergrade to varying degrees and may co-occur in some populations. The great similarity among these species has led to a maddening array of synonyms as new suites of species have been recognized and various new combinations have been made. In spite of the resultant confusion, these species are relatively distinct and are differentiated on the following characters: inflorescences generally greater than 3 cm diam. (*T. amoenum*) versus less than 2.5 cm diam. in the others; corollas usually 8–12 mm with calyx lobes 2–3 times the tube (*T. dichotomum*), versus corollas usually 5–8.5 mm with calyx lobes 3–6 times the tube; calyces 8–12+ mm with lobes exceeding the corollas by 3–6 mm (*T. olivaceum*), versus calyces 4–9 mm with lobes equaling the corollas or exceeding them by 1.5–2 mm (Isely 1998, Vincent and Isely 2012).

In Arizona, *Trifolium albopurpureum* is locally common in moist sites in the central region of the state.

2. Trifolium andinum Nuttall in J. Torrey and A. Gray, Fl. N. Amer. 1:314. 1838. * [of the Andes Mountains (though not native to the Andes)] INTERMOUNTAIN CLOVER Native

Trifolium andinum Nuttall var. *canone* S.L. Welsh & N.D. Atwood; *Trifolium andinum* Nuttall var. *navajoense* S.L. Welsh & N.D. Atwood; *Trifolium andinum* Nuttall var. *podocephalum* Barneby; *Trifolium andinum* Nuttall var. *wahwahense* S.L. Welsh & N.D. Atwood

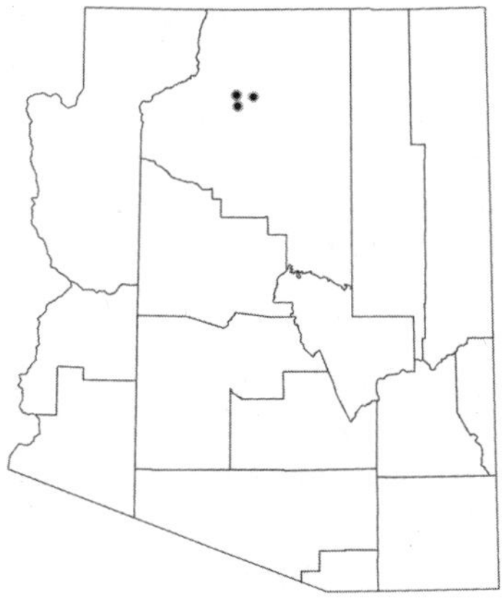

Herbs perennial, 5–15 cm, pubescent. **Stems** erect or ascending, growing in dense tufts, much-branched, often forming a mat, caudex woody. **Leaves** palmate; stipules oblanceolate, 0.5–1.4 cm, margins entire, apices acute-acuminate; petiole 0.5–4 cm; petiolules to 0.1 mm. **Leaflets** 3, blades oblanceolate, often folded, 0.4–1.5 × 0.2–0.4 cm, base cuneate, veins obscure, margins subentire to denticulate distally, apex acute, mucronulate, surfaces villous. **Inflorescences** terminal, usually formed of 2–3 nearly sessile heads, 15–25-flowered, globose, 0.8–2 × 0.6–1.5 cm; peduncles 0.1–1 cm, pseudopeduncle below inflorescences, 0.5–6 cm; involucres absent, distal stipules and leaves forming pseudoinvolucre; bracteoles broadly ovate, truncate, membranous, to 0.5 mm. **Pedicels** straight, (0–)1–2 mm. **Flowers** 10–15 mm; calyx tubular-campanulate, 6–9 mm, rough-hairy or glabrous, veins 10, tube 3–6 mm, lobes subequal, subulate, orifice open; corolla light purple, 10–15 mm, banner oblong, 9–13 × 3–4 mm, apex rounded or truncate; ovaries pubescent distally. **Fruits** ellipsoid, 4–5 mm. **Seeds** 1 or 2, brown, ovoid, 1.5–2 mm, smooth.

Flowering May–Jun. Shale or clay bluffs and hilltops, crevices of volcanic or limestone rock, pinyon-juniper belt; [1600–]2310–2680 m; Great Basin Conifer Woodland, Petran Montane Conifer Forest; Coconino Co.; Colorado, Nevada, New Mexico, Utah, Wyoming.

In Arizona, *Trifolium andinum* is localized on the Kaibab Plateau along the north rim of the Grand Canyon.

3. Trifolium dubium Sibthorp. Fl. Oxon. 231. 1794. * [from Latin, *dubi*, doubtful] LITTLE HOP CLOVER Adventive

Chrysaspis dubia (Sibthorp) Greene; *Trifolium minus* Smith

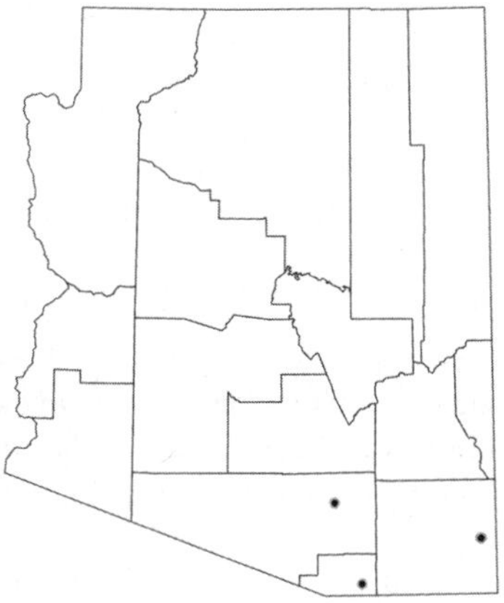

Plants annual, erect to prostrate, branched from base, 20–40 cm, glabrous or sparingly hairy. **Leaves** pinnate; stipules ovate, 0.3–0.5 cm, margins entire, apices acute; petiole to 1.5 cm; petiolules to 0.5 mm in lateral leaflets, 1–1.5 mm in terminal leaflets. **Leaflets** 3, obovate to elliptic-lanceolate, 1–2 × 0.5–1.5 cm, base cuneate, lateral veins prominent, ± parallel, ascending, margins dentate distally, apex rounded or retuse, surfaces glabrous or with scattered hairs. **Inflorescences** axillary or terminal, racemose, 5–20-flowered, ovoid or globose, 0.5–0.9 × 0.6 cm; peduncles 1–1.5 cm; involucres absent; bracteoles reduced to fringe of red bristles. **Pedicels** 0.2–0.5 mm, reflexed after anthesis. **Flowers** 2.8–3.5 mm; calyx campanulate, 1.5–2 mm, glabrous, veins 5, tube 0.8–1 mm, lobes unequal, nearly same length as or shorter than tube, upper (adaxial) 2 shorter, orifice open; corolla pale yellow becoming brown, 2.6–3.2 mm, not or slightly ribbed, banner persistent, spatulate, 2.6–3.2 × 2 mm, apex rounded to broadly acute. **Fruits** stipitate, ellipsoidal, 1.5–2 mm, short beaked. **Seeds** 1, yellow or pale brown, ellipsoid, 0.9–1 mm, surfaces smooth, glossy. $2n = 16, 28, 32$.

Flowering May–Oct. Disturbed ground, fields; [0–]800–1700[–1800] m; Plains Grassland, Madrean Evergreen Woodland; Cochise, Pima, Santa Cruz cos.; widely introduced and often naturalized across much of the US and parts of Canada; introduced also in s South America, Asia (e China), n, s Africa, Pacific Islands (Hawaii, New Zealand), Australia; native to central and southern Europe and western Asia.

Trifolium dubium is often confused with *Medicago lupulina* Linnaeus; the latter may be distinguished by its toothed stipules, deciduous corollas, and shiny, black fruits. Little hop clover may be the "shamrock" of Irish folk lore (Nelson 1991, Wyse Jackson 2014); other candidates include other species of *Trifolium* or species of *Medicago* or *Oxalis*.

Trifolium dubium is weakly adventive in the state with records from the University of Arizona campus in Tucson, Sharp Springs Cienega in Santa Cruz County, and Cave Creek in the Chiricahua Mountains, Cochise County.

4. Trifolium fragiferum Linnaeus. Sp. Pl. 2:772. 1753. * [from Latin, *frag*, strawberry] STRAWBERRY CLOVER Adventive

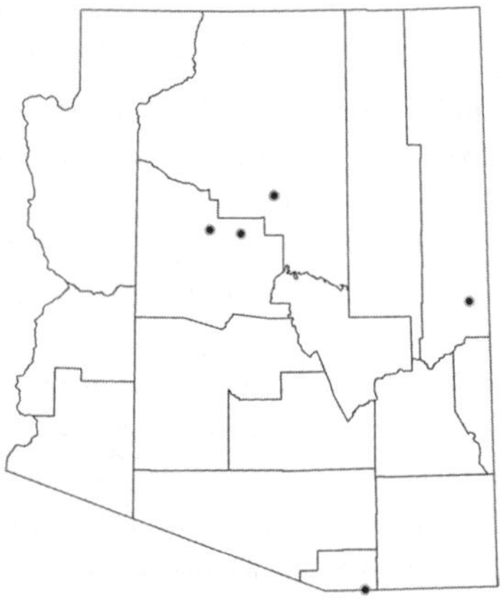

Plants perennial, prostrate or creeping, branched, 5–50 cm, pubescent or glabrous, rooting at nodes. **Leaves** palmate; stipules linear or lanceolate, dilated proximally, 1.3–2 cm, margins entire, apices subulate or acuminate; petiole 1–9 cm; petiolules 1 mm. **Leaflets** 3, blades ovate, obovate, or elliptic, 0.5–3 × 0.3–1.5 cm, base cuneate, veins prominent, thickened, curved, margins spinulose-denticulate, apex obtuse, often retuse, surfaces hairy below (abaxially), glabrous above (adaxially). **Inflorescences** axillary, erect or ascending, 10–30-flowered, globose, 0.8–2 × 0.8–2 cm; peduncles 3–17 cm; involucres cup-shaped, lanceolate-oblong bracts, 3–6 mm, distinct or united proximally; bracteoles linear or lanceolate, 2–3 mm, acuminate or distally cleft with two teeth. **Pedicels** straight, to 0.5 mm. **Flowers** 6–8 mm; calyx tubular, inflated in fruit, markedly asymmetric-2-lipped, 2.5–7 mm, usually pilose to woolly, sometimes glabrescent, veins 5–10, connected by reticulating lateral veins, tube 2–7 mm, lobes unequal, upper (adaxial) spreading in fruit, shorter than tube, lower (abaxial) equal to or longer than tube, straight, very unequal in fruit, orifice open; corolla white to pink, 5–8 mm, banner oblong-lancelate, 5–8 × 2 mm, apex emarginate to crenulate. **Fruits** ovoid, 2.5–3 mm, long-beaked. **Seeds** 1 or 2, tan or brown-spotted, globose-reniform, 1–1.4 mm, surfaces smooth, dull. $2n = 16$.

Flowering Apr–Nov. Meadows, fields, roadsides, sandy and saline soil; [0–]1290–2134 m; Semidesert Grassland, Petran Montane Conifer Forest; Apache, Coconino, Santa Cruz, Yavapai cos.; introduced and widely adventive or naturalized in the U.S. and in Canada (British Columbia), introduced also in South America, Pacific Islands (New Zealand), Australia; native to southern and central Europe and western Asia.

Trifolium fragiferum is cultivated for pastures and in mixtures with grass for hay and silage and, to a lesser extent, as groundcover in orchards and vineyards and as a green manure cover crop (St. John et al. 2010). In North America, it appears to have been introduced accidentally in the 1870s and intentionally as a crop around 1900 (Hollowell 1939). In Arizona this species is weakly adventive at a very few localities.

5. Trifolium gracilentum Torrey & A. Gray. Fl. N. Amer. 1:316. 1838. * [from Latin, *gracil*, slender] PINPOINT CLOVER Native

Trifolium denudatum Nuttall; *Trifolium exile* Greene; *Trifolium gracilentum* Torrey & A. Gray var. *exile* (Greene) P.B. Kennedy; *Trifolium gracilentum* Torrey & A. Gray var. *inconspicuum* Fernald; *Trifolium gracilentum* Torrey & A. Gray var. *reductum* Parish; *Trifolium inconspicuum* (Fernald) A. Heller

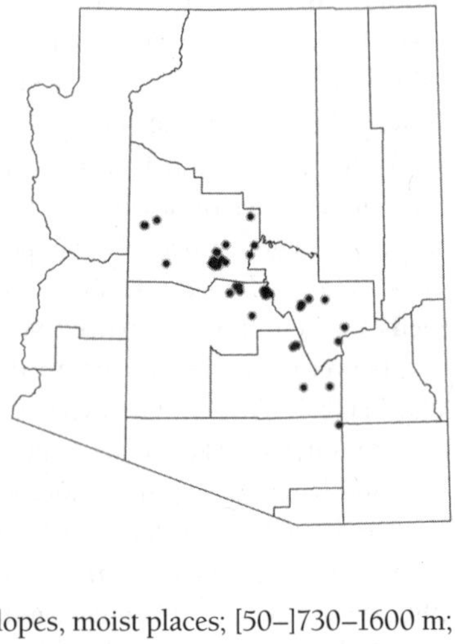

Plants annual, erect or ascending, branched, 2–60 cm, glabrous. **Leaves** palmate; stipules ovate-lanceolate, 0.5–2.5 cm, membranous, margins entire or slightly serrate, apices long-acuminate; petiole 0.5–9 cm; petiolules to 0.5 mm. **Leaflets** 3, blades obovate to obcordate, 0.5–2.5 × 0.2–1.5 cm, base cuneate, veins moderately thickened, margins setose-serrulate, apex rounded, shallowly retuse, surfaces glabrous. **Inflorescences** axillary or terminal, 10–25-flowered, globose becoming subglobose, 0.5–2 × 0.5–2 cm, rachis prolonged beyond flowers; peduncles 1–9 cm; involucres reduced to a narrow, membranous, dentate rim, to 0.5 mm; bracteoles low, cup-like, membranous, to 0.5 mm. **Pedicels** becoming dramatically reflexed, 3–4 mm. **Flowers** 5–7.5 mm; calyx narrowly campanulate, 4.5–6.5 mm, glabrous, veins 10, tube 2–2.5 mm, lobes unequal, long-triangular, margins green or purple, orifice open; corolla white, pink, or purple, 5–8 mm, banner ovate, 5–8 × 3–4 mm, apex broadly rounded, retuse or apiculate. **Fruits** ovoid-ellipsoid, 4–6 mm. **Seeds** 1 or 2, tan to brown, mitten-shaped, 1.1–1.5 mm, surface smooth. $2n = 16$.

Flowering Mar–Jun. Open, grassy areas, gravelly ridges, roadsides, adobe slopes, moist places; [50–]730–1600 m; Sonoran Desertscrub, Semidesert Grassland, Interior Chaparral, Great Basin Conifer Woodland; Gila, Maricopa, Pima, Pinal, Yavapai cos.; California, Nevada, Oregon, South Carolina, Washington; Mexico (Baja California).

Trifolium gracilentum is locally common in some areas of central Arizona. Abundance of *T. gracilentum* increases in response to burns (DiTomaso et al. 1999).

6. Trifolium gymnocarpon Nuttall in J. Torrey and A. Gray, Fl. N. Amer. 1:320. 1838. * [from Greek, *gymno*, naked, and *carpos*, fruit] NUTTALL'S CLOVER Native

Trifolium gymnocarpon Nuttall var. *plummerae* (S. Watson) J.S. Martin; *Trifolium nemorale* Greene; *Trifolium plummerae* S. Watson; *Trifolium subcaulescens* A. Gray

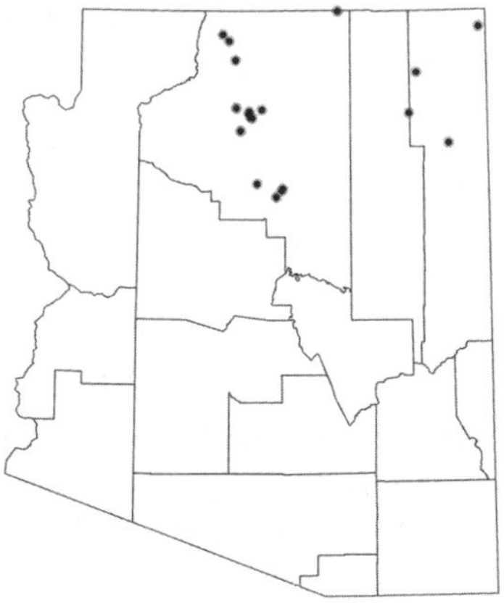

Plants perennial, growing in dense tufts, numerous short stems, branched from woody crown, 2–15 cm, pubescent. **Leaves** palmate; stipules ovate to lanceolate, 0.5–1.5 cm, margins entire or irregularly dentate, apices acute to acuminate; petiole 1.5–10 cm; petiolules to 0.5 mm. **Leaflets** 3–5, blades ovate, obovate, oblong, or elliptic, 0.8–3 × 0.2–1.8 cm, base cuneate, veins thickened distally, margins spinose-dentate, apex rounded or acute, surfaces pubescent, sometimes glabrous above (adaxially). **Inflorescences** terminal, 6–15-flowered, subglobose, 1–2 × 1.3–2.4 cm; peduncles 1–6.5 cm; involucres narrow, membranous, dentate rims, 0.5 mm; bracteoles ovate-triangular, to 0.8 mm. **Pedicels** erect, those of proximal flowers sometimes reflexed, 1–4 mm. **Flowers** 7.5–13 mm; calyx tubular-campanulate, 4–7.5 mm, strigose, veins 10, tube 2–3 mm, lobes subequal, narrowly triangular, orifice open; corolla pink-purple, often with whitish tips, sometimes nearly entirely whitish, 7–12 mm, banner oblong, 7–14 × 5–7 mm, apex rounded, usually retuse. **Fruits** ovoid, 4–5 mm. **Seeds** 1, tan to brown, mitten-shaped, 3–4.5 mm, surface roughened.

Flowering Apr–Jun. Rocky slopes, clay or gumbo soil on plains; [1500–]1950–2610[–3200] m; Great Basin Conifer Woodland, Petran Montane Conifer Forest; Apache, Coconino, Navajo cos.; California, Colorado, Idaho, Montana, New Mexico, Nevada, Oregon, Utah, Wyoming.

Two loosely differentiated varieties or subspecies of *Trifolium gymnocarpon* have sometimes been recognized: var. *plummerae* with 3–5 leaflets with upper (adaxial) surface pilose, and var. *gymnocarpon* with 3 leaflets with upper (adaxial) surface glabrous (Gillett 1972, Isely 1998). There is considerable overlap among characters, and other authors have rejected the distinction (Zohary and Heller 1984, Barneby 1989). In Arizona, *T. gymnocarpon* occurs as scattered populations in the northern part of the state. Plants may be locally abundant.

7. Trifolium hybridum Linnaeus. Sp. Pl. 2:766. 1753. * [from Latin, *hybrida*, a mongrel or hybrid] ALSIKE CLOVER Adventive

Amoria hybrida (Linnaeus) C. Presl; *Trifolium elegans* Savi; *Trifolium fistulosum* Gilibert

Plants perennial, erect or ascending, branched, 10–70 cm, glabrous or sparsely hairy. **Leaves** palmate; stipules obovate or lanceolate, 1–2.5 cm, margins entire or toothed, apices long-acuminate; petiole 1–21 cm; petiolules 1 mm. **Leaflets** 3, blades usually obovate or ovate, rarely rhombic, 1–4 × 1–2.5 cm, base cuneate, veins prominent, margins sharply serrate, apex broadly acute, obtuse, or emarginate, surfaces glabrous. **Inflorescences** terminal or axillary, with 10–70+ flowers, globose becoming subglobose, rachis not prolonged beyond flowers, 1–2.5 × 1–2.5 cm; peduncles 2–10 cm, glabrous or nearly so; involucres absent; bracteoles lanceolate, 1 mm. **Pedicels** strongly reflexed, 2–3.5 mm. **Flowers** 8–12 mm; calyx campanulate, 3.5–4 mm, glabrous or sparsely hairy, veins 10, tube 1.5–2 mm, lobes unequal, lanceolate-subulate, orifice open; corolla pale to dark pink, 7–10 mm, banner ovate-oblong, 6.5–10.5 × 4–6 mm, apex rounded-denticulate. **Fruits** stipitate, ellipsoid, 3–7 mm. **Seeds** 2–4, tan, brown, or red-brown, mitten-shaped, 1–1.4 mm, surfaces smooth, dull. $2n = 16$.

Flowering May–Oct. Roadsides, fields, meadows, waste places; [0–]1670–3050 m; Petran Montane Conifer Forest; Apache, Coconino, Gila cos.; native to Europe and w Asia; widely introduced and naturalized in the U.S. and Canada, introduced also in South America, Europe, elsewhere in Asia, Africa, Pacific Islands, Australia.

Trifolium hybridum is widely cultivated as a forage crop. It may cause dermatitis in humans (Hardin and Arena 1974) and is implicated as a cause of dermatitis and biliary fibrosis in horses (Fisher 1995); the connection between the clover and the diseases is inconclusive (Nation 1989). Several bird species are reported to feed upon the seeds (Graham 1941). This species is adventive or weakly naturalized in Arizona.

8. Trifolium longipes Nuttall in J. Torrey and A. Gray, Fl. N. Amer. 1:314. 1838. * [from Latin, *longi*, long, and *pes*, foot] LONGSTALK CLOVER

Plants perennial, erect or ascending, growing in dense tufts, unbranched to little branched distally, 5–46 cm, slightly pilose. **Leaves** palmate; stipules ovate to lanceolate, 1–5 cm, margins usually entire, sometimes slightly serrate proximally, apices acute or acuminate; petiole 1–13 cm; petiolules to 1 mm. **Leaflets** 3, blades linear, lanceolate to elliptic, or oblong, cauline sometimes ovate, 0.8–6 × 0.2–1.6 cm, base cuneate, veins slightly to very thickened, curving, margins entire or serrate, apex acute, surfaces glabrous below (abaxially), usually appressed-pubescent above (adaxially). **Inflorescences** terminal, 10–40-flowered, globose, ellipsoidal, obconical, or subglobose, 1–3.5 × 1–3.2 cm; peduncles 1–30 cm; involucres absent; bracteoles scale-like, 0.5 mm. **Pedicels** erect or reflexed in fruit, 0.5–2 mm. **Flowers** 10–19 mm; calyx tubular, 4–10 mm, hairy or glabrous, veins 5–10, tube 2.5–3 mm, lobes unequal, linear, shorter than or to 6 times length of tube, orifice open; corolla white, cream, pink, lavender, or purple, sometimes pale with purple tips, 10–18 mm, banner ovate to oblanceolate, 10–18 × 3–4 mm, apex usually not beaked (except subsp. *multipedunculatum*), usually obtuse, acute, or acuminate, rarely emarginate; ovaries pubescent distally. **Fruits** oblong to ellipsoid, 4–5 mm. **Seeds** 1–4, dark brown, flattened globose, 2–3 mm, surface smooth.

Subspecies 12 (3 native in the flora): w United States.

Trifolium longipes encompasses a morphologically diverse and taxonomically difficult suite of populations. Gillett (1969) united *T. longipes* with *T. rusbyi* and differentiated ten subspecies, with *T. neurophyllum* recognized as distinct. Zohary and Heller (1984) maintained *T. rusbyi* (with seven subspecies, including subsp. *neurophyllum*) as distinct from *T. longipes* (with five subspecies). Analyses of molecular data support the union of *T. longipes* and *T. rusbyi* (Ellison et al. 2006).

1. Plants with thickened roots, often fusiform, with crown of buried stems ____ **8b. Trifolium longipes** subsp. **pygmaeum**
1′ Plants usually rhizomatous, if rarely with roots ± thickened, then the crown of rhizomes above thickened portion, or with slender taproots bearing a crown of wiry, pilose stems.
   2. Calyces white-villous, veins and lobes green; corollas purple-violet or lavender ____ **8a. Trifolium longipes** subsp. **neurophyllum**
   2′ Calyces sparsely pilose or curly-pilose at sinuses, often glabrous proximally; corollas white, cream, or violet ____ **8c. Trifolium longipes** subsp. **reflexum**

8a. Trifolium longipes Nuttall subsp. **neurophyllum** (Greene) J.M. Gillett in J. Maschinski et al., S.W. Rare Endang. Pl. 1995, 262. 1996. * [from Greek, *neuro*, nerve or cord, and *phyllum*, leaf] WHITE MOUNTAIN CLOVER Native

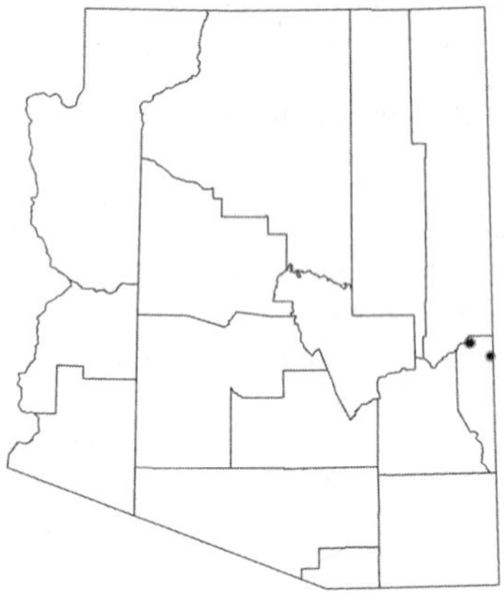

Trifolium neurophyllum Greene, Leafl. Bot. Observ. Crit. 1:154. 1905; *Trifolium longipes* Nuttall var. *neurophyllum* (Greene) J.S. Martin ex Isely; *Trifolium rusbyi* Greene subsp. *neurophyllum* (Greene) D. Heller & Zohary

Plants 10–46 cm; short-rhizomatous, becoming thickened woody roots. **Inflorescences** 35–40-flowered, hemispherical or subglobose, 2–3 cm wide. **Pedicels** strongly reflexed, stout, to 1 mm. **Flowers** 11–14 mm; calyx white-villous, veins and lobes green; corolla purple-violet or lavender, banner apex obtuse. $2n = 16$.

Flowering Jul–Aug. Open meadows, pine forests; 2300–3000 m; Petran Montane Conifer Forest; Apache, Greenlee cos.; endemic.

Subspecies *neurophyllum* as *Trifolium neurophyllum* is listed in NatureServe as a G2 plant of conservation concern. It is edemic to the mountains of east central Arizona.

8b. Trifolium longipes Nuttall subsp. **pygmaeum** (A. Gray) J.M. Gillett. Canad. J. Bot. 47:102. 1969. * [from Greek, *pygm*, dwarf] PYGMY CLOVER Native

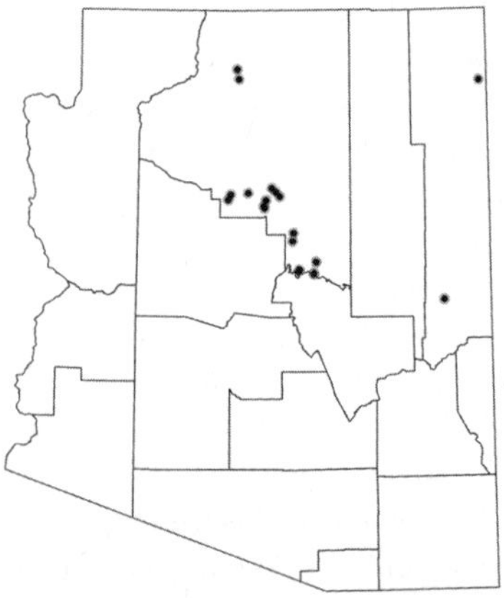

Trifolium longipes Nuttall var. *pygmaeum* A. Gray in J.C. Ives, Rep. Colorado R. 4:9. 1860; *Trifolium brachypus* (S. Watson) Blankinship; *Trifolium confusum* Rydberg; *Trifolium rusbyi* Greene

Plants 5–30 cm; roots thickened, often fusiform, with a crown of buried stems. **Inflorescences** 10–25-flowered, globose, 1.5–2.5 cm wide. **Pedicels** at least proximal, often all, strongly reflexed, slender, 1.5–2 mm. **Flowers** 11–12 mm; calyx curly-pubescent throughout or at sinuses; corolla purple or pink, banner apex acute or acuminate. *2n* = 16.

Flowering Apr–Aug. Ponderosa pine or Engelmann spruce forests, aspen groves, open meadows; [1800–]2070–2680[–3500] m; Petran Montane Conifer Forest, [Petran Subalpine Conifer Forest]; Apache, Coconino cos.; Colorado, New Mexico, Utah.

Trifolium longipes subsp. *pygmaeum* is locally abundant within its distribution in northern Arizona. The plants favor moist sites.

8c. Trifolium longipes Nuttall subsp. **reflexum** (A. Nelson) J.M. Gillett. Canad. J. Bot. 47:105. 1969. * [reflexed, bent] REFLEXED LONGSTALK CLOVER Native

Trifolium longipes Nuttall var. *reflexum* A. Nelson, Wyoming Agric. Exp. Sta. Bull. 28:94. 1896; *Trifolium rydbergii* Greene

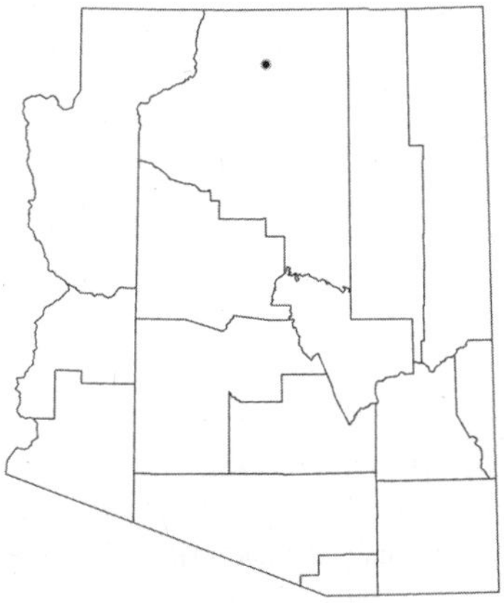

Plants 15–37 cm; rhizomatous. **Inflorescences** 10–20-flowered, hemispherical, obconical, or subglobose, 2–3 cm wide. **Pedicels** strongly reflexed, stout, to 1 mm. **Flowers** 13–15 mm; calyx slightly pilose or curly-pilose at sinuses, often glabrous proximally; corolla white, cream, or violet, banner apex acute or acuminate. *2n* = 24, 48.

Flowering May–Jul. Moist meadows; [1000–3600] m; Coconino Co.; Colorado, Idaho, Montana, New Mexico, Utah, Washington, Wyoming.

Trifolium longipes subsp. *reflexum* is documented in Arizona from a single specimen from the Kaibab National Forest (*Henrietta L. Zobel s.n.*, 12 Jun 1936). It may no longer be present in the state.

9. Trifolium macilentum Greene. Pittonia 3:223. 1897. * [from Latin, *maci*, leanness] LEAN CLOVER Native

Plants perennial, erect, ascending, or decumbent, unbranched or branched distally, 12–38 cm, glabrous. **Leaves** palmate; stipules lanceolate-elliptic on basal leaves, triangular-ovate on distal leaves, 0.8–3 cm, margins entire, apices acute or acuminate; petiole 1–15 cm; petiolules 1 mm. **Leaflets** 3, blades ovate, rhombic, or lanceolate, 1–4.5 × 0.3–2.5 cm, basal blades thin, base cuneate, veins prominent, margins serrate, those of basal leaflets setose to dentate, apex rounded or acute, surfaces glabrous. **Inflorescences** terminal or axillary, 20–40-flowered, ovoid or obovoid, 2–4 × 1.4–3.5 cm, rachis internodes between floral whorls elongated, rachis tip sometimes surpassing flowers; peduncles 4–15 cm; involucres absent; bracteoles minute, blunt. **Pedicels** reflexed, to 1 mm. **Flowers** 14–17 mm; calyx campanulate, 4–5.7 mm, sparsely hairy, veins 10, tube 2–3.5 mm, lobes unequal, narrowly triangular, orifice open; corolla violet or deep purple, 14–17 mm, banner ovate-oblong, 14–16 × 8–10 mm, apex rounded or slightly emarginate. **Fruits** oblong or clavate, 4–5 mm. **Seeds** 1–3, brown, flattened ovoid, 2 mm, surface smooth or slightly roughened, dull.

Flowering May–Jun. Dry hillsides; [1200–]2060–2130[–2800] m; Great Basin Conifer Woodland, Petran Montane Conifer Forest; Mohave Co.; Nevada, Utah.

Trifolium macilentum and morphologically similar species (*T. dedeckerae*, *T. kingii*, *T. productum*, and *T. rollinsii*) have been treated in diverse ways. Gillett (1972) recognized the latter four as distinct and considered *T. macilentum* a subspecies of *T. kingii*. Isely (1998) recognized *T. kingii*, *T. macilentum*, and *T. productum*, and considered *T. dedeckerae* and *T. rollinsii* to be varieties of *T. macilentum*. Barneby (1989) treated these as two species, *T. kingii* (with *T. productum* in synonymy) and *T. macilentum* (with *T. dedeckerae* and *T. rollinsii* as varieties). Zohary and Heller (1984) recognized *T. kingii*, with the other four taxa considered subspecies. As treated here, *T. macilentum* is known from extreme southwestern Utah, northwestern Arizona, and adjacent Nevada. The Arizona populations are localized and occur in the ecotone between pinyon-juniper woodland and ponderosa pine forest.

10. Trifolium microcephalum Pursh. Fl. Amer. Sept. 2:478. 1813. * [from Greek, *micro*, small, and *cephalo*, head] SMALLHEAD CLOVER Native

Lojaconoa microcephala (Pursh) Bobrov; *Trifolium microcephalum* Pursh var. *bipedale* Hooker; *Trifolium microcephalum* Pursh var. *lemmonii* Lojacono

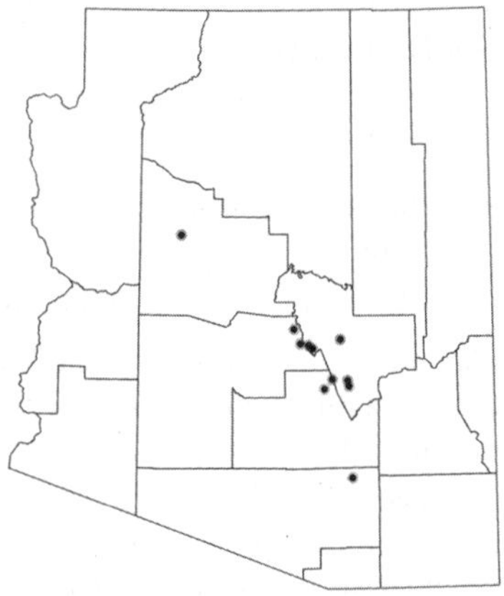

Plants annual, erect to ascending, branched, 3–55 cm, densely or rarely sparsely pubescent. **Leaves** palmate; stipules obliquely ovate, 0.4–1.2 cm, margins entire or slightly serrate, apices acuminate; petiole 1–5 cm; petiolules 0.5 mm. **Leaflets** 3, blades oblanceolate, obovate, or obcordate, 0.4–1.7 × 0.3–1.1 cm, base cuneate, veins fine or slightly thickened, margins setose, often dentate distally, apex usually retuse, rarely rounded, surfaces villous. **Inflorescences** axillary or terminal, 10–40-flowered, globose, 0.5–1.3 × 0.4–1.2 cm; peduncles 1–8 cm; involucres flattened or bowl-shaped, 4–10 mm, conspicuously villous, lobes 5–12, divided 1/2 their length, lanceolate-ovate, entire or slightly toothed proximally, acuminate, when flattened laterally, nearly completely hiding calyces; bracteoles absent. **Pedicels** absent. **Flowers** 4–6 mm; calyx campanulate-tubular, 3–5.5 mm, pubescent, veins 10, tube 1.8–2.5 mm, lobes ± equal, triangular-aristate, margins hyaline, wavy, orifice open; corolla white or pink to lavender, 4–6 mm, banner oblong, 3–6 × 1–2 mm, apex emarginate. **Fruits** broadly ellipsoid, 1.5–2 mm. **Seeds** 1, yellow, reddish mottled, oblong, 1.5 mm, surface smooth. $2n = 16$.

Flowering Mar–Jul. Meadows, roadsides, stream banks, forest clearings, grassy slopes; [0–]850–2300 m; Interior Chaparral, Madrean Evergreen Woodland, Great Basin Conifer Woodland, Petran Montane Conifer Forest; Gila, Maricopa, Pima, Pinal, Yavapai cos.; Alaska, California, Idaho, Montana, Nevada, Oregon, Washington; Mexico (Baja California); Canada (British Columbia).

Trifolium microcephalum is common in much of its range in California and north into Oregon and is rare in much of the rest of its range. Arizona populations are highly localized in moist sites within several biotic communities. It is closely allied with the Chilean *T. vernum* Philippi (Ellison et al. 2006), which it resembles greatly, and from which it is distinguished by its shorter calyx lobes (Zohary and Heller 1984). Graham (1941) indicates that valley quail feed upon the seeds of *T. microcephalum*.

11. Trifolium mucronatum Willdenow ex Sprengel. Syst. Veg. 3:208. 1826. * [from Latin, *mucro*, a sharp point] CUSP CLOVER

Trifolium fistulosum A.E. Vaughan; *Trifolium ortegae* Greene; *Trifolium wormskioldii* Lehmann var. *ortegae* (Greene) Barneby

Plants perennial, 5–65 cm, glabrous, stems erect, ascending, or decumbent, branched, solid or fistulose. **Leaves** palmate; stipules lanceolate or narrowly ovate, 1.5–2 cm, margins toothed (proximal stipules) or lacerate (distal stipules), apices acuminate to long-acuminate; petiole 1–10 cm; petiolules to 0.5 mm. **Leaflets** 3, blades linear, narrowly ovate, elliptic, obovate, oblanceolate, or lanceolate, 0.8–4 × 0.1–1.3 cm, base cuneate or rounded, veins fine or thickened distally, margins serrulate to setose or entire, apex acute, rarely rounded, often apiculate, surfaces glabrous. **Inflorescences** axillary or terminal, 8–50-flowered, subglobose or globose, 1.5–2.5 × 1.5–4 cm; peduncles 1.5–8 cm; involucres broadly campanulate, 10–22 mm, lobes 10+, incised 1/2 or more of their length, aristate to triaristate or irregular to bifurcate, sharply and acutely serrate or lacerate, veins prominent; bracteoles subulate or narrowly ovate, 0.5–3 mm. **Pedicels** straight, 0.5–1 mm. **Flowers** 12–20 mm; odor somewhat sweet; calyx greenish or whitish, tubular-conical, 5–11 mm, glabrous, veins 10, tube 2–3 mm, lobes green to purple, ± equal lower (abaxial) rarely longer than tube, orifice open; corolla usually reddish, purple, lavender, pink, or white, rarely bicolored, 10–18 mm, banner oblong to obovate-oblong, 10–18 × 2–5 mm, apex truncate or rounded. **Fruits** oblong, 3–6 mm. **Seeds** 2–4, olive, squarish or reniform, 1.3–1.5 mm, smooth. $2n = 16$.

Subspecies 3 (1 native in the flora): w United States, n Mexico.

The difficulty of distinguishing *Trifolium mucronatum* from *T. wormskioldii* is discussed under the latter species. Populations in the United States represent *T. mucronatum* subsp. *lacerum*; in Mexico, subsp. *lacerum* is found in the north, subsp. *mucronatum* (petals pink or reddish, flowers 1.5–1.7 cm, inflorescences 2.5–3.5 cm wide) is widespread, and subsp. *vaughanae* J.M. Gillett (petals lavender or white, flowers 1.2–1.4 cm, inflorescences 1.5–2 cm wide) is restricted to central Mexico (Guanajuato, San Luis Potosí) (Gillett 1980).

Trifolium involucratum Ortega (1797), which pertains here, is a later homonym of *T. involucratum* Lamarck (1778, = *T. cherleri* Linnaeus) and thus illegitimate.

11a. Trifolium mucronatum Willdenow ex Sprengel subsp. **lacerum** (Greene) J.M. Gillett. Canad. J. Bot. 58:1444. 1980. * [from Latin, *lacer*, mangled, torn] SPINYTOOTH CLOVER Native

Trifolium lacerum Greene, Erythea 2:182. 1894; *Trifolium arizonicum* Greene; *Trifolium involucratum* Ortega var. *arizonicum* (Greene) McDermott; *Trifolium oxyodon* Greene ex Rydberg; *Trifolium wormskioldii* Lehmann var. *arizonicum* (Greene) Barneby

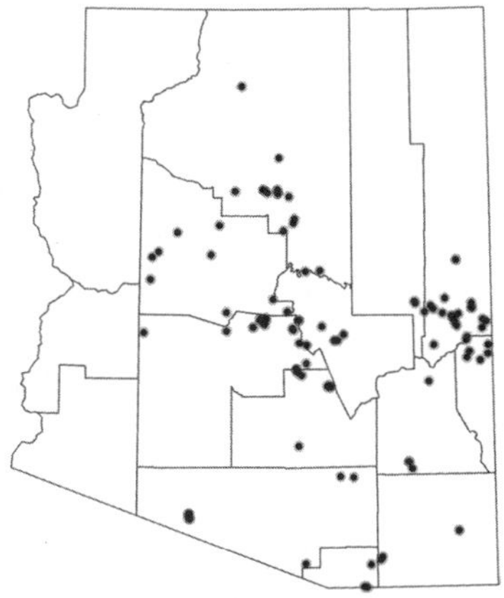

Plants perennial, 10–65 cm, stems erect or decumbent, sparsely branched; roots taproots, slender, rhizomes absent or short. **Leaves** palmate; stipules lanceolate to narrowly ovate, apices acuminate; petiole 1–3.5 cm; petiolules to 0.5 mm. **Leaflet** blades linear, elliptic, or obovate, 0.8–2.5 × 0.1–0.3 cm, base cuneate, veins thickened, margins entire or finely serrate to setose-spinulose, apex acute or apiculate. **Inflorescences** 8–20-flowered, subglobose becoming globose, 1.5–2 × 1.5–4 cm; peduncles 4–6 cm; involucre lobes 10+, lobed to 3/4 their length, triaristate to lacerate; bracteoles subulate or narrowly ovate, 2–3 mm. **Pedicels** straight, 0.5 mm. **Flowers** calyx 5–7 mm, tube 2–2.5 mm, lobes ± equal, lower (abaxial) lobe longest; corolla white, lavender, or purple, sometimes with different colors within inflorescence, 11–13 mm, banner narrowly [broadly obovate-]oblong, 11–13 × 4 mm, apex rounded. **Fruits** oblong, 3–6 mm. **Seeds** 2–4, olive, reniform. $2n = 16$.

Flowering Jun–Sep. Damp meadows, seeps, along streams in pine forests; 570–3050 m; Sonoran Desertscrub, Semidesert Grassland, Plains Grassland, Interior Chaparral, Madrean Evergreen Woodland, Petran Montane Conifer Forest, Petran Subalpine Conifer Forest; Apache, Cochise, Coconino, Gila, Graham, Greenlee, Maricopa, Navajo, Pima, Pinal, Santa Cruz, Yavapai cos.; California, Colorado, New Mexico, Texas; Mexico (Chihuahua, Coahuila, Sonora).

Trifolium mucronatum subsp. *lacerum* is widespread and often abundant in Arizona, ranging from moist sites within the Sonoran Desert to high elevation montane plant communities.

12. Trifolium pinetorum Greene. Erythea 2:182. 1894. * [from Latin, *pinet*, pine wood] PINE CLOVER Native

Trifolium longicaule Wooton & Standley; *Trifolium willdenovii* Sprengel var. *longicaule* (Wooton & Standly) L.D. Benson; *Trifolium wormskioldii* Lehmann var. *longicaule* (Wooton & Standley) L.D. Benson

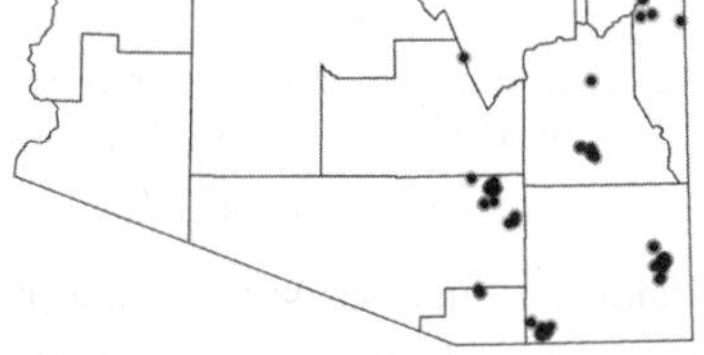

Plants perennial, prostrate to ascending, branched, 5–30 cm, glabrous or lightly pilose; roots stout, branched, rhizomes absent. **Leaves** palmate; stipules lanceolate-ovate, 0.7–1.3 cm, margins entire, apices acuminate; petiole 1–8.5 cm; petiolules 1–1.5 mm. **Leaflets** 3, blades oblanceolate or obovate, 0.5–2.9 × 0.4–1.3 cm, base cuneate, veins prominent, margins denticulate to spinulose, apex rounded or truncate, sometimes retuse, surfaces glabrous. **Inflorescences** axillary or terminal, 10–20-flowered, obconical, globose, or subglobose, 1.7–2.5 × 1–1.5 cm; peduncles 4–6 cm; involucres broadly campanulate, 6–8 mm, divided ± 3/4 their length, lobes 15–20, linear-lanceolate, entire, acuminate; bracteoles ovate, 0.5 mm. **Pedicels** erect to slightly reflexed, 1–2 mm. **Flowers** 10–16 mm; calyx campanulate, 6–10 mm, glabrous, veins 10, tube 2–2.3 mm, lobes ± equal, subulate, 3–5 mm, orifice open; corolla white to pale purple, keel petals with dark purple-red tips, 11–13 mm, banner oblong, 11–13 × 4–5 mm, apex retuse. **Fruits** short-stipitate, oblong, 3.5–5 mm. **Seeds** 1 or 2, olive-brown, mottled purple, oblong, 1.5 mm, surface smooth, semiglossy. $2n = 16$.

Flowering Jun–Oct. Disturbed areas in pine, fir, or spruce forests; (1260–)1610–2930 m; (Madrean Evergreen Woodland, Greast Basin Conifer Woodland), Petran Montane Conifer Forest, Petran Subalpine Conifer Forest; Apache, Cochise, Coconino, Graham, Greenlee, Navajo, Pima, Pinal, Santa Cruz, Yavapai cos.; New Mexico; Mexico (Chihuahua).

Trifolium pinetorum occupies disturbed areas in pine-fir-spruce forests and has adapted well to anthropogenic disturbances, such as roadsides and other graded areas (Gillett 1980). It is locally abundant at higher elevation sites in Arizona. The lowest elevation records are from the Grand Canyon.

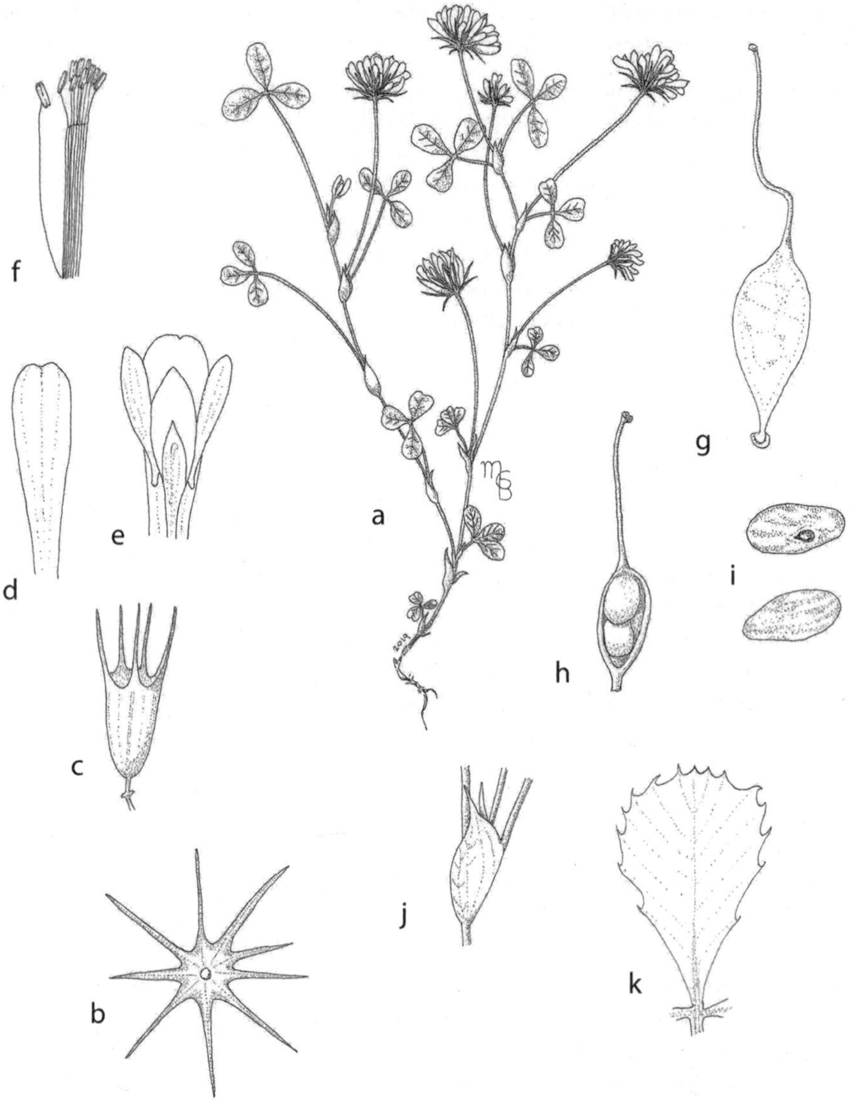

***Trifolium pinetorum*: a.** habit; **b.** involucre; **c.** calyx; **d.** petal; **e.** flower; **f.** stamens; **g.** fruit; **h.** fruit with seeds; **I.** seeds; **j.** stipules; **k.** leaflet. Illustrator: Marceline VandeWater.

13. Trifolium pratense Linnaeus. Sp. Pl. 2:768. 1753. * [from Latin, *pratum*, of meadows] RED CLOVER Naturalized

Trifolium pensylvanicum Willdenow

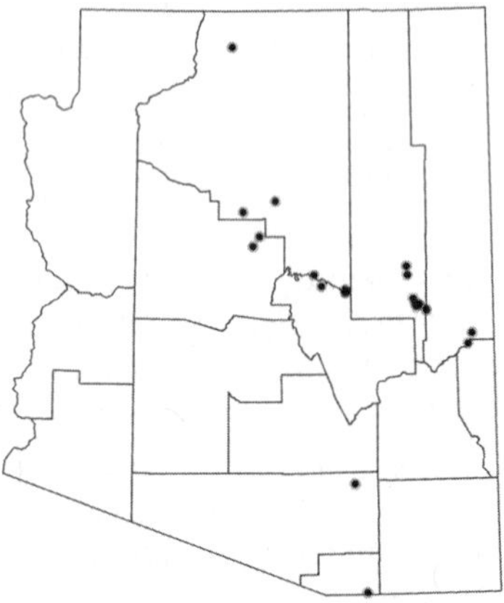

Plants perennial, erect, ascending, or decumbent, branched, 20–70 cm, pilose or glabrous. **Leaves** palmate; stipules broadly triangular, 1–2 cm, margins entire, sometimes ciliate, apices mucronate or bristle-like; petiole 1–9 cm; petiolules 1 mm. **Leaflets** 3, blades elliptic to obovate, 1.5–5 × 0.7–1.5 cm, base cuneate, veins prominent, margins subentire, apex obtuse, acute, or retuse, surfaces appressed-pubescent. **Inflorescences** terminal, solitary or paired, with 75–100+ flowers, globose or ovoid, 1.2–7.7 × 0.7–2.2 cm; involucres absent, pseudoinvolucres formed of stipules of distalmost leaves; bracteoles absent. **Pedicels** straight, 0–1 mm. **Flowers** 15–18 mm; calyx tubular-campanulate, 8–11 mm, hairy, veins 10, tube 3–4 mm, lobes upper (adaxial)

equaling tube, lateral and lower (abaxial) 2 times tube, orifice hairy, slightly closed; corolla usually rose-purple, rarely pink or white, 13–18 mm, banner elliptic-ovate, 10–13 × 3–5 mm, apex narrowly rounded. **Fruits** oblong, leathery distally, transversely dehiscent, 3 mm. **Seeds** 1 or 2, yellow, yellow-brown, or purple, ovoid, 1.6–2 mm, surfaces smooth, dull. $2n = 14$.

Flowering May–Sep. Fields, prairies, roadsides, disturbed areas; [0–](1190–)1910–2560 m; (Plains Grassland, Great Basin Conifer Woodland), Petran Montane Conifer Forest; Apache, Coconino, Gila, Navajo, Pima, Santa Cruz, Yavapai cos.; widely naturalized in the U.S., Canada, and Greenland; native to Eurasia; introduced also in s South America, s Africa, Pacific Islands (Hawaii, New Zealand), Australia.

Trifolium pratense is morphologically variable and numerous varieties have been recognized (Zohary and Heller 1984). Distinctions among these are slight and intergradation is common, perhaps due to the long-time cultivation of the species. In North America, the following varieties are sometimes recognized: var. *pratense* (stems decumbent to ascending, 20–40 cm, with dense, appressed, white hairs); var. *sativum* Schreber (stems mostly erect, 40–100 cm, sparsely hairy or glabrous); and var. *americanum* Harz (stems ascending to erect, 30–100 cm, with dense, spreading hairs).

Trifolium pratense is weakly naturalized in Arizona. The plants occur in scattered populations but can be locally abundant. Seeds and leaves of *T. pratense* are eaten by many species of wildlife, and the flowers provide bee forage for honey production (Graham 1941).

14. Trifolium repens Linnaeus. Sp. Pl. 2:767. 1753. * [from Latin, *repen*, creeping] WHITE CLOVER Naturalized

Trifolium saxicola Small

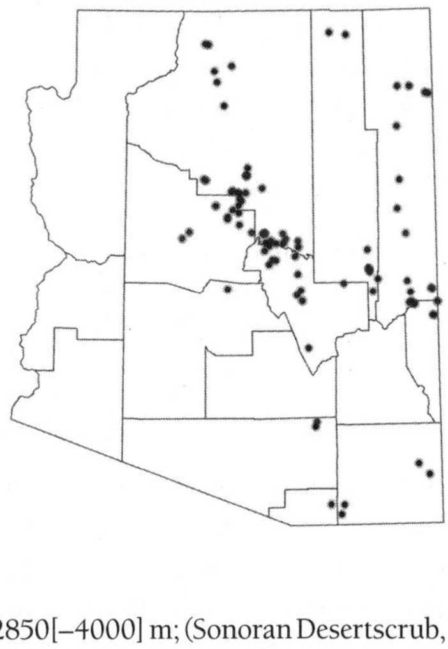

Plants perennial, creeping, branched, 10–40 cm, glabrous or glabrescent, rooting at nodes. **Leaves** palmate; stipules lanceolate, 0.9–1.3 cm, margins entire, apices short-subulate; petiole 5–20 cm; petiolules to 1 mm. **Leaflets** 3, blades obovate, obcordate, or orbiculate, 0.6–4 × 0.4–2.5 cm, base cuneate, veins moderately prominent, margins serrulate distally, apex emarginate, surfaces glabrous. **Inflorescences** axillary, erect, from prostrate stems, with 20–40 flowers, globose, 1.5–3.5 × 1.5–3.5 cm; peduncles 1.5–30 cm, glabrous or nearly so; involucres absent; bracteoles white, lanceolate, 1–2 mm. **Pedicels** strongly reflexed in fruit, elongate, 3–5 mm. **Flowers** 8–13 mm; calyx campanulate, 3–5 mm, glabrous, veins 6–10, tube 1.5–2.5 mm, lobes unequal to subequal, upper (adaxial) shorter than tube, triangular-lanceolate, orifice open; corolla white, often becoming pinkish in age, 4–12 mm, banner ovate-lanceolate or oblong, 4–12 × 1–4 mm, apex rounded. **Fruits** linear-oblong, 4–5 mm. **Seeds** 3 or 4, yellow, reddish brown, or light brown, ovoid-reniform, 1 mm, surfaces smooth, glossy. $2n = 16, 28, 32, 48, 64$.

Flowering Feb–Oct. Fields, lawns, roadsides, waste places; [0–](350–)1600–2850[–4000] m; (Sonoran Desertscrub, Great Basin Desertscrub, Semidesert Grassland, Plains Grassland), Interior Chaparral, Madrean Evergreen Woodland, Great Basin Conifer Woodland, Petran Montane Conifer Forest, Petran Subalpine Conifer Forest; Apache, Cochise, Coconino, Gila, Greenlee, Maricopa, Navajo, Pima, Santa Cruz, Yavapai cos.; widely naturalized in the U.S., Canada, and Greenland; native to Eurasia; introduced also in South America, Africa, Pacific Islands.

Trifolium repens may very well be the most important temperate pasture plant (Baker and Williams 1987) and has been considered the most important perennial pasture plant in North America (Piper 1924). It was introduced at least as early as the mid-1800s (Mack 2003) and spread so rapidly that it became known to Native Americans as "White Man's Foot Grass" (Strickland 1801). It is morphologically diverse; most material from the flora area represents var. *repens*, but some specimens fit within the circumscriptions given by Zohary and Heller (1984) of var. *giganteum* Lagrèze-Fossat, with inflorescences to 3.5 cm diameter and leaflets nearly 4 cm; others have smaller, pale-pink petals with hairy petioles and pedicels, and approach var. *biasolettii* (Steudel & Hochstetter) Ascherson & Graebner (*T. occidentale* Coombe).

Trifolium repens is naturalized in Arizona where it is extensively distributed and locally abundant. It is most commonly encountered at mid- and higher elevations in the state, particularly in the ponderosa pine zone. The seeds and leaves are eaten by a wide range of wildlife (Graham 1941).

15. Trifolium sonorense T.K. Ahlquist & Vincent. Phytoneuron 2018-1:1–5. 2018. * [of Sonora] SONORAN CLOVER Native

Plants perennial, prostrate, branched, 25–38+ cm, pubescent. **Leaves** palmate; stipules narrowly triangular, 0.7–1.7 cm, margins entire or finely serrulate, apices usually acute or acuminate, sometimes distally cleft with two teeth; petiole 0.3–5 cm; petiolules to 1 mm. **Leaflets** 3, blades obovate to obcordate, 0.9–1.6 × 0.7–1.2 cm, base cuneate, veins moderately thickened, margins serrate to weakly so, apex rounded or emarginate, mucronulate, surfaces slightly hairy below (abaxially), glabrous above (adaxially). **Inflorescences** axillary, 10–20-flowered, depressed-globose or globose, 0.8–1.6 × 0.9–1.6 cm, rachis prolonged beyond flowers; peduncles 2.8–5 cm; involucres a narrow rim, or slightly proximally united bracteoles of proximal flowers, 0.1–0.5 mm; bracteoles linear-triangular, 0.5 mm. **Pedicels** reflexed in fruit, 1–3 mm. **Flowers** 5.7–6.7 mm; calyx tubular, 3.3–6.5 mm, pubescent, veins 10, tube 0.8–1 mm, lobes subequal, narrowly triangular, orifice open; corolla white or pink, 5.3–6.5 mm, banner obovate, 5.3–6.5 × 2.8–3.8 mm, apex broadly rounded or broadly acute. **Fruits** obovoid, 4.3–4.4 mm. **Seeds** 2, brownish orange or olive-brown, globose or broadly reniform, 1.2–1.7 mm, surface smooth. $2n = 16$.

Flowering Aug–Oct. Dry stream banks, grassy places; 1500–1800 m; Madrean Evergreen Woodland; Cochise Co.; Mexico (Chihuahua, Jalisco, Sinaloa, Sonora).

Trifolium sonorense has long been equated with *T. amabile* Kunth, from which Ahlquist has shown it to be distinct by longer sepal lobes (more than 3 mm), keel petal claws (more than 1.7 mm), and filament tubes (more than 3.6 mm) (Ahlquist 2012, Ahlquist and Vincent 2018). *Trifolium sonorense* is known from western Mexico and southernmost Arizona, largely in the Sonoran Desert region, and in mountainous Jalisco. In Arizona the species is localized in the vicinity of the Huachuca Mountains in southern Cochise County. It is to be expected in adjacent Santa Cruz County.

16. Trifolium variegatum Nuttall in J. Torrey and A. Gray, Fl. N. Amer. 1:317. 1838. * [from Latin, *varigat*, varigated, variously marked] WHITETIP CLOVER Native

Trifolium dianthum Greene; *Trifolium geminiflorum* Greene; *Trifolium melananthum* Hooker & Arnott; *Trifolium pusillum* Greene; *Trifolium spinulosum* Douglas ex Hooker var. *triste* Torrey & A. Gray; *Trifolium subsalinum* Greene; *Trifolium tridentatum* Lindley var. *melananthum* (Hooker & Arnott) S. Watson; *Trifolium ultramontanum* Greene; *Trifolium variegatum* Nuttall var. *geminiflorum* (Greene) Vincent; *Trifolium variegatum* Nuttall var. *melananthum* (Hooker & Arnott) Greene

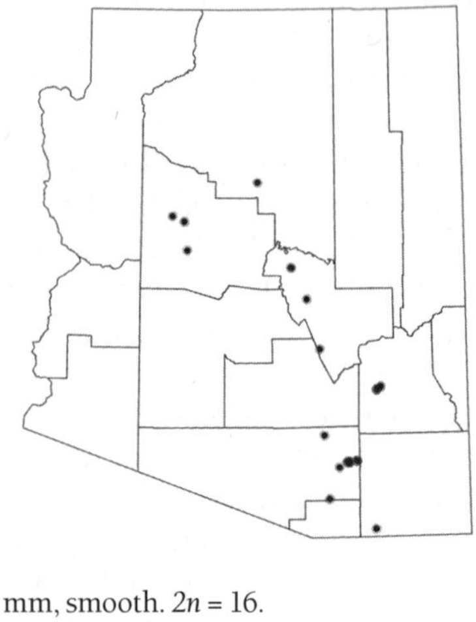

Herbs annual, 1.5–48+ cm, glabrous. **Stems** usually erect, ascending, or prostrate, branched. **Leaves** palmate; stipules lanceolate or ovate, 0.3–1 cm, margins lacerate, apices acute to acuminate; petiole 1–9 cm; petiolules to 0.8 mm. **Leaflets** 3, blades elliptic, obovate, oblanceolate, or obcordate, 0.2–5 × 0.1–1.2 cm, base cuneate, veins thickened, margins denticulate, apex acute or retuse, surfaces glabrous. **Inflorescences** terminal or axillary, with 1–20 flowers, subglobose or globose, 0.5–1.3 × 0.1–1 cm; peduncles 0.5–5 cm; involucres flat, rotate, not hiding flowers, 2–6 mm, incised 1/2 their length, lobes 2–7, lacerate; bracteoles absent. **Pedicels** straight, to 0.5 mm. **Flowers** 5–11 mm; calyx campanulate, 2.1–6.8 mm, not slit between upper (adaxial) lobes, glabrous, veins 10–20, tube 1.5–4 mm, lobes subequal, longer than tube, narrowly triangular, margins often without shoulders, entire, rarely with 1–3 lateral teeth, orifice open; corolla purple, usually white-tipped, 3–10 mm, banner obovate, 3–10 × 0.6–2.6 mm, apex emarginate. **Fruits** oblong or ellipsoid, 2–4 mm. **Seeds** 2, blackish brown, sometimes mottled, ovoid or ellipsoid, 1–1.6 mm, smooth. $2n = 16$.

Flowering Mar–Sep. Wet places, coastal hills, meadows; [0–]830–2300[–2500] m; (Sonoran Desertscrub), Semidesert Grassland, Interior Chaparral, Madrean Evergreen Woodland, (Petran Montane Conifer Forest); Cochise, Coconino, Gila, Graham, Maricopa, Pima, Yavapai cos.; Alaska, California, Idaho, Montana, Nevada, Oregon, Utah, Washington, Wyoming; Mexico (Baja California); Canada (British Columbia).

Trifolium variegatum is widespread and morphologically variable (Foster and Vincent 2018). Because of this variability, *T. variegatum* and related species were lumped into seven so-called phases by Isely (1998), who described these as ecotypic groups, some of which might deserve varietal or specific rank.

Cooper (1957) considered *T. variegatum* a good choice for cultivation in wet areas for grazing and hay production. Graham (1941) reports that quail species in California feed upon the seeds. *Trifolium variegatum* has a scattered distribution in Arizona, though the plants may be locally common.

17. Trifolium willdenovii Sprengel. Syst. Veg. 3:208. 1826. * [for Carl Ludwig Willdenow, 1765–1812, German botanist] THIMBLE CLOVER Native

Trifolium involucratum Willdenow, Sp. Pl. 3:1372. 1802, not Lamarck 1778; *Trifolium aciculare* Nuttall; *Trifolium nuttallii* Steudel; *Trifolium scabrellum* Greene; *Trifolium segetum* Greene; *Trifolium tridentatum* Lindley; *Trifolium tridentatum* Lindley var. *aciculare* (Nuttall) McDermott; *Trifolium tridentatum* Lindley var. *scabrellum* Greene; *Trifolium tridentatum* Lindley var. *segetum* (Greene) McDermott; *Trifolium tridentatum* Lindley var. *watsonii* (Lojacono) Jepson; *Trifolium trimorphum* Greene; *Trifolium watsonii* Lojacono

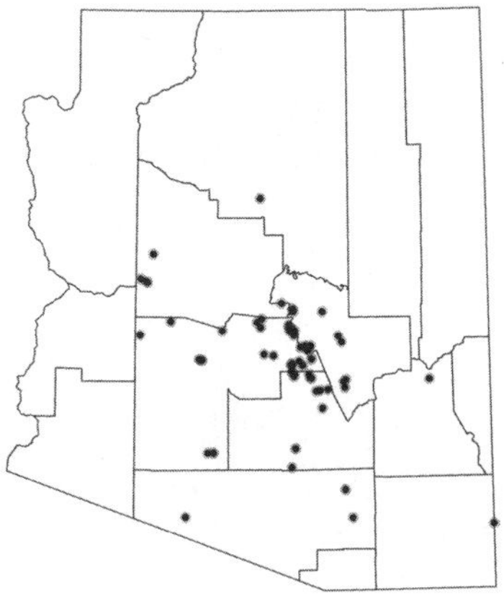

Plants annual, erect, branched, 10–60 cm, glabrous. **Leaves** palmate; stipules lanceolate or ovate, 1–2 cm, margins dentate or lacerate, apices acuminate; petiole 1–8 cm; petiolules 0.5 mm. **Leaflets** 3, blades linear, lanceolate, or elliptic, 1–5 × 0.2–1.5 cm, base cuneate, veins fine, margins serrate, apex acute, blunt, or retuse, surfaces glabrous. **Inflorescences** axillary or terminal, 10–50-flowered, globose or ovoid, 1–3.5 × 1–3 cm; peduncles 2–15 cm; involucres flattened or bowl-shaped, 3–9 mm, divided 1/4–1/3 their length, deeply lacerate, when flattened laterally, not hiding flowers except proximally; bracteoles absent. **Pedicels** erect, 0.5 mm. **Flowers** 12–20 mm; calyx tubular-campanulate, slit between upper (adaxial) lobes, 6–9 mm, glabrous, veins 10–15, tube 3–7 mm, lobes unequal, shorter than or equal to tube, triangular or subulate, distally cleft with three teeth or shouldered, orifice open; corolla usually white with purple spot, sometimes lavender or pink or all white, 10–18 mm, banner narrowly oblong, 12–20 × 3–4 mm, apex obtuse or retuse. **Fruits** ellipsoid, 3 mm. **Seeds** 1 or 2, yellow, mottled, globose-ellipsoid, 1.5–2.5 mm, surface smooth. $2n = 16$.

Flowering Mar–Jul. Roadsides, meadows, lawns, hillsides, stream margins, open oak chaparral; [0–]480–1680 m; Sonoran Desertscrub, Semidesert Grassland, Interior Chaparral, Madrean Evergreen Woodland; Cochise, Coconino, Gila, Graham, Maricopa, Pima, Pinal, Yavapai cos.; California, Idaho, Oregon, Washington; Mexico (Baja California); Canada (British Columbia).

Trifolium willdenovii was described by Zohary and Heller (1984) as one of the most variable clover species in North America. *Trifolium involucratum* Willdenow (1802) is the earliest name for this species, which is a later homonym of *T. involucratum* Lamarck (1778). The Arizona populations are localized but plants may be abundant. Plants are most common in riparian habitats and canyons but may also be found on rocky slopes. Records of *T. willdenovii* from Cochise and Coconino counties are outside the typical distribution for this species in Arizona.

18. Trifolium wormskioldii Lehmann. Sem. Hort. Bot. Hamburg 17. 1825. * [for Morten Wormskjold, 1783–1845, Danish botanist and explorer] COW CLOVER Native

Lupinaster wormskioldii (Lehmann) C. Presl; *Trifolium fendleri* Greene; *Trifolium fimbriatum* Lindley; *Trifolium heterodon* Torrey & A. Gray; *Trifolium spinulosum* Douglas ex Hooker

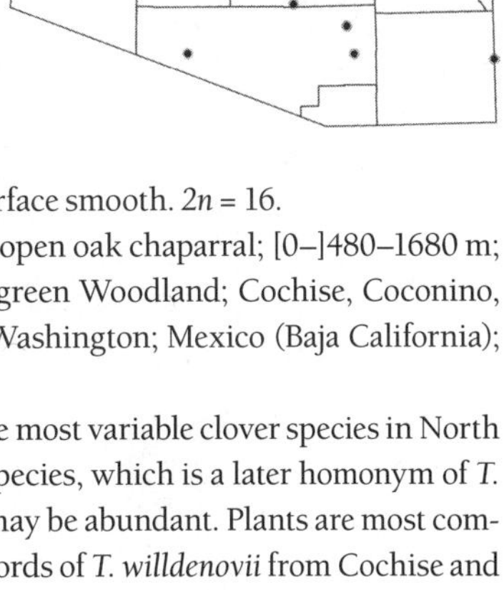

Herbs perennial, 10–40 cm, glabrous. **Stems** erect, ascending, or decumbent, branched. **Leaves** palmate; stipules ovate to lanceolate, 1.5–2 cm, margins entire (proximal stipules) or lacerate (distal stipules), apices acute to acuminate; petiole 1–11 cm; petiolules 0.5 mm. **Leaflets** 3, blades obovate to elliptic, 0.4–4.2 × 0.2–1.3 cm, base cuneate or rounded, veins fine or thickened distally, margins spinulose-serrate, apex usually acute to rounded, sometimes retuse or mucronate, surfaces glabrous. **Inflorescences** axillary or terminal, 20–50-flowered, subglobose or globose, 1.8–2.5 × 2–2.5 cm; peduncles 1.5–11 cm; involucres broadly bowl-shaped, 12–20 mm, incised 1/3–1/2 their length, lobes 10+, sharply and acutely serrate, veins prominent; bracteoles linear or ovate, 1–2 mm. **Pedicels** straight, 1.5–2 mm. **Flowers** odor somewhat unpleasant, 12–14 mm; calyx whitish, tubular, 5–6 mm, glabrous, veins 10, tube 2–3

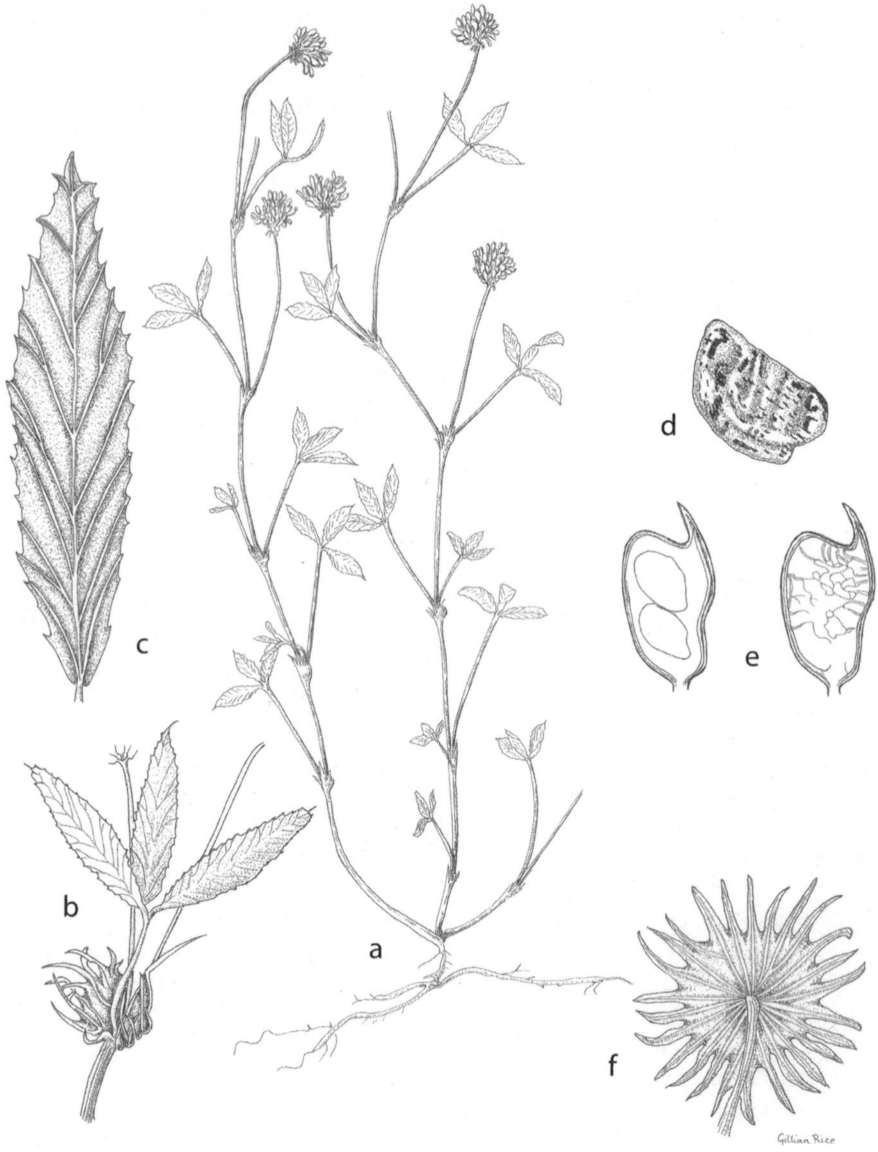

***Trifolium wormskioldii*: a. habit; b. leaf and stipules; c. leaflet; d. seed; e. fruit (longitudinal section); f. involucre. Illustrator: Gillian Rice.**

mm, lobes green to purple, subequal, lower (abaxial) lobes rarely longer than tube, orifice open; corolla usually magenta to purple, rarely white or bicolor, 10–12 mm, banner ovate-oblong, 12–14 × 2–3 mm, apex truncate or rounded, slightly retuse. **Fruits** oblong, 2–4 mm. **Seeds** 3 or 4, brown, sometimes mottled, subglobose or mitten-shaped, 1.5 mm, smooth. $2n = 16, 32$.

Flowering Mar–Sep. Saline flats, beaches, meadows, grassy areas, alluvial soil; [0–]730–2920 m; (Sonoran Desertscrub, Semidesert Grassland, Plains Grassland, Interior Chaparral), Madrean Evergreen Woodland, Petran Montane Conifer Forest, Petran Subalpine Conifer Forest; Apache, Coconino, Gila, Greenlee, Maricopa, Pima, Santa Cruz, Yavapai cos.; California, Colorado, Idaho, Nevada, New Mexico, Oregon, Utah, Washington, Wyoming; Mexico (Baja California); Canada (British Columbia).

Trifolium wormskioldii is largely tetraploid and is widespread in western North America. Distinguishing *T. wormskioldii* from closely allied species (especially *T. mucronatum*) is sometimes difficult, which led Barneby (1989) to synonymize *T. mucronatum* and *T. pinetorum* under the former name. *Trifolium wormskioldii* produces long, white rhizomes; *T. mucronatum* produces only short rhizomes at most, often only small, fibrous roots or taproots (Gillett 1980). The distinction of *T. pinetorum* from *T. wormskioldii* was supported by Ellison et al. (2006); those authors did not include material of *T. mucronatum* in their DNA analyses.

Trifolium wormskioldii rhizomes may have been a food resource for Native American groups in the Pacific Northwest (Turner and Kuhnlein 1982). The leaves are reported to be utilized for food by mule deer and rabbits (Graham 1941). In Arizona, *T. wormskioldii* is most frequently encountered at higher elevations.

VACHELLIA Wight & Arnott. Prodr. Fl. Ind. Orient. 1:272. 1834. * [named in honor of the Rev. G.H. Vachell, a collector of the flora of China]

John E. Ebinger and David S. Seigler

Shrubs or trees, not scandent, mostly erect. **Stems** lightly to highly branched; twigs usually rounded, glabrous to pubescent; short shoots usually present; stipular spines present, prickles absent. **Leaves** alternate; bipinnate, pinna pairs 1–numerous per leaf, mostly opposite; stipular spines woody, paired at the nodes, straight to curved, sometimes asymmetrical, in some species enlarged and inhabited by ants; petioles round, channeled above, petiolar gland present; rachis channeled above, sessile to stalked glands at various positions. **Leaflets** 1–numerous pairs per pinna, usually opposite, subsessile, membranous, glabrous to pubescent. **Inflorescence** loosely- to densely-flowered globose head to cylindrical spike, solitary to clustered in the leaf axis or on spur branches, rarely in short racemose clusters; peduncles short, glabrous to pubescent, not elongating in fruit; bracts present. **Pedicels** absent. **Flowers** mimosoid, radially symmetric, 5- or less commonly 4-merous; calyx cup-shaped, lobes triangular, glabrous to pubescent; corolla cup-shaped, membranous, lobes triangular, glabrous to pubescent, valvate in bud; nectariferous disk absent; stamens 20–120, 2–7 mm, exserted, yellow to gold or creamy white, rarely pink; anthers attached dorsally, lacking glands; ovary 1, sessile to short-stalked; style thread-like; stigma thread-like. **Fruits** erect to pendulous, usually dehiscing along both sutures, occasionally indehiscent, linear to oblong, straight to slightly curved, flattened to nearly circular in cross section, dry, stiffly papery to leathery or woody, glabrous to pubescent; stipe short; apex sometimes beaked. **Seeds** arranged in one or two rows, sometimes surrounded by pulp, ovoid to ellipsoid, sometimes flattened, usually smooth, pleurogram present. $x = 13$.

Species about 60 in the New World (3 native in the flora, 7 cultivated): s United States, Caribbean, Mexico, Central America, South America (to c Argentina and Chile); additional species in Old World.

Vachellia species have been employed for diverse uses. The wood of many species is important locally for construction, posts, wooden articles, and for fuel and charcoal. Pods of various species are utilized as food by wildlife and livestock and have sometimes been used as food by people. The flowers are suitable bee forage for honey production. The bark of some species has been used in tanning leather. Medicinal products and gums are also obtained from species of *Vachellia*. In Africa, the spiny stems are utilized for corrals to protect livestock from predators. Many species are well-suited for landscape use in dry regions and several make excellent shade trees.

SELECTED REFERENCES: Clarke et al. 1989, 1990, 2009; Ebinger et al. 2002; Ross 1979; Seigler and Ebinger 2005; Smit 2008.

1. Leaves with 17 to 48 pairs of pinnae; young foliage densely pubescent with yellow-green hairs ____ **7. Vachellia pennatula**
1′ Leaves with 1 to 13 pairs of pinnae; young foliage not densely pubescent with yellow-green hairs.
  2. Flowers 4-merous, inflorescence an elongated spike ____ **8. Vachellia rigidula**
  2′ Flower 5-merous, inflorescence globose.
    3. Pericarpic strip lining inside of each fruit valve.
      4. Pinnae more than 15 mm, many with more than 12 leaflets/pinna ____ **5. Vachellia karroo**
      4′ Pinnae less than 14 mm, consistently with fewer than 12 leaflets/pinna.
        5. Leaves with 2–8 pairs of pinnae; rachis more than 10 mm ____ **2. Vachellia constricta**
        5′ Leaves with 1–2(–3) pairs of pinnae; rachis less than 10 mm ____ **9. Vachellia vernicosa**
    3′ Pericarpic strip absent.
      6. Fruits, twigs, and petioles with pustular glands, glabrous ____ **6. Vachellia nebrownii**
      6′ Fruits, twigs, and petioles lacking pustular glands, some parts of the plant pubescent.
        7. Fruits 20–50 mm wide, densely puberulent ____ **3. Vachellia erioloba**
        7′ Fruits less that 20 mm wide, glabrous to pubescent with erect hairs, not puberulent.
          8. Petiolar gland at or just below the lowermost pinna pair; rachis 0–18 mm ____ **1. Vachellia bravoensis**
          8′ Petiolar gland near middle of petiole; rachis 20–55 mm ____ **4. Vachellia farnesiana**

1. Vachellia bravoensis (Isely) Seigler & Ebinger. Phytologia 87:146. 2005. * [named for the Rio Bravo (Rio Grande) which forms the border between Texas and Mexico] SCHAFFNER ACACIA; *HUISACHE CHINO, HUISACHILLO* Cultivated

Acacia schaffneri (S. Watson) F.J. Hermann var. *bravoensis* Isely, Sida 3:383. 1969

Shrub or small tree, erect, to 6 m. **Stems:** twigs red brown, slightly zigzagged, round, glabrous to lightly pubescent; short shoots present between the stipular spines; bark of limbs and trunk dark brown or black, rough and fissured. **Leaves:** pinnae 1–4 pairs per leaf, 7–28 mm; stipular spines straight, 1–15(–30) × 0.6–1.5 mm near the base, pubescent at least near the base; petiole 2–8 mm, usually densely pubescent; petiolar gland solitary, sessile, circular, 0.2–1.1 mm across, at or just below the first pinna pair; rachis 0–18 mm. **Leaflets** 10–24 pairs per pinna, oblong, oblique and obtuse at base, obtuse apex, 2–4.2(–5) × 0.6–1.1 mm, glabrous. **Inflorescence** a densely-flowered globose head, 8–11 mm across, 1–5 clustered on the short shoots. **Flowers** 5-merous, yellow; calyx 0.8–1.6 mm, the lobes glabrous to puberulent; corolla 2–3 mm, the lobes glabrous to puberulent; filaments 3.2–4.2 mm. **Fruits** dark brown or black, 70–160 × 6–9 mm, straight to slightly curved, nearly circular in cross section, usually constricted between the seeds, leathery, densely pubescent, glandular with minute, red, deciduous glands; pericarpic strip absent; stipe 4–10 mm; apex acute to obtuse. **Seeds** brown, arranged in single row, ovoid, slightly flattened, 6–9.1 × 5–6.5 mm.

Flowering Feb–Apr. s Texas; n Mexico (Coahuila, Hidalgo, Nuevo León, San Luis Potosí, Tamaulipas).

Vachellia bravoensis is locally common on plains and along arroyos in the Tamaulipan Thornscrub of southern Texas and northeastern Mexico, where it also occurs westward into the Chihuahuan Desert. It is frequently found on disturbed sites. Elevation ranges from 0 to 1500 m. The species is reported to be strongly cyanogenic (Miller n.d., Clark et al. 1989).

Occasionally planted in Arizona landscapes, *Vachellia bravoensis* is suitable for patios, street medians, and other small spaces. The canopy is relatively open in at least some individuals, with the leaves densely clustered along the stems, giving the tree a distinctive appearance. The leaves are deciduous in areas with winter freezing.

Vachellia bravoensis does best in deep soil. Plants in rocky soil or caliche may require extra irrigation to maintain an acceptable appearance. Established plants are drought resistant but do best with occasional deep irrigation during the growing season. Plants appear to be hardy to at least -12°C. Easily propagated by scarified seeds.

2. Vachellia constricta (Bentham) Seigler & Ebinger. Phytologia 87:152. 2005. * [the specific epithet refers to the constrictions in the fruits between the seeds] WHITETHORN ACACIA, MESCAT ACACIA; *MEZQUITILLO* Native, Cultivated

Acacia constricta Bentham in A. Gray, Pl. Wright. 1:66. 1852; *Acacia constricta* Bentham var. *paucispina* Wooton & Standley; *Acaciopsis constricta* (Bentham) Britton & Rose

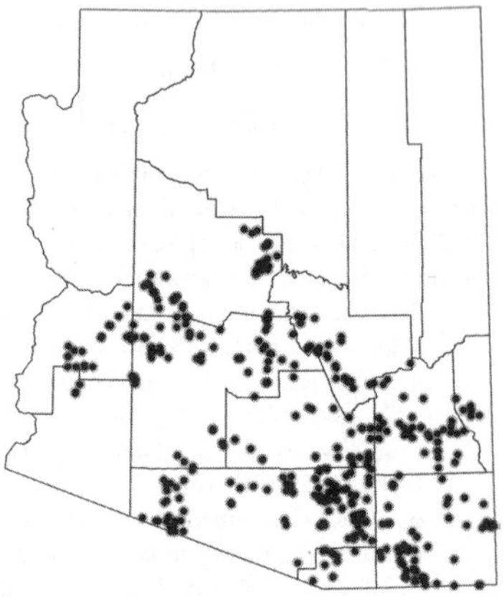

Shrub or small tree, erect, to 6 m. **Stems:** twigs red brown, not zigzagged, round, glabrous to pubescent; short shoots present between the stipular spines; bark gray and smooth, becoming rough and flaking on larger limbs and trunks. **Leaves:** pinnae 2–8 pairs per leaf, 5–12 mm; stipular spines straight, 5–15 (–38) × 0.8–1.2 mm near the base, usually glabrous; petiole 4–14 mm, lightly puberulent; petiolar gland solitary, usually sessile, circular, 0.2–0.7 mm across; rachis 10–35 mm. **Leaflets** 5–12 pairs per pinna, elliptic to oblong or oval, oblique and obtuse at base, acute to obtuse apex, 1.5–3.6 × 0.7–1.3 mm, mostly glabrous. **Inflorescence** a densely-flowered globose head, 5.5–9.5 mm across, 1–5 clustered on the short shoots. **Flowers** 5-merous, yellow; calyx 0.9–1.4 mm, glabrous; corolla, 1.7–2.4 mm, glabrous; filaments 3.5–5 mm. **Fruits** tan to red brown, 50–150 × 3.6–5.6 mm, straight to slightly curved, flattened, constricted between the seeds, papery, glabrous, without glands; a papery pericarpic strip lining the inside of each valve; stipe 4–8 mm; apex acuminate and beaked. **Seeds** mottled gray-black, arranged in a single row, oblong, flattened, 5–7.1 × 2.5–3.6 mm.

Flowering Apr–Oct. Arid sites; 400–2000 m; Chihuahuan Desertscrub, Sonoran Desertscrub, Semidesert Grassland, (Interior Chaparral); Cochise, Gila, Graham, Greenlee, La Paz, Maricopa, Pima, Pinal, Santa Cruz, Yavapai cos.; s New Mexico, w Texas; Mexico (s to Puebla and Oaxaca).

In Arizona, *Vachellia constricta* is widespread in a variety of soil types and habitats. In the Sonoran and Chihuahuan Deserts it is found from alluvial valleys and arroyos to rocky slopes. It is also found in semidesert grassland and

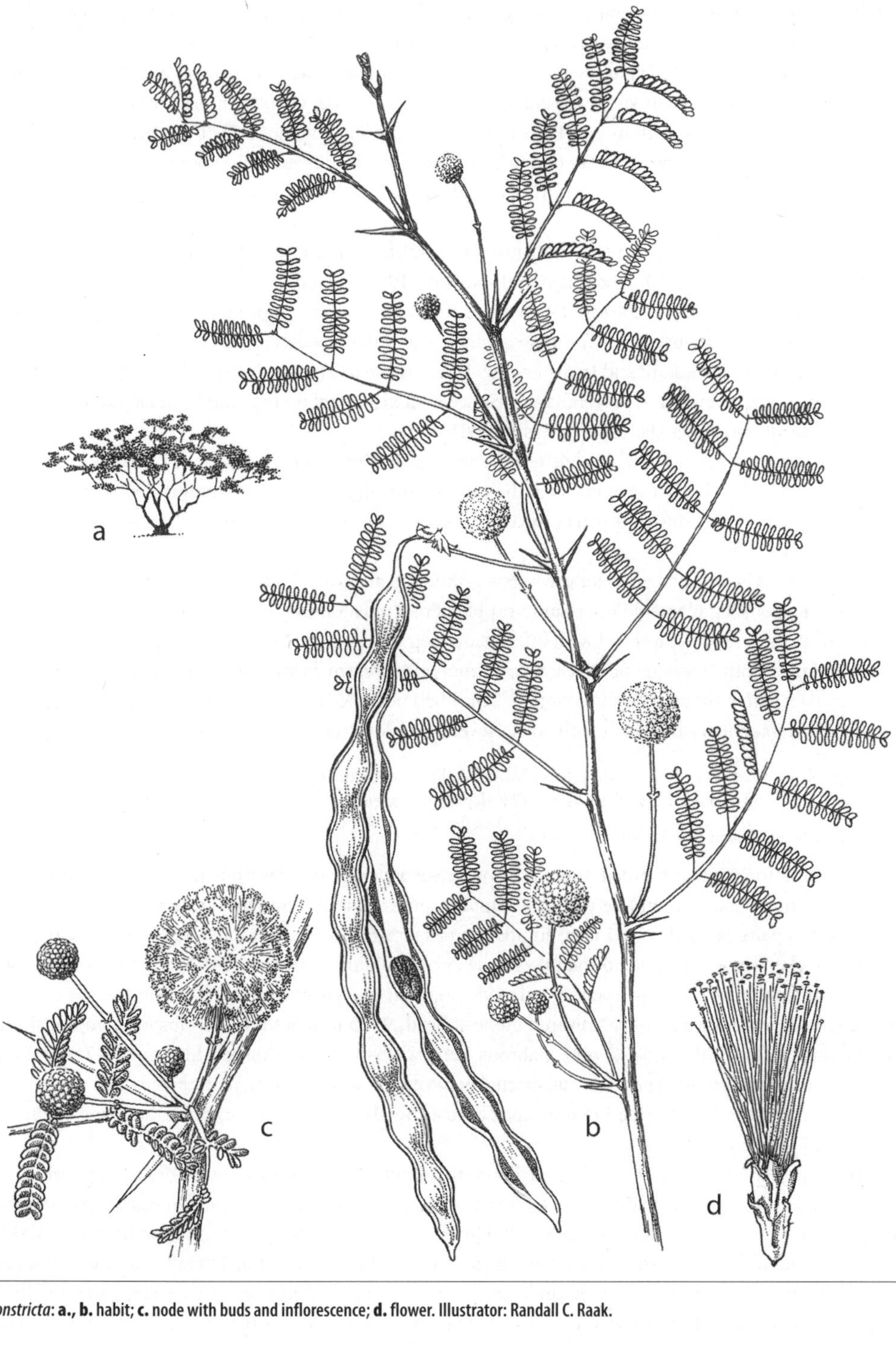

***Vachellia constricta*: a., b. habit; c. node with buds and inflorescence; d. flower. Illustrator: Randall C. Raak.**

occasionally in interior chaparral. Its western distribution in Arizona appears to be limited by a lack of sufficient summer rainfall, and the northern and altitudinal distribution is likely determined by a combination of winter cold, competition from other plants, and wildfire. Elsewhere, *V. constricta* is found in Sinaloan Thornscrub and Sinaloan Thornscrub.

Whitethorn acacia is often a major component of the vegetation in areas where it is present. In winter, the leafless, maroon to pinkish stems are conspicuous. The appearance of new leaves on whitethorn acacia in the spring is a good indicator that the danger of damaging freezes has past; however, in years with particularly dry winters, leaves may not appear until the onset of the summer rains. During extended drought, the plants may become leafless. Plants often flower in the spring in years with adequate rainfall, and again with summer rains. At least 11 mm of rain is required to initiate flowering (Bowers and Dimmitt 1994). In most areas, whitethorn acacia grows as a shrub 2–4 m tall but can develop into a small tree in favorable sites with adequate moisture. The degree of spininess varies greatly from plant to plant and even on a single stem, with many plants or stems having small spines and others bearing conspicuous long white spines. This species is long-lived, with a documented age of at least 72 years (Goldberg and Turner 1986). *Vachellia constricta* is tetraploid while the related *V. vernicosa* is diploid (Isely 1969). This species is reported to nodulate (Allen and Allen 1981).

The seeds of *Vachellia constricta* are eaten by cattle and wildlife, including quail, and native peoples ground them to make pinole (Vines 1960, Tull 1987, Cheatham and Johnston 1995, Powell 1998). This species is considered a good honey plant in Texas (Powell 1988). Parker (1972) states that *V. constricta* can produce hydrocyanic acid under some range conditions. The roots are used to make a medicinal tea, and the powdered pods have been employed in treating skin rashes (Moore 1989). Cheatham and Johnston (1995) report the use of powdered fruits and leaves to stop external bleeding and to treat diaper rash. The Seri Indians of Sonora, Mexico, treat diarrhea and stomach ache with a tea made from the mashed seeds and leaves (Felger and Moser 1985).

Whitethorn acacia is grown for landscape use in the state. It is ideal for small spaces. It makes an excellent patio tree, providing light shade to break the intensity of the summer sun and allowing maximum winter sun when it is deciduous. The plants are also well-suited for street medians and beneath utility lines due to their small stature. They also make an effective barrier.

Whitethorn acacia is recommended for use in revegetation in areas to which it is native. It is hardy to at least -18°C for brief periods. Established plants do not require supplemental irrigation within their natural distribution; however, growth rate and appearance are improved with infrequent deep watering during the growing season. The flower heads are attractive and pleasantly fragrant and will appear at intervals in warm weather with rainfall or irrigation. The plants produce minimal leaf litter, though the abundant pods may be a nuisance in some situations. When young, pruning can encourage the development of a tree form with one to several trunks. Propagation is by scarified seeds.

3. Vachellia erioloba (E. Meyer) P.J.H. Hurter in Mabberly, Pl. Book (ed.3) 1021. 2008. * [from the Greek *erios* wool and *lobos* lobe, referring to the fruit] GIRAFFE THORN Cultivated

Acacia erioloba E. Meyer, Comm. Pl. Afr. Austr. 1:171. 1836

Shrub or tree, erect, to 22 m, with broad, rounded crown. **Stems:** twigs often purplish, zigzagged, round, glabrous to rarely pubescent; short shoots present between the stipular spines; bark of limbs and trunk black, rough and fissured. **Leaves:** pinnae 1–5 pairs per leaf, 11–35 mm; stipular spines straight, 5–50(–100) × 1–5 mm near the base, glabrous; petiole 1–14 mm, glabrous; petiolar gland solitary, sessile, oval to slightly circular, 0.3–1.1 mm across; rachis 0–40 mm. **Leaflets** 7–17 pairs per pinna, oblong to rarely narrowly obovate, oblique and obtuse base, obtuse apex, 4–10 × 1.3–4 mm, glabrous. **Inflorescence** a densely-flowered globose head, 9–15 mm across, 1–4 clustered on the short shoots. **Flowers** 5-merous, yellow; calyx 1.5–2.5 mm, glabrous; corolla 2.5–3.6 mm, glabrous; filaments 5–7 mm. **Fruits** gray, 60–130 × 20–50 mm, curved, flattened, not constricted between the seeds, woody, densely puberulent with gray hairs, glandular; pericarpic strip absent; stipe 1–5 mm; apex obtuse. **Seeds** light to dark brown, arranged in one row, elliptic, flattened, 8–14 × 7–10 mm.

Flowering Apr–Oct. Africa (native to Angola, Botswana, Zimbabwe, Namibia, South Africa (Transvaal), Zambia).

Vachellia erioloba is a conspicuous, long-lived tree in areas of the Kalahari Desert and in adjacent semi-arid habitats where it often forms open savannah woodlands and may be the dominant species. It is found from 0 to 1500 m. This species varies considerably in size, sometimes growing as shrubs seldom more than 2 m tall, but more often developing into small to large trees. It typically grows in sandy or rocky soil on plains and along dry watercourses and floodplains. In the driest habitats, it is confined to watercourses where it has access to groundwater. Giraffe thorn is the largest tree in many areas where it is found. Nodulation has been reported for this species (Allen and Allen 1981).

In its native habitat, *Vachellia erioloba* provides food and shelter for a variety of wildlife species. The large gray pods are eagerly sought by wildlife and livestock (Palgrave 1983, Timberlake et al. 1999); however, Palgrave (1983) reports that the pods can be toxic in some seasons. Timberlake (1980) indicates that the seeds contain prussic acid and have caused livestock poisoning. Sociable weaverbirds (*Philetairus socius* Latham) often construct huge nests in the canopies of *V. erioloba*, and these sometimes become so heavy that they will cause large limbs to break (Ross 1975).

The wood of *Vachellia erioloba* is red brown, hard and heavy, and is an excellent fuel. It has been used for mine timbers and for building wagons (Palgrave 1983), as well as for posts, corrals, and in machinery (Timberlake 1980). It is resistant to wood boring insects and termites (Palgrave 1983). The tree has been severely overharvested in some areas of South Africa (Ross 1975). The species is deep-rooted. Roots, thought to be of this species, were found at a depth of 46 m in a borehole in Namibia (Ross 1975, Timberlake 1980). In areas that receive less than 250 mm of annual rainfall, *V. erioloba* is considered an indicator of underground water (Timberlake 1980). The flowers are considered good bee forage, and the seeds have been used as a coffee substitute (Timberlake 1980). Gum produced by this species is edible and has been eaten by local people (Palgrave 1983) and used to treat gonorrhea (Timberlake 1980). In Botswana, headache is treated with burned and crushed bark, and a powder made from the dry pods is used for ear infections (Timberlake 1980, Palgrave 1983). The roots have been employed as a remedy for cough (Timberlake 1980).

Giraffe thorn is uncommon in the Arizona nursery trade. It is a suitable tree for larger spaces, as the potential mature size is larger than some residential lots. Trunks of large individuals can exceed 1.5 m in diameter and the canopy may spread to over 25 m. The stout spines are a consideration when siting this species. The canopy provides dense shade, and on older plants may hang down unless pruned. The plants are typically evergreen in cultivation in Arizona.

Giraffe thorn does best in deep soil and is drought resistant when established. The slow growth rate of this species can be increased with periodic deep irrigation. Cultivated plants in Arizona have survived temperatures to -8°C without damage. Propagation is by scarified seeds.

4. Vachellia farnesiana (Linnaeus) Wight & Arnott. Prodr. Fl. Ind. Orient. 1:272. 1834. * [named for Odoardo Farnese, 1573–1626, Roman Catholic Cardinal] SWEET ACACIA; *HUISACHE*

Mimosa farnesiana Linnaeus, Sp. Pl. 1:521. 1753; *Acacia acicularis* Humboldt & Bonpland ex Willdenow; *Acacia edulis* Humboldt & Bonpland ex Willdenow; *Acacia farnesiana* (Linnaeus) Willdenow; *Acacia ferox* M. Martens & Galeotti; *Acacia lenticellata* F. Mueller; *Acacia pedunculata* Willdenow; *Acacia smallii* Isely; *Poponax farnesiana* (Linnaeus) Rafinesque; *Poponax venosa* Britton in Britton and Killip; *Vachellia densiflora* Alexander ex Small

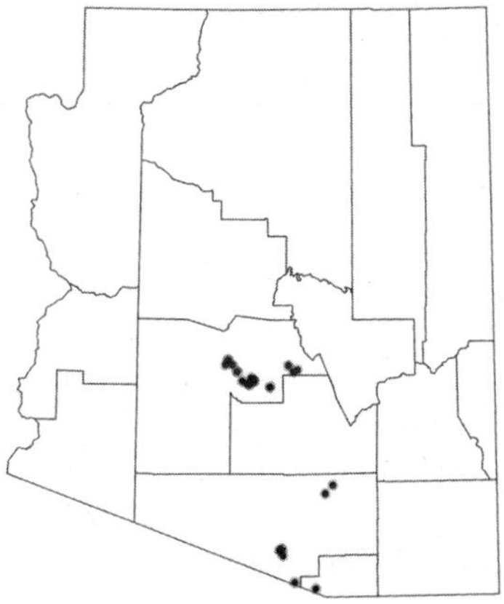

Shrub to small tree, erect, to 8 m. **Stems:** twigs red brown, slightly zigzagged, round, usually glabrous; short shoots present between the stipular spines; bark green to gray with conspicuous lenticels, becoming rough and fissured. **Leaves:** pinnae 2–6 pairs per leaf, 16–33 mm; stipular spines straight, 5–35 (–55) × 0.5–1.5 mm near the base, mostly glabrous; petioles 4–12(–17) mm, sparsely pubescent; petiolar gland solitary, sessile, circular to slightly elongated, 0.3–1 mm across, near middle of petiole; rachis 20–55 mm. **Leaflets** 8–19 pairs per pinna, oblong, oblique and obtuse at base, obtuse to broadly acute apex, 2.8–6.3 × 0.7–1.7 mm, glabrous to pubescent. **Inflorescence** a densely-flowered globose head, 6–10 mm across, 1–5 clustered on the short shoots. **Flowers** 5-merous, bright yellow; calyx 1.1–1.8 mm, glabrous; corolla 1.9–2.8 mm, glabrous; filaments 3.5–5.5 mm. **Fruits** red brown to nearly black, 40–90 × 9–18 mm, straight to slightly curved, nearly circular in cross section, not constricted between the seeds, leathery, glabrous, without glands; pericarpic strip absent; stipe 1–7 mm; apex obtuse to acuminate, sometimes beaked. **Seeds** olive green to dark brown, mostly arranged in two rows, ovoid to ellipsoid, slightly flattened, 4.2–8 × 3.5–5.8 mm.

Varieties 3 (1 native in the flora): widespread from southern United States and the Caribbean to southern South America and widely naturalized in dry to humid climates in parts of Africa, Asia and Australia.

There are three varieties. The native Arizona populations and all or most of the plants grown in horticulture in the state are var. *farnesiana* with fruits less than 100 mm and seeds in two rows or irregularly arranged; however, var. *minuta* (M.E. Jones) Seigler & Ebinger, with fruits 100–170 mm and seeds in one row may be present in horticulture. The third variety, var. *pinetorum* (F. J. Hermann) Seigler & Ebinger, occurs outside the region.

4a. Vachellia farnesiana (Linnaeus) Wight & Arnott var. **farnesiana** Native, Cultivated

Large shrub or small tree, erect, to 8 m. **Leaves** mostly more than 30 mm; pinnae 4 to 6 pairs per primary leaf, mostly more than 18 mm; usually lightly pubescent with erect hairs. **Leaflets** mostly more than 3 mm, mostly glabrous or nearly so beneath. **Fruits** less than 90 mm. **Seeds** arranged in two rows or irregularly arranged, imbedded in a sweet pulp.

Flowering Dec–May. Relatively dry, disturbed habitats; [0–]300–1220[–1500] m; Sonoran Desertscrub, Semidesert Grassland; Pima Co.; Texas, Florida; West Indies; Mexico; Central America; Caribbean; South America (s to Argentina).

Vachellia farnesiana var. *farnesiana* is native to Arizona in the Baboquivari Mountains, but in recent decades has become established and is spontaneously reproducing in other areas of eastern Pima County and Maricopa County. In these areas it can be found along roads and arroyos, sometimes well away from settlements, as escapees from cultivated plants. It is establishing in Pinal County as well. Cold and aridity apparently limit its distribution in Arizona. Across its wide range, var. *farnesiana* can be found in a variety of habitats from dry plains to periodically inundated areas with a shallow water table. It can form extensive populations in favorable habitats. *Vachellia farnesiana* var. *farnesiana* has become widely naturalized in many warm regions of the world including Africa, Australia, and India. Nodulation is widely reported in *V. farnesiana* (Allen and Allen 1981).

Vachellia farnesiana is best known outside of horticulture for the use of oils from the flowers in the manufacture of perfumes. This species is valued for a variety of purposes. The wood is used for fuel and to make charcoal, and for posts and woodworking; the foliage and pods are considered to be good forage; a high quality honey is produced from the nectar; the bark and pods have been used for making ink and in dyeing and tanning; glue has been made from the pods to mend broken pottery (Tull 1987, Powell 1988, Cheatham and Johnston 1995, Powell 1998). Tull (1987) indicates that the bark fiber has been used to make rope. Cheatham and Johnston (1995) report that the bark and flowers have been used to produce a yellow dye for leather and the flowers alone make a yellow dye for wool. Gum from the stems has been used as a substitute for gum Arabic (Tull 1987, Cheatham and Johnston 1995). Cheatham and Johnston (1995) report that the leaves are used as a condiment in southeast Asia. They also report the use of flowers as an aphrodisiac and as an insecticide. The Guarajío people of Mexico mix the flowers with grease to make a salve to treat bruises and headaches (Gentry 1942). The roots have reportedly been chewed to counter scorpion envenomation (Martin et al. 1998).

Cheatham and Johnston (1995) report extensive uses in folk medicine including boiling the green fruits and applying the liquid to sooth poison ivy and other skin irritations, using the ground leaves to make a poultice for treating wounds, and making various decoctions to treat diarrhea and intestinal disorders. They also report the following local uses: the Tarahumara make a decoction of the bark and spines to treat scorpion stings and insect bites; a decoction of the spines is also used to treat kidney disorders; the Tarahumara apply the flowers to bruises and the flowers mixed with grease are applied to the head to treat headache; the Pima Bajo of Sonora apply mashed bark to the forehead for headache; the Mexican Kickapoo boiled the roots and used the liquid to treat conjunctivitis, and a leaf decoction was taken to treat flu; the Huasteca boiled the bark to make a mouthwash to alleviate toothache, and a root decoction was employed for backache; a root decoction has been used in San Luis Potosí to treat tuberculosis; the Maya make a decoction of the heartwood to treat biliousness; the pods are used in Cuba to treat sore throat and conjunctivitis; strips of bark are applied in the West Indies for relief of arthritis; flowers are used to treat nervous disorders in the Yucatán; a root decoction was used in Venezuela to treat stomach cancer, and tumors were treated with plaster made from pulp; in the Philippines, a root decoction is used to treat leucorrhea; and Lodha women of India mixed powdered bark with banana as a contraceptive following menstruation.

Sweet acacia is popular and widely planted at lower elevations in Arizona where it is grown for shade and for its showy masses of fragrant flowers. It is suitable as a patio or street median tree. It commonly develops a vase- shaped form but requires pruning when young to remove basal sprouts and outer limbs which tend to droop with time. The stems are spiny.

Sweet acacia does best in deep soil with periodic deep watering during dry weather. Growth is rapid with sufficient irrigation. Native populations, and most cultivated sweet acacias grown in the state, flower in February and March. They are hardy to ca. -9°C. Some cultivated plants flower from November–January and may suffer freeze damage at -6° or -7°C. These plants may have originated from material introduced from a milder portion of the species' range. Sweet acacia can become partially deciduous in cold winters. The plants produce a fine leaf litter but the abundant pods can be a maintenance issue. Sweet acacia is sensitive to infections of cotton root rot or Texas root rot (*Phymatotrichopsis omnivora*), a fungus that can cause the rapid death of affected plants during hot weather. Scarified seeds germinate readily. *Vachellia farnesiana* should be considered potentially invasive beyond its natural distribution.

5. Vachellia karroo (Hayne) Banfi & Galasso. Atti Soc. Ital. Sci. Nat. Mus. Civico Storia Nat. Milano 149:149. 2007. * [named for the Karoo region of South Africa where this species is found] KAROO THORN, SWEET THORN Cultivated

Acacia karroo Hayne, Getreue Darstell. Gew. 10:pl. 33. 1827

Shrubs to trees, erect, to 12 m, crown rough to flattened. **Stems:** twigs green aging to brown, not zigzagged, round, glabrous; short shoots very small but usually present between the stipular spines; bark tan, smooth, becoming dark brown or black, rough and fissured on larger limbs and trunk. **Leaves:** pinnae 1–7(–13) pairs per leaf, 20–55 mm; stipular spines straight, 5–100(–250) × 1–3.5 mm near the base, glabrous; petiole 5–18 mm, glabrous; petiolar gland solitary, usually sessile, circular, 0.8–3.5 mm across; rachis 0–60 mm. **Leaflets** (8–)11–26 pairs per pinna, oblong to obovate, oblique and obtuse at base, obtuse apex, 4–9 × 1.2–2.5 mm, glabrous. **Inflorescence** a densely-flowered globose head, 7–12 mm across, 1–10 clustered in the leaf axils or in short terminal racemes; involucral bract at middle of peduncle. **Flowers** 5-merous, golden yellow; calyx 1.3–2.1 mm, glabrous; corolla 2–3.2 mm, glabrous; filaments 4–6 mm. **Fruit** brown, 40–150 × 5–9 mm, curved, flattened, slightly constricted between the seeds, leathery, glabrous, without glands; a papery pericarpic strip lining the inside of each valve; stipe 5–10 mm; apex acuminate, beaked. **Seeds** light brown, arranged in a single row, oblong, flattened, 5.3–9.4 × 3.5–6.5 mm.

Flowering Apr–Oct. Widespread in southern Africa.

Vachellia karroo commonly grows as a spiny shrub or small tree with a rounded to flat crown. In some areas it forms a large tree to 22 m tall (Ross 1975). Several distinctive forms are widely reported, but these are generally not distinguished taxonomically. One form is upright, spindly, and open and there are differences in bark color and other features, but these forms appear to grade into typical *V. karroo* (Ross 1975). This is the most widespread species of *Vachellia* in southern Africa, and it is found in a wide variety of habitats and soil types (Ross 1975). It is found from 0 to 1500 m in dry thornveld, scrub land, woodland, savanna, grassland, and coastal dunes. Sometimes it forms dense stands, and in some areas, it may be the only tree present. In southern Africa, the species increases greatly on overgrazed rangeland (Ross 1975). Nodulation has been observed on *V. karroo* (Allen and Allen 1981).

Vachellia karroo is widely used in regions where it is native. The wood is hard and heavy but not especially durable. The sapwood is yellow to pale brown in color and the heartwood is red brown (Timberlake et al. 1999). The wood has had many uses including fuel and charcoal, fencing, corrals, and making furniture (Timberlake 1980, Palgrave 1983). The wood has a density of 800–890 kg/m^3 (Timberlake et al. 1999). Cut trees often coppice (Ross 1975). The spines have occasionally been used as needles (Palgrave 1983). The leaves and pods are considered good livestock forage, and the trees are considered excellent for honey production (Timberlake 1980, Palgrave 1983). The bark was used in tanning leather and is employed in treating tulp (*Moraea* sp.) poisoning in cattle, while the inner bark is used to make durable rope (Timberlake 1980, Palgrave 1983). The bark yields a yellow or brown fabric dye (Timberlake et al. 1999). The high-quality gum produced by *V. karroo* is used in confectionary, pharmaceuticals, and as an adhesive (Timberlake 1980, Timberlake et al. 1999, Palgrave 1983). The roots are chewed as an aphrodisiac, and an infusion of the roots is used to treat gonorrhea and as an emetic (Timberlake 1980). Roots have been placed in poultry enclosures to reduce parasites (Timberlake et al 1999).

Karoo thorn is occasionally available from commercial nurseries in Arizona. It grows as a spiny shrub or small tree, typically with a single, low-branched trunk and a rounded crown. The abundant, stout white spines of some individuals of this species are formidable and can inflict serious injury. The spines are especially impressive when the trees are leafless. It can be planted to form an effective barrier. With pruning, it can be grown as a patio tree. Karoo thorn is also useful for erosion control (Jones and Sacamano 2000). The canopy provides dense shade. The flowers are fragrant.

Established trees do best with infrequent deep irrigation and are tolerant of a wide variety of soil types. Plants have a moderate growth rate with irrigation. The species prefers well-drained soil but is often found on clay soils (Timberlake 1980). Plants are typically evergreen but will become deciduous with extended drought or extreme cold. At least two forms have been grown in Arizona. One form is damaged at ca. -6°C, but the form available for landscaping appears to be hardy to at least -12°C. Leaf litter is not a problem in most landscape situations, though the pods may require removal. Karoo thorn is known to seed aggressively in some situations where it is native and should be considered potentially invasive. The lifespan for this species is 30–40 years (Timberlake et al. 1999). Scarified seeds are easily germinated. It can also be propagated from root suckers and will successfully transplant (Timberlake 1980).

6. Vachellia nebrownii (Burtt Davy) Seigler & Ebinger. Phytologia 92(1):95. 2010. * [named for the English botanist Nicholas E. Brown, 1849–1934] WATER ACACIA Cultivated

Acacia nebrownii Burtt Davy, Bull. Misc. Inform. Kew 1921:50. 1921

Shrubs or slender trees, erect, to 5 m. **Stems:** twigs red brown, sometimes slightly zigzagged, round, glabrous, with sessile pustular glands; short shoots present above the stipular spines; bark orange brown to gray and somewhat papery, green and smooth beneath. **Leaves:** pinnae 1–2 pairs per leaf, 3–17 mm; stipular spines straight to slightly deflexed, 1–70 × 1–3 mm near the base, glabrous; petiole 2–17 mm, with sessile pustular glands; petiolar gland solitary, short columnar, 0.2–0.9 mm long, apex oval to orbicular, 0.3–0.8 mm across; rachis 0–10 mm. **Leaflets** 2–6 pairs per pinna, oblong to orbicular, oblique and obtuse at base, acute apex, 2–7 × 1.1–4.5 mm, glabrous beneath. **Inflorescence** a densely-flowered globose head, 9–14 mm across, 1–4 clustered on the short shoots. **Flowers** 5-merous, yellow; calyx 1.1–1.9 mm, glabrous; corolla 2–3 mm, glabrous; filaments 6–8 mm. **Fruits** tan, 20–60 × 7–15 mm, curved, flattened, not constricted between the seeds, papery, glabrous, with pustular glands; pericarpic strip absent; stipe 1–5 mm; apex acute. **Seeds** brown, arranged in a single row, elliptic, flattened, 6–10 × 5–7 mm.

Flowering Apr–Oct. Botswana, Namibia, w Zimbabwe, Transvaal region of South Africa.

In habitat, *Vachellia nebrownii* is found on rocky outcrops, sand flats, pans, riverbanks, dry watercourses, and river valleys in dry scrub and woodland. It is frequently thicket-forming (Ross 1975). There are two distinct forms that are separated geographically (Timberlake et al. 1999). Nodulation is reported for this species (Allen and Allen 1981).

Vachellia nebrownii produces an edible gum (Timberlake 1980). The common name is derived from a belief that this species is an indicator of groundwater; however, this is probably due to occurrence over particular rock types that hold water (Timberlake 1980).

Water acacia makes an attractive shrub or small, multiple-stemmed patio tree for warmer locations. It can be grown as a spiny barrier and is suitable for narrow spaces where the spiny stems can be pruned to a sufficient height. The papery bark is interesting. Flowers appear at intervals during warm weather with sufficient moisture.

Plants are drought tolerant when established but periodic deep irrigation helps to maintain an acceptable appearance. The plants are deciduous with freezing or extended drought. This species appears to be hardy to ca. -5°C. Cultivated plants in Tucson have sustained severe damage at -7°C. Water acacia produces small quantities of fine leaf litter. Propagation is by scarified seeds.

7. Vachellia pennatula (Schlechtendal & Chamisso) Seigler & Ebinger. Phytologia 87:164. 2005. * [the specific epithet refers to the feathery pinnate leaves] FEATHER ACACIA; *ALGARROBO*

Inga pennatula Schlechtendal & Chamisso, Linnaea 5:593. 1830; *Acacia lanata* M. Martens & Galeotti; *Acacia pennatula* (Schlechtendal & Chamisso) Bentham; *Pithecellobium minutissimum* M.E. Jones; *Poponax pennatula* (Schlechtendal & Chamisso) Britton & Rose

Trees, erect, to 15 m, with a broad crown. **Stems:** twigs green, slightly zigzagged, round, densely pubescent with grayish to yellow-green hairs; short shoots absent; bark red brown to brown with granulated texture, becoming rough and fissured on larger limbs and trunks. **Leaves:** pinnae 17–48 pairs per leaf, 15–38 mm; stipular spines straight, 6–25(–40) × 1–3 mm near the base, densely pubescent at the base and sometimes throughout; petiole 3–8 mm, densely pubescent; petiolar gland solitary, sessile, circular to slightly elongated, 1–2.5 mm across; rachis 60–200 mm. **Leaflets** 24–60 pairs per pinna, linear, oblique and obtuse at base, obtuse apex, 0.8–3 × 0.4–0.7 mm, glabrous to lightly pubescent. **Inflorescence** a densely-flowered globose head, 9–16 mm across, 1–5(–10) clustered in the leaf axils. **Flowers** 5-merous, yellow; calyx 0.7–1.6 mm, puberulent; corolla 2–2.6 mm, puberulent; filaments 3–4.5 mm. **Fruits** dark red brown, 50–160 × 16–30 mm, straight, flattened, not constricted between the seeds, leathery, glabrous, without glands; pericarpic strip absent; stipe 1–3(–8)mm; apex obtuse. **Seeds** light brown, arranged in a single row, ovoid to ellipsoid, slightly flattened, 6.5–8.4 × 4.2–6 mm.

The plants in Arizona horticulture are var. *pennatula*, mostly or all from material originating in eastern Sonora, Mexico. A second variety, var. *parvicephala* (Seigler & Ebinger) Seigler & Ebinger, ranges from southern Mexico to northern South America (Seigler and Ebinger 1988).

Varieties 2 (1 cultivated in the flora): Mexico and Central America to northern South America.

7a. Vachellia pennatula (Schlechtendal & Chamisso) Seigler & Ebinger var. **pennatula** Cultivated

Twigs densely pubescent with erect, yellow-green hairs mostly more than 0.3 mm long. **Leaves**: petioles densely pubescent with erect, yellow-green hairs mostly more than 3 mm. **Inflorescence**: peduncles mostly more than 12 mm, densely pubescent with erect, yellow-green hairs; floral bracts 1.2–2 mm. **Fruits:** stipe 2–6(–8)mm.

Flowering Mar–June in Mexico and Central America, May–Dec in South America. Mexico and Central America to northern South America. Not native and rarely planted in Arizona.

Vachellia pennatula var. *pennatula* is found from 500–2500 m. It is found on slopes and in valleys at upper elevations in thornscrub and in tropical deciduous forest, oak woodland, and occasionally in lower elevation pine-oak woodland, most commonly on igneous soils. It is often abundant in disturbed habitats. Plants of var. *pennatula* begin flowering at four years of age, and mature trees may produce up to 100,000 seeds (Cházaro 1977). *Vachellia pennatula* is reported to fix nitrogen (Van Kessel et al. 1983).

Fruits of *Vachellia pennatula* are eaten by cattle (Cházaro 1977). The Guarajío people of Mexico ground the pods and seeds for flour, and the bark was used to dye leather (Gentry 1942). Standley (1922) states the bark has been used to treat indigestion.

Feather acacia is rarely grown in Arizona, where it is suited to warm locations and protected microclimates. The distinctive feathery leaves lend a graceful, tropical effect to the landscape. In Arizona, plants seldom exceed 5–7 m tall, making them suitable for use as a patio tree. This species should be carefully sited due to the spiny stems. The foliage is evergreen unless killed by freezing temperatures.

Feather acacia does best with regular irrigation during the growing season. Cultivated plants in Arizona are hardy to ca. -4°C and suffer severe damage below -6°C. Prune freeze-damaged wood after the danger of frost is past. The leaves produce fine litter though the thick pods can become a nuisance in some situations. Propagation is by scarified seeds.

8. Vachellia rigidula (Bentham) Seigler & Ebinger. Phytologia 87:166. 2005. * [the specific epithet refers to the rigid stems of this species] BLACKBRUSH ACACIA; *CHAPARRO PRIETO* Cultivated

Acacia rigidula Bentham, London J. Bot. 1:504. 1842; *Acaciopsis rigidula* (Bentham) Britton & Rose

Shrubs or small trees, erect, to 4 m. **Stems:** twigs gray, not zigzagged, round, glabrous to lightly puberulent; short shoots present between the stipular spines; bark gray and smooth, becoming fissured on large trunks. **Leaves**: pinnae 1 pair per leaf, 5–20 mm; stipular spines straight, 10–50(–80) × 1.2–2.1 mm near the base, glabrous; petiole 2–5(–9) mm, densely puberulent; petiolar gland solitary, sessile, circular, 0.3–0.6(–1.1) mm across; rachis absent. **Leaflets** (2–)3–5 pairs per pinna, oblong to elliptic, oblique and obtuse at base, obtuse apex, (4–)6–13(–16) × 3–8 mm, mostly glabrous. **Inflorescence** a loosely-flowered elongated spike, 10–35 mm, 1–4(–8) clustered on the short shoots. **Flowers** 4-merous, white; calyx 0.4–0.8 mm, puberulent; corolla 1.2–1.7 mm, glabrous; filaments 2.7–4 mm. **Fruit** brown or black, 40–100 × 3–6 mm, straight to slightly curved, flattened, constricted between the seeds, leathery, puberulent, without glands; a papery pericarpic strip lining the inside of each valve; stipe 3–7 mm; apex acuminate. **Seeds** dark brown, arranged in a single row, oblong, strongly flattened, 5.5–7.5 × 3.5–4.5 mm.

Flowering Feb–Jul. s Texas; n Mexico.

Vachellia rigidula is a conspicuous and often dominant element of the Tamaulipan Thornscrub of southern Texas and northeastern Mexico where it reaches the eastern margin of the Chihuahuan Desert. It is found at lower elevations on slopes, plains, bluffs, and along drainage channels and arroyos in sandy to clay soils as well as rocky substrates, especially limestone. It can also be found on disturbed sites and along fence rows. The plants can form impenetrable thickets. The dark green foliage is distinctive and stands out among the often paler adjacent vegetation. Its northern range along the Edwards Plateau of Texas is limited by winter freezing. The western distribution is determined by higher terrain and limited rainfall. To the south and east, higher rainfall and competition with more mesic vegetation likely limits distribution.

The spiny stems of *Vachellia rigidula* frequently cause flat tires in southern Texas. This species is considered a good honey plant (Powell 1988, Cheatham and Johnston 1995) and is useful for erosion control (Powell 1988). The Mexican Kickapoo use the stems for construction of huts, ramadas, and benches (Cheatham and Johnston 1995). A leaf decoction of *V. rigidula* or a root decoction of *V. rigidula* and *V. farnesiana* was taken by the Mexican Kickapoo for bloody diarrhea.

Though seldom cultivated in Arizona, blackbrush acacia makes an ideal spiny screen, hedge, or barrier plant. It is suitable as a patio or street median tree and affords excellent wildlife cover. The distinctive dark green foliage resembles that of *Ebenopsis ebano* and contrasts with the pale gray bark. Blackbrush acacia is usually a shrub 3–4 m tall, but in cultivation with adequate irrigation, it sometimes develops into a tree to 6 m tall. Some pruning may be required to encourage this plant to develop into a tree. Showy, cream colored flowers can appear in profusion in the spring. The leaves turn maroon following freezing and are deciduous with extreme cold. Growth rate is slow.

Established plants are drought tolerant in areas that receive at least 300 mm of annual rainfall, but growth rate and appearance are improved with occasional deep irrigation during hot weather. Cultivated plants in Arizona are hardy to at least -9°C. Blackbrush acacia tolerates a wide range of soil conditions including rocky soils and caliche. Litter is minimal, consisting mainly of the small leaflets. Scarified seeds germinate readily.

9. **Vachellia vernicosa** (Britton & Rose) Seigler & Ebinger. Phytologia 87:169. 2005. * [Latin *vernicose*, having a polished or glossy surface] CHIHUAHUAN WHITETHORN, VISCID ACACIA Native

Acaciopsis vernicosa Britton & Rose, N. Amer. Fl. 23:96. 1928; *Acacia constricta* Bentham var. *vernicosa* (Britton & Rose) L.D. Benson; *Acacia neovernicosa* Isely; *Acacia vernicosa* Standley, non *Acacia vernicosa* W. Fitzgerald

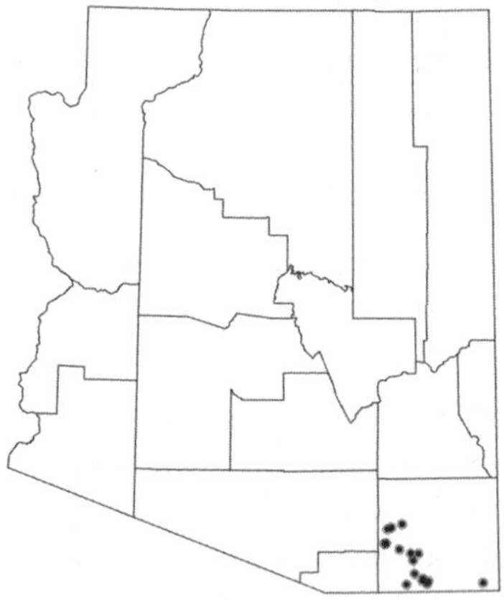

Shrubs, erect, to 3 m. **Stems:** twigs red brown, not zigzagged, round, glabrous and glutinous; short shoots present between the stipular spines; bark gray and often pink-tinged, smooth. **Leaves:** pinnae 1–2(–3) pairs per leaf, 4.5–13 mm; stipular spines straight to slightly reflexed, 3–11(–25) × 0.7–1.1 mm near the base, glabrous and glutinous; petiole 3–10 mm, glabrous and glutinous; petiolar gland solitary, sessile, circular, 0.1–0.4 mm across; rachis 0–9 mm. **Leaflets** 4–12 pairs per pinna, elliptic to oval, oblique and obtuse at base, obtuse apex, 1.1–2.7 × 0.5–0.9 mm, glabrous and glutinous. **Inflorescence** a densely-flowered globose head, 6–8 mm across, 1–5 clustered on the short shoots. **Flowers** 5-merous, pale yellow; calyx 0.6–1.2 mm, glabrous; corolla 1.4–2.1 mm, glabrous; filaments 2.7–4 mm. **Fruits** tan to red brown, 38–90 × 2–4 mm, straight to slightly curved, flattened, usually constricted between the seeds, papery, glabrous and glutinous, glandular with small, clear to reddish, sessile glands; a papery pericarpic strip lining the inside of each valve; stipe 8–12 mm; apex acuminate and usually beaked. **Seeds** mottled gray-brown, arranged in a single row, oblong, strongly flattened, 4.5–6.5 × 1.7–3 mm.

Flowering Mar–Sep. Disturbed, arid sites; [580–]1160–1590[–1800] m; Chihuahuan Desertscrub and degraded Semidesert Grassland; Cochise, Pima cos.; s New Mexico, sw Texas; n Mexico (Chihuahua, Coahuila, Durango, Querétaro, San Luis Potosí, Sonora, Tamaulipas, Zacatecas).

Vachellia vernicosa grows as a multiple-stemmed shrub with an almost skeletal appearance. The reddish to pink bark is particularly distinctive in winter when the plants are leafless. It is localized but often extremely abundant on calcareous soils in southeastern Arizona in Chihuahuan Desert vegetation. It can form almost impenetrable thickets and is frequently the tallest element in these communities. The species is also localized in the Sonoran Desert in northwest Sonora. Occasionally sympatric with the related *V. constricta* and distinguished by the viscid herbage and usually smaller leaves with mostly fewer pinnae. The seeds of *V. vernicosa* were ground into a coarse meal by native peoples (Tull 1987).

Chihuahuan whitethorn is an excellent shrub for the xeric garden where minimal shade is desired, though it is rarely grown. Its open form and sparsely branched, wand-like stems can best be appreciated against a wall. It is suitable for small spaces and hot exposures. When massed, the plants form an effective barrier due to the spiny stems. The colorful bark and fragrant flowers complement the growth habit.

Peak flowering occurs with summer rains and is enhanced with irrigation. The plants are drought tolerant once established in areas with 200 mm or more of yearly rainfall but will become leafless during extended drought. The leaves are also deciduous in winter. Chihuahuan whitethorn is hardy to -20°C for brief periods and can survive several consecutive days of subfreezing temperatures in habitat. The plants produce minimal litter except for the pods, which can be abundant in some years. Plants are easily grown from scarified seeds.

Taxa with potential for cultivation

Additional taxa of *Vachellia* have been introduced to Arizona horticulture but are generally not available from commercial sources. New introductions are likely to become available over time.

Vachellia gerrardii (Bentham) P. J. H. Hurter in Mabberley, Pl.-Book (ed. 3) 1021. 2008. GRAY-HAIRED ACACIA
Gray-haired acacia has been grown on a very limited basis in Arizona. It forms a tree to 15 m tall with a flat-topped canopy. It has gray to blackish outer bark that exfoliates in small flakes, ultimately becoming rough and fissured. The twigs have paired white stipular spines and bipinnate leaves. The foliage is deciduous with winter freezing. The globose flower heads are cream colored to yellowish. Cultivated plants in Arizona appear to be hardy to ca. -7°C. They do best with periodic irrigation.

Vachellia xanthophloea (Bentham) P. J. H. Hurter in Mabberley, Pl.-Book (ed. 3) 1021. 2008. FEVER TREE
Fever tree is rarely available in the state. It forms a tree 10–18 m tall and has powdery yellow or pale green bark. The twigs have paired white stipular spines and bipinnate leaves. Flowers are pale yellow and occur in globose heads. Fever tree is damaged at temperatures below ca. -5°C. The plants need regular deep irrigation to thrive.

VICIA Linnaeus. Sp. Pl. 2:734. 1753. * [Latin name for these plants] VETCH
Steven L. Broich

Herbs, annual, biennial, or perennial, unarmed, from rhizomes or taproots. **Stems** erect, sprawling, or climbing, angled or ridged, not winged, glabrous or pubescent, hairs unicellular, without glands. **Leaves** alternate, even-pinnate (or with odd number of leaflets but still lacking a terminal leaflet); rachis not winged, usually terminating in a simple or branched, sometimes mucronate, tendril; stipules present, inconspicuous or leaf-like, with or without nectariferous patch; petiolate. **Leaflets** 2–30, usually scattered, folded lengthwise, except in some *V. faba*, blade margins entire or serrate, surfaces glabrous or pubescent. **Inflorescences** 1–50-flowered, axillary racemes, erect or lax; bracts and bracteoles absent. **Flowers** papilionoid, chasmogamous or cleistogamous; calyx campanulate, symmetrical or oblique, base sometimes gibbous, lobes 5, equal or unequal; corolla white, cream, yellow, rose, lilac, blue, purple, lavender, or violet, wings and keel often lighter than banner; stamens 10,9 united and vexillary stamen free or partially united with others; anthers attached dorsally, uniform; ovary glabrous or pubescent; style cylindrical or compressed dorsally (perpendicular to the floral axis) or laterally (parallel to the floral axis), never twisted, evenly pubescent just proximal to stigma and/or tufted on lower (outer) side; ovules (1 or) 2–10+. **Fruits** legumes, sessile or stipitate, cylindrical or flattened, oblong, oblong-falcate, linear, or elliptic, dehiscent, valves twisted after dehiscence, without a septum, margins usually obscure, splitting along both margins, glabrous, glabrate, or pubescent. **Seeds** 2–15, usually brown or black, often spherical, sometimes angular. $x = 7$.

Species ca. 140 (4 native in the flora and 2 introduced): North America, n Mexico, s South America, Europe, Asia, n Africa, Mediterranean region; introduced in temperate regions worldwide.

Several species of *Vicia* are widely planted as cover crops and for green manure, for forage, and for erosion control (Allen and Allen 1981, Lewis et al. 2005). *Vicia faba*, fava bean or broadbean, is widely grown as a vegetable crop for human consumption and has also been used in folk medicine (Allen and Allen 1981, Duke 1981, Lewis et al. 2005). Some people are genetically disposed to develop fabism or favism, a type of anemia, that results from consuming the seeds or inhaling the pollen of *V. faba*. Nodulation is widely reported for *Vicia*. Seed descriptions in this treatment are from F. J. Hermann (1960).

1. Stipules with nectariferous patch on lower surfaces; racemes 0–1 cm ____ **5. Vicia sativa**
1′ Stipules without nectariferous patch; racemes 1–25 cm.
  2. Banners oblong, blade equal to claw in width, or banner blade wider than claw; calyx base symmetrical; styles cylindrical or dorsally compressed.
    3. Leaflets 8–18, blade surfaces glabrous or finely pubescent; racemes 3–9-flowered; flowers 12–25 mm, styles dorsally compressed ____ **1. Vicia americana**
    3′ Leaflets 6–8, lower surfaces of blade long-villous, upper surfaces glabrous; racemes 1 or 2-flowered; flowers 7–9 mm, styles cylindrical ____ **2. Vicia leucophaea**
  2′ Banner claws and blades equal in width separated by a narrower "waist"; calyx base symmetrical or gibbous; styles laterally compressed.
    4. Flowers 12–18 mm ____ **6. Vicia villosa**
    4′ Flowers 4–7 mm.
      5. Plants perennial; leaflets 10–16; stipules ligulate ____ **4. Vicia pulchella**
      5′ Plants annual; leaflets 7–10; stipules semisagittate ____ **3. Vicia ludoviciana** var. **ludoviciana**

1. Vicia americana Muhlenburg ex Willdenow. Sp. Pl. 3:1096. 1802. * [for North America, where the species is widespread] AMERICAN VETCH

Herbs perennial. **Stems** erect, trailing, or climbing, slender to stout, to 20 dm. **Leaves** 2–8 cm; tendrils simple or branched; stipules inconspicuous, semisagittate, without nectariferous patch. **Leaflets** 8–18, ovate or elliptic to linear, 6–55 × 1–14 mm, apex obtuse to truncate-emarginate or apiculate, surfaces glabrous or finely brown pubescent. **Inflorescences** racemes, 3–9-flowered, 2–8 cm. **Flowers** 12–25 mm; calyx base symmetrical, lobes subequal and shorter than tube; corolla usually bluish purple, rarely white, banner oblong, blade equal to claw in width, glabrous; style dorsally compressed, pubescent around and tufted on the outer (lower) side. **Fruits** stipitate, tawny to brown, oblong, 25–39 × 5–9 mm, oblique-tipped, glabrous or pubescent; stipe to 4–5 mm. **Seeds:** number not known, olive-brown to deep violet-brown, subglobose, 3–4 mm diam.; hilum encircling 1/4–1/3 circumference of seed.

Varieties 2 (2 native in the flora): North America, east Asia.

Gunn (1968) has shown that *Vicia americana* consists of a polymorphic assemblage of populations, in which leaflet size, shape, and vestiture is quite variable. Intergradation is widespread, but there are two relatively distinct entities on the continent.

This species is reported to be utilized for food by quail and mule deer in California (Graham 1941). *Vicia americana* is widely reported to nodulate.

1. Stems usually 4–20 dm, trailing or climbing; leaves with branched tendrils; leaflets more than 30 mm long and more than 4 mm wide; racemes 5–9-flowered ______ **1a. Vicia americana** var. **americana**
1′ Stems usually less than 4 dm, erect; leaves often with simple tendrils; leaflets less than 30 mm long and less than 4 mm wide; racemes 3–5-flowered______ **1b. Vicia americana** var. **minor**

1a. Vicia americana Muhlenberg ex Willdenow var. **americana** * Native

Vicia americana Muhlenberg ex Willdenow subsp. *oregana* (Nuttall) Abrams; *Vicia americana* Muhlenberg ex Willdenow var. *truncata* (Nuttall) W.H. Brewer ex S. Watson; *Vicia americana* Muhlenberg ex Willdenow var. *villosa* (Kellogg) F.J. Hermann; *Vicia californica* Greene; *Vicia californica* Greene var. *madrensis* Jepson; *Vicia oregana* Nuttall

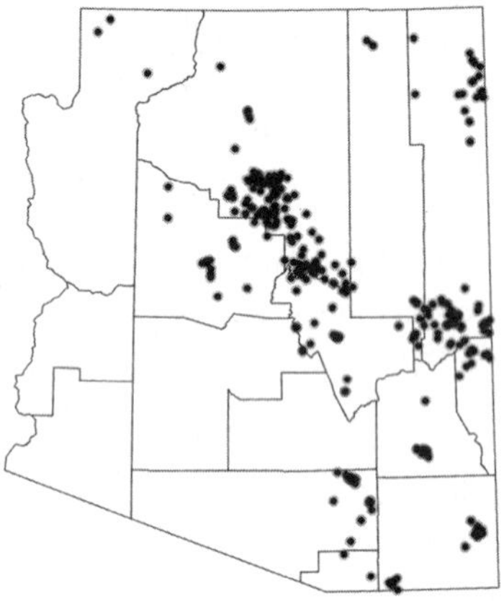

Stems trailing or climbing, usually 4–20 dm. **Leaves** with branched tendrils. **Leaflets** greater than 30 × 4 mm. **Inflorescences** racemes 5–9-flowered. $2n$ = 14, 28.

Flowering May–Aug. Ridges, slopes, canyons, stream banks, springs, moist woods, thickets, meadows, roadsides, waste areas in various soils; [0–](1030–)1500–3450 m; (Chihuahuan Desertscrub, Interior Chaparral), Madrean Evergreen Woodland, (Great Basin Conifer Woodland), Petran Montane Conifer Forest, Petran Subalpine Conifer Forest; Apache, Cochise, Coconino, Gila, Graham, Greenlee, Mohave, Navajo, Pima, Santa Cruz, Yavapai cos.; widespread in North America; Mexico (Baja California, Chihuahua, Coahuila, south to México).

Variety *americana* is the common, widespread variety of this species in Arizona. It is found in diverse habitats where sufficient moisture is available.

1b. Vicia americana Muhlenberg ex Willdenow var. **minor** Hooker. Fl. Bor.-Amer. 1:157. 1831. * [from Latin, *minor*, smaller] Native

Vicia americana Muhlenberg ex Willdenow var. *angustifolia* Nees; *Vicia americana* Muhlenberg ex Willdenow var. *linearis* (Nuttall) S. Watson; *Vicia americana* Muhlenberg ex Willdenow subsp. *minor* (Hooker) Piper & Beattie

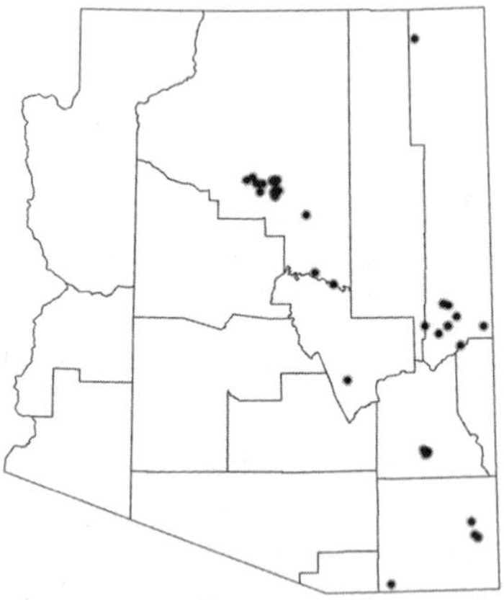

Stems erect, usually less than 4 dm. **Leaves** often with simple tendrils. **Leaflets** usually less than 30 × 4 mm. **Inflorescences** racemes 3–5-flowered. $2n$ = 14, 28.

Flowering May–Jul(–Sep). Slopes, canyons, meadows, stream banks, springs, forest edges, roadsides, and disturbed areas; [500–]1830–3200 m; (Interior Chaparral, Great Basin Conifer Woodland), Petran Montane Conifer Forest; Petran Subalpine Conifer Forest; Apache, Coconino, Cochise, Graham, Gila cos.; fairly widespread west of the Mississippi River.

Variety *minor* is less abundant in the state than is var. *americana*.

2. Vicia leucophaea Greene. Bot. Gaz. 6:217. 1881 (as *Vicia leucophoea*) * [from Greek, *leuco*, white, and *phae*, dusky] MOGOLLON VETCH Native

Herbs perennial. **Stems** erect, sprawling, or climbing, slender to robust, 1–8 dm. **Leaves** 1–4 cm; tendrils simple or branched; stipules foliose, semisagittate, without nectariferous patch. **Leaflets** 6 or 8, oblong-elliptic to linear, 7–25 × 1–6 mm, apex obtuse, lower surfaces long-villous, upper surfaces glabrous. **Inflorescences** racemes, 1 or 2-flowered, 1–3 cm. **Flowers** 7–9 mm; calyx base symmetrical, lobes subequal, longer than tube; corolla cream to white with purple keel tip, banner blades wider than claws, less than or equal to claw in length, glabrous; style cylindrical, with hairs in dense ring distinctly proximal to stigma. **Fruits** subsessile, pale brown, oblong, 25–40 × 4–6 mm, oblique-tipped, appressed-villous. **Seeds** 6–9, reddish brown, compressed-globose, 2.5 mm diam.; hilum encircling 1/4 circumference of seed.

Flowering Jul–Sep. Slopes, canyons, roadsides, disturbed areas; (1370–) 1600–2680 m; Madrean Evergreen Woodland, Petran Montane Conifer Forest; Apache, Cochise, Pima, Santa Cruz cos.; New Mexico; Mexico (Coahuila, Sonora).

Vicia leucophaea is endemic to four counties in eastern and southern Arizona, three counties in southwestern New Mexico, and in adjacent northern Mexico. Populations are localized.

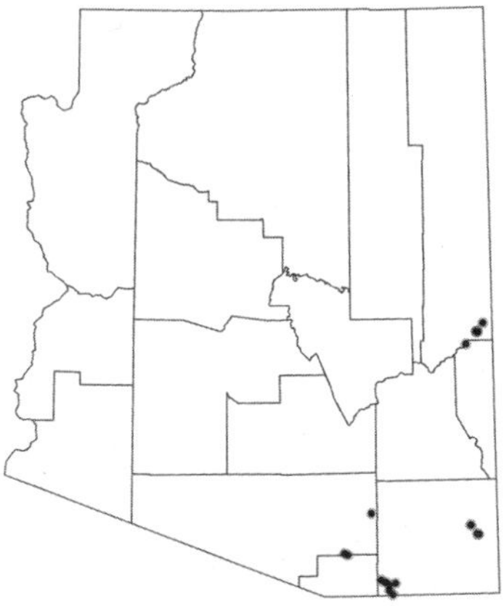

3. Vicia ludoviciana Nuttall ex Torrey & A. Gray. Fl. N. Amer. 1:271. 1838. * [of Louisiana] DEERPEA, LOUISIANA VETCH

Herbs annual or winter annual. **Stems** sprawling or climbing, slender to robust, 2–20 dm. **Leaves** 2–10 cm; tendrils branched; stipules much smaller than leaflets, semisagittate, without nectariferous patch. **Leaflets** 7–14, blades ovate to linear, 5–25 × 1–11 mm, apex acute to emarginate, surfaces glabrous or puberulent. **Inflorescences** racemes, 1–19-flowered, 1–15 cm, or flowers 1, ± shorter than subtending leaf rachis. **Flowers** 4–7 mm; calyx base symmetric, lobes subequal, shorter than or equal to tube; corolla deep blue to bluish purple or pinkish white to light lavender, banner blade and claw equal in width separated by a narrower "waist," glabrous; style compressed laterally, pubescent apically. **Fruits** yellow to brown, rhombic-oblong, 16–30 × 4–6 mm, oblique-tipped, glabrous; stipe to 1 mm. **Seeds** 4–8, brownish gray, mottled dark purple, subglobose to compressed-subglobose, 2.2–2.5 mm diam.; hilum encircling 1/4–1/3 circumference of seed.

Varieties 2 (1 native in the flora): North America (southern U.S., Mexico).

Lassetter (1984) has shown that the taxa traditionally treated as separate species (*Vicia exigua*, *V. leavenworthii*, *V. ludoviciana*, and *V. producta*) are morphologically confluent and part of one evolving species complex. Additional treatments of the variation within the complex have been given by Isely (1998) and Broich (2007). *Vicia ludoviciana* var. *ludoviciana* race "*producta*" is known to occur in Arizona (Isely 1998, Broich 2007).

3a. Vicia ludoviciana Nuttall ex Torrey & A. Gray var. **ludoviciana** * LOUISIANA VETCH, DEERPEA VETCH Native

Vicia exigua Nuttall; *Vicia leavenworthii* Nuttall ex Torrey & A. Gray var. *occidentalis* Shinners; *Vicia ludoviciana* Nuttall ex Torrey & A. Gray var. *laxiflora* Shinners; *Vicia ludoviciana* Nuttall ex Torrey & A. Gray var. *texana* (Torrey & A. Gray) Shinners; *Vicia producta* Rydberg; *Vicia texana* (Torrey & A. Gray) Small

Leaflets 7–10. **Inflorescences** 1–3(–5)-flowered. **Flowers** opening after peduncle and internode elongation, fruits not present when first open; corolla deep blue to bluish purple. $2n = 14$.

Flowering Mar–May. Sandy soil, dense to open woodlands, chaparrals, thickets, grasslands, ridges, rocky slopes, canyons, arroyos, roadsides, waste areas; (0–)220–1700(–2310)[–2500] m; Sonoran Desertscrub, Mohave Desertscrub, Semidesert Grassland, Interior Chaparral, Madrean Evergreen Woodland, Great Basin Conifer Woodland, (Petran Montane Conifer Forest); Apache, Cochise, Coconino, Gila, Graham, Greenlee, La Paz, Maricopa, Mohave, Navajo, Pima, Pinal, Santa Cruz, Yavapai cos.; Colorado, New Mexico, Oklahoma, Texas, Utah; Mexico (Baja California, Baja California Sur, Chihuahua, Coahuila, Nuevo León, Sinaloa, Sonora).

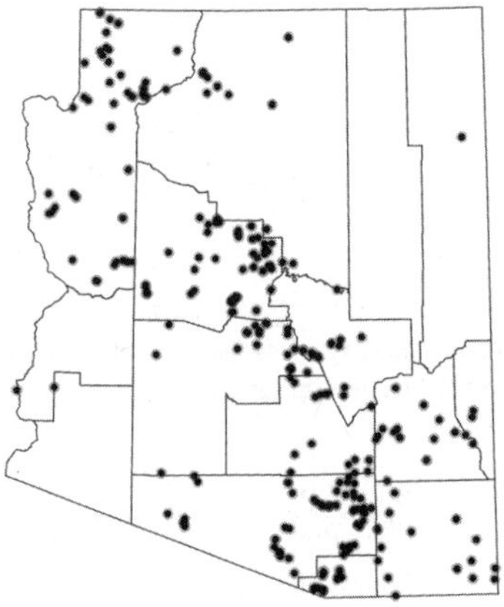

Vicia ludoviciana is found nearly statewide at lower elevations and can be locally common in years with sufficient winter-spring rainfall. The seeds are eaten by scaled quail and the plants are palatable to mule deer in experimental feeding trials (Graham 1941). Nodulation is reported for *V. ludoviciana* (Allen and Allen 1981).

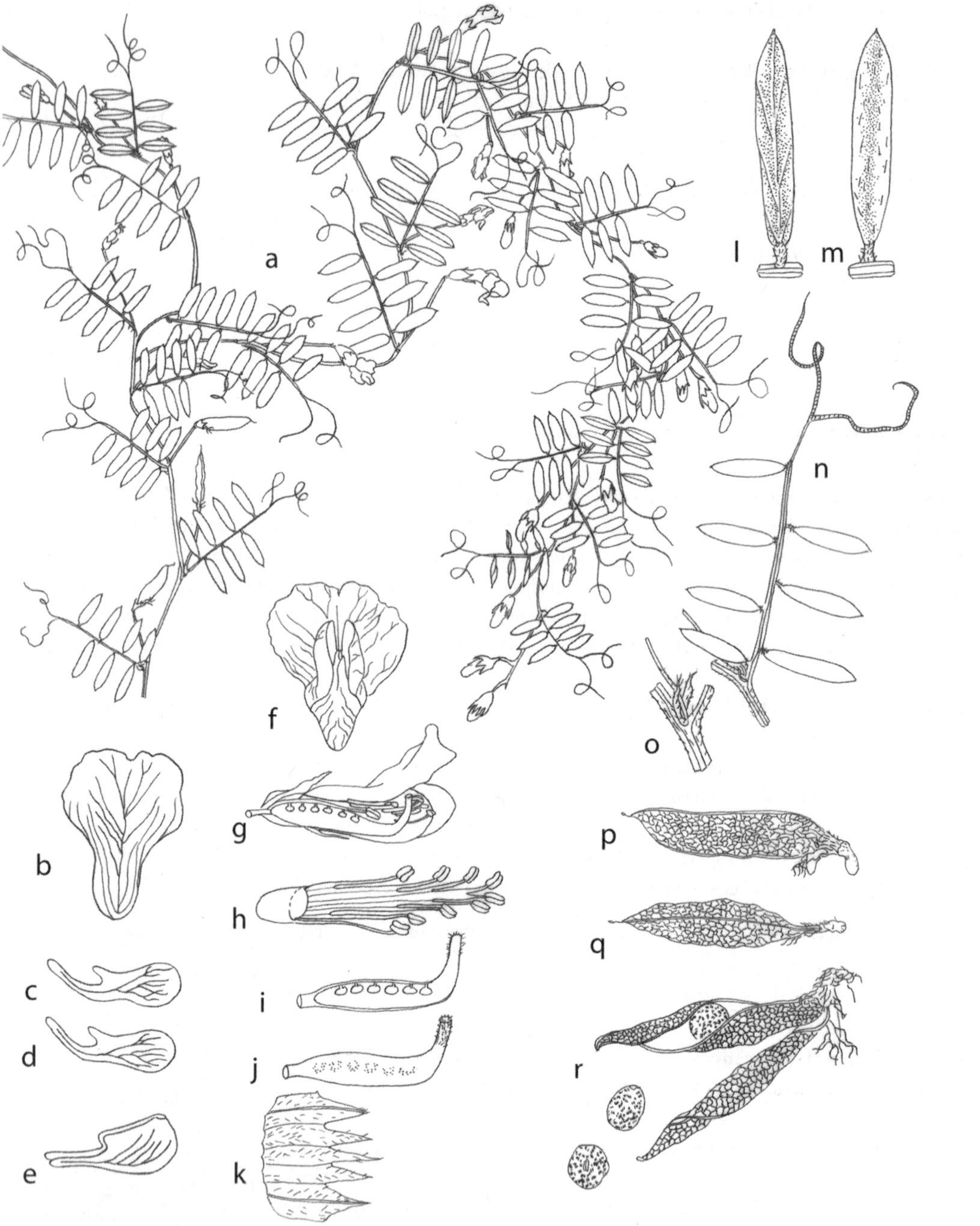

***Vicia ludoviciana* var. *ludoviciana*: a. habit; b., c., d., e. petals; f. flower; g. flower cutaway; h. stamens; i., j. pistil; k. calyx; l., m. leaflets; n. leaf; o. stem node; p. fruit (lateral view); q. fruit (top view); r. dehisced fruit with seeds. Illustrator: Margaret Pope.**

4. Vicia pulchella Kunth in A. von Humboldt et al. Nov. Gen. Sp. 6(fol.):390; 6(qto.):499; pl. 583. 1824. * [from Latin, *pulchell*, beautiful] BEAUTIFUL VETCH, SHOWY VETCH Native

Vicia melilotoides Wooton & Standley

Herbs perennial. **Stems** sprawling or climbing, slender, 2–15 dm. **Leaves** 4–11 cm; tendrils branched; stipules inconspicuous, ligulate, without nectariferous patch. **Leaflets** 10–16, elliptic-oblong to linear, 7–30 × 1–11 mm, apex obtuse, lower (abaxial) surfaces glabrous, upper (adaxial) surfaces often sparsely long-pubescent. **Inflorescences** racemes, 8–25-flowered, 2–15 cm. **Flowers** 4–7 mm; calyx base symmetrical, lobes equal, usually shorter than tube; corolla

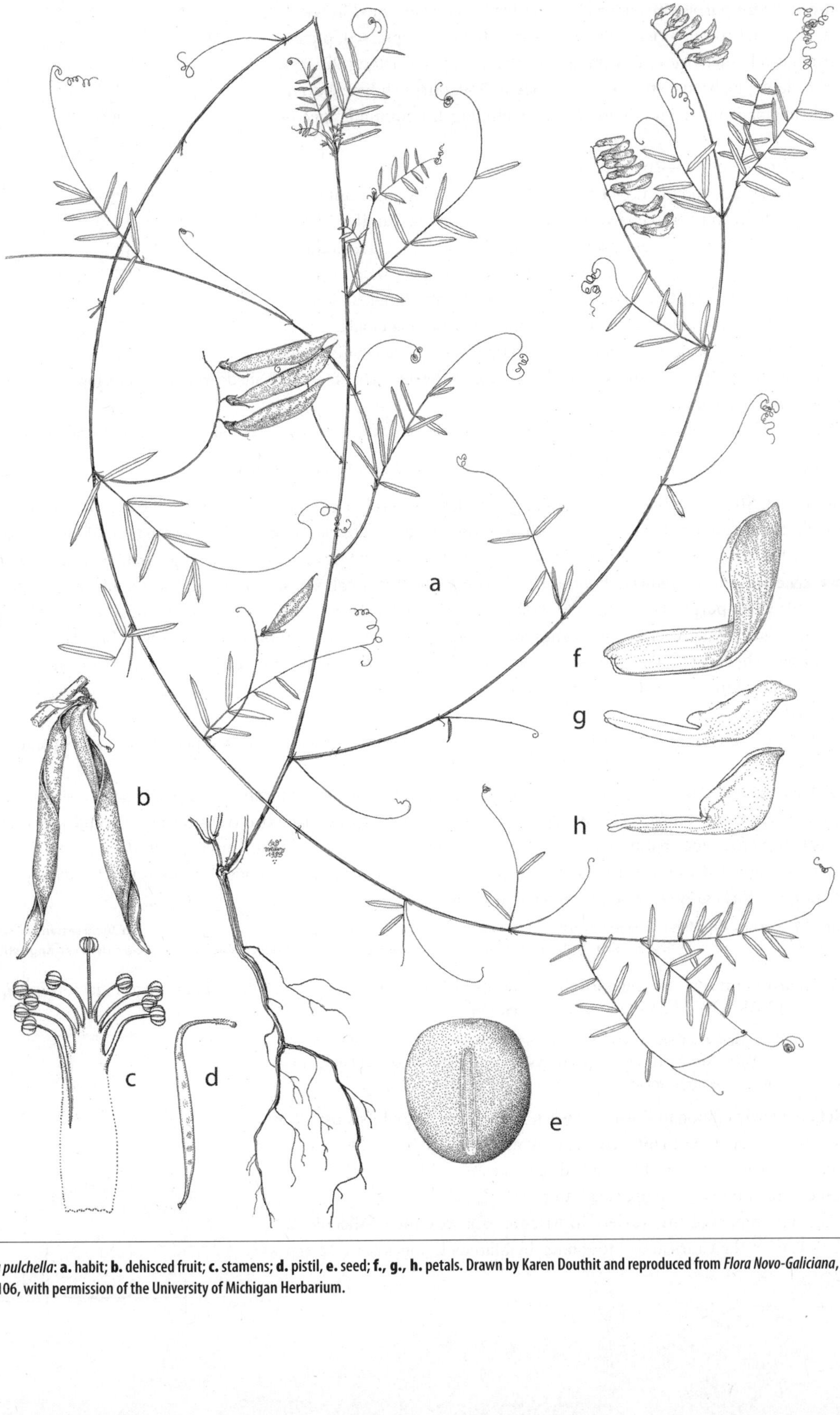

Vicia pulchella: **a.** habit; **b.** dehisced fruit; **c.** stamens; **d.** pistil, **e.** seed; **f., g., h.** petals. Drawn by Karen Douthit and reproduced from *Flora Novo-Galiciana*, vol. 5, fig. 106, with permission of the University of Michigan Herbarium.

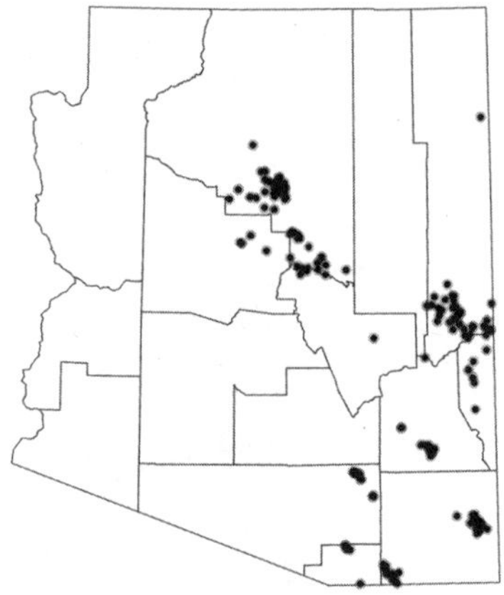

cream-white with purple-tipped keel, banner blade and claw equal in width separated by a narrower "waist," glabrous; style laterally compressed, pubescent around and tufted below. **Fruits** stipitate, reddish brown, oblong, 20–35 × 5–8 mm, acute, glabrous; stipe 1–2 mm. **Seeds** 5–9, purplish black, compressed-subglobose, 2–3 mm diam.; hilum encircling 1/4 circumference of seed.

Flowering (Jun–)Jul–Sep(–Oct). Ridges, slopes, canyons, stream banks, meadows, roadsides, and disturbed areas; (1010–)1500–3500 m; (Semidesert Grassland), Madrean Evergreen Woodland, Petran Montane Conifer Forest, Petran Subalpine Conifer Forest; Apache, Cochise, Coconino, Gila, Graham, Greenlee, Navajo, Pima, Santa Cruz, Yavapai cos.; New Mexico; Mexico (widespread from Chihuahua, Coahuila, Durango, Nuevo León, Sinaloa, Sonora, Tamaulipas, Zacatecas southward to Chiapas and Oaxaca); Guatemala.

Vicia pulchella is widespread at higher elevations in Arizona where it may be locally abundant. It thrives in a variety of soils and in diverse habitats in both shade and sun. At its lower elevational limits it is confined to moist sites.

Gunn (1979) has described an additional subspecies *mexicana* (Hemsley) C. R. Gunn native to central Mexico.

5. Vicia sativa Linnaeus. Sp. Pl. 2:736. 1753. * COMMONVETCH, SPRING VETCH

Herbs annual. **Stems** erect-ascending or climbing, slender to robust, 3–10 dm. **Leaves** 3–8 cm; tendrils simple or branched; stipules foliose, semisagittate, with nectariferous patch abaxially (lower side). **Leaflets** 8–14, ovate-oblong, elliptic, or linear, 15–30 × 5–15 mm, apex obtuse to truncate-emarginate, distinctly apiculate, surfaces hirsute. **Inflorescences** racemes, 2-flowered, subsessile. **Flowers** 10–30 mm; calyx base symmetrical, lobes subequal, ± equal to tube; corolla violet-purple, lavender, or whitish, banner blades wider that claws, less than or equal to claw, glabrous; style compressed below, evenly pubescent around and tufted abaxially. **Fruits** sessile, yellow to brown, or reddish brown to black, linear, 25–60 × 3–11 mm, oblique-tipped, glabrous or pubescent. **Seeds** 4–12, usually greenish gray to maroon or black, rarely yellowish white, globose or ± compressed, 3–5 mm diam.; hilum encircling 1/6–1/5 circumference of seed.

Varieties 7 (2 introduced in the flora): Europe, west Asia; introduced into temperate and subtropical areas worldwide.

Zohary and Plitmann (1979) have provided a detailed description of the morphological and genetic variation found within the *Vicia sativa* complex. Of the seven infraspecific taxa, they describe two (vars. *angustifolia* and *sativa*) cultivated as forage crops and that have been widely introduced into North America. Variety *angustifolia* is an aggressive colonizer of cultivated and disturbed habitats. Doves and mule deer have been observed to utilize *V. sativa* for food (Graham 1941). *Vicia sativa* is widely reported to nodulate.

1. Flowers 20–30 mm; calyx lobes usually equal to tube; legumes yellow to brown, pubescent ____ **5b. Vicia sativa** var. **sativa**
1′ Flowers 10–20 mm; calyx lobes usually slightly shorter than tube; legumes reddish brown to black, glabrous _ **5a. Vicia sativa** var. **angustifolia**

5a. Vicia sativa Linnaeus var. **angustifolia** Linnaeus Fl. Suec. (ed. 2):255. 1755. * [from Latin, *angust*, narrow, and *folia*, a leaf] NARROW-LEAVED VETCH Occasional

Vicia angustifolia Linnaeus, Amoen. Acad. 4:105. 1759; *Vicia angustifolia* Linnaeus var. *segetalis* (Thuillier) W.D.J. Koch; *Vicia angustifolia* Linnaeus var. *uncinata* (Desvaux) Rouy; *Vicia sativa* Linnaeus var. *nigra* Linnaeus

Leaflets narrowly elliptic to linear. **Flowers** 10–20 mm; calyx lobes usually slightly shorter than tube. **Fruits** reddish brown to black, glabrous. $2n = 12$.

Flowering May–Aug. Fields, roadsides, waste areas; [0–2000 m]; Petran Montane Conifer Forest; Cochise Co.; Europe.

The only report of this variety in Arizona is a specimen collected at Rustler's Park in the Chiricahua Mountains by Marcus E. Jones (*s.n.*, 22 Sep 1931).

5b. Vicia sativa Linnaeus var. **sativa** * Occasional

Vicia sativa Linnaeus var. *linearis* Lange

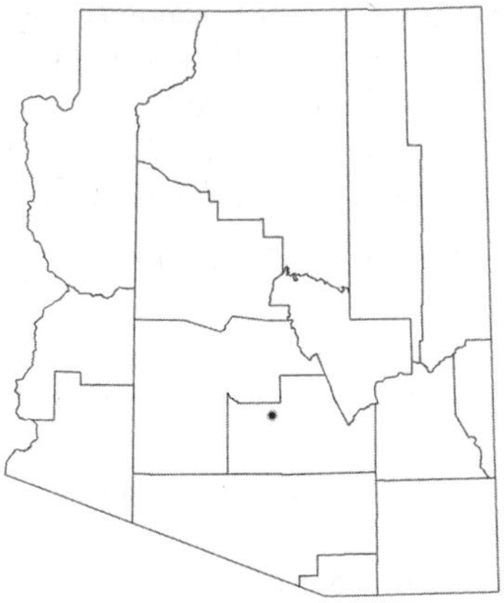

Leaflets usually ovate-oblong to narrowly elliptic, sometimes linear. **Flowers** 20–30 mm; calyx lobes usually equal to tube. **Fruits** yellow to brown, pubescent. $2n = 12$.

Flowering May–Aug. Fields, roadsides, waste areas; [0–1000 m]; Pinal Co.; possibly occurring in agricultural areas. Europe.

6. Vicia villosa Roth. Tent. Fl. Germ. 2(2):182. 1793. * [from Latin *villosa*, hairy] HAIRY VETCH, WINTER VETCH, WOOLLY VETCH

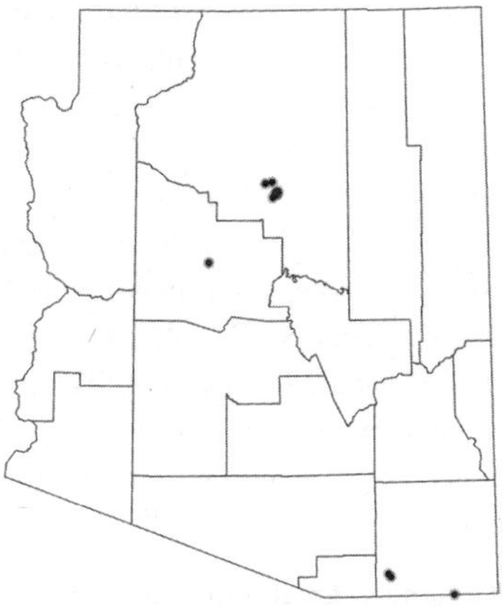

Herbs annual or biennial. **Stems** erect, sprawling, or climbing, slender, 2–13 dm. **Leaves** 4–9 cm; tendrils branched; stipules inconspicuous, semisagittate, without nectariferous patch. **Leaflets** 10–18, narrowly oblong to linear, 10–35 × 1–10 mm, apex obtuse to acute, or apiculate, surfaces glabrate to villous. **Inflorescences** racemes, with 10–20+ flowers, 3–19 cm. **Flowers** 12–18 mm; calyx base distinctly gibbous, lobes unequal, usually longer than tube; corolla violet to rose and white, banner blade and claw equal in width separated by a narrower "waist," blade 1/2 length of claw, glabrous; style laterally compressed, pubescent around and tufted on outer (lower) side. **Fruits** stipitate, tawny, oblong, 15–40 × 6–11 mm, oblique-tipped, glabrous; stipe to 2–3 mm. **Seeds** 2–8, black to blackish brown, globose to ovoid-oblong, 3–5 mm diam.; hilum encircling 1/7 circumference of seed.

Varieties 5 (2 introduced in the flora): Europe, central and southwest Asia, north Africa; introduced also into South America and Australia.

Arizona specimens are not distinguished to variety, so this distribution statement encompasses both varieties.

Flowering Apr–Sep. Slopes, meadows, stream banks, roadsides, vacant lots, and other disturbed areas; [0–]1280–3170 m; (Semidesert Grassland), Petran Montane Conifer Forest, Petran Subalpine Conifer Forest; Cochise, Coconino, Yavapai cos.; widely introduced and naturalized in North America; native to Europe.

Vicia villosa was introduced as a forage crop and has become widely established in North America. It can be distinguished from the similar *V. cracca* by its distinctly asymmetrical (gibbous) calyx base and by a banner blade that is half the length of the claw. Graham (1941) reports that some wildlife species will utilize this species for food. Nodulation is widely reported for *V. villosa*.

1. Plants villous; racemes with more than 20 flowers ____ **6b. Vicia villosa** var. **villosa**
1′ Plants glabrate; racemes 10–20-flowered ____ **6a. Vicia villosa** var. **glabrescens**

6a. Vicia villosa Roth var. **glabrescens** W.D.J. Koch. Syn. Fl. Germ. Helv. 194. 1835. * [from Latin, *glabr*, smooth] Weakly adventive

Vicia dasycarpa Tenore; *Vicia varia* Host; *Vicia villosa* Roth subsp. *varia* (Host) Corbière

Herbs glabrate. **Inflorescences** racemes, 10–20-flowered. **Flowers** 12–16 mm. $2n = 14$.

6b. Vicia villosa Roth var. **villosa** * Weakly Adventive

Herbs villous throughout. **Inflorescences** racemes, with 20+ flowers. **Flowers** 14–18 mm. $2n = 14$.

VIGNA Savi. Nuovo Giorn. Lett. Ser. 3, 8:113. 1824, name conserved. * [name commemorating Domenico Vigni, 1577(?)–1647, botanist and curator of the Pisa Botanic Garden] COWPEA

Alfonso Delgado-Salinas

Azukia Takahashi ex Ohwi; *Cadelium* Medikus; *Haydonia* R. Wilczek; *Liebrechtsia* De Wildeman; *Plectrotropis* Schumacher & Thonning; *Rudua* F. Maekawa; *Scytalis* E. Meyer; *Voandzeia* Thouars

Vines, or erect plants (cultivars), herbaceous to woody, annual or perennial; roots tuberous or fibrous. **Stems** prostrate to climbing, striate, covered with oblique to appressed, retrorse hairs (hairs not hooked at tip). **Leaves** alternate, 3-foliolate; stipules cordate-triangular, slightly auriculate or distinctly peltate; stipels lanceolate. **Leaflets** membranous or firm, margins entire to shallowly or rarely deeply lobed, glabrous or pubescent. **Inflorescences** pseudoracemes, axillary, with 2–many swollen floral nodes, glandular, flowers 2 per node; primary nodal bracts entire, deciduous; bracteoles shorter or longer than calyx, often deciduous. **Pedicels** slightly erect, arched to reflexed in fruit. **Flowers** papilionoid; calyces campanulate, with 4 lobes shorter or longer than tube; upper lobes usually entirely fused or almost so; corollas greenish yellow to bright yellow, white, pink, purple, or bicolored; banner symmetric or asymmetric, glabrous, often with appendages on the central inner face, slightly to strongly thickened at point of reflexion, apex emarginate; wing petals ovate to spatulate, subequal in length to other petals; keel apically expanded, not or slightly or fully incurved, sometimes with a prominent lateral appendage on the left-side lamina supporting left wing; stamens 10, 9 fused and 1 free, free stamen thickened at base; ovary with a nectary disk at base, linear, up to 20 ovules, style incurved, with a distal brush and often extending beyond the stigma into a conic or slender tip, stigma lateral. **Fruits** ascending, spreading, or pendulous, usually falcate, rarely oblong-straight; valves membranous or chartaceous, dehiscent, twisting. **Seeds** 20; 0.25–ca. 1 cm (in wild species), rounded to oval or oblong rectangular, hilum concave or protruding, with or without aril, epihilum present.

Species ca. 60 (1 cultivated in the flora). *Vigna* species occur mostly in the tropics and warmer temperate regions of Africa and Asia. There are less than 10 native to the New World (Delgado-Salinas et al. 2011, Isely 1998, Maxted et al. 2004, Tomooka et al. 2002).

Vigna aconitifolia (moth bean), *V. angularis* (Adzuki bean), *V. hosei* (Sarawak bean), *V. mungo* (black gram), *V. radiata* (mungbean), *V. umbellata* (rice bean), and *V. unguiculata* (cowpea) are agriculturally important grain legumes. Several species have become naturalized beyond their natural distributions. Nodulation is widely reported for the genus.

1. Vigna unguiculata (Linnaeus) Walpers. Repert. Bot. Syst. 1:779. 1843. * [from Latin, *unguis*, claw] BLACKEYE PEA, COWPEA, SOUTHERN PEA, CROWDER PEA, CAUPI, CATJANG, YARDLONG BEAN Cultivated

Dolichos unguiculatus Linnaeus, Sp. Pl. 2:725. 1753; *Dolichos catjang* Burman f.; *Dolichos sinensis* Linnaeus; *Phaseolus cylindricus* Linnaeus; *Vigna catjang* (Burman f.) Walpers; *Vigna sinensis* (Linnaeus) Savi ex Hasskarl; *Vigna unguiculata* (Linnaeus) Walpers subsp. *cylindrica* (Linnaeus) Verdcourt

Vines, or erect plants (cultivars), annual. **Stems** trailing or climbing, sometimes hollow, nodes in trailing stems streaked red, glabrous. **Leaves** with petioles 5–14 cm long, with longitudinal grooves and striate; rachis 2 cm; stipules triangular-lanceolate, 1.5 × 0.5 cm, produced at base, appendage shorter and narrower; stipels lanceolate. **Leaflets** 3, 4–15 × 3–8 cm, greyish green, ovate to lanceolate or rhombic-ovate, often lobed at base. **Inflorescences** long-pedunculate, (8–)15–25 cm, with 2–6 flowers on swollen nodes at apex; bracteoles lanceolate, 0.5 cm, deciduous. **Pedicels** 0.2 cm. **Flowers** with calyx campanulate, striate, 0.5 cm, 4-lobed, upper lobe triangular, shorter than laterals and lower lobe, triangular-lanceolate, 0.5 cm; corolla white or purple, rarely yellowish, 2.5 cm, banner with two parallel appendages on the inner face, wings spatulate, keel with short beak, slightly twisted towards the right; style shortly beaked, stigma with a short lobe-like appendage. **Fruits** linear to slightly falcate, nearly cylindrical, spreading or pendulous, 10–30 × 1 cm, white to yellowish-colored, occasionally streaked with rose or dark purple. **Seeds** 18–25, reniform or subglobose, 4–12 × 0.3–0.6 mm, light brown with red or purple tint and purple or pink eyes, or creamy-white with black eyes, or reddish black all over; hilum white, not raised.

Flowering Jun–Sep. Originally African, now cultivated nearly worldwide in tropical and subtropical regions. Sometimes escaping and becoming naturalized.

Vigna unguiculata is a worldwide cultivated species of African origin (Maxted et al. 2004). Cowpea was occasionally cultivated by native people in Arizona and is sometimes grown by home gardeners. Many varieties thrive in hot desert locations. The plants are drought resistant and are grown in spring with warm temperatures or during summer. Irrigation requirements vary with soil and environmental conditions. *Vigna unguiculata* is suitable as a warm season cover crop and green manure. Experiments in Arizona indicated that the plants substantially improved soil nitrogen levels (Wang and Nolte 2010). This species has become locally naturalized in California, Florida, Illinois, and Missouri. Nodulation is widely reported for *V. unguiculata*.

WISTERIA Nuttall. Gen. N. Amer. Pl. 2:115. 1818, name conserved. * [in honor of Dr. Caspar Wistar, 1761–1818, American scientist] WISTERIA

Lawrence R. Stritch

Diplonyx Rafinesque; *Glycine* Linnaeus; *Krauhnia* Steudel; *Phaseoloides* Duhamel; *Thyrsanthus* Elliott

Vines, woody, high-climbing or scandent shrubs with long-creeping rhizomes. **Stems** twining sinistrorsely (counter-clockwise) or dextrorsely (clockwise). **Leaves** alternate, odd pinnate; stipules and stipels linear-lanceolate, pubescent, quickly deciduous; leaflets petiolulate, opposite to subopposite. **Leaflets** elliptic, oblong-elliptic, or obovate, glabrous or villous, if villous, upper and lower surfaces densely villous during expansion, becoming glabrous with maturity, base wedge-shaped to rounded, margins entire, apex acuminate. **Inflorescence** a terminal raceme, axis pubescent; floral bracts lanceolate, pubescent, clavate glands or without glands, quickly deciduous; bracteoles absent. **Pedicels** pubescent, clavate glands or without glands, quickly deciduous. **Flowers** azure, purple, rose, or white, showy, papilionoid; calyx bell-shaped, two-lipped, pubescent, 5-lobed, the upper (adaxial) lip 2-lobed, lobes equal, the lower (abaxial) lip 3-lobed with the middle lobe long-acuminate; petals subequal, the keel incurved, petals of the keel fused at their apices, 1-clawed with an auricle to one side of the base, wing petals oblong-falcate, 1-clawed with a downward angled auricle to one side of the base, banner almost circular, reflexed near the middle with a two-auricled callosity near the middle or reflexed near base with a 2-auricled callosity at the base; stamens 10, 9 united, 1 free; ovary stipitate, glabrous or pubescent, with a disc surrounding the stipe, multi-ovulate. **Fruit** linear-oblong or oblong-elliptic to oblanceolate, rounded or flattened, torulose, glabrous or pubescent, 2-valved, explosively or tardily dehiscent. **Seeds** kidney-shaped flat, lens-shaped, black, black mottled with brown.

Species 8 (3 species and 1 hybrid cultivated in the flora): e United States and eastern Asia.

Wisteria japonica and *W. sinensis* are native in temperate eastern Asia; the two species and one ornamental hybrid have been introduced in eastern North America where they have become naturalized and invasive. *Wisteria japonica*, *W. sinensis*, and *W. ×formosa* are commonly cultivated around the world.

Wisteria bark has been used for textile fiber, and fragrances and essential oils are obtained from some species (Lewis et al. 2005). *Wisteria* are widely reported to be toxic. The seeds and fruits are particularly toxic and contain the glycoside wisterin (Lewis and Elvin-Lewis 1977, Schmutz and Hamilton 1979, Blackwell 1990). Nodulation is reported for the genus (Allen and Allen 1981).

Wisteria species and cultivars are occasionally cultivated in Arizona for their showy flowers. Plants are frequently grown on a trellis, ramada, or other structure, but can be trained to develop a shrubby habit. *Wisteria* grown in Arizona share similar horticultural requirements. The vines can grow to 10 m or more and are very long-lived. They are winter-deciduous. Showy masses of flowers are produced in spring. Many plants can be reluctant to flower at lower elevations in Arizona and may require cold temperatures to encourage flowering (Norem 2002). The plants seem to do well in full sun or partial shade and tolerate high summer temperatures. Results vary, and plants may thrive in some sites while the same species can languish in other sites. *Wisteria* require ample irrigation in Arizona to maintain an acceptable appearance and performance. These plants are noted for their fast growth.

Wisteria are generally hardy to -20° or -25°C and can survive successive days of subfreezing temperatures. Fallen leaves and flowers are maintenance considerations and, in most garden situations, considerable pruning is required to maintain the plants. At least in more humid climates, unmaintained plants are reported to damage structures. *Wisteria* can be propagated by seeds, though grafting, hardwood cuttings or air layering are used to maintain genetic uniformity of horticultural selections (Vines 1960).

SELECTED REFERENCES: Makino 1910 and 1911, Matsumara 1902, Rehder and Wilson 1916.

1. Ovary and fruit glabrous; banner reflexed near the middle; seeds kidney-shaped ____ **2. Wisteria frutescens**
1′ Ovary and fruit pubescent; banner reflexed at the base; seeds lens-shaped.
 2. Banner 2.1–2.4 × 2.1–2.3 cm ____ **3. Wisteria sinensis**
 2′ Banner 1.6–1.8 × 1.6–1.8 cm.
  3. Auricles of the banner callosity 0.7–0.8 mm ____ **4. Wisteria ×formosa**
  3′ Auricles of the banner callosity 1.1–1.2 mm ____ **1. Wisteria floribunda**

1. Wisteria floribunda (Willdenow) de Candolle. Prodr. 2:390. 1825. * [many flowers] JAPANESE WISTERIA Cultivated

Stems twining dextrorsely (clockwise). **Leaflets** (11–)13–17(–19), elliptic to lanceolate, to 11.2 × 2.6 cm. **Inflorescence** to 132 cm with the raceme flowering gradually from the base to the tip with flowers abscising from the base before flowers at the tip open. **Pedicels** to 2.8 cm. **Flowers** azure, purple, rose, or white; calyx with middle lobe of lower (abaxial) lip to 4 mm; corolla 1.5–2 cm; petals of the keel 1.4–1.5 × 0.7–0.8 cm, wing petals 1.4–1.5 × 0.7–0.8 cm, banner 1.6–1.8 × 1.6–1.7 cm, reflexed at base, auricles 1.1–1.2 mm, acuminate at apex; ovary pubescent. **Fruits** linear-oblong or oblong-elliptic to oblanceolate, flattened, 12–22(–30) cm, pubescent. **Seeds** lens-shaped.

Flowering late spring. Naturalized in Alabama, Arkansas, Florida, Georgia, Illinois, Kentucky, Louisiana, Maine, Mississippi, New Carolina, New Hampshire, Ohio, Pennsylvania, South Carolina, Tennessee, Virginia; Native to Japan.

Wisteria floribunda is endemic to Japan. It is common in moist deciduous forests and riparian habitats and can be found near sea level to 1200 m. It is extensively cultivated in Japan. *Wisteria floribunda* is a popular ornamental in the eastern United States, where it has become naturalized and invasive in habitats similar to its native environs. It is found on abandoned home sites, wooded edges of low fields, disturbed bottomland hardwood forests, and wet ground from sea level to 500m.

Wisteria floribunda differs from *W. sinensis*, with which it has been confused, in that its inflorescence expands simultaneously with or after the leaves expand, the raceme flowering gradually from the base to the tip with flowers abscising from the base before flowers at the tip open. Additionally, its banner is smaller, and the auricles are larger and acuminate. *Wisteria floribunda* differs from *W. ×formosa* in that the inflorescence expands simultaneous with or after the leaves have expanded, and the auricles of the banner callosity are larger and acuminate.

SELECTED REFERENCES: Sargent 1916, von Siebold and Zuccarini 1839, Wyman 1949, 1969, 1977.

2. Wisteria frutescens (Linnaeus) Poiret in Lamarck & Poiret, Tabl. Encycl. 3:674. 1823. * AMERICAN WISTERIA, KIDNEY-BEAN TREE, WOODY WISTERIA Cultivated

Glycine frutescens Linnaeus, Sp. Pl. 2:753. 1753; *Wisteria frutescens* (Linnaeus) Poiret var. *macrostachya* Torrey & A. Gray; *Wisteria macrostachya* (Torrey & A. Gray) Nuttall ex B.L. Robinson & Fernald; *Wisteria speciosa* Nuttall

Stems twining dextrorsely (clockwise). **Leaflets** 9–15, elliptic to lanceolate, 6 × 3 cm. **Inflorescence** 5–15(–22) cm with the raceme flowering almost all at once. **Pedicels** 5–8 mm. **Flowers** azure, purple or rarely white; calyx with middle lobe of lower (abaxial) lip to 3 mm; corolla 1.5–2 cm; petals of the keel 1.2–1.3 × 0.6–0.7 cm, wing petals 1.7–1.8 × 0.6 cm, banner 1.7–1.8 × 1.4–1.5 cm, reflexed near middle, auricles 0.7–0.8 mm, rounded at apex; ovary glabrous. **Fruit** linear-oblong, round in cross-section, 5–12(–15) cm, glabrous. **Seeds** kidney-shaped.

Flowering late spring to early summer. Native to Alabama, Arkansas, Connecticut, Florida, Georgia, Illinois, Indiana, Kentucky, Louisiana, Massachusetts, Michigan, Missouri, Mississippi, North Carolina, New Jersey, New York, Ohio, Pennsylvania, South Carolina, Tennessee, Texas, Virginia, West Virginia.

Across its native range in the eastern U.S., *Wisteria frutescens* is found from near sea level to 250 m. It is found in bottomland forests, river and stream banks, and bayheads particularly on the Atlantic and Gulf coastal plains and up the Mississippi embayment.

SELECTED REFERENCES: Duncan 1975, Robinson and Fernald 1916.

3. Wisteria sinensis (Sims) de Candolle. Prodr. 2:390 1825. * [of China] CHINESE WISTERIA Cultivated

Stems twining sinistrorsely (counterclockwise). **Leaflets** (7–)9–11(–13), ovate-elliptic to obovate, 9 × 3.5 cm. **Inflorescence** to 33 cm with the raceme flowering almost all at once. **Pedicels** to 3 cm. **Flowers** azure, purple, rose, or white; calyx with middle lobe of lower (abaxial) lip to 6 mm; corolla 2–2.5 cm; petals of the keel 1.7–1.8 × 0.7–0.8 cm, wing petals 1.7–1.8 × 0.7 cm, banner 2.1–2.4 × 2.1–2.3 cm, reflexed at base, auricles 0.7–0.8 mm, rounded at apex; ovary pubescent. **Fruits** linear-oblong or oblong-elliptic to oblanceolate, flattened, 8–21 cm, pubescent. **Seeds** lens-shaped.

Flowering late spring. Naturalized in Alabama, Arkansas, Connecticut, Florida, Georgia, Illinois, Kentucky, Louisiana, Maine, Massachusetts, Michigan, Mississippi, North Carolina, New York, Pennsylvania, South Carolina, Tennessee, Texas, Virginia; Native to China.

Wisteria sinensis is endemic to eastern and central China, where it is common in moist deciduous forests and riparian habitats from near sea level to 1000 m. It is extensively cultivated as an ornamental in China. It is also a popular ornamental in the eastern United States, where it has become naturalized and invasive in habitats similar to its native environs. It is found in abandoned home sites, wooded edges of low fields, disturbed bottomland hardwood forests, and wet ground from sea level to 500 m.

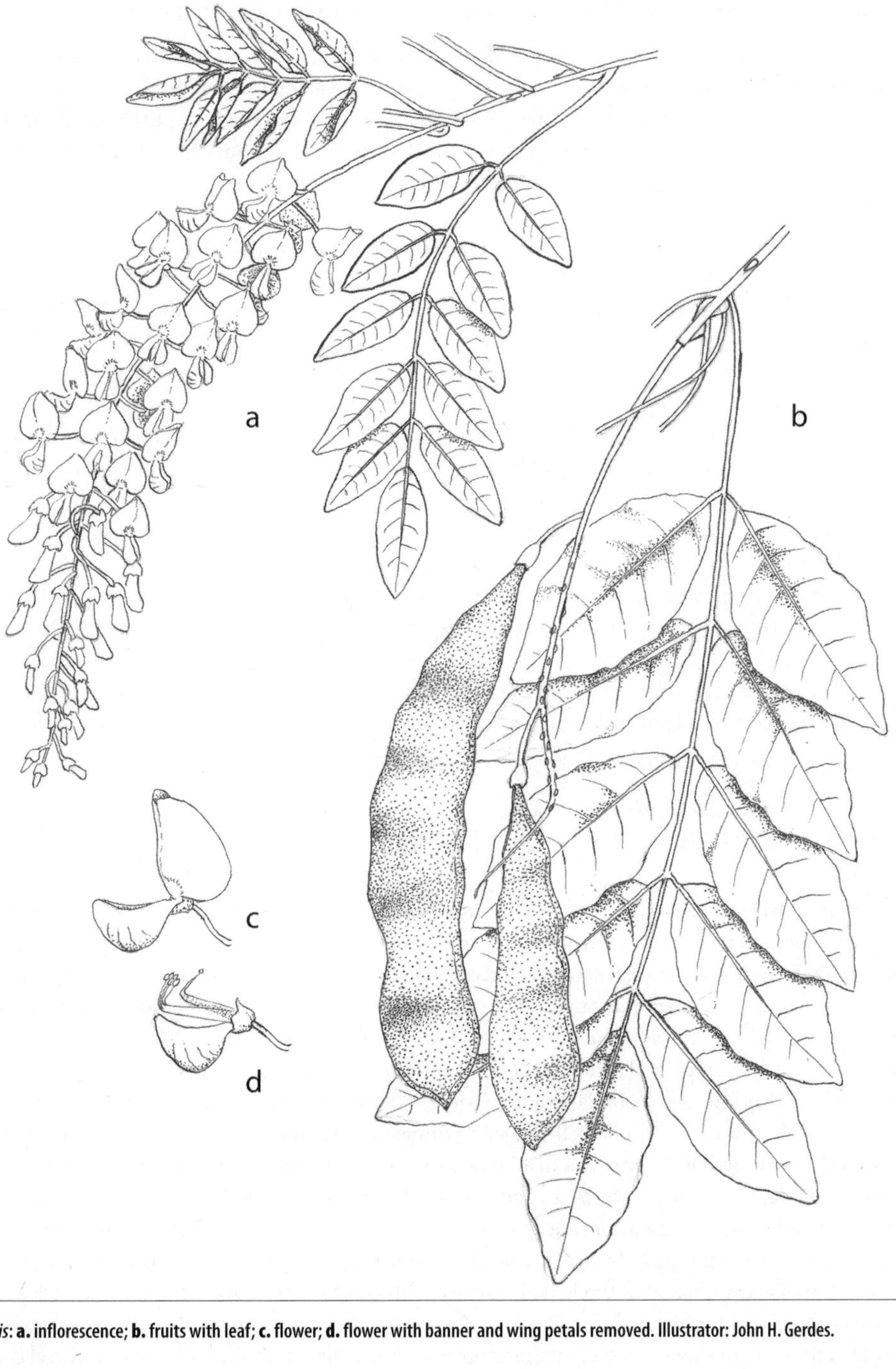

***Wisteria sinensis*: a. inflorescence; b. fruits with leaf; c. flower; d. flower with banner and wing petals removed. Illustrator: John H. Gerdes.**

The petals of *Wisteria sinensis* are a delicacy in China, where they are eaten steamed or fried. It is the only species of *Wisteria* whose petals are eaten (Wyman 1949).

In flowering condition, *Wisteria sinensis* differs from *W. floribunda* and *W. ×formosa* in that it has no scent whereas the latter two are very pleasantly scented; the inflorescence expands and flowers before the expansion of the leaves and the raceme flowers almost all at once from the base to the tip; the banner is larger; the auricles are smaller and rounded. *Wisteria sinensis* differs from *W. floribunda*, in number of leaflets (fewer in *W. sinensis*) and the direction of stem twining (sinistrorsely for *W. sinensis*, dextrorsely for *W. floribunda*).

SELECTED REFERENCES: Rehder 1926, Sargent 1916, Wyman 1949, 1969, 1977.

4. Wisteria ×formosa Rehder. J. Arnold Arbor. 3:36. 1921. * [of Formosa] HYBRID WISTERIA Cultivated

Stems twining sinistrorsely (counterclockwise) or rarely dextrorsely (clockwise). **Leaflets** (7–)9–13(–15), oblong-elliptic, 8.2 × 3.3 cm. **Inflorescence** to 36 cm with the raceme flowering gradually from the base to the tip with flowers abscising from the base before flowers at the tip open. **Pedicels** to 2.8 cm. **Flowers** azure, purple, or white; corolla 1.5–2 cm; petals of the keel 1.5–1.6 × 0.7–0.8 cm, wing petals 1.7–1.8 × 0.8–0.9 cm, banner 1.6–1.8 × 1.6–1.8 cm, reflexed at base, auricles 0.7–0.8 mm, rounded at apex; ovary pubescent. **Fruits** linear-oblong or oblong-elliptic to oblanceolate, flattened, 6–9.5 cm; pubescent. **Seeds** lens-shaped.

Flowering late spring. Naturalized in Alabama, Arkansas, Florida, Georgia, Kentucky, Louisiana, Michigan, North Carolina, South Carolina, Texas, Virginia.

As an ornamental, *Wisteria ×formosa* is superior to both parents. It has been sold extensively throughout the eastern United States and almost always under the name Chinese wisteria.

Wisteria ×formosa is similar to *W. floribunda* in having fragrant flowers and the size of the banner is the same. In both species the flowers abscise from the base before they open at the tip of the raceme. It is similar to *W. sinensis* in that the inflorescence expands and flowers before the leaves expand, and the auricles of the banner callosity are the same. The leaflet number is intermediate between that of the parents. The direction of twining of the stem can be that of either parent.

This hybrid has confused most botanists attempting to identify *Wisteria sinensis*, *W. floribunda*, and *W. ×formosa* itself. No major North American flora includes it in its keys despite the fact that Wyman (1949, 1969, and 1977) stated that it is probable that many Chinese wisterias being grown in gardens are actually *Wisteria ×formosa*. Isely (1990) states that morphological characters said to distinguish the two species do not correlate and notes the possibility of hybridization.

Wisteria ×formosa is naturalized extensively throughout the southeast United States. It is found on abandoned home sites, wooded edges of low fields, disturbed bottomland hardwood forests, and wet ground from sea level to 500 m.

SELECTED REFERENCES: Isely 1990, Rehder 1921, Wyman 1949, 1969, 1977.

ZAPOTECA H.M. Hernández. Ann. Missouri Bot. Gard. 73:755–763. 1986. * [named for the Zapotecan culture at the heart of the genus' diversity in central southern Mexico]

Philip Jenkins

Shrubs, rarely small trees, highly branched, erect, scandent, prostrate, or subprostrate, glabrous or hairy, unarmed. **Stems:** young branches rounded or 4-angled. **Leaves** even-bipinnate; stipules usually conspicuous, leafy, rarely spinescent, persistent; petioles rarely with nectariferous glands, usually with conspicuous channel on the upper (adaxial) side. **Leaflets** one to numerous pairs per pinna, opposite, sessile, usually membranous, rarely stiffly papery or leathery, usually glaucous beneath, glabrous or hairy, not overlapping. **Inflorescences:** hemispherical heads solitary and axillary or in terminal pseudopanicles (simple and compound), densely flowered; peduncles solitary or clustered at distant nodes. **Flowers** mimosoid, with bracts, sessile, radially symmetric, (4–)5(–6)-merous; calyx cup shaped, lobes toothed or with very small teeth, glabrous or hairy; corolla bell-shaped or funnel-shaped, membranous, petals valvate in bud, usually revolute at anthesis; nectariferous disk present in all flowers; stamens ca. 30–60; filaments long-exserted (19–43 mm), fused proximally, white, pink, red-purple, or a combination of two colors, staminal tube included; anthers attached dorsally, without glands; ovary 1, sessile or shortly stipitate; style thread-like, 15–56 mm in fertile pistils; stigma cup-shaped. **Fruits** with valves dehiscing elastically from apex to base, usually pendent, linear, straight, rarely slightly curved, flattened, translucent, margins thickened, usually with constrictions, dry, membranous to leathery. **Seeds** ovoid or rhomboid, rarely ellipsoid. $x = 13$.

Species 21 (1 native in the flora): most common in open areas in tropical deciduous forests, or in arid and semiarid, scrubby vegetation, United States (Arizona, Texas), Mexico, West Indies, South America (to n Argentina).

Zapoteca was most recently treated within *Calliandra*. Hernández separated the genera on general morphology, chromosome numbers, and reproductive features (Hernández 1986, 1989). Further studies by Barneby and Grimes (1996, 1997) and by Luckow et al. (2003) further separated *Zapoteca* from previously believed relationships and placed it with *Faidherbia* at a more basal position in tribe Ingeae (Lewis et al. 2005).

Flowering in *Zapoteca* is nocturnal. The flowers are pollinated by settling moth in the wet summer season.

SELECTED REFERENCE: Hernández 1989.

1. Zapoteca formosa (Kunth) H.M. Hernández. Ann. Missouri Bot. Gard. 73:755–763. 1986

Acacia formosa Kunth, Mimos. 102–105, pl. 32. 1822; *Anneslia formosa* (Kunth) Britton & Millspaugh; *Calliandra formosa* (Kunth) Bentham; *Feuilleea formosa* (Kunth) Kuntze

Shrubs, erect, to 5 m. **Stems** erect, sometimes drooping, usually slender, sometimes to 5 mm diam.; branches rounded to angular. **Leaves** 1–3(–4) pairs of pinnae; without glands; stipules triangular to linear-triangular, or lanceolate to ovate triangular, to 13 mm, glabrous to villous; petioles rounded with conspicuous channel above, 0.3–7.8 cm, glabrous to villous; rachis 0.3–6.8 cm; secondary rachis 0.6–12.5 cm. **Leaflets** (2–)3–15(–17) pairs per pinna, oblong to oblong-obovate or very widely obovate, proximal leaflets sometimes becoming elliptic or widely obovate, oblique at the base, or proximal leaflets sometimes becoming acute, rounded to acute at apex, usually mucronate, distal leaflets 4–59 × 2–36 mm, proximal ones smaller, mostly membranous, glabrous to pubescent, pilose or villous. **Inflorescences:** hemispherical heads solitary and axillary or capitula in terminal pseudopanicles (simple); peduncles 1–11 cm at anthesis. **Flowers** 5-merous; calyx 1.5–4 mm, glabrous, teeth triangular to oblong lanceolate; corolla bell-shaped, 3–5 mm, glabrous lobes elliptic to lanceolate; filaments 25–30 mm, white, greenish white, red purple, or white in the basal half and red-purple or pink in the distal half. **Fruits** to 13(–15) × 0.5–0.9 cm when mature, apex truncate to rounded or acute, usually with hook, thickly membranous, glabrous, glabrate, or densely pilose. **Seeds** brown or gray, rhomboid-ovoid, 4–8 × 3–6 mm.

Subspecies 7 (1 native in the flora): United States (Arizona); Mexico, West Indies, Central America, South America.

Although only one subspecies of *Zapoteca formosa* is recognized from Arizona, some individual collections from the Baboquivari Mountains appear intermediate between two subspecies. A key is included to differentiate the two closely related taxa, however, only *Z. formosa* subsp. *schottii* is described and certainly occurs in Arizona. Subspecies *rosei* replaces *schottii* from north-central Sonora, Mexico, southward. The two subspecies form a contiguous range on the Pacific slope of Mexico.

1. Leaflets 4–14 × 2–6(–7) mm; flower heads axillary; filaments white proximally and red-purple to pink distally; petioles 1.2–3.5 cm ________ **1a. Zapoteca formosa** subsp. **schottii**

1′ Leaflets 8–31 × 3–21 mm; flower heads axillary or in terminal pseudopanicles; filaments either wholly red-purple or white proximally and red-purple to pink distally; petioles 1–4.2(–5) cm ________ **Zapoteca formosa** subsp. **rosei**

1a. Zapoteca formosa (Kunth) H.M. Hernández subsp. **schottii** (Torrey ex S. Watson) H M. Hernández. Ann. Missouri Bot. Gard. 76:852. 1989. * [for Arthur Carl Victor Schott, 1814–1875, German-American topographical engineer, cartographer, botanist, geologist, artist] *TASOPOLO* Native

Calliandra schottii Torrey ex S. Watson, Proc. Amer. Acad. Arts 20:364. 1885; *Anneslia schottii* (Torrey ex S. Watson) Britton & Rose; *Calliandra schottii* Torrey ex S. Watson subsp. *schottii*

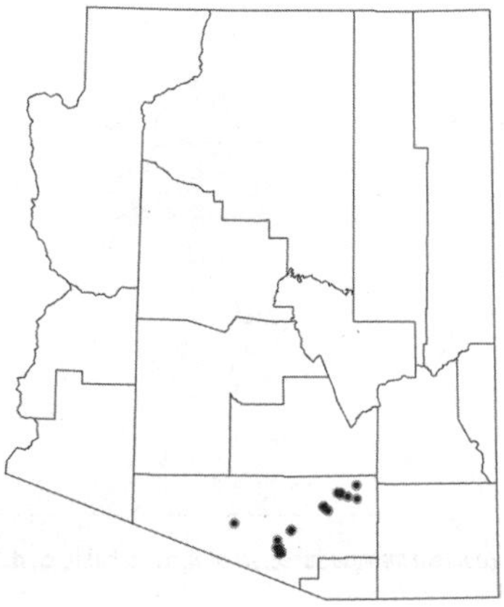

Shrubs to 1.5 (2) m. **Stems** slender, densely branched; branchlets glabrous to finely pilose. **Leaves:** 1–2 pairs of pinnae; stipules narrowly triangular to lanceolate, sometimes slightly curved, to 5 mm, glabrous to finely pubescent; petioles 1.2–3.5 cm; rachis 0.4–1.7 cm; secondary rachis 1.9–4.5 cm. **Leaflets** (3–)4–6(–7) pairs per pinna, oblong-obovate, 4–12(–14) × 2–6(–7) mm, base oblique, apex rounded to acute, usually mucronate, membranous, glabrous to finely pubescent. **Inflorescences:** hemispherical heads, axillary; peduncles 1.8–4.5(–6) cm. **Flowers:** calyx 1.5–2 mm, glabrous, the teeth triangular-lanceolate; corolla 3 mm; filaments white in basal half, pink to red-purple in the distal half. **Fruits** to 7 × 0.6 cm, glabrous, very rarely sparsely and minutely pilose. **Seeds** brown, rhomboid-ovoid, 5–8 × 3.5–5 mm.

Flowering, fruiting Jul–Oct, rarely May. Rocky slopes; Sonoran Desertscrub; 1100–1400 m; Pima Co (Santa Catalina Mountains to the Ajo Mountains); Mexico (n Sonora).

Zapoteca formosa subsp. *schottii* is infrequently encountered and is certainly not a common plant. It is easily overlooked in daylight when the stamens hang drooped and tangled and are rather inconspicuous. The individual inflorescences are functional for a single night and form an attractive pink ball reminiscent of *Calliandra eriophylla.*

A statement in Arizona Flora (Kearney and Peebles 1960) that the species is found in the Chiricahua Mountains (Cochise County) could not be verified.

Zapoteca formosa subsp. *schottii* is easily propagated from scarified seeds. Cultivated plants in Tucson have been damaged by overnight temperatures of ca. -6 °C but recover rapidly the following season. No specific uses are noted for

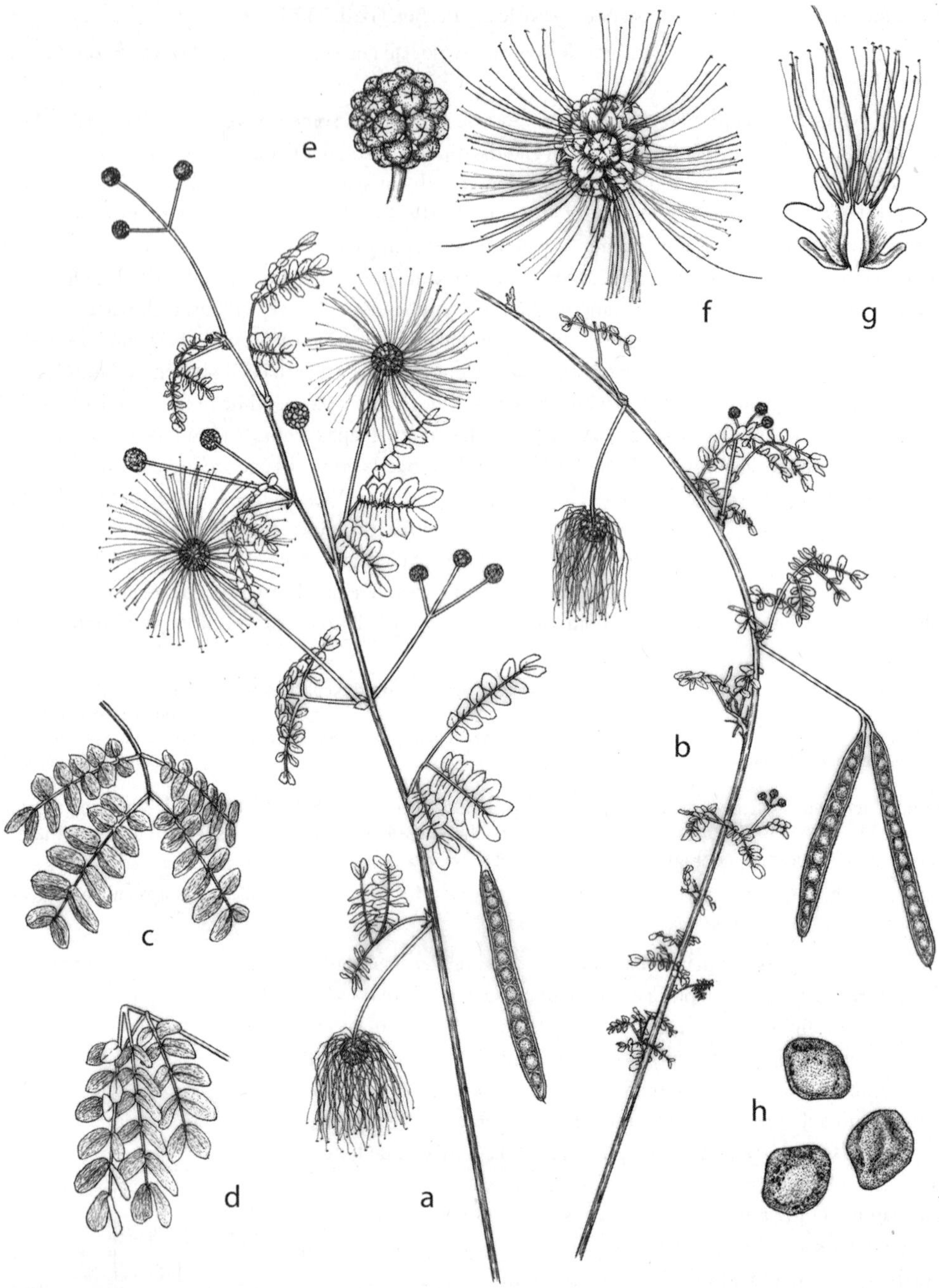

Zapoteca formosa subsp. *schottii*: **a., b.** habit; **c., d.** leaves; **e.** immature inflorescence; **f.** mature inflorescence; **g.** flower; **h.** seeds. Illustrator: Chris Bondante.

subsp. *schottii*; however, subsp. *formosa* is considered a nutritional forage plant for cattle in Venezuela (Newman et al. 1999).

ZORNIA J.F. Gmelin. Syst. Nat. 2(2):1076, 1096. 1791[1792]. * [for Johannes Zorn, 1739–1799, German pharmacist, botanist, and botanical illustrator]

Teague Embrey

Perennial herbs or subshrubs, glabrous to glandular-pubescent. **Leaves** even-pinnate; stipules leaf-like, paired, peltate. **Leaflets** 2 or 4; pellucid-dotted or not; stipels absent. **Inflorescences** in erect spikes, axillary and terminal, these subtended by stiff, leaf-like, paired bracts. **Pedicels** absent. **Flowers** papilionoid, bisexual, chasmogamous, small; calyx teeth 5, calyx yellow to yellow-orange; the banner nearly circular; 10 stamens united into a single bundle, anthers of two lengths, alternating; ovary stalked or not; style glabrous or pubescent. **Fruits** a strongly compressed loment, linear to oblong, with 2–15 segments, constricted, hispid, ciliate margined. **Seeds** 2–15 per fruit, ovoid, shiny or dull, dark brown or black. $x = 10$.

Approximately 75 species (1 native in the flora): North America including Mexico; Central America, West Indies, South America; South Africa, India; introduced in southeast Asia, Australia.

SELECTED REFERENCES: Mohlenbrock 1961, Mohlenbrock in preparation.

1. Zornia reticulata Smith in Rees, Cycl. 39:2. 1819. * [from Latin *reticulata*, reticulate, presumably in reference to the fruit surface] RETICULATE VIPERINA; *ZARZABOCOA DE DOS HOJAS* Native

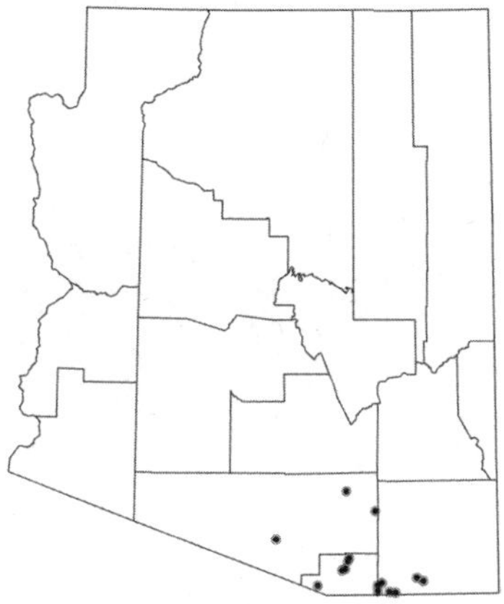

Herbaceous perennials, erect to ascending, to 0.5 m, sparsely hairy to strigose, multiple stems from a taproot. **Stems** wiry, slender. **Leaves:** 2 leaflets per leaf; 30 mm; stipules linear, 8–9 × 1–4 mm; petioles 10–15 mm. **Leaflets** ciliate-margined, midvein evident, 15–25 × 1.5–11 mm, linear acuminate to ovate, leaflets on lower portion of plant tending towards ovate, in some cases rotund, pellucid-dotted or not. **Inflorescence** a spike, 100–200[–250] × 10 mm; peduncles 20–30 mm; paired auriculate, acuminate floral bracts, 10–12 × 9–10 mm, pellucid dots present or not, glabrous to villous, generally 5-nerved, that partially hide the flower. **Flowers** solitary; calyx hairy, 4 × 2 mm; corolla yellow with purple streaking, glabrous, 9–10 × 6–8 mm; 10 stamens, anthers yellow; ovary stalked. **Fruit** a sessile loment, reticulate, covered with white tomentulose hairs and bearing retrorsely barbed, reddish prickles visible with the naked eye, 4–6(–7) pod segments, 10–15 × 4 mm. **Seeds** reniform, dark brown, smooth, 1 × 0.1 mm.

Flowering post-monsoonal rains (Aug–Sep). Hillsides, canyons, disturbed areas including roadsides; [250–]1100–1800[–2200] m; Semidesert Grassland, Plains Grassland, Madrean Evergreen Woodland; Cochise, Pima, Santa Cruz cos.; Mexico, Central America, South America, Caribbean.

Zornia reticulata occurs in scattered populations in southeastern Arizona. The plants may be locally abundant and often occur in sunny exposures. The flowers are showy. Reports of *Z. gemella* and *Z. leptophylla* from Arizona are apparently based on misidentified specimens of *Z. reticulata*.

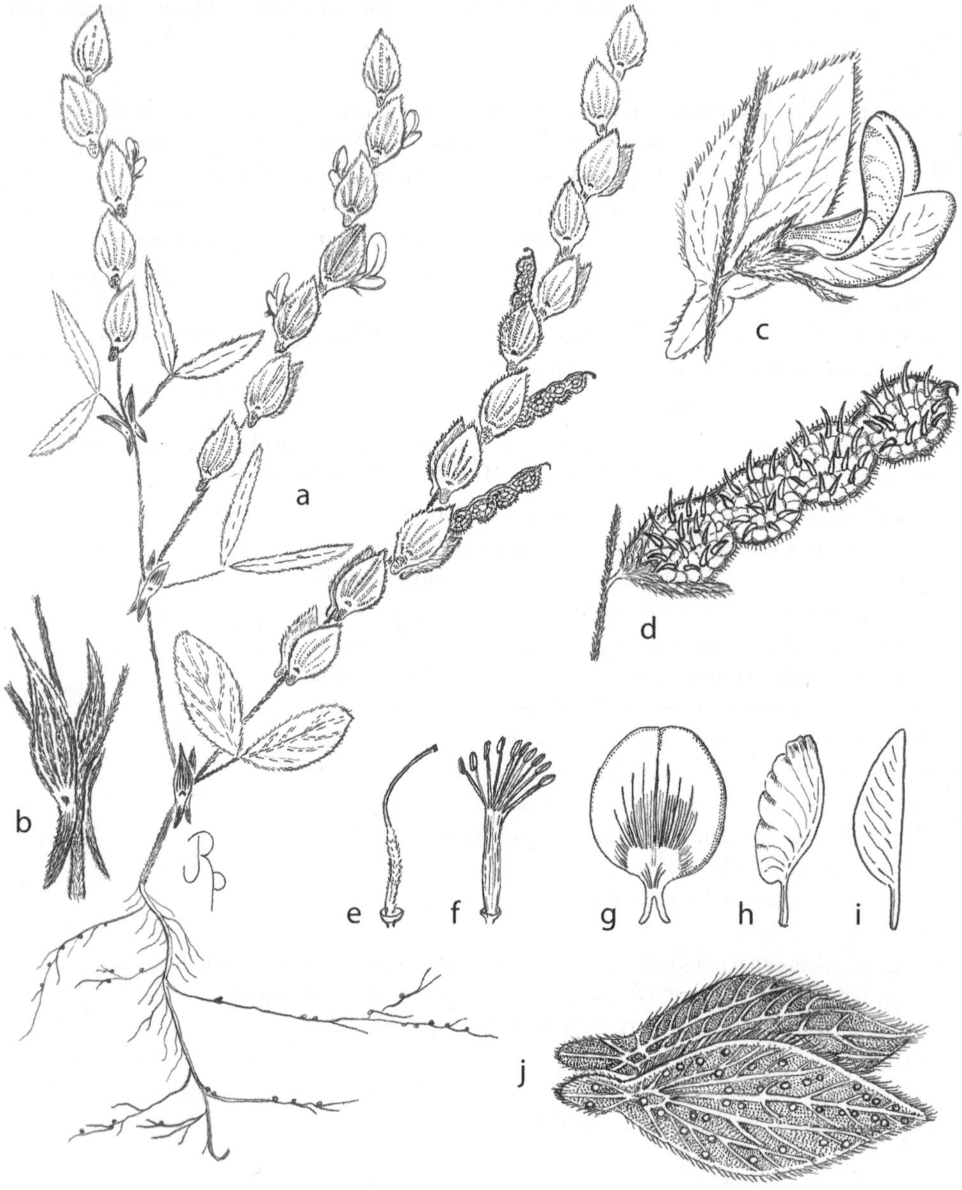

Zornia reticulata: **a.** habit; **b.** node with stipules; **c.** flower with bract; **d.** fruit; **e.** pistil; **f.** anthers; **g., h., i.** petals; **j.** bracts. Illustrator Joyce Riney Peters.

GLOSSARY

Throughout this volume, the authors and editors have simplified botanical terms where it was possible to do so without loss of precision or accuracy. We provide definitions here for remaining technical terms. Adj. = adjective form of the given term; alt. = alternative; opp. = opposite; pl. = plural; dim. = diminutive, e.g., the diminutive form of capitate (head-like or with head-like structures) is capitellate (with small head-like structures), and the diminutive form of pilose (covered with long soft hairs) is pilosulous (slightly covered with long soft hairs). Complete, illustrated glossaries are available from the New York Botanical Garden (as an online resource, nybg.org) and from the Royal Botanical Gardens, Kew (Beentje 2016).

Symbol

± : more or less.

A

Abaxial (opp. adaxial): facing away from the axis, e.g., the lower surface of a leaf, which faces away from the stem.

Acaulescent (opp. caulescent): lacking a well-developed stem above ground.

Acute: 1) pointed; 2) forming an angle of less than 90 degrees.

Acropetal: more developed/mature toward the base, less so toward the apex, e.g., of an inflorescence.

Acuminate: with a slender tapering point.

Adaxial (opp. abaxial): facing toward the axis, e.g., the upper surface of a leaf, which faces toward the stem.

Adventitious: produced in a non-standard position, e.g., of roots coming from a main stem.

Allopatric (opp. sympatric): separated in geographical space, e.g., of species.

Alveolate: pitted with angular depressions.

Amphitropical: disjunctly distributed on both sides of the equatorial tropics.

Anastomosing: forming an interconnected network, e.g., of veins.

Androecium: the male reproductive organs, i.e., the stamens.

Angular: not rounded, e.g., of a stem cross section.

Anthesis: the developmental stage in which a flower is receptive for pollination.

Apex (adj. apical): tip, or end.

Apiculum (adj. apiculate): a small point.

Appressed: laying close to or flat against.

Areole (adj. areolate): small, angular space on a surface, e.g., delimited by veins [as used here; defined otherwise for Cactaceae].

Aril (adj. arillate): a fleshy outgrowth on a seed that can partially or completely cover the seed, often brightly colored and thereby attracting birds or other animals that aid in seed dispersal.

Arista (adj. aristate): a stiff bristle or bristle-like appendage at the apex of a plant organ.

Ascending: directed upward.

Attenuate: narrowing gradually.

Auricle (adj. auriculate): ear-shaped appendage usually located at the base of a leaf or petal.

Axil (adj. axillary): the area in the upper region of the leaf node between the leaf and the stem.

Axis (pl. axes): 1) main stem, e.g., on which flowers are borne; 2) central column of an ovary.

B

Banner: upper petal, esp. of papilionoid flowers.

Barbate (dim. barbellate): having short, stiff, fine barbs, hairs, spines, etc.

Basifixed: attached at the base, of anthers or hairs.

Basionym: original name for a taxon, used in its first description.

Beak (adj. beaked): a projection that is thicker basally and tapers to a point.

Bi- : prefix, referring to two.

Biconvex: raised on both sides.

Bipinnate: doubly compound, i.e., a leaf in which primary leaflets are themselves pinnately compound.

Blade: flat portion of an organ, e.g., leaf or petal.

Brachyblast: short lateral stem in which the internodes are indistinguishable (also called a spur branch).

Bract: modified leaf at the base of a peduncle, pedicel, or just below a flower.

Bracteole: secondary bract, often small and paired at the distal end of the pedicel, just below the flower.

C

Caesalpinoid: flower form found in legume species outside the Mimosoid group and the subfamily Papilionoideae, in which the petals are free, the corolla is generally monosymmetric, the upper petal is inside the lateral petals in bud, and the stamens are free.

Caespitose: with multiple stems forming clusters or clumps.

Caliche: a calcium carbonate-rich layer of natural cement, often a few to several feet under the soil surface.

Callosity: irregular hardened surface outgrowth.

Callus (pl. calli): 1) hard protuberance; 2) hardened associated with a wound.

Calyx (pl. calyces): outermost whorl of a flower, i.e., the sepals.

Campanulate: bell-shaped.

Canaliculate: with longitudinal grooves.

Canescent: pale, often ashy in appearance due to the presences of abundant short hairs.

Capitate (dim. capitellate or subcapitate): head-like, spherical.

Capitulum (pl. capitula): congested, spherical inflorescence.

Cartilaginous: tough, hardened, but slightly elastic.

Cataphyll: small scale-like bract (as used here, usually derived from fused stipules).

Caudate: having a tapering, tail-like structure.

Caudex: persistent, often woody, stem at or just below ground.

Caulescent: with a stem/trunk above ground.

Chartaceous: stiffly papery.

Chasmogamy (adj. chasmogamous): flowers open prior to fertilization.

Cleistogamy (adj. cleistogamous): flowers remain closed, often involving self-pollination.

Chlorophyllous: green, containing chlorophyll, able to photosynthesize.

Cilia (adj. ciliate, dim. ciliolate): fine hair-like structures, usually along a margin.

Cinereous: ashy gray.

Clavate: club-shaped, distally thickened.

Claw: narrow proximal portion, of a petal.

Connate: fused or united, of like parts, e.g., petals or sepals.

Cordate: heart-shaped.

Coriaceous: tough and leathery.

Corolla: the second outermost whorl in a flower, i.e., the petals.

Corymb (adj. corymbiform): flat-topped inflorescence with individual flowers originating from different points along a common axis, with flowers that emerge lower on the axis having longer stalks, and those that emerge higher on the axis having shorter stalks.

Cotyledon: first leaf, seed leaf (note that these are enlarged and serve as nutritive tissue in mature seeds of legumes).

Crenate (dim. crenulate): with rounded irregular teeth, e.g., of a margin.

Crispate (alt. crisped): wavy or curled, e.g., of a margin.

Cupulate: shaped like a cup.

Cuneate: shaped like a wedge with the wide end distal-most.

Curvilinear: curved and of uniform thickness along the entire length.

Cyme: inflorescence in which the first flower formed is on the main stem axis while each subsequent flower is formed on a separate lateral stem.

D

Decumbent: lying along the ground, with the distal portion upright.
Decurved: gradually curved downward (often used as indistinguishable from recurved).
Decurrent: a part that extends down and is united with another part, e.g., of a leaf base.
Deflexed: abruptly bent downward.
Dehisce (adj. dehiscent, opp. adj. indehiscent): to open at maturity, releasing, e.g., seeds or pollen.
Deltate: shaped like a triangle with equal sides.
Deltoid: triangular in shape with a wide base, tapering to a point.
Diadelphous: with filaments united to form two groups of stamens, in legumes usually 9+1.
Diffuse: loosely spreading.
Dioecious: having male and female flowers on separate plants.
Distal (opp. proximal): away from the point of attachment.
Dorsal: the lower (abaxial), back, or outer surface, i.e., the side facing away from the axis.
Dorsiventral (opp. lateral): pertaining to the vertical plane, e.g., a dorsiventrally compressed fruit is short and wide in cross section.
Drupe: an indehiscent fleshy fruit with a single seed surrounded by a hard endocarp (e.g., a peach).

E

E-: prefix used to mean without.
Emarginate: with a notch at the apex, e.g., of a leaf.
Endemic: native and limited to a specific geographic region.
Endocarp: innermost layer of a multi-layered fruit wall.
Endosperm: nutritive tissue formed from double fertilization in seeds of flowering plants (note that legumes have very little endosperm remaining at seed maturity).
Erose: irregularly jagged, e.g., of a leaf margin.
Exocarp: outer-most layer of a multi-layered fruit wall.
Exserted: protruding out beyond other organs, e.g., an exserted stamen is not enclosed by the petals.
Extrafloral: not associated with a flower, e.g., extrafloral nectaries may be on petioles or leaf blades.
Extrorse (opp. introrse): outward, esp. of opening, e.g., anther dehiscence.

F

Falcate: sickle-shaped, curving to one side.
Fascicle (adj. fasciculate): a cluster, attached at a common point, e.g., of leaves.
Filiform: thread-like.
Fimbriate: fringed at the margins.
Flabellate: fan-shaped.
Foliolate: with leaflets, e.g., 3-foliolate, with 3 leaflets, unifoliolate, with 1 leaflet.
Free: not fused or united, e.g., of flower parts.
Funicule (alt. funiculus, adj. funicular): stalk that connects the ovule or seed to the placenta in the ovary.
Fusiform: thick in the middle and tapering at both ends.

G

Gibbous: with a swollen enlargement on one side at or near the base.
Glabrescent: becoming glabrous.
Glabrous (dim. glabrate): without hairs.
Gland (adj. glandular): area of secreted material, e.g., on a leaf or stem surface, or at the tip of a hair.
Glaucous: having a whitish or grayish waxy coating.
Globose: spherical.
Glutinous: slimy, sticky, or tacky.
Gum: water-soluble exudate associated with wounds.
Gynoecium: the female reproductive organs, i.e., the pistil(s).
Gynophore: stalk below an ovary or fruit, derived from the flower receptacle (not from the ovary).

H

Heterantherous: bearing anthers of more than one type or size.
Hilum (adj. hilar): scar on the seed coat where the funiculus was attached.
Hirsute (dim. hirtellous, hirsutulous): with rough or stiff hairs.
Hispid (dim. hispidulous): with long, stiff hairs.
Holotype: single specimen upon which a taxonomic name is based.
Humistrate: lying along the ground.
Hypanthium: cup of tissue surrounding the base of an ovary, with the sepals, petals, and stamens attached to the rim.

I

Imbricate: closely overlapping.
Incurved: gradually bent inward.
Indehiscent (opp. dehiscent): not opening at maturity.
Indumentum (alt. indument): covering of trichomes or hair-like structures on the surface of a plant structure.
Introrse (opp. extrorse): inward, esp. of opening, e.g., anther dehiscence.
Involucre: series of bracts below a congested inflorescence.
Involute (opp. revolute): rolled or curled inwards, e.g., towards the upper surface of a leaf.

K

Keel: 1) lower pair of petals in a papilionoid flower; 2) narrow surface ridge, e.g., of a seed.

L

Lamina (adj. laminar, alt. blade): flat portion of an organ, e.g., of a petal or leaf.
Lanceolate: shaped with a proximal wide portion, tapering to a point at the distal end, more narrow than ovate.
Lateral (opp. dorsiventral): pertaining to the horizontal plane, e.g., a laterally compressed fruit is taller than wide.
Lenticel: spongy pore allowing gas exchange, e.g., on woody stems.
Lenticular (alt. lentiform): domed on both sides, i.e., shaped like a lentil.
Liana: woody vine.
Lignescent: becoming woody in maturity.
Lineate: marked with lines.
Lineoles (adj. lineolate): small, fine lines.
Lobe (adj. lobate): a rounded surface or margin, e.g., of a leaf or petal.
Loment (adj. lomentaceous): an indehiscent fruit that separates into individual segments, each containing a single seed.
Lunate: shaped like a crescent or half-moon.
Lutescent: yellowish or becoming yellow.

M

-merous: referring to the numbers of the various structures that make up, e.g., a flower.
Medifixed: attached at or near the middle, e.g., of hairs or anthers.
Mesocarp: middle layer of tissue in a multi-layered fruit wall.
Mimosoid: flower form found in most members of the Mimosoid group, a clade within subfamily Caesalpinioideae, in which the corolla is radially symmetric, the petals are basally fused or sometimes free, the stamens are free or sometimes basally fused and are often numerous.
Monadelphous: with filaments fused into a single group of stamens.
Moniliform: cylindric with constrictions, like a string of beads.
Monophyletic: descending from a common ancestor, forming a clade, of a group of taxa.
Monosymmetric: referring to a flower whose structure can only be split in one plane to give a mirror image, i.e., zygomorphic, bilaterally symmetrical.
Mucilaginous: slimy.
Mucro (adj. mucronate, dim. adj. mucronulate): a short, sharp point.

N

Nodulation: the act of forming a small protuberance, e.g., as in many legume species that form root nodules, housing beneficial nitrogen-fixing bacteria.

Nyctinasty (adj. nyctinastic): movement of plant parts in response to light or darkness.

O

Ob- : prefix used to denote against or opposite of expected, e.g., an obovate leaf, in which the widest part is near the end.

Obcompressed: not used in this publication; in other literature, obcompressed may be used to describe dorsiventrally compressed *Astragalus* fruit.

Oblique: 1) uneven, e.g., of a leaf base; 2) at an angle, e.g., of a fruit that is not aligned on the symmetric plane.

Obsolete: rudimentary, not apparent.

Obtuse: blunt, having an angle between 90° and 180°, of a leaf base or apex.

Ochroleucous: a yellowish-white or cream color.

Orbicular: round or circular in shape.

Ovate: egg-shaped (two-dimensional), widest at the proximal end, e.g., of a leaf.

Ovoid: egg-shaped (three dimensional), e.g., of a fruit.

P

Palmate: with segments that originate from a single point, e.g., of venation, leaflets, or lobes.

Panicle (adj. paniculate): a type of compound inflorescence in which the main axis branches, secondary branches may branch again.

Papilionoid: flower form found in most members of the subfamily Papilionoideae, in which the upper petal (banner) is outside the others in bud, two lateral petals (wings) are outside the two lower petals (keel petals), which usually enclose the androecium and gynoecium.

Papilla (pl. papillae, adj. papillose): small, pimple-like protrusion on the surface.

Pedicel (adj. pedicellate): stem beneath a single flower.

Peduncle: stem beneath an inflorescence.

Pellucid: transparent or clear, e.g., of a gland.

Peltate: attached by a stalk at or near the center of the lower surface, e.g., of a scale or leaf.

Pericarp: mature fruit wall.

Petaloid: resembling a petal.

Petiole (adj. petiolar, petiolate): stalk, usually narrow, connecting leaf blade to stem.

Petiolule (adj. petiolulate): stalk, connecting leaflet to rachis in a compound leaf.

Phyllode: a leaf-like structure formed from a modified petiole or rachis.

Pilose (dim. pilosulose, pilosulous): covered with long soft hairs.

Pinna (pl. pinnae, adj. pinnular): leaflet, division of a compound leaf.

Pinnate: referring to a type of compound leaf having pinnae, usually in pairs, along a central axis.

Pith (adj. pithy): spongy tissue, e.g., at center of woody stem.

Plane: flat.

Pleurogram: a horseshoe-shaped mark on a seed.

Plicate: pleated, with parallel folds.

Plumose: feathery, with soft hairs.

Polyad: pollen assemblage, comprised of more than four grains.

Poricidal: having pores or openings from which material exits, e.g., of anthers.

Porrect: directed forward and outward.

Prickle: a hard-pointed unvascularized outgrowth of the epidermis.

Procumbent: lying along the ground or nearly so.

Prostrate: lying flat on the ground.

Protogynous: with stigma receptive prior to anther dehiscence.

Proximal (opp. distal): near to the point of attachment.

Pseudo- : prefix used to mean seeming or nearly.

Puberulent (alt. puberulous): minutely hairy.

Pubescent: with hairs.

Pulvinus (pl. pulvini, adj. pulvinate): swollen area of a leaf petiole or leaflet petiolule, can function to control leaf movement.

Punctate: dotted or pitted, e.g., with sunken glands.

Pungent: 1) with a sharp odor; 2) with a sharp end.

Pustule (adj. pustulate): blister.

Q

Quadrangular: with four angles.

Quadrate: square or nearly so.

R

Raceme (adj. racemose): inflorescence type in which the flowers are borne on pedicels along a central axis and the oldest flowers are at the bottom.

Rachis: 1) axis of compound leaf, above the petiole; 2) axis of an inflorescence, above the peduncle.

Radicle: root of a germinating seed.

Recurved: gradually bent downward (often used as indistinguishable from decurved).

Reflexed: abruptly bent downward.

Reniform: shaped like a kidney.

Resin (adj. resinous): water-insoluble exudate associated with wounds.

Resupinate: inverted or twisted 180 degrees, appearing upside down, e.g., of flowers.

Reticulate: 1) net-like, e.g., of veins; 2) involving network-like evolutionary history, e.g., from past hybridization events.

Reticulum: network of veins.

Retrorse: directed downward or backward.

Retuse: with a rounded notch at the apex.

Revolute (opp. involute): curled or rolled outwards, e.g., towards lower surface of a leaf.

Rhizome (adj. rhizomatous): a modified stem growing underground with roots and shoots developing from the nodes.

Rhombic: shaped as an equilateral parallelogram.

Rhomboid: nearly rhombic, with adjacent sides that are unequal.

Rootstock: underground roots and/or stems, often persisting when above ground tissue declines seasonally.

Rugose (dim. rugulose): creased or wrinkled.

S

Samara (adj. samaroid): indehiscent fruit having a flat wing-like structure.

Scandent: climbing or growing upward on a surface or another plant without benefit of tendrils.

Scape (adj. scapose): leafless stem, from ground level, bearing a flower or an inflorescence.

Scapiform: resembling a scape.

Scarification: the process of mechanically breaking the seed coat to break seed dormancy.

Scarious: dry, thin, membranous.

Sclerophyllous: with leaves that are small, thick, stiff, or leathery, often evergreen.

Secund: arranged on one side, one-sided.

Selenophyte (adj. selenophytic): a plant that concentrates selenium in its tissues.

Septum (pl. septa): wall-like tissue within ovary or fruit, separating locules.

Sericeous: with soft, silky hairs.

Serrate: having notches or teeth like a knife, e.g., of a margin.

Sessile: attached directly to another structure without a stalk.

Setose: with bristles.

Setaceous: bristle-like.

Sinus: 1) a recess, e.g., between lobes of a leaf or of a calyx; 2) a depression, e.g., on a seed.

Spatulate: shaped like a spatula or spoon.

Spheroid: resembling a sphere.

Spike: type of inflorescence in which flowers are borne directly on the axis, i.e., sessile, without pedicels.

Spine (adj. spinescent, spinose): sharp, pointed, often woody structure modified from a stipule or leaf.
Spur: 1) a tapering projection, e.g., of a perianth; 2) a short branch with minute internodes.
Staminode (pl. staminodia): sterile stamen, often small or incomplete.
Stipe: stalk below an ovary or fruit, derived from the ovary.
Stipel: stipule-like projection at the base of a leaflet, sometimes in pairs.
Stipitate: borne on a specialized stalk.
Stipule (adj. stipular): projection at the base of a leaf, at the junction with the stem, often paired on either side.
Stratification: the process of breaking seed dormancy, e.g., through cold treatments.
Striate: having fine, parallel lines created by shallow ridges or depressions.
Strigose (dim. strigillose, strigulose): with stiff hairs lying along the surface.
Sub-: prefix used to mean nearly, almost, or below.
Subopposite: almost opposite, i.e., closer to opposite than to alternate.
Subshrub: small perennial with both woody and herbaceous stems.
Subtend: lies just below, in reference to another part, e.g., a bract subtends a flower or pedicel.
Subulate: awl-shaped, narrow and tapering to a point.
Suffrutescent: similar to a subshrub, often with more woody stems.
Sulcus (adj. sulcate): groove.
Superficial (opp. subterranean): at or above ground surface.
Suture: a seam or fissure.
Sympatric (opp. allopatric): sharing geographic space, e.g., of species.

T

Tannin: generalized dark, water-soluble phenolic compound.
Thorn: a usually woody, modified stem that ends in a sharp point.
Tomentose (dim. tomentulose): covered densely with short soft hairs.
Torose (dim. torulose): cylindrical, with regularly spaced indentations.
Trichome: hair-like structure arising from the epidermis.
Trullate: trowel-shaped.
Truncate: ending abruptly, with a blunt end or apex.
Tubercle (adj. tuberculate): a small protuberance.
Turbinate: top- or turban-shaped.

U

Umbel: a type of inflorescence in which multiple pedicels or secondary peduncles attach to the main peduncle at the same point.

V

Valvate: 1) meeting at the edges of adjacent structures without overlapping, e.g., of petals or sepals; 2) referring to the valves of a fruit.
Valve: individual part of a capsule or pod that separates, along sutures, in fruits that dehisce.
Velutinous: with short dense soft hairs, velvety.
Venation: arrangement of veins, e.g., of a leaf or fruit.
Ventral: the upper (adaxial), front, or inner surface, i.e., the side facing towards the axis.
Vestiture: the assemblage of hairs, scales, or other projections from a surface.
Vexillary: referring to the upper, or banner, petal in a papilionoid flower.
Villous (dim. villosulous): with long soft hairs.
Viscid: sticky.

W

Whorl: set of similar organs, arranged around an axis, e.g., more than two leaves emanating from a node, or, e.g., each of the organ sets in a flower: sepals, petals, stamens, carpels.
Wing: 1) lateral petal of a papilionoid flower; 2) any flat extension on an organ.
Winged: with one or more wings, e.g., of fruits.

X

Xerophytic: adapted to dry habitats, drought resistant.
Xerophyllous: adapted to dry conditions, e.g., with leaves that are small, drought-deciduous.

REFERENCES

Adams, D.K. & A.C. Comrie. 1997. The North American monsoon. Bull. Amer. Meteor. Soc. 78:2197–2213.

Adams, J.L. 2002. Ground Stone Analysis: a Technological Approach. University of Utah Press, Salt Lake City, Utah, U.S.A.

Adorno, R. & P. Pautz. 1999. Álvar Núñez Cabeza de Vaca: His Account, His Life, and the Expedition of Pánfilo de Narváez. 3 vols. University of Nebraska Press, Lincoln, Nebraska, U.S.A.

Adovasio, J.M. & G.F. Fry. 1976. Prehistoric psychotropic drug use in northeastern Mexico and Trans-Pecos Texas. Econ. Bot. 30:94–96.

Ahlquist, T. 2012. A morphological analysis of the *Trifolium amabile* Kunth species complex in North America. M.S. thesis. Miami University, Oxford, Ohio, U.S.A.

Ahlquist, T. & M.A. Vincent. 2018. *Trifolium sonorense* (Fabaceae), a new species in the *T. amabile* species complex from Arizona and Mexico. Phytoneuron 2018-1:1–5.

Ajilvsgi, G. 1984. Wildflowers of Texas. Shearer Publishing, Austin, Texas, U.S.A.

Alexander, J.A. 2005. The taxonomic status of *Astragalus mokiacensis* (Fabaceae), a species endemic to the southwestern United States. Brittonia 57:320–323.

Allen, G.J. & M. Porter. 2000. Tribal delimitation and phylogenetic relationships of Loteae and Coronillae (Faboideae: Fabaceae) with special reference to *Lotus*: evidence from nuclear ribosomal ITS sequences. Amer. J. Bot. 87:1871–1881.

Allen, O.N. & E.K. Allen. 1981. The Leguminosae: A Source Book of Characteristics, Uses, and Nodulation. The University of Wisconsin Press, Madison, Wisconsin, U.S.A.

Allison, C.D., J.L. Turner, & J.C. Wenzel. 2016. Poisonous Plants of New Mexico Rangelands. New Mexico Cooperative Extension Service Circular 678. New Mexico State University, Las Cruces, New Mexico, U.S.A.

Allred, K.W. 2012. Flora Neomexicana I: Annotated Checklist. Published by author.

Anderson, E.A. 1953. Introgressive hybridization. Biol. Rev. 28:280–307.

Anderson, K. 1991. California Indian horticulture: Management and use of redbud by the Southern Sierra Miwok. J. Ethnobiol. 11(1):145–157.

Anonymous. 1882. The botanist in Arizona. Bot. Gaz. 7:8–9.

Anthony, R.G. & N.S. Smith. 1977. Ecological relationships between mule deer and white-tail deer in southeastern Arizona. Ecol. Monogr. 47(3):255–277.

Antisell, T. 1856. Synoptical tables of botanical localities. In: J.G. Parke. 1856. Report of explorations for railroad routes from San Francisco Bay to Los Angeles, California, west of the Coast Range, and from the Pimas Villages on the Gila to the Rio Grande, near the thirty-second parallel of north latitude (Pacific Railr. Rep. 7, part 3). House Exec. doc. 91, 33rd Congress, 2nd Session. A.O.P. Nicholson printer, Washington, D.C., U.S.A. 23–26.

Arizona Register of Big Trees, The. 2005 ed. Ironwood Press, Tucson, Arizona, U.S.A.

Aronson, J. 1990. Desert plants of use and charm from northern Chile. Desert Pl. 10(2):65–74, 79–86.

Aronson, J. & C. Saravia Toledo. 1992. *Caesalpinia paraguariensis* (Fabaceae): forage tree for all seasons. Econ. Bot. 46(2):121–132.

Asif, M. & A. Kumar. 2009. Anti-inflammatory activity of ethanolic extract of *Dalbergia sissoo* (Roxb.) bark. Malaysian J. Pharm. Sci. 7(1):39–50.

Austin, D.F. 2010. Baboquivari Mountain Plants: Identification, Ecology, & Ethnobotany. University of Arizona Press, Tucson, Arizona, U.S.A.

Australian Native Plant Society (Australia) (ANPSA). 2015. "*Hardenbergia violacea*." ANPSA, Ulverstone, Tasmania. Accessed 12 Mar 2017. http://anpsa.org.au/h-viol.html

Avicenna (Ibn Sīnā). 1976. Qānūn fí al-tibb. Latin (The Canon of Avicenna). (Gherardo da Cremona, Andrea Alpago, Benedictus Rinius, 'Abd al-Wāhid Jūzjānī, Niccolò Massa, Arnaldus de Villanova, & Armegandus Blasii, Latin Trans.) Pahlavi Commemorative Reprint Series. Imperial Organization for Social Services, Tehran, Iran. (Original work published 1025)

Awmack, C.S. & J.M. Lock. 2002. The genus *Alhagi* (Leguminosae: Papilionoideae) in the Middle East. Kew Bull. 57:435–443.

Baker, M.J. & W.M. Williams. 1987. White Clover. CAB International, Wallingford, U.K.

Baldwin, B.G., D. Goldman, D.J. Keil, R. Patterson, T.J. Rosatti, & D. Wilken, eds. 2012. The Jepson Manual: Vascular Plants of California. 2nd ed. University of California Press, Berkeley, California, U.S.A.

Ballenger, J.A. 1992. A biosystematic revision of the genus *Cercis* L. (Leguminosae) in North America. Ph.D. dissertation. Miami University, Oxford, Ohio, U.S.A.

Ballenger, J.A. & M.A. Vincent. In prep. *Cercis*. For: Flora of North America Editorial Committee, eds. 1993+. Flora of North America North of Mexico. Oxford University Press, New York, New York, U.S.A. and Oxford, U.K. Vols. 10–11.

Barneby, R.C. 1943. Miscellaneous diagnoses. Leafl. W. Bot. 3:193–194.

Barneby, R.C. 1944. Pugillus Astragalorum alter. Proc. Calif. Acad. Sci. IV. 25:147–170.

Barneby, R.C. 1945. Pugillus Astragalorum IV: The section *Diplocystium*. Leafl. W. Bot. 4:65–147.

Barneby, R.C. 1947. Pugillus Astragalorum VI: Notes of section *Drabellae*. Leafl. W. Bot. 5:1–9.

Barneby, R.C. 1949. Pugillus Astragalorum X: New species, varieties, and combinations. Amer. Midl. Naturalist 41:496–502.

Barneby, R.C. 1964a. Atlas of North American *Astragalus*. Mem. New York Bot. Gard. 13:1–1188.

Barneby, R.C. 1964b. *Sphaerophysa*. In: R.C. Barneby. 1964. Atlas of North American *Astragalus*, Part II. Mem. New York Bot. Gard. 13:1162–1164.

Barneby, R.C. 1977. *Dalea* Imagines: An Illustrated Revision of *Errazurizia* Philippi, *Psorothamnus* Rydberg, *Marina* Liebmann, and *Dalea* Lucanus emend. Barneby, including all species of Leguminosae tribe Amorpheae Borissova ever referred to *Dalea*. Mem. New York Bot. Gard. 27:1–892.

Barneby, R.C. 1989. Fabales. In: A. Cronquist, A.H. Holmgren, N.H. Holmgren, J.L. Reveal, & P.K. Holmgren, eds. 1989. Intermountain Flora: Vascular Plants of the Intermountain West, U.S.A. Volume 3B. Fabales. New York Botanical Garden, Bronx, New York, U.S.A.

Barneby, R.C. 1991. Sensitivae Censitae: A description of the genus *Mimosa* Linnaeus (Mimosaceae) in the New World. Mem. New York Bot. Gard. 65:1–835.

Barneby, R.C. 1998. Silk Tree, Guanacaste, Monkey's Earring: A Generic System for the Synandrous Mimosaceae of the Americas. Part III. *Calliandra*. Mem. New York Bot. Gard. 74(3):1–223.

Barneby, R.C. & J.W. Grimes. 1996. Silk Tree, Guanacaste, Monkey's Earring: A Generic System for the Synandrous Mimosaceae of the Americas. Part I. *Abarema*, *Albizia*, and Allies. Mem. New York Bot. Gard. 74(1):1–292.

Barneby, R.C. & J.W. Grimes. 1997. Silk Tree, Guanacaste, Monkey's Earring: A Generic System for the Synandrous Mimosaceae of the Americas. Part II. *Pithecellobium*, *Cojoba*, and *Zygia*. Mem. New York Bot. Gard. 74(2):1–149.

Bartram, E.B. 1922. Midwinter botanizing in southern Arizona. Bull. Torrey Bot. Club 49:237–250.

Batlle, I. & J. Tous. 1997. Carob tree. *Ceratonia siliqua* L. International Plant Genetic Resources Institute, Rome, Italy.

Becker, R. 1981. The Nutritive Value of *Prosopis* Pods. U.S. Department of Agriculture Agricultural Research Service, Western Regional Research Center, Berkeley, California, U.S.A.

Becker, R. 1983. Nutritional quality of the fruit from the chanar tree (*Geoffroea decorticans*). Ecol. Food Nutr. 13:91–97.

Beentje, H. 2016. The Kew Plant Glossary. 2nd ed. Royal Botanic Gardens, Kew, London, U.K.

Bell, W.H. & E.F. Castetter. 1937. The utilization of mesquite and screwbean by the aborigines in the American Southwest. University of New Mexico biological series, vol. 5, no. 2, University of New Mexico bulletin, whole no. 314, Ethnobiological studies in the American Southwest, 5 5, 2.

Belofsky, G., D. Percivill, K. Lewis, G.P. Tegos, & J. Ekart. 2004. Phenolic metabolites of *Dalea versicolor* that enhance antibiotic activity against model pathogenic bacteria. J. Nat. Prod. 67:481–484.

Bena, G. 2001. Molecular phylogeny supports the morphologically based taxonomic transfer of the "medicagoid" *Trigonella* species to the genus *Medicago* L. Pl. Syst. Evol. 229:217–236.

Benson, L. 1940. Taxonomic contributions. Amer. J. Bot. 27:186–190.

Benson, L. 1941. The mesquites and screw-beans of the United States. Amer. J. Bot. 28:748–754.

Benson, L. & R.A. Darrow. 1981. Trees and Shrubs of the Southwestern Deserts. 3rd ed. The University of Arizona Press, Tucson, Arizona, U.S.A.

Bentham, G. 1839. Plantas Hartwegianas Imprimis Mexicanas. G. Pamplin, London, U.K.

Bentham, G. 1844. The Botany of the Voyage of H.M.S. Sulphur. Smith, Elder, and Co., London, U.K.

Bentham, G. 1875. Revision of the suborder Mimoseae. Trans. Linn. Soc. London 30:335–664.

Bessega, C., L. Ferreyra, J.C. Vilardi, & B.O. Saidman. 2000. Unexpected low genetic differentiation among allopatric species of section *Algarobia* of *Prosopis* (Leguminosae). Genetica 109:255–266.

Bessega, C., J.C. Vilardi, & B.O. Saidman. 2006. Genetic relationships among American species of the genus *Prosopis* (Mimosoideae, Leguminosae) inferred from ITS sequences: Evidence for long-distance dispersal. J. Biogeogr. 33:1905–1915.

Beyra Matos, A., P. Herrera, G. Reyes, & L. Hernández. 2005. Revisión taxonómica del género *Galactia* P. Br. (Leguminosae-Papilionoideae) en Cuba. Rev. Acad. Colomb. Cienc. 29:467–494.

Bieda, S.W. III, C.L. Castro, S.L. Mullen, A.C. Comrie, & E. Pytlak. 2009. The relationship of transient upper-level troughs to variability of the North American Monsoon System. J. Clim. 22:4213–4227.

Bigelow, J.M. 1856. General description of the botanical character of the country. In: A.W. Whipple. 1856. Report of explorations for a railway route, near the thirty-fifth parallel of north latitude, from the Mississippi River to the Pacific Ocean (Pacific Railr. Rep. 3, part 5). House Exec. doc. 91, 33rd Congress, 2nd Session. A.O.P. Nicholson printer, Washington, D.C., U.S.A. 1–16.

Blackwell, W.H. 1990. Poisonous and Medicinal Plants. Prentice Hall, Englewood Cliffs, New Jersey, U.S.A.

Blohm, H. 1962. Poisonous Plants of Venezuela. Harvard University Press, Cambridge, Massachusetts, U.S.A.

Bongarten, B.C. 1992. Genetic variation in black locust within its native range. In: J.W. Hanover, P. Bloese, & K. Miller, eds. 1992. Black Locust: Biology, Culture and Utilization. Department of Forestry, Michigan State University, East Lansing, Michigan, U.S.A. 78–97.

Boughton, I.B. & W.T. Hardy. 1935. Mescalbean (*Sophora secundiflora*) poisonous for livestock. Bull. 519. Texas Agricultural Experiment Station, College Station, Texas, U.S.A.

Bourke, J.G. 1889. Notes on the cosmogony and theogony of the Mohave Indians of the Rio Colorado, Arizona. J. Amer. Folklore 2:169–189.

Bowers, J.E. & M.A. Dimmitt. 1994. Flowering phenology of six woody plants in the northern Sonoran Desert. Bull. Torrey Bot. Club 121:215–229.

Brenan, J.P.M. 1963. Notes on African Caesalpinioideae: The genus *Peltophoropsis* Chiov. and its relationship. Kew Bull. 17:203–209.

Brendel, F. 1879. Historical sketch of the science of botany in North America from 1635 to 1840. Amer. Naturalist 13:754–771.

Brendel, F. 1880. Historical sketch of the science of botany in North America from 1840 to 1858. Amer. Naturalist 14:25–38.

Brett, C.H. 1946. Insecticidal properties of the indigobush (*Amorpha fruticosa*). J. Agric. Res. 73(3):81–96.

Brewbaker, J.L. 1987. *Leucaena*, a multipurpose tree genus for tropical agroforestry. In: H.A. Steppler & P.K.R. Nair, eds. 1987. Agroforestry: A Decade of Development. International Council for Research in Agroforestry, Nairobi, Kenya. 289–323.

Britton, N.L. & J.N. Rose. 1930. Caesalpiniaceae. N. Amer. Fl. 23:201–349.

Broich, S.L. 2007. New combinations in North American *Lathyrus* and *Vicia* (Fabaceae: Faboideae: Fabeae). Madroño 54:63–71. Erratum 54:371.

Brouillet, L. 2008. The taxonomy of North American Loti (Fabaceae: Loteae): New names in *Acmispon* and *Hosackia*. J. Bot. Res. Inst. Texas 2(1):387–394.

Brouillet, L. 2012. *Acmispon maritimus* var. *brevivexillus*. In: B.G. Baldwin, D. Goldman, D.J. Keil, R. Patterson, T.J. Rosatti, & D. Wilken, eds. 2012. The Jepson Manual, Vascular Plants of California. 2nd ed. University of California Press, Berkeley, California, U.S.A. 620.

Brown, D.E., ed. 1982. Biotic communities of the American Southwest: United States & Mexico. Desert Pl. 4(1–4):3–341.

Brown, D.E., ed. 1994. Biotic Communities: Southwestern United States and Northwestern Mexico. University of Utah Press, Salt Lake City, Utah, U.S.A.

Brown, R.C. & M.L. Brown. 1972. Woody Plants of Maryland. University of Maryland, College Park, Maryland, U.S.A.

Bruneau, A., F. Forest, P.S. Herendeen, B.B. Klitgaard, and G.P. Lewis. 2001. Phylogenetic relationships in the Caesalpinioideae (Leguminosae) as inferred from chloroplast *trn*L intron sequences. Syst. Bot. 26:487–514.

Bruneau, A., M. Mercure, G.P. Lewis, & P. Herendeen. 2008. Phylogenetic patterns and diversification in the caesalpinioid legumes. Botany 86:697–718.

Bruneton, J. 1999. Pharmacognosy. 2nd ed. Lavoisier Publishing, Paris, France.

Buchmann, S.L. 1974. Buzz pollination of *Cassia quiedondilla* (Leguminosae) by bees of the genera *Centris* and *Melipona*. Bull. S. Calif. Acad. Sci. 73:171–173.

Buhrow, R. 1983. The wild beans of southwestern North America. Desert Pl. 5(2):67–88.

Burghardt, A.D. & S.M. Espert. 2007. Phylogeny of *Prosopis* (Leguminosae) as shown by morphological and biochemical evidence. Austral. Syst. Bot. 20:332–339.

Burkart, A. 1949. La posición sistemática del "chañar" y las especies del género *Geoffroea* (Leguminosae-Dalbergieae). Darwiniana 9(1):9–23.

Burkart, A. 1952. Las Leguminosas Argentinas, silvestres y cultivadas. Acme Agency, Buenos Aires, Argentina.

Burkart, A. 1971. El género *Galactia* (Legum.-Phaseoleae) en Sudamérica con especial referencia a la Argentina y países vecinos. Darwiniana 16:663–796.

Burkart, A. 1976. A monograph of the genus *Prosopis* (Leguminosae subfam. Mimosoideae). J. Arnold Arbor. 57:219–249, 450–525.

Burlage, H.M. 1968. Index of Plants of Texas with Reputed Medicinal and Poisonous Properties. Published by author, Austin, Texas, U.S.A.

Burlando, B., G. Pastorino, A. Salis, G. Damonte, M. Clericuzio, & L. Cornara. 2017. The bioactivity of *Hedysarum coronarium* extracts on skin enzymes and cell correlates with phenolic content. Pharm. Biol. 55(1):1984–1991. doi:10.1080/13880209.2017.1346691

Burrows, G.E. & R.J. Tyrl. 2013. Toxic Plants of North America. 2nd ed. John Wiley & Sons, Inc., Hoboken, New York, U.S.A.

Burton, J.C., R.L. Curly, & C.J. Martinez. 1974. Rhizobia inoculants for various leguminous species. Nitrogen Inform. Bull. 101:1–3.

California Department of Food and Agriculture (CDFA). n.d. "Integrated Pest Control/Weeds: Encyclopeedia." CDFA, Sacramento, California, U.S.A. Accessed 20 Dec 2010. http://www.cdfa.ca.gov/phpps/pc/weedinfo/sphaerophysa-salsula.htm

Calvillo-Canadell, L. & S.R.S. Cevallos-Ferriz. 2005. Diverse assemblage of Eocene and Oligocene Leguminosae from Mexico. Int. J. Pl. Sci. 166:671–692.

Calviño, A. & L. Galetto. 2010. Variation in sexual expression in relation to plant height and local density in the andromonoecious shrub *Caesalpinia gilliesii* (Fabaceae). Pl. Ecol. 209(1):37–45.

Camazine, S. & R.A. Bye. 1980. A study of the medical ethnobotany of the Zuni Indians of New Mexico. J. Ethnopharmacol. 2:365–368.

Campbell, T.N. 1958. Origin of the mescal bean cult. Amer. Anthropologist 60:156–160.

Cannon, H.L. 1971. The use of plant indicators in ground water surveys, geologic mapping, and mineral prospecting. Taxon 20(2/3):227–256.

Cardoso, D., L.P. de Queiroz, H.C. de Lima, E. Suganuma, C. van den Berg, & M. Lavin. 2013. A molecular phylogeny of the vetaireoid legumes underscores floral evolvability that is general to many early-branching papilionoid lineages. Amer. J. Bot. 100:403–421.

Carleton, A.M. 1986. Synoptic–dynamic character of "bursts" and "breaks" in the southwest U.S. summer precipitation singularity. J. Climatol. 6:605–623.

Carleton, A.M., D.A. Carpenter, & P.J. Weser. 1990. Mechanisms of interannual variability of the southwest United States summer rainfall maximum. J. Climate 3:999–1015.

Carter, A.M. 1974a. The genus *Cercidium* (Leguminosae: Caesalpinoideae) in the Sonoran Desert of Mexico and the United States. Proc. Calif. Acad. Sci. 40(2):17–57.

Carter, A.M. 1974b. Evidence for the hybrid origin of *Cercidium sonorae* (Leguminosae: Caesalpinioideae) of northwestern Mexico. Madroño 22:266–272.

Carter, A.M. & N.C. Rem. 1974. Pollen studies in relation to hybridization in *Cercidium* and *Parkinsonia* (Leguminosae: Caesalpinioideae). Madroño 22:303–311.

Castetter, E.F. 1935. Ethnobiological Studies in the American Southwest, I: Uncultivated Native Plants Used as Sources of Food. Univ. New Mexico Bull. no. 266, Biological Series, vol. 4, no. 1. University of New Mexico Press, Albuquerque, New Mexico, U.S.A. 1–62.

Castetter, E.F. & W.H. Bell. 1942. Pima and Papago Agriculture. The University of New Mexico Press, Albuquerque, New Mexico, U.S.A.

Castetter, E.F. & W.H. Bell. 1951. Yuman Indian Agriculture: Primitive Subsistence on the Lower Colorado and Gila Rivers. The University of New Mexico Press, Albuquerque, New Mexico, U.S.A.

Castetter, E.F. & M.E. Opler. 1936. Ethnobotanical studies in the American Southwest III. The Ethnobiology of the Chiricahua and Mescalero Apache. The Use of Plants for Foods, Beverages and Narcotics. Univ. New Mexico Bull. no. 297, Biol. Ser. 4(5). University of New Mexico Press, Albuquerque, New Mexico, U.S.A. 1–63.

Castetter, E.F. & R.M. Underhill. 1935. Ethnobiological studies in the American Southwest II. The Ethnobiology of the Papago Indians. Univ. New Mexico Bull. no. 275, Biological Series, vol. 4, no. 3. University of New Mexico Press, Albuquerque, New Mexico, U.S.A. 1–84.

Castro, C.L., T.B. McKee, & R.A. Pielke Sr. 2001. The relationship of the North American monsoon to tropical and North Pacific surface temperatures as revealed by observational analysis. J. Climate 14:4449–4473.

Castro, C.L., R.A. Pielke Sr., J.O. Adegoke, S.D. Schubert, & P.J. Pegion. 2007. Investigation of the summer climate of the contiguous U.S. and Mexico using the Regional Atmospheric Modeling System (RAMS). Part II: Model climate variability. J. Climate 20:3866–3887.

Ceolin, G.B. 2011. O gênero Galactia P. Browne (Leguminosae, Papilionoideae) no Brasil. Ph.D. dissertation. Universidade Federal do Rio Grande do Sul, Instituto de Biociências, Programa de Pós-Graduação em Botânica, Porto Alegre, Brazil.

Ceolin, G.B. & S.T.S. Miotto. 2013. Synopsis of the genus *Galactia* (Phaseoleae, Papilionoideae, Leguminosae) in Brazil. Phytotaxa 134:1–26.

Chadwick, H.W. & P.D. Dalke. 1965. Plant succession on dune sands in Fremont County, Idaho. Ecology 46(6):765–780.

Cházaro, M. 1977. El huizache, *Acacia pennatula* (Schlecht. & Cham.) Benth. Una invasora del Centro de Veracruz. Biotica 2:1–18.

Cheatham, S. & M.C. Johnston. 1995. The Useful Wild Plants of Texas, the Southeastern and Southwestern United States, the Southern Plains, and Northern Mexico. Useful Wild Plants, Inc., Austin, Texas, U.S.A.

Cheatham, S. & M.C. Johnston with L. Marshall. 2000. Useful Wild Plants of Texas, the Southeastern and Southwestern United States, the Southern Plains, and Northern Mexico. Volume 2. Useful Wild Plants, Inc., Austin, Texas, U.S.A

Chen, C.J., M.G. Mendenhall, & B.L. Turner. 1994. Taxonomy of *Thermopsis* (Fabaceae) in North America. Ann. Missouri Bot. Gard. 81:714–742.

Chen, D., D. Zhang, & M. Thulin. 2010. *Galactia* (Fabaceae). In: Z.Y. Wu, P.H. Raven, & D.Y. Hong, eds. 2010. Flora of China. Vol. 10 (Fabaceae). Science Press, Missouri Botanical Garden Press, St. Louis, Missouri, U.S.A. 199–200.

Chevallier, A. 1996. The Encyclopedia of Medicinal Plants. DK Publishing, New York, New York, U.S.A.

Chiang, L.C., W. Chiang, M.C. Liu, & C.C. Lin. 2003. *In vitro* antiviral activities of *Caesalpinia pulcherrima* and its related flavonoids. J. Antimicrob. Chemother. 52:194–198. doi:10.1093/jac/dkg291

Chittenden, F.J., ed. 1956. The Royal Horticultural Society Dictionary of Gardening: A Practical and Scientific Encyclopedia of Horticulture. 4 vols. Clarendon Press, Oxford, U.K.

Chittenden, N.H. 1901. Among the Cocopahs. The Land of Sunshine 14:196–204.

Clarke, H.D., D.S. Seigler, & J.E. Ebinger. 1989. *Acacia farnesiana* (Fabaceae: Mimosoideae) and related species from Mexico, the southwestern U.S., and the Caribbean. Syst. Bot. 14:549–564.

Clarke, H.D., D.S. Seigler, & J.E. Ebinger. 1990. *Acacia constricta* (Fabaceae: Mimosoideae) and related species from the southwestern U.S. and Mexico. Amer. J. Bot. 77:305–315.

Clarke, H.D., D.S. Seigler, & J.E. Ebinger. 2009. Taxonomic revision of the *Vachellia acuifera* species group (Fabaceae: Mimosoideaea) in the Caribbean. Syst. Bot. 34:84–101.

Cleland, J.B. 1966. The ecology of the Aboriginal in south and central Australia. In: B.C. Cotton, ed. 1966. Aboriginal Man in South and Central Australia. Government Printer, Adelaide, South Australia, Australia. 111–158.

Colladon, L.T.F. 1816. Histoire naturelle et médicale des Casses, et particulièrement de la Casse et des Sénés employés en médecine. Jean Martel, Montpellier, France.

Comrie, A.C. & E.C. Glenn. 1998. Principal components based regionalization of precipitation regimes across the southwest United States and northern Mexico, with an application to monsoon precipitation variability. Climate Res. 10:201–215.

Conn, J.S. 1976. The relationship of rock outcrops to the distribution of *Erythrina flabelliformis* Kearney in southern Arizona. M.S. thesis. The University of Arizona, Tucson, Arizona, U.S.A.

Cook, D., D.R. Gardner, S.T. Lee, J.A. Pfister, C.A. Stonecipher, & S.L. Welsh. 2016. A swainsonine survey of North American *Astragalus* and *Oxytropis* taxa implicated as locoweeds. Toxicon 118:104–111.

Cooper, C.S. 1957. A legume for native flood meadows:6. Establishment and maintenance of stands of white-tip clover (*T. variegatum*) in native flood meadows and its effect upon yields, vegetative and chemical composition of hay. Agron. J. 49:473–477.

Cooper, J.R., G.C. Cone, & H.W. Peirce. 1969. Geologic Map and Cross-Sections of Arizona 1969, scale 1:2,500,000. repository.azgs.az.gov/uri_gin/azgs/dlio/1083

Correll, D.S. & M.C. Johnston. 1970. Manual of the Vascular Plants of Texas. Texas Research Foundation, Renner, Texas, U.S.A.

Coulot, P. & P. Rabaute. 2013. Monographie des Leguminosae de France. Tome 3: Tribu des Trifolieae. Société Botanique du Centre-Ouest, Jarnac, France.

Coulter, T. 1835. Notes on Upper California. J. Roy. Geogr. Soc. 5:59–70.

Coville, F.V. 1895. The botanical explorations of Thomas Coulter in Mexico and California. Bot. Gaz. 20:519–531.

Cox, B.J. 1972. Biosystematics of *Lupinus lepidus-L. caespitosus* complex. Ph.D. dissertation. University of Missouri, Columbia, Missouri, U.S.A.

Craig, W.J. & T.T. Nguyen. 1984. Caffeine and theobromine levels in cocoa and carob products. J. Food Sci. 49:302–303.

Craighead, J.J., F.C. Craighead, Jr., & R.J. Davis. 1963. A Field Guide to Rocky Mountain Wildflowers from Northern Arizona and New Mexico to British Columbia. Houghton Mifflin Company, Boston, Massachusetts, U.S.A

Crosswhite, F.S. 1981. Dry country plants of the South Texas plains. Desert Pl. 2(3):141–179.

Cunningham, G.M., W.E. Mulham, P.L. Milthorpe, & J.H. Leigh. 1992. Plants of Western New South Wales. Inkata Press, Clayton, Victoria, Australia.

Curtin, L.S.M. 1947. Healing Herbs of the Upper Rio Grande. San Vicente Foundation, Inc., Santa Fe, New Mexico, U.S.A.

Curtin, L.S.M. 1949. By the Prophet of the Earth: Ethnobotany of the Pima. San Vicente Foundation, Inc. Reissued 1984. University of Arizona Press, Tucson, Arizona, U.S.A.

Cutright, P.R. & M.J. Broadhead. 1981. Elliott Coues: Naturalist and Frontier Historian. University of Illinois Press, Champaign, Illinois, U.S.A.

Dahmer, N., M.F. Simon, M.T. Schifino-Wittmann, C.E. Hughes, S.T.S. Miotto, & J.C. Giuliani. 2011. Chromosome numbers in the genus *Mimosa* L.: Cytotaxonomic and evolutionary implications. Pl. Syst. Evol. 291:211–220.

D'Antoni, H.L. & O.T. Solbrig. 1977. Algarrobos in South American cultures: Past and present. In: B.B. Simpson, ed. 1977. Mesquite: Its Biology in Two Desert Scrub Ecosystems. Dowden, Hutchinson and Ross, Stroudsburg, Pennsylvania, U.S.A. 189–200.

Darbyshire, S. & E. Small. 2018. Are *Melilotus albus* and *M. officinalis* conspecific? Genet. Resources Crop Evol. 65:1571–1580.

Darwin, C.R. 1859. On the Origin of Species. John Murray, London, U.K.

Davidson, B.R. & H.F. Davidson. 1993. Legumes, The Australian Experience: The Botany, Ecology and Agriculture of Indigenous and Immigrant Legumes. Research Studies Press Ltd., Somerset, U.K.

Davies, A.M.R., N.Maxted, & L.J.G. van der Maesen. 2007. A natural infrageneric classification for *Cicer* (Leguminosae, Cicereae). Blumea 52:379–400.

Davis, C.C., P.W. Fritsch, J. Li, & M.J. Donoghue. 2002. Phylogeny and biogeography of *Cercis* (Fabaceae): Evidence from nuclear ribosomal ITS and chloroplast *ndhF* sequence data. Syst. Bot. 27:289–302.

Delgado-Salinas, A., M. Thulin, R. Pasquet, N. Weeden, & M. Lavin. 2011. *Vigna* (Leguminosae) sensu lato: The Names and Identities of the American segregate genera. Amer. J. Bot. 98:1694–1715.

Delgado-Salinas, A. & L. Torres-Colín. 1995. Una nueva combinación en el género *Macroptilium* (Fabaceae: Phaseolinae). Anales Inst. Biol. Univ. Nac. Autón. México, Ser. Bot. 66(2):189–192.

Desert Harvesters. 2018. Eat Mesquite and More: A Cookbook for Sonoran Desert Foods and Living. RainSource Press, Tucson, Arizona, U.S.A.

de Souza Conceição, A., L. Paganucci de Queiroz, G.P. Lewis, M.J. Gomes de Andrade, P.R. Machado de Almeida, A.S. Schnadelbach, & C. van den Berg. 2009. Phylogeny of *Chamaecrista* Moench (Leguminosae-Caesalpinioideae) based on nuclear and chloroplast DNA regions. Taxon 58:1168–1180.

Devitt, J. 1992. Acacias: A traditional Aboriginal food source in central Australia. In: A.P.N. House & C.E. Harwood, eds. 1992. Australian Dry-zone Acacias for Human Food. CSIRO Publications, North Clayton, Victoria, Australia.

Diamond, A.R. 2016. New vascular plant county records from Alabama. Phytoneuron 2016-70:1–6.

Diggs Jr., G.M., B.L. Lipscomb, & R.J. O'Kennon. 1999. Shiners and Mahler's Illustrated Flora of North Central Texas. Sida, Botanical Miscellany, No. 16. Botanical Research Institute of Texas Press, Fort Worth, Texas. U.S.A.

Dimmitt, M.A. 1987. The hybrid palo verde 'Desert Museum': A new, superior tree for desert landscapes. Desert Pl. 8(3):99–103.

Dimmitt, M.A. 2000. Fabaceae (legume family). In: Phillips, S.J. & P.W. Comus, eds. 2000. A Natural History of the Sonoran Desert. Arizona-Sonoran Desert Museum Press, Tucson, Arizona, U.S.A., and University of California Press, Berkeley, California, U.S.A. 227–238.

DiTomaso, J.M. & E.A. Healy. 2007. Weeds of California and Other Western States. 2 vols. Publication 3488. University of California Agriculture and Natural Resources Communication Services, Oakland, California, U.S.A.

DiTomaso, J.M., G.B. Kyser, & M.S. Hastings. 1999. Prescribed burning for control of yellow starthistle (*Centaurea solstitialis*) and enhanced native plant diversity. Weed Sci. 47:233–242.

Dixon, J.S. 1934. A study of the life history and feeding habits of mule deer in California. Calif. Fish & Game 20:181–282, 315–354.

Dong, Y., D. Tang, N. Zhang, Y. Li, C. Zhang, L. Li, & M. Li. 2013. Phytochemicals and biological studies of plants in genus *Hedysarum*. Chem. Central J. 7:124. doi:10.1186/1752-153X-7-124

Douglas, M.W., R.A. Maddox, K. Howard, & S. Reyes. 1993. The Mexican monsoon. J. Climate 6:1665–1677.

Drummond, C.S., R.J. Eastwood, S.T.S. Miotto, & C.E. Hughes. 2012. Multiple continental radiations and correlates of diversification in *Lupinus* (Leguminosae): Testing for key innovation with incomplete taxon sampling. Syst. Biol. 61:443–460.

Dubey, H.D., R. Woodbury, & R.L. Rodríguez. 1972. New records of tropical legume nodulation. Bot. Gaz. 133(1):35–38.

Duke, J.A. 1981. Handbook of Legumes of World Economic Importance. Plenum Press, New York, New York, U.S.A.

Duke, J.A. & K.K. Wain. 1981. Medicinal Plants of the World. 3 vols. Computer index with more than 85,000 entries. Plant Genetics and Germplasm Institute, Agricultural Research Service, Beltsville, Maryland, U.S.A.

Duncan, W.H. 1975. Woody vines of the Southeastern United States. University of Georgia Press, Athens, Georgia, U.S.A.

Duncan, W.H. 1979. Changes in *Galactia* (Fabaceae) of the southeastern United States. Sida 8:170–180.

Duno de Stefano, R., G.C. Fernández-Concha, L.L. Can-Itza, & M. Lavin. 2010. The morphological and phylogenetic distinctions of *Coursetia greenmanii* (Leguminosae): Taxonomic and ecological implications. Syst. Bot. 35(2):289–295.

Ebinger, J.E., D.S. Seigler, & H.D. Clarke. 2002. Notes on the segregates of *Acacia farnesiana* (L.) Willd. (Fabaceae: Mimosoideae) and related species in North America. SouthW. Naturalist 47:86–91.

Egan, A.N. 2014. Species delimitation and recognition in the *Pediomelum megalanthum* complex (Fabaceae) via multivariate morphometrics. PhytoKeys 44:65–87. doi:10.3897/phytokeys.44.8750

Egan, A.N. & K.A. Crandall. 2008a. Divergence and diversification in North American Psoraleeae (Fabaceae) due to climate change. BMC Biol. 6:55.

Egan, A.N. & K.A. Crandall. 2008b. Incorporating gaps as phylogenetic characters across eight DNA regions: Ramifications for North American Psoraleeae (Leguminosae). Molec. Phylogen. Evol. 46:532–546.

Egan, A.N. & J.L. Reveal. 2009. A new combination in *Pediomelum* and a new genus, *Ladeania*, from western North American (Fabaceae, Psoraleeae). Novon 19:310–314.

Eisner, T. 1994. Chemical prospecting: A global imperative. Proc. Amer. Philos. Soc. 138:385–395.

Ellison, N.W., A. Liston, J.J. Steiner, W.M. Williams, & N.L. Taylor. 2006. Molecular phylogenetics of the clover genus (*Trifolium*-Leguminosae). Molec. Phylogen. Evol. 39:688–705.

Elmore, F.H. 1978. Ethnobotany of the Navajo. AMS Press, New York, New York, U.S.A.

Emory, W.H. 1848. Notes of a military reconnoissance, from Fort Leavenworth, in Missouri, to San Diego, in California, including part of the Arkansas, Del Norte, and Gila Rivers. House Exec. doc. 41, 30th Congress, 1st Session. Wendell and Van Benthuysen Printers, Washington, D.C., U.S.A.

Encarnación-Dimayuga, R., J. Agúndez-Espinoza, A. Garcia, G. Delgado, G.M. Molina-Salinas, & S. Said-Fernández. 2006. Two new cassane-type diterpenes from *Calliandra californica* with antituberculosis and cytotoxic activities. Pl. Med. 72(8):757–761.

Encarnación-Dimayuga, R., N. Ochoa, U. Anthoni, C. Christophersen, & P.H. Nielsen. 1994. Two new flavones from *Calliandra californica*. J. Nat. Prod. 57(9):1307–1309.

Everett, T.H. 1968. Living Trees of the World. Doubleday & Co., Inc., New York, New York, U.S.A.

Everett, T H. 1981. The New York Botanical Garden Illustrated Encyclopedia of Horticulture, vol 2. Garland Publishing, Inc., New York, New York, U.S.A. and London, U.K.

Ewan, J. 1936. Bibliography of the botany of Arizona. Amer. Midl. Naturalist 17:430–454.

Fantz, P.R. 1977. A monograph of the genus *Clitoria* (Leguminosae: *Glycineae*). Ph.D. dissertation. The University of Florida, Gainesville, Florida, U.S.A.

Fearing, O.S. 1959. A cytotaxonomic study of the genus *Cologania* and its relationship to *Amphicarpaea* (Leguminosae-Papilionoideae). Ph.D. dissertation. The University of Texas, Austin, Texas, U.S.A.

Feldman, I. & S.R. Feldman. 1987. Brush problems in Argentina's rangelands: I. Chañar [*Geoffroea decorticans*] in northwestern Santa Fe Province. Rangelands 9(6):251–254.

Felger, R.S. 1977. Mesquite in Indian cultures of southwest North America. In: B.B. Simpson, ed. 1977. Mesquite: Its Biology in Two Desert Scrub Ecosystems. Dowden, Hutchinson and Ross, Stroudsburg, Pennsylvania, U.S.A. 150–176.

Felger, R.S. 1979. Ancient crops for the 21st century. In: G. Ritchie, ed. 1979. New Agricultural Crops. AAAS Selected Symposium 38. Westview Press, Boulder, Colorado. 5–20.

Felger, R.S. 1990. Mesquite, a world food crop. Aridus 2(1):1–3.

Felger, R.S. 1992. Reflections on a desert legume trinity. Aridus 4(4):1–5, 7.

Felger, R.S. 2000. Flora of the Gran Desierto and Río Colorado of Northwestern Mexico. The University of Arizona Press Tucson, Arizona, U.S.A.

Felger, R.S. 2007. Living resources at the center of the Sonoran Desert: Native American plant and animal utilization. In: R.S. Felger & B. Broyles, eds. 2007. Dry Borders: Great Natural Reserves of the Sonoran Desert. University of Utah Press, Salt Lake City, Utah, U.S.A. 147–192.

Felger, R.S., M.B. Johnson, & M.F. Wilson. 2001. The Trees of Sonora, Mexico. Oxford University Press, Oxford, U.K.

Felger, R.S. & N. Logan. 2010. Mesquite—Food for the World. In: B. Lancaster. 2010. Eat Mesquite! Desert Harvesters, Tucson, Arizona, U.S.A. 4–7.

Felger, R.S. & N. Logan. 2018. The Desert Is Food. In: Desert Harvesters. 2018. Eat Mesquite and More: A Cookbook for Sonoran Desert Foods and Living. RainSource Press, Tucson, Arizona, U.S.A. 318–321.

Felger, R.S. & M.B. Moser. 1985. People of the Desert and Sea: Ethnobotany of the Seri Indians. The University of Arizona Press, Tucson, Arizona, U.S.A.

Felger, R.S. & S. Rutman. 2015. Ajo Peak to Tinajas Altas: A flora of southwestern Arizona. Part 14. Eudicots: Fabaceae—Legume Family. Phytoneuron 2015-58:1–83.

Felker, P. & P.R. Clark. 1980. Nitrogen fixation (acetylene reduction) and cross inoculation in 12 *Prosopis* (mesquite) species. Plant and Soil 57:177–186.

Felker, P. & P.R. Clark. 1981. Nodulation and nitrogen fixation (acetylene reduction) in desert ironwood (*Olneya tesota*). Oecologia 48:292–293.

Felker, P., N. Grados, G. Cruz, & D. Prokopiuk. 2003. Economic assessment of production of flour from *Prosopis alba* and *P. pallida* pods for human food applications. J. Arid Environm. 53:517–528.

Fischer, D.L. 2001. Early Southwest Ornithologists, 1528–1900. University of Arizona Press, Tucson, Arizona, U.S.A.

Fisher, C. 1995. Horse care: Perilous pasture plants. Rural Heritage 20:44–45.

Fisher, E.M. 1892. Revision of North American species of *Hoffmannseggia*. Contr. U.S. Natl. Herb. 1:143–150.

Fleet, R.R. & B.L. Young. 2000. Facultative mutualism between imported fire ants (*Solenopsis invicta*) and a legume (*Senna occidentalis*). SouthW. Naturalist 45:289–298.

Flint, R. & S. Flint. 2005. Documents of the Coronado Expedition, 1539–1542: "They Were Not Familiar with His Majesty, Nor Did They Wish to Be His Subjects." Southern Methodist University Press, Dallas, Texas, U.S.A.

Flores-Franco, G. 2013. Análisis Taxonómico del Género *Cologania* (Leguminosae). Maestría en Ciencias (Biología), Facultad de Ciencias, U.N.A.M., Ciudad México, México.

Food and Agriculture Organization of the United Nations (FAO). n.d. "FAOSTAT." FAO, Rome, Italy. Accessed 15 Jun 2020. http://www.fao.org/faostat

Foster, B.G. & M.A. Vincent. 2018. A new taxonomy for *Trifolium variegatum* and its relatives. Madroño 65:141–150.

Fox, W.E., K.W. Allred, & E.H. Roalson. 2010. A guide to the common locoweeds and milkvetches of New Mexico. Circular 557. New Mexico State University Cooperative Extension Service, Las Cruces, New Mexico, U.S.A.

Fritsch, P.W. & B.C. Cruz. 2012. Phylogeny of *Cercis* based on DNA sequences of nuclear ITS and four plastid regions: Implications for transatlantic historical biogeography. Molec. Phylogen. Evol. 62(3):816–825.

Fritsch, P.W., C.F. Nowell, L.S.T. Leatherman, W. Gong, B.C. Cruz, D.O. Burge, & A. Delgado-Salinas. 2018. Leaf adaptations and species boundaries in North American *Cercis*: Implications for the evolution of dry floras. Amer. J. Bot. 105(9):1577–1594.

Fritsch, P.W., A.M. Schiller, & K.W. Larson. 2009. Taxonomic implications of morphological variation in *Cercis canadensis* (Fabaceae) from Mexico and adjacent parts of Texas. Syst. Bot. 34:510–520.

Gagnon, E., A. Bruneau, C.E. Hughes, L. Paganucci de Queiroz, & G.P. Lewis. 2016. A new generic system for the pantropical *Caesalpinia* group (Leguminosae). PhytoKeys 71:1–160. doi:10.3897/phytokeys.71.9203

Gallagher, M.V. 1976. Contemporary ethnobotany among the Apache of the Clarkdale, Arizona area: Coconino and Prescott National Forests. Archeological Report No. 14. U.S. Department of Agriculture Forest Service Southwestern Region, Albuquerque, New Mexico, U.S.A.

Gandhi, K.N. 2015. Validation of *Pediomelum palmeri* (Fabaceae). Harvard Pap. Bot. 20(2):213.

Gandhi, K.N., M.A. Vincent, & J.L. Reveal. 2011. *Dermatophyllum*, the correct name for *Calia* (Fabaceae). Phytoneuron 57:1–4.

Genaust, H. 1976. Etymologisches Wörterbuch der botanischen Pflanzennamen. Birkhäuser Verlag, Basel, Switzerland.

Gentry, H.S. 1942. Río Mayo Plants: A study of the flora and vegetation of the valley of the Rio Mayo, Sonora. Publication 527. Carnegie Institution of Washington, Washington, D.C., U.S.A.

Ghane, S.G., V.H. Lokhande, M.L. Ahire, & T.D. Nikam. 2010. *Indigofera glandulosa* Wendl. (Barbada) a potential source of nutritious food: Underutilized and neglected legume in India. Genet. Resources Crop Evol. 57:147–153. doi:10.1007/s10722-009-9496-1

Gifford, E.W. 1933. The Cocopa. Univ. Calif. Publ. Amer. Archaeol. Ethnol. 31(5):257–334.

Gillett, J.M. 1965. Taxonomy of *Trifolium*: Five American species of section *Lupinaster* (Leguminosae). Brittonia 17:121–136.

Gillett, J.M. 1969. Taxonomy of *Trifolium* (Leguminosae). II. The *T. longipes* complex in North America. Canad. J. Bot. 47:93–113.

Gillett, J.M. 1971. Taxonomy of *Trifolium* (Leguminosae). III. *T. eriocephalum*. Canad. J. Bot. 49:395–405.

Gillett, J.M. 1972. Taxonomy of *Trifolium* (Leguminosae). IV. The American species of section *Lupinaster* (Adanson) Seringe. Canad. J. Bot. 50:1975–2007.

Gillett, J.M. 1980. Taxonomy of *Trifolium* (Leguminosae). V. The perennial species of section *Involucrarium*. Canad. J. Bot. 58:1425–1448.

Gillett, J.M. & N.L. Taylor. 2001. The World of Clovers. Iowa State University Press, Ames, Iowa, U.S.A.

Gilman, E.F. & D.G. Watson. 1994. *Sophora affinis* Eves-Necklace. Fact Sheet ST-591. Environmental Horticulture Department, Florida Cooperative Extension Service, Institute of Food and Agricultural Sciences, University of Florida, Gainesville, Florida, U.S.A.

Giusti, L. 1938. Plantas forrajeras cianogenéticas [*Astragalus bergii, Sesbania punicea*]. Jornada Agronomía y Veterinaria 1937:43–49.

Gleason, H.A. & A. Cronquist. 1991. Manual of Vascular Plants of Northeastern United States and Adjacent Canada. 2nd ed. New York Botanical Garden, Bronx, New York, U.S.A.

Goertzen, L.R. & E. Small. 1993. The defensive role of trichomes in black medick (*Medicago lupulina, Fabaceae*). Pl. Syst. Evol. 184:101–111.

Goldberg, D.E. & R.M. Turner. 1986. Vegetation change and plant demography in permanent plots in the Sonoran Desert. Ecology 67:695–712.

Gottsberger, G. & I. Silberbauer-Gottsberger. 1988. Evolution of flower structures and pollination in Neotropical *Cassiinae (Caesalpiniaceae)* species. Phyton (Horn) 28:293–320.

Govindarajulu, R., C.E. Hughes, P.J. Alexander, & C.D. Bailey. 2011. The complex evolutionary dynamics of ancient and recent polyploidy in *Leucaena* (Leguminosae; Mimosoideae). Amer. J. Bot. 98:2064–2076.

Govindarajulu, R., C.E. Hughes, & C.D. Bailey. 2011. Phylogenetic and population genetic analyses of diploid *Leucaena* (Leguminosae; Mimosoideae) reveal cryptic species diversity and patterns of divergent allopatric speciation. Amer. J. Bot. 98:2049–2063.

Graham, E.H. 1941. Legumes for Erosion Control and Wildlife. U.S. Department of Agriculture, Miscellaneous Publication No. 412. United States Government Printing Office, Washington, D.C., U.S.A.

Grantz, K., B. Rajagopalan, M. Clark, & E. Zagona. 2007. Seasonal shifts in the North American monsoon. J. Climate 20: 1923–1935.

Gray, A. 1885. How to reach the Grand Cañon. Science 5:516–517.

Gray, A., J. Torrey, G. Thurber, & G. Engelmann. 1861. Part IV: Botany. In: J.C. Ives & J.S. Newberry. 1861. Report upon the Colorado River of the West. Senate Exec. doc. [s.n.], 36th Congress, 1st Session. Government Printing Office, Washington, D.C., U.S.A.

Grear, J.W. 1978. A Revision of the New World Species of *Rhynchosia* (Leguminosae-Faboideae). Mem. New York Bot. Gard. 31(1):1–168.

Grether, R., A. Martínez-Bernal, & S.L. Camargo-Ricalde. 2007. *Mimosa* L. In: G. Calderón de Rzedowski & J. Rzedowski, eds. 2007. Flora del Bajío y de Regiones Adyacentes 150:140–183.

Grimes, J.W. 1990. A revision of the New World species of Psoraleeae (Leguminosae: Papilionoideae). Mem. New York Bot. Gard. 61:1–113.

Guinet, P. & H.M. Hernández. 1989. Pollen characters in the genera *Zapoteca* and *Calliandra* (Leguminosae, Mimosoideae), their systematic and phylogenetic relevance. Pollen & Spores 31:5–22.

Gunn, C.R. 1968. The *Vicia americana* complex (Leguminosae). Iowa State J. Sci. 42:171–214.

Gunn, C.R. 1979. Genus *Vicia* with notes about tribe Vicieae (Fabaceae) in Mexico and Central America. U.S. Department of Agriculture, Tech. Bull. No. 1601. United States Government Printing Office, Washington, D.C., U.S.A.

Gutterman, J.U. & V. Haridas. 2016. Avicin D for use in the treatment of Mantle Cell Lymphoma. Research Development Foundation. Internat. Patent No. WO2016134361.

Hajare, S.W., S. Chandra, S.K. Tandan, J. Sarma, J. Lal, & A.G. Telang. 2000. Analgesic and antipyretic activities of *Dalbergia sissoo* leaves. Indian J. Pharmacol. 32:357–360.

Hanausek, M., P. Ganesh, Z. Walaszek, C.J. Arntzen, T.J. Slaga, & J.U. Gutterman. 2001. Avicins, a family of triterpenoid saponins from *Acacia victoriae* (Bentham), suppress H-ras mutations and aneuploidy in a murine skin carcinogenesis model. Proc. Natl. Acad. Sci. 98:11551–11556.

Harborne, J.B. 1994. Phytochemistry of the Leguminosae. In: F.A. Bisby, J. Buckingham, & J.B. Harborne, eds. 1994. Phytochemical Dictionary of the Leguminosae. Chapman & Hall. London, U.K. xx-xxiii.

Hardin, J.W. & J.M. Arena. 1974. Human Poisoning from Native and Cultivated Plants. 2nd ed. Duke University Press, Durham, North Carolina, U.S.A.

Hardin, J.W. & C.F. Brownie. 1993. Plants Poisonous to Livestock and Pets in North Carolina. North Carolina Agricultural Research Service Bulletin No. 414 (revised). North Carolina State University, Raleigh, North Carolina, U.S.A.

Haridas, V., Z.-X. Xu, D. Kitchen, A. Jiang, P. Michels, & J.U. Gutterman. 2011. The anticancer plant triterpenoid, avicin D, regulates glucocorticoid receptor signaling: Implications for cellular metabolism. PLoS ONE 6: e28037.

Hart, C.R., T. Garland, C. Barr, B.B. Carpenter, & J.C. Raegor. 2003 (revised). Toxic Plants of Texas. Texas A&M Extension Publication, College Station, Texas, U.S.A.

Haston, E.M., G.P. Lewis, & J.A. Hawkins. 2005. A phylogenetic reappraisal of the *Peltophorum* group (Caesalpinieae: Leguminosae) based on the chloroplast *trnL-F*, *rbcL* and *rps16* sequence data. Amer. J. Bot. 92:1359–1371.

Hawkins, J.A. 1996. Systematics of *Parkinsonia* L. and *Cercidium* Tul. (Leguminosae: Caesalpinioideae). Ph.D. dissertation. University of Oxford, Oxford, U.K.

Hawkins, J.A., L.W. Olascoaga, C.E. Hughes, J.-L.R. Contreras Jiménez, & P. Mercado Ruaro. 1999. Investigation and documentation of hybridization between *Parkinsonia aculeata* and *Cercidium praecox* (Leguminosae: Caesalpinioideae). Pl. Syst. Evol. 216:49–68.

Hayden, J.D. 1969. Gyratory crushers of the Sierra Pinacate, Sonora. Amer. Antiquity 34:154–161.

Heenan, P.B., M.I. Dawson, & S.J. Wagstaff. 2004. The relationship of *Sophora* sect. *Edwardsia* (Fabaceae) to *Sophora tomentosa*, the type species of the genus *Sophora*, observed from DNA sequence data and morphological characters. Bot. J. Linn. Soc. 146:439–446.

Heintzelman, S.P. 1857. Report of July 1853. In: Indian Affairs on the Pacific, U.S. House of Representatives, 34th Congress, 3d session, House Executive Document No. 76. U.S. Government Printing Office, Washington, D.C., U.S.A. 34–54.

Hendricks, D.M. 1985. Arizona Soils. College of Agriculture, University of Arizona, Tucson, Arizona, U.S.A.

Herendeen, P.S., W.L. Crepet, & D.L. Dilcher. 1992. The fossil history of the Leguminosae: Phylogenetic and biogeographic implications. In: P.S. Herendeen & D.L. Dilcher, eds. 1992. Advances in Legume Systematics: Part 4 The Fossil Record. Royal Botanic Gardens, Kew, London, U.K. 303–316.

Hermann, F.J. 1960. Vetches of the U.S.: Native, Naturalized and Introduced. U.S. Department of Agriculture, Agriculture Handbook No. 168. Government Printing Office, Washington, D.C., U.S.A.

Hernández, H M. 1986. *Zapoteca*, a new genus of neotropical Mimosoideae. Ann. Missouri Bot. Gard. 73:755–763.

Hernández, H.M. 1989. Systematics of *Zapoteca* (Leguminosae). Ann. Missouri Bot. Gard. 76:781–862.

Hernández, H.M. 1991. Taxonomía, distribución geográfica y biología reproductiva de *Calliandra calothyrsus* (Leguminosae, Mimosoideae), una especie con potencial agroforestal. Anales Inst. Biol. Univ. Nac. Autón. México, Ser. Bot. 62(2):121–132.

Heyn, C.C. 1963. The Annual Species of *Medicago*. Magnes Press, The Hebrew University, Jerusalem, Israel.

Hitchcock, C.L. 1952. A revision of the North American species of *Lathyrus*. Univ. Wash. Publ. Biol. 15:1–104.

Hitchcock, C.L., A. Cronquist, M. Ownbey, & J.W. Thompson. 1961. Vascular plants of the Pacific Northwest, Vol. 3: Saxifragaceae to Ericaceae. University of Washington Press, Seattle, Washington, U.S.A.

Hocking, P.J. & A.A. Kortt. 1987. Growth and nutrient accumulation by fruits of the perennial legume, *Hardenbergia violacea*, with special reference to myrmecochory. New Phytol. 105:89–102.

Hodgson, W.C. 2001. Food Plants of the Sonoran Desert. University of Arizona Press, Tucson, Arizona, U.S.A.

Hoffmann, J.J., G. Wächter, & J.U. Gutterman. 2000. Antibacterial and antifungal flavanones from *Eysenhardtia texana*. Research Development Foundation. Internat. Patent No. WO2000020406.

Hollowell, E.A. 1939. Strawberry Clover. Leaflet No. 176. U.S. Department of Agriculture, Washington, D.C., U.S.A.

Holmgren, P.K., N.H. Holmgren, & L.C. Barnett, eds. 1990. Index Herbariorum, Part I: The Herbaria of the World, Ed. 8. New York Botanical Garden, Bronx, New York, U.S.A.

Hooker, W.J. 1833. Botanical Miscellany. John Murray, London, U.K.

Hopkins, M. 1942. *Cercis* in North America. Rhodora 44:192–211.

Howard, J.H. 1957. The mescal bean cult of the Central and Southern Plains: An ancestor of the peyote cult? Amer. Anthropol. 59:75–87.

Hsu, P.-J., J.S. Miller, & J.M. Berger. 2009. Bakuchiol, an antibacterial component of *Psoralidium tenuiflorum*. Natural Product Res. 23(8):781–788.

Hughes, C.E. 1998a. Monograph of *Leucaena* (Leguminosae-Mimosoideae). Syst. Bot. Monogr. 55:1–244.

Hughes, C.E. 1998b. *Leucaena*: A Genetic Resources Handbook. Tropical Forestry Paper 37. Oxford Forestry Institute, University of Oxford, Oxford, U.K.

Hughes, C.E., R. Govindarajulu, A. Robertson, D.L. Filer, S.A. Harris, & C.D. Bailey. 2007. Serendipitous backyard hybridization and the origin of crops. Proc. Natl. Acad. Sci. U.S.A. 104:14389–14394.

Hughes C.E., J.J. Ringelberg, G.P. Lewis, & S.A. Catalano. 2022. Disintegration of the genus *Prosopis* L. (Leguminosae, Caesalpinioideae, mimosoid clade). Phytokeys 205:147–189.

Hughes, C.E. & B.T. Styles. 1984. Exploration and seed collection of multiple-purpose dry zone trees in Central America. Int. Tree Crops J. 3(1):1–31.

Irwin, H.S. & R.C. Barneby. 1978. Monographic studies in *Cassia* (Leguminosae Caesalpinioideae). III. Sections *Absus* and *Grimaldia*. Mem. New York Bot. Gard. 30:1–300.

Irwin, H.S. & R.C. Barneby. 1981. Tribe 2. Cassieae Bronn. In: R.M. Polhill & P.H. Raven, eds. 1981. Advances in Legume Systematics: Part 1. Royal Botanic Gardens, Kew, London, U.K. 97–106.

Irwin, H.S. & R.C. Barneby. 1982. The American Cassiinae. Mem. New York Bot. Gard. 35:1–918.

Isely, D. 1954. Keys to sweet clovers (*Melilotus*). Proc. Iowa Acad. Sci. 61:119–131.

Isely, D. 1969. Legumes of the United States: I. Native *Acacia*. Sida 3:365–386.

Isely, D. 1972. Legumes of the U.S. VI. *Calliandra, Pithecellobium, Prosopis.* Madroño 21:273–298.

Isely, D. 1973. Leguminosae of the United States: I. Subfamily Mimosoideae. Mem. New York Bot. Gard. 25(1):1–152.

Isely, D. 1975. Leguminosae of the United States: II. Subfamily Caesalpinioideae. Mem. New York Bot. Gard. 25(2):1–228.

Isely, D. 1981. Leguminosae of the United States: III. Subfamily Papilionoideae: Tribes Sophorae, Podalyrieae, Loteae. Mem. New York Bot. Gard. 25(3):1–264.

Isely, D. 1983. New combinations and two new varieties in *Astragalus*, *Orophaca*, and *Oxytropis* (Leguminosae). Syst. Bot. 8:420–426.

Isely, D. 1990. Vascular Flora of the Southeastern United States: Volume 3, Part 2 Leguminosae (Fabaceae). The University of North Carolina Press, Chapel Hill, North Carolina, U.S.A.

Isely, D. 1998. Native and Naturalized Leguminosae (Fabaceae) of the United States (exclusive of Alaska and Hawaii). Monte L. Bean Life Science Museum, Brigham Young University, Provo, Utah, U.S.A.

Isely, D. & F.J. Peabody. 1984. *Robinia* (Leguminosae: Papilionoideae). Castanea 49:187–202.

Ives, J.C. 1861. Part I: General report. In: J.C. Ives & J.S. Newberry. 1861. Report upon the Colorado River of the West. Senate Exec. doc. [s.n.], 36th Congress, 1st Session. Government Printing Office, Washington, D.C., U.S.A. 19–131.

Jacobs, K.L., G.M. Garfin, & B.J. Morehouse. 2005. Climate science and drought planning: The Arizona experience. J. Amer. Water Resour. Assoc. 41(2):437–445.

Jaiwal, P.K. & R.P. Singh, eds. 2003. Improvement Strategies of Leguminosae Biotechnology. Kluwer Academic Publishers, Dordrecht, Netherlands.

James, L.F. & S.L. Welsh. 1992. Poisonous plants of North America. In: L.F. James, R.F. Keeler, E.M. Bailey, Jr., P.R. Cheeke, & M.P. Hegarty, eds. 1992. Poisonous Plants: Proceedings of the Third International Symposium. Iowa State University Press, Ames, Iowa, U.S.A. 94–103.

Jawad, J.T., D.S. Seigler, & J.E. Ebinger. 2000. A systematic treatment of *Acacia coulteri* (Fabaceae, Mimosoideae) and similar species in the New World. Ann. Missouri Bot. Gard. 87:528–548.

Jayatilake, G.S., D.R. Freeberg, Z. Liu, S.L. Richheimer, M.E. Blake, D.T. Bailey, V. Haridas, & J.U. Gutterman. 2003. Isolation and structures of avicins D and G: In vitro tumor-inhibitory saponins derived from *Acacia victoriae*. J. Nat. Prod. 66:779–783.

Jenkins, M.B., R.A. Virginia, & W.M. Jarrell. 1988. Rhizobial ecology of *Psorothamnus spinosus* in a Sonoran Desert arroyo. Pl. & Soil 105:113–120.

Jia, H. & S.R. Manchester. 2014. Fossil leaves and fruits of *Cercis* L. (Leguminosae) from the Eocene of western North America. Int. J. Pl. Sci. 175:601–612.

Johnson, H.B. & H.S. Mayeux. 1990. *Prosopis glandulosa* and the nitrogen balance of rangelands: Extent and occurrence of modulation. Oecologia 84:176–185.

Johnson, M.B., J.H. Brown, & D.A. Palzkill. 1993. Desert smoke tree—desert legume with a promising future. Aridus 5(1):1–5, 8.

Johnston, I.M. 1924. Taxonomic records concerning American spermatophytes. 1. *Parkinsonia* and *Cercidium*. Contr. Gray Herb. 70:61–68.

Johnston, I.M. 1940. Field notes of Charles Wright for 1849 and 1851–52 relating to collections from Texas, New Mexico and adjacent Sonora and Chihuahua. Gray Herbarium, Harvard University, Cambridge, Massachusetts, U.S.A.

Johnston, M.C. 1962. The North American mesquites *Prosopis* sect. *Algarobia* (Leguminosae). Brittonia 14:72–90.

Johnston, M.C. 1990. The Vascular Plants of Texas. 2nd ed. Published by author, Austin, Texas, U.S.A.

Jones, M.E. 1923. Revision of North American Species of *Astragalus*. Published by author, Salt Lake City, Utah, U.S.A.

Jones, W. & C. Sacamano. 2000. Landscape Plants for Dry Regions: More than 600 Species from around the World. Fisher Books, Cambridge, Massachusetts, U.S.A.

Karron, J.D. 1989. Breeding systems and levels of inbreeding depression in geographically restricted and widespread species of *Astragalus* (Fabaceae). Amer. J. Bot. 76:331–340.

Karron, J.D., Y.B. Linhart, C.A. Chaulk, & C.A. Robertson. 1988. Genetic structure of populations of geographically restricted and widespread species of *Astragalus* (Fabaceae). Amer. J. Bot. 75:1114–1119.

Kartesz, J.T. & K.N. Gandhi. 1992. *Pediomelum ockendonii* sp. nov.: Correct name for *P. palmeri* Grimes (Fabaceae). Sida 15(1):137–138.

Kearney, T.H. & R.H. Peebles. 1939. Arizona plants: New species, varieties, and combinations. J. Washington Acad. Sci. 29(11):474–492.

Kearney, T.H. & R.H. Peebles. 1942. Flowering Plants and Ferns of Arizona. U.S. Department of Agriculture, Miscellaneous Publication No. 423. Government Printing Office, Washington, D.C., U.S.A.

Kearney, T.H. & R.H. Peebles. 1960. Arizona Flora. University of California Press, Berkeley, California, U.S.A.

Keeler, R.F., L.F. James, J.L. Shupe, & K.R. Van Kampen. 1977. Lupine-induced crooked calf disease and a management method to reduce incidence. J. Range Managem. 30(2):97–102.

Keeler, R.F., A.E. Johnson, & R.L. Chase. 1986. Toxicity of *Thermopsis montana* in cattle. Cornell Veterin. 76(2):115–127.

Kelly, W. 1977. Cocopa Ethnography. Anthropological Papers of the University of Arizona, No. 29. University of Arizona Press, Tucson, Arizona, U.S.A.

Killeen, T.J., E. García E., & S.G. Beck, eds. 1993. Guía de Arboles de Bolivia. Herbario Nacional de Bolivia and Missouri Botanical Garden. Quipas S.R.L., La Paz, Bolivia.

Kim, J.H., S.Y. Kim, S.Y. Lee, & C.G. Jang. 2007. Antidepressant-like effects of *Albizzia julibrissin* in mice: Involvement of the 5-HT_{1A} receptor system. Pharmacol. Biochem. Behav. 87(1):41–47. doi:10.1016/j.pbb.2007.03.018

Kindscher, K. 1987. Edible Wild Plants of the Prairie: An Ethnobotanical Guide. University Press of Kansas, Lawrence, Kansas, U.S.A.

Kindscher, K. 1992. Medicinal Wild Plants of the Prairie: An Ethnobotanical Guide. University Press of Kansas, Lawrence, Kansas, U.S.A.

Kingsbury, J.M. 1964. Poisonous Plants of the United States and Canada. Prentice-Hall, Upper Saddle River, New Jersey, U.S.A.

Kingsolver J.M. & J.E. Decelle. 1979. Host associations of *Specularius impressithorax* (PIC) (Insecta: Coleoptera: Bruchidae) with species of *Erythrina* (Fabales: Fabaceae). Ann. Missouri Bot. Gard. 66:528–532.

Knaus, B.J., R.C. Cronn, & A. Liston. 2005. Genetic characterization of three varieties of *Astragalus lentiginosus* (Fabaceae). Brittonia 57:334–344.

Knight, A.P. & R.G. Walter. 2001. A Guide to Plant Poisoning of Animals in North America. Teton NewMedia, Jackson, Wyoming, U.S.A.

Kosterin, O.E. & V.S. Bogdanova. 2008. Relationship of wild and cultivated forms of *Pisum* L. as inferred from an analysis of three markers, of the plastid, mitochondrial and nuclear genomes. Genet. Resources Crop Evol. 55(5):735–755.

Kremen, C., R.L. Bugg, N. Nicola, S.A. Smith, R.W. Thorp, & N.M. Williams. 2002. Native bees, native plants, and crop pollination in California. Fremontia 30(3–4):41–50.

Krukoff, B.A. & R.C. Barneby. 1974. Conspectus of species of the genus *Erythrina*. Lloydia 37:332–459.

La Barre, W. 1957. Mescalism and peyotism. Amer. Anthropologist 59:708–711.

Lancaster, B. 2010. Eat Mesquite! Desert Harvesters, Tucson, Arizona, U.S.A.

Lang, J.M. & D. Isely. 1982. *Eysenhardtia* (Leguminosae: Papilionoideae). Iowa State J. Res. 56:393–417.

Lassen, P. 1989. A new delimitation of the genera *Coronilla*, *Hippocrepis*, and *Securigera* (Fabaceae). Willdenowia 19(1):49–62.

Lassetter, J.S. 1984. Taxonomy of the *Vicia ludoviciana* complex (Leguminosae). Rhodora 86:475–505.

Lau, C.P.Y., L. Ramsden, & R.M.K. Saunders. 2005. Hybrid origin of "*Bauhinia blakeana*" (Leguminosae: Caesalpinioideae), inferred using morphological, reproductive, and molecular data. Amer. J. Bot. 92:525–533.

Lavin, M. 1988. Systematics of *Coursetia* (Leguminosae-Papilionoideae). Syst. Bot. Monogr. 21:1–167.

Lavin, M. 1990. The genus *Sphinctospermum* (Leguminosae): Taxonomy and tribal relationships as inferred from a cladistic analysis of traditional data. Syst. Bot. 15:544–559.

Lavin, M. & M. Sousa S. 1995. Phylogenetic systematics and biogeography of the tribe Robinieae (Leguminosae). Syst. Bot. Monogr. 45:1–165.

Lavin, M., M.F. Wojciechowski, P. Gasson, C. Hughes, & E. Wheeler. 2003. Phylogeny of Robinioid legumes (Fabaceae) revisited: *Coursetia* and *Gliricidia* recircumscribed, and a biogeographical appraisal of the Caribbean endemics. Syst. Bot. 28:387–409.

Lee, S.G. & P. Felker. 1992. Influence of water/heat stress on flowering and fruiting of mesquite (*Prosopis glandulosa* var. *glandulosa*). J. Arid Environm. 23:309–319.

Legume Phylogeny Working Group (LPWG). 2013. Legume phylogeny and classification in the 21st century: Progress, prospects and lessons for other species-rich clades. Taxon 62:217–248.

Legume Phylogeny Working Group (LPWG). 2017. A new subfamily classification of the Leguminosae based on a taxonomically comprehensive phylogeny. Taxon 66(1):44–77.

Leonardis, R.F.J., ed. 1975. Libro del árbol, Tomo 2—Esencias Forestales Indígenas de la Argentina de Aplicación Industrial. 2nd ed. Celulosa Argentina, S.A., Buenos Aires, Argentina.

Lesins, K.A. & I. Lesins. 1979. Genus *Medicago* (Leguminosae): A Taxogenetic Study. Dr. W. Junk bv Publishers, The Hague, Netherlands.

Lewis, G. 2005. Tribe Caesalpineae. In: G. Lewis, B. Schrire, B. Mackinder, & M. Lock, eds. 2005. Legumes of the World. Royal Botanic Gardens, Kew, London, U.K. 127–161.

Lewis, G.P. & L. Rico Arce. 2005. Tribe Ingeae. In: G. Lewis, B. Schrire, B. MacKinder, & M. Lock, eds. 2005. Legumes of the World. Royal Botanic Gardens, Kew, London, U.K. 193–213.

Lewis, G., B. Schrire, B. MacKinder, & M. Lock, eds. 2005. Legumes of the World. Royal Botanic Gardens, Kew, London, U.K.

Lewis, M., W. Clark, et al. 2003. The Lewis and Clark Journals: An American Epic of Discovery, G. E. Mouton, ed. University of Nebraska Press, Lincoln, Nebraska, U.S.A.

Lewis, W.H. & M.P.F. Elvin-Lewis. 1977. Medical Botany: Plants Affecting Man's Health. John Wiley & Sons, Inc., Hoboken, New Jersey, U.S.A.

Lock, J.M. 1989. Legumes of Africa: A Checklist. Royal Botanic Gardens, Kew, London, U.K.

López, J.A., E.L. Little Jr., G.F. Ritz, J.S. Rombold, & W.J. Hahn. 1987. Arboles comunes del Paraguay. Cuerpo de Paz, Colección y Intercambio de Información, Asunción, Paraguay.

Louca, A. & A. Papas. 1973. The effect of different proportions of carob pod meal in the diet on the performance of calves and goats. Animal Production 17:139–146.

Luckow, M. 1993. Monograph of *Desmanthus* (Leguminosae-Mimosoideae). Syst. Bot. Monogr. 38:1–166.

Luckow, M. 1996. The cultivated species of *Cassia*, *Senna* and *Chamaecrista* (Leguminosae). Baileya 23:195–241.

Luckow, M., J.T. Miller, D.J. Murphy, & T. Livshultz. 2003. A phylogenetic analysis of the Mimosoideae (Leguminosae) based on chloroplast DNA sequence data. In: B.B. Klitgaard & A. Bruneau, eds. 2003. Advances in Legume Systematics: Part 10 Higher level Systematics. Royal Botanic Gardens, Kew, London, U.K. 197–220.

Lumholtz, C. 1912. New Trails in Mexico. Charles Scribner's Sons. Reprinted 1990 by University of Arizona Press, Tucson, Arizona, U.S.A.

Mabberley, D.J. 1987. The Plant Book. Cambridge University Press, Cambridge, U.K.

MacDougal, D.T. 1906. The delta of the Rio Colorado. Bull. Amer. Geogr. Soc. 38:1–16.

Mack, R.N. 2003. Plant naturalizations and invasions in the eastern United States:1634–1860. Ann. Missouri Bot. Gard. 90:77–90.

Maekawa, M., Hasen, & F. Kita. 1991. Identification of reciprocal translocations observed in several *Melilotus* species (subgenus *Eumelilotus*) by interspecific triple crossings. Euphytica 54:255–261.

Makino, T. 1910. Observations on the flora of Japan. Bot. Mag. (Tokyo) 24:290–305.

Makino, T. 1911. Observations on the flora of Japan. Bot. Mag. (Tokyo) 25:9–18.

Mali, R.G. & N.B. Mahale. 2008. Evaluation of *Rhynchosia minima* (Linn.) DC leaves for anthelmintic activity. Int. J. Pharm. Sci. Nanotechnology 1(2):191–194.

Manje, J.M. 1954. Unknown Arizona and Sonora, 1693–1721: Luz de Tierra Incógnito. Translated by Harry J. Karns. Arizona Silhouettes, Tucson, Arizona, U.S.A.

Marazzi, B., E. Conti, & P.K. Endress. 2007. Diversity in anthers and stigmas in the buzz-pollinated genus *Senna* (Leguminosae, Cassiinae). Int. J. Pl. Sci. 168:371–391.

Marazzi, B., E. Conti, M.J. Sanderson, M.M. McMahon, & J.L. Bronstein. 2013. Diversity and evolution of a trait mediating ant-plant interactions: Insights from extrafloral nectaries in *Senna* (Leguminosae). Ann. Bot. (Oxford) 111:1263–1275.

Marazzi, B. & P.K. Endress. 2008. Patterns and development of floral asymmetry in *Senna* (Leguminosae, Cassiinae). Amer. J. Bot. 95:22–40.

Marazzi, B., P.K. Endress, L.P. Queiroz, & E. Conti. 2006. Phylogenetic relationships within *Senna* (Leguminosae, Cassiinae) based on three chloroplast DNA regions: Patterns in the evolution of floral symmetry and extrafloral nectaries. Amer. J. Bot. 93:288–303.

Marazzi, B., R.H. Fortunato, P.K. Endress, & R. Spichiger. 2006. *Senna* (Cassiinae, Leguminosae) in Paraguay: Synopsis, occurrence, ecologic role and ethnobotany. Candollea 61:315–329.

Marazzi, B. & M.J. Sanderson. 2010. Large-scale patterns of diversification in the widespread legume genus *Senna* and the evolutionary role of extrafloral nectaries. Evolution 64:3570–3592.

Martin, C.W. & C.R. Hutchins. 1980. A Flora of New Mexico, Vol. 1. J. Kramer, Vaduz, Germany.

Martin, P.S., D. Yetman, M. Fishbein, P. Jenkins, T.R. Van Devender, & R.K. Wilson. 1998. Gentry's Río Mayo Plants: The Tropical Deciduous Forest & Environs of Northwest Mexico. The University of Arizona Press, Tucson, Arizona, U.S.A.

Martin, S.C. 1948. Mesquite seeds remain viable after 44 years. Ecology 29:393.

Martínez, M. 1959. Las plantas medicinales de México. 4th ed. Ediciones Botas, Mexico City, Mexico.

Martínez-Bernal, A. & R. Grether. 2006. *Mimosa* L. In: A. Novelo Retana, R. Medina Lemos, H. Ochoterena Booth, G.A. Salazar Chávez, & L.O. Alvarado-Cárdenas, eds. 2006. Flora del Valle de Tehuacán-Cuicatlán 44:42–99.

Martínez-Bernal, A., R. Grether, & R.M. González-Amaro. 2008. *Mimosa*. In: G. Castillo-Campos, M. Cházaro, A. Espejo-Serna, M.T. Mejía-Saulés, A. Lot, M. Nee, & J. Rzedowski, eds. 2008. Flora de Veracruz 147:9–127.

Mason, C.T., Jr. & P.B. Mason. 1987. A Handbook of Mexican Roadside Flora. The University of Arizona Press, Tucson, Arizona, U.S.A.

Matschat, C.H. 1935. Mexican Plants for American Gardens. Houghton Mifflin Co., Boston, Massachusetts, U.S.A.

Matsumara, J. 1902. A conspectus of the Leguminosae found growing wild, or cultivated in Japan, Loochoo, and Formosa. Bot. Mag. (Tokyo) 16:37–42, 61–86, 91–118.

Mauz, K. 2011. An Agreeable Landscape: Historical Botany and Plant Biodiversity of a Sonoran Desert Bottomland, 1855–1920. Sida, Bot. Misc. 35. Botanical Research Institute of Texas Press, Fort Worth, Texas, U.S.A.

Maxted, N. & M. Ambrose. 2001. Peas (*Pisum* L.). In: N. Maxted & S.J. Bennett, eds. 2001. Plant Genetic Resources of Legumes in the Mediterranean. Kluwer Academic Publishers, Dordrecht, Netherlands. 181–190.

Maxted, N., P. Mabuza-Dlamini, H. Moss, S. Padulosi, A. Jarvis, & L. Guarino. 2004. An ecogeographic study. African *Vigna*. International Plant Genetic Resources Institute, Rome, Italy.

McDermott, L.F. 1910. An Illustrated Key to the North American Species of *Trifolium*. Cunningham, Curtiss & Welch, San Francisco, California, U.S.A.

McDougall, W.B. 1973. Seed Plants of Northern Arizona. Museum of Northern Arizona Press, Flagstaff, Arizona, U.S.A.

McGrath-Hill, C.A. & I.M. Vicas. 1997. Case series of *Thermopsis* exposures. J. Toxicol. Clin. Toxicol. 35(6):659–665.

McMahon, M. & L. Hufford. 2004. Phylogeny of Amorpheae (Fabaceae: Papilionoideae). Amer. J. Bot. 91:1219–1230.

McVaugh, R. 1943. The travels of Thomas Coulter, 1824–1827. J. Wash. Acad. Sci. 33:65–70.

McVaugh, R. 1956. Edward Palmer: Plant Explorer of the American West. University of Oklahoma Press, Norman, Oklahoma, U.S.A.

McVaugh, R. 1987. Flora Novo-Galiciana: A Descriptive Account of the Plants of Western Mexico. Volume 5: Leguminosae. The University of Michigan Press, Ann Arbor, Michigan, U.S.A.

Mearns, E.A. 1892–93. Field books, Mexican boundary survey, vol. 3. On file, United States National Herbarium, library, Natural History Museum, Smithsonian Institution. Washington, D.C., U.S.A.

Meikle, R.D. 1977. Flora of Cyprus, Vol. 1. Royal Botanic Gardens, Kew, London, U.K.

Menon, S.R., V.K. Patel, L.A. Mitscher, P. Shih, S.P. Pillai, & D.M. Shankel. 1999. Structure-antimutagenic activity relationship study of plicatin B. J. Nat. Prod. 62(1):102–106.

Meyer, D. 1984. Processing, utilization and economics of mesquite pods as a raw material for the food industry. Ph.D. dissertation. Nr. 7688, Swiss Federal Institute of Technology, Zurich, Switzerland.

Meyer, D., R. Becker, & H. Neukom. 1982. Milling and separation of *Prosopis* pod components and their application in food products. In: H.W. Parker, ed. 1982. Mesquite Utilization. Texas Tech University, Lubbock, Texas, U.S.A. L1–L12.

Miller, J. n.d. "Fact Sheet: *Vachellia bravoensis*." Australian National Botanic Gardens, Canberra, Australia. Accessed 8 Nov 2018. http://www.anbg.gov.au/jmiller/factsheets/Vachellia/bravoensis.htm

Miller, R.H. 1967. *Crotalaria* Seed Morphology, Anatomy, and Identification. Agricultural Research Service, Technical Bulletin No. 1373. U.S. Dept. of Agriculture, Washington, D.C., U.S.A.

Miller, S.M. 1994. The role of *Acacia aneura* in animal production. In: R.C. Gutteridge & H.M. Shelton, eds. 1994. Forage Tree Legumes in Tropical Agriculture. CAB International, Wallingford, U.K.

Missbah El Idrissi, M., N. Aujjar, A. Belabed, Y. Dessaux, & A. Filali-Maltouf. 1996. Characterization of rhizobia isolated from the carob tree (*Ceratonia siliqua*). J. Appl. Microbiol. 80:165–173.

Mitscher, L.A., S. Drake, S.R. Gollapudi, & S.K. Okwute. 1987. A modern look at folkloric use of anti-infective agents. J. Nat. Prod. (Lloydia) 50(6):1025–1040.

Moerman, D.E. 1998. Native American Ethnobotany. Timber Press, Portland, Oregon, U.S.A.

Mohlenbrock, R.H. 1957. A revision of the genus *Stylosanthes*. Ann. Missouri Bot. Gard. 44(4):299–355.

Mohlenbrock, R.H. 1961. A Monograph of the Leguminous Genus *Zornia*. Webbia 16:1–141.

Mohlenbrock, R.H. In prep. *Zornia*. For: Flora of North America Editorial Committee, eds. 1993+. Flora of North America North of Mexico. Oxford University Press, New York, U.S.A., and Oxford, U.K. Vols. 10–11.

Moody, M. & J. Moody. 2009. Arizona mesquite company. Aridus 21(1):1–3.

Mooney, H.A., B.B. Simpson, & O.T. Solbrig. 1977. Phenology, morphology, physiology. In: B.B. Simpson, ed. 1977. Mesquite: Its Biology in Two Desert Scrub Ecosystems. Dowden, Hutchinson and Ross, Stroudsburg, Pennsylvania, U.S.A. 26–43.

Moore, D.B. & J.B. Cole. 2004. Arizona's local floras and plant lists: A bibliography with locations and maps. J. Arizona-Nevada Acad. Sci. 37:1–55.

Moore, M. 1989. Medicinal Plants of the Desert and Canyon West. Museum of New Mexico Press, Santa Fe, New Mexico, U.S.A.

Moré, M., A.N. Sérsic, & A.A. Cocucci. 2006. Specialized use of pollen vectors by *Caesalpinia gilliesii*, a legume shrub with brush-type flowers. Biol. J. Linn. Soc. 88:579–592.

Moreno Salazar, S.F., A.E. Verdugo. C.C. López, E.B. Martínez, T.E. Candelas, & R.E. Robles-Zapata. 2008. Activity of medicinal plants, used by native populations from Sonora, Mexico, against enteropathogenic bacteria. Pharm. Biol. 46(10–11):732–737.

Moser, M.B. & S.A. Marlett, compilers. 2010. "Comcaac quih yaza quih hant ihiip hac: Diccionario seri-español-inglés, 2nd ed." Editorial Unison, Hermosillo, Mexico, and Plaza y Váldás Editores, Mexico, D.F. Mexico. Accessed 9 Sep 2012. http://www.sil.org/mexico/seri/G004c-DiccionarioEd2-sei.pdf

Mujoo, K., V. Haridas, J.J. Hoffmann, G.A. Wächter, L.K. Hutter, Y. Lu, M.E. Blake, G.S. Jayatilake, D. Bailey, G.B. Mills, & J.U. Gutterman. 2001. Triterpenoid saponins from *Acacia victoriae* (Bentham) decrease tumor cell proliferation and induce apoptosis. Cancer Res. 61:5486–5490.

Musselman, L.J. 2007. Figs, Dates, Laurel, and Myrrh: Plants of the Bible and the Quran. Timber Press, Portland, Oregon, U.S.A.

Muyskens, M. 2006. The fluorescence of *Lignum nephriticum*: A flashback to the past and a simple demonstration of natural substance fluorescence. J. Chem. Educ. 83:765–768.

Nabhan, G.P. 1985. Gathering the Desert. University of Arizona Press, Tucson, Arizona, U.S.A.

Nabhan, G.P., J.W. Berry, & C.W. Weber. 1980. Wild beans of the Greater Southwest: *Phaseolus metcalfei* and *P. ritensis*. Econ. Bot. 34:68–85.

Nabhan, G.P. & R.S. Felger. 1978. Teparies in southwestern North America. Econ. Bot. 32:3–19.

Nation, P.N. 1989. Alsike clover poisoning: A review. Canad. Vet. J. 30:410–415.

National Research Council (NRC). 1977. *Leucaena*: Promising Forage and Tree Crop for the Tropics. The National Academies Press, Washington, D.C., U.S.A.

National Research Council (NRC). 1979. Tropical Legumes: Resources for the Future. The National Academies Press, Washington, D.C., U.S.A.

National Research Council (NRC). 1980. Firewood Crops: Shrub and Tree Species for Energy Production. The National Academies Press, Washington, D.C., U.S.A.

National Research Council (NRC). 1983. Firewood Crops: Shrub and Tree Species for Energy Production: Volume 2. The National Academies Press, Washington, D.C., U.S.A.

NatureServe. 2007. "NatureServe Explorer: An online encyclopedia of life." NatureServe, Arlington, Virginia, U.S.A. Accessed 15 Jul 2017. http://www.natureserve.org/explorer

Nelson, E.C. 1991. Shamrock: Botany and History of an Irish Myth. Boethius Press, Aberystwyth, Wales, U.K.

Newman, Y., H. Delgado, J. Zambrano, & G. Sthormes. 1999. *Zapoteca formosa* (Kunth) H. Hern. subsp. *formosa*: Nueva especie forrajera arbustiva natural para Venezuela. I. Estudio de fenología y contenido nutricional. Revista Fac. Agron. Univ. Zulia 16:196–203.

Newton, D.R. 2013. The Vascular Flora of the Eagletail Mountain Region. Desert Pl. 29(1):3–51.

Nichol, A.A. 1938. Experimental feeding of deer. Tech. Bull. 75. Univ. Arizona, Agric. Exp. Stat. 1–39.

Nielsen, I.C. 1992. Acacia. Flora Malesiana Ser. 1, 11:34–64.

Nilsen, E.T., M.R. Sharifi, & P.W. Rundel. 1984. Comparative water relations of phreatophytes in the Sonoran Desert of California. Ecology 65(3):767–778.

Niranjan, P.S., D. Singh, K. Prajapati, & S.K. Jain. 2010. Antidiabetic activity of ethanolic extract of *Dalbergia sissoo* L. leaves in alloxan-induced diabetic rats. Int. J. Curr. Pharm. Res. 2(2):24–27.

Nisson, N. n.d. "Tipu psyllid, *Platycorypha nigrivirga*." Center for Invasive Species Research, University of California, Riverside, California, U.S.A. Accessed 11 Jun 2012. http://cisr.ucr.edu/tipu_psyllid.html

Nokes, J. 1986. How to Grow Native Plants of Texas and the Southwest. Texas Monthly Press, Austin, Texas, U.S.A.

Norem, M.A. 2002. Mysterious wisteria. Aridus 14(1):1–4.

Nores, M.J., B.B. Simpson, P. Hick, A.M. Anton, & R.H. Fortunato. 2012. The phylogenetic relationships of four monospecific caesalpinioids (Leguminosae) endemic to southern South America. Taxon 61:790–802.

Ockendon, D.J. 1965. A taxonomic study of *Psoralea* subgenus *Pediomelum* (Leguminosae). SouthW. Naturalist 10:81–124.

Ockendon, D.J. 1966. *Psoralea palmeri*. SouthW. Naturalist 11:412.

Ohashi, H. 2013. New combinations in North American *Desmodium* (*Leguminosae*: Tribe *Desmodieae*). J. Jap. Bot. 88(3):166–175.

Orchard, A.E. & A.J.G. Wilson. 2001a. Flora of Australia. Volume 11A, Mimosaceae, *Acacia* part 1. ABRS/CSIRO, Melbourne, Australia.

Orchard, A.E. & A.J.G. Wilson. 2001b. Flora of Australia. Volume 11B, Mimosaceae, *Acacia* part 2. ABRS/CSIRO, Melbourne, Australia.

Ortiz, P.L. 1999. *Ceratonia* L. In: S. Castroviejo, ed. 1999. Flora Iberica, Vol. 7 (I), Plantas vasculares de la Península Ibérica e Islas Baleares. Real Jardín Botánico, Madrid, Spain.

Ottley, A.M. 1944. The American Loti with special consideration of a proposed new section, Simpeteria. Brittonia 5(2):81–123.

Owensby, C.E. 1980. Kansas Prairie Wildflowers. Iowa State University Press, Ames, Iowa, U.S.A.

Oxford English Dictionary. 2012. "Algarroba." Oxford University Press, Oxford, U.K. Accessed 2 Oct 2012. http://oxforddictionaries.com/words/the-oxford-english-dictionary

Palacios, R. 2006. Los mezquites Mexicanos: Biodiversidad y distribución geográfica. Bol. Soc. Argent. Bot. 41(1–2):99–121.

Palgrave, K.C. 1983. Trees of Southern Africa. Struik Publishers, Cape Town, South Africa.

Palomino, G., P. Martínez, C. Bernal, & M. Sousa. 1993. Diferencias cromosomicas entre algunas especies de los generos *Sophora* L. y *Styphnolobium* Schott. Ann. Missouri Bot. Gard. 80:284–290.

Palzkill, D.A. 1998. Natural asexual regeneration in woody desert plants—how common is it? Aridus 10(2):1–5.

Pammel, L.H. 1911. A Manual of Poisonous Plants. The Torch Press, Cedar Rapids, Iowa, U.S.A.

Parke, J.G. 1855. Report of explorations for that portion of a railroad route, near the thirty-second parallel of latitude, lying between Doña Ana, on the Rio Grande, and Pimas Villages, on the Gila. House doc. 129, 33rd Congress, 1st Session. A.O.P. Nicholson printer, Washington, D.C., U.S.A.

Parker, H.W., ed. 1982. Mesquite Utilization. Texas Tech University, Lubbock, Texas, U.S.A.

Parker, K.F. 1972. An Illustrated Guide to Arizona Weeds. The University of Arizona Press, Tucson, Arizona, U.S.A.

Pattie, J.O. 1988. Personal Narrative of James O. Pattie. Edited by Richard Batman. Mountain Press, Missoula, Montana, U.S.A.

Peattie, D.C. 1953. A Natural History of Western Trees. Houghton Mifflin, Boston, Massachusetts, U.S.A.

Peña, R.C., L. Iturriaga, G. Montenegro, & B.K. Cassels. 2000. Phylogenetic and biogeographic aspects of *Sophora* sect. *Edwardsia* (Papilionaceae). Pacific Sci. 54:159–167.

Piper, C.V. 1924. Book review: Native American Forage Plants. Agronomy J. 16:682.

Polhill, R.M. 1981. Dalbergieae. In: R.M. Polhill & P.H. Raven, eds. 1981. Advances in Legume Systematics: Part 1. Royal Botanic Gardens, Kew, London, U.K. 233–242.

Polhill, R.M. & P.H. Raven, eds. 1981. Advances in Legume Systematics: Part 1. Royal Botanic Gardens, Kew, London, U.K.

Polhill, R.M. & J.E. Vidal. 1981. Caesalpinieae. In: R.M. Polhill & P.H. Raven, eds. 1981 Advances in Legume Systematics, Part 1. Royal Botanic Gardens, Kew, London, U.K. 81–95.

Pooja, P. Sharma, K.C. Samanta, & V. Garg. 2010. Evaluation of nitric oxide and hydrogen peroxide scavenging activity of *Dalbergia sissoo* roots. Pharmacophore 1(2):78–81.

Powell, A.M. 1988. Trees and Shrubs of Trans-Pecos Texas: Including Big Bend and Guadalupe Mountains National Parks. Big Bend Natural History Association, Big Bend Park, Texas, U.S.A.

Powell, A.M. 1998. Trees and Shrubs of Trans-Pecos Texas and Adjacent Areas. University of Texas Press, Austin, Texas, U.S.A.

Powell, J.W. 1875. Exploration of the Colorado River of the West and its tributaries, explored in 1869, 1870, 1871, and 1872, under the direction of the Secretary of the Smithsonian Institution. Government Printing Office, Washington, D.C., U.S.A.

Powell, R.G., C.R. Smith Jr., & R.V. Madrigal. 1976. Antitumor activity of *Sesbania vesicaria, S. punicea* and *S. drummondii* seed extracts. Pl. Med. 30(1):1–8.

Pratt, R.C. & G.P. Nabhan. 1988. Evolution and diversity of *Phaseolus acutifolius* genetic resources. In: P. Gepts, ed. 1988. Genetic Resources of *Phaseolus* Beans. Kluwer Academic Publishers, Dordrecht, Netherlands. 404–440.

Radford, A.E., H.E. Ahles, & C.R. Bell. 1968. Manual of the Vascular Flora of the Carolinas. The University of North Carolina Press, Chapel Hill, North Carolina, U.S.A.

Ragasa, C.Y., J.G. Hofileña, & J.A. Rideout. 2002. New furanoid diterpenes from *Caesalpinia pulcherrima*. J. Nat. Prod. 65(8):1107–1110. doi:10.1021/np0201523

Ralphs, M.H., J.D. Graham, & L.F. James. 2002. A close look at locoweed poisonings on shortgrass prairies. Rangelands 24(2):30–34.

Randell B.R. 1989. Revision of the Cassiinae in Australia. 2. *Senna* Miller sect. *Psilorhegma* (J. Vogel) Irwin and Barneby. J. Adelaide Bot. Gard. 12:165–272.

Randell, B.R. & B.A. Barlow. 1998. *Senna*. Flora of Australia 12:89–138.

Raulston, J.C. 1990. Redbud—A comprehensive overview of the genus *Cercis* spotlighting the most promising species and cultivars. Amer. Nurseryman 171(5):39–51.

Rea, A.M. 1997. At the Desert's Green Edge: An Ethnobotany of the Gila River Pima. University of Arizona Press, Tucson, Arizona, U.S.A.

Rechinger, K.H. 1984. Flora Iranica 157: Papilionaceae II. Akademische Druck und Verlagsanstalt, Graz, Austria.

Record, S.J. & R.W. Hess. 1943. Timbers of the New World. Yale University Press, New Haven, Connecticut, U.S.A.

Rehder, A. 1921. New species, varieties and combinations from the herbarium and the collections of the Arnold Arboretum. J. Arnold Arbor. 3:12–50.

Rehder, A. 1926. Enumeration of the ligneous plants of northern China, III. J. Arnold Arbor. 7:151–227.

Rehder, A. & E.H. Wilson. 1916. *Wistaria*. In: C.S. Sargent, ed. 1916. Plantae Wilsonianae. Vol. 2. The University Press, Cambridge, Massachusetts, U.S.A. 509–515.

Reid, W.V., S.A. Laird, C.A. Meyer, R. Gámez, A. Sittenfeld, D.H. Janzen, M.A. Gollin, & C. Juma. 1993. Biodiversity Prospecting: Using Genetic Resources for Sustainable Development. World Resources Institute, Washington, D.C., U.S.A.

Rhodes, S., J. Beasley, & T. Ayers. 2011. Vascular Plants of Arizona: Fabaceae Part One: *Errazuria* [sic] Phillips, *Marina* Liebm., *Parryella* Torr. & A. Gray, and *Psorothamnus* Rydb. Canotia 7:1–13.

Rico-Arce, M. de L. 2007. A Checklist and Synopsis of American Species of *Acacia* (Leguminosae: Mimosoideae). Royal Botanic Gardens, Kew, London, U.K.

Rico Arce, M. de L. & S. Bachman. 2006. A taxonomic revision of *Acaciella* (Leguminosae, Mimosoideae). Anales Jard. Bot. Madrid 63:189–244.

Rico Arce, M. de L., S.L. Gale, & N. Maxted. 2008. A taxonomic study of *Albizia* (Leguminosae: Mimosoideae: Ingeae) in Mexico and Central America. Anales Jard. Bot. Madrid 65:255–305.

Riley-Hulting, E.T., A. Delgado-Salinas, & M. Lavin. 2004. Phylogenetic systematics of *Strophostyles* (Fabaceae): A North American temperate genus within a neotropical diversification. Syst. Bot. 29:627–653.

Rincón-Rosales, R., L. Lloret, E. Ponce, & E. Martinez-Romero. 2009. Rhizobia with different symbiotic efficiencies nodulate *Acaciella angustissima* in Mexico, including *Sinorhizobium chiapanecum* sp. nov. which has common symbiotic genes with *Sinorhizobium mexicanum*. FEMS Microbiol. Ecol. 67:103. doi:10.1111/j.1574-6941.2008.00590.x

Ritchie, E.A., K.M. Wood, D.S. Gutzler, & S. White. 2011. The influence of eastern Pacific tropical cyclone remnants on the southwestern United States. Mon. Wea. Rev. 139:192–210.

Ritter, N. 1917. Histology of *Astragalus mollissimus*. Kansas Univ. Sci. Bull. 10:197–208.

Roark, R.C. 1947. Some promising insecticidal plants. Econ. Bot. 1:437–445.

Robinson, B.L. & M.L. Fernald. 1916. Gray's New Manual of Botany. 7th ed. American Book Company, New York, New York, U.S.A.

Rogers, D.J. 1980. Lakota Names and Traditional Uses of Native Plants by Sicangu (Brule) People in the Rosebud Area, South Dakota. Rosebud Educational Society, Inc., St. Francis, South Dakota, U.S.A.

Rosales, M. 2014. Plant guide for *Hedysarum boreale* (Utah sweetvetch). USDA NRCS Upper Colorado Environmental Plant Center, Meeker, Colorado, U.S.A.

Ross, J.H., ed. 1975. Flora of Southern Africa, Vol. 16, Part 1. Botanical Research Institute, Department of Agricultural Technical Services, Republic of South Africa, Pretoria, South Africa.

Ross, J.H. 1979. A Conspectus of the African *Acacia* Species. Mem. Bot. Surv. South Africa 44: v+1–156.

Rothrock, J.T. 1878. Catalogue of plants collected in Nevada, Utah, Colorado, New Mexico, and Arizona. In: G.M. Wheeler. 1878. United States Geographical Surveys west of the One Hundredth Meridian, Vol. VI–Botany. Government Printing Office, Washington, D.C., U.S.A. 55–351.

Rozen, Jr., J.G. & B.L. Rozen. 1986. Bionomics of crepuscular bees associated with the plant *Psorothamnus scoparius* (Hymenoptera: Apoidea). J. New York Entomol. Soc. 94(4):472–479.

Rudd, V.E. 1955. The American Species of *Aeschynomene*. Contr. U.S. Natl. Herb. 32(1):1–172.

Rudd, V.E. 1956. A revision of the genus *Nissolia*. Contr. U.S. Natl. Herb. 32:173–206.

Russell, A.B., J.W. Hardin, L. Grand, & A. Fraser. 1997. "Poisonous Plants of North Carolina." North Carolina State University, Raleigh, North Carolina, U.S.A. Accessed 16 Oct 2011. http://www.ces.ncsu.edu/depts/hort/consumer/poison/Sesbapu.htm

Russell, F. 1908. The Pima Indians. 26th Annual Report, Smithsonian Institute Bureau of American Ethnology. Government Printing Office, Washington, D.C., U.S.A.

Rydberg, P.A. 1919. Fabaceae, tribe Psoraleae. Part 1. N. Amer. Fl. 24:1–64.

Rydberg, P.A. 1929. Galegeae, Subtribe 10. Astragalaneae. N. Amer. Fl. 24:251–462.

Rzedowski, J. 1988. Análisis de la distribución geográfica del complejo *Prosopis* (Leguminosae, Mimosoideae) en Norteamérica. Acta Bot. Mex. 3:7–19.

Sahle, M., J. Coleou, & C. Haas. 1992. Carob pod (*Ceratonia siliqua*) meal in geese diets. British Poultry Sci. 33:531–541.

St. John, L., D.G. Ogle, & D. Tilley. 2010. Plant Guide for Strawberry Clover (*Trifolium fragiferum*). USDA-Natural Resources Conservation Service, Aberdeen, Idaho Plant Materials Center, Aberdeen, Idaho, U.S.A.

Sánchez-Marroquín, A., L. García, & M. Méndez. 1958. Brazilin, antibacterial substance from *Haematoxylon brasiletto* Karst. Rev. Latinoam. Microbial. 1(3):225–232.

Santanu Saha, E.V.S. Subrahmanyam, K.S. Chandrashekar, & Shashidhara C. Shastry. 2011. *In vivo* study for anti-inflammatory activity of *Bauhinia variegata* L. leaves. Pharm. Crops 2:70–73.

Saravia Toledo, C. & E.M. Del Castillo. 1986. Regeneración de cepa en especias arbóreas del Chaco occidental Argentino. In: Proc. V. Reunion de Intercambio Tecnologico en Zonas Aridas y Semiaridas. La Rioja. E. Orientación Grafica, Buenos Aires, Argentina. 382–405.

Sargent, C.S. 1889. Notes upon some North American trees. V. *Parkinsonia*. Gard. & Forest 2:388.

Sargent, C.S. 1916. Plantae Wilsonianae. Vol. 2. The University Press, Cambridge, Massachusetts, U.S.A.

Saunders, R.M., R. Becker, D. Meyer, F.R. Del Valle, E. Marco, & M.E. Torres. 1986. Identification of commercial milling techniques to produce high sugar, high fiber, high protein and high galactomannan gum fractions from *Prosopis* pods. Forest Ecol. Managem. 16:169–180.

Schmitt, A., H. Telikepalli, & L.A. Mitscher. 1991. Plicatin B, the antimicrobial principle of *Psoralea juncea*. Phytochemistry 30(11):3569–3570.

Schmutz, E.M. & L.B. Hamilton. 1979. Plants That Poison: An Illustrated Guide for the American Southwest. Northland Press, Flagstaff, Arizona, U.S.A.

Schultes, R.E. & A. Hofmann. 1980. The Botany and Chemistry of Hallucinogens. Charles C. Thomas, Publisher, Springfield, Illinois, U.S.A.

Schulz, O.E. 1901. Monographie der Gattung *Melilotus*. Bot. Jahrb. 29:660–735.

Seigler, D.S. & J.E. Ebinger. 1988. *Acacia macracantha. A. pennatula*, and *A. cochliacantha* (Fabaceae: Mimosoideae) species complexes in Mexico. Syst. Bot. 13:7–15.

Seigler, D.S. & J.E. Ebinger. 2005. New combinations in the genus *Vachellia* (Fabaceae: Mimosoideae) from the New World. Phytologia 87:139–178.

Seigler, D.S. & J.E. Ebinger. 2018. New combinations in *Parasenegalia* and *Mariosousa* (Fabaceae: Mimosoideae). Phytologia 100(4):256–259.

Seigler, D.S., J.E. Ebinger, & J.T. Miller. 2006a. The genus *Senegalia* (Fabaceae: Mimosoideae) from the New World. Phytologia 88:38–93.

Seigler, D.S., J.E. Ebinger. & J.T. Miller. 2006b. *Mariosousa*, a new segregate genus from *Acacia* s.l. (Fabaceae, Mimosoideae) from Central and North America. Novon 16:413–420.

Seigler, D.S., J.E. Ebinger, C.W. Riggins, V. Terra, & J.T. Miller. 2017. *Parasenegalia* and *Pseudosenegalia* (Fabaceae): New genera of the Mimosoideae. Novon 25(2):180–205.

Sgariglia, M.A., J.R. Soberón, D.A. Sampietro, E.N. Quiroga, & M.A. Vattuone. 2011. Isolation of antibacterial components from infusion of *Caesalpinia paraguariensis* bark. A bio-guided phytochemical study. Food Chem. 126(2):395–404. doi:10.1016/j.foodchem.2010.10.104

Shaparenko, K.K. 1972. *Alhagi*. In: V.L. Komarov, ed. 1972. Flora of the U.S.S.R., Vol. XIII. *Leguminosae: Oxytropis, Hedysarum*. (R. Lavoott, Israel Program for Scientific Translations, Jerusalem, Trans.). Keter Press, Jerusalem, Israel. (Original work published 1948) 281–285.

Sharma, V. & G.P. Rajani. 2011. Evaluation of *Caesalpinia pulcherrima* Linn. for anti-inflammatory and antiulcer activities. Indian J. Pharm. 43(2):168–171. doi:10.4103/0253-7613.77354

Sheikh, M.I. 1989. Sissoo—The Versatile Rosewood, NFT Highlights 89-7. Nitrogen Fixing Tree Association, Waimanalo, Hawaii, U.S.A.

Sheppard, P.R., A.C. Comrie, G.D. Packin, K. Angersbach, & M.K. Hughes. 2002. The climate of the US southwest. Clim. Res. 21:219–238.

Sherbrooke, W.C. & J.C. Scheerens. 1979. Ant-visited extrafloral (calyx and foliar) nectaries and nectar sugars of *Erythrina flabelliformis* Kearney in Arizona. Ann. Missouri Bot. Gard. 66:472–481.

Shreter, A.I. 2010. "Medicinal Plants." (from: The Great Soviet Encyclopedia, 3rd Edition (1970–1979)). The Gale Group, Inc., Farmington Hills, Michigan. Accessed 20 Dec 2010. http://encyclopedia2.thefreedictionary.com/Medicinal+Plants

Shreve, F. 1951. Vegetation of the Sonoran Desert. In: F. Shreve & I.L. Wiggins. 1951. Vegetation and Flora of the Sonoran Desert, Vol. 1, Publ. 591. Carnegie Institution of Washington, Washington, D.C., U.S.A. 1–191.

Shreve, F. & I.L. Wiggins. 1964. Vegetation and Flora of the Sonoran Desert, Vol. 1. Stanford University Press, Stanford, California, U.S.A.

Simmons, M.H. 1981. Acacias of Australia, Volume 1. Nelson Publishers, Melbourne, Australia.

Simon, M.F., R. Grether, L.P. de Queiroz, T.E. Särkinen, V.F. Dutra, & C.E. Hughes. 2011. The evolutionary history of *Mimosa* (Leguminosae): Toward a phylogeny of the sensitive plants. Amer. J. Bot. 98(7):1201–1221.

Simpson, B.B., ed. 1977. Mesquite: Its Biology in Two Desert Scrub Ecosystems. Dowden, Hutchinson and Ross, Stroudsburg, Pennsylvania, U.S.A.

Simpson, B.B. 1989. Krameriaceae. Fl. Neotrop. Monogr. 49:1–108.

Simpson, B.B. 1998. A revision of *Pomaria* (Fabaceae) in North America. Lundellia 1:46–71.

Simpson, B.B. 1999. A revision of *Hoffmannseggia* (Fabaceae) in North America. Lundellia 2:14–54.

Simpson, B.B., L. Larkin, A. Weeks, & J. McDill. 2006. Phylogeny and biogeography of *Pomaria* (Caesalpinioideae: Leguminosae). Syst. Bot. 31:792–804.

Simpson, B.B. & G.P. Lewis. 2003. New combinations in *Pomaria* (Caesalpinioideae: Leguminosae). Kew Bull. 58:175–184.

Simpson, B.B., J.A. Tate, & A. Weeks. 2004a. Phylogeny and character evolution of *Hoffmannseggia* (Caesalpinieae: Caesalpinioideae: Leguminosae). Syst. Bot. 29:933–946.

Simpson, B.B., J.A. Tate, & A. Weeks. 2004b. The biogeography of *Hoffmannseggia* (Leguminosae, Caesalpinioideae, Caesalpinieae): A tale of many travels. J. Biogeogr. 31:1–13.

Simpson, B.B. & E.A. Ulibarri. 2006. A synopsis of the genus *Hoffmannseggia* (Leguminosae). Lundellia 9:7–33.

Sitgreaves, L. 1853. Report of an Expedition down the Zuñi and Colorado Rivers. Senate Exec. doc. 59, 32nd Congress, 2nd Session. Robert Armstrong public printer, Washington, D.C., U.S.A.

Small, E. 1985. Morphological differentiation in *Medicago sativa* s.l. in relation to ploidy. Canad. J. Bot. 63:1747–1752.

Small, E. 1987. Generic Changes in Trifolieae Subtribe Trigonellinae. In: C.H. Stirton, ed. 1987. Advances in Legume Systematics, Part 3. Royal Botanic Gardens, Kew, London, U.K. 159–181.

Small, E. 2011. Alfalfa and Relatives: Evolution and Classification of *Medicago*. NRC Research Press, Ottawa, Canada.

Small, E. & B. Brookes. 1984. Taxonomic circumscription and identification in the *Medicago sativa-falcata* (alfalfa) continuum. Econ. Bot. 38:83–96.

Small, E. & M. Jomphe. 1989. A synopsis of the genus *Medicago* (Leguminosae). Canad. J. Bot. 67:3260–3294.

Small, E., P. Lassen, & B. Brookes. 1987. An expanded circumscription of *Medicago* (Leguminosae, Trifolieae) based on explosive flower tripping. Willdenowia 16:415–437.

Small, J.K. 1933. Manual of the Southeastern Flora. University of North Carolina Press, Chapel Hill, North Carolina, U.S.A.

Smit, G.N. 2008. Field Guide to the Acacias of South Africa. Briza Publications, Pretoria, South Africa.

Smith, G.W., K.W. Allred, & D.E. Kiehl. 1992. Swainsonine content of New Mexico locoweeds. Proc., Western Sect., Amer. Soc. Animal Sci. 42:405–407.

Smith, M.C. & D.M. Sherman. 1994. Goat Medicine. John Wiley & Sons, Inc., Hoboken, New Jersey, U.S.A.

Smýkal, P., G. Kenicer, A.J. Flavell, J. Corander, O. Kosterin, R.J. Redden, R. Ford, C.J. Coyne, N. Maxted, M.J. Ambrose, & N.T.H. Ellis. 2011. Phylogeny, phylogeography and genetic diversity of the *Pisum* genus. Pl. Genet. Resources Charact. Utilz. 9:4–18.

Solbrig, O.T. & P.D. Cantino. 1975. Reproductive adaptations in *Prosopis* (Leguminosae: Mimosoideae). J. Arnold Arbor. 56:185–210.

Sousa, M. & V.E. Rudd. 1993. Revision del genero *Styphnolobium* (Leguminosae: Papilionoideae: Sophoreae). Ann. Missouri Bot. Gard. 80:270–283.

Souza, E.R. de, M.V. Krishnaraj, & L.P. de Queiroz. 2016. *Sanjappa*, a new genus in the tribe Ingeae (Leguminosae: Mimosoideae) from India. Rheedea 26(1):1–12.

Souza, E.R. de, G.P. Lewis, F. Forest, A.S. Schnadelbach, C. van den Berg, & L.P. de Queiroz. 2013. Phylogeny of Calliandra (Leguminosae: Mimosoideae) based on nuclear and plastid molecular markers. Taxon 62:1200–1219.

Spellenberg, R. 1976. Chromosome numbers and their cytotaxonomic significance for North American *Astragalus* (Fabaceae). Taxon 25:463–476.

Sprent, J.I. 2009. Legume Nodulation: A Global Perspective. Wiley-Blackwell, Oxford, U.K.

Standley, P.C. 1922. Trees and Shrubs of Mexico. Part 2: Fagaceae–Fabaceae. Contr. U.S. Natl. Herb. 23(2).

Starr, G. 1998. Chihuahuan Desert Bauhinias. Aridus 10(1):1–3.

Stevens, P.F. (2001 onwards). "Angiosperm Phylogeny Website. Version 14, July 2017 [and more or less continuously updated since]." Accessed Dec 2020. http://www.mobot.org/MOBOT/research/APweb/

Stevenson, G.A. 1969. An agronomic and taxonomic review of the genus *Melilotus* Mill. Canad. J. Pl. Sci. 49:1–20.

Stevenson, M.C. 1915. Ethnobotany of the Zuni Indians. 30th Annual Report, Smithsonian Institute Bureau of American Ethnology. Government Printing Office, Washington, D.C., U.S.A.

Straub, S.C.K. & J.J. Doyle. 2014. Molecular phylogenetics of *Amorpha* (Fabaceae): An evaluation of monophyly, species relationships, and polyploid origins. Molec. Phylogen. Evol. 76:49–66.

Strickland, W. 1801. Observations on the Agriculture of the United States of America. W. Bulmer and Co., London, U.K.

Stubbendieck, J. & E.C. Conard. 1989. Common Legumes of the Great Plains: An Illustrated Guide. University of Nebraska Press, Lincoln, Nebraska, U.S.A.

Suttie, J.M. 2000. Hay and Straw Conservation: For Small-scale Farming and Pastoral Conditions. Food and Agriculture Organization of the United Nations, Rome, Italy.

Symon, D.E. 1998. *Senna phyllodinea* a new combination in *Senna* (Caesalpiniaceae). J. Adelaide Bot. Gard. 18:101–102.

Szarek, S.R. & R.M. Woodhouse. 1978. Ecophysiological studies of Sonoran Desert plants: IV. Seasonal photosynthetic capacities of *Acacia greggii* and *Cercidium microphyllum*. Oecologia 37:221–229.

Taylor, N.L., ed. 1985. Clover Science and Technology. American Society of Agronomy, Madison, Wisconsin, U.S.A.

Téllez, O. 2001. *Tephrosia*. In: W.D. Stevens, C. Ulloa Ulloa, A. Pool, & O.M. Montiel, eds. 2001. Flora of Nicaragua. Monogr. Syst. Bot. Missouri Bot. Gard. 85:1063–1066.

Texas Big Tree Registry. n.d. "Lists." Texas A&M Forest Service, College Station, Texas, U.S.A. Accessed 21 Jun 2018. http://texasforestinfo.tamu.edu/BigTreeRegistry/Lists

Thomson, L. 1992. Australia's subtropical dry-zone *Acacia* species with human food potential. In: A.P.N. House & C.E. Harwood, eds. 1992. Australian Dry-zone Acacias for Human Food. CSIRO Publications, North Clayton, Victoria, Australia.

Thorn, K.A., A.M. Tinsley, C.W. Weber, & J.W. Berry. 1983. Antinutritional factors in legumes of the Sonoran Desert. Ecol. Food Nutr. 13(4):251–256.

Thulin, M. 1993. Fabaceae (Leguminosae). In: M. Thulin, ed. 1993. Flora of Somalia: Vol. 1, Pteridophyta; Gymnospermae; Angiospermae (Annonaceae–Fabaceae). Royal Botanic Gardens, Kew, London, U.K.

Tidestrom, I. & T. Kittell. 1941. A Flora of Arizona and New Mexico. Catholic University of America Press, Washington, D.C., U.S.A.

Timberlake, J. 1980. Handbook of Botswana Acacias. Ministry of Agriculture, Botswana, Gaborone, Botswana.

Timberlake, J., C. Fagg, & R. Barnes. 1999. A Field Guide to the *Acacias* of Zimbabwe. CBC Publishing, Harare, Zimbabwe.

Toft, C.A. & S.L. Welsh. 1972. A revision of the *Psoralea lanceolata* complex (Leguminosae). Great Basin Naturalist 32(2):76–87.

Tomooka, N., D.A. Vaughan, H. Moss, & N. Maxted. 2002. The Asian *Vigna*: Genus *Vigna* subgenus *Ceratotropis* genetic resources. Kluwer Academic Publishers, Dordrecht, Netherlands.

Tonne, P. 2000. Status report for Chihuahua scurfpea (*Pediomelum pentaphyllum;* Fabaceae). U.S. Fish and Wildlife Service, Albuquerque, New Mexico, U.S.A.

Torres, Y., G. Agostini, & N. Xena de Enrech. 1983. Revisión taxonómica del género *Galactia* P. Br. en Venezuela (Leguminosae, Faboideae, Phaseoleae). Acta Biol. Venez. 11:45–93.

Torrey, J. 1848. Appendix 2 [Botany]. In: W.H. Emory. 1848. Notes of a military reconnoissance, from Fort Leavenworth, in Missouri, to San Diego, in California, including part of the Arkansas, Del Norte, and Gila Rivers. House Exec. doc. 41, 30th Congress, 1st Session. Wendell and Van Benthuysen Printers, Washington, D.C., U.S.A. 135–186.

Torrey, J. 1853. Botany. In: L. Sitgreaves. 1853. Report of an Expedition down the Zuñi and Colorado Rivers. Senate Exec. doc. 59, 32nd Congress, 2nd Session. Robert Armstrong public printer, Washington, D.C., U.S.A. 155–178.

Torrey, J. 1856. Botanical report: List and description of the plants collected. In: J.G. Parke. 1856. Report of explorations for railroad routes from San Francisco Bay to Los Angeles, California, west of the Coast Range, and from the Pimas Villages on the Gila to the Rio Grande, near the thirty-second parallel of north latitude (Pacific Railr. Rep. 7, part 3). House Exec. doc. 91, 33rd Congress, 2nd Session. A.O.P. Nicholson printer, Washington, D.C., U.S.A. 7–22.

Torrey, J. 1857. Descriptions of the general botanical collections. In: A.W. Whipple. 1857. Report of explorations for a railway route, near the thirty-fifth parallel of north latitude, from the Mississippi River to the Pacific Ocean (Pacific Railr. Rep. 3, part 5). House Exec. doc. 91, 33rd Congress, 2nd Session. A.O.P. Nicholson printer, Washington, D.C., U.S.A. 59–161.

Torrey, J. 1859. Botany of the boundary. In: W.H. Emory. 1859. Report on the United States and Mexican Boundary Survey made under the direction of the Secretary of the Interior (vol. II, part 1). Senate Exec. doc. 108, 34th Congress, 1st Session. A.O.P. Nicholson printer, Washington, D.C., U.S.A. 29–236.

Travis, S.E., J. Maschinski, & P. Keim. 1996. An analysis of genetic variation in *Astragalus cremnophylax* var. *cremnophylax*, a critically endangered plant, using AFLP markers. Molec. Ecol. 5:735–745.

Trelease, S.F. & O.A. Beath. 1949. Selenium. The Champlain Printers, Burlington, Vermont, U.S.A.

Tull, D. 1987. A Practical Guide to Edible and Useful Plants: Including Recipes, Harmful Plants, Natural Dyes and Textile Fibers. Texas Monthly Press, Austin, Texas, U.S.A.

Turchi, G., G. Alagona, & V. Lubrano. 2009. Protective activity of plicatin B against human LDL oxidation induced in metal ion-dependent and -independent processes. Experimental and theoretical studies. Phytomedicine 16(11):1014–1026.

Turkington, R. & P.B. Cavers. 1979. The biology of Canadian weeds. 33. *Medicago lupulina* L. Canad. J. Pl. Sci. 59:99–110.

Turkington, R.A., P.B. Cavers, & E. Rempel. 1978. The biology of Canadian weeds. 29. *Melilotus alba* Desr. and *M. officinalis* (L.) Lam. Canad. J. Pl. Sci. 58:523–537.

Turner, B.L. 1956. New Names for Texas Leguminosae. Field & Lab. 24(1):15–17.

Turner, B.L. 1959. The Legumes of Texas. University of Texas Press, Austin, Texas, U.S.A.

Turner, B.L. 1992. Taxonomic overview of the genus *Cologania* (Fabaceae, Phaseoleae). Phytologia 73:281–301.

Turner, B.L. 2011. Systematics of the *Rhynchosia senna* complex. Lundellia 14:27–31.

Turner, N.J. & H.V. Kuhnlein. 1982. Two important "root" foods of the Northwest Coast Indians: Springbank clover (*Trifolium wormskioldii*) and Pacific silverweed (*Potentilla anserina* ssp. *pacifica*). Econ. Bot. 36:411–432.

Turner, R.M. 1990. Long-term vegetation change at a fully protected Sonoran Desert site. Ecology 71:464–477.

Turner, R.M., J.E. Bowers, & T.L. Burgess. 1995. Sonoran Desert Plants: An Ecological Atlas. The University of Arizona Press, Tucson, Arizona, U.S.A.

Ulubelen, A., W.F. McCaughey, & J.R. Cole. 1967. Proteinaceous antitumor substances from plants III. *Caesalpinia gilliesii* (Leguminosae). J. Pharm. Sci. 56(7):914–916.

Uphof, J.C.T. 1968. Dictionary of Economic Plants. 2nd ed. J. Cramer, Lehre, Germany.

Urban, A. 1990. Wildflowers and Plants of Central Australia. Southbank Editions, Port Melbourne, Victoria, Australia.

Usher, G. 1974. A Dictionary of Plants Used by Man. Constable and Company Ltd., London, U.K.

Vail, A.M. 1895. A study of the genus *Galactia* in North America. Bull. Torrey Bot. Club 22:500–511.

van der Maesen, L.J.G. 1972. *Cicer* L., a monograph of the genus, with special reference to the chickpea (*Cicer arietinum* L.), its ecology and cultivation. Mededelingen Landbouwhogeschool Wageningen. 72–10.

van der Maesen, L.J.G. 1987. Origin, History and Taxonomy of Chickpea. In: M.C. Saxena & K.B. Singh, eds. 1987. The Chickpea. CAB International/International Center for Agricultural Research in the Dry Areas, Wallingford, U.K. 11–34.

van der Maesen, L.J.G., N. Maxted, F. Javadi, S. Coles, & A.M.R. Davies. 2007. Taxonomy of the genus *Cicer* revisited. In: S.S. Yadav, R. Redden, W. Chen, & B. Sharma, eds. 2007. Chickpea Breeding and Management. CAB International, Wallingford, U.K. 14–46.

Van Kessel, C., J.P. Roskoski, T. Wood, & J. Montano. 1983. $^{15}N_2$ fixation and H_2 evolution by six species of tropical leguminous trees. Plant Physiology 72:909–910.

Vaughn, P.K. & D.B. Dunn. 1977. The *Lupinus latifolius* Agardh complex. Trans. Missouri Acad. Sci. 10:89–106.

Veloz-García, R.A., R. Marín-Martínez, R. Veloz-Rodríguez, C.I. Muñoz-Sánchez, L. Guevara-Olvera, R Miranda-López, M.M. González-Chavira, I. Torres-Pacheco, S.H. Guzmán-Maldonado, A. Cardador-Martínez, G. Loarca-Piña, & R.G. Guevara-González. 2004. Antimutagenic and antioxidant activities of cascalote (*Caesalpinia cacalaco*) phenolics. J. Sci. Food Agric. 84(13):1632–1638.

Veloz-García, R., R. Marín-Martínez, R. Veloz-Rodríguez, R. Rodríguez-Guerra, I. Torres-Pacheco, M.M. González-Chavira, J.L. Anaya-López, L. Guevara-Olvera, A.A. Feregrino-Pérez, G. Loarca-Piña, & R.G. Guevara-González. 2010. Antimicrobial activities of cascalote (*Caesalpinia cacalaco*) phenolics-containing extracts against fungus *Colletotrichum lindemuthianum*. Industr. Crops Prod. 31(1):134–138. doi:10.1016/j.indcrop.2009.09.013

Vestal, P.A. 1952. The Ethnobotany of the Ramah Navaho. Pap. Peabody Mus. Amer. Archeol. Ethnol. 40(4):1–94.

Vijayakumari, K., P. Siddhuraju, & K. Janardhanan. 1994. Nutritional assessment and chemical composition of the lesser known tree legume, Acacia leucophloea (Roxb.) Willd. Food Chem. 50:285–288.

Villiers, J.F. 2002. *Viguieranthus*. In: D.J. Du Puy, J.N. Labat, R. Rabevohitra, J.F. Villiers, J. Bosser, & J. Moat, eds. 2002. The Leguminosae of Madagascar. Royal Botanic Gardens, Kew, London, U.K. 271–275.

Vincent, M.A. & D. Isely. 2012. *Trifolium*. In: B.G. Baldwin, D. Goldman, D.J. Keil, R. Patterson, T.J. Rosatti, & D. Wilken, eds. 2012. The Jepson Manual, Vascular Plants of California. 2nd ed. University of California Press, Berkeley, California, U.S.A. 789–798.

Vines, R.A. 1960. Trees, Shrubs and Woody Vines of the Southwest. University of Texas Press, Austin, Texas, U.S.A.

von Siebold, P.F. & J.G. Zuccarini. 1839. Flora Japonica. Vol. 1. Leiden, Netherlands.

Wächter, G.A., J.J. Hoffmann, T. Furbacher, M.E. Blake, & B.N. Timmermann. 1999. Antibacterial and antifungal flavanones from *Eysenhardtia texana*. Phytochemistry 52:1469–1471.

Walpers, W.G. 1842. Repertorium Botanices Systematicae, Tom. 1. Sumtibus Friderici Hofmeister, Leipzig, Germany.

Walter, K.S. & H.J. Gillett, eds. 1998. 1997 IUCN Red List of Threatened Plants. Compiled by the World Conservation Monitoring Centre. IUCN - The World Conservation Union, Gland, Switzerland, and Cambridge, U.K.

Wang, G. & K. Nolte. 2010. Summer cover crop use in Arizona vegetable production systems. Arizona Cooperative Extension Publication AZ1519. The University of Arizona, Tucson, Arizona.

Wang, H., V. Haridas, J.U. Gutterman, & Z.-X. Xu. 2010. Natural triterpenoid avicins selectively induce tumor cell death. Commun. Integr. Biol. 3:205–208.

Ward, D.B. & D.W. Hall. 2004. Keys to the flora of Florida–10, *Galactia* (Leguminosae). Phytologia 86:65–74.

Washington State Noxious Weed Control Board (Washington State NWCB). 2008. "Swainsonpea, *Sphaerophysa salsula*." Washington State NWCB, Olympia, Washington. Accessed 20 Dec 2010. https://www.nwcb.wa.gov/weeds/swainsonpea

Watson, A.I., R.L. Holle, & R.E. López. 1994. Cloud-to-ground lightning and upper-air patterns during bursts and breaks in the Southwest monsoon. Mon. Wea. Rev. 122:1726–1739.

Watson, S. 1871. Botany. In: C. King, ed. 1871. Report of the Geological Exploration of the Fortieth Parallel. Vol. 5. Government Printing Office, Washington, D.C., U.S.A.

Watson, S. 1876. Descriptions of new species of plants, chiefly Californian, with revisions of certain genera. Proc. Amer. Acad. Arts 11:121–148.

Webster, G.T. 1955. Interspecific hybridization of *Melilotus alba* × *M. officinalis* using embryo culture. Agron. J. 47:138–142.

Wei, Z. & M.A. Vincent. 2010. Tribe Trifolieae. In: Z.Y. Wu, P.H. Raven, & D.Y. Hong, eds. 2010. Flora of China. Vol. 10 (Fabaceae). Science Press, Beijing, China, and Missouri Botanical Garden Press, St. Louis, Missouri, U.S.A. 547–559.

Weitemier, K.A. 2010. Phylogeographic patterns and intervarietal relationships within *Lupinus lepidus*: Morphological differences, genetic similarities. Dissertations and Theses. Paper 919.

Welsh, S.L. 1960. Legumes of the north-central states: Galegeae. Iowa State J. Sci. 35:111–250.

Welsh, S.L. 2007. North American Species of *Astragalus* Linnaeus (Leguminosae): A Taxonomic Revision. Brigham Young University, Provo, Utah, U.S.A.

Welsh, S.L., N.D. Atwood, S. Goodrich, & L.C. Higgins. 1993. A Utah Flora. Brigham Young University Press, Provo, Utah, U.S.A.

Welsh, S.L., N.D. Atwood, L.C. Higgins, & S. Goodrich, eds. 1987. A Utah Flora. Great Basin Naturalist Mem. 9(Art. 17). Brigham Young University, Provo, Utah, U.S.A.

Welsh, S.L., N.D. Atwood, S. Goodrich, & L.C. Higgins, eds. 2003. A Utah Flora. 3rd ed. Monte L. Bean Life Sciences Museum, Brigham Young University, Provo, Utah, U.S.A.

Welsh, S.L. & M. Licher. 2010. *Pediomelum* Rydberg (Leguminosae) in Arizona and two previously undescribed species. W. N. Amer. Naturalist 70:9–18.

Wheeler, G.M. 1872. Preliminary report of explorations in Nevada and Arizona. In: A.A. Humphreys. 1872. Preliminary report concerning explorations and surveys principally in Nevada and Arizona, 1871. Government Printing Office, Washington, D.C., U.S.A. 9–62.

Whibley, D.J.E. 1980. Acacias of South Australia. D.J. Woolman, Government Printer South Australia, Adelaide, South Australia, Australia.

Whisenant, S.G. & D.N Ueckert. 1982. Germination responses of *Eysenhardtia texana* and *Leucaena retusa*. J. Range Managem. 35(6):748–750.

Whiteside, L. 1981. The Carob Cookbook. Thorsons Publishers, Wellingborough, U.K.

Whiting, A.F. 1939. Ethnobotany of the Hopi. Mus. Northern Arizona Bull. 15:33, 80.

Whyte, R.O., G. Nilsson-Leissner, & H.C. Trumble. 1953. Legumes in Agriculture. FAO Agricultural Studies No. 21. Food and Agriculture Organization of the United Nations, Rome, Italy.

Wiener, M.A. 1990. Earth Medicine, Earth Food. Ballantine Books, New York, New York, U.S.A.

Wiggins, I.A. 1964. Flora of the Sonoran Desert. In: Shreve, F. & I.A. Wiggins. 1964. Vegetation and Flora of the Sonoran Desert. 2 vols. Stanford University Press, Stanford, California, U.S.A. 189–1740.

Wilbur, R.L. 1975. A revision of the North American genus *Amorpha* (Leguminosae––Psoraleae). Rhodora 77:337–409.

WildEarth Guardians. 2008. "Petition to list the Chihuahua Scurfpea *(Pediomelum pentaphyllum)* under the U.S. Endangered Species Act." WildEarth Guardians, Santa Fe, New Mexico, U.S.A. Accessed 6 Jun 2011. http://www.wildearthguardians.org/site/DocServer/petition_scurfpea.pdf?docID=624&AddInterest=1103

Windler D.R. 1971a. Systematic studies in *Crotalaria sagittalis* L. and related species in North America (Leguminosae). Diss. Abstr. Int., B 31(11):6460–6461.

Windler D.R. 1971b. New North American unifoliolate *Crotalaria* taxa (Leguminosae). Phytologia 21:257–266.

Windler D.R. 1973. Field and garden studies in *Crotalaria sagittalis* L. and related species. Phytologia 26:289–354.

Windler D.R. 1974a. Chromosome numbers for native North American unifoliolate species of *Crotalaria* (Leguminosae). Brittonia 26:172–176.

Windler D.R. 1974b. A systematic treatment of the native unifoliolate Crotalarias of North America (Leguminosae). Rhodora 76:151–204.

Wojciechowski, M.F., M. Lavin, & M.J. Sanderson. 2004. A phylogeny of Legumes (Leguminosae) based on analysis of the plastid *matK* gene resolves many well-supported subclades within the family. Amer. J. Bot. 91:1846–1862.

Woldemichael, G.M., M.P. Singh, W.M. Maiese, & B.N. Timmermann. 2003. Constituents of antibacterial extracts of *Caesalpinia paraguariensis* Burk. Z Naturforsch C. 58(1–2):70–75.

Wood, C.E. 1949. The American barbistyled species of *Tephrosia* (Leguminosae). Rhodora 51:193–231, 233–302, 305–364, 369–384.

Woodhouse, S.W. 1853. Report on the natural history. In: L. Sitgreaves. 1853. Report of an Expedition down the Zuñi and Colorado Rivers. Senate Exec. doc. 59, 32nd Congress, 2nd Session. Robert Armstrong public printer, Washington, D.C., U.S.A. 32–40.

Wunderlin, R.P. 1983. Revision of the arborescent Bauhinias (Fabaceae: Caesalpinioideae: Cercideae) native to Middle America. Ann. Missouri Bot. Gard. 70:95–127.

Wunderlin, R.P. 2010. Reorganization of the Cercideae (Fabaceae: Caesalpinioideae). Phytoneuron 2010-48:1–5.

Wyman, D. 1949. The wisterias. Arnoldia (Jamaica Plain) 9:17–28.

Wyman, D. 1969. Shrubs and Vines for American Gardens. Macmillan Publishing Company, New York, New York, U.S.A.

Wyman, D. 1977. Wyman's Gardening Encyclopedia. Macmillan Publishing Company, New York, New York, U.S.A.

Wyman, L.C. & S.K. Harris. 1941. Navajo Indian Medical Ethnobotany. University of New Mexico Press, Albuquerque, New Mexico, U.S.A.

Wyse Jackson, P. 2014. Ireland's Generous Nature. Missouri Botanical Garden Press, St. Louis, Missouri, U.S.A.

Xu, L., J. Xu, Q. Liu, R. Xie, & G. Wei. 2009. Genetic diversity of rhizobia isolated from *Sphaerophysa salsula* in several regions of northwestern China. Biodivers. Sci. 17(1):69–75.

Yakovlev, G.P. 1967. Systematical and geographical studies of the genus *Sophora* L. and allied genera. Trudy Leningradsk. Khim.-Farm. Inst. 21:42–62.

Yakovlev, G.P. 1979. Notes on the taxonomy of the genus *Alhagi* Gagneb. (Fabaceae). Bot. Zhurn. (Moscow & Leningrad) 64:1794–1799.

Zepeda, O. 1985. The Sand Papago Oral History Project. Division of Archaeology. Western Archaeological and Conservation Center. National Park Service, Tucson, Arizona.

Zohary, D. & M. Hopf. 1973. Domestication of pulses in the old world: Legumes were companions of wheat and barley when agriculture began in the Near East. Science 182:887–894.

Zohary, D. & U. Plitmann. 1979. Chromosome polymorphism, hybridization and colonization in the *Vicia sativa* group (Fabaceae). Pl. Syst. Evol. 131:143–156.

Zohary, M. & D. Heller. 1984. The Genus *Trifolium*. The Israel Academy of Sciences and Humanities, Jerusalem, Israel.

Zolfaghari, R. & M. Harden. 1982. Nutritional value of mesquite beans (*Prosopis glandulosa*). In: H.W. Parker, ed. 1982. Mesquite Utilization. Texas Tech University, Lubbock, Texas, U.S.A. K1–K16.

Zolfaghari, R., M. Harden, & L. Huffman. 1986. Some physical and chemical properties of honey mesquite pod (*Prosopis glandulosa*) and applications in food products. Cereal Chem. 63:104–108.

INDEX

The accepted scientific names are given in **bold** with main page number for taxonomic treatment also in **bold.** Taxonomic synonyms and names of plants casually mentioned are in *italics*. Common names are in all CAPS. Page numbers in *italics* note b/w illustrations while page numbers with an underline note a color figure.

ALPHABETICAL LIST OF GENERA